www.devbio.com

the website for

DEVELOPMENTAL BIOLOGY

SIXTH EDITION

This website exists to provide materials to supplement and enrich courses in developmental biology. The material here is loosely based on the theme "This is really interesting; it's too bad I can't put it into the textbook." The website contains:

1. Material to update the *Developmental Biology* textbook.
2. Studies and experiments deemed too specialized to put into the textbook.
3. Philosophical, sociological, and historical studies in developmental biology. These include ethical issues raised by new technologies.
4. Interviews with people in the field who have been influential in the "morphogenesis" of developmental biology.
5. Opinions (labeled as such) that can be used as a springboard for discussion.

The website will be updated with new information, videos, and interviews. The following directory summarizes the numbered references to the website, along with references to *Vade Mecum*, that you will find throughout each of the book's chapters.

(continued on inside back cover)

Developmental Biology

SIXTH EDITION

Developmental Biology

Sixth Edition

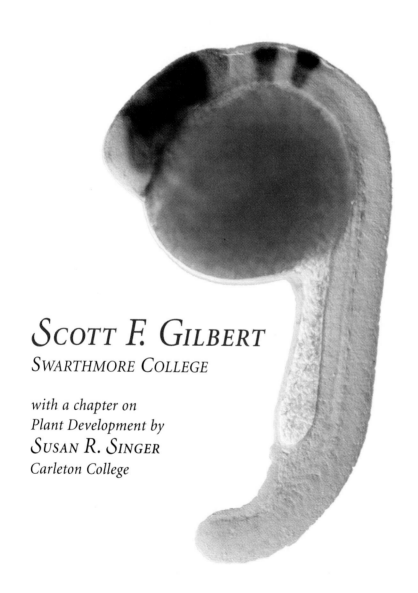

Scott F. Gilbert
Swarthmore College

with a chapter on
Plant Development by
Susan R. Singer
Carleton College

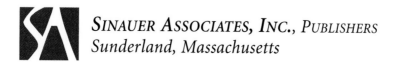

Sinauer Associates, Inc., Publishers
Sunderland, Massachusetts

The Front Cover

The garter snake *Thamnophis* has no limbs. The expression of certain genes during development prevent the formation of limb-forming regions and convert nearly all of its vertebrae into rib-bearing thoracic vertebrae. The embryos shown have been treated with Alcian Blue, a dye that stains cartilage. (See Chapter 22. Photograph courtesy of A. C. Burke.)

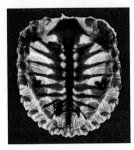

The Back Cover

The ribs of the red-eared slider turtle *Chrysemys* enter the dermis of its back. It is probable that these ribs are active in transforming that dermis into bone, thereby making the carapace of the turtle shell. The vertebrae form the central portion of the carapace. The embryos have been treated with Alizarin Red, a dye that stains bone. (See Chapter 14. Photograph courtesy of G. Loredo, A. Brukman, and S. F. Gilbert.)

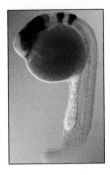

The Title Page

Three gene products that help specify the regions of the brain are localized to the regions where they function in a zebrafish embryo (shown at roughly 20 hours of development). The *otx-2* message translates into a protein that is involved in specifying the forebrain and midbrain. It is seen here in magenta stain. The *krx-20* message forms two distinct bands (purple); its protein is critical in specifying regions of the hindbrain. Engrailed protein (yellow) is seen at the midbrain-hindbrain boundary and in certain muscle cells along the midline of the embryo. (See Chapters 11 and 12. Photograph courtesy of G. Hauptmann and Springer-Verlag.)

Developmental Biology, 6th Edition

Sinauer Associates, Inc.
23 Plumtree Road/PO Box 407
Sunderland, MA 01375 U.S.A.
FAX: 413-549-1118
Email: publish@sinauer.com
www.sinauer.com

Library of Congress Cataloging-in-Publication Data

Gilbert, Scott F., 1949–
    Developmental biology / Scott F. Gilbert. -- 6th ed.
        p.   cm.
    Includes bibliographical references.
    ISBN 0-87893-243-7 (cloth)
    1. Embryology. 2. Developmental biology. I. Title.

QL955.G48 2000
571.8--dc21   00-027504

Printed in U.S.A.
5   4   3   2   1

To Daniel, Sarah, and David

Contents

part 2 Early embryonic development

CHAPTER 7 Fertilization: Beginning a new organism 185

CHAPTER 8 Early development in selected invertebrates 223

p a r t 3 *Later embryonic development*

part 4 *Ramifications of developmental biology*

Preface

The poet Vladimir Mayakowsky wrote that revolutions of content require revolutions of structure. And if any science has had a revolution in content, it is developmental biology. The structure of the Sixth Edition of *Developmental Biology* has changed significantly to account for several ongoing revolutions of content. It has also been rewritten to emphasize a core of developmental biology paradigms and principles. In the revision process, *Developmental Biology's* Sixth Edition actually became about 200 pages shorter than its predecessor. This was accomplished by putting much of the more advanced material, as well as nearly all the material now covered in introductory biology textbooks, onto the website that is integrated with the text.

The growth of developmental biology actually made it easier to organize chapters. Studies on phenomena that had been considered separate entities can be brought together into coherent stories. So this edition has rearranged its chapters to best introduce new biology students to the remarkable embryos and the ways we study them.

- There is now an entirely new introductory section of the textbook. The first six chapters, *Principles of Developmental Biology*, provide the basic conceptual background of the anatomical, experimental, genetic, cellular, evolutionary, and ecological approaches to developmental biology. The chapters build on one another, such that material presented in Chapter 1 is revisited throughout the other five chapters, but with different emphases.

- The Sixth Edition integrates the discussion of the early development of several organisms into concise units that detail cleavage, gastrulation, and axis formation simultaneously. This integration has only been made possible in recent years.

- A new chapter on a new topic, *Mechanisms of Plant Development*, was written for the Sixth Edition by Dr. Susan Singer. This chapter concisely and elegantly introduces the basic concepts and exciting new findings of plant developmental biology, and it introduces the student to *Arabidopsis* as a model organism for studying plant developmental genetics.

- Another new chapter, *Metamorphosis, Regeneration, and Aging*, highlights the new importance of these areas of developmental biology.

This reorganization was made possible and necessary by numerous revolutions in the science: Developmental biology is becoming a central organizing discipline in biology, relating cell and molecular biology, anatomy, ecology, evolution, and medicine to each other. It is an exceptionally interesting and busy time to be a developmental biologist, and you should be able to find questions that will excite, challenge, and be worthy of your intellectual talents. These revolutions include:

- *Molecular embryology.* New knowledge about how the inherited genes are expressed differently in different populations of cells is changing our views as to how organs are made. Conclusions that had been stable since the 1920s are now being revised, if not completely overthrown, by studies that use more refined techniques.

- *Biotechnology.* The combining of developmental biology with biotechnology is promising to regenerate spinal neurons and bones for the first time in human history. The altering of our development by embryonic stem cells, cloning, and even the enhancement of our genetic endowment is now theoretically possible.

- *Ecological developmental biology.* As human activity alters the environment, the possible effects of global warming, pesticides, and other chemicals have incited a new interest in the environmental regulation and disruption of development. Developmental biologists may soon find themselves at the forefront of conservation biology and ecological issues.

- *Clinical genetics.* During the past five years, there has been an integration of human embryology and medical genetics to create a new medical developmental biology that seeks to understand and treat the molecular bases of birth defects.

- *Evolutionary developmental biology.* The emergence of new phenotypes is made possible by changes in development. The regulatory genes that have long been thought to control the generation of novel structures are now being discovered.

- *Bioethics.* As a result of our new scientific abilities, there has been the emergence of an entire field of bioethics. The cloning of Dolly has brought both controversy and capital into developmental biology. Developmental biologists are suddenly being asked to discuss ethical and legal issues that they never before had to address.

The Sixth Edition of *Developmental Biology* covers all these issues, many of which are new to developmental biologists (and which certainly were not a part of our own training as professors). At the same time, this volume attempts to provide the fundamental set of techniques, paradigms, and models necessary for a student to understand the core of developmental biology—differentiation and morphogenesis.

In addition to the structural changes mentioned above, there have been many pedagogical changes in the textbook.

- The book has new and elegant *color illustrations.* Gastrulation aficionado Ray Keller (in his review of one of the chapters) told me, "Students should NOT read this material quickly, but too typical a scene is some poor bastard hunkered over this text at 2:30 A.M. with a cup of coffee, frantically scanning the figures to see if he or she can figure out what is happening." If such a scene really does occur, the wonderful illustrations of the J/B Woolsey studio should make that student's life much easier.

- This book is also graced with an outstanding collection of *color photographs.* I wish to thank all the scientists who contributed their micrographs and other photographic material to this volume. I especially wish to thank M. Danilchik, G. Müller, S. Paddock, S. Carroll, and G. Schatten for allowing us to use so many of their spectacular images.

- In this edition, each chapter ends with a *Snapshot Summary.* These summaries are to be used to see the forest as well as the trees. While they should be useful in reviewing the chapter, please do not confuse these summaries with the actual science—which consists of evidence and process.

- As mentioned above, the text is linked to a new *website* (www.devbio.com). Our website started as a student-operated site in 1994 and has become so successful that our server could not handle the load. We are on a new server and have greatly expanded our sites. Please see the endpapers for a directory of material on the website that indicates how it is keyed to the book's chapters.

- This edition also sees the debut of **Vade Mecum,** a CD-ROM created specifically for studying developmental biology. Mary Tyler and Ron Kozlowski have created a unique and remarkable resource for anyone teaching or studying animal development. Time-lapse videomicroscopy has always played a key role in developmental biology education, but these movies and videos depended on television or movie projectors and were therefore confined to classroom or laboratory times. *Vade Mecum* (from the Latin for "come with me") will enable you to see these videos whenever you want. Developmental biology is about change, and print cannot do the field justice. We are thrilled to include the *Vade Mecum* CD-ROM inside the front cover of this book. The directory on the endpapers shows the references to *Vade Mecum* within the text.

I hope this textbook will provide two gateways—one portal to the actual organisms, and the other to the research that studies them. It is my hope that the textbook will enable each person to appreciate both the development of the organism and the research papers in developmental biology. Developmental biologists find themselves within traditions of inquiry that extend back to the beginnings of human inquisitiveness. We are discovering the mechanisms by which male and female are distinguished, by which left and right are separated, by which caterpillars become butterflies, and by which organs are formed. We are beginning to have answers to questions that have perplexed human thought for as long as we have been human.

But these answers are not "facts," but conclusions based on the evidence available. Indeed, this book is not about "Facts." It is about evidence, interpretations, and attitude. In developmental biology, we have become used to many of our most cherished "facts" being overturned by new evidence. But obtaining such evidence is not at all easy. To say that something is true, one must prove that everything else is false, and this is determined by the techniques available at any time. Thus, primary embryonic induction, the mainstay of developmental biology since the 1920s, has recently been shown by molecular techniques to be neither primary nor an induction. The new studies do not invalidate the experiments and observations of our predecessors. Rather, they show how complex the subject is, and they have revitalized these areas of research. If any discipline should embrace change, it is developmental biology. As you read this book, probably the best advice to keep in mind is the instruction of Alfred North Whitehead, who wrote that every scientist should "Seek simplicity and distrust it."

Acknowledgments

This edition, like its earlier incarnations, has benefited enormously from the students of my embryology and developmental genetics courses and seminars. The extremely supportive staff and faculty of Swarthmore College have also played major roles in producing this work. The scientists who reviewed each chapter deserve special thanks. First among equals is the "Gang of Four" who took it upon themselves to read all six new introductory chapters. Everyone teaching or learning from this book owes thanks to Jessica Bolker, Carol Erikson, Margaret Saha, and Mary Tyler. Other people who read major parts of chapters include M. Susan Lindee, F. Anne McNabb, Kirsi Sainio, and Hannu Sariola. My profound thanks go to those who reviewed the individual chapters, contributing insight and information to their revision: Kathryn Anderson, Robert Arking, Philip Benfey, Bruce Bowerman, Marianne Bronner-Fraser, Karen Crawford, Diana Darnell, Ali Hemmati-Brivanlou, C. Drew Harvell, Laurinda Jaffe, Ray Keller, Anthony Samuel LaMantia, Jim Langeland, Marta Laskowski, Michael Levine, Kersti Linask, Craig Nelson, Lee Niswander, David Page, Rudy Raff, Richard Schultz, Kathleen Sulik, Billie Swalla, Rocky Tuan, and Greg Wray.

This book has shown that the management skills of Andy Sinauer, editor Carol Wigg, production coordinator Chris Small, and art director John Woolsey were greater than they ever thought possible. Their efforts have been heroic. With this edition we entered the brave new world of electronic photograph transmission, and David McIntyre and Mariah Peelle coped admirably with hundreds of images transmitted via the Internet. Jeff Johnson did a fine job with the striking cover image. Janice Holabird and Joan Gemme produced stunning chapter layouts and then worked unstintingly to incorporate many last-minute changes to the art and text. It is in part a testament to their flexibility and patience that this textbook could be updated right up until the day it went to press, in March of 2000. Very few publishers allow their authors the freedom to revise at this level, and I hope readers benefit from the work that went into the final polishing.

In particular, this book could not have been started or completed without the enthusiastic support of my wife, Anne Raunio, and the patience of our friends. My deep thanks to you all.

SCOTT GILBERT
MARCH 2000

PART I Principles of Developmental Biology

chapter **1** — *Developmental biology: The anatomical tradition*

BETWEEN FERTILIZATION AND BIRTH, the developing organism is known as an embryo. The concept of an embryo is a staggering one, and forming an embryo is the hardest thing you will ever do. To become an embryo, you had to build yourself from a single cell. You had to respire before you had lungs, digest before you had a gut, build bones when you were pulpy, and form orderly arrays of neurons before you knew how to think. One of the critical differences between you and a machine is that a machine is never required to function until after it is built. Every animal has to function as it builds itself.

For animals, fungi, and plants, the sole way of getting from egg to adult is by developing an embryo. The embryo mediates between genotype and phenotype, between the inherited genes and the adult organism. Whereas most of biology studies adult structure and function, developmental biology finds the study of the transient stages leading up to the adult to be more interesting. Developmental biology studies the initiation and construction of organisms rather than their maintenance. It is a science of becoming, a science of process. To say that a mayfly lives but one day is profoundly inaccurate to a developmental biologist. A mayfly may be a winged *adult* for only a day, but it spends the other 364 days of its life as an aquatic juvenile under the waters of a pond or stream.

The questions asked by developmental biologists are often questions about becoming rather than about being. To say that XX mammals are usually females and XY mammals are usually males does not explain sex determination to a developmental biologist, who wants to know *how* the XX genotype produces a female and *how* the XY genotype produces a male. Similarly, a geneticist might ask how globin genes are transmitted from one generation to the next, and a physiologist might ask about the function of globins in the body. But the developmental biologist asks how it is that the globin genes become expressed only in red blood cells and how they become active only at specific times in development. (We don't know the answers yet.)

Developmental biology is a great field for scientists who want to integrate different levels of biology. We can take a problem and study it on the molecular and chemical levels (e.g., How are globin genes transcribed, and how do the factors activating their transcription interact with one another on the DNA?), on the cellular and tissue levels (Which cells are able to make globin, and how does globin mRNA leave the nucleus?), on the organ and organ system levels (How do the capillaries form in each tissue, and how are they instructed to branch and connect?), and even at the ecological and evolutionary levels (How do differences in globin gene activation enable oxygen to flow from mother to fetus, and how do environmental factors trigger the differentiation of more red blood cells?).

Developmental biology is one of the fastest growing and most exciting fields in biology, creating a framework that integrates molecular biology, physiology, cell biology, anatomy, cancer research, neurobiology, immunology, ecology, and evolutionary biology. The study of development has become essential for understanding any other area of biology.

The Questions of Developmental Biology

According to Aristotle, the first embryologist known to history, science begins with wonder: "It is owing to wonder that people began to philosophize, and wonder remains the beginning of knowledge." The development of an animal from an egg has been a source of wonder throughout history. The simple procedure of cracking open a chick egg on each successive day of its 3-week incubation provides a remarkable experience as a thin band of cells is seen to give rise to an entire bird. Aristotle performed this procedure and noted the formation of the major organs. Anyone can wonder at this remarkable—yet commonplace—phenomenon, but the scientist seeks to discover how development actually occurs. And rather than dissipating wonder, new understanding increases it.

Multicellular organisms do not spring forth fully formed. Rather, they arise by a relatively slow process of progressive change that we call **development**. In nearly all cases, the development of a multicellular organism begins with a single cell—the fertilized egg, or **zygote**, which divides mitotically to produce all the cells of the body. The study of animal development has traditionally been called **embryology**, from that stage of an organism that exists between fertilization and birth. But development does not stop at birth, or even at adulthood. Most organisms never stop developing. Each day we replace more than a gram of skin cells (the older cells being sloughed off as we move), and our bone marrow sustains the development of millions of new red blood cells every minute of our lives. In addition, some animals can regenerate severed parts, and many species undero metamorphosis (such as the transformation of a tadpole into a frog, or a caterpillar into a butterfly). Therefore, in recent years it has become customary to speak of **developmental biology** as the discipline that studies embryonic and other developmental processes.

Development accomplishes two major objectives: it generates cellular diversity and order within each generation, and it ensures the continuity of life from one generation to the next. Thus, there are two fundamental questions in developmental biology: How does the fertilized egg give rise to the adult body, and how does that adult body produce yet another body? These two huge questions have been subdivided into six general questions scrutinized by developmental biologists:

- **The question of differentiation**. A single cell, the fertilized egg, gives rise to hundreds of different cell types—muscle cells, epidermal cells, neurons, lens cells, lymphocytes, blood cells, fat cells, and so on (Figure 1.1). This generation of cellular diversity is called **differentiation**. Since each cell of the body (with very few exceptions) contains the same set of genes, we need to understand how this same set of genetic instructions can produce different types of cells. How can the fertilized egg generate so many different cell types?

- **The question of morphogenesis**. Our differentiated cells are not randomly distributed. Rather, they are organized into intricate tissues and organs. These organs are arranged in a given way: the fingers are always at the tips of our hands, never in the middle; the eyes are always in our heads, not in our toes or gut. This creation of ordered form is called **morphogenesis**. How can the cells form such ordered structures?

- **The question of growth**. How do our cells know when to stop dividing? If each cell in our face were to undergo just one more cell division, we would be considered horribly malformed. If each cell in our arms underwent just one more round of cell division, we could tie our shoelaces without bending over. Our arms are generally the same size on both sides of the body. How is cell division so tightly regulated?

- **The question of reproduction**. The sperm and egg are very specialized cells. Only they can transmit the instructions for making an organism from one generation to the next. How are these cells set apart to form the next generation, and what are the instructions in the nucleus and cytoplasm that allow them to function this way?

- **The question of evolution**. Evolution involves inherited changes in development. When we say that today's one-toed horse had a five-toed ancestor, we are saying that changes in the development of cartilage and muscles occurred over many generations in the embryos of the horse's ancestors. How do changes in development create new body forms? Which heritable changes are possible, given the constraints imposed by the necessity of the organism to survive as it develops?

- **The question of environmental integration**. The development of many organisms is influenced by cues from the environment. Certain butterflies, for instance, inherit the ability to produce different wing colors based on the temperature or the amount of daylight experienced by the caterpillar before it undergoes metamorphosis. How is the development of an organism integrated into the larger context of its habitat?

Anatomical Approaches to Developmental Biology

A field of science is defined by the questions it seeks to answer, and most of the questions in developmental biology

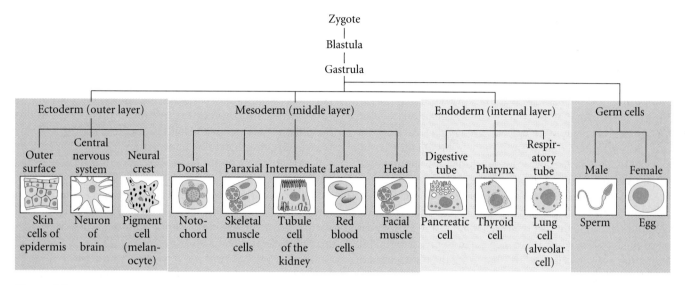

Figure 1.1
Some representative differentiated cell types of the vertebrate body. The progeny of the fertilized egg must diversify into hundreds of cell types. The cell types are organized according to the germ layers from which they arise. The germ cells (precursors of the sperm and egg) are shown as not arising from any of the germ layers.

have been bequeathed to it through its embryological heritage. There are numerous strands of embryology, each predominating during a different era. Sometimes they are very distinct traditions, and sometimes they blend. We can identify three major ways of studying embryology:

- Anatomical approaches
- Experimental approaches
- Genetic approaches

While it is true that anatomical approaches gave rise to experimental approaches, and that genetic approaches built on the foundations of the earlier two approaches, all three traditions persist to this day and continue to play a major role in developmental biology. Chapter 3 of this text discusses experimental approaches, and Chapters 4 and 5 examine the genetic approaches in greater depth. In recent years, each of these traditions has become joined with molecular genetics to produce a vigorous and multifaceted science of developmental biology.

But the basis of all research in developmental biology is the changing anatomy of the organism. What parts of the embryo form the heart? How do the cells that form the retina position themselves the proper distance from the cells that form the lens? How do the tissues that form the bird wing relate to the tissues that form the fish fin or the human hand?

There are several strands that weave together to form the anatomical approaches to development. The first strand is **comparative embryology**, the study of how anatomy changes during the development of different organisms. For instance, a comparative embryologist may study which tissues form the nervous system in the fly or in the frog. The second strand, based on the first, is **evolutionary embryology**, the study of how changes in development may cause evolutionary changes and of how an organism's ancestry may constrain the types of changes that are possible. The third anatomical approach to developmental biology is **teratology**, the study of birth defects. These anatomical abnormalities may be caused by mutant genes or by substances in the environment that interfere with development. The study of abnormalities is often used to discover how normal development occurs. The fourth anatomical approach is **mathematical modeling**, which seeks to describe developmental phenomena in terms of equations. Certain patterns of growth and differentiation can be explained by interactions whose results are mathematically predictable. The revolution in graphics technology has enabled scientists to model certain types of development on the computer and to identify mathematical principles upon which those developmental processes are based.

Comparative Embryology

The first known study of comparative developmental anatomy was undertaken by Aristotle in the fourth century B.C.E. He noted the different ways that animals are born: from eggs (**oviparity**, as in birds, frogs, and most invertebrates), by live birth (**viviparity**, as in eutherian mammals), or by producing an egg that hatches inside the body (**ovoviviparity**, as in certain reptiles and sharks). Aristotle also identified the two major cell division patterns by which embryos are formed: the **holoblastic** pattern of cleavage (in which the entire egg is divided into smaller cells, as it is in frogs and mammals) and the **meroblastic** pattern of cleavage (as in chicks, wherein only part of the egg is destined to become the embryo, while the other portion—the yolk—serves as nutrition). And should anyone want to know who first figured out the functions of the placenta and the umbilical cord, it was Aristotle.

After Aristotle, there was remarkably little progress in embryology for the next two thousand years. It was only in 1651 that William Harvey concluded that all animals—even mammals—originate from eggs. *Ex ovo omnia* ("All from the egg") was the motto on the frontispiece of his *On the Generation of Living Creatures*, and this precluded the spontaneous generation of animals from mud or excrement. This statement was not made lightly, for Harvey knew that it went against the views of Aristotle, whom Harvey still venerated. (Aristotle had thought that menstrual fluid formed the material of the embryo, while the semen acted to give it form and animation.) Harvey also was the first to see the **blastoderm** of the chick embryo—that small region of the egg that contains the yolk-free cytoplasm that gives rise to the embryo—and he was the first to notice that "islands" of blood cells form before the heart does. Harvey also suggested that the amnionic fluid might function as a shock absorber for the embryo.

As might be expected, embryology remained little but speculation until the invention of the microscope allowed detailed observations. In 1672, Marcello Malpighi published the first microscopic account of chick development. Here, for the first time, the neural groove (precursor of the neural tube), the muscle-forming somites, and the first circulation of the arteries and veins—to and from the yolk—were identified (Figure 1.2).

Epigenesis and preformation

With Malpighi begins one of the great debates in embryology—the controversy over whether the organs of the embryo are formed de novo ("from scratch") at each generation, or whether the organs are already present, but in miniature form, within the egg (or sperm). The first view is called **epigenesis,** and it was supported by Aristotle and Harvey. The second view is called **preformation,** and it was reinvigorated with support from Malpighi. Malpighi showed that the unincubated* chick egg already had a great deal of structure. This observation provided him with reasons to question epigenesis. According to the preformationist view, all the organs of the adult were prefigured in miniature within the sperm or (more usually) the egg. Organisms were not seen to be "developed," but rather "unrolled."

The preformationist hypothesis had the backing of eighteenth-century science, religion, and philosophy (Gould 1977; Roe 1981, Pinto-Correia 1997). First, because all organs were prefigured, embryonic development merely required the growth of existing structures, not the formation of new ones. No extra mysterious force was needed for embryonic develop-

ment. Second, just as the adult organism was prefigured in the germ cells, another generation already existed in a prefigured state within the germ cells of the first prefigured generation. This corollary, called *embôitment* (encapsulation), ensured that the species would always remain constant. Although certain microscopists claimed to see fully formed human miniatures within the sperm or egg, the major proponents of this hypothesis—Albrecht von Haller and Charles Bonnet—knew that organ systems develop at different rates and that embryonic structures need not be in the same place as those in the newborn.

The preformationists had no cell theory to provide a lower limit to the size of their preformed organisms (the cell theory arose in the mid-1800s), nor did they view humankind's tenure on Earth as potentially infinite. Rather, said Bonnet (1764), "Nature works as small as it wishes," and the human species existed in that finite time between Creation and Resurrection. This view was in accord with the best science of its time, conforming to the French mathematician-philosopher René Descartes's principle of the infinite divisibility of a mechanical nature initiated, but not interfered with, by God. It also conformed to Enlightenment views of the Deity. The scientist-priest Nicolas Malebranche saw in preformationism the fusion of the rule-giving God of Christianity with Cartesian science (Churchill 1991; Pinto-Correia 1997).[†]

The embryological case for epigenesis was revived at the same time by Kaspar Friedrich Wolff, a German embryologist working in St. Petersburg. By carefully observing the development of chick embryos, Wolff demonstrated that the embryonic parts develop from tissues that have no counterpart in the adult organism. The heart and blood vessels (which, according to preformationism, had to be present from the beginning to ensure embryonic growth) could be seen to develop anew in each embryo. Similarly, the intestinal tube was seen to arise by the folding of an originally flat tissue. This latter observation was explicitly detailed by Wolff, who proclaimed (1767), "When the formation of the intestine in this manner has been duly weighed, almost no doubt can remain, I believe, of the truth of epigenesis." However, to explain how an organism is created anew each generation, Wolff had to postulate an unknown force, the *vis essentialis* ("essential force"), which, acting like gravity or magnetism, would organize embryonic development.

A reconciliation of sorts was attempted by the German philosopher Immanuel Kant (1724–1804) and his colleague,

---

*As was pointed out by Maître-Jan in 1722, the egg examined by Malpighi may technically be called "unincubated," but as it was left sitting in the Bolognese sun in August, it was not unheated.

[†]Preformation was a conservative theory, emphasizing the lack of change between generations. Its principal failure was its inability to account for the variations known by the limited genetic evidence of the time. It was known, for instance, that matings between white and black parents produced children of intermediate skin color, an impossibility if inheritance and development were solely through either the sperm or the egg. In more controlled experiments, the German botanist Joseph Kölreuter (1766) had produced hybrid tobacco plants having the characteristics of both species. Moreover, by mating the hybrid to either the male or female parent, Kölreuter was able to "revert" the hybrid back to one or the other parental type after several generations. Thus, inheritance seemed to arise from a mixture of parental components.

biologist Johann Friedrich Blumenbach (1752–1840). Attempting to construct a scientific theory of racial descent, Blumenbach postulated a mechanical, goal-directed force called the *Bildungstrieb* ("development force"). Such a force, he said, was not theoretical, but could be shown to exist by experimentation. A hydra, when cut, regenerates its amputated parts by rearranging existing elements (see Chapter 18). Some purposive organizing force could be observed in operation, and this force was a property of the organism itself. This *Bildungstrieb* was thought to be inherited through the germ cells. Thus, development could proceed through a predetermined force inherent in the matter of the embryo (Cassirer 1950;

Figure 1.2
Depictions of chick developmental anatomy. (A) Dorsal view (looking "down" at what will become the back) of a 2-day chick embryo, as depicted by Marcello Malpighi in 1672. (B) Ventral view (looking "up" at the prospective belly) of a chick embryo at a similar stage, seen through a dissecting microscope and rendered by F. R. Lillie in 1908. (C) Eduard d'Alton's depiction of a later stage 2-day chick embryo in Pander (1817). (D) Modern rendering of a 3-day chick embryo. Details of the anatomy will be discussed in later chapters. (A from Malpighi 1672; B from Lillie 1908; C from Pander 1817, courtesy of Ernst Mayr Library of the Museum of Comparative Zoology, Harvard; D after Carlson 1981.)

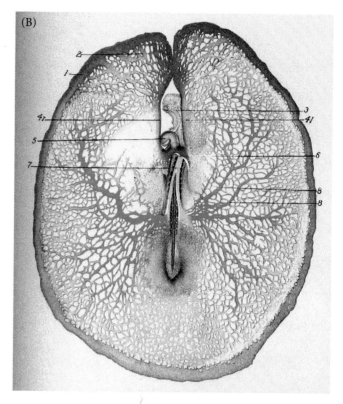

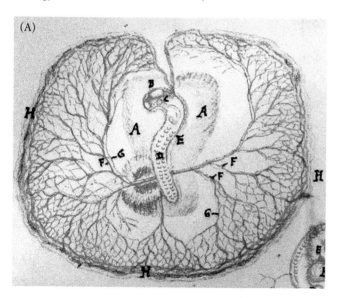

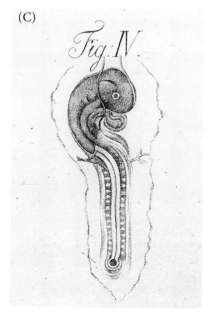

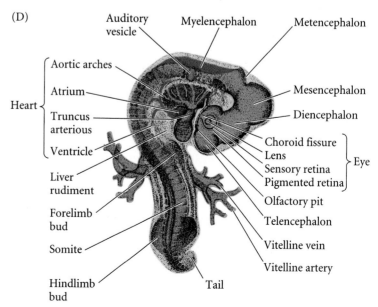

Lenoir 1980). Moreover, this force was believed to be susceptible to change, as demonstrated by the left-handed variant of snail coiling (where left-coiled snails can produce right-coiled progeny). In this hypothesis, wherein epigenetic development is directed by preformed instructions, we are not far from the view held by modern biologists that most of the instructions for forming the organism are already present in the egg.

Naming the parts:
The primary germ layers and early organs

The end of preformationism did not come until the 1820s, when a combination of new staining techniques, improved microscopes, and institutional reforms in German universities created a revolution in descriptive embryology. The new techniques enabled microscopists to document the epigenesis of anatomical structures, and the institutional reforms provided audiences for these reports and students to carry on the work of their teachers. Among the most talented of this new group of microscopically inclined investigators were three friends (born within a year of each other) who came from the Baltic region and who studied in northern Germany. The work of Christian Pander, Karl Ernst von Baer, and Heinrich Rathke transformed embryology into a specialized branch of science (and allowed the term *embryology* to be used to describe their work).

Pander studied the chick embryo for less than two years (before becoming a paleontologist), but in those 15 months, he discovered the three germ layers,* the specific regions of the embryo that give rise to the specific organ systems (see Figure 1.1).

- The **ectoderm** generates the outer layer of the embryo. It produces the surface layer (epidermis) of the skin and forms the nerves.
- The **endoderm** becomes the innermost layer of the embryo and produces the digestive tube and its associated organs (including the lungs).
- The **mesoderm** becomes sandwiched between the ectoderm and endoderm. It generates the blood, heart, kidney, gonads, bones, and connective tissues.

These three layers are found in the embryos of all **triploblastic** (three-layer) organisms. Some phyla, such as the porifera (sponges), cnidarians (sea anemones, hydra, jellyfish), and ctenophores (comb jellies) lack a true mesoderm and are considered **diploblastic** animals.

Pander (1817) also made observations that weighted the balance in favor of epigenesis. The germ layers, he noted, did not form their organs independently. Rather, each germ layer

"is not yet independent enough to indicate what it truly is; it still needs the help of its sister travelers, and therefore, although already designated for different ends, all three influence each other collectively until each has reached an appropriate level." Pander had discovered the tissue interactions that we now call **induction**. No tissue is able to construct organs by itself; it must interact with other tissues. (We will discuss the principles of induction more thoroughly in Chapter 6.) Thus, Pander felt that preformation could not be true, since the organs come into being through interactions between simpler structures.

Interestingly, the glory of Pander's book is its engravings; the artist, Eduard d'Alton, drew details for which the vocabulary had not yet been invented. Today we can look at these drawings and see the four regions of the embryonic chick brain, even though these regions had not yet been separately defined or given names (Figure 1.2C; see Churchill 1991). The ability to make precise observations has been among the greatest skills of embryologists, and even today modern developmental biologists looking at gene expression patterns are "rediscovering" regions of the embryo that were observed by embryologists a century ago.

Rathke looked at the development of frogs, salamanders, fish, birds, and mammals, and emphasized the similarities in the development of all these vertebrate groups. During his 40 years of embryological research, he described for the first time the vertebrate **pharyngeal arches** (Figure 1.3), which become the gill apparatus of fish but become the mammalian jaws and ears, (among other things, as we will see in Figure 1.14), the formation of the vertebrate skull, and the origin of the reproductive, excretory, and respiratory systems. He also studied the development of invertebrates, especially the crayfish. He is memorialized today in the name Rathke's pouch, the embryonic rudiment of the glandular portion of the pituitary. That

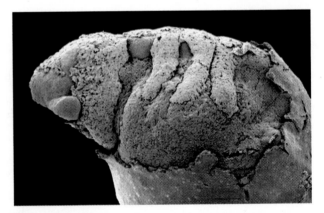

Figure 1.3
Pharyngeal arches (also called branchial arches and gill arches) in the embryo of the salamander *Ambystoma mexicanum*. The surface ectoderm has been removed to permit the easy visualization of these arches (highlighted) as they form. (Photograph courtesy of P. Falck and L. Olsson.)

---

*From the same root as *germination:* the Latin *germen,* meaning "sprout" or "bud." The names of the three germ layers are from the Greek: ectoderm from *ektos* ("outside") plus *derma* (skin); mesoderm from *mesos* ("middle"), and endoderm from *endon* ("within").

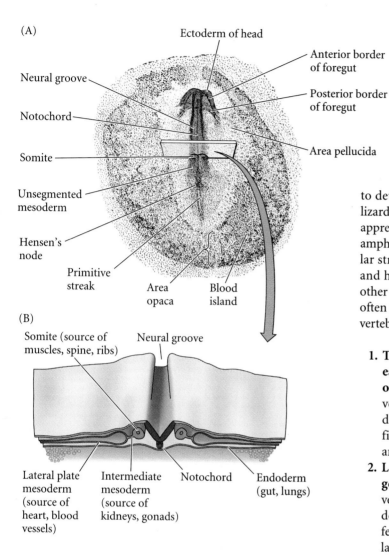

(A)

Ectoderm of head

Neural groove

Notochord

Somite

Unsegmented mesoderm

Hensen's node

Primitive streak

Anterior border of foregut

Posterior border of foregut

Area pellucida

Area opaca

Blood island

(B)

Somite (source of muscles, spine, ribs)

Neural groove

Lateral plate mesoderm (source of heart, blood vessels)

Intermediate mesoderm (source of kidneys, gonads)

Notochord

Endoderm (gut, lungs)

Figure 1.4
The notochord in the chick embryo. (A) Dorsal view of the 24-hour chick embryo. (B) A cross-section through the trunk shows the notochord and developing neural tube. By comparing Figures 1.2 and 1.4, you should see the remarkable changes between days 1, 2, and 3 of chick egg incubation. (A after Patten 1951.)

to determine the genus to which they belong. They may be lizards, small birds, or even mammals." Figure 1.5 allows us to appreciate his quandary. All vertebrate embryos (fish, reptiles, amphibians, birds, and mammals) begin with a basically similar structure. From his detailed study of chick development and his comparison of chick embryos with the embryos of other vertebrates, von Baer derived four generalizations (now often referred to as "von Baer's laws"), stated here with some vertebrate examples:

1. **The general features of a large group of animals appear earlier in development than do the specialized features of a smaller group.** All developing vertebrates appear very similar shortly after gastrulation. It is only later in development that the special features of class, order, and finally species emerge. All vertebrate embryos have gill arches, notochords, spinal cords, and primitive kidneys.

2. **Less general characters are developed from the more general, until finally the most specialized appear.** All vertebrates initially have the same type of skin. Only later does the skin develop fish scales, reptilian scales, bird feathers, or the hair, claws, and nails of mammals. Similarly, the early development of the limb is essentially the same in all vertebrates. Only later do the differences between legs, wings, and arms become apparent.

3. **The embryo of a given species, instead of passing through the adult stages of lower animals, departs more and more from them.**[†] The visceral clefts of embryonic birds and mammals do not resemble the gill slits of adult fish in detail. Rather, they resemble the visceral clefts of *embryonic* fish and other *embryonic* vertebrates. Whereas fish preserve and elaborate these clefts into true gill slits, mammals convert them into structures such as the eustachian tubes (between the ear and mouth).

4. **Therefore, the early embryo of a higher animal is never like a lower animal, but only like its early embryo.** Human embryos never pass through a stage equivalent to an adult fish or bird. Rather, human embryos initially share characteristics in common with fish and avian embryos. Later, the mammalian and other embryos diverge, none of them passing through the stages of the others.

he could see such a structure using the techniques available at that time is testimony to his remarkable powers of observation and his steady hand.

Karl Ernst von Baer extended Pander's studies of the chick embryo. He discovered the **notochord**, the rod of dorsalmost mesoderm that separates the embryo into right and left halves and which instructs the ectoderm above it to become the nervous system (Figure 1.4). He also discovered the mammalian egg, that long-sought cell that everyone believed existed but no one had yet seen.*

The four principles of Karl Ernst von Baer

In 1828, von Baer reported, "I have two small embryos preserved in alcohol, that I forgot to label. At present I am unable

---

*von Baer could hardly believe that he had at last found it when so many others—Harvey, de Graaf, von Haller, Prevost, Dumas, and even Purkinje—had failed. "I recoiled as if struck by lightening ... I had to try to relax a while before I could work up enough courage to look again, as I was afraid I had been deluded by a phantom. Is it not strange that a sight which is expected, and indeed hoped for, should be frightening when it eventually materializes?"

[†]von Baer formulated these generalizations prior to Darwin's theory of evolution. "Lower animals" would be those appearing earlier in life's history.

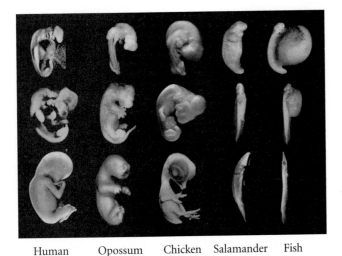

| Human | Opossum | Chicken | Salamander (axolotl) | Fish (gar) |

Figure 1.5
The similarities and differences between different vertebrate embryos as they proceed through development. They each begin with a basically similar structure, although they acquire this structure at different ages and sizes. As they develop, they become less like each other. (Adapted from Richardson et al. 1998; photograph courtesy of M. Richardson.)

Von Baer also recognized that there is a common pattern to all vertebrate development: the three germ layers give rise to different organs, and this derivation of the organs is constant whether the organism is a fish, a frog, or a chick.

> WEBSITE 1.1 **The reception of von Baer's principles.** The acceptance of von Baer's principles and their interpretation over the past hundred years has varied enormously. Recent evidence suggests that one important researcher in the 1800s even fabricated data when his own theory went against these postulates.

Fate mapping the embryo

By the late 1800s, the cell had been conclusively demonstrated to be the basis for anatomy and physiology. Embryologists, too, began to base their field on the cell. One of the most important programs of descriptive embryology became the tracing of **cell lineages**: following individual cells to see what they become. In many organisms, this fine a resolution is not possible, but one can label groups of cells to see what that *area* of the embryo will become. By bringing such studies together, one can construct a **fate map**. These diagrams "map" the larval or adult structure onto the region of the embryo from which it arose. Fate maps are the bases for experimental embryology, since they provide researchers with information on which portions of the embryo normally become which larval or adult structures. Fate maps of some embryos at the early gastrula stage are shown in Figure 1.6. Fate maps have been generated in several ways.

OBSERVING LIVING EMBRYOS. In certain invertebrates, the embryos are transparent, have relatively few cells, and the daughter cells remain close to one another. In such cases, it is actually possible to look through the microscope and trace the descendants of a particular cell into the organs they generate. This type of study was performed about a century ago by Edwin G. Conklin. In one of these studies, he took eggs of the tunicate *Styela partita*, a sea squirt that resides in the waters off the coast of Massachusetts, and he patiently followed the fates of each cell in the embryo until they differentiated into particular structures (Figure 1.7; Conklin 1905). He was helped in this endeavor by the peculiarity of the *Styela* egg, wherein the different cells contain different pigments. For example, the muscle-forming cells always had a yellow color. Conklin's fate map was confirmed by cell removal experiments. Removal of the B4.1 cell (which should produce all the tail musculature), for example, resulted in the absence of tail muscles (Reverberi and Minganti 1946).

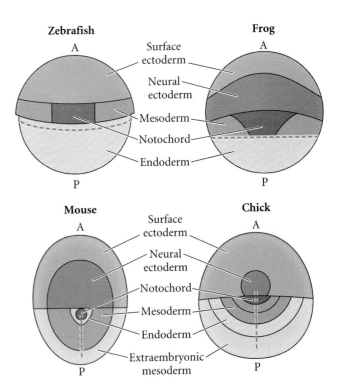

Figure 1.6
Fate maps of different vertebrate classes at the early gastrula stage. All views are dorsal surface views (looking "down" on the embryo at what will be its back). Despite the different appearances of these adult animals, their fate maps show numerous similarities among the embryos. The cells that will form the notochord occupy a central dorsal position, while the precursors of the neural system lie immediately anterior to it. The neural ectoderm is surrounded by less dorsal ectoderm, which will form the skin. A indicates the anterior end of the embryo, P the posterior end. The dashed green lines indicate the site of ingression—the path cells will follow as they migrate from the exterior to the interior of the embryo.

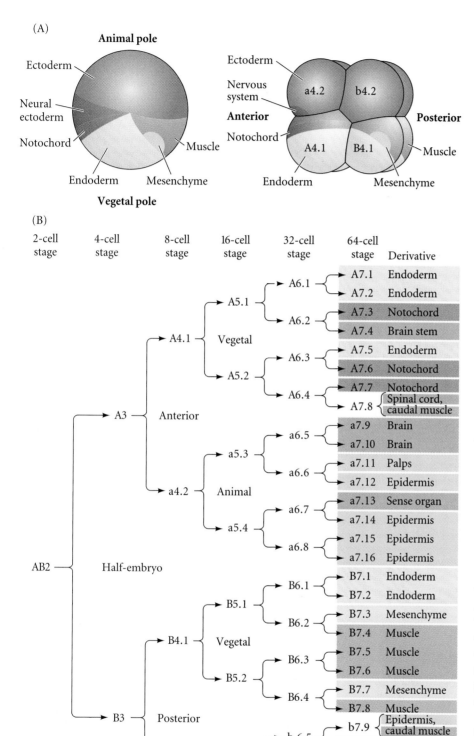

(A)

Animal pole

Ectoderm

Neural ectoderm

Notochord

Endoderm   Mesenchyme   Muscle

Vegetal pole

Ectoderm

Nervous system

Anterior   a4.2   b4.2   **Posterior**

Notochord

A4.1   B4.1

Endoderm   Mesenchyme   Muscle

(B)

2-cell stage	4-cell stage	8-cell stage	16-cell stage	32-cell stage	64-cell stage	Derivative
					A7.1	Endoderm
			A5.1	A6.1	A7.2	Endoderm
		A4.1 Vegetal		A6.2	A7.3	Notochord
					A7.4	Brain stem
	A3 Anterior		A5.2	A6.3	A7.5	Endoderm
					A7.6	Notochord
				A6.4	A7.7	Notochord
					A7.8	Spinal cord, caudal muscle
		a4.2 Animal	a5.3	a6.5	a7.9	Brain
					a7.10	Brain
				a6.6	a7.11	Palps
					a7.12	Epidermis
			a5.4	a6.7	a7.13	Sense organ
AB2 Half-embryo					a7.14	Epidermis
				a6.8	a7.15	Epidermis
					a7.16	Epidermis
			B5.1	B6.1	B7.1	Endoderm
		B4.1 Vegetal			B7.2	Endoderm
				B6.2	B7.3	Mesenchyme
					B7.4	Muscle
			B5.2	B6.3	B7.5	Muscle
					B7.6	Muscle
	B3 Posterior			B6.4	B7.7	Mesenchyme
					B7.8	Muscle
		b4.2 Animal	b5.3	b 6.5	b7.9	Epidermis, caudal muscle
					b7.10	Epidermis
				b6.6	b7.11	Epidermis
					b7.12	Epidermis
			b5.4	b6.7	b7.13	Epidermis
					b7.14	Epidermis
				b6.8	b7.15	Epidermis
					b7.16	Epidermis

Figure 1.7
Fate map of the tunicate embryo. (A) The 1-cell embryo (left), shown shortly before the first cell division, with the fate of the cytoplasmic regions indicated. The 8-cell embryo on the right shows these regions after three cell divisions. (B) A linear version of the fate map, showing the fates of each cell of the embryo. (A after Nishida 1987 and Reverberi and Minganti 1946; B after Conklin 1905 and Nishida 1987.)

WEBSITE **1.2 Conklin's art and science.** The plates from Conklin's remarkable 1905 paper are online. Looking at them, one can see the precision of his observations and how he constructed his fate map of the tunicate embryo.

VADE MECUM **The compound microscope.** The compound microscope has been the critical tool of developmental anatomists. Mastery of microscopic techniques allows one to enter an entire world of form and pattern. **[Click on Microscope]**

VITAL DYE MARKING. Most embryos are not so accommodating as to have cells of different colors. Nor do all embryos have as few cells as tunicates. In the early years of the twentieth century, Vogt (1929) traced the fates of different areas of amphibian eggs by applying vital dyes to the region of interest. Vital dyes will stain cells but not kill them. He mixed the dye with agar and spread the agar on a microscope slide to dry. The ends of the dyed agar would be very thin. He cut chips from these ends and placed them onto a frog embryo. After the dye stained the cells, the agar chip was removed, and cell movements within the embryo could be followed (Figure 1.8).

RADIOACTIVE LABELING AND FLUORESCENT DYES. A variant of the dye marking technique is to make one area of the embryo radioactive. To do this, a donor embryo is usually grown in a solution containing radioactive thymidine. This base will be incorporated into the DNA of the dividing embryo. A second embryo (the host embryo) is grown under normal conditions. The region of interest is cut

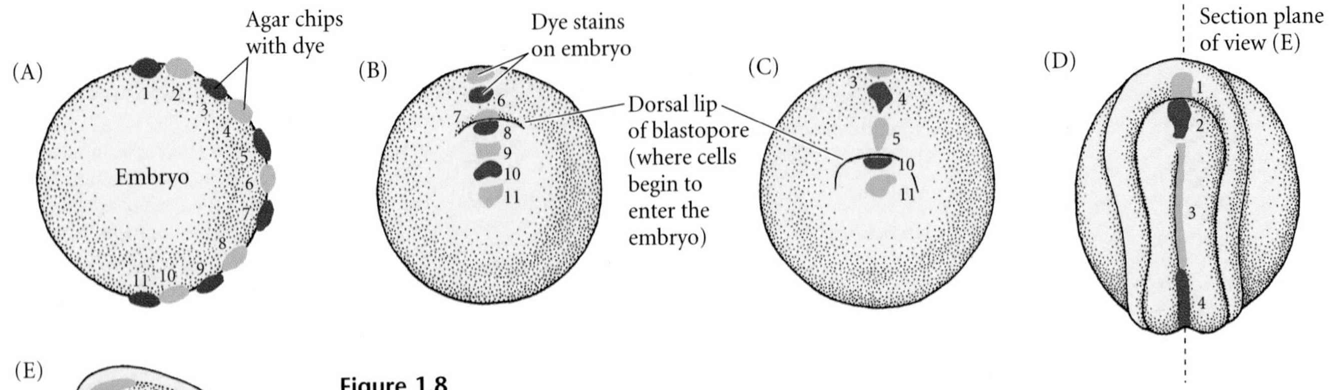

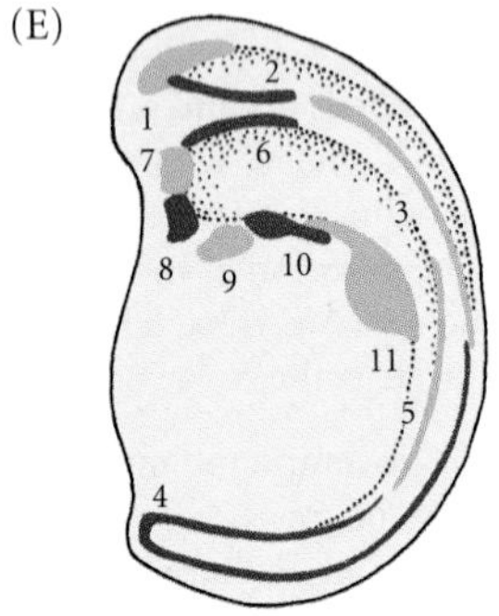

(A) Agar chips with dye

(B) Dye stains on embryo

Dorsal lip of blastopore (where cells begin to enter the embryo)

(C)

(D) Section plane of view (E)

Embryo

(E)

Figure 1.8
Vital dye staining of amphibian embryos. (A) Vogt's method for marking specific cells of the embryonic surface with vital dyes. (B–D) Dorsal surface views of stain on successively later embryos. (E) Newt embryo dissected in a medial sagittal section to show the stained cells in the interior. (After Vogt 1929.)

out from the host embryo and replaced by a radioactive graft from the donor embryo. After some time, the host embryo is sectioned for microscopy. The cells that are radioactive will be the descendants of the cells of the graft, and can be distinguished by **autoradiography**. Fixed microscope slides containing the sectioned tissues are dipped into photographic emulsion. The high-energy electrons from the radioactive thymidine will reduce the silver ions in the emulsion (just as light would). The result is a cluster of dark silver grains di-

rectly above the radioactive region. In this manner, the fates of different regions of the chick embryo have been determined (Rosenquist 1966).

One of the problems with vital dyes and radioactive labels is that they become diluted at each cell division. One way around this problem was the creation of fluorescent dyes that were extremely powerful and could be injected into individual cells. Fluorescein-conjugated dextran, for example, could be injected into a single cell of an early embryo. The descendants of that cell could then be seen by examining the embryo under ultraviolet light (Figure 1.9). More recently, diI, a powerfully fluorescent molecule that becomes incorporated into lipid membranes, has also been used to follow the fates of cells and their progeny.

GENETIC MARKING. The problems with radioactive and vital dye marking include their dilution over many cell divisions

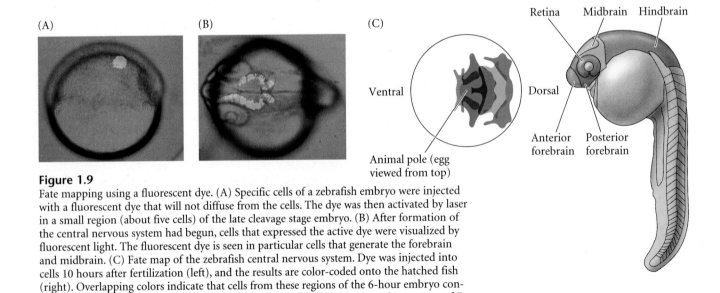

Figure 1.9
Fate mapping using a fluorescent dye. (A) Specific cells of a zebrafish embryo were injected with a fluorescent dye that will not diffuse from the cells. The dye was then activated by laser in a small region (about five cells) of the late cleavage stage embryo. (B) After formation of the central nervous system had begun, cells that expressed the active dye were visualized by fluorescent light. The fluorescent dye is seen in particular cells that generate the forebrain and midbrain. (C) Fate map of the zebrafish central nervous system. Dye was injected into cells 10 hours after fertilization (left), and the results are color-coded onto the hatched fish (right). Overlapping colors indicate that cells from these regions of the 6-hour embryo contribute to two or more regions. (A, B from Kozlowski et al. 1998; photographs courtesy of E. Weinberg. C after Woo and Fraser 1995.)

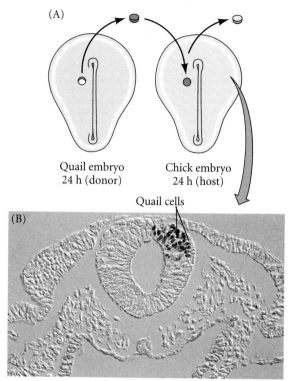

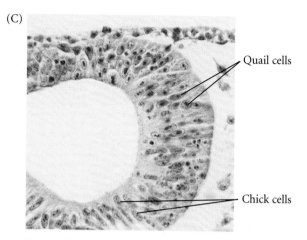

(A)

Quail embryo
24 h (donor)

Chick embryo
24 h (host)

Quail cells

(B)

Chick embryo with region of quail cells on the neural tube

(C)

Quail cells

Chick cells

Figure 1.10
Genetic markers as cell lineage tracers. (A) Grafting experiment wherein the cells from a particular region of a 1-day quail embryo have been placed into a similar region of a 1-day chick embryo. (B) After several days, the quail cells can be seen by using an antibody to quail-specific proteins. This region of the 3-day embryo produces cells that populate the neural tube. (C) Chick and quail cells can also be distinguished by the heterochromatin of their nuclei. Results of an experiment similar to (A) but using the heterochromatin of the nuclei as markers. The quail cells have a single large nucleolus (staining purple), distinguishing them from the diffuse nuclei of the chick. (After Darnell and Schoenwolf 1997; photographs courtesy of the authors.)

and the laborious preparation of the slides. One permanent way of marking cells is to create mosaic embryos having different genetic constitutions. One of the best examples of this technique is the construction of chimeric embryos, consisting, for example, of a graft of quail cells inside a chick embryo. Chick and quail develop in a very similar manner (especially during early embryonic development), and a graft of quail cells will become integrated into a chick embryo and participate in the construction of the various organs. The substitution of quail cells for chick cells can be performed on an embryo while it is still inside the egg, and the chick that hatches will have quail cells in particular sites, depending upon where the graft was placed. The quail cells differ from the chick's in two important ways. First, the quail heterochromatin in the nucleus is concentrated around the nucleoli, making the quail nucleus easily distinguishable from chick nuclei. Second, there are cell-specific antigens that are quail-specific and can be used to find individual quail cells, even if they are in a large population of chick cells. In this way, fine-structure maps of the chick brain and skeletal system can be made (Figure 1.10; Le Douarin 1969; Le Douarin and Teillet 1973).

VADE MECUM **Histotechniques.** Most cells must be stained in order to see them; different dyes stain different types of molecules. Instructions on staining cells to observe particular structures (such as the nucleus) are given here. **[Click on Histotechniques]**

Cell migration

One of the most important contributions of fate maps has been their demonstration of extensive cell migration during development. Mary Rawles (1940) showed that the pigment cells (**melanocytes**) of the chick originate in the **neural crest**, a transient band of cells that joins the neural tube to the epidermis. When she transplanted small regions of neural crest-containing tissue from a pigmented strain of chickens into a similar position in an embryo from an unpigmented strain of chickens, the migrating pigment cells entered the epidermis and later entered the feathers (Figure 1.11A). Ris (1941) used similar techniques to show that while almost all of the external pigment of the chick embryo came from the migrating neural crest cells, the pigment of the retina formed in the retina itself and was not dependent on the migrating neural crest cells. By using radioactive marking techniques, Weston (1963) demonstrated that the migrating neural crest cells gave rise to the melanocytes, and also to the peripheral neurons and the epinephrine-secreting adrenal medulla (Figure 1.11B,C). This pattern was confirmed in chick-quail hybrids, in which the quail neural crest cells produced their own pigment and pattern in the chick feathers. More recently, fluorescent dye labeling has followed the movements of individual neural crest cells as they form their pigment, adrenal, and neuronal lineages (see Chapter 13).

(A)   (B)   (C)

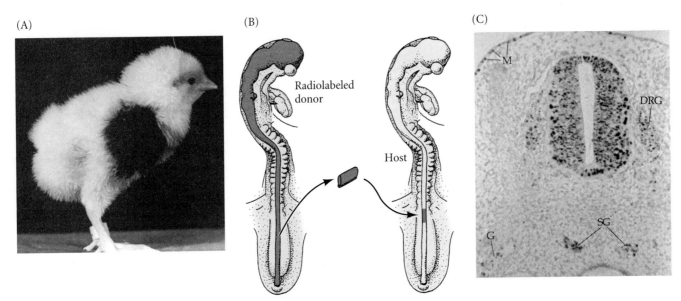

Figure 1.11
Neural crest cell migration. (A) Chick resulting from the transplantation of a trunk neural crest region from an embryo of a pigmented strain of chickens into the same region of an embryo of an unpigmented strain. The neural crest cells that gave rise to the pigment migrated into the wing epidermis and feathers. (B) Technique for following neural crest cells using radioactive tissue. (C) Autoradiograph showing locations of neural crest cells that have migrated from the radioactive donor cells. These cells form melanoblasts (M), sympathetic neural ganglia (SG), dorsal root ganglia (DRG), and glial cells (G). (A, original photograph from the archives of B. H. Willier; B after Weston 1963; C courtesy of J. Weston.)

In addition to the travels of pigment cells, other wide-scale migrations include those of the primordial germ cells (which migrate from the yolky cells to the gonads and form the sperm and eggs) and the blood cell precursors (which undergo several migrations to colonize the liver and bone marrow).

Evolutionary Embryology

Charles Darwin's theory of evolution restructured comparative embryology and gave it a new focus. After reading Johannes Müller's summary of von Baer's laws in 1842, Darwin saw that embryonic resemblances would be a very strong argument in favor of the genetic connectedness of different animal groups. "Community of embryonic structure reveals community of descent," he would conclude in *On the Origin of Species* in 1859.

Larval forms had been used for taxonomic classification even before Darwin. J. V. Thompson, for instance, had demonstrated that larval barnacles were almost identical to larval crabs, and he therefore counted barnacles as arthropods, not molluscs (Figure 1.12; Winsor 1969). Darwin, an expert on barnacle taxonomy, celebrated this finding: "Even the illustrious Cuvier did not perceive that a barnacle is a crustacean, but a glance at the larva shows this in an unmistakable manner." Darwin's evolutionary interpretation of von Baer's

laws established a paradigm that was to be followed for many decades, namely, that relationships between groups can be discovered by finding common embryonic or larval forms. Kowalevsky (1871) would soon make a similar type of discovery (publicized in Darwin's *Descent of Man*) that tunicate larvae have notochords and form their neural tubes and other

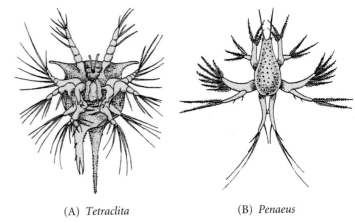

(A) *Tetraclita*   (B) *Penaeus*

Figure 1.12
Nauplius larvae of (A) a barnacle (*Tetraclita*, seen in ventral view) and (B) a shrimp (*Penaeus*, seen in dorsal view). The shrimp and barnacle share a similar larval stage despite their radical divergence in later development. (After Müller 1864.)

organs in a manner very similar to that of the primitive chordate *Amphioxus*. The tunicates, another enigma of classification schemes (formerly placed, along with barnacles, among the molluscs), thereby found a home with the chordates.

Darwin also noted that embryonic organisms sometimes make structures that are inappropriate for their adult form but that show their relatedness to other animals. He pointed out the existence of eyes in embryonic moles, pelvic rudiments in embryonic snakes, and teeth in embryonic baleen whales.

Darwin also argued that adaptations that depart from the "type" and allow an organism to survive in its particular environment develop late in the embryo.* He noted that the differences between species within genera become greater as development persists, as predicted by von Baer's laws. Thus, Darwin recognized two ways of looking at "descent with modification." One could emphasize the common descent by pointing out embryonic similarities between two or more groups of animals, or one could emphasize the modifications by showing how development was altered to produce structures that enabled animals to adapt to particular conditions.

Embryonic homologies

One of the most important distinctions made by the evolutionary embryologists was the difference between analogy and homology. Both terms refer to structures that appear to be similar. **Homologous** structures are those organs whose underlying similarity arises from their being derived from a common ancestral structure. For example, the wing of a bird and the forelimb of a human are homologous. Moreover, their respective parts are homologous (Figure 1.13). **Analogous** structures are those whose similarity comes from their performing a similar function, rather than their arising from a common ancestor. Therefore, for example, the wing of a butterfly and the wing of a bird are analogous. The two types of wings share a common function (and therefore are both called wings), but the bird wing and insect wing did not arise from an original ancestral structure that became modified through evolution into bird wings and butterfly wings.

Homologies must be made carefully and must always refer to the level of organization being compared. For instance, the bird wing and the bat wing are homologous as forelimbs, but not as wings. In other words, they share a common underlying structure of forelimb bones because birds and mammals share a common ancestry. However, the bird wing developed independently from the bat wing. Bats descended from a long line of nonwinged mammals, and the structure of the bat wing is markedly different from that of a bird wing.

*Moreover, as first noted by Weismann (1875), larvae must have their own adaptations to help them survive. The adult viceroy butterfly mimics the monarch butterfly, but the viceroy caterpillar does not resemble the beautiful larva of the monarch. Rather, the viceroy larva escapes detection by resembling bird droppings (Begon et al. 1986).

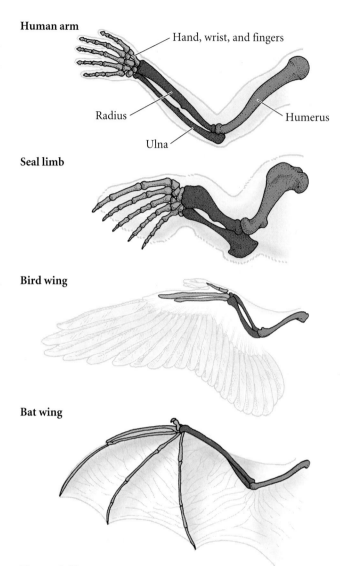

Human arm
Hand, wrist, and fingers
Radius
Humerus
Ulna

Seal limb

Bird wing

Bat wing

Figure 1.13
Homologies of structure among a human arm, a seal forelimb, a bird wing, and a bat wing; homologous supporting structures are shown in the same color. All four are homologous as forelimbs and were derived from a common tetrapod ancestor. The adaptations of bird and bat forelimbs to flight, however, evolved independently of each other, after the two lineages diverged from their common ancestor. Therefore, as wings they are not homologous, but analogous.

One of the most celebrated cases of embryonic homology is that of the fish gill cartilage, the reptilian jaw, and the mammalian middle ear (reviewed in Gould 1990). First, the gill arches of jawless (agnathan) fishes became modified to form the jaw of the jawed fishes. In the jawless fishes, a series of gills opened behind the jawless mouth. When the gill slits became supported by cartilaginous elements, the first set of these gill supports surrounded the mouth to form the jaw. There is ample evidence that jaws are modified gill supports. First, both these sets of bones are made from neural crest cells. (Most other bones come from mesodermal tissue.) Second,

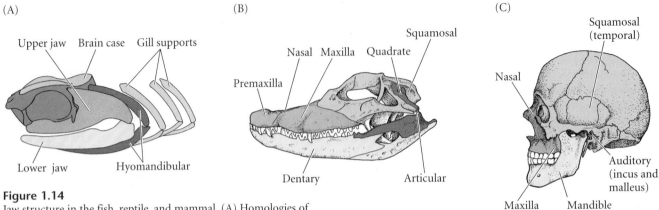

Figure 1.14
Jaw structure in the fish, reptile, and mammal. (A) Homologies of the jaws and gill arches as seen in the skull of the Paleozoic shark *Cobeledus aculentes*. (B) Lateral view of an alligator skull. The articular portion of the lower jaw articulates with the quadrate bone of the skull. (C) Lateral view of a human skull, showing the junction of the lower jaw with the squamosal (temporal) region of the skull. In mammals, the quadrate becomes internalized to form the incus of the middle ear. The articular bone retains its contact with the quadrate, becoming the malleus of the middle ear. (A after Zangerl and Williams 1975.)

both structures form from upper and lower bars that bend forward and are hinged in the middle. Third, the jaw musculature seems to be homologous to the original gill support musculature. Thus, the vertebrate jaw appears to be homologous to the gill arches of jawless fishes.

But the story does not end here. The upper portion of the second embryonic arch supporting the gill became the hyomandibular bone of jawed fishes. This element supports the skull and links the jaw to the cranium (Figure 1.14A). As vertebrates came up onto land, they had a new problem: how to hear in a medium as thin as air. The hyomandibular bone happens to be near the otic (ear) capsule, and bony material is excellent for transmitting sound. Thus, while still functioning as a cranial brace, the hyomandibular bone of the first amphibians also began functioning as a sound transducer (Clack 1989). As the terrestrial vertebrates altered their locomotion, jaw structure, and posture, the cranium became firmly attached to the rest of the skull and did not need the hyomandibular brace. The hyomandibular bone then seems to have become specialized into the stapes bone of the middle ear. What had been this bone's secondary function became its primary function.

The original jaw bones changed also. The first embryonic arch generates the jaw apparatus. In amphibians, reptiles, and birds, the posterior portion of this cartilage forms the quadrate bone of the upper jaw and the articular bone of the lower jaw. These bones connect to each other and are responsible for articulating the upper and lower jaws. However, in mammals, this articulation occurs at another region (the dentary and squamosal bones), thereby "freeing" these bony ele-

ments to acquire new functions. The quadrate bone of the reptilian upper jaw evolved into the mammalian incus bone of the middle ear, and the articular bone of the reptile's lower jaw has become our malleus. This latter process was first described by Reichert in 1837, when he observed in the pig embryo that the mandible (jawbone) ossifies on the side of Meckel's cartilage, while the posterior region of Meckel's cartilage ossifies, detaches from the rest of the cartilage, and enters the region of the middle ear to become the malleus (Figure 1.14B,C). Thus, the middle ear bones of the mammal are homologous to the posterior lower jaw of the reptile and to the gill arches of agnathan fishes. Chapter 22 will detail more recent information concerning the relationship of development to evolution.

Medical Embryology and Teratology

While embryologists could look at embryos to describe the evolution of life and how different animals form their organs, physicians became interested in embryos for more practical reasons. About 2% of human infants are born with a readily observable anatomical abnormality (Thorogood 1997). These abnormalities may include missing limbs, missing or extra digits, cleft palate, eyes that lack certain parts, hearts that lack valves, and so forth. Physicians need know the causes of these birth defects in order to counsel parents as to the risk of having another malformed infant. In addition, the different birth defects can tell us how the human body is normally formed. In the absence of experimental data on human embryos, we often must rely on nature's "experiments" to learn how the human body becomes organized.* Some birth defects are pro-

---

*The word "monster," used frequently in textbooks prior to the mid-twentieth century to describe malformed infants, comes from the Latin *monstrare*, "to show or point out." This is also the root of our word "demonstrate." It was realized by Meckel (of jaw cartilage fame) that syndromes of congenital anomalies *demonstrated* certain principles about normal development. Parts of the body that were affected together must have some common developmental origin or mechanism that was being affected.

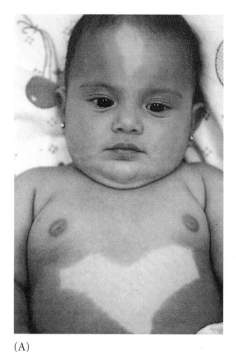

(A)

(B)

Figure 1.15
Developmental anomalies caused by genetic mutation. (A) Piebaldism in a human infant. This genetically produced condition results in sterility, anemia and underpigmented regions of the skin and hair, along with defective development of gut neurons and the ear. Piebaldism is caused by a mutation in the *KIT* gene. The Kit protein is essential for the proliferation and migration of neural crest cells, germ cell precursors, and blood cell precursors. (B) A piebald mouse with a mutation of the *Kit* gene. Mice provide important models for studying human developmental diseases. (Photographs courtesy of R. A. Fleischman.)

Developmental biologists and clinical geneticists often study human syndromes (and determine their causes) by studying animals that display the same syndrome. These are called **animal models** of the disease; the mouse model for piebaldism is shown in Figure 1.15B. It has a phenotype very similar to that of the human condition, and it is caused by a mutation in the *Kit* gene of the mouse.*

Abnormalities due to exogenous agents (certain chemicals or viruses, radiation, or hyperthermia) are called **disruptions**. The agents responsible for these disruptions are called **teratogens** (Greek, "monster-formers"), and the study of how environmental agents disrupt normal development is called **teratology**. In 1961, Lenz and McBride independently accumulated evidence that thalidomide, prescribed as a mild sedative to many pregnant women, caused an enormous increase in a previously rare syndrome of congenital anomalies. The most noticeable of these anomalies was phocomelia, a condition in which the long bones of the limbs are deficient or absent (Figure 1.16A). Over 7000 affected infants were born to women who took this drug, and a woman need only have taken one tablet to produce children with all four limbs deformed (Lenz 1962, 1966; Toms 1962). Other abnormalities induced by the ingestion of thalidomide included heart defects, absence of the external ears, and malformed intestines.

Nowack (1965) documented the period of susceptibility during which thalidomide caused these abnormalities. The drug was found to be teratogenic only during days 34–50 after the last menstruation (about 20 to 36 days postconception). The specificity of thalidomide action is shown in Figure 1.16B. From day 34 to day 38, no limb abnormalities are seen. During this period, thalidomide can cause the absence or deficiency of ear components. Malformations of upper limbs are seen before those of the lower limbs, since the arms form slightly before the legs during development. The only animal

duced by mutant genes or chromosomes, and some are produced by environmental factors that impede development.

Abnormalities caused by genetic events (gene mutations, chromosomal aneuploidies and translocations) are called **malformations**. Malformations often appear as **syndromes** (from the Greek, "running together"), where several abnormalities are seen concurrently. For instance, a human malformation called piebaldism, shown in Figure 1.15A, is due to a dominant mutation in a gene (*KIT*) on the long arm of chromosome 4 (Halleban and Moellmann 1993). The syndrome includes anemia, sterility, unpigmented regions of the skin and hair, deafness, and the absence of the nerves that cause peristalsis in the gut. The common feature underlying these conditions is that the *KIT* gene encodes a protein that is expressed in the neural crest cells and in the precursors of blood cells and germ cells. The Kit protein enables these cells to proliferate. Without this protein, the neural crest cells—which generate the pigment cells, certain ear cells, and the gut neurons—do not multiply as much as they should (resulting in underpigmentation, deafness, and gut malformations), nor do the precursors of the blood cells (resulting in anemia) or the germ cells (resulting in sterility).

*The mouse *Kit* and human *KIT* genes are considered homologous by their structural similarities and their presumed common ancestry. Human genes are usually italicized and written in all capitals. Mouse genes are italicized, but only the first letter is usually capitalized. Gene products—proteins—are not italicized. If the protein has no standard biochemical or physiological name, it is usually represented with the name of the gene in roman type, with the first letter capitalized. These rules are frequently bent, however. One is reminded of Cohen's (1982) dictum that "Academicians are more likely to share each other's toothbrush than each other's nomenclature."

(A)

(B)

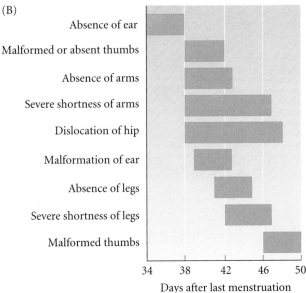

Figure 1.16
Developmental anomalies caused by an environmental agent. (A) Phocomelia, the lack of proper limb development, was the most visible of the birth defects that occurred in many children whose mothers took the drug thalidomide during pregnancy. (B) Thalidomide disrupts different structures at different times of human development. (Photograph © Deutsche Presse/Archive Photos; B after Nowack 1965.)

models for thalidomide, however, are primates, and we still do not know the mechanisms by which thalidomide causes human developmental disruptions. Thalidomide was withdrawn from the market in November 1961, but it is beginning to be prescribed again, this time as a potential anti-tumor and anti-autoimmunity drug (Raje and Anderson 1999).

The integration of anatomical information about congenital malformations with our new knowledge concerning the genes responsible for development has had a revolutionary effect and is currently restructuring medicine. This integration is allowing us to discover the genes responsible for inherited malformations, and it permits us to identify the steps in development being disrupted by teratogens. We will see examples of this integration throughout this text, and Chapter 21 will detail some of the remarkable new discoveries in teratology.

Mathematical Modeling of Development

Developmental biology has been described as the last refuge of the mathematically incompetent scientist. This phenomenon, however, is not going to last. While most embryologists have been content trying to analyze specific instances of development or even formulating some general principles of embryology, some researchers are now seeking the *laws* of development. The goal of these investigators is to base embryology on formal mathematical or physical principles (see Held 1992; Webster and Goodwin 1996). Pattern formation and growth are two areas in which such mathematical modeling

has given biologists interesting insights into some underlying laws of animal development.

The mathematics of organismal growth

Most animals grow by increasing their volume while retaining their proportions. Theoretically, an animal that increases its weight (volume) twofold will increase its length only 1.26 times (as $1.26^3 = 2$). W. K. Brooks (1886) observed that this ratio was frequently seen in nature, and he noted that the deep-sea arthropods collected by the *Challenger* expedition increased about 1.25 times between molts. In 1904, Przibram and his colleagues performed a detailed study of mantises and found that the increase of size between molts was almost exactly 1.26 (see Przibram 1931). Even the hexagonal facets of the arthropod eye (which grow by cell expansion, not by cell division) increased by that ratio.

D'Arcy Thompson (1942) similarly showed that the spiral growth of shells (and fingernails) can be expressed mathematically ($r = a^\theta$), and that the ratio of the widths between two whorls of a shell can be calculated by the formula $r = e^{2\pi\cot\theta}$ (Figure 1.17; Table 1.1). Thus, if a whorl were 1 inch in breadth at one point on a radius and the angle of the spiral were 80°, the next whorl would have a width of 3 inches on the same radius. Most gastropod (snail) and nautiloid molluscs have an angle of curvature between 80° and 85°.* Lower-angle curvatures are seen in some shells (mostly bivalves) and are common in teeth and claws.

Such growth, in which the shape is preserved because all components grow at the same rate, is called **isometric**

---

*If the angle were 90°, the shell would form a circle rather than a spiral, and growth would cease. If the angle were 60°, however, the next whorl would be 4 feet on that radius, and if the angle were 17°, the next whorl would occupy a distance of some 15,000 miles!

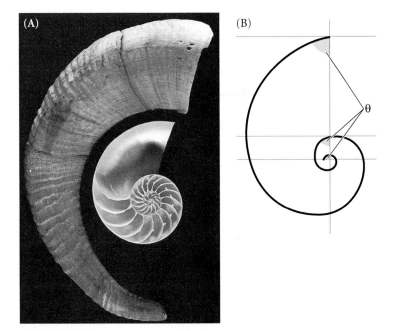

Table 1.1 Constant angle of an equiangular spiral and the ratio of widths between whorls

Constant angle	Ratio of widths[a]
90°	1.0
89°8′	1.1
86°18′	1.5
83°42′	2.0
80°5′	3.0
75°38′	5.0
69°53′	10.0
64°31′	20.0
58°5′	50.0
53°46′	10^2
42°17′	10^3
34°19′	10^4
28°37′	10^5
24°28′	10^6

Source: From Thompson 1942.

[a]The ratio of widths is calculated by dividing the width of one whorl by the width of the next larger whorl.

Figure 1.17
Equiangular spiral growth patterns. (A) A ram's horn and the shell of a chambered nautilus both show equiangular spiral growth. The nautilus shell (below) is cut in cross section. (B) René Descartes' analysis of an equiangular spiral, showing that if the curve cuts each radius vector at a constant angle (symbolized θ), then the curve grows continuously without ever changing its shape. (B after Thompson 1942.)

growth. In many organisms, growth is not a uniform phenomenon. It is obvious that there are some periods in an organism's life during which growth is more rapid than in others. Physical growth during the first 10 years of person's existence is much more dramatic than in the 10 years following one's graduation from college. Moreover, not all parts of the body grow at the same rate. This phenomenon of the different growth rates of parts within the same organism is called **allometric growth** (or **allometry**). Human allometry is depicted in Figure 1.18. Our arms and legs grow at a faster rate than our torso and head, such that adult proportions differ markedly from those of infants. Julian Huxley (1932) likened

Figure 1.18
Allometry in humans. The embryo's head is exceedingly large in proportion to the rest of the body. After the embryonic period, the head grows more slowly than the torso, hands, and legs. Human allometry has been represented in Western art only since the Renaissance. Before that, children resembled little adults. (After Moore 1983.)

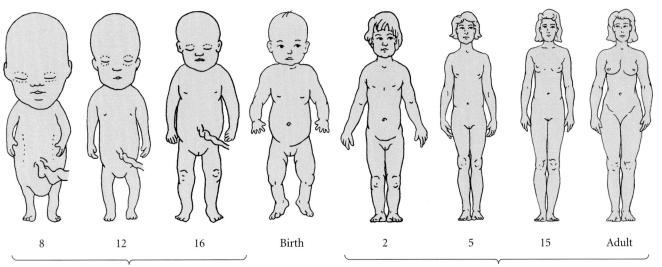

| 8 | 12 | 16 | Birth | 2 | 5 | 15 | Adult |

Weeks after fertilization   Age in years

Figure 1.19
Male specimens of the fiddler crab, *Uca pugnax*. Allometric growth occurs only in one of the male's claws. In females (not shown), both claws retain isometric growth. (Photograph courtesy of Swarthmore College Marine Biology laboratory.)

allometry to putting money in the bank at two different continuous interest rates.

The formula for allometric growth (or for comparing moneys invested at two different interest rates) is $y = bx^{a/c}$,

where a and c are the growth rates of two body parts, and b is the value of y when $x = 1$. If $a/c > 1$, then that part of the body represented by a is growing faster than that part of the body represented by c. In logarithmic terms (which are much easier to graph), $\log y = \log b + (a/c)\log x$.

One of the most vivid examples of allometric growth is seen in the male fiddler crab, *Uca pugnax*. In small males, the two claws are of equal weight, each constituting about 8% of the crab's total weight. As the crab grows larger, its chela (the large crushing claw) grows even more rapidly, eventually constituting about 38% of the crab's weight (Figure 1.19). When these data are plotted on double logarithmic plots (the body mass on the x axis, the chela mass on the y axis), one obtains a straight line whose slope is the a/c ratio. In the male *Uca pugnax* (whose name is derived from the huge claw), the a/c ratio is 6:1. This means that the mass of the chela increases six times faster than the mass of the rest of the body. In females of the species, the claw remains about 8% of the body weight throughout growth. It is only in the males (who use the claw for defense and display) that this allometry occurs.

The mathematics of patterning

One of the most important mathematical models in developmental biology has been that formulated by Alan Turing (1952), one of the founders of computer science (and the mathematician who cracked the German "Enigma" code during World War II). He proposed a model wherein two homogeneously distributed solutions would interact to produce stable patterns during morphogenesis. These patterns would represent regional differences in the concentrations of the two

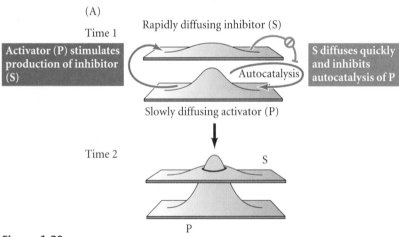

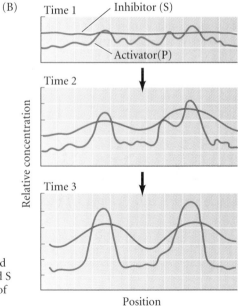

Figure 1.20
Reaction-diffusion (Turing model) system of pattern generation. Generation of periodic spatial heterogeneity can come about spontaneously when two reactants, S and P, are mixed together under the conditions that S inhibits P, P catalyzes production of both S and P, and S diffuses faster than P. (A) The conditions of the reaction-diffusion system yielding a peak of P and a lower peak of S at the same place. (B) The distribution of the reactants is initially random, and their concentrations fluctuate over a given average. As P increases locally, it produces more S, which diffuses to inhibit more peaks of P from forming in the vicinity of its production. The result is a series of P peaks ("standing waves") at regular intervals.

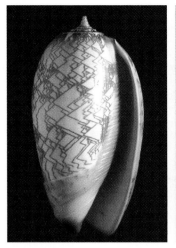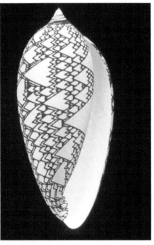

Figure 1.21
A photograph of the snail *Oliva porphyria* (left), and a computer model of the same snail (right) in which the growth parameters of the shell and its pigmentation pattern were both mathematically generated. (From Meinhardt 1998; computer image courtesy of D. Fowler, P. Prusinkiewicz, and H. Meinhardt.)

substances. Their interactions would produce an ordered structure out of random chaos.

Turing's **reaction-diffusion model** involves two substances. One of them, substance S, inhibits the production of the other, substance P. Substance P promotes the production of more substance P as well as more substance S. Turing's mathematics show that if S diffuses more readily than P, sharp waves of concentration differences will be generated

for substance P (Figure 1.20). These waves have been observed in certain chemical reactions (Prigogine and Nicolis 1967; Winfree 1974).

The reaction-diffusion model predicts alternating areas of high and low concentrations of some substance. When the concentration of such a substance is above a certain threshold level, a cell (or group of cells) may be instructed to differentiate in a certain way. An important feature of Turing's model is that particular chemical wavelengths will be amplified while all others will be suppressed. As local concentrations of P increase, the values of S form a peak centering on the P peak, but becoming broader and shallower because of S's more rapid diffusion. These S peaks inhibit other P peaks from forming. But which of the many P peaks will survive? That depends on the size and shape of the tissues in which the oscillating reaction is occurring. (This pattern is analogous to the harmonics of vibrating strings, as in a guitar. Only certain resonance vibrations are permitted, based on the boundaries of the string.)

The mathematics describing which particular wavelengths are selected consist of complex polynomial equations. Such functions have been used to model the spiral patterning of slime molds, the polar organization of the limb, and the pigment patterns of mammals, fish, and snails (Figures 1.21 and 1.22; Kondo and Asai 1995; Meinhardt 1998). A computer simulation based on a Turing reaction-diffusion system can successfully predict such patterns, given the starting shapes and sizes of the elements involved.

One way to search for the chemicals predicted by Turing's model is to find genetic mutations in which the ordered struc-

(A)

(B)

(C)

(D)

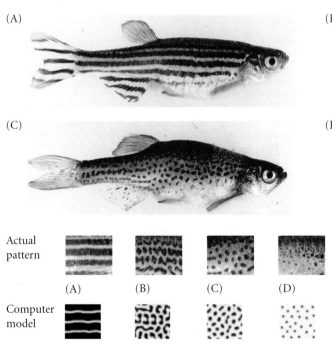

Actual pattern

(A) (B) (C) (D)

Computer model

Rate of pigment synthesis ⟶

Figure 1.22
Pigment patterns of zebrafish homozygous for the wild-type allele (A) and for three different mutant alleles (B–D) of the *leopard* gene. Computer simulations of the pigment patterns are shown in the bottom row. Changing a single parameter of the reaction-diffusion equation results in changes in the pattern. As the values of this parameter become larger, the stripes break into spots and the spots get smaller and less dense. (From Asai et al. 1999; photographs courtesy of S. Kondo.)

ture of a pattern has been altered. The wild-type alleles of these genes may be responsible for generating the normal pattern. Such a candidate is the *leopard* gene of zebrafish (Asai et al. 1999). Zebrafish usually have five parallel stripes along their flanks. However, in the different mutations, the stripes are broken into spots of different sizes and densities. Figure 1.22 shows fish homozygous for four different alleles of the *leopard* gene. If the *leopard* gene encodes an enzyme that catalyzes one of the reactions of the reaction-diffusion system, the different mutations of this gene may change the kinetics of synthesis or degradation. Indeed, all the mutant patterns (and those of their heterozygotes) can be computer-generated by changing a single parameter in the reaction-diffusion equa-

tion. The cloning of this gene should enable further cooperation between theoretical biology and developmental anatomy.

WEBSITE **1.3 The mathematical background of pattern formation.** The equations modeling pattern formation are a series of partial derivatives depicting rates of synthesis, degradation, and diffusion of the activator and inhibitor molecules.

WEBSITE **1.4 How does the zebra get its stripes?** No one knows for sure, but adding the Turing equations to what's known about equine embryology allows one to model how each of the three known zebra species acquired its unique striping pattern.

Principles of Development: Developmental Anatomy

1. Organisms must function as they form their organs. They have to use one set of structures while constructing others.

2. The main question of development is, How does the egg becomes an adult? This question can be broken down into the component problems of differentiation (How do cells become different from one another and from their precursors?), morphogenesis (How is ordered form is generated?), growth (How is size regulated?), reproduction (How does one generation create another generation?), and evolution (How do changes in developmental processes create new anatomical structures?).

3. Epigenesis happens. New organisms are created de novo each generation from the relatively disordered cytoplasm of the egg.

4. Preformation is not in the anatomical structures, but in the instructions to form them. The inheritance of the fertilized egg includes the genetic potentials of the organism.

5. The preformed nuclear instructions include the ability to respond to environmental stimuli in specific ways.

6. The ectoderm gives rise to the epidermis, nervous system, and pigment cells.

7. The mesoderm generates the kidneys, gonads, bones, heart, and blood cells.

8. The endoderm forms the lining of the digestive tube and the respiratory system.

9. Karl von Baer's principles state that the general features of a large group of animals appear earlier in the embryo than do the specialized features of a smaller group. As each embryo of a given species develops, it diverges from the adult forms of other species. The early embryo of a "higher" animal species is not like the adult of a "lower" animal.

10. Labeling cells with dyes shows that some cells differentiate where they form, while others migrate from their original sites and differentiate in their new locations. Migratory cells include neural crest cells and the precursors of germ cells and blood cells.

11. "Community of embryonic structure reveals community of descent" (Charles Darwin).

12. Homologous structures in different species are those organs whose similarity is due to their sharing a common ancestral structure. Analogous structures are those organs whose similarity comes from their serving a similar function (but which are not derived from a common ancestral structure).

13. Congenital anomalies can be caused by genetic factors (mutations, aneuploidies, translocations) or by environmental agents (certain chemicals, certain viruses, radiation).

14. Syndromes consists of sets of developmental abnormalities that "run together."

15. Organs that are linked in developmental syndromes share either a common origin or a common mechanism of formation.

16. If growth is isometric, a twofold change in weight will cause a 1.26-fold expansion in length.

17. Allometric growth can create dramatic changes in the structure of organisms.

18. Complex patterns may be self-generated by reaction-diffusion events, wherein the activator of a local phenomenon stimulates the production of more of itself as well as the production of a more diffusible inhibitor.

Literature Cited

Aristotle. ca. 400 B.C.E. *The Generation of Animals*. A. L. Peck (trans.); G. P. Goold, (ed). Harvard University Press, Cambridge, MA, 1990.

Asai, R., E. Taguchi, Y. Kume, M. Saito and S. Kondo. 1999. Zebrafish *Leopard* gene as a component of the putative reaction-diffusion system. *Mech. Dev.* 89: 87–92.

Begon, M., J. L. Harper and C. R. Townsend. 1986. *Ecology: Individuals, Populations, and Communities*. Blackwell Scientific, Oxford.

Bonnet, C. 1764. *Contemplation de la Nature*. Marc-Michel Ray, Amsterdam.

Brooks, W. K. 1886. Report in the Stomatopoda collected by H.M.S. *Challenger. Challenger Reports* 16: 1–114.

Carlson, B. M. 1981. *Patten's Foundations of Embryology*. McGraw-Hill, New York.

Cassirer, E. 1950. Developmental mechanics and the problem of cause in biology. *In* E. Cassirer (ed.), *The Problem of Knowledge*. Yale University Press, New Haven.

Churchill, A. 1991. The rise of classical descriptive embryology. *In* S. F. Gilbert (ed.), *A Conceptual History of Modern Embryology*, Plenum Press, New York, pp. 1–29.

Clack, J. A. 1989. Discovery of the earliest known tetrapod stapes. *Nature* 342: 425–427.

Cohen, M. M. Jr. 1982. *The Child with Multiple Birth Defects*. Raven, New York.

Conklin, E. G. 1905. The organization and cell lineage of the ascidian egg. *J. Acad. Nat. Sci. Phila.* 13: 1–119.

Darnell, D. K. and G. C. Schoenwolf. 1997. Modern techniques for labeling in avian and murine embryos. *In* G. P. Daston (ed.), *Molecular and Cellular Methods in Developmental Toxicology*. CRC Press, Boca Raton, FL, pp. 231–272.

Darwin, C. 1859. *On the Origin of Species by Means of Natural Selection, or the Preservation of Favoured Races in the Struggle for Life*. John Murray, London.

Darwin, C. 1874. *The Descent of Man, and Selection in Relation to Sex*. 2nd Ed. John Murray, London.

Giebel, L. B. and R. A. Spritz. 1991. Mutation of the *KIT* (mast/stem cell growth factor receptor) protooncogene in human piebaldism. *Proc. Nat. Acad. Sci. USA* 88: 8696–8699.

Gould, S. J. 1977. *Ontogeny and Phylogeny*. Belknap Press, Cambridge, MA.

Gould, S. J. 1990. An earful of jaw. *Nat. Hist.* 1990(3): 12–23.

Harvey, W. 1651. *Exercitationes de generatione animalium: quibus accedunt quaedum de partu, de membranis ac humoribus uteri et de conceptione*. London.

Held, L. I., Jr. 1992. *Models for Embryonic Periodicity*. Karger, New York.

Huxley, J. S. 1932. *Problems of Relative Growth*. Dial Press, New York.

Kölreuter, J. G. 1766. Vorläufige Nachricht von einigen das feschlecht der Planzen betreffenden Versuchen und Beobachtung, nebst Fortsetzungen, 1, 2, 3.

Kondo, S. and R. Asai. 1995. A reaction-diffusion wave on the skin of the marine angelfish *Pomacanthus*. *Nature* 376: 765–768.

Kowalevsky, A. 1871. Weitere Studien II. Die Entwicklung der einfachen Ascidien. *Arch. Micr. Anat.* 7: 101–130.

Kozlowski, D. J., T. Muramaki, R. K. Ho and E. S. Weinberg. 1998. Regional cell movement and tissue patterning in the zebrafish embryo revealed by fate mapping with caged fluorescein. *Biochem. Cell Biol.* 75: 551–562.

Le Douarin, N. M. 1969. Particularités du noyau interphasique chez la Caille japonaise (*Coturnix coturnix japonica*). Utilisation de ces particularités comme "marquage biologique" dans les recherches sur les interactions tissulaires et les migrations cellulaires au cours de l'ontogenèse. *Bull. Biol. Fr. Belg.* 103: 435–452.

Le Douarin, N. M. and M.-A. Teillet. 1973. The migration of neural crest cells to the wall of the digestive tract in avian embryo. *J. Embryol. Exp. Morphol.* 30: 31–48.

Lenoir, T. 1980. Kant, Blumenbach, and vital materialism in German biology. *Isis* 71: 77–108.

Lenz, W. 1962. Thalidomide and congenital abnormalities. *Lancet* 1: 45 (reported in a symposium in 1961.)

Lenz, W. 1966. Malformations caused by drugs in pregnancy. *Am. J. Dis. Child.* 112: 99–l06.

Lillie, F. R. 1908. *The Embryology of the Chick*. Henry Holt, New York.

Maître-Jan, A. 1722. *Observations sur la formation du poluet*. L. d'Houdry, Paris.

Malpighi, M. 1672. De Formatione Pulli in Ovo (London). Reprinted in H. B. Adelmannm, *Marcello Malpighi and the Evolution of Embryology*. Cornell University Press, Ithaca, NY, 1966.

McBride, W. G. 1961. Thalidomide and congenital abnormalities. *Lancet* 2: 1358.

Meinhardt, M. 1998. *The Algorhythmic Beauty of Sea Shells*. Springer, Berlin.

Moore, K. L. 1983. *The Developing Human*. 3rd Ed. Saunders, Philadelphia.

Müller, F. 1864. *Für Darwin*. Engelmann, Leipzig.

Nishida, H. 1987. Cell lineage analysis in ascidian embryos by intracellular injection of a tracer enzyme. III. Up to the tissue-restricted stage. *Dev. Biol.* 121: 526–541.

Nowack, E. 1965. Die sensible Phase bei der Thalidomide-Embryopathie. *Humangenetik* 1: 516–536.

Olsson, L. and J. Hanken. 1996. Cranial neural crest cell migration and chondrogenic fate in the oriental fire-bellied toad *Bombina orientalis*: Defining the ancestral pattern of head development in anuran amphibians. *J. Morphol.* 229: 105–120.

Pander, C. 1817. *Beiträge zur Entwickelungseschichte des Hünchens im Eye*. Brönner, Würzburg.

Patten, B. M. 1951. *The Early Embryo of the Chick*. 4th Ed. McGraw-Hill, New York.

Pinto-Correia, C. 1997. *The Ovary of Eve*. University of Chicago Press, Chicago.

Prigogine, I. and G. Nicolis. 1967. On symmetry-breaking instabilities in dissipative systems. *J. Chem. Phys.* 46: 3542–3550.

Przibram, H. 1931. *Connecting Laws in Animal Morphology*. University of London Press, London.

Raje, N. and K. Anderson. 1999. Thalidomide: A revival story. *New Engl. J. Med.* 341: 1606–1609.

Rawles, M. E. 1940. The pigment forming potency of early chick blastoderm. *Proc. Natl. Acad. Sci. USA* 26: 86–94.

Reichert, C. B. 1837. Entwicklungsgeschichte der Gehörknöchelchen der sogenannte Meckelsche Forsatz des Hammers. *Müller's Arch. Anat. Phys. Wissensch. Med.* 177–188.

Reverberi, G. and A. Minganti. 1946. Fenomeni di evocazione nello sviluppo dell'uovo di Ascidie. Risultati dell'indagine spermentale sull'ouvo di Ascidiella aspersa e di Ascidia malaca allo stadio di 8 blastomeri. *Pubbl. Staz. Zool. Napoli* 20: 199–252.

Richardson, M. K., J. Hanken, L. Selwood, G. M. Wright, R. J. Richards, C. Pieau and A. Raynaud. 1998. Haeckel, embryos, and evolution. *Science* 280: 983–984.

Ris, H. 1941. An experimental study of the origins of melanophores in birds. *Physiol. Zool.* 14: 48–66.

Roe, S. 1981. *Matter, Life, and Generation: Eighteenth-Century Embryology and the Haller-Wolff Debate*. Cambridge University Press, Cambridge.

Rosenquist, G. C. 1966. A radioautographic study of labeled grafts in the chick blastoderm. Development from primitive streak stages to stage 12. *Contrib. Embryol. Carnegie Inst.* 38: 71–110.

Thompson, D. W. 1942. *On Growth and Form*. Cambridge University Press, Cambridge.

Thorogood, P. 1997. The relationship between genotype and phenotype: Some basic concepts. *In* P. Thorogood (ed.), *Embryos, Genes, and Birth Defects*. Wiley, New York, pp. 1–16.

Toms, D. A. 1962. Thalidomide and congenital abnormalities. *Lancet* 2: 400.

Turing, A. M. 1952. The chemical basis of morphogenesis. *Philos. Trans. R. Soc. Lond.* [B] 237: 37–72.

Vogt, W. 1929. Gestaltungsanalyse am Amphibienkeim mit örtlicher Vitalfärbung. II. Teil Gastrulation und Mesodermbildung bei Urodelen und Anuren. *Wilhelm Roux Arch. Entwicklungsmech. Org.* 120: 384–706.

von Baer, K. E. 1828. *Entwicklungsgeschichte der Thiere: Beobachtung und Reflexion.* Bornträger, Konigsberg.

Webster, G. and B. Goodwin. 1996. *Form and Transformation: Generative and Relational Principles in Biology.* Cambridge University Press, Cambridge.

Weismann, A. 1875. Über den Saison-Dimorphismus der Schmetterlinge. *In Studien zur Descendenz-Theorie.* Engelmann, Leipzig.

Weston, J. 1963. A radiographic analysis of the migration and localization of trunk neural crest cells in the chick. *Dev. Biol.* 6: 274–310.

Winfree, A. T. 1974. Rotating chemical reactions. *Sci. Am.* 230(6): 82–95.

Winsor, M. P. 1969. Barnacle larvae in the nineteenth century: A case study in taxonomic theory. *J. Hist. Med. Allied Sci.* 24: 294–309.

Wolff, K. F. 1767. De formatione intestinorum praecipue. *Novi Commentarii Academine Scientarum Imperialis Petropolitanae* 12: 403–507.

Woo, K. and S. E. Fraser. 1995. Order and coherence in the fate map of the zebrafish embryo. *Development* 121: 2595–2609.

Zangerl, R. and M. E. Williams. 1975. New evidence on the nature of the jaw suspension in Paleozoic anacanthus sharks. *Paleontology* 18: 333–341.

<table>
<tr><td>c h a p t e r</td><td>2</td><td>Life cycles and the evolution
of developmental patterns</td></tr>
</table>

TRADITIONAL WAYS OF CLASSIFYING catalog animals according to their adult structure. But, as J. T. Bonner (1965) pointed out, this is a very artificial method, because what we consider an individual is usually just a brief slice of its life cycle. When we consider a dog, for instance, we usually picture an adult. But the dog is a "dog" from the moment of fertilization of a dog egg by a dog sperm. It remains a dog even as a senescent dying hound. Therefore, the dog is actually the entire life cycle of the animal, from fertilization through death.

The life cycle has to be adapted to its environment, which is composed of nonliving objects as well as other life cycles. Take, for example, the life cycle of *Clunio marinus*, a small fly that inhabits tidal waters along the coast of western Europe. Females of this species live only 2–3 hours as adults, and they must mate and lay their eggs within this short time. To make matters even more precarious, egg laying is confined to red algae mats that are exposed only during the lowest ebbing of the spring tide. Such low tides occur on four successive days shortly after the new and full moons (i.e., at about 15-day intervals). Therefore, the life cycle of these insects must be coordinated with the tidal rhythms as well as the daily rhythms such that the insects emerge from their pupal cases during the few days of the spring tide and at the correct hour for its ebb (Beck 1980; Neumann and Spindler 1991).

One of the major triumphs of descriptive embryology was the idea of a generalizable life cycle. Each animal, whether an earthworm, an eagle, or a beagle, passes through similar stages of development. The major stages of animal development are illustrated in Figure 2.1. The life of a new individual is initiated by the fusion of genetic material from the two gametes—the sperm and the egg. This fusion, called **fertilization**, stimulates the egg to begin development. The stages of development between fertilization and hatching are collectively called **embryogenesis**.

The Circle of Life: The Stages of Animal Development

Throughout the animal kingdom, an incredible variety of embryonic types exist, but most patterns of embryogenesis are variations on five themes:

1. Immediately following fertilization, **cleavage** occurs. Cleavage is a series of extremely rapid mitotic divisions wherein the enormous volume of zygote cytoplasm is divided into numerous smaller cells. These cells are called **blastomeres**,

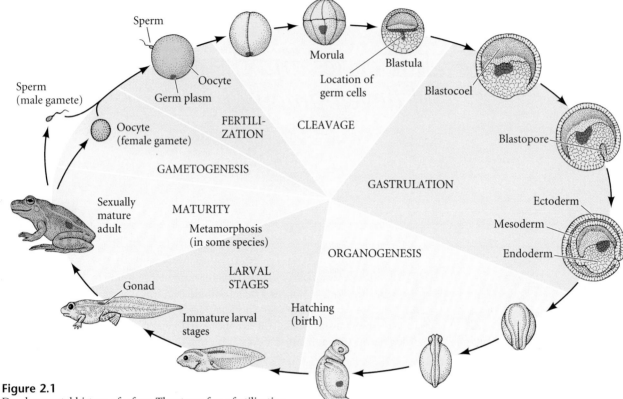

Figure 2.1
Developmental history of a frog. The stages from fertilization through hatching (birth) are known collectively as embryogenesis. The region set aside for producing germ cells is shown in bright purple. Gametogenesis, which is completed in the sexually mature adult, begins at different times during development, depending on the species. (The sizes of the varicolored wedges shown here are arbitrary and do not correspond to the proportion of the life cycle spent in each stage.)

and by the end of cleavage, they generally form a sphere known as a **blastula**.

2. After the rate of mitotic division has slowed down, the blastomeres undergo dramatic movements wherein they change their positions relative to one another. This series of extensive cell rearrangements is called **gastrulation,** and the embryo is said to be in the **gastrula** stage. As a result of gastrulation, the embryo contains three **germ layers**: the ectoderm, the endoderm, and the mesoderm.

3. Once the three germ layers are established, the cells interact with one another and rearrange themselves to produce tissues and organs. This process is called **organogenesis**. Many organs contain cells from more than one germ layer, and it is not unusual for the outside of an organ to be derived from one layer and the inside from another. For example, the outer layer of skin comes from the ectoderm, while the inner layer (the dermis) comes from the mesoderm. Also during organogenesis, certain

cells undergo long migrations from their place of origin to their final location. These migrating cells include the precursors of blood cells, lymph cells, pigment cells, and gametes. Most of the bones of our face are derived from cells that have migrated ventrally from the dorsal region of the head.

4. As seen in Figure 2.1, in many species a specialized portion of egg cytoplasm gives rise to cells that are the precursors of the **gametes** (the sperm and egg). The gametes and their precursor cells are collectively called **germ cells**, and they are set aside for reproductive function. All the other cells of the body are called **somatic cells**. This separation of somatic cells (which give rise to the individual body) and germ cells (which contribute to the formation of a new generation) is often one of the first differentiations to occur during animal development. The germ cells eventually migrate to the gonads, where they differentiate into gametes. The development of gametes, called **gametogenesis**, is usually not completed until the organism has become physically mature. At maturity, the gametes may be released and participate in fertilization to begin a new embryo. The adult organism eventually undergoes senescence and dies.

5. In many species, the organism that hatches from the egg or is born into the world is not sexually mature. Indeed, in most animals, the young organism is a **larva** that may

look significantly different from the adult. Larvae often constitute the stage of life that is used for feeding or dispersal. In many species, the larval stage is the one that lasts the longest, and the adult is a brief stage solely for reproduction. In the silkworm moths, for instance, the adults do not have mouthparts and cannot feed. The larvae must eat enough for the adult to survive and mate. Indeed, most female moths mate as soon as they eclose from their pupa, and they fly only once—to lay their eggs. Then they die.

The Frog Life Cycle

Figure 2.1 uses the development of a frog to show a representative life cycle. Let us look at this life cycle in a bit more detail. First, in most frogs, gametogenesis and fertilization are seasonal events for this animal, because its life depends upon the plants and insects in the pond where it lives and on the temperature of the air and water. A combination of photoperiod (hours of daylight) and temperature tells the pituitary gland of the female frog that it is spring. If the frog is mature, the pituitary gland secretes hormones that stimulate the ovary to make estrogen. Estrogen is a hormone that can instruct the liver to make and secrete the yolk proteins, which are then transported through the blood into the enlarging eggs in the ovary.* The yolk is transported into the bottom portion of the egg (Figure 2.2A).

Since the bottom half of the egg usually contains the yolk, it divides more slowly (because the large yolk deposits interfere with cleavage). This portion is the **vegetal** hemisphere of the egg. Conversely, the upper half of the egg usually has less yolk and divides faster. This upper portion is called the **animal** hemisphere of the egg.†

Another ovarian hormone, progesterone, signals the egg to resume its meiotic division. This is necessary because the egg had been "frozen" in the metaphase of its first meiosis. When it has completed this first meiotic division, the egg is released from the ovary and can be fertilized. In many species, the eggs are enclosed in a jelly coat that acts to enhance their size (so they won't be as easily eaten), to protect them against bacteria, and to attract and activate sperm.

---

*As we will see in later chapters, there are numerous ways by which the synthesis of a new protein can be induced. Estrogen stimulates the production of vitellogenin protein in two ways. First, it uses **transcriptional regulation** to make new vitellogenin mRNA. Before estrogen stimulation, no vitellogenin message can be seen in the liver cells. After stimulation, there are over 50,000 vitellogenin mRNA molecules in these cells. Estrogen also uses **translational regulation** to stabilize these particular messages, increasing their half-life from 16 hours to 3 weeks. In this way, more protein can be translated from each message.

†The terms *animal* and *vegetal* reflect the movements of cells seen in some embryos (such as those of frogs). The cells derived from the upper portion of the egg are actively mobile (hence, animated), while the yolk-filled cells were seen as being immobile (hence, like plants).

Sperm also occur on a seasonal basis. The male leopard frogs make their sperm in the summer, and by the time they begin hibernation in autumn, they have all the sperm that are to be available for the following spring's breeding season. In most species of frogs, fertilization is external. The male frog grabs the female's back and fertilizes the eggs as the female frog releases them (Figure 2.2B). *Rana pipiens* usually lays around 2500 eggs, while the bullfrog, *Rana catesbiana*, can lay as many as 20,000. Some species lay their eggs in pond vegetation, and the jelly adheres to the plants and anchors the eggs (Figure 2.2C). Other species float their eggs into the center of the pond without any support.

Fertilization accomplishes several things. First, it allows the egg to complete its second meiotic division, which provides the egg with a haploid **pronucleus**. The egg pronucleus and the sperm pronucleus will meet in the egg cytoplasm to form the diploid zygotic nucleus. Second, fertilization causes the cytoplasm of the egg to move such that different parts of the cytoplasm find themselves in new locations (Figure 2.2D). Third, fertilization activates those molecules necessary to begin cell cleavage and development (Rugh 1950). The sperm and egg die quickly unless fertilization occurs.

During cleavage, the volume of the frog egg stays the same, but it is divided into tens of thousands of cells (Figure 2.2E–H). The animal hemisphere of the egg divides faster than the vegetal hemisphere does, and the cells of the vegetal hemisphere become progressively larger the more vegetal the cytoplasm. A fluid-filled cavity, the **blastocoel**, forms in the animal hemisphere (Figure 2.2H). This cavity will be important for allowing cell movements to occur during gastrulation.

Gastrulation in the frog begins at a point on the embryo surface roughly 180 degrees opposite the point of sperm entry with the formation of a dimple, called the **blastopore**. Cells migrate through the blastopore and toward the animal pole (Figure 2.3A,B). These cells become the dorsal mesoderm. The blastopore expands into a circle (Figure 2.3C), and cells migrating through this circle become the lateral and ventral mesoderm. The cells remaining on the outside become the ectoderm, and this outer layer expands vegetally to enclose the entire embryo. The large yolky cells that remain at the vegetal hemisphere (until they are encircled by the ectoderm) become the endoderm. Thus, at the end of gastrulation, the ectoderm (the precursor of the epidermis and nerves) is on the outside of the embryo, the endoderm (the precursor of the gut lining) is on the inside of the embryo, and the mesoderm (the precursor of connective tissue, blood, skeleton, gonads, and kidneys) is between them.

Organogenesis begins when the notochord—a rod of mesodermal cells in the most dorsal portion of the embryo—tells the ectodermal cells above it that they are not going to become skin. Rather, these dorsal ectoderm cells are to form a tube and become the nervous system. At this stage, the embryo is called a **neurula**. The neural precursor cells elon-

(A)

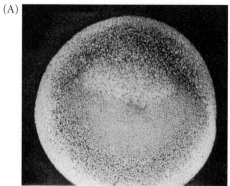

Figure 2.2
Early development of the frog *Xenopus laevis*. (A) As the egg matures, it accumulates yolk (here stained yellow and green) in the vegetal cytoplasm. (B) Frogs mate by amplexus, the male grasping the female around the belly and fertilizing the eggs as they are released. (C) A newly laid clutch of eggs. The brown area of each egg is the pigmented animal cap. The white spot in the middle of the pigment is where the egg's nucleus resides. (D) A 2-cell embryo near the end of its first cleavage. (E) An 8-cell embryo. (F) Early blastula. Note that the cells get smaller, but the volume of the egg remains the same. (G) Late blastula. (H) Cross section of a late blastula, showing the blastocoel (cavity). (A–G courtesy of Michael Danilchik and Kimberly Ray; H courtesy of J. Heasman.)

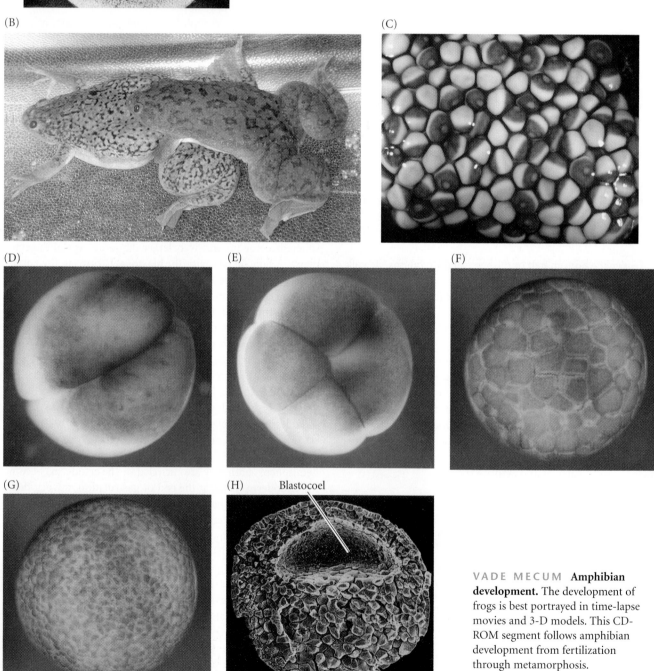

(B)

(C)

(D)

(E)

(F)

(G)

(H)   Blastocoel

VADE MECUM **Amphibian development.** The development of frogs is best portrayed in time-lapse movies and 3-D models. This CD-ROM segment follows amphibian development from fertilization through metamorphosis.
[Click on Amphibian]

Figure 2.3
Continued development of *Xenopus laevis*. (A) Gastrulation begins with an invagination, or slit, in the future dorsal side of the embryo. (B) This slit, the dorsal blastopore lip, as seen from the ventral surface (bottom) of the embryo. (C) The slit becomes a circle—the blastopore—and future mesoderm cells migrate into the interior of the embryo along the blastopore edges. The ectoderm (future skin and nerves) migrates down the outside of the embryo. The remaining part, the yolk-filled endoderm, is eventually encircled. (D) Neural folds begin to form on the dorsal surface. (E) A groove can be seen where the bottom of the neural tube will be. (F) The neural folds are coming together at the dorsal midline, creating a neural tube. (G) Cross section of the *Xenopus* embryo at the neurula stage. (H) A prehatching tadpole, as the protrusions of the forebrain begin to induce eyes to form. (I) A mature tadpole, having swum away from the egg mass and feeding independently. (Photographs courtesy of Michael Danilchik and Kimberly Ray.)

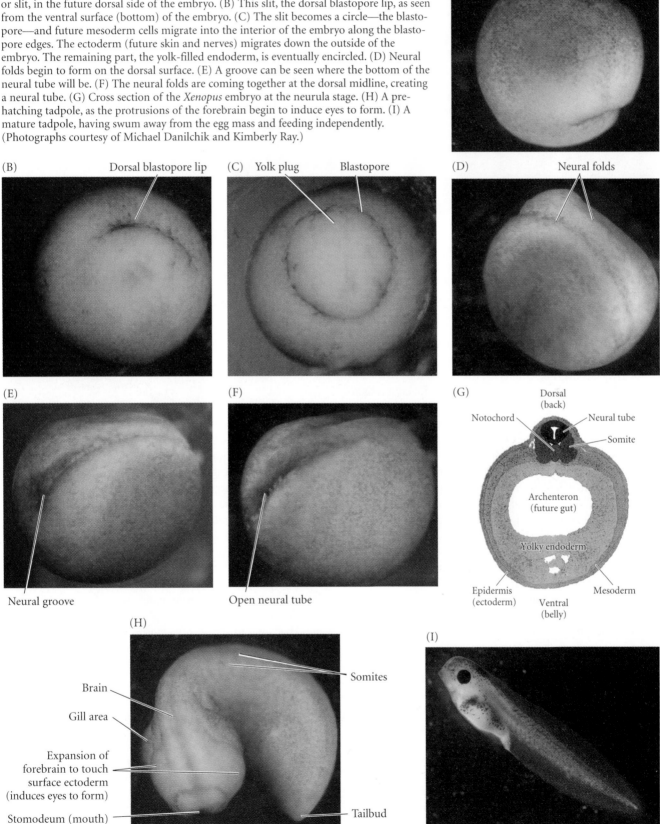

(A)

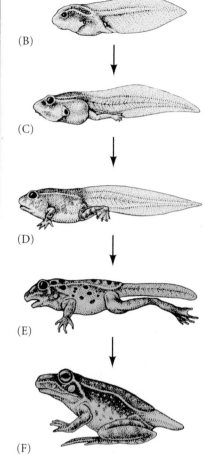

(B)

(C)

(D)

(E)

(F)

Figure 2.4
Metamorphosis of the frog *Rana*. (A) Huge changes are obvious when one contrasts the tadpole and the adult bullfrog. Note especially the differences in jaw structure and limbs. (B) Premetamorphic tadpole. (C) Prometamorphic tadpole, showing hindlimb growth. (D) Onset of metamorphic climax as forelimbs emerge. (E, F) Climax stages. (Photograph © Patrice Ceisel/Stock, Boston.)

gate, stretch, and fold into the embryo (Figure 2.3A–D), forming the **neural tube**. The future back epidermal cells cover them. The cells that had connected the neural tube to the epidermis become the **neural crest cells**. The neural crest cells are almost like a fourth germ layer. They give rise to the pigment cells of the body (the melanocytes), the peripheral neurons, and the cartilage of the face. Once the neural tube has formed, it induces changes in its neighbors, and organogenesis continues. The mesodermal tissue adjacent to the notochord becomes segmented into **somites**, the precursors of the frog's back muscles, spinal cord, and dermis (the inner portion of the skin). These somites appear as blocks of mesodermal tissue (Figure 2.3F,G). The embryo develops a mouth and an anus, and it elongates into the typical tadpole structure. The neurons make their connections to the muscles and to other neurons, the gills form, and the larva is ready to hatch from its egg jelly. The hatched tadpole will soon feed for itself once the yolk supply given it by its mother is exhausted (Figure 2.3H).

Metamorphosis of the tadpole larva into an adult frog is one of the most striking transformations in all of biology (Figure 2.4). In amphibians, metamorphosis is initiated by hormones from the tadpole's thyroid gland, and these changes prepare an aquatic organism for a terrestrial existence. (The mechanisms by which thyroid hormones accomplish these changes will be discussed in Chapter 18.) In anurans (frogs and toads), the metamorphic changes are most striking, and almost every organ is subject to modification. The changes in form are very obvious. For locomotion, the hindlimbs and forelimbs differentiate as the paddle tail recedes. The cartilaginous skull of the tadpole is replaced by the predominantly bony skull of the young frog. The horny teeth the tadpole uses to tear up

pond plants disappear as the mouth and jaw take a new shape, and the fly-catching tongue muscle of the frog develops. Meanwhile the large intestine characteristic of herbivores shortens to suit the more carnivorous diet of the adult frog. The gills regress, and the lungs enlarge.

As metamorphosis ends, the development of the first germ cells begins. In *Rana pipiens*, egg development lasts 3 years. At that time, the frog is sexually mature and can produce offspring of her own. The speed of metamorphosis is carefully keyed to environmental pressures. In temperate regions, for instance, metamorphosis must occur before the pond becomes frozen. A *Rana pipiens* frog can burrow into the mud and survive the winter; its tadpole cannot.

WEBSITE **2.1 Immortal animals.** Imagine a multicellular animal that acquires immortality by reverting back to its larval form instead of growing old. That seems to be what the marine hydrant *Turritopsis* does.

WEBSITE **2.2 The human life cycle.** The human animal provides a fascinating life cycle to study. Here are some websites that speculate about (A) when is an embryo or fetus "human"? (B) how might the strange way the human brain develops necessitate childhood? and (C) do humans undergo metamorphosis?

The Evolution of Developmental Patterns in Unicellular Protists

Every living organism develops. Development can be seen even among the unicellular organisms. Moreover, by studying the development of unicellular protists, we can see the simplest forms of cell differentiation and sexual reproduction.

Control of developmental morphogenesis: The role of the nucleus

A century ago, it had not yet been proved that the nucleus contained hereditary or developmental information. Some of the best evidence for this theory came from studies in which unicellular organisms were fragmented into nucleate and anucleate pieces (reviewed in Wilson 1896). When various protists were cut into fragments, nearly all the pieces died. However, the fragments containing nuclei were able to live and to regenerate entire complex cellular structures.

Nuclear control of cell morphogenesis and the interaction of nucleus and cytoplasm are beautifully demonstrated in studies of *Acetabularia*. This enormous single cell (2–4 cm long) consists of three parts: a cap, a stalk, and a rhizoid (Figure 2.5A; Mandoli 1998). The rhizoid is located at the base of the cell and holds it to the substrate. The single nucleus of the cell resides within the rhizoid. The size of *Acetabularia* and the location of its nucleus allow investigators to remove the nucleus from one cell and replace it with a nucleus from another cell. In the 1930s, J. Hämmerling took advantage of these unique features and exchanged nuclei between two morphologically distinct species, *A. mediterranea** and *A. crenulata*. As Figure 2.5A shows, these two species have very different cap structures. Hämmerling found that when he transferred

the nucleus from one species into the stalk of another species, the newly formed cap eventually assumed the form associated with the *donor* nucleus (Figure 2.5B). Thus, the nucleus was seen to control *Acetabularia* development.

The formation of a cap is a complex morphogenic event involving the synthesis of numerous proteins, which must be accumulated in a certain portion of the cell and then assembled into complex, species-specific structures. The transplanted nucleus does indeed direct the synthesis of its species-specific cap, but it takes several weeks to do so. Moreover, if the nucleus is removed from an *Acetabularia* cell early in development, before it first forms a cap, a normal cap is formed weeks later, even though the organism will eventually die. These studies suggest that (1) the nucleus contains information specifying the type of cap produced (i.e., it contains the genetic information that specifies the proteins required for the pro-

*After a recent formal name change, this species is now called *Acetabularia acetabulum*. For the sake of simplicity, however, we will use Hämmerling's designations here.

(A)

(B)

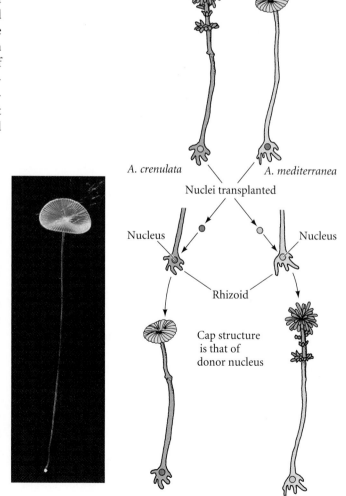

Figure 2.5

(A) *Acetabularia crenulata* (left) and *A. mediterranea* (right). Each individual is a single cell. The rhizoid contains the nucleus. (B) Effect of exchanging nuclei between two species of *Acetabularia*. Nuclei were transplanted into enucleated rhizoid fragments. *A. crenulata* structures are darker, *A. mediterranea* structures lighter green. (Photographs courtesy of S. Berger.)

duction of a certain type of cap), and (2) material containing this information enters the cytoplasm long before cap production occurs. This information in the cytoplasm is not used for several weeks.

One current hypothesis proposed to explain these observations is that the nucleus synthesizes a stable mRNA that lies dormant in the cytoplasm until the time of cap formation. This hypothesis is supported by an observation that Hämmerling published in 1934. Hämmerling fractionated young *Acetabularia* into several parts (Figure 2.6). The portion with the nucleus eventually formed a new cap, as expected; so did the apical tip of the stalk. However, the intermediate portion of the stalk did not form a cap. Thus, Hämmerling postulated (nearly 30 years before the existence of mRNA was known) that the instructions for cap formation originated in the nucleus and were somehow stored in a dormant form near the tip of the stalk. Many years later, researchers established that nucleus-derived mRNA does accumulate in the tip of the stalk, and that the destruction of this mRNA or the inhibition of protein synthesis in this region prevents cap formation (Kloppstech and Schweiger 1975; Garcia and Dazy 1986).

It is clear from the preceding discussion that nuclear transcription plays an important role in the formation of the *Acetabularia* cap. But note that the cytoplasm also plays an essential role in cap formation. The mRNAs are not translated for weeks, even though they are in the cytoplasm. Something in the cytoplasm controls when the message is utilized. Hence, the expression of the cap is controlled not only by nuclear transcription, but also by the translation of the cytoplasmic RNA. In this unicellular organism, "development" is controlled at both the transcriptional and translational levels.

WEBSITE **2.3 Protist differentiation.** Three of the most remarkable areas of protist development concern the control of sex type in fission yeast, the transformation of *Naegleria* amoebae into streamlined, flagellated cells, and the cortical inheritance of the cell surface in paramecia.

Unicellular protists and the origins of sexual reproduction

Sexual reproduction is another invention of the protists that has had a profound effect on more complex organisms. It should be noted that sex and reproduction are two distinct and separable processes. **Reproduction** involves the creation of new individuals. **Sex** involves the combining of genes from two different individuals into new arrangements. Reproduction in the absence of sex is characteristic of organisms that reproduce by fission (i.e., splitting into two); there is no sorting of genes when an amoeba divides or when a hydra buds off cells to form a new colony.

Sex without reproduction is also common among unicellular organisms. Bacteria are able to transmit genes from one individual to another by means of sex pili. This transmission

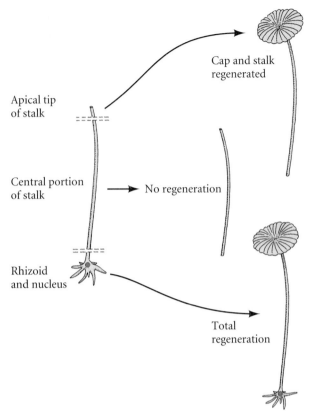

Figure 2.6
Regenerative ability of different fragments of *A. mediterranea*.

is separate from reproduction. Protists are also able to reassort genes without reproduction. Paramecia, for instance, reproduce by fission, but sex is accomplished by **conjugation**. When two paramecia join together, they link their oral apparatuses and form a cytoplasmic connection through which they can exchange genetic material (Figure 2.7). Each macronucleus (which controls the metabolism of the organism) degenerates, while each micronucleus undergoes meiosis to produce eight haploid micronuclei, of which all but one degenerate. The remaining micronucleus divides once more to form a stationary micronucleus and a migratory micronucleus. Each migratory micronucleus crosses the cytoplasmic bridge and fuses with ("fertilizes") the stationary micronucleus, thereby creating a new diploid nucleus in each cell. This diploid nucleus then divides mitotically to give rise to a new micronucleus and a new macronucleus as the two partners disengage. Therefore, no reproduction has occurred, only sex.

The union of these two distinct processes, sex and reproduction, into **sexual reproduction** is seen in unicellular eukaryotes. Figure 2.8 shows the life cycle of *Chlamydomonas*. This organism is usually haploid, having just one copy of each chromosome (like a mammalian gamete). The individuals of each species, however, are divided into two mating types: plus and minus. When a plus and a minus meet, they join their cytoplasms, and their nuclei fuse to form a diploid zygote.

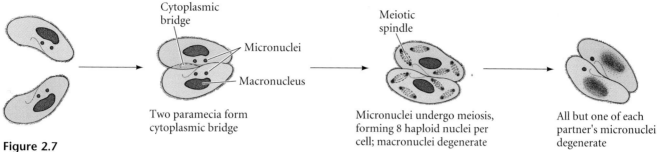

Figure 2.7
Conjugation across a cytoplasmic bridge in paramecia. Two paramecia can exchange genetic material, leaving each with genes that differ from those with which they started. (After Strickberger 1985.)

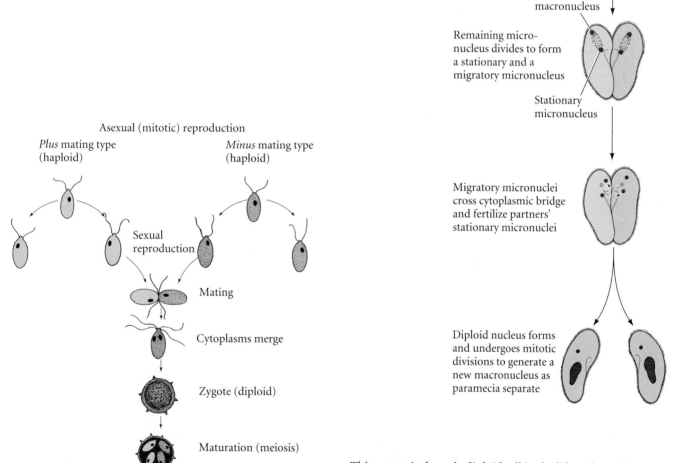

Figure 2.8
Sexual reproduction in *Chlamydomonas*. Two strains, both haploid, can reproduce asexually when separate. Under certain conditions, the two strains can unite to produce a diploid cell that can undergo meiosis to form four new haploid organisms. (After Strickberger 1985.)

This zygote is the only diploid cell in the life cycle, and it eventually undergoes meiosis to form four new *Chlamydomonas* cells. This is true sexual reproduction, for chromosomes are reassorted during the meiotic divisions and more individuals are formed. Note that in this protist type of sexual reproduction, the gametes are morphologically identical; the distinction between sperm and egg has not yet been made.

In evolving sexual reproduction, two important advances had to be achieved. The first was the mechanism of meiosis (Figure 2.9), whereby the diploid complement of chromosomes is reduced to the haploid state (discussed in detail in Chapter 19). The second was a mechanism whereby the two

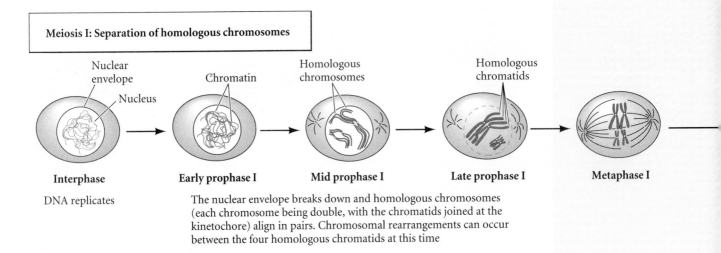

Meiosis I: Separation of homologous chromosomes

| Interphase | Early prophase I | Mid prophase I | Late prophase I | Metaphase I |

DNA replicates

The nuclear envelope breaks down and homologous chromosomes (each chromosome being double, with the chromatids joined at the kinetochore) align in pairs. Chromosomal rearrangements can occur between the four homologous chromatids at this time

Figure 2.9
Summary of meiosis. The DNA and associated proteins replicate during interphase. During prophase, the nuclear envelope breaks down and homologous chromosomes (each chromosome is double, with the chromatids joined at the kinetochore) align in pairs. Chromosomal rearrangements can occur between the four homologous chromatids at this time. After the first metaphase, the two original homologous chromosomes are segregated into different cells. During the second meiotic division the kinetochore splits, thereby leaving each new cell with one copy of each chromosome.

different mating types could recognize each other. In *Chlamydomonas*, recognition occurs first on the flagellar membranes (Figure 2.10; Bergman et al. 1975; Goodenough and Weiss 1975). The flagella of two individuals twist around each other, enabling specific regions of the cell membranes to come together. These specialized regions contain mating type-specif-ic components that enable the cytoplasms to fuse. Following flagellar agglutination, the plus individuals initiate fusion by extending a fertilization tube. This tube contacts and fuses with a specific site on the minus individual. Interestingly, the mechanism used to extend this tube—the polymerization of the protein actin—is also used to extend processes of sea urchin eggs and sperm. In Chapter 7, we will see that the recognition and fusion of sperm and egg occur in an amazingly similar manner.

Unicellular eukaryotes appear to possess the basic elements of the developmental processes that characterize the more complex organisms: protein synthesis is controlled such that certain proteins are made only at certain times and in certain places; the structures of individual genes and chromo-

Figure 2.10
Two-step recognition in mating *Chlamydomonas*. (A) Scanning electron micrograph (7000×) of mating pair. The interacting flagella twist around each other, adhering at the tips (arrows). (B) Transmission electron micrograph (20,000×) of a cytoplasmic bridge connecting the two organisms. The microfilaments extend from the donor (lower) cell to the recipient (upper) cell. (From Goodenough and Weiss 1975 and Bergman et al. 1975, courtesy of U. Goodenough.)

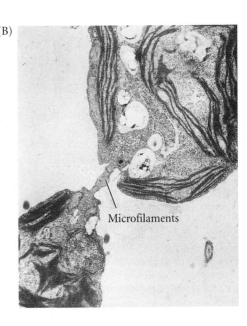

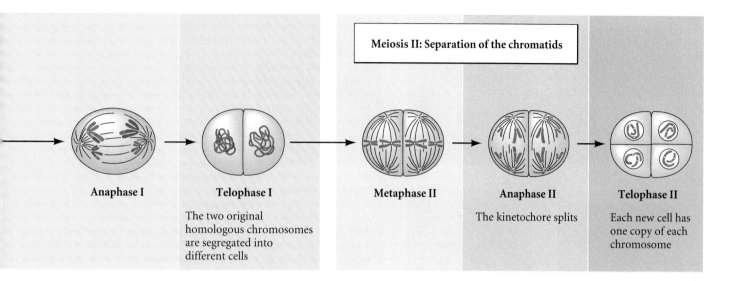

Meiosis II: Separation of the chromatids

Anaphase I

Telophase I

The two original homologous chromosomes are segregated into different cells

Metaphase II

Anaphase II

The kinetochore splits

Telophase II

Each new cell has one copy of each chromosome

somes are as they will be throughout eukaryotic evolution; mitosis and meiosis have been perfected; and sexual reproduction exists, involving cooperation between individual cells. Such intercellular cooperation becomes even more important with the evolution of multicellular organisms.

Multicellularity: The Evolution of Differentiation

One of evolution's most important experiments was the creation of multicellular organisms. There appear to be several paths by which single cells evolved multicellular arrangements; we will discuss only two of them here (see Chapter 22 for a fuller discussion). The first path involves the orderly division of the reproductive cell and the subsequent differentiation of its progeny into different cell types. This path to multicellularity can be seen in a remarkable series of multicellular organisms collectively referred to as the family Volvocaceae, or the volvocaceans (Kirk 1999).

The Volvocaceans

The simpler organisms among the volvocaceans are ordered assemblies of numerous cells, each resembling the unicellular protist *Chlamydomonas*, to which they are related (Figure 2.11A). A single organism of the volvocacean genus *Gonium* (Figure 2.11B), for example, consists of a flat plate of 4 to 16 cells, each with its own flagellum. In a related genus, *Pandorina*, the 16 cells form a sphere (Figure 2.11C); and in *Eudorina*, the sphere contains 32 or 64 cells arranged in a regular pattern (Figure 2.11D). In these organisms, then, a very important developmental principle has been worked out: the ordered division of one cell to generate a number of cells that are organized in a predictable fashion. As occurs during cleavage in most animal embryos, the cell divisions by which a single

volvocacean cell produces an organism of 4 to 64 cells occur in very rapid sequence and in the absence of cell growth.

The next two genera of the volvocacean series exhibit another important principle of development: the differentiation of cell types within an individual organism. The reproductive cells become differentiated from the somatic cells. In all the genera mentioned earlier, every cell can, and normally does, produce a complete new organism by mitosis. In the genera *Pleodorina* and *Volvox*, however, relatively few cells can reproduce. In *Pleodorina californica* (Figure 2.11E), the cells in the anterior region are restricted to a somatic function; only those cells on the posterior side can reproduce. In *P. californica*, a colony usually has 128 or 64 cells, and the ratio of the number of somatic cells to the number of reproductive cells is usually 3:5. Thus, a 128-cell colony typically has 48 somatic cells, and a 64-cell colony has 24.

In *Volvox*, almost all the cells are somatic, and very few of the cells are able to produce new individuals. In some species of *Volvox*, reproductive cells, as in *Pleodorina*, are derived from cells that originally look and function like somatic cells before they enlarge and divide to form new progeny. However, in other members of the genus, such as *V. carteri*, there is a complete division of labor: the reproductive cells that will create the next generation are set aside during the division of the original cell that is forming a new individual. The reproductive cells never develop functional flagella and never contribute to motility or other somatic functions of the individual; they are entirely specialized for reproduction. Thus, although the simpler volvocaceans may be thought of as colonial organisms (because each cell is capable of independent existence and of perpetuating the species), in *V. carteri* we have a truly multicellular organism with two distinct and interdependent cell types (somatic and reproductive), both of which are required for perpetuation of the species (Figure

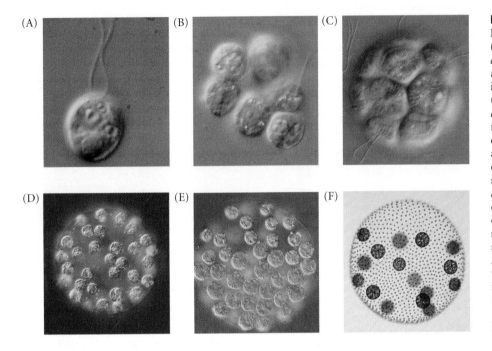

(A) (B) (C)

(D) (E) (F)

Figure 2.11
Representatives of the order Volvocales. (A) The unicellular protist *Chlamydomonas reinhardtii.* (B) *Gonium pectorale,* with 8 *Chlamydomonas*-like cells in a convex disc. (C) *Pandorina morum.* (D) *Eudorina elegans.* (E) *Pleodorina californica.* Here, all 64 cells are originally similar, but the posterior ones dedifferentiate and redifferentiate as asexual, reproductive cells called gonidia, while the anterior cells remain small and biflagellate, like *Chlamydomonas.* (F) *Volvox carteri.* Here, cells destined to become gonidia are set aside early in development and never have somatic characteristics. The smaller somatic cells resemble *Chlamydomonas.* All but *Chlamydomonas* are members of the family Volvocaceae. Complexity increases from the single-celled *Chlamydomonas* to the multicellular *Volvox.* (Photographs courtesy of D. Kirk.)

2.11F). Although not all animals set aside the reproductive cells from the somatic cells (and plants hardly ever do), this separation of germ cells from somatic cells early in development is characteristic of many animal phyla and will be discussed in more detail in Chapter 19.

WEBSITE **2.4** *Volvox* **cell differentiation.** The pathways leading to germ cells or somatic cells are controlled by genes that cause cells to follow one or the other fate. Mutations can prevent the formation of one of these lineages.

Sidelights & Speculations

Sex and Individuality in Volvox

SIMPLE AS IT IS, *Volvox* shares many features that characterize the life cycles and developmental histories of much more complex organisms, including ourselves. As already mentioned, *Volvox* is among the simplest organisms to exhibit a division of labor between two completely different cell types. As a consequence, it is among the simplest organisms to include death as a regular, genetically regulated part of its life history.

Death and Differentiation
Unicellular organisms that reproduce by simple cell division, such as amoebae, are potentially immortal. The amoeba you see today under the microscope has no dead ancestors. When an amoeba divides, neither of the two resulting cells can be considered either ances-

tor or offspring; they are siblings. Death comes to an amoeba only if it is eaten or meets with a fatal accident, and when it does, the dead cell leaves no offspring.

Death becomes an essential part of life, however, for any multicellular organism that establishes a division of labor between somatic (body) cells and germ (reproductive) cells. Consider the life history of *Volvox carteri* when it is reproducing asexually (Figure 2.12). Each asexual adult is a spheroid containing some 2000 small, biflagellated somatic cells along its periphery and about 16 large, asexual reproductive cells, called **gonidia**, toward one end of the interior. When mature, each gonidium divides rapidly 11 or 12 times. Certain of these divisions are asymmetrical and produce the 16 large cells that will become a new set of goni-

dia in the next generation. At the end of cleavage, all the cells that will be present in an adult have been produced from the gonidium. But the embryo is "inside out": it is now a hollow sphere with its gonidia on the outside and the flagella of its somatic cells pointing toward the interior. This predicament is corrected by a process called **inversion**, in which the embryo turns itself right side out by a set of cell movements that resemble the gastrulation movements of animal embryos (Figure 2.13). Clusters of bottle-shaped cells open a hole at one end of the embryo by producing tension on the interconnected cell sheet (Figure 2.14). The embryo everts through this hole and then closes it up. About a day after this is done, the juvenile *Volvox* are enzymatically released from the parent and swim away.

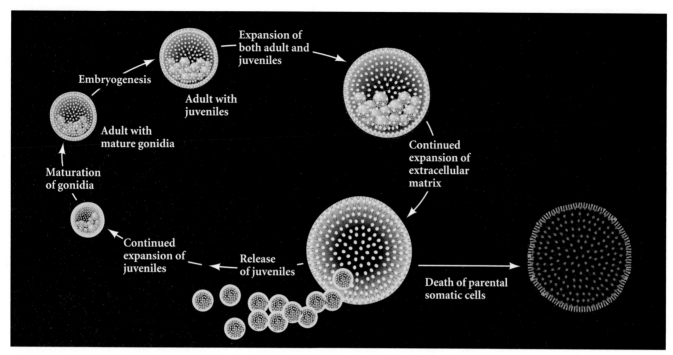

Figure 2.12
Asexual reproduction in *V. carteri*. When reproductive cells (gonidia) are mature, they enter a cleavage-like stage of embryonic development to produce juveniles within the adult. Through a series of cell movements resembling gastrulation, the embryonic *Volvox* invert and are eventually released from the parent. The somatic cells of the parent, lacking the gonidia, undergo senescence and die, while the juvenile colonies mature. The entire asexual cycle takes 2 days. (After Kirk 1988.)

Figure 2.13
Inversion of embryos of *V. carteri*. A–E are scanning electron micrographs of whole embryos. F–J are sagittal sections through the center of the embryo, visualized by differential interference microscopy. Before inversion, the embryo is a hollow sphere of connected cells. When the "bottle cells" change their shape, a hole (the phialopore) opens at the apex of the embryo (A, B, F, G). Cells then curl around and rejoin at the bottom (C–E, H–J). (From Kirk et al. 1982; photographs courtesy of D. Kirk.)

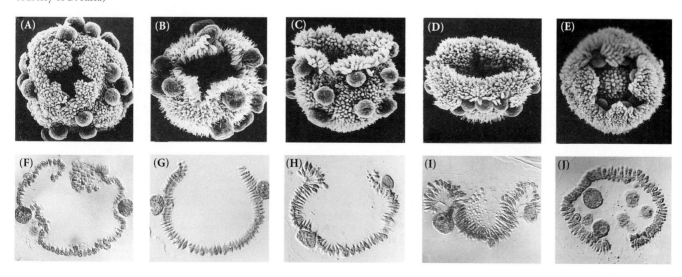

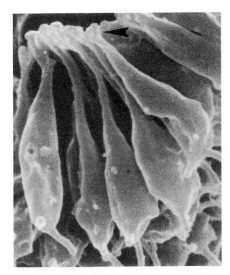

Figure 2.14
"Bottle cells" near the opening of the phialopore in a *V. carteri* embryo. These cells remain tightly interconnected through cytoplasmic bridges near their elongated apices, thereby creating the tension that causes the curvature of the interconnected cell sheet. (From Kirk et al. 1982; photograph courtesy of D. Kirk.)

What happens to the somatic cells of the "parent" *Volvox* now that its young have "left home"? Having produced offspring and being incapable of further reproduction, these somatic cells die. Actually, these cells commit suicide, synthesizing a set of proteins that cause the death and dissolution of the cells that make them (Pommerville and Kochert 1982). Moreover, in death, the cells release for the use of others, including their own offspring, all the nutrients that they had stored during life. "Thus emerges," notes David Kirk, "one of the great themes of life on planet Earth: 'Some die that others may live.'"

In *V. carteri*, a specific gene, *regA*, plays a central role in regulating cell death has been identified (Kirk 1988; Kirk et al. 1999). In laboratory strains possessing mutations of this gene, somatic cells abandon their suicidal ways, gain the ability to reproduce asexually, and become potentially immortal (Figure 2.15). The fact that such mutants have never been found in nature indicates that cell death most likely plays an important role in the survival of *V. carteri* under natural conditions.

Enter Sex
Although *V. carteri* reproduces asexually much of the time, in nature it reproduces sexually once each year. When it does, one generation of individuals passes away and a new and genetically different generation is produced. The naturalist Joseph Wood Krutch (1956, pp. 28–29) put it more poetically:

The amoeba and the paramecium are potentially immortal. … But for Volvox, death seems to be as inevitable as it is in a mouse or in a man. Volvox must die as Leeuwenhoek saw it die because it had children and is no longer needed. When its time comes it drops quietly to the bottom and joins its ancestors. As Hegner, the Johns Hopkins zoologist, once wrote, "This is the first advent of inevitable natural death in the animal kingdom and all for the sake of sex." And he asked: "Is it worth it?"

For *Volvox carteri*, it most assuredly is worth it. *V. carteri* lives in shallow temporary ponds that fill with spring rains but dry out in the heat of late summer. During most of that time, *V. carteri* swims about, reproducing asexually. These asexual volvoxes would die in minutes once the pond dried up. *V. carteri* is able to survive by turning sexual shortly before the pond dries up, producing dormant zygotes that survive the heat and drought of late summer and the cold of winter. When rain fills the pond in spring, the zygotes break their dormancy and hatch out a new generation of individuals to reproduce asexually until the pond is about to dry up once more.

How do these simple organisms predict the coming of adverse conditions with sufficient accuracy to produce a sexual generation just in time, year after year? The stimulus for switching from the asexual to the sexual mode of reproduction in *V. carteri* is known to be a 30-kDa sexual inducer protein. This protein is so powerful that concentrations as low as 6×10^{-17} M cause gonidia to undergo a modified pattern of embryonic development that results in the production of eggs or sperm, depending on the genetic sex of the individual (Sumper et al. 1993). The sperm are released and swim to a female, where they fertilize eggs to produce the dormant zygotes (Figure 2.16).

What is the source of this sexual inducer protein? Kirk and Kirk (1986) discovered that the sexual cycle could be initiated by heating dishes of *V. carteri* to temperatures that might be expected in a shallow pond in late summer. When this was done, the somatic cells of the asexual volvoxes produced the sexual inducer protein. Since the amount of sexual inducer protein secreted by one individual is sufficient to initiate sexual development in over 500 million asexual volvoxes, a single inducing volvox can convert the entire pond to sexuality. This discovery explained an observation made nearly 90 years ago that "in the full blaze of Nebraska sunlight, *Volvox* is able to appear, multiply, and riot in sexual reproduction in pools of rainwater of scarcely a fortnight's duration" (Powers 1908). Thus, in temporary ponds formed by spring rains and dried up by summer's heat, *Volvox* has found a means of survival: it uses that heat to induce the formation of sexual individuals whose mating produces zygotes capable of surviving conditions that kill the adult organism. We see, too, that development is critically linked to the ecosystem in which the organism has adapted to survive. ■

(A) 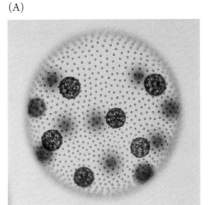   (B)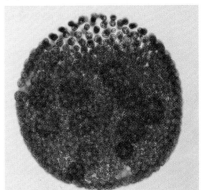

Figure 2.15
Mutation of a single gene (called *somatic regenerator A*) abolishes programmed cell death in *V. carteri*. A newly hatched *Volvox* carrying this mutation (A) is indistinguishable from the wild-type spheroid. However, shortly before the time when the somatic cells of wild-type spheroids begin to die, the somatic cells of this mutant redifferentiate as gonidia (B). Eventually, every cell of the mutant will divide to form (regenerate) a new spheroid that will repeat this potentially immortal developmental cycle. (Photographs courtesy of D. Kirk.)

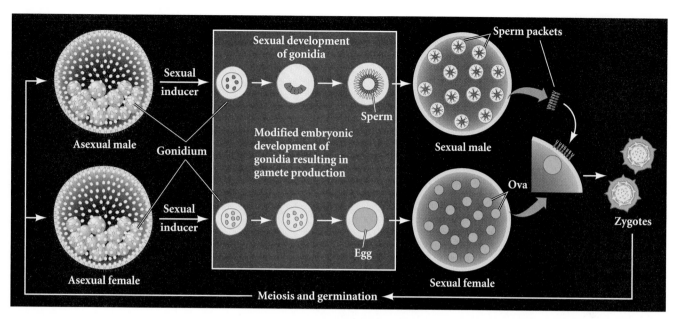

Figure 2.16
Sexual reproduction in *V. carteri*. Males and females are indistinguishable in their asexual phase. When the sexual inducer protein is present, the gonidia of both mating types undergo a modified embryogenesis that leads to the formation of eggs in the females and sperm in the males. When the gametes are mature, sperm packets (containing 64 or 128 sperm each) are released and swim to the females. Upon reaching a female, the sperm packet breaks up into individual sperm, which can fertilize the eggs. The resulting zygote has tough cell walls that can resist drying, heat, and cold. When spring rains cause the zygote to germinate, it undergoes meiosis to produce haploid males and females that reproduce asexually until heat induces the sexual cycle again.

Although all the volvocaceans, like their unicellular relative *Chlamydomonas*, reproduce predominantly by asexual means, they are also capable of sexual reproduction, which involves the production and fusion of haploid gametes. In many species of *Chlamydomonas*, including the one illustrated in Figure 2.10, sexual reproduction is **isogamous** ("the same gametes"), since the haploid gametes that meet are similar in size, structure, and motility. However, in other species of *Chlamydomonas*—as well as many species of colonial volvocaceans—swimming gametes of very different sizes are produced by the different mating types. This pattern is called **heterogamy** ("different gametes"). But the larger volvocaceans have evolved a specialized form of heterogamy, called **oogamy**, which involves the production of large, relatively immotile eggs by one mating type and small, motile sperm by the other (see Sidelights and Speculations). Here we see one type of gamete specialized for the retention of nutritional and developmental resources and the other type of gamete specialized for the transport of nuclei. Thus, the volvocaceans include the simplest organisms that have distinguishable male and female members of the species and that have distinct developmental pathways for the production of eggs or sperm. In all the volvocaceans, the fertilization reaction resembles that of *Chlamydomonas* in that it results in the production of a dormant diploid zygote, which is capable of surviving harsh environmental conditions. When conditions allow the zygote to germinate, it first undergoes meiosis to produce haploid offspring of the two different mating types in equal numbers.

Differentiation and Morphogenesis in Dictyostelium: Cell Adhesion

THE LIFE CYCLE OF *DICTYOSTELIUM*. Another type of multicellular organization derived from unicellular organisms is found in *Dictyostelium discoideum*.* The life cycle of this fascinating organism is illustrated in Figure 2.17A. In its asexual cycle, solitary haploid amoebae (called myxamoebae or "social amoebae" to distinguish them from amoeba species that always remain solitary) live on decaying logs, eating bacteria and reproducing by binary fission. When they have exhausted their food supply, tens of thousands of these myxamoebae join together to form moving streams of cells that converge at a central point. Here they pile atop one another to produce a conical mound called a tight aggregate. Subsequently, a tip arises at the top of this mound, and the tight aggregate bends over to produce the migrating slug (with the tip at the front). The slug (often given the more dignified title of pseudoplasmodium or grex) is usually 2–4 mm long and is encased in a slimy sheath. The grex begins to migrate (if the environment is dark and moist) with its anterior tip slightly raised. When it

---

*Though colloquially called a "cellular slime mold," *Dictyostelium* is not a mold, nor is it consistently slimy. It is perhaps best to think of *Dictyostelium* as a social amoeba.

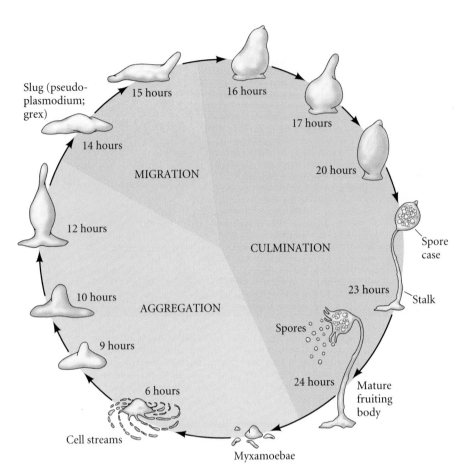

Slug (pseudo-plasmodium; grex)

15 hours

16 hours

17 hours

14 hours

MIGRATION

20 hours

12 hours

CULMINATION

Spore case

23 hours

Stalk

10 hours

AGGREGATION

Spores

9 hours

24 hours

Mature fruiting body

6 hours

Cell streams

Myxamoebae

Figure 2.17
Life history of *Dictyostelium discoideum*. Haploid spores give rise to myxamoebae, which can reproduce asexually to form more haploid myxamoebae. As the food supply diminishes, aggregation occurs at central points, and a migrating pseudoplasmodium is formed. Eventually it stops moving and forms a fruiting body that releases more spores. The times refer to hours since nutrient starvation began the developmental sequence.

reaches an illuminated area, migration ceases, and the grex differentiates into a fruiting body composed of spore cells and a stalk. The anterior cells, representing 15–20% of the entire cellular population, form the tubed stalk. This process begins as some of the central anterior cells, the prestalk cells, begin secreting an extracellular coat and extending a tube through the grex. As the prestalk cells differentiate, they form vacuoles and enlarge, lifting up the mass of prespore cells that had made up the posterior four-fifths of the grex (Jermyn and Williams 1991). The stalk cells die, but the prespore cells, elevated above the stalk, become spore cells. These spore cells disperse, each one becoming a new myxamoeba.

In addition to this asexual cycle, there is a possibility for sex in *Dictyostelium*. Two myxamoebae can fuse to create a giant cell, which digests all the other cells of the aggregate. When it has eaten all its neighbors, it encysts itself in a thick wall and undergoes meiotic and mitotic divisions; eventually, new myxamoebae are liberated.

Dictyostelium has been a wonderful experimental organism for developmental biologists because initially identical cells are differentiated into one of two alternative cell types, spore and stalk. It is also an organism wherein individual cells come together to form a cohesive structure composed of differentiated cell types, akin to tissue formation in more complex organisms. The aggregation of thousands of myxamoebae into a single organism is an incredible feat of organization that invites experimentation to answer questions about the mechanisms involved.

VADE MECUM **Slime mold.** The life cycle of *Dictyostelium*—the remarkable aggregation of myxamoebae, the migration of the slug, and the truly awesome culmination of the stalk and fruiting body—can best be viewed through movies.
[Click on Slime Mold]

AGGREGATION OF DICTYOSTELIUM CELLS. The first question is, What causes the myxamoebae to aggregate? Time-lapse microcinematography has shown that no directed movement occurs during the first 4–5 hours following nutrient starvation. During the next 5 hours, however, the cells can be seen moving at about 20 μm/min for 100 seconds. This movement ceases for about 4 minutes, then resumes. Although the movement is directed toward a central point, it is not a simple radial movement. Rather, cells join with one another to form streams; the streams converge into larger streams, and eventually all streams merge at the center. Bonner (1947) and Shaffer (1953) showed that this movement is due to chemotaxis: the cells are guided to aggregation centers by a soluble substance. This substance was later identified as cyclic adenosine 3′,5′-monophosphate (cAMP) (Konijn et al. 1967; Bonner et al. 1969), the chemical structure of which is shown in Figure 2.18A.

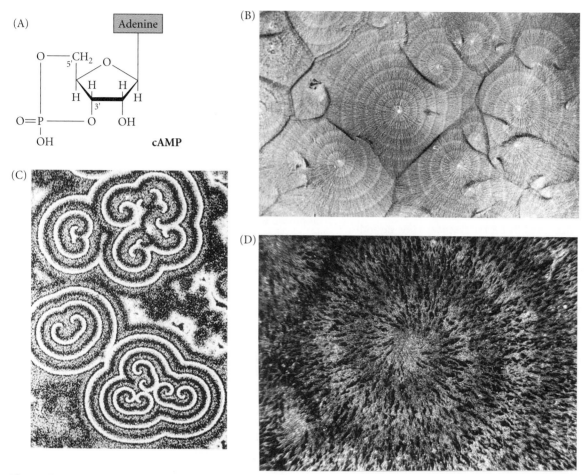

Figure 2.18

Chemotaxis of *Dictyostelium* myxamoebae due to spiral waves of cAMP. (A) Chemical structure of cAMP. (B) Visualization of several cAMP "waves." Central cells secrete cAMP at regular intervals, and each pulse diffuses outward as a concentric wave. Waves are charted by saturating filter paper with radioactive cAMP and placing it on an aggregating colony. The cAMP from the secreting cells dilutes the radioactive cAMP. When the radioactivity on the paper is recorded (by placing it over X-ray film), the regions of high cAMP concentration in the culture appear lighter than those of low cAMP concentration. (C, D) Spiral waves of myxamoebae moving toward the initial source of cAMP. (C) This digitally processed dark-field photomicrograph shows about 10^7 cells. Because moving and nonmoving cells scatter light differently, the photograph reflects cell movement. The bright bands are composed of elongated migrating cells; the dark bands are cells that have stopped moving and have rounded up. (D) As cells form streams, the spiral of movement can still be seen moving toward the center. (B from Tomchick and Devreotes 1981; C, D from Siegert and Weijer 1989.)

Aggregation is initiated as each of the cells begins to synthesize cAMP. There are no dominant cells that begin the secretion or control the others. Rather, the sites of aggregation are determined by the distribution of myxamoebae (Keller and Segal 1970; Tyson and Murray 1989). Neighboring cells respond to cAMP in two ways: they initiate a movement toward the cAMP pulse, and they release cAMP of their own

(Robertson et al. 1972; Shaffer 1975). After this, the cell is unresponsive to further cAMP pulses for several minutes. The result is a rotating spiral wave of cAMP that is propagated throughout the population of cells (Figure 2.18B–D). As each wave arrives, the cells take another step toward the center.*

The differentiation of individual myxamoebae into either stalk (somatic) or spore (reproductive) cells is a complex matter. Raper (1940) and Bonner (1957) demonstrated that the anterior cells normally become stalk, while the remaining, pos-

---

*The biochemistry of this reaction involves a receptor that binds cAMP. When this binding occurs, specific gene transcription takes place, motility toward the source of the cAMP is initiated, and adenyl cyclase enzymes (which synthesize cAMP from ATP) are activated. The newly formed cAMP activates the cell's own receptors, as well as those of its neighbors. The cells in the area remain insensitive to new waves of cAMP until the bound cAMP is removed from the receptors by another cell surface enzyme, phosphodiesterase (Johnson et al. 1989). The mathematics of such oscillation reactions predict that the diffusion of cAMP should initially be circular. However, as cAMP interacts with the cells that receive and propagate the signal, the cells that receive the front part of the wave begin to migrate at a different rate than the cells behind them (see Nanjundiah 1997, 1998). The result is the rotating spiral of cAMP and migration seen in Figure 2.18. Interestingly, the same mathematical formulas predict the behavior of certain chemical reactions and the formation of new stars in rotating spiral galaxies (Tyson and Murray 1989).

terior cells are usually destined to form spores. However, surgically removing the anterior part of the slug does not abolish its ability to form a stalk. Rather, the cells that now find themselves at the anterior end (and which originally had been destined to produce spores) now form the stalk (Raper 1940). Somehow a decision is made so that whichever cells are anterior become stalk cells and whichever are posterior become spores. This ability of cells to change their developmental fates according to their location within the whole organism and thereby compensate for missing parts is called **regulation**. We will see this phenomenon in many embryos, including those of mammals.

CELL ADHESION MOLECULES IN *DICTYOSTELIUM*. How do individual cells stick together to form a cohesive organism? This problem is the same one that embryonic cells face, and the solution that evolved in the protists is the same one used by embryos: developmentally regulated cell adhesion molecules.

While growing mitotically on bacteria, *Dictyostelium* cells do not adhere to one another. However, once cell division stops, the cells become increasingly adhesive, reaching a plateau of maximum cohesiveness around 8 hours after starvation. The initial cell-cell adhesion is mediated by a 24,000-Da (24-kDa) glycoprotein that is absent in myxamoebae but appears shortly after division ceases (Figure 2.19; Knecht et al. 1987; Loomis 1988). This protein is synthesized from newly transcribed mRNA and becomes localized in the cell membranes of the myxamoebae. If myxamoebae are treated with antibodies that bind to and mask this protein, they will not stick to one another, and all subsequent development ceases.

Once this initial aggregation has occurred, it is stabilized by a second cell adhesion molecule. This 80-kDa glycoprotein is also synthesized during the aggregation phase. If it is defective or absent in the cells, small slugs will form, and their fruiting bodies will be only about one-third the normal size. Thus, the second cell adhesion system seems to be needed for retaining a large enough number of cells to form large fruiting bodies (Müller and Gerisch 1978; Loomis 1988). In addition, a third cell adhesion system is activated late in development, while the slug is migrating. This protein appears to be important in the movement of the prestalk cells to the apex of the mound (Ginger et al. 1998). Thus, *Dictyostelium* has evolved three developmentally regulated systems of cell-cell adhesion that are necessary for the morphogenesis of individual cells into a coherent organism. As we will see in subsequent chapters, metazoan cells also use cell adhesion molecules to form the tissues and organs of the embryo.

Dictyostelium is a "part-time multicellular organism" that does not form many cell types (Kay et al. 1989), and the more complex multicellular organisms do not form by the aggregation of formerly independent cells. Nevertheless, many of the principles of development demonstrated by this "simple" organism also appear in embryos of more complex phyla (see Loomis and Insall 1999). The ability of individual cells to

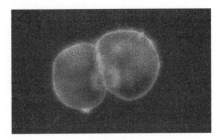

Figure 2.19
Dictyostelium cells synthesize an adhesive 24-kDa glycoprotein shortly after nutrient starvation. These *Dictyostelium* cells were stained with a fluorescently labeled antibody that binds to the 24-kDa glycoprotein and were then observed under ultraviolet light. This protein is not seen on myxamoebae that have just stopped dividing. However, as shown here—10 hours after cell division has ceased—individual myxamoebae have this protein in their cell membranes and are capable of adhering together. (Photograph courtesy of W. Loomis.)

sense a chemical gradient (as in the myxamoeba's response to cAMP) is very important for cell migration and morphogenesis during animal development. Moreover, the role of cell surface proteins in cell cohesiveness is seen throughout the animal kingdom, and differentiation-inducing molecules are beginning to be isolated in metazoan organisms.

DIFFERENTIATION IN *DICTYOSTELIUM*. Differentiation into stalk cell or spore cell reflects another major phenomenon of embryogenesis: the cell's selection of a developmental pathway. Cells often select a particular developmental fate when alternatives are available. A particular cell in a vertebrate embryo, for instance, can become either an epidermal skin cell or a neuron. In *Dictyostelium*, we see a simple dichotomous decision, because only two cell types are possible. How is it that a given cell becomes a stalk cell or a spore cell? Although the details are not fully known, a cell's fate appears to be regulated by certain diffusible molecules. The two major candidates are differentiation-inducing factor (DIF) and cAMP. DIF appears to be necessary for stalk cell differentiation. This factor, like the sex-inducing factor of *Volvox*, is effective at very low concentrations (10^{-10} *M*); and, like the *Volvox* protein, it appears to induce differentiation into a particular type of cell. When added to isolated myxamoebae or even to prespore (posterior) cells, it causes them to form stalk cells. The synthesis of this low molecular weight lipid is genetically regulated, for there are mutant strains of *Dictyostelium* that form only spore precursors and no stalk cells. When DIF is added to these mutant cultures, stalk cells are able to differentiate (Kay and Jermyn 1983; Morris et al. 1987), and new prestalk-specific mRNAs are seen in the cell cytoplasm (Williams et al. 1987). While the mechanisms by which DIF induces 20% of the grex cells to become stalk tissue are still controversial (see Early et al. 1995), DIF may act by releasing calcium ions from intracellular compartments within the cell (Shaulsky and Loomis 1995).

Evidence and Antibodies

BIOLOGY, like any other science, does not deal with Facts; rather, it deals with evidence. Several types of evidence will be presented in this book, and they are not equivalent in strength. As an example, we will use the analysis of cell adhesion in *Dictyostelium*. The first, and weakest, type of evidence is **correlative evidence.** Here, correlations are made between two or more events, and there is an inference that one event causes the other. As we have seen, fluorescently labeled antibodies to a certain 24-kDa glycoprotein do not bind to dividing myxamoebae, but they do find this protein in myxamoeba cell membranes soon after the cells stop dividing and become competent to aggregate (see Figure 2.19). Thus, there is a correlation between the presence of this cell membrane glycoprotein and the ability to aggregate.

Correlative evidence gives a starting point to investigations, but one cannot say with certainty that one event causes the other based solely on correlations. Although one might infer that the synthesis of this 24-kDa glycoprotein caused the adhesion of the cells, it is also possible that cell adhesion caused the cells to synthesize this new glycoprotein, or that cell adhesion and the synthesis of the glycoprotein are separate events initiated by the same underlying cause. The simultaneous occurrence of the two events could even be coincidental, the events having no relationship to each other.*

*In a tongue-in-cheek letter spoofing such correlative inferences, Sies (1988) demonstrated a remarkably good correlation between the number of storks seen in West Germany from 1965 to 1980 and the number of babies born during those same years.

How, then, does one get beyond mere correlation? In the study of cell adhesion in *Dictyostelium*, the next step was to use the antibodies that bound to the 24-kDa glycoprotein to block the adhesion of myxamoebae. Using a technique pioneered by Gerisch's laboratory (Beug et al. 1970), Knecht and co-workers (1987) isolated the antibodies' antigen-binding sites (the portions of the antibody molecule that actually recognize the antigen). This was necessary because the whole antibody molecule contains two antigen-binding sites and would therefore artificially crosslink and agglutinate the myxamoebae. When these antigen-binding fragments (called Fab fragments) were added to aggregation-competent cells, the cells could not aggregate. The antibody fragments inhibited the cells' adhering together, presumably by binding to the 24-kDa glycoprotein and blocking its function. This type of evidence is called **loss-of-function evidence.** While stronger than correlative evidence, it still does not make other inferences impossible. For instance, perhaps the antibodies killed the cells (as might have been the case if the 24-kDa glycoprotein were a critical transport channel). This would also stop the cells from adhering. Or perhaps the 24-kDa glycoprotein has nothing to do with adhesion itself but is necessary for the real adhesive molecule to function (perhaps, for example, it stabilizes membrane proteins in general). In this case, blocking the glycoprotein would similarly cause the inhibition of cell aggregation. Thus, loss-of-function evidence must be bolstered by many controls demonstrating that the agents causing the loss of function specifically knock out the particular function and nothing else.

The strongest type of evidence is **gain-of-function evidence.** Here, the initiation of the first event causes the second event to happen even in instances where neither event usually occurs. For instance, da Silva and Klein (1990) and Faix and co-workers (1990) obtained such evidence to show that the 80-kDa glycoprotein of *Dictyostelium* is an adhesive molecule. They isolated the gene for the 80-kDa protein and modified it in a way that would cause it to be expressed all the time. They then placed it back into well-fed, dividing myxamoebae, which do not usually express this protein and are not usually able to adhere to one another. The presence of this protein on the cell membrane of these dividing cells was confirmed by antibody labeling. Moreover, the treated cells now adhered to one another even in the proliferative stages (when they normally do not). Thus, they had gained an adhesive function solely upon expressing this particular glycoprotein on their cell surfaces. This gain-of-function evidence is more convincing than other types of evidence. Similar experiments have recently been performed on mammalian cells to demonstrate the presence of particular cell adhesion molecules in the developing embryo.

Evidence must also be taken together. "Every scientist," writes Fleck (1979), "knows just how little a single experiment can prove or convince. To establish proof, an entire system of experiments and controls is needed." Science is a communal endeavor, and it is doubtful that any great discovery is the achievement of a single experiment, or of any individual. Correlative, loss-of-function, and gain-of-function evidence must consistently support each other to establish and solidify a conclusion. ■

Although DIF stimulates myxamoebae to become prestalk cells, the differentiation of prespore cells is most likely controlled by the continuing pulses of cAMP. High concentrations of cAMP initiate the expression of prespore-specific mRNAs in aggregated myxamoebae. Moreover, when slugs are placed in a medium containing an enzyme that destroys extracellular cAMP, the prespore cells lose their differentiated characteristics (Figure 2.20; Schaap and van Driel 1985; Wang et al. 1988a,b).

WEBSITE **2.5 The control of *Dictyostelium* cell type.** Whether a myxamoeba becomes a spore cell or a stalk cell depends on the stage in the cell cycle at which it stops moving as well as on what chemical cues are in its immediate environment.

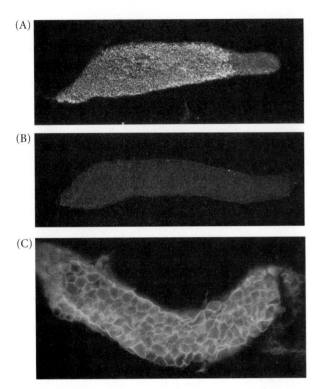

Figure 2.20
Chemicals controlling differentiation in *Dictyostelium*. (A) and (B) show the effects of placing *Dictyostelium* slugs into a medium containing enzymes that destroy extracellular cAMP. (A) Control grex stained for the presence of a prespore-specific protein (white regions). (B) Similar grex stained after treatment with cAMP-degrading enzymes. No prespore-specific product is seen. (C) Higher magnification of a slug treated with DIF (in the absence of ammonia). The stain used here binds to the cellulose wall of the stalk cells.

Developmental Patterns among the Metazoa

Since the remainder of this book concerns the development of **metazoans**—multicellular animals* that pass through embryonic stages of development—we will present an overview of their developmental patterns here. Figure 2.21 illustrates the major evolutionary trends of metazoan development. The most striking pattern is that life has not evolved in a straight line; rather, there are several branching evolutionary paths. We can see that metazoans belong to one of three major branches: Diploblasts, protostomes, and deuterostomes.

---

*Plants undergo equally complex and fascinating patterns of embryonic and postembryonic development. However, plant development differs significantly from that of animals, and the decision was made to focus this text on the development of animals. Readers who wish to discover some of the differences are referred to Chapter 20, which provides an overview of plant life cycles and the patterns of angiosperm (seed plant) development.

Sponges develop in a manner so different from that of any other animal group that some taxonomists do not consider them metazoans at all, and call them "parazoans." A sponge has three major types of somatic cells, but one of these, the **archeocyte,** can differentiate into all the other cell types in the body. Individual cells of a sponge passed through a sieve can reaggregate to form new sponges. Moreover, in some instances, such reaggregation is species-specific: if individual sponge cells from two different species are mixed together, each of the sponges that re-forms contains cells from only one species (Wilson 1907). In these cases, it is thought that the motile archeocytes collect cells from their own species and not from others (Turner 1978). Sponges contain no mesoderm, so the Porifera have no true organ systems; nor do they have a digestive tube or circulatory system, nerves, or muscles. Thus, even though they pass through an embryonic and a larval stage, sponges are very unlike most metazoans (Fell 1997). However, sponges do share many features of development (including gene regulatory proteins and signaling cascades) with all the other animal phyla, suggesting that they share a common origin (Coutinho et al. 1998).

Diploblasts

Diploblastic animals are those who have ectoderm and endoderm, but no true mesoderm. These include the cnidarians (jellyfish and hydras) and the ctenophores (comb jellies).

Cnidrians and ctenophores constitute the Radiata, so called because they have radial symmetry, like that of a tube or a wheel. In these animals, the mesoderm is rudimentary, consisting of sparsely scattered cells in a gelatinous matrix.

Protostomes and deuterostomes

Most metazoans have bilateral symmetry and three germ layers. The animals of these phyla, known collectively as the Bilatera, are classified as either protostomes or deuterostomes. All Bilateria are thought to have descended from a primitive type of flatworm. These flatworms were the first to have a true mesoderm (although it was not hollowed out to form a body cavity), and they may have resembled the larvae of certain contemporary coelenterates.

There are two divisions of bilaterian phyla, the protostomes and the deuterostomes. **Protostomes** (Greek, "mouth first"), which include the mollusc, arthropod, and worm phyla, are so called because the mouth is formed first, at or near the opening to the gut, which is produced during gastrulation. The anus forms later at another location. The **coelom,** or body cavity, of these animals forms from the hollowing out of a previously solid cord of mesodermal cells.

There are two major branches of the protostomes. The **ecdysozoa** includes those animals that molt. Its major constituent is Arthropoda, a phylum containing insects, arachnids, mites, crustaceans, and millipedes. The second major group of protostomes are the **lophotrochozoa**. They are char-

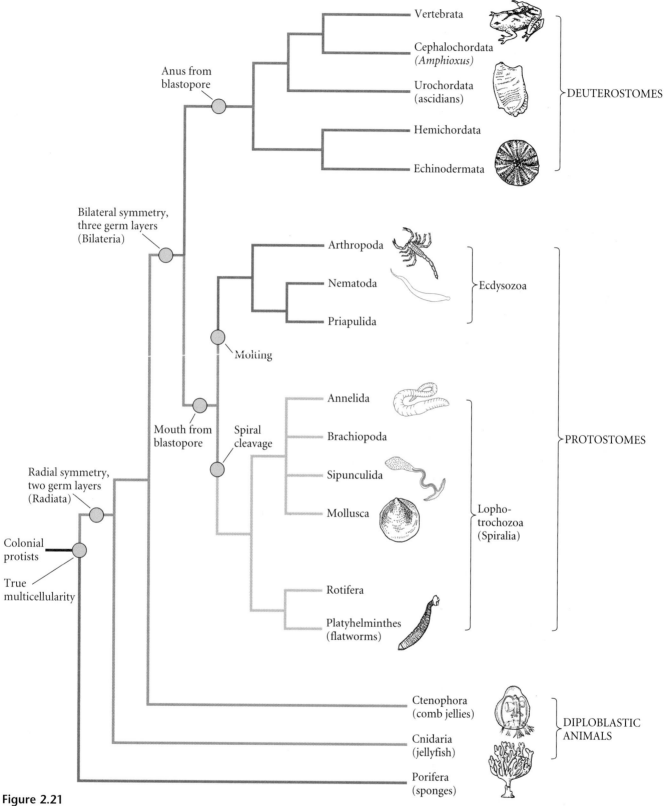

Figure 2.21

Major evolutionary divergences in extant animals. Other models of evolutionary relation-ships among the phyla are possible. This grouping of metazoa is based on embryonic, mor-phological, and molecular criteria. (Based on J. R. Garey, personal communication.)

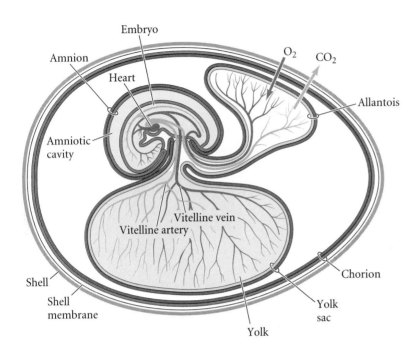

Embryo

Amnion

Heart

Amniotic
cavity

O_2 CO_2

Allantois

Vitelline vein

Vitelline artery

Shell

Shell
membrane

Chorion

Yolk
sac

Yolk

Figure 2.22
Diagram of the amniote egg of the chick, showing the membranes enfolding the 7-day chick embryo. The yolk is eventually surrounded by the yolk sac, which allows the entry of nutrients into the blood vessels. The chorion is derived in part from the ectoderm and extends from the embryo to the shell (where it will exchange oxygen and carbon dioxide and absorb calcium from the shell). The amnion provides the fluid medium in which the embryo grows, and the allantois collects nitrogenous wastes that would be dangerous to the embryo. Eventually the endoderm becomes the gut tube and encircles the yolk.

VADE MECUM **Egg development.** The anatomy of the amniote egg is seen in video. **[Click on Chick-Early]**

acterized by a common type of cleavage (spiral), a common larval form, and a distinctive feeding apparatus. These phyla include annelids, molluscs, and flatworms.

Phyla in the **deuterostome** lineage include the chordates and echinoderms. Although it may seem strange to classify humans, fish, and frogs in the same group as starfish and sea urchins, certain embryological features stress this kinship. First, in deuterostomes ("mouth second"), the mouth opening is formed after the anal opening. Also, whereas protostomes generally form their body cavities by hollowing out a solid mesodermal block (**schizocoelous** formation of the body cavity), most deuterostomes form their body cavities from mesodermal pouches extending from the gut (**enterocoelous** formation of the body cavity). It should be mentioned that there are many exceptions to these generalizations.

The evolution of organisms depends on inherited changes in their development. One of the greatest evolutionary advances—the **amniote egg**—occurred among the deuterostomes. This type of egg, exemplified by that of a chicken (Figure 2.22), is thought to have originated in the amphibian ancestors of reptiles about 255 million years ago. The amniote egg allowed vertebrates to roam on land, far from existing ponds. Whereas most amphibians must return to water to lay their eggs, the amniote egg carries its own water and food supplies. It is fertilized internally and contains yolk to nourish the developing embryo. Moreover, the amniote egg contains four sacs: the yolk sac, which stores nutritive proteins; the amnion, which contains the fluid bathing the embryo; the allantois, in which waste materials from embry-

onic metabolism collect; and the chorion, which interacts with the outside environment, selectively allowing materials to reach the embryo.* The entire structure is encased in a shell that allows the diffusion of oxygen but is hard enough to protect the embryo from environmental assaults and dehydration. A similar development of egg casings enabled arthropods to be the first terrestrial invertebrates. Thus, the final crossing of the boundary between water and land occurred with the modification of the earliest stage in development: the egg.

Embryology provides an endless assortment of fascinating animals and problems to study. In this text, we will use but a small sample of them to illustrate the major principles of animal development. This sample is an incredibly small collection. We are merely observing a small tidepool within our reach, while the whole ocean of developmental phenomena lies before us.

After a brief outline of the experimental and genetic approaches to developmental biology, we will investigate the early stages of animal embryogenesis: fertilization, cleavage, gastrulation, and the establishment of the body axes. Later chapters will concentrate on the genetic and cellular mechanisms by which animal bodies are constructed. Although an attempt has been made to survey the important variations throughout the animal kingdom, a certain deuterostome chauvinism may be apparent. (For a more comprehensive survey of the diversity of animal development across the phyla, see Gilbert and Raunio 1997.)

---

*In mammals, the chorion is modified to form the placenta—another example of the modification of development to produce evolutionary change.

Principles of Development: Life Cycles and Developmental Patterns

1. The life cycle can be considered a central unit in biology. The adult form need not be paramount. In a sense, the life cycle is the organism.

2. The basic life cycle consists of fertilization, cleavage, gastrulation, germ layer formation, organogenesis, metamorphosis, adulthood, and senescence.

3. Reproduction need not be sexual. Some organisms, such as *Volvox* and *Dictyostelium*, exhibit both asexual reproduction and sexual reproduction.

4. Cleavage divides the zygote into numerous cells called blastomeres.

5. In animal development, gastrulation rearranges the blastomeres and forms the three germ layers.

6. Organogenesis often involves interactions between germ layers to produce distinct organs.

7. Germ cells are the precursors of the gametes. Gametogenesis forms the sperm and the eggs.

8. There are three main ways to provide nutrition to the developing embryo: (1) supply the embryo with yolk; (2) form a larval feeding stage between the embryo and the adult; or (3) create a placenta between the mother and the embryo.

9. Life cycles must be adapted to the nonliving environment and interwoven with other life cycles.

10. Don't regress your tail until you've formed your hindlimbs.

11. There are several types of evidence. Correlation between phenomenon A and phenomenon B does not imply that A causes B or that B causes A. Loss-of-function data (if A is experimentally removed, B does not occur) suggests that A causes B, but other explanations are possible. Gain-of-function data (if A happens where or when it does not usually occur, then B also happens in this new time or place) is most convincing.

12. Protostomes and deuterostomes represent two different sets of variations on development. Protostomes form the mouth first, while deuterostomes form the anus first.

Literature Cited

Alley, K. E. and M. D. Barnes. 1983. Birthdates of trigeminal motor neurons and metamorphic reorganization of the jaw myoneural system in frogs. *J. Comp. Neurol.* 218: 395–405.

Beck, S. D. 1980. *Insect Photoperiodism*, 2nd Ed. Academic Press, New York.

Bergman, K., U. W. Goodenough, D. A. Goodenough, J. Jawitz and H. Martin. 1975. Gametic differentiation in *Chlamydomonas reinhardtii*. II. Flagellar membranes and the agglutination reaction. *J. Cell Biol.* 67: 606–622.

Beug, H., G. Gerisch, S. Kempff, V. Riedel and G. Cremer. 1970. Specific inhibition of cell contact formation in *Dictyostelium* by univalent antibodies. *Exp. Cell Res.* 63: 147–158.

Bonner, J. T. 1947. Evidence for the formation of cell aggregates by chemotaxis in the development of the slime mold *Dictyostelium discoideum*. *J. Exp. Zool.* 106: 1–26.

Bonner, J. T. 1957. A theory of the control of differentiation in the cellular slime molds. *Q. Rev. Biol.* 32: 232–246.

Bonner, J. T. 1965. *Size and Cycle*. Princeton University Press, Princeton, NJ.

Bonner, J. T., D. S. Berkley, E. M. Hall, T. M. Konijn, J. W. Mason, G. O'Keefe and P. B. Wolfe. 1969. Acrasin, acrasinase, and the sensitivity to acrasin in *Dictyostelium discoideum*. *Dev. Biol.* 20: 72–87.

Coutinho, C., J. Seack, G. de Vyler, R. Borojevic and W. E. G. Müller. 1998. Origin of the metazoan body plan: Characterization and functional testing of the promoter of the homeobox gene *EmH-3* from the freshwater sponge *Ephydatia muelleri* in mouse 3T3 cells. *Biol. Chem.* 379: 1243–1251.

Currie, J. and W. M. Cowan. 1974. Evidence for the late development of the uncrossed retinothalamic projections in the frog *Rana pipiens*. *Brain Res.* 71: 133–139.

da Silva, A. M. and C. Klein. 1990. Cell adhesion transformed *D. discoideum* cells: Expression of gp80 and its biochemical characterization. *Dev. Biol.* 140: 139–148.

Early, A., T. Abe and J. Williams. 1995. Evidence for positional differentiation of prestalk cells and for a morphogenetic gradient in *Dictyostelium*. *Cell* 83: 91–99.

Faix, J., G. Gerisch and A. A. Noegel. 1990. Constitutive overexpression of the contact A glycoprotein enables growth-phase cells of *Dictyostelium discoideum* to aggregate. *EMBO J.* 9: 2709–2716.

Fell, P. E. 1997. Porifera: The sponges. *In* S. F. Gilbert and A. M. Raunio (eds.), *Embryology: Constructing the Organism*. Sinauer Associates, Sunderland, MA, pp. 39–54.

Fleck, L. 1979. *Genesis and Development of a Scientific Fact*. Translated by F. Bradley and T. J. Trenn. University of Chicago Press, Chicago.

Forehand, C. J. and P. B. Farel. 1982. Spinal cord development in anuran larvae. I. Primary and secondary neurons. *J. Comp. Neurol.* 209: 386–394.

Garcia, E. and A.-C. Dazy. 1986. Spatial distribution of poly(A)$^+$ RNA and protein synthesis in *Acetabularia mediterranea*. *Biol. Cell* 58: 23–29.

Gilbert, S. F. and A. M. Raunio (eds.). 1997. *Embryology: Constructing the Organism*. Sinauer Associates, Sunderland, MA.

Ginger, R. S., L. Drury, C. Baader, N. V. Zhukovskaya and J. G. Williams. 1998. A novel *Dictyostelium* cell surface protein important in both cell aggregation and cell sorting. *Development* 125: 3343–3352.

Goodenough, U. W. and R. L. Weiss 1975. Gametic differentiation in *Chlamydomonas reinhardtii*. III. Cell wall lysis and microfilament associated mating structure activation in wildtype and mutant strains. *J. Cell Biol.* 67: 623–637.

Grobstein, P. 1987. On beyond neuronal specificity: Problems in going from cells to networks and from networks to behavior. *In* P. Shinkman (ed.), *Advances in Neural and Behavioral Development*, Vol. 3. Ablex, Norwood, NJ, pp. 1–58.

Hämmerling, J. 1934. Über formbildendend Substanzen bei *Acetabularia mediterranea*, ihre räumliche und zeitliche Verteilung und ihre Herkunft. *Wilhelm Roux Arch. Entwicklungsmech. Org.* 131: 1–82.

Harwood, A. J., N. A. Hopper, M.-N. Simon, D. M. Driscoll, M. Veron and J. G. Williams. 1992. Culmination in *Dictyostelium* is regulated by the cAMP-dependent protein kinase. *Cell* 69: 615–624.

Herre, E. A. 1993. Population structure and the evolution of virulence in nematode parasites of fig wasps. *Science* 259: 1442–1445.

Hoskins, S. G. and P. Grobstein. 1984. Thyroxine induces the ipsilateral retinothalamic projection in *Xenopus laevis. Nature* 307: 730–733.

Jermyn, K. A. and J. Williams. 1992. An analysis of culmination in *Dictyostelium* using prestalk and stalk-specific cell autonomous markers. *Development* 111: 779–787.

Johnson, R. L. and 7 others. 1989. G-protein-linked signal transduction systems control development in *Dictyostelium. Development* [Suppl.]: 75–81.

Kay, R. R. and K. A. Jermyn. 1983. A possible morphogen controlling differentiation in *Dictyostelium. Nature* 303: 242–244.

Kay, R. R., M. Berks and D. Traynor. 1989. Morphogen hunting in *Dictyostelium. Development* [Suppl.]: 81–90.

Keller, E. F. and L. A. Segal. 1970. Initiation of slime mold aggregation viewed as an instability. *J. Theor. Biol.* 26: 399–415.

Kirk, D. L. 1988. The ontogeny and phylogeny of cellular differentiation in *Volvox. Trends Genet.* 4: 32–36.

Kirk, D. L. 1999. Evolution of multicellularity in the Volvocine lineage. *Curr. Opin. Plant Biol.* 2: 496–501.

Kirk, D. L. and M. M. Kirk. 1986. Heat shock elicits production of sexual inducer in *Volvox. Science* 231: 51–54.

Kirk, D. L., G. I. Viamontes, K. J. Green and J. L. Bryant, Jr. 1982. Integrated morphogenetic behavior of cell sheets: *Volvox* as a model. *In* S. Subtelny and P. B. Green (eds.), *Developmental Order: Its Origin and Regulation.* Alan R. Liss, New York, pp. 247–274.

Kirk, M. M. and 7 others. 1999. reg A, a *Volvox* gene that plays a central role in germ-soma differentiation, encodes a novel regulatory protein. *Development* 126: 639–647.

Kloppstech, K. and H. G. Schweiger. 1975. Polyadenylated RNA from *Acetabularia. Differentiation* 4: 115–123.

Knecht, D. A., D. Fuller and W. F. Loomis. 1987. Surface glycoprotein gp24 involved in early adhesion of *Dictyostelium discoideum. Dev. Biol.* 121: 277–283.

Krutch, J. W. 1956. *The Great Chain of Life.* Houghton Mifflin, Boston.

Loomis, W. F. 1988. Cell-cell adhesion in *Dictyostelium discoideum. Dev. Genet.* 9: 549–559.

Loomis, W. F. and R. H. Insall. 1999. A cell for all seasons. *Nature* 401: 440–441.

Mandoli, D.F. 1998. What ever happened to *Acetabularia?* Bringing a once-classic model system into the age of molecular genetics. *Int. Rev. Cytol.* 182: 1–67.

Morris, H. R., G. W. Taylor, M. S. Masento, K. A. Jermyn and R. R. Kay. 1987. Chemical structure of the morphogen differentiation-inducing factor from *Dictyostelium discoideum. Nature* 328: 811–814.

Müller, K. and G. Gerisch. 1978. A specific glycoprotein as the target of adhesion blocking Fab in aggregating *Dictyostelium* cells. *Nature* 274: 445–447.

Nanjundiah, V. 1997. Models for pattern formation in the Dictyostelid slime molds. *In Dictyostelium: A Model System for Cell and Developmental Biology.* Universal Academy Press, India, pp. 305–322.

Nanjundiah, V. 1998. Cyclic AMP oscillations in *Dictyostelium discoideum*: Models and observations. *Biophys. Chem.* 72: 1–8.

Neumann, D. and K.-D. Spindler. 1991. Circasemilunar control of imaginal disc development in *Clunio marinus*: Temporal switching point, temperature-compensated developmental time, and ecdysteroid profile. *J. Insect Physiol.* 37: 101–109.

Pommerville, J. and G. Kochert. 1982. Effects of senescence on somatic cell physiology in the green alga *Volvox carteri. Exp. Cell Res.* 14: 39–45.

Powers, J. H. 1908. Further studies on *Volvox*, with description of three new species. *Trans. Am. Micros. Soc.* 28: 141–175.

Raper, K. B. 1940. Pseudoplasmodium formation and organization in *Dictyostelium discoideum. J. Elisha Mitchell Sci. Soc.* 56: 241–282.

Robertson, A., D. J. Drage and M. H. Cohen. 1972. Control of aggregation in *Dictyostelium discoideum* by an external periodic pulse of cyclic adenosine monophosphate. *Science* 175: 333–335.

Rugh, R. 1950. *The Frog: Its Reproduction and Development.* McGraw-Hill, New York.

Schaap, P. and R. van Driel. 1985. The induction of post-aggregative differentiation in *Dictyostelium discoideum* by cAMP. Evidence for involvement of the cell surface cAMP receptor. *Exp. Cell Res.* 159: 388–398.

Shaffer, B. M. 1953. Aggregation in cellular slime molds: In vitro isolation of acrasin. *Nature* 171: 975.

Shaffer, B. M. 1975. Secretion of cyclic AMP induced by cyclic AMP in the cellular slime mold *Dictyostelium discoideum. Nature* 255: 549–552.

Shaulsky, G. and W. F. Loomis. 1995. Mitochondrial DNA replication but no nuclear DNA replication during development. *Proc. Natl. Acad. Sci. USA* 92: 5660–5663.

Siegert, F. and C. J. Weijer. 1989. Digital image processing of optical density wave propagation in *Dictyostelium discoideum* and analysis of the effects of caffeine and ammonia. *J. Cell Sci.* 93: 325–335.

Sies, H. 1988. A new parameter for sex education. *Nature* 332: 495.

Strickberger, M. W. 1985. *Genetics*, 3rd Ed. Macmillan, New York.

Sumper, M., E. Berg, S. Wenzl and K. Godl. 1993. How a sex pheromone might act at a concentration below 10^{-16} *M. EMBO J.* 12: 831–836.

Tomchick, K. J. and P. N. Devreotes. 1981. Adenosine 3′,5′ monophosphate waves in *Dictyostelium discoideum. Science* 212: 443–446.

Turner, R. S. Jr. 1978. Sponge cell adhesions. *In* D. R. Garrod (ed.), *Specificity of Embryological Interactions.* Chapman and Hall, London, pp. 199–232.

Tyson, J. J. and J. D. Murray. 1989. Cyclic AMP waves during aggregation of *Dictyostelium* amoebae. *Development* 106: 421–426.

Wang, M., R. J. Aerts, W. Spek and P. Schaap. 1988a. Cell cycle phase in *Dictyostelium discoideum* is correlated with the expression of cyclic AMP production, detection and degradation. Involvement of cAMP signaling in cell sorting. *Dev. Biol.* 125: 410–416.

Wang, M., R. van Driel, and P. Schaap. 1988b. Cyclic AMP-phosphodiesterase induces dedifferentiation of prespore cells in *Dictyostelium discoideum* slugs: Evidence that cyclic AMP is the morphogenetic signal for prespore differentiation. *Development* 103: 611–618.

Williams, J. G. and 7 others. 1987. Direct induction of *Dictyostelium* pre-stalk gene expression of DIF provides evidence that DIF is a morphogen. *Cell* 49: 185–192.

Wilson, E. B. 1896. *The Cell in Development and Inheritance.* Macmillan, New York.

Wilson, H. V. 1907. On some phenomena of coalescence and regeneration in sponges. *J. Exp. Zool.* 5: 245–258.

chapter 3 ## *Principles of experimental embryology*

DESCRIPTIVE EMBRYOLOGY AND EVOLUTIONARY EMBRYOLOGY both had their roots in anatomy. At the end of the nineteenth century, however, the new biological science of physiology made inroads into embryological research. The questions of "what?" became questions of "how?" A new generation of embryologists felt that embryology should not merely be a guide to the study of anatomy and evolution, but should answer the question, "How does an egg become an adult?" Embryologists were to study the mechanisms of organ formation (morphogenesis) and differentiation. This new program was called *Entwicklungsmechanik*, often translated as "causal embryology," "physiological embryology," or "developmental mechanics." Its goals were to find the molecules and processes that caused the visible changes in embryos. Experimentation was to supplement observation in the study of embryos, and embryologists were expected to discover the properties of the embryo by seeing how the embryonic cells responded to perturbations and disruptions. Wilhelm Roux (1894), one of the founders of this branch of embryology, saw it as a grand undertaking:

> *We must not hide from ourselves the fact that the causal investigation of organisms is one of the most difficult, if not the most difficult, problem which the human intellect has attempted to solve ... since every new cause ascertained only gives rise to fresh questions regarding the cause of this cause.*

In this chapter, we will discuss three of the major research programs in experimental embryology. The first concerns how forces outside the embryo influence its development. The second concerns how forces within the embryo cause the differentiation of its cells. The third looks at how the cells order themselves into tissues and organs.

WEBSITE **3.1 Establishing experimental embryology.** The foundations of *Entwicklungsmechanik* were laid by a group of young investigators who desired a more physiological approach to embryology. These scientists disagreed with one another concerning the mechanisms of development, but they cooperated to secure places to perform and publish their research.

Environmental Developmental Biology

The developing embryo is not isolated from its environment. In numerous instances, environmental cues are a fundamental part of the organism's life cycle. Moreover, removing or altering these environmental parameters can alter development.

49

Environmental sex determination

SEX DETERMINATION IN AN ECHIUROID WORM: *BONELLIA.*
When developmental mechanics was first formulated, some of
the obvious variables to manipulate were the temperature and
media in which embryos were developing. These early studies
initiated several experimental programs on the effects of the
environment on development. For instance, Baltzer (1914)
showed that the sex of the echiuroid worm *Bonellia viridis* depended on where the *Bonellia* larva settled. The female
Bonellia worm is a marine, rock-dwelling animal, with a body
about 10 cm long (Figure 3.1). She has a proboscis that can extend over a meter in length. The male *Bonellia*, however, is
only 1–3 mm long and resides within the uterus of the female,
fertilizing her eggs. Baltzer showed that if a *Bonellia* larva settles on the seafloor, it becomes a female. However, should a
larva land on a female's proboscis (which apparently emits
chemical signals that attract the larva), it enters the female's
mouth, migrates into her uterus, and differentiates into a
male. Thus, if a larva lands on the seafloor, it becomes female;
if it settles on a proboscis, it becomes male. Baltzer (1914) and
Leutert (1974) were able to duplicate this phenomenon in the
laboratory, incubating larvae in either the absence or presence
of adult females (Figure 3.2).

SEX DETERMINATION IN A VERTEBRATE: *ALLIGATOR.* Recent research has shown that the effects of the environment on development can have important consequences. Such research
has shown that the sex of the alligators, crocodiles, and many
other reptiles depends not on chromosomes, but on tempera-

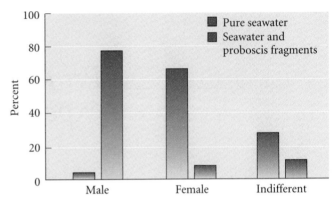

Figure 3.2
In vitro analysis of *Bonellia* sex determination. Larval *Bonellia* were
placed either in normal seawater or in seawater containing fragments of the female proboscis. A majority of the animals cultured
in the presence of the proboscis fragments became males, whereas
in their absence, most became females. (After Leutert 1974.)

ture. After studying the sex determination of the Mississippi
alligator both in the laboratory and in the field, Ferguson and
Joanen (1982) concluded that sex is determined by the temperature of the egg during the second and third weeks of incubation. Eggs incubated at 30°C or below during this time
period produce female alligators, whereas those eggs incubated at 34°C or above produce males. (At 32°C, 87% of the
hatchlings were female.) Moreover, nests constructed on lev-
ees (close to 34°C) give rise to males, whereas nests built in
wet marshes (close to 30°C) produce females. These findings
are obviously important to wildlife managers and farmers
who wish to breed this species; but they also raise questions of
environmental policy. The shade of buildings or the heat of
thermal effluents can have dramatic effects on the sex ratios of
reptiles. We will discuss the mechanisms of temperature-dependent sex determination further in Chapter 17.

> **WEBSITE 3.2 The hazards of environmental sex deter-
> mination.** Ferguson and Joanen (1982) speculate that tem-
> perature-dependent sex determination may be have been
> responsible for the extinction of the dinosaurs. The depen-
> dence on temperature for sex determination may also be
> dangerous for reptilian species in our present era of climate
> change.

Adaptation of embryos and larvae to their environments

NORMS OF REACTION. Another program of environmental developmental biology concerns how the embryo adapts to its
particular environment. August Weismann (1875) pioneered
the study of larval adaptations, and recent research in this area
has provided some fascinating insights into how an organism's
development is keyed to its environment. Weismann noted
that butterflies that hatched during different seasons were colored differently, and that this season-dependent coloration

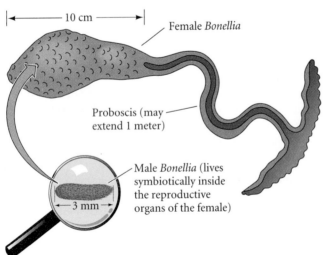

Figure 3.1
Sexual dimorphism in *Bonellia viridis*. The body of the mature female is about 10 cm in length, but the proboscis can extend up to a
meter. The body of the symbiotic male is a minute 1–3 mm in
length. While the body of the adult female is buried in the ocean
sediments, her proboscis extends out of the sediments, where it can
be used for feeding or attracting larvae.

Figure 3.3
Two morphs of *Araschnia levana*, the European map butterfly. The summer morph is represented at the top, the spring morph at the bottom. In this species, the phenotypic differences are elicited by differences in day length and temperature during the larval period. These morphs represent the range of the *Arachina* reaction norm. (Photographs courtesy of H. F. Nijhout.)

could be mimicked by incubating larvae at different temperatures. Phenotypic variations caused by environmental differences are often called **morphs**. One example of such seasonal variation is the European map butterfly, *Araschnia levana*, which has two seasonal phenotypes so different that Linnaeus classified them as two different species (van der Weele 1995). The spring morph is bright orange with black spots, while the summer form is mostly black with a white band (Figure 3.3). The change from spring to summer morph is controlled by changes in both day length and temperature during the larval period. When researchers experimentally mimic spring conditions, summer caterpillars can give rise to "spring" butterflies (Koch and Buchmann 1987; Nijhout 1991).

Another dramatic example of seasonal change in development occurs in the moth *Nemoria arizonaria*. This moth has a fairly typical insect life cycle. Eggs hatch in the spring, and the caterpillars feed on young oak flowers (catkins). These larvae metamorphose in the late spring, mate in the summer, and produce another brood of caterpillars on the oak trees. These caterpillars eat the oak leaves, metamorphose, and mate. Their eggs overwinter to start the cycle over again next spring. What is remarkable is that the caterpillars that hatch in the spring look nothing like their progeny that hatch in the summer (Figure 3.4). The caterpillars that hatch in the spring

and eat oak catkins (flowers) are yellow-brown, rugose, and beaded, resembling nothing else but an oak catkin. They are magnificently camouflaged against predation. But what of the caterpillars that hatch in the summer, after all the catkins are gone? They, too, are well camouflaged, resembling year-old oak twigs. What controls this difference? By doing reciprocal feeding experiments, Greene (1989) was able to convert spring caterpillars into summer morphs by feeding them oak leaves. The reciprocal experiment did not turn the summer morphs into catkin-like caterpillars. Thus, it appears that the catkin form is the "default state" and that something induces the twiglike morphology. That something is probably a tannin that is concentrated in oak leaves as they mature.

Embryologists have emphasized that what gets inherited is not a deterministic genotype, but rather a genotype that encodes a potential range of phenotypes. The environment is often able to select the phenotype that is adaptive for that season or habitat. This continuous range of phenotypes expressed by a single genotype across a range of environmental

(A)

(B)

Figure 3.4
Two morphs of *Nemoria arizonaria*. (A) Caterpillars that hatch in the spring eat oak catkins and develop a cuticle that resembles these flowers. (B) Caterpillars that hatch in the summer (after the catkins are gone) eat oak leaves. These caterpillars develop a cuticle that resembles young oak twigs. (Photographs courtesy of E. Greene.)

Figure 3.5
The photolyase reaction. Photolyase repairs thymidine dimers caused by ultraviolet radiation. The bonds (red lines) linking the two thymidines (colored molecules) are shorter than the normal spacing between the bases and therefore distort the DNA in that area.

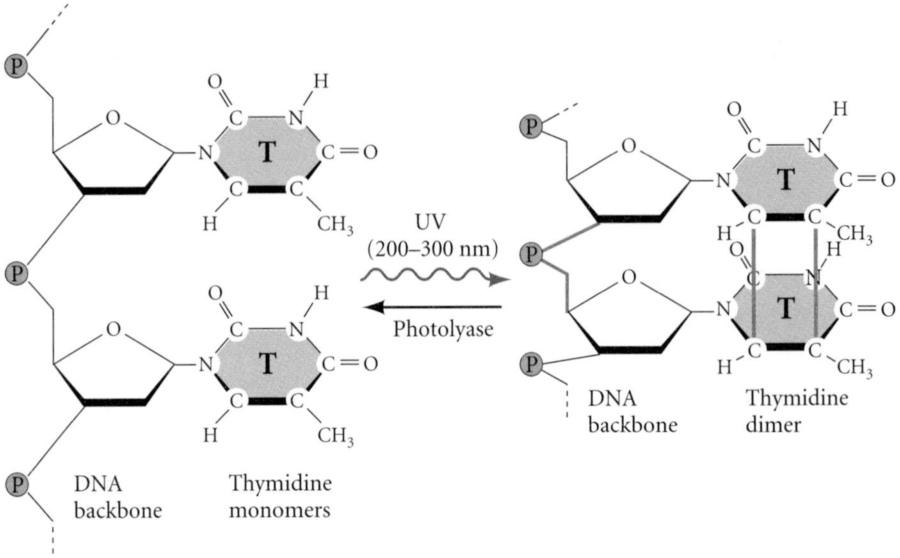

conditions is called the **reaction norm** (Woltereck 1909; Schmalhausen 1949; Stearns et al. 1991; Schlichting and Pigliucci 1998). The reaction norm is thus a property of the genome and is also subject to selection. Different genotypes will be expected to differ in the direction and amount of plasticity that they are able to express (Gotthard and Nylin 1995; Via et al. 1995).

PROTECTION OF THE EGG BY SUNSCREENS AND REPAIR ENZYMES. The survival of embryos in their environments poses major problems. Indeed, as Darwin clearly noted, most eggs and embryos fail to survive. A sea urchin may broadcast tens of thousands of eggs into the seawater, but only one or two of the resulting embryos will become adult urchins. Most become food for other organisms. Moreover, if the environment changes, embryonic survival may be increased or decreased dramatically. For instance, many eggs and early embryos lie in direct sunlight for long periods. If we lie in the sun for hours without sunscreen, we get sunburn from the ultraviolet rays of the sun; this radiation is harmful to our DNA. How can eggs survive all those hours of constant exposure to the sun (often on the same beaches where we sun ourselves)? First, it seems that many eggs have evolved natural sunscreens. The eggs of many marine organisms possess high concentrations of mycosporine amino acid pigments, which absorb ultraviolet radiation (UV-B). Moreover, just like our melanin pigment, these pigments can be induced by exposure to UV-B radiation (Jokiel and York 1982; Siebeck 1988). The eggs of tunicates are very resistant to UV-B radiation, and much of this resistance comes from extracellular coats that are enriched with mycosporine compounds (Mead and Epel 1995). Similarly, Adams and Shick (1996) experimentally modulated the amount of mycosporine amino acids in sea urchin eggs and found that embryos from eggs with more of these compounds were better protected from UV damage than embryos with

less. So some of the beautiful pigments found in marine eggs have a very practical function.

VADE MECUM   **Sea urchins and UV radiation.** This segment presents data documenting the protection of sea urchin embryos by mycosporine amino acids. This type of research is linking developmental biology with ecology and conservation biology.
[Click on Sea Urchin-UV]

The possibility exists that the recent global decline of amphibian populations may be caused by increasing amounts of UV-B reaching the Earth's surface. Populations of amphibians in widely scattered locations have been drastically reduced in the past decade, and some of these species (such as the golden toad of Costa Rica) have recently become extinct. While no single cause for these declines has been identified (and the extinctions may be due to the convergence of more than one cause), the fact that they have occurred in undisturbed areas and throughout the world has prompted considerations of global phenomena (see Phillips 1994).

Blaustein and his colleagues (1994) have looked at levels of photolyase, an enzyme that repairs UV damage to DNA by excising and replacing damaged thymidine residues (Figure 3.5), in amphibian eggs and oocytes. Levels of photolyase varied 80-fold among the tested species and correlates with the site of egg laying. Those eggs more exposed to the sun had higher levels of photolyase (Table 3.1) These levels also correlated with whether or not the species was suffering population decline. The highest photolyase levels were in those species (such as the Pacific tree frog, *Hyla regilla*) whose populations were not seen to be in decline. The lowest levels were seen in those species (such as the Western toad, *Bufo boreas*, and the Cascades frog, *Rana cascadae*) whose populations had declined dramatically.

Table 3.1 Photolyase activity correlated with exposure of eggs to ultraviolet radiation in 10 amphibian species

Species	Photolyase activity[a]	Mode of egg-laying	Exposure to sunlight
Plethodon dunni	<0.1	Eggs hidden	None
Xenopus laevis	0.1	Eggs laid in laboratory[b]	Limited
Triturus granulosa	0.2	Eggs hidden	Limited
Rana variegatus	0.3	Eggs hidden	None
Plethodon vehiculum	0.5	Eggs hidden	None
Ambystoma macrodactylum	0.8	Eggs often laid in open water	Some
Ambystoma gracile	1.0	Eggs often laid in open water	Some
Bufo boreas	1.3	Eggs laid in open, often in shallow water	High
Rana cascadae	2.4	Eggs laid in open shallow water	High
Hyla regilla	7.5	Eggs laid in open shallow water	High

Source: After Blaustein et al. 1994.

[a]Specific activity of photolyase, 10^{11} thymidine dimers separated per hour per μg. The values are averages of 6–8 assays.

[b]In nature, *Xenopus laevis* eggs are laid under vegetation with limited exposure to sunlight.

Blaustein and his colleagues tested whether or not UV-B could be a factor in lowering the hatching rate of amphibian eggs. At two field sites, they divided the eggs of each of three amphibian species into three groups (Figure 3.6). The first group developed without any sun filter. The second group developed under a filter that allowed UV-B to pass through. The third group developed under a filter that blocked UV-B from reaching the eggs. For *Hyla regilla*, the filters had no effect, and hatching success was excellent under all three conditions. For *Rana cascadea* and *Bufo boreas*, however, the UV-B blocking filter raised the percentage of eggs hatched from about 60% to close to 80%.

The environmental programs of experimental embryology were a major part of the discipline when *Entwicklungsmechanik* was first established. However, it soon became obvious that experimental variables could be better controlled in the laboratory than in the field, and that a scientist could do many more experiments in the laboratory. Thus, field experimentation in embryology dwindled in the first decades of the twentieth century (see Nyhart 1995). However, with our increasing concern about the environment, this area of developmental biology has become increasingly important. Other recent work in this field will be detailed in Chapter 21.

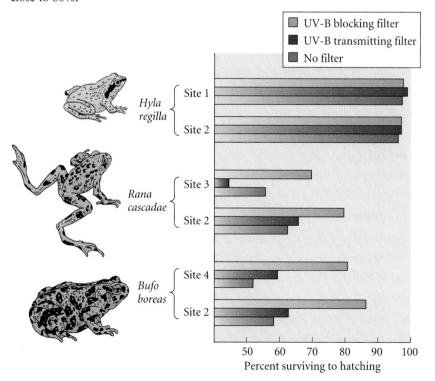

■ UV-B blocking filter
■ UV-B transmitting filter
■ No filter

Hyla regilla
{ Site 1
{ Site 2

Rana cascadae
{ Site 3
{ Site 2

Bufo boreas
{ Site 4
{ Site 2

50 60 70 80 90 100
Percent surviving to hatching

Figure 3.6
Hatching success rates in three amphibian species in the field. At each of two sites, eggs were placed in enclosures that were unshielded, shielded with an acetate screen that admitted UV-B radiation, or shielded with a Mylar™ screen that blocked UV-B radiation. Eggs of the tree frog *Hyla regilla* hatched successfully under all three conditions. Eggs of the frog *Rana cascadae* and the toad *Bufo boreas* hatched significantly better when protected from UV-B radiation. (After Blaustein et al. 1994.)

The Developmental Mechanics of Cell Specification

An embryo's environment may be a tide pool, a pond, or a uterus. As we saw above, the embryo interacts with its environment, and its developmental trajectory can be guided by information from its surroundings. On a smaller scale, the environment of an embryonic cell consists of the surrounding tissues within the embryo, and the fate of that cell (for instance, whether it becomes part of the skin or part of the lens) often depends upon its interactions with other components of its immediate "ecosystem."

Thus, a second research program of experimental embryology studies how interactions between embryonic cells generate the embryo. The development of specialized cell types is called **differentiation** (Table 3.2). These overt changes in cellular biochemistry and function are preceded by a process involving the **commitment** of the cell to a certain fate. At this point, even though the cell or tissue does not appear phenotypically different from its uncommitted state, its developmental fate has become restricted. The process of commitment can be divided into two stages (Harrison 1933; Slack 1991). The first stage is a labile phase called **specification**. The fate of a cell or a tissue is said to be *specified* when it is capable of differentiating autonomously when placed in a neutral environment such as a petri dish or test tube. (The environment is neutral with respect to the developmental pathway.) At this stage, the commitment is still capable of being reversed. The second stage of commitment is **determination**. A cell or tissue is said to be *determined* when it is capable of differentiating autonomously even when placed into another region of the embryo. If it is able to differentiate according to its original fate even under these circumstances, it is assumed that the commitment is irreversible.*

Autonomous Specification

Three basic modes of commitment have been described (Table 3.3; Davidson 1991). The first is called **autonomous specification**. In this case, if a particular blastomere is removed from an embryo early in its development, that isolated blastomere will produce the same cells that it would have made if it were still part of the embryo (Figure 3.7). Moreover, the embryo from which that cell is taken will lack those cells (and only those cells) that would have been produced by the missing blastomere. Autonomous specification gives rise to a pattern of development referred to as **mosaic development**, since the embryo appears to be constructed like a tile mosaic of independent self-differentiating parts. Invertebrate embryos, especially those of molluscs, annelids, and tunicates, often use autonomous specification to determine the fate of their cells. In these embryos, **morphogenetic determinants** (certain proteins or messenger RNAs) are placed in different regions of the egg cytoplasm and are apportioned to the different cells as the embryo divides. These morphogenetic determinants specify the cell type.

Autonomous specification was first demonstrated in 1887 by a French medical student, Laurent Chabry. Chabry desired to know the causes of birth defects, and he reasoned that such malformations might be caused by the lack of certain

Table 3.2 Some differentiated cell types and their major products

Cell type	Differentiated cell product	Specialized function of cell
Keratinocyte (epidermal cell)	Keratin	Protection against abrasion, desiccation
Erythrocyte (red blood cell)	Hemoglobin	Transport of oxygen
Lens cell	Crystallins	Transmission of light
B lymphocyte	Immunoglobulins	Antibody synthesis
T lymphocyte	Cell-surface antigens (lymphokines)	Destruction of foreign cells; regulation of immune response
Melanocyte	Melanin	Pigment production
Pancreatic islet cells	Insulin	Regulation of carbohydrate metabolism
Leydig cell (♂)	Testosterone	Male sexual characteristics
Chondrocyte (cartilage cell)	Chondroitin sulfate; type II collagen	Tendons and ligaments
Osteoblast (bone-forming cell)	Bone matrix	Skeletal support
Myocyte (muscle cell)	Muscle actin and myosin	Contraction
Hepatocyte (liver cell)	Serum albumin; numerous enzymes	Production of serum proteins and numerous enzymatic functions
Neurons	Neurotransmitters (acetylcholine epinephrine, etc.)	Transmission of electrical impulses
Tubule cell (♀) of hen oviduct	Ovalbumin	Egg white proteins for nutrition and protection of embryo
Follicle cell (♀) of insect ovary	Chorion proteins	Eggshell proteins for protection of embryo

*This irreversibility of commitment is only with regard to normal development. As Dolly and other cloned animals have recently shown, the nucleus of a differentiated cell can be reprogrammed experimentally to give rise to any cell type in the body. We will discuss this in detail in the next chapter.

Normal development of *Patella*

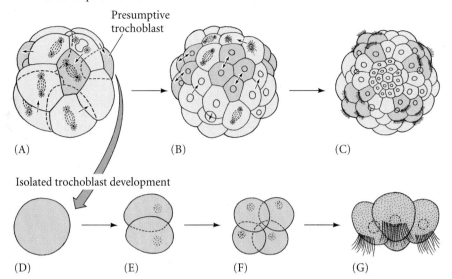

Presumptive
trochoblast

(A)   (B)   (C)

Isolated trochoblast development

(D)   (E)   (F)   (G)

Figure 3.7
Autonomous specification (mosaic development). (A–C) Differentiation of trochoblast (ciliated) cells of the mollusc *Patella*. (A) 16-cell stage seen from the side; the presumptive trochoblast cells are shaded. (B) 48-cell stage. (C) Ciliated larval stage, seen from the animal pole. (D–G) Differentiation of a *Patella* trochoblast cell isolated from the 16-cell stage and cultured in vitro. (E, F)Results of the first and second divisions in culture. (G) Ciliated products of (F). Even in isolated culture, the cells divide and become ciliated at the correct time. (After Wilson 1904.)

cells. He decided to perform experiments on tunicate embryos, since they have relatively large cells and were abundant in a nearby bay. This was a fortunate choice, because tunicate embryos develop rapidly into larvae with relatively few cells and cell types (Chabry 1887; Fischer 1991). Chabry set out to produce specific malformations by isolating or lancing specific blastomeres of the cleaving tunicate embryo. He discovered that each blastomere was responsible for producing a particular set of larval tissues (Figure 3.8). In the absence of particular blastomeres, the larva lacked just those structures normal-

ly formed by those cells. Moreover, he observed that when particular cells were isolated from the rest of the embryo, they formed their characteristic structure apart from the context of the other cells. Thus, each of the tunicate cells appeared to be developing autonomously.*

Recent studies have confirmed that when particular cells of the 8-cell tunicate embryo are removed, the embryo lacks those structures normally produced by the missing cells, and the isolated cells produce these structures away from the embryo. J. R. Whittaker provided dramatic biochemical confirmation of the cytoplasmic segregation of the morphogenetic determinants responsible for this pattern. Whittaker (1973) stained blastomeres for the presence of the enzyme acetylcholinesterase. This enzyme is found only in muscle tissue and is involved in enabling larval muscles to respond to repeated nerve impulses. From the cell lineage studies of Conklin and others (see Chapter 1), it was known that only one pair of blastomeres (the posterior vegetal pair, B4.1) in the 8-cell tunicate embryo is capable of producing tail muscle tissue. (As discussed in Chapter 1, the B4.1 blastomere pair contains the yellow crescent cytoplasm that correlates with muscle determination.) When Whittaker re-

---

Table 3.3 Modes of cell type specification and their characteristics

I. *Autonomous specification*

  Characteristic of most invertebrates.

  Specification by differential acquisition of certain cytoplasmic molecules present in the egg.

  Invariant cleavages produce the same lineages in each embryo of the species. Blastomere fates are generally invariant.

  Cell type specification precedes any large-scale embryonic cell migration.

  Produces "mosaic" ("determinative") development: cells cannot change fate if a blastomere is lost.

II. *Conditional specification*

  Characteristic of all vertebrates and few invertebrates.

  Specification by interactions between cells. Relative positions are important.

  Variable cleavages produce no invariant fate assignments to cells.

  Massive cell rearrangements and migrations precede or accompany specification.

  Capacity for "regulative" development: allows cells to acquire different functions.

III. *Syncytial specification*

  Characteristic of most insect classes.

  Specification of body regions by interactions between cytoplasmic regions prior to cellularization of the blastoderm.

  Variable cleavage produces no rigid cell fates for particular nuclei.

  After cellularization, conditional specification is most often seen.

Source: After Davidson 1991.

---

*This was not the answer Chabry expected, nor the one he had hoped to find. In nineteenth-century France, conservatives favored preformationist views, which were interpreted to support hereditary inequalities between members of a community. What you were was determined by your lineage. Liberals, especially Socialists, favored epigenetic views, which were interpreted to indicate that everyone started off with an equal hereditary endowment, and that no one had a "right" to a higher position than any other person. Chabry, a Socialist who hated the inherited rights of the aristocrats, took pains not to extrapolate his data to anything beyond tunicate embryos (see Fischer 1991).

moved these two cells and placed them in isolation, they produced muscle tissue that stained positively for the presence of acetylcholinesterase (Figure 3.9). When he transferred some of the yellow crescent cytoplasm of the B4.1 (muscle-forming) blastomere into the b4.2 (ectoderm-forming) blastomere of an 8-cell tunicate embryo, the ectoderm-forming blastomere generated muscle cells as well as its normal ectodermal progeny (Figure 3.10; Whittaker 1982).

Conditional specification

THE PHENOMENON OF CONDITIONAL SPECIFICATION. A second mode of commitment involves interactions with neighboring cells. In this type of specification, each cell originally has the ability to become many different cell types. However, the interactions of the cell with other cells restricts the fate of one or both of the participants. This mode of commitment is sometimes called **conditional specification**, because the fate of a cell depends upon the conditions in which the cell finds

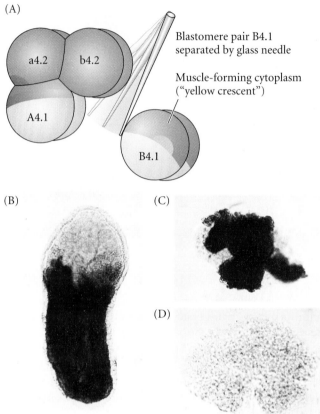

(A)

Blastomere pair B4.1 separated by glass needle

Muscle-forming cytoplasm ("yellow crescent")

(B)

(C)

(D)

Figure 3.9
Acetylcholinesterase in the progeny of the muscle lineage blastomeres (B4.1) isolated from a tunicate embryo at the 8-cell stage. (A) Diagram of the isolation procedure. (B) Localization of acetylcholinesterase in the tail muscles of an intact tunicate larva. The presence of the enzyme is demonstrated by the dark staining. The same dark staining is seen in the progeny of the B4.1 blastomere pair (C), but not in the remaining 6/8 of the embryo (D) when incubated for the length of time it normally takes to form a larva. (From Whittaker 1977; photographs courtesy of J. R. Whittaker.)

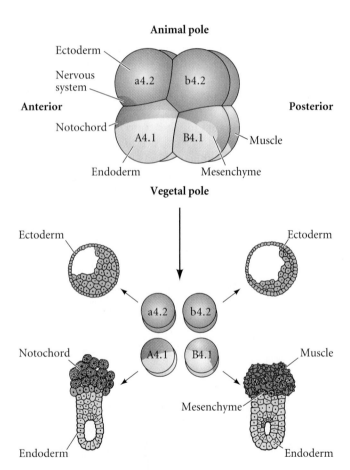

Figure 3.8
Autonomous specification in the early tunicate embryo. When the four blastomere pairs of the 8-cell embryo are dissociated, each forms structures that it would have formed if it had remained in the embryo. (The fate map of the tunicate shows that the left and right sides produce identical cell lineages.) (After Reverberi and Minganti 1946.)

itself. If a blastomere is removed from an early embryo that uses conditional specification, the remaining embryonic cells alter their fates so that the roles of the missing cells can be taken over. This ability of the embryonic cells to change their fates to compensate for the missing parts is called **regulation** (Figure 3.11). The isolated blastomere can also give rise to a wide variety of cell types (and sometimes generates cell types that the cell would normally not have made if it were part of the embryo). Thus, conditional specification gives rise to a pattern of embryogenesis called **regulative development.***

---

*Sydney Brenner (quoted in Wilkins 1993) has remarked that animal development can proceed according to either the American or the European plan. Under the European plan (autonomous specification), you are what your progenitors were. Lineage is important. Under the American plan (conditional specification), the cells start off undetermined, but with certain biases. There is a great deal of mixing, lineages are not critical, and one tends to becomes what one's neighbors are.

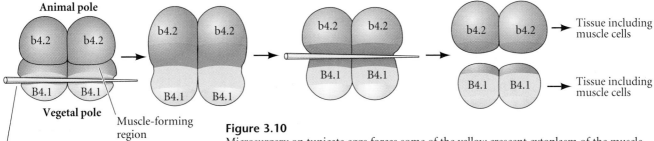

Animal pole

b4.2 | b4.2

B4.1 | B4.1

Vegetal pole

Muscle-forming region

Needle pushes muscle-forming cytoplasm into animal cells

→ Tissue including muscle cells

→ Tissue including muscle cells

Figure 3.10
Microsurgery on tunicate eggs forces some of the yellow crescent cytoplasm of the muscle-forming B4.1 blastomeres to enter the b4.2 (skin- and nerve-producing) blastomere pair. Pressing the B4.1 blastomeres with a glass needle causes the regression of the cleavage furrow. The furrow will re-form at a more vegetal position where the cells are cut with a needle. The new furrow will thereby separate the cells in such a way that the b4.2 blastomeres receive some of the muscle-forming ("yellow crescent") B4.1 cytoplasm. These modified b4.2 cells produce muscle cells as well as their normal ectodermal progeny. (After Whittaker 1983.)

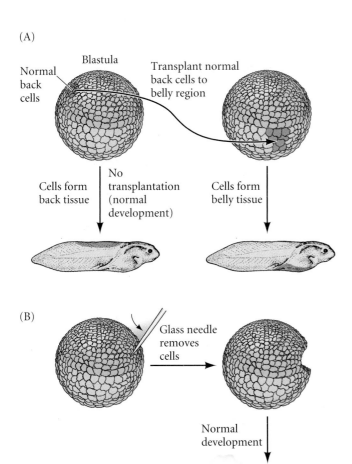

(A)

Blastula

Normal back cells

Transplant normal back cells to belly region

Cells form back tissue

No transplantation (normal development)

Cells form belly tissue

(B)

Glass needle removes cells

Normal development

Figure 3.11
Conditional specification. (A) What a cell becomes depends upon its position in the embryo. Its fate is determined by interactions with neighboring cells. (B) If cells are removed from the embryo, the remaining cells can regulate and compensate for the missing part.

Regulative development is seen in most vertebrate embryos, and it is obviously critical in the development of identical twins. In the formation of such twins, the cleavage-stage cells of a single embryo divide into two groups, and each group of cells produces a fully developed individual (Figure 3.12).

The research leading to the discovery of conditional specification began with the testing of a hypothesis claiming that there was no such thing. In 1883, August Weismann proposed the first testable model of cell specification, the **germ plasm theory**. Based on the scant knowledge of fertilization available at that time, Weismann boldly proposed that the sperm and egg provided equal chromosomal contributions, both quantitatively and qualitatively, to the new organism. Moreover, he postulated that the chromosomes carried the inherited poten-

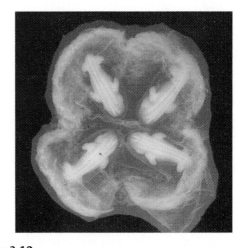

Figure 3.12
In the early developmental stages of many vertebrates, the separation of the embryonic cells into two parts can create twins. This phenomenon occurs sporadically in humans. However, in the nine-banded armadillo, *Dasypus novemcinctus*, the original embryo always splits into four separate groups of cells, each of which forms its own embryo. (Photograph courtesy of K. Benirschke.)

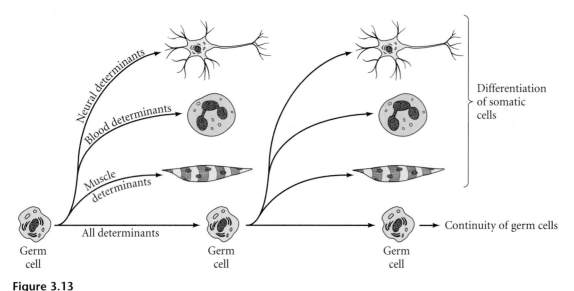

Figure 3.13
Weismann's theory of inheritance. The germ cell gives rise to the differentiating somatic cells of the body (indicated in color), as well as to new germ cells. Weismann hypothesized that only the germ cells contained all the inherited determinants. The somatic cells were each thought to contain a subset of the determinants. The types of determinants found in the nucleus would determine the cell type. (After Wilson 1896.)

tials of this new organism.* However, not all the determinants on the chromosomes were thought to enter every cell of the embryo. Instead of dividing equally, the chromosomes were hypothesized to divide in such a way that different chromosomal determinants entered different cells. Whereas the fertilized egg would carry the full complement of determinants, certain somatic cells would retain the "blood-forming" determinants while others would retain the "muscle-forming" determinants (Figure 3.13). Only in the nuclei of those cells destined to become gametes (the germ cells) were all types of determinants

*Embryologists were thinking in these terms some 15 years before the rediscovery of Mendel's work. Weismann (1892, 1893) also speculated that these nuclear determinants of inheritance functioned by elaborating substances that became active in the cytoplasm!

thought to be retained. The nuclei of all other cells would have only a subset of the original determinant types.

In postulating this model, Weismann had proposed a hypothesis of development that could be tested immediately. Based on the fate map of the frog embryo, Weismann claimed that when the first cleavage division separated the future right half of the embryo from the future left half, there would be a separation of "right" determinants from "left" determinants in the resulting blastomeres. The testing of this hypothesis pioneered three of the four major techniques involved in experimental embryology:

- The **defect experiment**, wherein one destroys a portion of the embryo and then observes the development of the impaired embryo.
- The **isolation experiment**, wherein one removes a portion of the embryo and then observes the development of the partial embryo and the isolated part.

Figure 3.14
Roux's attempt to show mosaic development. Destroying (but not removing) one cell of a 2-cell frog embryo results in the development of only one-half of the embryo.

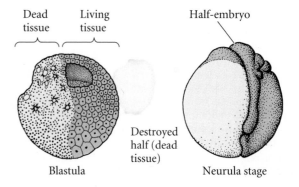

- The **recombination experiment**, wherein one observes the development of the embryo after replacing an original part with a part from a different region of the embryo.
- The **transplantation experiment**, wherein one portion of the embryo is replaced by a portion from a different embryo. This fourth technique was used by some of the same scientists when they first constructed fate maps of early embryos (see Chapter 1).

One of the first scientists to test Weismann's hypothesis was Wilhelm Roux, a young German embryologist. In 1888, Roux published the results of a series of defect experiments in which he took 2- and 4-cell frog embryos and destroyed some of the cells of each embryo with a hot needle. Weismann's hypothesis predicted the formation of right or left half-embryos. Roux obtained half-blastulae, just as Weismann had predicted (Figure 3.14). These developed into half-neurulae having a complete right or left side, with one neural fold, one ear pit, and so on. He therefore concluded that the frog embryo was a mosaic of self-differentiating parts and that it was likely that each cell received a specific set of determinants and differentiated accordingly.

Nobody appreciated Roux's work and the experimental approach to embryology more than Hans Driesch. Driesch's goal was to explain development in terms of the laws of physics and mathematics. His initial investigations were similar to those of Roux. However, while Roux's studies were *defect* experiments that answered the question of how the remaining blastomeres of an embryo would develop when a subset of them was destroyed, Driesch (1892) sought to extend this research by performing *isolation* experiments. He separated sea urchin blastomeres from each other by vigorous shaking (or, later, by placing them in calcium-free seawater). To Driesch's surprise, each of the blastomeres from a 2-cell embryo developed into a complete larva. Similarly, when Driesch separated the blastomeres from 4- and 8-cell embryos, some of the isolated cells produced entire pluteus larvae (Figure 3.15).

Here was a result drastically different from the predictions of Weismann or Roux. Rather than self-differentiating into its future embryonic part, each isolated blastomere regulated its development so as to produce a complete organism. Moreover, these experiments provided the first experimentally observable instance of regulative development.

Driesch confirmed regulative development in sea urchin embryos by performing an intricate *recombination* experiment. In sea urchin eggs, the first two cleavage planes are meridional, passing through both the animal and vegetal poles, whereas the third division is equatorial, dividing the embryo into four upper and four lower cells. Driesch (1893) changed the direction of the third cleavage by gently compressing early embryos between two glass plates, thus causing the third division to be meridional like the preceding two. After he released the pressure, the fourth division was equatorial. This procedure reshuffled the nuclei, causing a nucleus that normally would be in the region destined to form endoderm to now be in the presumptive ectoderm region. Some nuclei that would normally have produced dorsal structures were now found in the ventral cells (Figure 3.16). If segregation of nuclear determinants had occurred (as had been proposed by Weismann and Roux), the resulting embryo should have been strangely disordered. However, Driesch obtained normal larvae from these embryos. He concluded, "The relative position of a blastomere within the whole will probably in a general way determine what shall come from it."

Figure 3.15
Driesch's demonstration of regulative development. (A) An intact 4-cell sea urchin embryo generates a normal pluteus larva. (B) When one removes the 4-cell embryo from its fertilization envelope and isolates each of the four cells, each cell can form a smaller, but normal, pluteus larva. (All larvae are drawn to the same scale.) Note that the four larvae derived in this way are not identical, despite their ability to generate all the necessary cell types. Such variations are also seen in adult sea urchins formed in this way (Marcus 1979). (Photograph courtesy of G. Watchmaker.)

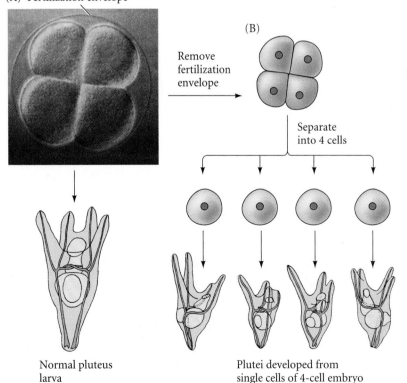

(A) Fertilization envelope

(B)

Remove fertilization envelope

Separate into 4 cells

Normal pluteus larva

Plutei developed from single cells of 4-cell embryo

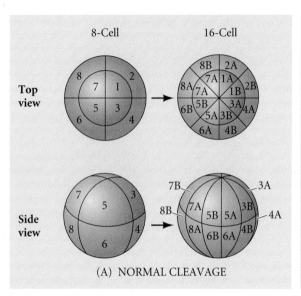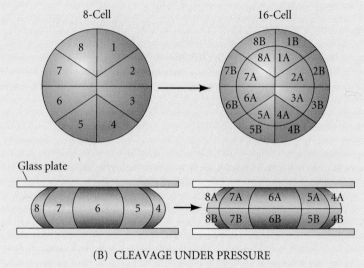

Figure 3.16
Driesch's pressure-plate experiment for altering the distribution of nuclei. (A) Normal cleavage in 8- to 16-cell sea urchin embryos, seen from the animal pole (upper sequence) and from the side (lower sequence). (B) Abnormal cleavage planes formed under pressure, as seen from the animal pole and from the side. (After Huxley and deBeer 1934.)

The consequences of these experiments were momentous, both for embryology and for Driesch personally. First, Driesch had demonstrated that the prospective potency of an isolated blastomere (those cell types it was possible for it to form) is greater than its prospective fate (those cell types it would normally give rise to over the unaltered course of its development). According to Weismann and Roux, the prospective potency and the prospective fate of a blastomere should be identical. Second, Driesch concluded that the sea urchin embryo is a "harmonious equipotential system" because all of its potentially independent parts functioned together to form a single organism. Third, he concluded that the fate of a nucleus depended solely on its location in the embryo. Driesch (1894) hypothesized a series of events wherein development proceeded by the interactions of the nucleus and cytoplasm:

> *Insofar as it contains a nucleus, every cell, during development, carries the totality of all primordia; insofar as it contains a specific cytoplasmic cell body, it is specifically enabled by this to respond to specific effects only. …When nuclear material is activated, then, under its guidance, the cytoplasm of its cell that had first influenced the nucleus is in turn changed, and thus the basis is established for a new elementary process, which itself is not only the result but also a cause.*

This strikingly modern concept of nuclear-cytoplasmic interaction and nuclear equivalence eventually caused Driesch to abandon science. Because the embryo could be subdivided into parts that were each capable of re-forming the entire organism, he could no longer envision it as a physical machine. In other words, Driesch had come to believe that development could not be explained by physical forces. Harking back to Aristotle, he invoked a vital force, *entelechy* ("internal goal-directed force"), to explain how development proceeds. Essentially, he believed that the embryo was imbued with an internal psyche and wisdom to accomplish its goals despite the obstacles embryologists placed in its path. Unable to explain his results in terms of the physics of his day, Driesch renounced the study of developmental physiology and became a philosophy professor, proclaiming **vitalism** (the doctrine that living things cannot be explained by physical forces alone) until his death in 1941. Others, especially Oscar Hertwig (1894), were able to incorporate Driesch's experiments into a more sophisticated experimental embryology.*

---

*Driesch also became an outspoken opponent of the Nazis, and was one of the first non-Jewish professors to be forcibly retired when Hitler came to power (Harrington 1996). Hertwig used Driesch's experiments and some of his own to strengthen within embryology a type of materialistic philosophy called **wholist organicism**. This philosophy embraces the views that (1) the properties of the whole cannot be predicted solely from the properties of the component parts, and (2) the properties of the parts are informed by their relationship to the whole. As an analogy, the meaning of a sentence obviously depends on the meanings of its component parts, words. However, the meaning of each word depends on the entire sentence. In the sentence, "The party leaders were split on the platform," the possible meanings of each noun and verb are limited by the meaning of the entire sentence and by their relationships to other words within the sentence. Similarly, the phenotype of a cell in the embryo depends on its interactions within the entire embryo. The opposite materialist view is **reductionism**, which maintains that the properties of the whole can be known if all the properties of the parts are known. Embryology has traditionally espoused wholist organicism as its ontology (model of reality) while maintaining a reductionist methodology (experimental procedures) (Needham 1943; Haraway 1976; Hamburger 1988; Gilbert and Faber 1996).

Table 3.4 Experimental procedures and results of Roux and Dreisch

Investigator	Organism	Type of experiment	Conclusion	Interpretation concerning potency and fate
Roux (1888)	Frog (*Rana fusca*)	Defect	Mosaic development (autonomous)	Prospective potency equals prospective fate
Driesch (1892)	Sea urchin (*Echinus microtuberculatus*)	Isolation	Regulative development (conditional)	Prospective potency is greater than prospective fate
Dreisch (1893)	Sea urchin (*Echinus* and *Paracentrotus*)	Recombination	Regulative development (conditional)	Prospective potency is greater than prospective fate

VADE MECUM Sea urchin development. Roux's and Dreisch's experiments manipulated normal development. Normal sea urchin development is seen here in video and labeled photographs.
[Click on Sea Urchin]

The differences between Roux's experiments and those of Driesch are summarized in Table 3.4. The difference between isolation and defect experiments and the importance of the interactions among blastomeres were highlighted in 1910, when J. F. McClendon showed that isolated frog blastomeres behave just like separated sea urchin cells. Therefore, the mosaic-like development of the first two frog blastomeres in Roux's study was an artifact of the defect experiment. Something in or on the dead blastomere still informed the live cells that it existed. Therefore, even though Weismann and Roux pioneered the study of developmental physiology, their proposition that differentiation is caused by the segregation of nuclear determinants was soon shown to be incorrect.

THE INFLUENCE OF NEIGHBORING CELLS. Driesch referred to the embryo as an "harmonious equipotential system" because each of the composite cells had surrendered most of its potential in order to form part of a single complete organism. Each cell could have become a complete animal on its own, yet didn't. What made the cells cooperate instead of becoming autonomous entities? Recent evidence suggests that the "harmonious equipotential system" is the result of negative induction events that mutually restrict the fates of neighboring cells. Jon Henry and colleagues in Rudolf Raff's laboratory (1989) showed that if one isolates pairs of cells from the animal cap of a 16-cell sea urchin embryo, those cells can give rise to both ectodermal and mesodermal components. However, their capacity to form mesoderm is severely restricted if they are aggregated with other animal cap pairs. Thus, the presence of neighbor cells, even of the same kind, restricts the potencies of both partners (Figure 3.17). Ettensohn and McClay (1988) and Khaner and Wilt (1990, 1991) showed that potency is also restricted when a cell is combined with its neighbors along the animal-vegetal axis. First, they demonstrated that the number

of skeletal mesoderm cells is fixed and can be regulated by changes in the cells that normally produce the gut endoderm. If all the 60 skeletal mesoderm cells of the sea urchin *Lytechinus variegatus* are removed from the early gastrula, an equal number of skeletal cells is produced from cells that would otherwise have become part of the gut. If one removes 20 skeletal precursor cells, about 20 gut precursor cells become skeletal mesoderm. Thus, the skeletal mesoderm cells have a restrictive influence, preventing the formation of new skeletal mesoderm cells from the gut precursors.

MORPHOGEN GRADIENTS. Cell fates may be specified by neighboring cells, but cell fates can also be specified by specific amounts of soluble molecules secreted at a distance from the target cells. Such a soluble molecule is called a **morphogen**, and a morphogen may specify more than one cell type by forming a **concentration gradient**. The concept of morphogen gradients had been used to model another phe-

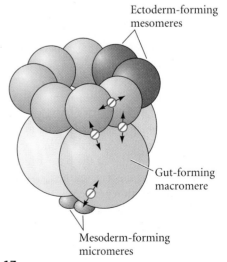

Figure 3.17
Summary of inhibitory interactions in the sea urchin blastula. Double-headed arrows illustrate the mutually restrictive interactions between adjacent cells. (After Henry et al. 1989.)

nomenon of regulative development: **regeneration**. It had been known since the 1700s that when hydras and planarian flatworms were cut in half, the head half would regenerate a tail from the wound site, while the tail half would regenerate a head. Allman (1864) had called attention to the fact that this phenomenon indicated a polarity in the organization of the hydra. It was not until 1905, however, that Thomas Hunt Morgan (1905, 1906) realized that such polarity indicated an important principle in development. He pointed out that if the head and tail were both cut off a flatworm, leaving only the medial segment, this segment would regenerate a head from the former anterior end and a tail from the former posterior end— never the reverse (Figure 3.18A,B). Moreover, if the medial segment were sufficiently small, the regenerating portions would be abnormal (Figure 3.18C). Morgan postulated a gradient of anterior-producing materials concentrated in the head region. The middle segment would be told what to regenerate at both ends by the concentration gradient of these materials. If the piece were too small, however, the gradient would not be sensed within the segment. (It is possible that there are actually two gradients in the flatworm, one to instruct the formation of a head and one to instruct the production of a tail. Regeneration will be discussed in more detail in Chapter 18.)

VADE MECUM   **Flatworm regeneration.** You should see it for yourself. Flatworms are easy to obtain, and cutting the animal in half does nothing more than what the animal does to itself. Here are videos and easy instructions for experimenting with these fascinating animals. **[Click on Flatworm]**

In the 1930s through the 1950s, gradient models were used to explain conditional cell specification in sea urchin and amphibian embryos (Hörstadius and Wolsky 1936; Hörstadius 1939; Toivonen and Saxén 1955). In the 1960s, these gradient models were extended to explain how cells might be told their position along an embryonic axis (Lawrence 1966; Stumpf 1966; Wolpert 1968, 1969). In such models, a soluble substance—the morphogen—is posited to diffuse from its site of synthesis (source) to its site of degradation (sink). Wolpert (1968) illustrated this type of positional information using "the French flag analogy." Imagine a row of "flag cells," each of which is capable of differentiating into a red, white, or blue cell. Then imagine a morphogen whose source is on the left-hand edge of the blue stripe, and whose sink is at the other end of the flag, on the right-hand edge of the red stripe. A concentration gradient is thus formed, being highest at one end of the "flag tissue" and lowest at the other. The specification of what type of cell any of the multipotential cells in this tissue will become is accomplished by the concentration of the morphogen. Cells sensing a high concentration of the morphogen become blue. Then there is a threshold of morphogen concentration below which cells become white.

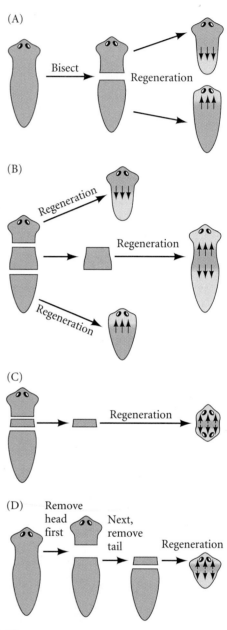

Figure 3.18
Flatworm regeneration and its limits. (A) If a flatworm is cut in two, the anterior portion of the bottom half regenerates a head, while the posterior of the upper half regenerates a tail. The same tissue can generate a head (if it is at the anterior portion of the tail piece) or a tail (if it at the posterior portion of the head piece). (B) If a flatworm is cut into three pieces, the middle piece will regenerate a head from its anterior end and a tail from its posterior end. (C) However, if the middle piece is too thin, there is no morphogen gradient within it, and regeneration is abnormal. (D) If the second cut is delayed, an equally thin middle section forms a normal worm, since the time lag has allowed an anterior–posterior gradient to become established. (After Gosse 1969.)

As the declining concentration of morphogen falls below another threshold, the cells become red (Figure 3.19).

Other tissues may use the same gradient system, but respond to the gradient in a different way. If cells that would normally become the middle segment of a *Drosophila* leg are removed from the leg-forming area of the larva and placed into the region that will become the tip of the fly's antenna, they differentiate into *claws*. These cells retain their committed status as leg cells, but respond to the positional information of their environment. Thereby, they became leg tip cells—claws. This phenomenon, said Wolpert, is analogous to reciprocally transplanting portions of American and French flags into each other. The segments will retain their identity (French or American), but will be positionally specified (develop colors) appropriate to their new positions.

WEBSITE **3.3 Receptor gradients.** In addition to a gradient of morphogen, there can also be a gradient of those molecules that recognize the morphogen. The interplay of morphogen gradients and the gradients of molecules that interpret them can give rise to interesting developmental patterns.

The molecules involved in establishing such gradients are beginning to be identified. For a diffusible molecule to be considered a morphogen, it must be demonstrated that cells respond directly to that molecule and that the differentiation of those cells depends upon the concentration of that molecule. One such system currently being analyzed concerns the ability of different concentrations of the protein **activin** to specify different fates in the frog *Xenopus*. In the *Xenopus* blastula, the cells in the middle of the embryo become mesodermal by responding to activin (or an activin-like compound) produced in the vegetal hemisphere. Makoto Asashima and his colleagues (Fukui and Asashima 1994; Ariizumi and Asashima 1994) have shown that the animal cap of the *Xenopus* embryo (which normally becomes ectoderm, but which can be induced to form mesoderm if transplanted into other regions within the embryo) responds differently to different concentrations of activin. If left untreated in saline solution, animal cap blastomeres form an epidermis-like mass of cells. However, if exposed to small amounts of activin, they form ventral mesodermal tissue—blood and connective tissue. Progressively higher concentrations of activin will cause the animal cap cells to develop into other types of mesodermal cells: muscles, notochord cells, and heart cells (Figure 3.20).

John Gurdon's laboratory has shown that these animal cap cells respond to activin by changing the expression of particular genes (Figure 3.21; Gurdon et al. 1994,1995). Gurdon and his colleagues placed activin-releasing beads or control beads into "sandwiches" of *Xenopus* animal cap cells. They found that cells exposed to little or no activin failed to express any of the genes associated with mesodermal tissues. These cells differentiated into ectoderm. Higher concentrations of

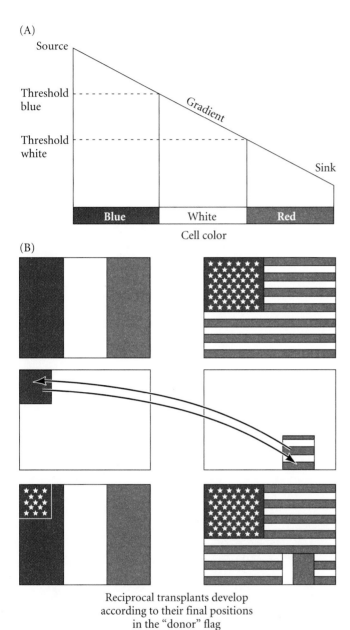

Reciprocal transplants develop according to their final positions in the "donor" flag

Figure 3.19
The French flag analogy for the operation of a gradient of positional information. (A) In this model, positional information is delivered by a gradient of a diffusible morphogen from a source to a sink. The thresholds indicated on the left are cellular properties that enable the gradient to be interpreted. For example, cells becomes blue at one concentration of the morphogen, but as the concentration declines below a certain threshold, cells become white. Where the concentration falls below another threshold, cells become red. The result is a pattern of three colors. (B) An important feature of this model is that a piece of tissue transplanted from one region of an embryo to another retains its identity (as to its origin), but differentiates according to its new positional instructions. This phenomenon is indicated schematically by reciprocal "grafts" between the flag of the United States of America and the French flag. (After Wolpert 1978.)

Figure 3.20
Activin (or a closely related compound such as Nodal) is thought to be responsible for converting animal hemisphere cells into mesoderm. When animal cap cells were removed from *Xenopus* blastulae and placed in saline solutions containing activin, the activin conferred different fates on the cells at different concentrations. (After Fukui and Asashima 1994.)

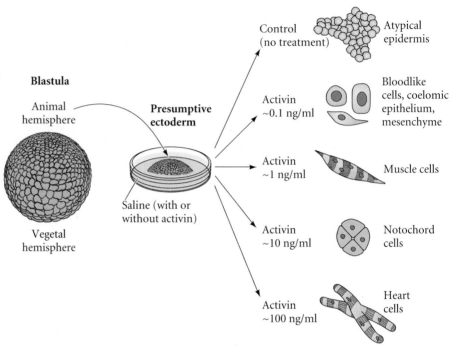

activin turned on genes such as *Brachyury*, which are responsible for instructing cells to become mesoderm. Still higher concentrations of activin caused the cells to express genes such as *goosecoid*, which are associated with the most dorsal mesodermal structure, the notochord. The expression of the *Brachyury* and *goosecoid* genes has been correlated with the number of activin receptors on each cell that are bound by activin. Each cell has about 500 activin receptors. If about 100 of them are bound, this activates *Brachyury* expression, and the tissue becomes ventrolateral mesoderm, such as blood and connective tissue cells. If about 300 of these receptors are occupied, the cell turns on its *goosecoid* gene and differentiates into a more dorsal mesodermal cell type such as notochord (Figure 3.22; Dyson and Gurdon 1998; Shimizu and Gurdon 1999).

WEBSITE **3.4 Demonstrating a morphogen.** It takes a lot of work to show that a particular chemical functions as a morphogen. This website discusses some of the controls used by Gurdon's laboratory to make certain that activin was functioning as a morphogen.

MORPHOGENETIC FIELDS. One of the most interesting ideas to come from experimental embryology has been that of the **morphogenetic field**. A morphogenetic field can be described as a group of cells whose position and fate are specified with

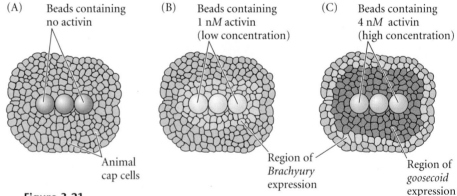

Figure 3.21
A gradient of activin causes different gene expression in *Xenopus* animal cap cells. The mRNAs from the *Brachyury* and *goosecoid* genes can be monitored by hybridization techniques that will be discussed in the next chapter. The cells containing these mRNAs appear darker than the cells not expressing them. Beads containing activin induce the expression (transcription of mRNA) of the *Brachyury* gene at distances removed from the beads. (A) Beads containing no activin did not elicit *Brachyury* gene expression. (B) Beads containing 1 nM activin elicited *Brachyury* expression near the beads. (C) Beads containing 4 nM activin elicited *Brachyury* gene expression several cell diameters away from the beads. However, *goosecoid* expression is seen where the concentration of activin is higher, near the source. Thus, it appears that *Brachyury* gene expression is induced at particular concentrations of activin, and that *goosecoid* is induced at higher concentrations of activin. (After Gurdon et al. 1994, 1995.)

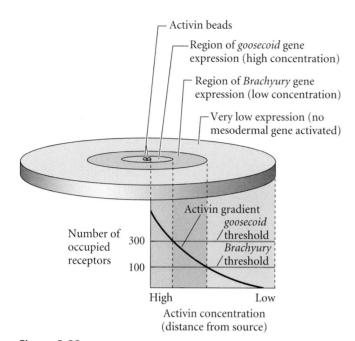

Figure 3.22
Interpretation of activin gradient by *Xenopus* animal cap cells. High concentrations of activin activate the *goosecoid* gene, while lower concentrations activate the *Brachyury* gene. This pattern correlates with the number of activin receptors occupied on the individual cells. A threshold value appears to exist that determines whether a cell will express *goosecoid*, *Brachyury*, or neither gene. (After Gurdon et al. 1998.)

respect to the same set of boundaries (Weiss 1939; Wolpert 1977). The general fate of a morphogenetic field is determined; thus, a particular field of cells will give rise to its particular organ (forelimb, eye, heart, etc.) even when transplanted to a different part of the embryo. However, the the individual cells *within* the field are not committed, and the cells of the field can regulate their fates to make up for missing cells in the field (Huxley and De Beer 1934; Opitz 1985; De Robertis et al. 1991). Moreover, as described earlier (i.e., the case of presumptive *Drosophila* leg cells transposed to the tip region of the presumptive antennal field), if cells from one field are placed within another field, they can use the positional cues of their new location, even if they retain their organ-specific commitment.

One of the first morphogenetic fields identified was the limb field. The mesodermal cells that give rise to a vertebrate limb can be identified by (1) removing certain groups of cells and observing that a limb does not develop in their absence (Detwiler 1918; Harrison 1918), (2) transplanting these cells to new locations and observing that they form a limb in this new place (Hertwig 1925), and (3) marking groups of cells with dyes or radioactive precursors and observing that their descendants partake in limb development (Rosenquist 1971). Figure 3.23 shows the prospective forelimb area in the tailbud

stage of the salamander *Ambystoma maculatum*. The center of this disc normally gives rise to the limb itself. Adjacent to it are the cells that will form the peribrachial flank tissue and the shoulder girdle. However, if all these cells are extirpated from the embryo, a limb will still form, albeit somewhat later, from an additional ring of cells that surrounds this area (and which would not normally form a limb). If this last ring of cells is included in the extirpated tissue, no limb will develop. This larger region, representing all the cells in the area capable of forming a limb, is called the **limb field**.

When it first forms, the limb field has the ability to regulate for lost or added parts. In the tailbud stage of *Ambystoma*, any half of the limb disc is able to regenerate the entire limb when grafted to a new site (Harrison 1918). This potential can also be shown by splitting the limb disc vertically into two or more segments and placing thin barriers between the segments to prevent their reunion. When this is done, each segment develops into a full limb. The regulative ability of the limb bud has recently been highlighted by a remarkable experiment of nature. In several ponds in the United States, numerous multilegged frogs and salamanders have been found (Figure 3.24). The presence of these extra appendages has been linked to the infestation of the larval abdomen by parasitic trematode worms. The eggs of these worms apparently split the limb buds in several places while the tadpoles were first forming these structures (Sessions and Ruth 1990; Sessions et al. 1999). Thus, like an early sea urchin embryo, the limb field represents a "harmonious equipotential system" wherein a cell can be instructed to form any part of the limb.

The morphogenetic field has been referred to as a "field of organization" (Spemann 1921) and as a "cellular ecosystem" (Weiss 1923, 1939). The ecosystem metaphor is quite appropriate, in that recent studies have shown that there are webs of

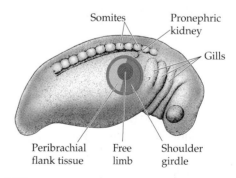

Figure 3.23
Prospective forelimb field of the salamander *Ambystoma maculatum*. The central area contains cells destined to form the limb per se (the free limb). The cells surrounding the free limb give rise to the peribrachial flank tissue and the shoulder girdle. The ring of cells outside these regions usually is not included the limb, but can form a limb if the more central tissues are extirpated. (After Stocum and Fallon 1983.)

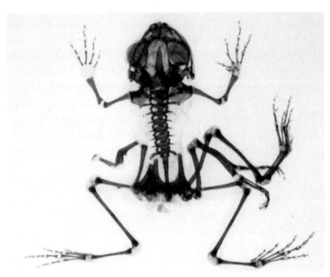

Figure 3.24
The regulative ability of the limb field as demonstrated by an experiment of nature. This multilimbed Pacific tree frog, *Hyla regilla*, is the result of infestation of the developing limb buds in the tadpole stage by trematode cysts. In this picture of the adult frog's skeleton, the cartilage is stained blue; the bones are stained red. (Courtesy of S. Sessions.)

interactions among the cells in different regions of a morphogenetic field. The molecular connections among the various cells of such fields are now being studied in the limb, eye, and heart fields of several vertebrates, as well as in the imaginal discs that form the eyes, antennae, legs, wings, and balancers of insects.

> WEBSITE **3.5 Rediscovery of the morphogenetic field.** The morphogenetic field was one of the most important concepts of embryology during the early twentieth century. This concept was eclipsed by research on the roles of genes in development, but it is being "rediscovered" as a consequence of those developmental genetic studies.

Syncytial specification

Many insects also use a third means, known as **syncitial specification**, to commit cells to their fates. Here, interactions occur not between cells, but between parts of one cell. In early embryos of these insects, cell division is not complete. Rather, the nuclei divide within the egg cytoplasm. This creates many nuclei in the large egg cell. A cytoplasm that contains many nuclei is called a **syncytium**. The egg cytoplasm, however, is not uniform. Rather, the anterior of the egg cytoplasm is markedly different from the posterior. In *Drosophila*, for instance, the anteriormost portion of the egg contains an mRNA that encodes a protein called Bicoid. The posteriormost portion of the egg contains an mRNA that encodes a protein called Nanos. When the egg is laid and fertilized, these two mRNAs are translated into their respective proteins. The

concentration of Bicoid protein is highest in the anterior and declines toward the posterior; that of Nanos protein is highest in the posterior and declines as it diffuses anteriorly. Thus, the long axis of the *Drosophila* egg is spanned by two opposing gradients—one of Bicoid protein coming from the anterior, and one of Nanos protein coming from the posterior. The Bicoid and Nanos proteins form a coordinate system based on their ratios, such that each region of the embryo will be distinguished by a different ratio of the two proteins. As the nuclei divide and enter different regions of the egg cytoplasm, they will be instructed by these ratios as to their position along the anterior–posterior axis. Those nuclei in regions containing high amounts of Bicoid and little Nanos will be instructed to activate those genes necessary for producing the head. Those nuclei in regions with slightly less Bicoid but with a small amount of Nanos will be instructed to activate those genes that generate the thorax. Those nuclei in regions that have little or no Bicoid and plenty of Nanos will be instructed to form the abdominal structures (Figure 3.25; Nüsslein-Volhard et al. 1987). The mechanisms of syncytial specification will be detailed in Chapter 9.

No embryo uses only autonomous, conditional, or syncytial mechanisms to specify its cells. One finds autonomous specification even in a "regulative embryo" such as the sea urchin, and the nervous system and some musculature of the "autonomously developing" tunicate have been shown to come from regulative interactions between its cells. Insects such as *Drosophila* use all three modes of specification to commit their cells to particular fates. Later chapters will detail the mechanisms by which cell fates are committed in these species.

Morphogenesis and Cell Adhesion

A body is more than a collection of randomly distributed cell types. Development involves not only the differentiation of cells, but also their organization into multicellular arrangements such as tissues and organs. When we observe the detailed anatomy of a tissue such as the neural retina of the eye, we see an intricate and precise arrangement of many types of cells. How can matter organize itself so as to create a complex structure such as a limb or an eye?

There are five major questions for embryologists who study morphogenesis:

1. **How are tissues formed from populations of cells?** For example, how do neural retina cells stick to other neural retina cells and not become integrated into the pigmented retina or iris cells next to them? How are the various cell types within the retina (the three distinct layers of photoreceptors, bipolar neurons, and ganglion cells) arranged so that the retina is functional?

2. **How are organs constructed from tissues?** The retina of the eye forms at a precise distance behind the cornea and the lens. The retina would be useless if it developed be-

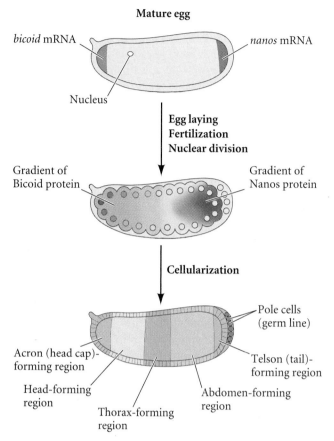

Mature egg

bicoid mRNA

nanos mRNA

Nucleus

**Egg laying
Fertilization
Nuclear division**

Gradient of
Bicoid protein

Gradient of
Nanos protein

Cellularization

Pole cells
(germ line)

Acron (head cap)-
forming region

Telson (tail)-
forming region

Head-forming
region

Abdomen-forming
region

Thorax-forming
region

Figure 3.25
Syncytial specification in the fruit fly *Drosophila melanogaster*.
Anterior–posterior specification originates from gradients within
the egg cell. *Bicoid* mRNA is stabilized in the most anterior portion
of the egg, while *Nanos* mRNA is tethered to the posterior end. (The
anterior can be recognized by the micropyle on the shell; this structure permits sperm to enter.) When the egg is laid and fertilized,
these two mRNAs are translated into proteins. The Bicoid protein
forms a gradient that is highest at the anterior end, and the Nanos
protein forms a gradient that is highest at the posterior end. These
two proteins form a coordinate system based on their ratios. Each
position along the axis is thus distinguished from any other position. When nuclear division occurs, each nucleus is given its positional information by the *ratio* of these proteins. The proteins forming these gradients will activate the transcription of the genes that
specify the segmental identities of the larva and the adult fly.

hind a bone or in the middle of the kidney. Moreover,
neurons from the retina must enter the brain to innervate
the regions of the brain cortex that analyze visual information. All these connections must be precisely ordered.

3. **How do organs form in particular locations, and how
   do migrating cells reach their destinations?** Eyes develop only in the head and nowhere else. What stops an eye
   from forming in some other area of the body? Some
   cells—for instance, the precursors of our pigment cells,
   germ cells, and blood cells—must travel long distances to
   reach their final destinations. How are cells instructed to

travel along certain routes in our embryonic bodies, and
how are they told to stop once they have reached their appropriate destinations?

4. **How do organs and their cells grow, and how is their
   growth coordinated throughout development?** The cells
   of all the tissues in the eye must grow in a coordinated
   fashion if one is to see. Some cells, including most neurons, do not divide after birth. In contrast, the intestine is
   constantly shedding cells, and new intestinal cells are regenerated each day. The mitotic rate of this tissue must be
   carefully regulated. If the intestine generated more cells
   than it sloughed off, it could produce tumorous outgrowths. If it produced fewer cells than it sloughed off, it
   would soon become nonfunctional. What controls the
   rate of mitosis in the intestine?

5. **How do organs achieve polarity?** If one were to look at a
   cross section of the fingers, one would see a certain organized collection of tissues—bone, cartilage, muscle, fat,
   dermis, epidermis, blood, and neurons. Looking at a cross
   section of the forearm, one would find the same collection of tissues. But they are arranged very differently in
   different parts of the arm. How is it that the same cell
   types can be arranged in different ways in different parts
   of the same structure?

All these questions concern aspects of cell behavior. There
are two major types of cell arrangements in the embryo: **epithelial cells**, which are tightly connected to one another in
sheets or tubes, and **mesenchymal cells**, which are unconnected to one another and which operate as independent
units. Morphogenesis is brought about through a limited
repertoire of variations in cellular processes within these two
types of arrangements: (1) the direction and number of cell
divisions; (2) cell shape changes; (3) cell movement; (4) cell
growth; (5) cell death; and (6) changes in the composition of
the cell membrane or secreted products. We will discuss the
last of these considerations here.

WEBSITE **3.6 How morphogenetic behaviors work.**
Although the repertoire of morphogenetic behaviors is
small, cells can do a great deal with this limited set of instructions. This website illustrates the epithelial and mesenchymal changes that effect development.

Differential cell affinity

Many of the answers to our questions about morphogenesis
involve the properties of the cell surface. The cell surface looks
pretty much the same in all cell types, and many early investigators thought that the cell surface was not even a living part
of the cell. We now know that each type of cell has a different
set of proteins in its surfaces, and that some of these differences
are responsible for forming the structure of the tissues and organs during development. Observations of fertilization and
early embryonic development made by E. E. Just (1939) sug-

gested that the cell membrane differed among cell types, but the modern analysis of morphogenesis began with the experiments of Townes and Holtfreter in 1955. Taking advantage of the discovery that amphibian tissues become dissociated into single cells when placed in alkaline solutions, they prepared single-cell suspensions from each of the three germ layers of amphibian embryos soon after the neural tube had formed. Two or more of these single-cell suspensions could be combined in various ways, and when the pH was normalized, the cells adhered to one another, forming aggregates on agar-coated petri dishes. By using embryos from species having cells of different sizes and colors, Townes and Holtfreter were able to follow the behavior of the recombined cells (Figure 3.26).

The results of their experiments were striking. First, they found that reaggregated cells become spatially segregated. That is, instead of the two cell types remaining mixed, each cell type sorts out into its own region. Thus, when epidermal (ectodermal) and mesodermal cells are brought together to form a mixed aggregate, the epidermal cells move to the periphery of the aggregate and the mesodermal cells move to the inside. In no case do the recombined cells remain randomly mixed, and in most cases, one tissue type completely envelops the other.

Second, the researchers found that the final positions of the reaggregated cells reflect their embryonic positions. The mesoderm migrates centrally with respect to the epidermis, adhering to the inner epidermal surface (Figure 3.27A). The mesoderm also migrates centrally with respect to the gut or endoderm (Figure 3.27B). However, when the three germ layers are mixed together, the endoderm separates from the ecto-

derm and mesoderm and is then enveloped by them (Figure 3.27C). In its final configuration, the ectoderm is on the periphery, the endoderm is internal, and the mesoderm lies in the region between them. Holtfreter interpreted this finding in terms of **selective affinity**. The inner surface of the ectoderm has a positive affinity for mesodermal cells and a negative affinity for the endoderm, while the mesoderm has positive affinities for both ectodermal and endodermal cells. Mimicry of normal embryonic structure by cell aggregates is also seen in the recombination of epidermis and neural plate cells (Figures 3.26 and 3.27D). The presumptive epidermal cells migrate to the periphery as before; the neural plate cells migrate inward, forming a structure reminiscent of the neural tube. When axial mesoderm (notochord) cells are added to a suspension of presumptive epidermal and presumptive neural cells, cell segregation results in an external epidermal layer, a centrally located neural tissue, and a layer of mesodermal tissue between them (Figure 3.27E). Somehow, the cells are able to sort out into their proper embryonic positions.

Such selective affinities were also noted by Boucaut (1974), who injected individual cells from specific germ layers into the body cavity of amphibian gastrulae. He found that these cells migrated back to their appropriate germ layer. Endodermal cells found positions in the host endoderm, whereas ectodermal cells were found only in host ectoderm. Thus, selective affinity appears to be important for imparting positional information to embryonic cells.

The third conclusion of Holtfreter and his colleagues was that selective affinities change during development. This should be expected, because embryonic cells do not retain a

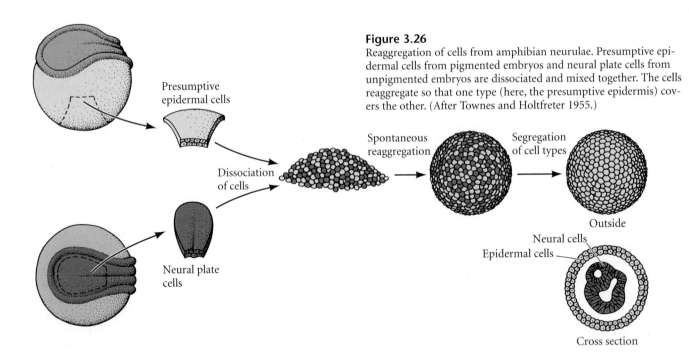

Figure 3.26
Reaggregation of cells from amphibian neurulae. Presumptive epidermal cells from pigmented embryos and neural plate cells from unpigmented embryos are dissociated and mixed together. The cells reaggregate so that one type (here, the presumptive epidermis) covers the other. (After Townes and Holtfreter 1955.)

Presumptive epidermal cells

Neural plate cells

Dissociation of cells

Spontaneous reaggregation

Segregation of cell types

Outside

Neural cells

Epidermal cells

Cross section

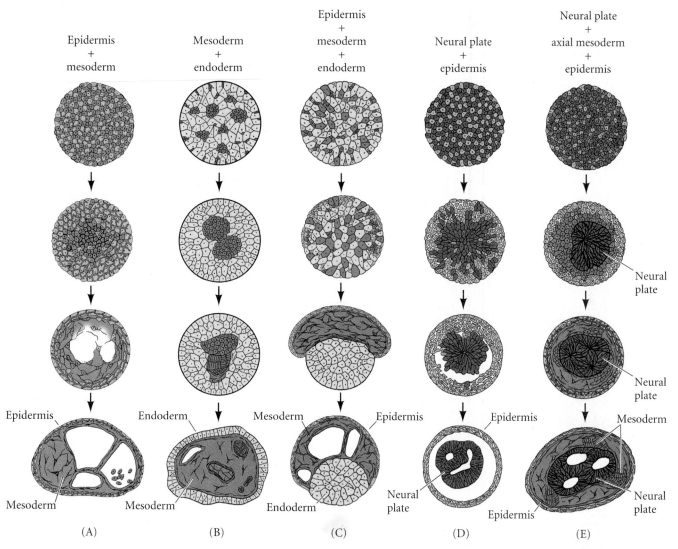

Epidermis + mesoderm

Mesoderm + endoderm

Epidermis + mesoderm + endoderm

Neural plate + epidermis

Neural plate + axial mesoderm + epidermis

(A) (B) (C) (D) (E)

Figure 3.27
Sorting out and reconstruction of spatial relationships in aggregates of embryonic amphibian cells. (After Townes and Holtfreter 1955.)

single stable relationship with other cell types. For development to occur, cells must interact differently with other cell populations at specific times. Such changes in cell affinity are extremely important in the processes of morphogenesis.

The reconstruction of aggregates from cells of later embryos of birds and mammals was accomplished by the use of the protease trypsin to dissociate the cells from one another (Moscona 1952). When the resulting single cells were mixed together in a flask and swirled so that the shear force would break any nonspecific adhesions, the cells sorted themselves out according to their cell type. In so doing, they reconstructed the organization of the original tissue (Moscona 1961; Giudice 1962). Figure 3.28 shows the "reconstruction" of skin tissue from a 15-day embryonic mouse. The skin cells are sep-

arated by proteolytic enzymes and then aggregated in a rotary culture. The epidermal cells of each aggregate migrate to the periphery, and the dermal cells migrate toward the center. In 72 hours, the epidermis has been reconstituted, a keratin layer has formed, and interactions between these tissues form hair follicles in the dermal region. Such reconstruction of complex tissues from individual cells is called **histotypic aggregation**.

The thermodynamic model of cell interactions

Cells, then, do not sort randomly, but can actively move to create tissue organization. What forces direct cell movement during morphogenesis? In 1964, Malcolm Steinberg proposed the **differential adhesion hypothesis**, a model that explained patterns of cell sorting based on thermodynamic principles. Using cells derived from trypsinized embryonic tissues, Steinberg showed that certain cell types always migrate centrally when combined with some cell types, but migrate peripherally when combined with others. Figure 3.29 illustrates

(A) Epidermis   Dermis   Primary
                           hair follicle

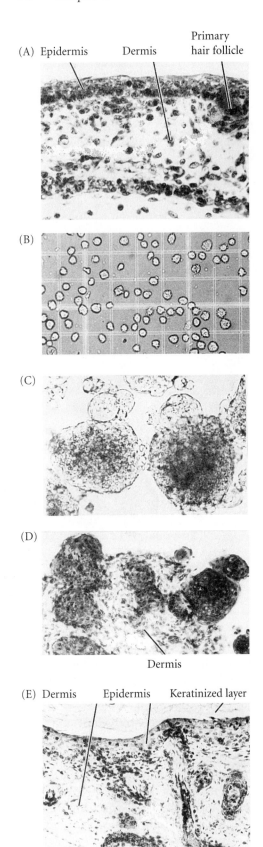

(B)

(C)

(D)

Dermis

(E) Dermis   Epidermis   Keratinized layer

Hair follicles

Figure 3.28
Reconstruction of skin from a suspension of skin cells from a 15-day embryonic mouse. (A) Section through intact embryonic skin, showing epidermis, dermis, and primary hair follicle. (B) Suspension of single skin cells from both the dermis and the epidermis. (C) Aggregates after 24 hours. (D) Section through an aggregate, showing migration of epidermal cells to the periphery. (E) Further differentiation of aggregates (72 hours), showing reconstituted epidermis and dermis, complete with hair follicles and keratinized layer. (From Monroy and Moscona 1979; photographs courtesy of A. Moscona.)

the interactions between pigmented retina cells and neural retina cells. When single-cell suspensions of these two cell types are mixed together, they form aggregates of randomly arranged cells. However, after several hours, the pigmented retina cells are no longer seen on the periphery of the aggregates, and after 2 days, two distinct layers are seen, with the pigmented retina cells lying internal to the neural retina cells. Moreover, such interactions form a hierarchy (Steinberg 1970). If the final position of one cell type, A, is internal to a second cell type, B, and the final position of B is internal to a third cell type, C, then the final position of A will always be internal to C. For example, pigmented retina cells migrate internally to neural retina cells, and heart cells migrate internally to pigmented retina cells. Therefore, heart cells migrate internally to neural retina cells.

(A)   (C)

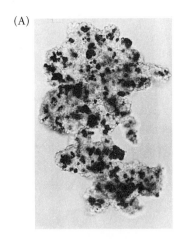

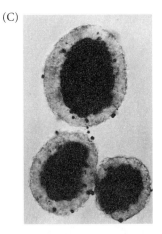

(B)

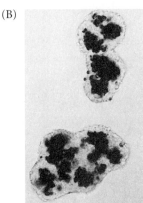

Figure 3.29
Aggregates formed by mixing 7-day-old chick embryo neural retina (unpigmented) cells with pigmented retina (dark) cells. (A) 5 hours after the single-cell suspensions are mixed, aggregates of randomly distributed cells are seen. (B) At 19 hours, the pigmented retina cells are no longer seen on the periphery. (C) At 2 days, a great majority of the pigmented retina cells are located in a central internal mass, surrounded by the neural retina cells. (The scattered pigmented cells are probably dead cells.) (From Armstrong 1989; photographs courtesy of P. B. Armstrong.)

This observation led Steinberg to propose that cells interact so as to form an aggregate with the smallest interfacial free energy. In other words, the cells rearrange themselves into the most thermodynamically stable pattern. If cell types A and B have different strengths of adhesion, and if the strength of A-A connections is greater than the strength of A-B or B-B connections, sorting will occur, with the A cells becoming central. On the other hand, if the strength of A-A connections is less than or equal to the strength of A-B connections, then the aggregate will remain as a random mix of cells. Finally, if the strength of A-A connections is far greater than the strength of A-B connections—in other words, if A and B cells show essentially no adhesivity toward one another—then A cells and B cells will form separate aggregates. According to this hypothesis, the early embryo can be viewed as existing in an equilibrium state until some change in gene activity changes the cell surface molecules. The movements that result seek to restore the cells to a new equilibrium configuration.

All that is needed for sorting to occur is that cell types differ in the strengths of their adhesion. In 1996, Foty and his colleagues in Steinberg's laboratory demonstrated that this was indeed the case: the cell types that had greater surface cohesion sorted within those cells that had less surface tension (Figure 3.30; Foty et al. 1996). In the simplest form of this model, all cells could have the same type of "glue" on the cell surface. The amount of this cell surface product, or the cellular architecture that allows the substance to be differentially distributed across the surface, could cause a difference in the number of stable contacts made between cell types. In a more specific version of this model, the thermodynamic differences could be caused by different types of adhesion molecules (see Moscona 1974). When Holtfreter's studies were revisited using modern techniques, Davis and colleagues (1997) found that the tissue surface tensions of the individual germ layers were precisely those required for the sorting patterns observed both in vitro and in vivo.

WEBSITE **3.7 Demonstrating the thermodynamic model.** The original in vivo evidence for the thermodynamic model of cell adhesion came from studies of limb regeneration. This website goes into some of the details of these experiments and how they are interpreted.

Cadherins and cell adhesion

Recent evidence shows that boundaries between tissues can indeed be created both by (1) different cell types having different types of cell adhesion molecules and (2) different cell types having different amounts of cell adhesion molecules. There are several classes of molecules that can mediate cell adhesion. The major cell adhesion molecules appear to be the **cadherins**. As their name suggests, they are *ca*lcium-dependent *adh*esion molecules. Cadherins are critical for establishing and maintaining intercellular connections, and they appear to be crucial

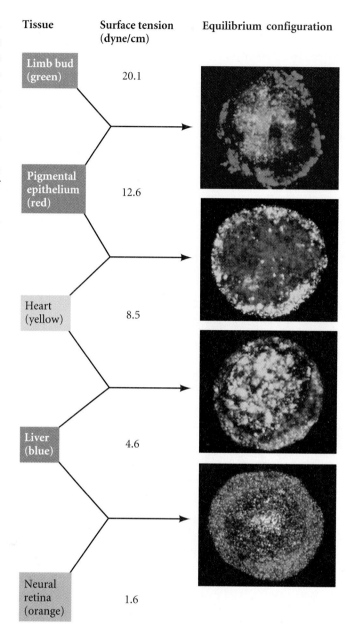

Tissue	Surface tension (dyne/cm)	Equilibrium configuration
Limb bud (green)	20.1	
Pigmental epithelium (red)	12.6	
Heart (yellow)	8.5	
Liver (blue)	4.6	
Neural retina (orange)	1.6	

Figure 3.30
Hierarchy of cell sorting in order of decreasing surface tensions. The equilibrium configuration reflected the strength of cell cohesion, with the cell types having the more cell cohesion segregating inside the cells with less cohesion. The images were obtained by sectioning the aggregates and assigning colors to the cell types by computer. The black areas represent cells whose signal was edited out in the program of image optimization. (From Foty et al. 1996; photograph courtesy of M. S. Steinberg and R. A. Foty.)

to the spatial segregation of cell types and to the organization of animal form (Takeichi 1987). Cadherins interact with other cadherins on adjacent cells, and they are anchored into the cell by a complex of proteins called **catenins** (Figure 3.31). The cadherin–catenin complex forms the classic adherens junctions that connect epithelial cells together. Moreover, since the

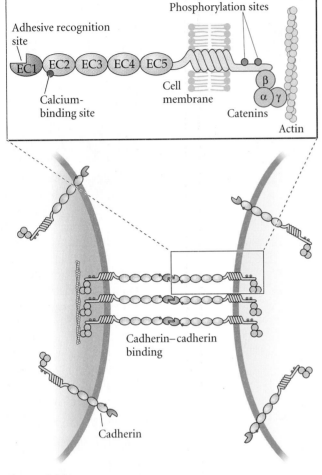

Figure 3.31
Schematic representation of cadherin-mediated cell adhesion. Cadherins are associated with three types of catenins. The catenins can become associated with the actin microfilament system within the cell. (After Takeichi 1991.)

catenins bind to the actin cytoskeleton of the cell, they integrate the epithelial cells together into a mechanical unit.

In vertebrate embryos, several major cadherin classes have been identified:

- **E-cadherin** (epithelial cadherin, also called uvomorulin and L-CAM) is expressed on all early mammalian embryonic cells, even at the 1-cell stage. Later, this molecule is restricted to epithelial tissues of embryos and adults.
- **P-cadherin** (placental cadherin) appears to be expressed primarily on the trophoblast cells (those placental cells of the mammalian embryo that contact the uterine wall) and on the uterine wall epithelium (Nose and Takeichi 1986). It is possible that P-cadherin facilitates the connection of the embryo to the uterus, since P-cadherin on the uterine cells is seen to contact P-cadherin on the trophoblast cells of mouse embryos (Kadokawa et al. 1989).
- **N-cadherin** (neural cadherin) is first seen on mesodermal cells in the gastrulating embryo as they lose their E-cadherin expression. It is also highly expressed on the cells of the developing central nervous system (Figure 3.32; Hatta and Takeichi 1986).

Figure 3.32
Localization of two different cadherins during the formation of the mouse neural tube. (A) Double immunofluorescent staining was used to localize E-cadherin (B) and N-cadherin (C) in the same transverse section of an 8.5-day embryonic mouse hindbrain. Antibodies to E-cadherin were labeled with one type of fluorescent dye (which fluoresces under one set of wavelengths), while antibodies to N-cadherin were marked with a second type of fluorescent dye (which emits its color at other wavelengths). Photographs taken at the different wavelengths reveal that the outer ectoderm expresses predominantly E-cadherin, while the invaginating neural plate ceases E-cadherin expression and instead expresses N-cadherin. (D) When the neural tube has formed, it expresses N-cadherin, the epidermis expresses E-cadherin, and the neural crest cells between them express neither. (B, C photographs by K. Shimamura and H. Matsunami, courtesy of M. Takeichi; D after Rutishauser et al. 1988.)

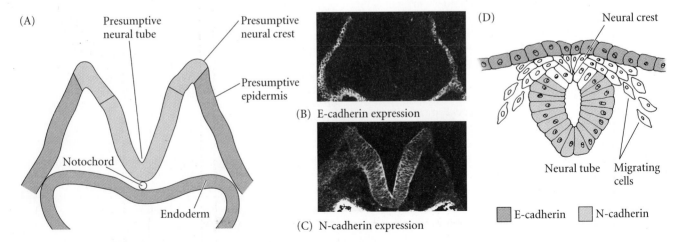

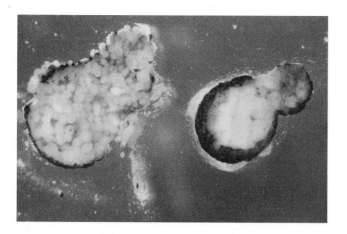

Figure 3.33
The importance of cadherins for maintaining cohesion between developing cells can be demonstrated by interfering with their production. When oocytes are injected with an antisense oligonucleotide against a maternally inherited cadherin mRNA (thus preventing the synthesis of the cadherin), the inner cells of the resulting embryo disperse when the animal cap is removed (left). In control embryos (right), the inner cells remain together. (After Heasman et al. 1994; photograph courtesy of J. Heasman.)

- **EP-cadherin** (C-cadherin) has been found to be critical for maintaining adhesion between the blastomeres of the *Xenopus* blastula and is required for the normal movements of gastrulation (Figure 3.33; Heasman et al. 1994; Lee and Gumbiner 1995).
- **Protocadherins** are calcium-dependent adhesion proteins that differ from the classic cadherins in that they lack connections to the cytoskeleton through catenins. Protocadherins have been found to be very important in separating the notochord from the other mesodermal tissues during *Xenopus* gastrulation (Chapter 10).

Cadherins join cells together by binding to the same type of cadherin on another cell. Thus, cells with E-cadherin stick best to other cells with E-cadherin, and they will sort out from cells containing N-cadherin in their membranes. This pattern is called **homophilic binding**. Cells expressing N-cadherin readily sort out from N-cadherin-negative cells in vitro, and univalent (Fab) antibodies against cadherins will convert a three-dimensional, histotypic aggregate of cells into a single layer of cells (Takeichi et al. 1979). Moreover, when activated E-cadherin genes are added to and expressed in cultured mouse fibroblasts (mesenchymal cells that usually do not express this protein), E-cadherin is seen on their cell surfaces, and the treated fibroblasts become tightly connected to one another (Nagafuchi et al. 1987). In fact, these cells begin acting like epithelial cells. The sorting out of cells can be explained by the amounts and types of cadherins on their cell surfaces. Fibroblasts made to express E-cadherin adhere to other E-cadherin-bearing fibroblasts, while fibroblasts made to express P-cadherin stick to other fibroblasts expressing P-cadherin (Takeichi 1987; Nose et al. 1988).

> **WEBSITE** **3.8 Cadherins: Functional anatomy**. The cadherin molecule has several functional domains that mediate its activities, and the mechanisms of homophilic adhesion are currently being resolved.

These adhesion patterns may have important consequences in the embryo. In the gastrula of the frog *Xenopus*, the neural tube expresses N-cadherin, while the epidermis expresses E-cadherin. Normally, these two tissues separate from each other such that the neural tube is inside the body and the epidermis covers the body (see Figure 3.32). If the epidermis is experimentally manipulated to remove its E-cadherin, the epidermal epithelium cannot hold together. If the epidermis is made to express N-cadherin, or if the neural cells are made to lose it, the neural tube will not separate from the epidermis (Figure 3.34; Detrick et al. 1990; Fujimori et al. 1990).

The amount of cadherin can also mediate the formation of embryonic structures. This was first shown to be a possibility when Steinberg and Takeichi (1994) collaborated on an experiment using two cell lines that were identical except that they synthesized different amounts of P-cadherin. When these two groups of cells were mixed, the cells that expressed more cadherin had a higher surface cohesion and sorted out within the lower-expressing group of cells. Recent studies show that differences in the degree of cell adhesion may be critical in the

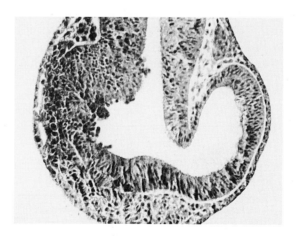

Figure 3.34
The importance of N-cadherin in the separation of neural and epidermal ectoderm. At the 4-cell stage, the blastomeres that form the left side of the *Xenopus* embryo were injected with an mRNA for N-cadherin that lacks the extracellular region of the cadherin. This mutation blocks N-cadherin function. During neurulation, the cells with the mutant protein did not form a coherent layer distinct from the epidermis. (From Kintner et al. 1992; photograph courtesy of C. Kintner.)

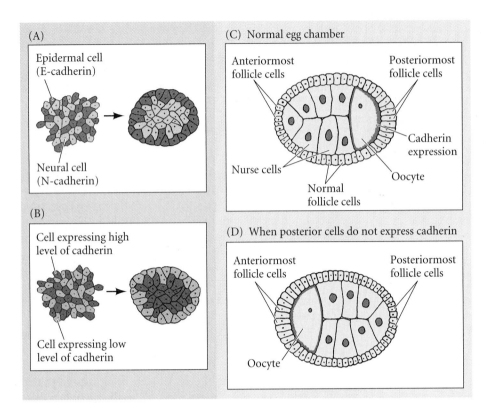

Figure 3.35
Cell sorting out in vivo: *Drosophila* oocytes. (A, B) Molecular bases for sorting. (A) Cells having different types of cadherins can sort from each other. (B) Cells having different amounts of cadherins can sort from each other. In both cases, those with the highest surface tension sort inward from those with less surface tension. (C, D) Sorting out in the *Drosophila* ovary caused by differential E-cadherin expression. (C) Normal egg chamber, with the oocyte positioned posteriorly and the nurse cells more anteriorly. This pattern is a consequence of the posterior follicle cells and the oocyte expressing higher levels of E-cadherin than any other cell in the field. (D) When the anterior follicle cells are induced to express more E-cadherin, the oocyte is positioned at the anterior pole. This disturbs axis formation in the fly embryo. (After Peifer 1998.)

development of the fruit fly embryo. Within the *Drosophila* ovary, the developing egg, or **oocyte,** is always found at the most posterior side of the egg chamber, or **follicle.** The oocyte's **nurse cells** (which export messenger RNA and ribosomes into the oocyte) are found more anteriorly (Figure 3.35). This pattern reflects the distribution of E-cadherin in these cells. Although all follicle cells, nurse cells, and the oocyte express E-cadherin, the oocyte and the posterior follicle cells express it at far higher levels than the other cells (Godt and Tepass 1998; González-Reyes and St. Johnston 1998). Moreover, when E-cadherin was experimentally removed from the oocyte and nurse cells (or from the follicle cells), the position of the oocyte became random.

> WEBSITE **3.9 Other cell adhesion molecules.** There are more types of cell adhesion molecules than cadherins. This website looks at some of the other cell adhesion and substrate adhesion molecules that have been discovered.

During development, the cadherins often work with other adhesion systems. For instance, one of the most critical times in a mammal's life is when the embryo is passing through the uterus. If development is to continue, the embryo must adhere to the uterus and embed itself in the uterine wall. That is why the first differentiation event in mammalian development distinguishes the **trophoblast** cells (the outer cells that bind to the uterus) from the **inner cell mass** (those cells that will generate the adult organism). This process occurs as the embryo travels down from the upper regions of the oviduct on its way to the uterus. The trophoblast cells are endowed with several adhesion molecules to anchor the embryo to the uterine wall. First, they contain both E-cadherins and P-cadherins (Kadokawa et al. 1989), and these cadherins recognize similar cadherins on the uterine cells. Second, they have receptors (the integrin proteins) for the collagen and the heparan sulfate glycoproteins of the uterine wall (Farach et al. 1987; Carson et al. 1988; 1993; Cross et al. 1994). Third, the trophoblast cells also have a modified glycosyltransferase enzyme that extends out from the membrane and that can bind to specific carbohydrate residues on uterine glycoproteins (Dutt et al. 1987). For something as important as the implantation of the mammalian embryo, it is not surprising that several cell adhesion systems appear to be working together.

As the psalmist said, "I am fearfully and wonderfully made." The questions of morphogenesis remain some of the most fascinating of all developmental biology. Think, for example, of the thousands of specific connections made by the millions of cells within the human brain; or ponder the mechanisms by which the heart chambers form on the correct sides and become connected to the appropriate arteries and veins. These and other questions will be specifically addressed in later chapters.

Principles of Development: Experimental Embryology

1. There are norms of reaction that describe an embryo's inherited ability to develop a range of phenotypes. The environment can play a role in selecting which phenotype is expressed. (Examples include temperature-dependent sex determination and seasonal phenotypic changes in caterpillars and butterflies.)

2. Developing organisms are adapted to the ecological niches in which they develop. (Examples include the ability of frog eggs exposed to sunlight to repair DNA damage.)

3. Before cells overtly differentiate into the many cell types of the body, they undergo a "covert" commitment to a certain fate. This commitment is first labile (the specification step) but later becomes irreversible (the determination step).

4. In autonomous specification, removal of a blastomere from an embryo causes the absence in the embryo of those tissues formed by that blastomere. This mechanism of specification produces a mosaic pattern of development. (Examples include early snail and tunicate embryos.)

5. In autonomous specification, morphogenetic determinants are apportioned to different blastomeres during cell cleavage. (An example is the yellow crescent cytoplasm that is found in the muscle-forming cells of tunicate embryos.)

6. In conditional specification, the removal of a blastomere from the embryo can be compensated for by the other cells' changing their fates. Each cell can potentially give rise to more cell types than it normally does. This produces a regulative pattern of development wherein cell fates are determined relatively late. (Examples include frog and mammalian embryos.)

7. In conditional specification, the fate of a cell often depends upon its neighbors ("whom it meets").

8. In conditional specification, groups of cells can have their fates determined according to a concentration gradient of morphogen. The cells specified by such a morphogen can constitute a field.

9. In syncytial specification, the fates of cells can be determined by gradients of morphogens within the egg cytoplasm.

10. Different cell types can sort themselves into regions by means of cell surface molecules such as cadherins. These molecules can be critical in patterning cells into tissues and organs.

Literature Cited

Adams, N. L. and J. M. Shick. 1996. Mycosporine-like amino acids provide protection against ultraviolet radiation in eggs of the green sea urchin *Strongylocentrotus doebachiensis*. *Photochem. Photobiol.* 64: 149–158.

Allman, J. A. 1864. Reproductive systems in the Hydroidaea. In *Report of the Thirty-third Meeting of the British Association, 1869.*

Ariizumi, T. and M. Asashima. 1994. In vitro control of the embryonic form of *Xenopus laevis* by activin A: Time- and dose-dependent inducing properties of activin-treated ectoderm. *Dev. Growth Diff.* 36: 499–507.

Armstrong, P. B. 1989. Cell sorting out: The self-assembly of tissues in vitro. *CRC Crit. Rev. Biochem. Mol. Biol.* 24: 119–149.

Baltzer, F. 1914. Die Bestimmung und der Dimorphismus des Geschlechtes bei *Bonellia*. *Sber. Phys.-Med. Ges. Würzb.* 43: 1–4.

Blaustein, A. R., P. D. Hoffman, D. G. Hokit, J. M. Kiesecker, S. C. Walls and J. B. Hays. 1994. UV repair and resistance to solar UV-B in amphibian eggs: A link to population declines? *Proc. Natl. Acad. Sci. USA* 91: 1791–1795.

Boucaut, J. C. 1974. Étude autoradiographique de la distribution de cellules embryonnaires isolées, transplantées dans le blastocèle chez *Pleurodeles waltii* Michah (Amphibien, Urodele). *Ann. Embryol. Morphol.* 7: 7–50.

Carson, D. D., J.-P. Tang and S. Gay. 1988. Collagens support embryo attachment and outgrowth in vitro: Effects of the Arg-Gly-Asp sequence. *Dev. Biol.* 127: 368–375.

Carson, D. D., J.-P. Tang and J. Julian. 1993. Heparan sulfate proteoglycan (perlecan) expression by mouse embryos during acquisition of attachment competence. *Dev. Biol.* 155: 97–106.

Chabry, L. M. 1887. Contribution a l'embryologie normale tératologique des ascidies simples. *J. Anat. Physiol. Norm. Pathol.* 23: 167–321.

Cross, J. C., Z. Werb and S. J. Fisher. 1994. Implantation and the placenta: Key pieces of the development puzzle. *Science* 266: 1508–1518.

Davidson, E. H. 1991. Spatial mechanisms of gene regulation in metazoan embryos. *Development* 113: 1–26.

Davis, G. S., H. M. Phillips and M. S. Steinberg. 1997. Germ-layer surface tension and "tissue affinities" in *Rana pipiens*: Quantitative measurement. *Dev. Biol.* 192: 630–644.

De Robertis, E. A., E. M. Morita and K. W. Y. Cho. 1991. Gradient fields and homeobox genes. *Development* 112: 669–678.

Detrick, R. J., D. Dickey and C. R. Kintner. 1990. The effects of N-cadherin misexpression on morphogenesis in *Xenopus* embryos. *Neuron* 4: 493–506.

Driesch, H. 1892. The potency of the first two cleavage cells in echinoderm development: Experimental production of partial and double formations. *In* B. H. Willier and J. M. Oppenheimer (eds.), *Foundations of Experimental Embryology*. Hafner, New York, 1974.

Driesch, H. 1893. Zur Verlagerung der Blastomeren des Echinideneies. *Anat. Anz.* 8: 348–357.

Driesch, H. 1894. *Analytische Theorie de organischen Entwicklung*. W. Engelmann, Leipzig.

Dutt, A., J.-P. Tang and D. D. Carson. 1987. Lactosaminoglycans are involved in uterine epithelial cell adhesion in vitro. *Dev. Biol.* 119: 27–37.

Dyson, S. and J. B. Gurdon. 1998. The interpretation of position in a morphogen gradient as revealed by occupancy of activin receptors. *Cell* 93: 557–568.

Ettensohn, C. A. and D. R. McClay. 1988. Cell lineage conversion in the sea urchin embryo. *Dev. Biol.* 125: 396–409.

Farach, M. C., J. P. Tang, G. L. Decker and D. D. Carson. 1987. Heparin/heparan sulfate is involved in attachment and spreading of mouse embryos in vitro. *Dev. Biol.* 123: 401–410.

Ferguson, M. W. J. and T. Joanen. 1983. Temperature of egg incubation determines sex in *Alligator mississippiensis*. *Nature* 296: 850–853.

Fischer, J.-L. 1991. Laurent Chabry and the beginnings of experimental embryology in France. *In* S. Gilbert (ed.), *A Conceptual History of Modern Embryology*. Plenum, New York, pp. 31–41.

Foty, R. A., C. M. Pfleger, G. Forgacs and M. S. Steinberg. 1996. Surface tensions of embryonic cells predict their mutual envelopment behavior. *Development* 122: 1611–1620.

Fujimori, T., S. Miyatani and M. Takeichi. 1990. Ectopic expression of N-cadherin perturbs histogenesis in *Xenopus* embryos. *Development* 110: 97–104.

Fukui, A. and M. Asashima. 1994. Control of cell differentiation and morphogenesis in amphibian development. *Int. J. Dev. Biol.* 38: 257–266.

Gilbert. S. F. and M. Faber. 1996. Looking at embryos: The visual and conceptual aesthetics of embryology. *In* A. I. Tauber (ed.), *The Elusive Synthesis: Aesthetics and Science*. Kluwer, Dordrecht, pp. 125–151.

Giudice, G. 1963. Restitution of whole larvae from disaggregated cells of sea urchin embryos. *Dev. Biol.* 5: 402–411.

Godt, D. and U. Tepass. 1998. *Drosophila* oocyte localization is mediated by differential cadherin-based adhesion. *Nature* 395: 387–391.

González-Reyes, A. and D. St. Johnston. 1998. The *Drosophila* AP axis is polarised by the cadherin-mediated positioning of the oocyte. *Development* 125: 3635–3644.

Gosse, R. J. 1969. *Principles of Regeneration*. Academic Press, New York.

Gotthard, K. and S. Nylin. 1995. Adaptive plasticity and plasticity as an adaptation: A selective review of plasticity in animal morphology and life history. *Oikos* 74: 3–17.

Greene, E. 1989. A diet-induced developmental polymorphism in a caterpillar. *Science* 243: 643–646.

Gurdon, J. B., P. Harger, A. Mitchell and P. Lemaire. 1994. Activin signalling and responses to a morphogen gradient. *Nature* 371: 487–493.

Gurdon, J. B., A. Mitchell and D. Mahony. 1995. Direct and continuous assessment by cells of their position in a morphogen gradient. *Nature* 376: 520–521.

Gurdon, J. B., S. Dyson and D. St. Johntson. 1998. Cells' perception of position in a concentration gradient. *Cell* 95: 159–163.

Hamburger, V. 1988. *The Heritage of Experimental Embryology: Hans Spemann and the Organizer*. Oxford University Press, New York, p. vii.

Haraway, D. J. 1976. *Crystals, Fabrics and Fields: Metaphors of Organicism in Twentieth-Century Biology*. Yale University Press, New Haven.

Harrington, A. 1996. *Reenchanted Science: Holism in German Culture from Wilhelm II to Hitler*. Princeton University Press, Princeton, NJ.

Harrison, R. G. 1918. Experiments on the development of the forelimb of *Ambystoma*, a self-differentiating equipotential system. *J. Exp. Zool.* 25: 413–461.

Harrison, R. G. 1933. Some difficulties of the determination problem. *Am. Nat.* 67: 306–321.

Hatta, K. and M. Takeichi. 1986. Expression of N-cadherin adhesion molecules associated with early morphogenetic events in chick development. *Nature* 320: 447–449.

Heasman, J., D. Ginsberg, K. Goldstone, T. Pratt, C. Yoshidanaro and C. Wylie. 1994. A functional test for maternally inherited cadherin in *Xenopus* shows its importance in cell adhesion at the blastula stage. *Development* 120: 49–57.

Henry, J. J., S. Amemiya, G. A. Wray and R. A. Raff. 1989. Early inductive interactions are involved in restricting cell fates of mesomeres in sea urchin embryos. *Dev. Biol.* 136: 140–153.

Hertwig, O. 1894. *Zeit- und Streitfragen der Biologie I. Präformation oder Epigenese? Grundzüge einer Entwicklungstheorie der Organismen*. Gustav Fischer, Jena.

Hertwig, O. 1925. Haploidkernige Transplante als Organisatoran diploidkeniger Extremitaten be Triton. *Anat. Anz.* [Suppl.] 60: 112–118.

Hörstadius, S. 1939. The mechanics of sea urchin development studied by operative methods. *Biol. Rev.* 14: 132–179.

Hörstadius, S. and A. Wolsky. 1936. Studien Über die Determination der Bilateralsymmetrie des jungen Seeigelkeimes. *Wilhelm Roux Arch. Entwicklungsmech. Org.* 135: 69–113.

Huxley, J. and G. R. de Beer. 1934. *The Elements of Experimental Embryology*. Cambridge University Press, Cambridge.

Jokiel, P. L. and R. H. York, Jr. 1982. Solar ultraviolet photobiology of the reef coral *Pocillopora damicornis* and symbiotic zooxanthellae. *Bull. Mar. Sci.* 32: 301–315.

Just, E. E. 1939. *The Biology of the Cell Surface*. Blackiston, Philadelphia.

Kadokawa, Y., I. Fuketa, A. Nose, M. Takeichi and N. Nakatsuji. 1989. Expression of E- and P-cadherin in mouse embryos and uteri during the periimplantation period. *Dev. Growth Diff.* 31: 23–30.

Khaner, O. and F. Wilt. 1990. The influence of cell interactions and tissue mass on differentiation of sea urchin mesomeres. *Development* 109: 625–634.

Khaner, O. and F. Wilt. 1991. Interactions of different vegetal cells with mesomeres during early stages of sea urchin development. *Development* 112: 881–890.

Kintner, C. 1993. Regulation of embryonic cell adhesion by the cadherin cytoplasm domain. *Cell* 69: 225–236.

Koch, P. B. and D. Buchmann. 1987. Hormonal control of seasonal morphs by the timing of ecdysteroid release in *Araschnia levana* (Nymphalidae: Lepidoptera). *J. Insect Physiol.* 36: 159–164.

Lawrence, P. A. 1966. Gradients in the insect segment: The orientation of hairs in the milkweed bug *Oncopeltus fasciatus*. *J. Exp. Zool.* 44: 602–620.

Lee, C.-H. and B. M. Gumbiner. 1995. Disruption of gastrulation movements in *Xenopus* by a dominant-negative mutant for C-cadherin. *Dev. Biol.* 171: 363–373.

Leutert, T. R. 1974. Zur Geschlechtsbestimmung und Gametogenese von *Bonellia vividis* Rolando. *J. Embryol. Exp. Morphol.* 32: 169–193.

Marcus, N. H. 1979. Developmental aberrations associated with twinning in laboratory-reared sea urchins. *Dev. Biol.* 70: 274–277.

McClendon, J. F. 1910. The development of isolated blastomeres of the frog's egg. *Am. J. Anat.* 10: 425–430.

Mead, K. S. and D. Epel. 1995. Beakers versus breakers: How fertilization in the laboratory differs from fertilization in nature. *Zygote* 3: 95–99.

Monroy, A. and A. A. Moscona. 1979. *Introductory Concepts in Developmental Biology*. University of Chicago Press, Chicago.

Morgan, T. H. 1905 An attempt to analyze the phenomenon of polarity in *Tubularia*. *J. Exp. Zool.* 1: 589–591.

Morgan, T. H. 1906. "Polarity" considered as a phenomenon of gradation of materials. *J. Exp. Zool.* 2: 495–506.

Moscona, A. A. 1952. Cell suspension from organ rudiments of chick embryos. *Exp. Cell Res.* 3: 535–539.

Moscona, A. A. 1961. Rotation-mediated histogenetic aggregation of dissociated cells: A quantifiable approach to cell interaction in vitro. *Exp. Cell Res.* 22: 455–475.

Moscona, A. A. 1974. Surface specification of embryonic cells: Lectin receptors, cell recognition, and specific cell ligands. *In* A. Monroy (ed.), *The Cell Surface in Development*. Wiley, New York, pp. 67–99.

Nagafuchi, A., Y. Shirayoshi, K. Okazaki, K. Yasuda and M. Takeichi. 1987. Transformation of cell adhesion properties of exogenously introduced E-cadherin cDNA. *Nature* 329: 341–343.

Needham, J. 1943. *Time: The Refreshing River*. Allen and Unwin, London.

Nijhout, H. F. 1991. *The Development and Evolution of Butterfly Wing Patterns*. Smithsonian Institution Press, Washington, D.C.

Nose, A. and M. Takeichi. 1986. A novel cadherin adhesion molecule: Its expression patterns associated with implantation and organogenesis of mouse embryos. *J. Cell Biol.* 103: 2649–2658.

Nose, A., A. Nagafuchi and M. Takeichi. 1988. Expressed recombinant cadherins mediate cell sorting in model systems. *Cell* 54: 993–1001.

Nüsslein-Volhard, C., H. G. Fröhnhofer and R. Lehmann. 1987. Determination of anterioposterior polarity in *Drosophila. Science* 238: 1675–1681.

Nyhart, L. K. 1995. *Biology Takes Form: Animal Morphology and the German Universities, 1800–1900.* University of Chicago Press, Chicago.

Opitz, J. M. 1985. The developmental field concept. *Am. J. Med. Genet.* 21: 1–11.

Peifer, M. 1998. Birds of a feather flock together. *Nature* 395: 324–325.

Phillips, K. 1994. *Tracking the Vanishing Frogs: An Ecological Mystery.* St. Martin's Press, New York.

Reverberi, G. and A. Minganti. 1946. Fenomeni di evocazione nello sviluppo dell'uovo di Ascidie. Risultati dell'indagine spermentale sull'ouvo di Ascidiella aspersa e di Ascidia malaca allo stadio di 8 blastomeri. *Pubbl. Staz. Zool. Napoli* 20: 199–253. (Quoted in G. Reverberi,. *Experimental Embryology of Marine and Freshwater Invertebrates.* North-Holland, Amsterdam, 1971, p. 537.)

Rosenquist, G. C. 1971. The origin and movement of the limb-bud epithelium and mesenchyme in the chick embryo as determined by radioautographic mapping. *J. Embryol. Exp. Morphol.* 25: 85–96.

Roux, W. 1888. Contributions to the developmental mechanics of the embryo. On the artificial production of half-embryos by destruction of one of the first two blastomeres and the later development (postgeneration) of the missing half of the body. *In* B. H. Willier and J. M. Oppenheimer (eds.) 1974, *Foundations of Experimental Embryology.* Hafner, New York, pp. 2–37.

Roux, W. 1894. Einleitung zum Archiv für Entwicklungsmechanik. *Arch. Embryol.* 1: 1–142. English translation reprinted in J. M. Maienschein (ed.), *Defining Biology*, 1986. Harvard University Press, Cambridge, MA, pp. 107–148.

Rutishauser, U., A. Acheson, A. Hall, D. M. Mann and J. Sunshine. 1988. The neural cell adhesion molecule (N-CAM) as a regulator of cell-cell interactions. *Science* 240: 53–57.

Schlichting, C. D. and M. Pigliucci. 1998. *Phenotypic Evolution: A Reaction Norm Perspective.* Sinauer Associates, Sunderland, MA.

Schmalhausen, I. I. 1949. *Factors of Evolution: The Theory of Stabilizing Selection.* University of Chicago Press, Chicago.

Sessions, S. and S. B. Ruth. 1990. Explanation for naturally occurring supernumerary limbs in amphibians. *J. Exp. Zool.* 254: 38–47.

Sessions, S. K., R. A. Franssen and V. C. Horner. 1999. Morphological elves from multilegged frogs: Are retinoids to blame? *Science* 284: 800–802.

Shimizu, and J. B. Gurdon. 1999. A quantitative analysis of signal transduction for activin receptor to nucleus and its relevance to morphogen gradient interpretation. *Proc. Natl. Acad. Sci. USA* 96: 6791–6796.

Siebeck, O. 1988. Experimental investigation of UV tolerance in hermatypic corals. *Mar. Ecol. Prog. Ser.* 43: 95–103.

Slack, J. M. W. 1991. *From Egg to Embryo: Regional Specification in Early Development.* Cambridge University Press, New York.

Spemann, H. 1921. Die Erzeugung tierischer Chimaeren durch heteroplastiche embryonale Transplantation zwischen *Triton cristatus* u. *taeniatus. Wilhelm Roux Arch. Entwicklungsmech. Org.* 48, 533–570.

Stearns, S. C., G. de Jong and R. A. Newman. 1991. The effects of phenotypic plasticity on genetic correlations. *Trends Ecol. Evol.* 6: 122–126.

Steinberg, M. S. 1964. The problem of adhesive selectivity in cellular interactions. *In* M. Locke (ed.), *Cellular Membranes in Development.* Academic Press, New York, pp. 321–434.

Steinberg, M. S. 1970. Does differential adhesion govern self-assembly processes in histogenesis? Equilibrium configurations and the emergence of a hierarchy among populations of embryonic cells. *J. Exp. Zool.* 173: 395–434.

Steinberg, M. S. and M. Takeichi. 1994. Experimental specification of cell sorting, tissue spreading and specific spatial patterning by quantitative differences in cadherin expression. *Proc. Natl. Acad. Sci. USA* 91: 206–209.

Stocum, D. L. and J. F. Fallon. 1982. Control of pattern formation in urodele limb ontogeny: A review and a hypothesis. *J. Embryol. Exp. Morphol.* 69: 7–36.

Stumpf, H. 1966. Mechanism by which cells estimate their location within the body. *Nature* 212: 430–431.

Takeichi, M. 1987. Cadherins: A molecular family essential for selective cell–cell adhesion and animal morphogenesis. *Trends Genet.* 3: 213–217.

Takeichi, M. 1991. Cadherin cell adhesion receptors as a morphogenetic regulator. *Science* 251: 1451–1455.

Takeichi, M., H. S. Ozaka, K. Tokunaga and T. S. Okada. 1979. Experimental manipulation of cell surface to affect cellular recognition mechanisms. *Dev. Biol.* 70: 195–205.

Toivonen, S. and L. Saxén. 1955. The simultaneous inducing action of liver and bone marrow of the guinea pig in implantation and explantation experiments of *Triturus. Exp. Cell Res. Supp.* 3: 346–357.

Townes, P. L. and J. Holtfreter. 1955. Directed movements and selective adhesion of embryonic amphibian cells. *J. Exp. Zool.* 128: 53–120.

van der Weele, C. 1995. *Images of Development: Environmental Causes in Ontogeny.* Elinkwijk, Utrecht.

Via, S., R. Gomulkiewicz, G. De Jong, S. M. Scheiner, C. D. Schlichting and P. H. Van Tienderen. 1995. Adaptive phenotypic plasticity: Consensus and controversy. *Trends Ecol. Evol.* 10: 212–217.

Weismann, A. 1875. Über den Saison-Dimorphismus der Schmetterlinge. *In. Studien zur Descendenz-Theorie.* Engelmann, Leipzig.

Weismann, A. 1892. *Essays on Heredity and Kindred Biological Problems.* Translated by E. B. Poulton, S. Schoenland and A. E. Shipley. Clarendon, Oxford.

Weismann, A. 1893. *The Germ-Plasm: A Theory of Heredity.* Translated by W. Newton Parker and H. Ronnfeld. Walter Scott Ltd., London.

Weiss, P. 1923. *Naturwissenschaft* 11: 669 (quoted in Weiss 1939).

Weiss, P. 1939. *Principles of Development.* Holt, New York.

Whittaker, J. R. 1973. Segregation during ascidian embryogenesis of egg cytoplasmic information for tissue-specific enzyme development. *Proc. Natl. Acad. Sci. USA* 70: 2096–2100.

Whittaker, J. R. 1977. Segregation during cleavage of a factor determining endodermal alkaline phosphatase development in ascidian embryos. *J. Exp. Zool.* 202: 139–153.

Whittaker, J. R. 1982. Muscle cell lineage can change the developmental expression in epidermal lineage cells of ascidian embryos. *Dev. Biol.* 93: 463–470.

Wilkins, A. S. 1993. *Genetic Analysis of Animal Development,* 2nd Ed. Wiley-Liss, New York.

Wilson, E. B. 1896. *The Cell in Development and Inheritance.* Macmillan, New York.

Wilson, E. B. 1904. Experimental studies on germinal location. *J. Exp. Zool.* 1: 1–72.

Wolpert, L. 1968. The French flag problem: A contribution to the discussion on pattern formation and regulation. *In* C. H. Waddington (ed.), *Towards a Theoretical Biology.* Edinburgh University Press, Edinburgh, pp. 125–133.

Wolpert, L. 1969. Positional information and the spatial pattern of cellular differentiation. *J. Theor. Biol.* 25: 1–47.

Wolpert, L. 1977. *The Development of Pattern and Form in Animals.* Carolina Biological, Burlington, NC.

Wolpert, L. 1978. Pattern formation in biological development. *Sci. Am.* 239(4): 154–164.

Woltereck, R. 1909. Weitere experimentelle Untersuchungen über Artveränderung, speziell über das Wesen quantitativer Artunderscheide bei Daphniden. *Versuch. Deutsch. Zool. Ges.* 1909: 110–173.

4 *Genes and development: Techniques and ethical issues*

c h a p t e r

The secrets that engage me—that sweep me away—are generally secrets of inheritance: how the pear seed becomes a pear tree, for instance, rather than a polar bear.

CYNTHIA OZICK (1989)

The future is already here. It's just not evenly distributed yet.

WILLIAM GIBSON (1999)

"BETWEEN THE CHARACTERS that furnish the data for the theory, and the postulated genes, to which the characters are referred, lies the whole field of embryonic development." Here Thomas Hunt Morgan noted in 1926 that the only way to get from genotype to phenotype is through developmental processes.

In the early twentieth century, embryology and genetics were not considered separate sciences. They diverged in the 1920s, when Morgan redefined genetics as the science studying the *transmission* of traits, as opposed to embryology, the science studying the *expression* of traits. Within the past decade, however, the techniques of molecular biology have effected a rapprochement of embryology and genetics. In fact, the two fields have become linked to a degree that makes it necessary to discuss molecular genetics early in this text. Problems in animal development that could not be addressed a decade ago are now being solved by a set of techniques involving nucleic acid synthesis and hybridization. This chapter seeks to place these new techniques within the context of the ongoing dialogues between genetics and embryology.

The Embryological Origins of the Gene Theory

Nucleus or cytoplasm: Which controls heredity?

Mendel called them *bildungsfähigen Elemente*, "form-building elements"; we call them genes. It is in Mendel's term, however, that we see how closely intertwined were the concepts of inheritance and development in the nineteenth century. Mendel's observations, however, did not indicate where these hereditary elements existed in the cell, or how they came to be expressed. The gene theory that was to become the cornerstone of modern genetics originated from a controversy within the field of physiological embryology (see Chapter 3). In the late 1800s, a group of scientists began to study the mechanisms by which fertilized eggs give rise to adult organisms. Two young American embryologists, Edmund Beecher Wilson and Thomas Hunt Morgan (Figure 4.1), became part of this group of "physiological embryologists," and each became a partisan in the controversy over which of the two compartments of the fertilized egg—the nucleus or the cytoplasm—controls inheritance. Morgan allied himself with those embryologists who thought the control of development lay within the cytoplasm, while Wilson allied himself with Theodor Boveri, one of the biologists who felt that the nucleus contained the instructions for development. In fact, Wilson (1896, p. 262) declared that the processes of meiosis, mitosis, fertilization, and unicellular regeneration (only from the fragment containing the nucleus; see Chapter 2)

(A)

(B)

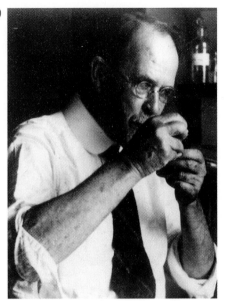

Figure 4.1
(A) E. B. Wilson (1856–1939; shown here around 1899), an embryologist whose work on early embryology and sex determination greatly advanced the chromosomal hypotheses of development. (Wilson was also acknowledged to be among the best amateur cellists in the country.) (B) Thomas Hunt Morgan (1866–1945), who brought the gene theory out of embryology. This photo—taken in 1915, as the basic elements of the gene theory were coming together—shows Morgan using a hand lens to sort fruit flies. (A courtesy of W. N. Timmins; B courtesy of G. Allen.)

"converge to the conclusion that the chromatin is the most essential element in development."* He did not shrink from the consequences of this belief. Years before the rediscovery of Mendel or the gene theory, Wilson (1895, p. 4) noted, "Now, chromatin is known to be closely similar, if not identical with, a substance known as nuclein ... which analysis shows to be a tolerably definite chemical composed of a nucleic acid (a complex organic acid rich in phosphorus) and albumin. And thus we reach the remarkable conclusion that inheritance may, perhaps, be effected by the physical transmission of a particular chemical compound from parent to offspring."

Some of the major support for the chromosomal hypothesis of inheritance was coming from the embryological studies of Theodor Boveri (Figure 4.2A), a researcher at the Naples

Zoological Station. Boveri fertilized sea urchin eggs with large concentrations of their sperm and obtained eggs that had been fertilized by two sperm. At first cleavage, these eggs formed four mitotic poles and divided into four cells instead of two (see Chapter 7). Boveri then separated the blastomeres and demonstrated that each cell developed abnormally, and in a different way, as a result of each of the cells having different types of chromosomes. Thus, Boveri claimed that each chromosome had an individual nature and controlled different vital processes.

Adding to Boveri's evidence, E. B. Wilson (1905) and Nettie Stevens (1905a,b; Figure 4.2B) demonstrated a critical correlation between nuclear chromosomes and organismal development: XO or XY embryos became male; XX embryos became female. Here was a nuclear property that correlated with development. Eventually, Morgan began to obtain mutations that correlated with sex and with the X chromosome, and he began to view the genes as being physically linked to one another on the chromosomes. The embryologist Morgan

---

*Notice that Wilson was writing about form-building units in chromatin in 1896—before the rediscovery of Mendel's paper or the founding of the gene theory. For further analysis of the interactions between Morgan and Wilson that led to the gene theory, see Gilbert 1978, 1987 and Allen 1986.

(A)  (B)

Figure 4.2
Chromosomal uniqueness was shown by Boveri and Stevens. (A) Theodor Boveri (1862–1915), whose work, Wilson said, "accomplished the actual amalgamation between cytology, embryology, and genetics—a biological achievement which … is not second to any of our time" (Wilson 1896). This photograph was taken in 1908, when Boveri's chromosomal and embryological studies were at their zenith. (B) Nettie M. Stevens (1861–1912), who trained with both Boveri and Morgan, seen here in 1904 when she was a postdoctoral student pursuing the research that correlated the number of X chromosomes with sexual development. (A from Baltzer 1967; B courtesy of the Carnegie Institute of Washington.)

had shown that nuclear chromosomes are responsible for the development of inherited characters.*

> WEBSITE **4.1 The embryological origin of the gene theory.** The emergence of the gene theory from embryological research is a fascinating story and complements the history of genetics that begins with Mendel's experiments.

The split between embryology and genetics

Morgan's evidence provided a material basis for the concept of the gene. Originally, this type of genetics was seen as being part of embryology, but by the 1930s, genetics became its own discipline, developing its own vocabulary, journals, societies, favored research organisms, professorships, and rules of evidence. Hostility between embryology and genetics also emerged. Geneticists believed that the embryologists were old-fashioned and that development would be completely explained as the result of gene expression. Conversely, the embryologists regarded the geneticists as uninformed about how organisms actually developed and felt that genetics was irrelevant to embryological questions. Embryologists such as Frank Lillie (1927), Ross Granville Harrison (1937), Hans Spemann (1938), and Ernest E. Just (1939) (Figure 4.3) claimed that there could be no genetic theory of development until at least three major challenges had been met by the geneticists:

1. Geneticists had to explain how chromosomes—which were thought to be identical in every cell of the organism—produce different and changing types of cell cytoplasms.

2. Geneticists had to provide evidence that genes control the early stages of embryogenesis. Almost all the genes known at the time affected the final modeling steps in development (eye color, bristle shape, wing venation in *Drosophila*). As Just said (quoted in Harrison 1937), embryologists were interested in how a fly forms its back, not in the number of bristles on its back.

3. Geneticists had to explain phenomena such as sex determination in certain invertebrates (and vertebrates such as reptiles), in which the environment determines sexual phenotype.

The debate became quite vehement. In rhetoric reflecting the political anxieties of the late 1930s, Harrison (1937) warned:

> *Now that the necessity of relating the data of genetics to embryology is generally recognized and the Wanderlust of geneticists is beginning to urge them in our direction, it may not be inappropriate to point out a danger of this threatened invasion. The prestige of success enjoyed by the gene theory might easily become a hindrance to the understanding of development by directing our attention solely to the genom, whereas cell movements, differentiation, and in fact all of developmental processes are actually effected by cytoplasm. Already we have theories that refer the processes of development to gene action and regard the whole performance as no more than the realization of the potencies of genes. Such theories are altogether too one-sided.*

---

*Morgan's evidence for nuclear control of development went against his expectations; until 1910, he was the leading proponent of the cytoplasm. Wilson was one of Morgan's closest friends, and Morgan considered Stevens his best graduate student at that time. Both were against Morgan on this issue. Even though they disagreed, Morgan wholeheartedly supported Stevens's request for research funds, saying that her qualifications were the best possible. Wilson wrote an equally laudatory letter of support, even though she would be a rival researcher (see Brush 1978).

(A)

(C)

Figure 4.3
Embryologists attempted to keep genetics from "taking over" their field in the 1930s. (A) Frank Lillie headed the Marine Biology Laboratory at Woods Hole and was a leader in fertilization research and reproductive endocrinology. (B) Hans Spemann (left) and Ross Harrison (right) perfected the transplantation operations used to discover when the body and limb axes are determined. They argued that geneticists had no mechanism for explaining how the same nuclear genes could create different cell types during development. (C) Ernest E. Just made critical discoveries on fertilization. He spurned genetics and emphasized the role of the cell membrane in determining the fates of cells. (A courtesy of V. Hamburger; B courtesy of T. Horder; C courtesy of the Marine Biology Laboratory, Woods Hole.)

(B)

Until geneticists could demonstrate the existence of inherited variants during early development, and until geneticists had a well-documented theory for how the same chromosomes could produce different cell types, embryologists generally felt no need to ground their science in gene action.

Early attempts at developmental genetics

Some scientists, however, felt that neither embryology nor genetics was complete without the other. There were several attempts to synthesize the two disciplines, but the first successful reintegration of genetics and embryology came in the late 1930s from two embryologists, Salome Gluecksohn-Schoenheimer (now S. Gluecksohn Waelsch) and Conrad Hal Waddington (Figure 4.4). Both were trained in European embryology and had learned genetics in the United States from Morgan's students. Gluecksohn-Schoenheimer and Waddington attempted

to find mutations that affected early development and to discover the processes that these genes affected. Gluecksohn-Schoenheimer (1938, 1940) showed that mutations in the *Brachyury* genes of the mouse caused the aberrant development of the posterior portion of the embryo, and she traced the effects of these mutant genes to the notochord, which would normally have helped induce the dorsal axis.*

At the same time, Waddington (1939) isolated several genes that caused wing malformations in fruit flies (*Drosophila*). He, too, analyzed these mutations in terms of how the genes might affect the developmental primordia that give rise to these structures. The *Drosophila* wing, he correctly claimed, "appears favorable for investigations on the developmental action of genes." Thus, one of the main objections

---

*Gluecksohn-Schoenheimer's observations took 60 years to be confirmed by DNA hybridization. However, when the *Brachyury* (T-locus) gene was cloned and its expression detected by the in situ hybridization technique (discussed later in this chapter), Wilkinson and co-workers (1990) found that "the expression of the T gene has a direct role in the early events of mesoderm formation and in the morphogenesis of the notochord." While a comprehensive history of early developmental genetics needs to be written, more information on its turbulent origins can be found in Oppenheimer 1981; Sander 1986; Gilbert 1988, 1991, 1996; Burian et al. 1991; Harwood 1993; Keller 1995; and Morange 1996.

(A)

(B)

Figure 4.4
Two of the founders of developmental genetics. (A) Salome Gluecksohn-Schoenheimer (now S. Gluecksohn-Waelsch; b. 1907) received her doctorate in Spemann's laboratory. Fleeing Hitler's Germany, she brought her embryological acumen to Leslie Dunn's genetics laboratory in the United States. (B) Conrad Hal Waddington (1905–1975) did not believe in the distinction between genetics and embryology and sought to find mutations that were active during development. (A courtesy of S. Gluecksohn-Waelsch; B from Waddington 1948.)

of embryologists to the genetic model of development—that genes appear to be working only on the final modeling of the embryo and not on its major outlines—was countered.

WEBSITE **4.2 Creating developmental genetics.** The resynthesis of embryology and genetics into developmental genetics did not come easily. These websites look at some of the ways different researchers attempted to join the two fields together. The backgrounds of these scientists figure prominently in their particular syntheses.

Evidence for Genomic Equivalence

The other major objection to a genetically based embryology still remained: How could nuclear genes direct development when they were the same in every cell type? The existence of this **genomic equivalence** was not so much proved as assumed (because every cell is the mitotic descendant of the fertilized egg), so one of the first problems of developmental genetics was to determine whether every cell of an organism indeed had the same set of genes, or **genome**, as every other cell.

Metaplasia

The first evidence for genomic equivalence came from embryologists studying the regeneration of excised tissues. The study of salamander eye regeneration demonstrated that even adult differentiated cells can retain their potential to produce other cell types. Therefore, the genes for these other cell types' products must still be present in the cells, though not normally expressed. In the salamander eye, removal of the neural retina promotes its regeneration from the pigmented retina, and if the lens is removed, a new lens can be formed from the cells of the dorsal iris. The regeneration of lens tissue from the iris (called Wolffian regeneration after the person who first observed it, in 1894) has been intensively studied. Yamada and

his colleagues (Yamada 1966; Dumont and Yamada 1972) found that a series of events leads to the production of a new lens from the iris (Figure 4.5). The nuclei of the dorsal side of the iris begin to synthesize enormous numbers of ribosomes, their DNA replicates, and a series of mitotic divisions ensues. The pigmented iris cells then begin to dedifferentiate by throwing out their melanosomes (the pigmented granules that give the eye its color) These melanosomes are ingested by macrophages that enter the wound site. The dorsal iris cells continue to divide, forming a globe of dedifferentiated tissue in the region of the removed lens. These cells then start synthesizing the products of differentiated lens cells, the crystallin proteins. These crystallins are made in the same order as in normal lens development. Once a new lens has formed, the cells on the dorsal side of the iris cease mitosis (Mikhailov et al. 1997).

These events are not the normal route by which the vertebrate lens is formed. As mentioned in Chapter 3, the lens normally develops from a layer of head epithelial cells induced by the underlying retinal precursor cells. The formation of the lens by the differentiated cells of the iris is an example of **metaplasia** (or **transdifferentiation**), the transformation of one differentiated cell type into another (Okada 1991). The salamander iris, then, has not lost any of the genes that are used to differentiate the cells of the lens.

Amphibian cloning: The restriction of nuclear potency

The ultimate test of whether the nucleus of a differentiated cell has undergone any irreversible functional restriction is to have that nucleus generate every other type of differentiated cell in the body. If each cell's nucleus is identical to the zygote nucleus, then each cell's nucleus should be **totipotent** (capable of directing the entire development of the organism) when transplanted into an activated enucleated egg. Before such an experiment could be done, however, three techniques for

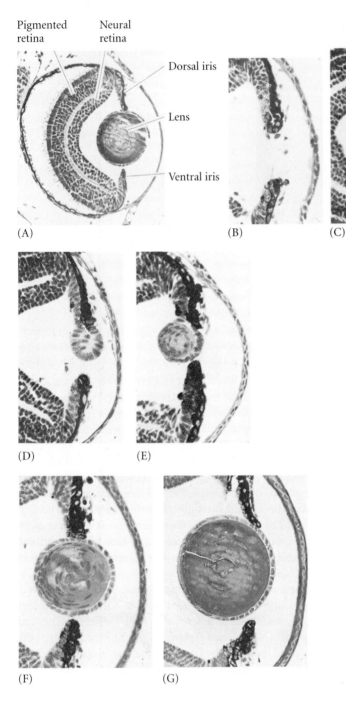

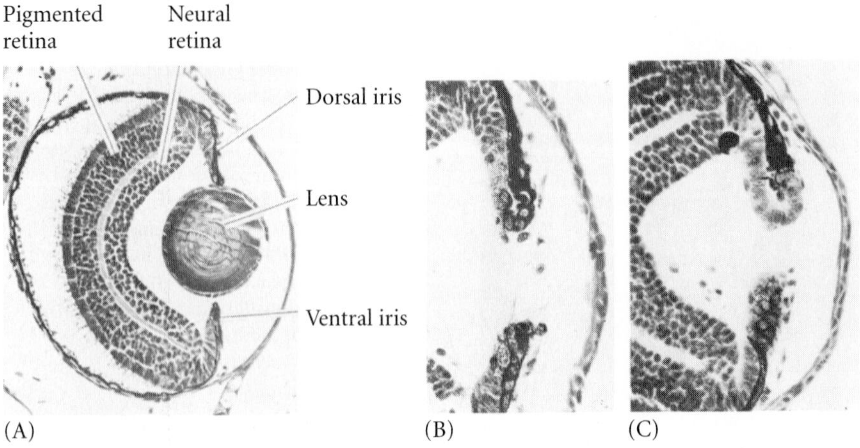

Pigmented retina   Neural retina

Dorsal iris

Lens

Ventral iris

(A)   (B)   (C)

(D)   (E)

(F)   (G)

Figure 4.5
Wolffian regeneration of the salamander lens from the dorsal margin of the iris. (A) Normal eye of the larval-stage newt *Notophthalmus viridescens*. (B–G) Regeneration of the lens, seen on days 5, 7, 9, 16, 18, and 30, respectively. The new lens is complete at day 30. (From Reyer 1954; photographs courtesy of R. W. Reyer.)

glass needle, the egg undergoes all the cytological and biochemical changes associated with fertilization. The internal cytoplasmic rearrangements of fertilization occur, and the completion of meiosis takes place near the animal pole of the cell. The meiotic spindle can easily be located as it pushes away the pigment granules at the animal pole, and puncturing the oocyte at this site causes the spindle and its chromosomes to flow outside the egg (Figure 4.6). The host egg is now considered both activated (the fertilization reactions necessary to initiate development have been completed) and enucleated. The transfer of a nucleus into the egg is accomplished by disrupting a donor cell and transferring the released nucleus into the oocyte through a micropipette. Some cytoplasm accompanies the nucleus to its new home, but the ratio of donor to recipient cytoplasm is only $1:10^5$, and the donor cytoplasm does not seem to affect the outcome of the experiments. In 1952, Briggs and King, using these techniques, demonstrated that blastula cell nuclei could direct the development of complete tadpoles when transferred into the oocyte cytoplasm.

What happens when nuclei from more advanced developmental stages are transferred into activated enucleated oocytes? King and Briggs (1956) found that whereas most blastula nuclei could produce entire tadpoles, there was a dramatic decrease in the ability of nuclei from later stages to direct development to the tadpole stage (Figure 4.7). When nuclei from the somatic cells of tailbud-stage tadpoles were used as donors, normal development did not occur. However, nuclei from the germ cells of tailbud-stage tadpoles (which could give rise to a complete organism after fertilization) were capable of directing normal development in 40% of the blastulae that developed (Smith 1956). Thus, most somatic cells appeared to lose their ability to direct development as they became determined and differentiated.

transplanting nuclei into eggs had to be perfected: (1) a method for enucleating host eggs without destroying them; (2) a method for isolating intact donor nuclei; and (3) a method for transferring such nuclei into the host egg without damaging either the nucleus or the oocyte.

All three techniques were developed in the 1950s by Robert Briggs and Thomas King. First, they combined the enucleation of the host egg with its activation. When an oocyte from the leopard frog (*Rana pipiens*) is pricked with a clean

Figure 4.6
Procedure for transplanting blastula nuclei into activated enucleated *Rana pipiens* eggs. The relative dimensions of the meiotic spindle have been exaggerated to show the technique. "Freddy," the handsome and mature *R. pipiens* in the photograph, was derived in this way by M. DiBerardino and N. Hoffner Orr. (After King 1966; photograph courtesy of M. DiBerardino.)

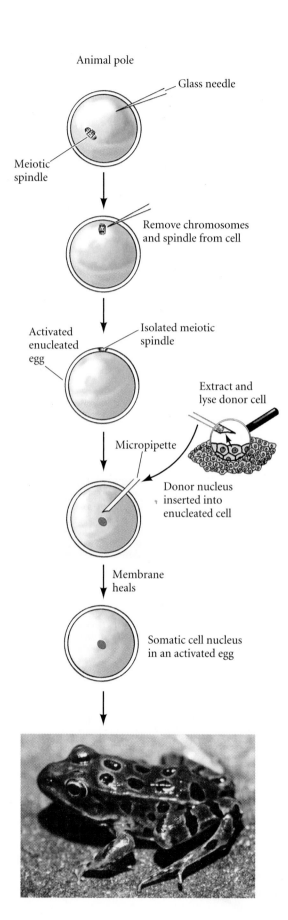

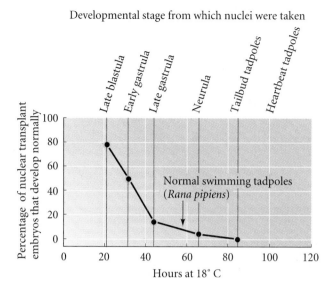

Figure 4.7

Percentage of successful nuclear transplants as a function of the developmental age of the donor nucleus. The abscissa represents the developmental stage at which a donor nucleus (from *R. pipiens*) was isolated and inserted into an activated enucleated oocyte. The ordinate shows the percentage of those transplants capable of producing blastulae that could then direct development to the swimming tadpole stage. (After McKinnell 1978.)

Amphibian cloning: The pluripotency of somatic cells

Is it possible that some differentiatied cell nuclei differ from others in their ability to direct development? John Gurdon and his colleagues, using slightly different methods of nuclear transplantation on the frog *Xenopus*, obtained results suggesting that the nuclei of some differentiated cells can remain totipotent. Gurdon, too, found a progressive loss of potency with increasing developmental age, although *Xenopus* cells retained their potencies for a longer period than did the cells of *Rana* (Figure 4.8). The exceptions to this rule, however, proved very interesting. Gurdon transferred nuclei from the intestinal endoderm of feeding *Xenopus* tadpoles into activated enucleated eggs. These donor nuclei contained a genetic marker (one nucleolus per cell instead of the usual two) that distinguished them from host nuclei. Out of 726 nuclei transferred, only 10 (1.4%) promoted development to the feeding tadpole stage. Serial transplantation (transplanting an intestinal nucleus into an egg and, when the egg had become a blastula, transferring the nuclei of the blastula cells into several more eggs) increased the yield to 7% (Gurdon 1962). In some instances, nuclei from tadpole intestinal cells were capable of generating all the cell lineages—neurons, blood cells, nerves, and so forth—of a living tadpole. Moreover, seven of these tadpoles (from two original nuclei) metamorphosed into fertile adult frogs (Gurdon and Uehlinger 1966); these two nuclei were totipotent (Figure 4.9).

Wild-type female donor
of enucleated eggs

Albino parents
of nucleus donor

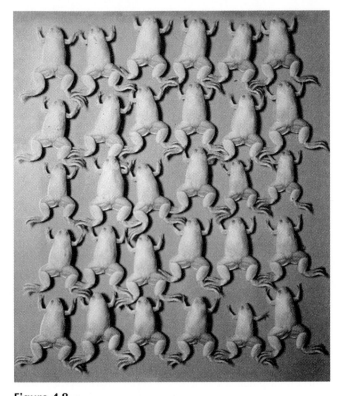

Figure 4.8
A clone of *Xenopus laevis* frogs. The nuclei for all the members of this clone came from a single individual—a female tailbud-stage tadpole whose parents (upper panel) were both marked by albino genes. The nuclei (containing these defective pigmentation genes) were transferred into activated enucleated eggs from a wild-type female (upper panel). The resulting frogs were all female and albino (lower panel). (Photographs courtesy of J. Gurdon.)

King and his colleagues criticized these experiments, pointing out that (1) not enough care was taken to make certain that primordial germ cells, which can migrate through to the gut, were not used as sources of nuclei, and (2) the intestinal epithelial cells of such a young tadpole may not qualify as a truly differentiated cell type because such cells of feeding tadpoles still contain yolk platelets (Di Berardino and King 1967; McKinnell 1978; Briggs 1979). To answer these criti-

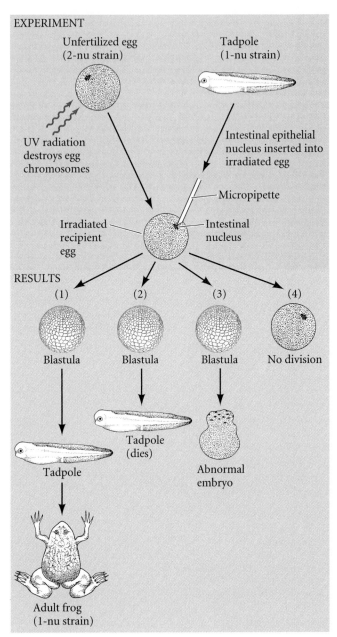

Figure 4.9
Procedure used to obtain mature frogs from the intestinal nuclei of *Xenopus* tadpoles. The wild-type egg (with two nucleoli per nucleus; 2-nu) is irradiated to destroy the maternal chromosomes, and an intestinal nucleus from a marked (1-nu) tadpole is inserted. In some cases, there is no cell division; in some cases, the embryo is arrested in development; but in other cases, an entire new frog, with a 1-nu genotype, is formed. (After Gurdon 1968, 1977.)

cisms, Gurdon and his colleagues cultured epithelial cells from adult frog foot webbing. These cells were shown to be differentiated; each of them contained a specific keratin, the characteristic protein of adult skin cells. When nuclei from these

(A)

(B)

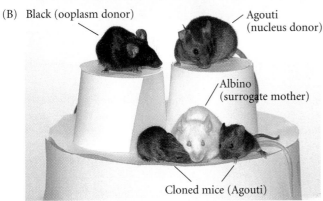

Black (ooplasm donor)

Agouti (nucleus donor)

Albino (surrogate mother)

Cloned mice (Agouti)

(C)

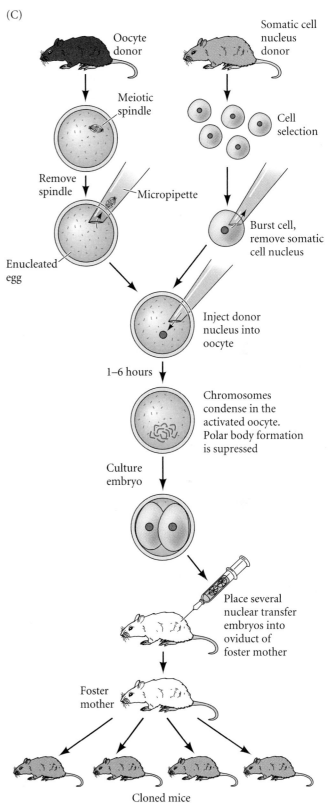

Oocyte donor

Somatic cell nucleus donor

Meiotic spindle

Cell selection

Remove spindle

Micropipette

Burst cell, remove somatic cell nucleus

Enucleated egg

Inject donor nucleus into oocyte

1–6 hours

Chromosomes condense in the activated oocyte. Polar body formation is supressed

Culture embryo

Place several nuclear transfer embryos into oviduct of foster mother

Foster mother

Cloned mice

Figure 4.10
Cloned mammals, whose nuclei came from adult somatic cells. (A) Dolly, the adult sheep on the left, was derived by fusing a mammary gland cell nucleus with an enucleated oocyte, which was then implanted in a surrogate mother (of a different breed of sheep) who gave birth to Dolly. Dolly has since produced a lamb (Bonnie, at right) by normal reproduction. (B) Cloned mice and their "parents." The upper left black mouse is the oocyte donor, while the upper right brown (agouti) mouse is the nucleus donor. The white mouse in the lower row is the mouse into whose uterus the resulting embryos were implanted. The two agouti mice beside her are cloned mice, derived from the injection of the agouti nucleus into the oocyte of the black-furred parent (C) Procedure used for cloning mice. (A photograph by Roddy Field, © Roslin Institute; B courtesy of T. Wakayama and R. Yanagimachi.)

cells were transferred into activated, enucleated *Xenopus* oocytes, none of the first-generation transfers progressed further than the formation of the neural tube shortly after gastrulation. By serial transplantation, however, numerous tadpoles were generated (Gurdon et al. 1975). Although these tadpoles all died prior to feeding, they showed that a single differentiated cell nucleus still retained incredible potencies.

Cloning mammals

In 1997, Ian Wilmut announced that a sheep had been cloned from a somatic cell nucleus from an *adult* female sheep. This was the first time that an adult vertebrate had been successfully cloned from another adult. To do this, Wilmut and his colleagues (1997) took cells from the mammary gland of an adult (6-year-old) pregnant ewe and put them into culture. The culture medium was formulated to keep the nuclei in these cells at the resting stage of the cell cycle (G_0). They then obtained oocytes (the maturing egg cell) from a different strain of sheep and removed their nuclei. The fusion of the donor cell and the enucleated oocyte was accomplished by bringing the two cells together and sending electrical pulses through them. The electric pulses destabilized the cell membranes, allowing the cells to fuse together. Moreover, the same pulses that fused the cells activated the egg to begin development. The resulting embryos were eventually transferred into the uteri of pregnant sheep. Of the 434 sheep oocytes originally used in this experiment, only one survived: Dolly (Figure 4.10A). DNA analysis confirmed that the nuclei of Dolly's cells were derived from the strain of sheep from which the donor nucleus was taken (Ashworth et al. 1998; Signer et al. 1998). Thus, it appears that the nuclei of adult somatic cells can be totipotent. No genes necessary for development have been lost or mutated in a way that would make them nonfunctional.

This result has been confirmed in cows (Kato et al. 1998) and mice (Wakayama et al. 1998). In mice, somatic cell nuclei from the cumulus cells of the ovary were injected directly into enucleated oocytes. These re-nucleated oocytes were able to develop into mice at a frequency of 2.5% (Figure 4.10B,C). Interestingly, nuclei from other somatic cells (such as neurons or Sertoli cells) that are similarly blocked at the G_0 stage did not generate any live mice. Cumulus cell nuclei from cows have also directed the complete development of oocytes into mature cows (Kato et al. 1998).

Sidelights & Speculations

Why Clone Mammals?

Given that we already knew from amphibian studies in the 1960s that nuclei were pluripotent, why clone mammals? Many of the reasons are medical and commercial, and there are good reasons why these techniques were first developed by pharmaceutical companies rather than at universities. Cloning is of interest to some developmental biologists who study the relationships between the nucleus and cytoplasm during fertilization or who study aging (and the loss of totipotency that appears to accompany it), but cloned mammals are of special interest to those people concerned with protein pharmaceuticals. Protein drugs such as human insulin, protease inhibitor, and clotting factors are difficult to manufacture. Due to immunological rejection problems, the human proteins are usually much better tolerated by patients than proteins from other animals. So the problem becomes how to obtain large amounts of the human protein. One of the most efficient ways of producing these proteins is to insert the human genes encoding them into the oocyte DNA of sheep, goats, or cows. Animals containing a gene from another individual (often of a different species)—a **transgene**—are called **transgenic animals**. A transgenic female sheep or cow might not only contain the gene for the human protein, but might also be able to express the gene in her mammary tissue and thereby secrete the protein in her milk (Figure 4.11; Prather 1991). Thus, shortly after the announcement of Dolly, the same laboratory announced the birth of Polly (Schnieke et al. 1997). Polly was cloned from transgenic fetal sheep fibroblasts that contained the gene for human clotting factor IX, a gene whose function is deficient in hereditary hemophilia.

Producing transgenic sheep, cows, or goats is not an efficient undertaking. Only 20% of the treated eggs survive the technique. Of these, only about 5% express the human gene. And of those transgenic animals expressing the human gene, only half are female, and only a small percentage of these actually secrete a high level of the protein into their milk. (And it often takes years for them to first produce milk). Moreover, after several years of milk production, they die, and their offspring are usually not as good at secreting the human protein as the originals. Cloning would enable pharmaceutical companies to make numerous copies of such an "elite transgenic animal," all of which should produce high yields of the human protein in their milk. The medical importance of such a technology would be great, since such proteins could become much cheaper for the patients who need them for survival. The economic incentives for cloning are therefore enormous (Meade 1997).

Why Not Clone Humans?

Attorney John Robertson (1998a,b) has listed several reasons why cloning of humans should be legal. First, it would provide infertile couples with a chance to have a genetically related child. The government, Robertson says, should not interfere with the ability of any couple to have a genetically related child, and if cloning technologies are available, they should be used toward that end. Second, cloning could provide a source of body parts for transplantation (liver, pancreas, etc.) that would not be immunologically rejected. Third, in Robertson's view, no harm would come to the individual or society. Fourth, the techniques for cloning are not that different from other techniques of assisted reproduction, and the clone is merely a "late-born twin."

In 1997, the Society for Developmental Biology, representing some of the developmental biologists who pioneered cloning technology, argued against each of these positions and passed a voluntary moratorium on human cloning for 5 years. This moratorium was subsequently adopted by the Federation of Societies of Experimental Biology, and it has served as the basis for the federal proscription of human cloning. Some of the reasons cited by scientists and ethicists against Robertson's human cloning arguments are:

1. Cloning is not a good technique for giving couples a chance to have genetically related children,* for scientific reasons that include the following: (a) The resulting child would be genetically related to only one member of the couple (which could cause enormous problems with divorce and custody laws). (b) With a success rate so far from 100%, a woman could not produce enough oocytes for the procedure to have a good chance of success. (c) This high failure rate predicts that many cloned fetuses would abort or be born malformed. As Marie DiBerardino and Robert McKinnell, two of the pioneers of cloning, have pointed out, "In nearly 100% of the cases, [human cloning] would give rise to abnormal embryos and fetuses, almost all of which would abort. And what an atrocity if an abnormal child were born!" (DiBerardino and McKinnell 1997).

2. Moral issues loom tremendous against the "spare parts" argument. A person

*In earlier times, adoption of a niece or nephew was a fairly common means of raising a genetically related child. K. E. von Baer, for instance, was raised by his childless aunt and uncle.

cannot "own" another person; a clone might not wish to be used for spare parts. Scenarios that involve the creation of "brain-dead" clones kept "in storage" for spare parts are not acceptable to most of us (see Krauthammer 1998). In addition, using cloning in this manner would extend the already great social inequalities caused by differential access to medical care. Those with enough resources could use cloned body parts to extend their lives to an unprecedented duration, creating a situation wherein economic advantage becomes an overwhelming biological advantage.

3. No one knows the community or personal harm in being raised by a genetically identical parent. Genetic uniqueness has always been assumed. (Or, as one student remarked, it's bad enough being told what to do by a regular parent.)

4. Biologically speaking, clones are not "late-born twins." First, they fail the genetic definition of identical twins, since their mitochondrial genes (derived from the eggs of different women) would be different. Second, they fail the

embryological criteria for twins in that they are not derived from the same zygote and do not share the same uterus.

5. The government already sets limits on procreation rights. (You cannot legally marry your sibling or your child, for instance.) It does so by consent of the government. A ban on human cloning would not be a new role for the government.

The Society for Developmental Biology and other scientific societies have taken the stand that researchers should be allowed to study the early human embryo in attempts to understand diseases and development and to enhance fertility. They also do not wish to block pharmaceutical companies from developing their products. However, these groups feel that we are not technically or morally prepared to start cloning human beings. ■

WEBSITE **4.3 The human cloning debate.** The debate surrounding human cloning has become more intense as the techniques for cloning mammals have circumvented many of the problems that were originally seen to preclude such work. This site details some of this ongoing controversy and includes the statement from the Society for Developmental Biology.

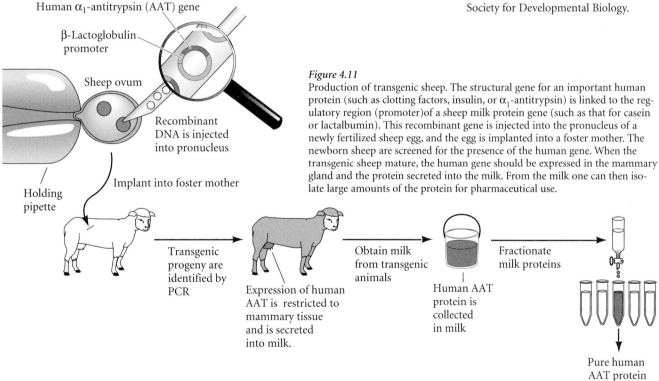

Figure 4.11
Production of transgenic sheep. The structural gene for an important human protein (such as clotting factors, insulin, or α_1-antitrypsin) is linked to the regulatory region (promoter) of a sheep milk protein gene (such as that for casein or lactalbumin). This recombinant gene is injected into the pronucleus of a newly fertilized sheep egg, and the egg is implanted into a foster mother. The newborn sheep are screened for the presence of the human gene. When the transgenic sheep mature, the human gene should be expressed in the mammary gland and the protein secreted into the milk. From the milk one can then isolate large amounts of the protein for pharmaceutical use.

The Exception to the Rule: Immunoglobulin Genes

HILE THE RULE is that the genome is the same in every cell in the body, some white blood cells that function as part of the immune system provide exceptions to that rule. The B lymphocyte ("B cell") is able to synthesize proteins called **immunoglobulins** that can function as antibodies. For decades, immunologists puzzled over how the immune system could possibly generate so many different types of antibodies. Could all the 10^7 different types of antibody proteins be encoded in the genome? This would take up an enormous amount of chromosomal space. Moreover, how could the immune system "know" how to make an antibody to some foreign molecule (**antigen**) that isn't even found outside the laboratory? They eventually discovered that the genome of the B cell does not contain DNA encoding for any of the antibody proteins. Rather, the DNA is rearranged during the development of the B cell to create the antibody-encoding genes. Moreover, while the mammalian organism has the ability to synthesize over 10 million different types of antibody proteins, each B cell can synthesize only one.

All immunoglobulins secreted from the B cells have a very similar structure. Each consists of two pairs of polypeptide subunits. There are two identical **heavy chains** and two identical **light chains**; the chains are linked together by disulfide bonds (Figure 4.12). The specificity of the immunoglobulin molecule (i.e., whether it will bind to a poliovirus, an *E. coli* cell, or some other antigen) is determined by the amino acid sequence of the **variable regions** at the amino-terminal ends of the heavy and light chains. The variable regions of the immunoglobulin molecule are attached to **constant regions** that give the antibody its effector properties needed for inactivating the antigen.

The genes that encode the immunoglobulin heavy and light chains are organized in segments. Mammalian light chain genes contain three segments (Figure 4.12). The first gene segment, *V*, encodes the first 97 amino acids of the light chain variable region. There are about 300 different *V* sequences linked tandemly on the mouse genome. The second

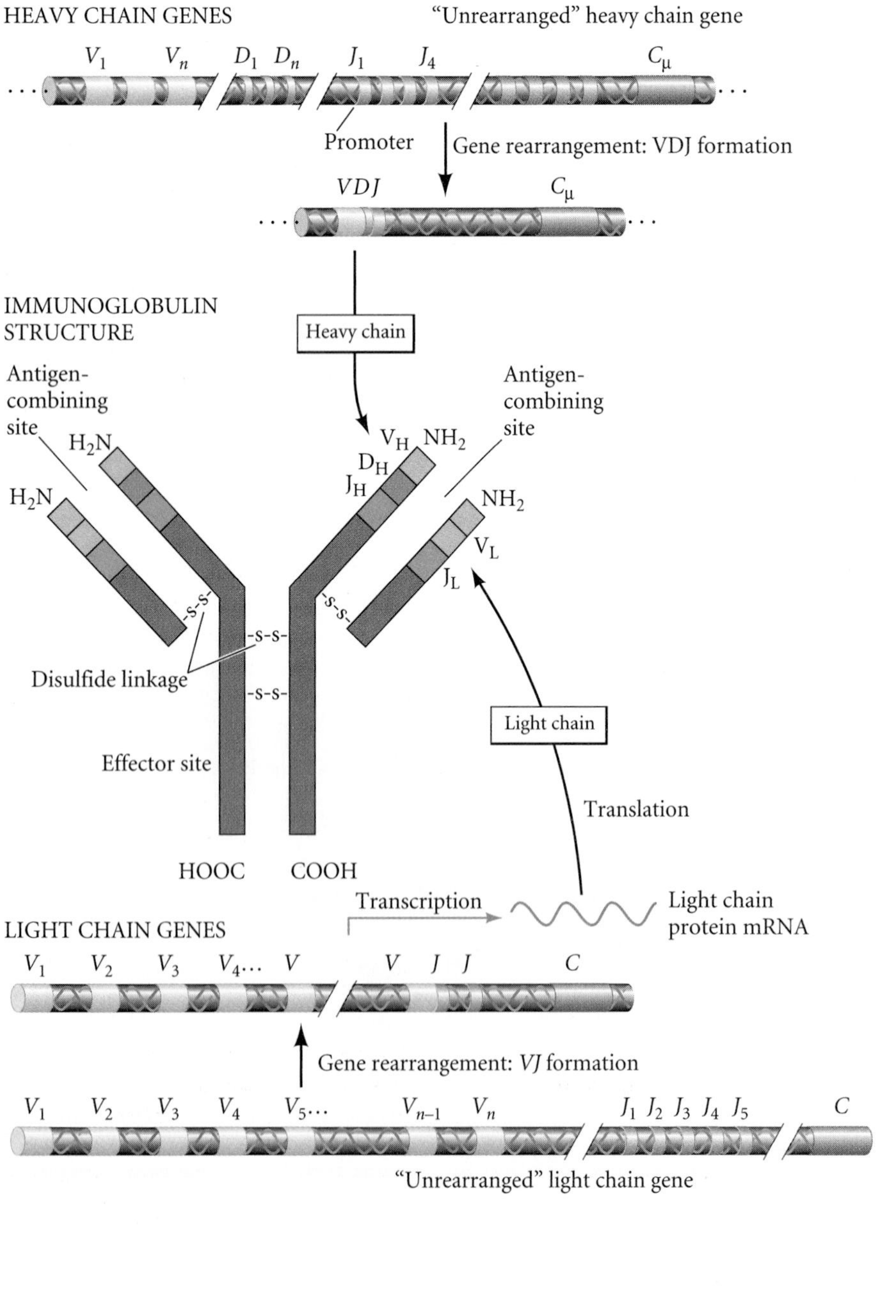

Figure 4.12
(Center) Structure of a typical immunoglobulin (antibody) protein. Two identical heavy chains and two identical light chains are connected by disulfide linkages. The antigen-binding site is composed of the variable regions of the heavy and light chains, whereas the effector site of the antibody is determined by the amino acid sequence of the heavy chain constant region. (Bottom) Rearrangement of the light chain genes during B lymphocyte differentiation. While the developing B cell is still maturing in the bone marrow, one of the 300 or more *V* gene segments combines with one of the five *J* gene segments and moves closer to the constant (*C*) gene segment. (Top) Rearrangement of the heavy chain genes. A heavy chain gene contains three segments (*V*, *D*, and *J*) that come together to form the variable region, and a constant region.

segment, *J*, consists of 4 or 5 possible DNA sequences for the last 15–17 residues of the variable region. The third segment, *C*, encodes the constant region of the light chain. During B cell differentiation, which occurs as the B cells are maturing in the bone marrow, one of the 300 *V* segments and one of the 5 *J* segments combine to form the variable region of the antibody gene. This is done by moving a *V* segment sequence next to a *J* segment sequence, a rearrangement that eliminates the intervening DNA (Hozumi and Tonegawa 1976).

The heavy chain genes contain even more segments than the light chain genes. Heavy chain genes include a *V* segment (200 different sequences for the first 97 amino acids), a *D* segment (10–15 different sequences encoding 3–14 amino acids), and a *J* segment (4 sequences for the last 15–17 amino acids of the variable region). The next segment, *C*, codes for the constant region. The heavy chain variable region is formed by adjoining one *V* segment and one *D* segment to one *J* segment (Figure 4.12). This *VDJ* variable region sequence is now adjacent to the first constant region of the heavy chain genes—the C_μ region, which is specific for antibodies that can be inserted into the plasma membrane.

Thus, an immunoglobulin molecule is formed from two genes created during the antigen-independent stage of B lymphocyte development. About 10^3 different light chain genes and about 10^4 different heavy chain genes can be formed. Since each is formed independently of the other, about 10^7 types of immunoglobulins can be created from the union of the light chain and the heavy chain within a cell. Each cell makes only one of these 10^7 antibody types.

The B cell is not the only cell type that alters its genome during differentiation. The other major cell type of the immune system, the T cell, also recombines and deletes a portion of its genome in the construction of its antigen receptor (Fujimoto and Yamagishi 1987). The enzymes responsible for mediating these DNA recombination events appear to be the same in the B and T cell lineages. Called **recombinases** (Agrawal et al. 1998), these two proteins recognize the signal regions of DNA immediately upstream from the recombinable DNA segments and form a complex there that initiates the double-stranded breaks. Moreover, the genes for these recombinase enzymes are active only in pre-B cells and pre-T cells, where the genes are being recombined.* Mutations that eliminate the function of either of the recombinases lead to severe immunodeficiency syndromes that are manifest at birth (Schwarz et al. 1996; Villa et al. 1998). ■

WEBSITE **4.4 Antibody formation.** How the immunoglobulin genes eventually produce antibody proteins is a fascinating story and full of exceptions to rules. More details are given here about DNA rearrangements.

*Recombinase proteins were once thought to be found solely in lymphocytes, but there is evidence (Chun et al. 1991; Matsuoka et al. 1991) that recombination events and recombinases exist In brain tissue as well. It is not known what their function might be in neural cells, but it is fascinating to speculate that some of the receptors that bind a nerve cell axon to its specific target might be made by the recombination of several gene regions.

Differential Gene Expression

If the genome is the same in all somatic cells within an organism (with the exception of the above-mentioned lymphocytes), how do the cells become different from one another? If every cell in the body contains the genes for hemoglobin and insulin proteins, how are the hemoglobin proteins made only in the red blood cells, the insulin proteins made only in certain pancreas cells, and neither made in the kidney or nervous system? Based on the embryological evidence for genomic equivalence (and on bacterial models of gene regulation), a consensus emerged in the 1960s that cells differentiate through **differential gene expression**. The three postulates of differential gene expression are as follows:

1. Every cell nucleus contains the complete genome established in the fertilized egg. In molecular terms, the DNAs of all differentiated cells are identical.
2. The unused genes in differentiated cells are not destroyed or mutated, and they retain the potential for being expressed.
3. Only a small percentage of the genome is expressed in each cell, and a portion of the RNA synthesized in the cell is specific for that cell type.

The first two postulates have already been discussed. The third postulate—that only a small portion of the genome is active in making tissue-specific products—was first tested in insect larvae. Fruit fly larvae have certain cells whose chromosomes become **polytene**. These chromosomes, beloved by *Drosophila* geneticists, undergo DNA replication in the absence of mitosis and therefore contain 512 (2^9), 1024 (2^{10}), or even more parallel DNA double helices instead of just one (Figure 4.13A,B). These cells do not undergo mitosis, and they grow by expanding to about 150 times their original volume. Beermann (1952) showed that the banding patterns of polytene chromosomes were identical throughout the larva, and that no loss or addition of any chromosomal region was seen when different cell types were compared. However, he and others showed that in different tissues, different regions of these chromosomes were making organ-specific RNA. In certain cell types, particular regions of the chromosomes would loosen up, "puff" out, and transcribe mRNA. In other cell types, these regions would be "silent," but other regions would puff out and synthesize mRNA.

The idea that the genes of chromosomes were differentially expressed in different cell types was confirmed using DNA–RNA hybridization (Figure 4.13C). This technique involves annealing single-stranded pieces of RNA and DNA to allow complementary strands to form double-stranded hybrids. While some mRNAs from one cell type were also found in other cell types (as expected for mRNAs encoding enzymes concerned with cell metabolism), many mRNAs were found to be specific for a particular type of cell and were not ex-

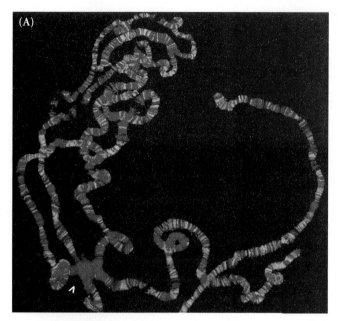

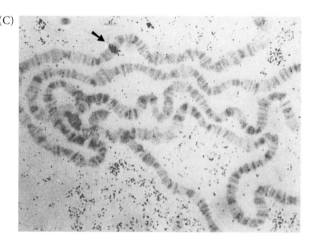

Figure 4.13
Polytene chromosomes. (A) Polytene chromosomes from the salivary gland cells of *Drosophila melanogaster*. The four chromosomes are connected at their centromere regions, forming a dense chromocenter (arrowhead). The DNA has been stained red with propidium iodide stain. The yellow stain represents the binding of a particular protein to the DNA. This protein is involved in compartmentalizing the regions of the chromosome so that the activation of one gene does not cause the activation of its neighbors. (B) Electron micrograph of a small region of a *Drosophila* polytene chromosome. The bands (dark) are highly condensed compared with the interband (lighter) regions. (C) Hybridization of a yolk protein mRNA with the polytene chromosome of a larval *Drosophila* salivary gland. The dark grains (arrow) show where the radioactive yolk protein message has bound to the chromosomes. Note that the gene for the yolk protein is present in the salivary gland chromosomes, even though yolk protein is not synthesized there. (A courtesy of U. K. Laemmli; B from Burkholder 1976, courtesy of G. D. Burkholder; C from Barnett et al. 1980; photograph courtesy of P. C. Wensink.)

pressed in other cell types, even though the genes encoding them were present (Wetmur and Davidson 1968). Thus, differential gene expression was shown to be the way a single genome derived from the fertilized egg could generate the hundreds of different cell types in the body. The question then became, How does this differential gene expression occur? The answers to that question will be the topic of the next chapter. To understand the results that will be presented there, however, one must become familiar with some of the techniques of molecular biology that are being applied to the study of development. These include techniques to determine the spatial and temporal location of specific mRNAs, as well as techniques to determine the functions of these messages.

WEBSITE **4.5 DNA isolation techniques.** The basic techniques of DNA analysis—gene cloning, sequencing, and Southern blotting—are discussed in most introductory biology books. This site gives a review of these procedures.

RNA Localization Techniques

Northern blotting

We can determine the temporal and spatial locations where RNAs are expressed by running an RNA blot (often called a **Northern blot**). First, one extracts the RNA from embryos at different stages of development, or from different organs of the same embryo. The investigator then places the RNA samples side by side at one end of a gel and runs an electrical current through the gel. The smaller the RNA, the faster it moves through the gel. Thus, different RNAs are separated by their sizes. This technique is called **electrophoresis**.* The investigator then transfers the separated RNAs to a nitrocellulose paper or nylon membrane filter, and incubates the RNA-containing filter in a solution containing a radioactive single-stranded DNA fragment from a particular gene (Figure 4.14A). This radioactively labeled DNA probe binds only to regions of the fil-

---

*Given the same charge-to-mass ratio, smaller RNA fragments obtain a faster velocity than larger ones when propelled by the same energy (i.e., larger fragments will move more slowly than smaller fragments). This is a function of the kinetic energy equation, $E = 1/2mv^2$. Solving for velocity, we find that velocity is inversely proportional to the square of the mass.

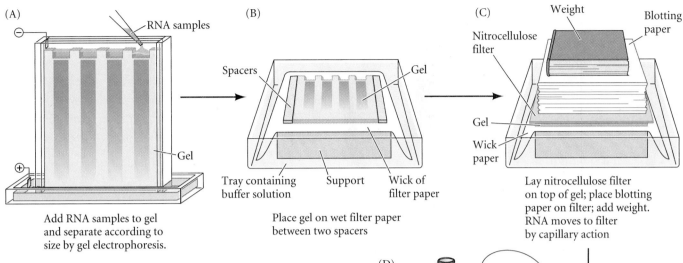

(A) Add RNA samples to gel and separate according to size by gel electrophoresis.

(B) Place gel on wet filter paper between two spacers

(C) Lay nitrocellulose filter on top of gel; place blotting paper on filter; add weight. RNA moves to filter by capillary action

Figure 4.14
Developmental Northern blotting. (A–E) Procedure for Northern blotting. (A) RNA is isolated from various tissues and is separated by size using gel electrophoresis. (B) The gel is then placed on a paper wick, which absorbs an ionic solution from a trough. (C) A filter that traps RNA is placed above the gel, and blotting paper is placed above that. Capillary action draws the solution through the gel, trapping the RNA on the filter. (D) The filter is incubated with radioactive single-stranded DNA complementary to the mRNA of interest. (E) After washing off unbound DNA, autoradiography localizes the mRNA in the samples that contain it. (F) Drawing of a developmental Northern blot showing the presence of *Pax6* mRNA in the eye, brain, and pancreas of a mammalian embryo. (F after Ton et al. 1991.)

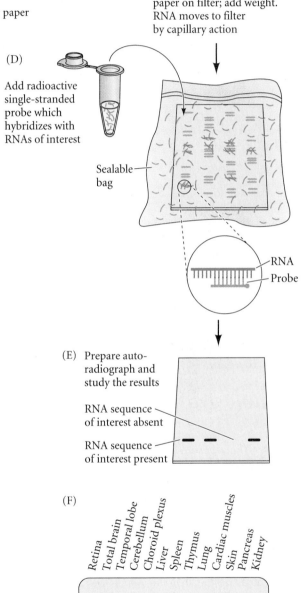

(D) Add radioactive single-stranded probe which hybridizes with RNAs of interest

(E) Prepare auto-radiograph and study the results

RNA sequence of interest absent

RNA sequence of interest present

(F)

ter where the complementary RNA is located. If the mRNA for that gene is present in a sample, the labeled DNA will bind to it and can be detected by autoradiography. X-ray film is placed above the filter and incubated in the dark. The localized radioactivity in the probe reduces the silver in the X-ray film, and grains form. The resulting spots, which appear directly above the places where the radioactive DNA has bound, appear black when viewed directly. Autoradiographs of this type, in which RNAs from several stages or tissues are compared simultaneously, are called **developmental Northern blots**.

Figure 4.14F shows a developmental Northern blot used to investigate the expression of *Pax6* expression in the mammalian embryo. Pax6 is a protein that is critical for normal eye development; mutations in this gene give small eyes (in heterozygous mice) or no eyes or nose (in mice or humans homozygous for the loss-of-function mutation). The Northern blot shows that this gene is expressed in the embryo in the brain, eyes, and pancreas, but in no other tissue.

In situ hybridization

But where was this gene expressed in the eye? Northern blot analysis (which uses mRNA extracted from pieces of tissue that have been removed from the embryo) can give only an

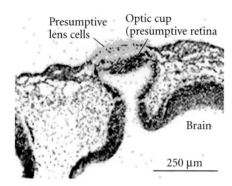

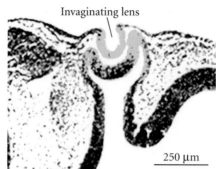

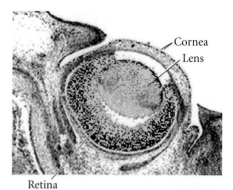

Presumptive lens cells / Optic cup (presumptive retina / Brain / 250 µm

Invaginating lens / 250 µm

Cornea / Lens / Retina

Figure 4.15
In situ hybridization showing the expression of the *Pax6* gene in the developing mouse eye. Transverse microscopic sections were taken through the developing heads of 9-, 10-, and 15-day embryonic mice. At this time, the optic cup is touching the outer ectoderm and inducing it to form a lens. The radioactive *Pax6* DNA probe binds only where *Pax6* mRNA is localized, and can be visualized by developing the photographic emulsion. The locations where the probe has bound, depicted here as yellow dots (by computer imaging), show that the *Pax6* message is expressed in both the presumptive lens ectoderm and the optic stalk, which forms the retina and optic nerve. (From Grindley et al. 1995; photograph courtesy of R. E. Hill.)

approximate location and time for gene expression. A more detailed map of gene expression patterns can be obtained by using a process called **in situ hybridization**. Instead of using a DNA probe to seek mRNA on a filter, the probe is hybridized with the mRNA in the organ itself. Embryos or organs are fixed to preserve their structure and to prevent the RNA from being degraded, then sectioned for microscopy and placed on a slide. When the radioactive DNA probe is added, it binds only where the complementary mRNA is present. After any unbound probe is washed off, the slide is covered with a transparent photographic emulsion for autoradiography. By means of computer-mediated bright-field imaging, the reduced silver grains can be shown in a color that contrasts with the background stain. Thus, we can visualize those cells (or even regions within cells) that have accumulated a specific type of mRNA. Figure 4.15 shows an in situ hybridization for *Pax6* mRNA in mice. One can see that *Pax6* mRNA is found in the region where the presumptive retina meets the presumptive lens tissue. As development proceeds, it is seen in the developing retina, lens, and cornea of the eye.*

*The role that *Pax6* is playing in this process will be discussed in Chapter 6.

Sidelights & Speculations

Whole-Mount In Situ Hybridization

Working with radioactive probes and emulsions necessitates the use of finely sliced microscopic sections. A more recent technique for in situ hybridization uses probes that bind colored reagents. This technique allows us to look at entire organs (and organisms) without sectioning them, thereby observing large regions of gene expression.

Figure 4.16 shows an in situ hybridization performed on a whole chick embryos that have been fixed without being sectioned. It has also been permeabilized by lipid and protein solvents so that the probe can get in and out of its cells. The probe used in this experi-

ment recognizes the mRNA encoding *Pax6* in the chick embryo. This probe is labeled not with a radioactive isotope, but with a modified UTP. To create this probe, a region of the cloned *Pax6* gene was transcribed into mRNA, but with two important modifications. First, in addition to regular UTP, the nucleotide mix also contained UTP conjugated with digoxigenin. This does not interfere with the coding properties of the resulting mRNA, but does make it recognizably different from any other RNA in the cell. (Digoxigenin is a compound made by particular groups of plants and is not found in animal cells.) Second, the gene was reversed with re-

spect to its promoter. By manipulating how a cloned piece of DNA gets into its vector, one can clone a gene in its reversed orientation. An "antisense" mRNA can be transcribed from the reversed gene and used as a probe, since it will recognize the "sense" mRNA in the cell.

The digoxigenin-labeled probe is then incubated with the embryo. After several hours, numerous washes remove any probe that has not bound to the embryo. Then the embryo is incubated in a solution containing an antibody against digoxigenin. The only places where digoxigenin should exist is where the probe has bound (i.e., where it recognized its

(A)

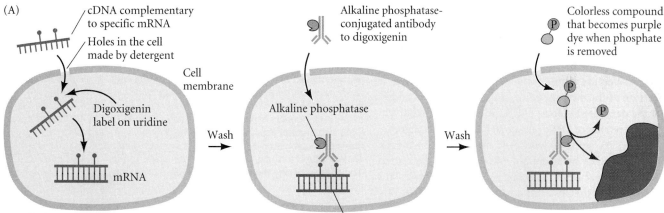

cDNA complementary
to specific mRNA

Holes in the cell
made by detergent

Cell
membrane

Digoxigenin
label on uridine

mRNA

Alkaline phosphatase-
conjugated antibody
to digoxigenin

Wash

Alkaline phosphatase

Colorless compound
that becomes purple
dye when phosphate
is removed

Wash

1. Add digoxigenin-labeled probe
   complementary to RNA of interest

2. Add alkaline phosphatase-conjugated
   antibody that binds to digoxigenin

3. Add chemical that becomes a dark
   purple dye when phosphate is removed;
   dye colors the cell.

(B)

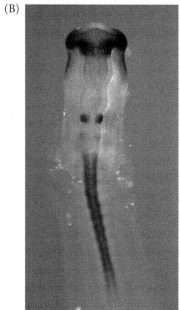

Figure 4.16
Whole-mount in situ hybridization localizing *Pax6* gene expression in early chick embryos.
(A) Schematic of procedure whereby a digoxigenin-labeled probe hybridizes to a specific mRNA.
Alkaline phosphatase-conjugated antibodies to digoxigenin recognize the digoxigenin labeled
probe. The enzyme is able to convert a colorless compound into a dark purple precipitate. (B)
Pax6 mRNA can be seen to accumulate in the roof of the brain region that will form the eyes, as
well as in the ectoderm that will form lenses. More caudal expression of this gene in the nervous
system is also seen. (After Li et al. 1994; photograph courtesy of O. Sundin.)

RNA), so the antibody sticks there. After repeated washes to remove all the unbound antibody, the embryo is incubated in a different antibody solution. This second antibody, which is covalently linked to an enzyme such as alkaline phosphatase, recognizes and binds to the first antibody. If the first antibody has bound to the probe, the second antibody will find it. After any excess antibody is washed away, the presence of the second antibody is assayed by using a dye that is activated by the phosphatase enzyme attached to that antibody. A dark blue precipitate forms in every cell that contains the second antibody. And the second antibody should be present only where the particular mRNA encoding chick *Pax6* is present. This message is found in the roof of the brain region that will form the eyes, as well as in the head ectoderm that will form the lenses. It is also expressed in other, more caudal, regions of the neural tube. ■

The polymerase chain reaction

The **polymerase chain reaction** (PCR) is a method of in vitro gene cloning* that can generate enormous quantities of a specific DNA fragment from a small amount of starting material (Saiki et al. 1985). It can be used for cloning a specific gene or for determining whether a specific gene is actively transcribing RNA in a particular organ or cell type. The standard methods of cloning use living microorganisms to amplify recombinant DNA. PCR, however, can amplify a single region of a DNA molecule several million times in a few hours, and can do it in a test tube. This technique has been extremely useful

in cases in which there is very little nucleic acid to study. Preimplantation mouse embryos, for instance, have very little mRNA, and we cannot obtain millions of such embryos to study. If we wanted to know whether the preimplantation mouse embryo contained the mRNA for a particular protein, it would be very difficult to find out using standard methods. We would have to lyse thousands of mouse embryos to get enough mRNA. However, the PCR technique allows us to find this message in a few embryos by specifically amplifying only that message millions of times (Rappolee et al. 1988).

The use of PCR for finding rare mRNAs is illustrated in Figure 4.17. First, the mRNAs from a group of cells are purified and converted into complementary DNA (cDNA) using an enzyme called **reverse transcriptase**. By using DNA polymerase and S1 nuclease, this population of single-stranded

*Gene cloning—making numerous copies of the same DNA sequence—should not be confused with organism cloning (making numerous genetically identical copies of the same organism).

cDNAs is then made into double-stranded cDNAs. Next a specific cDNA is targeted for amplification. In preparation for this step, the cDNA double helices are separated, or denatured, by heating them. Two small oligonucleotide **primers** that are complementary to a portion of the message being looked for are then added to the denatured cDNA. Oligonucleotides are relatively short stretches of DNA (about 20 bases). If the oligonucleotides recognize sequences in the cDNA, this means that the mRNA being sought was present in the source of the original DNA. The oligonucleotide primers are made so that they hybridize to opposite strands at opposite ends of the targeted sequence. (If we are trying to isolate the gene or mRNA for a specific protein of

Figure 4.17
The polymerase chain reaction (PCR) technique for determining whether a particular type of mRNA is present. First, all the mRNA is converted to double-stranded DNA by reverse transcriptase and DNA polymerase. This DNA is denatured, and two sets of primers are added. If the specific sequence is present, the primers will hybridize to its opposite ends. (Specific primers are made based on the sequence we are looking for. If we know only the sequence of the protein encoded by the message, a set of different primers is made, each possibly being complementary to the DNA.) Using thermostable DNA polymerase from *T. aquaticus* (*Taq* polymerase), each DNA strand synthesizes its complement. These strands are then denatured, and the primers are hybridized to them, starting the cycle again. In this way, the number of new strands having the sequence between the two primers increases exponentially.

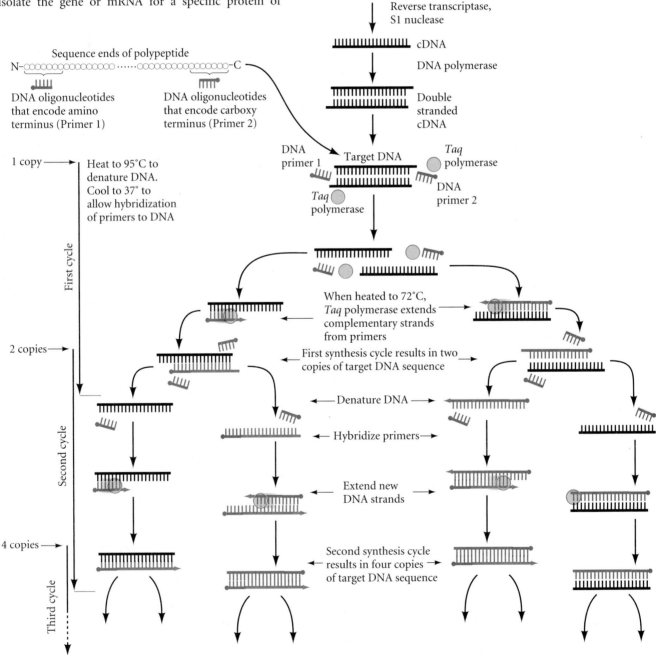

known sequence, we can synthesize oligonucleotides that are complementary to the amino end and the carboxyl end of the protein.) The 3′ ends of these primers face each other, so that replication is through the target DNA. Once the first primer has hybridized with the DNA, DNA polymerase can be used to synthesize a new strand.

The DNA polymerase used in this process is not normal *E. coli* DNA polymerase, however; it is DNA polymerase from bacteria such as *Thermus aquaticus* or *Thermococcus littoralis*. These bacteria live in hot springs (such as those in Yellowstone National Park) or in submarine thermal vents, where the temperature reaches nearly 90°C. These DNA polymerases can withstand temperatures near boiling. PCR takes advantage of this evolutionary adaptation. Once the second strand is made, it is heat-denatured from its complement. (The temperatures used would inactivate the *E. coli* DNA polymerase, but the thermostable polymerases are not damaged.) The second primer is added, and now both strands can synthesize new DNA. Repeated cycles of denaturation and synthesis amplify this region of DNA in geometric fashion. After 20 such rounds, that specific region has been amplified 2^{20} (a little more than a million) times. When the DNA is subjected to electrophoresis, the presence of such an amplified fragment is easily detected. Its presence shows that in the original messenger RNA population there was an mRNA with the sequence of interest.

Determining the Function of Genes during Development

Transgenic cells and organisms

While it is important to know the sequence of a gene and its temporal-spatial pattern of expression, what's really crucial is to know the functions of that gene during development. Recently developed techniques have enabled us to study gene function by moving certain genes into and out of embryonic cells.

> WEBSITE **4.6 Bioinformatics.** Information about gene regulation and developmental pathways may soon be modeled on computers. Accessibility to this information may enable researchers to design experiments that have higher chances of success.

INSERTING NEW DNA INTO A CELL. Cloned pieces of DNA can be isolated, modified (if so desired), and placed into cells by several means. One very direct technique is **microinjection**, in which a solution containing the cloned gene is injected very carefully into the nucleus of a cell (Capecchi 1980). This is an especially useful technique for injecting genes into newly fertilized eggs, since the haploid nuclei of the sperm and egg are relatively large (Figure 4.18). In **transfection**, DNA is incorporated directly into cells by incubating them in a solution that makes them "drink" it in. The chances of a DNA fragment being incorporated into the chromosomes in this way are rel-

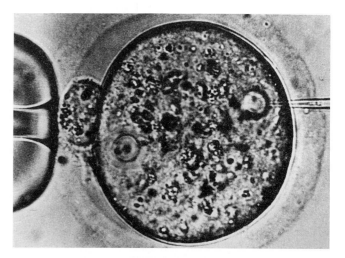

Figure 4.18
Insertion of new DNA into embryonic cells. Here, DNA (from cloned genes) is injected into the a pronucleus of a mouse egg. (From Wagner et al. 1981; photograph courtesy of T. E. Wagner.)

atively small, however, so the DNA of interest is usually mixed with another gene, such as a gene encoding resistance to a particular antibiotic, that enables those rare cells that incorporate the DNA to survive under culture conditions that will kill all the other cells (Perucho et al. 1980; Robins et al. 1981). Another technique is **electroporation**, in which a high-voltage pulse "pushes" the DNA into the cells.

A more "natural" way of getting genes into cells is to put the cloned gene into a **transposable element** or **retroviral vector**. These are naturally occurring mobile regions of DNA that can integrate themselves into the genome of an organism. Retroviruses are RNA-containing viruses. Within a host cell, they make a DNA copy of themselves (using their own virally encoded reverse transcriptase); the copy then becomes double-stranded and integrates itself into a host chromosome. The integration is accomplished by two identical sequences (long terminal repeats) at the ends of the retroviral DNA. Retroviral vectors are made by removing the viral packaging genes (needed for the exit of viruses from the cell) from the center of a mouse retrovirus. This extraction creates a vacant site where other genes can be placed. By using the appropriate restriction enzymes, researchers can insert an isolated gene (such as a gene isolated by PCR) and insert it into a retroviral vector. These retroviral vectors infect mouse cells with an efficiency approaching 100%. Similarly, in *Drosophila*, new genes can be carried into a fly via **P elements**. These DNA sequences are naturally occurring transposable elements that can integrate like viruses into any region of the *Drosophila* genome. Moreover, they can be isolated, and cloned genes can be inserted into the center of the P element. When the recombined P element is injected into a *Drosophila* oocyte, it can integrate into the DNA and provide the embryo with the new gene (Spradling and Rubin 1982).

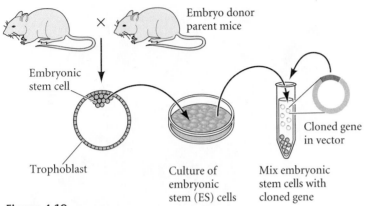

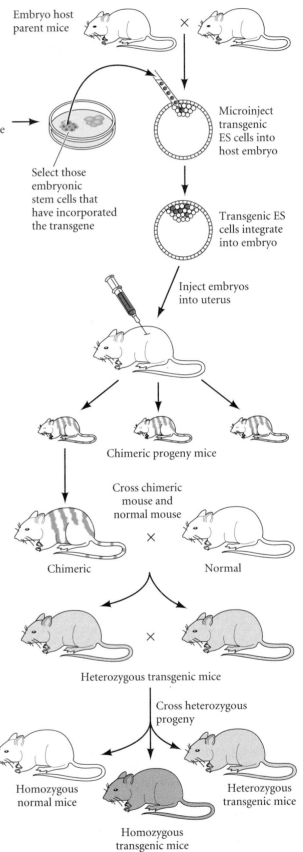

Figure 4.19
Production of transgenic mice. Embryonic stem cells from one mouse are cultured and their genome altered by the addition of a cloned gene. These transgenic cells are selected and then injected into the early stages of a host mouse embryo. Here, the transgenic embryonic stem cells integrate with the host's embryonic stem cells. The embryo is placed into the uterus of a pregnant mouse and develops into a chimeric mouse. If the donor stem cells have contributed to the germ line, and the chimeric mouse is crossed with a wild-type mouse, some of the progeny will be heterozygous for the added allele. By mating heterozygotes, a strain of transgenic mice can be generated that is homozygous for the added allele. The added gene (the transgene) can be from any eukaryotic source.

CHIMERIC MICE. The techniques described above have been used to transfer genes into every cell of the mouse embryo (Figure 4.19). During early mouse development, there is a stage when only two cell types are present: the outer trophoblast cells, which will form the fetal portion of the placenta, and the inner cell mass, whose cells will give rise to the embryo itself. These inner cells are the cells whose separation can lead to twins (Chapters 3 and 11), and if an inner cell mass blastomere of one mouse is transferred into the embryo of a second mouse, that donor cell can contribute to every organ of the host embryo.

Inner cell mass blastomeres can be isolated from the embryo and cultured in vitro; such cultured cells are called **embryonic stem cells (ES cells)**. ES cells retain their totipotency, and each of them can contribute to all organs if injected into a host embryo (Gardner 1968; Moustafa and Brinster 1972). Moreover, once in culture, these cells can be treated as described in the preceding section so that they will incorporate new DNA. A treated ES cell (the entire cell, not just the DNA) can then be injected into another early-stage mouse embryo, and will integrate into the host embryo. The result is a **chimeric mouse.*** Some of this mouse's cells will be derived from the host embryonic stem cells, but some portion of its cells will be derived from the treated embryonic stem cell. If

*It is critical to note the difference between a chimera and a hybrid. A **hybrid** results from the union of two different genomes within the same cell: the offspring of an *AA* genotype parent and an *aa* genotype parent is an *Aa* hybrid. A **chimera** results when cells of different genetic constitution appear in the same organism. The term is apt: it refers to a mythical beast with a lion's head, a goat's body, and a serpent's tail.

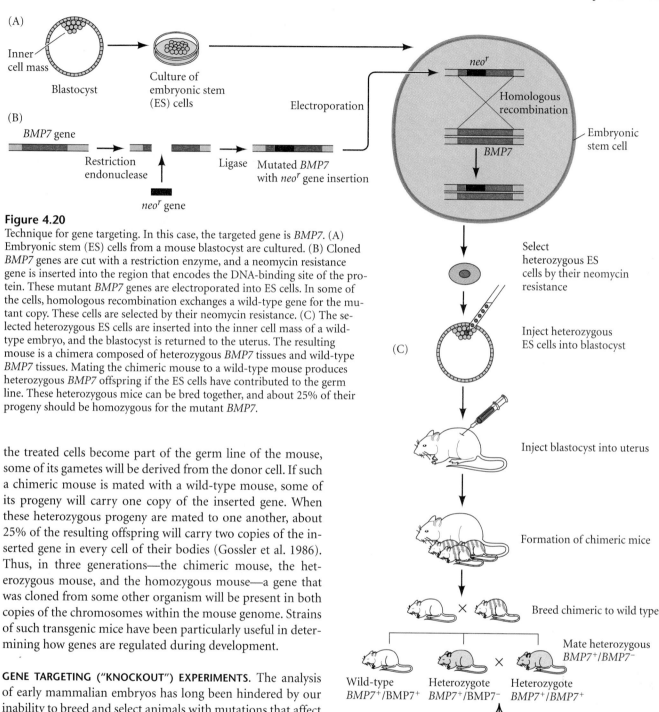

Figure 4.20

Technique for gene targeting. In this case, the targeted gene is *BMP7*. (A) Embryonic stem (ES) cells from a mouse blastocyst are cultured. (B) Cloned *BMP7* genes are cut with a restriction enzyme, and a neomycin resistance gene is inserted into the region that encodes the DNA-binding site of the protein. These mutant *BMP7* genes are electroporated into ES cells. In some of the cells, homologous recombination exchanges a wild-type gene for the mutant copy. These cells are selected by their neomycin resistance. (C) The selected heterozygous ES cells are inserted into the inner cell mass of a wild-type embryo, and the blastocyst is returned to the uterus. The resulting mouse is a chimera composed of heterozygous *BMP7* tissues and wild-type *BMP7* tissues. Mating the chimeric mouse to a wild-type mouse produces heterozygous *BMP7* offspring if the ES cells have contributed to the germ line. These heterozygous mice can be bred together, and about 25% of their progeny should be homozygous for the mutant *BMP7*.

the treated cells become part of the germ line of the mouse, some of its gametes will be derived from the donor cell. If such a chimeric mouse is mated with a wild-type mouse, some of its progeny will carry one copy of the inserted gene. When these heterozygous progeny are mated to one another, about 25% of the resulting offspring will carry two copies of the inserted gene in every cell of their bodies (Gossler et al. 1986). Thus, in three generations—the chimeric mouse, the heterozygous mouse, and the homozygous mouse—a gene that was cloned from some other organism will be present in both copies of the chromosomes within the mouse genome. Strains of such transgenic mice have been particularly useful in determining how genes are regulated during development.

GENE TARGETING ("KNOCKOUT") EXPERIMENTS. The analysis of early mammalian embryos has long been hindered by our inability to breed and select animals with mutations that affect early embryonic development. This block has been circumvented by the techniques of **gene targeting** (or, as it is sometimes called, **gene knockout**). These techniques are similar to those that generate transgenic mice, but instead of *adding* genes, gene targeting *replaces* wild-type alleles with mutant ones. As an example, we will look at the gene knockout of bone morphogenetic protein 7 (BMP7). Bone morphogenetic proteins are involved in numerous developmental interactions whereby one set of cells interacts with other neighboring cells to alter their properties. BMP7 has been implicated as a protein that prevents cell death and promotes cell division in sev-

eral developing organs. Dudley and his colleagues (1995) used gene targeting to find the function of BMP7 in the development of the mouse. First, they isolated the *BMP7* gene, cut it at one site with a restriction enzyme, and inserted a bacterial gene for neomycin resistance into that site (Figure 4.20). In

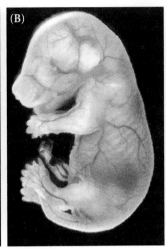

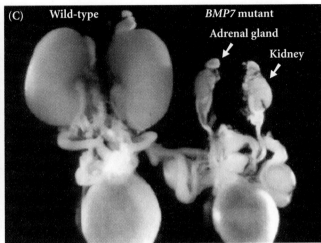

Figure 4.21
Morphological analysis of *BMP7* knockout mice. A wild-type (A) and a homozygous BMP7-deficient mouse (B) at day 17 of their 21-day gestation. The BMP7-deficient mouse lacks eyes. The kidneys of these mice at day 19 of gestation are shown in (C). The kidney of the BMP7-deficient mouse is severely atrophied. Microscopic sections revealed the death of the cells that would otherwise have formed the nephrons. (From Dudley et al. 1995; photographs courtesy of E. Robertson.)

other words, they mutated the *BMP7* gene by inserting into it a large piece of foreign DNA, destroying the ability of the BMP7 protein to function. These mutant *BMP7* genes were electroporated into ES cells that were sensitive to neomycin. Once inside the nucleus of an ES cell, the mutated *BMP7* gene may replace a normal allele of *BMP7* by a process called homologous recombination. In this process, the enzymes involved in DNA repair and replication incorporate the mutant gene in the place of the normal copy. It's a rare event, but such cells can be selected by growing the ES cells in neomycin. Most

of the cells are killed by the drug, but the ones that have acquired resistance from the incorporated gene survive. The resulting cells have one normal *BMP7* gene and one mutated *BMP7* gene. These heterozygous ES cells were then microinjected into mouse blastocysts, where they were integrated into the cells of the embryo. The resulting mice were chimeras composed of wild-type cells from the host embryo and heterozygous *BMP7*-containing cells from the donor ES cells. The chimeras were mated to wild-type mice, producing progeny that were heterozygous for the *BMP7* gene. These heterozygous mice were then bred with each other, and about 25% of their progeny carried two copies of the mutated *BMP7* gene. These homozygous mutant mice lacked eyes and kidneys (Figure 4.21). In the absence of BMP7, it appears that many of the cells that normally form these two organs stop dividing and die. In this way, gene targeting can be used to analyze the roles of particular genes during mammalian development.

Sidelights & Speculations

Human Somatic and Germ Line Gene Therapy

Embryonic Stem Cells

In 1998, two laboratories (Gearhart 1998; Thomson et al. 1998) announced that they had derived human embryonic stem cells. In some instances, these cells were derived from inner cell masses of embryos that were not implanted into infertility patients. In other cases, they were generated from germ cells derived from spontaneously aborted fetuses. In both instances, the embryonic stem cells were pluripotent, since they were able to

differentiate in culture to form more restricted stem cells that produced neurons and blood when injected into immunodeficient mice. The hope is that human ES cells can be used to produce new neurons for patients with degenerative brain disorders or spinal cord injuries, and to produce new blood cells for people with anemias (Figure 4.22A). Such therapy has already worked in mice, where ES cells were cultured in conditions causing them to form glial stem cells. These glial stem

cells were transplanted into mice that had a genetic deficiency of glial function, and cured the defect (Figure 4.22B; Brüstle et al. 1999). Similarly, neural stem cells derived from mouse ES cells were able to divide and differentiate into functional neurons when injected into a damaged rodent nervous system (McDonald et al. 1999). But laboratory mice are inbred, and humans are not. To get around the problem of host rejection, human ES cells might have to be modified, or doctors

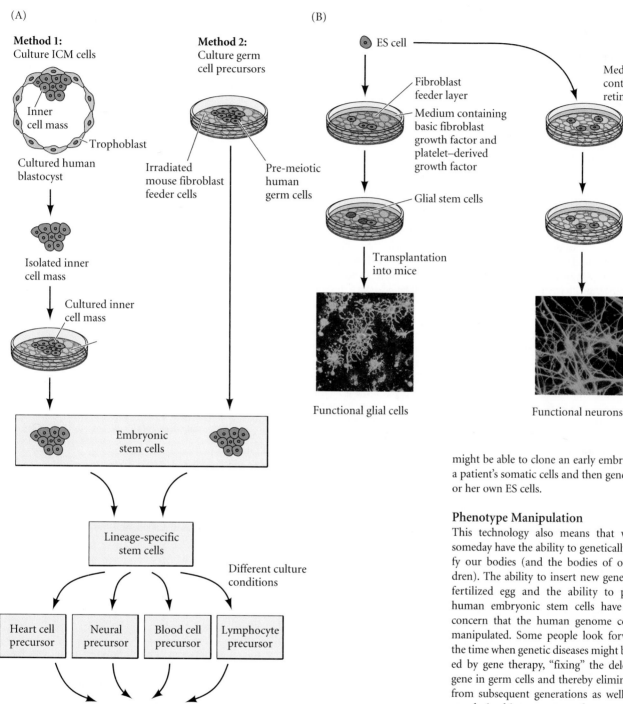

(A)

Method 1:
Culture ICM cells

Inner cell mass

Trophoblast

Cultured human blastocyst

Isolated inner cell mass

Cultured inner cell mass

Method 2:
Culture germ cell precursors

Irradiated mouse fibroblast feeder cells

Pre-meiotic human germ cells

Embryonic stem cells

Lineage-specific stem cells

Different culture conditions

Heart cell precursor | Neural precursor | Blood cell precursor | Lymphocyte precursor

Transplantation therapy

(B)

ES cell

Fibroblast feeder layer

Medium containing basic fibroblast growth factor and platelet–derived growth factor

Glial stem cells

Transplantation into mice

Functional glial cells

Medium containing retinoic acid

Neural stem cells

Functional neurons

Figure 4.22
ES cell therapeutics. (A) Human embryonic stem (ES) cells can be derived from the inner cell mass of blastomeres or from human germ cells before they initiate meiosis. These cells can differentiate in culture to form the more restricted stem cells for neural, blood, heart muscle, or lymphocyte lineages. (B) The differentiation of ES cells into lineage-restricted (neuronal and glial) stem cells can be accomplished by altering the media in which the ES cells grow. (A after Gearhart 1998; B photographs from Brüstle et al. 1999 and Wickelgren 1999, courtesy of O. Brüstle and J. W. McDonald.)

might be able to clone an early embryo from a patient's somatic cells and then generate his or her own ES cells.

Phenotype Manipulation

This technology also means that we may someday have the ability to genetically modify our bodies (and the bodies of our children). The ability to insert new genes into a fertilized egg and the ability to produce human embryonic stem cells have caused concern that the human genome could be manipulated. Some people look forward to the time when genetic diseases might be treated by gene therapy, "fixing" the deleterious gene in germ cells and thereby eliminating it from subsequent generations as well. Other people (and in some cases, the same people) worry that this ability to manipulate germ line genes will result in misguided attempts to enhance human physical or mental abilities.

There is relatively little concern about genetic therapy being applied to somatic cells. This type of gene therapy is currently being tested at several medical centers. In such cases, a stem cell (such as a blood stem cell) is cultured, given the new gene, and reinserted into the body (Anderson 1998; Gage 1998; Ye et al. 1999). This procedure is akin to standard medical treatment wherein the individual is treated. Germ line gene therapy, in contrast, would seek to eliminate "bad" genes from the population.

When would germ line genetic manipulation be justified? Wivel and Walters (1993) see two possible cases. First, such manipulation may be justified when both parents are afflicted with a recessive autosomal disorder, so that 100% of their offspring would be expected to have it. This is an exceptionally rare situation. More common is the case in which both parents are heterozygotes for a recessive genetic disorder. These parents have a 75% chance of having a phenotypically normal child, and screening can be carried out during pregnancy, followed by selective abortion if the fetus is found to be homozygous for the mutant allele. Germ line genetic modification is seen as an alternative to screening and selective abortion. Wivel and Walters view such monogenic deficiency diseases as Lesch-Nyhan syndrome, Tay-Sachs disease, and metachromatic leukodystrophy as candidates for this type of genetic therapy.

The arguments against germ line gene therapy come from several sources. Danks (1994) can see very little medical good coming from germ line gene therapy. First, he finds that there are very few candidates for such a procedure. The case of two homozygous parents is extraordinarily rare. Moreover, in the case of the two heterozygous parents, the procedures that would be used for germ line therapy would include in vitro fertilization, a technique that discards several viable embryos and is seen by many as equivalent to abortion. Indeed, the same techniques used to identify the affected offspring of two heterozygous parents would also be able to identify those embryos that are not at risk. "Are we to propose discarding those ova that are shown to be unaffected in order to correct an affected one and reimplant it? Surely one should discard the affected and reimplant one of the unaffected."

Danks also argues against the view that germ line gene therapy would result in the correction of genetic disease in future generations. It might even lead to more mutations. No one can target a gene to a particular location in the egg nucleus. So a new gene added to a cell could become inserted within a normal gene and mutate it. To achieve heritable correction, it will be necessary to achieve replacement of the defective gene by the inserted gene, and this has not yet been done. Zanjani and Anderson (1999) have proposed that somatic gene therapy in utero might be a worthwhile alternative to germ line gene therapy.

There are important social issues also at stake in germ line gene therapy. In the recent symposium on Engineering the Human Genome (UCLA 1998), James D. Watson was explicit about using gene therapy techniques for enhancement: "I mean, if we could make better human beings by knowing how to add genes, why shouldn't we do it?" Critics and commentators (Haraway 1997; Silver 1997; Rifkin 1998) claim that if this technology were successful, it would cause a large gap between the haves and have-nots. Those who could afford it could have "prettier," healthier, perhaps even smarter children. Those who couldn't afford it would be left out of this bright future. Thus, genetic engineering of the germ line could exacerbate social stratification, reify economic differences into biological ones, and create a genetic underclass.* Moreover, what we think of today as being a good trait might not turn out to be so good in a future environment.

Another reason for caution concerns individuality. Until this moment in history, each person could see himself or herself as an individual whose genetic capabilities were the products of a chance union of a particular sperm and egg. But what if you knew that your good looks and abilities were given to you by your parents through the insertion of their choice of particular alleles into your zygote chromatin? What if you knew that you could have been taller or thinner if your parents had been richer when you were conceived? As Stan Lee wrote in 1963, "With great power comes great responsibility." ∎

WEBSITE **4.7 ES cells and genetic engineering.** This website can keep one current on debates and policies in these areas.

*If this sounds like science fiction, remember that we have been just been discussing cloning mammals—something that *was* science fiction until 1997.

Determining the function of a message: Antisense RNA

Another method for determining the function of a gene is to use "antisense" copies of its message to block the function of that message. Antisense RNA allows developmental biologists to determine the function of genes during development and to analyze the action of genes that would otherwise be inaccessible for genetic analysis.

Antisense messages can be generated by cloning DNA into vectors that have promoters at both ends of the inserted gene. When incubated with a particular RNA polymerase and nucleotide triphosphates, the promoter will initiate transcription of the message "in the wrong direction." In so doing, it synthesizes a transcript that is complementary to the natural one (Figure 4.23A). This complementary transcript is called **antisense RNA** because it is the complement of the original ("sense") message. When large amounts of antisense RNA are injected or transfected into cells containing the normal mRNA from the same gene, the antisense RNA binds to the normal message, and the resulting double-stranded nucleic acid is degraded. (Cells have enzymes to digest double-stranded nucleic acids in the cytoplasm.) This causes a functional deletion of the message, just as if there were a deletion mutation for that gene.

The similarities between the phenotypes produced by a loss-of-function mutation and by antisense RNA treatment were seen when antisense RNA was made to the *Krüppel* gene of *Drosophila*. *Krüppel* is critical for forming the thorax and abdomen of the fly. If this gene is absent, fly larvae die because they lack thoracic and anterior abdominal segments (Figure 4.23B). A similar situation is created when large amounts of antisense RNA against the *Krüppel* message are injected into early fly embryos (Rosenberg et al. 1985).

WEBSITE **4.8 RNA interference.** Soaking the nematode *C. elegans* in a solution containing double-stranded RNA will knock out the expression of that gene not only in the soaked animal, but also in the progeny of that animal.

Figure 4.23
Use of antisense RNA to examine the roles of genes in development. (A) An antisense message (in this case, to the *Krüppel* gene of *Drosophila*) is produced by placing the cloned cDNA fragment for the *Krüppel* message between two strong promoters (where RNA polymerases bind to initiate transcription). The two promoters are in opposite orientation with respect to the *Krüppel* cDNA. In this case, the T3 promoter is in normal orientation and the T7 promoter is reversed. These promoters are recognized by different RNA polymerases (from the T3 and T7 bacteriophages, respectively). T3 polymerase enables the transcription of "sense" mRNA, whereas T7 polymerase transcribes antisense transcripts. (B) Result of injecting the *Krüppel* antisense message into an early embryo (syncytial blastoderm stage) of *Drosophila* before the normal *Krüppel* message is produced. The central figure is a wild-type embryo just prior to hatching. Above it is a mutant caused by the lack of *Krüppel* genes. Below it is a wild-type embryo that was injected with *Krüppel* antisense message at the early embryo stage. Both the mutant and the antisense-treated embryos lack thoracic and anterior abdominal segments. (B after Rosenberg et al. 1985.)

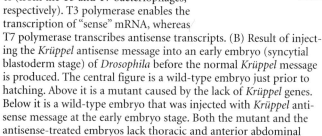

(A)

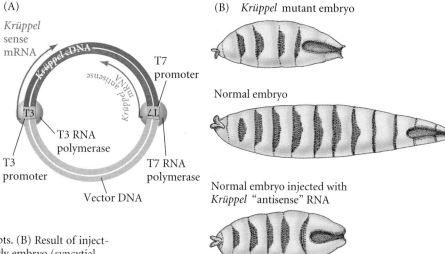

(B) *Krüppel* mutant embryo

Normal embryo

Normal embryo injected with *Krüppel* "antisense" RNA

Identifying the Genes for Human Developmental Anomalies

One cannot experiment on human embryos, nor can one selectively breed humans to express a particular mutant phenotype. How, then, can we find the genes that are involved in normal human development and whose mutations cause congenital (inborn) developmental malformations? Two approaches have revolutionized human embryology within the past five years.

The first approach is **positional gene cloning** (Collins 1992; Scambler 1997). Here, pedigree analysis highlights a region of the genome where a particular mutant gene is thought to reside. By making a DNA map of that area, one can hope to find a region of DNA that differs between people who have the mutation and people who lack it. Then, by sequencing that region of the genome, the gene can be located. As an example, we will look at the discovery of the human Aniridia gene. Ton and his colleagues (1991) attempted to find the gene whose absence caused defects in normal eye development. People heterozygous for this gene, *Aniridia*, lack the irises of their eyes. Fetuses homozygous for this condition are stillborn and have no eyes or nose. Using DNA from individuals with aniridia having chromosomal breaks in this region, Ton and co-workers found a region of chromosome 11 that was present in wild-type individuals but completely or partially absent in individuals with aniridia. Moreover, this region of DNA contained a region that could be a gene (i.e., it had a promoter site, and sequences connoting introns and exons). To determine whether this sequence was a gene that was active in eye development, they did a Northern blot, using this region of DNA as a probe. As Figure 4.14F shows, this probe found complementary mRNA primarily in the brain and eye regions of the body. This fragment of DNA contained a gene that was indeed expressed in the brain and eye. Sequencing this fragment showed that it was the human *PAX6* gene: the human homologue of the *Pax6* gene already known in mice. This was confirmed through in situ hybridization (Figure 4.24).

The second approach is **candidate gene mapping**. This approach is similar to positional cloning, but here, one starts with a correlation between the genetic mapping of a particular syndrome and the genetic mapping for a particular gene. As an example, we will take another condition that produces small eyes: Waardenburg syndrome type 2. This autosomal dominant condition is characterized by deafness, heterochromatic (multicolored) irises, and a white forelock (Figure 4.25). By analyzing the pedigrees of several families whose members had this condition, Hughes and her colleagues (1994) showed that it is caused by a mutation in a gene on the small arm of chromosome 3, between bands 12.3 and 14.4. A strikingly similar condition is found in mice, in which it is called *microphthalmia*. Mutations of the microphthalmia gene (*mi*) cause a dominant syndrome involving deafness, a white patch of fur, and eye abnormalities (Figure 4.25). Could Waardenburg syndrome type 2 be caused by mutations in the human equivalent of the microphthalmia gene? The mouse *mi* gene was cloned, and was found (by sequencing and in situ hybridization) to encode a DNA-binding protein that is expressed in the pigment cells of the eyes, ears, and hair follicles of embryonic mice (Hemesath et al. 1994; Hodgkinson et al. 1994; Nakayama et al. 1998). DNA-binding proteins are important because they often act to regulate the transcription of other genes (as will be detailed in the next two chapters).

(A)

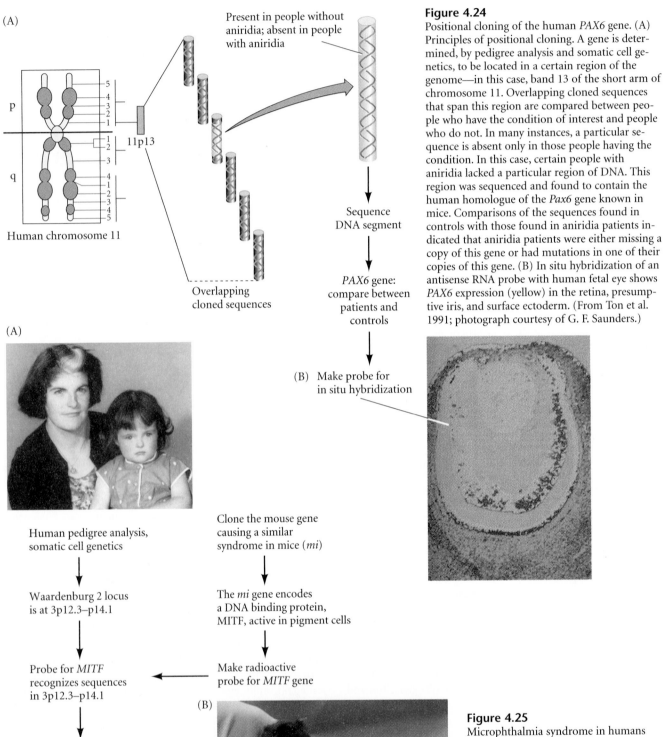

Human chromosome 11

Present in people without aniridia; absent in people with aniridia

11p13

Overlapping cloned sequences

Sequence DNA segment

PAX6 gene: compare between patients and controls

(B) Make probe for in situ hybridization

Figure 4.24
Positional cloning of the human *PAX6* gene. (A) Principles of positional cloning. A gene is determined, by pedigree analysis and somatic cell genetics, to be located in a certain region of the genome—in this case, band 13 of the short arm of chromosome 11. Overlapping cloned sequences that span this region are compared between people who have the condition of interest and people who do not. In many instances, a particular sequence is absent only in those people having the condition. In this case, certain people with aniridia lacked a particular region of DNA. This region was sequenced and found to contain the human homologue of the *Pax6* gene known in mice. Comparisons of the sequences found in controls with those found in aniridia patients indicated that aniridia patients were either missing a copy of this gene or had mutations in one of their copies of this gene. (B) In situ hybridization of an antisense RNA probe with human fetal eye shows *PAX6* expression (yellow) in the retina, presumptive iris, and surface ectoderm. (From Ton et al. 1991; photograph courtesy of G. F. Saunders.)

(A)

Human pedigree analysis, somatic cell genetics

Waardenburg 2 locus is at 3p12.3–p14.1

Probe for *MITF* recognizes sequences in 3p12.3–p14.1

Mutations in *MITF* gene found in human subjects with Waardenburg syndrome 2 and not in people without this syndrome.

Clone the mouse gene causing a similar syndrome in mice (*mi*)

The *mi* gene encodes a DNA binding protein, MITF, active in pigment cells

Make radioactive probe for *MITF* gene

(B)

Figure 4.25
Microphthalmia syndrome in humans and mice. (A) Human patients with Waardenburg syndrome type 2. (B) Mice with *microphthalmia* mutation. The relationship between these two syndromes was ascertained by showing that patients with Waardenburg syndrome type 2 had mutations in a gene that is homologous with the mouse *mi* gene. (A photograph from Partington 1959; B photograph courtesy of D. Fischer.)

The mouse *mi* gene was used to make a probe to look for similar genes in the human genome. Indeed, the probe found a human homologue of the mouse microphthalmia gene, and this homologous sequence (*MITF*) mapped to the exact same region as Waardenburg syndrome type 2 (Tachibana et al. 1994; Tassabehji et al. 1994). When Waardenburg syndrome patients were studied, it was found that they each had mutations of the *MITF* gene. Thus, Waardenburg syndrome type 2 was correlated with mutations in a DNA-binding protein encoded by the human *MITF* locus.

Candidate gene and positional gene mapping techniques have brought together medical embryology and medical genetics. This fusion has enabled scientists and physicians to understand normal human development and the causes of many malformations.

Appendix 1 lists some of the developmental anomalies whose genetic causes have been elucidated by these techniques.

WEBSITE **4.9 OMIM.** The clinical studies involving Waardenburg syndrome type 2 and aniridia are detailed in the Online McKusick's Mendelian Inheritance in Man (OMIM). This site connects you to the OMIM sites.

Principles of Development: Genes and Development

1. Development connects genotype and phenotype.

2. Nuclear genes are not lost or mutated during development. The genome of every cell is equivalent.

3. The exceptions to the rule of genomic equivalence are the lymphocytes. During differentiation, these cells rearrange their DNA to create new immunoglobulin and antigen receptor genes.

4. Genomic equivalence is implied by metaplasia, in which one differentiated cell type becomes another differentiated cell type. An example is the transdifferentiation of the salamander dorsal iris into a lens when the lens is removed.

5. The ability of nuclei from differentiated cells to direct the development of complete adult organisms has recently confirmed the principle of genomic equivalence.

6. The cloning of human beings, as well as regenerating damaged organs or enhancing physical abilities, may soon be possible through cloning technology and the use of embryonic stem cells.

7. Only a small percentage of the genome is expressed in any particular cell.

8. Polytene chromosomes, in which the DNA has replicated without separating (as in larval *Drosophila* salivary glands), show regions where DNA is being transcribed. Different cell types show different regions of DNA being transcribed.

9. Northern blots, in situ hybridization, and the polymerase chain reaction can show which cells are transcribing particular genes.

10. The functions of a gene often can be ascertained by antisense mRNA, transgenic expression, or (in the case of mammals) gene knockouts.

11. Knowledge of gene activity in humans can be obtained by candidate gene mapping or positional cloning.

Literature Cited

Agrawal, A., Q. M. Eastman and D. G. Schatz. 1998. Transposition mediated by *RAG1* and *RAG2* and its implications for the evolution of the immune system. *Nature* 394: 744–751.

Allen, G. E. 1986. T. H. Morgan and the split between embryology and genetics, 1910–1935. *In* T. J. Horder, J. A. Witkowski and C. C. Wylie (eds.), *A History of Embryology.* Cambridge University Press, New York, pp. 113–146.

Anderson, H. F. 1998. Human gene therapy. *Nature* 392: 25–30.

Ashworth, D. and 10 others. 1998. DNA satellite analysis of Dolly. *Nature* 394: 329.

Baltzer, F. 1967. *Theodor Boveri: Life and Work of a Great Biologist.* D. Rudnick (trans.) University of California Press, Berkeley.

Barnett, T., C. Pachl, J. P. Gergen and P. C. Wensink. 1980. The isolation and characterization of *Drosophila* yolk protein genes. *Cell* 21: 729–738.

Beermann, W. 1952. Chromomerenkonstanz und spezifische Modifikationen der Chromosomenstruktur in der Entwicklung und Organdifferenzierung von *Chironomus tentans*. *Chromosoma* 5: 139–198.

Briggs, R. 1979. Genetics of cell type determination. *Int. Rev. Cytol.* [Suppl.] 9: 107–127.

Briggs, R. and T. J. King. 1952. Transplantation of living nuclei from blastula cells into enucleated frogs' eggs. *Proc. Natl. Acad. Sci. USA* 38: 455–464.

Brush, S. 1978. Nettie Stevens and the discovery of sex determination. *Isis* 69: 132–172.

Brüstle, O. and 7 others. 1999. Embryonic stem cell-derived glial precursors: A source of myelinating transplants. *Science* 285: 754–756.

Burian, R., J. Gayon and D. T. Zallen. 1991. Boris Ephrussi and the synthesis of genetics and embryology. *In* S. Gilbert (ed.), *A Conceptual History of Modern Embryology.* Plenum, New York, pp. 207–227.

Burkholder, G. D. 1976. Whole mount electron microscopy of polytene chromosome from *Drosophila melanogaster. Can. J. Genet. Cytol.* 18: 67–77.

Capecchi, M. R. 1980. High efficiency transformation by direct microinjection of DNA into cultured mammalian cells. *Cell* 22: 479–488.

Chun, J. J. M., D. G. Schatz, M. A. Oettinger, R. Jaenisch and D. Baltimore. 1991. The recombination activating gene 1 (*RAG-1*) is present in the murine central nervous system. *Cell* 64: 189–200.

Collins, F. S. 1992. Positional cloning: Let's not call it reverse anymore. *Nature Genet.* 1: 3–6.

DiBerardino, M. A. and T. J. King. 1967. Development and cellular differentiation of neural nuclear transplants of known karyotypes. *Dev. Biol.* 15: 102–128.

DiBerardino, M. A. and R. G. McKinnell. 1997. Backward compatible. *The Sciences* (Sept/Oct 1997): 32–37.

Dudley, A. T., K. M. Lyons and E. J. Robertson. 1995. A requirement for bone morphogenetic protein 7 during development of the mammalian kidney and eye. *Genes Dev.* 9: 2795–2807.

Dumont, J. N. and T. Yamada. 1972. Dedifferentiation of iris epithelial cells. *Dev. Biol.* 29: 385–401.

Fujimoto, S. and H. Yamagishi. 1987. Isolation of an excision product of T cell receptor α-chain gene rearrangements. *Nature* 327: 242–244.

Gage, F. H. 1998. Cell therapy. *Nature* 392: 18–24.

Gardner, R. L. 1968. Mouse chimeras obtained by the injection of cells into the blastocyst. *Nature* 220: 596–597.

Gearhart, J. 1998. New potential for human embryonic stem cells. *Science* 282: 1061–1062.

Gilbert, S. F. 1978. The embryological origins of the gene theory. *J. Hist. Biol.* 11: 307–351.

Gilbert, S. F. 1987. In friendly disagreement: Wilson, Morgan, and the embryological origins of the gene theory. *Am. Zool.* 27: 797–806.

Gilbert, S. F. 1988. Cellular politics: Ernest Everett Just, Richard B. Goldschmidt, and the attempts to reconcile embryology and genetics. *In* R. Rainger, K. R. Benson and J. Maienschein (eds.), *The American Development of Biology.* University of Pennsylvania Press, Philadelphia, pp. 311–346.

Gilbert, S. F. 1991. Induction and the origins of developmental genetics. *In* S. F. Gilbert (ed.), *A Conceptual History of Modern Embryology.* Plenum, New York, pp. 181–206.

Gilbert, S. F. 1996. Enzyme adaptation and the entrance of molecular biology into embryology. *In* S. Sarkar (ed.), *The Molecular Philosophy and History of Molecular Biology: New Perspectives.* Kluwer Academic Publishers, Dordrecht, pp. 101–124.

Glucksohn-Schoenheimer, S. 1938. The development of two tailless mutants in the house mouse. *Genetics* 23: 573–584.

Glucksohn-Schoenheimer, S. 1940. The effect of an early lethal (*t⁰*) in the house mouse. *Genetics* 25: 391–400.

Gossler, A., T. Doetschman, R. Korn, E. Serfling and R. Kemler. 1986. Transgenesis by means of blastocyst-derived stem cell lines. *Proc. Natl. Acad. Sci. USA* 83: 9065–9069.

Grindley, J. C., D. R., Davidson and R. E. Hill. 1995. The role of *Pax-6* in eye and nasal development. *Development* 121: 1433–1442.

Gurdon, J. B. 1962. The developmental capacity of nuclei taken from intestinal epithelial cells of feeding tadpoles. *J. Embryol. Exp. Morphol.* 10: 622–640.

Gurdon, J. B. 1968. Transplanted nuclei and cell differentiation. *Sci. Am.* 219(6): 24–35.

Gurdon, J. B. 1977. Egg cytoplasm and gene control in development. *Proc. R. Soc. Lond.* B 198: 211–247.

Gurdon, J. B. and V. Uehlinger. 1966. "Fertile" intestinal nuclei. *Nature* 210: 1240–1241.

Gurdon, J. B., R. A. Laskey and O. R. Reeves. 1975. The developmental capacity of nuclei transplanted from keratinized cells of adult frogs. *J. Embryol. Exp. Morphol.* 34: 93–112.

Haraway, D. 1997. *Modest_Witness@ Second_Millenium.* Routledge, New York.

Harrison, R. G. 1937. Embryology and its relations. *Science* 85: 369–374.

Harwood, J. 1993. *Styles of Scientific Thought: The German Genetics Community 1900–1934.* University of Chicago Press, Chicago.

Hemesath, T. J. and 9 others. 1994. *Microphthalmia*, a critical factor in melanocyte development, defines a discrete transcription factor family. *Genes Dev.* 8: 2770–2780.

Hodgkinson, C. A., K. J. Moore, A. Nakayama, E. Steingrimsson, N. G. Copeland, N. A. Jenkins and H. Arnheiter. 1994. Mutations at the mouse *microphthalmia* locus are associated with defects in a gene encoding a novel basic helix-loop-helix-zipper protein. *Cell* 74: 395–404.

Hozumi, N. and S. Tonegawa. 1976. Evidence for somatic rearrangement of immunoglobulin genes coding for variable and constant regions. *Proc. Natl. Acad. Sci. USA* 73: 3628–3632.

Hughes, A. E., V. E. Newton, X. Z. Liu and A. P. Read. 1994. A gene for Waardenburg Syndrome type II maps close to the human homologue of the *microphthalmia* gene at chromosome 3 p12–p14.1. *Nature Genet.* 7: 509–512.

Just, E. E. 1939. *The Biology of the Cell Surface.* Blakiston, Philadelphia.

Kato, Y. and 7 others. 1998. Eight calves cloned from somatic cells of a single adult. *Science* 282: 2095–2098.

Keller, E. F. 1995. *Refiguring Life: Metaphors of Twentieth-Century Biology.* Colorado University Press.

King, T. J. 1966. Nuclear transplantation in amphibia. *Methods Cell Physiol.* 2: 1–36.

King, T. J. and Briggs, R. 1956. Serial transplantation of embryonic nuclei. *Cold Spring Harbor Symp. Quant. Biol.* 21: 271–289.

Krauthammer. C. 1998. Of headless mice . . . and men. *Time*, Jan. 19, p. 76.

Lee, S. 1963. *The Amazing Spiderman.* Marvel Comics, New York.

Li, H.-S., J.-M. Yang, R. D. Jacobson, D. Pasko and O. Sundin. 1994. *Pax-6* is first expressed in a region of ectoderm anterior to the early neural plate: Implications for stepwise determination of the lens. *Dev. Biol.* 162: 181–194.

Lillie, F. R. 1927. The gene and the ontogenetic process. *Science* 64: 361–368.

Matsuoka, M. and 8 others. 1991. Detection of somatic DNA recombination in the transgenic mouse brain. *Science* 254: 81–86.

McDonald, J. W. and 7 others. 1999. Transplanted embryonic stem cells survive, differentiate, and promote recovery in injured rat spinal cord. *Nature Med.* 5: 1410–1412.

McKinnell, R. G. 1978. *Cloning: Nuclear Transplantation in Amphibia.* University of Minnesota Press, Minneapolis.

Meade, H. M. 1997. *Dairy* gene. *The Sciences* (Sept./Oct.): 20–25.

Mendel, G. 1866. Versuche über Pflanzen-hybriden. *Verh. Naturf. Vereines* (Brünn) 4: 42.

Mikhailov, A. T., V. N. Simirskii, K. S. Aleinikova and N. A. Gorgolyuk. 1997. Developmental patterns of crystallin expression during lens fiber differentiation in amphibians. *Int. J. Dev. Biol.* 41: 883–891.

Morange, M. 1996. Construction of the developmental gene concept. The crucial years: 1960–1980. *Biol. Zent. bl.* 115: 132–138.

Morgan, T. H. 1926. *The Theory of the Gene.* Yale University Press, New Haven.

Moustafa, L. A. and R. L. Brinster. 1972. Induced chimaerism by transplanting embryonic cells into mouse blastocysts. *J. Exp. Zool.* 181: 193–202.

Nakayama, A., M.-T. T. Nguyen, C. C. Chen, K. Opdecamp, C. A. Hodgkinson and H. Arnheiter. 1998. Mutations in *microphthalmia*, the mouse homolog of the human deafness gene *MITF*, affect neuroepithelial and neural crest-derived melanocytes differently. *Mech. Dev.* 70: 155–166.

Okada, T. S. 1991. *Transdifferentiation.* Oxford University Press, New York.

Oppenheimer, J. M. 1981. Walter Landauer and developmental genetics. *In* S. Subtelny and U. K. Abbott (eds.), *Levels of Genetic Control in Development.* Alan R. Liss, New York, pp. 1–14.

Orr, N. H., M. A. DiBerardino and R. G. McKinnell. 1986. The genome of frog erythrocytes displays centuplicate replications. *Proc. Natl. Acad. Sci. USA* 83: 1369–1374.

Partington, M. W. 1959. Mother and daughter show the white forelock of Waardenburg's syndrome. *Arch. Dis. Childhood* 34: 154–157.

Perucho, M., D. Hanahan and M. Wigler. 1980. Genetic and physical linkage of exogenous sequences in transformed cells. *Cell* 22: 309–317.

Prather, R. S. 1991. Nuclear transplantation and embryo cloning in mammals. *Int. Lab. Anim. Res. News* 33: 62–68.

Rappolee, D. A., C. A. Brenner, R. Schultz, D. Mark and Z. Werb. 1988. Developmental expression of *PDGF*, *TGF-α* and *TGF-β* genes in preimplantation mouse embryos. *Science* 241: 1823–1825.

Reyer, R. W. 1954. Regeneration in the lens in the amphibian eye. *Q. Rev. Biol.* 29: 1–46.

Rifkin, J. 1998. *The Biotech Century: Harnessing the Gene and Remaking the World.* Tarcher/Putnam, New York.

Robertson, J. A. 1998a. Human cloning and the challenge of regulation. *New Engl. J. Med.* 339: 119–122.

Robertson, J. A. 1998b. Liberty, identity, and human cloning. *Texas Law Rev.* 77: 1371–1456.

Robins, D. M., S. Ripley, A. S. Henderson and R. Axel. 1981. Transforming DNA integrates into the host chromosome. *Cell* 23: 29–39.

Rosenberg, U. B., A. Preiss, E. Seifert, H. Jäckle and D. C. Knipple. 1985. Production of phenocopies by *Krüppel* antisense RNA injection into *Drosophila* embryos. *Nature* 313: 703–706.

Saiki, R. K., S. Scharf, F. Faloona, K. B. Mullis, G. T. Horn, H. A. Erlich and N. Arnheim. 1985. Enzymatic amplification of β-globin genomic sequences and restriction site analysis for diagnosis of sickle cell anemia. *Science* 230: 1350–1354.

Sander, K. 1986. The role of genes in ontogenesis: Evolving concepts from 1883 to 1983 as perceived by an insect embryologist. *In* T. J. Horder, J. A. Witkowski and C. C. Wylie (eds.), *A History of Embryology.* Cambridge University Press, New York, pp. 363–395.

Scambler, P. J. 1997. Positional cloning and analysis of loci implicated in human birth defects. *In* P. Thorogood (ed.), *Embryos, Genes, and Birth Defects.* John Wiley, New York, pp. 33–47.

Schnieke, A. and 8 others. 1997. Human factor IX transgenic sheep produced by transfer of nuclei from transfected fetal fibroblasts. *Science* 278: 2130–2134.

Schwarz, K. and 11 others. 1996. Rag mutations in human B cell-negative SCID. *Science* 274: 97–99.

Signer, E. N., Y. E. Dubrova, A. J. Jeffreys, C. Wilde, L. M. B. Finch, M. Wells and M. Peaker. 1998. DNA fingerprinting Dolly. *Nature* 394: 329–330.

Silver, L. M. 1997. *Remaking Eden: Cloning and Beyond in the Brave New World.* Avon, New York.

Smith, L. D. 1956. Transplantation of the nuclei of primordial germ cells into enucleated eggs of *Rana pipiens. Proc. Natl. Acad. Sci. USA* 54: 101–107.

Spemann, H. 1938. *Embryonic Development and Induction.* Yale University Press, New Haven.

Spradling, A. C. and G. M. Rubin. 1982. Transposition of cloned P elements into *Drosophila* germ line chromosomes. *Science* 218: 341–347.

Stevens, N. M. 1905a. *Studies in Spermatogenesis with Especial Reference to the "Accessory Chromosome."* Carnegie Institute of Washington, Washington, D.C.

Stevens, N. M. 1905b. A study of the germ cells of *Aphis rosae* and *Aphis oenotherae. J. Exp. Zool.* 2: 371–405, 507–545.

Tachibana, M and 9 others. 1994. Cloning of *MITF*, the human homolog of the mouse *microphthalmia* gene, and the assignment to human chromosome 3, region p14.1–p12.4. *Hum. Mol. Genet.* 3: 553–557.

Tassabehji, M., V. E. Newton and A. P Read. 1994. Waardenburg syndrome type 2 caused by mutations in the human *microphthalmia* (*MITF*) gene. *Nature Genet.* 8: 251–255.

Thomson, J. A., J. Itskovitz-Eldor, S. S. Shapiro, M. A. Waknitz, J. J. Swiegiel, V. S. Marshall and J. M. Jones. 1998. Embryonic stem cell lines derived from human blastocysts. *Science* 282: 1145–1147.

Ton, C. T. and 14 others. 1991. Positional cloning and characterization of a paired-box and homeobox-containing gene from the *aniridia* region. *Cell* 67: 1059–1074.

Villa, A. and 11 others. 1998. Partial V(D)J recombination activity leads to Omenn syndrome. *Cell* 93: 885–896.

Waddington, C. H. 1939. Preliminary notes on the development of wings in normal and mutant strains of *Drosophila. Proc. Natl. Acad. Sci. USA* 25: 299–307.

Waddington, C. H. 1948. *The Scientific Attitude.* Pelican Books, New York.

Wagner, T. E., P. Hoppe, J. D. Jollick, D. R. Scholl, R. L. Hodinka and J. B. Gault. 1981. Microinjection of rabbit β-globin gene into zygotes and its subsequent expression in adult mice and offspring. *Proc. Natl. Acad. Sci. USA* 78: 6376–6380.

Wakayama, T., A. C. F. Perry, M. Zuccotti, K. R. Johnson and R. Yanagimachi. 1998. Full-term development of mice from enucleated oocytes injected with cumulus cell nuclei. *Nature* 394: 369–374.

Wetmur, J. G. and N. Davidson. 1968. Kinetics of renaturation of DNA. *J. Mol. Biol.* 31: 349–370.

Wickelgren, I. 1999. Rat spinal cord function partially restored. *Science* 286: 1826–1827.

Wilkinson, D. G., S. Bhatt and B. G. Herrmann. 1990. Expression pattern of the mouse *T* gene and its role in mesoderm formation. *Nature* 343: 657–659.

Wilmut, I., A. E. Schnieke, J. McWhir, A. J. Kind and K. H. S. Campbell. 1997. Viable offspring from fetal and adult mammalian cells. *Nature* 385: 810–814.

Wilson, E. B. 1895. *An Atlas of the Fertilization and Karyogenesis of the Ovum.* Macmillan, New York.

Wilson, E. B. 1896. *The Cell in Development and Inheritance.* Macmillan, New York.

Wilson, E. B. 1905. The chromosomes in relation to the determination of sex in insects. *Science* 22: 500–502.

Wivel, N. A. and L. Walters. 1993. Germ-line gene modification and disease prevention: Some medical and ethical perspectives. *Science* 262: 533–538.

Yamada, T. 1966. Control of tissue specificity: The pattern of cellular synthetic activities in tissue transformation. *Am. Zool.* 6: 21–31.

Ye, X. and 8 others. 1999. Regulated delivery of therapeutic proteins after in vivo somatic cell gene transfer. *Science* 283: 88–91.

Zanjani, E. D. and W. F. Anderson. 1999. Prospects for in utero human gene therapy. *Science* 285: 2084–2088.

chapter 5

The genetic core of development: Differential gene expression

DIFFERENT CELL TYPES make different sets of proteins, even though their genomes are identical. Each human being has roughly 150,000 genes in each nucleus, but each cell uses only a small subset of these genes. Moreover, different cell types use different subsets of these genes. Red blood cells make globins, lens cells make crystallins, melanocytes make melanin, and endocrine glands make their specific hormones. **Developmental genetics** is the discipline that examines how the genotype is transformed into the phenotype, and the major paradigm of developmental genetics is *differential gene expression from the same nuclear repertoire*. The regulation of gene expression can be accomplished at several levels:

- **Differential gene transcription**, regulating which of the nuclear genes are transcribed into RNA
- **Selective nuclear RNA processing**, regulating which of the transcribed RNAs (or which parts of such a nuclear RNA) enter into the cytoplasm to become messenger RNAs
- **Selective messenger RNA translation**, regulating which of the mRNAs in the cytoplasm become translated into proteins
- **Differential protein modification**, regulating which proteins are allowed to remain or function in the cell

Some genes (such as those coding for the globin proteins of hemoglobin) are regulated at each of these levels. This chapter will discuss the mechanisms of cell differentiation: how different genes from the same genetic repertoire are activated or repressed at particular times and places to cause cells to become different from one another.

Differential Gene Transcription

Anatomy of the gene: Exons and introns

There are two fundamental differences distinguishing most eukaryotic genes from most prokaryotic genes. First, eukaryotic genes are contained within a complex of DNA and protein called **chromatin**. The protein component constitutes about half the weight of chromatin and is composed largely of **nucleosomes**. The nucleosome is the basic unit of chromatin structure. It is composed of an octamer of **histone** proteins (two molecules each of histones H2A-H2B and histones H3-H4) wrapped with two loops containing approximately 140 base pairs of DNA (Figure 5.1; Kornberg

109

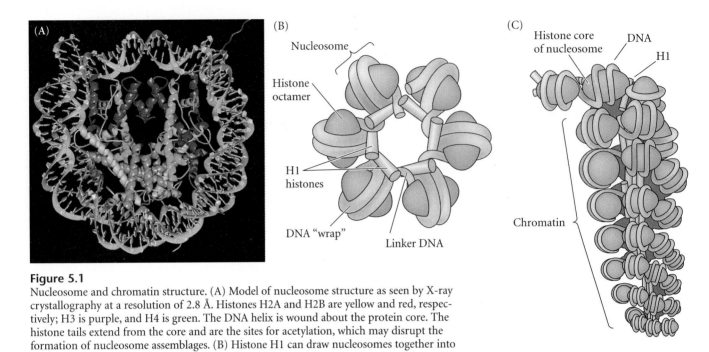

Figure 5.1
Nucleosome and chromatin structure. (A) Model of nucleosome structure as seen by X-ray crystallography at a resolution of 2.8 Å. Histones H2A and H2B are yellow and red, respectively; H3 is purple, and H4 is green. The DNA helix is wound about the protein core. The histone tails extend from the core and are the sites for acetylation, which may disrupt the formation of nucleosome assemblages. (B) Histone H1 can draw nucleosomes together into compact forms. About 140 base pairs of DNA encircle each histone octamer, and about 60 base pairs of DNA link the nucleosomes together. (C) Model for the arrangement of nucleosomes in the highly compacted solenoidal chromatin structure. (A from Luger et al. 1997, photograph courtesy of the authors; B, C after Wolfe 1993.)

and Thomas 1974). Chromatin can thus be visualized as a string of nucleosome beads linked by ribbons of DNA. While classic geneticists have likened genes to "beads on a string," molecular geneticists liken genes to "string on the beads." Most of the time, the nucleosomes are themselves wound into tight "solenoids" that are stabilized by histone H1. Histone H1 is found in the 60 or so base pairs of "linker" DNA between the nucleosomes (Weintraub 1984). This H1-dependent conformation of nucleosomes inhibits the transcription of genes in somatic cells by packing adjacent nucleosomes together into tight arrays that prohibit the access of transcription factors and RNA polymerases to the genes (Thoma et al. 1979; Schlissel and Brown 1984). It is generally thought, then, that the "default" condition of chromatin is a repressed state, and that tissue-specific genes become activated by local interruption of this repression (Weintraub 1985).

> WEBSITE **5.1 Displacing nucleosomes.** Transcription can occur even in a region of nucleosomes. If the promoter is accessible to RNA polymerase and transcription factors, the presence of nucleosomes will not inhibit the elongation of the message.

The second difference is that eukaryotic genes are not co-linear with their peptide products. Rather, the single nucleic acid strand of eukaryotic mRNA comes from noncontiguous regions on the chromosome. Between the regions of DNA coding for a protein—**exons**—are intervening sequences—**introns**—that have nothing whatsoever to do with the amino

acid sequence of the protein.* The structure of a typical eukaryotic gene can be represented by the human β-globin gene, shown in Figure 5.2. This gene consists of the following elements:

1. A **promoter** region, which is responsible for the binding of RNA polymerase and for the subsequent initiation of transcription. The promoter region of the human β-globin gene has three distinct units and extends from 95 to 26 base pairs before ("upstream from")† the transcription initiation site (i.e., from –95 to –26).

2. The **transcription initiation site**, which for human β-globin is ACATTTG. This site is often called the **cap sequence** because it represents the 5′ end of the RNA, which will receive a "cap" of modified nucleotides soon after it is transcribed. The specific cap sequence varies among genes.

3. The **translation initiation site**, ATG. This codon (which becomes AUG in the mRNA) is located 50 base pairs after the transcription initiation site in the human β-globin gene (although this distance differs greatly among differ-

---

*The term *exon* refers to a nucleotide sequence whose RNA "exits" the nucleus. It has taken on the functional definition of a protein-encoding nucleotide sequence. Leader sequences and 3′ untranslated sequences are also derived from exons, even though they are not translated into protein.

†By convention, upstream, downstream, 5′, and 3′ directions are specified in relation to the RNA. Thus, the promoter is upstream of the gene, near its 5′ end.

(A)

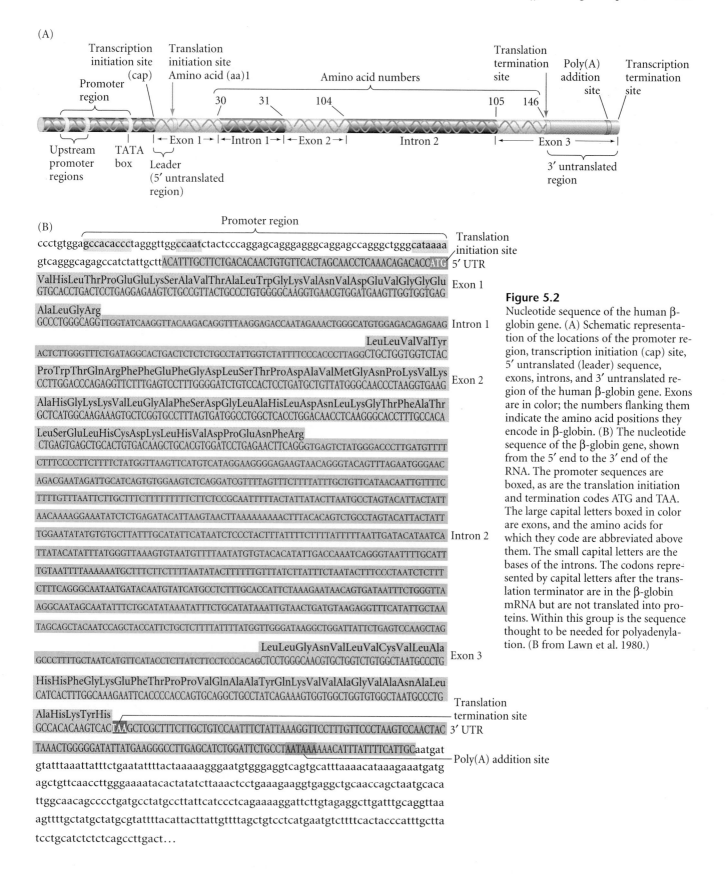

Figure 5.2
Nucleotide sequence of the human β-globin gene. (A) Schematic representation of the locations of the promoter region, transcription initiation (cap) site, 5′ untranslated (leader) sequence, exons, introns, and 3′ untranslated region of the human β-globin gene. Exons are in color; the numbers flanking them indicate the amino acid positions they encode in β-globin. (B) The nucleotide sequence of the β-globin gene, shown from the 5′ end to the 3′ end of the RNA. The promoter sequences are boxed, as are the translation initiation and termination codes ATG and TAA. The large capital letters boxed in color are exons, and the amino acids for which they code are abbreviated above them. The small capital letters are the bases of the introns. The codons represented by capital letters after the translation terminator are in the β-globin mRNA but are not translated into proteins. Within this group is the sequence thought to be needed for polyadenylation. (B from Lawn et al. 1980.)

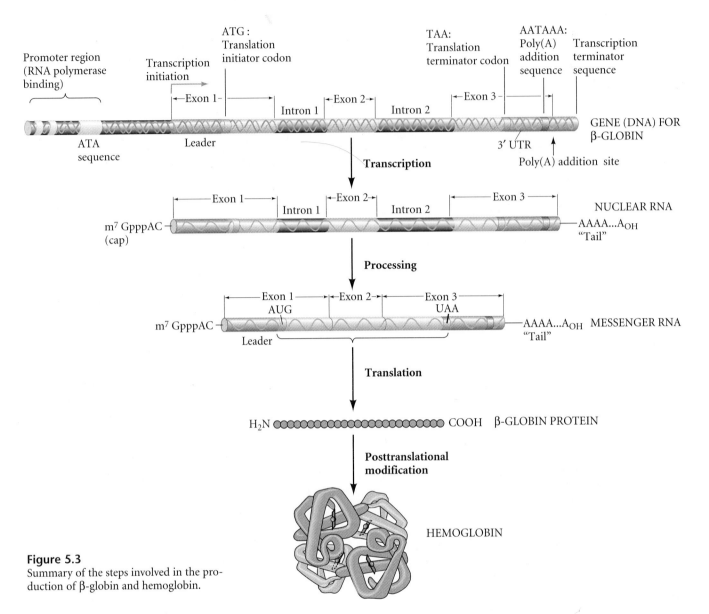

Figure 5.3
Summary of the steps involved in the production of β-globin and hemoglobin.

ent genes). The intervening sequence of 50 base pairs between the initiation points of transcription and translation is the **5′ untranslated region**, often called the 5′ UTR or **leader sequence**. The 5′ UTR can determine the rate at which translation is initiated.

4. The first **exon**, which contains 90 base pairs coding for amino acids 1–30 of human β-globin.

5. An **intron** containing 130 base pairs with no coding sequences for the globin protein. The structure of this intron is important in enabling the RNA to be processed into messenger RNA and exit from the nucleus.

6. An exon containing 222 base pairs coding for amino acids 31–104.

7. A large intron—850 base pairs—having nothing to do with the globin protein structure.

8. An exon containing 126 base pairs coding for amino acids 105–146.

9. A **translation termination codon**, TAA. This codon becomes UAA in the mRNA. The ribosome dissociates at this codon, and the protein is released.

10. A **3′ untranslated region** that, (3′ UTR) although transcribed, is not translated into protein. This region includes the sequence AATAAA, which is needed for **polyadenylation**: the placement of a "tail" of some 200 to 300 adenylate residues on the RNA transcript. This **poly(A) tail** (1) confers stability on the mRNA, (2) allows the mRNA to exit the nucleus, and (3) permits the mRNA to be translated into protein. The poly(A) tail is inserted into the RNA about 20 bases downstream of the AAUAAA sequence. Transcription continues beyond the AATAAA site for about 1000 nucleotides before being terminated.

The original nuclear RNA transcript for such a gene contains the capping sequence, the 5′ untranslated region, the exons, the introns, and the 3′ untranslated region (Figure 5.3). In addition, both its ends become modified. A cap consisting of methylated guanosine is placed on the 5′ end of the RNA in opposite polarity to the RNA itself. This means that there is no free 5′ phosphate group on the nuclear RNA. The 5′ cap is necessary for the binding of mRNA to the ribosome and for subsequent translation (Shatkin 1976). The 3′ terminus is usually modified in the nucleus by the addition of a poly(A) tail. These adenylate residues are put together enzymatically and are added to the transcript; they are not part of the gene sequence. Both the 5′ and 3′ modifications may protect the RNA from exonucleases that would otherwise digest the mRNA (Sheiness and Darnell 1973; Gedamu and Dixon 1978). The modifications thus stabilize the message and its precursor.

> WEBSITE **5.2 Structure of the 5′ cap.** The capping and methylation of the 5′ end are critical steps in the synthesis of mRNA. If the cap is missing or unmethylated, translation may fail to occur.

Anatomy of the gene: Promoters and enhancers

In addition to the protein-encoding region of the gene, there are regulatory sequences that can be on either end of the gene (or even within it). These sequences—the promoters and enhancers—are necessary for controlling where and when a particular gene is transcribed.

Promoters are the sites where RNA polymerase binds to the DNA to initiate transcription. Promoters of genes that synthesize messenger RNAs (i.e., genes that encode proteins*) are typically located immediately upstream from the site where the RNA polymerase initiates transcription. Most of these promoters contain the sequence TATA, where RNA polymerase will be bound (Figure 5.4). This site, known as the **TATA box,** is usually about 30 base pairs upstream from the site where the first base is transcribed. Eukaryotic RNA polymerases, however, will not bind to this naked DNA sequence. Rather, they require additional protein factors to bind efficiently to the promoter. The protein-encoding genes are transcribed by RNA polymerase II, and at least six nuclear proteins have been shown to be necessary for the proper initiation of transcription by RNA polymerase II (Buratowski et al. 1989; Sopta et al. 1989). These proteins are called **basal transcription factors.** The first of these, **TFIID,**† recognizes the TATA box through one of its subunits, **TATA-binding protein** (**TBP**). TFIID serves as the foundation of the transcription initiation complex, and it also serves to

---

*There are several types of RNA that do *not* encode proteins. These include the ribosomal RNAs and transfer RNAs (which are used in protein synthesis) and the small nuclear RNAs (which are used in RNA processing). In addition, there are regulatory RNAs (such as *Xist* and *lin-4*, which we will discuss later in this chapter) that are involved in regulating gene expression (and which do not encode proteins).

†**TF** stands for transcription factor; **II** indicates that the factor was first found to be needed for RNA polymerase II; and the letter designations refer to the active fractions from the phosphocellulose columns used to purify these proteins.

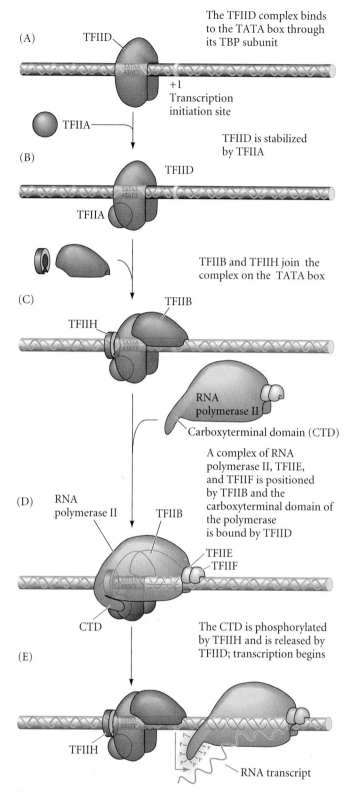

Figure 5.4
The formation of the active eukaryotic initiation complex. The diagrams represent the complexes formed on the TATA box by the transcription factors and RNA polymerase II. (A) The TFIID complex binds to the TATA box through its TBP subunit. (B) TFIID is stabilized by TFIIA. (C) TFIIB and TFIIH join the complex on the TATA box while TFIIE and TFIIF associate with RNA polymerase II. (D) RNA polymerase is positioned by TFIIB, and its carboxy-terminal domain (CTD) is bound by TFIID. (E) The CTD is phosphorylated by TFIIH and is released by TFIID. The RNA polymerase II is now competent to transcribe mRNA from the gene.

Figure 5.5
Model of TAF stabilization of TBP.
(A) A minimal complex near the promoter, containing TBP on the TATA box of the promoter and two upstream sites occupied by two transcription factors, Sp1 and NTF-1. TAF 250 is bound to the TBP, but this complex is not stable enough to activate transcription. (B) Certain TAFs bind to these proteins, forming bridges that stabilize the TBP on the promoter. (After Chen et al. 1995.)

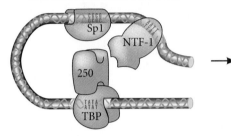

(A) A minimal complex of TBP and a TAF fails to activate transcription (Sp1 and NTF cannot associate with TBP)

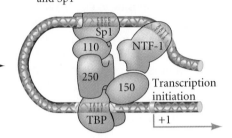

(B) Addition of the p110 TAF and the p150 TAF allows activation by both NTF and Sp1

keep nucleosomes from forming in this region. Once TFIID is stabilized by TFIIA, it becomes able to bind **TFIIB**. Once TFIIB is in place, RNA polymerase can bind to this complex. Other transcription factors (TFIIE, F, and H) are then used to release RNA polymerase from the complex so that it can transcribe the gene, and to unwind the DNA helix so that the RNA polymerase will have a free template from which to transcribe.

In addition to these basal transcription factors, which are found in each nucleus, there is also a set of transcription factors called **TBP-associated factors**, or **TAFs** (Figure 5.5; Buratowski 1997; Lee and Young 1998), which can stabilize the TBP. This function is critical for gene transcription, for if the TBP is not stabilized, it can fall off the small TATA sequence. The TAFs are bound by **upstream promoter elements** on the DNA. These DNA sequences are near the TATA sequence, and usually upstream from it. These TAFs need not be in every cell of the body, however. **Cell-specific transcription factors** (such as the Pax6 and microphthalmia proteins mentioned in Chapter 4) can also activate the gene by stabilizing the transcription initiation complex. They can do so by binding to the TAFs, by binding directly to other factors such as TFIIB, or (as we will see soon) by destabilizing nucleosomes.

WEBSITE **5.3 Promoter struction and the mechanisms of transcription complex assembly.** Getting RNA polymerase to a promoter is not an easy task. The transcriptional initiation complex is a major protein complex that must be created at each round of transcription. The delineation of the different parts of promoters is worked out via mutations and transgenes.

An **enhancer** is a DNA sequence that can activate the utilization of a promoter, controlling the efficiency and rate of transcription from that particular promoter. Enhancers can activate only *cis*-linked promoters (i.e., promoters on the same chromosome*), but they can do so at great distances (some as great as 50 kilobases away from the promoter). Moreover, enhancers do not need to be on the 5′ (upstream) side of the gene. They can also be at the 3′ end, in the introns, or even on the complementary DNA strand (Maniatis et al. 1987). The human β-globin gene has an enhancer in its 3′ UTR, roughly 700 base pairs downstream from the AATAAA

site. This sequence is necessary for the temporal- and tissue-specific expression of the β-globin gene in adult red blood cell precursors (Trudel and Constantini 1987). Like promoters, enhancers function by binding specific regulatory proteins called transcription factors.

Enhancers can regulate the temporal and tissue-specific expression of any differentially regulated gene, but different types of genes normally have different enhancers. In the pancreas, for instance, the exocrine protein genes (for the digestive proteins chymotrypsin, amylase, and trypsin) have enhancers different from that of the gene for the endocrine protein insulin. These enhancers both lie in the 5′ flanking sequences of their genes. Walker and colleagues (1983) created transgenes by placing flanking regions from the genes for chymotrypsin and insulin onto the gene for bacterial chloramphenicol acetyltransferase (CAT), an enzyme that is not found in mammalian cells. CAT activity is easy to assay in mammalian cells, so the bacterial *CAT* gene can be used as a **reporter gene** to tell investigators whether a particular enhancer is functioning. The researchers then transfected the transgenes into (1) ovary cells (which do not secrete either insulin or chymotrypsin), (2) an insulin-secreting cell line, and (3) a chymotrypsin-secreting cell line, and measured the activity of CAT in each of these cells. As shown in Figure 5.6, neither enhancer sequence caused the enzyme to be made in the ovarian cells. In the insulin-secreting cell, however, the 5′ flanking region from the insulin gene enabled the *CAT* gene to be expressed, but the 5′ flanking region of the chymotrypsin gene did not. Conversely, when the clones were placed into the exocrine pancreatic cell line, the chymotrypsin 5′ flanking sequence allowed *CAT* expression, while the insulin enhancer did not. The enhancers for 10 different exocrine proteins share a 20-base-pair consensus sequence, suggesting that these sim-

---

*Cis- and *trans*-regulatory elements are so named by analogy with *E. coli* genetics. There, *cis*-factors are regulatory elements that reside on the same strand of DNA (*cis*-, meaning "on the same side as"), while *trans*-elements are those that could be supplied from another chromosome (*trans*-, meaning "on the other side of"). *Cis*-regulatory elements now refers to those DNA sequences that regulate a gene on the same stretch of DNA (i.e., the promoters and enhancers). *Trans*-regulatory factors are soluble molecules whose genes are located elsewhere in the genome and which bind to the *cis*-regulatory elements. They are usually transcription factors.

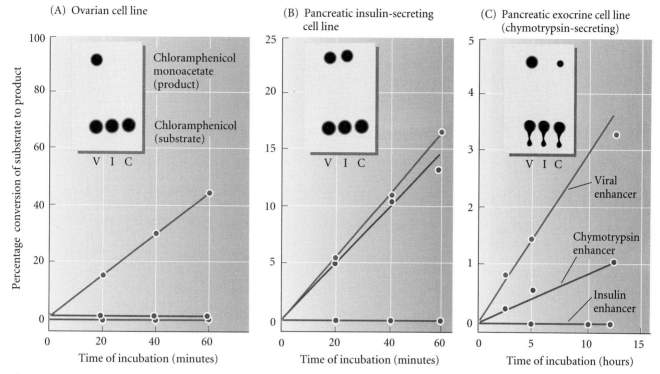

Figure 5.6

Tissue specificity of pancreatic gene enhancers. The 5′ flanking regions of the insulin gene (open circles) and the chymotrypsin gene (shaded circles) were each inserted next to the gene for bacterial CAT. As a positive control, the enhancer from Rous sarcoma virus (solid circles), which appears to operate in all cell types, was also placed next to the *CAT* gene. The three clones were transfected into three types of cells: (A) an ovarian cell line making neither insulin nor chymotrypsin, (B) an insulin-secreting cell line, or (C) a chymotrypsin-secreting cell line. CAT activity was assayed on the cell lysates. The inserts show typical autoradiographs of the CAT assay, in which radioactive chloramphenical (the *substrate* of the CAT reaction) can be separated from chloramphenicol monoacetate (the *product* of the CAT reaction). The results show that different enhancers function in different cell types. (After Walker et al. 1983.)

ilar sequences play a role in activating an entire set of genes specifically expressed in the exocrine cells of the pancreas (Boulet et al. 1986). Thus, the expression of genes in exocrine and in endocrine cells of the pancreas appears to be controlled by different enhancers.

By taking an enhancer from one gene and fusing it to another gene, it has been shown that enhancers can direct the expression of any gene sequence. For instance, the β-**galactosidase** gene from *E. coli* (the *lacZ* gene) can be used as a reporter gene and placed onto an enhancer that normally directs a particular mouse gene to become expressed in muscles. If the resulting transgene is injected into a newly fertilized mouse egg and gets incorporated into its DNA, the β-galactosidase gene will be expressed in the muscle cells. By staining for the presence of β-galactosidase, the expression pattern of that muscle-specific gene can be seen (Figure 5.7A). Similarly, if the gene

for **green fluorescent protein** (**GFP**, a reporter protein that is usually made only in jellyfish) is placed on the enhancer of genes encoding the crystallin proteins of the eye lens, GFP expression is seen solely in the lens (Figure 5.7B).

Enhancers are critical in the regulation of normal development. Over the past decade, six generalizations that emphasize their importance for differential gene expression have been made:

1. Most genes require enhancers for their transcription.
2. Enhancers are the major determinant of differential transcription in space (cell type) and time.
3. The ability of an enhancer to function while far from the promoter means that there can be multiple signals to determine whether a given gene is transcribed. A given gene can have several enhancer sites linked to it, and each enhancer can be bound by more than one transcription factor.
4. The interaction between the proteins bound to the enhancer sites and the transcription initiation complex assembled at the promoter is thought to regulate transcription. The mechanism of this association is not fully known, nor do we comprehend how the promoter integrates all these signals.
5. Enhancers are modular. There are various DNA elements that regulate temporal and spatial gene expression, and these can be mixed and matched. As we will see, the enhancers for endocrine hormones such as insulin and for lens-specific proteins such as crystallins both have sites that bind Pax6 protein. But Pax6 doesn't tell the lens to

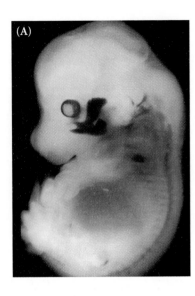

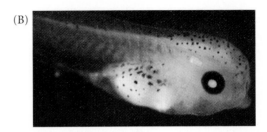

Figure 5.7
The genetic elements regulating cell type-specific transcription can be identified by fusing reporter genes to the enhancer regions of the genes found in particular cell types. (A) The enhancer region of the muscle-specific protein Myf-5 is fused to a β-galactosidase reporter gene and is incorporated into a mouse embryo. When stained for β-galactosidase activity (darkly staining region), the 13.5-day mouse embryo shows that the reporter gene is expressed in the eye muscles, forelimb muscles, neck muscles, and segmented myotomes (which give rise to the back musculature). (B) The gene for green fluorescent protein (GFP) is fused to a lens crystallin enhancer in *Xenopus tropicalis*. The result is the placement of green fluorescent protein in the tadpole lens. (A courtesy of A. Patapoutian and B. Wold; B from Offield et al., in press, courtesy of R. Grainger.)

make insulin or the pancreas to make crystallins, because there are other transcription factor proteins that also must bind. It is the combination of transcription factors that causes particular genes to be transcribed.

6. A gene can have several enhancer elements, each turning it on in a different set of cells

7. Enhancers can also be used to inhibit transcription. In some cases, the same transcription factors that activate the transcription of one gene can be used to repress the transcription of other genes. These "negative enhancers" are also called **silencers**.

Transcription factors

Transcription factors are proteins that bind to enhancer or promoter regions and interact to activate or repress the transcription of a particular gene. Most transcription factors can bind to specific DNA sequences. These proteins can be grouped together in families based on similarities in structure (Table 5.1). The transcription factors within such a family share a common framework structure in their DNA-binding sites, and slight differences in the amino acids at the binding site can alter the sequence of the DNA to which the factor binds.

WEBSITE **5.4 Families of transcription factors.** There are several families of transcription factors grouped together by their structural similarities and their mechanisms of action. Homeodomain transcription factors are important in specifying anterior-posterior axes, and hormone receptors mediate the effects of hormones to the genes.

Transcription factors have three major domains. The first is a **DNA-binding domain** that recognizes a particular DNA se-

Table 5.1 Some major transcription factor families and subfamilies

Family	Representative transcription factors	Some functions
Homeodomain:		
Hox	Hoxa-1, Hoxb-2, etc.	Axis formation
POU	Pit-1, Unc-86, Oct-2	Pituitary development; neural fate
LIM	Lim-1, Forkhead	Head development
Pax	Pax1, 2, 3, etc.	Neural specification; eye development
Basic helix-loop-helix (bHLH)	MyoD, achaete, daughterless	Muscle and nerve specification; *Drosophila* sex determination
Basic leucine zipper (bZip)	C/EBP, AP1	Liver differentiation; fat cell specification
Zinc finger:		
Standard	WT1, Krüppel, Engrailed	Kidney, gonad, and macrophage development; *Drosophila* segmentation
Nuclear hormone receptors	Glucocorticoid receptor, estrogen receptor, testosterone receptor, retinoic acid receptors	Secondary sex determination; craniofacial development; limb development
Sry-Sox	Sry, SoxD, Sox2	Bend DNA; mammalian primary sex determination; ectoderm differentiation

quence. The second is a ***trans*-activating domain** that activates or suppresses the transcription of the gene whose promoter or enhancer it has bound. Usually, this *trans*-activating domain enables the transcription factor to interact with proteins involved in binding RNA polymerase (such as TFIIB or TFIIE; see Sauer et al. 1995). In addition, there may be a **protein-protein interaction domain** that allows the transcription factor's activity to be modulated by TAFs or other transcription factors.

EXAMPLES OF TRANSCRIPTION FACTORS: MITF AND PAX6. As examples of transcription factors, we can look at the Pax6 and microphthalmia proteins mentioned in Chapter 4. The microphthalmia (MITF) protein is necessary for the production of pigment cells and their pigments. There are three functionally important domains of the MITF protein. First, MITF has a protein-protein interaction domain that enables it to dimerize with another MITF protein (Ferré-D'Amaré et al. 1993). This homodimer (two microphthalmia proteins bound together) forms the functional protein that can bind to DNA and activate the transcription of certain genes (Figure 5.8). The second region, the DNA-binding domain, is close to the amino-termi-

nal end of the protein and contains numerous basic amino acids that make contact with the DNA (Hemesath et al. 1994; Steingrímsson et al. 1994). This assignment was confirmed by the discovery of various human and mouse mutations that map within the DNA-binding site for MITF and prevent the attachment of the MITF protein to the DNA. Sites for MITF binding have been found in the promoter regions for three pigment-cell-specific proteins of the tyrosinase family (Bentley et al. 1994; Yasumoto et al. 1997). Without MITF, these proteins are not synthesized properly (Figure 5.9) and the melanin pigment is not made. These promoters all contain the same 11-base-pair sequence, including the core sequence CATGTG.

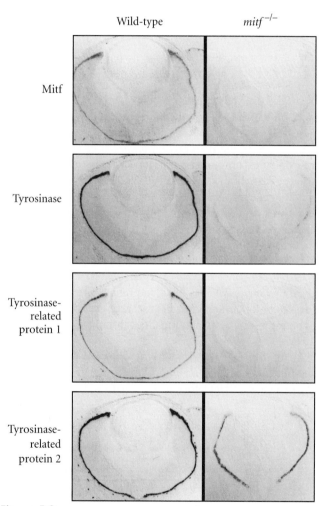

Figure 5.9
Mitf is required for the transcription of pigmention genes. Serial sections of the eye in 15.5-day mouse embryos are shown. In the wild-type mouse embryo, in situ hybridization reveals the presence of Mitf (dark staining) in the retinal pigment epithelial layer, as well as revealing the transcription of the tyrosinase gene and two related genes. In the *mi* mutant embryo (*mitf*⁻/⁻), no Mitf is present, nor are the genes for tyrosinase or tyrosinase-related protein 1 transcribed. The transcription of the tyrosinase-related protein 2 gene is significantly reduced in the mutant. (From Nakayama et al. 1998; photographs courtesy of H. Arnheiter.)

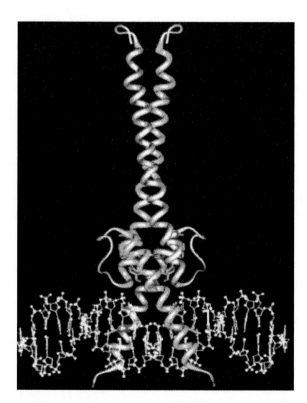

Figure 5.8
Three-dimensional model of the MITF dimer binding to its promoter element in DNA. The amino termini are located at the bottom of the figure and form the DNA-binding domain. The protein-protein interaction domain is located immediately above it, and the carboxyl end of the molecule is thought to be the *trans*-activating domain that binds the p300/CBP co-activator protein. (After Steingrímsson et al. 1994; photograph courtesy of Dr. N. Jenkins.)

The third functional region of MITF is its *trans*-activating domain. This domain includes a long stretch of amino acids in the center of the protein. When the MITF dimer is bound to its target element in a promoter or enhancer, the *trans*-activating region is able to bind a TAF called p300/CBP. The p300/CPB protein is a **histone acetyltransferase** enzyme that can transfer acetyl groups to each histone in the nucleosomes (Ogryzko et al. 1996; Price et al. 1998). This acetylation of the nucleosomes destabilizes them and allows the genes for pigment-forming enzymes to be expressed. Recent discoveries have shown that numerous transcription factors operate by recruiting histone acetyltransferases. It is thought that once the nucleosomes are destabilized, other transcription factors and RNA polymerase can bind more easily to the DNA in that region.

> WEBSITE **5.5 Histone acetylation.** Histone acetylation is a critical step in clearing the way for the transcription initiation complex. Derepressed chromatin is characterized by acetylated histones.

The Pax6 transcription factor, which is needed for mammalian eye, nervous system, and pancreas development, contains two potential DNA-binding domains. The major DNA-binding site of the Pax6 protein resides at its amino-terminal end, and these amino acids interact with a specific 20–26-base-pair sequence of DNA (Figure 5.10; Xu et al. 1995). Such Pax6-binding sequences have been found in the enhancers of vertebrate lens crystallin genes and in the genes expressed in the endocrine cells of the pancreas (insulin, glucagon, and somatostatin) (Cvekl and Piatigorsky 1996; Andersen 1999). When Pax6 binds to a particular site in an enhancer or promoter, it can either activate or repress that gene. The *trans*-activating domain of Pax6 is rich in proline, threonine, and serine. Mutations in this region cause severe nervous system, pancreatic, and optic abnormalities in humans (Glaser et al. 1994).

The use of Pax6 by different organs demonstrates the modular nature of transcriptional regulatory units. Figure 5.11 shows two gene regulatory regions that use Pax6. The first is

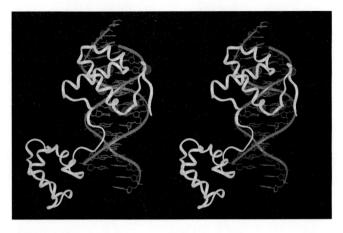

Figure 5.10
Stereoscopic model of Pax6 binding to its enhancer element in DNA. The DNA-binding region (the "paired domain") is in yellow; the DNA is in blue. The red dots indicate the sites of loss-of-function mutations in the *Pax6* gene that give rise to nonfunctional Pax6 proteins. It is worth trying to cross your eyes to get the central three-dimensional figure. (From Xu et al. 1995; photograph courtesy of S. O. Pääbo.)

that of the chick *δ1* lens crystallin gene. This gene has a promoter containing a site for TBP binding and an upstream site that binds Sp1, a general transcriptional activator found in all cells. The gene also has an enhancer in its third intron that controls the time and place of crystallin gene expression. This enhancer has two Pax6-binding sites. The crystallin gene will not be expressed unless Pax6 is present in the nucleus and bound to these enhancer sites. As mentioned in Chapter 4, Pax6 is present during early development in the central nervous system and head surface ectoderm of the chick. Moreover, this enhancer has a site for another transcription factor, the Sox2 protein. Sox2 is not usually found in the outer ectoderm, but it appears in those outer ectodermal cells that will become lens by virtue of their being induced by the optic vesicle evaginating

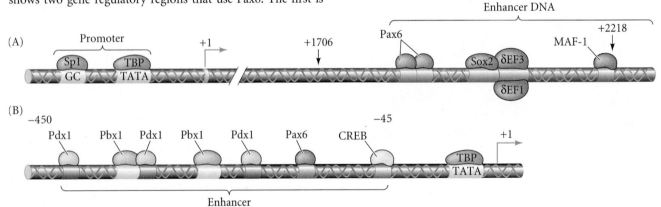

Figure 5.11
Modular transcriptional regulatory regions using Pax6 as an activator. (A) Promoter and enhancer of the chick lens *δ1* crystallin gene. Pax6 interacts with Sox2 and Maf to activate this gene. (B) Enhancer of the rat somatostatin gene. Pax6 activates this gene by cooperating with the Pdx1 transcription factor. (A after Cvekl and Piatigorsky 1996; B after Andersen et al. 1999.)

from the brain (Kamachi et al. 1998). Thus, only those cells that contain both Sox2 and Pax6 can express the lens crystallin gene. In addition, there is a third site that can bind either an activator (the δEF3 protein) or a repressor (the δEF1 protein) of transcription. It is thought that the repressor may be critical in preventing crystallin expression in the nervous system. Thus, enhancers function in a combinatorial manner, wherein several transcription factors work together to promote or inhibit transcription.

Another set of regulatory regions that use Pax6 are the enhancers regulating the transcription of the insulin, glucagon, and somatostatin genes of the pancreas (Figure 5.11B). Here, Pax6 is essential for gene expression, and it works in cooperation with other transcription factors such as Pdx1 (specific for the pancreatic region of the endoderm) and Pbx1 (Andersen et al. 1999; Hussain and Habener 1999). In the absence of Pax6 (as in the homozygous small eye mutation in mice and rats), the endocrine parts of the pancreas do not develop properly, and the production of these proteins from those cells is deficient (Sander et al. 1997). One can see that the genes for spe-

cific proteins use numerous transcription factors in various combinations. In this way, transcription factors can regulate the timing and place of gene expression.

There are other genes that are activated by Pax6 binding, and one of them is the *Pax6* gene itself. Pax6 protein can bind to the *Pax6* gene promoter (Plaza et al. 1993). This means that once the *Pax6* gene is turned on, it will be continue to be expressed, even if the signal that originally activated it is no longer given.

Within the past decade, our knowledge of transcription factors has progressed enormously, giving us a new, dynamic view of gene expression. The gene itself is no longer seen as an independent entity controlling the synthesis of proteins. Rather, the gene both directs and is directed by protein synthesis. Natalie Angier (1992) has written, "A series of new discoveries suggests that DNA is more like a certain type of politician, surrounded by a flock of protein handlers and advisers that must vigorously massage it, twist it and, on occasion, reinvent it before the grand blueprint of the body can make any sense at all."

Sidelights & Speculations

Studying DNA Regulatory Elements

Identifying DNA regulatory elements How do we know that a particular DNA fragment binds a transcription factor? One of the simplest ways is to perform a **gel mobility shift assay**. The basis for this assay is gel electrophoresis. Fragments of DNA can be placed into a depression at one end of a gel, and an electrical current run through the gel. The DNA will be moved toward the negative pole, and the distance it travels in a given time will depend upon its mass and conformation. Larger fragments will run more slowly than smaller fragments. If the fragments are incubated in a solution of Pax6 protein before they are placed in the gel, two things can happen. If the Pax6 does not recognize a sequence in a DNA fragment, the DNA will not be bound, and the fragment will migrate as it normally would in the gel. Alternatively, if the Pax6 protein does recognize a sequence in the DNA, it will bind to it. This will increase the mass of the fragment and cause it to run more slowly in the gel. Figure 5.12A shows such a gel used to find the binding site of Pax6 protein.

The results of this procedure are often confirmed by a **DNase protection assay**. If a

DNA-binding protein such as Pax6 finds its target sequence in a DNA fragment, it will bind to it. If the fragment is then placed into a solution of DNase I (which cleaves DNA

Figure 5.12
Procedures for determining the DNA-binding sites of transcription factors. (A) Gel mobility shift assay. A DNA fragment containing the Pax6 recognition element changes its mobility in the gel as Pax6 protein binds to it. Lanes 1 and 3 show the position of a DNA fragment containing a Pax6 binding site. When Pax6 is added to the fragment, it moves more slowly. Lanes 2 and 4 show the position of a similar fragment whose Pax6 binding site is mutated and does not bind Pax6. (B) A DNase footprint of the intron between the third and fourth exons of the chick δ1 crystallin gene. DNA is labeled at one end and is subjected to varying concentrations of DNase I. This DNase will randomly cleave the DNA. The resulting fragments are run on an electrophoretic gel and autoradiographed. No cleavage is seen in the region where a bound protein (Pax6) has prevented DNase from binding. There are two sites (heavy bars) where Pax6 binding is able to protect the DNA from digestion with DNase. They are enhancers. (A after Beimesche et al. 1999; B from Cvekl et al. 1995)

randomly), the transcription factor will protect that specific region of DNA from being cleaved. Figure 5.12B shows the results of one such assay for Pax6 binding.

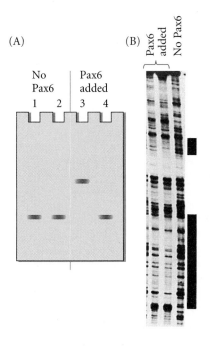

Reporter genes such as the *lacZ* gene mentioned above are also used to determine the function of various DNA fragments. If a sequence of DNA is thought to contain an enhancer region, it can be fused to a reporter gene and injected into an egg by various means (see Chapter 4). (This can be done only in certain species, such as fruit flies or mice, for which techniques of gene insertion have been established.) If the tested sequence is able to direct the expression of the reporter gene in the appropriate tissues, it is assumed to contain an enhancer.

Using enhancers to activate genes

The ability of an enhancer from one gene to activate other genes has been used by scientists to find new enhancers and the genes regulated by them. To do this, one makes an **enhancer trap**. The trap consists of a reporter gene (such as the *lacZ* gene for *E. coli* β-galactosidase or the jellyfish gene for green fluorescent protein, GFP) fused to a relatively weak eukaryotic promoter. The weak promoter will not initiate the transcription of the reporter gene without the help of an enhancer. This recombinant enhancer trap is then introduced into an egg or oocyte, where it integrates randomly into the genome. If the reporter gene becomes expressed, that means that it has come within the domain of an active enhancer (Figure 5.13). By isolating this activated region of the genome in wild-type flies or mice, the normal gene activated by this enhancer can be discovered (O'Kane and Gehring 1987).

One of the most powerful uses of genetic technology has been to activate regulatory genes such as *Pax6* in new places. Using *Drosophila* embryos, Halder and his colleagues (1995) in Walter Gehring's laboratory placed a gene encoding the yeast GAL4 transcriptional activator protein downstream from an enhancer that was known to function in the antennal imaginal discs (those parts of the *Drosophila* larva that become the adult antennae). In other words, the gene for the GAL4 transcription factor was placed next to an enhancer for genes normally expressed in the developing antennae. Therefore, GAL4 should also be expressed in antennal tissue. Halder and his colleagues then constructed a second transgenic fly, placing the cDNA for the *Drosophila Pax6* gene downstream from a sequence composed of five GAL4-binding sites. The GAL4 protein should be made only in a particular group of cells destined to become the anten-

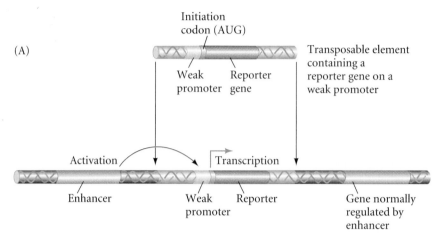

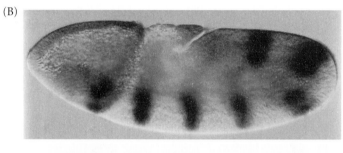

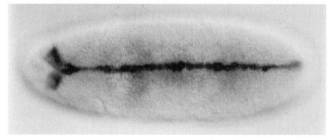

Figure 5.13
Enhancer trap technique. (A) A reporter gene is fused to a weak promoter that cannot direct transcription on its own. The recombinant gene is injected into the nucleus of the egg and integrates randomly into the genome. If it integrates near an enhancer, the reporter gene will become expressed when that enhancer is activated, showing the normal expression pattern of a gene normally associated with that enhancer. (B) Reporter gene expression (dark regions) in *Drosophila* embryos injected with an enhancer trap. These expression patterns demonstrated the presence of enhancers that are active in the development of the insect nervous system and which were unrecognized before this procedure. (Photographs courtesy of Y. Hiromi.)

nae, and when that protein was made, it should cause the transcription of the *Pax6* gene in those particular cells (Figure 5.14A). In flies in which the *Pax6* gene was expressed in the incipient antennal cells, part of the antennae gave rise to eyes (Figure 5.14B). Pax6 in *Drosophila* and frogs (but not in mice) is able to turn several developing tissue types into eyes (Chou et al. 1999). It appears that in *Drosophila*, Pax6 not only activates those genes that are necessary for the construction

of eyes, but also represses those genes that are used to construct other organs. ∎

WEBSITE **5.6 Enhancers and cancers.** Enhancer trapping may occur accidentally during development and place a new gene next to a particular regulatory element. When growth factor genes are placed next to the genes involved in making antibodies, the results are leukemias.

(A)

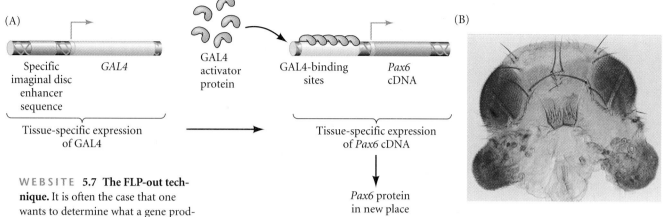

Specific imaginal disc enhancer sequence

GAL4

GAL4 activator protein

GAL4-binding sites

Pax6 cDNA

Tissue-specific expression of GAL4

Tissue-specific expression of *Pax6* cDNA

Pax6 protein in new place

(B)

WEBSITE 5.7 The FLP-out technique. It is often the case that one wants to determine what a gene product does when it is expressed in a cell type that does not usually express that gene. This is especially important when attempting to understand the interactions between embryonic cells. The FLP-out technique has greatly facilitated our ability to cause particular genes to be expressed in "the wrong" cell type.

Figure 5.14
Targeted expression of the *Pax6* gene in a *Drosophila* non-eye imaginal disc. (A) A strain of *Drosophila* was constructed wherein the gene for the yeast GAL4 transcription factor was placed downstream from an enhancer sequence that stimulates gene expression in the imaginal discs for mouth parts. Usually, the yeast protein can find no sequence to activate. However, if the embryo also contains a transgene that carries the DNA for *Pax6* downstream from the GAL4-binding sites, that *Pax6* DNA will be expressed in whichever imaginal disc the GAL4 protein is made. (B) *Drosophila* ommatidia (compound eye) emerging from the mouth parts of a fruit fly in which *Pax6* DNA was expressed in the antennal discs. (Photograph courtesy of W. Gehring and G. Halder.)

CASCADES OF TRANSCRIPTION FACTORS. We now know that the tissue-specific expression of a particular gene, such as the gene encoding the δ1 crystallin of the lens, is the result of the presence of a particular constellation of transcription factors in the nucleus. In the case of this lens crystalline, the Pax6 and Sox2 transcription factors are especially important. It is important to see that this combinatorial mode of operation means that none of these transcription factors has to be cell type-specific.

But how do the transcription factors themselves get to be expressed in a tissue-specific manner? In many cases, the genes for transcription factors are activated by other transcription factors. Let us look again at the *Pax6* gene as an example. Not only does the Pax6 protein regulate other genes (such as the ones for insulin and crystallins); it is itself regulated. The regulatory regions of the mouse *Pax6* gene were discovered by taking regions from its 5′ flanking sequence and introns and fusing them to a β-galactosidase reporter gene. This transgene was then microinjected into newly fertilized mouse pronuclei, and

(A)

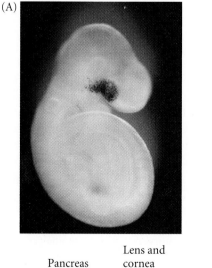

Figure 5.15
Regulatory regions of the mouse *Pax6* gene. (A) A sequence from the upstream enhancer of the murine *Pax6* gene directs the expression of the *lacZ* reporter transgene in the surface ectoderm overlying the optic cup, as shown by the dark staining in this area. (B) Map of the enhancer sites of the murine *Pax6* gene, based on reporter gene studies. There are two major promoters (in exons 0 and 1, respectively) initiating transcription of the same RNA, and the translation initiation codon is in exon 4. (Exon 5a is an alternatively spliced exon, discussed in the text.) Regions A–D are enhancers that activate the *Pax6* gene in the pancreas, lens, neural tube, and retina, respectively. The B region expression pattern is shown in (A). (A from Williams et al. 1998; B after Kammandel et al. 1998.)

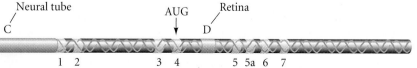

Pancreas   Lens and cornea   Neural tube   AUG   Retina

Enhancers: A   B   C   D

(B)

Exons: 0   1 2   3 4   5 5a 6 7

the resulting embryos were stained for β-galactosidase (Figure 5.15A; Kammandel et al. 1998; Williams et al. 1998). The results are summarized in Figure 5.15B. An enhancer farthest upstream from the promoter contains the regions necessary for *Pax6* expression in the pancreas, while a second enhancer activates *Pax6* expression in surface ectoderm (lens, cornea, and conjunctiva). A third enhancer resides in the leader sequence, and it contained the sequences that direct *Pax6* expression in the neural tube. A fourth enhancer sequence, located in an intron shortly after the translation initiation sequence, determines the expression of *Pax6* in the retina. The search is on now for those transcription factors that activate the gene for the Pax6 transcription factor. Wilhelm Roux (1894) described this situation eloquently in his manifesto for experimental embryology when he stated that the causal analysis of development may be the greatest problem the human intellect has attempted to solve, "since every new cause ascertained only gives rise to fresh questions concerning the cause of this cause."

Silencers

Some sequences act specifically to block transcription. These silencer domains are useful in restricting the transcription of a particular gene to a particular group of cells or for regulating the timing of the gene's expression. For example, the fetal mouse liver makes serum albumin, but only after a certain stage of gut development. At first, the endodermal cells that will form the liver do not transcribe this albumin gene. However, when the endodermal tube contacts the cardiac mesoderm (which is in the process of forming a heart), the heart precursors are able to instruct the endodermal tube to begin forming the liver and to start transcribing liver-specific genes (Le Douarin 1964; Gualdi et al. 1996). This contact is thought to release a transcription factor bound to a silencer region in the serum albumin gene. This silencer site is occupied prior to the contact of the endoderm with the cardiac mesoderm, but it becomes vacant in the endodermal tube immediately after contact with the heart-forming cells (Figure 5.16; Gualdi et al. 1996).

Another example of a silencer is found in certain neural genes. There is a sequence found in certain promoters that prevents the promoter's activation in any tissue except neurons. This sequence was given the name **neural restrictive silencer element** (**NRSE**), and it has been found in several mouse genes whose expression is limited to the nervous system: those for synapsin I, sodium channel type II, brain-derived neurotrophic factor, Ng-CAM, and L1. The protein that binds to the NRSE is a zinc finger transcription factor called neural restrictive silencer factor (NRSF). NRSF appears to be expressed in every cell that is not a mature neuron (Chong et al. 1995; Schoenherr and Anderson 1995). Thus, the down-regulation of NRSF seems to be a key event in allowing the expression of several genes that are critical to neural function.

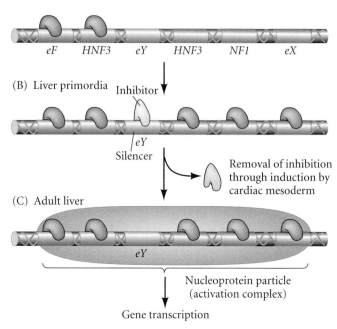

(A) General gut endoderm

eF   *HNF3*   *eY*   *HNF3*   *NF1*   *eX*

(B) Liver primordia   Inhibitor

eY
Silencer

Removal of inhibition through induction by cardiac mesoderm

(C) Adult liver

eY

Nucleoprotein particle (activation complex)

Gene transcription

Figure 5.16
The importance of silencers in liver-specific gene transcription. (A) In the early digestive tube endoderm, most of the transcription factors are not bound to their sites on the enhancer for serum albumin. (B) As endoderm development proceeds, the sites on the enhancer become occupied by five proteins whose presence is essential for activating the gene, and one protein, bound to the silencer (site *eY*), that can inhibit transcription. (C) As the liver forms, the inhibitory protein is no longer found on the enhancer, and the serum albumin gene is transcribed. Interestingly, this change may take place shortly after the association of the pre-liver endodermal region with heart-forming tissue. At this time, the chromatin in this region clumps together to form a nucleoprotein activation complex that spans 180 base pairs of DNA and activates the albumin promoter. (After Gualdi et al. 1996.)

To test the hypothesis that NRSE was necessary in the normal repression of neural genes in non-neural cells, *lacZ* transgenes were made by fusing a β-galactosidase gene (*lacZ*) with part of the *L1* neural cell adhesion gene. (L1 is a protein whose function is critical for brain development, as we will see in later chapters.) In one case, the *L1* gene, from its promoter through the fourth exon, was fused to the *lacZ* sequences. A second transgene was made just like the first, except that the NRSE had been deleted from the *L1* promoter. The two transgenes were separately inserted into the pronuclei of fertilized oocytes, and the resulting transgenic mice were analyzed for the expression of β-galactosidase (Kallunki et al. 1995, 1997). In the embryos receiving the complete transgene (which included the NSRE), expression was seen only in the nervous system (Figure 5.17A). However, in those mice whose transgene lacked the NRSE, expression was seen in the heart, limb

(A)

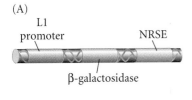

(B)

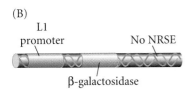

Figure 5.17
Analysis of β-galactosidase staining patterns in 11.5-day embryonic mice containing (A) a transgene composed of the *L1* promoter, a portion of the *L1* gene, and a bacterial *lacZ* gene fused to the second exon (which contains the NRSE region), or (B) a similar transgene, but lacking the NRSE sequence. The dark areas reveal the presence of β-galactosidase (the lacZ product). (From Kallunki et al. 1997.)

mesenchyme and limb ectoderm, kidney mesoderm, ventral body wall, and cephalic mesenchyme (Figure 5.17B).

Locus control regions in globin genes

There are some regions of DNA called **locus control regions (LCRs)**, which function as "super-enhancers." These LCRs establish an "open" chromatin configuration, inhibiting the normal repression of transcription over an area spanning several genes. The mechanism by which the LCR opens up the chromatin is not yet known.

One of the best-studied LCRs is that regulating the tissue-specific expression of the β-globin family of genes in humans, mice, and chicks. In many species, including chicks and humans, the embryonic or fetal hemoglobin differs from that found in adult red blood cells (Figure 5.18). Human hemoglobin consists largely of four globin chains of two different types and four molecules of heme. Human embryonic hemoglobin has two zeta (ζ) globin chains and two epsilon (ε) globin chains (and four molecules of heme). During the second month of human gestation, ζ- and ε-globin synthesis abruptly ceases, while alpha (α) and gamma (γ) globin synthesis increases. The association of two γ-globin chains with two α-globin chains produces fetal hemoglobin ($\alpha_2\gamma_2$). (The physiological importance of the γ-globin chain in fetal hemoglobin is examined in Chapter 14.) At 3 months gestation, the beta (β) globin and delta (δ) globin genes begin to be active, and their products slowly increase, while γ-globin levels gradually decline. This switchover is greatly accelerated after birth, and fetal hemoglobin is replaced by adult hemoglobin ($\alpha_2\beta_2$). The normal adult hemoglobin profile is 97% $\alpha_2\beta_2$, 2–3% $\alpha_2\delta_2$, and 1% $\alpha_2\gamma_2$.

A schematic diagram of human hemoglobin types and the genes that code for them is shown in Figure 5.19. In humans, the ζ- and α-globin genes are located on chromosome 16, but the ε-, γ-, δ-, and β-globin genes (known as the β-globin gene family) are linked together, in order of appearance, on chromosome 11. It appears, then, that there is a mechanism that directs the sequential switching of the chromosome 11 genes from embryonic, to fetal, to adult globins.

The discovery of the human globin LCR came from studies of the genetic disease **β-thalassemia**. This anemic condition results from a lack of β-globin and can be caused in several ways. The usual causes of β-thalassemia involve deletions or mutations in either the coding region of the β-globin gene or its promoter. However, in some patients, there is a

Figure 5.18
Percentages of hemoglobin chain types as a function of human developmental stage. (After Karlsson and Nieuhaus 1985.)

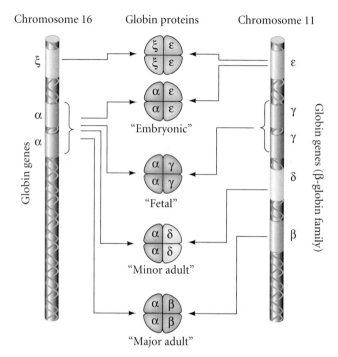

Chromosome 16   Globin proteins   Chromosome 11

Figure 5.19
Diagram of the human β-globin family of genes on chromosome 11. The erythroid-specific LCR region is located 6 to 22 kilobases upstream of the ε-globin gene. The five DNase I-hypersensitive sites within this region are marked by arrows. Downstream from the ε-globin (embryonic) gene are two nearly identical γ-globin (fetal) genes. These are followed by the adult δ- and β-globin genes.

deletion in a region *upstream* from the β-globin gene family, while the genes themselves are normal. Moreover, without this upstream region, the β-globin family DNA was found to be DNase I-insensitive (van der Ploeg et al. 1980; Kioussis et al. 1983). **DNase I** treatment is used to see whether the DNA in chromatin is accessible to transcription factors. If the DNA in chromatin is not digested by DNase I, it means that DNase I cannot reach it, and therefore, transcription factors couldn't reach it either. Promoters are usually DNase I-sensitive in the cells where they function; and they are usually DNase I-insensitive in those cells where they are not active (Weintraub and Groudine 1976; Stalder et al. 1980). So it appeared that there was a region of DNA upstream from the β-globin gene cluster that was responsible for "opening up" the chromatin of the genes, making them accessible to transcription factors. This region of DNA was termed the β-globin locus control region.

The locus control region for the β-globin gene complex is located far upstream from its most 5′ member (ε). This LCR contains four sites that are DNase I-hypersensitive only in erythroid precursor cells. Sites are said to be DNase-hypersensitive when the DNA in the chromatin can be digested there with only small amounts of DNase I. In most instances, a site is thought to be DNase I hypersensitive when it lacks nucleo-

somes (Elgin 1988). These sites in the LCR are therefore within nucleosome coils in most cells' nuclei, but in the precursors of the red blood cells, this DNA is exposed. The entire LCR is necessary for activating high levels of erythroid cell-specific transcription of the entire β-globin gene family on human chromosome 11 (Grosveld et al. 1987). Deletion or mutation of the LCR causes the silencing of all these genes. Conversely, if the LCR is placed adjacent to a gene that is not usually expressed in red blood cells (such as the T cell-specific *thy-1* gene) and then transfected into erythroid precursor cells, the additional gene will be expressed in the red blood cells. This effect is specific for red blood cell precursors, since only they have the appropriate transcription factors to bind to the LCR (Blom van Assendelft et al. 1989; Fiering et al. 1993). If any of the globin genes are separated from the LCR, they are repressed, even in the erythroid cells that would normally be transcribing them. The locus control region is crammed with binding sites for transcription factors. As Gary Felsenfeld (1992) observed, "The domains look as though they were put together by an overenthusiastic student determined to construct a powerful *cis*-acting element." He suggested that one of the functions of the LCR is to loop around to one of the globin promoter regions during DNA replication and bind to it in a manner that prevents nucleosomes from forming on that promoter. Indeed, the globin promoters are not DNase I-hypersensitive (i.e., accessible to transcription factors) except in the presence of the LCR.

It is not known why the genes closest to the LCR are transcribed earlier than the genes farther from it. Interestingly, experiments have shown that the distance between the LCR and the globin genes does affect their activation (Hanscombe et al. 1991; Dillon et al. 1997). When linked closely to the LCR, the human β-globin gene becomes expressed in transgenic mouse *embryonic* cells. Its correct activation (in *adult* red cells only) is restored only when it is placed farther away from the LCR. Similarly, the human γ-globin gene is repressed earlier (like the normal β-globin gene) when it is placed farther from the LCR. These findings suggest that the interaction between the LCR and the globin genes is polarized: those globin genes closest to the LCR are turned on earliest, while those genes more distal are turned on later. Presumably, there is physical contact between the LCR and the gene-specific promoters and enhancers.

LCRs have been found for other genes, such as the human growth hormone locus, the macrophage-specific lysozyme gene, and the CD2 and CD4 genes expressed in T lymphocytes (see Kioussis and Festenstein 1997; Fraser and Grosveld 1998).

WEBSITE **5.8 Further mechanisms of transcriptional regulation.** These three websites cover (1) DNase I hypersensitivity, (2) the mechanisms by which LCRs may reglate the temporal expression of tandemly linked genes, and (3) the association of active genes with the nuclear matrix.

Figure 5.20
Methylation of globin genes in human embryonic blood cells. The activity of the globin genes correlates inversely with the methylation of their promoters. (After Mavilio et al. 1983.)

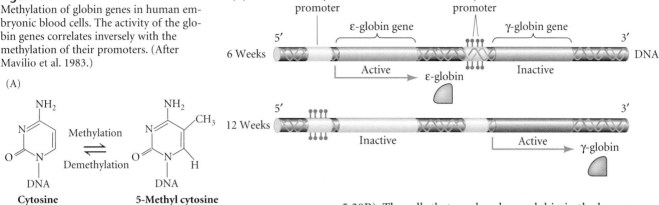

Methylation Pattern and the Control of Transcription

DNA methylation and gene activity

How does a pattern of transcription become stable? How can a lens cell continue to remain a lens cell and not activate muscle-specific genes? How can cells undergo rounds of mitosis and still maintain their differentiated characteristics? The answer appears to be **DNA methylation**. The promoters of inactive genes become methylated at certain cytosine residues, and the resulting methylcytosine stabilizes nucleosomes and prevents transcription factors from binding.

It is often assumed that a gene contains exactly the same nucleotides whether it is active or inactive; that is, that a β-globin gene in a red blood cell precursor has the same nucleotides as a β-globin gene in a fibroblast or retinal cell of the same animal. There is, however, a subtle difference in the DNA. In 1948, R. D. Hotchkiss discovered a "fifth base" in DNA, 5-methylcytosine. In vertebrates, this base is made enzymatically after DNA is replicated, at which time about 5% of the cytosines in mammalian DNA are converted to 5-methylcytosine (Figure 5.20A). This conversion can occur only when the cytosine residue is followed by a guanosine. Recent studies have shown that the degree to which the cytosines of a gene are methylated can control the level of the gene's transcription. Cytosine methylation appears to be a major mechanism of transcriptional regulation in vertebrates. However, *Drosophila*, nematodes, and perhaps most invertebrates do not methylate their DNA.

In vertebrates, the presence of methylated cytosines in the promoter of a gene correlates with the repression of transcription from that gene. In developing human and chick red blood cells, the DNA of the globin promoters is almost completely unmethylated, whereas the same promoters are highly methylated in cells that do not produce globin. Moreover, the methylation pattern changes during development (Figure 5.20B). The cells that produce hemoglobin in the human embryo have unmethylated promoters for the genes encoding the ε-globins of embryonic hemoglobin. These promoters become methylated in the fetal tissue (van der Ploeg and Flavell 1980; Groudine and Weintraub 1981; Mavilio et al. 1983). Similarly, when the fetal globin gives way to adult globin, the γ-globin gene promoters become methylated. The correlation between methylated cytosines and transcriptional repression has been confirmed experimentally. By adding transgenes to cells and giving them different patterns of methylation, Busslinger and co-workers (1983) showed that methylation in the promoter or enhancer of a gene correlates extremely well with the repression of gene transcription. In vertebrate development, the absence of DNA methylation correlates well with the tissue-specific expression of many genes.

Possible mechanisms by which methylation represses gene transcription

How is methylation involved in repressing genes? One hypothesis is that methylated DNA stabilizes nucleosomes. Here, DNA methylation is linked to histone deacetylation. Whereas acetylated histones are relatively unstable and cause the nucleosomes to disperse (see above), deacetylated histones form a stable nucleosome. The protein MeCP2 selectively binds to methylated regions of DNA; it also binds to histone deacetylases. Thus, when MeCP2 binds to methylated DNA it can stabilize the nucleosomes in that particular region of chromatin (Keshet et al. 1986; Jones et al. 1998; Nan et al. 1998). Methylated DNA may also be preferentially bound by histone H1, the histone that associates nucleosomes into higher-order folded complexes (McArthur and Thomas 1996). These patterns are maintained through cell division by the enzyme DNA (cytosine-5)-methyltransferase. During replication, one strand of the DNA (the template strand) retains the methylation pattern, while the newly synthesized strand does not. However, the enzyme DNA (cytosine-5)-methyltransferase has a strong preference for DNA that has one methylated strand, and when it sees a methyl-CpG on one side of the DNA, it methylates the new C on the other side (Gruenbaum et al. 1982; Bestor and Ingram 1983).

Genomic Imprinting

IT IS USUALLY ASSUMED that the genes one inherits from one's father and the genes one inherits from one's mother are equivalent. In fact, the basis for Mendelian ratios (and the Punnett square analyses used to teach them) is that it does not matter whether the genes came from the sperm or from the egg. But in some cases, it *does* matter. In certain mutations of mice and humans, a severe or lethal condition arises if the mutant gene is derived from one parent, but that same mutant gene has no deleterious effects if inherited from the other parent. For instance, in mice, the gene for insulin-like growth factor II (*Igf-2*) on chromosome 7 is active in early embryos only on the chromosome transmitted from the father. Conversely, the gene (*Igf-2r*) for a protein that binds this growth factor, located on chromosome 17, is active only in the chromosome transmitted from the mother (Barlow et al. 1991; DeChiara et al. 1991; Bartolomei and Tilghman 1997). The Igf-2r protein acts to bind and degrade excess Igf-2. A mouse pup that inherits a deletion of the *Igf-2r* gene from its father is normal, but if the same deletion is inherited from the mother, the fetus experiences a 30% increase in growth and dies late in gestation.*

In humans, the loss of a particular segment of the long arm of chromosome 15 results in different phenotypes, depending on whether the loss is in the male- or the female-derived chromosome (Figure 5.21). If the chromosome with the defective or missing segment comes from the father, the child is born with Prader-Willi syndrome, a disease associated with mild mental retardation, obesity, small

*The increase in growth is caused by the excess of Igf-2. The lethality is probably due to lysosomal defects, since the Igf-2 receptor also serves to target lysosomal enzymes into that organelle.

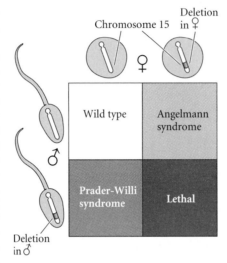

Figure 5.21
Inheritance patterns for Prader-Willi and Angelman syndrome. The region in the long arm of chromosome 15 contains the genes whose absence causes these syndromes. However, they are imprinted in reverse fashion. In Prader-Willi syndrome, the paternal genes are active, while in Angelman syndrome, the maternal genes are active.

gonads, and short stature. If the defective or missing segment comes from the mother, the child has Angelman syndrome, characterized by severe mental retardation, seizures, lack of speech, and inappropriate laughter (Knoll et al. 1989; Nicholls et al. 1998). Angelman syndrome is thought to occur through the loss of the gene for E6-AP ubiquitin protein ligase (*UBE3A*), a gene that is found in this 4-kilobase deletion (see Jiang et al. 1998).

These differences involve methylation. In the primordial germ cells that give rise to the sperm and egg, all methylation differences are wiped out. The DNA is almost entirely unmethylated (Monk et al. 1987; Driscoll and Migeon 1990). However, as the germ cells develop into sperm or eggs, their genes undergo extensive methylation. Moreover, the pattern of methylation on a given gene can differ between egg and sperm. These gene-specific methylation differences can be seen in the chromosomes of embryonic cells (Sanford et al. 1987; Chaillet et al. 1991; Kafri et al. 1992). Thus, methylation differences may reveal whether a gene came from the father or from the mother. Methylation also determines whether the gene will be active or inactive. Thus, the *Igf-2* gene is methylated in the mouse egg and unmethylated in the mouse sperm. The expression of *UBE3A* in the brain comes only from the unmethylated maternally derived gene. The paternal gene for *UBE3A* in the brain is methylated (Zeschingk et al. 1997). This ability to mark a gene as coming either from the father or the mother is called **genomic imprinting**. Imprinting adds additional information to the inherited genome, information that may regulate spatial and temporal gene activity. It also provides a reminder that the organism cannot be explained solely by its genes. One needs knowledge of developmental parameters as well as genetic ones. ▪

WEBSITE **5.9 Imprinting in humans and mice.** Insulin-like growth factor II and one of its receptors are imprinted in the mouse. Their expression may be controlled by an adjacent region whose RNA product does not leave the nucleus. In humans, different methylation patterns effect the transmission of certain diseases, and unmethylating γ-globin genes may be a cure for certain blood diseases.

Transcriptional Regulation of an Entire Chromosome: Dosage Compensation

In *Drosophila* and mammals, females are characterized as having two X chromosomes per cell, while males are characterized as having a single X chromosome per cell. Unlike the Y chromosome, the X chromosome contains thousands of genes that are essential for cell activity. Yet, despite the female's cells having double the number of X chromosomes the male's have, male and female cells contain approximately equal amounts of X chromosome-encoded gene products. This equalization is

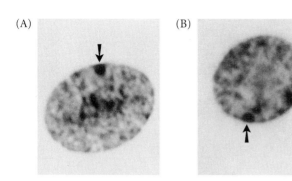

Figure 5.22
Nuclei of human oral epithelial cells stained with Cresyl violet. (A) Cell from a normal XX female, showing a single Barr body (arrow). (B) Cell from a female with three X chromosomes. Two Barr bodies can be seen, and only one X chromosome per cell is active. (Photographs courtesy of M. L. Barr.)

called **dosage compensation**. The transcription rates of the X chromosomes are altered so that male and female cells transcribe the same amount of RNAs from their X chromosomes. In *Drosophila*, both X chromosomes in the female are active, but there is increased transcription from the male's X chromosome, so that the single X chromosome of male cells produces as much product as the two X chromosomes in female cells (Lucchesi and Manning 1987). This is accomplished by the binding of particular transcription factors to hundreds of sites along the male X chromosome (Kuroda et al. 1991).

In mammals, X chromosome dosage compensation occurs through the inactivation of one X chromosome in each female cell. Thus, each mammalian somatic cell, whether male or female, has only one functioning X chromosome. This phenomenon is called **X chromosome inactivation**. The chromatin of the inactive X chromosome is converted into **heterochromatin**—chromatin that remains condensed throughout most of the cell cycle and replicates later than most of the other chromatin (the **euchromatin**) of the nucleus. The heterochromatic (inactive) X chromosome can often be seen on the nuclear envelope of female cells and is referred to as a **Barr body** (Figure 5.22; Barr and Bertram 1949). X chromosome inactivation must occur early in development. Using a mutated X chromosome that would not inactivate, Tagaki and Abe (1990) showed that the expression of two X chromosomes per cell in mouse embryos leads to ectodermal cell death and the absence of mesoderm formation, eventually causing embryonic death at day 10 of gestation.

The early inactivation of one X chromosome per cell has important phenotypic consequences. One of the earliest analyses of X chromosome inactivation was performed by Mary Lyon (1961), who observed coat color patterns in mice. If a mouse is heterozygous for an autosomal gene controlling hair pigmentation, then it resembles one of its two parents, or has a color intermediate between the two. In either case, the mouse is a single color. But if a female mouse is heterozygous

for a pigmentation gene on the X chromosome, a different result is seen: patches of one parental color alternate with patches of the other parental color (Figure 5.23). Lyon proposed the following hypothesis to account for these results:

1. Very early in the development of female mammals, both X chromosomes are active.
2. As development proceeds, one X chromosome is turned off in each cell.
3. This inactivation is random. In some cells, the paternally derived X chromosome is inactivated; in other cells, the maternally derived X chromosome is shut off.
4. This process is irreversible. Once an X chromosome has been inactivated, the same X chromosome is inactivated in all that cell's progeny. Since X inactivation happens relatively early in development, an entire region of cells derived from a single cell may all have the same X chromosome inactivated. Thus, all tissues in female mammals are mosaics of two cell types.

The Lyon hypothesis of X chromosome inactivation provides an excellent account of differential gene inactivation at the level of transcription. Some interesting exceptions to the general rules further show its importance. First, X chromosome inactivation holds true only for somatic cells, not germ cells. In female germ cells, the inactive X chromosome is reactivated shortly before the cells enter meiosis (Gartler et al. 1973; Migeon and Jelalian 1977). Thus, in early oocytes, both X chromosomes are unmethylated (and active). In each generation, X chromosome inactivation has to be established anew.

Second, there are some exceptions to the rule of randomness in the inactivation pattern. For instance, the first X chromosome inactivation in the mouse is seen in the fetal portion of the placenta, where the paternally derived X chromosome is specifically inactivated (Tagaki 1974). Third, X chromosome inactivation does not extend to every gene on the human X chromosome. There are a few genes (such as that encoding steroid sulfatase) on both arms of the X chromosome that "escape" X inactivation (Brown et al. 1997).

The fourth exception really ends up proving the rule. There are a few male mammals with coat color patterns we would not expect to find unless the animals exhibited X chromosome inactivation. Male calico and tortoiseshell cats are among these examples. These orange and black spotted coat patterns are normally seen in females and are thought to result from random X chromosome inactivation.* But rare males exhibit these coat patterns as well. How can this be? It turns out that these cats are XXY. The Y chromosome makes

---

*Although the terms calico and tortoiseshell are sometimes used synonymously, calico cats usually have white patches as well (where no pigment is found). If the cat has an allele for orange pigment on one X chromosome and an allele for black pigment on the other X chromosome, X inactivation will give patches of orange and black fur.

(A)

AT FERTILIZATION

Female zygote with
two X chromosomes

Paternal X
chromosome

Maternal X
chromosome

EARLY CLEAVAGE

Both X chromosomes
active in all cells

IMPLANTATION

Random inactivation
of one X chromosome
in all cells of the inner
cell mass, which will
form the embryo

Barr bodies

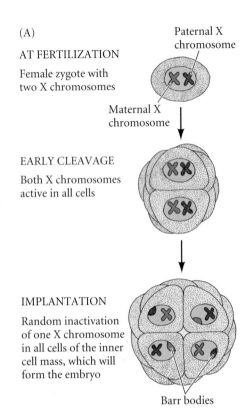

(B)

Figure 5.23
X chromosome inactivation in mammals. (A) Schematic diagram illustrating
random X chromosome inactivation. The inactivation is believed to occur at
about the time of implantation. (B) A calico cat, with orange and black alleles of a pigmentation gene on the X chromosome. The regions of different
color correspond to one or the other X chromosome being active. (Photograph courtesy of R. Loredo and G. Loredo.)

them male, but one X chromosome undergoes inactivation,
just as in females, so there is only one active X per cell
(Centerwall and Benirschke 1973). Thus, these cats have cells
with a Barr body and random X chromosome inactivation. It
is clear, then, that one mechanism for transcription-level control of gene regulation is to make a large number of genes heterochromatic and thus transcriptionally inert.

W E B S I T E **5.10 Chromosome elimination and diminution.** The inactivation or the elimination of entire chromosomes is not uncommon among invertebrates and is sometimes used as a mechanism of sex determination.

Sidelights & Speculations

The Mechanisms of X Chromosome Inactivation

THE MECHANISMS of X chromosome inactivation are still poorly understood,
but new research is giving us some indication of the factors that may be involved
in initiating and maintaining a heterochromatic X chromosome.

Initiation of X Chromosome
Inactivation: The *Xist* Gene
In 1991, Brown and her colleagues found an
RNA transcript that was made solely from
the *inactive* X chromosome of humans. This

transcript, *XIST*, does not encode a protein.
Rather, It stays within the nucleus and interacts with the inactive X chromatin, forming
an *XIST*-Barr body complex (Brown et al.
1992). A similar situation exists in the mouse,
in which the transcript of the *Xist* gene is seen
to coat the inactive X chromosome (Figure
5.24; Borsani et al. 1991; Brockdorrf et al.
1992).

The *Xist* gene is an excellent candidate for
the initiator of X inactivation. First, the transcripts from the *Xist* gene are seen in mouse

embryos prior to X chromosome inactivation, which would be expected if this gene
plays a role in initiating inactivation (Kay et
al. 1993). Second, knocking out one *Xist*
locus in an XX cell prevents X inactivation
from occurring on that particular chromosome (Penny et al. 1996). Third, the transfer
of a 450-kilobase segment containing the
mouse *Xist* gene into an autosome of male
embryonic stem cells causes the random inactivation of either that autosome or the endogenous X chromosome (Lee et al. 1996).

(A)

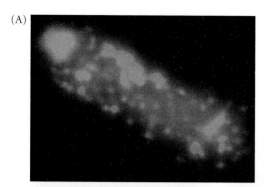

Figure 5.24

Xist RNA associates with the inactive X chromosome that made it. (A) Mouse *Xist* RNA (red) on an inactive X chromosome (blue) in metaphase. (In humans, *XIST* RNA is not seen to coat metaphase X chromosomes.) (B–D) XX murine embryonic stem cells undergoing differentiation and X chromosome inactivation. *Xist* RNA is stained light blue; mRNA from the X-linked phosphoglycerate kinase (*Pgk*) gene is stained red; the background DNA is stained blue. (B) XX cell before differentiating shows both X chromosomes transcribing *Pgk* and *Xist*. (C) As X chromosome inactivation begins, the *Xist* RNA is stabilized on one of the chromosomes. This chromosome no longer transcribes the *Pgk* gene. (D) As the cell finishes differentiating, X chromosome inactivation is complete. One X chromosome continues to make *Xist*. The other chromosome no longer makes *Xist*, but continues transcribing *Pgk*. (A courtesy of R. Jaenisch; B–D from Sheardown et al. 1997, courtesy of N. Brockdorff.)

(B)

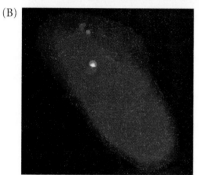

(C)

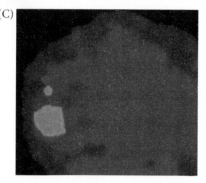

(D)

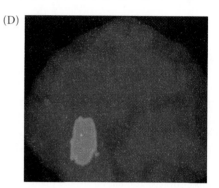

The autosome is "counted" as an X chromosome. Fourth, *Xist* appears to be involved in "choosing" which X chromosome is inactivated. Female mice heterozygous for a deletion of a particular region of the *Xist* gene will preferentially inactivate the wild-type chromosome (Marahrens et al. 1998). *Xist* expression is needed only for the initiation of X chromosome inactivation; once inactivation occurs, *Xist* transcription is dispensable (Brown and Willard 1994).

It is still not known what *Xist* RNA does to inactivate the chromosome. The *Xist* RNA works only in *cis*; that is, on the chromosome that made it. In the preimplantation mouse embryo, both X chromosomes synthesize *Xist* RNA, but this RNA is quickly degraded. When the cells begin to differentiate, the *Xist* RNA is stabilized on one of the two X chromosomes (Figure 5.24B–D; Sheardown et al. 1997; Panning et al. 1997). *Xist* RNA appears to be critical in the counting, selection, and intrachromosomal spreading of X chromosome inactivation.

Maintaining X Chromosome Inactivation

Once *Xist* initiates the inactivation of an X chromosome, the silencing of that chromosome is maintained in at least two ways. The first way involves methylation. The *Xist* locus on the active X chromosome becomes methylated, while the active *Xist* gene (on the inactive X chromosome) remains unmethy-

lated (Norris et al. 1994). Conversely, the promoter regions of numerous genes are methylated on the inactive X chromosome and unmethylated on the active X chromosome (Wolf et al. 1984; Keith et al. 1986; Migeon et al. 1991). This pattern may be reflected in the observation that the inactive X chromosomes of humans and mice have hardly any acetylated histone 4 (Figure 5.25; Jeppesen and Turner 1993). Conversely, the inactive X

(A)

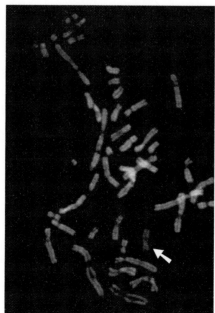

chromosome becomes associated with a higher concentration of the histone 2A variant macroH2A1 (Costanzi and Pehrson 1998). These nucleosome changes may create the heterochromatin that is characteristic of the Barr body.

We are still ignorant of the mechanisms by which the *Xist* transcript regulates the state of the chromatin and by which the spreading of inactivation occurs. We still do

(B)

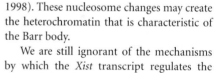

Attributes of the inactive X chromosome
Xist RNA production
Xist/Barr body complex
Histone H4 hypoacetylation
Histone macroH2A1 concentration
Nuclear envelope association
Late replication
Heterochromatin
Methylated promoters

Figure 5.25

The inactive X chromosome of human female cells contains underacetylated histone H4. (A) Chromosomes from a human female fibroblast cell stained green with fluorescent antibody to acetylated histone H4. While all the other chromosomes are stained green, the inactive X is not and appears red (arrow). (B) List of attributes characterizing the inactive X chromosome. (From Jeppesen and Turner 1993; photograph courtesy of the authors.)

not understand the ways in which *Xist* transcription is linked to DNA methylation. We do not yet know how the choice between the two X chromosomes is originally made, nor how the *Xist* RNA is transcribed from a region surrounded by inactivated genes.* There is still much to learn about dosage compensation in mammals, as well as in other animals that have chromosomal sex determination. ■

WEBSITE **5.11 The medical importance of X chromosome inactivation.** When a woman is heterozygous for an X-linked trait, she can often supply the missing product to the cells expressing the mutant allele. In other cases, she may remain mosaic. Sometimes random X inactivation is such that she can express the disease like a male.

*Lee and Lu (1999) provide a fascinating clue to *Xist* regulation. They have shown that *Xist* may be regulated by a natural antisense RNA that binds to it. This RNA (called *Tsix*) is transcribed from the X chromosome that remains active.

Differential RNA Processing

The regulation of gene expression is not confined to the differential transcription of DNA. Even if a particular RNA transcript is synthesized, there is no guarantee that it will create a functional protein in the cell. To become an active protein, the RNA must be (1) processed into a messenger RNA by the removal of introns, (2) translocated from the nucleus to the cytoplasm, and (3) translated by the protein-synthesizing apparatus. In some cases, the synthesized protein is not in its mature form and (4) must be posttranslationally modified to become active. Regulation can occur at any of these steps during development.

The essence of differentiation is the production of different sets of proteins in different types of cells. In bacteria, differential gene expression can be effected at the levels of transcription, translation, and protein modification. In eukaryotes, however, another possible level of regulation exists—namely, control at the level of RNA processing and transport. There are two major ways in which differential RNA processing can regulate development. The first involves the "censoring" of which nuclear transcripts are processed into cytoplasmic messages. Here, different cells can select different nuclear transcripts to be processed and sent to the cytoplasm as messenger RNA. The same pool of nuclear transcripts can thereby give rise to different populations of cytoplasmic mRNAs in different cell types (Figure 5.26A). The second mode of differential RNA processing is the splicing of the mRNA precursors into messages for different proteins by using different combinations of potential exons. If an mRNA precursor had five potential exons, one cell might use exons 1, 2, 4, and 5; a different cell might utilize exons 1, 2, and 3; and yet another cell type might use yet another combination (Figure 5.26B). Thus, one gene can create a family of related proteins.

Control of early development by nuclear RNA selection

In the late 1970s, numerous investigators found that mRNA was not the primary transcript from the genes. Rather, the genes transcribed **nuclear RNA** (**nRNA**), sometimes called heterogeneous nuclear RNA (hnRNA) or pre-messenger RNA (pre-mRNA). This nRNA is usually many times longer than the messenger RNA because the nuclear RNA contains introns that get spliced out during the passage from nucleus to cytoplasm. Originally, investigators thought that whatever RNA was transcribed in the nucleus was processed into cytoplasmic mRNA. But studies of sea urchins showed that different cell types could be *transcribing* the *same* type of nuclear RNA, but *processing* different subsets of this population into mRNA in different types of cells (Kleene and Humphreys 1977, 1985). Wold and her colleagues (1978) showed that sequences present in sea urchin blastula *messenger* RNA, but absent in gastrula and adult tissue mRNA, were nonetheless present in the *nuclear* RNA of the gastrula and adult tissues.

(A) RNA selection

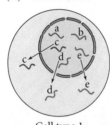

Cell type 1

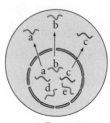

Cell type 2

(B) Differential splicing

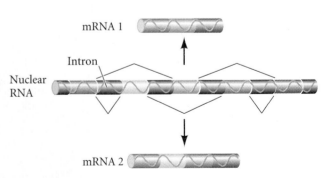

Figure 5.26
Roles of differential RNA processing during development. (A) RNA selection, whereby the same nuclear RNAs are made in two cell types, but the set that becomes cytoplasmic messenger RNAs is different. (B) Differential splicing, whereby the same nuclear RNA is spliced into different proteins by selectively removing possible exons.

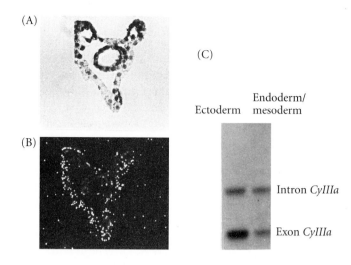

(A)

(B)

(C)

Ectoderm   Endoderm/ mesoderm

Intron *CyIIIa*

Exon *CyIIIa*

Figure 5.27
Regulation of ectoderm-specific gene expression by RNA process-ing. (A, B) *CyIIIa* mRNA is seen by autoradiography to be present only in the ectoderm. (A) Phase contrast micrograph and (B) in situ hybridization using a probe that binds to a *CyIIIa* exon. (C) The *CyIIIa* gene transcript is found in both ectoderm and endoderm/ mesoderm. The left-hand column represents RNA isolated from the gastrula ectodermal tissue; the right-hand column represents RNA isolated from endodermal and mesodermal tissues. The upper band is the RNA bound by a probe that binds to an intron sequence (which should be found only in the nucleus) of *CyIIIa*. The lower band represents the RNA bound by a probe complementary to an exon sequence. The presence of the intron indicates that the *CyIIIa* nuclear RNA is being made in both groups of cells, even if the mRNA is only seen in the ectoderm. (From Gagnon et al. 1992, courtesy of R. and L. Angerer.)

More genes are transcribed in the nucleus than are al-lowed to become mRNAs in the cytoplasm. This censoring of RNA transcripts has been confirmed for specific messages by probing for the introns and exons of specific genes. Gagnon and his colleagues (1992) performed such an analysis on the transcripts from the *CyIIIa* genes of the sea urchin *Strongylocentrotus purpuratus*. These genes encode calcium-binding and actin proteins, respectively, that are expressed only in a particular part of the ectoderm of the sea urchin larva. Using probes that bound to an exon (which is included in the mRNA) and to an intron (which is not included in the mRNA), they found that these genes were being transcribed not only in the ectodermal cells, but also in the mesoderm and endoderm. The analysis of the *CyIIIa* gene showed that the concentration of introns was the same in both the gastrula ec-toderm and in the mesoderm/endoderm samples, suggesting that this gene was being transcribed at the same rate in the nu-clei of all cell types, but was made into cytoplasmic mRNA only in ectodermal cells (Figure 5.27). The unprocessed nRNA for CyIIIa is degraded while still in the nuclei of the endoder-mal and mesodermal cells.

WEBSITE **5.12 Differential nRNA censoring.** Studies of differential nRNA censoring overturned the paradigm that differential gene transcription was the ultimate means of regulating embryonic differentiation. It "freed" embryol-ogy from microbiological ways of thinking about gene ex-pression.

WEBSITE **5.13 An inside-out gene.** Some RNAs stay in the nucleus to function. In one interesting case, the exons go outside the nucleus to be degraded, while the introns stay and help construct the nucleolus.

Creating families of proteins through differential nRNA splicing

The average vertebrate nRNA consists of relatively short exons (averaging about 140 bases) separated by introns that are usu-ally much longer. Most mammalian nRNAs contain numer-ous exons. By splicing together different sets of exons, differ-ent cells can make different types of mRNAs, and hence, dif-ferent proteins. Whether a sequence of RNA is recognized as an exon or as an intron is a crucial step in gene regulation. What is an intron in one cell's nucleus may be an exon in an-other cell's nucleus.

Alternative nRNA splicing is based on determining which sequences can be spliced out as introns. This can occur in sev-eral ways (Figure 5.28). Cells can differ in their ability to rec-ognize the 5' splice site (at the beginning of the intron) or the 3' splice site (at the end of the intron). Or some cells could fail to recognize a sequence as an intron at all, retaining it within the message. The splicing of nRNA is mediated through a complex called a **spliceosome**, made up of **small nuclear RNAs (snRNA)** and proteins, that assembles at a splice site. Whether a spliceosome recognizes the splice sites depends on certain factors in the nucleus that can interact with those sites and compete or cooperate with the proteins that direct spliceosome formation. The 5' splice site is normally recog-nized by small nuclear RNA U1 (U1 snRNA) and splicing fac-tor 2 (SF2; also known as alternative splicing factor). The choice of alternative 3' splice sites is often controlled by which splice site can best bind a protein called U2AF. The spliceo-some forms when the 5' and 3' splice sites are brought togeth-er and the intervening RNA is cut out.

WEBSITE **5.14 The mechanism of differential nRNA splicing.** Differential nRNA splicing depends on the assem-bly of the nucleosome and upon the ratio of certain pro-teins in the nucleus of the cell.

Differential RNA processing has been found to control the alternative forms of expression of genes encoding over 100 proteins. The deletion of certain potential exons in some cells but not in others enables one gene to create a family of close-ly related proteins. Instead of one gene-one polypeptide, one can have one gene-one family of proteins. For instance, alter-native RNA splicing enables the α-tropomyosin gene to en-code brain, liver, skeletal muscle, smooth muscle, and fibrob-last forms of this protein (Figure 5.29; Breitbart et al. 1987). The nuclear RNA for α-tropomyosin contains 11 potential

CONSTITUTIVE SPLICING

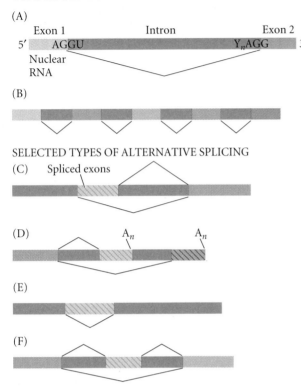

SELECTED TYPES OF ALTERNATIVE SPLICING

Figure 5.28
Schematic diagram of alternative nRNA splicing. Exons are represented as shaded boxes, alternatively spliced exons are represented by hatched boxes, and introns are represented by broad lines. By convention, the path of splicing is shown by fine V-shaped lines. (A) The exon-intron borders, showing the consensus sequences at the 5′ and 3′ ends of the intron. Y represents pyrimidine. (B) The splicing of an nRNA that has five exons. (C–F) Alternative splicing by (C) alternative 5′ splice sites, (D) alternative 3′ splice sites (in some cases, this would provide different termini to the mRNA, and both sites would need a polyadenylation sequence, shown here as A$_n$), (E) a splice/no splice decision, and (F) exon inclusion/exon skipping. (After Horowitz and Krainer 1995.)

DNA-binding site and enables another to be used (Epstein et al. 1994). The two forms of Pax 6 appear to be made at similar rates in all the cells expressing the *Pax6* gene. If the human gene for *PAX6* is mutated such that the 5′ splicing site becomes more efficient, the Pax6 isoform containing the amino acids encoded by the alternatively spliced exon is made in excess. The eyes of people with this mutation have defects in their lenses, corneas, and pupils.

If you think that differential splicing means that certain genes with dozens of introns can create thousands of different related proteins, you are probably correct. Proteins derived from the neurexin genes, for example, are found on the cell surfaces of developing neurons, and they may be important in specifying the connections that these neurons make.* These genes can be alternatively spliced at several different sites, creating hundreds of proteins from the same gene (Ullrich et al. 1995; Ichtchenko et al. 1995).

---

*The neurexins are cell recognition molecules and appear to be involved in neuron-neuron adhesion and recognition. The venom of the black widow spider works by binding to neurexins, causing massive neurotransmitter release (Rosenthal and Meldolesi 1989).

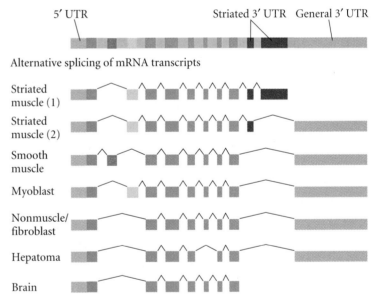

Figure 5.29
Alternative RNA splicing to form a family of rat α-tropomyosin proteins. The α-tropomyosin gene is represented on top. The numbers correspond to the amino acids encoded by the exons. The thin lines represent the sequences that become introns and are spliced out to form the mature mRNA. Constitutive exons (found in all tropomyosins) are green. Those expressed only in smooth muscle are red; those expressed only in striated muscle are purple. Those that are variously expressed are yellow. Note that in addition to the internal possibilities of exons, there are two different alternative 3′ ends possible. (After Breitbart et al. 1987.)

exons, but different sets of exons are used in different cells. Such different proteins encoded by the same gene are called **splicing isoforms** of the protein.

In some instances, alternatively spliced RNAs yield proteins that play similar, yet distinguishable, roles in the same cell. The nuclear RNA for the Pax6 transcription factor is actually spliced to yield two types of Pax6 proteins. In about half the mRNAs, there is an added exon that interrupts one major

Sidelights & Speculations

Differential nRNA Processing and Drosophila *Sex Determination*

S EX DETERMINATION in *Drosophila* is regulated by a cascade of RNA processing events (see Baker et al. 1987; Mac-Dougall et al. 1995). As we will see in Chapter 17, the development of the sexual phenotype in *Drosophila* is mediated by the ratio of X chromosomes to autosomes (non-sex chromosomes). The X chromosomes produce transcription factors that activate the *Sex-lethal (Sxl)* gene,* while the autosomes produce transcription factors that compete for these activators. When the X-to-autosome ratio is 1 (i.e., when there are two X chromosomes per diploid cell), the *Sxl* gene is active, and the embryo develops into a female fly. When the ratio is 0.5 (i.e., when the fly is XY with only one X chromosome per diploid cell), the *Sxl* gene is not active, and the embryo develops into a male (Figure 5.30).

But what is the Sxl protein doing to determine sex? It is acting as a differential splicing factor on the nRNA of the *transformer (tra)* gene. Throughout the larval period, the *tra* gene actively synthesizes a transcript that is processed into either a general mRNA (found in both females and males) or a female-specific mRNA. The female-specific message is made when Sxl binds to the nRNA and inhibits spliceosome formation on the general 3′ splice site of the first intron (Sosnowski et al. 1989; Valcárcel et al. 1993). Instead, the spliceosome forms on another, less efficient 3′ site and allows splicing to occur there. As a result, the female form of the *transformer* mRNA lacks a portion of RNA that is found in the general form. And that is a crucial difference, for this portion contains a translational stop codon (UGA)

*The name of this gene, *Sex-lethal*, comes from the deadly decoupling of the dosage compensation mechanism that arises when this gene is mutated. When this happens, the fly will become male, even if it is XX. However, its two X chromosomes will be instructed to transcribe their genes at the higher (male) rate. This will create regulatory defects that will kill the embryo (Cline 1986).

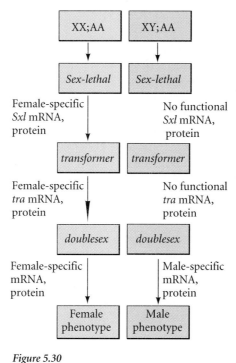

Figure 5.30
Sex determination in *Drosophila*. This simplified scheme shows that the X-to-autosome ratio is monitored by the *Sex-lethal* gene. If this gene is active, it processes the *transformer* nRNA into a functional female-specific message. In the presence of the female-specific Transformer protein, the *doublesex* gene transcript is processed in a female-specific fashion. The female-specific Doublesex protein is a transcription factor that leads to the production of the female phenotype. If the *transformer* gene does not make a female-specific product (i.e., if the *Sex-lethal* gene is not activated), the *doublesex* transcript is spliced in the male-specific manner, leading to the formation of a male-specific Doublesex protein. This is a transcription factor that generates the male phenotype.

that causes the message to make a small, nonfunctional, protein. Therefore, the general transcript has no bearing on sex determination (Belote et al. 1989). However, in the fe-

male-specific message, the UGA codon is spliced out during mRNA formation and does not interfere with the translation of the message. In other words, the female *transformer* transcript is the only functional transcript of this gene.

The Transformer protein is, itself, an alternative splicing factor that regulates the splicing of the nuclear transcript of the *doublesex (dsx)* gene. This gene is needed for the production of either sexual phenotype, and mutations of *dsx* can reverse the expected sexual phenotype, causing XX embryos to become males or XY embryos to become females. During pupation, the *doublesex* gene makes a transcript that can be processed in two alternative ways. It can generate a female-specific mRNA or a male-specific mRNA (Nagoshi et al. 1988). In females and males, the first three exons of the *doublesex* mRNA are the same (Tian and Maniatis 1992, 1993). But if the Transformer protein is present, it converts a weak 3′ splicing site into a strong site (a more efficient binder of U2AF), and exon 4 is retained, resulting in female-specific *doublesex* mRNA. If Transformer protein is not present, U2AF will not bind to this 3′ site, and exon 4 will not be included, resulting in the male-specific *doublesex* message. The Doublesex proteins made by the male and female mRNAs are both transcription factors, and they recognize the same sequence of DNA. However, while the female Doublesex protein activates female-specific enhancers (such as those on the genes encoding yolk proteins), the male Doublesex protein inhibits transcription from those same enhancers (Coschigano and Wensink 1993; Jursnich and Burtis 1993). Conversely, the female Doublesex protein can inhibit transcription from genes that are otherwise activated by the male Doublesex protein. Research into *Drosophila* sex determination shows that differential RNA processing plays enormously important roles throughout development. ■

Control of Gene Expression at the Level of Translation

Once the RNA has reached the cytoplasm, there is still no guarantee that it will be translated. The control of gene expression at the level of translation can occur by many means; some of the most important of these are described below.

Differential mRNA longevity

The longer an mRNA persists, the more protein can be translated from it. If a message with a relatively short half-life were selectively stabilized in certain cells at certain times, it would make large amounts of its particular protein only at those times and places. The stability of a message is often dependent upon the length of its poly(A) tail. This, in turn, appears to depend upon sequences in the 3′ untranslated region. Certain 3′ UTR sequences allow longer poly(A) tails than others. If these regions are experimentally traded, the half-lives of the resulting mRNAs will be altered: usually long-lived messages will decay rapidly, while normally short-lived mRNAs will remain around longer (Shaw and Kamen 1986; Wilson and Treisman 1988; Decker and Parker 1994). In some instances, messenger RNAs can be selectively stabilized at specific times in specific cells. The mRNA for casein, the major protein of milk, has a half-life of 1.1 hours in rat mammary gland tissue. However, during periods of lactation, the presence of the hormone prolactin increases this half-life to 28.5 hours (Figure 5.31; Guyette et al. 1979).

WEBSITE **5.15 Mechanism of mRNA translation and degradation.** Translation is a complex process involving the initiation, elongation, and termination of protein synthesis. It has numerous points at which regulation can occur. Similarly, the degradation of mRNA is a tightly regulated event.

Selective inhibition of mRNA translation

Some of the most remarkable cases of translational regulation of gene expression occur in the oocyte. The oocyte often makes and stores mRNAs that will be used only after fertilization occurs. These messages stay in a dormant state until they are actived by ionic signals (to be discussed in Chapter 7) that spread through the egg during ovulation or sperm binding. Table 5.3 gives a partial list of some of the mRNAs that are stored in the oocyte cytoplasm. Some of these stored mRNAs are for proteins that will be needed during cleavage, when the embryo makes enormous amounts of chromatin, cell membranes, and cytoskeletal components. Some of them encode cyclin proteins that regulate the timing of early cell division (Rosenthal et al. 1980; Standart et al. 1986). Indeed, in many species (including sea urchins and *Drosophila*), maintenance of the normal rate and pattern of early cell divisions does not require a nucleus; rather, it requires continued protein synthesis from stored maternal mRNAs (Wagenaar and Mazia 1978;

Edgar et al. 1989). Other stored messages encode proteins that determine the fates of cells. These include the *bicoid* and *nanos* messages that provide positional information in the *Drosophila* embryo, as we saw in Chapter 3, and the *glp-1* mRNA of the nematode *C. elegans*.

WEBSITE **5.16 The discovery of stored mRNAs.** The existence of maternally transcribed mRNAs stored in the oocyte was one of the first discoveries of molecular embryology. Even before gene cloning became available, the identity of several of these mRNAs was known.

In some instances, these messages are prevented from being translated by the binding of some inhibitory protein. For instance, the smaug protein binds to the 3′ UTR of the *Drosophila nanos* message, and this prevents the translation of *nanos* mRNA in the oocyte until fertilization (Smibert et al. 1996). (At that time, the Nanos protein will become critical for determining which part of the fly will be its abdomen.) In other instances, the translatability of the mRNA is regulated by the length of its poly(A) tail. In oocytes, having a short poly(A) tail does not lead to the degradation of the message; however, such messages are not translated. In the *Drosophila* oocyte, the *bicoid* message remains untranslated until signals at fertilization allow the Cortex and Grauzone proteins to add poly(A) residues to the *bicoid* mRNA (Sallés et al. 1994; Lieberfarb et al. 1996). At that point, the *bicoid* message becomes translatable (and its product determines which part of the embryo becomes the head and thorax). Other organisms use ingenious ways of regulating the translatability of their

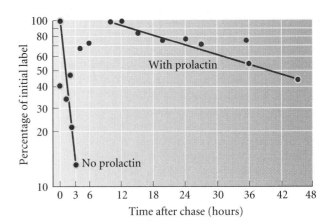

Figure 5.31
Degradation of casein mRNA in the presence and absence of prolactin. Cultured mammary cells were given radioactive RNA precursors (pulse) and after a given time were washed and given nonradioactive precursors (chase). The casein RNA synthesized during the pulse time was then isolated and its radioactive labelmeasured. In the absence of prolactin, the newly synthesized casein mRNA decayed rapidly, with a half-life of 1.1 hours. When the same experiment was done in a medium containing prolactin, the half-life was extended to 28.5 hours. (After Guyette et al. 1979.)

Table 5.3 Some mRNAs stored in oocyte cytoplasm and translated at or near fertilization

mRNAs encoding	Function(s)	Organism(s)
Cyclins	Cell division regulation	Sea urchin, clam, starfish, frog
Actin	Cell movement and contraction	Mouse, starfish
Tubulin	Formation mitotic spindles, cilia, flagella	Clam, mouse
Small subunit of ribonucleotide reductase	DNA synthesis	Sea urchin, clam, starfish
Hypoxanthine phosphoribosyl-transferase	Purine synthesis	Mouse
Vg1	Mesodermal determination(?)	Frog
Histones	Chromatin formation	Sea urchin, frog, clam
Cadherins	Blastomere adhesion	Frog
Metalloproteinases	Implantation in uterus	Mouse
Growth factors	Cell growth; uterine cell growth(?)	Mouse
Sex determination factor FEM-3	Sperm formation	*C. elegans*
PAR gene products	Segregate morphogenetic determinants	*C. elegans*
SKN-1 morphogen	Blastomere fate determination	*C. elegans*
Hunchback morphogen	Anterior fate determination	*Drosophila*
Caudal morphogen	Posterior fate determination	*Drosophila*
Bicoid morphogen	Anterior fate determination	*Drosophila*
Nanos morphogen	Posterior fate determination	*Drosophila*
GLP-1 morphogen	Anterior fate determination	*C. elegans*
Germ cell-less protein	Germ cell determination	*Drosophila*
Oskar protein	Germ cell localization	*Drosophila*
Ornithine transcarbamylase	Urea cycle	Frog
Elongation factor 1α	Protein synthesis	Frog
Ribosomal proteins	Protein synthesis	Frog, *Drosophila*

Sources: Compiled from numerous sources, including Raff 1980; Shiokawa et al. 1983; Rappoll et al. 1988; Brenner et al. 1989; Standart 1992.

messages. The oocyte of the tobacco hornworm moth makes some of its mRNAs without their methylated 5′ caps. In this state, they cannot be efficiently translated. However, at fertilization, a methyltransferase completes the formation of the caps, and these mRNAs can be translated (Kastern et al. 1982).

One of the most remarkable ways of regulating the translation of a specific message is seen in *Caenorhabditis elegans*. This nematode lives up to its name, having evolved a particularly elegant solution to the problem of controlling larval gene expression (Lee et al. 1993; Wightman et al. 1993). It makes a naturally occuring antisense mRNA to one of its own messages. High levels of the LIN-14 transcription factor are im-

portant in the development of early larval organs. Thereafter, the LIN-14 protein is no longer seen, although *lin-14* messages can be detected throughout development. *C. elegans* is able to inhibit the synthesis of LIN-14 from these messages by activating the *lin-4* gene. The *lin-4* gene does not encode a protein. Rather, it encodes two small RNAs (the most abundant being 25 nucleotides long, the other continuing for 40 more nucleotides) that are complementary to an imperfectly repeated site in the *lin-14* 3′ UTR. Figure 5.32 shows a hypothetical sketch of what might be happening. It appears that the binding of the *lin-4* transcripts to the *lin-14* mRNA 3′ UTR does not signal the destruction of the message, but rather prevents the message from being translated.

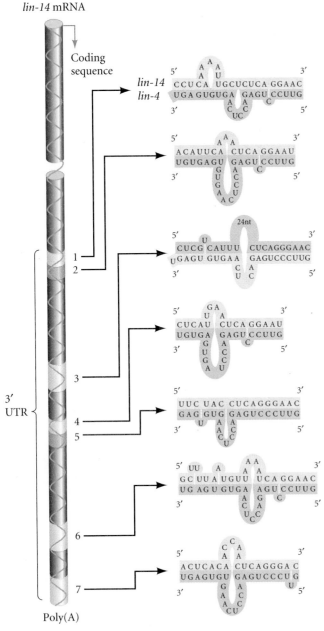

Figure 5.32
Hypothetical model of the regulation of *lin-14* mRNA translation by the *lin-4* mRNAs. The *lin-4* gene does not produce an mRNA. Rather, it produces small RNAs that are complementary to a repeated sequence in the 3′ UTR of the *lin-14* mRNA, which bind to it and prevent its translation. (After Wickens and Takayama 1995.)

Control of RNA expression by cytoplasmic localization

Not only is the time of mRNA translation regulated, but so is the place of RNA expression. Just like the selective repression of mRNA translation, the selective localization of messages is also often accomplished through their 3′ UTRs, and it is also often performed in oocytes. Rebagliati and colleagues (1985) showed

that there are certain mRNAs in *Xenopus* embryos that are selectively transported to the vegetal pole of the frog oocyte (Figure 5.33). After fertilization, these messages make proteins that are found only in the vegetal blastomeres. In *Drosophila*, the *bicoid* and *nanos* messages are each localized to different ends of the oocyte. The 3′ UTR of the *bicoid* mRNA allows this message to bind to the microtubules through its association with two other proteins (swallow and staufen). If the *bicoid* 3′ UTR is attached to some other message, that mRNA also will be bound to the anterior pole of the oocyte (Driever and Nüsslein-Volhard 1988a,b; Ferrandon et al. 1994). The 3′ UTR of the *nanos* message similarly allows it to be transported to the posterior pole of the egg, where it will be bound to the cytoskeleton (Gavis and Lehmann 1994). As we saw in Chapter 3, this localization allows the Bicoid protein to form a gradient wherein the highest amount of it is at the anterior pole, while the Nanos protein forms a gradient with its peak at the posterior pole (Figure 5.34). The ratio of these two proteins will eventually determine the anterior-posterior axis of the *Drosophila* embryo and adult. (The ability of cells to be specified by gradients of proteins is a critical phenomenon and is discussed in more detail in Chapters 3 and 9.)

WEBSITE **5.17 Other examples of translational regulation of gene expression.** There are numerous other fascinating examples wherein mRNA is selectively translated under different conditions. The 1:1 ratio of α- and β-globins in adult blood comes from the differential translation of the respective globin messages. The production of heme for the hemoglobin is also regulated at the level of translation.

Epilogue: Posttranslational Gene Regulation

When a protein is synthesized, the story is still not over. Once a protein is made, it becomes part of a larger level of organization. For instance, it may become part of the structural framework of the cell, or it may become involved in one of the myriad enzymatic pathways for the synthesis or breakdown of cellular metabolites. In any case, the individual protein is now part of a complex "ecosystem" that integrates it into a relationship with numerous other proteins. Thus, several changes can still take place that determine whether or not the protein will be active. Some newly synthesized proteins are inactive without the cleaving away of certain inhibitory sections. This is what happens when insulin is made from its larger protein precursor. Some proteins must be "addressed" to their specific intracellular destinations in order to function. Proteins are often sequestered in certain regions, such as membranes, lysosomes, nuclei, or mitochondria. Some proteins need to assemble with other proteins to form a functional unit. The hemoglobin protein, the microtubule, and the ribosome are all examples of numerous proteins joining together to form a

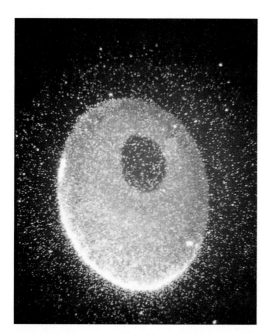

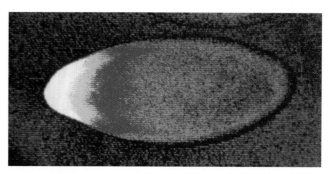

Figure 5.34
Control of anterior-posterior polarity in *Drosophila* embryos by maternal mRNA. The bicoid message is tethered to the cytoskeleton in the anterior (left) region of the egg. When the egg is fertilized, it synthesizes the Bicoid protein. Thus, a gradient of Bicoid protein extends from its peak (represented by yellow) at the most anterior portion of the *Drosophila* egg about one-third of the way into the egg (orange and red). (Photograph courtesy of C. Nüsslein-Volhard.)

Figure 5.33
Localization of *Vg1* mRNA to the vegetal portion of the *Xenopus* oocyte. the white crescent at the bottom of the egg represents the tethered *Vg1* message. The black area is the haploid nucleus of the oocyte. At fertilization, the *Vg1* message is translated into an inactive protein. Moreover, if that protein is processed to its active form, it can deliver a TGF-β signal of the kind required for organizer formation.(Photograph courtesy of D. Melton.)

functional unit. And some proteins are not active unless they bind an ion such as calcium, or are modified by the covalent addition of a phosphate or acetate group. This last type of protein modification will become very important in the next chapter, since many important proteins in embryonic cells are just sitting there until some signal activates them. We turn next to how the embryo develops by activating certain proteins in specific cells.

Principles of Development: Developmental Genetics

1. Differential gene expression from genetically identical nuclei creates different cell types. Differential gene expression can occur at the levels of gene transcription, nuclear RNA processing, mRNA translation, and protein modification.

2. Genes are usually repressed. Activation of a gene often means inhibiting its repressor. This leads to thinking in double and triple negatives: Activation is often the inhibition of the inhibitor; repression is the inhibition of the inhibitor of the inhibitor.

3. Eukaryotic genes contain promoter sequences to which RNA polymerase can bind to initiate transcription. The eukaryotic RNA polymerases are bound by a series of proteins called basal transcription factors.

4. Eukaryotic genes expressed in specific cell types contain enhancer sequences that regulate their transcription in time and space.

5. Specific transcription factors can recognize specific sequences of DNA in the promoter and enhancer regions. They activate or repress transcription from the genes to which they have bound.

6. Enhancers work in a combinatorial fashion. The binding of several transcription factors can act to promote or inhibit transcription from a certain promoter. In some cases transcription is activated only if *both* factor A *and* factor B are present, while in other cases, transcription is activated if *either* factor A *or* factor B is present.

7. A gene encoding a transcription factor can keep itself activated if the transcription factor it encodes also activates its own promoter. Thus, a transcription factor gene can have one set of enhancer sites to initiate its activation and a second set of enhancer sites (that bind the encoded transcription factor) to maintain its activation.

8. Often, the same transcription factors that are used during the differentiation of a particular cell type are also used to activate the genes for that cell type's specific products. For instance, Pax6 is needed both for the differentiation of the lens and for the transcription of the lens crystallin genes, and Mitf is needed for pigment cell differentiation and for the transcription of the genes whose products catalyze the synthesis of melanin.

9. Enhancers can act as silencers to suppress the transcription of a gene in inappropriate cell types.

10. Locus control regions may function by making relatively large portions of a chromosome accessible to transcription factors.

11. Transcription factors act in different ways to regulate RNA synthesis. Some transcription factors stabilize RNA polymerase binding to the DNA, some disrupt nucleosomes, and some increase the efficiency of transcription.

12. Transcription correlates with a lack of methylation on the promoter and enhancer regions of genes. Methylation differences can account for examples of genomic imprinting, wherein a gene transmitted through the sperm is expressed differently from the same gene transmitted through the egg.

13. Dosage compensation enables the X chromosome-derived products of males (which have one X chromosome per cell in fruit flies and mammals) to equal the X chromosome-derived products of females (which have two X chromosomes per cell). This compensation is accomplished at the level of transcription, either by accelerating transcription from the lone X chromosome in males (*Drosophila*) or by inactivating a large portion of one of the two X chromosomes in females (mammals).

14. X chromosome inactivation in placental mammals is generally random and involves the activation of the *Xist* gene on the chromosome that will be inactivated.

15. Differential RNA selection can allow certain transcripts to enter the cytoplasm while preventing other transcripts from leaving the nucleus.

16. Differential RNA splicing can create a family of related proteins by causing different regions of the nRNA to be read as exons and introns. What is an exon in one set of circumstances may be an intron in another.

17. Some messages are translated only at certain times. The oocyte, in particular, uses translational regulation to set aside certain messages that it transcribes during egg development but uses only after the egg is fertilized. This activation is often accomplished either by the removal of inhibitory proteins or by the polyadenylation of the message.

18. Many messenger RNAs are localized to particular regions of the oocyte or other cells. This localization appears to be regulated by the 3′ untranslated region of the mRNA.

Literature Cited

Andersen, F. G., J. Jensen, R. S. Heller, H. V. Petersen, L.-I. Larsson, O. D. Madsen and P. Serup. 1999. Pax6 and Pdx1 form a functional complex on the rat somatostain gene upstream enhancer. *FEBS Lett.* 445: 315–320.

Angier, N. 1992. A first step in putting genes into action: Bend the DNA. *New York Times,* Aug. 4, 1992, pp. C1, C7.

Baker, B., R. N. Nagoshi and K. C. Burtin. 1987. Molecular genetic aspects of sex determination in *Drosophila. BioEssays* 6: 66–70.

Barlow, D. P., R. Stoger, B. G. Herrmann, K. Saito and N. Schweifer. 1991. The mouse insulin-like growth factor type-2 receptor is imprinted and closely linked to the *Tme* locus. *Nature* 349: 84–87.

Barr, M. L. and E. G. Bertram. 1949. A morphological distinction between neurones of the male and female, and the behavior of the nucleolar satellite during accelerated nucleoprotein synthesis. *Nature* 163: 676.

Bartolomei, M. S. and S. M. Tilghman. 1997. Genomic imprinting in mammals. *Annu. Rev. Genet.* 31: 493–525.

Beimesche, S. and 8 others. 1999. Tissue-specific transcriptional activity of a pancreatic islet cell-specific enhancer sequence/Pax6 binding site determined in normal adult tissues in vivo

using transgenic mice. *Mol. Endocrin.* 13: 718–728.

Belote, J. M., M. McKeown, R. T. Boggs, R. Ohkawa and B. A. Sosnowski. 1989. Molecular genetics of *transformer,* a genetic switch controlling sexual differentiation in *Drosophila. Dev. Genet.* 10: 143–155.

Bentley, N. J., T. Eisen and C. R. Goding. 1994. Melanocyte-specific expression of the human tyrosinase promoter: Activation by the *microphthalmia* gene product and the role of the initiator. *Mol. Cell. Biol.* 14: 7996–8006.

Bestor, T. H. and V. M. Ingram. 1983. Two DNA methyltransferases from murine erythroleukemia cells: Purification, sequence specificity, and mode of interaction with DNA. *Proc. Natl. Acad. Sci. USA* 82: 2674–2678.

Blom van Assendelft, G., O. Hanscombe, F. Grosveld and D. R. Greaves. 1989. The β-globin dominant control region activates homologous and heterologous promoters in a tissue-specific manner. *Cell* 56: 969–977.

Borsani, G. and 13 others. 1991. Characterization of a murine gene expressed from the inactive X chromosome. *Nature* 351: 325–329.

Boulet, A. M., C. R. Erwin and W. J. Rutter. 1986. Cell-specific enhancers in the rat exocrine pancreas. *Proc. Natl. Acad. Sci. USA* 83: 3599–3603.

Breitbart, R. A., A. Andreadis and B. Nadal-Ginard. 1987. Alternative splicing: A ubiquitous mechanism for the generation of multiple protein isoforms from single genes. *Annu. Rev. Biochem.* 56: 481–495.

Brockdorrf, N. and 7 others. 1992. The product of the mouse *Xist* gene is a 15-kb inactive X-specific transcript containing no conserved ORF and located in the nucleus. *Cell* 71: 515–526.

Brown, C. J. and H. F. Willard. 1994. The human X-inactivation centre is not required for maintenance of X-chromosome inactivation. *Nature* 368: 154–156.

Brown, C. J. A. Ballabio, J. L. Rupert, R. G. Lafreniere, M. Grompe, R. Tonlorenzi and H. F. Willard. 1991a. A gene from the region of the human X inactivation center is expressed exclusively from the inactive X chromosome. *Nature* 349: 38–45.

Brown, C. J. and 9 others. 1991b. Localization of the X chromosome inactivation center on the human X chromosome. *Nature* 349: 82–85.

Brown, C. J., B. D. Hendrich, J. L. Rupert, R. G. Lafreniere, Y. Xing, J. Lawrence and H. F. Willard. 1992. The human *XIST* gene: Analysis of a 17-kb inactive X-specific RNA that contains conserved repeats and is highly localized within the nucleus. *Cell* 71: 527–542.

Brown, C. J., L. Carrel and H. F. Willard. 1997. Expression of genes from the human active and inactive X chromosomes. *Am. J. Hum. Genet.* 60: 1333–1343.

Buratowski, S. 1997. Multiple TATA-binding factors come back into style. *Cell* 91: 13-15.

Buratowski, S., S. Hahn, L. Guarente and P. A. Sharp. 1989. Five initiation complexes in transcription initiation by RNA polymerase II. *Cell* 56: 549–561.

Busslinger, M., J. Hurst and R. A. Flavell. 1983. DNA methylation and the regulation of globin gene expression. *Cell* 34: 197–206.

Centerwall, W. R. and K. Benirschke. 1973. Male tortoiseshell and calico (T-C) cats. *J. Hered.* 64: 272–278.

Chaillet, J. R., T. F. Vogt, D. R. Beier and P. Leder. 1991. Parental-specific methylation of an imprinted transgene is established during gametogenesis and progressively changes during embryogenesis. *Cell* 66: 77–83.

Chen, J.-L., L. D. Attardi, C. P. Verrijzer, K. Yokomori and R. Tjian. 1995. Assembly of recombinant TFIID reveals differential cofactor requirements for distinct transcriptional activators. *Cell* 79: 93–105.

Chong, J. A. and 9 others. 1995. REST: a mammalian silencer protein that restricts sodium channel gene expression to neurons. *Cell* 80: 949–957.

Chou, R. L., C. R. Altmann, R. A. Lang and A. Hemmati-Brivanlou. 1999. Pax6 induces ectopic eyes in a vertebrate. *Development* 126: 4213–4222.

Cline, T. 1986. A female-specific lethal lesion in an X-linked positive regulator of the *Drosophila* sex determination gene, *Sex-lethal*. *Genetics* 113: 641–663.

Coschigano, K. T. and P. Wensink. 1993. Sex-specific transcriptional regulation by the male and female doublesex proteins of *Drosophila*. *Genes Dev.* 7: 42–55.

Costanzi, C. and J. R. Pehrson. 1998. Histone macroH2A1 is concentrated in the inactive X chromosome of female mammals. *Nature* 393: 599–601.

Cvekl, A. and J. Piatigorsky. 1996. Lens development and crystallin gene expression: Many roles for Pax-6. *BioEssays* 18: 621–630.

Cvekl, A., C. M. Sax, X. Li, J. B. McDermott and J. Piatigorsky. 1995. Pax-6 and lens-specific transcription of the chicken δ1-crystallin gene. *Proc. Natl. Acad. Sci. USA* 92: 4681–4685.

DeChiara, T. M., E. J. Robertson and A. Efstratiadis. 1991. Parental imprinting of the mouse insulin-like growth factor II gene. *Cell* 64: 849–859.

Decker, C. J. and R. Parker. 1995. Mechanisms of mRNA degradation in eukaryotes. *Trends Biochem.* 19: 336–340.

Dillon, N., T. Trimborn, J. Strouboulis, P. Fraser and F. Grosveld. 1997. The effect of distance on long-range chromatin interactions. *Mol. Cell* 1: 131–139.

Driever, W. and C. Nüsslein-Volhard. 1988a. The bicoid protein determines position in the *Drosophila* embryo in a concentration-dependent manner. *Cell* 54: 95–105.

Driever, W. and C. Nüsslein-Volhard. 1988b. A gradient of bicoid protein in *Drosophila*. *Cell* 54: 83–93.

Driscoll, D. J. and B. R. Migeon. 1990. Sex difference in methylation of single-copy genes in human meiotic germ cells: Implications for X chromosome inactivation, parental imprinting, and the origin of PGC mutations. *Somat. Cell Mol. Genet.* 16: 267–268.

Edgar, B., F. Sprenger, R. J. Duronio, P. Leopold and P. O'Farrell. 1994. MPF regulation during the embryonic cell cycles of *Drosophila*. *Genes Dev.* 8: 440–453.

Elgin, S. C. R. 1988. The formation and function of DNase-I hypersensitivity sites in the process of gene activation. *J. Biol. Chem.* 263: 9259–9262.

Epstein, J., T. Cai, T. Glaser, L. Jepeal and R. Maas. 1994. Identification of a Pax paired domain recognition sequence and evidence for DNA-dependent conformational changes. *J. Biol. Chem.* 269: 8355–8361.

Felsenfeld, G. 1992. Chromatin as an essential part of the transcriptional mechanism. *Nature* 355: 219–225.

Ferrandon, D., L. Elphick, C. Nüsslein-Volhard and D. St. Johnston. 1994. Staufen protein associates with the 3′ UTR of *bicoid* to form particles that move in a microtubule-dependent manner. *Cell* 79: 1221–1232.

Ferré-D'Amaré, A. R., G. C. Predergast, E. B. Ziff and S. K. Burley. 1993. Recognition by Max of its cognate DNA through a dimeric bHLH/Z domain. *Nature* 363: 38–45.

Fiering, S., C. G. Kim, E. M. Epner and M. Groudine. 1993. An "in-out" strategy using gene targeting and FLP recombinase for the functional dissection of complex DNA regulatory elements: Analysis of the β-globin locus control region. *Proc. Natl. Acad. Sci. USA* 90: 8469–8473.

Fraser, P. and F. Grosveld. 1998. Locus control regions, chromatin activation, and transcription. *Curr. Opin. Cell Biol.* 10: 361–365.

Gagnon, M. L., L. M. Angerer and R. C. Angerer. 1992. Posttranscriptional regulation of ectoderm-specific gene expression in early sea urchin embryos. *Development* 114: 457– 467.

Gartler, S. M., R. M. Liskay and N. Grant. 1973. Two functional X chromosomes in human fetal oocytes. *Exp. Cell Res.* 82: 464–466.

Gavis, E. R. and R. Lehmann. 1994. Translational regulation of *nanos* by RNA localization. *Nature* 369: 315–318.

Gedamu, L. and G. H. Dixon. 1978. Effect of enzymatic decapping on protamine messenger RNA translation in wheat-germ S-30. *Biochem. Biophys. Res. Comm.* 85: 114–125.

Glaser, T., L. Jepeal, J. G. Edwards, S. R. Young, J. Favor and R. L. Maas. 1994.. *PAX6* gene dosage effects in a family with congenital cataracts,

aniridia, anophthalmia, and central nervous system defects. *Nature Genet.* 8: 463–471.

Grosveld, F., G. Blom van Assendelft, D. R. Greaves and G. Kollins. 1987. Position-dependent high-level expression of the human β-globin gene in transgenic mice. *Cell* 51: 975–985.

Groudine, M. and H. Weintraub. 1981. Activation of globin genes during chick development. *Cell* 24: 393–401.

Gruenbaum, Y., H. Ceder and A. Razin. 1982. Substrate and sequence specificity of a eukaryotic DNA methylase. *Nature* 295: 620–622.

Gualdi, R., P., Bossard, M. Zheng, Y. Hamada, J. R. Coleman and K. S. Zaret. 1996. Hepatic specification of the gut endoderm in vitro: Cell signaling and transcriptional control. *Genes Dev.* 10: 1670–1682.

Guyette, W. A., R. J. Matusik and J. M. Rosen. 1979. Prolactin-mediated transcriptional and post-transcriptional control of casein gene expression. *Cell* 17: 1013–1023.

Halder, G., P. Callaerts and W. J. Gehring. 1995. Induction of ectopic eyes by targeted expression of the *eyeless* gene in *Drosophila*. *Science* 267: 1788–1792.

Hanscombe, O., D. Whyall, P. Fraser, N. Yannoutsos, D. Greaves, N. Dillon and F. Grosveld. 1991. Importance of globin gene order for correct developmental expression. *Genes Dev.* 5: 1387–1395.

Hemesath, T. J. and 9 others. 1994. Microphthalmia, a critical factor in melanocyte development, defines a discrete transcription factor family. *Genes Dev.* 8: 2770–2780.

Horowitz, D. S. and A. R. Krainer. 1995. Mechanisms for selecting 5′ splice sites in mammalian pre-mRNA splicing. *Trends Genet.* 10: 100–106.

Hotchkiss, R. D. 1948. The quantitative separation of purines, pyrimidines, and nucleosides by paper chromatography. *J. Biol. Chem.* 175: 315–332.

Hussain, M. A. and J.-F. Habener. 1999. Glucagon gene transcription activation mediated by synergistic interactions of Pax6 and Cdx2 with the p300 co-activator. *J. Biol. Chem.* 274: 28950–28957.

Ichtchenko, K., Y. Hata, T. Nguyen, B. Ullrich, M. Missler, C. Moomaw and T. C. Südhof. 1995. Neuroglian 1: A splice-site specific ligand for β-neurexins. *Cell* 81: 435–443.

Jeppesen, P. and B. M. Turner. 1993. The inactive X chromosome in female mammals is distinguished by a lack of histone H4 acetylation, a cytogenetic marker for gene expression. *Cell* 74: 281–289.

Jiang, Y.-H., T.-F. Tsai, J. Bressler and A. L. Beaudet. 1998. Imprinting in Angelman and Prader-Willi syndromes. *Curr. Opin. Genet. Dev.* 8: 334–342.

Jones, P. L. and 7 others. 1998. Methylated DNA and MeCP2 recruit histone deacetylase to repress transcription. *Nature Genet.* 19: 187–191.

Jursnich, V. A. and K. C. Burtis. 1993. A positive role in differentiation of the male doublesex protein of *Drosophila*. *Dev. Biol.* 155: 235–249.

Kafri, T. and 7 others. 1992. Developmental pattern of gene-specific DNA methylation in the mouse embryo and germ line. *Genes Dev.* 6: 705–715.

Kallunki, P., S. Jenkinson, G. M. Edelman and F. S. Jones. 1995. Silencer elements modulate the expression of the gene for neuron-glia cell adhesion molecule, Ng-CAM. *J. Biol. Chem.* 270: 21291–21298.

Kallunki, P., G. M. Edelman and F. S. Jones. 1997. Tissue-specific expression of the L1 cell adhesion molecule is modulated by the neural restrictive silencer element. *J. Cell Biol.* 138: 1343–1355.

Kamachi, Y., M. Uchikawa, J. Collignon, R. Lovell-Badge and H. Kondoh. 1998. Involvement of Sox1, 2, and 3 in the early and subsequent molecular events of lens induction. *Development* 125: 2521–2532.

Kammandel, B., K. Chowdhury, A. Stoykova, S. Aparicio, S. Brenner and P. Gruss. 1998. Distinct *cis*-essential modules direct the time-space pattern of *Pax6* gene activity. *Dev. Biol.* 205: 79–97.

Karlsson, S. and A. W. Nieuhaus. 1985. Developmental regulation of human globin genes. *Annu. Rev. Biochem.* 54: 1071–1108.

Kastern, W. H., M. Swindlehurst, C. Aaron, J. Hooper and S. J. Berry. 1982. Control of mRNA translation in oocytes and developing embryos of giant moths. I. Functions of the 5′ terminal "cap" in the tobacco hornworm *Manduca sexta*. *Dev. Biol.* 89: 437–449.

Kay, G. F., G. D. Penny, D. Patel, A. Ashworth, N. Brockdorrf and S. Rastan. 1993. Expression of *Xist* during mouse development suggests a role in the initiation of X chromosome inactivation. *Cell* 72: 171–182.

Keith, D. H., J. Singersam and A. D. Riggs. 1986. Active X-chromosome DNA is unmethylated at eight CCGG sites clustered in a guanine-plus-cytosine-rich island at the 5′ end of the gene for phosphoglycerate kinase. *Mol. Cell Biol.* 6: 4122–4125.

Keshet, I., J. Lieman-Hurwitz and H. Cedar. 1986. DNA methylation affects the formation of active chromatin. *Cell* 44: 535–543.

Kioussis, D. and R. Festenstein. 1997. Locus control regions: Overcoming heterochromatin-induced gene inactivation in mammals. *Curr. Opin. Genet. Dev.* 7: 614–619.

Kioussis, D., E. Vanin, T. de Lange, R. A. Flavell and F. G. Grosveld. 1983. Beta-globin gene inactivation by DNA translocation in gamma-beta thalassemia. *Nature* 306: 662–666.

Kleene, K. C. and T. Humphreys. 1977. Similarity of hnRNA sequences in blastula and pluteus stage sea urchin embryos. *Cell* 12: 143–155.

Kleene, K. C. and T. Humphreys. 1985. Transcription of similar sets of rare maternal RNAs and rare nuclear RNAs in sea urchin blastulae and adult coelomocytes. *J. Embryol. Exp. Morphol.* 85: 131–149.

Knoll, J. H. M., R. D. Nicholls, R. E. Magenis, J. M. Graham, Jr., M. Lalande and S. A. Latt. 1989. Angelman and Prader-Willi syndromes share a common chromosome 15 deletion but differ in the parental origin of the deletion. *Am. J. Med. Genet.* 32: 285–290.

Kornberg, R. D. and J. D. Thomas. 1974. Chromatin structure: Oligomers of histones. *Science* 184: 865–868.

Kuroda, M. I., M. J. Kernan, R. Kreber, B. Ganetzky and B. S. Baker. 1991. The maleless protein associates with the X chromosome to regulate dosage compensation in *Drosophila*. *Cell* 66: 935–947.

Lawn, R. M., A. Efstratiadis, C. O'Connell and T. Maniatis. 1980. The nucleotide sequence of the human β-globin gene. *Cell* 21: 647–651.

Le Douarin, N. 1964. Etude expérimentale de l'organogenèse du tube digestif et du foie chez l'embryon de poulet. *Bull. Biol. Fr. Belg.* 98: 543–676.

Lee, J. T. and N. Lu. 1999. Targeted mutagenesis of *Tsix* leads to nonrandom X inactivation. *Cell* 99: 47–57.

Lee, J. T., W. M. Strauss, J. A. Dausman and R. Jaenisch. 1996. A 450-kb transgene displays properties of the mammalian X-inactivation center. *Cell* 86: 83–95.

Lee, R. C., R. L. Feinbaum and V. Ambros. 1993. The *C. elegans* heterochromatic gene *lin-4* encodes small RNAs with antisense complementarity to *lin-15*. *Cell* 75: 843–855.

Lee, T. I. and R. A. Young. 1998. Regulation of gene expression by TBP-associated proteins. *Genes Dev.* 12: 1398–1408.

Lieberfarb, M. E., T. Chu, C. Wreden, W. Theurkauf, J. P. Gergen and S. Strickland. 1996. Mutations that perturb poly(A)-dependent maternal mRNA activation block the initiation of development. *Development* 122: 579–588.

Lucchesi, J. C. and J. E. Manning. 1987. Gene dosage and compensation in *Drosophila melanogaster*. *Adv. Genet.* 24: 371–429.

Luger , K., A. W. Mäder, R. K. Richmond, D. F. Sargent and T. J. Richmond. 1997. Crystal structure of the nucleosome core particle at 2.8Å resolution. *Nature* 389: 251–260.

Lyon, M. F. 1961. Gene action in the X chromosome of the mouse (*Mus musculus* L.) *Nature* 190: 372–373.

MacDougall, C., D. Harbison and M. Bownes. 1995. The developmental consequences of alternative splicing in sex determination and differentiation in *Drosophila*. *Dev. Biol.* 172: 353–376.

Maniatis, T., S. Goodbourn and J. A. Fischer. 1987. Regulation of inducible and tissue-specific gene expression. *Science* 236: 1237–1245.

Marahrens, Y., J. Loring and R. Jaenisch. 1998. Role of the *Xist* gene in X chromosome choosing. *Cell* 92: 657–665.

Mavilio, F. and 9 others. 1983. Molecular mechanisms for human hemoglobin switching: Selective undermethylation and expression of globin genes in embryonic, fetal, and adult ery-throblasts. *Proc. Natl. Acad. Sci. USA* 80: 6907–6911.

McArthur, M. and J. O. Thomas. 1996. A preference of histone H1 for methylated DNA. *EMBO J.* 15: 1705–1715.

Migeon, B. R. and K. Jelalian. 1977. Evidence for two active X chromosomes in germ cells of female before meiotic entry. *Nature* 269: 242–243.

Migeon, B. R., M. M. Holland, D. J. Driscoll and J. C. Robinson. 1991. Programmed demethylation in CpG islands during human fetal development. *Somatic Cell Mol. Genet.* 17: 159–168.

Monk, M., M. Boubelik and S. Lehnert. 1987. Temporal and regional changes in DNA methylation in the embryonic, extraembryonic, and germ cell lineages during mouse embryo development. *Development* 99: 371–382.

Moore, K. L. 1977. *The Developing Human*. Saunders, Philadelphia.

Nagoshi, R. N., M. McKeown, K. C. Burtis, J. M. Belote and B. S. Baker. 1988. The control of alternative splicing at genes regulating sexual differentiation in *D. melanogaster*. *Cell* 53: 229–236.

Nakayama, A., M.-T., T. Nguyen, C. C. Chen, K. Opdecamp, C. A. Hodgkinson and H. Arnheiter. 1998. Mutations in *microphthalmia*, the mouse homolog of the human deafness gene *MITF*, affect neuroepithelial and neural crest-derived melanocytes differently. *Mech. Dev.* 70: 155–166.

Nan, X., H.-H. Ng, C. A. Johnson, C. D. Laherty, B. M. Turner, R. N. Eisenman and A. Bird. 1998. Transcriptional repression by the methyl-CpG-binding protein MeCP2 involves a histone deacetylase complex. *Nature* 393: 386–389.

Nicholls, R. D. 1998. Imprinting in Prader-Willi and Angelman syndromes. *Trends Genet.* 14: 194–200.

Norris, D. P., D. Patel, G. F. Kay, G. D. Penny, N. Brockdorff, S. A. Sheardown and S.Rastan. 1994. Evidence that random and imprinted *Xist* expression is controlled by preemptive methylation. *Cell* 77: 41–51.

Offield, M., F. N. Hirsch and R. M. Grainger. In press. The use of *Xenopus tropicalis* transgenic lines for studying lens developmental timing in living embryos.

Ogryzko, V. V., R. L. Schlitz, V. Russanova, B. H. Howard and Y. Nakatani. 1996. The transcriptional coactivators p300 and CBP are histone acetyltransferases. *Cell* 87: 953–959.

O'Kane, C. J. and W. J. Gehring. 1987. Detection in situ of genomic regulatory elements in *Drosophila*. *Proc. Natl. Acad. Sci. USA* 84: 9123–9127.

Panning, B., J. Dausman and R. Jaenisch. 1997. X chromosome inactivation is mediated by *Xist* stabilization. *Cell* 90: 907–916.

Penny, G. D., G. F. Kay, S. A. Sheardown, S. Rastan and N. Brockdorff. 1996. Requirement for *Xist* in X chromosome inactivation. *Nature* 379: 131–137.

Plaza, S., C. Dozier and S. Saule. 1993. Quail *PAX6* (*PAX-QNR*) encodes a transcription factor able to bind and transactivate its own promoter. *Cell Growth Diff.* 4: 1041–1050.

Price, E. R. and 7 others. 1998. Lineage-specific signaling in melanocytes: c-Kit stimulation recruits p300/CBP to microphthalmia. *J. Biol. Chem.* 273: 33042–33047.

Rebagliati, M. R., D. L. Weeks, R. P. Harvey and D. A. Melton. 1985. Identification and cloning of localized maternal RNAs from *Xenopus* eggs. *Cell* 42: 769–777.

Rosenthal, E., T. Hunt and J. V. Ruderman. 1980. Selective translation of mRNA controls the pattern of protein synthesis during early development of the surf clam, *Spisula solidissima. Cell* 20: 487–495.

Rosenthal, L. and J. Meldolesi. 1989. α-Latrotoxin and related toxins. *Pharm. Ther.* 42: 115–134.

Roux, W. 1894. The problems, methods, and scope of developmental mechanics. *In* W. M. Wheeler (trans.), *Biological Lectures of the Marine Biology Laboratory, Woods Hole.* Ginn, Boston, pp. 149–190.

Sallés, F. J., M. E. Lieberfarb, C. Wreden, J. P. Gergen and S. Strickland. 1994. Coordinate initiation of *Drosophila* development by regulated polyadenylation of maternal messenger RNAs. *Science* 266: 1996–1999.

Sander, M., A. Neubüser, H. Ee, G. R. Martin and M. S. German. 1997. Genetic analysis reveals that Pax6 is required for normal transcription of pancreatic hormone genes and islet development. *Genes Dev.* 11: 1662–1673.

Sanford, J. P., H. J. Clark, V. M. Chapman and J. Rossant. 1987. Differences in DNA methylation during oogenesis and spermatogenesis and their persistence during early embryogenesis in the mouse. *Genes Dev.* 1: 1039–1046.

Sauer, F., J. D. Fondell, Y. Ohkuma, R. G. Roeder and H. Jäckle. 1995. Control of transcription by Krüppel through interactions with TFIIB and TFIIE. *Nature* 375: 162–165.

Schlissel, M. S. and D. D. Brown. 1984. The transcriptional regulation of *Xenopus* 5S RNA genes in chromatin: The roles of active stable transcription complex and histone H1. *Cell* 37: 903–913.

Schoenherr, C. J. and D. J. Anderson. 1995. The neuron-restrictive silencer factor (NRSF): A coordinate repressor of multiple neuron-specific genes. *Science* 267: 1360–1363.

Shatkin, A. J. 1976. Capping of eucaryotic mRNAs. *Cell* 9: 645–653.

Shaw, G. and R. Kamen. 1986. A conserved AU sequence from the 3′ untranslated region of GM-CSF mRNA mediates selective mRNA degradation. *Cell* 46: 659–667.

Sheardown, S. A. and 9 others. 1997. Stabilization of *Xist* RNA mediates initiation of X chromosome inactivation. *Cell* 91: 99–107.

Sheiness, D. and J. E. Darnell. 1973. Polyadenylic segment in mRNA becomes shorter with age. *Nature New Biol.* 241: 265–268.

Smibert, C. A., J. E. Wilson, K. Kerr and P. M. Macdonald. 1996. Smaug protein represses

translation of unlocalized *nanos* mRNA in the *Drosophila* embryo. *Genes Dev.* 10: 2600–2609.

Sopta, M., Z. F. Burton and J. Greenblatt. 1989. Structure and associated DNA helicase activity of a general transcription factor that binds to RNA polymerase II. *Nature* 341: 410–415.

Sosnowski, B. A., J. M. Belote and M. McKeown. 1989. Sex-specific alternative splicing of RNA from the *transformer* gene results from sequence-specific splice site blockage. *Cell* 58: 449–459.

Stalder, J., A. Larsen, J. D. Engel, M. Dolan, M. Groudine and H. Weintraub. 1980. Tissue-specific DNA cleavages in the globin chromatin domain introduced by DNase I. *Cell* 20: 451–460.

Standart, N., T. Hunt and J. V. Ruderman. 1986. Differential accumulation of ribonucleotide reductase subunits in clam oocytes: The large subunit is stored as a polypeptide, the small subunit as untranslated mRNA. *J. Cell Biol.* 103: 2129–2136.

Steingrímsson, E. and 10 others. 1994. Molecular basis of mouse *microphthalmia* (*mi*) mutations helps explain their developmental and phenotypic consequences. *Nature Genet.* 8: 256–263.

Tagaki, N. 1974. Differentiation of X chromosomes in early female mouse embryos. *Exp. Cell Res.* 86: 127–135.

Tagaki, N. and K. Abe. 1990. Detrimental effects of two active X chromosomes on early mouse development. *Development* 109: 189–201.

Thoma, F., T. Koller and A. Klug. 1979. Involvement of histone H1 in the organization of the nucleosome and of the salt-dependent superstructures of chromatin. *J. Cell Biol.* 83: 403–427.

Tian, M. and T. Maniatis. 1992. Positive control of pre-mRNA splicing in vitro. *Science* 256: 237–240.

Tian, M. and T. Maniatis. 1993. A splicing enhancer complex controls alternative splicing of *doublesex* pre-mRNA. *Cell* 74: 105–115.

Trudel, M. and F. Constantini. 1987. A 3′ enhancer contributes to the stage-specific expression of the human β-globin gene. *Genes Dev.* 1: 954–961.

Ullrich, B., Y. A. Uskaryov and T. C. Südhof. 1995. Cartography of neurexins: More than 1000 isoforms generated by alternative splicing and expressed in distinct subsets of neurons. *Neuron* 14: 497–507.

Valcárcel, J., R. Singh, P. D. Zamore and M. R. Greene. 1993. The protein Sex-lethal antagonizes the splicing factor U2AF to regulate alternative splicing of *transformer* pre-mRNA. *Nature* 362: 171–175.

van der Ploeg, L. H. T. and R. D. Flavell. 1980. DNA methylation in the human γ-δ-β globin locus in erythroid and non-erythroid cells. *Cell* 19: 947–958.

van der Ploeg, L. H., A. Konings, M. Oort, D. Roos, L. Bernini and R. A. Flavell. 1980. Gamma-beta thalassemia studies showing that deletion of the gamma- and delta-genes influences beta-globin expression in man. *Nature* 283: 637–642.

Wagenaar, E. B. and D. Mazia. 1978. The effect of emetine on the first cleavage division of the sea urchin, *Strongylocentrotus purpuratus. In* E. R. Dirksen, D. M. Prescott and L. F. Fox (eds.), *Cell Reproduction: In Honor of Daniel Mazia.* Academic Press, New York, pp. 539–545.

Walker, M. D., T. Edlund, A. M. Boulet and W. J. Rutter. 1983. Cell-specific expression controlled by the 5′ flanking region of the insulin and chymotrypsin genes. *Nature* 306: 557–561.

Weintraub, H. 1984. Histone H1-dependent chromatin superstructures and the suppression of gene activity. *Cell* 38: 17–27.

Weintraub, H. 1985. Assembly and propagation of repressed and derepressed chromosomal states. *Cell* 42: 705–711.

Weintraub, H. and M. Groudine. 1976. Chromosomal subunits in active genes have an altered configuration . *Science* 193: 848–856.

Wickens, M. and K. Takayama. 1995. Deviants—or emissaries. *Nature* 367: 17–18.

Wightman, B., I. Ha and G. Ruvkun. 1993. Posttranslational regulation of the heterochronic gene *lin-14* by *lin-4* mediates temporal pattern formation in *C. elegans. Cell* 75: 855–862.

Williams, S. C., C. R. Altmann, R. L. Chow, A. Hemmati-Brivanlou and R. A. Lang. 1998. A highly conserved lens transcriptional control element from the *Pax-6* gene. *Mech. Dev.* 73: 225–229.

Wilson, T. and R. Treisman. 1988. Removal of poly(A) and consequent degradation of *c-fos* mRNA facilitated by 3′ AU-rich sequences. *Nature* 336: 396–399.

Wold, B. J., W. H. Klein, B. R. Hough-Evans, R. J. Britten and E. H. Davidson. 1978. Sea urchin embryo mRNA sequences expressed in nuclear RNA of adult tissues. *Cell* 14: 941–950.

Wolf, S. F., S. Dintgis, D. Toniolo, G. Persico, K. D. Lunnen, J. Axelman and B. R.Migeon. 1 1984. Complete concordance between glucose-6-phosphate dehydrogenase activity and hypomethylation of 3′ CpG clusters: Implication for X chromosome dosage compensation. *Nucleic Acids Res.* 12: 9333–9348.

Wolfe, S. L. 1993. *Molecular and Cellular Biology.* Wadsworth, Belmont, CA.

Xu, W., M. AS. Rould, S. Jun, C. Desplan and Pabo, C. O. 1995. Crystal structure of a paired domain-DNA complex at 2.5 Å resolution reveals structural basis for *Pax* developmental mutations. *Cell* 80: 639–650.

Yasumoto, K.-I., K. Yokoyama, K. Takahashi, Y. Tomita and S. Shibihara. 1997. Functional analysis of *microphthalmia*-associated transcription factor in pigment cell-specific transcription of human tyrosinase family genes. *J. Biol. Chem.* 272: 503–509.

Zeschingk, M., B. Schmitz, B. Dittrich, K. Buiting, B. Horsthemke and W. Doerfler. 1997. Imprinted segments in the human genome: Different DNA methylation patterns in the Prader-Willi/Angelman syndrome region as determined by the genomic sequencing method. *Hum. Mol. Genet.* 6: 387–395.

chapter 6 Cell-cell communication in development

THE FORMATION OF ORGANIZED BODIES has been one of the great sources of wonder for humankind. Indeed, the "miracle of life" seems just that—matter has become organized in such a way that it lives.* While each organism starts off as a single cell, the progeny of that cell form complex structures—tissues and organs—that are themselves integrated into larger systems. Probably no one better recognizes how remarkable life actually is than the developmental biologists who get to study how all this complexity arises. In the past decade, developmental biologists have started to answer some of the most important questions of natural science: We have begun to understand how organs form.

Induction and Competence

Organs are complex structures composed of numerous types of tissues. In the vertebrate eye, for example, light is transmitted through the transparent corneal tissue and focused by the lens tissue (the diameter of which is controlled by muscle tissue), eventually impinging on the tissue of the neural retina. The precise arrangement of tissues in this organ cannot be disturbed without impairing its function. Such coordination in the construction of organs is accomplished by one group of cells changing the behavior of an adjacent set of cells, thereby causing them to change their shape, mitotic rate, or fate. This kind of interaction at close range between two or more cells or tissues of different history and properties is called proximate interaction, or **induction**.† There are at least two components to every inductive interaction. The first component is the inducer: the tissue that produces a signal (or signals) that changes the cellular behavior of the other tissue. The second component, the tissue being induced, is the responder.

---

*The twelfth-century rabbi and physician Maimonides (1190) framed the question of morphogenesis beautifully when he noted that the pious persons of his day believed that an angel of God had to enter the womb to form the organs of the embryo. How much more powerful a miracle would life be, he asked, if the Deity had made matter such that it could generate such remarkable order without a matter-molding angel having to intervene in every pregnancy? The idea of an angel was still part of the embryology of the Renaissance. The problem addressed today is the secular version of Maimonides' question: How can matter alone construct the organized tissues of the embryo?

†Often, these inductions are called "secondary" inductions, whereas the tissue interactions that generate the neural tube are called "primary embryonic induction." However, there is no difference in the molecular nature of "primary" and "secondary" inductions. Primary embryonic induction will be detailed separately in Chapters 10 and 11.

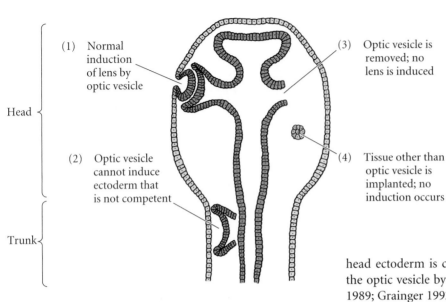

(1) Normal induction of lens by optic vesicle

(2) Optic vesicle cannot induce ectoderm that is not competent

(3) Optic vesicle is removed; no lens is induced

(4) Tissue other than optic vesicle is implanted; no induction occurs

Head

Trunk

Figure 6.1
Ectodermal competence and the ability to respond to the optic vesicle inducer in *Xenopus*. (1) The optic vesicle is able to induce lenses in the anterior portion of the ectoderm, but not in the presumptive trunk and abdomen (2). If the optic vesicle is removed (3), the surface ectoderm forms either an abnormal lens or no lens at all. (4) Most other tissues are not able to substitute for the optic vesicle.

head ectoderm is competent to respond to the signals from the optic vesicle by producing a lens* (Figure 6.1; Saha et al. 1989; Grainger 1992).

This ability to respond to a specific inductive signal is called **competence** (Waddington 1940). Competence is not a passive state, but an actively acquired condition. For example, in the developing chick and mammalian eye, the Pax6 protein

Not all tissues can respond to the signal being produced by the inducer. For instance, if the optic vesicle (presumptive retina) of *Xenopus laevis* is placed in an ectopic location (i.e., in a different place from where it normally forms) underneath the head ectoderm, it will induce that ectoderm to form lens tissue. Only the optic vesicle appears to be able to do this; therefore, it is an inducer. However, if the optic vesicle is placed beneath ectoderm in the flank or abdomen of the same organism, that ectoderm will not be able to respond. Only the

*When describing lens induction, one has to be careful to mention which species one is studying because there are numerous species-specific differences. In some species, induction will not occur at certain temperatures. In other species, the entire ectoderm can respond to the optic vesicle by forming lenses. These species-specific differences have made this area very difficult to study (Jacobson and Sater 1988; Saha et al. 1989, 1991).

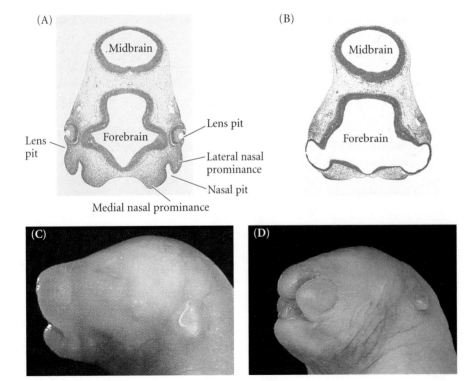

(A)

Midbrain

Forebrain

Lens pit

Lens pit

Lateral nasal prominance

Nasal pit

Medial nasal prominance

(B)

Midbrain

Forebrain

(C)

(D)

Figure 6.2
Induction of optic and nasal structures by Pax6 in the rat embryo. (A, B) Histology of wild-type (A) and homozygous *Pax6* mutant (B) embryos at day 12 of gestation shows induction of lenses and retinal development in the wild-type embryo, but neither lens nor retina in the mutant. Similarly, neither the nasal pit nor the medial nasal prominence is induced in the mutant rats. (C) Newborn wild-type rats show prominent nose as well as (closed) eyes. (D) Newborn *Pax6* mutant rats show neither eyes nor nose. (From Fujiwara et al. 1994; photographs courtesy of M. Fujiwara.)

appears to be important in making the ectoderm competent to respond to the inductive signal from the optic vesicle. Pax6 expression is seen in the head ectoderm, which can respond to the optic vesicle by forming lenses, and it is not seen in other regions of the surface ectoderm (See Figure 4.16; Li et al. 1994). Moreover, the importance of Pax6 as a **competence factor** was demonstrated by recombination experiments using embryonic rat eye tissue (Fujiwara et al. 1994). The homozygous Pax6-mutant rat has a phenotype similar to the homozygous Pax6-mutant mouse (see Chapter 4), lacking eyes and nose. It has been shown that part of this phenotype is due to the failure of lens induction (Figure 6.2). But which is the defective component—the optic vesicle or the surface ectoderm? When head ectoderm from Pax6-mutant rat embryos was combined with a wild-type optic vesicle, no lenses were formed. However, when the head ectoderm from wild-type rat

embryos was combined with a Pax6-mutant optic vesicle, lenses formed normally (Figure 6.3). Therefore, Pax6 is needed for the surface ectoderm to respond to the inductive signal from the optic vesicle. The inducing tissue does not need it. It is not known how Pax6 becomes expressed in the anterior ectoderm of the embryo, although it is thought that its expression is induced by the anterior regions of the neural plate. Competence to respond to the optic vesicle inducer can be conferred on ectodermal tissue by incubating it next to anterior neural plate tissue (Henry and Grainger 1990; Li et al. 1994; Zygar et al. 1998).

Thus, there is no single inducer of the lens. Studies on amphibians suggest that the first inducers may be the pharyngeal endoderm and heart-forming mesoderm that underlie the lens-forming ectoderm during the early- and mid-gastrula stages (Jacobson 1963, 1966). The anterior neural plate may produce the next signals, including a signal that promotes the synthesis of Pax6 in the anterior ectoderm (Zygar et al. 1998; Figure 6.4). Thus, the optic vesicle appears to be *the* inducer, but the anterior ectoderm has already been induced by at least two other factors. (The situation is like that of the player who kicks *the* "winning goal" of a soccer match.) The optic vesicle appears to secrete two induction factors, one of which is BMP4 (see the discussion below), a protein that induces the transcription of the Sox2 and Sox3 transcription factors (and another, as yet unidentified, signal that induces the appearance of the L-Maf transcription factor; Ogino and Yasuda 1998). The combination of Pax6, Sox2, Sox3, and L-Maf ensures the production of the lens.

Cascades of induction: Reciprocal and sequential inductive events

Another feature of induction is the reciprocal nature of many inductive interactions. Once the lens has formed, it can then induce other tissues. One of these responding tissues is the optic vesicle it-

Optic vesicles	Surface ectoderm	Lens induction
Wild-type	Wild-type	Yes
Pax6⁻/Pax6⁻	Wild-type	Yes
Wild-type	Pax6⁻/Pax6⁻	No
Pax6⁻/Pax6	Pax6⁻/Pax6⁻	No

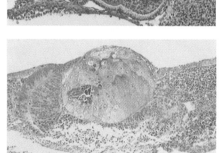

Lens

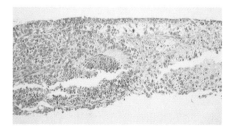

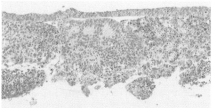

Figure 6.3
Recombination experiments showing that the induction deficiency of Pax6-deficient rats is caused by the inability of the surface ectoderm to respond to the optic vesicle. (Photographs courtesy of M. Fujiwara.)

(A)

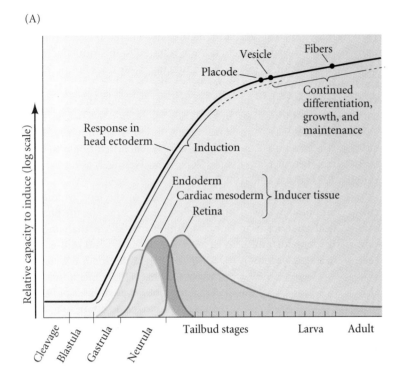

Figure 6.4

Lens induction in embryonic amphibians. (A) The additive effects of inducers, as shown by transplantation and extirpation (removal) experiments on the salamander *Tarichosa torosa*. The ability to produce lens tissue is first induced by pharyngeal endoderm, then by cardiac mesoderm, and finally by the optic vesicle. The competence of the lens ectoderm to respond to these inducers increases logarithmically from the early gastrula through the tailbud larval stages. (B) Sequence of induction postulated by similar experiments performed on embryos of the frog *Xenopus laevis*. Unidentified inducers (possibly from the pharyngeal endoderm and heart-forming mesoderm) cause the synthesis of the Otx-2 transcription factor in the head ectoderm during the late gastrula stage. As the neural folds rise, inducers from the anterior neural plate (including the region that will form the retina) induce *Pax6* expression in the anterior ectoderm that can form lens tissue. Expression of the Pax6 transcription factor may constitute the competence of the surface ectoderm to respond to the optic vesicle during the late neurula stage. The optic vesicle secretes factors (probably of the BMP family) that induce the synthesis of the Sox transcription factors and initiate observable lens formation. (A after Jacobson 1966; B after Grainger 1992.)

(B)

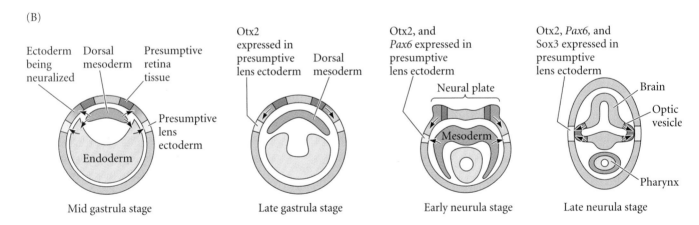

self. Now the inducer becomes the induced. Under the influence of factors secreted by the lens, the optic vesicle becomes the optic cup, and the wall of the optic cup differentiates into two layers, the pigmented retina and the neural retina (Figure 6.5; Cvekl and Piatigorsky 1996). Such interactions are called **reciprocal inductions**.

At the same time, the lens is also inducing the ectoderm above it to become the cornea. Like the lens-forming ectoderm, the cornea-forming ectoderm has achieved a particular competence to respond to inductive signals, in this case the signals from the lens (Meier 1977). Under the influence of the lens, the corneal ectodermal cells become columnar and secrete multiple layers of collagen. Mesenchymal cells from the neural crest use this collagen matrix to enter the area and se-

crete a set of proteins (including the enzyme hyaluronidase) that further differentiate the cornea. A third signal, the hormone thyroxine, dehydrates the tissue and makes it transparent (see Hay 1980; Bard 1990). Thus, there are sequential inductive events, and multiple causes for each induction.

Instructive and permissive interactions

Howard Holtzer (1968) distinguished two major modes of inductive interaction. In **instructive interaction**, a signal from the inducing cell is necessary for initiating new gene expression in the responding cell. Without the inducing cell, the responding cell would not be capable of differentiating in that particular way. For example, when the optic vesicle is experimentally placed under a new region of the head ectoderm and

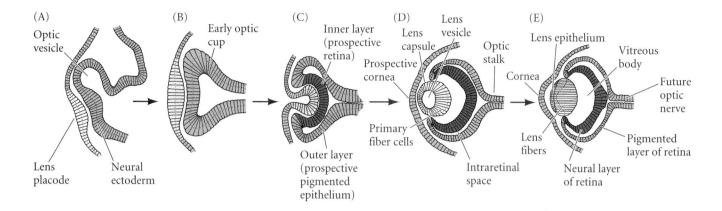

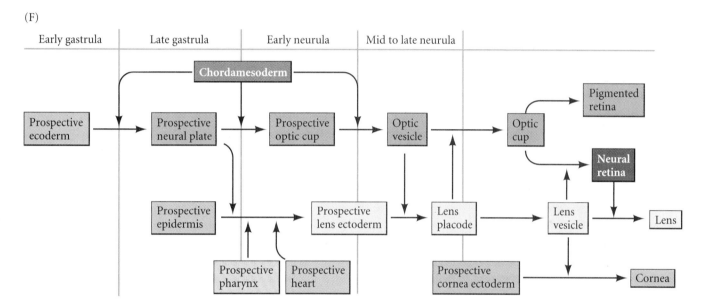

Figure 6.5
Schematic diagram of the induction of the mouse lens. (A) At embryonic day 9, the optic vesicle extends toward the surface ectoderm from the forebrain. The lens placode (the prospective lens) appears as a local thickening of the surface ectoderm near the optic vesicle. (B) By the middle of day 9, the lens placode has enlarged and the optic vesicle has formed an optic cup. (C) By the middle of day 10, the central portion of the lens-forming ectoderm invaginates, while the two layers of the retina become distinguished. (D) By the middle of day 11, the lens vesicle has formed, and by day 13 (E), the lens consists of anterior cuboidal epithelial cells and elongating posterior fiber cells. The cornea develops in front of the lens. (F) Summary of some of the inductive interactions during eye development. (A–E after Cvekl and Piatigorsky 1996.)

causes that region of the ectoderm to form a lens, that is an instructive interaction. Wessells (1977) has proposed three general principles characteristic of most instructive interactions:

1. In the presence of tissue A, responding tissue B develops in a certain way.

2. In the absence of tissue A, responding tissue B does not develop in that way.

3. In the absence of tissue A, but in the presence of tissue C, tissue B does not develop in that way.

The second type of induction is **permissive interaction**. Here, the responding tissue contains all the potentials that are to be expressed, and needs only an environment that allows the expression of these traits.* For instance, many tissues need a solid substrate containing fibronectin or laminin in order to develop. The fibronectin or laminin does not alter the type of cell that is to be produced, but only enables what has been determined to be expressed.

---

*It is easy to distinguish permissive and instructive interactions by an analogy with a more familiar situation. This textbook is made possible by both permissive and instructive interactions. The reviewers can convince me to change the material in the chapters. This is an instructive interaction, as the information in the book is changed from what it would have been. However, the information in the book could not be expressed without permissive interactions with the publisher and printer.

Table 6.1 Some epithelial-mesenchymal interactions

Organ	Epithelial component	Mesenchymal component
Cutaneous structures (hair, feathers, sweat glands, mammary glands)	Epidermis (ectoderm)	Dermis (mesoderm)
Limb	Epidermis (ectoderm)	Mesenchyme (mesoderm)
Gut organs (liver, pancreas, salivary glands)	Epithelium (endoderm)	Mesenchyme (mesoderm)
Pharyngeal and respiratory associated organs (lungs, thymus, thyroid)	Epithelium (endoderm)	Mesenchyme (mesoderm)
Kidney	Ureteric bud epithelim (mesoderm)	Mesenchyme (mesoderm)
Tooth	Jaw epithelium (ectoderm)	Mesenchyme (neural crest)

Epithelial-mesenchymal interactions

Some of the best-studied cases of induction are those involving the interactions of sheets of epithelial cells with adjacent mesenchymal cells. These interactions are called **epithelial-mesenchymal interactions**. Epithelia are sheets or tubes of connected cells; they can originate from any germ layer. Mesenchyme refers to loosely packed, unconnected cells. Mesenchymal cells are derived from the mesoderm or neural crest. All organs consist of an epithelium and an associated mesenchyme, so epithelial-mesenchymal interactions are among the most important phenomena in nature. Some examples are listed in Table 6.1.

REGIONAL SPECIFICITY OF INDUCTION. Using the induction of cutaneous structures as our examples, we will look at the properties of epithelial-mesenchymal interactions. The first of these properties is the regional specificity of induction. Skin is composed of two main tissues: an outer epidermis, an epithelial tissue derived from ectoderm, and a dermis, a mesenchymal tissue derived from mesoderm. The chick epidermis signals the underlying dermal cells to form condensations (probably by secreting Sonic hedgehog and TGF-β2 proteins, which will be discussed below), and the condensed dermal mesenchyme responds by secreting factors that cause the epidermis to form regionally specific cutaneous structures (Figure 6.6; Nohno et al. 1995, Ting-Berreth and Chuong 1996). These structures can be the broad feathers of the wing, the narrow feathers of the thigh, or the scales and claws of the feet. Researchers can separate the embryonic epithelium and mesenchyme from each other and recombine them in different ways (Saunders et al. 1957). As Figure 6.7 demonstrates, the dermal mesenchyme is responsible for the regional specificity of induction in the competent epidermal epithelium. The same type of epithelium develops cutaneous structures according to the region from which the mesenchyme was taken. Here, the mesenchyme plays an instructive role, calling into play different sets of genes in the responding epithelial cells.

GENETIC SPECIFICITY OF INDUCTION. The second property of epithelial-mesenchymal interactions is the genetic specificity of induction. Whereas the mesenchyme may instruct the epithelium as to what sets of genes to activate, the responding epithelium can comply with these instructions only so far as its genome permits. This property was discovered through experiments involving

(A)

(B)

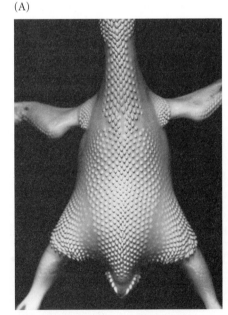
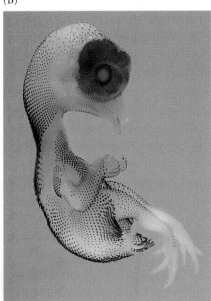

Figure 6.6
(A) Feather tracts on the dorsum of a day 9 chick embryo. Note that each feather primordium is located between the primordia of adjacent rows. (B) In situ hybridization of a day 10 chick embryo shows Sonic hedgehog expression (dark spots) in the ectoderm of the developing feathers and scales. (A courtesy of P. Sengal; B courtesy of W.-S. Kim and J. F. Fallon.)

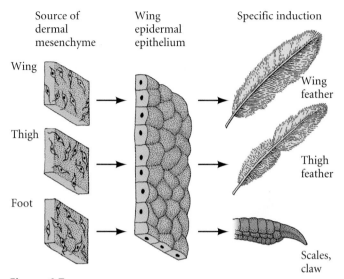

Figure 6.7
Regional specificity of induction. When cells from different regions of the dermis (mesenchyme) are recombined with the epidermis (epithelium) in the chick, the type of cutaneous structure made by the epidermal epithelium is determined by the original source of the mesenchyme. (After Saunders 1980.)

the transplantation of tissues from one species to another. In one of the most dramatic examples of interspecific induction, Hans Spemann and Oscar Schotté (1932) transplanted flank ectoderm from an early frog gastrula to the region of a newt gastrula destined to become parts of the mouth. Similarly, they placed presumptive flank ectodermal tissue from a newt gastrula into the presumptive oral regions of frog embryos. The structures of the mouth region differ greatly between salamander and frog larvae. The salamander larva has club-shaped balancers beneath its mouth, whereas the frog tadpole produces mucus-secreting glands and suckers (Figure 6.8). The frog tadpole also has a horny jaw without teeth, whereas the salamander has a set of calcareous teeth in its jaw. The larvae resulting from the transplants were chimeras. The salamander larvae had froglike mouths, and the frog tadpoles had salamander teeth and balancers. In other words, the mesodermal cells instructed the ectoderm to make a mouth, but the ectoderm responded by making the only kind of mouth it "knew" how to make, no matter how inappropriate.*

Thus, the instructions sent by the mesenchymal tissue can cross species barriers. Salamanders respond to frog signals, and chick tissue responds to mammalian inducers. The response of the epithelium, however, is species-specific. So, whereas organ type specificity (e.g., feather or claw) is usually controlled by the mesenchyme within a species, species specificity is usually controlled by the responding epithelium. As

*Spemann is reported to have put it this way: "The ectoderm says to the inducer, 'you tell me to make a mouth; all right, I'll do so, but I can't make your kind of mouth; I can make my own and I'll do that'" (quoted in Harrison 1933).

we will see in Chapters 21 and 22, large evolutionary changes can be brought about by changing the response to a particular inducer.

WEBSITE **6.1 Hen's teeth.** Some inductive events between species can bring forth lost structures. Mouse molar mesenchyme may be able to induce teeth in the bird jaw.

Paracrine Factors

How are the signals between inducer and responder transmitted? While studying the mechanisms of induction that produce the kidney tubules and teeth, Grobstein (1956) and others (Saxén et al. 1976; Slavkin and Bringas 1976) found that some inductive events could occur despite a filter separating the epithelial and mesenchymal cells. Other inductions, however, were blocked by the filter. The researchers therefore concluded that some of the inductive molecules were soluble factors that could pass through the small pores of the filter, and that other inductive events required physical contact between the epithelial and mesenchymal cells (Figure 6.9). When cell membrane proteins on one cell surface interact with receptor proteins on adjacent cell surfaces, these events are called **juxtacrine inter-**

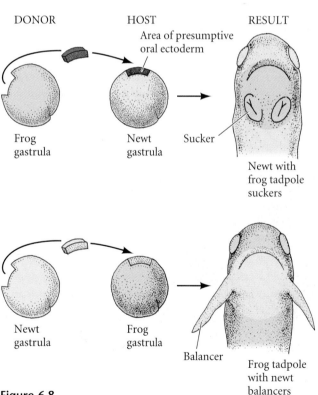

Figure 6.8
Genetic specificity of induction. Reciprocal transplantation between the presumptive oral ectoderm regions of salamander and frog gastrulae leads to newts with tadpole suckers and tadpoles with newt balancers. (After Hamburgh 1970.)

Figure 6.9

Mechanisms of inductive interaction. (A) Paracrine induction. Presumptive mouse lens ectoderm and mesenchyme were placed on a filter. Retinal tissue was placed beneath it. After 3 days, a lens had developed from the surface ectoderm. In the absence of a signal from the retinal tissue, the surface ectoderm would have become epidermal. (B-D) Paracrine and juxtracrine modes of signaling. (B) Paracrine modes of signaling involve the secretion of diffusible molecules from one cell for reception by another cell nearby. (C) In some cases, the paracrine signal can come from an extracellular matrix protein secreted by a cell. (D) In juxtracine interactions, contact is made between a signaling molecule on the surface of one cell and its receptor on another cell. (A from Muthukkarapan 1965, photograph courtesy of R. Auerbach; B–D after Grobstein 1956.)

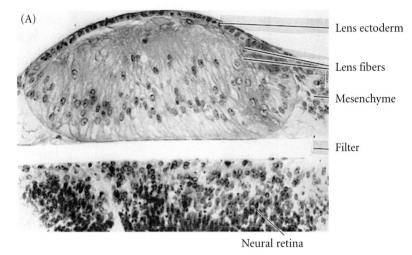

(A)

Lens ectoderm

Lens fibers

Mesenchyme

Filter

Neural retina

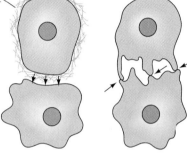

(B)

(C)

Matrix

(D)

Paracrine factors

Diffusion of inducers from one cell to another

Matrix of one cell induces change in another cell

Contact (arrows) between the inducing and responding cells

actions (since the cell membranes are juxtaposed). When proteins synthesized by one cell can diffuse over small distances to induce changes in neighboring cells, the event is called a **paracrine interaction**, and the diffusible proteins are called **paracrine factors** or **growth and differentiation factors** (**GDFs**). We will consider paracrine interactions first and then return to juxtracrine interactions later in the chapter.

Whereas *endocrine* factors (hormones) travel through the blood to exert their effects, *paracrine* factors are secreted into the immediate spaces around the cell producing them.* These proteins are the "inducing factors" of the classic experimental embryologists. During the past decade, developmental biologists have discovered that the induction of numerous organs is actually effected by a relatively small set of paracrine factors. The embryo inherits a rather compact "tool kit" and uses many of the same proteins to construct the heart, the kidneys, the teeth, the eyes, and other organs. Moreover, the same proteins are utilized throughout the animal kingdom; the factors active in creating the *Drosophila* eye or heart are very similar to those used in generating mammalian organs. Many of these paracrine factors can be grouped into four major families on the basis of their structures. These families are the fibroblast growth factor (FGF) family, the Hedgehog family, the Wingless (Wnt) family, and the TGF–β superfamily.

---

*There is considerable debate as to how far paracrine factors can operate. Activin, for instance, can diffuse over many cell diameters and can induce different sets of genes at different concentrations (Gurdon et al. 1994, 1995). The Vg1, BMP4, and Nodal proteins, however, probably work only on their adjacent neighbors (Jones et al. 1996; Reilly and Melton 1996). These factors may induce the expression of other short-range factors from these neighbors, and a cascade of paracrine inductions can be initiated.

In addition to endocrine, paracrine, and juxtracrine regulation, there is also *autocrine* regulation. Autocrine regulation occurs when the same cells that secrete paracrine factors also respond to them. In this case, the cell synthesizes a molecule for which it has its own receptor. Although autocrine regulation is not common, it is seen in placental cytotrophoblast cells; these cells synthesize and secrete platelet-derived growth factor, whose receptor is on the cytotrophoblast cell membrane (Goustin et al. 1985). The result is the explosive proliferation of that tissue.

The fibroblast growth factors

The **fibroblast growth factor** (**FGF**) family currently has over a dozen structurally related members. FGF1 is also known as acidic FGF; FGF2 is sometimes called basic FGF; and FGF7 sometimes goes by the name of keratinocyte growth factor. Over a dozen distinct FGF genes are known in vertebrates, and they can generate hundreds of protein isoforms by varying their RNA splicing or initiation codons in different tissues (Lappi 1995). FGFs can activate a set of receptor tyrosine kinases called the **fibroblast growth factor receptors** (**FGFRs**). As we will discuss later in this chapter, receptor tyrosine kinases are proteins that extend through the cell membrane (Figure 6.10A). On the extracellular side is the portion of the protein that binds the paracrine factor. On the intracellular side is a dormant tyrosine kinase (i.e., a protein that can phosphorylate another protein by splitting ATP). When the FGF receptor binds an FGF (and only when it binds an FGF), the dormant kinase is activated, and it phosphorylates certain proteins within the responding cell. The proteins are now activated and can perform new functions. FGFs are associated with several developmental functions, including angiogenesis

(A)

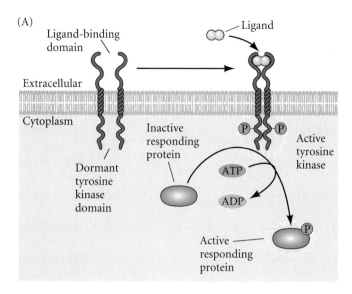

(B)

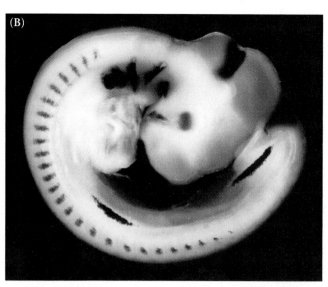

Figure 6.10

FGF expression. (A) Structure of a receptor tyrosine kinase. The dormant tyrosine kinase is activated by the binding of FGF by the extracellular portion of the receptor protein. This enzyme activity phosphorylates specific tyrosine residues of certain proteins. (B) FGF8 expression in the 3-day chick embryo, shown by in situ hybridization. FGF8 expression (dark areas) is seen in the distalmost limb bud ectoderm, in the somitic mesoderm (the segmented blocks of cells along the anterior-posterior axis), in the branchial arches of the neck, at the boundary between the midbrain and hindbrian, and in the tail. (B courtesy of E. Laufer, C.-Y. Yeo, and C. Tabin.)

(blood vessel formation), mesoderm formation, and axon extension. While FGFs can often substitute for one another, their expression patterns give them separate functions. FGF2 is especially important in angiogenesis, and FGF8 is important for the development of the midbrain and limbs (Figure 6.10B; Crossley et al. 1996).

> WEBSITE **6.2 FGF binding.** The binding of FGFs to their receptors is a complex acrobatic act involving an interesting cast of cell surface molecules. Glycoproteins play a major supporting role in this event.

The Hedgehog family

The **Hedgehog** proteins constitute a family of paracrine factors that are often used by the embryo to induce particular cell types and to create boundaries between tissues. Vertebrates have at least three homologues of the Drosophila *hedgehog* gene: *sonic hedgehog* (*shh*), *desert hedgehog* (*dhh*), and *indian hedgehog* (*ihh*). Desert hedgehog is expressed in the Sertoli cells of the testes, and mice homozygous for a null allele of *dhh* exhibit defective spermatogenesis. Indian hedgehog is expressed in the gut and in cartilage and is important in postnatal bone growth (Bitgood and McMahon 1995; Bitgood et al. 1996).

Sonic hedgehog* is the most widely used of the three vertebrate homologues. Made by the notochord, it is processed so that only the amino-terminal two-thirds of the molecule is secreted. This peptide is responsible for patterning the neural tube such that motor neurons are formed from the ventral neurons and sensory neurons are formed from the dorsal neurons (see Chapter 12; Yamada et al. 1993). Sonic hedgehog is also responsible for patterning the somites so that the portion of the somite closest to the notochord becomes the cartilage of the spine (Fan and Tessier-Lavigne 1994; Johnson et al. 1994). As we will see in later chapters, Sonic hedgehog has been shown to mediate the formation of the left-right axis in chicks, to initiate the anterior-posterior axis in limbs, to induce the regionally specific differentiation of the digestive tube, and to induce feather formation (see Figures 6.11 and 6.6). Sonic hedgehog often works with other paracrine factors, such as Wnt and FGF proteins. In the developing tooth, Sonic hedgehog, FGF4, and other paracrine factors are concentrated in the region where cell interactions are creating the cusps of the teeth (see Figure 13.9; Vaahtokari et al. 1996a).

> WEBSITE **6.3 Functions of the Hedgehog family.** While Sonic hedgehog is used to induce and specify numerous tissues in the embryo, Desert hedgehog and Indian hedgehog are used postnatally to regulate bone growth and sperm production.

---

*Yes, it is named after the Sega Genesis character. The original *hedgehog* gene was found in *Drosophila*, in which genes are named after their mutant phenotype. The loss-of-function *hedgehog* mutation in *Drosophila* causes the fly embryo to be covered with pointy denticles on its cuticle. Hence, it looks like a hedgehog. The vertebrate *hedgehog* genes were discovered by searching chick gene libraries with probes that would find sequences similar to that of the fruit fly *hedgehog* gene. Riddle and his colleagues in Cliff Tabin's laboratory (1993) discovered three genes homologous to the *Drosophila hedgehog*. Two were named after species of hedgehogs, the third was named after the cartoon character.

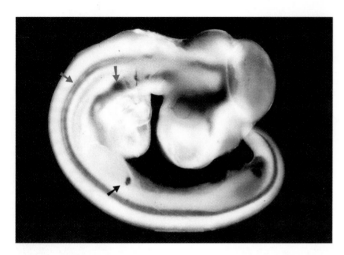

Figure 6.11
The *sonic hedgehog* gene is shown by in situ hybridization to be expressed in the 3-day embryonic chick nervous system (red arrow), gut (blue arrow), and limb bud (black arrow). (Photograph courtesy of C. Tabin.)

The Wnt family

The **Wnt**s constitute a family of cysteine-rich glycoproteins. There are at least 15 members of this family in vertebrates. Their name comes from fusing the name of the *Drosophila* segment polarity gene *wingless* with the name of one of its vertebrate homologues, *integrated*. While Sonic hedgehog is important in patterning the ventral portion of the somites (causing the cells to become cartilage), Wnt1 appears to be active in inducing the dorsal cells of the somites to become muscle (McMahon and Bradley 1990; Stern et al. 1995). Wnt proteins also are critical in establishing the polarity of insect and vertebrate limbs, and they are used in several steps of urogenital system development (Figure 6.12).

> WEBSITE 6.4 **Wnts: An ancient family.** The biochemistry of the Wnt proteins and the mechanism of their actions is a fascinating tale of theme and variations. The Wnt family may be one of the oldest group of signaling molecules in the animal kingdom.

Figure 6.12
Wnt proteins play several roles in the development of the urogenital organs. Wnt4 is necessary for kidney development and for female sex determination. (A) Whole-mount in situ hybirdization of Wnt4 expression in a 14-day embryonic male urogenital rudiment. Expression (dark blue staining) is seen in the mesenchyme that condenses to form the kidney's nephrons. (B) The urogenital rudiment of a wild-type newborn female mouse. (C) The urogenital rudiment of a targeted knockout of the Wnt4 genes in a female mouse shows that the kidney fails to develop. The ovary also starts synthesizing testosterone and becomes surrounded by a modified male duct system. (Photographs courtesy of J. Perasaari and S. Vainio.)

The TGF-β superfamily

There are over 30 structurally related members of the **TGF-β superfamily**,* and they regulate some of the most important interactions in development (Figure 6.13). The proteins encoded by TGF-β superfamily genes are processed such that the carboxy-terminal region contains the mature peptide. These peptides are dimerized into homodimers (with themselves) or heterodimers (with other TGF-β peptides) and are secreted from the cell. The TGF-β superfamily includes the TGF-β family, the activin family, the bone morphogenetic proteins (BMPs), the Vg1 family, and other proteins, including glial-derived neurotrophic factor (necessary for kidney and enteric neuron differentiation) and Müllerian inhibitory factor (which is involved in mammalian sex determination).

TGF-β family members TGF-β1, 2, 3, and 5 are important in regulating the formation of the extracellular matrix between cells and for regulating cell division (both positively and negatively). TGF-β1 increases the amount of extracellular matrix epithelial cells make (both by stimulating collagen and

---

*TGF stands for *transforming growth factor*. The designation *superfamily* is often given when each of the different classes of molecules constitutes a "family." The members of a superfamily all have similar structures, but are not as close as the molecules within a family are to one another.

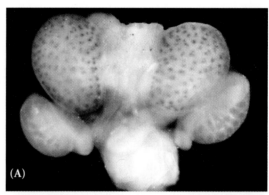

(A)

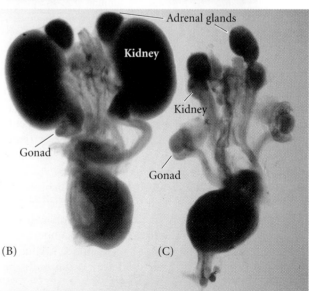

Adrenal glands

Kidney

Kidney

Gonad

Gonad

(B)   **(C)**

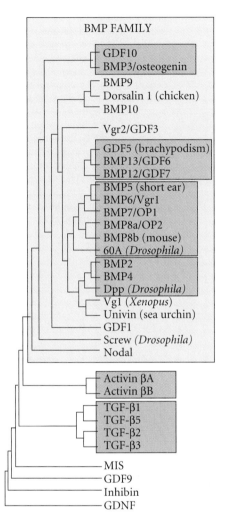

BMP FAMILY

GDF10
BMP3/osteogenin
BMP9
Dorsalin 1 (chicken)
BMP10
Vgr2/GDF3
GDF5 (brachypodism)
BMP13/GDF6
BMP12/GDF7
BMP5 (short ear)
BMP6/Vgr1
BMP7/OP1
BMP8a/OP2
BMP8b (mouse)
60A *(Drosophila)*
BMP2
BMP4
Dpp *(Drosophila)*
Vg1 *(Xenopus)*
Univin (sea urchin)
GDF1
Screw *(Drosophila)*
Nodal

Activin βA
Activin βB
TGF-β1
TGF-β5
TGF-β2
TGF-β3
MIS
GDF9
Inhibin
GDNF

Figure 6.13
Relationships among members of the TGF-β superfamily. (After Hogan 1996.)

fibronectin synthesis and by inhibiting matrix degradation). TGF-βs may be critical in controlling where and when epithelia can branch to form the ducts of kidneys, lungs, and salivary glands (Daniel 1989; Hardman et al. 1994; Ritvos et al. 1995). The effects of the individual TGF–β family members are difficult to sort out, because members of the TGF–β family appear to function similarly and can compensate for losses of the others when expressed together. Moreover, targeted deletions of the *Tgf-β1* gene in mice are difficult to interpret, since the mother can supply this factor through the placenta and milk (Letterio et al. 1994).

The members of the **BMP** family were originally discovered by their ability to induce bone formation; hence, they are the *bone morphogenetic proteins*. Bone formation, however, is only one of their many functions, and they have been found to regulate cell division, apoptosis (programmed cell death), cell migration, and differentiation (Hogan 1996). BMPs can

be distinguished from other members of the TGF-β superfamily by their having seven, rather than nine, conserved cysteines in the mature polypeptide. The BMPs include proteins such as Nodal (responsible for left-right axis formation) and BMP4 (important in neural tube polarity, eye development, and cell death; see Figure 4.21). (As it turns out, BMP1 is not a member of the family; it is a protease.) The *Drosophila* Decapentaplegic protein is homologous to the vertebrate BMP4, and human BMP4 can replace the *Drosophila* homologue, rescuing those flies deficient in Dpp (Padgett et al. 1993).

Other paracrine factors

Although most of the paracrine factors are members of the above-mentioned four families, some have few or no close relatives. Factors such as epidermal growth factor, hepatocyte growth factor, neurotrophins, and stem cell factor are not in the above-mentioned families, but each plays important roles during development. In addition, there are numerous factors involved almost exclusively with developing blood cells: erythropoietin, the cytokines, and the interleukins. These factors will be discussed when we detail blood cell formation in Chapter 14.

Cell Surface Receptors and Their Signal Transduction Pathways

The paracrine factors are inducer proteins. We now turn to the molecules involved in the *response* to induction. These molecules include the receptors in the membrane of the responding cell, which binds the paracrine factor, and the cascade of interacting proteins that transmit a signal through a pathway from the bound receptor to the nucleus. These pathways between the cell membrane and the genome are called **signal transduction pathways**. Several types of signal transduction pathways have been discovered, and we will outline some of the major ones here. As you will see, they appear to be variations on a common and rather elegant theme: Each receptor spans the cell membrane and has an extracellular region, a transmembrane region, and a cytoplasmic region. When a ligand (the paracrine factor) binds its receptor in the extracellular region, the ligand induces a conformational change in the receptor's structure. This shape change is transmitted through the membrane and changes the shape of the cytoplasmic domains. The conformational change in the cytoplasmic domains gives them enzymatic activity—usually a kinase activity that can use ATP to phosphorylate proteins, including the receptor molecule itself. The active receptor can now catalyze reactions that phosphorylate other proteins, and this phosphorylation activates their latent activities in turn. Eventually, the cascade of phosphorylation activates a dormant transcription factor, which activates (or represses) a particular set of genes.

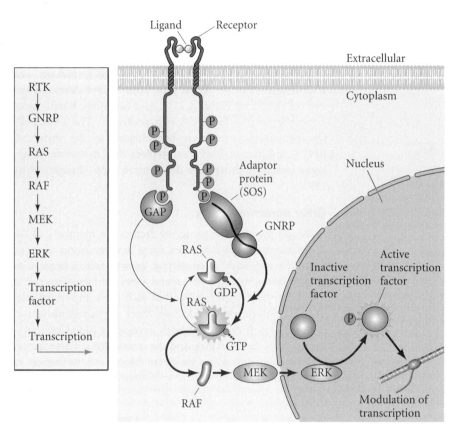

Figure 6.14
The widely used RTK signal transduction pathway. The receptor tyrosine kinase is dimerized by the ligand, which causes the autophosphorylation of the receptor. The adaptor protein recognizes the phosphorylated tyrosines on the RTK and activates an intermediate protein, GNRP, which activates the Ras G protein by allowing the phosphorylation of the GDP-bound Ras. At the same time, the GAP protein stimulates the hydrolysis of this phosphate bond, returning Ras to its inactive state. The active Ras activates protein kinase C (PKC), which in turn phosphorylates a series of kinases. Eventually, the activated ERK kinase alters gene expression in the responding cell by phosphorylating certain transcription factors (which can then enter the nucleus to change the types of genes transcribed) and certain translation factors (which alter the level of protein synthesis). In many cases, this pathway is reinforced by the release of calcium ions. A simplified version of the pathway is depicted on the left.

The RTK pathway

The RTK signal transduction pathway was one of the first pathways to unite various areas of developmental biology. Researchers studying *Drosophila* eyes, nematode vulvae, and human cancers found that they were all studying the same genes. The RTK-Ras pathway begins at the cell surface, where a **receptor tyrosine kinase** (**RTK**) binds its specific ligand. Ligands that bind to RTKs include the fibroblast growth factors, epidermal growth factors, platelet-derived growth factors, and stem cell factor. Each RTK can bind only one or a small set of these ligands. (Stem cell factor, for instance, will bind to only one RTK, the Kit protein.) The RTK spans the cell membrane, and when it binds its ligand, it undergoes a conformational change that enables it to dimerize with another RTK. This conformational change activates the latent kinase activity of each RTK, and these receptors phosphorylate each other on particular tyrosine residues. Thus, the binding of the ligand to the receptor causes the autophosphorylation of the cytoplasmic domain of the receptor.

The phosphorylated tyrosine on the receptor is then recognized by an adaptor protein (Figure 6.14). The adaptor protein serves as a bridge that links the phosphorylated RTK to a powerful intracellular signaling system. While binding to the phosphorylated RTK through one of its cytoplasmic domains, the adaptor protein also activates a **G protein**. Normally, the G protein is in an inactive, GDP–bound state. The activated receptor stimulates the adaptor protein to activate the **guanine nucleotide releasing factor** This protein exchanges a phosphate from a GTP to transform the bound GDP into GTP. The GTP-bound G protein is an active form that transmits the signal. After delivering the signal, the GTP on the G protein is hydrolyzed back into GDP. This catalysis is greatly stimulated by the complexing of the Ras protein with the **GTPase-activating protein** (**GAP**). In this way, the G protein is returned to its inactive state, where it can await further signaling. One of the major G proteins is called Ras; mutations in the *RAS* gene account for a large proportion of human tumors (Shih and Weinberg 1982), and the mutations of *RAS* that make it oncogenic all inhibit the binding of the GAP protein. Without the GAP protein, Ras protein cannot catalyze GTP well, and so remains in its active configuration (Cales et al. 1988; McCormick 1989).

The active G protein associates with a kinase called Raf. The G protein recruits the inactive Raf protein to the cell membrane, where it becomes active (Leevers et al. 1994; Stokoe et al. 1994). The Raf protein is a kinase that activates the MEK protein by phosphorylating it. MEK is itself a kinase, which activates ERK by phosphorylation. And ERK is a kinase that can enter the nucleus and phosphorylate certain transcription factors. This pathway is critical in numerous developmental processes.

(A)

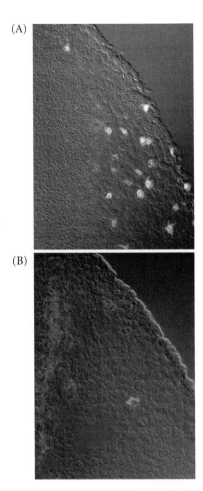

(B)

Figure 6.15
Activation of the Mitf transcription factor through the binding of stem cell factor by the Kit RTK protein. The information received at the cell membrane is sent to the nucleus by the RTK signal transduction pathway. (A, B) Demonstration that Kit protein and Mitf are present in the same cells. Antibodies to these proteins stain the Kit protein (red) and Mitf (green). The overlap is yellow or yellow-green. They are both present in the migrating melanocyte precursor cells (melanoblasts). (A) is from wild-type mouse embryo at day 10.5. (B) is from a *Mitf*-mutant embryo at the same day. The lack of melanoblasts in the mutant is due to the relative absence of Mitf. (C) Signal transduction pathway leading from the cell membrane to the nucleus. When the Kit protein binds the Steel paracrine factor, Kit dimerizes and becomes phosphorylated. This phosphorylation is used to activate the Ras G protein, which activates the chain of kinases that will phosphorylate the Mitf protein. Once phosphorylated, Mitf can bind the cofactor p300/CBP, acetylate the nucleosome histones, and initiate transcription of the genes for melanocyte development. (A, B from Nakayama et al. 1998; photographs courtesy of H. Arnheiter; C after Price et al. 1998.)

(C)

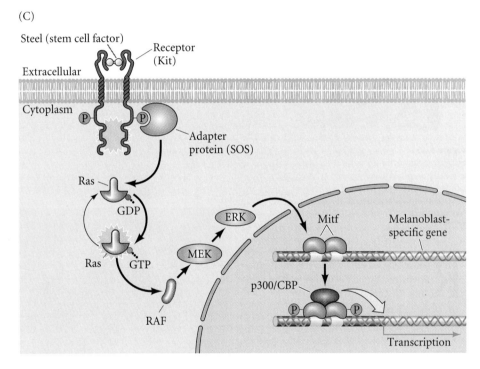

In the migrating neural crest cells of humans and mice, the RTK pathway is important in activating the microphthalmia transcription factor (Mitf) to produce the pigment cells. We have been following the Mitf transcription factor for the past two chapters. It is transcribed in the pigment-forming melanoblast cells that migrate from the neural crest into the skin and in the melanin-forming cells of the pigmented retina. But we have not yet discussed what proteins signal this transcription factor to become active. The clue lay in two mouse mutants whose phenotypes resemble those of mice homozygous for microphthalmia mutations. Like those mice, homozygous *White* mice and homozygous *Steel* mice are white because their pigment cells have failed to migrate. Perhaps all three genes (*Mitf*, *Steel*, and *White*) are on the same

developmental pathway. In 1990, several laboratories demonstrated that the *Steel* gene encodes a paracrine protein called **stem cell factor** (see Witte 1990). Stem cell factor binds to and activates the **Kit** receptor tyrosine kinase encoded by the *White* gene (Spritz et al. 1992; Wu et al. 2000). The binding of stem cell factor to the Kit RTK dimerizes the Kit protein, causing it to become phosphorylated. The phosphorylated Kit activates the pathway whereby phosphorylated ERK is able to phosphorylate the Mitf transcription factor (Hsu et al. 1997; Hemesath et al. 1998). Only the phosphorylated form of Mitf is able to bind the p300/CBP coactivator protein that enables it to activate transcription of the genes encoding tyrosinase and other proteins of the melanin-formation pathway (Figure 6.15; Price et al. 1998).

The RTK Pathway and Cell-to-Cell Induction

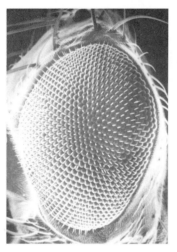

Figure 6.16
Scanning electron micrograph of a compound eye in *Drosophila*. Each facet is a single ommatidium. A sensory bristle projects from each ommatidium. (Photograph courtesy of T. Venkatesh.)

R ECENT RESEARCH into the development of *Drosophila* and *Caenorhabditis elegans* has shown that induction does indeed occur on the cell-to-cell level. Some of the best-studied examples involve the formation of the retinal photoreceptors in the *Drosophila* eye and the formation of the vulva in *C. elegans*. Remarkably, the signal transduction pathways involved turn out to be the same in both cases; only the

targeted transcription factors are different. In both cases, an epidermal growth factor-like inducer activates the RTK pathway.

Photoreceptor induction in *Drosophila*

The *Drosophila* retina consists of about 800 units called **ommatidia** (Figure 6.16). Each ommatidium is composed of 20 cells arranged in a precise pattern. Eight of those cells are photoreceptors; the rest are lens cells. The eye develops in the flat epithelial layer of the eye imaginal disc of the larva. There are no cells directly above or below this layer, so the interactions are confined to neighboring cells in the same plane. The differentiation of these randomly arranged epithelial cells into the retinal photoreceptors and their surrounding lens tissue occurs during the last (third) larval stage. An indentation forms at the posterior

Figure 6.17
Differentiation of photoreceptors in the *Drosophila* compound eye. The morphogenetic furrow (arrow) crosses the disc from posterior (left) to anterior (right). (A) Confocal micrograph of a triple-labeled late larval eye/antennal imaginal disc, showing *hairy* expression in green ahead of the morphogenetic furrow (arrow). Within the furrow, the Ci protein (red) is expressed as a consequence of the Hedgehog signal. (It will activate the decapentaplegic gene.) The neural specific protein, 22C10, is stained blue in the differentiating photoreceptors behind the morphogenetic furrow. (The blue horizontal line of staining is Bolweg's nerve.) (B) Behind the furrow, the photoreceptor cells differentiate in a defined sequence. The first photoreceptor cell to differentiate is R8. R8 appears to induce the differentiation of R2 and R5, and a cascade of induction continues until the R7 photoreceptor is differentiated. (Photograph courtesy of N. Brown, S. Paddock, and S. Carroll; B after Tomlinson 1988.)

(A)

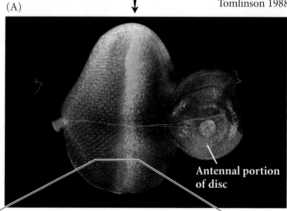

Antennal portion of disc

(B)

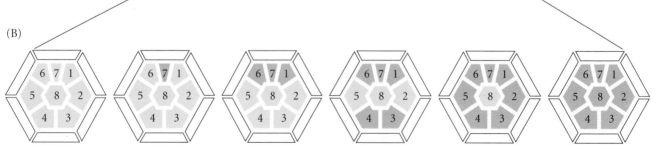

Later differentiation (posterior to morphogenetic furrow)

Early differentiation (entering morphogenetic furrow)

margin of the imaginal disc, and this **morphogenetic furrow** begins to travel forward toward the anterior of the epithelium (Figure 6.17). The movement of the furrow depends on the interactions between two paracrine factors, Hedgehog and Decapentaplegic. Hedgehog is expressed by the cells immediately posterior to the furrow (i.e., those that have just differentiated), and it induces the expression of the Decapentaplegic protein within the furrow (Heberlein et al. 1993; Ma et al. 1993). Thus, as retinal cells begin to differentiate behind the furrow, they secrete the Hedgehog protein, which drives the furrow anteriorly (Brown et al. 1995).

VADE MECUM *Drosophila* **imaginal discs.** Imaginal discs are groups of larval cells from which adult structures form. You can easily dissect imaginal discs from the *Drosophila* larva to see the magnificent eye-antenna disc for yourself. This segment shows how it is done. **[Click on Fruit Fly]**

As the morphogenetic furrow passes through a region of cells, those cells begin to differentiate in a specific order. The first cell to differentiate is the central (R8) photoreceptor (Chen and Chien 1999). (It is not yet known how the furrow instructs certain cells to become R8 photoreceptors, but it is possible that the Decapentaplegic and Hedgehog proteins in the furrow region induce R8 determination.) The R8 cell is thought to induce the cell before it and the cell after it (with respect to the furrow) to become the R2 and R5 photoreceptors, respectively. The R2 and R5 photoreceptors are functionally equivalent, so the signal from R8 is probably the same to both cells (Tomlinson and Ready 1987). Signals from these cells induce four more adjacent cells to become the R3, R4, and then the R1 and R6 photoreceptors. Last, the R7 photoreceptor appears. The other cells around these photoreceptors become the lens cells. Lens determination is the "default" condition if the cells are not induced.

A series of mutations has been found that blocks some of the steps of this induction cascade. Mutations in the **sevenless (sev)** gene or in the **bride of sevenless (boss)** gene can each prevent the R7 cell from differentiating into a photoreceptor. (It becomes a lens cell instead.) Analysis of these mutations has shown that they affect the inductive process. The *sev* gene is required in the R7 cell itself. If mosaic embryos are made such that some of the cells of

the eye imaginal disc are heterozygous (normal) and some are homozygous for the sevenless mutation, the R7 photoreceptor develops only if the R7 precursor cell has the wild-type *sev* allele (Basler and Hafen 1989; Bowtell et al. 1989). Antibodies to the Sevenless protein have found it in the cell membrane, and the sequence of the *sev* gene suggests that it encodes a transmembrane protein with a tyrosine kinase site in its cytoplasmic domain (Banerjee et al. 1987; Hafen et al. 1987). This finding is consistent with the protein's being a receptor for some signal—an RTK.

The signal that tells the R7 precursor to differentiate into an R7 photoreceptor comes from a protein encoded by the wild-type *bride-of-sevenless* gene. Flies homozygous for the *boss* mutation also lack R7 photoreceptors. Genetic mosaic studies wherein some of the cells of the eye imaginal disc are normal and some are homozygous for the *boss* mutation show that the wild-type *boss* gene is not needed in the R7 precursor cell itself. Rather, the R7 photoreceptor differentiates only if the wild-type *boss* gene is expressed in the R8 cell. Thus, the *boss* gene encodes some protein whose existence in the R8 cell is necessary for the differentiation of the R7 cell. In fact, the Boss protein is the ligand for the Sev RTK. The Boss protein probably works in a juxtacrine fashion, and its extracellular domain binds to the the extracellular domain of the Sev protein (Reinke and Zipursky 1988; Hart et al. 1993). In *Drosophila* eyes, the RTK cascade initiated by the binding of Boss to Sev activates the Sevenless-in-Absentia (Sina) transcription factor, whose activity is necessary for the differentiation of photoreceptor R7 (Carthew and Rubin 1990; Dickson et al. 1992). Once R7 is induced, it reciprocally induces the expression of opsin proteins in the R8 cell (Chou et al. 1999). A summary of some of the cell-to-cell inductions in the *Drosophila* retina (Figure 6.18) shows that individual cells are able to induce other individual cells to create the precise arrangement of cells in particular tissues.

WEBSITE **6.5 Eye formation: A conserved pathway.** The proteins of the sevenless pathway are seen not only in ommatidial development, but in vertebrate eye development as well. This appears to be a remarkably conserved pathway for photoreceptor differentiation. Moreover, there appear to be "safeguards" preventing the R8 cell from inducing R7 differentiation in other ommatidial cells.

Figure 6.18
Summary of genes known to be involved in the induction of *Drosophila* photoreceptors. For development to continue beyond the differentiation of the R8, R2, and R5 photoreceptors, the *rough* gene (*ro*) must be present in both the R2 and R5 cells. For the differentiation of the R7 photoreceptor, the *sevenless* gene (*sev*) has to be active in the R7 precursor cell, while the *bride of sevenless* gene (*boss*) must be active in the R8 photoreceptor. (After Rubin 1989.)

Vulval Induction in *Caenorhabditis elegans*

Most *C. elegans* individuals are hermaphrodites. In their early development, they are male, and the gonad produces sperm, which is stored for later use. As they grow older, they develop ovaries. The eggs "roll" through the region of sperm storage, are fertilized inside the nematode, and then pass out of the body through the vulva.

The vulva of *Caenorhabditis elegans* represents a case in which one inductive signal generates a variety of cell types. This organ forms during the larval stage from six cells called the **vulval precursor cells (VPCs)**. The cell connecting the overlying gonad to the vulval precursor cells is called the **anchor cell**. The anchor cell secretes the LIN-3 protein, a relative of epidermal growth factor (EGF) and the Boss protein (Hill and Sternberg 1992). If the anchor cell is destroyed (or if the *lin-3* gene is mutated), the VPCs will not form a vulva; they will instead become part of the hypodermis (skin) (Kimble 1981). The six VPCs influenced by the anchor cell form an **equivalence group**. Each member of this group is competent to become induced by the anchor cell and can assume any of three fates, depending on its proximity to the anchor cell (Figure 6.19). The cell directly beneath the anchor cell divides to form the central vulval cells. The two cells flanking that central cell divide to become the lateral vulval cells, while the three cells farther away from the anchor cell generate hypodermal cells. If the anchor cell is destroyed, all six cells of the equivalence group divide once and contribute to the hypodermal tissue. If the three central VPCs are destroyed, the three outer cells, which normally form hy-

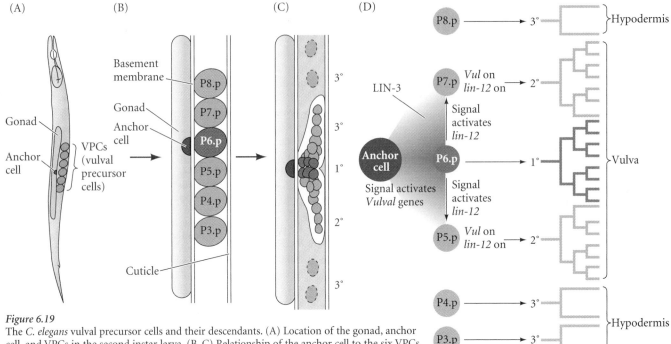

Figure 6.19
The *C. elegans* vulval precursor cells and their descendants. (A) Location of the gonad, anchor cell, and VPCs in the second instar larva. (B, C) Relationship of the anchor cell to the six VPCs and their subsequent lineages. 1° lineages result in the central vulval cells; 2° lineages constitute the lateral vulval cells; 3° lineages generate hypodermal cells. (C) Outline of the vulva in the fourth instar larva. The circles represent the positions of the nuclei. (D) Model for the determination of vulval cell lineages in *C. elegans*. The LIN-3 signal from the anchor cell causes the determination of the P6.p cell to generate the central vulval lineage. Lower concentrations of LIN-3 cause the P5.p and P7.p cells to form the lateral vulval lineages. The P6.p (central lineage) cell also secretes a short-range juxtacrine signal that induces the neighboring cells to activate the LIN-12 protein. This signal prevents the P5.p and P7.p cells from generating the primary, central vulval cell lineage. (After Katz and Sternberg 1996.)

podermal cells, generate vulval cells instead. The LIN-3 protein is received by the LET-23 receptor tyrosine kinase on the VPCs, and the signal is transferred to the nucleus through the RTK-Ras pathway. The target of the kinase cascade is the LIN-31 protein (Tan et al. 1998). When this protein is phosphorylated in the nucleus, it loses its inhibitory protein partner and is able to function as a transcription factor, promoting vulval cell fates.

Two major mechanisms coordinate the formation of the vulva through this induction (Figure 6.19D; Katz and Sternberg 1996):

1. The LIN-3 protein forms a concentration gradient. Here, the VPC closest to the anchor cell (i.e., the P6.p cell) receives the highest concentration of LIN-3 protein and generates the central vulval cells. The two VPCs adjacent to it (P5.p and P7.p) receive a lower amount of LIN-3 and become the lateral vulval cells. The VPCs farther away from the anchor cell do not receive enough LIN-3 to have an effect, so they become hypodermis (Katz et al. 1995).

2. In addition to forming the central vulval lineage, the VPC closest to the anchor cell also signals laterally to the two adjacent cells and instructs them to generate the lateral vulval lineages. These lateral cells do not instruct the peripheral VPCs to do anything, so they become hypodermis (Koga and Oshima 1995; Simske and Kim 1995). This lateral inhibition of the "secondary" vulval precursor cells by the "primary" VPC is accomplished through the LIN-12 proteins (which are discussed more thoroughly below; Sternberg 1988).

Both of these mechanisms function during normal development. As Kenyon (1995) notes, "together they could produce the ever-perfect tiny vulvae that *C. elegans* is so famous for." ■

The Smad pathway

Members of the TFG-β superfamily of paracrine factors activate members of the **Smad** family of transcription factors (Figure 6.20; Heldin et al. 1997). The TGF–β ligand binds to a type II TGF-β receptor, which allows that receptor to bind to a type I TGF-β receptor. Once the two receptors are in close contact, the type II receptor phosphorylates a serine or threonine on the type I receptor, thereby activating it. The activated type I receptor can now phosphorylate the Smad proteins. (Researchers named the Smad proteins by eliding the names of the

first identified members of this family: the *C. elegans* Sma protein and the *Drosophila* Mad protein.) Smads 1 and 5 are activated by the BMP family of TGF-β factors, while the receptors binding activin and the TGF-β family phosphorylate Smads 2 and 3. These phosphorylated Smads bind to Smad 4 and form the transcription factor complex that will enter the nucleus. In vertebrates, the TGF-β superfamily ligand **Nodal** appears to activate the Smads pathway in those cells responsible for the formation of the mesoderm and for specifying the left-right axis in vertebrates (Graff et al. 1996; Nomura and Li 1998).

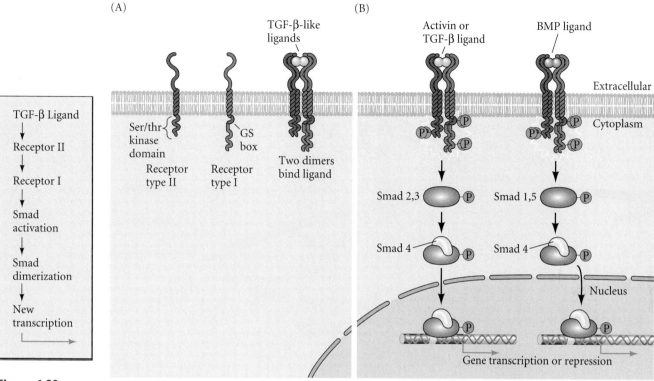

Figure 6.20
The Smad pathway activated by TGF-β superfamily ligands. (A) An activation complex is formed by the binding of the ligand by the type I and type II receptors. This allows the type II receptor to phosphorylate the type I receptor on particular serine or threonine residues (of the "GS box"). The phosphorylated type I receptor protein can now phosphorylate the Smad proteins. (B) Those receptors that bind TGF-β family proteins or members of the activin family phosphorylate Smads 2 and 3. Those receptors that bind to BMP family proteins phosphorylate Smads 1 and 6. These Smads can complex with Smad 4 to form active transcription factors. A simplified version of the pathway is shown at the left.

The JAK-STAT pathway

Another important pathway transducing information on the cell membrane to the nucleus is the JAK-STAT pathway. Here the set of transcription factors consists of the STAT (*signal transducers and activators of transcription*) proteins (Ihle 1995, 1996). STATs are phosphorylated by certain receptor tyrosine kinases, including fibroblast growth factor receptors and the JAK family of tyrosine kinases. The JAK-STAT pathway is extremely important in the differentiation of blood cells and in the activation of the casein gene during milk production (Briscoe et al. 1994; Groner and Gouilleux 1995). The role of this pathway in casein production is shown in Figure 6.21. Here, the endocrine factor prolactin binds to the extracellular regions of prolactin receptors, causing them to dimerize. A JAK protein kinase is bound to each of the receptors (in their respective cytoplasmic regions), and these JAK proteins are now brought together, where they can phosphorylate the receptors at several sites. The receptors are now activated and

have their own protein kinase activity. Therefore, the JAK proteins convert a receptor into a receptor tyrosine kinase. The activated receptors can now phosphorylate particular inactive STATs and cause them to dimerize. These dimers are the active form of the STAT transcription factors, and they are translocated into the nucleus, where they bind to specific regions of DNA. In this case, they bind to the upstream promoter elements of the casein gene, causing it to be transcribed.

The STAT pathway is very important in the regulation of human fetal bone growth. Mutations that prematurely activate the STAT pathway have been implicated in some severe forms of dwarfism such as the lethal **thanatophoric dysplasia**, wherein the growth plates of the rib and limb bones fail to proliferate. The short-limbed newborn dies because its ribs cannot support breathing. The genetic lesion resides in the gene encoding fibroblast growth factor receptor 3 (FGFR3) (Figure 6.22; Rousseau et al. 1994; Shiang et al. 1994). This protein is expressed in the cartilage precursor cells—known as **chondrocytes**—in the growth plates of the long bones. Normally, the FGFR3 protein is activated by a fibroblast growth factor, and it signals the chondrocytes to stop dividing and begin differentiating into cartilage. This signal is mediated by the STAT1 protein, which is phosphorylated by the activated FGFR3 and then translocated into the nucleus. Inside the nucleus, this transcription factor activates the genes encoding a cell cycle inhibitor, the p21 protein (Su et al. 1997). The mutations causing thanatophoric dwarfism result in a gain-of-function phenotype, wherein the mutant FGFR3

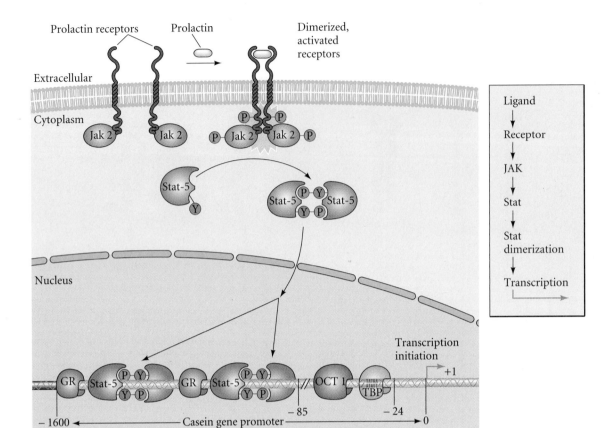

Figure 6.21
A STAT pathway: the casein gene activation pathway activated by prolactin. The casein gene is activated during the last (lactogenic) phase of mammary gland development, and its signal is the secretion of the hormone prolactin from the anterior pituitary gland. Prolactin causes the dimerization of prolactin receptors in the mammary duct epithelial cells. A particular JAK protein (Jak2) is "hitched" to these receptors. When the receptors are dimerized, the JAK proteins phosphorylate each other and the dimerized receptors, activating the dormant kinase activity of the receptors. The activated receptors add a phosphate group to a tyrosine residue (Y) of a particular STAT protein (in this case, Stat5). This allows Stat5 to dimerize, be translocated into the nucleus, and bind to particular regions of DNA. In combination with other transcription factors (which presumably have been waiting for its arrival), the STAT protein activates transcription of the casein gene. GR is the glucocorticoid receptor, OCT1 is a general transcription factor, and TBP is the set of proteins responsible for binding RNA polymerase. A simplified diagram is shown to the right. (For details, see Groner and Gouilleux 1996.)

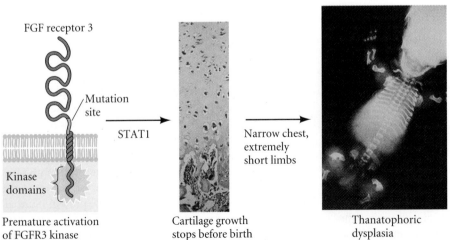

Figure 6.22
A mutation in the gene for FGFR3 causes the premature constitutive activation of the STAT pathway and the production of phosphorylated Stat1 protein. This transcription factor activates genes that cause the premature termination of chondrocyte cell division. The result is a condition of failed bone growth that results in the death of the newborn infant, since the thoracic cage cannot expand to take breaths. (From Gilbert-Barness and Opitz 1996.)

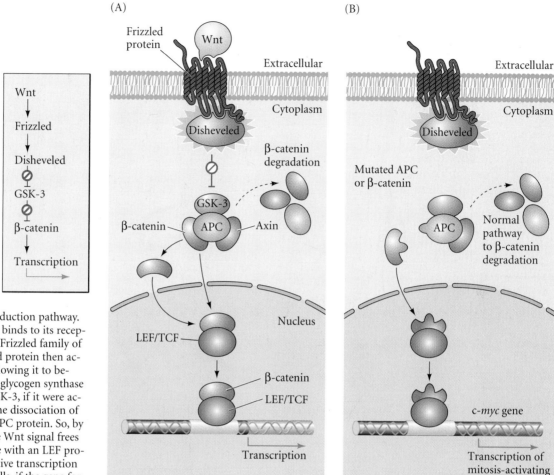

(A)   (B)

Figure 6.23
The Wnt signal transduction pathway. (A) The Wnt protein binds to its receptor, a member of the Frizzled family of proteins. The Frizzled protein then activates Disheveled, allowing it to become an inhibitor of glycogen synthase kinase 3 (GSK-3). GSK-3, if it were active, would prevent the dissociation of β-catenin from the APC protein. So, by inhibiting GSK-3, the Wnt signal frees β-catenin to associate with an LEF protein to become an active transcription factor. (B) In adult cells, if the gene for APC or β-catenin is mutated such that they cannot bind together, β-catenin is constitutively allowed into the nucleus. This causes it to activate certain cell division genes and initiate tumors. (B after Pennisi 1998.)

is active constitutively—that is, without the need to be activated by an FGF (Deng et al. 1996; Webster and Donoghue 1996). This causes the chondrocytes to stop proliferating shortly after they are formed, and the bones fail to grow. Mutations that activate FGFR3 to a lesser degree produce achondroplasic (short-limbed) dwarfism, the most prevalent human dominant syndrome.

> WEBSITE **6.6 FGFR mutations.** Mutations of the human FGF receptors have been associated with several skeletal malformation syndromes, including syndromes wherein skull cartilage, rib cartilage, or limb cartilage fails to grow or differentiate.

The Wnt pathway

Members of the Wnt family of paracrine factors interact with transmembrane receptors of the **Frizzled** family. In most in-

stances, the binding of Wnt by the Frizzled protein causes the Frizzled protein to activate the Disheveled protein. Once the Disheveled protein is activated, it inhibits the activity of the glycogen synthase kinase-3 enzyme. GSK-3, if it were active, would prevent the dissociation of the β-catenin protein from the APC protein, which targets β-catenin for degradation. However, when the Wnt signal is given and GSK-3 is inhibited, β-catenin can dissociate from the APC protein and enter the nucleus. Once inside the nucleus, it can form a heterodimer with an LEF or TCF DNA-binding protein, becoming a transcription factor. This complex binds to and activates the Wnt-responsive genes (Figure 6.23A; Behrens et al. 1996; Cadigan and Nusse 1997).

This model is undoubtedly an oversimplification, because different cells use this pathway in different ways (see Cox and Peifer 1998). Moreover, its components can have more than one function in the cell. In addition to being part of the Wnt signal transduction cascade, GSK-3 is also a metabolic enzyme regulating glycogen metabolism. The β-catenin protein was first recognized as being part of the cell adhesion complex on

(A)

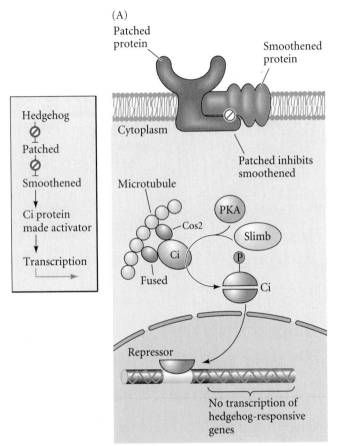

(B)

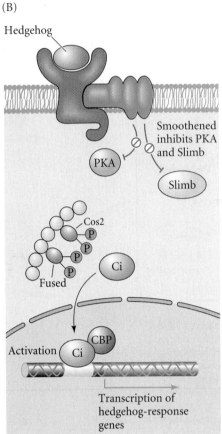

Figure 6.24
The Hedgehog signal transduction pathway. The Patched protein in the cell membrane is an inhibitor of the Smoothened protein. (A) In the absence of Hedgehog binding to Patched, the Ci protein is tethered to the microtubules (by the Cos2 and Fused proteins). This allows the PKA and Slimb proteins to cleave Ci into a transcriptional repressor that blocks the transcription of particular genes. (B) When Hedgehog binds to Patched, its conformation changes, releasing the inhibition of the Smoothened protein. Smoothened releases Ci from the microtubules (probably by adding more phosphates to the Cos2 and Fused proteins) and inactivates the cleavage proteins PKA and Slimb. The Ci protein enters the nucleus, binds a CBP protein and acts as a transcriptional activator of particular genes. (After Johnson and Scott 1998.)

the cell surface before it was also found to be a transcription factor. The APC protein also functions as a tumor suppressor in adults. The transformation of normal colon cells into colon cancer is thought to occur when the APC gene is mutated and can no longer keep the β-catenin protein out of the nucleus (Figure 6.23B; Korinek et al. 1997; He et al. 1998). Once in the nucleus, β-catenin can bind with another transcription factor and activate genes for cell division.

One principle that is readily seen in the Wnt pathway (and which is also evident in the Hedgehog pathway) is that activation is often accomplished by inhibiting an inhibitor. Thus, the GSK-3 protein is an inhibitor that is itself repressed by the Wnt signal.

The Hedgehog pathway

Members of the Hedgehog protein family function by binding to a receptor called Patched. The Patched protein, however, is not a signal transducer. Rather, it is bound to a signal transducer, the Smoothened protein. The Patched protein prevents the Smoothened protein from functioning. In the absence of Hedgehog binding to Patched, the Smoothened protein is inactive, and the Cubitus interruptus (Ci) protein is tethered to the microtubules of the responding cell. While on the microtubules, it is cleaved in such a way that a portion of it enters the nucleus and acts as a transcriptional repressor. This portion of the Ci protein binds to the promoters and enhancers of particular genes and acts as an inhibitor of transcription. When Hedgehog binds to the Patched protein, the Patched protein's shape is altered such that it no longer inhibits Smoothened. The Smoothened protein acts (probably by phosphorylation) to release the Ci protein from the microtubules and to prevent its being cleaved. The intact Ci protein can now enter the nucleus, where it acts as a transcriptional *activator* of the same genes it used to repress (Figure 6.24; Aza-Blanc et al. 1997).

The Hedgehog pathway is extremely important in limb and neural differentiation in vertebrates. When mice were made homozygous for a mutant allele of Sonic hedgehog, they had major limb abnormalities as well as cyclopia—a single eye in the center of the forehead (Chiang et al. 1996). The verte-

Figure 6.25
Head of a cyclopic lamb born of a ewe who had eaten *Veratrum californicum* early in pregnancy. The cerebral hemispheres fused, forming only one central eye and no pituitary gland. The jervine alkaloid made by this plant inhibits cholesterol synthesis, which is needed for Hedgehog production and reception. (Photograph courtesy of L. James and the USDA Poisonous Plant Laboratory.)

brate homologues of the Ci protein in *Drosophila* are the Gli proteins. Severe truncations of the human *GLI3* gene produce a nonfunctional protein that gives rise to Grieg cephalopolysyndactyly, a condition involving a high forehead and extra digits. A less severe truncation retains the DNA binding domain of the GLI3 protein, but deletes the activion region. Thus, this mutant GLI3 protein can act only as a repressor. This protein is found in patients with Pallister-Hall syndrome, a much more severe syndrome (indeed, lethal soon after birth) involving not only extra digits, but also poor develop-

ment of the pituitary gland, hypothalamus, anus, and kidneys (see Shin et al. 1999). While mutations that inactivate the Hedgehog pathway can cause malformations, mutations that activate the pathway ectopically can cause cancers. If the pathway is activated in somatic tissues such that it can be constitutively turned on, it can cause basal cell carcinomas of the basal cell layer of the epidermis. Heritable mutations of the patched gene cause basal cell nevus syndrome, an autosomal dominant condition characterized by both developmental anomalies (fused fingers, rib and facial abnormalities) and multiple malignant tumors such as basal cell carcinoma (Hahn et al. 1996; Johnson et al. 1996).

One remarkable feature of the Hedgehog signal transduction pathway is the importance of cholesterol. First, cholesterol is critical for the catalytic cleavage of Sonic hedgehog protein. Only the amino-terminal portion of the protein is functional and secreted. Second, the Patched protein that binds the Sonic hedgehog protein also needs cholesterol in order to function. It has recently been found (Kelley et al. 1996; Roessler et al. 1996) that some human cyclopia syndromes are caused by mutations in genes that encode either Sonic hedgehog or the enzymes that synthesize cholesterol. Moreover, certain chemicals that induce cyclopia do so by interfering with the cholesterol biosynthetic enzymes (Figure 6.25; Beachy et al. 1997; Cooper et al. 1998). Environmental factors that cause developmental anomalies are called **teratogens** (from the Greek, meaning "monster-former"), and they will discussed in more detail in Chapter 21. Two teratogens known to cause cyclopia in vertebrates are jervine and cyclopamine. Both substances are found in the plant *Veratrum californicum* (Keeler and Binns et al. 1968), and both block the synthesis of cholesterol.

Sidelights & Speculations

The Nature of Human Syndromes

As WE HAVE SEEN, human infants are sometimes born with congenital malformations that range from life-threatening to relatively benign. **Congenital** means "at birth," and this term reflects the fact that these malformations are errors of development. Often these malformations are linked into syndromes (see Chapter 1).

Pleiotropy
The production of several effects by one gene is called **pleiotropy**. For instance, in humans, heterozygosity for *MITF* causes a condition

called Waardenburg syndrome type 2, as we saw in Chapter 4 (Figure 4.25). This syndrome involves iris defects, pigmentation abnormalities, deafness, and inability to produce the normal number of mast cells (a type of blood cell). The skin pigment, the iris of the eye, the inner ear tissue, and the mast cells of the blood are not related to one another in such a way that the absence of one would produce the absence of the others. Rather, all four parts of the body independently use the MITF protein as a transcription factor. This type of pleiotropy has been called **mosaic pleiotropy**,

because the affected organ systems are separately affected by the abnormal gene function.

While the eye pigment, body pigment, and mast cell features of Waardenburg syndrome type 2 are separate events, other parts of the syndrome are not. For instance, the failure of MITF expression in the pigmented retina prevents this structure from fully differentiating. This in turn causes a malformation of the choroid fissure of the eye, resulting in the drainage of vitreous humor fluid. Without this fluid, the eye fails to enlarge (hence the name *microphthalmia*, which

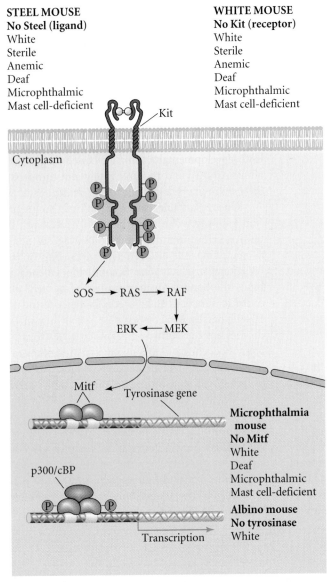

STEEL MOUSE
No Steel (ligand)
White
Sterile
Anemic
Deaf
Microphthalmic
Mast cell-deficient

WHITE MOUSE
No Kit (receptor)
White
Sterile
Anemic
Deaf
Microphthalmic
Mast cell-deficient

Kit

Cytoplasm

SOS → RAS → RAF

ERK ← MEK

Mitf

Tyrosinase gene

Microphthalmia mouse
No Mitf
White
Deaf
Microphthalmic
Mast cell-deficient

p300/cBP

Albino mouse
No tyrosinase
White

Transcription

Figure 6.26
Phenotypes of mice with mutations along the pigment cascade. The Steel and White mice have mutations in the genes for stem cell factor (ligand) and c-kit protein (receptor), respectively. These proteins activate the Mitf transcription factor through the pathway shown in Figure 6.15. *Mitf* mutations give rise to the *microphthalmia* phenotype, which contains a subset of those anomalies seen upstream in the pathway. The albino mutation is farther down on the pathway and contains a subset of those conditions found in the *Mitf* mutants.

means "small eye"). This phenomenon, in which several developing tissues are affected by the mutation even though they do not express the mutated gene, is called **relational pleiotropy** (see Gruneberg 1938).

Genetic heterogeneity

Another important feature of syndromes is that mutations in different genes can produce the same phenotype. If the genes are part of the same signal transduction pathway, a mutation in any of them can give a similar result.

The phenomenon whereby mutations in different genes produce similar phenotypes is called **genetic heterogeneity**. For example, cyclopia can be produced by mutations in the *Sonic hedgehog* gene or by mutations in cholesterol synthesis genes. Since they are in the same pathway, mutations in one gene generate a phenotype similar or identical to mutations in the other genes. Similarly, as we saw above, mutations in the stem cell factor (*Steel*) gene produce a syndrome resembling that produced by mutations in the gene for its recep-

tor, the Kit protein (*White*). Since mutations of either of these genes prevent Mitf from being activated, they produce a phenotype similar to that of the Mitf-deficient mouse (Figures 6.26 and 1.15B). But here, there is a difference. Since the Kit and stem cell factor proteins are also used by migrating germ cells and blood cell precursors (which do not use Mitf), mice that are mutant for Kit and stem cell factor also have fewer gametes and blood cells. Albinism, which is produced by a loss-of-function mutation of the tyrosinase gene, also gives a white phenotype, but it does not preclude mast cell function or fertility.

Phenotypic heterogeneity

Not only can different mutations produce the same phenotype, but the same mutation can produce a different phenotype in different individuals (Wolf 1995, 1997; Nijhout and Paulsen 1997). This phenomenon is called **phenotypic heterogeneity** (see Wolf 1995, 1997). It is caused by the existence of the signal transduction pathways and the integration of genes into complex networks. Genes are not autonomous entities. Rather, they interact with other genes and gene products to make pathways and networks. For instance, Bellus and colleagues (1996) have analyzed the phenotypes derived from the same mutation in the gene for FGFR3 in ten independent human families. These phenotypes range from relatively mild anomalies to potentially lethal malformations. Similarly, Freire-Maia (1975) reported that within one family, a mutant gene affecting limb development caused phenotypes ranging from severe phocomelia (lack of limb development shortly after the most proximal bone, ending in a "flipper-like stump") to a mild abnormality of the thumb. The severity of a mutant gene's effect often depends on the other genes in the pathway.

Mechanisms of dominance

Whether a syndrome is dominant or recessive can now be explained at a molecular level. First, it must be recognized that there are many syndromes that are referred to as dominant only because the homozygous condition is lethal to the embryo and the fetus is never born. Therefore, the homozygous condition never exists. Second, there are at least three ways of achieving a "dominant" phenotype.

The first mechanism of dominance is **haploinsufficiency**. This merely means that one copy of the gene (the haploid condition) is not enough to produce the required amount of product for normal development. For example, individuals with Waardenburg syndrome type 2 have roughly half the wild-type amount of MITF. This is not enough for full pigment cell proliferation, mast cell differentiation, or inner ear development. Thus, an aberrant phenotype results when only one of the two copies of this gene is absent or nonfunctional.

The second mechanism of dominance is **gain-of-function mutations**. Thus, thanatophoric dysplasia (as well as milder forms of dwarfism such as achondroplasia) results from a mutation causing the FGF receptor to be constitutively active. This activity is enough to cause an anomalous phenotype to develop.

The third mechanism of dominance is a **dominant negative allele**. This situation can occur when the active form of the protein is a multimer and all proteins of the multimer have to be wild-type in order for the protein to function. A dominant negative allele is the cause of Marfan syndrome, a disorder of the extracellular matrix. Marfan syndrome is associated with joint and connective tissue anomalies, not all of which are necessarily disadvantageous. Increased height, disproportionately long limbs and digits, and mild to moderate joint laxity are characteristic of this syndrome. However, patients with Marfan syndrome also experience vertebral column deformities, myopia, loose lenses, and (most importantly) aortic problems that may lead to aneurysm (bursting of the aorta) later in life. The mutation is in the gene for fibrillin, a secreted glycoprotein that forms multimeric microfibrils in elastic connective tissue. The presence of even small amounts of mutant fibrillin prohibits the association of wild-type fibrillin into microfibrils. Eldadah and colleagues (1995) have shown that when a mutant human gene for fibrillin is transfected into fibroblast cells that already contain two wild-type genes, the incorporation of fibrillin into the matrix is inhibited.

We are beginning to understand the molecular bases for many of the inherited congenital malformation syndromes in humans. This understanding constitutes a critically important synthesis of developmental biology, human genetics, and pediatric medicine. ▧

The Cell Death Pathways

"To be, or not to be: that is the question." While we all are poised at life-or-death decisions, this existential dichotomy is exceptionally stark for embryonic cells. Programmed cell death, called **apoptosis**,* is a normal part of development. In the nematode *C. elegans*, in which we can count the number of cells, exactly 131 cells die according to the normal developmental pattern. All the cells of this nematode are "programmed" to die unless they are actively told not to undergo apoptosis. In humans, as many as 10^{11} cells die in each adult each day and are replaced by other cells. (Indeed, the mass of cells we lose each year through normal cell death is close to our entire body weight!) Within the uterus, we were constantly making and destroying cells, and we generated about three times as many neurons as we eventually ended up with when we were born. Lewis Thomas (1992) has aptly noted,

> By the time I was born, more of me had died than survived. It was no wonder I cannot remember; during that time I went through brain after brain for nine months, finally contriving the one model that could be human, equipped for language.

Apoptosis is necessary not only for the proper spacing and orientation of neurons, but also for generating the middle ear space, the vaginal opening, and the spaces between our fingers and toes (Saunders and Fallon 1966; Roberts and Miller 1998; Rodriguez et al. 1997). Apoptosis prunes away unneeded structures, controls the number of cells in particular tissues, and sculpts complex organs.

Different tissues use different signals for apoptosis. One of the signals often used in vertebrates is bone morphogenetic protein 4 (BMP4). Some tissues, such as connective tissue, respond to BMP4 by differentiating into bone. Others, such as the frog gastrula ectoderm, respond to BMP4 by differentiating into skin. Still others, such as neural crest cells and tooth primordia, respond by degrading their DNA and dying. In the developing tooth, for instance, numerous growth and differentiation factors are secreted by the enamel knot. After the cusp has grown, the enamel knot synthesizes BMP4 and shuts itself down by apoptosis (see Chapter 13; Vaahtokari et al. 1996b).

In other tissues, the cells are "programmed" to die, and they will remain alive only if some growth or differentiation factor is present to "rescue" them. This happens during the development of mammalian red blood cells. The red blood cell precursors in the mouse liver need the hormone erythropoietin in order to survive. If they do not receive it, they undergo apoptosis. The erythropoietin receptor works through the JAK-STAT pathway, activating the Stat5 transcription factor. In this way, the amount of erythropoietin present can determine how many red blood cells enter the circulation.

One of the pathways for apoptosis was largely delineated through genetic studies of *C. elegans*. It was found that the proteins encoded by the ***ced-3*** and ***ced-4*** genes were essential for apoptosis, but that in the cells that did not undergo apoptosis, those genes were turned off by the product of the ***ced-9*** gene (Figure 6.27A; Hengartner et al. 1992). The CED-4 protein is a protease activating factor that activates CED-3, a protease that initiates the destruction of the cell. Mutations that inactivate the CED-9 protein cause numerous cells that would normally survive to activate their *ced-3* and *ced-4* genes and die. This leads to the death of the entire embryo. Conversely, gain-of-function mutations of *ced-9* cause CED-9 protein to be made

---

*Apoptosis (both "p"s are pronounced) comes from the Greek word for the natural processes of leaves falling from trees or petals from flowers. It is active and can be evolutionarily selected. The other type of cell death, *necrosis*, is a pathological death caused by external factors such as inflammation or toxic injury.

Figure 6.27
Apoptosis pathways in nematodes and mammals. (A) In *C. elegans*, the CED-4 protein is a protease activating factor that can activate the CED-3 protease. The CED-3 protease initiates the cell destruction events. CED-9 can inhibit CED-4 (and CED-9 can be inhibited upstream by EGL-1). (B) In mammals, a similar pathway exists, and appears to function in a similar manner. In this hypothetical scheme for the regulation of apoptosis in mammalian neurons, Bcl-x_L (a member of the Bcl-2 family) binds Apaf-1 and prevents it from activating the precursor of caspase-9. The signal for apoptosis allows another protein (here, Bik) to inhibit the binding of Apaf-1 to Bcl-x_L. Now, Apaf-1 is able to bind to the caspase-9 precursor and cleave it. Caspase-9 now dimerizes and activates caspase-3, which initiates apoptosis. (C) In addition, there are other pathways, such as the one initiated by the CD95 protein in the cell membranes of lymphocytes. The same colors are used to represent homologous proteins. (After Adams and Cory 1998.)

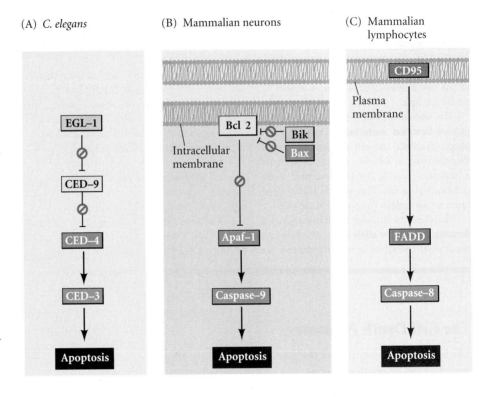

(A) *C. elegans*

(B) Mammalian neurons

(C) Mammalian lymphocytes

in cells that would otherwise die. Thus, the *ced-9* gene appears to be a binary switch that regulates the choice between life and death on the cellular level. It is possible that every cell in the nematode embryo is poised to die, and those cells that survive are rescued by the activation of the *ced-9* gene.

The CED-3 and CED-4 proteins form the center of the apoptosis pathway that is common to all animals studied. The trigger for apoptosis can be a developmental cue such as a particular molecule (such as BMP4 or glucocorticoids) or the loss of adhesion to a matrix. Either type of cue can activate the CED-3 or CED-4 proteins or inactivate the CED-9 molecules. In mammals, the homologues of the CED-9 protein are members of the **Bcl-2** family of genes. This family includes *Bcl-2*, *Bcl-X*, and similar genes. The functional similarities are so strong that if an active human *BCL-2* gene is placed into *C. elegans* embryos, it prevents normally occurring cell deaths in the nematode embryos (Vaux et al. 1992). In vertebrate red blood cell development (mentioned above), the Stat5 transcription factor activated by erythropoietin functions by binding to the promoter of the *Bcl-X* gene, where it activates the synthesis of that anti-apoptosis protein (Socolovsky et al. 1999).

The mammalian homologue of CED-4 is called **Apaf-1** (*a*poptotic *p*rotease *a*ctivating *f*actor-1), and it participates in the cytochrome *c*-dependent activation of the mammalian CED-3 homologues, the proteases **caspase-9** and **caspase-3** (Shaham and Horvitz 1996; Cecconi et al. 1998; Yoshida et al. 1998). The activation of the caspases causes the autodigestion

of the cell. Caspases are strong proteases, and they digest the cell from within. The cellular proteins are cleaved and the DNA is fragmented.*

While apoptosis-deficient nematodes deficient for CED-4 are viable (despite their having 15% more cells than wild-type worms), mice with loss-of-function mutations for either caspase-3 or caspase-9 die around birth from massive cell overgrowth in the nervous system (Figure 6.28; Kuida et al. 1996, 1998; Jacobson et al. 1997). Mice homozygous for targeted deletions of *Apaf-1* have severe craniofacial abnormalities, brain overgrowth, and webbing between their toes.

In mammals, there is more than one pathway to apoptosis. The apoptosis of the lymphocytes, for instance, is not affected by the deletion of *Apaf-1* or *caspase-9*, and works by a separate pathway initiated by the CD95 protein (Figure 6.27B,C). Different caspases may be functioning in different cell types to mediate the apoptotic signals (Hakem et al. 1998; Kuida et al. 1998).

---

*There is some evidence (see Barinaga 1998a,b; Saudou et al. 1998) that activation of the apoptosis pathway in adult neurons may be responsible for the pathology of Alzheimer's disease and stroke. Fragmentation of DNA is one of the major ways that apoptosis is recognized, and it is seen in regions of the brain affected by these diseases. Its fragmentation into specific sized pieces (protected and held together by nucleosomes) may be caused by the digestion of poly(ADP-ribose) polymerase (PARP) by caspase-3 (Lazebnik et al. 1994). PARP recognizes and repairs DNA breaks. For more on other pathways leading to apoptosis, see Green 1998.

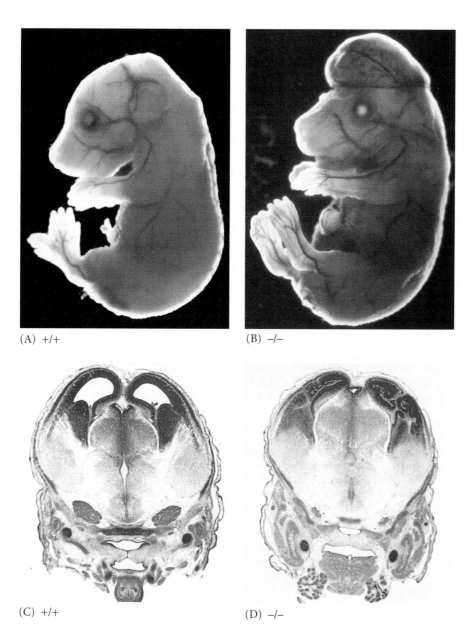

(A) +/+

(B) –/–

(C) +/+

(D) –/–

Figure 6.28
Disruption of normal brain development by blocking apoptosis. In mice in which *caspase-9* or *Apaf-1* has been knocked out, normal neural apoptosis fails to occur. In caspase 9-deficient mice, the overproliferation of brain neurons is obvious on a morphological level. (A) 16-day embryonic wild-type mouse. (B) Caspase 9-knockout mouse of the same age. The enlarged brain protrudes above the face, and the limbs are still webbed. This effect is confirmed by cross-sections through the forebrain at day 13.5 in (C) a normal mouse and (D) a caspase 9-knockout mouse. The knockout exhibits thickened ventricle walls and the near-obliteration of the ventricles. (From Kuida et al. 1998.)

WEBSITE **6.7 The uses of apoptosis.** Apoptosis is used for numerous processes throughout development. This website explores the role of apoptosis in such phenomena as *Drosophila* germ cell development and the eyes of blind cave fish.

Juxtacrine Signaling

In juxtacrine interactions, proteins from the inducing cell interact with receptor proteins of adjacent responding cells. The inducer does not diffuse from the cell producing it. There are three types of juxtacrine interactions. In the first type, a protein on one cell binds to its receptor on the adjacent cell. We saw this type of juxtacrine interaction when we discussed the interaction between the Bride of sevenless protein and its receptor, Sevenless. In the second type, a receptor on one cell binds to its ligand on the extracellular matrix secreted by another cell. In the third type, the signal is transmitted directly from the cytoplasm of one cell through small conduits into the cytoplasm of an adjacent cell.

The Notch pathway: Juxtaposed ligands and receptors

While most known regulators of induction are diffusible proteins, some inducing proteins remain bound to the inducing cell surface. In one such pathway, cells expressing the **Delta**, **Jagged**, or **Serrate** proteins in their cell membranes activate neighboring cells that contain the **Notch** protein in their cell

(A)

(B)

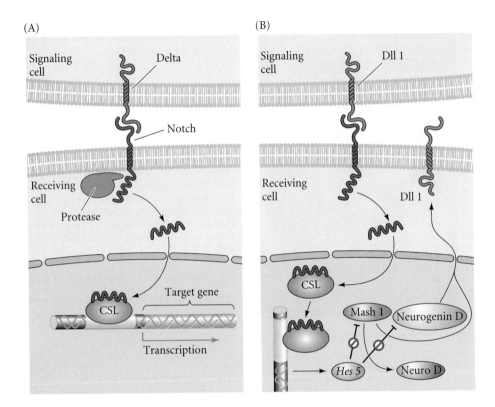

Figure 6.29

Mechanism of Notch activity. (A) Model for the activation of Notch. A ligand (Delta, Jagged, or Serrate protein) on one cell binds to the extracellular domain of the Notch protein on an adjacent cell. This binding causes a shape change in the intracellular domain of Notch, which activates a protease. The protease cleaves Notch and allows the intracellular region of the Notch protein to enter the nucleus and activate a transcription factor of the CSL family (such as Suppressor of hairless or CBF1). The activated CSL can then transcribe its target genes. (B) Inhibition of neurogenesis in the mouse. Delta family gene *Delta-1* (*Dll1*) signals Notch family genes *Notch-1* (and probably *Notch-3*). These proteins activate a CSL transcription factor (RBP), which transcribes the *Hes5* gene. Hes5 protein inhibits the *Mash1* gene, which would otherwise activate the neural determining gene *NeuroD*. Thus, neural determination is blocked. Hes5 also inhibits another gene (probably for neurogenin) that would otherwise activate the production of Delta-1. (A after Schroeder et al. 1998; B after Blaschuk and ffrench-Constant 1998.)

membranes. Notch extends through the cell membrane, and its external surface contacts Delta, Jagged, or Serrate proteins extending out from an adjacent cell. When complexed to one of these ligands, Notch undergoes a conformational change that enables it to be cut by a protease. The cleaved portion enters the nucleus and binds to a dormant transcription factor of the CSL family. When bound to the Notch protein, the CSL transcription factors activate their target genes (Figure 6.29A; Lecourtois and Schweisguth 1998; Schroeder et al. 1998; Struhl and Adachi 1998).

Notch proteins are extremely important receptors in the nervous system. In both the vertebrate and *Drosophila* ner-

vous system, the binding of Delta to Notch tells the receiving cell not to become neural (Figure 6.29B; Chitnis et al. 1995; Wang et al. 1998). In the vertebrate eye, the interactions between Notch and its ligands seem to regulate which cells become optic neurons and which become glial cells (Dorsky et al. 1997; Wang et al. 1998).

WEBSITE **6.8 Notch mutations.** Mutations in Notch proteins can cause nervous system abnormalities in humans. Humans have more than one *Notch* gene and more than one ligand. Their interactions may be critical in neural development.

Cell-Cell Interactions and Chance in the Determination of Cell Types

THE DEVELOPMENT OF THE VULVA in *C. el-egans* offers several examples of induction on the cellular level. We have already discussed the reception of the EGF-like LIN-3 signal by the cells of the equivalence group that can form the vulva. But before this induction occurs, there is an earlier interaction that forms the anchor cell. The formation of the anchor cell is mediated by the *lin-12* gene, the *C. elegans* homologue of the *Notch* gene. In wild-type *C. elegans* hermaphrodites, two adjacent cells, Z1.ppp and Z4.aaa, have the potential to become the anchor cell. They interact in a manner that causes one of them to become the anchor cell while the other one becomes the precursor of the uterine tissue. In recessive *lin-12* mutants, both cells become anchor cells, while in dominant mutations, both cells become uterine precursors (Greenwald et al. 1983). Studies using genetic mosaics and cell ablations have shown that this decision is made in the second larval stage, and that the *lin-12* gene only needs to function in that cell destined to become the uterine precursor cell. The presumptive anchor cell does not need it. Seydoux and Greenwald (1989) speculate that these two cells originally synthesize both the signal for uterine differentiation (the LAG-2 protein, homologous to Delta in *Drosophila*) and the receptor for this molecule (the LIN-12 protein, homologous to Notch) (Figure 6.30; Wilkinson et al. 1994). During a particular time in larval development, the cell that, by chance, is secreting more LAG-2 causes its neighbor to cease its production of this differentiation signal and to increase its production of LIN-12 protein. The cell secreting LAG-2 becomes the gonadal anchor cell, while the cell receiving the signal through its LIN-12 protein becomes the ventral uterine precursor cell. Thus, the two cells are thought to determine each other prior to their respective differentiation events.

The anchor cell/ventral uterine precursor decision illustrates two important aspects of

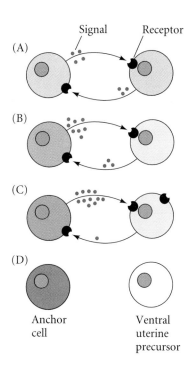

Signal   Receptor

(A)

(B)

(C)

(D)

Anchor   Ventral
cell   uterine
                precursor

Figure 6.30
Model for the generation of two cell types (anchor cell and ventral uterine precursor) from two equivalent cells (Z1.ppp and Z4.aaa). (A) The cells start off as equivalent, producing fluctuating amounts of signal (arrow) and receptor (inverted arrow). The *lag-2* gene is thought to encode the signal; the *lin-12* (*Notch*) gene is thought to encode the receptor. Reception of the signal turns down LAG-2 (Delta) production and up-regulates LIN-12. (B) A stochastic (chance) event causes one cell to produce more LAG-2 than the other cell at some particular critical time. This stimulates more LIN-12 production in the neighboring cell. (C) This difference is amplified, since the cell producing more LIN-12 produces less LAG-2. (D) Eventually, just one cell is delivering the LAG-2 signal, and the other cell is receiving it. The signaling cell becomes the anchor cell; the receiving cell becomes the ventral uterine precursor. (After Greenwald and Rubin 1992.)

determination in two originally equivalent cells. First, the initial difference between the two cells is created by chance. Second, this initial difference is reinforced by feedback. Such a mechanism is also seen in the determination of which of the originally equivalent epidermal cells of the insect embryo will generate the neurons of the peripheral nervous system. Here, the choice is between becoming a skin (hypodermal) cell or a neural precursor cell (a **neuroblast**). The *Notch* gene of *Drosophila*, like its *C. elegans* homologue, *lin-12*, channels a bipotential cell into one of two alternative paths. The LIN-12 protein will be used again during vulva formation. It is activated by the primary vulval lineage to stop the lateral vulval cells from forming the central vulval phenotype (Figure 6.19).

Soon after gastrulation, a region of about 1800 ectodermal cells lies along the ventral midline of the *Drosophila* embryo. These

cells, known as neurogenic ectodermal cells, all have the potential to form the ventral nerve cord of the insect. About one-quarter of these cells will become neuroblasts, while the rest will become the precursors of the hypodermis. The cells that give rise to neuroblasts are intermingled with those that become hypodermal precursors. Thus, each neurogenic ectoderm cell can give rise to either hypodermal or neural precursor cells (Hartenstein and Campos-Ortega 1984). In the absence of *Notch* gene transcription in the embryo, these cells develop exclusively into neuroblasts, rather than into a mixture of hypodermal and neural precursor cells (Artavanis-Tsakonis et al. 1983; Lehmann et al. 1983). These embryos die, having a gross excess of neural cells at the expense of the ventral and head hypodermis (Poulson 1937; Hoppe and Greenspan 1986).

Heitzler and Simpson (1991) have proposed that the Notch protein, like LIN-12, serves as a receptor for intercellular signals involved in differentiating equivalent cells.

Moreover, they provided evidence that Delta is the ligand for Notch. Genetic mosaics show that whereas Notch is needed in the cells that are to become hypodermis, Delta is needed in the cells that *induce* the hypodermal phenotype.

Greenwald and Rubin (1992) have proposed a model based on the LIN-12 hypothesis to explain the spacing of neuroblasts in these proneural clusters of epidermal and neural precursors (Figure 6.31). Initially, all the cells have equal potentials and produce the same signals. However, when one of the cells, by chance, produces more signal (say, Delta protein), it activates the receptors on adjacent cells and reduces their signaling level. Since the signaling levels on adjacent cells are lowered, the neighbors of those low-signaling cells will tend to become high-level signalers. In this way, a spacing of neuroblasts is produced. ■

Figure 6.31
Model to explain the spacing pattern of neuroblasts among initially equivalent neurogenic ectodermal cells. (A) A field of equivalent cells, all of which signal and receive equally. (B) A chance event causes one of the cells (darker shading) to produce more signal. The cells surrounding it receive this higher amount of signal and reduce their own signaling level (lighter shading). (C) The rest of the pattern is now constrained. Those cells that have down-regulated their own signaling (in response to the events in B) are less likely to express more signal than their neighboring cells. The cells surrounded by down-regulated signalers are more likely to become signalers. (D, E) The fates of the cells throughout the field become specified as the amplification of the signal creates populations of signalers surrounded by populations of receivers. In the case of the neurogenic-cells, the signal is thought to be the Delta protein, and the receiver is the Notch protein. (After Greenwald and Rubin 1992.)

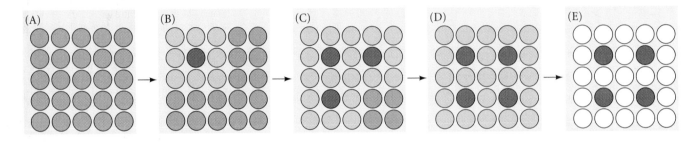

The extracellular matrix as a source of critical developmental signals

PROTEINS AND FUNCTIONS OF THE EXTRACELLULAR MATRIX. The **extracellular matrix** consists of macromolecules secreted by cells into their immediate environment. These macromolecules form a region of noncellular material in the interstices between the cells. The extracellular matrix is a critical region for much of animal development. Cell adhesion, cell migration, and the formation of epithelial sheets and tubes all depend on the ability of cells to form attachments to extracellular matrices. In some cases, as in the formation of epithelia, these attachments have to be extremely strong. In other instances, as when cells migrate, attachments have to be made, broken, and made again. In some cases, the extracellular matrix merely serves as a permissive substrate to which cells can adhere, or upon which they can migrate. In other cases, it provides the directions for cell movement or the signal for a developmental event.

Extracellular matrices are made up of collagen, proteoglycans, and a variety of specialized glycoprotein molecules, such as fibronectin and laminin. These large glycoproteins are responsible for organizing the matrix and cells into an ordered structure. **Fibronectin** is a very large (460-kDa) glycoprotein dimer synthesized by numerous cell types. One function of fibronectin is to serve as a general adhesive molecule, linking cells to one another and to other substrates such as collagen and proteoglycans. Fibronectin has several distinct binding sites, and their interaction with the appropriate molecules results in the proper alignment of cells with their extracellular matrix (Figure 6.32). As we will see in later chapters, fibronectin also has an important role in cell migration. The "roads" over which certain migrating cells travel are paved with this protein. Fibronectin paths lead germ cells to the gonads and lead heart cells to the midline of the embryo. If chick embryos are injected with antibodies to fibronectin, the heart-forming cells fail to reach the midline, and two separate hearts develop (Heasman et al. 1981; Linask and Lash 1988).

Laminin and **type IV collagen** are major components of a type of extracellular matrix called the **basal lamina** (Figure 6.33). This basal lamina is characteristic of the closely knit sheets that surround epithelial tissue. The adhesion of epithelial cells to laminin (upon which they sit) is much greater than the affinity of mesenchymal cells for fibronectin (to which they must bind and release if they are to migrate). Like fibronectin, laminin plays a role in assembling the extracellular matrix, promoting cell adhesion and growth, changing cell shape, and permitting cell migration (Hakamori et al. 1984).

INTEGRINS, THE CELL RECEPTORS FOR EXTRACELLULAR MATRIX MOLECULES. The ability of a cell to bind to adhesive glycoproteins depends on its expressing a cell membrane receptor for the cell-binding site of these large molecules. The main fi-

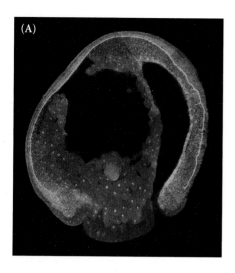

Figure 6.32
Fibronectin in the developing frog embryo. (A) Fluorescent antibodies to fibronectin show fibronectin deposition as a green band in the *Xenopus* embryo during gastrulation. The fibronectin will orient the mesoderm movements of the cells. (B) Structure and binding domains of fibronectin. The rectangles represent protease-resistant domains. The fibroblast-binding domain consists of two units, the RGD site and the high-affinity site, both of which are essential for cell binding. Avian neural crest cells have another site that is necessary for them to migrate on a fibronectin substrate. Other regions of fibronectin enable it to bind to collagen, heparin, and other molecules of the extracellular matrix. (A courtesy of M. Marsden and D. W. DeSimone; B after Dufour et al. 1988.)

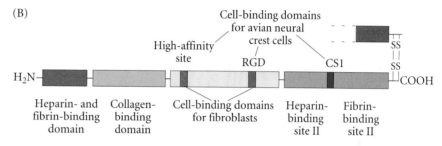

bronectin receptors were identified by using antibodies that block the attachment of cells to fibronectin (Chen et al. 1985; Knudsen et al. 1985). The fibronectin receptor complex was found not only to bind fibronectin on the outside of the cell, but also to bind cytoskeletal proteins on the inside of the cell. Thus, the fibronectin receptor complex appears to span the cell membrane and unite two types of matrices. On the outside of the cell, it binds to the fibronectin of the extracellular matrix; on the inside of the cell, it serves as an anchorage site for the actin microfilaments that move the cell (Figure 6.34). Horwitz and co-workers (1986; Tamkun et al. 1986) have

called this family of receptor proteins **integrins** because they integrate the extracellular and intracellular scaffolds, allowing them to work together. On the extracellular side, integrins bind to the sequence arginine-glycine-aspartate (RGD), found in several adhesive proteins in extracellular matrices, includ-

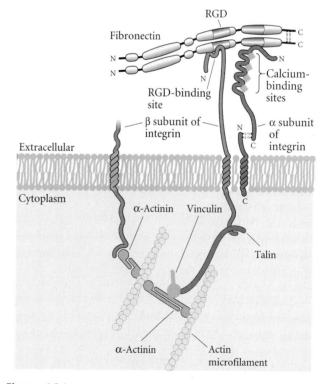

Figure 6.34
Speculative diagram relating the binding of cytoskeleton to the extracellular matrix through the integrin molecule. (After Luna and Hitt 1992.)

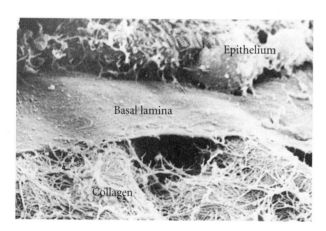

Figure 6.33
Location and formation of extracellular matrices in the chick embryo. The scanning electron micrograph shows the extracellular matrix at the junction of the epithelial cells (above) and mesenchymal cells (below). The epithelial cells synthesize a tight, glycoprotein-based basal lamina, while the mesenchymal cells secrete a loose reticular lamina made primarily of collagen. (Courtesy of R. L. Trelsted.)

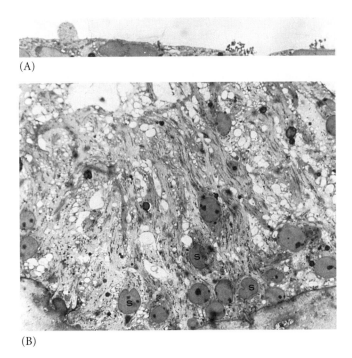

(A)

(B)

Figure 6.35
Role of the extracellular matrix in cell differentiation. Light micrographs of rat Sertoli testis cells grown for two weeks (A) on tissue culture plastic dishes and (B) on dishes coated with basal lamina. The two photographs were taken at the same magnification, 1200×. (From Hadley et al. 1985; photographs courtesy of M. Dym.)

ing fibronectin, vitronectin (found in the basal lamina of the eye), and laminin (Ruoslahti and Pierschbacher 1987). On the cytoplasmic side, integrins bind to talin and α-actinin, two proteins that connect to actin microfilaments. This dual binding enables the cell to move by contracting the actin microfilaments against the fixed extracellular matrix.

Bissell and her colleagues (1982; Martins-Green and Bissell 1995) have proposed that the extracellular matrix is capable of inducing specific gene expression in developing tissues, especially those of the liver, testis, and mammary gland, in which the induction of specific transcription factors depends on cell-substrate binding (Figure 6.35; Liu et al. 1991; Streuli et al. 1991; Notenboom et al. 1996). Often, the presence of bound integrin prevents the activation of genes that specify apoptosis (Montgomery et al. 1994; Frisch and Ruoslahti 1997). The chondrocytes that produce the cartilage of our vertebrae and limbs can survive and differentiate only if they are surrounded by an extracellular matrix and are joined to that matrix through their integrins (Hirsch et al. 1997). If chondrocytes from the developing chick sternum are incubated with antibodies that block the binding of integrins to the extracellular matrix, they shrivel up and die. While the mechanisms by which bound integrins inhibit apoptosis remain controversial (see Howe et al. 1998), the extracellular matrix is obviously an important source of signals that can be transduced into the nucleus to produce specific gene expression.

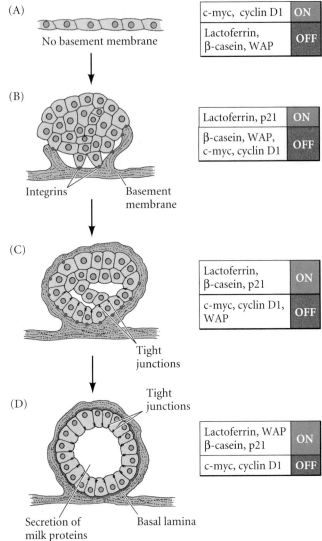

Figure 6.36
Basement membrane-directed gene expression in mammary gland tissue. (A) Mouse mammary gland tissue divides when placed on tissue culture plastic. Cell division genes are on, and the genes capable of synthesizing the differentiated products of the mammary gland (lactoferrin, casein, whey acidic protein) are off. (B) When presented with basement membrane that contains laminin, the genes for cell division proteins are turned off, while the gene inhibiting cell division (p21) and the gene for lactoferrin are turned on. (C, D) Mammary gland cells wrap the basement membrane about them, forming a secretory epithelium. The genes for casein and whey protein are sequentially activated. (After Bissell, personal communication.)

Some of the genes induced by matrix attachment are being identified. When plated onto tissue culture plastic, mouse mammary gland cells will divide (Figure 6.36). Indeed, the genes for cell division (c-*myc*, *cyclinD1*) are expressed, while the genes for the differentiated products of the mammary gland (casein, lactoferrin, whey acidic protein) are not expressed. If the same cells are plated onto plastic coated with a laminin-containing basement membrane, the cells stop dividing and the differentiated genes of the mammary gland are expressed. This happens only after the integrins of the mammary gland cells bind to the laminin of the extracellular basement membrane. Then the gene for lactoferrin is expressed, as is the gene for p21, a cell division inhibitor. The c-*myc* and *cyclinD1* genes become silent. Eventually, all the genes for the developmental products of the mammary gland are expressed and the cell division genes remain turned off. At this time, the mammary cells will have enveloped themselves in the basement membrane, forming a secretory epithelium reminiscent of the mammary gland tissue. The binding of integrins to laminin is essential for the transcription of the casein gene, and the integrins act in concert with prolactin (see Figure 6.21) to activate that gene's expression (Roskelley et al. 1994; Muschler et al. 1999).

Recent studies have shown that the binding of integrins to the extracellular matrix can stimulate the RTK-Ras pathway. When an integrin on the cell membrane of one cell binds to the fibronectin or collagen secreted by a neighboring cell, the integrin can activate the tyrosine kinase cascade through an adaptor protein-like complex that connects the integrin to the Ras G protein (Figure 6.37A; Wary et al. 1998). Cadherins

and other cell adhesion molecules can also transmit signals by "hijacking" the FGF receptors (Williams et al. 1994b; Clark and Brugge 1995). Cadherins, for example, are able to bind to the intramembrane region of FGF receptors and thereby dimerize these receptors just like the normal FGF ligands (Figure 6.37B; Williams et al. 1994a; Doherty et al. 1995).

Direct transmission of signals through gap junctions

Throughout this chapter we have been discussing the reception of signals by cell membrane receptors. In these mechanisms, the receptor is in some manner altered by binding the ligand, so that its cytoplasmic domain transmits the signal into the cell. Another mechanism transmits small, soluble signals directly through the cell membrane, for the membrane is not continuous in all places. There are regions called **gap junctions** that serve as communication channels between adjacent cells (Figure 6.38A,B). Cells so linked are said to be "coupled," and small molecules (molecular weight <1500) and ions can freely pass from one cell to the other. In most embryos, at least some of the early blastomeres are connected by gap junctions. The ability of cells to form gap junctions with some cells and not with others creates physiological "compartments" within the developing embryo (Figure 6.38C).

The gap junction channels are made of **connexin** proteins. In each cell, six identical connexins in the membrane group together to form a transmembrane channel containing a central pore. The channel complex of one cell connects to the channel complex of another cell, enabling the cytoplasms of both cells to be joined (Figure 6.38D). Different types of connexin proteins have separate, but overlapping, roles in normal development. For example, connexin-43 is found throughout the developing mouse embryo, in nearly every tis-

Figure 6.37

Two types of activation by cell adhesion molecules. (A) Cell-substrate adhesion molecules such as integrins may transmit a signal from the cytoplasmic portion of the integrin protein to the Ras G protein through a cascade involving caveolin and Fyn proteins. (B) The FGF receptors may be "hijacked" by cell adhesion molecules and dimerized. They may be brought together by the interaction of opposite cell adhesion molecules, or the "crosslinking" of FGF receptors by the apposing cell membrane may activate their kinase domains. (A after Wary et al.1998.)

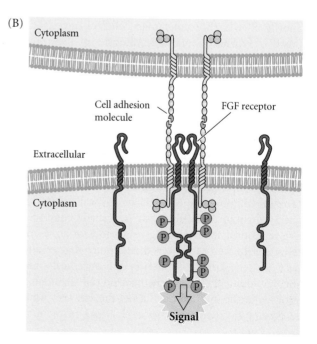

(A)

Fibronectin
RGD
RGD-binding site
Calcium-binding sites
β subunit of integrin
α subunit of integrin
Extracellular
Cytoplasm
Caveolin
Fyn
Shc
Kinase
SOS
GRB2
Ras
Signal

(B)

Cytoplasm
Cell adhesion molecule
FGF receptor
Extracellular
Cytoplasm
Signal

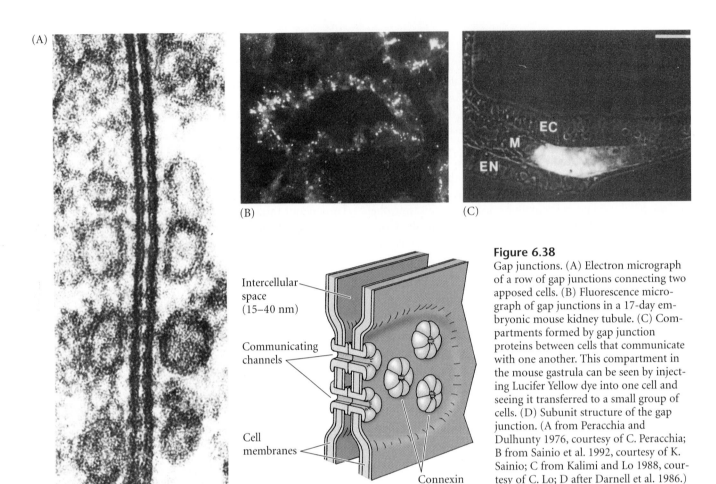

(A)

(B)

(C)

Intercellular space (15–40 nm)

Communicating channels

Cell membranes

Connexin proteins

(D)

Figure 6.38
Gap junctions. (A) Electron micrograph of a row of gap junctions connecting two apposed cells. (B) Fluorescence micrograph of gap junctions in a 17-day embryonic mouse kidney tubule. (C) Compartments formed by gap junction proteins between cells that communicate with one another. This compartment in the mouse gastrula can be seen by injecting Lucifer Yellow dye into one cell and seeing it transferred to a small group of cells. (D) Subunit structure of the gap junction. (A from Peracchia and Dulhunty 1976, courtesy of C. Peracchia; B from Sainio et al. 1992, courtesy of K. Sainio; C from Kalimi and Lo 1988, courtesy of C. Lo; D after Darnell et al. 1986.)

sue. However, if the connexin-43 genes are knocked out by gene targeting, the mouse embryo will still develop. It appears that most of the functions of the connexin-43 protein can be taken over by other related connexin proteins—but not all. Shortly after birth, connexin-43-deficient mice take gasping breaths, turn bluish, and die. Autopsies of these mice show that the right ventricle—the chamber that pumps blood through the pulmonary artery to the lungs—is filled with tissue that occludes the chamber and obstructs blood flow (Reaume et al. 1995; Huang et al. 1998). Although loss of the connexin-43 protein can be compensated for in many tissues, it appears to be critical for normal heart development.

The importance of gap junctions in development has been demonstrated in amphibian and mammalian embryos (Warner et al. 1984). When antibodies to connexins were microinjected into one specific cell of an 8-cell *Xenopus* blastula, the progeny of that cell, which are usually coupled through gap junctions, could no longer pass ions or small molecules from cell to cell. Moreover, the tadpoles that resulted from such treated blastulae showed defects specifically relating to the developmental fate of the injected cell (Figure 6.39). The progeny of the injected cell did not die, but they were unable to develop normally (Warner et al. 1984). In the mouse em-

bryo, the first eight blastomeres are also connected to one another by gap junctions. Although loosely associated with one another, these eight cells move together to form a compacted embryo. If compaction is inhibited by antibodies against connexins, the treated blastomeres continue to divide, but further development ceases (Lo and Gilula 1979; Lee et al. 1987). If antisense RNA to connexin messages is injected into one of the blastomeres of a normal mouse embryo, that cell will not form gap junctions and will not be included in the embryo (Bevilacqua et al. 1989).

WEBSITE **6.9 Connexin mutations.** Mutations in human connexin proteins cause congenital malformations of the heart and ear. In many cases, one connexin can substitute for another, but when the connexins cannot compensate, a mutant phenotype results.

Cross-Talk between Pathways

We have been representing the major signal transduction pathways as if they were linear chains, through which information flows in a single conduit. However, these pathways are just the major highways of information flow. Between them, avenues

Figure 6.39
Developmental effects of gap junctions. Section through *Xenopus* tadpoles in which one of the blastomeres at the 8-cell stage was injected with (A) a control antibody or (B) an antibody against connexins. The side formed by the injected blastomere lacks its eye and has abnormal brain morphology. (From Warner et al. 1984; photographs courtesy of A. E. Warner.)

(A)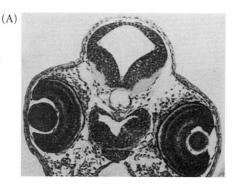

(B)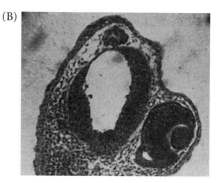

and streets connect one pathway with another. (This may be why there are so many steps between the cell surface and the nucleus. Each step is a potential regulatory point as well as a potential intersection.) This **cross-talk** can be seen in numerous tissues, wherein two signaling pathways reinforce each other. We must remember that a cell has numerous receptors and is constantly receiving many signals simultaneously.

In some cells, gene transcription requires two signals. This pattern is seen during lymphocyte differentiation, for which two signals are needed, each one producing one of the two peptides of a transcription factor needed for the production of interleukin 2 (IL-2, also known as T cell growth factor). One peptide, c-Fos, is produced by the binding of the T cell receptor to an antigen (Figure 6.40). This signal activates the Ras pathway, creating a transcription factor, Elk-1, that activates the *c-fos* gene to synthesize c-Fos. The second signal comes from the B7 glycoprotein on the surface of the cell presenting the antigen. This signal activates a second cascade of kinases, eventually producing c-Jun. The two peptides, c-Fos and c-Jun, can join to make the AP-1 protein, a transcription factor that binds to the *IL-2* enhancer and activates its expression (Li et al. 1996).

We also have seen that one receptor can activate several different pathways. The fibroblast growth factors, for instance, can activate the RTK pathway, the STAT pathway, or even a third pathway that involves lipid turnover and increases the levels of calcium ions in the cell.

Figure 6.40
Two signals are needed to effect the differentiation of the T lymphocyte. The first signal comes from the receptors that bind the antigen. The second signal comes from the binding of the CD28 protein to the B7 protein on the surface of the antigen-presenting cell. The first signal directs the synthesis of one subunit of the AP-1 transcription factor. The second signal directs the synthesis of the other subunit. The two subunits, c-Fos and c-Jun, form the AP-1 transcription factor, which can activate T cell-specific enhancers such as that regulating interleukin 2 production.

Coda

In 1782, the French essayist Denis Diderot posed the question of morphogenesis in the fevered dream of a noted physicist. This character could imagine that the body was formed from myriads of "tiny sensitive bodies" that collected together to

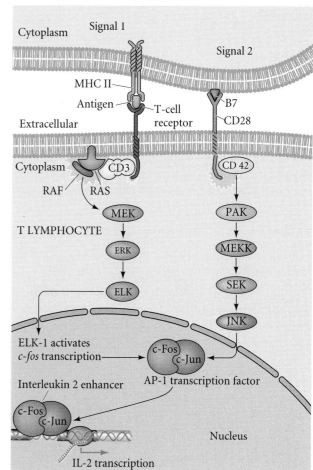

form an aggregate, but he could not envision how this aggregate could become an animal. A hundred years later, Charles Darwin wrote that the eye must have evolved by small useful variations, but he did not know the mechanisms by which the intricate coordination of eye development occurred.

Recent studies have shown that this coordination of cells to form organs and organisms is due to the molecules on the embryonic cell surfaces. Inducers and their competent responders interact with one another to instruct and coordinate the further development of the component parts. In subsequent chapters, we will look more closely at some of these morphogenetic interactions. As we continue our study of animal development, we will find that induction and competence provide the core of morphogenesis. Moreover, we will find these signal transduction pathways wherever we look, both throughout the animal kingdom and within each developing embryo. We are now at the stage where we can begin our study of early embryogenesis and see the integration of the organismal, genetic, and cellular processes of animal development.

Principles of Development: Cell-Cell Communication

1. Inductive tissue interaction involves inducer and responding tissues.

2. The ability to respond to inductive signals depends upon the competence of the responding cells.

3. Reciprocal induction occurs when the two interacting tissues are both inducers and are competent to respond to each other's signals.

4. Cascades of inductive events are responsible for organ formation.

5. Regionally specific inductions can generate different structures from the same tissue.

6. The ability to respond to inducers is determined by the genetic state of the responding tissue.

7. Juxtacrine interactions are inductions that occur between the cell membranes of adjacent cells or between a cell membrane and an extracellular matrix secreted by another cell.

8. Paracrine interactions occur when a cell or tissue secretes proteins that induce changes in neighboring cells.

9. Paracrine factors are inducing proteins that bind to cell membrane receptors in competent responding cells.

10. Competent cells respond to paracrine factors through signal transduction pathways. Competence is the ability to bind and to respond to the inducers, and it is often the result of a prior induction.

11. Signal transduction pathways begin with the paracrine or juxtacrine factor causing a conformational change in its cell membrane receptor. The new shape results in enzymatic activity in the cytoplasmic domains of the receptor protein. This allows the receptor to phosphorylate other cytoplasmic proteins, thereby activating a dormant kinase activity. Eventually, a transcription factor (or set of factors) is activated that activates or represses specific gene activity.

12. Pleiotropy is the phenomenon of many phenotypic changes being caused by one mutation. Mosaic pleiotropy results when the mutant gene is used in different parts of the body and each part is separately altered. Relational pleiotropy occurs when a particular defect caused by the mutant gene affects other parts of the body that do not express the gene.

13. Genetic heterogeneity results when multiple genes are needed to create a particular phenotype. Often mutant genes for a paracrine factor cause syndromes similar to those generated by mutant genes for the factor's receptor.

14. Phenotypic heterogeneity results when the same mutation produces different phenotypic effects in different individuals. It is caused by the interactions between gene products.

15. Dominant mutations (in which only one mutant gene of the diploid pair is necessary to produce an abnormal phenotype) can be caused by haploinsufficiency, gain-of-function mutations, or dominant negative alleles.

16. Programmed cell death is one possible response to inductive stimuli. Apoptosis is a critical part of life.

17. There is cross-talk between signal transduction pathways, which allows the cell to respond to multiple inputs simultaneously.

Literature Cited

Adams, J. M. and S. Cory. 1998. The Bcl-2 protein family: Arbiters of cell survival. *Science* 281: 1322–1326.

Artavanis-Tsakonis, S., M. A. T. Muskavitch and Y. Yedvobnick. 1983. Molecular cloning of *Notch*, a locus affecting neurogenesis in *Drosophila melanogaster. Proc. Natl. Acad. Sci. USA* 80: 1977–1981.

Aza-Blanc, P., F. A. Ramirez-Weber, M. P. Laget and T. B. Kornberg. 1997. Proteolysis thatis inhibited by hedgehog targets Cubitus interruptus protein to the nucleus and converts it to a repressor. *Cell* 89: 1043–1053.

Banerjee, U., P. J. Renfranz, J. A. Pollock and S. Benzer. 1987. Molecular characterization and expression of *sevenless*, a gene involved in neuronal pattern formation in the *Drosophila* eye. *Cell* 49: 281–291.

Bard, J. B. L. 1990. *Morphogenesis: The Cellular and Molecular Processes of Developmental Anatomy.* Cambridge University Press, Cambridge.

Barinaga, M. 1998a. Is apoptosis key in Alzheimer's disease? *Science* 281: 1303–1304.

Barinaga, M. 1998b. Stroke-damaged neurons may commit cellular suicide. *Science* 281: 1302–1303.

Basler, K. and E. Hafen. 1989. Ubiquitous expression of *sevenless*: Position-dependent specification of cell fate. *Science* 243: 931–934.

Beachy, P. A. and 7 others. Multiple roles of cholesterol in hedgehog protein biosynthesis and signaling. *Cold Spring Harbor Symp. Quant. Biol.* 62: 191–204.

Behrens, J., J. P. von Kries, M. Kühl, L. Bruhn, D. Wedlich, R. Grosschedl and W. Birchmeier. 1996. Functional interaction of β-catenin with the transcription factor LRF-1. *Nature* 382: 638–642.

Bellus, G. A., K. Gaudenz, E. H. Zackai, L. A. Clarke, J. Szabo, C. A. Francomano and M. Muenke. 1996. Identical mutations in three fibroblast growth factor receptor genes in autosomal dominant craniosynostosis syndromes. *Nature Genet.* 14: 174–176.

Bevilacqua, A., R. Loch-Caruso and Erickson, R. P. 1989. Abnormal development and dye coupling produced by antisense RNA to gap junction protein in mouse preimplantation embryos. *Proc. Natl. Acad. Sci. USA* 86: 5444–5448.

Binns, W., L. F. James and J. L. Shupe. 1964. Toxicosis of *Veratrum californicum* in ewes and its relationship to a congenital deformity in lambs. *Ann. NY Acad. Sci.* 111: 571–576.

Bissell, M. J., H. G. Hall and G. Parry. 1982. How does the extracellular matrix direct gene expression? *J. Theor. Biol.* 99: 31–68.

Bitgood, M. J. and A. P. McMahon. 1995. Hedgehog and BMP genes are coexpressed at many diverse sites of cell-cell interaction in the mouse embryo. *Dev. Biol.*172:126–158.

Bitgood, M. J., L. Shen and A. P. McMahon. 1996. Sertoli cell signalling by desert hedgehog regulates the male germline. *Curr. Biol.* 6: 298–304.

Blaschuk, K. and ffrench-Constant, C. 1998. Developmental Neurobiology: Notch is tops in the developing brain. *Curr. Biol.* 8: R334–R337.

Bowtell, D. D. L., M. A. Simon and G. M. Rubin. 1989. Ommatidia in the developing *Drosophila* eye require and can respond to *sevenless* for only a restricted period. *Cell* 56: 931–936.

Briscoe, J., D. Guschin and M. Müller. 1994. Just another signalling pathway. *Curr. Biol.* 4: 1033–1036.

Brown, N. L., C. A. Sattler, S. W. Paddock and S. B. Carroll. 1995. Hairy and Emc negatively regulate morphogenetic furrow progression in the *Drosophila* eye. *Cell* 80: 879–887.

Cadigan, K. M. and R. Nusse. 1997. Wnt signaling: A common theme in animal development. *Genes Dev.* 24: 3286–3306.

Cales, C., J. F. Hancock, C. J. Marshall and A. Hall. 1988. The cytoplasmic protein GAP is implicated as a target for regulation by the ras gene product. *Nature* 332: 548–551.

Carthew, R. W. and G. M. Rubin. 1990. *Seven-in-absentia*, a gene required for the specification of R7 cell fate in the *Drosophila* eye. *Cell* 63: 561–577.

Cecconi, F., G. Alvarez-Bolado, B. I. Meyer, K. A. Roth and P. Gruss. 1998. Apaf-1 (CED-4 homologue) regulates programmed cell death in mammalian development. *Cell* 94: 727–737.

Chen, K. C. and C. T. Chien. 1999. Negative regulation of *atonal* in proneural cluster formation in *Drosophila* R8 photoreceptors. *Proc. Natl. Acad. Sci. USA* 96: 5055–5060.

Chen, W. T., E. Hasegawa, T. Hasegawa, C. Weinstock and K. M. Yamada. 1985. Development of cell-surface linkage complexes in cultured fibroblasts. *J. Cell Biol.* 100: 1103–1114.

Chiang, C., L. T. Ying, E. Lee, K. E. Young, J. L. Corden, H. Westphal and P. A. Beachy. 1996. Cyclopia and defective axial patterning in mice lacking Sonic hedgehog gene function. *Nature* 383: 407–413.

Chitnis, A., D. Henrique, J. Lewis, D. Ish-Horowicz and C. Kintner. 1995. Primary neurogenesis in *Xenopus* embryos regulated by a homologue of the *Drosophila* gene *Delta. Nature* 375: 761–766.

Chou, W. H. and 7 others. 1999. Patterning of the R7 and R8 photoreceptor cells of *Drosophila*: Evidence for induced and default cell-fate specification. *Development* 126: 607–616.

Clark, E. A. and J. S. Brugge. 1995. Integrin and signal transduction pathways: The road taken. *Science* 268: 233–239.

Cooper, M. K., J. A. Porter, K. E. Young and P. A. Beachy. 1998. Teratogen-mediated inhibition of target tissue response to hedgehog signaling. *Science* 280: 1603–1607.

Cox, R. T. and M. Peifer. 1998 Wingless signaling: The inconvenient complexities of life. *Curr. Biol.* 8: R140–R144.

Crossley, P. H., G. Monowada, C. A. MacArthur and G. Martin. 1996. Roles for FGF8 in the induction, initiation, and maintenance of the chick limb development. *Cell* 84: 127–136.

Cvekl, A. and J. Piatigorsky. 1996. Lens development and crystallin gene expression: Many roles for Pax-6. *BioEssays* 18: 621–630.

Daniel, C. W. 1989. TGF-β1 induced inhibition of mouse mammary ductal growth: Developmental specificity and characterization. *Dev. Biol.* 134: 20–30.

Darnell, J., H. Lodish and D. Baltimore. 1986. *Molecular Cell Biology.* Scientific American Books, New York.

Deng, C., A. Wynshaw–Boris, F. Zhou, A. Kuo and P. Leder. 1996. Fibroblast growth factor receptor-3 is a negative regulator of bone growth. *Cell* 84: 911–921.

Dickson, B., F. Sprenger, D. Morrison and E. Hafen. 1992. Raf functions downstream of Ras1 in the Sevenless signal transduction pathway. *Nature* 360: 600–603.

Diderot, D. 1782. *D'Alembert's Dream.* Reprinted in J. Barzun and R. H. Bowen (eds.), *Rameau's Nephew and Other Works* (1956). Doubleday, Garden City, NY, p. 114.

Dorsky, R. I., W. S. Chang, D. H. Rapaport and W. A. Harris. 1997. Regulation of neuronal diversity in the *Xenopus* retina by Delta signalling. *Nature* 385: 67–74.

Dufour, S., J.-L. Duband, M. J. Humphries, M. Obara, K. M. Yamada and J. P. Thiery. 1988. Attachment, spreading and locomotion of avian neural crest cells are mediated by multiple adhesion sites on fibronectin molecules. *EMBO J.* 7: 2661–2671.

Eldadah, Z. A., T. Brenn, H. Furthmayr and H. C. Dietz. 1995. Expression of a mutant human fibrillin allele upon a normal human or murine genetic background recapitulates a Marfan cellular phenotype. *J. Clin. Invest.* 95: 874–880.

Fan, C. M. and M. Tessier-Lavigne. 1994. Patterning of mammalian somites by surface ectoderm and notochord: Evidence for sclerotome induction by a hedgehog homolog. *Cell* 79: 1175–1186.

Freire-Maia, N. 1975. A heterozygote expression of a "recessive" gene. *Hum. Hered.* 25: 302–304.

Frisch, S. M. and E. Ruoslahti.1997. Integrins and anoikis. *Curr. Opin. Cell Biol.* 9: 701–706.

Fujiwara, M., T. Uchida, N. Osumi-Yamashita and K. Eto. 1994. Uchida rat (rSey): A new mutant rat with craniofacial abnormalities resembling those of the mouse Sey mutant. *Differentiation* 57: 31–38.

Gilbert-Barness, E. and J. M. Opitz. 1996. Abnormal bone development: Histopathology and skeletal dysplasias. *In* M. E. Martini-Neri, G. Neri and J. M. Opitz (eds.), *Gene Regulation and Fetal Development*. March of Dimes Birth Defects Foundation Original Article Series 30: (1). Wiley-Liss, New York, pp. 103–156.

Goustin, A. S. and 9 others. 1986. Coexpression of the *sis* and *myc* proto-oncogenes in developing human placenta suggests autocrine control of trophoblast growth. *Cell* 41: 301–312.

Graff, J. M., A. Bansal and D. A. Melton. 1996. *Xenopus* Mad proteins transduce distinct subsets of signals for the TGF-β superfamily. *Cell* 85: 479–487.

Grainger, R. M. 1992. Embryonic lens induction: Shedding light on vertebrate tissue determination. *Trends Genet.* 8: 349–356.

Green, D. R. 1998. Apoptotic pathways: Roads to ruin. *Cell* 94: 695–698.

Greenwald, I. and G. M. Rubin. 1992. Making a difference: The role of cell-cell interactions in establishing separate identities for equivalent cells. *Cell* 68: 271–281.

Greenwald, I., P. W. Sternberg and H. R. Horvitz. 1983. The *lin-12* locus specifies cell fates in *Caenorhabditis elegans. Cell* 34: 435–444.

Grobstein, C. 1956. Trans-filter induction of tubules in mouse metanephrogenic mesenchyme. *Exp. Cell Res.* 10: 424–440.

Groner, B. and F. Gouilleux. 1995. Prolactin-mediated gene activation in mammary epithelial cells. *Curr. Opin. Genet. Dev.* 5: 587–594.

Gruneberg, H. 1938. An analysis of the "pleiotropic" effects of a new lethal mutation in the rat (*Mus norwegicus*). *Proc. R. Soc. Lond.* B 125: 123–144.

Gurdon, J. B., P. Harger, A. Mitchell and P. Lemaire. 1994. Activin signalling and response to a morphogen gradient. *Nature* 371: 487–492.

Gurdon, J. B., A. Mitchell and D. Mahony. 1995. Direct and continuous assessment by cells of their position in a morphogen gradient. *Nature* 376: 520–521.

Hadley, M. A., S. W. Byers, C. A. Suárez-Quian, H. Kleinman and M. Dym. 1985. Extracellular matrix regulates Sertoli cell differentiation, testicular cord formation, and germ cell development in vitro. *J. Cell Biol.* 101: 1511–1512.

Hafen, E., K. Basler, J. E. Edstrom and G. M. Rubin. 1987. *sevenless*, a cell-specific homeotic gene of *Drosophila*, encodes a putative transmembrane receptor with a tyrosine kinase domain. *Science* 236: 55–63.

Hahn, H. and 20 others. 1996. Mutations of the human homolog of *Drosophila patched* in the nevoid basal cell carcinoma syndrome. *Cell* 85: 841–851.

Hakamori, S., M. Fukuda, K. Sekiguchi and W. G. Carter. 1984. Fibronectin, laminin, and other extracellular glycoproteins. *In* K. A. Picz and A. H. Reddi (eds.), *Extracellular Matrix Biochemistry*. Elsevier, New York, pp. 229–276.

Hakem, R. and 15 others. 1998. Differential requirement for caspase-9 in apoptotic pathways in vivo. *Cell* 94: 339–352.

Hamburgh, M. 1970. *Theories of Differentiation.* Elsevier, New York.

Hardman, P., E. Landels, A. S. Woolf and B. S. Spooner. 1994. TGF-β1 inhibits growth and branching morphogenesis in embryonic mouse submandibular and sublingual glands in vitro. *Dev. Growth Diff.* 36: 567–577.

Harrison, R. G. 1933. Some difficulties of the determination problem. *Am. Nat.* 67: 306–321.

Hart, A. C., H. Krämer and S. L. Zipursky. 1993. Extracellular domain of the boss transmembrane ligand acts as an antagonist of the *sev* receptor. *Nature* 361: 732–736.

Hartenstein, V. and J. A. Campos-Ortega. 1984. Early neurogenesis in wild-type *Drosophila melanogaster. Wilhelm Roux Arch. Dev. Biol.* 193: 308–326.

Hay, E. D. 1980. Development of the vertebrate cornea. *Int. Rev. Cytol.* 63: 263–322.

He, T.-C. and 8 others. 1998. Identification of c-*MYC* as a target of the APC pathway. *Science* 281: 1509–1512.

Heasman, J., R. D. Hines, A. P. Swan, V. Thomas and C. C. Wylie. 1981. Primordial germ cells of *Xenopus* embryos: The role of fibronectin in their adhesion during migration. *Cell* 27: 437–447.

Heberlein, U., T. Wolff and G. M. Rubin. 1993. The TGF-β homolog dpp and the segment polarity gene *hedgehog* are required for propagation of a morphogenetic wave in the *Drosophila* retina. *Cell* 75: 913–926.

Heitzler, P. and P. Simpson. 1991. The choice of cell fate in the epidermis of *Drosophila. Cell* 64: 1083–1092.

Heldin, C.-H., K. Miyazono and P. ten Dijke. 1997. TGF-β signaling from cell membrane to nucleus through SMAD proteins. *Nature* 390: 465–471.

Hengartner, M. O., R. E. Ellis and H. R. Horvitz. 1992. *Caenorhabditis elegans* gene *ced-9* protects from programmed cell death. *Nature* 356: 494–499.

Henry, J. J. and R. M. Grainger. 1990. Early tissue interaction leading to embryonic lens formation in *Xenopus laevis. Dev. Biol.* 141: 149–163.

Hill, R. J. and P. W. Sternberg. 1992. The gene *lin-3* encodes an inductive signal for vulval development in *C. elegans. Nature* 358: 470–476.

Hirsch, M. S., L. E. Lunsford, V. Trinkaus-Randall and K. K. H. Svoboda. 1997. Chondrocyte survival and differentiation in situ are integrin medicated. *Dev. Dynam.* 210: 249–263.

Hogan, B. L. M. 1996. Bone morphogenesis proteins: Multifunctional regulators of vertebrate development. *Genes Dev.* 10: 1580–1594.

Holtzer, H. 1968. Induction of chondrogenesis: A concept in terms of mechanisms. *In* R. Gleischmajer and R. E. Billingham (eds.), *Epithelial-Mesenchymal Interactions.* Williams & Wilkins, Baltimore, pp. 152–164.

Hoppe, P. E. and R. J. Greenspan. 1986. Local function of the *Notch* gene for embryonic ectodermal pathway choice in *Drosophila. Cell* 46: 773–783.

Horwitz, A., K. Duggan, R. Greggs, C. Decker and C. Buck. 1986. The cell substrate attachment (CSAT) antigen has properties of a receptor for laminin and fibronectin. *J. Cell Biol.* 101: 2134–2144.

Horwitz, A., K. Duggan, R. Greggs, C. Decker and C. Buck. 1986. The cell substrate attachment (CSAT) antigen has properties of a receptor for laminin and fibronectin. *J. Cell Biol.* 101: 2134–2144.

Howe, A., A. E. Aplin, S. K. Alahari and R. L. Juliano. 1998. Integrin signaling and cell growth control. *Curr. Opin. Cell Biol.* 10: 220–231.

Hsu, Y.-R. and 10 others. 1997. The majority of stem cell factor exists as monomer under physiological conditions. *J. Biol. Chem.* 272: 6406–6416.

Huang, G. Y., A. Wessels, B. R. Smith, K. K. Linask, J. L. Ewart and C. W. Lo. 1998. Alteration in connexin 43 gap junction gene dosage impairs conotruncal heart development. *Dev. Biol.* 198: 32–44.

Ihle, J. N. 1996a. Cytokine receptor signalling. *Nature* 377: 591–594.

Ihle, J. N. 1996b. STATs: Signal transducers and activators of transcription. *Cell* 84: 331–334.

Jacobson, A. G. 1963. The determination and positioning of the nose, lens and ear. I. Interactions within the ectoderm, and between ectoderm and underlying tissues. *J. Exp. Zool.* 154: 273–283.

Jacobson, A. G. 1966. Inductive processes in embryonic development. *Science* 152: 25–34.

Jacobson, A. G. and A. K. Sater. 1988. Features of embryonic induction. *Development* 104: 341–359.

Jacobson, M. D., M. Weil and M. C. Raff. 1997. Programmed cell death in animal development. *Cell* 88: 347–354.

Jernvall, J. 1995. Mammalian molar cusp patterns: Developmental mechanisms of diversity. *Acta Zool. Fennica* 198: 1–61.

Johnson, R. L. and M. P. Scott. 1998. New players and puzzles in the hedgehog signaling pathway. *Curr. Opin. Genet. Dev.* 8: 450–456.

Johnson, R. L., E. Laufer, R. D. Riddle and C. Tabin. 1994. Ectopic expression of *Sonic hedgehog* alters dorsal-ventral patterning of somites. *Cell* 79: 1165–1173.

Johnson, R. L. and 10 others. 1996. Human homolog of *patched*, a candidate gene for the basal cell nevus syndrome. *Science* 272: 1668–1671.

Jones, C. M., N. Armes and J. C. Smith. 1996. Signalling by TGF-β family members: Short-range efects of Xnr-2 and BMP4 contrast with the long-range effects of activin. *Curr. Biol.* 6: 1468–1475.

Kalimi, G. H. and C. Lo. 1988. Communication compartments in the gastrulating mouse embryo. *J. Cell Biol.* 107: 241–256.

Katz, W. S. and P. W. Sternberg. 1996. Intercellular signalling in *Caenorhabditis elegans* vulval pattern formation. *Semin. Cell Dev. Biol.* 7: 175–183.

Katz, W., R. J. Hill, T. R. Clandenin and P. W. Sternberg. 1995. Different levels of the *C. elegans* growth factor LIN-3 promote distinct vulval precursor fates. *Cell* 82: 297–307.

Kauffmann, R. C., S. H. Li, P. A. Gallagher, J. J. Zhang and R. W. Carthew. 1996. Ras1 signaling and transcriptional competence in the R7 cell of *Drosophila. Genes Dev.* 10: 2167–2178.

Keeler, R. F. and W. Binns. 1968. Teratogenic compounds of *Veratrum californicum* (Durand). V. Comparison of cyclopian effects of steroidal alkaloids from the plan and structurally related compounds from other sources. *Teratology* 1: 5–10.

Kelley, R. I. and 7 others. 1996. Holoprosencephaly in RSH/Smith-Lemli-Opitz syndrome: Does abnormal cholesterol metabolism affect the function of Sonic hedghehog? *Am. J. Med. Genet.* 66: 478–484.

Kenyon, C. 1995. A perfect vulva every time: Gradients and signaling cascades in *C. elegans. Cell* 82: 171–174.

Kimble, J. 1981. Alterations in cell lineage following laser ablation of cells in the somatic gonad of *Caenorhabditis elegans. Dev. Biol.* 87: 286–300.

Knudsen, K., A. F. Horwitz and C. Buck. 1985. A monoclonal antibody identifies a glycoprotein complex involved in cell-substratum adhesion. *Exp. Cell Res.* 157: 218–226.

Koga, M. and Y. Ohshima. 1995. Mosaic analysis of the *let-23* gene function in vulval induction of *Caenorhabditis elegans. Development* 121: 2655–2666.

Korinek, V. and 7 others. 1997. Constitutive transcriptional activation by a β-catenin-Tcf complex in APC/colon carcinoma. *Science* 275: 1784–1786.

Kuida, K. and 7 others. 1996. Decreased apoptosis in the brain and premature lethality in CPP32-deficient mice. *Nature* 384: 368–372.

Kuida, K. and 8 others. 1998. Reduced apoptosis and cytochrome *c*-mediated caspase activation in mice lacking caspase 9. *Cell* 94: 325–337.

Lappi, D. A. 1995. Tumor targeting through fibroblast growth factor receptors. *Semin. Cancer Biol.* 6: 279–288.

Lazebnik, Y. A., S. H. Kaufmann, S. Desnoyers, G. G. Poirer and W. C. Earnshaw. 1994. Cleavage of poly(ADP-ribose) polymerase by a proteinase with properties like ICE. *Nature* 371: 346–347.

Lecourtois, M. and F. Schweisguth. 1998. Indirect evidence for Delta-dependent intercellular processing of Notch in *Drosophila* embryos. *Curr. Biol.* 8: 771–774.

Lee, S., N. B. Gilula and A. E. Warner. 1987. Gap junctional communication and compaction during preimplantation stages of mouse development. *Cell* 51: 851–860.

Leevers, S. J., H. F. Paterson and C. J. Marshall. 1994. Requirement for Ras in Raf activation is overcome by targeting Raf to the plasma membrane. *Nature* 369: 411–414.

Lehmann, R., F. Jimenez, U. Dietrich and J. A. Campos-Ortega. 1983. On the phenotype and development of mutants of early neurogenesis in *Drosophila melanogaster. Wilhelm Roux Arch. Dev. Biol.* 192: 62–74.

Letterio, J. J., A. G. Geiser, A. B. Kulkarni, A. B. Roche, N. S. Sporn and A. B. Roberts. 1994. Maternal rescue of TGF-β1-null mice. *Science* 264: 1936–1938.

Li, H.-S., J.-M. Yang, R. D. Jacobson, D. Pasko and O. Sundin. 1994. Pax-6 is first expressed in a region of ectoderm anterior to the early neural plate: Implications for stepwise determination of the lens. *Dev. Biol.* 162: 181–194.

Li, W., C. D. Whaley, A. Mondino and D. L. Mueller. 1996. Blocked signal transduction to the ERK and JNK protein kinases in anergic CD4$^+$ T cells. *Science* 271: 1272–1274.

Linask, K. L. and J. W. Lash. 1988. A role for fibronectin in the migration of avian precardiac cells. I. Dose-dependent effects of fibronectin antibody. *Dev. Biol.* 129: 315–323.

Liu, J.-K., M. C. Di Persio and K. S. Zaret. 1991. Extracellular signals that regulate liver transcription factors during hepatic differentiation in vitro. *Mol. Cell Biol.* 11: 773–784.

Lo, C. and N. B. Gilula. 1979. Gap junctional communication in the preimplantation mouse embryo. *Cell* 18: 399–409.

Luna, E. J. and A. L. Hitt. 1992. Cytoskeleton-plasma membrane interactions. *Science* 258: 955–964.

Ma, C., Y. Zhou, P. A. Beachy and K. Moses. 1993. The segment polarity gene hedgehog is required for progression of the morphogenetic furrow in the developing *Drosophila* retina. *Cell* 75: 927–938.

Maimonides (Moshe ben Maimon). 1190. *A Guide for the Perplexed* (M. Friedländer, trans.). Dover, New York, 1956.

Martins-Green, M. and M. J. Bissell. 1995. Cell-ECM interactions in development. *Semin. Dev. Biol.* 6: 149–159.

McCormick, F. 1989. Ras GTPase activating protein: Signal transmitter and signal terminator. *Cell* 56: 5–8.

McMahon, A. P. and A. Bradley. 1990. The *Wnt-1 (int-1)* proto-oncogene is required for the development of a large region of the mouse brain. *Cell* 62: 1073–1086.

Meier, S. 1977. Initiation of corneal differentiation prior to cornea-lens association. *Cell Tissue Res.* 184: 255–267.

Mlodzik, M., Y. Hiromi, U. Weber, C. S. Goodman and G. M. Rubin. 1990. The *Drosophila* *seven-up* gene, a member of the steroid receptor gene superfamily, controls photoreceptor cell fates. *Cell* 60: 211–224.

Montgomery, A. M. P., R. A. Reisfeld and D. A. Cheresh. 1994. Integrin αvβ3 rescues melanoma cells from apoptosis in a three-dimensional dermal collagen. *Proc. Natl. Acad. Sci. USA* 91: 8856–8860.

Muschler, J., A. Lochter, C. D. Roskelley, P. Yurchenko and M. J. Bissell. 1999. Division of labor among the α6β4 integrins, β1 integrins, and an E3 laminin receptor to signal morphogenesis and β-casein expression in mammary epithelial cells. *Mol. Biol. Cell* 10: 2817–2828.

Muthukkarapan, V. R. 1965. Inductive tissue interaction in the development of the mouse lens in vitro. *J. Exp. Zool.* 159: 269–288.

Nakayama, A., M.-T. Nguyen, C. C. Chen, K. Opdecamp, C. A. Hodgkinson and H. Arnheiter. 1998. Mutations in *microphthalmia*, the mouse homolog of the human deafness gene *MITF*, affect neuroepithelial and neural crest-derived melanocytes differently. *Mech. Dev.* 70: 155–166.

Nijhout, H. F. and S. M. Paulsen. 1997. Developmental models and polygenic characters. *Amer. Nat.* 149: 394-405.

Nohno, T. W., Y. Kawakami, H. Ohuchi, A. Fujiwara, H. Yoshioka and S. Noji. 1995. Involvement of the *sonic hedgehog* gene in chick feather formation. *Biochem. Biophys. Res. Comm.* 206: 33–39.

Nomura, M. and E. Li. 1998. Smad2 role in mesoderm formation, left-right patterning, and craniofacial development. *Nature* 393: 786–790.

Notenboom, R. G. E., P. A. J. de Poer, A. F. M. Moorman and W. H. Lamers. 1996. The establishment of the hepatic architecture is a prerequisite for the development of a lobular pattern of gene expression. *Development* 122: 321–332.

Ogino, H. and K. Yasuda. 1998. Induction of lens differentiation by activation of a bZIP transcription factor, L-maf. *Science* 280: 115–118.

Padgett, R. W., J. M. Wozney and W. M. Gelbart. 1993. Human BMP sequences can confer normal dosal-ventral patterning in the *Drosophila* embryo. *Proc. Natl. Acad. Sci. USA* 90: 2905–2809.

Pennisi, E. 1998. How a growth control path takes a wrong turn to cancer. *Science* 281: 1439–1441.

Peracchia, C. and A. F. Dulhunty. 1976. Low resistance junctions in crayfish: Structural changes with functional uncoupling. *J. Cell Biol.* 70: 419–439.

Placzek, M., M. Tessier-Lavigne, T. Yamada, T. Jessell and J. Dodd. 1990. Mesodermal control of neural cell identity: Floor plate induction by the notochord. *Science* 250: 985–988.

Poulson, D. F. 1937. Chromosomal deficiencies and the embryonic development of *Drosophila melanogaster. Proc. Natl. Acad. Sci. USA* 23: 133–137.

Price, E. R. and 7 others. 1998. Lineage-specific signaling in melanocytes: c-Kit stimulation recruits p300/CBP to microphthalmia. *J. Biol. Chem.* 273: 17983–17986

Reaume, A. G. and 8 others. 1995. Cardiac malformation in neonatal mice lacking connexin43. *Science* 267: 1831–1834.

Reilly, K. M. and D. A. Melton. 1996. Short-range signaling by candidate morphogens of the TGF-β family and evidence for a relay mechanism of induction. *Cell* 86: 743–754.

Reinke, R. and A. L. Zipursky. 1988. Cell-cell interaction in the *Drosophila* retina: The *bride of sevenless* gene is required in photoreceptor cell R8 for R7 cell development. *Cell* 55: 321–330.

Riddle, R. D., R. L. Johnson, E. Laufer and C. Tabin. 1993. *Sonic hedgehog* mediates the polarizing activity of the ZPA. *Cell* 75: 1401–1416.

Ritvos, O., T. Tuuri, M. Erämaa, K. Sainio, Hilden, L. Saxén and S. F. Gilbert.1995. Activin disrupts epithelial branching morphogenesis in developing murine kidney, pancreas, and salivary gland. *Mech. Dev.* 50: 229–246.

Roberts, D. S. and S. A. Miller. 1998. Apoptosis in cavitation of middle ear space. *Anatomical Rec.* 251: 286–289.

Rodrigez, I, K. Araki, K. Khatib, J.-C. Martinou and P. Vassalli. 1997. Mouse vaginal opening is an apoptosis-dependent process which can be prevented by the overexpression of Bcl2. *Dev. Biol.* 184: 115–121.

Roessler, E. and 7 others. 1996. Mutations in the human *Sonic hedgehog* gene cause holoprosencephaly. *Nat. Genet.* 14: 357–360.

Roskelley, C. D., P. Y. Desprez and M. J. Bissell. 1994. Extracellular matrix-dependent tissue-specific gene expression in mammary epithelial cells requires both physical and biochemical signal transduction. *Proc. Natl. Acad. Sci. USA* 91: 12378–12382.

Rousseau, F. and 7 others. 1994. Mutations in the gene encoding fibroblast growth factor receptor-3 in achondroplasia. *Nature* 371: 252–254.

Rubin, G. M. 1989. Development of the *Drosophila* retina: Inductive events studied at single cell resolution. *Cell* 57: 519–520.

Ruoslahti, E. and M. D. Pierschbacher. 1987. New perspectives in cell adhesion: RGD and integrins. *Science* 238: 491–497.

Saha, M. S., C. L. Spann and R. M. Grainger. 1989. Embryonic lens induction: More than meets the optic vesicle. *Cell Diff. Dev.* 28: 153–172.

Sainio, K., S. F. Gilbert, E. Lehtonen, M. Nishi, N. M. Kumar, N. B. Gilula and L. Saxén. 1992. Differential expression of gap junction mRNAs and proteins in the developing murine kidney and in experimentally induced nephric mesenchymes. *Development* 115: 827–837.

Saudou, F., S. Finkbeiner, D. Devys and M. E. Greenberg. 1998. Huntingtin acts in the nucleus to induce apoptosis but death does not correlate with the formation of intranuclear inclusions. *Cell* 95: 55–66.

Saunders, J. W., Jr. 1980. *Developmental Biology.* Macmillan, New York.

Saunders, J. W., Jr. and J. F. Fallon. 1966. Cell death and morphogenesis. *In* M. Locke (ed.) *Major Problems of Developmental Biology.* Academic Press, New York, pp. 289–314.

Saunders, J. W., Jr., J. M. Cairns and M. T. Gasseling. 1957. The role of the apical ectodermal ridge of ectoderm in the differentiation of the morphological structure of and inductive specificity of limb parts of the chick. *J. Morphol.* 101: 57–88.

Saxén, L., E. Lehtonen, M. Karkinen-Jääskeläinen, S. Nordling and J. Wartiovaara. 1976. Are morphogenetic tissue interactions mediated by transmissable signal substances or through cell contacts. *Nature* 259: 662–663.

Schroeder, E. H., J. A. Kisslinger and R. Kopan. 1998. Notch-1 signalling requires ligand-induced proteolytic release of intracellular domain. *Nature* 393: 382–386.

Seydoux, G. and I. Greenwald. 1989. Cell autonomy of *lin-12* function in a cell fate decision in *C. elegans. Cell* 57: 1237–1246.

Shaham, S. and H. R. Horvitz. 1996. An alternatively spliced *C. elegans ced-4* RNA encodes a novel cell death inhibitor. *Cell* 86: 201–208.

Shiang, R. and 7 others. 1994. Mutations in the transmembrane domain of FGFR3 cause the most common genetic form of dwarfism, achondroplasia. *Cell* 78: 335–342.

Shih, C. and R. A. Weinberg. 1982. Isolation of a transforming sequence from a human bladder carcinoma cell line. *Cell* 29: 161–169.

Shin, S. H., P. Kogerman, E. Lindström, R. Toftgård and L. G. Biesecker. 1999. *GLI3* mutations in human disorders mimic Cubitus interruptus protein functions and localization. *Proc. Natl. Acad. Sci. USA* 96: 2880–2884.

Simske, J. S. and S. K. Kim. 1996. Sequential signaling during *Caenorhabditis elegans* vulval induction. *Nature* 375: 142–146.

Slavkin, H. C. and P. Bringas, Jr. 1976. Epithelial mesenchymal interactions during odontogenesis. IV. Morphological evidence for direct heterotypic cell-cell contacts. *Dev. Biol.* 50: 428–442.

Spemann, H. and O. Schotté. 1932. Über xenoplatische Transplantation als Mittel zur Analyse der embryonalen Induktion. *Naturwissenschaften* 20: 463–467.

Socolovsky, M., A. E. J. Fllon, S. Wang, C. Brugnara and H. F. Lodish. 1999. Fetal anemia and apoptosis of red cell progenitors in Stat5a$^{-/-}$5b$^{-/-}$ mice: A direct role for Stat5 in Bcl-X$_L$ induction. *Cell* 98: 181–191.

Spritz, R. A., L. B. Gielbel and S. A. Holmes. 1992. Dominant negative and loss-of-function mutations of the *c-kit* (mast/Stem cell growth factor receptor) proto-oncogene in human piebaldism. *Am. J. Hum. Genet.* 50: 261–269.

Stern, H. M., A. M. C. Brown and S. D. Hauschka. 1995. Myogenesis in paraxial mesoderm: Preferential induction by dorsal neural tube and by cells expressing Wnt-1. *Development* 121: 3675–3686.

Sternberg, P. W. 1988. Lateral inhibition during vulval induction in *Caenorhabditis elegans. Nature* 335: 551–554.

Stokoe, D., S. G. Macdonald, K. Cadwallader, M. Symons and J. F. Hancock. 1994. Activation of raf as well as recruitment to the plasma membrane. *Science* 264: 1463–1467.

Streuli, C. H., N. Bailey and M. J. Bissell. 1991. Control of mammary epithelial differentiation: Basement membrane induces tissue specific gene expression in the absence of cell-cell interactions and morphological polarity. *J. Cell Biol.* 115: 1383–1396.

Struhl, G. and A. Adachi. 1998. Nuclear access and action of Notch in vivo. *Cell* 93: 382–386.

Su, W.-C. S. and 8 others. 1997. Activation of Stat1 by mutant fibroblast growth factor receptor in thanatophoric dysplasia type II dwarfism. *Nature* 386: 288–292.

Tamkun, J. W., D. W. DeSimone, D. Fonda, R. S. Patel, C. Buck, A. F. Horwitz and R. O. Hynes. 1986. Structure of integrin, a glycoprotein involved in transmembrane linkage between fibronectin and actin. *Cell* 46: 271–282.

Tan, P. B., M. R. Lackner and S. K. Kim. 1998. MAP kinase signaling specificity mediated by the LIN-1/Ets/LIN31 WH transcription factor complex during *C. elegans* vulval induction. *Cell* 93: 569–580.

Thomas, L. 1992. *The Fragile Species.* Macmillan, New York

Ting-Berreth, S. A. and C.-M. Chuong. 1996. Local delivery of TGFβ2 can substitute for placode epithelium to induce mesenchymal condensation during skin morphogenesis. *Dev. Biol.* 179: 347–359.

Tomlinson, A. 1988. Cellular interactions in the developing *Drosophila* eye. *Development* 104: 183–193.

Tomlinson, A. and D. F. Ready. 1987. Cell fate in the *Drosophila* ommatidium. *Dev. Biol.* 123: 264–276.

Vaahtokari, A., T. Aberg, J. Jernvall, S. Keränen and I. Thesleff. 1996a. The enamel knot as a signalling center in the developing mouse tooth. *Mech. Dev.* 54: 39–43.

Vaahtokari, A., T. Aberg and I. Thesleff. 1996b. Apoptosis in the developing tooth: Association with an embryonic signaling center and suppression by EGF and FGF-4. *Development* 122: 121–129.

Vaux, D. L., I. L. Weissman and S. K. Kim. 1992. Prevention of programmed cell death in *Caenorhabditis elegans* by human *bcl-2. Science* 258: 1955–1957.

Waddington, C. H. 1940. *Organisers and Genes.* Cambridge University Press, Cambridge.

Wang, N., J. P. Butler and D. E. Ingber. 1993. Mechanotransduction across the cell surface and through the cytoskeleton. *Science* 260: 1124–1127.

Wang, S. and 7 others. 1998. Notch receptor activation inhibits oligodendrocyte differentiation. *Neuron* 21: 63–76.

Warner, A. E., S. C. Guthrie and N. B. Gilula. 1984. Antibodies to gap junctional protein selectively disrupt junctional communication in the early amphibian embryo. *Nature* 311: 127–131.

Wary, K. K., A. Mariotti, C. Zurzolo and F. Giancotti. 1998. A requirement for caveolin-1 and associated kinase Fyn in integrin signaling and anchorage-dependent cell growth. *Cell* 94: 625–634.

Webster, M. K. and D. J. Donoghue. 1996. Constitutive activation of fibroblast growth factor receptor 3 by the transmembrane domain point mutation found in achondroplasia. *EMBO J.* 15: 520–527.

Wessells, N. K. 1977. *Tissue Interaction and Development*. Benjamin Cummings, Menlo Park, CA.

Wilkinson, H. A., K. Fitzgerald and I. Greenwald. 1994. Reciprocal changes in expression of the receptor *lin-12* and its ligand *lag-2* prior to commitment in a *C. elegans* cell fate decision. *Cell* 79: 1187–1198.

Williams, E. J., J. Furness, F. S. Walsh and P. Doherty. 1994a. Activation of the FGF receptor underlies neurite outgrowth stimulated by L1, N-CAM, and N-cadherin. *Neuron* 13: 583–594.

Williams, E. J., F. S. Walsch and P. Doherty. 1994b. Tyrosine kinase inhibitors can differentially inhibit integrin-dependent and CAM-dependent neurite outgrowth. *J. Cell Biol.* 124: 1029–1037.

Witte, O. N. 1990. *Steel* locus defines new multipotent growth factor. *Cell* 63: 5–6.

Wolf, U. 1995. The genetic contribution to the phenotype. *Hum. Genet.* 95: 127–148.

Wolf, U. 1997. Identical mutations and phenotypic variation. *Hum. Genet.* 100: 305–321.

Wu, M. and 7 others. 2000. c-Kit triggers dual phosphorylation, which couples activation and degradation of the essential melanocyte factor Mi. *Genes. Dev.* 14: 301–312.

Yamada, T., S. L. Pfaff, T. Edlund and T. M. Jessell. 1993. Control of cell pattern in the neural tube: Motor neuron induction by diffusible factors from notochord and floor plate. *Cell* 73: 673–686.

Yoshida, H. and 7 others. 1998. Apaf1 is required for mitochondrial pathways of apoptosis and brain development. *Cell* 94: 739–750.

Zygar, C. A., T. L. Cook, Jr. and R. M. Grainger. 1998. Gene activation during early stages of lens induction in *Xenopus*. *Development* 125: 3509–3519.

PART II Early Embryonic Development

7 *Fertilization: Beginning a new organism*

Urge and urge and urge,
Always the procreant urge of the world.
Out of the dimness opposite equals
*   advance,*
Always substance and increase,
*   always sex,*
Always a knit of identity,
*   always distinction,*
Always a breed of life.
                    WALT WHITMAN (1855)

The final aim of all love intrigues, be
they comic or tragic, is really of more
importance than all other ends in
human life. What it turns upon is noth-
ing less than the composition of the next
generation.

A. SCHOPENHAUER
(QUOTED BY C. DARWIN, 1871)

FERTILIZATION IS THE PROCESS whereby two sex cells (gametes) fuse together to create a new individual with genetic potentials derived from both parents. Fertilization accomplishes two separate ends: sex (the combining of genes derived from the two parents) and reproduction (the creation of new organisms). Thus, the first function of fertilization is to transmit genes from parent to offspring, and the second is to initiate in the egg cytoplasm those reactions that permit development to proceed.

Although the details of fertilization vary from species to species, conception generally consists of four major events:

1. Contact and recognition between sperm and egg. In most cases, this ensures that the sperm and egg are of the same species.
2. Regulation of sperm entry into the egg. Only one sperm can ultimately fertilize the egg. This is usually accomplished by allowing only one sperm to enter the egg and inhibiting any others from entering.
3. Fusion of the genetic material of sperm and egg.
4. Activation of egg metabolism to start development.

Structure of the Gametes

A complex dialogue exists between egg and sperm. The egg activates the sperm metabolism that is essential for fertilization, and the sperm reciprocates by activating the egg metabolism needed for the onset of development. But before we investigate these aspects of fertilization, we need to consider the structures of the sperm and egg—the two cell types specialized for fertilization.

Sperm

It is only within the past century that the sperm's role in fertilization has been known. Anton van Leeuwenhoek, the Dutch microscopist who co-discovered sperm in 1678, first believed them to be parasitic animals living within the semen (hence the term *spermatozoa*, meaning "sperm animals"). He originally assumed that they had nothing at all to do with reproducing the organism in which they were found, but he later came to believe that each sperm contained a preformed embryo. Leeuwenhoek (1685) wrote that sperm were seeds (both *sperma* and *semen* mean "seed") and that the female merely provided the nutrient soil in which the seeds were planted. In this, he was returning to a notion of procreation promulgated by Aristotle 2000 years earlier. Try as he might, Leeuwenhoek was continually disappointed in his attempts to

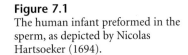

Figure 7.1
The human infant preformed in the sperm, as depicted by Nicolas Hartsoeker (1694).

find the preformed embryo within the spermatozoa. Nicolas Hartsoeker, the other co-discoverer of sperm, drew a picture of what he hoped to find: a preformed human ("homunculus") within the human sperm (Figure 7.1). This belief that the sperm contained the entire embryonic organism never gained much acceptance, as it implied an enormous waste of potential life. Most investigators regarded the sperm as unimportant. (See Pinto-Correia 1997 for details of this remarkable story.)

WEBSITE **7.1 Leeuwenhoek and images of homunculi.** Scholars in the 1600s thought that either the sperm or the egg carried the rudiments of the adult body. Moreover, these views became distorted by contemporary commentators and later historians.

The first evidence suggesting the importance of sperm in reproduction came from a series of experiments performed by Lazzaro Spallanzani in the late 1700s. Spallanzani demonstrated that filtered toad semen devoid of sperm would not fertilize eggs. He concluded, however, that the viscous fluid retained by the filter paper, and not the sperm, was the agent of fertilization. He, like many others, felt that the spermatic "animals" were parasites.

The combination of better microscopic lenses and the cell theory led to a new appreciation of spermatic function. In 1824, J. L. Prevost and J. B. Dumas claimed that sperm were not parasites, but rather the active agents of fertilization. They noted the universal existence of sperm in sexually mature males and their absence in immature and aged individuals. These observations, coupled with the known absence of spermatozoa in the sterile mule, convinced them that "there exists an intimate relation between their presence in the organs and the fecundating capacity of the animal." They proposed that the sperm entered the egg and contributed materially to the next generation.

These claims were largely disregarded until the 1840s, when A. von Kolliker described the formation of sperm from cells within the adult testes. He ridiculed the idea that the semen could be normal and yet support such an enormous number of parasites. Even so, von Kolliker denied that there was any physical contact between sperm and egg. He believed that the sperm excited the egg to develop, much as a magnet communicates its presence to iron. It was only in 1876 that Oscar Hertwig and Herman Fol independently demonstrated

sperm entry into the egg and the union of the two cells' nuclei. Hertwig had sought an organism suitable for detailed microscopic observations, and he found that the Mediterranean sea urchin, *Toxopneustes lividus*, was perfect. Not only was it common throughout the region and sexually mature throughout most of the year, but its eggs were available in large numbers and were transparent even at high magnifications. After mixing sperm and egg suspensions together, Hertwig repeatedly observed a sperm entering an egg and saw the two nuclei unite. He also noted that only one sperm was seen to enter each egg, and that all the nuclei of the embryo were derived from the fused nucleus created at fertilization. Fol made similar observations and detailed the mechanism of sperm entry. Fertilization was at last recognized as the union of sperm and egg, and the union of sea urchin gametes remains one of the best-studied examples of fertilization.

WEBSITE **7.2 The origins of fertilization research.** Studies by Hertwig, Fol, Boveri, and Auerbach integrated cytology with genetics. The debates over meiosis and nuclear structure were critical in these investigations of fertilization.

Each sperm consists of a haploid nucleus, a propulsion system to move the nucleus, and a sac of enzymes that enable the nucleus to enter the egg. Most of the sperm's cytoplasm is eliminated during maturation, leaving only certain organelles that are modified for spermatic function (Figure 7.2). During the course of sperm maturation, the haploid nucleus becomes very streamlined, and its DNA becomes tightly compressed. In front of this compressed haploid nucleus lies the **acrosomal vesicle**, or **acrosome**, which is derived from the Golgi apparatus and contains enzymes that digest proteins and complex sugars; thus, it can be considered a modified secretory vesicle. These stored enzymes are used to lyse the outer coverings of the egg. In many species, such as sea urchins, a region of globular actin molecules lies between the nucleus and the acrosomal vesicle. These proteins are used to extend a fingerlike **acrosomal process** from the sperm during the early stages of fertilization. In sea urchins and several other species, recognition between sperm and egg involves molecules on the acrosomal process. Together, the acrosome and nucleus constitute the head of the sperm.

The means by which sperm are propelled vary according to how the species has adapted to environmental conditions. In some species (such as the parasitic roundworm *Ascaris*), the sperm travel by the amoeboid motion of lamellipodial extensions of the cell membrane. In most species, however, each sperm is able to travel long distances by whipping its **flagellum**. Flagella are complex structures. The major motor portion of the flagellum is called the **axoneme**. It is formed by microtubules emanating from the centriole at the base of the sperm nucleus (Figures 7.2 and 7.3). The core of the axoneme consists of two central microtubules surrounded by a row of nine doublet microtubules. Actually, only one microtubule of

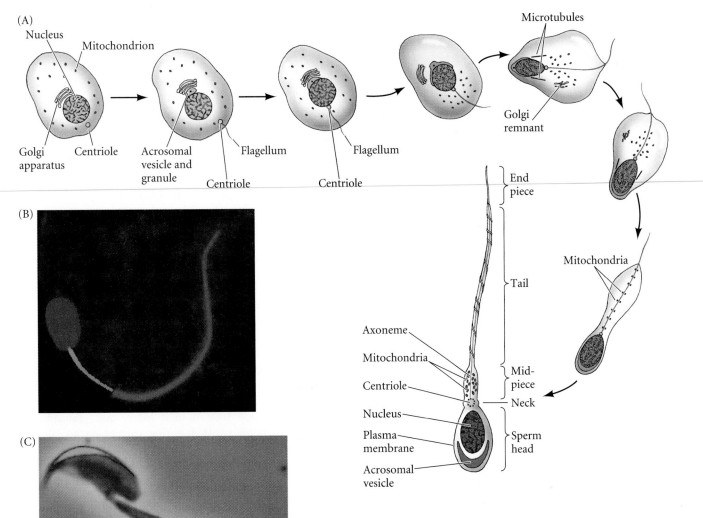

Figure 7.2
The modification of a germ cell to form a mammalian sperm. (A)
The centriole produces a long flagellum at what will be the posterior end of the sperm, and the Golgi apparatus forms the acrosomal vesicle at the future anterior end. The mitochondria (hollow dots) collect around the flagellum near the base of the haploid nucleus and become incorporated into the midpiece of the sperm. The remaining cytoplasm is jettisoned, and the nucleus condenses. The size of the mature sperm has been enlarged relative to the other stages. (B) Mature bull sperm. The DNA is stained blue with DAPI; the mitochondria are stained green, and the tubulin of the flagellum is stained red. (C) Acrosome of mouse sperm, stained green by GFP. A construct whereby the GFP gene was combined to the proacrosin promoter caused GFP to accumulate in the acrosome. (A after Clermont and Leblond 1955; B from Sutovsky et al. 1996, courtesy of G. Schatten; C courtesy of K.-S. Kim and G. L. Gerton.)

each doublet is complete, having 13 protofilaments; the other is C-shaped and has only 11 protofilaments (Figure 7.3B). A three-dimensional model of a complete microtubule is shown in Figure 7.3C. Here we can see the 13 interconnected protofilaments, which are made exclusively of the dimeric protein tubulin.

Although tubulin is the basis for the structure of the flagellum, other proteins are also critical for flagellar function. The force for sperm propulsion is provided by **dynein**, a protein that is attached to the microtubules (Figure 7.3B). Dynein hydrolyzes molecules of ATP and can convert the released chemical energy into the mechanical energy that propels the sperm. This energy allows the active sliding of the outer doublet microtubules, causing the flagellum to bend (Ogawa et al. 1977; Shinyoji et al. 1998). The importance of dynein can be seen in individuals with the genetic syndrome called the Kartagener triad. These individuals lack dynein on all their ciliated and flagellated cells, rendering these structures immotile. Males with this disease are sterile (immotile sperm), are susceptible to bronchial infections (immotile respiratory cilia), and have a 50% chance of having the heart on the right side of the body (Afzelius 1976). Another important flagellar

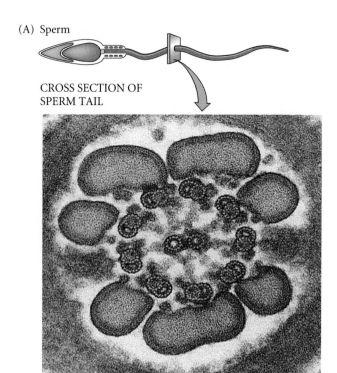

CROSS SECTION OF
SPERM TAIL

(B)

Plasma membrane

Radial spoke

Spoke head

Nexin

Subfiber A

Subfiber B

Central singlet
microtubule

Inner dynein arm

Outer dynein arm

AXONEME

MICROTUBULE
DOUBLET

(C)

"A"
MICROTUBULE

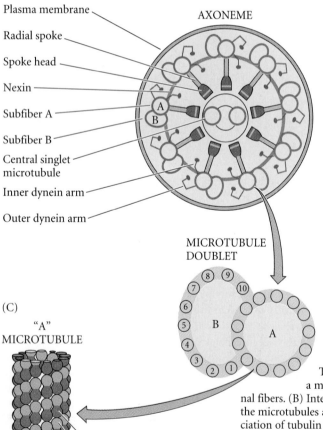

α-Tubulin monomer

β-Tubulin monomer

Tubulin
dimer

protein appears to be histone H1. This protein is usually found inside the nucleus, where it folds the chromatin into tight clusters. However, Multigner and colleagues (1992) found that this same protein stabilizes the flagellar microtubules so that they do not disassemble.

The "9 + 2" microtubule arrangement with the dynein arms has been conserved in axonemes throughout the eukaryotic kingdoms, suggesting that this arrangement is extremely well suited for transmitting energy for movement. The ATP needed to whip the flagellum and propel the sperm comes from rings of mitochondria located in the neck region of the sperm (see Figure 7.2). In many species (notably mammals), a layer of dense fibers has interposed itself between the mitochondrial sheath and the axoneme. This fiber layer stiffens the sperm tail. Because the thickness of this layer decreases toward the tip, the fibers probably prevent the sperm head from being whipped around too suddenly. Thus, the sperm has undergone extensive modification for the transport of its nucleus to the egg.

The differentiation of mammalian sperm is not completed in the testes. After being expelled into the lumen of the seminiferous tubules, the sperm are stored in the epididymis, where they acquire the ability to move. Motility is achieved through changes in the ATP-generating system (possibly through modification of dynein) as well as changes in the plasma membrane that make it more fluid (Yanagimachi 1994). The sperm released during ejaculation are able to move, yet they do not yet have the capacity to bind to and fertilize an egg. These final stages of sperm maturation (called capacitation) do not occur until the sperm has been inside the female reproductive tract for a certain period of time.

The egg

All the material necessary for the beginning of growth and development must be stored in the mature egg (the **ovum**). Whereas the sperm has eliminated most of its cytoplasm, the developing egg (called the **oocyte** before it reaches the stage of meiosis at which it is fertilized) not only conserves its material, but is actively involved in accumulating more. The meiotic divisions that form the oocyte conserve its cytoplasm (rather than giving half of it away), and the oocyte either synthesizes

Figure 7.3
The motile apparatus of the sperm. (A) Cross section of the flagellum of a mammalian spermatozoon, showing the central axoneme and the external fibers. (B) Interpretive diagram of the axoneme, showing the "9 + 2" arrangement of the microtubules and other flagellar components. The schematic diagram shows the association of tubulin protofilaments into a microtubule doublet. The first ("A") portion of the doublet is a normal microtubule comprising 13 protofilaments. The second ("B") portion of the doublet contains only 11 (occasionally 10) protofilaments. The dynein arms contain the ATPases that provide the energy for flagellar movement. (C) A three-dimensional model of an "A" microtubule. The α- and β-tubulin subunits are similar but not identical, and the microtubule can change size by polymerizing or depolymerizing tubulin subunits at either end. (A, photograph courtesy of D. M. Phillips; B after De Robertis et al. 1975 and Tilney et al. 1973.)

or absorbs proteins, such as yolk, that act as food reservoirs for the developing embryo. Thus, birds' eggs are enormous single cells, swollen with their accumulated yolk. Even eggs with relatively sparse yolk are comparatively large. The volume of a sea urchin egg is about 200 picoliters (2×10^{-4} mm^3, more than 10,000 times the volume of the sperm) (Figure 7.4). So, while sperm and egg have equal haploid nuclear components, the egg also has a remarkable cytoplasmic storehouse that it has accumulated during its maturation. This cytoplasmic trove includes the following:*

- **Proteins.** It will be a long while before the embryo is able to feed itself or obtain food from its mother. The early embryonic cells need a supply of energy and amino acids. In many species, this is accomplished by accumulating yolk proteins in the egg. Many of the yolk proteins are made in other organs (liver, fat body) and travel through the maternal blood to the egg.
- **Ribosomes and tRNA.** The early embryo needs to make many of its own proteins, and in some species, there is a burst of protein synthesis soon after fertilization. Protein synthesis is accomplished by ribosomes and tRNA, which exist in the egg. The developing egg has special mechanisms to synthesize ribosomes, and certain amphibian oocytes produce as many as 10^{12} ribosomes during their meiotic prophase.
- **Messenger RNA.** In most organisms, the instructions for proteins made during early development are already packaged in the oocyte. It is estimated that the eggs of sea urchins contain 25,000 to 50,000 different types of mRNA. This mRNA, however, remains dormant until after fertilization (see Chapter 5).
- **Morphogenetic factors.** Molecules that direct the differentiation of cells into certain cell types are present in the egg. They appear to be localized in different regions of the egg and become segregated into different cells during cleavage (see Chapter 8).
- **Protective chemicals.** The embryo cannot run away from predators or move to a safer environment, so it must come equipped to deal with threats. Many eggs contain ultraviolet filters and DNA repair enzymes that protect them from sunlight; some eggs contain molecules that potential predators find distasteful; and the yolk of bird eggs even contains antibodies.

WEBSITE **7.3 The egg and its environment.** The laboratory is not where most eggs are found. Eggs have evolved remarkable ways to protect themselves in particular environments.

*The contents of the egg vary greatly from species to species. The synthesis and placement of these materials will be addressed in Chapter 19, when we discuss the differentiation of germ cells.

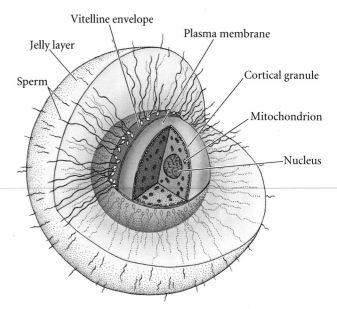

Figure 7.4
Structure of the sea urchin egg during fertilization. The drawing also shows the relative sizes of egg and sperm. (After Epel 1977.)

Within this enormous volume of cytoplasm resides a large nucleus. In some species (e.g., sea urchins), the nucleus is already haploid at the time of fertilization. In other species (including many worms and most mammals), the egg nucleus is still diploid, and the sperm enters before the meiotic divisions are completed. The stage of the egg nucleus at the time of sperm entry in different species is illustrated in Figure 7.5.

Enclosing the cytoplasm is the egg **plasma membrane**. This membrane must regulate the flow of certain ions during fertilization and must be capable of fusing with the sperm plasma membrane. Outside the plasma membrane is the **vitelline envelope** (Figure 7.6), which forms a fibrous mat around the egg. This envelope contains at least eight different glycoproteins and is often involved in sperm-egg recognition (Correia and Carroll 1997). It is supplemented by extensions of membrane glycoproteins from the plasma membrane and by proteinaceous vitelline posts that adhere the vitelline envelope to the membrane (Mozingo and Chandler 1991). The vitelline envelope is essential for the species-specific binding of sperm. In mammals, the vitelline envelope is a separate and thick extracellular matrix called the **zona pellucida**. The mammalian egg is also surrounded by a layer of cells called the **cumulus** (Figure 7.7), which is made up of the ovarian follicular cells that were nurturing the egg at the time of its release from the ovary. Mammalian sperm have to get past these cells to fertilize the egg. The innermost layer of cumulus cells, immediately adjacent to the zona pellucida, is called the **corona radiata.**

Lying immediately beneath the plasma membrane of the egg is a thin shell (about 5 μm) of gel-like cytoplasm called the **cortex**. The cytoplasm in this region is stiffer than the internal cytoplasm and contains high concentrations of globular actin

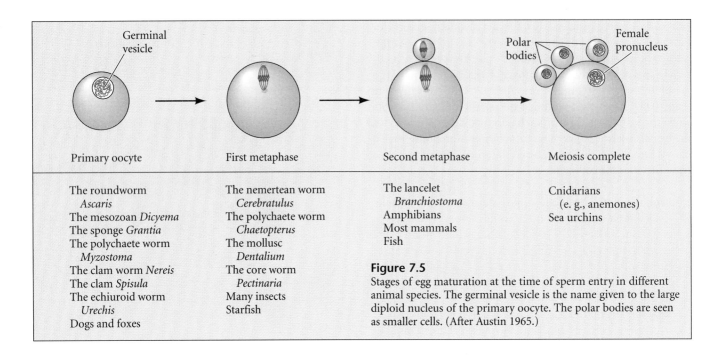

Primary oocyte	First metaphase	Second metaphase	Meiosis complete
The roundworm *Ascaris* The mesozoan *Dicyema* The sponge *Grantia* The polychaete worm *Myzostoma* The clam worm *Nereis* The clam *Spisula* The echiuroid worm *Urechis* Dogs and foxes	The nemertean worm *Cerebratulus* The polychaete worm *Chaetopterus* The mollusc *Dentalium* The core worm *Pectinaria* Many insects Starfish	The lancelet *Branchiostoma* Amphibians Most mammals Fish	Cnidarians (e. g., anemones) Sea urchins

Figure 7.5

Stages of egg maturation at the time of sperm entry in different animal species. The germinal vesicle is the name given to the large diploid nucleus of the primary oocyte. The polar bodies are seen as smaller cells. (After Austin 1965.)

molecules. During fertilization, these actin molecules polymerize to form long cables of actin known as **microfilaments**. Microfilaments are necessary for cell division, and they also are used to extend the egg surface into small projections called **microvilli**, which may aid sperm entry into the cell (see Figure 7.6B; also see Figure 7.19). Also within the cortex are the **cortical granules** (see Figures 7.4 and 7.6B). These membrane-bound structures, which are homologous to the acrosomal vesicle of the sperm, are Golgi-derived organelles containing proteolytic enzymes. However, whereas each sperm contains one acrosomal vesicle, each sea urchin egg contains approximately 15,000 cortical granules. Moreover, in addition to digestive enzymes, the cortical granules contain mucopolysaccharides, adhesive glycoproteins, and hyalin protein. The enzymes and mucopolysaccharides are active in preventing other sperm from entering the egg after the first sperm has entered, and the hyalin and adhesive glycoproteins surround the early embryo and provide support for the cleavage-stage blastomeres.

(A)

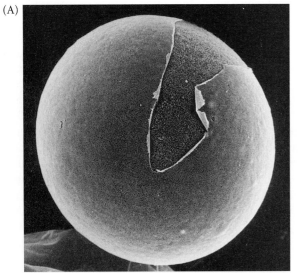

(B)

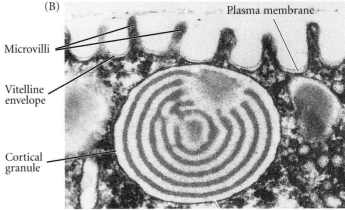

Figure 7.6

The sea urchin egg cell surface. (A) Scanning electron micrograph of an egg before fertilization. The plasma membrane is exposed where the vitelline envelope has been torn. (B) Transmission electron micrograph of an unfertilized egg, showing microvilli and plasma membrane, which are closely covered by the vitelline envelope. A cortical granule lies directly beneath the plasma membrane. (From Schroeder 1979; photographs courtesy of T. E. Schroeder.)

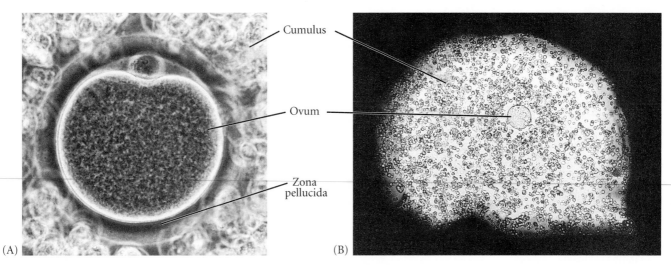

Cumulus

Ovum

Zona pellucida

(A)

(B)

Figure 7.7
Hamster eggs immediately before fertilization. (A) The hamster egg, or ovum, is encased in the zona pellucida. This, in turn, is surrounded by the cells of the cumulus. A polar body cell, produced during meiosis, is also visible within the zona pellucida. (B) At lower magnification, a mouse oocyte is shown surrounded by the cumulus. Colloidal carbon particles (India ink) are excluded by the hyaluronidate matrix. (Photographs courtesy of R. Yanagimachi.)

Many types of eggs also have an **egg jelly** outside the vitelline envelope (see Figure 7.4). This glycoprotein meshwork can have numerous functions, but most commonly is used either to attract or to activate sperm. The egg, then, is a cell specialized for receiving sperm and initiating development.

VADE MECUM **Gametogenesis.** Stained sections of testis and ovary illustrate the process of gametogenesis, the streamlining of developing sperm, and the remarkable growth of the egg as it stores nutrients for its long journey. You can see this in movies and labeled photographs that take you at each step deeper into the mammalian gonad. **[Click on Gametogenesis]**

Recognition of Egg and Sperm

The interaction of sperm and egg generally proceeds according to five basic steps (Figure 7.8; Vacquier 1998):

1. The chemoattraction of the sperm to the egg by soluble molecules secreted by the egg
2. The exocytosis of the acrosomal vesicle to release its enzymes
3. The binding of the sperm to the extracellular envelope (vitelline layer or zona pellucida) of the egg
4. The passing of the sperm through this extracellular envelope
5. Fusion of egg and sperm cell plasma membranes

Sometimes steps 2 and 3 are reversed (as in mammalian fertilization) and the sperm binds to the egg before releasing the contents of the acrosome. After these five steps are accomplished, the haploid sperm and egg nuclei can meet, and the reactions that initiate development can begin.

In many species, the meeting of sperm and egg is not a simple matter. Many marine organisms release their gametes into the environment. That environment may be as small as a tide pool or as large as an ocean. Moreover, it is shared with other species that may shed their sex cells at the same time. These organisms are faced with two problems: How can sperm and eggs meet in such a dilute concentration, and how can sperm be prevented from trying to fertilize eggs of another species? Two major mechanisms have evolved to solve these problems: species-specific attraction of sperm and species-specific sperm activation.

Sperm attraction: Action at a distance

Species-specific sperm attraction has been documented in numerous species, including cnidarians, molluscs, echinoderms, and urochordates (Miller 1985; Yoshida et al. 1993). In many species, sperm are attracted toward eggs of their species by **chemotaxis**, that is, by following a gradient of a chemical secreted by the egg. In 1978, Miller demonstrated that the eggs of the cnidarian *Orthopyxis caliculata* not only secrete a chemotactic factor but also regulate the timing of its release. Developing oocytes at various stages in their maturation were fixed on microscope slides, and sperm were released at a certain distance from the eggs. Miller found that when sperm were added to oocytes that had not yet completed their second meiotic division, there was no attraction of sperm to eggs. However, after the second meiotic division was finished and the eggs were ready to be fertilized, the sperm migrated toward them. Thus, these oocytes control not only the type of sperm they attract, but also the time at which they attract them.

The mechanisms of chemotaxis differ among species (see Metz 1978; Ward and Kopf 1993). One chemotactic molecule,

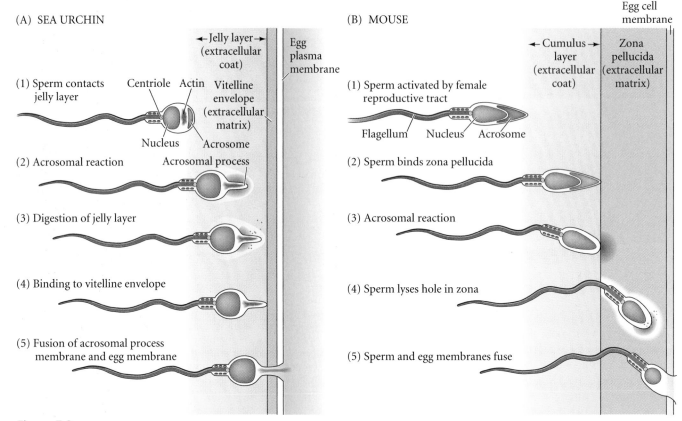

(A) SEA URCHIN

← Jelly layer → (extracellular coat)

Egg plasma membrane

(1) Sperm contacts jelly layer — Centriole, Actin, Nucleus, Acrosome, Vitelline envelope (extracellular matrix)

(2) Acrosomal reaction — Acrosomal process

(3) Digestion of jelly layer

(4) Binding to vitelline envelope

(5) Fusion of acrosomal process membrane and egg membrane

(B) MOUSE

Egg cell membrane

← Cumulus → layer (extracellular coat)

Zona pellucida (extracellular matrix)

(1) Sperm activated by female reproductive tract — Flagellum, Nucleus, Acrosome

(2) Sperm binds zona pellucida

(3) Acrosomal reaction

(4) Sperm lyses hole in zona

(5) Sperm and egg membranes fuse

Figure 7.8
Summary of events leading to the fusion of egg and sperm plasma membranes in the sea urchin (A) and the mouse (B). (A) Sea urchin fertilization is external. (1) The sperm is activated by and chemotactically attracted to the egg. (2, 3) The egg jelly causes the acrosomal reaction to occur, allowing the acrosomal process to form and release proteolytic enzymes. (4) The sperm adheres to the vitelline envelope and lyses a hole in it. (5) The sperm adheres to the egg plasma membrane and fuses with it. The sperm pronucleus can now enter the egg cytoplasm. (B) Mammalian fertilization is internal. (1) The contents of the female reproductive tract capacitate, attract, and activate the sperm. (2) The acrosome-intact sperm binds to the zona pellucida, which constitutes a thicker envelope than that of sea urchins. (3) The acrosomal reaction occurs on the zona pellucida. (4) The sperm digests a hole in the zona pellucida. (5) The sperm adheres to the egg, and their plasma membranes fuse.

a 14-amino acid peptide called **resact,** has been isolated from the egg jelly of the sea urchin *Arbacia punctulata* (Ward et al. 1985). Resact diffuses readily in seawater and has a profound effect at very low concentrations when added to a suspension of *Arbacia* sperm (Figure 7.9). When a drop of seawater con-

taining *Arbacia* sperm is placed on a microscope slide, the sperm generally swim in circles about 50 μm in diameter. Within seconds after a minute amount of resact is injected into the drop, sperm migrate into the region of the injection and congregate there. As resact continues to diffuse from the

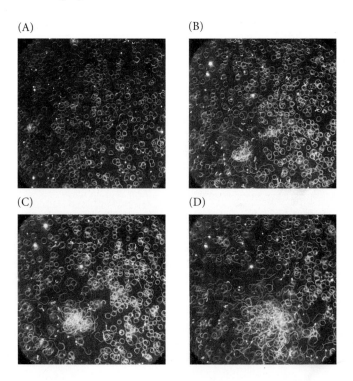

(A)   (B)

(C)   (D)

Figure 7.9
Sperm chemotaxis in *Arbacia*. One nanoliter of a 10-nM solution of resact is injected into a 20-μl drop of sperm suspension. The position of the micropipette is indicated in (A). (A) A 1-second photographic exposure showing sperm swimming in tight circles before the addition of resact. (B–D) Similar 1-second exposures showing migration of sperm to the center of the resact gradient 20, 40, and 90 seconds after injection. (From Ward et al. 1985; photographs courtesy of V. D. Vacquier.)

area of injection, more sperm are recruited into the growing cluster. Resact is specific for *A. punctulata* and does not attract sperm of other species. *A. punctulata* sperm have receptors in their plasma membranes that bind resact (Ramarao and Garbers 1985; Bentley et al. 1986) and can swim up a concentration gradient of this compound until they reach the egg.

Resact also acts as a **sperm-activating peptide**. Sperm-activating peptides cause dramatic and immediate increases in mitochondrial respiration and sperm motility (Tombes and Shapiro 1985; Hardy et al. 1994). The sperm receptor for resact is a transmembrane protein, and when it binds resact on the extracellular side, a conformational change on the cytoplasmic side activates the receptor's enzymatic activity. This activates the mitochondrial ATP-generating apparatus as well as the dynein ATPase that stimulates flagellar movement in the sperm (Shimomura et al. 1986; Cook and Babcock 1993).

The acrosomal reaction in sea urchins

A second interaction between sperm and egg is the **acrosomal reaction.** In most marine invertebrates, the acrosomal reaction has two components: the fusion of the acrosomal vesicle with the sperm plasma membrane (an exocytosis that results in the release of the contents of the acrosomal vesicle) and the exten-

sion of the acrosomal process (Colwin and Colwin 1963). The acrosomal reaction in sea urchins is initiated by contact of the sperm with the egg jelly. Contact with egg jelly causes the exocytosis of the sperm's acrosomal vesicle and the release of proteolytic enzymes that can digest a path through the jelly coat to the egg surface (Dan 1967; Franklin 1970; Levine et al. 1978). The sequence of these events is outlined in Figure 7.10.

In sea urchins, the acrosomal reaction is thought to be initiated by a fucose-containing polysaccharide in the egg jelly that binds to the sperm and allows calcium to enter into the sperm head (Schackmann and Shapiro 1981; Alves et al. 1997; Vacquier and Moy 1997). The exocytosis of the acrosomal vesicle is caused by the calcium-mediated fusion of the acrosomal membrane with the adjacent sperm plasma membrane (Figures 7.10 and 7.11). The egg jelly factors that initiate the acrosomal reaction in sea urchins are often highly specific to each species* (Summers and Hylander 1975).

The second part of the acrosomal reaction involves the extension of the acrosomal process (see Figure 7.10). This protrusion arises through the polymerization of globular actin molecules into actin filaments (Tilney et al. 1978).

---

*Such exocytotic reactions are seen in the release of insulin from pancreatic cells and in the release of neurotransmitters from synaptic terminals. In all cases, there is a calcium-mediated fusion between the secretory vesicle and the cell membrane. Indeed, the similarity of acrosomal vesicle exocytosis and synaptic vesicle exocytosis may actually be quite deep. Studies of acrosomal reactions in sea urchins and mammals (Florman et al. 1992; González-Martínez et al. 1992) suggest that when the receptors for the sperm-activating ligands bind these molecules, they cause a depolarization of the membrane that would open voltage-dependent calcium ion channels in a manner reminiscent of synaptic transmission. The proteins that dock the cortical granules of the egg to the plasma membrane also appear to be homologous to those used in the axon tip (Bi et al. 1995).

Figure 7.10
Acrosomal reaction in sea urchin sperm. (A–C) The portion of the acrosomal membrane lying directly beneath the sperm plasma membrane fuses with the plasma membrane to release the contents of the acrosomal vesicle. (D) The actin molecules assemble to produce microfilaments, extending the acrosomal process outward. Actual photographs of the acrosomal reaction in sea urchin sperm are shown below the diagrams. (After Summers and Hylander 1974; photographs courtesy of G. L. Decker and W. J. Lennarz.)

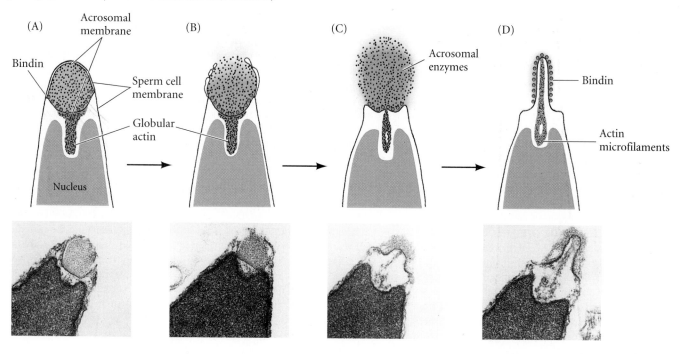

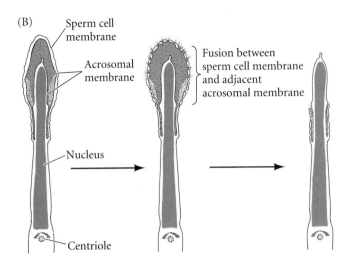

(A)

(B)

Sperm cell membrane

Acrosomal membrane

Nucleus

Centriole

Fusion between sperm cell membrane and adjacent acrosomal membrane

Figure 7.11
Acrosomal reaction in hamster sperm. (A) Transmission electron micrograph of hamster sperm undergoing the acrosomal reaction. The acrosomal membrane can be seen to form vesicles. (B) Interpretive diagram of electron micrographs showing the fusion of the acrosomal and cell membranes in the sperm head. (A from Meizel 1984, courtesy of S. Meizel; B after Yanagimachi and Noda 1970.)

Sidelights & Speculations

Action at a Distance: Mammalian Gametes

IT IS VERY DIFFICULT to study the interactions that might be occurring between mammalian gametes prior to sperm-egg contact. One obvious reason for this is that mammalian fertilization occurs inside the oviducts of the female. While it is relatively easy to mimic the conditions surrounding sea urchin fertilization (using either natural or artificial seawater), we do not yet know the components of the various natural environments that mammalian sperm encounter as they travel to the egg. A second reason for this difficulty is that the sperm population ejaculated into the female is probably very heterogeneous, containing spermatozoa at different stages of maturation. Of the 280×10^6 human sperm normally ejaculated into the vagina, only about 200 reach the ampullary region of the oviduct, where fertilization takes place (Ralt et al. 1991). Since fewer than 1 in 10,000 sperm get close to the egg, it is difficult to assay those molecules that might enable the sperm to swim toward the egg and become activated. There is a great deal of controversy concerning the mechanisms underlying the translocation of mammalian sperm to the oviduct, the possibility that the egg may be attracting the sperm through chemotaxis, and the capacitation and hyperactivation reactions that appear necessary for some species' sperm to bind with the egg.

Translocation and Capacitation

The reproductive tract of female mammals plays a very active role in the mammalian fertilization process. While sperm motility is required for mouse sperm to encounter the egg once it is in the oviduct, sperm motility is probably a minor factor in getting the sperm into the oviduct in the first place. Sperm are found in the oviducts of mice, hamsters, guinea pigs, cows, and humans within 30 minutes of sperm deposition in the vagina, a time "too short to have been attained by even the most Olympian sperm relying on their own flagellar power" (Storey 1995). Rather, the sperm appear to be transported to the oviduct by the muscular activity of the uterus.

By whatever means, mammalian sperm pass through the uterus and oviduct, interacting with the cells and secretions of the female reproductive tract as they do so. These interactions are critical for their ability to interact with the egg. Newly ejaculated mammalian sperm are unable to undergo the acrosomal reaction without residing for some time in the female reproductive tract (Chang 1951; Austin 1952). The set of physiological changes that allow the sperm to be competent to fertilize the egg is called **capacitation**. The requirement for capacitation varies from species to species (Gwatkin 1976). Capacitation can be mimicked in vitro by incubating

sperm in tissue culture media (containing calcium ions, bicarbonate, and serum albumin) or in fluid from the oviducts. Sperm that are not capacitated are "held up" in the cumulus and so do not reach the egg (Austin 1960; Corselli and Talbot 1987).

As mentioned above (and contrary to the opening scenes of the *Look Who's Talking* movies), "the race is not always to the swiftest." Although some human sperm reach the ampullary region of the oviduct within a half hour after intercourse, those sperm may have little chance of fertilizing the egg. Wilcox and colleagues (1995) found that nearly all human pregnancies result from sexual intercourse during a 6-day period ending on the day of ovulation. This means that the fertilizing sperm could have taken as long as 6 days to make the journey. Eisenbach (1995) has proposed a hypothesis wherein capacitation is a transient event, and sperm are given a relatively brief window of competence in which they can successfully fertilize the egg. As the sperm reach the ampulla, they acquire competence, but if they stay around too long, they lose it. Sperm may also have different survival rates depending on their location within the reproductive tract, and this may allow some sperm to arrive late but with better chance of success than those that have arrived days earlier.

The molecular changes that account for capacitation are still unknown, but there are four sets of molecular changes that may be important. First, the fluidity of the sperm plasma membrane is altered by the removal of cholesterol by albumin proteins found in the female reproductive tract. If serum albumin is experimentally preloaded with cholesterol, it will not permit capacitation to occur in vitro. Second, particular proteins or carbohydrates on the sperm surface are lost during capacitation (Lopez et al. 1985; Wilson and Oliphant 1987). It is possible that these compounds block the recognition sites for the proteins that bind to the zona pellucida. Third, the membrane potential of the sperm becomes more negative as potassium ions leave the sperm. This change in membrane potential may allow calcium channels to be opened and permit calcium to enter the sperm. Calcium and bicarbonate ions may be critical in activating cAMP production and in facilitating the membrane fusion events of the acrosomal reaction (Visconti et al. 1995; Arnoult et al. 1999). Fourth, protein phosphorylation occurs (Galantino-Homer et al.

1997). However, it is still uncertain whether these events are independent of one another and to what extent each of them causes sperm capacitation (Figure 7.12).

There may be an important connection between sperm translocation and capacitation. Timothy Smith (1998) and Susan Suarez (1998) have documented that before entering the ampulla of the oviduct (where mammalian fertilization occurs), the uncapacitated sperm bind actively to the membranes of the oviduct cells in the narrow passage (isthmus) preceding it (Figure 7.13). This binding is temporary and appears to be broken when the sperm become capacitated. Moreover, the life span of the sperm is significantly lengthened by this binding, and its capacitation is slowed down. This restriction of sperm entry into the ampulla, the slowing down of capacitation, and the expansion of sperm life span may have very important consequences. First, this binding may function as a block to polyspermy by preventing many sperm from reaching the egg at the same time. If the isthmus is excised in cows, a much higher rate of polyspermy results. Second, slowing the rate

of sperm capacitation and extending the active life of sperm may maximize the probability of there being some sperm in the ampulla to meet the egg if ejaculation does not occur at the same time as ovulation.

Hyperactivation and Chemotaxis

Different regions of the female reproductive tract may secrete different, regionally specific molecules. These factors may influence sperm motility as well as capacitation. For instance, when sperm of certain mammals (especially hamsters, guinea pigs, and some strains of mice) pass from the uterus into the oviducts, they become **hyperactivated**, swimming at higher velocities and generating greater force than before. Suarez and co-workers (1991) have shown that while this behavior is not conducive to traveling in low-viscosity fluids, it appears to be extremely well suited for linear sperm movement in the viscous fluid that sperm might encounter in the oviduct.

In addition to increasing the activity of sperm, soluble factors in the oviduct may also provide the directional component of sperm movement. There has been speculation that the ovum (or, more likely, the ovarian follicle in which it developed) may secrete chemotactic substances that attract the sperm toward the egg during the last stages of sperm migration (see Hunter 1989). Ralt and colleagues (1991) tested this hypothesis using follicular fluid from human follicles whose eggs were being used for in vitro fertilization. Performing an experiment similar to the one described earlier with sea urchins, they microinjected a drop of follicular fluid into a larger drop of sperm suspension. When they did this, some of the sperm changed their direction to migrate toward the source of follicular fluid. Microinjection of other solutions did not have this effect. These studies did not rule out the possibility that the effect was due to a general stimulation of sperm movement or metabo-

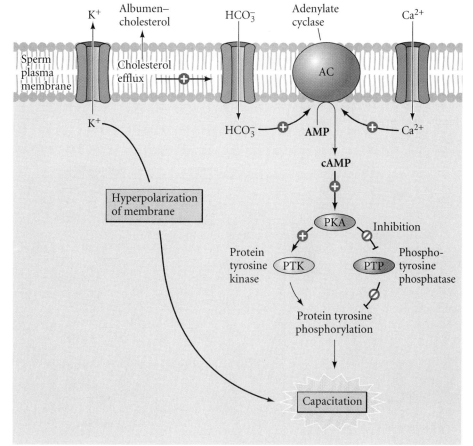

Figure 7.12
Hypothetical model for mammalian sperm capacitation. The efflux of potassium (whose cause we do not know) results in a change in the resting potential of the sperm cell membrane. The removal of cholesterol by albumin stimulates ion channels that enable calcium and bicarbonate ions to enter the sperm. These promote the activity of adenylate cyclase, which makes cAMP from AMP. The rise in cAMP activates protein kinase A, causing it to activate the protein tyrosine kinases (while inactivating the protein phosphatases). The kinases phosphorylate proteins that are essential for capacitation. (After Visconti and Kopf 1998.)

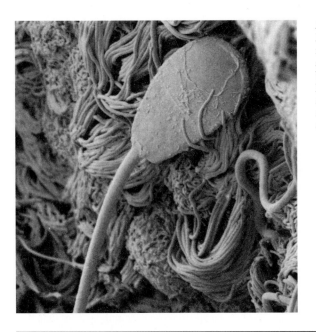

Figure 7.13
Mammalian sperm in the female reproductive tract. Bull sperm adhering to the membranes of the oviduct epithelial cells prior to entering the ampulla. (From Lefebvre et al. 1995; photograph courtesy of S. Suarez.)

lism. However, these investigations uncovered a fascinating correlation: the fluid from only about half the follicles tested showed a chemotactic effect, and in nearly every case, the egg was fertilizable if, and only if, the fluid showed chemotactic ability ($P < 0.0001$). It is possible, then, that like certain invertebrate eggs, the human egg secretes a chemotactic factor only when it is capable of being fertilized.

The female reproductive tract, then, is not a passive conduit through which the sperm race, but a highly specialized set of tissues that regulate the timing of sperm capacitation and access to the egg.

WEBSITE **7.4 In vitro fertilization.** In vitro fertilization has either mimicked or circumvented the ionic conditions of capacitation. In this manner, human sperm can be induced to fertilize eggs in plastic dishes.

Species-specific recognition in sea urchins

Once the sea urchin sperm has penetrated the egg jelly, the acrosomal process of the sperm contacts the surface of the egg (Figure 7.14A). A major species-specific recognition step occurs at this point. The acrosomal protein mediating this recognition is called **bindin**. In 1977, Vacquier and co-workers isolated this nonsoluble 30,500-Da protein from the acrosome of *Strongylocentrotus purpuratus* and found it to be capable of binding to dejellied eggs of the same species (Figure 7.14B; Vacquier and Moy 1977). Further, its interaction with eggs is relatively species-specific (Glabe and Vacquier 1977; Glabe and Lennarz 1979): bindin isolated from the acrosomes

of *S. purpuratus* binds to its own dejellied eggs, but not to those of *Arbacia punctulata*. Using immunological techniques, Moy and Vacquier (1979) demonstrated that bindin is located specifically on the acrosomal process—exactly where it should be for sperm-egg recognition (Figure 7.15).

Biochemical studies have shown that the bindins of closely related sea urchin species are indeed different.* This finding im-

*Bindin is probably the fastest evolving protein known. Closely related species may have near-identity of every other protein, but their bindins may have diverged significantly. For more information on bindin evolution, see Chapter 22.

(A)

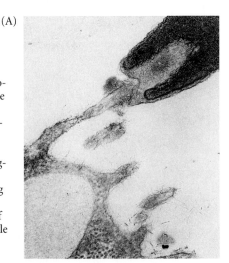

(B)

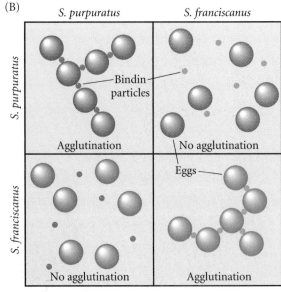

Figure 7.14
Species-specific binding of acrosomal process to egg cell surface in sea urchins. (A) Actual contact of a sea urchin sperm acrosomal process with an egg microvillus. (B) In vitro model of species-specific binding. The agglutination of dejellied eggs by bindin was measured by adding bindin aggregates to a plastic well containing a suspension of eggs. After 2–5 minutes of gentle shaking, the wells were photographed. Each bindin bound to and agglutinated only eggs from its own species. (A from Epel 1977, photograph courtesy of F. D. Collins and D. Epel; B based on photographs of Glabe and Vacquier 1977.)

(A)

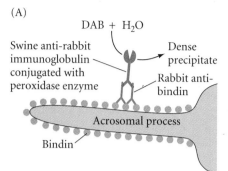

Swine anti-rabbit
immunoglobulin
conjugated with
peroxidase enzyme

DAB + H$_2$O

Dense
precipitate

Rabbit anti-
bindin

Acrosomal process

Bindin

(B) DAB precipitate

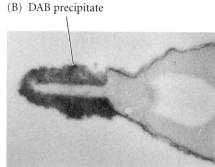

(C) Vitelline membrane
of egg

Acrosomal
process

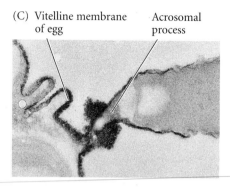

Figure 7.15
Localization of bindin on the acrosomal process. (A) Immunochemical technique used to localize bindin. Rabbit antibody was made to the bindin protein, and this antibody was incubated with sperm that had undergone the acrosomal reaction. If bindin was present, the rabbit antibody would remain bound to the sperm. After any unbound antibody was washed off, the sperm were treated with swine antibody that had been covalently linked to peroxidase enzymes. The swine antibody bound to the rabbit antibody, placing peroxidase molecules wherever bindin was present. Peroxidase catalyzes the formation of a dark precipitate from diaminobenzidine (DAB) and hydrogen peroxide. Thus, this precipitate formed only where bindin was present. (B) Localization of bindin to the acrosomal process after the acrosomal reaction (33,200×). (C) Localization of bindin to the acrosomal process at the junction of the sperm and the egg. (B and C from Moy and Vacquier 1979; photographs courtesy of V. D. Vacquier.)

plies the existence of species-specific bindin receptors on the egg, vitelline envelope, or plasma membrane. Such receptors were also suggested by the experiments of Vacquier and Payne (1973), who saturated sea urchin eggs with sperm. As seen in Figure 7.16A, sperm binding does not occur over the entire egg surface. Even at saturating numbers of sperm (approximately 1500), there appears to be room on the ovum for more sperm heads, implying a limiting number of sperm-binding sites. The bindin receptor on the egg has recently been isolated (Giusti et al. 1997; Stears and Lennarz 1997). This 350-kDa protein may have several regions that interact with bindin. At least one of these sites recognizes only the bindin of the same species. The other site or sites appear to recognize a general bindin structure and can recognize the bindin of many species. The bindin receptors are thought to be aggregated into complexes on the egg cell surface,

and hundreds of these complexes may be needed to tether the sperm to the egg (Figure 7.16B). Thus, species-specific recognition of sea urchin gametes occurs at the levels of sperm attraction, sperm activation, and sperm adhesion to the egg surface.

WEBSITE **7.5 The Lillie-Loeb dispute.** In the early 1900s, fertilization research was framed by a dispute between F. R. Lillie and Jacques Loeb, who disagreed over whether the sperm recognized the egg through soluble factors or through cell-cell interactions.

(A)

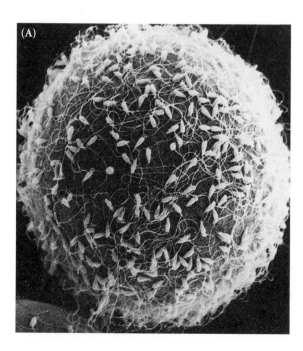

(B)

Figure 7.16
Bindin receptors on the egg. (A) Scanning electron micrograph of sea urchin sperm bound to the vitelline envelope of an egg. Although this egg is saturated with sperm, there appears to be room on the surface for more sperm, implying the existence of a limited number bindin receptors. (B) Binding of *S. purpuratus* sperm to polystyrene beads that have been coated with purified bindin receptor protein. (A courtesy of C. Glabe, L. Perez, and W. J. Lennarz; B from Foltz et al. 1993.)

Gamete binding and recognition in mammals

ZP3: THE SPERM-BINDING PROTEIN OF THE MOUSE ZONA PELLU-CIDA. The zona pellucida in mammals plays a role analogous to that of the vitelline envelope in invertebrates. This glycoprotein matrix, which is synthesized and secreted by the growing oocyte, plays two major roles during fertilization: it binds the sperm, and it initiates the acrosomal reaction after the sperm is bound (Saling et al. 1979; Florman and Storey 1982; Cherr et al. 1986). The binding of sperm to the zona is relatively, but not absolutely, species-specific. (Species-specific gamete recognition is not a major problem when fertilization occurs internally.)

The binding of mouse sperm to the mouse zona pellucida can be inhibited by first incubating the sperm with zona glycoproteins. Bleil and Wassarman (1980, 1986, 1988) isolated an 83-kDa glycoprotein, **ZP3**, from the mouse zona that was the active competitor for binding in this inhibition assay. The other two zona glycoproteins they found, ZP1 and ZP2, failed to compete for sperm binding (Figure 7.17). Moreover, they found that radiolabeled ZP3 bound to the heads of mouse sperm with intact acrosomes. Thus, ZP3 is the specific glycoprotein in the mouse zona pellucida to which the sperm bind. ZP3 also initiates the acrosomal reaction after sperm have bound to it. The mouse sperm can thereby concentrate its proteolytic enzymes directly at the point of attachment at the zona pellucida.

The molecular mechanism by which the zona pellucida and the mammalian sperm recognize each other is presently being studied. The current hypothesis of mammalian gamete binding postulates a set of proteins on the sperm capable of recognizing specific carbohydrate regions of ZP3 (Figure 7.18A; Florman et al. 1984; Florman and Wassarman 1985; Wassarman 1987; Saling 1989). Removal of these threonine- or serine-linked carbohydrate groups from ZP3 abolishes its ability to bind sperm. Several proteins have been identified on the sperm cell surface that specifically bind to the ZP3 carbohydrates. Moreover, the deletion of these proteins from the sperm can inhibit or eliminate sperm-zona binding (see Kopf 1998).

WEBSITE **7.6 Zona-binding proteins.** There are numerous proteins on the sperm that bind to ZP3 on the zona pellucida. These proteins are important in mediating the activation signal to the sperm and for initiating the acrosomal reaction.

INDUCTION OF THE MAMMALIAN ACROSOMAL REACTION BY ZP3. Unlike the sea urchin acrosomal reaction, the acrosomal reaction in mammals occurs only after the sperm has bound to the zona pellucida (Figure 7.8). The mouse sperm acrosomal reaction is induced by the crosslinking of ZP3 with the receptors for it on the sperm membrane (Endo et al. 1987; Leyton and Saling 1989). This crosslinking opens calcium channels to increase the concentration of calcium in the sperm (Leyton and Saling 1992; Florman et al. 1998). The mechanism by which ZP3 induces the opening of the calcium channels and the subsequent exocytosis of the acrosome remains controversial, but it

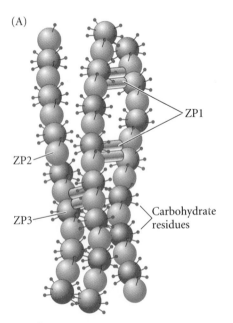

(A)

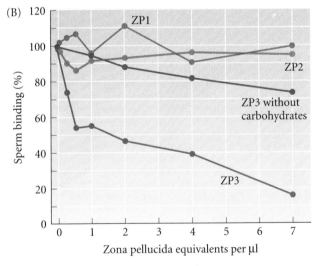

(B)

Figure 7.17
Mouse ZP3 as the zona protein that binds sperm. (A) Diagram of the fibrillar structure of the mouse zona pellucida. The major strands of the zona are composed of repeating dimers of proteins ZP2 and ZP3. These strands are occasionally crosslinked together by ZP1, forming a meshlike network. (B) Inhibition assay showing the specific decrease of mouse sperm binding to zonae pellucidae when sperm and zonae were first incubated with increasingly large amounts of the glycoprotein ZP3. The importance of the carbohydrate portion of ZP3 is also indicated by this graph. (A after Wassarman 1989; B after Bleil and Wassarman 1980 and Florman and Wassarman 1985.)

(A)

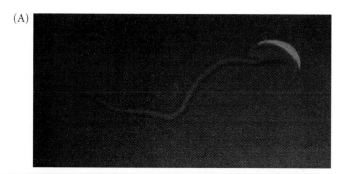

(B)

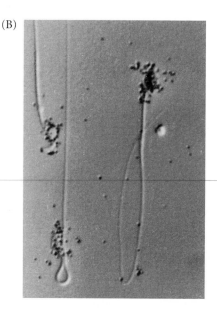

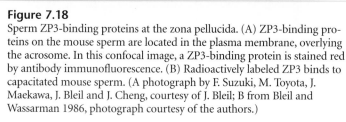

Figure 7.18
Sperm ZP3-binding proteins at the zona pellucida. (A) ZP3-binding proteins on the mouse sperm are located in the plasma membrane, overlying the acrosome. In this confocal image, a ZP3-binding protein is stained red by antibody immunofluorescence. (B) Radioactively labeled ZP3 binds to capacitated mouse sperm. (A photograph by F. Suzuki, M. Toyota, J. Maekawa, J. Bleil and J. Cheng, courtesy of J. Bleil; B from Bleil and Wassarman 1986, photograph courtesy of the authors.)

may involve the receptor's activating a cation channel (for sodium, potassium, or calcium), which would change the resting potential of the sperm plasma membrane. The calcium channels in the membrane would be sensitive to this change in membrane potential, allowing calcium to enter the sperm.

The difference between the acrosomal reaction in sea urchins and mammals may be due to the thickness of the extracellular envelopes surrounding the egg. In the sea urchin, the vitelline envelope is very thin and porous. Once a sperm has bound there, it is very close to the egg plasma membrane, and, indeed, the bindin receptor may extend through the vitelline envelope. In mammals, however, the zona pellucida is a very thick matrix, so the sperm is far removed from the egg. By undergoing the acrosomal reaction directly on the zona, the sperm is able to concentrate its proteolytic enzymes to lyse a hole in this envelope. Indeed, sperm that undergo the acrosomal reaction before they reach the zona pellucida are unable to penetrate it (Florman et al. 1998).

SECONDARY BINDING OF SPERM TO THE ZONA PELLUCIDA.
During the acrosomal reaction, the anterior portion of the sperm plasma membrane is shed from the sperm (see Figure 7.11). This region is where the ZP3-binding proteins are located, and yet the sperm must still remain bound to the zona in order to lyse a path through it. In mice, it appears that secondary binding to the zona is accomplished by proteins in the inner acrosomal membrane that bind specifically to ZP2 (Bleil et al. 1988). Whereas acrosome-intact sperm will not bind to ZP2, acrosome-reacted sperm will. Moreover, antibodies against the ZP2 glycoprotein will not prevent the binding of acrosome-intact sperm to the zona, but will inhibit the attachment of acrosome-reacted sperm. The structure of the zona consists of repeating units of ZP3 and ZP2, occasionally crosslinked by ZP1 (Figure 7.18). It appears that the acrosome-reacted sperm transfer their binding from ZP3 to the adjacent

ZP2 molecules. After a mouse sperm has entered the egg, the egg cortical granules release their contents. One of the proteins released by these granules is a protease that specifically alters ZP2 (Moller and Wassarman 1989). This inhibits other acrosome-reacted sperm from moving closer toward the egg.

In guinea pigs, secondary binding to the zona is thought to be mediated by the protein PH-20. Moreover, when this inner acrosomal membrane protein was injected into adult male or female guinea pigs, 100% of them became sterile for several months (Primakoff et al. 1988). The blood sera of these sterile guinea pigs had extremely high concentrations of antibodies to PH-20. The antiserum from guinea pigs sterilized in this manner not only bound specifically to PH-20, but also blocked sperm-zona adhesion in vitro. The contraceptive effect lasted several months, after which fertility was restored. These experiments show that the principle of immunological contraception is well founded.

Gamete Fusion and the Prevention of Polyspermy

Fusion of the egg and sperm plasma membranes

Recognition of sperm by the vitelline envelope or zona pellucida is followed by the lysis of that portion of the envelope or zona in the region of the sperm head by the acrosomal enzymes (Colwin and Colwin 1960; Epel 1980). This lysis is followed by the fusion of the sperm plasma membrane with the plasma membrane of the egg.

The entry of a sperm into a sea urchin egg is illustrated in Figure 7.19. Sperm-egg binding appears to cause the extension of several microvilli to form the **fertilization cone** (Summers et al. 1975; Schatten and Schatten 1980, 1983). Homology between the egg and the sperm is again demonstrated, because the transitory fertilization cone, like the acrosomal process,

(A)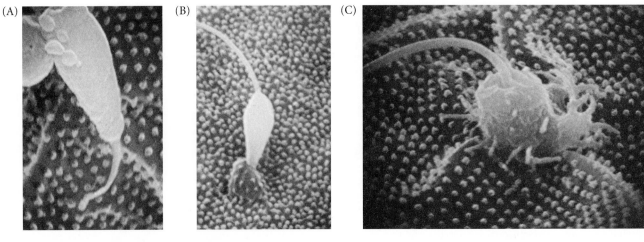

(B)

(C)

(D)

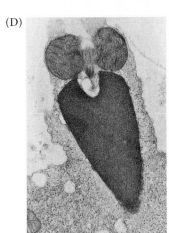

Figure 7.19
Scanning electron micrographs of the entry of sperm into sea urchin eggs. (A) Contact of sperm head with egg microvillus through the acrosomal process. (B) Formation of fertilization cone. (C) Internalization of sperm within the egg. (D) Transmission electron micrograph of sperm internalization through the fertilization cone. (A–C from Schatten and Mazia 1976, photographs courtesy of G. Schatten; D, photograph courtesy of F. J. Longo.)

appears to be extended by the polymerization of actin. The sperm and egg plasma membranes then join together, and material from the sperm membrane can later be found on the egg membrane (Gundersen et al. 1986). The sperm nucleus and tail pass through the resulting cytoplasmic bridge, which is widened by the actin polymerization. A similar process occurs during the fusion of mammalian gametes (Yanagimachi and Noda 1970; Figure 7.20).

In the sea urchin, all regions of the egg plasma membrane are capable of fusing with sperm. In several other species, certain regions of the membrane are specialized for sperm recognition and fusion (Vacquier 1979). Fusion is an active process, often mediated by specific "fusogenic" proteins. It seems that bindin plays a second role as a fusogenic protein. Glabe (1985) has shown that sea urchin bindin will cause phospholipid vesicles to fuse together and that, like viral fusogenic proteins, bindin contains a long stretch of hydrophobic amino acids near its amino terminus. This region is able to fuse phospholipid vesicles (Ulrich et al. 1998).

In mammals, the **fertilin** proteins in the sperm plasma membrane are essential for sperm membrane-egg membrane fusion (Primakoff et al. 1987; Blobel et al. 1992; Myles et al. 1994). Mouse fertilin is localized to the posterior plasma mem-

brane of the sperm head (Hunnicut et al. 1997). It adheres the sperm to the egg by binding to the $\alpha6\beta1$ integrin protein on the egg plasma membrane (Evans et al. 1997; Chen and Sampson 1999). Moreover, like sea urchin bindin (to which it is not structurally related), fertilin has a hydrophobic region that could potentially mediate the union of the two membranes (Almeida et al. 1995). Thus, fertilin appears to bind the sperm plasma membrane to the egg plasma membrane and then to fuse the two of them together. Mice homozygous for mutant fertilin have sperm with several defects, one of them being the inability to fuse with the egg plasma membrane (Cho et al. 1998). When the membranes are fused, the sperm nucleus, mitochondria, centriole, and flagellum can enter the egg.

The prevention of polyspermy

As soon as one sperm has entered the egg, the fusibility of the egg membrane, which was so necessary to get the sperm inside the egg, becomes a dangerous liability. In sea urchins, as in most animals studied, any sperm that enters the egg can provide a haploid nucleus and a centriole to the egg. In normal **monospermy**, in which only one sperm enters the egg, a haploid sperm nucleus and a haploid egg nucleus combine to form the diploid nucleus of the fertilized egg (zygote), thus restoring the chromosome number appropriate for the species. The centriole, which is provided by the sperm, will divide to form the two poles of the mitotic spindle during cleavage.

The entrance of multiple sperm—**polyspermy**—leads to disastrous consequences in most organisms. In the sea urchin, fertilization by two sperm results in a triploid nucleus, in which each chromosome is represented three times rather than twice. Worse, since each sperm's centriole divides to form the two poles of a mitotic apparatus, instead of a bipolar mi-

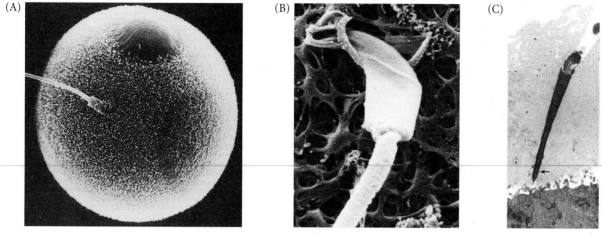

(A) (B) (C)

(D)

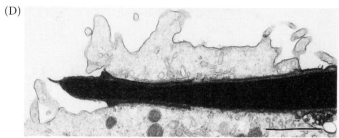

(E) ❶

Nucleus

Equatorial segment
of acrosome

Mitochondrion

Inner
acrosomal
membrane

Centriole

Cortical granules

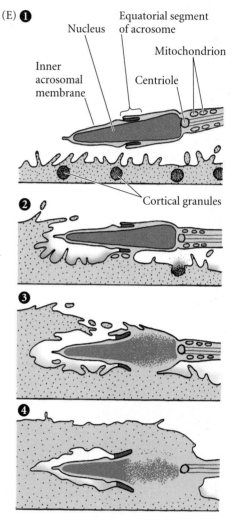

❷

❸

❹

Figure 7.20
Entry of sperm into golden hamster egg. (A) Scanning electron micrograph of sperm fusing with egg. The "bald" spot (without microvilli) is where the polar body has budded off. (B) Close-up of sperm-zona binding. (C) Transmission electron micrograph showing the sperm head passing through the zona. (D) Transmission electron micrograph of the hamster sperm fusing parallel to the egg plasma membrane. (E) Diagram of the fusion of the sperm acrosome and plasma membranes with the egg microvilli. (A–D from Yanagimachi and Noda 1970, Yanagimachi 1994; photographs courtesy of R. Yanagimachi.)

totic spindle separating the chromosomes into two cells, the triploid chromosomes may be divided into as many as four cells. Because there is no mechanism to ensure that each of the four cells receives the proper number and type of chromosomes, the chromosomes would be apportioned unequally. Some cells receive extra copies of certain chromosomes and other cells lack them. Theodor Boveri demonstrated in 1902 that such cells either die or develop abnormally (Figure 7.21).

Species have evolved ways to prevent the union of more than two haploid nuclei. The most common way is to prevent the entry of more than one sperm into the egg. The sea urchin egg has two mechanisms to avoid polyspermy: a fast reaction, accomplished by an electric change in the egg plasma membrane, and a slower reaction, caused by the exocytosis of the cortical granules (Just 1919).

THE FAST BLOCK TO POLYSPERMY. The **fast block to polyspermy** is achieved by changing the electric potential of the egg plasma membrane. This membrane provides a selective barrier between the egg cytoplasm and the outside environ-

ment, and the ionic concentration of the egg differs greatly from that of its surroundings. This concentration difference is especially significant for sodium and potassium ions. Seawater has a particularly high sodium ion concentration, whereas the egg cytoplasm contains relatively little sodium. The reverse is

(A)

Oocyte

Sperm centrosomes

Sperm pronuclei

Egg pronucleus

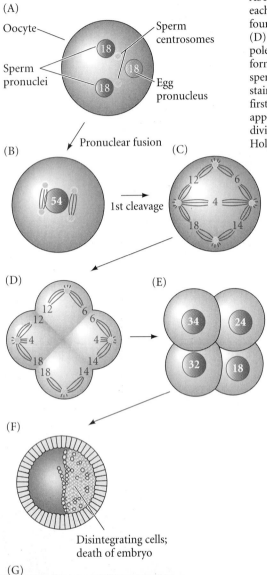

Pronuclear fusion

(B)

(C)

1st cleavage

(D)

(E)

(F)

Disintegrating cells; death of embryo

(G)

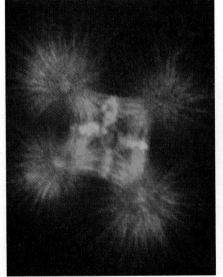

(H)

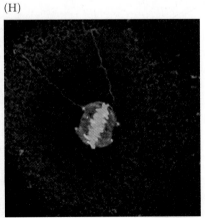

Figure 7.21
Aberrant development in a dispermic sea urchin egg. (A) Fusion of three haploid nuclei, each containing 18 chromosomes, and the division of the two sperm centrioles to form four mitotic poles. (B, C) The 54 chromosomes randomly assort on the four spindles. (D) At anaphase of the first division, the duplicated chromosomes are pulled to the four poles. (E) Four cells containing different numbers and types of chromosomes are formed, thereby causing (F) the early death of the embryo. (G) First metaphase of a dispermic sea urchin egg akin to (D). The microtubules are stained green, and the DNA is stained red. The triploid DNA is being split into four cells. (H) Human dispermic egg at first mitosis. The four centrioles are stained yellow, while the microtubules of the spindle apparatus (and of the two sperm tails) are stained red. The three sets of chromosomes divided by these four poles are stained blue. (A–F after Boveri 1907; G courtesy of J. Holy; H from Simerly et al. 1999, photograph courtesy of G. Schatten.)

the case with potassium ions. This condition is maintained by the plasma membrane, which steadfastly inhibits the entry of sodium ions into the oocyte and prevents potassium ions from leaking out into the environment. If we insert an electrode into an egg and place a second electrode outside it, we can measure the constant difference in charge across the egg plasma membrane. This **resting membrane potential** is generally about 70 mV, usually expressed as –70 mV because the inside of the cell is negatively charged with respect to the exterior.

Within 1–3 seconds after the binding of the first sperm, the membrane potential shifts to a positive level, about +20 mV (Longo et al. 1986). This change is caused by a small influx of sodium ions into the egg (Figure 7.22A). Although sperm can fuse with membranes having a resting potential of –70 mV, they cannot fuse with membranes having a positive resting potential, so no more sperm can fuse to the egg. It is not known whether the increased sodium permeability is due to the *binding* of the first sperm or to the *fusion* of the first sperm with the egg (Gould and Stephano 1987, 1991; McCulloh and Chambers 1992).

The importance of sodium ions and the change in resting potential was demonstrated by Laurinda Jaffe and colleagues. They found that polyspermy can be induced if sea urchin eggs are artificially supplied with an electric current that keeps their membrane potential negative. Conversely, fertilization can be prevented entirely by artificially keeping the membrane potential of eggs positive (Jaffe 1976). The fast block to polyspermy can also be circumvented by lowering the concentration of sodium ions in the water (Figure 7.22B–D). If the supply of sodium ions is not sufficient to cause the positive shift in membrane potential, polyspermy occurs (Gould-Somero et al. 1979; Jaffe 1980). It is not known how the change in membrane potential acts on the sperm to block secondary fertilization. Most likely, the sperm carry a voltage-sensitive component (possibly a positively charged fusogenic protein), and the insertion of this component into the egg plasma membrane could be regulated by the electric charge across the membrane (Iwao and Jaffe 1989). An electric block to polyspermy also occurs in frogs (Cross and Elinson 1980), but probably not in most mammals (Jaffe and Cross 1983).

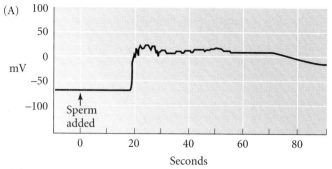

(A)

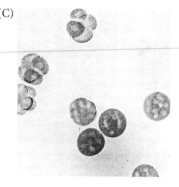

Figure 7.22
Membrane potential of sea urchin eggs before and after fertilization. (A) Before the addition of sperm, the potential difference across the egg plasma membrane is about −70 mV. Within 1–3 seconds after the fertilizing sperm contacts the egg, the potential shifts in a positive direction. (B, C) *Lytechinus* eggs photographed during first cleavage. (B) Control eggs developing in 490 m*M* Na⁺. (C) Polyspermy in eggs fertilized in 120 m*M* Na⁺ (choline was substituted for sodium). (D) Table showing the rise of polyspermy with decreasing sodium ion concentration. (From Jaffe 1980; photographs courtesy of L. A. Jaffe.)

(D)

Na⁺ (m*M*)	Percentage of polyspermic eggs
490	22
360	26
120	97
50	100

THE SLOW BLOCK TO POLYSPERMY. The eggs of sea urchins (and many other animals) have a second mechanism to ensure that multiple sperm do not enter the egg cytoplasm (Just 1919). The fast block to polyspermy is transient, since the membrane potential of the sea urchin egg remains positive for only about a minute. This brief potential shift is not sufficient to prevent polyspermy, which can still occur if the sperm bound to the vitelline envelope are not somehow removed (Carroll and Epel 1975). This removal is accomplished by the **cortical granule reaction**, a slower, mechanical block to polyspermy that becomes active about a minute after the first successful sperm-egg attachment.

Directly beneath the sea urchin egg plasma membrane are about 15,000 cortical granules, each about 1 µm in diameter (see Figure 7.6B). Upon sperm entry, these cortical granules fuse with the egg plasma membrane and release their contents into the space between the plasma membrane and the fibrous mat of vitelline envelope proteins. Several proteins are released by this cortical granule exocytosis. The first are proteases. These en-

zymes dissolve the protein posts that connect the vitelline envelope proteins to the cell membrane, and they clip off the bindin receptor and any sperm attached to it (Vacquier et al. 1973; Glabe and Vacquier 1978). Mucopolysaccharides released by the cortical granules produce an osmotic gradient that causes water to rush into the space between the plasma membrane and the vitelline envelope, causing the envelope to expand and become the **fertilization envelope** (Figures 7.23 and 7.24). A third pro-

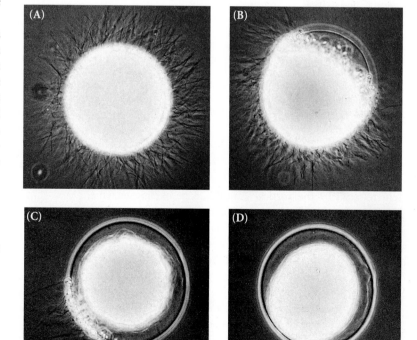

Figure 7.23
Formation of the fertilization envelope and removal of excess sperm. To create these photographs, sperm were added to sea urchin eggs, and the suspension was fixed in formaldehyde to prevent further reactions. (A) At 10 seconds after sperm addition, sperm are seen surrounding the egg. (B, C) At 25 and 35 seconds after insemination, a fertilization envelope is forming around the egg, starting at the point of sperm entry. (D) The fertilization envelope is complete, and excess sperm have been removed. (From Vacquier and Payne 1973; photographs courtesy of V. D. Vacquier.)

(A)

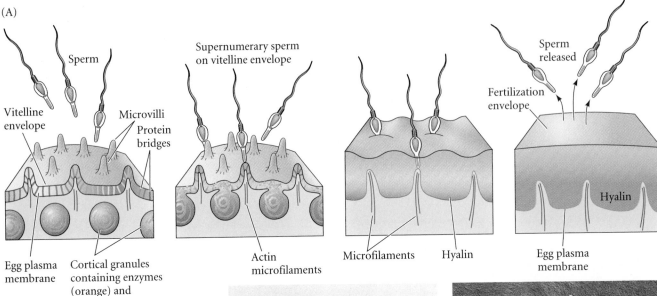

Figure 7.24
Cortical granule exocytosis. (A) Schematic diagram showing the events leading to the formation of the fertilization envelope and the hyaline layer. As cortical granules undergo exocytosis, they release proteases that cleave the proteins linking the vitelline envelope to the cell membrane. Mucopolysaccharides released by the cortical granules form an osmotic gradient, thereby causing water to enter and swell the space between the vitelline envelope and the plasma membrane. Other enzymes released from the cortical granules harden the vitelline envelope (now the fertilization envelope) and release sperm bound to it. (B, C) Transmission and scanning electron micrographs of the cortex of an unfertilized sea urchin egg. (D, E) Transmission and scanning electron micrographs of the same region of a recently fertilized egg, showing the raising of the fertilization envelope and the points at which the cortical granules have fused with the plasma membrane of the egg (arrows in D). (A after Austin 1965; B–E from Chandler and Heuser 1979, courtesy of D. E. Chandler.)

tein released by the cortical granules, a peroxidase enzyme, hardens the fertilization envelope by crosslinking tyrosine residues on adjacent proteins (Foerder and Shapiro 1977; Mozingo and Chandler 1991). As shown in Figure 7.23, the fertilization envelope starts to form at the site of sperm entry and continues its expansion around the egg. As it forms, bound sperm are released from the envelope. This process starts about 20 seconds after sperm attachment and is complete by the end of the first minute of fertilization. Finally, a fourth cortical granule protein, hyalin, forms a coating around the egg (Hylander and Summers 1982). The egg extends elongated mi-

crovilli whose tips attach to this **hyaline layer**. This layer provides support for the blastomeres during cleavage.

WEBSITE 7.7 **Building the egg's extracellular matrix.** In sea urchins, the cortical granules secrete not only hyalin but a number of proteins that construct the extracellular matrix of the embryo. This is a highly coordinated process that results in sequential layers.

VADE MECUM **Sea urchin fertilization.** The remarkable reactions that prevent polyspermy in a fertilized sea urchin egg can be seen in the raising of the fertilization envelope. This segment contains movies of this event shown in real-time.
[**Click on Sea Urchin**]

In mammals, the cortical granule reaction does not create a fertilization envelope, but its ultimate effect is the same.

Released enzymes modify the zona pellucida sperm receptors such that they can no longer bind sperm (Bleil and Wassarman 1980). During this process, called the **zona reaction,** both ZP3 and ZP2 are modified. Florman and Wassarman (1985) have proposed that the cortical granules of mouse eggs contain an enzyme that clips off the terminal sugar residues of ZP3, thereby releasing bound sperm from the zona and preventing the attachment of other sperm. Cortical granules of mouse eggs have been found to contain N-acetylglucosaminidase enzymes capable of cleaving N-acetylglucosamine from ZP3 carbohydrate chains. N-acetylglucosamine is one of the carbohydrate groups that sperm can bind to, and Miller and co-workers (1992, 1993) have demonstrated that when the N-acetylglucosamine residues are removed at fertilization, ZP3 will no longer serve as a substrate for the binding of other sperm. ZP2 is clipped by the cortical granule proteases and loses its ability to bind sperm as well (Moller and Wassarman 1989). Thus, once a sperm has entered the egg, other sperm can no longer initiate or maintain their binding to the zona pellucida and are rapidly shed.

Figure 7.25
Wave of calcium release across sea urchin eggs during fertilization. A sea urchin egg is preloaded with a dye that fluoresces when it binds calcium. When the sperm fuses with the egg, a wave of calcium release can be seen, beginning at the site of sperm entry and propagating across the egg. The wave takes 30 seconds to traverse the egg. (Photograph courtesy of G. Schatten.)

CALCIUM AS THE INITIATOR OF THE CORTICAL GRANULE REACTION. The mechanism of the cortical granule reaction is similar to that of the acrosomal reaction. Upon fertilization, the intracellular calcium ion concentration of the egg increases greatly. In this high-calcium environment, the cortical granule membranes fuse with the egg plasma membrane, releasing their contents (see Figure 7.24). Once the fusion of the cortical granules begins near the point of sperm entry, a wave of cortical granule exocytosis propagates around the cortex to the opposite side of the egg.

In sea urchins and mammals, the rise in calcium concentration responsible for the cortical granule reaction is not due to an influx of calcium into the egg, but rather comes from within the egg itself. The release of calcium from intracellular storage can be monitored visually using calcium-activated luminescent dyes such as aequorin (isolated from luminescent jellyfish) or fluorescent dyes such as fura-2. These dyes emit light when they bind free calcium ions. When a sea urchin egg is injected with dye and then fertilized, a striking wave of calcium release propagates across the egg (Figure 7.25). Starting at the point of sperm entry, a band of light traverses the cell (Steinhardt et al. 1977; Gilkey et al. 1978; Hafner et al. 1988). The calcium ions do not merely diffuse across the egg from the point of sperm entry. Rather, the release of calcium ions starts at one end of the cell and proceeds actively to the other end. The entire release of calcium ions is complete in roughly 30 seconds

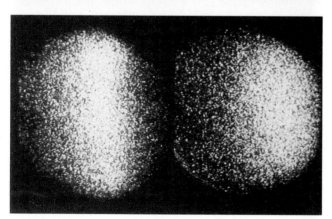

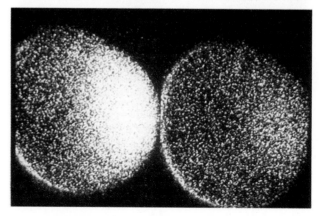

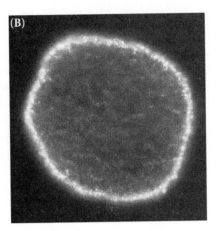

Figure 7.26
Endoplasmic reticulum surrounding cortical granules in sea urchin eggs. (A) The endoplasmic reticulum has been stained with osmium-zinc iodide to allow visualization by transmission electron microscopy. The cortical granule is seen to be surrounded by the endoplasmic reticulum. (B) An entire egg stained with fluorescent antibodies to calcium-dependent calcium release channels. The antibodies show these channels in the cortical endoplasmic reticulum. (A from Luttmer and Longo 1985, courtesy of S. Luttmer; B from McPherson et al. 1992, courtesy of F. J. Longo.)

in sea urchin eggs, and the free calcium ions are resequestered shortly after they are released. If two sperm enter the egg cytoplasm, calcium ion release can be seen starting at the two separate points of entry on the cell surface (Hafner et al. 1988).

Several experiments have demonstrated that calcium ions are directly responsible for propagating the cortical granule reaction, and that these calcium ions are stored within the egg itself. The drug A23187 is a calcium ionophore (a compound that transports free calcium ions across lipid membranes, allowing these cations to traverse otherwise impermeable barriers). Placing unfertilized sea urchin eggs into seawater containing A23187 causes the cortical granule reaction and the elevation of the fertilization envelope. Moreover, this reaction occurs in the absence of any calcium ions in the surrounding water. Therefore, A23187 must be causing the release of calcium ions already sequestered in organelles within the egg (Chambers et al. 1974; Steinhardt and Epel 1974).

The calcium ions responsible for the cortical granule reaction are stored in the endoplasmic reticulum of the egg (Eisen and Reynolds 1985; Terasaki and Sardet 1991). In sea urchins and frogs, this reticulum is pronounced in the cortex and surrounds the cortical granules (Figure 7.26; Gardiner and Grey 1983; Luttmer and Longo 1985). In *Xenopus*, the cortical endoplasmic reticulum becomes ten times more abundant during the maturation of the egg and disappears locally within a minute after the wave of cortical granule exocytosis occurs in any region of the cortex. Once initiated, the release of calcium is self-propagating. Free calcium is able to release sequestered calcium from its storage sites, thus causing a wave of calcium ion release and cortical granule exocytosis.

WEBSITE **7.8 Blocks to polyspermy.** Theodore Boveri's analysis of polyspermy is a classic of experimental and descriptive biology. E. E. Just's delineation of the fast and slow blocks was a critical paper in embryology. Both papers are reprinted here, along with commentaries.

The Activation of Egg Metabolism

Although fertilization is often depicted as merely the means to merge two haploid nuclei, it has an equally important role in initiating the processes that begin development. These events happen in the cytoplasm and occur without the involvement of the nuclei.*

The mature sea urchin egg is a metabolically sluggish cell that is activated by the sperm. This activation is merely a stimulus, however; it sets into action a preprogrammed set of metabolic events. The responses of the egg to the sperm can be divided into "early" responses, which occur within seconds of the cortical reaction, and "late" responses, which take place several minutes after fertilization begins (Table 7.1; Figure 7.27).

Early responses

As we have seen, contact between sea urchin sperm and egg activates the two major blocks to polyspermy: the fast block, initiated by sodium influx into the cell, and the slow block, initiated by the intracellular release of calcium ions. The activation of all eggs appears to depend on an increase in the concentration of free calcium ions within the egg. Such an increase can occur in two ways: calcium ions can enter the egg from outside, or calcium ions can be released from the endoplasmic reticulum within the egg. Both mechanisms are used to different degrees in different species. In snails and worms, much of the calcium probably enters the egg from outside, while in fishes, frogs, sea urchins, and mammals, most of the calcium ions probably come from the endoplasmic reticulum. In both cases, a wave of calcium ions sweeps across the egg, beginning at the site of sperm-egg fusion (Jaffe 1983; Terasaki and Sardet 1991).

*In certain salamanders, this developmental function of fertilization has been totally divorced from the genetic function. The silver salamander (*Ambystoma platineum*) is a hybrid subspecies consisting solely of females. Each female produces an egg with an unreduced chromosome number. This egg, however, cannot develop on its own, so the silver salamander mates with a male Jefferson salamander (*A. jeffersonianum*). The sperm from the male Jefferson salamander only stimulates the egg's development; it does not contribute genetic material (Uzzell 1964). For details of this complex mechanism of procreation, see Bogart et al. 1989.

Table 7.1 Events of sea urchin fertilization

Event	Approximate time postinsemination[a]
EARLY RESPONSES	
Sperm-egg binding	0 seconds
Fertilization potential rise (fast block to polyspermy)	within 1 sec
Sperm-egg membrane fusion	within 6 sec
Calcium increase first detected	6 sec
Cortical vesicle exocytosis (slow block to polyspermy)	15–60 sec
LATE RESPONSES	
Activation of NAD kinase	starts at 1 min
Increase in NADH and NADPH	starts at 1 min
Increase in O_2 consumption	starts at 1 min
Sperm entry	1–2 min
Acid efflux	1–5 min
Increase in pH (remains high)	1–5 min
Sperm chromatin decondensation	2–12 min
Sperm nucleus migration to egg center	2–12 min
Egg nucleus migration to sperm nucleus	5–10 min
Activation of protein synthesis	starts at 5–10 min
Activation of amino acid transport	starts at 5–10 min
Initiation of DNA synthesis	20–40 min
Mitosis	60–80 min
First cleavage	85–95 min

Main sources: Whitaker and Steinhardt 1985; Mohri et al. 1995.

[a]Approximate times based on data from *S. purpuratus* (15–17°C), *L. pictus* (16–18°C), *A. punctulata* (18–20°C), and *L. variegatus* (22–24°C). The timing of events within the first minute is best known for *Lytechinus variegatus*, so times are listed for that species.

The presence of calcium ions is essential for activating the development of the embryo. If the calcium-chelating chemical EGTA is injected into the sea urchin egg, there is no cortical granule reaction, no change in membrane resting potential, and no reinitiation of cell division (Kline 1988). Conversely, eggs can be activated artificially in the absence of sperm by procedures that release free calcium into the oocyte. Steinhardt and Epel (1974) found that injection of micromolar amounts of the calcium ionophore A23187 into a sea urchin egg elicits most of the responses characteristic of a normally fertilized egg. The elevation of the fertilization envelope, a rise of intracellular pH, a burst of oxygen utilization, and increases in protein and DNA synthesis are all generated in their proper order. In most of these cases, development ceases before the first mitosis because the egg is still haploid and lacks the sperm centriole needed for division. Calcium release activates a series of metabolic reactions (Figure 7.27). One of these is the activation of the enzyme NAD⁺ kinase, which converts NAD⁺ to NADP⁺ (Epel et al. 1981). This change may have important consequences for lipid metabolism, since NADP⁺ (but not NAD⁺) can be used as a coenzyme for lipid biosynthesis. Thus, the conversion of NAD⁺ to NADP⁺ may be important in the construction of the many new cell membranes required during cleavage. Another effect of calcium release involves oxygen consumption. A burst of oxygen reduction (to hydrogen peroxide) is seen during fertilization, and much of this "respiratory burst" is used to crosslink the fertilization envelope. The enzyme responsible for this reduction of oxygen is also NADPH-dependent (Heinecke and Shapiro 1989). Lastly, NADPH helps regenerate glutathione and ovothiols, which may be crucial for scavenging free radicals that could otherwise damage the DNA of the egg and early embryo (Mead and Epel 1995).

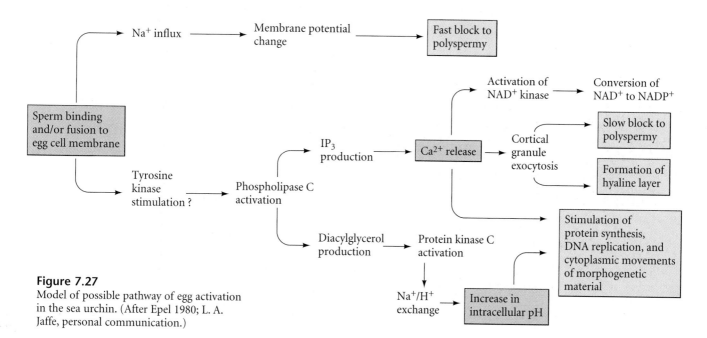

Figure 7.27
Model of possible pathway of egg activation in the sea urchin. (After Epel 1980; L. A. Jaffe, personal communication.)

The Activation of Gamete Metabolism

If calcium ion release is necessary for the activation of the oocyte, how does the sperm cause it to occur? We do not really know. As one investigator (Berridge 1993) stated, "Just how the sperm triggers the explosive release of calcium in the egg is still something of a mystery." Recent data suggest that the production of **inositol 1,4,5-trisphosphate** (**IP$_3$**) is the primary mechanism for releasing calcium ions from their intracellular storage.

The IP$_3$ pathway is shown in Figure 7.28. The membrane phospholipid **phosphatidylinositol 4,5-bisphosphate** (**PIP$_2$**) is split by the enzyme **phospholipase C** (**PLC**) to yield two active compounds: IP$_3$ and **diacylglycerol** (**DAG**). IP$_3$ is able to release calcium ions into the cytoplasm by opening the calcium ion channels of the endoplasmic reticulum.

DAG activates protein kinase C, which in turn activates a protein that exchanges sodium ions for hydrogen ions, raising the pH of the egg (Swann and Whitaker 1986; Nishizuka 1986). This Na$^+$/H$^+$ exchange pump also needs calcium ions for activity. The result of PLC activation is therefore the liberation of calcium ions and the alkalinization of the egg, and both of the proteins it creates, IP$_3$ and DAG, are involved in the initiation of development.

IP$_3$ is formed at the site of sperm entry in sea urchin eggs, and can be detected within seconds of their being fertilized. The inhibition of IP$_3$ synthesis prevents calcium release (Lee and Shen 1998), while injected IP$_3$ can release sequestered calcium ions in sea urchin eggs, leading to the cortical granule reaction (Whitaker and Irvine 1984; Busa et al. 1985).

Moreover, these IP$_3$-mediated effects can be thwarted by preinjecting the egg with calcium-chelating agents (Turner et al. 1986).

IP$_3$-responsive calcium channels have been found in the egg endoplasmic reticulum. The IP$_3$ formed at the site of sperm entry is thought to bind to the IP$_3$ receptors of these channels, effecting a local release of calcium (Ferris et al. 1989; Furuichi et al. 1989; Terasaki and Sardet 1991). Once released, the calcium ions can diffuse directly, or they can facilitate the release of more calcium ions by binding to calcium-sensitive receptors located in the cortical endoplasmic reticulum (McPherson et al. 1992). The binding of calcium ions to these receptors releases more calcium, and this released calcium binds to more receptors, and so on. The resulting wave of calcium release is propagated

Figure 7.28
The roles of inositol phosphates in initiating calcium release from the endoplasmic reticulum and the initiation of development. Phospholipase C splits PIP$_2$ into IP$_3$ and DAG. The IP$_3$ releases calcium from the endoplasmic reticulum, and the DAG, with assistance from the released calcium ions, activates the sodium/hydrogen exchange pump in the membrane.

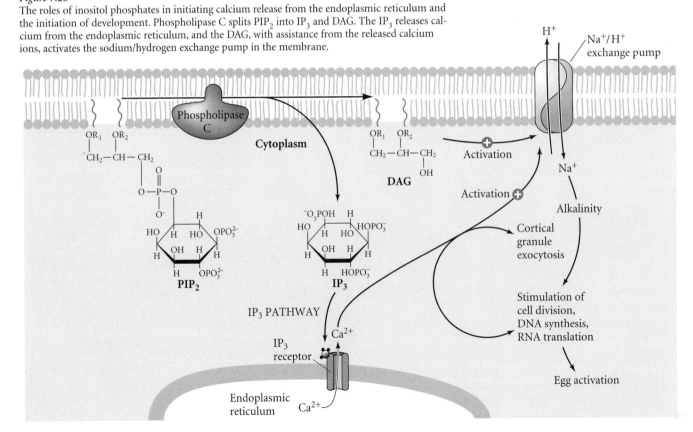

throughout the cell, starting at the point of sperm entry; and the cortical granules, which fuse with the cell membrane in the presence of high calcium concentrations, respond in a wave of exocytosis that follows the calcium ions. Mohri and colleagues (1995) have shown that IP_3-released calcium is both necessary and sufficient for initiating the wave of calcium release.

IP_3 is similarly found to release calcium ions in vertebrate eggs. As in sea urchins, waves of IP_3 are thought to mediate calcium release from sites within the endoplasmic reticulum (Lechleiter and Clapham 1992; Miyazaki et al. 1992; Ayabe et al. 1995). Blocking the IP_3 receptor in hamster eggs prevents the release of calcium at fertilization. Xu and colleagues (1994) found that blocking the

IP_3-mediated calcium release blocks every aspect of sperm-induced egg activation, including cortical granule exocytosis, mRNA recruitment, and cell cycle resumption.

The question then becomes, what initiates the production of IP_3? In other words, what activates the phospholipase C enzymes? This question has not been easy to address, since (1) there are numerous types of PLC, (2) they can be activated through different pathways, and (3) different species can use different mechanisms to activate them. Results from recent studies of sea urchin eggs suggest that the active PLC is a member of the γ family of PLCs and is activated by a protein tyrosine kinase (Carroll et al. 1997, 1999; Williams et al. 1998).

WEBSITE **7.9 Biochemistry of egg activation.** Inhibitors of phospholipase Cγ activity prevent the calcium ion flux across the echinoderm egg. This PLC is activated by protein kinases. This movie shows the evidence for this inhibition.

The next question, then, is, what protein tyrosine kinase is activated? This question has not yet been satisfactorily answered. According to one model, the sperm receptor protein crosses the egg plasma membrane and has a protein tyrosine kinase activity in its cytoplasmic domain (Figure 7.29A). This structure would make it a classic receptor tyrosine kinase (see Chapter 5). However, the sequence of the putative bindin receptor reveals neither transmembrane nor kinase domains (Just and Lennarz 1997). According to a second model, the bindin receptor is linked to a protein tyrosine kinase and can activate the kinase, perhaps as a consequence of receptor crosslinking by the sperm (Figure 7.29B; see Giusti et al. 1999). A third possibility is that the activation of the IP_3 pathway is caused not by the binding of sperm and egg, but by the fusion of the sperm and egg plasma membranes. McCulloh and Chambers (1992) have electrophysiological evidence that sea urchin egg activation does not occur until after sperm and egg cytoplasms are joined. They suggest that the egg-activating components are located on the sperm plasma membrane or in the cytoplasm. It is even possible that when the fusion of gamete membranes occurs, the sperm receptor tyrosine kinases (activated by the egg jelly to initiate the acrosomal reaction) activate the IP_3 cascade for calcium release in the egg (see Gilbert 1994). In this scenario, shown in Figure 7.29C, bindin serves for cell-cell ad-

ACTIVATION PRIOR TO SPERM FUSION

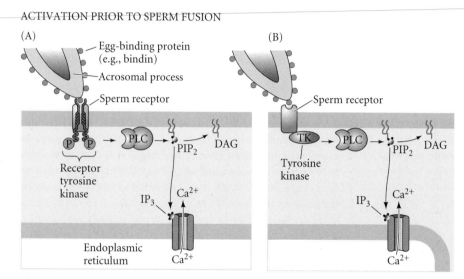

ACTIVATION AFTER SPERM FUSION

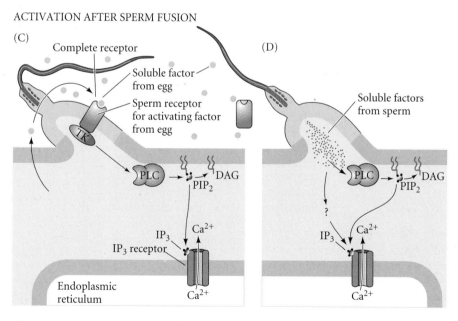

Figure 7.29
Possible mechanisms of egg activation. (A) The bindin receptor in the egg plasma membrane has tyrosine kinase activity. The tyrosine kinase activates PLC. (B) The bindin receptor activates a cytoplasmic tyrosine kinase. (C) An activated tyrosine kinase in the sperm plasma membrane activates the egg pathways. (D) Soluble activator pathways activate PLC or directly release calcium from the endoplasmic reticulum.

hesion and membrane fusion, but not for signaling. Rather, the egg "activates itself" through the sperm.

Still another possibility is that the agent active in releasing the sequestered calcium comes from the sperm cytosol (Figure 17.29D). Support for the sperm's carrying an activating molecule comes from the clinical procedure of intracytoplasmic sperm injection (ICSI) used to treat infertility when a man's sperm count is low. Here, a single intact human sperm is injected directly into the cytoplasm of the egg. This results in egg activation, the formation of a male pronucleus, and normal embryonic development (Van Steirtinghem 1994). Kimura and colleagues (1998) have shown that the isolated head of a mouse sperm is capable of activating the mouse oocyte, and that the active portion of the sperm head appears to be the proteins surrounding the haploid nucleus. It is not known what role these perinuclear components may play in the normal physiology of egg activation.

Late responses

Shortly after the calcium ion levels rise in a sea urchin egg, its intracellular pH also increases.* The rise in intracellular pH begins with a second influx of sodium ions, which causes a 1:1 exchange between sodium ions from the seawater and hydrogen ions from the egg. This loss of hydrogen ions causes the pH to rise (Shen and Steinhardt 1978). It is thought that the pH increase and the calcium ion elevation act together to stimulate new protein synthesis and DNA synthesis (Winkler et al. 1980; Whitaker and Steinhardt 1982; Rees et al. 1995). If one experimentally elevates the pH of an unfertilized egg to a level similar to that of a fertilized egg, DNA synthesis and nuclear envelope breakdown ensue, just as if the egg were fertilized (Miller and Epel 1999).

The late responses of fertilization brought about by these ionic changes include the activation of DNA synthesis and protein synthesis. In sea urchins, a burst of protein synthesis usually occurs within several minutes after sperm entry. This protein synthesis does not depend on the synthesis of new messenger RNA; rather, it utilizes mRNAs already present in the oocyte cytoplasm (Figure 7.30; see Table 5.2). These messages include mRNAs encoding proteins such as histones, tubulins, actins, and morphogenetic factors that are utilized during early development. Such a burst of protein synthesis can be induced by artificially raising the pH of the cytoplasm using ammonium ions (Winkler et al. 1980).

*Again, species-to-species variation is rampant. In the much smaller egg of the mouse, there is no elevation of pH after fertilization. Similarly in the mouse, there is no dramatic increase in protein synthesis immediately following fertilization (Ben-Yosef et al. 1996).

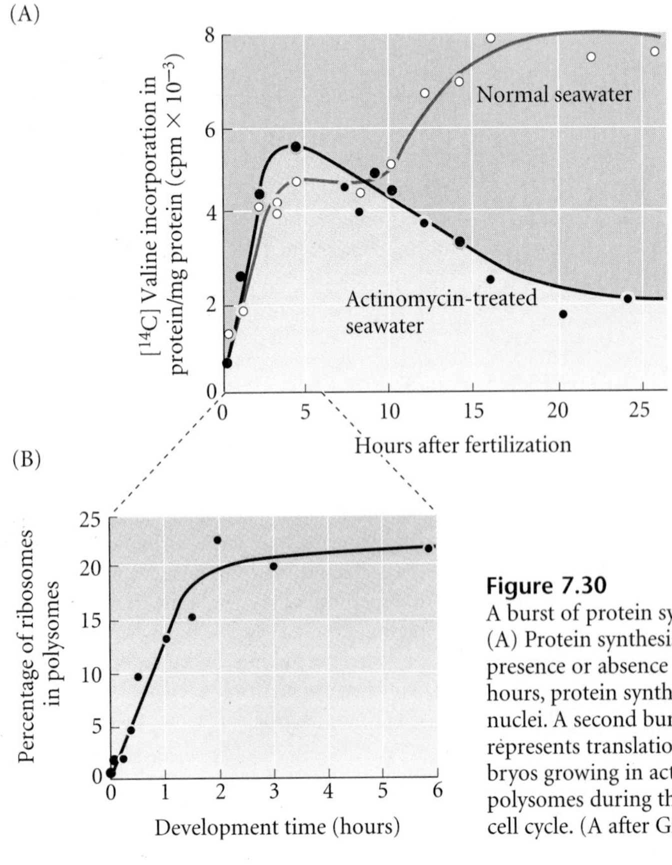

Fusion of the Genetic Material

Fusion of genetic material in sea urchins

In sea urchins, the sperm nucleus enters the egg perpendicular to the egg surface. After fusion of the sperm and egg plasma membranes, the sperm nucleus and its centriole separate from the mitochondria and the flagellum. The mitochondria and the flagellum disintegrate inside the egg, so very few, if any, sperm-derived mitochondria are found in developing or adult organisms (Dawid and Blackler 1972; Sutovsky et al. 1999). In mice, it is estimated that only 1 out of every 10,000 mitochondria is sperm-derived (Gyllensten et al. 1991). Thus, although each gamete contributes a hap-

(A)

(B)

Figure 7.30
A burst of protein synthesis at fertilization uses mRNAs stored in the oocyte cytoplasm. (A) Protein synthesis in embryos of the sea urchin *Arbacia punctulata* fertilized in the presence or absence of actinomycin D, an inhibitor of transcription. For the first few hours, protein synthesis occurs without any new transcription from the zygote or embryo nuclei. A second burst of protein synthesis occurs during mid-blastula stages. This burst represents translation of newly transcribed messages, and therefore is not seen in embryos growing in actinomycin. (B) Increase in the percentage of ribosomes recruited into polysomes during the first hours of sea urchin development, especially during the first cell cycle. (A after Gross et al. 1964; B after Humphreys 1971.)

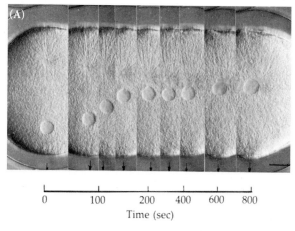

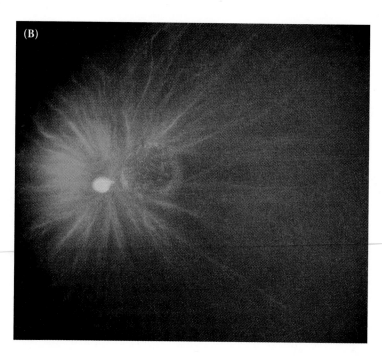

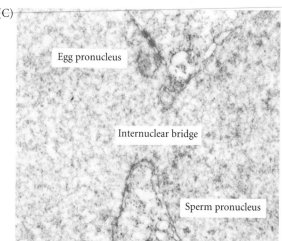

Figure 7.31
Nuclear events in the fertilization of the sea urchin. (A) Sequential photographs showing the migration of the egg pronucleus and the sperm pronucleus toward each other in an egg of *Clypeaster japonicus*. The sperm pronucleus is surrounded by its aster of microtubules. (B) The microtubules (stained with fluorescent antibodies to tubulin) radiate from the centrosome associated with the (smaller) male pronucleus and reach toward the female pronucleus. The two pronuclei migrate toward each other on these microtubular processes. (The pronuclear DNA is stained blue by Hoechst dye.) (C) Fusion of pronuclei in the sea urchin egg. (A from Hamaguchi and Hiramoto 1980, courtesy of the authors; B from Holy and Schatten 1991, courtesy of J. Holy; C courtesy of F. J. Longo.)

loid genome to the zygote, the mitochondrial genome is transmitted primarily by the maternal parent. Conversely, in almost all animals studied (the mouse being the major exception), the centrosome needed to produce the mitotic spindle of the subsequent divisions is derived from the sperm centriole (Sluder et al. 1989, 1993).

The egg nucleus, once it is haploid, is called the **female pronucleus**. Once inside the egg, the sperm nucleus decondenses to form the **male pronucleus**. The sperm nucleus undergoes a dramatic transformation. The nuclear envelope vesiculates into small packets, thereby exposing the compact sperm chromatin to the egg cytoplasm (Longo and Kunkle 1978; Poccia and Collas 1997). The proteins holding the sperm chromatin in its

condensed, inactive state are exchanged for other proteins derived from the egg cytoplasm. This exchange permits the decondensation of the sperm chromatin. In sea urchins, decondensation appears to be initiated by the phosphorylation of two sperm-specific histones that bind tightly to the DNA. This process begins when the sperm comes into contact with a glycoprotein in the egg jelly that elevates the level of cAMP-dependent protein kinase activity. These protein kinases phosphorylate several of the basic residues of the sperm-specific histones and thereby interfere with their binding to DNA (Garbers et al. 1980, 1983; Porter and Vacquier 1986). This loosening is thought to facilitate the replacement of the sperm-specific histones with other histones that have been stored in the oocyte cytoplasm (Poccia et al. 1981; Green and Poccia 1985). Once decondensed, the DNA can begin transcription and replication.

After the sea urchin sperm enters the egg cytoplasm, the male pronucleus rotates 180° so that the sperm centriole is between the sperm pronucleus and the egg pronucleus. The sperm centriole then acts as a microtubule organizing center, extending its own microtubules and integrating them with egg microtubules to form an aster.* These microtubules extend throughout the egg and contact the female pronucleus, and the two pronuclei migrate toward each other (Hamaguchi and Hiramoto 1980; Bestor and Schatten 1981). Their fusion forms the diploid **zygote nucleus** (Figure 7.31). The initiation of DNA synthesis can occur either in the pronuclear stage (during migration) or after the formation of the zygote nucleus.

---

*When Oscar Hertwig observed this radial array of sperm asters forming in his newly fertilized sea urchin eggs, he called it "the sun in the egg," and he thought it the happy indication of a successful fertilization (Hertwig 1877). More recently, Simerly and co-workers (1994) found that certain types of human male infertility are due to defects in the centriole's ability to form these microtubular asters. This deficiency causes the failure of pronuclear migration and the cessation of further development.

Fusion of genetic material in mammals

In mammals, the process of pronuclear migration takes about 12 hours, compared with less than 1 hour in the sea urchin. The mammalian sperm enters almost tangentially to the surface of the egg rather than approaching it perpendicularly, and it fuses with numerous microvilli (see Figure 7.20). The mammalian sperm nucleus also breaks down as its chromatin decondenses and is then reconstructed by coalescing vesicles. The DNA of the sperm nucleus is bound by basic proteins called protamines, and these nuclear proteins are tightly compacted through disulfide bonds. In the egg cytoplasm, glutathione reduces these disulfide bonds and allows the uncoiling of the sperm chromatin (Calvin and Bedford 1971; Kvist et al. 1980; Perreault et al. 1988). The mammalian male pronucleus enlarges while the oocyte nucleus completes its second meiotic division (Figure 7.32A). The centrosome (new centriole) accompanying the male pronucleus produces its asters (largely from proteins stored in the oocyte) and contacts the female pronucleus. Then each pronucleus migrates toward the other, replicating its DNA as it travels. Upon meeting, the two nuclear envelopes break down (Figure 7.32B). However, instead of producing a common zygote nucleus (as happens in sea urchin fertilization), the chromatin condenses into chromosomes that orient themselves on a common mitotic spindle (Figure 7.32C). Thus, a true diploid nucleus in mammals is first seen not in the zygote, but at the 2-cell stage.

WEBSITE **7.10 Sperm decondensation.** The formation of the male pronucleus involves changes in both the chromatin and the nuclear envelope. The nuclear annulus may be critical to uncoiling the DNA.

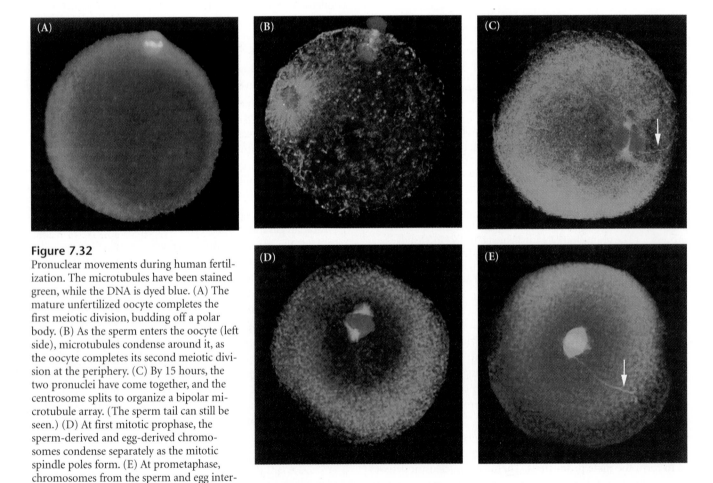

Figure 7.32
Pronuclear movements during human fertilization. The microtubules have been stained green, while the DNA is dyed blue. (A) The mature unfertilized oocyte completes the first meiotic division, budding off a polar body. (B) As the sperm enters the oocyte (left side), microtubules condense around it, as the oocyte completes its second meiotic division at the periphery. (C) By 15 hours, the two pronuclei have come together, and the centrosome splits to organize a bipolar microtubule array. (The sperm tail can still be seen.) (D) At first mitotic prophase, the sperm-derived and egg-derived chromosomes condense separately as the mitotic spindle poles form. (E) At prometaphase, chromosomes from the sperm and egg intermix on the metaphase equator and a mitotic spindle initiates the first mitotic division. (From Simerly et al. 1995; photographs courtesy of G. Schatten.)

The Nonequivalence of Mammalian Pronuclei

It is generally assumed that males and females carry equivalent haploid genomes. Indeed, one of the fundamental tenets of Mendelian genetics is that genes derived from the sperm are functionally equivalent to those derived from the egg. However, recent studies show that in mammals, the sperm-derived genome and the egg-derived genome may be functionally different and play complementary roles during certain stages of development. The first evidence for this nonequivalence came from studies of a human tumor called a **hydatidiform mole**, which resembles placental tissue. A majority of such moles have been shown to arise from a haploid sperm fertilizing an egg in which the female pronucleus is absent. After entering the egg, the sperm chromosomes duplicate themselves, thereby restoring the diploid chromosome number. Thus, the entire genome is derived from the sperm (Jacobs et al. 1980; Ohama et al. 1981). Here we see a situation in which the cells survive, divide, and have a normal chromosome number, but development is abnormal. Instead of forming an embryo, the egg becomes a mass of placenta-like cells. Normal development does not occur when the entire genome comes from the male parent.

Evidence for the nonequivalence of mammalian pronuclei also comes from attempts to get ova to develop in the absence of sperm. The ability to develop an embryo without spermatic contribution is called **parthenogenesis** (Greek, "virgin birth"), and the eggs of many invertebrates and some vertebrates are capable of developing normally in the absence of sperm (see Chapters 2 and 19). Mammals, however, do not exhibit parthenogenesis. Placing mouse oocytes in a culture medium that artificially activates the oocyte while suppressing the formation of the second polar body produces diploid mouse eggs whose inheritance is derived from the egg alone (Kaufman et al. 1977). These cells divide to form embryos with spinal cords, muscles, skeletons, and organs, including beating hearts. However, development does not continue, and by day 10 or 11 (halfway through the mouse's gestation), the parthenogenetic embryos deteriorate and become grossly disorganized. Neither human nor mouse development can be completed solely with egg-derived chromosomes.

The hypothesis that male and female pronuclei are different also gains support from pronuclear transplantation experiments (Surani and Barton 1983; Surani et al. 1986; McGrath and Solter 1984). Either male or female pronuclei of recently fertilized mouse eggs can be removed and added to other recently fertilized eggs. (The two pronuclei can be distinguished at this stage because the female pronucleus is the one beneath the polar bodies.) Thus, zygotes with two male or two female pronuclei can be constructed. Although embryonic cleavage occurs, neither of these types of eggs develops to birth, whereas some control eggs (containing one male pronucleus and one female pronucleus from different zygotes) undergoing such transplantation develop normally (Table 7.2). Moreover, the bimaternal or bipaternal embryos cease development at the same time as parthenogenetic embryos. Thus, although the two pronuclei are equivalent in many animals, in mammals there are important functional differences between them.

The reason for these embryonic deaths is that in some cells, only the maternally derived allele of certain genes is active, while in other cells, only the paternally derived allele of those genes is functional. (In most genes, of course, the male-derived and female-derived alleles are equivalent and are activated to the same degree in every cell. We are dealing here with exceptions to that Mendelian rule.) For instance, insulin-like growth factor II (Igf-2) promotes the growth of embryonic and fetal organs. In embryonic mice, the paternally derived allele of *Igf-2* is active throughout the embryo, whereas the maternally derived allele is usually inactive (except in a few neural cells). Thus, if a mouse inherits a mutant *Igf-2* allele from its mother, the mouse will develop to a normal size (since the maternally derived allele is not expressed), but if the same mutant allele is inherited from its father, the mouse will have stunted growth (DeChiara et al. 1991). The opposite pattern of allele expression is found for one of the receptors of Igf-2. Here, the paternal gene for the receptor is poorly transcribed, while the maternal allele is active (Barlow et al. 1991). The differences between the active and inactive alleles are caused by differential methylation of DNA in the egg and sperm nuclei (see Chapter 5). Because certain developmentally important genes are active only if they come from the sperm and others are active only if they come from the egg, maternal and paternal pronuclei are both necessary for the completion of mammalian development.

Table 7.2 Pronuclear transplantation experiments

Class of reconstructed zygotes	Operation	Number of successful transplants	Number of progeny surviving
Bimaternal		339	0
Bipaternal		328	0
Control		348	18

Source: McGrath and Solter 1984.

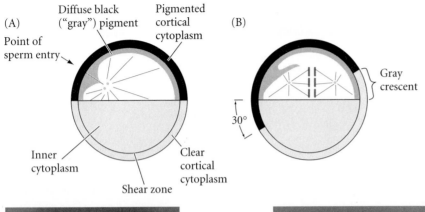

(A)

Point of sperm entry

Diffuse black ("gray") pigment

Pigmented cortical cytoplasm

(B)

Inner cytoplasm

Shear zone

Clear cortical cytoplasm

Gray crescent

30°

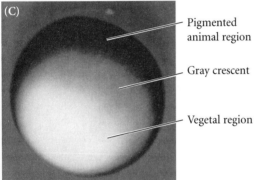

(C)

Pigmented animal region

Gray crescent

Vegetal region

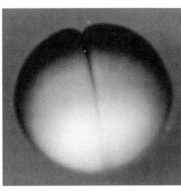

Figure 7.33
Reorganization of cytoplasm in the newly fertilized frog egg. (A) Schematic cross section of an egg midway through the first cleavage cycle. The egg has radial symmetry about its animal-vegetal axis. The sperm has entered at one side, and the sperm nucleus is migrating inward. The cortex represented is like that of *Rana*, with a heavily pigmented animal hemisphere and an unpigmented vegetal hemisphere. (B) About 80% of the way into first cleavage, the cortical cytoplasm rotates 30° relative to the internal cytoplasm. This rotation is important in that gastrulation will begin in that region opposite the point of sperm entry where the greatest displacement of cytoplasm occurs. (C) Gray crescent of *R. pipiens* immediately after the cortical rotation exposes the different pigment beneath the heavily pigmented cortical cytoplasm. (D) First cleavage furrow bisects the gray crescent. (A, B after Gerhart et al. 1989; C, D photographs courtesy of R. P. Elinson.)

Rearrangement of the Egg Cytoplasm

Fertilization can initiate radical displacements of the egg's cytoplasmic materials. While these cytoplasmic movements are not obvious in mammalian or sea urchin eggs, there are several species in which these rearrangements of oocyte cytoplasm are crucial for cell differentiation later in development. In the eggs of tunicates, as we will see in Chapter 8, cytoplasmic rearrangements are particularly obvious because of the differing pigmentation of the different regions of the egg. Such cytoplasmic movements are also easy to see in amphibian eggs. In frogs, a single sperm can enter anywhere on the animal hemisphere of the egg; when it does, it changes the cytoplasmic pattern of the egg. Originally, the egg is radially symmetrical about the animal-vegetal axis. After sperm entry, however, the cortical (outer) cytoplasm shifts about 30° toward the point of sperm entry, relative to the inner cytoplasm (Manes and Elinson 1980; Vincent et al. 1986). In some frogs (such as *Rana*), a region of the egg that was formerly covered by the dark cortical cytoplasm of the animal hemisphere is now exposed (Figure 7.33). This underlying cytoplasm, located near the equator on the side opposite the point of sperm entry, contains diffuse pigment granules and therefore appears gray. Thus, this region has been referred to as the **gray crescent** (Roux 1887; Ancel and Vintenberger 1948). As we will see in subsequent chapters, the gray crescent marks the region where gastrulation is initiated in amphibian embryos.

In frogs such as *Xenopus*, in which no gray crescent appears, dye labeling confirms that the cortical cytoplasm rotates

30° relative to the internal, subcortical cytoplasm (Vincent et al. 1986). The motor for these cytoplasmic movements in amphibian eggs appears to be an array of parallel microtubules that form between the cortical and inner cytoplasm of the vegetal hemisphere parallel to the direction of cytoplasmic rotation. These microtubular tracks are first seen immediately before the rotation commences, and they disappear when rotation ceases (Figure 7.34; Elinson and Rowning 1988). Treating the egg with colchicine or ultraviolet radiation at the beginning of rotation stops the formation of these microtubules, thereby inhibiting the cytoplasmic rotation. Using antibodies that bind to the microtubules, Houliston and Elinson (1991a) found that these tracks are formed from both sperm- and egg-derived microtubules, and that the sperm centriole directs the polymerization of the microtubules so that they grow into the vegetal region of the egg. Upon reaching the vegetal cortex, the microtubules angle away from the point of sperm entry, toward the vegetal pole. The off-center position of the sperm centriole as it initiates microtubule polymerization provides the directionality to the rotation. The motive force for the rotation may be provided by the ATPase **kinesin**. Like dynein and myosin, kinesin is able to attach to fibers and produce energy through ATP hydrolysis. This ATPase is located on the vegetal microtubules and the membranes of the cortical endoplasmic reticulum (Houliston and Elinson 1991b).

The movement of the cortical cytoplasm with respect to the inner cytoplasm causes profound changes within the inner cytoplasm. Danilchik and Denegre (1991) have labeled yolk platelets with Nile blue and watched their movement by fluo-

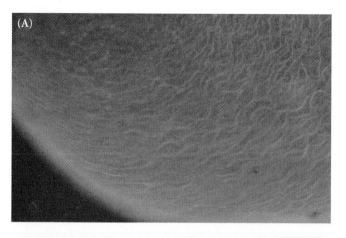

(A)

(B)

Figure 7.34
Parallel arrays of microtubules extend along the vegetal hemisphere along the future dorsal-ventral axis. These arrays are seen here in the second half of the first cell cycle using fluorescent antibodies to tubulin. (A) View looking at the vegetal hemisphere. (B) Higher magnification of the parallel microtubule arrays. Prior to cytoplasmic rotation (about midway through the first cell cycle), no array of microtubules can be seen, and at the end of cytoplasmic rotation, the microtubules depolymerize. (Photographs courtesy of R. P. Elinson.)

tive ventral side. What had been a radially symmetrical embryo is now a bilaterally symmetrical embryo. As we will see in Chapter 10, these cytoplasmic movements initiate a cascade of events that determine the dorsal-ventral axis of the frog. Indeed, the parallel microtubules that allow these rearrangements to stretch along what will become the dorsal-ventral axis of the frog (Klag and Ubbels 1975; Gerhart et al. 1983).

> VADE MECUM **Amphibian fertilization.** The shifting of cytoplasm in response to fertilization in the amphibian egg is often illustrated in models. Here you will see 3-D models of this event compiled into a movie.
> **[Click on Amphibian]**

Preparation for cleavage

The increase in intracellular free calcium ions that activates DNA and protein synthesis also sets in motion the apparatus for cell division. The mechanisms by which cleavage is initiated probably differ among species, depending on the stage of meiosis at which fertilization occurs. However, in all species studied, the rhythm of cell divisions is regulated by the synthesis and degradation of a protein called **cyclin**. As we will see in Chapters 8 and 19, cyclin keeps cells in metaphase, and the breakdown of cyclin enables the cells to return to interphase. In addition to their other activities, calcium ions appear to initiate the degradation of cyclin (Watanabe et al. 1991; Tokumoto et al. 1997). Once the cyclin is degraded, the cycles of cell division can begin anew.

Cleavage has a special relationship to the egg regions established by the cytoplasmic movements described above. In tunicate embryos, the first cleavage bisects the egg, with its established cytoplasmic pattern, into mirror-image duplicates. From that stage on, every division on one side of the cleavage furrow has a mirror-image division on the opposite side.

rescent microscopy (the bound dye fluoresces red). During the middle part of the first cell cycle, a mass of central egg cytoplasm flows from the presumptive ventral (belly) to the future dorsal (back) side of the embryo (Figure 7.35). By the end of first division, the cytoplasm of the prospective dorsal side of the embryo is distinctly different from that of the prospec-

Figure 7.35
Cytoplasmic rearrangement in the frog *Xenopus laevis*. (A) The unfertilized egg is radially symmetrical. (B) Cytoplasmic movements are seen as the egg starts cleaving, 90 minutes after fertilization. (C) The cytoplasm of the future dorsal side (right) becomes different from the future ventral side (left), where the sperm entered. (Photographs courtesy of M. V. Danilchik.)

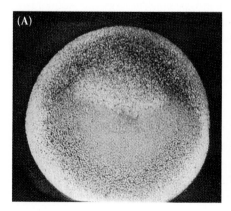

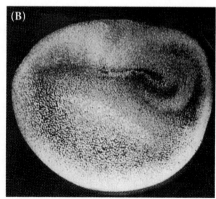

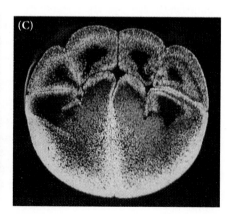

Similarly, the gray crescent is bisected by the first cleavage furrow in amphibian eggs (Figure 7.33D). Thus, the position of the first cleavage is not random, but tends to be specified by the point of sperm entry and the subsequent rotation of the egg cytoplasm. The coordination of cleavage plane and cytoplasmic rearrangements is probably mediated through the microtubules of the sperm aster (Manes et al. 1978; Gerhart et al. 1981; Elinson 1985).

Toward the end of the first cell cycle, then, the cytoplasm is rearranged, the pronuclei have met, DNA is replicating, and new proteins are being translated. The stage is set for the development of a multicellular organism.

Snapshot Summary: Fertilization

1. Fertilization accomplishes two separate activities: sex (the combining of genes derived from two parents) and reproduction (the creation of a new organism).

2. The events of conception usually include: (1) contact and recognition between sperm and egg; (2) regulation of sperm entry into the egg; (3) fusion of genetic material from the two gametes; and (4) activation of egg metabolism to start development.

3. The sperm head consists of a haploid nucleus and an acrosome. The acrosome is derived from the Golgi apparatus and contains enzymes needed to digest extracellular coats surrounding the egg. The neck of the sperm contains the mitochondria and the centriole that generates the microtubules of the flagellum. Energy for flagellar motion comes from mitochondrial ATP and a dynein ATPase in the flagellum.

4. The egg contains a haploid nucleus, and an enlarged cytoplasm storing ribosomes, mRNAs, and nutritive proteins. Other mRNAs and proteins, used as morphogenetic factors, are also stored in the egg. Cortical granules lie beneath the egg's plasma membrane. Many eggs also contain protective agents needed for survival in their particular environment.

5. Surrounding the egg plasma membrane is an extracellular layer often used in sperm recognition. In most animals, this extracellular layer is the vitelline envelope. In mammals, it is the much thicker zona pellucida.

6. In many species, eggs secrete diffusible molecules that attract and activate the sperm.

7. In sea urchins, the acrosome reaction is initiated by compounds in the egg jelly. The acrosomal vesicle undergoes exocytosis to release its enzymes. Globular actin polymerizes to extend the acrosomal process. Bindin on the acrosomal process is recognized by a protein complex on the sea urchin egg surface.

8. In mammals, sperm must be capacitated in the female reproductive tract before they are capable of fertilizing the egg.

9. Mammalian sperm bind to the zona pellucida before undergoing the acrosome reaction. In the mouse, this binding is mediated by ZP3 (zona protein 3) and one or many sperm proteins that recognize it. The mammalian acrosome reaction is initiated on the zona pellucida, and the acrosomal enzymes are concentrated there.

10. Fusion between sperm and egg is mediated by protein molecules whose hydrophobic groups can merge the sperm and egg plasma membranes. In sea urchins, bindin may mediate gamete fusion. In mammals, fertilin proteins in the sperm bind to integrins in the egg and allow the membranes to fuse.

11. Polyspermy results when two sperm fertilize the egg. It is usually lethal, since it results in three sets of chromosomes divided among four cells.

12. There are often two blocks to polyspermy. The fast block is electrical and is mediated by sodium ions: the egg membrane resting potential rises, and sperm can no longer fuse with the egg. The slow block is physical and is mediated by calcium ions. A wave of calcium ions propagates from the point of sperm entry, causing the cortical granules to fuse with the egg cell membrane. The released contents of the granules cause the vitelline membrane to rise and to harden into the fertilization envelope.

13. In mammals, blocks to polyspermy include the modification of the zona proteins by the contents of the cortical granules. Sperm can no longer bind to the zona.

14. Inositol 1,4,5-triphosphate (IP_3) is believed to be responsible for releasing calcium ions from storage in the endoplasmic reticulum. DAG (diacylglycerol) is thought to initiate the rise in egg pH. The free calcium ions, supported by the alkalization of the egg, activate egg metabolism, protein synthesis, and DNA synthesis.

15. The male pronucleus and the female pronucleus migrate toward each other, replicating DNA as they move.

16. In sea urchins, the two pronuclei merge and a diploid zygote nucleus is formed. In mammals, the pronuclei disintegrate as they approach each other, and their chromosomes gather around a common metaphase plate.

17. Some genes are transmitted differently depending on whether they are from the egg or the sperm. Methylation differences determine if these genes are to be expressed in the early embryo.

18. Microtubular changes cause cytoplasmic movements. These rearrangements of cytoplasm can be critical in specifying which portions of the egg are going to develop into which organs.

Literature Cited

Afzelius, B. A. 1976. A human syndrome caused by immotile cilia. *Science* 193: 317–319.

Almeida, E. A. C. and 10 others. 1995. Mouse egg integrin α 6 β 1 functions as a sperm receptor. *Cell* 81: 1095–1104.

Alves, A.-P., B. Mulloy, J. A., Diniz and P. A. S. Mourao. 1997. Sulfated polysaccharides from the egg jelly layer are species-specific inducers of acrosome reactions in sperms of sea urchins. *J. Biol. Chem.* 272: 6965–6971.

Amos, L. A. and A. Klug. 1974. Arrangement of subunits in flagellar microtubules. *J. Cell Sci.* 14: 523–549.

Ancel, P. and P. Vintenberger. 1948. Recherches sur le determinisme de la symmetrie bilatérale dans l'oeuf des amphibiens. *Bull. Biol. Fr. Belg.* [Suppl.] 31: 1–182.

Arnoult, C., I. G. Kazam, P.E. Visconti, G. Kopf, M. Villaz and H. Florman. 1999. Control of the low-voltage-activated calcium channel of mouse sperm by egg ZP3 and by membrane hyperpolarization during capacitation. *Proc. Natl. Acad. Sci. USA* 96: 6757–6762.

Austin, C. R. 1952. The "capacitation" of mammalian sperm. *Nature* 170: 327.

Austin, C. R. 1960. Capacitation and the release of hyaluronidase. *J. Reprod. Fertil.* 1: 310–311.

Austin, C. R. 1965. *Fertilization.* Prentice-Hall, Englewood Cliffs, NJ.

Ayabe, T., G. S. Kopf and R. M. Schultz. 1995. Regulation of mouse egg activation: Presence of ryanodine receptors and effects of microinjected ryanodine and cyclic ADP ribose on uninseminated and inseminated eggs. *Development* 121: 2233–2244.

Barlow, D. P., R. Stöger, B. G. Herrmann, K. Saito and N. Schweifer. 1991. The mouse insulin-like growth factor type-2 receptor is imprinted and closely linked to the *Tme* locus. *Nature* 349: 84–87.

Bentley, J. K., H. Shimomura and D. L.Garbers. 1986. Retention of a functional resact receptor in isolated sperm plasma membranes. *Cell* 45: 281–288.

Ben-Yosef, D., Y., Oron and R. Shalgi. 1996. Intracellular pH in rat eggs is not affected by fertilization and the resulting calcium oscillations. *Biol. Reprod.* 55: 461–468.

Berridge, M. J. 1993. Inositol triphosphate and calcium signalling. *Nature* 361: 315–325.

Bestor, T. M. and G. Schatten. 1981. Anti-tubulin immunofluorescence microscopy of microtubules present during the pronuclear movements of sea urchin fertilization. *Dev. Biol.* 88: 80–91.

Bi, G.-Q., J. M. Alderton and R. A. Steinhardt. 1995. Calcium-mediated exocytosis is required for cell membrane resealing. *J. Cell Biol.* 131: 1747–1758.

Bleil, J. D. and P. M. Wassarman. 1980. Mammalian sperm and egg interaction: Identification of a glycoprotein in mouse-egg zonae pellucidae possessing receptor activity for sperm. *Cell* 20: 873–882.

Bleil, J. D. and P. M.Wassarman. 1986. Autoradiographic visualization of the mouse egg's sperm receptor bound to sperm. *J. Cell Biol.* 102: 1363–1371.

Bleil, J. D. and P. M. Wassarman. 1988. Galactose at the nonreducing terminus of O-linked oligosaccharides of mouse egg zona pellucida glycoprotein ZP3 is essential for the glycoprotein's sperm receptor activity. *Proc. Natl. Acad. Sci. USA* 85: 6778–6782.

Bleil, J. D., J. M. Greve and P. M. Wassarman. 1988. Identification of a secondary sperm receptor in the mouse egg zona pellucida: Role in maintenance of binding of acrosome-reacted sperm to eggs. *Dev. Biol.* 28: 376–385.

Blobel, C. P, T. G. Wolfsberg, C. W. Turck, D. G. Myles, P. Primakoff and J. M. White. 1992. A potential fusion peptide and an integrin domain in a protein active in sperm-egg fusion. *Nature* 356: 248–251.

Bogart, J. P., R. P. Elinson and L. E. Licht. 1989. Temperature and sperm incorporation in polyploid salamanders. *Science* 246: 1032–1034.

Boveri, T. 1902. On multipolar mitosis as a means of analysis of the cell nucleus. (Translated by S. Gluecksohn-Waelsch.) *In* B. H. Willier and J. M. Oppenheimer (eds.), *Foundations of Experimental Embryology.* Hafner, New York, 1974.

Boveri, T. 1907. Zellenstudien VI. Die Entwicklung dispermer Seeigeleier. Ein Beiträge zur Befruchtungslehre und zur Theorie des Kernes. *Jena Z. Naturwiss.* 43: 1–292.

Busa, W. B., J. E. Ferguson, S. K. Joseph, J. R. Williamson and R. Nuccitelli. 1985. Activation of frog (*Xenopus laevis*) eggs by inositol triphosphate. I. Characterization of Ca^{2+} release from intracellular stores. *J. Cell Biol.* 100: 677–682.

Calvin, H. I. and J. M. Bedford. 1971. Formation of disulfide bonds in the nucleus and accessory structures of mammalian spermatozoa during maturation in the epididymis. *J. Reprod. Fertil.* [Suppl.] 13: 65–75.

Carroll, D. J., C. S. Ramarao, L. Mehlmann, S. Roche, M. Terasaki and L. A. Jaffe. 1997. Calcium release at fertilization in starfish eggs is mediated by phospholipase Cγ. *J. Cell Biol.* 138: 1303–1311.

Carroll, D. J., D. T. Albay, M. Terasaki, L. A. Jaffe and K. R. Foltz. 1999. Identification of PLCγ-dependent and independent events during fertilization of sea urchin eggs. *Dev. Biol.* 206: 232–247.

Carroll, E. J. and D. Epel. 1975. Isolation and biological activity of the proteases released by sea urchin eggs following fertilization. *Dev. Biol.* 44: 22–32.

Chambers, E. L., B. C. Pressman and B. Rose. 1974. The activity of sea urchin eggs by the divalent ionophores A23187 and X-537A. *Biochem. Biophys. Res. Comm.* 60: 126–132.

Chandler, D. E. and J. Heuser. 1979. Membrane fusion during secretion: Cortical granule exocytosis in sea urchin eggs as studied by quick-freezing and freeze fracture. *J. Cell Biol.* 83: 91–l08.

Chang, M. C. 1951. Fertilizing capacity of spermatozoa deposited into the fallopian tubes. *Nature* 168: 697–698.

Chen, H. and N. S. Sampson. 1999. Mediation of sperm-egg fusion: Evidence that mouse egg α6β1 integrin is the receptor for sperm fertilinβ. *Chem. Biol.* 6: 1–10.

Cherr, G. N., H. Lambert, S. Meizel and D. F.Katz. 1986. In vitro studies of the golden hamster sperm acrosomal reaction: Completion on zona pellucida and induction by homologous zonae pellucidae. *Dev. Biol.* 114: 119–131.

Cho, C, D. O. Bunch, J.-E. Faure, E. H. Goulding, E. M. Eddy, P. Primakoff and D. G. Myles. 1998. Fertilization defects in sperm from mice lacking fertilin β. *Science* 281: 1857–1859.

Clermont, Y. and C. P. Leblond. 1955. Spermiogenesis of man, monkey, and other animals as shown by the "periodic acid-Schiff" technique. *Am. J. Anat.* 96: 229–253.

Colwin, A. L. and L. H. Colwin. 1963. Role of the gamete membranes in fertilization in *Saccoglossus kowalevskii* (Enteropneustra). I. The acrosome reaction and its changes in early stages of fertilization. *J. Cell Biol.* 19: 477–500.

Colwin, L. H. and A. L. Colwin. 1960. Formation of sperm entry holes in the vitelline membrane of *Hydroides hexagonis* (Annelida) and evidence of their lytic origin. *J. Biophys. Biochem. Cytol.* 7: 315–320.

Cook, S. P. and D. F. Babcock. 1993. Selective modulation by cGMP of the K^+ channel activated by speract. *J. Biol. Chem.* 268: 22402–22407.

Correia, L. M. and E. J. Carroll Jr. 1997. Characterization of the vitelline envelope of the sea urchin *Strongylocentratus purpuratus.* *Dev. Growth Diff.* 39: 69–85.

Corselli, J. and P. Talbot. 1987. In vivo penetration of hamster oocyte-cumulus complexes using physiological numbers of sperm. *Dev. Biol.* 122: 227–242.

Cross, N. L. and R. P. Elinson. 1980. A fast block to polyspermy in frogs mediated by changes in the membrane potential. *Dev. Biol.* 75: 187–198.

Dan, J. C. 1967. Acrosome reaction and lysins. *In* C. B. Metz and A. Monroy (eds.), *Fertilization*, vol. 1. Academic Press, New York, pp. 237–367.

Danilchik, M. V. and J. M. Denegre. 1991. Deep cytoplasmic rearrangements during early development in *Xenopus laevis*. *Development* 111: 845–857.

Dawid, I. B. and A. W. Blackler. 1972. Maternal and cytoplasmic inheritance of mitochondria in *Xenopus*. *Dev. Biol.* 29: 152–161.

DeChiara, T. M., E. J. Robertson and A. Efstradiatis. 1991. Parental imprinting of the mouse insulin-like growth factor II gene. *Cell* 64: 849–859.

De Robertis, E. D. P., F. A. Saez and E. M. F. De Robertis. 1975. *Cell Biology*, 6th Ed. Saunders, Philadelphia.

Eisen, A. and G. T. Reynolds. 1985. Sources and sinks for the calcium release during fertilization of single sea urchin eggs. *J. Cell Biol.* 100: 1522–1527.

Eisenbach, M. 1995. Sperm changes enabling fertilization in mammals. *Curr. Opin. Endocrinol. Diabetes* 2: 468–475.

Elinson, R. P. 1985. Changes in levels of polymeric tubulin associated with activation and dorsoventral polarization of the frog egg. *Dev. Biol.* 109: 224–233.

Elinson, R. P. and B. Rowning. 1988. A transient array of parallel microtubules in frog eggs: Potential tracks for a cytoplasmic rotation that specifies the dorso-ventral axis. *Dev. Biol.* 128: 185–197.

Endo, Y. G., G. S. Kopf and R. M. Schultz. 1987. Effects of phorbol ester on mouse eggs: Dissociation of sperm receptor activity from acrosome reaction-inducing activity of the mouse zona pellucida protein, ZP3. *Dev. Biol.* 123: 574–577.

Epel, D. 1977. The program of fertilization. *Sci. Am.* 237(5): 128–138.

Epel, D. 1980. Fertilization. *Endeavour* N.S. 4: 26–31.

Epel, D., C. Patton, R. W. Wallace and W. Y. Cheung. 1981. Calmodulin activates NAD kinase of sea urchin eggs: An early response. *Cell* 23: 543–549.

Evans, J. P., G. S. Kopf and R. M. Schultz. 1997. Characterization of the binding of recombinant mouse sperm fertilin β subunit to mouse eggs. *Dev. Biol.* 187: 79–83.

Ferris, C. D., R. L. Huganir, S. Supattapone and S. H. Snyder. 1989. Purified inositol 1,4,5-trisphosphate receptor mediates calcium flux in reconstituted lipid vesicles. *Nature* 342: 87–89.

Florman, H. M. and B. T. Storey. 1982. Mouse gamete interactions: The zona pellucida is the site of the acrosome reaction leading to fertilization in vitro. *Dev. Biol.* 91: 121–130.

Florman, H. M. and P. M. Wassarman. 1985. O-linked oligosaccharides of mouse egg ZP3 account for its sperm receptor activity. *Cell* 41: 313–324.

Florman, H. M., K. B. Bechtol and P. M. Wassarman. 1984. Enzymatic dissection of the functions of the mouse egg's receptor for sperm. *Dev. Biol.* 106: 243–255.

Florman, H. M., M. E. Corron, T. D.-H. Kim and D. F. Babcock. 1992. Activation of voltage-dependent calcium channels of mammalian sperm is required for zona pellucida-induced acrosomal exocytosis. *Dev. Biol.* 152: 304–314.

Florman, H. M., C. Arnoult, I. Kazam, C. Li and C. M. B. O'Toole. 1998. A perspective on the control of mammalian fertilization by egg-activated ion channels in sperm: A tale of two channels. *Biol. Reprod.* 59: 12–17.

Foerder, C. A. and B. M. Shapiro. 1977. Release of ovoperoxidase from sea urchin eggs hardens fertilization membrane with tyrosine crosslinks. *Proc. Natl. Acad. Sci. USA* 74: 4214–4218.

Fol, H. 1877. Sur le commencement de l'hémogénie chez divers animaux. *Arch. Zool. Exp. Gén.* 6: 145–169.

Foltz, K. R., J. S. Partin and W. J. Lennarz. 1993. Sea urchin egg receptor for sperm: Sequence similarity of beinding domain and hsp 70. *Science* 259: 1421–1425.

Franklin, L. E. 1970. Fertilization and the role of the acrosomal reaction in non-mammals. *Biol. Reprod.* [Suppl.] 2: 159–177.

Furuichi, T., S. Yoshikawa, A. Miyawaki, K. Wada, N. Maeda and K. Mikoshiba. 1989. Primary structure and functional expression of the inositol 1,4,5-trisphosphate-binding protein P400. *Nature* 342: 32–38.

Galantino-Homer, H. L., P. E. Visconti and G. S. Kopf. 1997. Regulation of protein tyrosine kinase phosphorylation during bovine capacitation by a cyclic adenosine 3',5'-monophosphate-dependent pathway. *Biol. Reprod.* 56: 707–719.

Garbers, D. L., D. J. Tubb and G. S. Kopf. 1980. Regulation of sea urchin sperm cAMP-dependent protein kinases by an egg associated factor. *Biol. Reprod.* 22: 526–532.

Garbers, D. L., G. S. Kopf, D. J. Tubb and G. Olson. 1983. Elevation of sperm adenosine 3':5'-monophosphate concentrations by a fucose sulfate-rich complex associated with eggs. I. Structural characterization. *Biol. Reprod.* 29: 1211–1220.

Gardiner, D. M. and R. D. Grey. 1983. Membrane junctions in *Xenopus* eggs: Their distribution suggests a role in calcium regulation. *J. Cell Biol.* 96: 1159–1163.

Gerhart, J., G. Ubbels, S. Black, K. Hara and M. Kirschner. 1981. A reinvestigation of the role of the grey crescent in axis formation in *Xenopus laevis*. *Nature* 292: 511–517.

Gerhart, J., S. Black, R. Gimlich and S. Scharf. 1983. Control of polarity in the amphibian egg. In W. R. Jeffery and R. A. Raff (eds.), *Time, Space, and Pattern in Embryonic Development*. Alan R. Liss, New York, pp. 261–287.

Gerhart, J., M. Danilchik, T. Doniach, S. Roberts, B. Rowning and R. Stewart. 1989. Cortical rotation of the *Xenopus* egg: Consequences for the anterioposterior pattern of embryonic dorsal development. *Development* 1989 [Suppl.]: 37–51.

Gilbert, S. F. 1994. *Developmental Biology*, 4th Ed. Sinauer Associates, Sunderland, MA.

Giles, R. E., H. Blanc, H. M. Cann and D. C. Wallace. 1980. Maternal inheritance of human mitochondrial DNA. *Proc. Natl. Acad. Sci. USA* 77: 6715–6719.

Gilkey, J. C., L. F. Jaffe, E. G. Ridgway and G. T. Reynolds. 1978. A free calcium wave traverses the activating egg of the medaka, *Oryzias latipes*. *J. Cell Biol.* 76: 448–467.

Giusti, A, F., K. M. Hoang and K. R. Foltz. 1997. Surface localization of the sea urchin egg receptor for sperm. *Dev. Biol.* 184: 10–24.

Giusti, A. F., D. J. Carroll, Y. A. Abassi and K. R. Folty. 1999. Evidence that a starfish egg Src family tyrosine kinase associates with PLC-γ1 SH2 domains at fertilization. *Dev. Biol.* 208: 189–199.

Glabe, C. G. 1985. Interaction of the sperm adhesive protein, bindin, with phospholipid vesicles. II. Bindin induces the fusion of mixed-phase vesicles that contain phosphatidylcholine and phosphatidylserine in vitro. *J. Cell Biol.* 100: 800–807.

Glabe, C. G. and W. J. Lennarz. 1979. Species-specific sperm adhesion in sea urchins: A quantitative investigation of bindin-mediated egg agglutination. *J. Cell Biol.* 83: 595–604.

Glabe, C. G. and V. D. Vacquier. 1977. Species-specific agglutination of eggs by bindin isolated from sea urchin sperm. *Nature* 267: 836–838.

Glabe, C. G. and V. D. Vacquier. 1978. Egg surface glycoprotein receptor for sea urchin sperm bindin. *Proc. Natl. Acad. Sci. USA* 75: 881–885.

González-Martínez, M. T., A. Guerrero, E. Morales, L. de la Torre and A. Darszon. 1992. A depolarization can trigger Ca^{2+} uptake and the acrosome reaction when preceded by a hyperpolarization in *L. pictus* sea urchin sperm. *Dev. Biol.* 150: 193–202.

Gould, M. and J. L. Stephano. 1987. Electrical response of eggs to acrosomal protein similar to those induced by sperm. *Science* 235: 1654–1657.

Gould, M. and J. L. Stephano. 1991. Peptides from sperm acrosomal protein that activate development. *Dev. Biol.* 146: 509–518.

Gould-Somero, M., L. A. Jaffe and L. Z. Holland. 1979. Electrically mediated fast polyspermy block in eggs of the marine worm, *Urechis caupo*. *J. Cell Biol.* 82: 426–440.

Green, G. R. and E. L. Poccia. 1985. Phosphorylation of sea urchin sperm H1 and H2B histones precedes chromatin decondensation and H1 exchange during pronuclear formation. *Dev. Biol.* 108: 235–245.

Gross, P. R., L. I. Malkin and W. Moyer. 1964. Templates for the first proteins of embryonic development. *Proc. Natl. Acad. Sci. USA* 51: 407–414.

Gundersen, G. G., L. Medill and B. M. Shapiro. 1986. Sperm surface proteins are incorporated into the egg membrane and cytoplasm after fertilization. *Dev. Biol.* 113: 207–217.

Gwatkin, R. B. L. 1976. Fertilization. *In* G. Poste and G. L. Nicolson (eds.), *The Cell Surface in Animal Embryogenesis and Development.* Elsevier North-Holland, New York, pp. 1–53.

Gyllensten, U., D. Wharton, A. Josefson and A. Wilson. 1991. Paternal inheritance of mitochondrial DNA in mice. *Nature* 352: 255–258.

Hafner, M., C. Petzelt, R. Nobiling, J. B. Pawley, D. Kramp and G. Schatten. 1988. Wave of free calcium at fertilization in the sea urchin egg visualized with Fura-2. *Cell Motil. Cytoskel.* 9: 271–277.

Hamaguchi, M. S. and Y. Hiramoto. 1980. Fertilization process in the heart-urchin, *Clypaester japonicus,* observed with a differential interference microscope. *Dev. Growth Diff.* 22: 517–530.

Hardy, D. M., T. Harumi and D. L. Garbers. 1994. Sea urchin sperm receptors for egg peptides. *Semin. Dev. Biol.* 5: 217–224.

Hartsoeker, N. 1694. *Essai de dioptrique.* Paris.

Heinecke, J. W. and B. M. Shapiro. 1989. Respiratory oxygen burst of fertilization. *Proc. Natl. Acad. Sci. USA* 86: 1259–1263.

Hertwig, O. 1877. Beiträge zur Kenntniss der Bildung, Befruchtung, und Theilung des theirischen Eies. *Morphol. Jahr.* 1: 347–452.

Holy, J. and G. Schatten. 1991. Spindle pole centrosomes of sea urchin embryos are partially composed of material recruited from maternal stores. *Dev. Biol.* 147: 343–353.

Houliston, E. and R. P. Elinson. 1991a. Evidence for the involvement of microtubules, endoplasmic reticulum, and kinesin in cortical rotation of fertilized frog eggs. *J. Cell Biol.* 114: 1017–1028.

Houliston, E. and R. P. Elinson. 1991b. Patterns of microtubule polymerization relating to cortical rotation in *Xenopus laevis* eggs. *Development* 112: 107–117.

Humphreys, T. 1971. Measurements of messenger RNA entering polysomes upon fertilization in sea urchins. *Dev. Biol.* 26: 201–208.

Hunnicut, G. R., D. E. Koppel and D. G. Myles. 1997. Analysis of the process of localization of fertilin to the sperm posterior head plasma membrane domain during sperm maturation in the epididymis. *Dev. Biol.* 191: 146–159.

Hunter, R. H. F. 1989. Ovarian programming of gamete progression and maturation in the female genital tract. *Zool. J. Linn. Soc.* 95: 117–124.

Hylander, B. L. and R. G. Summers. 1982. An ultrastructural and immunocytochemical localization of hyaline in the sea urchin egg. *Dev. Biol.* 93: 368–380.

Iwao, Y. and L. A. Jaffe. 1989. Evidence that the voltage-dependent component in the fertilization process is contributed by the sperm. *Dev. Biol.* 134: 446–451.

Jacobs, P. A., C. M. Wilson, J. A. Sprenkle, N. B. Rosenshein and B. R. Migeon. 1980. Mechanism of origin of complete hydatidiform moles. *Nature* 286: 714–717.

Jaffe, L. A. 1976. Fast block to polyspermy in sea urchins is electrically mediated. *Nature* 261: 68–71.

Jaffe, L. A. 1980. Electrical polyspermy block in sea urchins: Nicotine and low sodium experiments. *Dev. Growth Diff.* 22: 503–507.

Jaffe, L. A. and N. L. Cross. 1983. Electrical properties of vertebrate oocyte membranes. *Biol. Reprod.* 30: 50–54.

Jaffe, L. F. 1983. Sources of calcium in egg activation: A review and hypothesis. *Dev. Biol.* 99: 265–277.

Just, E. E. 1919. The fertilization reaction in *Echinarachinus parma. Biol. Bull.* 36: 1–10.

Just, M. L. and W. J. Lennarz. 1997. Reexamination of the sequence of the sea urchin egg receptor for sperm: Implications with respect to its properties. *Dev. Biol.* 184: 25–30.

Kaufman, M. H., S. C. Barton and M. A. H. Surani. 1977. Normal postimplantation development of mouse parthenogenetic embryos to the forelimb bud stage. *Nature* 265: 53–55.

Kimura, Y., R. Yanagimachi, S. Kuretake, H. Bortiewicz, A. C. F. Perry and H. Yanagimachi. 1998. Analysis of mouse oocyte activation suggests the involvement of sperm perinuclear material. *Biol. Reprod.* 58: 1407–1415.

Klag, J. J. and G. A. Ubbels. 1975. Regional morphological and cytochemical differentiation of the fertilized egg of *Discoglossus pictus* (Anura). *Differentiation* 3: 15–20.

Kline, D. 1988. Calcium-dependent events at fertilization of the frog egg: Injection of a calcium buffer blocks ion channel opening, exocytosis, and formation of pronuclei. *Dev. Biol.* 126: 346–361.

Kopf, G. S. 1998. Acrosome reaction. *In* E. Knobil and J. D. Neill (eds.), *Encyclopedia of Reproduction,* Vol. 1. Academic Press, San Diego, pp. 17–27.

Kvist, U., B. A. Afzelius and L. Nilsson. 1980. The intrinsic mechanism of chromatin decondensation and its activation in human spermatozoa. *Dev. Growth Diff.* 22: 543–554.

Lechleiter, J. D. and D. E. Clapham. 1992. Molecular mechanisms of intracellular calcium excitability in *X. laevis* oocytes. *Cell* 69: 283–294.

Lee, S.-J. and S. S. Shen. 1998. The calcium transient in sea urchin eggs during fertilization requires the production of inositol 1,4,5-trisphosphate. *Dev. Biol.* 193: 195–208.

Leeuwenhoek, A. van. 1685. Letter to the Royal Society of London. Quoted in E. G. Ruestow 1983, Images and ideas: Leeuwenhoek's perception of the spermatozoa. *J. Hist. Biol.* 16: 185–224.

Lefebvre, R., P. J. Chenoweth, M. Drost, C. T. LeClear, M. MacCubbin, J. T. Dutton and S. S. Suarez. 1995. Characterization of the oviductal sperm receptor in cattle. *Biol. Reprod.* 53: 1066–1074.

Levine, A. E., K. A. Walsh and E. J. B. Fodor. 1978. Evidence of an acrosin-like enzyme in sea urchin sperm. *Dev. Biol.* 63: 299–307.

Leyton, L. and P. Saling. 1989. Evidence that aggregation of mouse sperm receptors by ZP3 triggers the acrosome reaction. *J. Cell Biol.* 108: 2163–2168.

Leyton, L., P. Leguen, D. Bunch and P. M. Saling. 1992. Regulation of mouse gametic interaction by a sperm tyrosine kinase. *Proc. Natl. Acad. Sci. USA* 93: 1164–1169.

Longo, F. J. and M. Kunkle. 1978. Transformation of sperm nuclei upon insemination. *In* A. A. Moscona and A. Monroy (eds.), *Current Topics in Developmental Biology,* Vol. 12. Academic Press, New York, pp. 149–184.

Longo, F. J., J. W. Lynn, D. H. McCulloh and E. L.Chambers. 1986. Correlative ultrastructural and electrophysiological studies of sperm-egg interactions of the sea urchin *Lytechinus variegatus. Dev. Biol.* 118: 155–167.

Lopez, L. C., E. M. Bayna, D. Litoff, N. L. Shaper, J. H. Shaper and B. D. Shur. 1985. Receptor function of mouse sperm surface galactosyltransferase during fertilization. *J. Cell Biol.* 101: 1501–1510.

Luttmer, S. and F. J. Longo. 1985. Ultrastructural and morphometric observations of cortical endoplasmic reticulum in *Arbacia, Spisula,* and mouse eggs. *Dev. Growth Diff.* 27: 349–359.

Manes, M. E. and R. P. Elinson. 1980. Ultraviolet light inhibits gray crescent formation in the frog egg. *Wilhelm Roux Arch. Dev. Biol.* 189: 73–77.

Manes, M. E., R. P. Elinson and F. D. Barbieri. 1978. Formation of the amphibian gray crescent: Effects of colchicine and cytochalasin-B. *Wilhelm Roux Arch. Dev. Biol.* 185: 99–104.

McCulloh, D. H. and E. L. Chambers. 1992. Fusion of membranes during fertilization. *J. Gen. Physiol.* 99: 137–175.

McGrath, J. and D. Solter. 1984. Completion of mouse embryogenesis requires both the maternal and paternal genome. *Cell* 37: 179–183.

McPherson, S. M., P. S. McPherson, L. Mathews, K. P. Campbell and F. J. Longo. 1992. Cortical localization of a calcium release channel in sea urchin eggs. *J. Cell Biol.* 116: 1111–1121.

Mead, K. S. and D. Epel. 1995. Beakers and breakers: How fertisation in the laboratory differs from fertisation in nature. *Zygote* 3: 95–99.

Meizel, S. 1984. The importance of hydrolytic enzymes to an exocytotic event, the mammalian sperm acrosome reaction. *Biol. Rev.* 59: 125–157.

Metz, C. B. 1978. Sperm and egg receptors involved in fertilization. *Curr. Top. Dev. Biol.* 12: 107–148.

Miller, B. S. and D. Epel. 1999. The roles of changes in NADPH and pH during fertilization and artificial activation of the sea urchin egg. *Dev. Biol.* 216: 394–405.

Miller, D. J., M. B. Macek and B. D. Shur. 1992. Complementarity between sperm surface β-1,4-galactosyltransferase and egg-coat ZPE mediates sperm-egg binding. *Nature* 357: 589–593.

Miller, D. J., X. Gong, G. Decker and B. D. Shur. 1993. Egg cortical granule N-acetylglucosaminidase is required for the mouse zona block to polyspermy. *J. Cell Biol.* 123: 1431–1440.

Miller, R. L. 1978. Site-specific agglutination and the timed release of a sperm chemoattractant by the egg of the leptomedusan, *Orthopyxis caliculata. J. Exp. Zool.* 205: 385–392.

Miller, R. L. 1985. Sperm chemo-orientation in the metazoa. *In* C. B. Metz, Jr. and A. Monroy (eds.), *Biology of Fertilization*, Vol. 2. Academic Press, New York, pp. 275–337.

Miyazaki, S.-I., M. Yuzaki, K. Nakada, H. Shirakawa, S. Nakanishi, S. Nakade and K. Mikoshiba. 1992. Block of Ca^{2+} wave and Ca^{2+} oscillation by antibody to the inositol 1,4,5-trisphosphate receptor in fertilized hamster eggs. *Science* 257: 251–255.

Mohri, T., P. I. Ivonnet and E. L. Chambers. 1995. Effect of sperm-induced activation current and increase of cytosolic Ca^{2+} by agents that modify the mobilization of $[Ca^{2+}]I$. Heparin and pentosan polysulfate. *Dev. Biol.* 172: 139–157.

Moller, C. C. and P. M. Wassarman. 1989. Characterization of a proteinase that cleaves zona pellucida glycoprotein ZP2 following activation of mouse eggs. *Dev. Biol.* 132: 103–112.

Moy, G. W. and V. D. Vacquier. 1979. Immunoperoxidase localization of bindin during the adhesion of sperm to sea urchin eggs. *Curr. Top. Dev. Biol.* 13: 31–44.

Mozingo, N. M. and D. E. Chandler. 1991. Evidence for the existence of two assembly domains within the sea urchin fertilization envelope. *Dev. Biol.* 146: 148–157.

Multigner, L., J. Gagnon, A. van Dorsselaer and D. Job. 1992. Stabilization of sea urchin flagellar microtubules by histone H1. *Nature* 360: 33–39.

Myles, D. G., L. H. Kimmel, C. P. Blobel, J. M. White and P. Primakoff. 1994. Identification of a binding site in the disintegrin domain of fertilin required for sperm-egg fusion. *Proc. Natl. Acad. Sci. USA* 91: 4195–4198.

Nilsson, L. 1990. *Syntyyuusi Ihminen.* Otava, Helsinki.

Nishizuka, Y. 1986. Studies and perspectives of protein kinase C. *Science* 233: 305–312.

Ogawa, K., T. Mohri and H. Mohri. 1977. Identification of dynein as the outer arms of sea urchin sperm axonemes. *Proc. Natl. Acad. Sci. USA* 74: 5006–5010.

Ohama, K. and 7 others. 1981. Dispermic origin of XY hydatidiform moles. *Nature* 292: 551–552.

Perreault, S. D., R. R. Barbee and V. L. Slott. 1988. Importance of glutathione in the acquisition and maintenance of sperm nuclear decondensing activity in maturing hamster oocytes. *Dev. Biol.* 125: 181–187.

Pinto-Correia, C. 1997. *The Ovary of Eve: Eggs and Sperm and Preformation.* University of Chicago Press, Chicago.

Poccia, D. and P. Collas. 1997. Nuclear envelope dynamics during male pronuclear development. *Dev. Growth Diff.* 39: 541–550.

Poccia, D., J. Salik and G. Krystal. 1981. Transitions in histone variants of the male pronucleus following fertilization and evidence for a maternal store of cleavage-stage histones in the sea urchin egg. *Dev. Biol.* 82: 287–297.

Porter, D. C. and V. D. Vacquier. 1986. Phosphorylation of sperm histone H1 is induced by the egg jelly layer in the sea urchin *Strongylocentrotus purpuratus. Dev. Biol.* 116: 203–212.

Prevost, J. L. and J. B. Dumas. 1824. Deuxieme mémoire sur la génération. *Ann. Sci. Nat.* 2: 129–149.

Primakoff, P., H. Hyatt and J. Tredick-Kline. 1987. Identification and purification of a sperm cell surface protein with a potential role in sperm-egg membrane fusion. *J. Cell Biol.* 104: 141–149.

Primakoff, P., W. Lathrop, L. Woolman, A. Cowan and D. Myles. 1988. Fully effective contraception in male and female guinea pigs immunized with the sperm protein PH-20. *Nature* 335: 543–547.

Ralt, D. and 8 others. 1991. Sperm attraction to a follicular factor(s) correlates with human egg fertilizability. *Proc. Natl. Acad. Sci. USA* 88: 2840–2844.

Ramarao, C. S. and D. L. Garbers. 1985. Receptor-mediated regulation of guanylate cyclase activity in spermatozoa. *J. Biol. Chem.* 260: 8390–8397.

Rees, B. B., C. Patton, J. L Grainger and D. Epel. 1995. Protein synthesis increases after fertilization of sea urchin eggs in the absence of an increase in intracellular pH. *Dev. Biol.* 169: 683–698.

Roux, W. 1887. Beiträge zur Entwicklungsmechanik des Embryo. *Arch. Mikrosk. Anat.* 29: 157–212.

Saling, P. M. 1989. Mammalian sperm interaction with extracellular matrices of the egg. *Oxford Rev. Reprod. Biol.* 11: 339–388.

Saling, P. M., J. Sowinski and B. T. Storey. 1979. An ultrastructural study of epididymal mouse spermatozoa binding to zonae pellucida in vitro: Sequential relationship to acrosome reaction. *J. Exp. Zool.* 209: 229–238.

Schackmann, R. W. and B. M. Shapiro. 1981. A partial sequence of ionic changes associated with the acrosome reaction of *Strongylocentrotus purpuratus. Dev. Biol.* 81: 145–154.

Schatten, G. and D. Mazia. 1976. The penetration of the spermatozoan through the sea urchin egg surface at fertilization: Observations from the outside on whole eggs and from the inside on isolated surfaces. *Exp. Cell Res.* 98: 325–337.

Schatten, G. and H. Schatten. 1983. The energetic egg. *The Sciences* 23(5): 28–35.

Schatten, H. and G. Schatten. 1980. Surface activity at the plasma membrane during sperm incorporation and its cytochalasin-B sensitivity: Scanning electron micrography and time-lapse video microscopy during fertilization of the sea urchin *Lytechinus variegatus. Dev. Biol.* 78: 435–449.

Schroeder, T. E. 1979. Surface area change at fertilization: Resorption of the mosaic membrane. *Dev. Biol.* 70: 306–327.

Shen, S. S. and R. A. Steinhardt. 1978. Direct measurement of intracellular pH during metabolic depression of the sea urchin egg. *Nature* 272: 253–254.

Shimomura, H., L. J. Dangott and D. L. Garbers. 1986. Covalent coupling of a resact analogue to guanylate cyclase. *J. Biol. Chem.* 261: 15778–15782.

Shinyoji, C., H. Higuchi, M. Yoshimura, E. Katayama and T. Yanagida. 1998. Dynein arms are oscillating force generators. *Nature* 393: 711–714.

Simerly, C. and 7 others. 1995. The paternal inheritance of the centrosome, the cell's microtubule-organizing center, in humans, and the implications for infertility. *Nature Med.* 1: 47–52.

Simerly, C. and 7 others. 1999. Biparental inheritance of β-tubulin during human fertilization: Molecular reconstitution of functional zygote centrosomes in inseminated human oocytes and in cell-free extracts nucleated by human sperm. *Mol. Cell Biol.* 10: 2955–2969.

Sluder, G., F. J. Miller, K. K. Lewis, E. D. Davison and C. L. Reider. 1989. Centrosome inheritance in starfish zygotes: Selective loss of the maternal centrosome after fertilization. *Dev. Biol.* 131: 567–579.

Sluder, G., F. J. Miller and K. Lewis. 1993. Centrosome inheritance in starfish zygotes II. Selective suppression of the maternal centrosome during meiosis. *Dev. Biol.* 155: 58–67.

Smith, T. T. 1998. The modulation of sperm function by the oviductal epithelium. *Biol. Reprod.* 58: 1102–1104.

Stears, R. L. and W. J. Lennarz. 1997. Mapping sperm binding domains on the sea urchin egg receptor for sperm. *Dev. Biol.* 187: 200–208.

Steinhardt, R. A. and D. Epel. 1974. Activation of sea urchin eggs by a calcium ionophore. *Proc. Natl. Acad. Sci. USA* 71: 1915–1919.

Steinhardt, R., R. Zucker and G. Schatten. 1977. Intracellular calcium release at fertilization in the sea urchin egg. *Dev. Biol.* 58: 185–197.

Storey, B. T. 1995. Interactions between gametes leading to fertilization: The sperm's eye view. *Reprod. Fertil. Dev.* 7: 927–942.

Suarez, S. S. 1998. The oviductal sperm reservoir in mammals: Mechanisms of formation. *Biol. Reprod.* 58: 1105–1107.

Suarez, S. S., D. F. Katz, D. H. Owen, J. B. Andrew and R. L. Powell. 1991. Evidence for the function of hyperactivated motility in sperm. *Biol. Reprod.* 44: 375–381.

Summers, R. G. and B. L. Hylander. 1974. An ultrastructural analysis of early fertilization in the sand dollar, *Echinarachnius parma*. *Cell Tissue Res.* 150: 343–368.

Summers, R. G. and B. L. Hylander. 1975. Species-specificity of acrosome reaction and primary gamete binding in echinoids. *Exp. Cell Res.* 96: 63–68.

Summers, R. G., B. L. Hylander, L. H. Colwin and A. L. Colwin. 1975. The functional anatomy of the echinoderm spermatozoon and its interaction with the egg at fertilization. *Am. Zool.* 15: 523–551.

Surani, M. A. H. and S. C. Barton. 1983. Development of gynogenetic eggs in the mouse: Implications for parthenogenetic embryos. *Science* 222: 1034–1037.

Surani, M. A. H., S. C. Barton and M. L.Norris. 1986. Nuclear transplantation in the mouse: Heritable differences between parental genomes after activation of the embryonic genome. *Cell* 45: 127–137.

Sutovsky, P., C. S. Navara and G. Schatten. 1996. Fate of the sperm mitochondria and the incorporation, conversion, and disassembly of the sperm tail structures during bovine fertilization. *Biol. Reprod.* 55: 1195–1205.

Sutovsky, P., R. D. Moreno, J. Ramalho-Santos, T. Dominko, C. Simerly and G. Schatten. 1999. Ubiquitin tag for sperm mitochondria. *Nature* 402: 371–372.

Swann, K. and M.Whitaker. 1986. The part played by inositol trisphosphate and calcium in the propagation of the fertilization wave in sea urchin eggs. *J. Cell Biol.* 103: 2333–2342.

Terasaki, M. and C. Sardet. 1991. Demonstration of calcium uptake and release by sea urchin egg cortical endoplasmic reticulum. *J. Cell Biol.* 115: 1031–1037.

Tilney, L. G., J. Bryan, D. J. Bush, K. Fujiwara, M. S. Mooseker, D. B. Murphy and D. H. Snyder. 1973. Microtubules: Evidence for 13 protofilaments. *J. Cell Biol.* 59: 267–275.

Tilney, L. G., D. P. Kiehart, C. Sardet and M. Tilney. 1978. Polymerization of actin. IV. Role of Ca^{2+} and H^+ in the assembly of actin and in membrane fusion in the acrosomal reaction of echinoderm sperm. *J. Cell Biol.* 77: 536–560.

Tokumoto, T., M. Yamashita, M. Tokumoto, Y. Katsu, R. Horiguchi, H. Kajiura and Y. Nagahama. 1997. Initiation of cyclin B degradation by the 26S proteasome upon egg activation. *J. Cell Biol.* 138: 1313–1322.

Tombes, R. M. and B. M. Shapiro. 1985. Metabolite channeling: A phosphocreatine shuttle to mediate high-energy phosphate transport between sperm mitochondrion and tail. *Cell* 41: 325–334.

Turner, P. R., L. A. Jaffe and A. Fein. 1986. Regulation of cortical vesicle exocytosis in sea urchin eggs by inositol 1,4,5-trisphosphate and GTP-binding protein. *J. Cell Biol.* 102: 70–77.

Ulrich, A. S., M. Otter, C. G. Glabe and D. Hoekstra. 1998. Membrane fusion is induced by a distinct peptide sequence of the sea urchin fertilization protein bindin. *J. Biol. Chem.* 273: 16748–16755.

Uzzell, T. M. 1964. Relations of the diploid and triploid species of the *Ambystoma jeffersonianum* complex. *Copeia* 1964: 257–300.

Vacquier, V. D. 1979. The interaction of sea urchin gametes during fertilization. *Am. Zool.* 19: 839–849.

Vacquier, V. D. 1998. Evolution of gamete recognition proteins. *Science* 281: 1995–1998.

Vacquier, V. D. and G. W. Moy. 1977. Isolation of bindin: The protein responsible for adhesion of sperm to sea urchin eggs. *Proc. Natl. Acad. Sci. USA* 74: 2456–2460.

Vacquier, V. D. and G. W. Moy. 1997. The fucose sulfate polymer of egg jelly binds to sperm REJ and is the inducer of the sea urchin sperm acrosome reaction. *Dev. Biol.* 192: 125–135.

Vacquier, V. D. and J. E. Payne. 1973. Methods for quantitating sea urchin sperm in egg binding. *Exp. Cell Res.* 82: 227–235.

Vacquier, V. D., M. J. Tegner and D. Epel. 1973. Protease release from sea urchin eggs at fertilization alters the vitelline layer and aids in preventing polyspermy. *Exp. Cell Res.* 80: 111–119.

Van Steirtinghem, A. 1994. IVF and micromanipulation techniques for male-factor fertility. *Curr. Opin. Obstet. Gynaecol.* 6: 173–177.

Vincent, J. P., G. F. Oster and J. C.Gerhart. 1986. Kinematics of gray crescent formation in *Xenopus* eggs: The displacement of subcortical cytoplasm relative to the egg surface. *Dev. Biol.* 113: 484–500.

Visconti, P. E. and 7 others. 1995. Capacitation of mouse spermatozoa. II. Protein tyrosine phosphorylation and capacitation are regulated by a cAMP-dependent pathway. *Development* 121: 1139–1150.

von Kolliker, A. 1841. Beiträge zur Kenntnis der Geschlectverhältnisse und der Samenflüssigkeit wirbelloser Thiere, nebst einem Versuch Über Wesen und die Bedeutung der sogenannten Samenthiere. Berlin.

Ward, C. R. and G. S. Kopf. 1993. Molecular events mediating sperm activation. *Dev. Biol.* 158: 9–34.

Ward, G. E., C. J. Brokaw, D. L. Garbers and V. D. Vacquier. 1985. Chemotaxis of *Arbacia punctulata* spermatozoa to resact, a peptide from the egg jelly layer. *J. Cell Biol.* 101: 2324–2329.

Wassarman, P. M. 1987. The biology and chemistry of fertilization. *Science* 235: 553–560.

Wassarman, P. M. 1989. Fertilization in mammals. *Sci. Am.* 256(6): 78–84.

Watanabe, N., T. Hunt, Y. Ikawa and N. Sagata. 1991. Independent inactivation of MPF and cytostatic factor (Mos) upon fertilization of *Xenopus* eggs. *Nature* 352: 247–249.

Whitaker, M. and R. F. Irvine. 1984. Inositol 1,4,5-trisphosphate microinjection activates sea urchin eggs. *Nature* 312: 636–639.

Whitaker, M. and R. Steinhardt. 1982. Ionic regulation of egg activation. *Q. Rev. Biophys.* 15: 593–667.

Wilcox, A. J., C. R. Weinberg and D. D. Baird. 1995. Timing of sexual intercourse in relation to ovulation: Effects on the probability of conception, survival of pregnancy, and the sex of the baby. *N. Engl. J. Med.* 333: 1517–121.

Williams, C. J., L. M. Mehlmann, L. A. Jaffee, G. S. Kopf and R. M. Schultz. 1998. Evidence that G_γ family G proteins do not function in mouse egg activation at fertilization. *Dev. Biol.* 198: 116–127.

Wilson, W. L. and G. Oliphant. 1987. Isolation and biochemical characterization of the subunits of the rabbit sperm acrosome stabilizing factor. *Biol. Reprod.* 37: 159–169.

Winkler, M. M., R. A. Steinhardt, J. L. Grainger and L. Minning. 1980. Dual ionic controls for the activation of protein synthesis at fertilization. *Nature* 287: 558–560.

Xu, Z., G. S. Kopf and R. M. Schultz. 1994. Involvement of inositol-1,4,5-trisphosphate-mediated Ca^{2+} release in early and late events of mouse egg activation. *Development* 120: 1851–1859.

Yanagimachi, R. 1994. Mammalian fertilization. *In* E. Knobil and J. D. Neill (eds.), *The Physiology of Reproduction*, 2nd Ed. Raven Press, New York.

Yanagimachi, R. and Y. D. Noda. 1970. Electron microscope studies of sperm incorporation into the golden hamster egg. *Am. J. Anat.* 128: 429–462.

Yoshida, M., K. Inabar and M. Morisawa. 1993. Sperm chemotaxis during the process of fertilization in the ascidians *Ciona savignyi* and *Ciona intestinalis*. *Dev. Biol.* 157: 497–507.

8 Early development in selected invertebrates

REMARKABLE AS IT IS, fertilization is but the initiating step in development. The zygote, with its new genetic potential and its new arrangement of cytoplasm, now begins the production of a multicellular organism. Between these events of fertilization and the events of organ formation are two critical stages: cleavage and gastrulation. During cleavage, rapid cell divisions divide the cytoplasm of the fertilized egg into numerous cells. These cells then undergo dramatic displacements during gastrulation, a process whereby they move to different parts of the embryo and acquire new neighbors (see Chapter 2). During cleavage and gastrulation, the major axes of the embryo are determined, and the cells begin to acquire their respective fates.

While cleavage always precedes gastrulation, axis formation can begin as early as oocyte formation. It can be completed during cleavage (as is the case with *Drosophila*) or extend all the way through gastrulation (as it does in *Xenopus*). There are three axes that need to be specified: the anterior-posterior (head-anus) axis, the dorsal-ventral (back-belly) axis, and the left-right axis. Different species specify these axes at different times, using different mechanisms.

An Introduction to Early Developmental Processes

Cleavage

After fertilization, the development of a multicellular organism proceeds by a process called **cleavage**, a series of mitotic divisions whereby the enormous volume of egg cytoplasm is divided into numerous smaller, nucleated cells. These cleavage-stage cells are called **blastomeres**. In most species (mammals being the chief exception), the rate of cell division and the placement of the blastomeres with respect to one another is completely under the control of the proteins and mRNAs stored in the oocyte by the mother. The zygotic genome, transmitted by mitosis to all the new cells, does not function in early-cleavage embryos. Few, if any, mRNAs are made until relatively late in cleavage, and the embryo can divide properly even when chemicals are used experimentally to inhibit transcription. During cleavage, however, cytoplasmic volume does not increase. Rather, the enormous volume of zygote cytoplasm is divided into increasingly smaller cells. First the egg is divided in half, then quarters, then eighths, and so forth. This division of egg cytoplasm without increasing its volume is accomplished by abolishing the growth period between cell divisions (that is, the G_1 and G_2 phases of the cell cycle). Meanwhile, the cleavage of nuclei occurs at a rapid

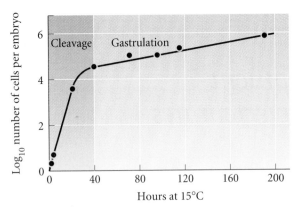

Figure 8.1
Rate of formation of new cells during the early development of the frog *Rana pipiens*. (After Sze 1953.)

rate never seen again (not even in tumor cells). A frog egg, for example, can divide into 37,000 cells in just 43 hours. Mitosis in cleavage-stage *Drosophila* embryos occurs every 10 minutes for over 2 hours and in just 12 hours forms some 50,000 cells. This dramatic increase in cell number can be appreciated by comparing cleavage with other stages of development. Figure 8.1 shows the logarithm of cell number in a frog embryo plotted against the time of development (Sze 1953). It illustrates a sharp discontinuity between cleavage and gastrulation.

One consequence of this rapid cell division is that the ratio of cytoplasmic to nuclear volume gets increasingly smaller as cleavage progresses. In many types of embryos (such as those of *Xenopus* and *Drosophila*, but not those of *C. elegans* or mammals), this decrease in the cytoplasmic to nuclear volume ratio is crucial in timing the activation of certain genes. For example, in the frog *Xenopus laevis*, transcription of new messages is not activated until after 12 divisions. At that

time, the rate of cleavage decreases, the blastomeres become motile, and nuclear genes begin to be transcribed. This stage is called the **mid-blastula transition**. It is thought that some factor in the egg is being titrated by the newly made chromatin, because the time of this transition can be changed by experimentally altering the ratio of chromatin to cytoplasm in the cell (Newport and Kirschner 1982a,b; Edgar et al. 1986). Thus, cleavage begins soon after fertilization and ends shortly after the stage when the embryo achieves a new balance between nucleus and cytoplasm.

From fertilization to cleavage

The transition from fertilization to cleavage is caused by the activation of **mitosis promoting factor** (**MPF**). MPF was first discovered as the major factor responsible for the resumption of meiotic cell divisions in the ovulated frog egg. It continues to play a role after fertilization, regulating the biphasic cell cycle of early blastomeres. Blastomeres generally progress through a cell cycle consisting of just two steps: M (mitosis) and S (DNA synthesis) (Figure 8.2). Gerhart and co-workers (1984) showed that MPF undergoes cyclical changes in its level of activity in mitotic cells. The MPF activity of early blastomeres is highest during M and undetectable during S. Newport and Kirschner (1984) demonstrated that DNA replication (S) and mitosis (M) are driven solely by the gain and loss of MPF activity. Cleaving cells can be experimentally

Figure 8.2
Cell cycles of somatic cells and early blastomeres. (A) The simple biphasic cell cycle of the early amphibian blastomeres has only two states, S and M. Cyclin synthesis allows progression to M (mitosis), while cyclin degradation allows cells to pass into S (synthesis) phase. (B) Cell cycle of a typical somatic cell. Mitosis (M) is followed by an "interphase" stage. This latter period is subdivided into G_1, S (synthesis), and G_2 phases. Cells that are differentiating are usually taken "out" of the cell cycle and are in an extended G_1 phase called G_0. The cyclins and their respective kinases responsible for the progression through the cell cycle are shown at their point of cell cycle regulation. (B after Nigg 1995.)

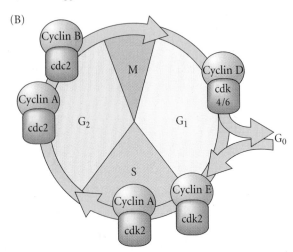

Table 8.1 Karyokinesis and cytokinesis

Process	Mechanical agent	Major protein composition	Location	Major disruptive drug
Karyokinesis	Mitotic spindle	Tubulin microtubules	Central cytoplasm	Colchicine, nocodazole[a]
Cytokinesis	Contractile ring	Actin microfilaments	Cortical cytoplasm	Cytochalasin B

[a]Because colchicine has been found to independently inhibit several membrane functions, including osmoregulation and the transport of ions and nucleosides, nocodazole has become the major drug used to inhibit microtubule-mediated processes (see Hardin 1987).

trapped in S phase by incubating them in an inhibitor of protein synthesis. When MPF is microinjected into these cells, they enter M. Their nuclear envelope breaks down and their chromatin condenses into chromosomes. After an hour, MPF is degraded and the chromosomes return to S phase.

What causes this cyclic activity of MPF? Mitosis-promoting factor contains two subunits. The large subunit is called **cyclin B**. It is this component that shows a periodic behavior, accumulating during S and then being degraded after the cells have reached M (Evans et al. 1983; Swenson et al. 1986). Cyclin B is often encoded by mRNAs stored in the oocyte cytoplasm, and if the translation of this message is specifically inhibited, the cell will not enter mitosis (Minshull et al. 1989). The presence of cyclin B depends upon its synthesis and its degradation. Cyclin B regulates the small subunit of MPF, the **cyclin-dependent kinase**. This kinase activates mitosis by phosphorylating several target proteins, including histones, the nuclear envelope lamin proteins, and the regulatory subunit of cytoplasmic myosin. This brings about chromatin condensation, nuclear envelope depolymerization, and the organization of the mitotic spindle.

Without cyclin, the cyclin-dependent kinase will not function. The presence of cyclin is controlled by several proteins that ensure its periodic synthesis and degradation. In most species studied, the regulators of cyclin (and thus, of MPF) are stored in the egg cytoplasm. Therefore, the cell cycle is independent of the nuclear genome for numerous cell divisions. These early divisions tend to be rapid and synchronous. However, as the cytoplasmic components are used up, the nucleus begins to synthesize them. The embryo now enters the mid-blastula transition, in which several new phenomena are added to the biphasic cell divisions of the embryo. First, the growth stages (G_1 and G_2) are added to the cell cycle, permitting the cells to grow. Before this time, the egg cytoplasm was being divided into smaller and smaller cells, but the total volume of the organism remained unchanged. *Xenopus* embryos add those phases to the cell cycle shortly after the twelfth cleavage. *Drosophila* adds G_2 during cycle 14 and G_1 during cycle 17 (Newport and Kirschner 1982a; Edgar et al. 1986). Second, the synchronicity of cell division is lost, as different cells synthesize different regulators of MPF. Third, new mRNAs are transcribed. Many of these messages encode proteins that will become necessary for gastrulation. If transcription is blocked, cell division will occur at normal rates and at normal times in many species, but the embryo will not be able to initiate gastrulation.

WEBSITE **8.1 Regulating the cell cycle.** Cyclins and the cyclin-dependent kinases are critical in regulating cell division and integrating it with the activities of the nuclear envelope and DNA synthesis.

The cytoskeletal mechanisms of mitosis

Cleavage is actually the result of two coordinated processes. The first of these cyclic processes is **karyokinesis**—the mitotic division of the nucleus. The mechanical agent of this division is the mitotic spindle, with its **microtubules** composed of **tubulin** (the same type of protein that makes up the sperm flagellum). The second process is **cytokinesis**—the division of the cell. The mechanical agent of cytokinesis is a **contractile ring** of **microfilaments** made of **actin** (the same type of protein that extends the egg microvilli and the sperm acrosomal process). Table 8.1 presents a comparison of these agents of cell division. The relationship and coordination between these two systems during cleavage is depicted in Figure 8.3A, in which a sea urchin egg is shown undergoing first cleavage. The mitotic spindle and contractile ring are perpendicular to each other, and the spindle is internal to the contractile ring. The contractile ring creates a **cleavage furrow**, which eventually bisects the plane of mitosis, thereby creating two genetically equivalent blastomeres.

The actin microfilaments are found in the cortex of the egg rather than in the central cytoplasm. Under the electron microscope, the ring of microfilaments can be seen forming a distinct cortical band 0.1 μm wide (Figure 8.3B). This contractile ring exists only during cleavage and extends 8–10 μm into the center of the egg. It is responsible for exerting the force that splits the zygote into blastomeres; for if it is disrupted, cytokinesis stops. Schroeder (1973) has proposed a model of cleavage wherein the contractile ring splits the egg like an "intercellular purse-string," tightening about the egg as cleavage continues. This tightening of the microfilamentous ring creates the cleavage furrow. Microtubules are also seen near the cleavage furrow (in addition to their role in creating the mitotic spindles), since they are needed to bring membrane material to the site of membrane addition (Figure 8.4; Danilchik et al. 1998).

Although karyokinesis and cytokinesis are usually coordinated, they are sometimes separated by natural or experi-

(A)

Microfilaments
(contractile ring)

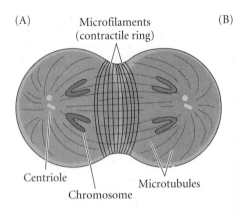

Centriole

Chromosome

Microtubules

(B)

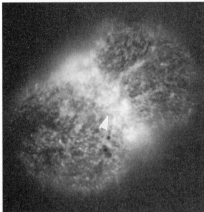

Figure 8.3
Role of microtubules and microfilaments in cell division. (A) Diagram of first-cleavage telophase. The chromosomes are being drawn to the centrioles by microtubules while the cytoplasm is pinched in by the contraction of microfilaments. (B) Localization of actin microfilaments in the cleavage furrow. Fluorescent labeling of the actin microfilaments shows the contractile ring in the first-cleavage furrow (arrowhead) of a telophase sea urchin egg. (C) Fluorescent labeling of tubulin shows the microtubular asters of a sea urchin egg during telophase of first cleavage. (B from Bonder et al. 1988; C from White et al. 1988; photographs courtesy of the authors.)

(C)

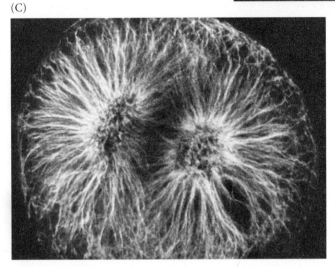

mental conditions. In insect eggs, karyokinesis occurs several times before cytokinesis takes place. Another way to produce this state is to treat embryos with the drug cytochalasin B. This drug inhibits the formation and organization of microfilaments in the contractile ring, thereby stopping cleavage without stopping karyokinesis (Schroeder 1972).

Patterns of embryonic cleavage

In 1923, embryologist E. B. Wilson reflected on how little we knew about cleavage: "To our limited intelligence, it would seem a simple task to divide a nucleus into equal parts. The cell, manifestly, entertains a very different opinion." Indeed, different organisms undergo cleavage in distinctly different ways. The pattern of embryonic cleavage particular to a species is determined by two major parameters: the amount and distribution of yolk protein within the cytoplasm, and factors in the egg cytoplasm that influence the angle of the mitotic spindle and the timing of its formation.

The amount and distribution of yolk determines where cleavage can occur and the relative size of the blastomeres. When one pole of the egg is relatively yolk-free, the cellular divisions occur there at a faster rate than at the opposite pole.

The yolk-rich pole is referred to as the **vegetal pole**; the yolk concentration in the **animal pole** is relatively low. The zygote nucleus is frequently displaced toward the animal pole. In general, yolk inhibits cleavage. Figure 8.5 provides a classification of cleavage types and shows the influence of yolk on cleavage symmetry and pattern.

At one extreme are the eggs of sea urchins, mammals, and snails. These eggs have sparse, equally spaced yolk and are thus **isolecithal** (Greek, "equal yolk"). In these species, cleavage is **holoblastic** (Greek *holos*, "complete"). meaning that the cleavage furrow extends through the entire egg. These embryos must have some other way of obtaining food. Most will generate a voracious larval form, while mammals get their nutrition from the placenta.

At the other extreme are the eggs of insects, fishes, reptiles, and birds. Most of their cell volumes are made up of yolk. The yolk must be sufficient to nourish these animals. Zygotes containing large accumulations of yolk undergo **mer-**

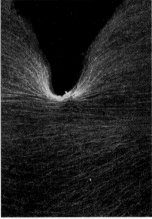

Figure 8.4
Microtubular array just ahead of the first-cleavage furrow of a dividing *Xenopus* egg. The microtubules are stained with fluorescent antibodies to tubulin. (From Danilchik et al. 1998; photograph courtesy of M. V. Danilchik.)

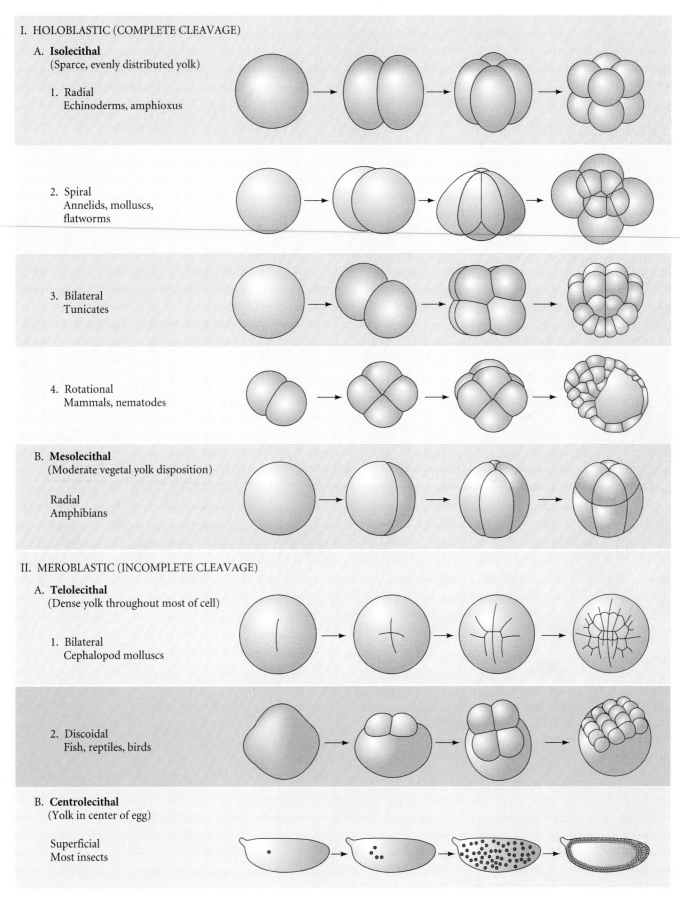

I. HOLOBLASTIC (COMPLETE CLEAVAGE)

A. **Isolecithal**
(Sparce, evenly distributed yolk)

1. Radial
Echinoderms, amphioxus

2. Spiral
Annelids, molluscs,
flatworms

3. Bilateral
Tunicates

4. Rotational
Mammals, nematodes

B. **Mesolecithal**
(Moderate vegetal yolk disposition)

Radial
Amphibians

II. MEROBLASTIC (INCOMPLETE CLEAVAGE)

A. **Telolecithal**
(Dense yolk throughout most of cell)

1. Bilateral
Cephalopod molluscs

2. Discoidal
Fish, reptiles, birds

B. **Centrolecithal**
(Yolk in center of egg)

Superficial
Most insects

Figure 8.5
Summary of the main patterns of cleavage.

oblastic cleavage, wherein only a portion of the cytoplasm is cleaved. The cleavage furrow does not penetrate into the yolky portion of the cytoplasm. The eggs of insects have their yolk in the center (i.e., they are **centrolecithal**), and the divisions of the cytoplasm occur only in the rim of cytoplasm around the periphery of the cell (i.e., **superficial** cleavage). The eggs of birds and fishes have only one small area of the egg that is free of yolk (**telolecithal** eggs), and therefore, the cell divisions occur only in this small disc of cytoplasm, giving rise to the **discoidal** pattern of cleavage. These are general rules, however, and closely related species can evolve different patterns of cleavage in a different environment.

However, the yolk is just one factor influencing a species' pattern of cleavage. There are also inherited patterns of cell division that are superimposed upon the constraints of the yolk. This can readily be seen in isolecithal eggs, in which very little yolk is present. In the absence of a large concentration of yolk, four major cleavage types can be observed: radial holoblastic, spiral holoblastic, bilateral holoblastic, and rotational holoblastic cleavage. We will see examples of these cleavage patterns below when we take a more detailed look at the early development of four different invertebrate groups.

> VADE MECUM **Cleavage patterns.** The quantities of yolk within an egg influence the pattern of cleavage. Eggs from various organisms commonly studied in developmental biology illustrate this, and are seen in labeled photographs and time-lapse movies.
> [Click on Sea Urchin; Fruit Fly; Chick-Early; and Amphibian]

Gastrulation

Gastrulation is the process of highly coordinated cell and tissue movements whereby the cells of the blastula are dramatically rearranged. The blastula consists of numerous cells, the positions of which were established during cleavage. During gastrulation, these cells are given new positions and new neighbors, and the multilayered body plan of the organism is established. The cells that will form the endodermal and mesodermal organs are brought inside the embryo, while the cells that will form the skin and nervous system are spread over its outside surface. Thus, the three germ layers—outer ectoderm, inner endoderm, and interstitial mesoderm—are first produced during gastrulation. In addition, the stage is set for the interactions of these newly positioned tissues.

The movements of gastrulation involve the entire embryo, and cell migrations in one part of the gastrulating embryo must be intimately coordinated with other movements occurring simultaneously. Although the patterns of gastrulation vary enormously throughout the animal kingdom, there are only a few basic types of cell movements. Gastrulation usually involves some combination of the following types of movements (Figure 8.6):

- **Invagination.** The infolding of a region of cells, much like the indenting of a soft rubber ball when it is poked.
- **Involution.** The inturning or inward movement of an expanding outer layer so that it spreads over the internal surface of the remaining external cells.
- **Ingression.** The migration of individual cells from the surface layer into the interior of the embryo.
- **Delamination.** The splitting of one cellular sheet into two more or less parallel sheets.
- **Epiboly.** The movement of epithelial sheets (usually of ectodermal cells) that spread as a unit, rather than individually, to enclose the deeper layers of the embryo.

As we look at gastrulation in different types of embryos, we should keep in mind the following questions (Trinkaus 1984):

- **What is the unit of migration?** Is migration dependent on the movements of individual cells, or are the cells part of a migrating sheet or region?
- **Is the spreading or folding of a cell sheet due to intrinsic factors within the sheet or to extrinsic forces stretching or distorting it?** It is essential to know the answer to this question if we are to understand how the various cell

Invagination:
Infolding of cell sheet into embryo

Example:
Sea urchin
endoderm

Involution:
Inturning of cell sheet over the basal surface of an outer layer

Example:
Amphibian
mesoderm

Ingression:
Migration of individual cells into the embryo

Example:
Sea urchin mesoderm, *Drosophila* neuroblasts

Delamination:
Splitting or migration of one sheet into two sheets

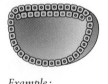

Example:
Mammalian and bird hypoblast formation

Epiboly:
The expansion of one cell sheet over other cells

Example:
Ectoderm formation in amphibians, sea urchins, and tunicates

Figure 8.6
Types of cell movements during gastrulation. The gastrulation of any particular organism is an ensemble of several of these movements.

movements of gastrulation are integrated. For instance, do involuting cells pull epibolizing cells down toward them, or are the two movements independent?

- **Is there active spreading of the whole tissue, or does the leading edge expand and drag the rest of a cell sheet passively along?**
- **Are changes in cell shape and motility during gastrulation the consequence of changes in cell surface properties,** such as adhesiveness to the substrate or to other cells?

Contrary to expectations, some regional migrational properties may be totally controlled by cytoplasmic factors that are independent of cellularization. F. R. Lillie (1902) was able to parthenogenetically activate eggs of the annelid *Chaetopterus* and suppress their cleavage. Many of the events of early development occurred even in the absence of cells. The cytoplasm of the zygote separated into defined regions, and cilia differentiated in the appropriate parts of the egg. Moreover, the outermost clear cytoplasm migrated down over the vegetal regions in a manner specifically reminiscent of the epiboly of animal hemisphere cells during normal development. This occurred at precisely the time that epiboly would have taken place during normal gastrulation. Thus, epiboly may be (at least in some respects) independent of the cells that form the migrating region.

VADE MECUM **Gastrulation movements.** The rearrangement of germ layers is best illustrated by viewing the living organism and 3-D models. These segments show time-lapse and real-time movies of gastrulation in several organisms, in which color coding of germ layers has been superimposed on the living embryo as well as on 3-D models. **[Click on Sea Urchin; Fruit Fly; Chick-Mid; and Amphibian]**

Axis Formation

Some of the most important phenomena in development concern the formation of embryonic axes (Figure 8.7). Embryos must develop three very important axes that are the foundations of the body: the anterior-posterior axis, the dorsal-ventral axis, and the right-left axis. The **anterior-posterior** (or **anteroposterior**) **axis** is the line extending from head to tail (or mouth to anus in those organisms that lack heads and tails). The **dorsal-ventral** (**dorsoventral**) **axis** is the line extending from back (dorsum) to belly (ventrum). For instance, in vertebrates, the neural tube is a dorsal structure. In insects, the neural cord is a ventral structure. The **right-left axis** is a line between the two lateral sides of the body. Although we may look symmetrical, recall that in most of us, the heart and liver are in the left half of the body only. Somehow, the embryo knows that some organs go on one side and other organs go on the other.

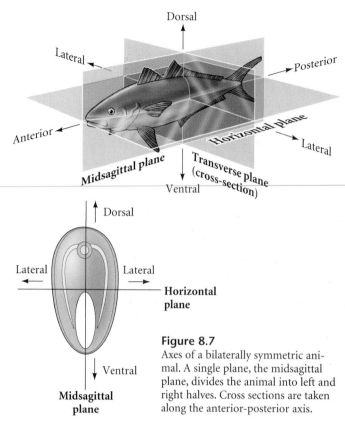

Figure 8.7
Axes of a bilaterally symmetric animal. A single plane, the midsagittal plane, divides the animal into left and right halves. Cross sections are taken along the anterior-posterior axis.

In this chapter, we will look at how four selected invertebrates—a sea urchin, a tunicate, a snail, and a nematode—undergo cleavage, gastrulation, axis specification, and cell fate determination. These four invertebrates were chosen because they have been important **model systems** for developmental biologists. In other words, they can be studied easily in the laboratory, and they have special properties that allow their mechanisms of development to be readily observed. They also represent a wide variety of cleavage types, patterns of gastrulation, and ways of specifying axes and cell fates.*

The Early Development of Sea Urchins

Cleavage in Sea Urchins

Sea urchins exhibit **radial holoblastic cleavage**. The first and second cleavages are both meridional and are perpendicular to each other. That is to say, the cleavage furrows pass through the animal and vegetal poles. The third cleavage is

---

*However, model systems—by their very ability to develop in the laboratory—preclude our asking certain questions concerning the relationship of development to its habitat. These questions will be addressed in Chapter 21.

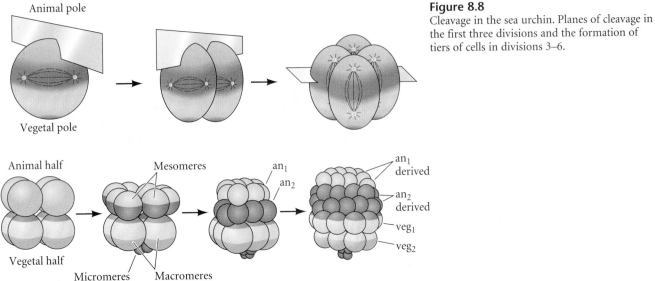

Figure 8.8
Cleavage in the sea urchin. Planes of cleavage in the first three divisions and the formation of tiers of cells in divisions 3–6.

equatorial, perpendicular to the first two cleavage planes, and separates the animal and vegetal hemispheres from one another (Figures 8.8 and 8.9). The fourth cleavage, however, is very different from the first three. The four cells of the animal tier divide meridionally into eight blastomeres, each with the same volume. These cells are called **mesomeres**. The vegetal tier, however, undergoes an *unequal* equatorial cleavage to produce four large cells, the **macromeres**, and four smaller **micromeres** at the vegetal pole (8.10; Summers et al. 1993). As the 16-cell embryo cleaves, the eight mesomeres divide to produce two "animal" tiers, an_1 and an_2, one staggered above the other. The macromeres divide meridionally, forming a tier of eight cells below an_2. The micromeres also divide, albeit somewhat later, producing a small cluster beneath the

larger tier. All the cleavage furrows of the sixth division are equatorial, and the seventh division is meridional, producing a 128-cell blastula.

Blastula formation

The **blastula** stage of sea urchin development begins at the 128-cell stage. Here the cells form a hollow sphere surrounding a central cavity, or **blastocoel** (Figure 8.11A). By this time, all the cells are the same size, the micromeres having slowed down their cell division. Every cell is in contact with the proteinaceous fluid of the blastocoel on the inside and with the hyaline layer on the outside. At this time, tight junctions unite the once loosely connected blastomeres into a seamless epithelial sheet that completely encircles the blastocoel (Figure

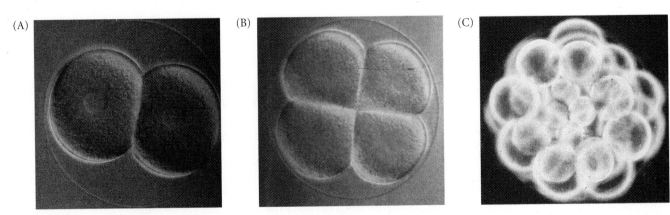

Figure 8.9
Cleavage in the sea urchin. (A–C) Photomicrographs of live embryos of the sea urchin *Lytechinus pictus*, looking down upon the animal pole. (A) The 2-cell stage. (B) The 4-cell stage. (C) The 32-cell stage, shown without the fertilization membrane to reveal the animal pole mesomeres, the central macromeres, and the vegetal micromeres, which angle into the center. (Photographs courtesy of G. Watchmaker.)

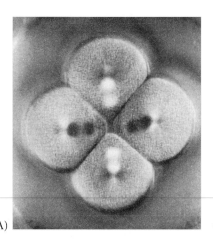

(A)

(B)

Figure 8.10
Formation of the micromeres during the fourth division of sea urchin embryos. The vegetal poles of the embryos are viewed from below. (A) The location and orientation of the mitotic spindle at the bottom portion of the vegetal cells is shown by viewing the living embryo with polarized light. (B) Cleavage through these asymmetrically placed spindles has produced micromeres and macromeres. (From Inoué 1982; photographs courtesy of S. Inoué.)

8.11B; Dan-Sohkawa and Fujisawa 1980). As the cells continue to divide, the blastula remains one cell layer thick, thinning out as it expands. This is accomplished by the adhesion of the blastomeres to the hyaline layer and by an influx of water that expands the blastocoel (Dan 1960; Wolpert and Gustafson 1961; Ettensohn and Ingersoll 1992).

These rapid and invariant cell cleavages last through the ninth or tenth cell division, depending upon the species. After that time, there is a mid-blastula transition, when the synchrony of cell division ends, new genes become expressed, and many of the nondividing cells develop cilia on their outer surfaces; Masuda and Sato 1984). The ciliated blastula begins to rotate within the fertilization envelope. Soon afterward, differences are seen in the cells. The cells at the vegetal pole of the blastula begin to thicken, forming a **vegetal plate**. The cells of the animal half synthesize and secrete a hatching enzyme that digests the fertilization envelope (Lepage et al. 1992; Figure 8.11C). The embryo is now a free-swimming hatched blastula.

Fate maps and the determination of sea urchin blastomeres

CELL FATE DETERMINATION. The fate map of the sea urchin embryo was originally created by observing each of the cell layers and what its descendants became. More recent investigations have refined these maps by following the fates of individual cells injected with fluorescent dyes such as diI (see Chapter 1). These studies have shown that by the 60-cell stage, most of the embryonic cell fates are specified, but that the cells are not irreversibly committed. In other words, particular blastomeres consistently produce the same cell types in each embryo, but these cells remain pluripotent and can give rise to other cell types if experimentally placed in a different part of the embryo.

A fate map of the 60-cell sea urchin embryo is shown in Figure 8.12 (Logan and McClay 1999; Wray 1999). The animal half of the embryo consistently gives rise to the ectoderm—the larval skin and its neurons. The veg_1 layer produces cells that can enter into either the ectodermal or endodermal organs. The veg_2 layer gives rise to cells that can populate three different structures—the endoderm, the coelom (body wall), and secondary mesenchyme (pigment cells, immunocytes, and muscle cells). The first tier of micromeres produces the primary mesenchyme cells that form the larval skeleton, while the second tier of micromeres contributes cells to the coelom (Logan and McClay 1997, 1999).

Although the early blastomeres have consistent fates in the larva, most of these fates are achieved by conditional specification. The only cells whose fates are determined au-

(A)

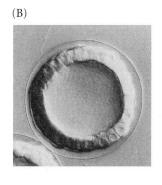

(B)

(C)

Figure 8.11
Sea urchin blastulae. (A) Formation of a blastocoel as cell division continues. (B) Soon after the rapid divisions of cleavage end, the previously rounded cells unite to form a true epithelium. The fertilization envelope can still be seen. As cilia develop, the blastula rotates within that envelope. (C) The vegetal plate thickens, while the animal hemisphere cells secrete hatching enzyme and allow the blastula to hatch from the fertilization envelope. (From Wray 1997; photographs courtesy of G. Wray.)

(A)

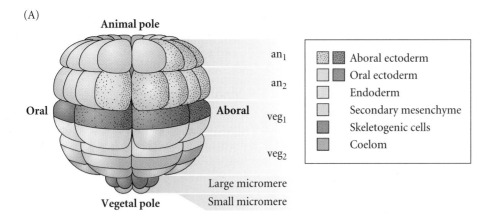

Animal pole

an₁

an₂

Oral · Aboral

veg₁

veg₂

Large micromere

Vegetal pole · Small micromere

		Aboral ectoderm
		Oral ectoderm
		Endoderm
		Secondary mesenchyme
		Skeletogenic cells
		Coelom

Figure 8.12
Fate map and cell lineage of the sea urchin *Strongylocentrotus purpuratus.* (A) The 60-cell embryo is shown, with the left side facing the viewer. Blastomere fates are segregated along the animal-vegetal axis of the egg. (B) Cell lineage map of the embryo. For simplicity, only one-quarter of the embryo is shown beyond second cleavage. The veg₁ tier gives rise to both ectodermal and endodermal lineages, and the coelom comes from two sources: the second tier of micromeres, and some veg₂ cells. (After Wray 1999.)

(B)

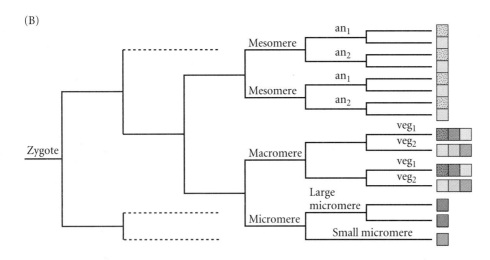

tonomously are the skeletogenic micromeres. If these micromeres are isolated from the embryo and placed in test tubes, they will still form skeletal spicules. Moreover, if these micromeres are transplanted into the animal region of the blastula, not only will their descendants form skeletal spicules, but the transplanted micromeres will alter the fates of nearby cells by inducing a secondary site for gastrulation. Cells that would normally have produced ectodermal skin cells will be respecified as endoderm and will produce a secondary gut (Figure 8.13; Hörstadius 1973; Ransick and Davidson 1993). The micromeres appear to produce a signal that tells the cells adjacent to them to become endoderm and induces them to invaginate into the embryo. Their ability to reorganize the embryonic cells is so pronounced that if the isolated micromeres are recombined with an isolated animal cap (the top two animal tiers), the animal cap cells will generate endoderm, and a more or less normal larva will develop (Figure 8.14; Hörstadius 1939).

In a normal embryo, the veg₂ cells become specified by the micromeres, and they, in turn, help specify the veg₁ layer. Without the veg₂ layer, the veg₁ cells are able to produce endoderm, but the endoderm is not specified as foregut, midgut,

or hindgut. Thus, there appears to be a cascade wherein the vegetal pole micromeres induce the cells above them to become the veg₂ cells, and the veg₂ cells induce the cells above them to assume the veg₁ fates. Thus, the micromeres undergo autonomous specification to become skeletogenic mesenchyme, and these micromeres produce the initial signals that specify the other tiers of cells.

The identities of the signaling molecules involved in this process are just now becoming known (Sherwood and McClay 1997). The molecule responsible for specifying the micromeres (and their ability to induce the neighboring cells) appears to be β-catenin. As we saw in Chapter 6, β-catenin is a transcription factor that is often activated by the Wnt pathway, and several pieces of evidence suggest it for this role. First, during normal sea urchin development, β-catenin accumulates in the nuclei of those cells fated to become endoderm and mesoderm (Figure 8.15A). This accumulation is autonomous and can occur even if the micromere precursors are separated from the rest of the embryo. Second, this accumulation appears to be responsible for specifying the vegetal half of the embryo. Treating sea urchin embryos with lithium chloride causes the accumulation of β-catenin in all their cells, and

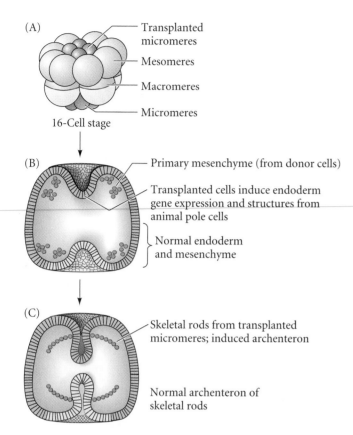

Figure 8.13
Ability of the micromeres to induce a secondary axis in sea urchin embryos. (A) Micromeres are transplanted from the vegetal pole of a 16-cell embryo into the animal pole of a host 16-cell embryo. (B) The transplanted micromeres invaginate into the blastocoel to create a new set of primary mesenchyme cells, and they induce the animal cells next to them to become vegetal plate endoderm cells. (C) The transplanted micromeres differentiate into skeletal cables while the induced animal cap cells form a secondary archenteron. Meanwhile, gastrulation proceeds normally from the original vegetal plate of the host. (After Ransick and Davidson 1993.)

this treatment transforms the presumptive ectoderm into endoderm. Conversely, experimental procedures that inhibit β-catenin accumulation in the vegetal cell nuclei prevent the formation of endoderm and mesoderm (Figure 8.15B,C; Logan et al. 1998; Wikramanayake et al. 1998). It is possible that the levels of nuclear β-catenin accumulation help to determine the mesodermal and endodermal fates of the vegetal cells. Third, β-catenin is essential for giving the micromeres their inductive ability.* Experiments described above demonstrated that the micromeres were able to induce a second embryonic axis when transplanted to the animal hemisphere. However,

---

*They probably activate the genes necessary for producing the inducing signal, and a Notch ligand (such as Delta) may be one of the signal's components.

---

micromeres from embryos in which β-catenin was prevented from entering the nucleus were unable to induce the animal hemisphere cells to form endoderm, and a second axis was not formed (Logan et al. 1998).

It appears that the specification of cell fates in sea urchins involves a cascade initiated by the micromeres. It has been hypothesized (Davidson 1989) that the signal from the micromeres induces a post-translational modification in some factor in the veg_2 layer. Similarly, the signal from the veg_2 layer probably modifies some protein (possibly a transcription factor) in the veg_1 layer. In this way, the different tiers are assigned different fates.

AXIS SPECIFICATION. In the sea urchin blastula, the cell fates line up along the animal-vegetal axis established in the egg cytoplasm prior to fertilization. The animal-vegetal axis also appears to structure the future anterior-posterior axis, with the vegetal region sequestering those maternal components necessary for posterior development (Boveri 1901; Maruyama et al. 1985).

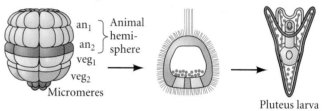

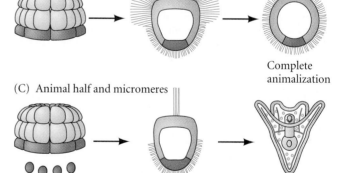

Figure 8.14
Ability of the micromeres to induce presumptive ectodermal cells to acquire other fates. (A) Normal development of the 64-cell sea urchin embryo, showing the fates of the different layers. (B) An isolated animal hemisphere becomes a ciliated ball of ectodermal cells. (C) When an isolated animal hemisphere is combined with isolated micromeres, a recognizable pluteus larva is formed, with all the endoderm derived from the animal hemisphere. (After Hörstadius 1939.)

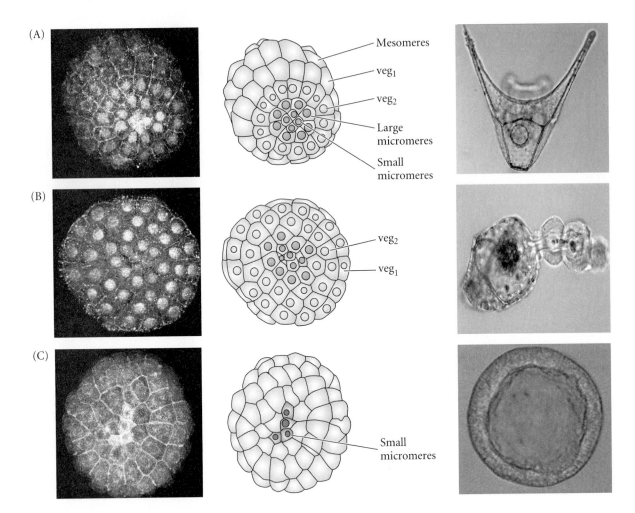

Figure 8.15
The role of β-catenin in specifying the vegetal cells of the sea urchin embryo. β-catenin is stained by a fluorescently labeled antibody. (A) During normal development, β-catenin accumulates predominantly in the micromeres and somewhat less in the veg_2 tier cells. (B) When lithium chloride treatment permits β-catenin to accumulate in the nuclei of all blastula cells (probably by blocking the GSK-3 enzyme of the Wnt pathway), the animal cells become specified as endoderm and mesoderm. (C) When β-catenin is prevented from entering into nuclei (and remains in the cytoplasm), the vegetal cell fates are not specified, and the entire embryo develops as a ciliated ectodermal ball. (After Logan et al. 1998; photographs courtesy of D. McClay.)

In most sea urchins, the dorsal-ventral and left-right axes are specified after fertilization, but the manner of their specification is not well understood. Since the first cleavage plane can be either parallel, perpendicular, or oblique with respect to the eventual dorsal-ventral axis, it is probable that the dorsal-ventral axis is not specified until the 8-cell stage, when there are cell boundaries that correspond to these positions (Kominami 1983; Henry et al. 1992). Interestingly, in those sea urchins that bypass the larval stage to develop directly into juveniles, the dorsal-ventral axis is specified maternally in the egg cytoplasm (Henry and Raff 1990).

WEBSITE **8.2 Sea urchin cell specification.** The specification of sea urchin cells was one of the first major research projects in experimental embryology and remains a fascinating area. This site looks at how new studies amplify some of the earliest investigations into cell differentiation.

Sea Urchin Gastrulation

The late sea urchin blastula consists of a single layer of about 1000 cells that form a hollow ball, somewhat flattened at the vegetal end. The blastomeres, derived from different regions

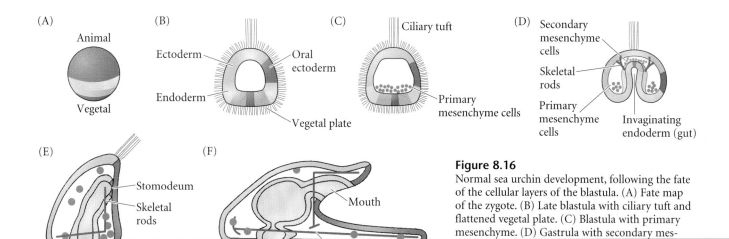

Figure 8.16
Normal sea urchin development, following the fate of the cellular layers of the blastula. (A) Fate map of the zygote. (B) Late blastula with ciliary tuft and flattened vegetal plate. (C) Blastula with primary mesenchyme. (D) Gastrula with secondary mesenchyme. (E) Prism-stage larva. (F) Pluteus larva. Fates of the zygote cytoplasm can be followed through the color pattern. (Courtesy of D. McClay.)

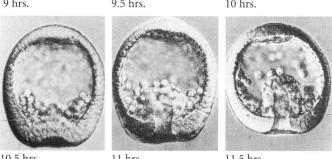

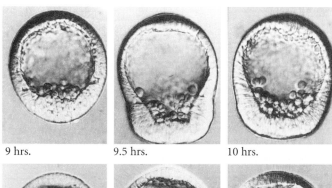

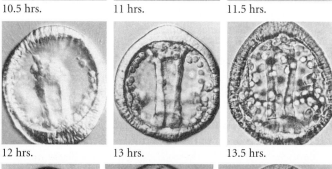

of the zygote, have different sizes and properties. Figures 8.16 and 8.17 show the fates of the various regions of the blastula as it develops through gastrulation to the **pluteus larva** stage characteristic of sea urchins. The fate of each cell layer can be seen through its movements during gastrulation.

Ingression of primary mesenchyme

FUNCTION OF PRIMARY MESENCHYME CELLS. Shortly after the blastula hatches from its fertilization envelope, the vegetal side of the spherical blastula begins to thicken and flatten (Figure 8.17, 9 hours). At the center of this flat vegetal plate, a cluster of small cells begins to change. These cells begin extending and contracting long, thin (30×5 μm) processes

Figure 8.17
Entire sequence of gastrulation in *Lytechinus variegatus*. The time shows the length of development at 25°C. (Photographs courtesy of J. Morrill; pluteus larva courtesy of G. Watchmaker.)

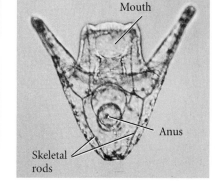

(A)

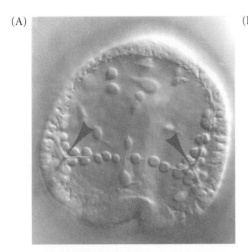

(B)

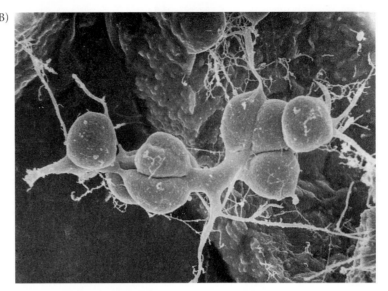

(C)

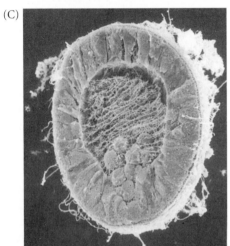

(D)

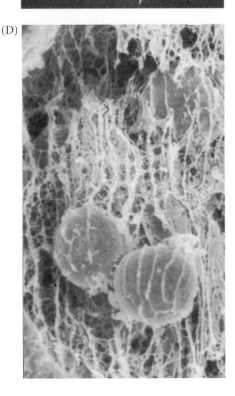

Figure 8.18
Formation of syncytial cables by mesenchyme cells of the sea urchin. (A) Primary mesenchyme cells in the early gastrula align and fuse to lay down the matrix of the calcium carbonate spicule (arrows). (B) Scanning electron micrograph (SEM) of spicules formed by the fusing of primary mesenchyme cells into syncytial cables. (C) SEM of primary mesenchyme cells enmeshed in the extracellular matrix of early *Strongylocentrotus* gastrula. (D) Gastrula-stage mesenchyme cell migration. The extracellular matrix fibrils of the blastocoel lie parallel to the animal-vegetal axis and are intimately associated with the primary mesenchyme cells. (A from Ettensohn 1990; B from Morrill and Santos 1985; C, D from Cherr et al. 1992; all photographs courtesy of the authors.)

called **filopodia** from their inner surfaces. The cells then dissociate from the epithelial monolayer and ingress into the blastocoel (Figure 8.17, 9–10 hours). These cells, derived from the micromeres, are called the **primary mesenchyme.** They will form the larval skeleton, so they are sometimes called the **skeletogenic mesenchyme.** At first the cells appear to move randomly along the inner blastocoel surface, actively making and breaking filopodial connections to the wall of the blastocoel. Eventually, however, they become localized within the prospective ventrolateral region of the blastocoel. Here they fuse into syncytial cables, which will form the axis of the calcium carbonate spicules of the larval skeleton (Figure 8.18).

IMPORTANCE OF EXTRACELLULAR LAMINA INSIDE THE BLASTOCOEL. The ingression of the micromere descendants into the blastocoel is caused by these cells losing their affinity for their neighbors and for the hyaline membrane and acquiring a strong affinity for a group of proteins that line the blastocoel. This model was first proposed by Gustafson and Wolpert (1967) and was confirmed in 1985, when Rachel Fink and David McClay measured the strengths of sea urchin blastomere adhesion to the hyaline layer, to the basal lamina lining the blastocoel, and to other blastomeres. Originally, all the cells of the blastula are connected on their outer surface to the hyaline layer and on their inner surface to a basal lamina se-

Table 8.2 Affinities of mesenchymal and nonmesenchymal cells to cellular and extracellular components[a]

	Dislodgment force (in dynes)		
Cell type	Hyaline	Gastrula cell monolayers	Basal lamina
16-cell-stage micromeres	5.8×10^{-5}	6.8×10^{-5}	4.8×10^{-7}
Migratory-stage mesenchyme cells	1.2×10^{-7}	1.2×10^{-7}	1.5×10^{-5}
Gastrula ectoderm and endoderm	5.0×10^{-5}	5.0×10^{-5}	5.0×10^{-7}

Source: After Fink and McClay 1985.

[a]Tested cells were allowed to adhere to plates containing hyaline, extracellular basal lamina, or cell monolayers. The plates were inverted and centrifuged at various strengths to dislodge the cells. The dislodgement force is calculated from the centrifugal force needed to remove the test cells from the substrate.

creted by the cells. On their lateral sides, each cell has another cell for a neighbor. Fink and McClay found that the prospective ectoderm and endoderm cells (descendants of the mesomeres and macromeres, respectively) bind tightly to one another and to the hyaline layer, but adhere only loosely to the basal lamina (Table 8.2). The micromeres originally display a similar pattern of binding. However, the micromere pattern changes at gastrulation. Whereas the other cells retain their tight binding to the hyaline layer and to their neighbors, the primary mesenchyme precursors lose their affinity for these structures (which drops to about 2% of its original value) while their affinity for components of the basal lamina and extracellular matrix (such as fibronectin) increases a hundredfold. This change in affinity causes the micromeres to release their attachments to the external hyaline layer and their neighboring cells and, drawn in by the basal lamina, migrate up into the blastocoel (Figures 8.18, 8.19). These changes in

affinity have been correlated with changes in cell surface molecules that occur during this time (Wessel and McClay 1985).

As shown in Figure 8.18, there is a heavy concentration of extracellular material around the ingressing primary mesenchyme cells (Galileo and Morrill 1985; Cherr et al. 1992). Once inside the blastocoel, the primary mesenchyme cells appear to migrate along the extracellular matrix of the blastocoel wall, extending their filopodia in front of them (Galileo and Morrill 1985; Karp and Solursh 1985). Several proteins (including fibronectin and a particular sulfated glycoprotein) are necessary to initiate and maintain this migration (Wessel et al. 1984; Sugiyama 1972; Lane and Solursh 1991; Berg et al. 1996).

But these guidance cues cannot be sufficient, since the migrating cells "know" when to stop their movement and form spicules near the equator of the blastocoel. The primary mesenchyme cells arrange themselves in a ring at a specific position along the animal-vegetal axis. At two sites near the future ventral side of the larva, many of these primary mesenchyme cells cluster together, fuse with each other, and initiate spicule formation (Figure 8.18A; Hodor and Ettensohn 1998). If a labeled micromere from another embryo is injected into the blastocoel of a gastrulating sea urchin embryo, it migrates to the correct location and contributes to the forma-

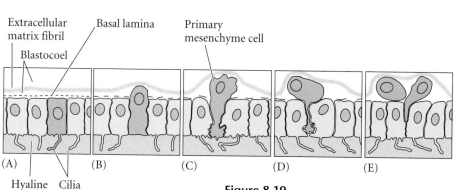

Figure 8.19

Ingression of primary mesenchyme cells. (A–E) Interpretative diagrams depicting changes in adhesive interactions in the presumptive primary mesenchyme cells (color). These cells lose their affinities for hyaline and for their neighboring blastomeres while gaining an affinity for the basal lamina. Nonmesenchymal blastomeres retain their original high affinities for hyaline and neighboring cells. (F) Scanning electron micrograph montage showing the ingression of the primary mesenchyme cells of *Lytechinus variegatus*. (F courtesy of J. B. Morrill and D. Flaherty.)

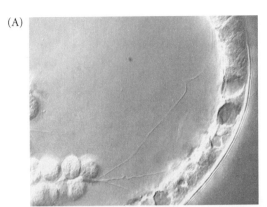

(A)

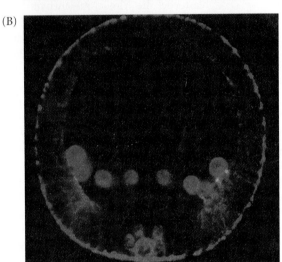

(B)

Figure 8.20

Placement of the primary mesenchyme cells. (A) Nomarski videomicrograph showing a long, thin filopodium from a primary mesenchyme cell extending to the ectodermal wall of the gastrula, as well as a shorter filopodium extending inward from the ectoderm. The mesenchymal filopodia extend through the extracellular matrix and directly contact the cell membrane of the ectodermal cells. (B) The localization of the micromeres to form the calcium carbonate skeleton is determined by the ectodermal cells. The primary mesenchyme cells are stained green, while β-catenin is stained red. The primary mesenchyme cells appear to accumulate in those regions characterized by high β–catenin concentrations. (A from Miller et al. 1995; photographs courtesy of J. R. Miller and D. McClay.)

tion of the embryonic spicules (Figure 8.20; Ettensohn 1990). It is thought that this positional information is provided by the prospective ectodermal cells and their basal laminae (von Übisch 1939; Harkey and Whiteley 1980). Only the primary mesenchyme cells (and not other cell types or latex beads) are capable of responding to these patterning cues (Ettensohn and McClay 1986). Miller and colleagues (1995) have reported the existence of extremely fine (0.3-μm diameter) filopodia on the skeleton-forming mesenchyme. These filopodia are not thought to function in locomotion; rather, they appear to explore and sense the blastocoel wall and may be responsible for picking up dorsal-ventral and animal-vegetal patterning cues from the ectoderm (Figure 8.20; Malinda et al. 1995).

First stage of archenteron invagination

As the ring of primary mesenchyme cells leaves the vegetal region of the blastocoel, important changes are occurring in the

Figure 8.21

Invagination of the vegetal plate. (A) Vegetal plate invagination in *Lytechinus variegatus*, seen by scanning electron micrography of the external surface of the early gastrula. The blastopore is clearly visible. (B) The hyaline layer consists of inner and outer laminae. Microvilli from the vegetal plate cells extend through the hyaline layer, and their cytoplasm contains secretory vesicles that store a chondroitin sulfate proteoglycan (CSPG). (C) The storage granules secrete CSPG into the inner lamina of the hyaline layer. The CSPG absorbs water and swells the inner lamina, while the outer lamina, to which it is attached, does not swell. This causes the bending of the hyaline layer and its attached epithelium inward. (A from Morrill and Santos 1985, courtesy of J. B. Morrill; B and C after Lane et al. 1993.)

(A)

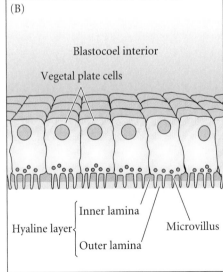

(B)

Blastocoel interior

Vegetal plate cells

Hyaline layer { Inner lamina

Outer lamina

Microvillus

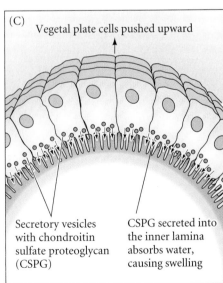

(C)

Vegetal plate cells pushed upward

Secretory vesicles with chondroitin sulfate proteoglycan (CSPG)

CSPG secreted into the inner lamina absorbs water, causing swelling

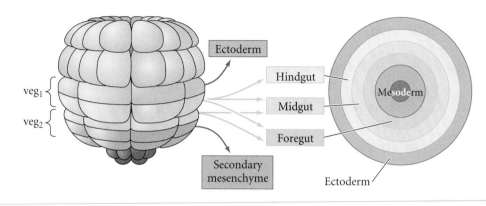

Figure 8.22
Fate map of the vegetal plate of the sea urchin embryo, looking "upward" at the vegetal surface. The central portion becomes the secondary mesenchyme cells, while the concentric layers around it become the foregut, midgut, and hindgut, respectively. The place where the endoderm meets the ectoderm marks the anus. The secondary mesenchyme and foregut come from the veg_2 layer, the midgut comes from veg_1 and veg_2 cells, and the foregut (and the ectoderm in contact with it) comes from the veg_1 layer. (After Logan and McClay 1999.)

cells that remain at the vegetal plate. These cells remain bound to one another and to the hyaline layer of the egg, and they move to fill the gaps caused by the ingression of the primary mesenchyme. Moreover, the vegetal plate bends inward and invaginates about one-fourth to one-half the way into the blastocoel (Figure 8.21A; also see Figure 8.17, 10.5–11.5 hours). Then invagination suddenly ceases. The invaginated region is called the **archenteron** (primitive gut), and the opening of the archenteron at the vegetal region is called the **blastopore**.

Lane and co-workers (1993) have provided evidence that the mechansim of this invagination is similar to that of the buckling produced by heating a bimetallic strip. The hyaline layer is actually made up of two layers, an outer lamina made primarily of hyalin protein and an inner lamina composed of fibropellin proteins (Hall and Vacquier 1982; Bisgrove et al. 1991). Fibropellins are stored in secretory granules within the oocyte, and are secreted from those granules after cortical granule exocytosis releases the hyalin protein. By the blastula stage, the fibropellins have formed a meshlike network over the embryo surface. At the time of invagination, the vegetal plate cells (and only those cells) secrete a chondroitin sulfate proteoglycan into the inner lamina of the hyaline layer directly beneath them. This hygroscopic (water-absorbing) molecule swells the inner lamina, but not the outer lamina. This causes the vegetal region of the hyaline layer to buckle (Figure 8.21B,C). Slightly later, a second force arising from the movements of epithelial cells adjacent to the vegetal plate may facilitate this invagination by drawing the buckled layer inward (Burke et al. 1991).

At the stage when the skeletogenic mesenchyme cells begin ingressing into the blastocoel, the fates of the vegetal plate cells have already been specified (Figure 8.22; Ruffins and Ettensohn 1996). The endodermal cells adjacent to the micromere-derived mesenchyme become foregut, migrating the farthest distance into the blastocoel. The next layer of endodermal cells becomes midgut, and the last circumferential row to invaginate forms the hindgut and anus.

Second and third stages of archenteron invagination

The invagination of the vegetal cells occurs in three discrete stages. After a brief pause, the second phase of archenteron formation begins. During this time, the archenteron extends dramatically, sometimes tripling its length. In this process of extension, the wide, short gut rudiment is transformed into a long, thin tube (see Figure 8.17, 12 hours; Figure 8.23). To accomplish this extension, the cells of the archenteron rearrange

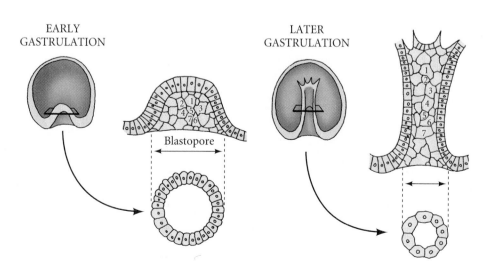

Figure 8.23
Cell rearrangement during the extension of the archenteron in sea urchin embryos. In this species, the early archenteron has 20 to 30 cells around its circumference. Later in gastrulation, the archenteron has a circumference made by only 6 to 8 cells. Fluorescently labeled clones can be seen to stretch apically. (After Hardin 1990.)

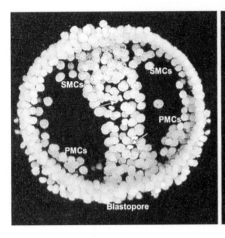
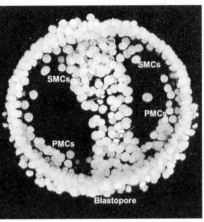

Figure 8.24
Stereo pair showing *Lytechinus* gastrula during the dispersal of the secondary mesenchyme cells. Convergent extension has occured, and a few mitotic spindles (amall arrows) can be seen. (PMC, primary mesenchyme cell; SMC, secondary mesenchyme cell.) If you cross your eyes, a 3 dimensional "central image" will form. (From Martins et al. 1998; photographs courtesy of the authors.)

themselves by migrating over one another and by flattening themselves (Ettensohn 1985; Hardin and Cheng 1986). This phenomenon, wherein cells intercalate to narrow the tissue and at the same time move it forward, is called **convergent extension**. Moreover, cell division continues, producing more endodermal and secondary mesenchyme cells as the archenteron extends (Figure 8.24; Martins et al. 1998).

In at least some species of sea urchins, a third stage of archenteron elongation occurs. This last phase is initiated by the tension provided by secondary mesenchyme cells, which form at the tip of the archenteron and remain there (see Figure 8.17, 13 hours; Figure 8.25). Filopodia are extended from these cells through the blastocoel fluid to contact the inner surface of the blastocoel wall (Dan and Okazaki 1956; Schroeder 1981). The filopodia attach to the wall at the junctions between the blastoderm cells and then shorten, pulling up the archenteron. Hardin (1988) ablated the secondary mesenchyme cells with a laser, with the result that the archenteron could elongate to only about two-thirds of the normal length. If a few secondary mesenchyme cells were left, elongation continued, although at a slower rate. The secondary mesenchyme cells, then, play an essential role in pulling the archenteron up to the blastocoel wall during the last phase of invagination.

But can the secondary mesenchyme filopodia attach to any part of the blastocoel wall, or is there a specific target in the animal hemisphere that must be present for attachment to occur? Is there a region of the blastocoel wall that is already committed to becoming the ventral side of the larva? Studies by Hardin and McClay (1990) show that there is a specific "target" site for the filopodia that differs from other regions of the animal hemisphere. The filopodia extend, touch the blastocoel wall at random sites, and then retract. However, when the filopodia contact a particular region of the wall, they remain attached there, flatten out against this region, and pull the archenteron toward it. When Hardin and McClay poked in the other side of the blastocoel wall so that the contacts were made most readily with that region, the filopodia continued to extend and retract after touching it. Only when the filopodia found their "target" did they cease these movements. If the gastrula was constricted so that filopodia never reached the target area, the secondary mesenchyme cells continued to explore until they eventually moved off the archenteron and found the target tissue as freely migrating cells. There appears, then, to be a target region on what is to become the ventral side of the larva that is recognized by the secondary mesenchyme cells, and which positions the archenteron in the region where the mouth will form.

(A)

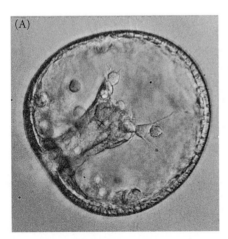

(B)

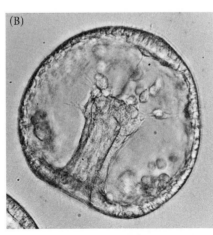

Figure 8.25
Mid-gastrula stage of the sea urchin *Lytechinus pictus*, showing filopodial extensions of secondary mesenchyme extending from the archenteron tip to the blastocoel wall. (A) Mesenchyme cells extending filopodia from the tip of the archenteron. (B) Filopodial cables connecting the blastocoel wall to the archenteron tip. The tension of the cables can be seen as they pull on the blastocoel wall at the point of attachment. (Photographs courtesy of C. Ettensohn.)

(A) VIEW FROM ANIMAL POLE

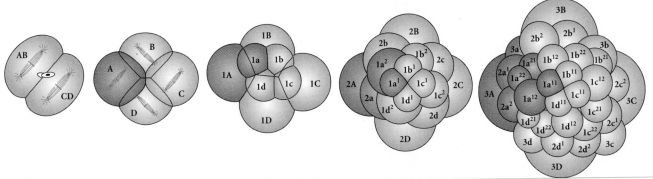

(B) SIDE VIEW

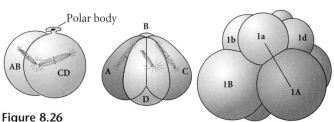

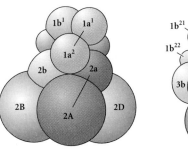

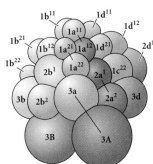

Figure 8.26
Spiral cleavage of the mollusc *Trochus* viewed (A) from the animal pole and (B) from one side. In (B), the cells derived from the A blastomere are shown in color. The mitotic spindles, sketched in the early stages, divide the cells unequally and at an angle to the vertical and horizontal axes.

As the top of the archenteron meets the blastocoel wall in the target region, the secondary mesenchyme cells disperse into the blastocoel, where they proliferate to form the mesodermal organs (see Figure 8.17, 13.5 hours). Where the archenteron contacts the wall, a mouth is eventually formed. The mouth fuses with the archenteron to create a continuous digestive tube. Thus, as is characteristic of deuterostomes, the blastopore marks the position of the anus.

> VADE MECUM **Sea urchin development.** The CD-ROM provides an excellent review of sea urchin development as well as containing questions on the fundamentals of echinoderm cleavage and gastrulation.
> [Click on Sea Urchin]

The Early Development of Snails

Cleavage in Snail Eggs

Spiral holoblastic cleavage is characteristic of several animal groups, including annelid worms, some flatworms, and most molluscs. It differs from radial cleavage in numerous ways. First, the cleavage planes are not parallel or perpendicular to the animal-vegetal axis of the egg; rather, cleavage is at oblique angles, forming a "spiral" arrangement of daughter blastomeres. Second, the cells touch one another at more places

than do those of radially cleaving embryos. In fact, they assume the most thermodynamically stable packing orientation, much like that of adjacent soap bubbles. Third, spirally cleaving embryos usually undergo fewer divisions before they begin gastrulation, making it possible to follow the fate of each cell of the blastula. When the fates of the individual blastomeres from annelid, flatworm, and mollusc embryos were compared, many of the same cells were seen in the same places, and their general fates were identical (Wilson 1898). Blastulae produced by radial cleavage have no blastocoel and are called **stereoblastulae**.

Figures 8.26 and 8.27 depict the cleavage of mollusc embryos. The first two cleavages are nearly meridional, producing four large macromeres (labeled A, B, C, and D). In many species, the blastomeres are different sizes (D being the largest), a characteristic that allows them to be individually identified. In each successive cleavage, each macromere buds off a small micromere at its animal pole. Each successive quartet of micromeres is displaced to the right or to the left of its sister macromere, creating the characteristic spiral pattern. Looking down on the embryo from the animal pole, the upper ends of the mitotic spindles appear to alternate clockwise and counterclockwise. This causes alternate micromeres to form obliquely to the left and to the right of their macromere. At the third cleavage, the A macromere gives rise to two daughter cells, macromere 1A and micromere 1a. The B, C, and D cells behave similarly, producing the first quartet of micromeres. In most species, the micromeres are to the right of their macromeres (looking down on the animal pole). At the

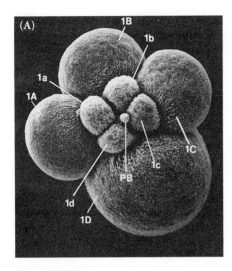
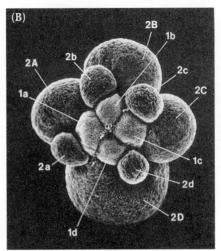
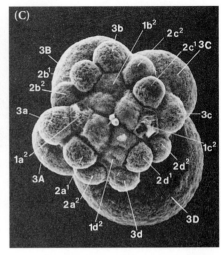

Figure 8.27
Spiral cleavage of the snail *Ilyanassa*. The D blastomere is larger than the others, allowing the identification of each cell. Cleavage is dextral. (A) 8-cell stage. PB is a polar body. (B) Mid-fourth cleavage (12-cell embryo). The macromeres have already divided into large and small spirally oriented cells; 1a–d have not divided yet. (C) 32-cell embryo. (From Craig and Morrill 1986; photographs courtesy of the authors.)

Figure 8.28
Looking down on the animal pole of left-coiling and right-coiling snails. The origin of sinistral and dextral coiling can be traced to the orientation of the mitotic spindle at the second cleavage. The left-coiling (A) and right-coiling (B) snails develop as mirror images of each other. (After Morgan 1927.)

fourth cleavage, macromere 1A divides to form macromere 2A and micromere 2a; and micromere 1a divides to form two more micromeres, $1a^1$ and $1a^2$. Further cleavage yields blastomeres 3A and 3a from macromere 2A, and micromere $1a^2$ divides to produce cells $1a^{21}$ and $1a^{22}$.

The orientation of the cleavage plane to the left or to the right is controlled by cytoplasmic factors within the oocyte. This was discovered by analyzing mutations of snail coiling. Some snails have their coils opening to the right of their shells (**dextral** coiling), whereas other snails have their coils opening to the left (**sinistral** coiling). Usually, the direction of coiling is the same for all members of a given species. Occasionally, though, mutants are found. For instance, in species in which the coils open on the right, some individuals will be found with coils that open on the left. Crampton (1894) analyzed the

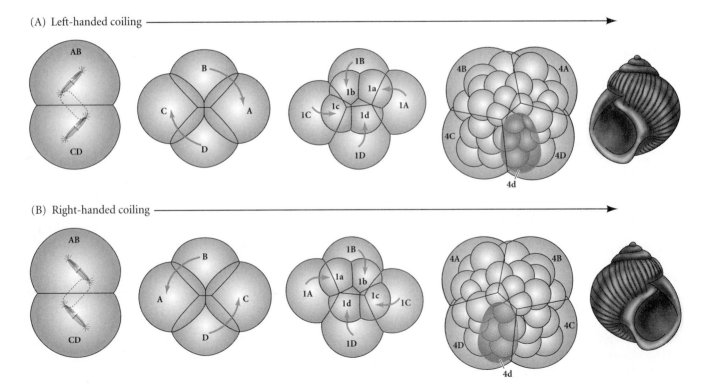

(A) Left-handed coiling

(B) Right-handed coiling

embryos of such aberrant snails and found that their early cleavage differed from the norm. The orientation of the cells after the second cleavage was different in the sinistrally coiling snails owing to a different orientation of the mitotic apparatus (Figure 8.28). All subsequent divisions in left-coiling embryos are mirror images of those in dextrally coiling embryos. In Figure 8.28, one can see that the position of the 4d blastomere (which is extremely important, as its progeny will form the mesodermal organs) is different in the two types of spiraling embryos. Eventually, two snails are formed, with their bodies on different sides of the coil opening.

The direction of snail shell coiling is controlled by a single pair of genes (Sturtevant 1923; Boycott et al. 1930). In the snail *Limnaea peregra*, most individuals are dextrally coiled. Rare mutants exhibiting sinistral coiling were found and mated with wild-type snails. These matings showed that there is a right-coiling allele *D*, which is dominant to the left-coiling allele *d*. However, the direction of cleavage is determined not by the genotype of the developing snail, but by the genotype of the snail's mother. A *dd* female snail can produce only sinistrally coiling offspring, even if the offspring's genotype is *Dd*. A *Dd* individual will coil either left or right, depending on the genotype of its mother. Such matings produce a chart like this:

		Genotype	Phenotype
DD ♀ × *dd* ♂	→	*Dd*	All right-coiling
DD ♂ × *dd* ♀	→	*Dd*	All left-coiling
Dd × *Dd*	→	1*DD*:2*Dd*:1*dd*	All right-coiling

The genetic factors involved in snail coiling are brought to the embryo by the oocyte cytoplasm. It is the genotype of the ovary in which the oocyte develops that determines which orientation cleavage will take. When Freeman and Lundelius (1982) injected a small amount of cytoplasm from dextrally coiling snails into the eggs of *dd* mothers, the resulting embryos coiled to the right. Cytoplasm from sinistrally coiling snails did not affect the right-coiling embryos. These findings confirmed that the wild-type mothers were placing a factor into their eggs that was absent or defective in the *dd* mothers.

WEBSITE **8.3 Alfred Sturtevant and the genetics of snail coiling.** By a masterful thought experiment, Sturtevant demonstrated the power of applying genetics to embryology. To do this, he brought Mendelian genetics into the study of the oocyte.

Sidelights & Speculations

Adaptation by Modifying Embryonic Cleavage

EVOLUTION IS CAUSED by the hereditary alteration of embryonic development. Sometimes we are able to identify a modification of embryogenesis that has enabled the organism to survive in an otherwise inhospitable environment. One such modification, discovered by Frank Lillie in 1898, is brought about by altering the typical pattern of spiral cleavage in the unionid family of clams.

Unlike most clams, *Unio* and its relatives live in swift-flowing streams. Streams create a problem for the dispersal of larvae: because the adults are sedentary, free-swimming larvae would always be carried downstream by the current. These clams, however, have adapted to this environment by effecting two changes in their development. The first alters embryonic cleavage. In the typical cleavage of molluscs, either all the macromeres are equal in size or the 2D blastomere is the largest cell at that embryonic stage. However, the division of *Unio* is such that the 2d blastomere gets the largest amount of cytoplasm (Figure 8.29).

Figure 8.29
Formation of glochidium larva by the modification of spiral cleavage. After the 8-cell embryo is formed (A), the placement of the mitotic spindle causes most of the D cytoplasm to enter the 2d blastomere (B). This large 2d blastomere divides (C) so as to eventually give rise to the large "bear-trap" shell of the larva (D). (After Raff and Kaufman 1983.)

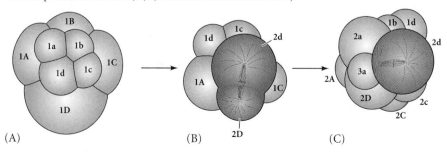
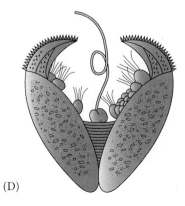

This cell divides to produce most of the larval structures, including a gland capable of producing a large shell. The resulting larvae (called **glochidia**) resemble tiny bear traps; they have sensitive hairs that cause the valves of the shell to snap shut when they are touched by the gills or fins of a wandering fish. They attach themselves to the fish and "hitchhike" with it until they are ready to drop off and metamorphose into adult clams. In this manner, they can spread upstream.

In some species, glochidia are released from the female's brood pouch and merely wait for a fish to come wandering by. Some other species, such as *Lampsilis ventricosa*, have increased the chances of their larvae finding a fish by yet another modification of their development (Welsh 1969). Many clams develop a thin mantle that flaps around the shell and surrounds the brood pouch. In

Figure 8.30
Phony fish atop the unionid clam *Lampsilis ventricosa*. The "fish" is actually the brood pouch and mantle of the clam. (Photograph courtesy of J. H. Welsh.)

some unionids, the shape of the brood pouch (marsupium) and the undulations of the mantle mimic the shape and swimming behavior of a minnow. To make the deception all the better, they develop a black "eyespot" on one end and a flaring "tail" on the other. The "fish" seen in Figure 8.30 is not a fish at all, but the brood pouch and mantle of the clam beneath it. When a predatory fish is lured within range, the clam discharges the glochidia from the brood pouch. Thus, the modification of existing developmental patterns has permitted unionid clams to survive in challenging environments. ■

The polar lobe: Cell determination and axis formation

Molluscs provide some of the most impressive examples of mosaic development, in which the blastomeres are specified autonomously, and of cytoplasmic localization, wherein morphogenetic determinants are placed in a specific region of the oocyte (see Chapter 3). Mosaic development is widespread throughout the animal kingdom, especially in protostomal organisms such as annelids, nematodes, and molluscs, all of which initiate gastrulation at the future anterior end after only a few cell divisions. Moreover, the cytoplasmic factors responsible for specification are actively moved to one pole of the cell so that a blastomere containing these factors can restrict their transmission to only one of its two daughter cells. The fate of the two daughter cells is thus changed by which one of them gets the morphogenetic determinant.

E. B. Wilson demonstrated that mosaic development characterizes the early snail embryo (see Figure 3.7). He also was able to demonstrate that such development is predicated on the segregation of specific morphogenetic determinants into specific blastomeres. Certain spirally cleaving

embryos (mostly in the mollusc and annelid phyla) extrude a bulb of cytoplasm immediately before first cleavage (Figure 8.31). This protrusion is called the **polar lobe**. In certain species of snails, the region uniting the polar lobe to the rest of the egg becomes a fine tube. The first cleavage splits the zygote asymmetrically, so that the polar lobe is connected only to the CD blastomere. In several species, nearly one-third of the total cytoplasmic volume is present in this anucleate lobe, giving it the appearance of another cell. This three-lobed structure is often referred to as the **trefoil-stage embryo**

Figure 8.31
Cleavage in the mollusc *Dentalium*. Extrusion and reincorporation of the polar lobe occur twice. (After Wilson 1904.)

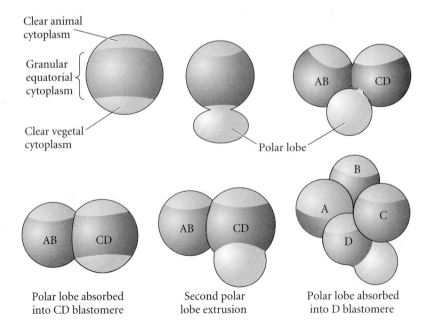

Clear animal cytoplasm

Granular equatorial cytoplasm

Clear vegetal cytoplasm

Polar lobe

Polar lobe absorbed into CD blastomere

Second polar lobe extrusion

Polar lobe absorbed into D blastomere

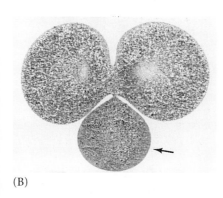

(A)

(B)

Figure 8.32
Polar lobes of molluscs. (A) Scanning electron micrograph of the extending polar lobe in the uncleaved egg of *Buccinum undatum*. The surface ridges are confined to the polar lobe region. (B) Section through first-cleavage, or trefoil-stage, embryo of *Dentalium*. The arrow points to the large polar lobe. (Photographs courtesy of M. R. Dohmen.)

(Figure 8.32). The CD blastomere then absorbs the polar lobe material, but extrudes it again prior to second cleavage. After this division, the polar lobe is attached only to the D blastomere, which absorbs its material. Thereafter, no polar lobe is formed.

Wilson (1904) showed that if one removes the polar lobe at the trefoil stage, the remaining cells divide normally. However, instead of producing a normal **trochophore** (snail) larva, they produce an incomplete larva, wholly lacking its mesodermal organs—muscles, mouth, shell gland,* and foot. Moreover, Wilson demonstrated that the same type of abnormal embryo can be produced by removing the D blastomere from the 4-cell embryo. Wilson concluded that the polar lobe cytoplasm contains the mesodermal determinants, and that these determinants give the D blastomere its mesoderm-forming capacity. Wilson also showed that the localization of the mesodermal determinants is established shortly after fertilization, thereby demonstrating that a specific cytoplasmic region of the egg, destined for inclusion in the D blastomere, contains whatever factors are necessary for the special cleavage rhythms of the D blastomere and for the differentiation of the mesoderm.

The morphogenetic determinants sequestered within the polar lobe are probably located in the cytoskeleton or cortex, not in the diffusible cytoplasm of the embryo. Evidence for this localization came from the studies of A. C. Clement

*The shell gland is an ectodermal organ formed through induction by mesodermal cells. Without the mesoderm, no cells are present to induce the competent ectoderm. Here we see an example of limited induction within a mosaic embryo. For more information on the formation of snail embryos, see Collier 1997.

(1968). When he separated the animal hemisphere of the egg of the snail *Ilyanassa obsoleta* from the vegetal hemisphere, the animal hemisphere formed ectodermal organs that resembled those formed by lobeless embryos. Clement then took embryos that had begun resorbing their second polar lobe and placed them into gelatin slabs. He centrifuged the embedded embryos, forcing the fluid, yolky cytoplasm from the vegetal part of the cell into the animal hemisphere. By centrifuging these embryos in a second, viscous medium, he caused the separation of the animal and vegetal hemispheres. The animal halves from such centrifuged embryos did not develop any more mesodermal or endodermal structures than those of uncentrifuged eggs. Thus, the determinants of the polar lobe were not transferred to the animal hemisphere in the fluid contents of the vegetal hemisphere. Van den Biggelaar (1977) obtained similar results when he removed the cytoplasm from the polar lobe with a micropipette. Cytoplasm from other regions of the cell flowed into the polar lobe, replacing the portion that he had removed. The subsequent development of these embryos was normal. In addition, when he added the soluble polar lobe cytoplasm to the B blastomere, duplications of structures were not seen (Verdonk and Cather 1983). Therefore, the diffusible part of the cytoplasm does not contain the morphogenetic determinants. They probably reside in the nonfluid cortical cytoplasm or on the cytoskeleton.

Clement (1962) also analyzed the further development of the D blastomere in order to observe the further appropriation of these determinants. The development of the D blastomere is illustrated in Figure 8.27. This macromere, having received the contents of the polar lobe, is larger than the other three. When one removes the D blastomere or its first or second macromere derivatives (1D or 2D), one obtains an incomplete larva, lacking heart, intestine, velum (the ciliated border of the larva), shell gland, eyes, and foot. When one removes the 3D blastomere (after the division of the 2D cell to form the 3D and 3d blastomeres), one obtains an almost normal embryo, having eyes, foot, velum, and some shell gland, but no heart or intestine (Figure 8.33). Therefore, some of the morphogenetic determinants originally present in the D blastomere must have been apportioned to the 3d cell. After the 4d cell is given off (by the division of the 3D blastomere), removal of the D derivative (the 4D cell) produces no qualitative difference in development. In fact, all the essential determinants for heart and intestine formation are now in the 4d blastomere, and removal of that cell results in a heartless and gutless larva (Clement 1986). The 4d blastomere is responsible for forming (at its next division) the two **mesentoblasts**, the cells that give rise to both the mesodermal (heart) and endodermal (intestine) organs.

(A)

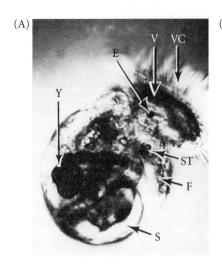

(B)

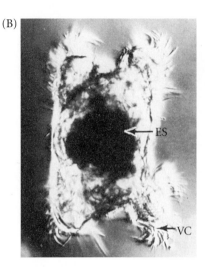

Figure 8.33
Importance of the polar lobe in the development of *Ilyanassa*. (A) Normal larva. (B) Abnormal larva, typical of those produced when the polar lobe of the D blastomere is removed. (E, eye; F, foot; S, shell; ST, statocyst, a balancing organ; V, velum; VC, velar cilia; Y, residual yolk; ES, everted stomodeum; DV, disorganized velum.) (From Newrock and Raff 1975; photographs courtesy of K. Newrock.)

The material in the polar lobe is also responsible for organizing the dorsal-ventral (back-belly) polarity of the embryo. When polar lobe material is forced to pass into the AB blastomere as well as into the CD blastomere, twin larvae are formed that are joined at their ventral surfaces (Figure 8.34; Guerrier et al. 1978; Henry and Martindale 1987).

Thus, experiments have demonstrated that the nondiffusible polar lobe cytoplasm is extremely important in normal mollusc development for a number of reasons:

- It contains the determinants for the proper cleavage rhythm and cleavage orientation of the D blastomere.
- It contains certain determinants (those entering the 4d blastomere and hence leading to the mesentoblasts) for mesodermal and intestinal differentiation.
- It is responsible for permitting the inductive interactions (through the material entering the 3d blastomere) leading to the formation of the shell gland and eye.
- It contains determinants needed for specifying the dorsal-ventral axis of the embryo.

Although the polar lobe is clearly important for normal snail development, we still do not know the mechanisms of its effects. There appear to be no major differences in mRNA or protein synthesis between lobed and lobeless embryos (Brandhorst and Newrock 1981; Collier 1983, 1984). One possible clue has been provided by Atkinson (1987), who has observed differentiated cells of the velum, digestive system, and shell gland within lobeless embryos. Thus, lobeless embryos can produce these cells, but they appear unable to organize them into functional tissues and organs. Tissues of the digestive tract can be found, but they are not connected; muscle cells are scattered around the lobeless larva, but are not organized into a functional muscle tissue. Thus, the developmental functions of the polar lobe may be very complex.

Fate map of *Ilyanassa obsoleta*

As detailed above, fate maps of the snail *Ilyanassa* have been made means of ablation experiments. More recently, Joanne Render (1997) has constructed a more detailed fate map by injecting specific micromeres with beads containing the fluorescent dye Lucifer Yellow. The fluorescence is maintained over the period of embryogenesis and can be seen in the larval tissue derived from the injected cells. Figure 8.35 shows the new fate map of *Ilyanassa obsoleta*. The second quartet micromeres (2a–d) generally contribute to the shell-forming mantle, the velum, the mouth, and the heart. The third quartet micromeres (3a–d) generate large regions of the foot, velum, esophagus, and heart. The 4d cell, the mesentoblast, contributes to the larval kidney, heart, retractor muscles, and intestine.

WEBSITE **8.4 Modifications of cell fate in spiralian eggs.** Within the gastropods, differences in the timing of cell fate result in significantly different body plans. Furthermore, the leeches and nemerteans have modified the spiralian cleavage pattern to produce new types of body plans.

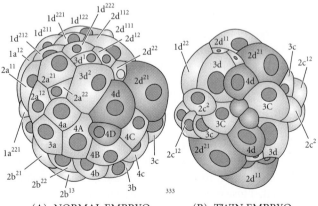

(A) NORMAL EMBRYO   (B) TWIN EMBRYO

Figure 8.34
Formation of twin embryos by the suppression of polar lobe formation in *Dentalium*. (A) Normal embryo at sixth-cleavage stage. (B) Twin embryos formed when low concentrations of cytochalasin are used to inhibit polar lobe formation and the polar lobe material is distributed to both the AB and CD blastomeres. (After Guerrier et al. 1978.)

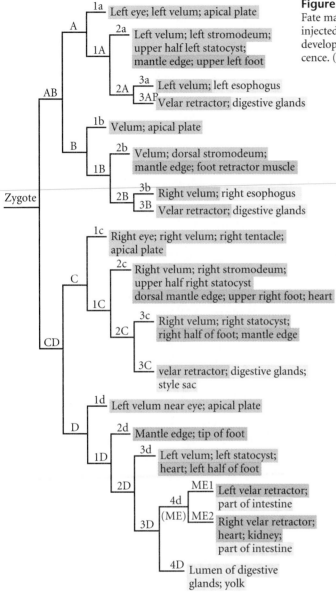

- A — 1A
  - 1a — Left eye; left velum; apical plate
  - 2a — Left velum; left stromodeum; upper half left statocyst; mantle edge; upper left foot
  - 2A
    - 3a — Left velum; left esophogus
    - 3AP — Velar retractor; digestive glands
- B — 1B
  - 1b — Velum; apical plate
  - 2b — Velum; dorsal stromodeum; mantle edge; foot retractor muscle
  - 2B
    - 3b — Right velum; right esophogus
    - 3B — Velar retractor; digestive glands
- C — 1C
  - 1c — Right eye; right velum; right tentacle; apical plate
  - 2c — Right velum; right stromodeum; upper half right statocyst dorsal mantle edge; upper right foot; heart
  - 2C
    - 3c — Right velum; right statocyst; right half of foot; mantle edge
    - 3C — velar retractor; digestive glands; style sac
- D — 1D
  - 1d — Left velum near eye; apical plate
  - 2d — Mantle edge; tip of foot
  - 2D
    - 3d — Left velum; left statocyst; heart; left half of foot
    - 3D
      - 4d (ME)
        - ME1 — Left velar retractor; part of intestine
        - ME2 — Right velar retractor; heart; kidney; part of intestine
      - 4D — Lumen of digestive glands; yolk

Figure 8.35
Fate map of the snail *Ilyanassa obsoleta*. Beads containing Lucifer Yellow were injected into individual blastomeres at the 32-cell stage. When the embryos developed into larvae , their descendants could be identified by their fluorescence. (After Render 1998.)

Gastrulation in Snails

The snail stereoblastula is relatively small, and its cell fates have already been determined by the D series of macromeres. Gastrulation is accomplished primarily by epiboly, wherein the micromeres at the animal cap multiply and "overgrow" the vegetal macromeres. Eventually, the micromeres will cover the entire embryo, leaving a small slit at the vegetal pole (Figure 8.36; Collier 1997).

Early Development in Tunicates

Tunicate Cleavage

Ascidians, members of the tunicate subphylum, are fascinating animals for several reasons, but the foremost is that they are invertebrate chordates. They have a notochord as larvae (and therefore are chordates), but they lack vertebrae. As larvae, they are free-swimming tadpoles; but when the tadpole undergoes metamorphosis, it sticks to the sea floor, its nerve cord and notochord degenerate, and it secrete a cellulose tunic (which gave the name "tunicates" to these creatures). These animals are characterized by **bilateral holoblastic cleavage**, a pattern found primarily in tunicates (Figure 8.37). The most striking phenomenon in this type of cleavage is that the first cleavage plane establishes the earliest axis of symmetry in the embryo, separating the embryo into its future right and left sides. Each successive division orients itself to this plane of symmetry, and the half-embryo formed on one side of the first cleavage plane is the mirror image of the half-embryo on the other side. The second cleavage is meridional, like the first, but unlike the first division, it does not pass through the center of the egg. Rather, it creates two large anterior cells (the A

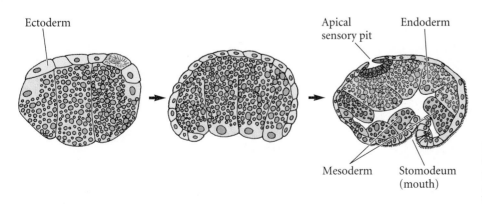

Ectoderm

Apical sensory pit   Endoderm

Mesoderm   Stomodeum (mouth)

Figure 8.36
Gastrulation in *Crepidula*. The ectoderm undergoes epiboly from the animal pole and envelops the other cells of the embryo. (After Conklin 1898.)

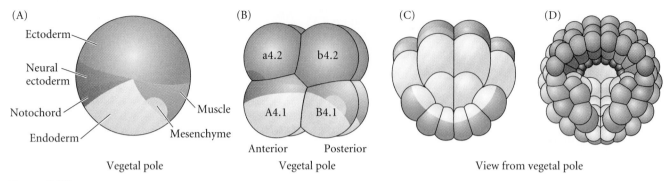

Figure 8.37
Bilateral symmetry in the egg of the tunicate *Styela partita*. (A) Uncleaved egg. The regions of cytoplasm destined to form particular organs are labeled here and coded by color throughout the diagrams. (B) 8-cell embryo, showing the blastomeres and the fates of various cells. The embryo can be viewed as two 4-cell halves; from here on, each division on the right side of the embryo has a mirror-image division on the left. (C, D) Views of later embryos from the vegetal pole. (A after Balinsky 1981.)

and D blastomeres) and two smaller posterior cells (blastomeres B and C). Each side now has a large and a small blastomere. During the next three divisions, differences in cell size and shape highlight the bilateral symmetry of these embryos. At the 32-cell stage, a small blastocoel is formed, and gastrulation begins. The cell lineages of the tunicate *Styela partita* are shown in Figure 1.7.

The tunicate fate map

As mentioned in Chapter 3, early tunicate cells are specified autonomously, each cell acquiring a specific type of cytoplasm that will determine its fate. In tunicates such as *Styela partita*, the different regions of cytoplasm have distinct pigmentation, and the cell fates can easily be seen to correspond to the type of cytoplasm taken up by each cell. These cytoplasmic regions are apportioned to the egg during fertilization. In the unfertilized egg of *Styela partita*, a central gray cytoplasm is enveloped by a cortical layer containing yellow lipid inclusions (Figure 8.38A). During meiosis, the breakdown of the nucleus releases a clear substance that accumulates in the animal hemisphere of the egg. Within 5 minutes of sperm entry, the inner clear and cortical yellow cytoplasms migrate into the vegetal (lower) hemisphere of the egg. As the male pronucleus migrates from the vegetal pole to the equator of the cell along the future posterior side of the embryo, the yellow lipid inclusions migrate with it. This migration forms a **yellow crescent**, extending from the vegetal pole to the equator (Figure 8.38B–D); this region will produce most of the tail muscles of the tunicate larva. The movement of these cytoplasmic regions depends on microtubules that are generated by the sperm centriole and on the wave of calcium ions that

Figure 8.38
Cytoplasmic rearrangement in the fertilized egg of *Styela partita*. (A) Before fertilization, yellow cortical cytoplasm surrounds the gray yolky inner cytoplasm. (B) After sperm entry (in the vegetal hemisphere of the oocyte), the yellow cortical cytoplasm and the clear cytoplasm derived from the breakdown of the oocyte nucleus stream vegetally toward the sperm. (C) As the sperm pronucleus migrates animally toward the newly formed egg pronucleus, the yellow and clear cytoplasms move with it. (D) The final positions of the clear and yellow cytoplasms mark the locations where cells give rise to mesenchyme and muscles, respectively. (After Conklin 1905.)

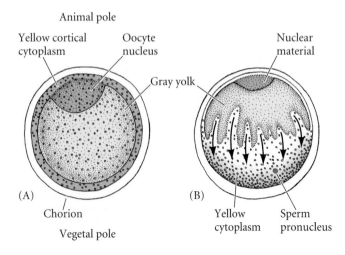

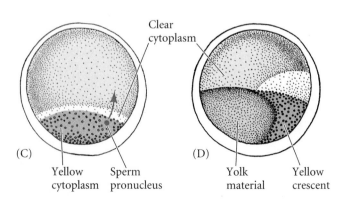

contracts the animal pole cytoplasm (Sawada and Schatten 1989; Speksnijder et al. 1990; Roegiers et al. 1995).

Edwin Conklin (1905) took advantage of the differing coloration of these regions of cytoplasm to follow each of the cells of the tunicate embryo to its fate in the larva (Figure 1.7). He found that cells receiving clear cytoplasm become ectoderm; those containing yellow cytoplasm give rise to mesodermal cells; those that incorporate slate-gray inclusions become endoderm; and light gray cells become the neural tube and notochord. The cytoplasmic regions are localized bilaterally around the plane of symmetry, so they are bisected by the first cleavage furrow into the right and left halves of the embryo. The second cleavage causes the prospective mesoderm to lie in the two posterior cells, while the prospective neural ectoderm and chordamesoderm (notochord) will be formed from the two anterior cells. The third division further partitions these cytoplasmic regions such that the mesoderm-forming cells are confined to the two vegetal posterior blastomeres, and the chordamesoderm cells are likewise restricted to the two vegetal anterior cells.

WEBSITE **8.5 The experimental analysis of tunicate cell specification.** Researchers analyzing tunicate development are using biochemical and molecular probes to find the morphogenetic determinants that are segregated to different regions of the egg cytoplasm.

Autonomous and conditional specification of tunicate blastomeres

As mentioned in Chapter 3, the autonomous specification of tunicate blastomeres was one of the first observations in the field of experimental embryology (Chabry 1888). Reverberi and Minganti (1946) extended this analysis in a series of isolation experiments, and they, too, observed the self-differentiation of each isolated blastomere and of the remaining embryo. The results of one of these experiments are shown in see Figure 3.8. When the 8-cell embryo is separated into its four doublets (the right and left sides being equivalent), mosaic determination is the rule. The animal posterior pair of blastomeres gives rise to the ectoderm, and the vegetal posterior pair produces endoderm, mesenchyme, and muscle tissue, just as expected from the fate map.

From the cell lineage studies of Conklin and others, it was known that only one pair of blastomeres (posterior vegetal; B4.1) in the 8-cell embryo is capable of producing tail muscle tissue. These cells contain the **yellow crescent** cytoplasm (Figure 8.39). When yellow crescent cytoplasm is transferred from the B4.1 (muscle-forming) blastomere to the b4.2 (ectoderm-forming) blastomere of an 8-cell tunicate embryo, the ectoderm-forming blastomere generates muscle cells as well as its normal ectodermal progeny (Whittaker 1982; Figure 3.10). Moreover, cytoplasm from the yellow crescent area of the fertilized egg can cause the a4.2 blastomere to express muscle-specific proteins (Nishida 1992a). Conversely, Tung and colleagues

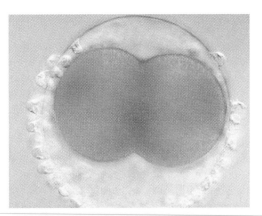
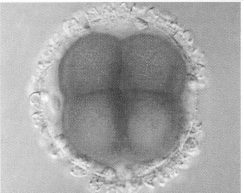
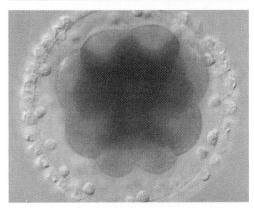
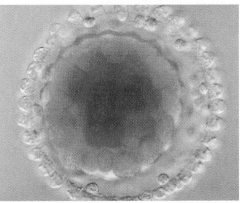

Figure 8.39
Cytoplasmic segregation in the egg of *Styela partita*. The yellow crescent, originally seen in the vegetal pole, becomes segregated into the B4.1 blastomere pair and thence into the muscle cells. (Photographs courtesy of J. R. Whittaker.)

(1977) showed that when larval cell nuclei are transplanted into enucleated tunicate egg fragments, the newly formed cells show the structures typical of the egg regions providing the cytoplasm, not of those cells providing the nuclei. We can conclude, then, that certain determinants that exist in the cytoplasm cause the formation of certain tissues. These morphogenetic determinants appear to work by selectively activating (or inactivating) specific genes. The determination of the blastomeres and the activation of certain genes are controlled by the spatial localization of the morphogenetic determinants within the egg cytoplasm.

Conditional specification, however, also plays an important role in tunicate development. One example of this process involves the development of neural cells. The nerve-producing cells are generated from both the animal and the vegetal anterior cells, yet neither the anterior or posterior cell of each half can produce them alone. When these anterior pairs are reunited, though, the brain and palp tissues arise. Ortolani (1959) has shown that this region of ectoderm is not determined for "neuralness" until the 64-cell stage, right before gastrulation. Similarly, while most muscles form autonomously from the yellow crescent material of the B4.1 blastomere, the most posterior muscle cells form through conditional specification by cell interactions with the descendants of the A4.1 and b4.2 blastomeres (Nishida 1987, 1992a,b). Thus, although most tissues are determined autonomously by segregation of the egg cytoplasm, certain tissues in tunicate embryos have a conditional determination by cell-cell interaction.

Specification of the embryonic axes

The axes of the tunicate larva are among its earliest committments. Indeed, all of its embryonic axes are determined by the cytoplasm of the zygote prior to first cleavage. The first axis to be determined is the dorsal-ventral axis, which is defined by the cap of cytoplasm at the vegetal pole. This cap defines the future dorsal side of the larva and the site where gastrulation is initiated (Bates and Jeffery 1987). When small regions of vegetal pole cytoplasm were removed from the zygote (between the first and second waves of zygote cytoplasmicmovement), the eggs neither gastrulated nor formed a dorsal-ventral axis.

The second axis to appear is the anterior-posterior axis, which is also determined during the migration of the oocyte cytoplasm. The yellow crescent forms in the region of the egg that will become the posterior side of the larva. When roughly 10% of the cytoplasm from this posterior vegetal region of the egg was removed after the second wave of cytoplasmic movement, most of the embryos failed to form an anterior-posterior axis. Rather, these embryos developed into radially symmetrical larvae with anterior fates (Figure 8.40). This posterior vegetal cytoplasm (PVC) was "dominant" to other cytoplasms in that when it was transplanted into the anterior vegetal region of zygotes that had had their own PVC removed, the anterior of the cell became the new posterior, and the axis was reversed (Nishida 1994). The left-right axis is specified as a consequence of these first two axes, and the first cleavage divides the embryo into its future right and left sides.

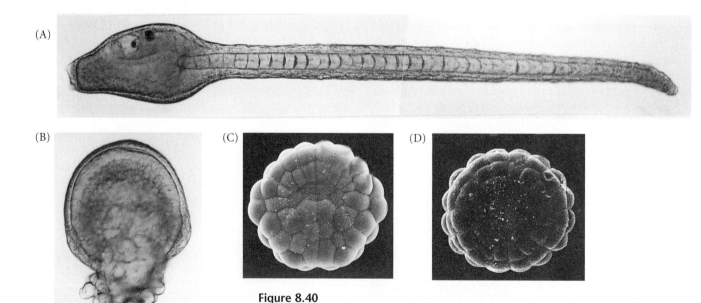

Figure 8.40
Comparison of normal tunicate embryos and embryos from which posterior vegetal cytoplasm has been removed. (A) Wild-type larva. (B) Radially symmetrical larva from egg in which PVC was removed. The larva has no anterior-posterior axis. It consistd of an outer epidermal layer, a central notochordal mass, and an intermediately placed endodermal layer. (C) Vegetal view of normal 76-cell embryo. (D) Vegetal view of radially symmetrical embryo whose PVC was removed. (Photographs from H. Nishida 1994; courtesy of H. Nishida.)

Gastrulation in Tunicates

Tunicate gastrulation is characterized by the invagination of the endoderm, the involution of the mesoderm, and the epiboly of the ectoderm. About 4–5 hours after fertilization, the vegetal (endoderm) cells assume a wedge shape, expanding their apical margins and contracting near their vegetal margins (Figure 8.41). The A8.1 and B8.1 pairs of blastomeres appear to lead this invagination into the center of the embryo. This invagination forms a blastopore whose lips will become the mesodermal cells. The presumptive notochord cells are now on the anterior portion of the blastopore lip, while the presumptive tail muscle cells (from the yellow crescent) are on the posterior lip of the blastopore. The lateral lips of the blastopore comprise those cells that will become mesenchyme.

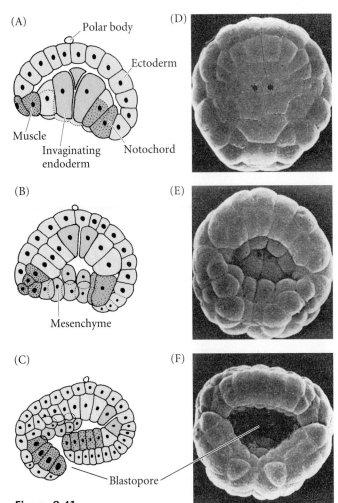

Figure 8.41
Gastrulation in the tunicate. Cross-sections (A–C) and scanning electron micrographs viewed from the vegetal pole (D–F) illustrate the invagination of the endoderm (E, D), the involution of the mesoderm (B, E), and the epiboly of the ectoderm (C, F). Cell fate is colored as in Figure 8.37. (From Satoh 1978; Jeffery and Swalla 1998; photographs courtesy of N. Satoh.)

The second step of gastrulation involves the involution of the mesoderm. The presumptive mesoderm cells involute over the lips of the blastopore, and by migrating upon the basal surfaces of the ectodermal cells, move inside the embryo. The ectodermal cells then flatten and epibolize over the mesoderm and endoderm, eventually covering the embryo. After gastrulation is completed, the embryo elongates along its anterior-posterior axis. The dorsal ectodermal cells that are the precursors of the neural tube invaginate into the embryo and are enclosed by neural folds. This forms the neural tube, which will form a brain anteriorly and a spinal chord posteriorly. Meanwhile, the presumptive notochord cells on the right and left sides of the embryo migrate to the midline and interdigitate to form the notochord, a single row of 40 cells. The muscle cells of the tail differentiate on either side of the neural tube and notochord (Jeffery and Swalla 1997). Thus, a tadpole is formed and will now search for a substrate on which to settle and metamorphose into an adult tunicate.

Early Development of the Nematode *Caenorhabditis elegans*

Why *C. elegans*?

Our ability to analyze development requires appropriate organisms. Sea urchins have long been a favorite organism of embryologists because their gametes are readily obtainable in large numbers, their eggs and embryos are transparent, and fertilization and development can occur under laboratory conditions. But sea urchins are difficult to rear in the laboratory for more than one generation, making their genetics difficult to study. Geneticists, on the other hand (at least those working with multicellular eukaryotes), favor *Drosophila*. Its rapid life cycle, its readiness to breed, and the polytene chromosomes of the larva (which allow gene localization) make the fruit fly superbly suited for hereditary analysis. But *Drosophila* development is complex and difficult to study.

A research program spearheaded by Sydney Brenner (1974) was set up to identify an organism wherein it might be possible to identify each gene involved in development as well as to trace the lineage of every single cell. The researchers settled upon *Caenorhabditis elegans*, a small (1 mm long), free-living soil nematode (Figure 8.42A). It has a rapid period of embryogenesis (about 16 hours), which it can accomplish in a petri dish, and relatively few cell types. Moreover, its predominant adult form is hermaphroditic, with each individual producing both eggs and sperm. These roundworms can reproduce either by self-fertilization or by cross-fertilization with the infrequently occurring males. The body of an adult *C. elegans* hermaphrodite contains exactly 959 somatic cells, whose entire lineage has been traced through its transparent cuticle (Figure 8.42B; Sulston and Horvitz 1977; Kimble and Hirsch 1979; Sulston et al. 1983). Furthermore, unlike verte-

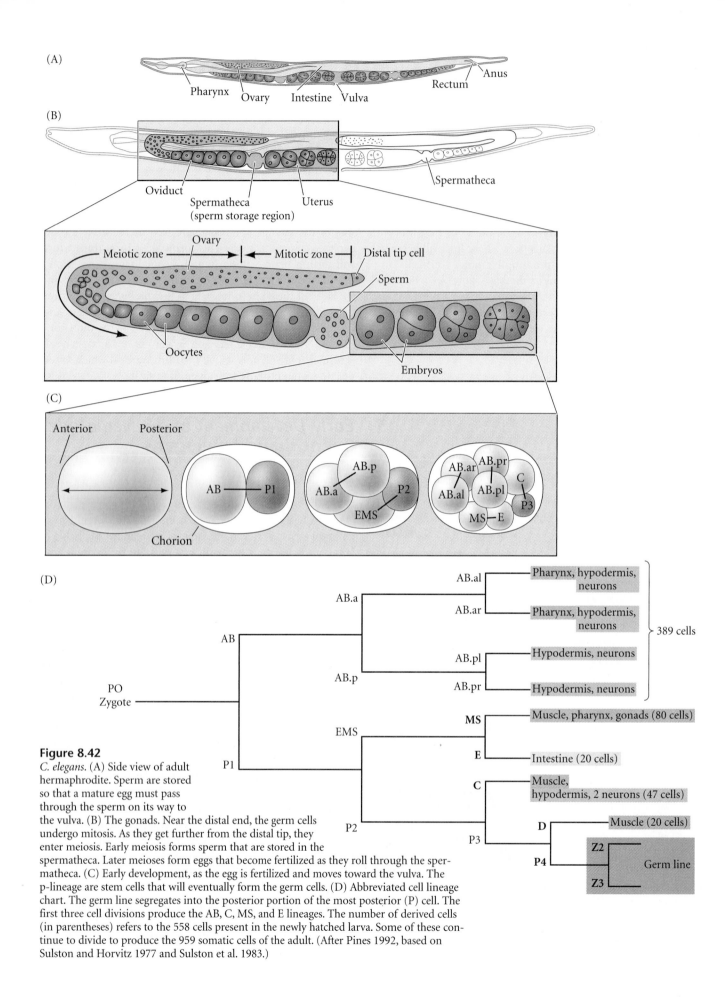

Figure 8.42

C. elegans. (A) Side view of adult hermaphrodite. Sperm are stored so that a mature egg must pass through the sperm on its way to the vulva. (B) The gonads. Near the distal end, the germ cells undergo mitosis. As they get further from the distal tip, they enter meiosis. Early meiosis forms sperm that are stored in the spermatheca. Later meioses form eggs that become fertilized as they roll through the sper-matheca. (C) Early development, as the egg is fertilized and moves toward the vulva. The p-lineage are stem cells that will eventually form the germ cells. (D) Abbreviated cell lineage chart. The germ line segregates into the posterior portion of the most posterior (P) cell. The first three cell divisions produce the AB, C, MS, and E lineages. The number of derived cells (in parentheses) refers to the 558 cells present in the newly hatched larva. Some of these continue to divide to produce the 959 somatic cells of the adult. (After Pines 1992, based on Sulston and Horvitz 1977 and Sulston et al. 1983.)

brate cell lineages, the cell lineage of *C. elegans* is almost entirely invariant from one individual to the next. There is little room for randomness (Sulston et al. 1983). *C. elegans* also has a small number of genes for a multicellular organism—about 19,000—and its genome has been entirely sequenced, the first ever for a multicellular organism (*C. elegans* Sequencing Consortium 1999).

Cleavage and Axis Formation in C. elegans

Rotational cleavage of the C. elegans egg

The *C. elegans* zygote exhibits **rotational holoblastic cleavage** (Figure 8.42C). During early cleavage, each asymmetrical division produces one **founder cell** (denoted AB, MS, E, C, and D), which produces differentiated descendants, and one **stem cell** (the P1–P4 lineage). In the first cell division, the cleavage furrow is located asymmetrically along the anterior-posterior axis of the egg, closer to what will be the posterior pole. It forms a founder cell (AB) and a stem cell (P1). During the second division, the anterior founder cell (AB) divides equatorially (longitudinally; 90° to the anterior-posterior axis), while the P1 cell divides meridionally (transversely) to produce another founder cell (EMS) and a posterior stem cell (P2). The stem cell lineage always undergoes meridional division to produce (1) an anterior founder cell and (2) a posterior cell that will continue the stem cell lineage.

The descendants of each founder cell divide at specific times in ways that are nearly identical from individual to individual. In this way, the exactly 558 cells of the newly hatched larva are generated. The descendants of the founder cells can be observed through the transparent cuticle and are named according to their positions relative to their sister cells. For instance, ABal is the "left-hand" daughter cell of the Aba cell, and ABa is the "anterior" daughter cell of the AB cell.

Anterior-posterior axis formation

The elongated axis of the *C. elegans* egg defines the future anterior-posterior axis of the nematode's body. The decision as to which end will become the anterior and which the posterior seems to reside with the position of sperm pronucleus. When it enters the oocyte cytoplasm, the centriole associated with the sperm pronucleus initiates cytoplasmic movements that push the male pronucleus to the nearest end of the oblong oocyte. This end becomes the posterior pole (Goldstein and Hird 1996).

A second anterior-posterior asymmetry seen shortly after fertilization is the migration of the **P-granules**. P-granules are ribonucleoprotein complexes that probably function in specifying the germ cells. Using fluorescent antibodies to a component of the P-granules, Strome and Wood (1983) discovered that shortly after fertilization, the randomly scattered P-granules move toward the posterior end of the zygote, so that they enter only the blastomere (P1) formed from the posterior cytoplasm (Figure 8.43). The P-granules of the P1 cell remain in the posterior of the P1 cell and are thereby passed to the P2 cell when P1 divides. During the division of P2 and P3, however, the P-granules become associated with the nucleus that enters the P3 cytoplasm. Eventually, the P-granules will reside in the P4 cell, whose progeny become the sperm and eggs of the adult. The localization of the P-granules requires microfilaments, but can occur in the absence of microtubules. Treating the zygote with cytochalasin D (a microfilament inhibitor) prevents the segregation of these granules to the posterior of the cell, whereas demecolcine (a colchicine-like microtubule inhibitor) fails to stop this movement (Strome and Wood 1983). The partitioning of the P-granules and the orientation of the mitotic spindles are both deficient in those embryos whose mothers were deficient in any of the *par* (partition-defective) genes. The proteins encoded by these genes are found in the cortex of the embryo and appear to interact with the actin cytoskeleton (Kemphues et al. 1988; Kirby et al. 1990; Bowerman 1999).

WEBSITE **8.6 P-granule migration.** Movies of P-granule migration under natural and experimental conditions were taken in the laboratory of Susan Strome. They show P-granule segregation to the P-lineage blastomeres except when perturbed by mutations or chemicals that inhibit microfilament function.

WEBSITE **8.7 The PAR proteins.** The polarity of cell divisions and the distribution of the morphogenetic determinants are specified by the position of the PAR proteins. These proteins are critical in coordinating cell division and cytoplasmic localization.

Formation of the dorsal-ventral and right-left axes

The dorsal-ventral axis of the nematode is seen in the division of the AB cell. As the AB cell divides, it becomes longer than the eggshell is wide. This causes the cells to slide, resulting in one AB daughter cell being anterior and one being posterior (hence their names, ABa and ABp, respectively). This squeezing also causes the ABp cell to take a position above the EMS cell that results from the division of the P1 blastomere. The ABp cell defines the future dorsal side of the embryo, while the EMS cell, the precursor of the muscle and gut cells, marks to future ventral surface of the embryo. The left-right axis is specified later, at the 12-cell stage, when the MS blastomere (from the division of the EMS cell) contacts half the "granddaughters" of the ABa cell, distinguishing the right side of the body from the left side (Evans et al. 1994).

Control of blastomere identity

C. elegans demonstrates both the conditional and autonomous modes of cell specification. Both modes can be seen if the first two blastomeres are experimentally separated (Priess and Thomson 1987). The P1 cell develops autonomously without the presence of AB. It makes all the cells that it would

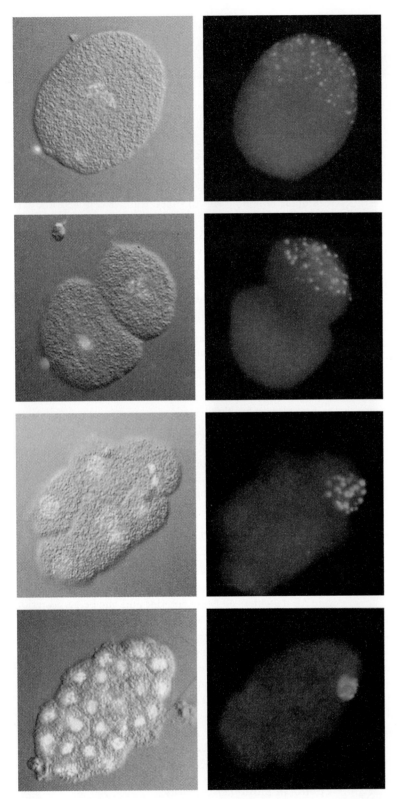

Figure 8.43
Segregation of the P-granules into the germ line lineage of the *C. elegans* embryo. The left column shows the cell nuclei (the DNA is stained blue by Hoescht dye), while the right column shows the same embryos stained for P-granules. At each successive division, the P-granules enter into the P-lineage blastomere, the one that will form the germ cells. (Photographs courtesy of S. Strome.)

normally make, and the result is the posterior half of an embryo. However, the AB cell, in isolation, makes only a fraction of the cell types that it would normally make. For instance, the resulting ABa blastomere fails to make the anterior pharyngeal muscles that it would have made in an intact embryo. Therefore, the specification of the AB blastomere is conditional, and it needs the descendants of the P1 cell to interact with it.

Autonomous specification

The determination of the P1 lineages appears to be autonomous, with the cell fates determined by internal cytoplasmic factors rather than by interactions with neighboring cells. It is thought that protein factors might determine cell fate by entering the nuclei of the appropriate blastomeres and activating or repressing specific fate-determining genes. Have any transcription factors been found in autonomously determined cell lineages? While the P-granules of *C. elegans* are localized in a way consistent with a role as a morphogenetic determinant, they do not enter the nucleus, and their role in development is still unknown. However, the SKN-1, PAL-1, and PIE-1 proteins are thought to encode transcription factors that act intrinsically to determine the fates of cells derived from the four P1-derived somatic founder cells, MS, E, C, and D.

The **SKN-1** protein is a maternally expressed polypeptide that may control the fate of the EMS blastomere, the cell that generates the posterior pharynx. After first cleavage, only the posterior blastomere, P1, has the ability to autonomously produce pharyngeal cells when isolated. After P1 divides, only EMS is able to generate pharyngeal muscle cells in isolation (Priess and Thomson 1987). Similarly, when the EMS cell divides, only one of its progeny, MS, has the intrinsic ability to generate pharyngeal tissue. These findings suggest that pharyngeal cell fate may be determined autonomously by maternal factors residing in the cytoplasm that are parceled out to these particular cells. Bowerman and his co-workers (1992a,b, 1993) found maternal effect mutants lacking pharyngeal cells, and isolated a mutation in the *skn-1* gene. Embryos from homozygous *skn-1*-deficient mothers lack both pharyngeal mesoderm and endoderm derivatives of EMS (Figure 8.44). Instead of making the normal intestinal and pharyngeal structures, these embryos seem to make extra hypodermal (skin) and body wall tissue where their intestine and pharynx should be. In other words, EMS appears to be respecified as C. Only those

Wild-type   *skn-1* mutant

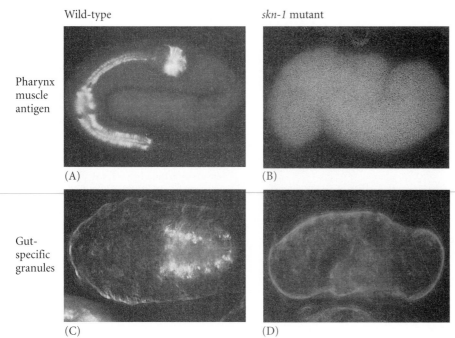

Pharynx
muscle
antigen

(A)   (B)

Gut-
specific
granules

(C)   (D)

Figure 8.44
Deficiencies of intestine and pharynx in *skn-1* mutants of *C. elegans*. Embryos derived from wild-type females (A, C) and females homozygous for mutant *skn-1* (B, D) were tested for the presence of pharyngeal muscles (A, B) and gut-specific granules (C, D). A pharyngeal muscle-specific antibody labels the pharynx musculature of those embryos derived from wild-type females (A), but does not bind to any structure in the embryos from *skn-1* mutant females (B). Similarly, the birefringent gut granules characteristic of embryonic intestines (C) are absent from embryos derived from the *skn-1* mutant females (D). (From Bowerman et al. 1992a; photographs courtesy of B. Bowerman.)

cells that are destined to form pharynx or intestine are affected by this mutation. Moreover, the protein encoded by the *skn-1* gene has a DNA-binding site motif similar to that seen in the bZip family of transcription factors (Blackwell et al. 1994).

A second possible transcription factor, **PAL-1**, is also required for the differentiation of the P1 lineage. PAL-1 activity is needed for the normal development of the somatic descendants of the P2 blastomere. Thus, embryos lacking PAL-1 have no somatic cell types derived from the C and D stem cells (Hunter and Kenyon 1996). PAL-1 is regulated by the MEX-3 protein, an RNA-binding protein that appears to inhibit the translation of the *pal-1* mRNA. Wherever MEX-3 is expressed, PAL-1 is absent. Thus, in *mex-3*-deficient mutants, PAL-1 is seen in every blastomere. SKN-1 also inhibits PAL-1 (thereby preventing it from becoming active in the EMS cell).

A third putative transcription factor, **PIE-1**, is necessary for germ line fate. PIE-1 appears to inhibit both SKN-1 and PAL-1 function in the P2 and subsequent germ line cells (Hunter and Kenyon 1996). Mutations of the maternal *pie-1* gene result in germ line blastomeres adopting somatic fates, with the P2 cell behaving similarly to a wild-type EMS blastomere. The localization and the genetic properties of PIE-1 suggest that it represses the establishment of somatic cell fate and preserves the totipotency of the germ cell lineage (Mello et al. 1996; Seydoux et al. 1996).

WEBSITE **8.8 Mechanisms of cytoplasmic localization in *C. elegans*.** Analyses of *C. elegans* mutations have isolated genes whose protein products are essential in specifying cell fate. In some instances, these proteins are involved in the placement of the morphogenetic determinants.

Conditional specification

As we saw above, the *C. elegans* embryo uses both autonomous and conditional modes of specification. Conditional specification can be seen in the development of the endoderm cell lineage. At the 4-cell stage, the EMS cell requires a signal from its neighbor (and sister), the P2 blastomere. Usually, the EMS cell divides into an MS cell (which produces mesodermal muscles) and an E cell (which produces the intestinal endoderm). If the P2 cell is removed at the early 4-cell stage, the EMS cell will divide into two MS cells, and endoderm will not be produced. If the EMS cell is recombined with the P2 blastomere, however, it will form endoderm; it will not do so, however, when combined with ABa, ABp, or both AB derivatives (Figure 8.45; Goldstein 1992).

The P2 cell produces a signal that interacts with the EMS cell and instructs the EMS daughter that is next to it to become the E cell. This message is transmitted through the Wnt signaling cascade (Figure 8.46; Rocheleau et al. 1997; Thorpe et al. 1997). The P2 cell produces the *C. elegans* homologue of a Wnt protein, the MOM-2 peptide. The MOM-2 peptide is received in the EMS cell by the MOM-5 protein, the *C. elegans* version of the Wnt receptor protein, Frizzled. The result of this signaling cascade is to down-regulate the expression of the *pop-1* gene in the EMS daughter destined to become the E cell. In *pop-1*-deficient embryos, both EMS daughter cells become E cells (Lin et al. 1995).

The P2 cell is also critical in giving the signal that distinguishes ABp from its sister, ABa (Figure 8.48). ABa gives rise to neurons, hypodermis, and the anterior pharynx cells, while ABp makes only neurons and hypodermal cells. However, if one experimentally reverses their positions, their fates are similarly reversed, and a normal embryo is formed. In other

(A)

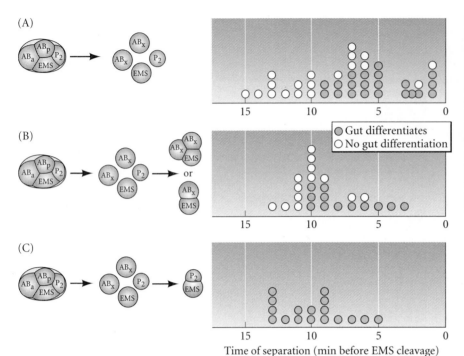

(B)

(C)

● Gut differentiates
○ No gut differentiation

Time of separation (min before EMS cleavage)

Figure 8.45
Results of isolation and recombination experiments, showing that cellular interactions are required for the EMS cell to form intestinal lineage determinants. (A) When isolated shortly after its formation, the EMS blastomere cannot produce gut-specific granules. If left in place for longer periods, it can. (B) If the EMS cell is recombined with either or both derivatives of the AB blastomere, it will not form gut-specific granules. (C) If recombined with the P2 blastomere, the EMS cell will give rise to gut-specific structures. (After Goldstein 1992.)

words, ABa and ABp are equivalent cells whose fate is determined by their positions within the embryo (Priess and Thomson 1987). Transplantation and genetic studies have shown that ABp becomes different from ABa through its inter-

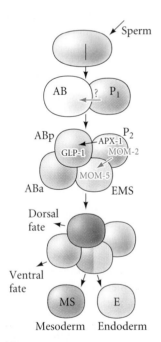

Figure 8.46
Cell-cell signaling in the 4-cell embryo of *C. elegans*. The P2 cell produces two signals: (1) the juxtacrine protein APX-1 (Delta), which is bound by GLP-1 (Notch) on the ABp cell, and (2) the paracrine protein MOM-2 (Wnt), which is bound by the MOM-5 (Frizzled) protein on the EMS cell. (After Han 1998.)

action with the P2 cell. In an unperturbed embryo, both ABa and ABp contact the EMS blastomere, but only ABp contacts the P2 cell. If the P2 cell is killed at the early 4-cell stage, the ABp cell does not generate its normal complement of cells (Bowerman et al. 1992a,b). Contact between ABp and P2 is essential for the specification of ABp cell fates, and the ABa cell can be made into an ABp-type cell if it is forced into contact with P2 (Hutter and Schnabel 1994; Mello et al. 1994).

Moreover, these studies show that this interaction is mediated by the GLP-1 protein on the ABp cell and the APX-1 (anterior pharynx excess) protein on the P2 blastomere. In embryos whose mothers have mutant *glp-1*, ABp is transformed into an ABa cell (Hutter and Schnabel 1994; Mello et al. 1994). The GLP-1 protein is a member of a widely conserved family called the Notch proteins, which serve as cell membrane receptors in many cell-cell interactions, and it is seen on both the ABa and ABp cells (Evans et al. 1994).* As mentioned in Chapter 5, one of the most important ligands for Notch proteins such as GLP-1 is another cell surface protein called Delta. In *C. elegans*, the Delta-like protein is APX-1, and it is found on the P2 cell (Mango et al. 1994; Mello et al. 1994). This APX-1 signal breaks the symmetry between ABa and ABp, since it stimulates the GLP-1 protein solely on the AB descendant that it touches, namely, the ABp blas-

*The GLP-1 protein is localized in the ABa and ABp blastomeres, but the maternally encoded *glp-1* mRNA is found throughout the embryo. Evans and colleagues (1994) have postulated that there might be some translational determinant in the AB blastomere that enables the *glp-1* message to be translated in its descendants. The *glp-1* gene is also active in regulating postembryonic cell-cell interactions. It is used later by the distal tip cell of the gonad to control the number of germ cells entering meiosis; hence the name GLP, "germ line proliferation."

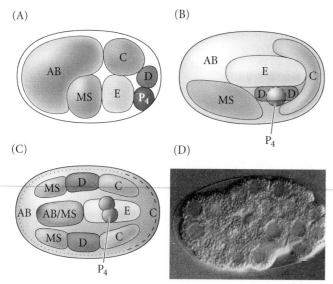

Figure 8.47
Gastrulation in *C. elegans*. (A) Positions of founder cells and their descendants at the 26-cell stage, just prior to gastrulation. (B) 102-cell stage, after the migration of the E, P4, and D descendants. (C) Positions of the cells near the end of gastrulation. The dotted and dashed lines represent regions of the hypodermis contributed by AB and C, respectively. (D) Early gastrulation, as the two E cells start moving inward. (After Schierenberg 1998; photograph courtesy of E. Schierenberg.)

tomere. In doing this, the P2 cell initiates the dorsal-ventral axis of *C. elegans*, and it confers on the ABp blastomere a fate different from that of its sister cell.

Gastrulation in *C. elegans*

Gastrulation in *C. elegans* starts extremely early, just after the generation of the P4 cell in the 24-cell embryo (Skiba and

Schierenberg 1992). At this time, the two daughters of the E cell (Ea and Ep) migrate from the ventral side into the center of the embryo. There, they will divide to form a gut consisting of 20 cells. There is a very small and transient blastocoel prior to the movement of the Ea and Ep cells, and their inward migration creates a tiny blastopore. The next cell to migrate through this blastopore is the P4 cell, the precursor of the germ cells. It migrates to a position beneath the gut primordium. The mesodermal cells move in next: the descendants of the MS cell migrate inward from the anterior side of the blastopore, and the C- and D-derived muscle precursors enter from the posterior side. These cells flank the gut tube on the left and right sides (Figure 8.47; Schierenberg 1997). Finally, about 6 hours after fertilization, the AB-derived cells that contribute to the pharynx are brought inside, while the **hypoblast** (hypodermal precursor) cells move ventrally by epiboly, eventually closing the blastopore. During the next 6 hours, the cells move and organize into organs, and the ball-shaped embryo stretches out to become a worm (see Priess and Hirsch 1986; Schierenberg 1997). This hermaphroditic worm will have 558 somatic cells. An additional 115 cells will have formed, but undergone apoptosis (see Chapter 6). After four molts, this worm will be a sexually mature adult, containing 959 somatic cells, as well as numerous sperm and eggs.

Coda

This chapter has described early embryonic development in four invertebrate species, each of which develops in a different pattern. The largest group of animals on this planet, however, is another invertebrate group—the insects. We probably know more about the development of one particular insect, *Drosophila melanogaster*, than any other organism. The next chapter details the early development of this particularly well-studied creature.

Snapshot Summary: Early Invertebrate Development

1. During cleavage, most cells do not grow. Rather, the volume of the oocyte is cleaved into numerous cells. The major exceptions to this rule are mammals.

2. The blastomere cell cycle is governed by the synthesis and degradation of cyclin. Cyclin synthesis promotes the formation of MPF, and MPF promotes mitosis. Degradation of cyclin brings the cell back to the S phase. The G phases are added at the midblastula transition.

3. "Blast" vocabulary: A blastomere is a cell derived from cleavage in an early embryo. A blastula is an embryonic structure composed of blastomeres. The cavity in the blastula is the blastocoel. If the blastula lacks a blastocoel, it is a stereo-

blastula. A mammalian blastula is called a blastocyst (in Chapter 11), and the invagination where gastrulation begins is the blastopore.

4. The movements of gastrulation include invagination, involution, ingression, delamination, and epiboly.

5. Three axes are the foundations of the body: the anterior-posterior axis (head to tail or mouth to anus), the dorsal-ventral axis (back to belly), and the right-left axis (between the two lateral sides of the body).

6. In all four invertebrates described here, cleavage is holoblastic. In the sea urchin, cleavage is radial; in the snail, spiral; in the tunicate, bilateral; and in the nematode, rotational.

7. In the tunicate, snail, and nematode, gastrulation occurs when there are relatively few cells, and the blastopore becomes the mouth. This is the protostome mode of gastrulation.

8. Body axes in these species are established in different ways. In some, such as the sea urchin and tunicate, the axes are established at fertilization through determinants in the egg cytoplasm. In other species, such as the nematode and snail, the axes are established by cell interactions later in development.

9. In the sea urchin, gastrulation occurs only after thousands of cells have formed, and the blastopore becomes the anus. This is the deuterostome mode of gastrulation, and is characteristic only of echinoderms and chordates.

10. In sea urchins, cell fates are determined by signaling. The micromeres constitute a major signaling center. β-catenin is important for the inducing capacity of the micromeres.

11. Differential cell adhesion is important in regulating sea urchin gastrulation. The micromeres delaminate first from the vegetal plate. They form the primary mesenchyme which becomes the skeletal rods of the pluteus larva. The vegetal plate invaginates to form the endodermal archenteron, with a tip of secondary mesenchyme cells. The archenteron elongates by convergent extension and is guided to the future mouth region by the secondary mesenchyme.

12. Snails exhibit spiral cleavage and form stereoblastulae, having no blastocoels. The direction of the spiral cleavage is regulated by a factor encoded by the mother and placed into the oocyte. Spiral cleavage can be modified by evolution, and adaptations of spiral cleavage have allowed some molluscs to survive in otherwise harsh conditions.

13. The polar lobe of certain molluscs contains the determinants for mesoderm and endoderm. These will enter the D blastomere.

14. The tunicate fate map is identical on its right and left sides. The yellow cytoplasm contains muscle-forming determinants; these act autonomously. The nervous system of tunicates is formed conditionally, by interactions between blastomeres.

15. The soil nematode *Caenorhabditis elegans* was chosen as a model organism because it has a small number of cells, a small genome, is easily bred and maintained, has a short lifespan, can be genetically manipulated, and has a cuticle through which one can see cell movements.

16. In the early divisions of the *C. elegans* zygote, one daughter cell becomes a founder cell (producing differentiated descendants) and the other becomes a stem cell (producing other founder cells and the germ line).

17. Blastomere identity in *C. elegans* is regulated by both autonomous and conditional specification.

Literature Cited

Atkinson, J. W. 1987. An atlas of light micrographs of normal and lobeless larvae of the marine gastropod *Ilyanassa obsoleta*. *Int. J. Invert. Reprod. Dev.* 9: 169–178.

Balinsky, B. I. 1981. *Introduction to Embryology*, 5th Ed. Saunders, Philadelphia.

Bates, W. R. and W. R. Jeffery. 1988. Localization of axial determinants in the vegetal pole region of ascidian eggs. *Dev. Biol.* 124: 65–76.

Berg, L. K., S. W. Chen and G. M. Wessel. 1996. An extracellular matrix molecule that is selectively expressed during development is important for gastrulation in the sea urchin embryo. *Development* 122: 703–713.

Bisgrove, B. W., M. E. Andrews and R. A. Raff. 1991. Fibropellins, products of an EGF repeat-containing gene, form a unique extracellular matrix structure that surrounds the sea urchin embryo. *Dev. Biol.* 146: 89–99.

Blackwell. T. K., B. Bowerman, J. R. Priess and H. Weintraub. 1994. Formation of a monomeric DNA binding domain by SKN bZIP and homeodomain elements. *Science* 266: 621–628.

Bonder, E. M., D. J. Fishkind, J. H. Henson, N. M. Cotran and D. A. Begg. 1988. Actin in cytokinesis: Formation of the contradtile apparatus. *Zool. Sci.* 5: 699–701.

Boveri, T. 1901. Die Polaritat von Ovocyte, Ei, und Larve des *Strongylocentrotus lividus*. *Zool. Jahrb. Abt. Anat. Ontog. Tiere* 14: 630–653.

Bowerman, B. 1999. Maternal control of polarity and patterning during embryogenesis in the nematode *Caenorhabditis elegans*. *In* S. A. Moody (ed.), *Cell Lineage and Determination*. Academic Press, New York, pp. 97–118.

Bowerman, B., B. A. Eaton and J. R. Priess. 1992a. *skn-1*, a maternally expressed gene required to specify the fate of ventral blastomeres in the early *C. elegans* embryo. *Cell* 68: 1061–1075.

Bowerman, B., F. E. Tax, J. H. Thomas and J. R. Priess. 1992b. Cell interactions involved in the development of the bilaterally symmetrical intestinal valve cells during embryogenesis of *Caenorhabditis elegans*. *Development* 116: 1113–1122.

Bowerman, B., B. W. Draper, C. C. Mello and J. R. Priess. 1993. The maternal gene *skn-1* encodes a protein that is distributed unequally in early *C. elegans* embryos. *Cell* 74: 443–452.

Boycott, A. E., C. Diver, S. L. Garstang and F. M. Turner. 1930. The inheritance of sinistrality in *Limnaea peregra* (Mollusca: Pulmonata). *Phil. Trans. R. Soc. Lond.* [B] 219: 51–131.

Brandhorst, B. P. and K. M. Newrock. 1981. Post-transcriptional regulation of protein synthesis in *Ilyanassa* embryos and isolated polar lobes. *Dev. Biol.* 83: 250–254.

Brenner, S. 1974. The genetics of *Caenorhabditis elegans*. *Genetics* 77: 71–94.

Burke, R. D., R. L. Myers, T. L. Sexton and C. Jackson. 1991 Cell movements during the initial phase of gastrulation in the sea urchin embryo. *Dev. Biol.* 146: 542–558.

C. elegans Sequencing Consortium. 1999. Genome sequence of the nematode *C. elegans*: A platform for investigating biology. *Science* 282: 2012–2018. www.sanger.ac.uk; genome.wustl.edu/gsc/gschmpg.html

Chabry, L. M. 1888. Contribution a l'embryologie normale tératologique des ascidies simples. *J. Anat. Physiol. Norm. Pathol.* 23: 167–321.

Cherr, G. N., R. G. Summers, J. D. Baldwin and J. B. Morrill. 1992. Preservation and visualization of the sea urchin blastocoelic extracellular matrix. *Microsc. Res. Tech.* 22: 11–22.

Clement, A. C. 1962. Development of *Ilyanassa* following removal of the D micromere at successive cleavage stages. *J. Exp. Zool.* 149: 193–215.

Clement, A. C. 1968. Development of the vegetal half of the *Ilyanassa* egg after removal of most of the yolk by centrifugal force, compared with the development of animal halves of similar visible composition. *Dev. Biol.* 17: 165–186.

Clement, A. C. 1986. The embryonic value of the micromeres in *Ilyanassa obsoleta*, as determined by deletion experiments. III. The third quartet cells and the mesentoblast cell, 4d. *Int. J. Invert. Reprod. Dev.* 9: 155–168.

Collier, J. R. 1983. The biochemistry of molluscan development. *In* N. H. Verdonk, J. A. M. van den Biggelaar and A. S. Tompa (eds.), *The Mollusca*, vol. 3. Academic Press, New York, pp. 215–252.

Collier, J. R. 1984. Protein synthesis in normal and lobeless gastrulae of *Ilyanassa obsoleta*. *Biol. Bull.* 167: 371–378.

Collier, J. R. 1998. Gastropods, the snails. *In* S. F. Gilbert and A. Raunio (eds.), *Embryology: Constructing the Organism*. Sinauer Associates, Sunderland, MA, pp. 189–218.

Conklin, E. G. 1897. The embryology of *Crepidula*. *J. Morphol.* 13: 3–209.

Conklin, E. G. 1905. The orientation and cell-lineage of the ascidian egg. *J. Acad. Nat. Sci. Phila.* 13: 5–119.

Craig, M. M. and J. B. Morrill. 1986. Cellular arrangements and surface topography during early development in embryos of *Ilyanassa obsoleta*. *Int. J. Invert. Reprod. Dev.* 9: 209–228.

Crampton, H. E. 1894. Reversal of cleavage in a sinistral gastropod. *Ann. NY Acad. Sci.* 8: 167–170.

Dan, K. 1960. Cytoembryology of echinoderms and amphibia. *Int. Rev. Cytol.* 9: 321–368.

Dan, K. and K. Okazaki. 1956. Cyto-embryological studies of sea urchins. III. Role of secondary mesenchyme cells in the formation of the primitive gut in sea urchin larvae. *Biol. Bull.* 110: 29–42.

Danilchik, M. V., W. C. Funk, E. E. Brown and K. Larkin. 1998. Requirement for microtubules in new membrane formation during cytokinesis of *Xenopus* embryos. *Dev. Biol.* 194: 47–60.

Dan-Sohkawa, M. and H. Fujisawa. 1980. Cell dynamics of the blastulation process in the starfish, *Asterina pectinifera*. *Dev. Biol.* 77: 328–339.

Davidson, E. H. 1989. Lineage-specific gene expression and the regulative capacities of the sea urchin embryo: A proposed mechanism. *Development* 105: 421–445.

Edgar, B., C. P. Kiehle and G. Schubiger. 1986. Cell cycle control by the nucleo-cytoplasmic ratio in early *Drosophila* development. *Cell* 44: 365–372.

Elinson, R. P. 1987. Change in developmental patterns: Embryos of amphibians with large eggs. *In* R. A. Raff and E. C. Raff (eds.), *Development as an Evolutionary Process*. Alan R. Liss, New York, pp. 1–21.

Ettensohn, C. A. 1985. Gastrulation in the sea urchin embryo is accompanied by the rearrangement of invaginating epithelial cells. *Dev. Biol.* 112: 383–390.

Ettensohn, C. A. 1990. The regulation of primary mesenchyme cell patterning. *Dev. Biol.* 140: 261–271.

Ettensohn, C. A. and E. P. Ingersoll. 1992. Morphogenesis of the sea urchin embryo. *In* E. F. Rossomondo and S. Alexander (eds.), *Morphogenesis*. Marcel Dekker, New York, pp. 189–262.

Ettensohn, C. A. and D. R. McClay. 1986. The regulation of primary mesenchyme cell migration in the sea urchin embryo: Transplantations of cells and latex beads. *Dev. Biol.* 117: 380–391.

Evans, T., E. Rosenthal, J. Youngblom, D. Distel and T. Hunt. 1983. Cyclin: A protein specified by maternal mRNA in sea urchin eggs that is destroyed at each cleavage division. *Cell* 33: 389–396.

Evans, T. C., S. L. Crittenden, V. Kodoyianni and J. Kimble. 1994. Translational control of maternal glp-1 mRNA establishes an asymmetry in the *C. elegans* embryo. *Cell* 77: 183–194.

Fink, R. D. and D. R. McClay. 1985. Three cell recognition changes accompany the ingression of sea urchin primary mesenchyme cells. *Dev. Biol.* 107: 66–74.

Freeman, G. and J. W. Lundelius. 1982. The developmental genetics of dextrality and sinistrality in the gastropod *Limnea peregra*. *Wilhelm Roux Arch. Dev. Biol.* 191: 69–83.

Galileo, D. S. and J. B. Morrill. 1985. Patterns of cells and extracellular material of the sea urchin *Lytechinus variegatus* (Echinodermata; Echinoidea) embryo, from hatched blastula to late gastrula. *J. Morphol.* 185: 387–402.

Gerhart, J. C., M. Wu and M. Kirschner. 1984. Cell dynamics of an M-phase-specific cytoplasmic factor in *Xenopus laevis* oocytes and eggs. *J. Cell Biol.* 98: 1247–1255.

Goldstein, B. 1992. Induction of gut in *Caenorhabditis elegans* embryos. *Nature* 357: 255–258.

Goldstein, B. and S. N. Hird. 1996. Specification of the anterioposterior axis in *Caenorhabditis elegans*. *Development* 122: 1467–1474.

Guerrier, P., J. A. M. van den Biggelaar, C. A. M. Dongen and N. H. Verdonk. 1978. Significance of the polar lobe for the determination of dorsoventral polarity in *Dentalium vulgare* (da Costa). *Dev. Biol.* 63: 233–242.

Gustafson, T. and L. Wolpert. 1967. Cellular movement and contact in sea urchin morphogenesis. *Biol. Rev.* 42: 442–498.

Hall, H. G. and V. D. Vacquier. 1982. The apical lamina of the sea urchin embryo: Major glycoprotein associated with the hyaline layer. *Dev. Biol.* 89: 168–178.

Han, M. 1998. Gut reaction to Wnt signaling in worms. *Cell* 90: 581–584.

Hardin, J. D. 1987. Archenteron elongation in the sea urchin embryo is a microtubule independent process. *Dev. Biol.* 121: 253–262.

Hardin, J. D. 1988. The role of secondary mesenchyme cells during sea urchin gastrulation studied by laser ablation. *Development* 103: 317–324.

Hardin, J. D. 1990. Context-dependent cell behaviors during gastrulation. *Semin. Dev. Biol.* 1: 335–345.

Hardin, J. D. and L. Y. Cheng. 1986. The mechanisms and mechanics of archenteron elongation during sea urchin gastrulation. *Dev. Biol.* 115: 490–501.

Hardin, J. D. and D. R. McClay. 1990. Target recognition by the archenteron during sea urchin gastrulation. *Dev. Biol.* 142: 87–105.

Harkey, M. A. and A. M. Whiteley. 1980. Isolation, culture and differentiation of echinoid primary mesenchyme cells. *Wilhelm Roux Arch. Dev. Biol.* 189: 111–122.

Henry, J. J. and M. Q. Martindale. 1987. The organizing role of the D quadrant as revealed through the phenomenon of twinning in the polychaete *Chaetopterus variopedatus*. *Wilhelm Roux Arch. Dev. Biol.* 196: 449–510.

Henry, J. J. and R. A. Raff. 1990. Evolutionary change in the process of dorsoventral axis determination in the direct developing sea urchin, *Heliocidaris erythrogramma*. *Dev. Biol.* 141: 55–69.

Henry, J. J., K. M. Klueg and R. A. Raff. 1992. Evolutionary dissociation between cleavage, cell lineage, and embryonic axes in sea urchin embryos. *Development* 114: 931–938.

Hodor, P. G. and C. A. Ettensohn. 1998. The dynamics and regulation of mesenchymal cell fusion in the sea urchin. *Dev. Biol.* 199: 111–124.

Hörstadius, S. 1939. The mechanics of sea urchin development, studied by operative methods. *Biol. Rev.* 14: 132–179.

Hörstadius, S. 1973. *Experimental Embryology of Echinoderms*. Clarendon Press, Oxford.

Hunter, C. P. and C. Kenyon. 1996. Spatial and temporal controls target pal-1 blastomere-specification activity to a single blastomere in *C. elegans* embryos. *Cell* 87: 217–226.

Hutter, H. and R. Schnabel. 1994. glp-1 and inductions establishing embryonic axes in *C. elegans*. *Development* 120: 2051–2064.

Inoué, S. 1982. The role of self-assembly in the generation of biologic form. *In* S. Subtelny and B. P. Green (eds.), *Developmental Order: Its Origin and Regulation*. Alan R. Liss, New York, pp. 35–76.

Jeffery, W. R. and B. J. Swalla. 1998. Tunicates. *In* S. F. Gilbert and A. M. Raunio, (eds.) *Embryology: Constructing the Organism*. Sinauer Associates, Sunderland, MA, pp. 331–364.

Karp, G. C. and M. Solursh. 1985. Dynamic activity of the filopodia of sea urchin embryonic cells and their role in directed migration of the primary mesenchyme in vitro. *Dev. Biol.* 112: 276–283.

Kemphues, K. J., J. R. Priess, D. G. Morton and N. Cheng. 1988. Identification of genes required for cytoplasmic localization in early *C. elegans* embryos. *Cell* 52: 311–320.

Kimble, J. and D. Hirsch. 1979. The postembryonic cell lineages of the hermaphrodite and male gonads in *Caenorhabditis elegans*. *Dev. Biol.* 70: 396–418.

Kirby, C. M., M. Kusch and K. Kemphues. 1990. Mutations in the *par* genes of *Caenorhabditis elegans* affect cytoplasmic reorganization during the first cell cycle. *Dev. Biol.* 142: 203–215.

Kominami, T. 1983. Establishment of embryonic axes in larvae of the starfish *Asterina pectinifera*. *J. Embryol. Exp. Morphol.* 75: 87–100.

Lane, M. C. and M. Solursh. 1991. Primary mesenchyme cell migration requires a chondroitin sulfate/dermatan sulfate proteoglycan. *Dev. Biol.* 143: 389–398.

Lane, M. C., M. A. R. Koehl, F. Wilt and R. Keller. 1993. A role for regulated secretion of apical matrix during epithelial invagination in the sea urchin. *Development* 117: 1049–1060.

Lepage, T., C. Sardet and C. Gache. 1992. Spatial expression of the hatching enzyme gene in the sea urchin embryo. *Dev. Biol.* 150: 23–32.

Lillie, F. R. 1898. Adaptation in cleavage. In *Biological Lectures of the Marine Biological Laboratory of Woods Hole*. Ginn, Boston, pp. 43–68.

Lillie, F. R. 1902. Differentiation without cleavage in the egg of the annelid *Chaetopterus pergamentaceous*. *Wilhelm Roux Arch. Entwicklungsmech. Org.* 14: 477–499.

Lin, R., S. Thompson and J. R. Priess. 1995. *pop-1* encodes an HMG box protein required for the specification of a mesoderm precursor in early *C. elegans* embryos. *Cell* 83: 599–609.

Logan, C. Y. and D. R. McClay. 1997. The allocation of early blastomeres to the ectoderm and endoderm is variable in the sea urchin embryo. *Development* 124: 2213–2223.

Logan, C. Y. and D. R. McClay. 1999. Lineages that give rise to endoderm and mesoderm in the sea urchin embryo. *In* S. A. Moody (ed.), *Cell Lineage and Determination*. Academic Press, New York, pp. 41–58.

Logan, C. Y., J. R. Miller, M. J. Ferkowicz and D. R. McClay. 1998. Nuclear β-catenin is required to specify vegetal cell fates in the sea urchin embryo. *Development* 126: 345–358.

Lutz, B. 1947. Trends towards non-aquatic and direct development in frogs. *Copeia* 4: 242–252.

Malinda, K. M., G.W. Fisher and C. A. Ettensohn. 1995. Four-dimensional microscopic analysis of the filopodial behavior of primary mesenchyme cells during gastrulation in the sea urchin embryo. *Dev. Biol.* 172: 552–566.

Mango, S. E., C. J. Thorpe, P. R. Martin, S. H. Chamberlain and B. Bowerman. 1994. Two maternal genes, *apx-1* and *pie-1*, are required to distinguish the fates of equivalent blastomeres in early *C. elegans* embryos. *Development* 120: 2305–2315.

Martins, G. G., R. G. Summers and J. B. Morrill. 1998. Cells are added to the archenteron during and following secondary invagination in the sea urchin *Lytechinus variegatus*. *Dev. Biol.* 198: 330–342.

Maruyama, Y. K., Y. Nakeseko and S. Yagi. 1985. Localization of cytoplasmic determinants responsible for primary mesenchyme formation and gastrulation in the unfertilized eggs of the sea urchin *Hemicentrotus pulcherrimus*. *J. Exp. Zool.* 236: 155–163.

Masuda, M. and H. Sato. 1984. Asynchronization of cell divisions is concurrently related with ciliogenesis in sea urchin blastulae. *Dev. Growth. Diff.* 26: 281–294.

Mello, C. C., B. W. Draper and J. R. Priess. 1994. The maternal genes *apx-1* and *glp-1* and establishment of dorsal-ventral polarity in the early *C. elegans* embryo. *Cell* 77: 95–106.

Mello, C. C., C. Schubert, B. Draper, W. Zhang, R. Lobel and J. R. Priess. 1996. The PIE-1 protein and germline specification in *C. elegans* embryos. *Nature* 382: 710–712.

Miller, J. R., S. E. Fraser and D. R. McClay. 1995. Dynamics of thin filopodia during sea urchin gastrulation. *Development* 121: 2505–2511.

Minshull, J., J. J. Blow and T. Hunt. 1989. Translation of cyclin mRNA is necessary for extracts of activated *Xenopus* eggs to enter mitosis. *Cell* 56:497–956.

Morgan, T. H. 1927. *Experimental Embryology*. Columbia University Press, New York.

Morrill, J. B. and L. L. Santos. 1985. A scanning electron micrographical overview of cellular and extracellular patterns during blastulation and gastrulation in the sea urchin, *Lytechinus variegatus*. *In* R. H. Sawyer and R. M. Showman (eds.), *The Cellular and Molecular Biology of Invertebrate Development*. University of South Carolina Press, Columbia, SC. pp. 3–33.

Newport, J. W. and M. W. Kirschner. 1982a. A major developmental transition in early *Xenopus* embryos: I. Characterization and timing of cellular changes at midblastula stage. *Cell* 30: 675–686.

Newport, J. W. and M. W. Kirschner. 1982b. A major developmental transition in early *Xenopus* embryos: II. Control of the onset of transcription. *Cell* 30: 687–696.

Newport, J. W. and M. W. Kirschner. 1984. Regulation of the cell cycle during *Xenopus laevis* development. *Cell* 37: 731–742.

Newrock, K. M. and R. A. Raff. 1975. Polar lobe specific regulation of translation in embryos of *Ilyanassa obsoleta*. *Dev. Biol.* 42: 242–261.

Nigg, E. A. 1995. Cyclin-dependent protein kinases: Key regulators of the eukaryotic cell cycle. *BioEssays* 17: 471–480.

Nishida, H. 1987. Cell lineage analysis in ascidian embryos by intracellular injection of a tracer enzyme. III. Up to the tissue restricted stage. *Dev. Biol.* 121: 526–541.

Nishida, H. 1992a. Determination of developmental fates of blastomeres in ascidian embryos. *Dev. Growth Diff.* 34: 253–262.

Nishida, H. 1992b. Regionality of egg cytoplasm that promotes muscle differentiation in embryo of the ascidian *Halocynthia roretzi*. *Development* 116: 521–529.

Nishida, H. 1994. Localization of determinants for formation of the anterior-posterior axis in eggs of the ascidian *Halocynthia roretzi*. *Development* 120: 3093–3104.

Ortolani, G. 1959. Richerche sulla induzione del sistema nervoso nelle larve delle Ascidie. *Boll. Zool.* 26: 341–348.

Pines, M. (ed.). 1992. *From Egg to Adult*. Howard Hughes Medical Institute, Bethesda, MD, pp. 30–38.

Priess, R. A. and D. I. Hirsch. 1986. *Caenorhabditis elegans* morphogenesis: The role of the cytoskeleton in elongating the embryo. *Dev. Biol.* 117: 156–173.

Priess, R. A. and J. N.Thomson. 1987. Cellular interactions in early *C. elegans* embryos. *Cell* 48: 241–250.

Raff, R. A. and T. C. Kaufman. 1983. *Embryos, Genes, and Evolution: The Developmental-Genetic Basis of Evolutionary Change*. Macmillan, New York.

Ransick, A. and E. H. Davidson. 1993. A complete second gut induced by transplanted micromeres in the sea urchin embryo. *Science* 259: 1134–1138.

Render, J. 1997. Cell fate maps in the *Ilyanassa obsoleta* embryo beyond the third division. *Dev. Biol.* 189: 301–310.

Reverberi, G. and A. Minganti. 1946. Fenomeni di evocazione nello sviluppo dell'uovo di Ascidie. Risultati dell'indagine spermentale sull'uovo di *Ascidiella aspersa* e di *Ascidia malaca* allo stadio di 8 blastomeri. *Pubbl. Staz. Zool. Napoli* 20: 199–252.

Rocheleau, C. E. and 8 others. 1998. Wnt signaling and an APC-related gene specify endoderm in early *C. elegans* embryos. *Cell* 90: 707–716.

Roegiers, F., A. McDougall and C. Sardet. 1995. The sperm entry point defines the orientation of the calcium-induced contraction wave that directs the first phase of cytoplasmic reorganization in the ascidian egg. *Development* 121: 3457–3466.

Ruffins, S. W. and C. A. Ettensohn. 1996. A fate map of the vegetal plate of the sea urchin (*Lytechinus variegatus*) mesenchyme blastula. *Development* 122: 253–263.

Satoh, N. 1978. Cellular morphology and architecture during early morphogenesis of the ascidian egg: An SEM study. *Biol. Bull.* 155: 608-614.

Sawada, T. and G. Schatten. 1989. Effects of cytoskeletal inhibitors on ooplasmic segregation and microtubule organization during fertilization and early development in the ascidian *Molgula occidentalis*. *Dev. Biol.* 132: 331–342.

Schierenberg, E. 1998. Nematodes, the round-worms. *In* S. F. Gilbert and A. M. Raunio (eds.), *Embryology: Constructing the Organism*. Sinauer Associates, Sunderland, MA, pp. 131–148.

Schroeder, T. E. 1972. The contractile ring. II. Determining its brief existence, volumetric changes, and vital role in cleaving *Arbacia* eggs. *J. Cell Biol.* 53: 419–434.

Schroeder, T. E. 1973. Cell constriction: Contractile role of microfilaments in division and development. *Am. Zool.* 13: 687–696.

Schroeder, T. 1981. Development of a "primitive" sea urchin (*Eucidaris tribuloides*): Irregularities in the hyaline layer, micromeres, and primary mesenchyme. *Biol. Bull.* 161: 141–151.

Seydoux, G., C. C. Mello, J. Pettitt, W. B. Wood, J. R. Priess and A. Fire. 1996. Repression of gene expression in the embryonic germ lineage of *C. elegans*. *Nature* 382: 713–716.

Sherwood, D. R. and D. R. McClay. 1998. Identification and localization of a sea urchin Notch homologue: Insights into vegetal plate regionalization and Notch receptor regulation. *Development* 124: 3363–3374.

Skiba, F. and E. Schierenberg. 1992. Cell lineage, developmental timing, and spatial pattern formation in embryos of free-living soil nematodes. *Dev. Biol.* 151: 597–610.

Speksnijder, J. E., C. Sardet and L. F. Jaffe. 1990. The activation wave of calcium in the ascidian egg and its role in ooplasmic segregation. *J. Cell Biol.* 110: 1589–1598.

Strome, S. and W. B. Wood. 1983. Generation of asymmetry and segregation of germ-like granules in early *Caenorhabditis elegans* embryos. *Cell* 35: 15–25.

Sturtevant, M. H. 1923. Inheritance of direction of coiling in *Limnaea*. *Science* 58: 269–270.

Sugiyama, K. 1972. Occurrence of mucopolysaccharides in the early development of the sea urchin embryo and its role in gastrulation. *Dev. Growth Diff.* 14: 62–73.

Sulston, J. and H. R. Horvitz. 1977. Postembryonic cell lineages of the nematode *Caenorhabditis elegans*. *Dev. Biol.* 56: 110–156.

Sulston, J. E., J. Schierenberg, J. White and N. Thomson. 1983. The embryonic cell lineage of the nematode *Caenorhabditis elegans*. *Dev. Biol.* 100: 64–119.

Summers, R. G., J. B. Morrill, A. Leith, M. Marko, D. W. Piston and A. T. Stonebraker. 1993. A stereometric analysis of karyogenesis, cytokinesis, and cell arrangements during and following fourth cleavage period in the sea urchin, *Lytechinus variegatus*. *Dev. Growth Diff.* 35: 41–58.

Swenson, K. L., K. M. Farrell and J. V. Ruderman. 1986. The clam embryo protein cyclin A induces entry into M phase and the resumption of meiosis in *Xenopus* oocytes. *Cell* 47: 861–870.

Sze, L. C. 1953. Changes in the amount of deoxyribonucleic acid in the development of *Rana pipiens*. *J. Exp. Zool.* 122: 577–601.

Thorpe, C. J., A. Schlesinger, J. C. Carter and B. Bowerman. 1998. Wnt signaling polarizes an early *C. elegans* blastomere to distinguish endoderm from mesoderm. *Cell* 90: 695–705.

Trinkaus, J. P. 1984. *Cells into Organs: The Forces that Shape the Embryo*, 2nd Ed. Prentice-Hall, Englewood Cliffs, NJ.

Tung, T. C., S. C. Wu, Y. F. Yel, K. S. Li and M. C. Hsu. 1977. Cell differentiation in ascidians studied by nuclear transplantation. *Scientia Sinica* 20: 222–233.

van den Biggelaar, J. A. M. 1977. Developtment of dorsoventral polarity and mesentoblast determination in *Patella vulgata*. *J. Morphol.* 154: 157–186.

van den Biggelaar, J. A. M. and P. Guerrier. 1979. Dorsoventral polarity and mesentoblast determination as concomitant results of cellular interactions in the mollusc *Patella vulgata*. *Dev. Biol.* 68: 462–471.

Verdonk, N. H. and J. N. Cather. 1983. Morphogenetic determination and differentiation. *In* N. H. Verdonk, J. A. M. van den Biggelaar and A. S. Tompa (eds.), *The Mollusca*. Academic Press, New York, pp. 215–252.

von Übisch, L. 1939. Keimblattchimarenforschung an Seeigellarven. *Biol. Rev. Cambr. Phil. Soc.* 14: 88–103.

Welsh, J. H. 1969. Mussels on the move. *Nat. Hist.* 78: 56–59.

Wessel, G. M. and D. R. McClay. 1985. Sequential expression of germ layer specific molecules in the sea urchin embryo. *Dev. Biol.* 111: 451–463.

Wessel, G. M., R. B. Marchase and D. R. McClay. 1984. Ontogeny of the basal lamina in the sea urchin embryo. *Dev. Biol.* 103: 235–245.

White, J. C., W. B. Amos and M. Fordham. 1988. An evaluation of confocal versus conventional imaging of biological structures by fluorescence light microscopy. *J. Cell Biol.* 105: 41–48.

Whittaker, J. R. 1982. Muscle cell lineage can change the developmental expression in epidermal lineage cells of ascidian embryos. *Dev. Biol.* 93: 463–470.

Wikramanayake, A. H., L. Huang and W. H. Klein. 1998. β-catenin is essential for patterning the maternally specified animal-vegetal axis in the sea urchin embryo. *Proc. Nat. Acad. Sci. USA* 95: 9343–9348.

Wilson, E. B. 1898. Cell lineage and ancestral reminiscences. *In Biological Lectures of the Marine Biological Laboratory of Woods Hole*. Ginn, Boston, pp. 21–42.

Wilson, E. B. 1904. Experimental studies on germinal localization. I. The germ regions of the egg of *Dentalium*. II. Experiments on the cleavage-mosaic in *Patella* and *Dentalium*. *J. Exp. Zool.* 1: 1–72.

Wilson, E. B. 1923. *The Physical Basis of Life*. Yale University Press, New Haven. p. 10.

Wolpert, L. and T. Gustafson. 1961. Studies in the cellular basis of morphogenesis of the sea urchin embryo: The formation of the blastula. *Exp. Cell Res.* 25: 374–382.

Wray, G. A. 1997. Echinoderms. *In* S. F. Gilbert and A. Raunio (eds.), *Embryology: Constructing the Organism*. Sinauer Associates, Sunderland, MA. pp. 309–329.

Wray, G. A. 1999. Introduction to sea urchins. *In* S. A. Moody (ed.), *Cell Lineage and Determination*. Academic Press, New York. pp. 3–9.

chapter 9

The genetics of axis specification in Drosophila

THANKS LARGELY TO THE STUDIES by Thomas Hunt Morgan's laboratory during the first decade of the twentieth century, we know more about the genetics of *Drosophila* than about any other multicellular organism. The reasons for this have to do with both the flies and the people who first studied them. *Drosophila* is easy to breed, hardy, prolific, tolerant of diverse conditions, and the polytene chromosomes of its larvae are easy to identify (see Chapter 4). The techniques for breeding and identifying mutants are easy to learn. Moreover, the progress of *Drosophila* genetics was aided by the relatively free access of every scientist to the mutants and the techniques of every other researcher. Mutants were considered the property of the entire scientific community, and Morgan's laboratory established the database and exchange network whereby anyone could obtain them.

Undergraduates (starting with Calvin Bridges and Alfred Sturtevant) played important roles in *Drosophila* research, which achieved its original popularity as a source of undergraduate research projects. As historian Robert Kohler noted (1994), "Departments of biology were cash poor but rich in one resource: cheap, eager, renewable student labor." The *Drosophila* genetics program was "designed by young persons to be a young person's game," and the students set the rules for *Drosophila* research: "No trade secrets, no monopolies, no poaching, no ambushes."

But *Drosophila* was a difficult organism on which to study embryology. Although Jack Schultz and others attempted to relate the genetics of *Drosophila* to its development, the fly embryos proved too complex and too intractable to study, being neither large enough to manipulate experimentally nor transparent enough to observe. It was not until the techniques of molecular biology allowed researchers to identify and manipulate the genes and RNAs of the insect that its genetics could be related to its development. And when that happened, a revolution occurred in the field of biology. The merging of our knowledge of the molecular aspects of *Drosophila* genetics with our knowledge of its development built the foundations on which are current sciences of developmental genetics and evolutionary developmental biology are based.

Early Drosophila *Development*

In the last chapter, we discussed the specification of early embryonic cells by their acquisition of different cytoplasmic determinants that had been stored in the oocyte. The cell membranes establish the region of cytoplasm incorporated into each new blastomere, and it is thought that the morphogenetic determinants then direct differential gene expression in these blastomeres. During *Drosophila* development, however, cellular membranes do not form until after the thirteenth nuclear division. Prior to this time, all the nuclei share a common cytoplasm, and material can diffuse throughout the embryo. In these embryos, the specification of cell types along anterior-posterior and dorsal-ventral axes is accomplished by the interactions of cytoplasmic materials *within* the single, multinucleated cell. Moreover, the initiation of the anterior-posterior and dorsal-ventral differences is controlled by the position of the egg within the mother's ovary. Whereas the sperm entry site may fix the axes in ascidians and nematodes, the fly's anterior-posterior and dorsal-ventral axes are specified by interactions between the egg and its surrounding follicle cells.

WEBSITE **9.1** *Drosophila* **fertilization.** Fertilization of Drosophila can only occur in the region of the oocyte that will become the anterior of the embryo. Morover, the sperm tail appears to stay in this region.

Cleavage

Most insect eggs undergo **superficial cleavage**, wherein a large mass of centrally located yolk confines cleavage to the cytoplasmic rim of the egg. One of the fascinating features of this cleavage type is that cells do not form until after the nuclei have divided. Cleavage in a *Drosophila* egg is shown in Figure 9.1. The zygote nucleus undergoes several mitotic divisions within the central portion of the egg. In *Drosophila*, 256 nuclei are produced by a series of eight nuclear divisions averaging 8 minutes each. The nuclei then migrate to the periphery of the egg, where the mitoses continue, albeit at a progressively slower rate. During the ninth division cycle, about five nuclei reach the surface of the posterior pole of the embryo. These nuclei become enclosed by cell membranes and generate the pole cells that give rise to the gametes of the adult. Most of the other nuclei arrive at the periphery of the embryo at cycle 10 and then undergo four more divisions at progressively slower rates. During these stages of nuclear division, the embryo is called a **syncytial blastoderm**, meaning that all the cleavage nuclei are contained within a common cytoplasm. No cell membranes exist other than that of the egg itself.

Although the nuclei divide within a common cytoplasm, this does not mean that the cytoplasm is itself uniform. Karr and Alberts (1986) have shown that each nucleus within the syncytial blastoderm is contained within its own little territory of cytoskeletal proteins. When the nu-

clei reach the periphery of the egg during the tenth cleavage cycle, each nucleus becomes surrounded by microtubules and microfilaments. The nuclei and their associated cytoplasmic islands are called **energids**. Figure 9.2 shows the nuclei and their essential microfilament and microtubule domains in prophase of the twelfth mitotic division.

Following cycle 13, the oocyte plasma membrane folds inward between the nuclei, eventually partitioning off each somatic nucleus into a single cell (Figure 9.3). This process creates the **cellular blastoderm**, in which all the cells are arranged in a single-layered jacket around the yolky core of the egg (Turner and Mahowald 1977; Foe and Alberts 1983). Like any other cell formation, the formation of the cellular blastoderm involves a delicate interplay between microtubules and microfilaments. The first phase of blastoderm cellularization is characterized by the invagination of cell membranes and their underlying actin microfilament network into the regions between the nuclei to form furrow canals. This process can be inhibited by drugs that block microtubules. After the furrow canals have passed the level of the nuclei, the second phase of cellularization occurs. Here, the rate of invagination increases, and the actin-membrane complex begins to constrict at what will be the basal end of the cell (Schejter and Wieschaus 1993; Foe et al. 1993). In *Drosophila*, the cellular blastoderm consists of approximately 6000 cells and is formed within 4 hours of fertilization.

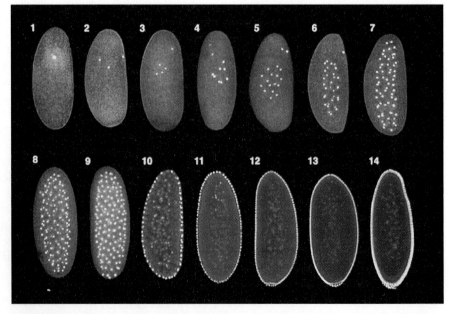

Figure 9.1

Superficial cleavage in a *Drosophila* embryo. The early divisions occur centrally. The numbers refer to the cell cycle. At the tenth cell cycle (512-nucleus stage 2 hours after fertilization), the pole cells form in the posterior, and the nuclei and their cytoplasmic islands ("energids") migrate to the periphery of the cell. This creates the syncytial blastoderm. After cycle 13, the oocyte membranes ingress between the nuclei to form the cellular blastoderm. (Laser confocal images of stained chromatin courtesy of W. Baker and G. Schubiger.)

(A)

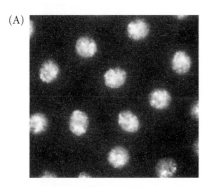

(B)

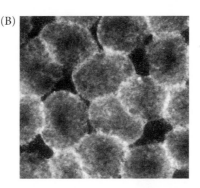

(C)

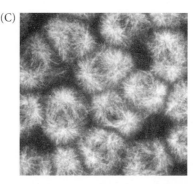

Figure 9.2
Localization of the cytoskeleton around nuclei in the syncytial blastoderm of *Drosophila*. A *Drosophila* embryo entering the mitotic prophase of its twelfth division was sectioned and triple-stained. (A) The nuclei were localized by a dye that binds to DNA. (B) Microfilaments were identified using a fluorescent antibody to actin. (C) Microtubules were recognized by a fluorescent antibody to tubulin. Cytoskeletal domains can be seen surrounding each nucleus. (From Karr and Alberts 1986; photographs courtesy of T. L. Karr.)

The midblastula transition

After the nuclei reach the periphery, the time required to complete each of the next four divisions becomes progressively longer. While cycles 1–10 are each 8 minutes long, cycle 13, the last cycle in the syncytial blastoderm, takes 25 minutes to complete. Cycle 14, in which the *Drosophila* embryo forms cells (i.e., after 13 divisions), is asynchronous. Some groups of cells complete this cycle in 75 minutes, whereas other groups of cells take 175 minutes (Figure 9.4; Foe 1989). Transcription from the nuclei (which begins around the eleventh cycle) is greatly enhanced at this stage. This slowdown of nuclear division and the concomitant increase in RNA transcription is often referred to as the midblastula transition (see Chapter 8). Such a transition is also seen in the embryos of numerous vertebrate and invertebrate phyla. The control of this mitotic slowdown (in *Xenopus*, sea urchin, starfish, and *Drosophila* embryos) appears to be effected by the ratio of chromatin to cytoplasm (Newport and Kirschner 1982; Edgar et al. 1986a). Edgar and his colleagues compared the early development of wild-type *Drosophila* embryos with that of a haploid mutant. These haploid *Drosophila* embryos have half the wild-type quantity of chromatin at each cell division. Hence a haploid embryo at the eighth cell cycle has the

(A)

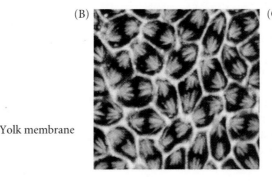

Egg surface

Mitotic spindle

Cleavage furrow
Aster
Nucleus

Furrow canal

Microtubules

Yolk membrane

Figure 9.3
Formation of the cellular blastoderm in *Drosophila*. (A) Developmental series showing the progressive cellularization. (B) Confocal fluorescence photomicrographs of nuclei dividing during the cellularization of the blastoderm. While there are no cell boundaries, actin (green) can be seen forming regions within which each nucleus divides. The microtubules of the mitotic apparatus are stained red with antibodies against tubulin. (C) Cross section during cellularization. As the cells form, the domain of the actin expands into the egg. (A after Fullilove and Jacobson 1971; B and C from Sullivan et al. 1993; photographs courtesy of E. Theurkauf and W. Sullivan.)

(B)

(C)

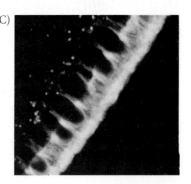

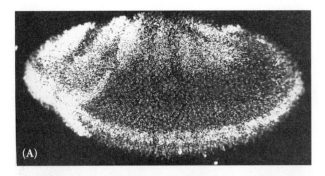

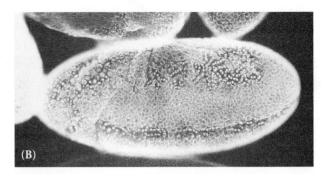

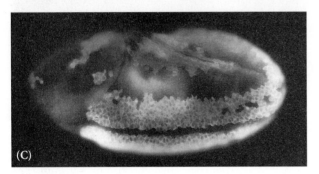

Figure 9.4

Differences in regional rates of cell division in *Drosophila* embryos. (A) Expression of the *string* gene correlates with cell division. In this example, a late stage 14 embryo is stained with a radioactive nucleotide sequence that specifically recognizes and binds *string* mRNA (seen here as the white dots in the autoradiograph). Different regions of the embryo are seen as capable of mitosis. (B) A slightly older embryo is stained with fluorescent antibodies to tubulin to show the microtubules of the mitotic spindles. A comparison of the fluorescence photomicrograph and the autoradiograph obtained from the binding of the radioactive probe shows that only those cells capable of dividing synthesize *string* mRNA. (C) Antibodies to the cyclin A protein show that it is degraded after mitosis and is not seen in those regions containing string protein. (From Edgar and O'Farrell 1989; photographs courtesy of B. A. Edgar.)

same amount of chromatin that a wild-type embryo has at cell cycle 7. The investigators found that whereas wild-type embryos formed their cellular blastoderm immediately after the thirteenth division, the haploid embryos underwent an extra, fourteenth, division before cellularization. Moreover, the lengths of cycles 11–14 in wild-type embryos corresponded to those of cycles 12–15 in the haploid embryos. Thus, the haploid embryos follow a pattern similar to that of the wild-type embryos—only they lag by one cell division.

> WEBSITE **9.2 The regulation of *Drosophila* cleavage.** The control of the cell cycle in *Drosophila* is a story of how the zygote nucleus gradually takes control from the mRNAs and proteins stored in the oocyte cytoplasm.

> WEBSITE **9.3 The early development of other insects.** *Drosophila* is a highly derived species. There are other insect species that develop in ways very different from the "standard" fruit fly.

Gastrulation

At the time of midblastula transition, gastrulation begins. The first movements of *Drosophila* gastrulation segregate the presumptive mesoderm, endoderm, and ectoderm. The prospective mesoderm—about 1000 cells constituting the ventral midline of the embryo—folds inward to produce the **ventral furrow** (Figure 9.5). This furrow eventually pinches off from the surface to become a ventral tube within the embryo. It then flattens to form a layer of mesodermal tissue beneath the ventral ectoderm. The prospective endoderm invaginates as two pockets at the anterior and posterior ends of the ventral furrow.

The pole cells are internalized along with the endoderm. At this time, the embryo bends to form the cephalic furrow.

The ectodermal cells on the surface and the mesoderm undergo convergence and extension, migrating toward the ventral midline to form the **germ band**, a collection of cells along the ventral midline that includes all the cells that will form the trunk of the embryo. The germ band extends posteriorly and, perhaps because of the egg case, wraps around the top (dorsal) surface of the embryo (Figure 9.5D). Thus, at the end of germ band formation, the cells destined to form the most posterior larval structures are located immediately behind the future head region. At this time, the body segments begin to appear, dividing the ectoderm and mesoderm. The germ band then retracts, placing the presumptive posterior segments into the posterior tip of the embryo (Figure 9.5E).

While the germ band is in its extended position, several key morphogenetic processes occur: organogenesis, segmentation, and the segregation of the imaginal discs* (Figure 9.5e). In addition, the nervous system forms from two regions of ventral ectoderm. As described in Chapter 6, neuroblasts differentiate from this neurogenic ectoderm within each segment (and also from the nonsegmented region of the head ectoderm). Therefore, in insects like *Drosophila*, the nervous system is located ventrally, rather than being derived from a dorsal neural tube as in vertebrates.

The general body plan of *Drosophila* is the same in the embryo, the larva, and the adult, each of which has a distinct head end and a distinct tail end, between which are repeating

*Imaginal discs are those cells set aside to produce the adult structures. The details of imaginal disc differentiation will be discussed in Chapter 18. For more information on *Drosophila* developmental anatomy, see Bate and Martinez-Arias 1993; Tyler and Schetzer 1996; and Schwalm 1997.

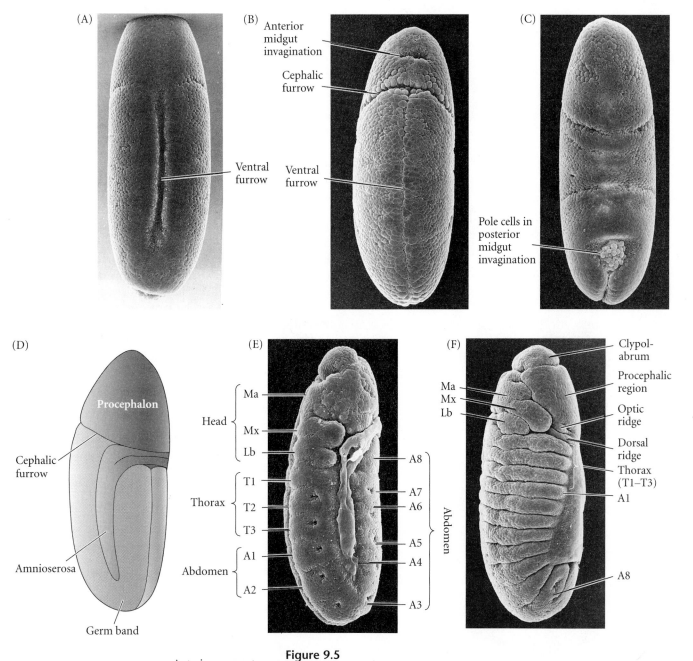

(A)

(B) Anterior midgut invagination
Cephalic furrow
Ventral furrow
Ventral furrow

(C) Pole cells in posterior midgut invagination

(D) Procephalon
Cephalic furrow
Amnioserosa
Germ band

(E) Head { Ma, Mx, Lb }
Thorax { T1, T2, T3 }
Abdomen { A1, A2 }
A8, A7, A6, A5, A4, A3 } Abdomen

(F) Ma, Mx, Lb
Clypol-abrum
Procephalic region
Optic ridge
Dorsal ridge
Thorax (T1–T3)
A1
A8

(G) Anterior segment

Figure 9.5

Gastrulation in *Drosophila*. (A) Ventral furrow beginning to form as cells flanking the ventral midline invaginate. (B) Closing of ventral furrow, with mesodermal cells placed internally and surface ectoderm flanking the ventral midline. (C) Dorsal view of a slightly older embryo, showing the pole cells and posterior endoderm sinking into the embryo (D) Diagram of the dorsolateral view of *Drosophila* embryo at fullest germ band extension, just prior to segmentation. The cephalic furrow separates the future head region (procephalon) from the germ band that will form the thorax and abdomen. (E) Lateral view, showing fullest extension of germ band and the beginnings of segmentation. Subtle indentations mark the incipient segments along the germ band: Ma, Mx, and Lb correspond to the mandibular, maxillary, and labial head segments; T1–T3, the thoracic segments; A1–A8, the abdominal segments. (F) Germ band reversing direction. The true segments are now visible, as well as the other territories of the dorsal head, such as the clypolabrum, procephalic region, optic ridge, and dorsal ridge. (G) Newly hatched first-instar larva. (Photographs courtesy of F. R. Turner. D after Campos-Ortega and Hartenstein 1985.)

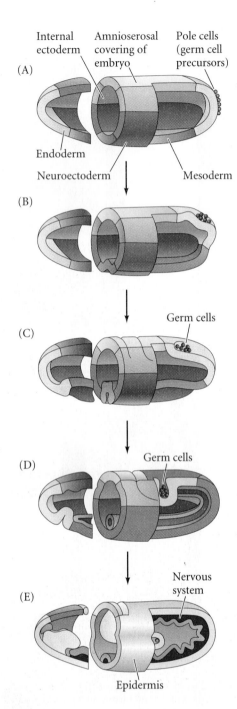

Internal ectoderm
Amnioserosal covering of embryo
Pole cells (germ cell precursors)

(A)

Endoderm

Neuroectoderm
Mesoderm

(B)

Germ cells

(C)

Germ cells

(D)

Nervous system

(E)

Epidermis

Figure 9.6
Schematic representation of gastrulation in *Drosophila*. (A) and (B) are surface and cut-away views showing the fates of the tissues immediately prior to gastrulation. (C) shows the beginning of gastrulation as the ventral mesoderm invaginates into the embryo. (D) corresponds to Figure 9.5A, while (E) corresponds to 9.5B and C. In (E), the neuroectoderm is largely differentiated into the nervous system and the epidermis. (After Campos-Ortega and Hartenstein 1985.)

and embryology have led to a detailed model describing how a segmented pattern is generated along the anterior-posterior axis and how each segment is differentiated from the others.

The anterior-posterior and dorsal-ventral axes of *Drosophila* form at right angles to one another, and they are both determined by the position of the oocyte within the follicle cells of the ovary. The rest of this chapter is divided into three main parts. The first part concerns how the anterior-posterior axis is specified and how it determines the identity of each segment. The second part concerns how the dorsal-ventral axis is specified by the interactions between the oocyte and its surrounding follicle cells. The third part concerns how embryonic tissues are specified to become particular organs by their placement along these two axes.

VADE MECUM **Drosophila development.** The CD-ROM contains some remarkable time-lapse sequences of *Drosophila* development, including cleavage and gastrulation. This segment also provides access to the fly life cycle. The color coding superimposed on the germ layers allows you to readily understand tissue movements.
[Click on Fruit Fly]

Figure 9.7
Comparison of larval and adult segmentation in *Drosophila*. The three thoracic segments can be distinguished by their appendages: T1 (prothorax) has legs only; T2 (mesothorax) has wings and legs; T3 (metathorax) has halteres and legs.

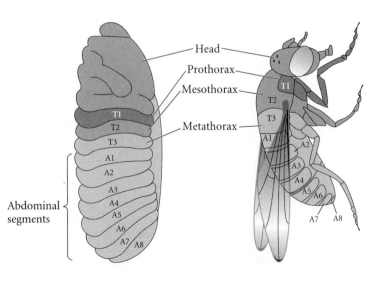

Head
Prothorax
Mesothorax
Metathorax
Abdominal segments

segmental units (Figure 9.7). Three of these segments form the thorax, while another eight segments form the abdomen. Each segment of the adult fly has its own identity. The first thoracic segment, for example, has only legs; the second thoracic segment has legs and wings; and the third thoracic segment has legs and halteres (balancers). Thoracic and abdominal segments can also be distinguished from each other by differences in the cuticle. How does this pattern arise? During the past decade, the combined approaches of molecular biology, genetics,

The Origins of Anterior-Posterior Polarity

The anterior-posterior polarity of the embryo, larva, and adult has its origin in the anterior-posterior polarity of the egg (Figure 9.8). The **maternal effect genes** expressed in the mother's ovaries produce messenger RNAs that are placed in different regions of the egg. These messages encode transcriptional and translational regulatory proteins that diffuse through the syncytial blastoderm and activate or repress the expression of certain zygotic genes. Two of these proteins, **Bicoid** and **Hunchback**, regulate the production of anterior structures, while another pair of maternally specified proteins, **Nanos** and **Caudal**, regulates the formation of the posterior parts of the embryo. Next, the zygotic genes regulated by these maternal factors are expressed in certain broad (about three segments wide), partially overlapping domains. These genes are called **gap genes** (because mutations in them cause gaps in the segmentation pattern), and they are among the first genes transcribed in the

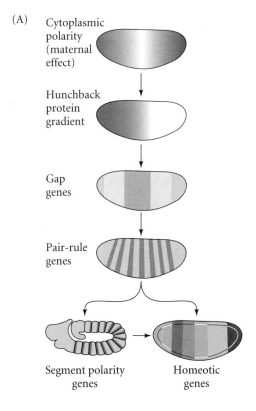

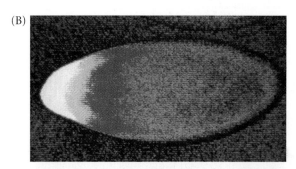

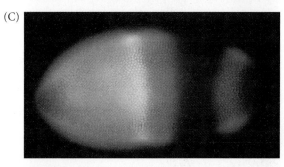

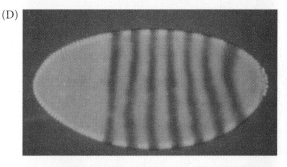

Figure 9.8
Generalized model of *Drosophila* anterior-posterior pattern formation. (A) The pattern is established by maternal effect genes that form gradients and regions of morphogenetic proteins. These morphogenetic determinants create a gradient of Hunchback protein that differentially activates the gap genes, which define broad territories of the embryo. The gap genes enable the expression of the pair-rule genes, each of which divides the embryo into regions about two segments wide. The segment polarity genes then divide the embryo into segment-sized units along the anterior-posterior axis. Together, the actions of these genes define the spatial domains of the homeotic genes that define the identities of each of the segments. In this way, periodicity is generated from nonperiodicity, and each segment is given a unique identity. (B) Maternal effect genes. The anterior axis is specified by the gradient of Bicoid protein (yellow through red). (C) Gap gene protein expression and overlap. The domain of Hunchback protein (orange) and the domain of Krüppel protein (green) overlap to form a region containing both transcription factors (yellow). (D) Products of the *fushi tarazu* pair-rule gene form seven bands across the embryo. (E) Products of the segment polarity gene *engrailed*, seen here at the extended germ band stage. (B courtesy of C. Nüsslein-Vollhard; C courtesy of C. Rushlow and M. Levine; D courtesy of T. Karr; E courtesy of S. Carroll and S. Paddock.)

embryo. Differing concentrations of the gap gene proteins cause the transcription of **pair-rule genes**, which divide the embryo into periodic units. The transcription of the different pair-rule genes results in a striped pattern of seven vertical bands perpendicular to the anterior-posterior axis. The pair-rule gene proteins activate the transcription of the **segment polarity genes**, whose mRNA and protein products divide the embryo into 14 segment-wide units, establishing the periodicity of the embryo. At the same time, the protein products of the gap, pair-rule, and segment polarity genes interact to regulate another class of genes, the homeotic selector genes, whose transcription determines the developmental fate of each segment.

The Maternal Effect Genes

Embryological evidence of polarity regulation by oocyte cytoplasm

Classic embryological experiments demonstrated that there are at least two "organizing centers" in the insect egg, one in the anterior of the egg and one in the posterior. For instance, Klaus Sander (1975) found that if he ligated the egg early in development, separating the anterior from the posterior region, one half developed into an anterior embryo and one half developed into a posterior embryo, but neither half contained the middle segments of the embryo. The later in development the ligature was made, the fewer middle segments were missing. Thus, it appeared that there were indeed gradients emanating from the two poles during cleavage, and that these gradients interacted to produce the positional information determining the identity of each segment. Moreover, when the RNA of the anterior of insect eggs was destroyed (by either ultraviolet light or RNase), the resulting embryos lacked a head and thorax. Instead, these embryos developed two abdomens and telsons (tails) with mirror-image symmetry: telson-abdomen-abdomen-telson (Figure 9.9; Kalthoff and Sander 1968; Kandler-Singer and Kalthoff 1976). Thus, Sander's laboratory postulated the existence of a gradient at both ends of the egg, and hypothesized that the egg sequesters an RNA that generates a gradient of anterior-forming material.

WEBSITE 9.4 **Evidence for gradients in insect development.** The original evidence for gradients in insect development came from studies providing evidence for two "organization centers" in the egg, one located anteriorly and one located posteriorly.

The molecular model: Protein gradients in the early embryo

In the late 1980s, the gradient hypothesis was united with a genetic approach to the study of *Drosophila* embryogenesis. If there were gradients, what were the morphogens whose concentrations changed over space? What were the genes that

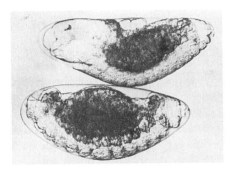

Figure 9.9
Normal and irradiated embryos of the midge *Smittia*. The normal embryo (top) shows a head on the left and abdominal segments on the right. The UV-irradiated embryo has no head region, but has abdominal segments at both ends. (From Kalthoff 1969; photographs courtesy of K. Kalthoff.)

shaped these gradients? And did these morphogens act by activating or inhibiting certain genes in the areas where they were concentrated? Christiane Nüsslein-Volhard led a research program that addressed these questions. The researchers found found that one set of genes encoded gradient morphogens for the anterior part of the embryo, another set of genes encoded morphogens responsible for organizing the posterior region of the embryo, and a third set of genes encoded proteins that produced the terminal regions at both ends of the embryo (Figure 9.10; Table 9.1). This work resulted in a Nobel Prize for Nüsslein-Volhard and her colleague, Eric Wieschaus, in 1995.

WEBSITE 9.5 **Christiane Nüsslein-Volhard and the molecular approach to development.** The research that revolutionized developmental biology had to wait for someone to synthesize molecular biology, embryology, and *Drosophila* genetics.

The anterior-posterior axis of the *Drosophila* embryo appears to be patterned before the nuclei even begin to function. The nurse cells of the ovary deposit mRNAs in the developing oocyte, and these mRNAs are apportioned to different regions of the cell. In particular, four maternal messenger RNAs are critical to the formation of the anterior-posterior axis:

- *bicoid* and *hunchback* mRNAs, whose protein products are critical for head and thorax formation
- *nanos* and *caudal* mRNAs, whose protein products are critical for the formation of the abdominal segments

The *bicoid* mRNAs are located in the anterior portion of the unfertilized egg, and are tethered to the anterior microtubules. The *nanos* messages are bound to the cytoskeleton in the posterior region of the unfertilized egg. The *hunchback* and *caudal* mRNAs are distributed throughout the oocyte. Upon fertilization, these mRNAs can be translated into proteins. At the ante-

Table 9.1 Maternal effect genes that effect the anterior-posterior polarity of the *Drosophila* embryo

Gene	Mutant phenotype	Proposed function and structure
ANTERIOR GROUP		
bicoid (bcd)	Head and thorax deleted, replaced by inverted telson	Graded anterior morphogen; contains homeodomain; represses caudal
exuperantia (exu)	Anterior head structures deleted	Anchors bicoid mRNA
swallow (swa)	Anterior head structures deleted	Anchors bicoid mRNP
POSTERIOR GROUP		
nanos (nos)	No abdomen	Posterior morphogen; represses hunchback
tudor (tud)	No abdomen, no pole cells	Localization of Nanos
oskar (osk)	No abdomen, no pole cells	Localization of Nanos
vasa (vas)	No abdomen, no pole cells; oogenesis defective	Localization of Nanos
valois (val)	No abdomen, no pole cells; cellularization defective	Stabilization of the Nanos localization complex
pumilio (pum)	No abdomen	Helps Nanos protein bind hunchback message
caudal (cad)	No abdomen	Activates posterior terminal genes
TERMINAL GROUP		
torso (tor)	No termini	Possible morphogen for termini
trunk (trk)	No termini	Transmits Torsolike signal to Torso
fs(1)Nasrat[fs(1)N]	No termini; collapsed eggs	Transmits Torsolike signal to Torso
fs(1)polehole[fs(1)ph]	No termini; collapsed eggs	Transmits *Torsolike* signal to *Torso*

Source: After Anderson 1989.

rior pole, the *bicoid* RNA is translated into Bicoid protein, which forms a gradient highest at the anterior. At the posterior pole, the *nanos* message is translated into Nanos protein, which forms a gradient highest at the posterior. Bicoid protein inhibits the translation of the *caudal* RNA, allowing Caudal protein to be synthesized only in the posterior of the cell. Conversely, Nanos protein, in conjunction with Pumilio protein, binds to *hunchback* RNA, preventing its translation in the posterior portion of the embryo. Bicoid also elevates the level of Hunchback protein in the anterior of the embryo by binding to the enhancers of the *hunchback* gene and stimulating its transcription. The result of these interactions is the creation of four protein gradients in the early embryo (Figure 9.11):

• An anterior-to-posterior gradient of Bicoid protein
• An anterior-to-posterior gradient of Hunchback protein
• A posterior-to-anterior gradient of Nanos protein
• A posterior-to-anterior gradient of Caudal protein

The Bicoid, Hunchback, and Caudal proteins are transcription factors whose relative concentrations can activate or repress particular zygotic genes. The stage is now set for the activation of zygotic genes in those nuclei that were busily dividing while this gradient was being established.

Evidence that the Bicoid gradient constitutes the anterior organizing center

In *Drosophila*, the phenotype of the bicoid mutant provides valuable information about the function of gradients. Instead of having anterior structures (acron, head, and thorax) followed by abdominal structures and a telson, the structure of the bicoid mutant is telson-abdomen-abdomen-telson (Figure 9.12). It would appear that these embryos lack whatever substances are needed for the formation of anterior structures. Moreover, one could hypothesize that the substance that these mutants lack is the one postulated by Sander and Kalthoff to turn on genes for the anterior structures and turn off genes for the telson structures (compare Figures 9.9 and 9.12).

Further studies have strengthened the view that the product of the wild-type *bicoid* gene is the morphogen that controls anterior development. First, *bicoid* is a maternal effect gene. Messenger RNA from the mother's *bicoid* genes is placed in the embryo by the mother's ovarian cells (Figure 9.13A; Frigerio et al. 1986; Berleth et al. 1988). The *bicoid* RNA is strictly localized in the anterior portion of the oocyte (Figure 9.13B), where the anterior cytoskeleton anchors it through the message's 3' untranslated region (Ferrandon et al. 1997; Macdonald and Kerr 1998). This mRNA is dormant until fertilization, at which time

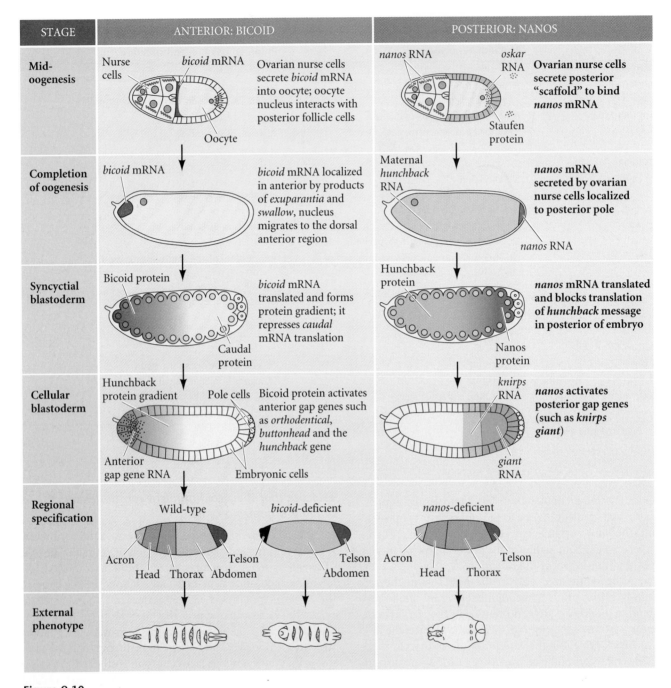

Figure 9.10
Three independent genetic pathways interact to form the anterior-posterior axis of the *Drosophila* embryo. In each case, the initial asymmetry is established during oogenesis, and the pattern is organized by maternal proteins soon after fertilization. The realization of the pattern comes about when the localized maternal proteins activate or repress specific zygotic genes in different regions of the embryo. (After St. Johnston and Nüsslein-Volhard 1992.)

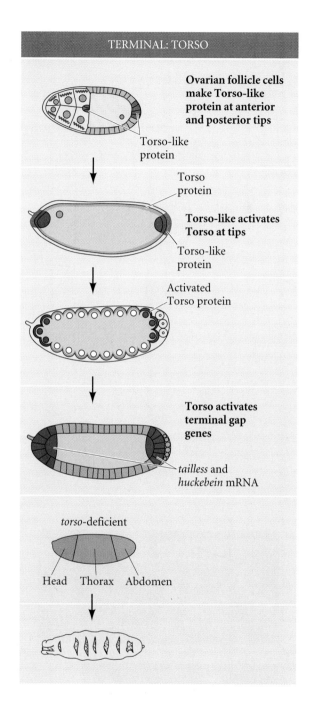

TERMINAL: TORSO

Ovarian follicle cells make Torso-like protein at anterior and posterior tips

Torso-like protein

Torso protein

Torso-like activates Torso at tips

Torso-like protein

Activated Torso protein

Torso activates terminal gap genes

tailless and *huckebein* mRNA

torso-deficient

Head Thorax Abdomen

it receives a longer polyadenylate tail and can be translated. Driever and Nüsslein-Volhard (1988b) have shown that when Bicoid protein is translated from this RNA during early cleavage, it forms a gradient, with the highest concentration in the anterior of the egg and the lowest in the posterior third of the egg. Moreover, this protein soon becomes concentrated in the embryonic nuclei in the anterior portion of the embryo (Figure 9.13C–E; see also Figure 5.35).

WEBSITE **9.6 Mechanism of *bicoid* mRNA localization.** One of the most critical steps in *Drosophila* pattern formation is the binding of the *bicoid* mRNA to the anterior microtubules. Several genes are involved in this process, wherein the *bicoid* message forms a complex with several proteins.

Further evidence that Bicoid protein is the anterior morphogen came from experiments that altered the steepness of the gradient. Two genes, *exuperantia* and *swallow*, are responsible for keeping the *bicoid* message at the anterior pole of the egg. In their absence, the *bicoid* message diffuses farther into the posterior of the egg, and the gradient of Bicoid protein is less steep (Driever and Nüsslein-Volhard 1988a). The phenotype produced by these two mutants is similar to that of *bicoid*-deficient embryos, but less severe. These embryos lack their most anterior structures and have an extended mouth and thoracic region. Thus, by altering the gradient of Bicoid protein, one correspondingly alters the fate of the embryonic regions.

Confirmation that the Bicoid protein is crucial for initiating head and thorax formation came from experiments in which purified *bicoid* RNA was injected into early-cleavage embryos (Figure 9.14; Driever et al. 1990). When injected into the anterior of *bicoid*-deficient embryos (whose mothers lacked *bicoid* genes), the *bicoid* RNA rescued the embryos and caused them to have normal anterior-posterior polarity. Moreover, any location in an embryo where the *bicoid* message was injected became the head. If *bicoid* RNA was injected into the center of an embryo, that middle region became the head, and the regions on either side of it became thorax structures. If a large amount of *bicoid* RNA was injected into the posterior end of a wild-type embryo (with its own endogenous *bicoid* message in its anterior pole), two heads emerged, one at either end.

The next question then emerged: How might a gradient in Bicoid protein control the determination of the anterior-posterior axis? Recent evidence suggests that Bicoid acts in two ways to specify the anterior of the *Drosophila* embryo. First, it acts as a repressor of posterior formation. It does this by binding to and suppressing the translation of *caudal* RNA, which is found throughout the egg and early embryo. The Caudal protein is critical in specifying the posterior domains of the embryo, and it activates the genes responsible for the invagination of the hindgut (Wu and Lengyel 1998). The Bicoid protein binds to a specific region of the *caudal* message's 3′ untranslated region, thereby preventing the translation of this message in the anterior section of the embryo (Figure 9.15; Dubnau and Struhl 1996; Rivera-Pomar et al. 1996). This suppression is necessary, for if Caudal protein is made in the anterior, the head and thorax are not properly formed.

Figure 9.11
A model of anterior-posterior pattern generation by the *Drosophila* maternal effect genes. (A) The *bicoid, nanos, hunchback,* and *caudal* messenger RNAs are placed in the oocyte by the ovarian nurse cells. The *bicoid* message is sequestered anteriorly. The *nanos* message is sent to the posterior pole. (B) Upon translation, the Bicoid protein gradient extends from anterior to posterior, and the Nanos protein gradient extends from posterior to anterior. Nanos inhibits the translation of the *hunchback* message (in the posterior), while Bicoid prevents the translation of the *caudal* message (in the anterior). This inhibition results in opposing Caudal and Hunchback gradients. The Hunchback gradient is secondarily strengthened by the transcription of the *hunchback* gene in the anterior nuclei (since Bicoid acts as a transcription factor to activate *hunchback* transcription). (C) Parallel interactions whereby translational gene regulation establishes the anterior-posterior patterning of the *Drosophila* embryo. (C after Macdonald and Smibert 1996.)

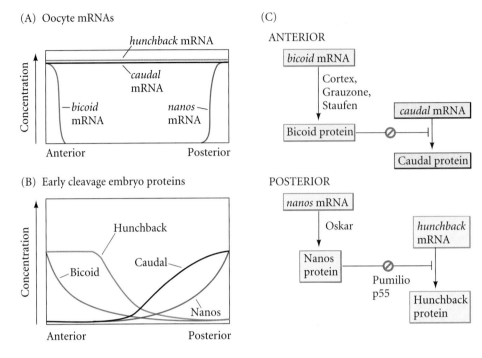

(A) Oocyte mRNAs

(B) Early cleavage embryo proteins

(C)

Second, the Bicoid protein functions as a transcription factor. Bicoid protein enters the nuclei of early-cleavage embryos, where it activates the *hunchback* gene. The transcription of *hunchback* is seen only in the anterior half of the embryo—the region where Bicoid protein is found. Mutants deficient in maternal and zygotic Hunchback protein lack mouthparts and thorax structures. In the late 1980s, two laboratories independently demonstrated that Bicoid protein binds to and activates the *hunchback* gene (Driever and Nüsslein-Volhard 1989; Struhl et al. 1989). The Hunchback protein derived from the synthesis of new *hunchback* mRNA joins the Hunchback protein synthesized by the translation of maternal messages in the anterior of the embryo. The Hunchback protein, also a transcription factor, represses abdominal-specific genes, thereby allowing the region of *hunchback* expression to form the head and thorax.

The Hunchback protein also works with Bicoid in generating the anterior pattern of the embryo. Based on two pieces of evidence, Driever and co-workers (1989) predicted that at least one other anterior gene besides *hunchback* must be activated by Bicoid. First, deletions of *hunchback* produce only some of the defects seen in the bicoid mutant phenotype. Second, as we saw in the *swallow* and *exuperantia* experiments, only moderate levels of Bicoid protein are needed to activate thorax formation (i.e., *hunchback* gene expression), but head formation requires higher concentrations. Driever

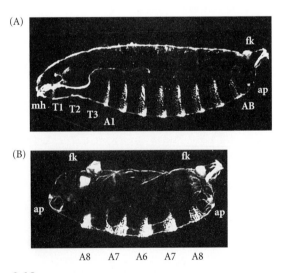

Figure 9.12
Phenotype of a strongly affected embryo from a female deficient in the bicoid gene. (A) Wild-type cuticle pattern. (B) Bicoid mutant. The head and thorax have been replaced by a second set of posterior telson structures. Abbreviations: fk, filzkörper; ap, anal plates (both telson structures); T1–T3, thoracic segments; A1, A8, the two terminal abdominal segments; mh, cs, head structures. (From Driever et al. 1990; photographs courtesy of W. Driever.)

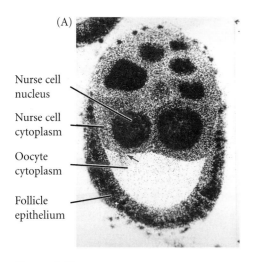

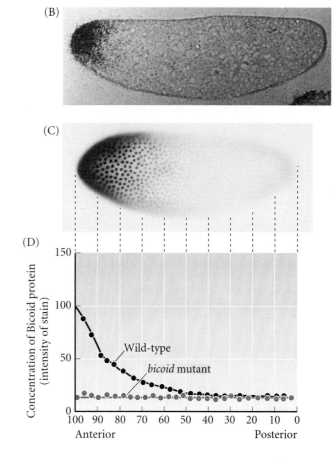

Figure 9.13
Gradient of Bicoid protein in the early *Drosophila* embryo. (A) *bicoid* mRNA passing into the oocyte from the nurse cells during oogenesis. (B) Localization of *bicoid* mRNA to the anterior tip of the embryo. (C) Gradient of Bicoid protein shortly after fertilization. Note that the concentration is greatest anteriorly and trails off posteriorly. Notice also that Bicoid protein is concentrated in the nuclei of the embryo. (D) Densitometric scan of the Bicoid protein gradient. The upper curve represents the gradient of Bicoid protein in wild-type embryos. The lower curve represents Bicoid protein in embryos of bicoid mutant mothers. (A from Stephanson et al. 1988; B from Kaufman et al. 1990; C and D from Driever and Nüsslein-Volhard 1988b; photographs courtesy of the authors.)

Figure 9.14
Schematic representation of the experiments demonstrating that the *bicoid* gene encodes the morphogen responsible for head structures in *Drosophila*. The phenotypes of *bicoid*-deficient and wild-type embryos are shown at the sides. When *bicoid*-deficient embryos are injected with *bicoid* mRNA, the point of injection forms the head structures. When the posterior pole of an early-cleavage wild-type embryo is injected with *bicoid* mRNA, head structures form at both poles. (After Driever et al. 1990.)

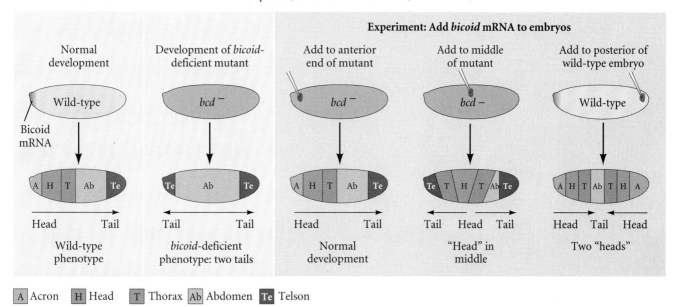

Figure 9.15
Gradient of Caudal protein in the syncitial blastoderm of a wild-type *Drosophila* embryo. The protein (stained darkly) enters the nuclei and helps specify posterior fates. Compare with the complementary gradient of Bicoid protein in Figure 9.13. (From Macdonald and Struhl 1986; photograph courtesy of the authors.)

and co-workers (1989) predicted that the promoters of such a head-specific gap gene would have low-affinity binding sites for Bicoid protein. This gene would be activated only at extremely high concentrations of Bicoid protein—that is, near the anterior tip of the embryo. Since then, three gap genes of the head have been discovered that are dependent on very high concentrations of Bicoid protein for their expression (Cohen and Jürgens 1990; Finkelstein and Perrimon 1990; Grossniklaus et al. 1994). The *buttonhead, empty spiracles,* and *orthodenticle* genes are needed to specify the progressively anterior regions of the head. In addition to needing high Bicoid levels for activation, these genes also require the presence of Hunchback protein to be transcribed (Simpson-Brose et al. 1994; Reinitz et al. 1995). The Bicoid and Hunchback proteins act synergistically at the enhancers of these "head genes" to promote their transcription.

The posterior organizing center: Localizing and activating Nanos

The posterior organizing center is defined by the activities of the *nanos* gene (Lehmann and Nüsslein-Volhard 1991; Wang and Lehmann 1991; Wharton and Struhl 1991). The *nanos* RNA is produced by the ovarian nurse cells and is transported into the posterior region of the egg (farthest away from the nurse cells). The *nanos* message is bound to the cytoskeleton in the posterior region of the egg through its 3' UTR and its association with the products of several other genes (*oskar, valois, vasa, staufen,* and

tudor).* If *nanos* or any other of these maternal effect genes are absent in the mother, no embryonic abdomen forms (Lehmann and Nüsslein-Volhard 1986; Schüpbach and Wieschaus 1986).

The *nanos* message is dormant in the unfertilized egg, as it is repressed by the binding of the Smaug protein to its 3' UTR (Smibert et al. 1996). At fertilization, this repression is removed, and Nanos protein can be synthesized. The Nanos protein forms a gradient that is highest at the posterior end. Nanos functions by inactivating *hunchback* mRNA translation (Figure 9.16, see also Figure 9.11; Tautz 1988). In the anterior of the cleavage-stage embryo, the *hunchback* message is bound in its 3' UTR by the Pumilio protein, and the message can be translated into Hunchback protein. In the posterior of the early embryo, however, the bound Pumilio can be joined by the Nanos protein. Nanos binds to Pumilio and deadenylates the *hunchback* mRNA, preventing its translation (Barker et al. 1992; Wreden et al. 1997). The *hunchback* RNA is initially present throughout the embryo, although more can be made from zygotic nuclei if they are activated by Bicoid protein. Thus, the combination of Bicoid and Nanos proteins causes a

---

*Like the placement of the *bicoid* message, the location of the *nanos* message is determined by its 3' untranslated region. If the *bicoid* 3' UTR is experimentally placed on the protein-encoding region of *nanos* RNA, the *nanos* message gets placed in the anterior of the egg. When the RNA is translated, the Nanos protein inhibits the translation of *hunchback* and *bicoid* mRNAs, and the embryo forms two abdomens—one in the anterior of the embryo and one in the posterior (Gavis and Lehmann 1992). The localization of *nanos* RNA is ultimately dependent on interactions between the oocyte and the neighboring follicle cells that localize the *oskar* message to the posterior pole.

Figure 9.16
Control of *hunchback* mRNA translation by Nanos. In the anterior of the embryo, Pumilio protein binds to the Nanos Response Element (NRE) in the 3' UTR of the *hunchback* message, and the message is polyadenylated normally. This polyadenylated message can be translated into Hunchback protein. In the posterior of the embryo, where Nanos protein is found, Nanos binds to Pumilio to cause the deadenylation of the *hunchback* message. This prevents the translation of the *hunchback* message. (After Wreden et al. 1997.)

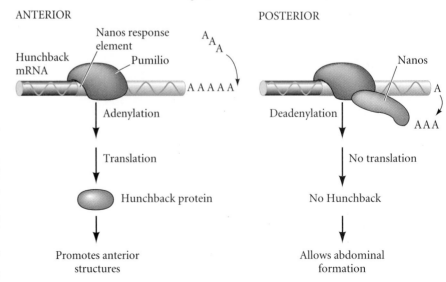

(A)

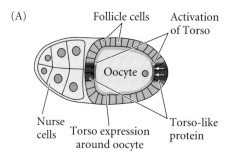

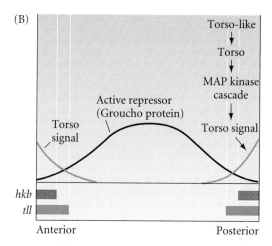

Figure 9.17
Formation of the unsegmented poles by *torso* signaling. (A) Torso-like protein is expressed by the follicle cells at the poles of the oocyte. Torso protein is expressed around the entire oocyte. Torso-like activates torso protein at the poles (see Casanova et al. 1995). (B) Inactivation of the transcriptional suppression of *huckebein* (*hkb*) and *tailless* (*tll*) genes. The *torso* signal antagonizes the Groucho protein. Groucho acts as a repressor of *tailless* and *huckebein* expression. The gradient of Torso protein is thought to provide the information that allows *tailless* to be expressed further into the embryo than *huckebein*. (A after Gabay et al. 1997; B after Paroush et al. 1997.)

gradient of Hunchback protein across the egg. The Bicoid protein activates *hunchback* gene transcription in the anterior part of the embryo, while the Nanos protein inhibits the translation of *hunchback* RNA in the posterior part of the embryo.

The terminal gene group

In addition to the anterior and posterior morphogens, there is third set of maternal genes whose proteins generate the extremes of the anterior-posterior axis. Mutations in these terminal genes result in the loss of the unsegmented extremities of the organism: the **acron** and the most anterior head segments and the **telson** (tail) and the most posterior abdominal segments (Degelmann et al. 1986; Klingler et al. 1988). A critical gene here appears to be **torso**, a gene encoding a receptor tyrosine kinase. The embryos of mothers with mutations of the *torso* gene have neither acron nor telson, suggesting that the two termini of the embryo are formed through the same pathway. The *torso* RNA is synthesized by the ovarian cells, deposited in the oocyte, and translated after fertilization. The transmembrane Torso protein is not spatially restricted to the ends of the egg, but is evenly distributed throughout the plasma membrane (Casanova and Struhl 1989). Indeed, a dominant mutation of *torso*, which imparts constitutive activity to the receptor, converts the entire anterior half of the embryo into an acron and the entire posterior half into a telson. Thus, Torso must normally be activated only at the ends of the egg.

Stevens and her colleagues (1990) have shown that this is the case. Torso protein is activated by the follicle cells only at the two poles of the oocyte. Two pieces of evidence suggest that the activator of the Torso protein is probably the **Torso-like** protein: first, loss-of-function mutations in the *torso-like* gene create a phenotype almost identical to that produced by torso mutants, and second, ectopic expression of Torso-like causes the activation of the Torso protein in the new location. The *torso-like* gene is usually expressed only in the anterior and posterior follicle cells, and the secreted Torso-like protein can cross the perivitelline space to activate the Torso protein in the egg membrane (Martin et al. 1994; Furriols et al. 1998). In this manner, the Torso-like protein activates the Torso protein in the anterior and posterior regions of the oocyte membrane. The end products of the RTK-kinase cascade activated by the Torso protein diffuse into the cytoplasm at both ends of the embryo (Figure 9.17; Gabay et al. 1997; see Chapter 6). These kinases are thought to inactivate a transcriptional inhibitor of the *tailless* and *huckebein* gap genes (Paroush et al. 1997). These two genes then specify the termini of the embryo. The distinction between the anterior and posterior termini depends on the presence of Bicoid. If the terminal genes act alone, the terminal regions differentiate into telsons. However, if Bicoid is also present, the region forms an acron (Pignoni et al. 1992).

The anterior-posterior axis of the embryo is therefore specified by three sets of genes: those that define the anterior organizing center, those that define the posterior organizing center, and those that define the terminal boundary region. The anterior organizing center is located at the anterior end of the embryo and acts through a gradient of Bicoid protein that functions as a transcription factor to activate anterior-specific gap genes and as a translational repressor to suppresses posterior-specific gap genes. The posterior organizing center is located at the posterior pole and acts translationally through the Nanos protein to inhibit anterior formation and transcriptionally through the Caudal protein to activate those genes that form the abdomen. The boundaries of the acron and telson are defined by the product of the torso gene, which is activated at the tips of the embryo. The activation of those genes responsible for constructing the posterior is performed by Caudal, a protein whose synthesis (as we have seen above) is inhibited in the anterior portion of the embryo. The next step in development will be to use these gradients of transcription factors to activate specific genes along the anterior-posterior axis.

Figure 9.18
Segments and parasegments. A and P represent the anterior and posterior compartments of the segments. The parasegments are shifted one compartment forward. Ma, Mx, and Lb represent three of the head segments (mandibular, maxillary, and labial), the T segments are thoracic, and the A segments are abdominal. The parasegments are numbered 1 through 14. The bars below the map show the boundaries of gene expression observed by the in situ hybridization of radioactive cDNA from the pair-rule gene *fushi tarazu* (*ftz*). (After Martinez-Arias and Lawrence 1985.)

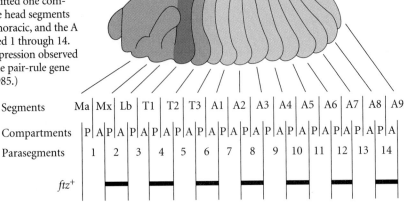

Segments	Ma	Mx	Lb	T1	T2	T3	A1	A2	A3	A4	A5	A6	A7	A8	A9
Compartments	P	A P	A P	A P	A P	A P	A P	A P	A P	A P	A P	A P	A P	A P	A
Parasegments	1	2	3	4	5	6	7	8	9	10	11	12	13	14	

ftz^+

The Segmentation Genes

The process of cell fate commitment in *Drosophila* appears to have two steps: specification and determination (Slack 1983). Early in development, the fate of a cell depends on environmental cues, such as those provided by the protein gradients mentioned above. This specification of cell fate is flexible and can still be altered in response to signals from other cells. Eventually, the cells undergo a transition from this loose type of commitment to an irreversible determination. At this point, the fate of a cell becomes cell-intrinsic.* The transition from specification to determination in *Drosophila* is mediated by the **segmentation genes**. These genes divide the early embryo into a repeating series of segmental primordia along the anterior-posterior axis. Mutations in segmentation genes cause the embryo to lack certain segments or parts of segments. Often these mutations affect **parasegments**, regions of the embryo that are separated by mesodermal thickenings and ectodermal grooves. The segmentation genes divide the embryo into 14 parasegments (Martinez-Arias and Lawrence 1985). The parasegments of the embryo do not become the segments of the larva or adult; rather, they include the posterior part of an anterior segment and the anterior portion of the segment behind it (Figure 9.18). While the segments are the major anatomical divisions of the larval and adult body plan, they are built according to rules that use the parasegment as the basic unit of construction.

There are three classes of segmentation genes, which are expressed sequentially (see Figure 9.8). The transition from an embryo characterized by gradients of morphogens to an embryo with distinct units is accomplished by the products of the

*Aficionados of information theory will recognize that the process by which the anterior-posterior information in morphogenetic gradients is transferred to discrete and different parasegments represents a transition from analog to digital specification. Specification is analog, determination digital. This process enables the transient information of the gradients in the syncytial blastoderm to be stabilized so that it can be utilized much later in development (Baumgartner and Noll 1990).

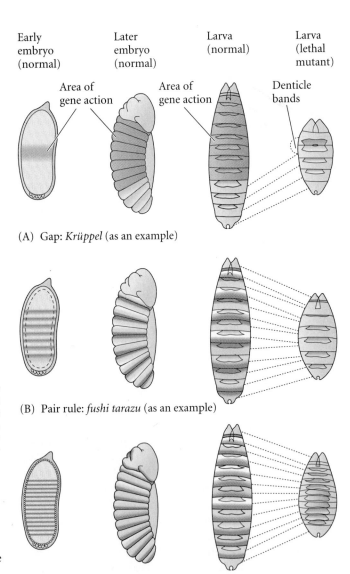

Early embryo (normal) | Later embryo (normal) | Larva (normal) | Larva (lethal mutant)

Area of gene action

Area of gene action

Denticle bands

(A) Gap: *Krüppel* (as an example)

(B) Pair rule: *fushi tarazu* (as an example)

(C) Segment polarity: *engrailed* (as an example)

Figure 9.19
Three types of segmentation gene mutations. The left panel shows the early-cleavage embryo, with the region where the particular gene is normally transcribed in wild-type embryos shown in color. These areas are deleted as the mutants develop.

gap genes. The gap genes are activated or repressed by the maternal effect genes, and they divide the embryo into broad regions, each containing several parasegment primordia. The *Krüppel* gene, for example, is expressed primarily in parasegments 4–6, in the center of the *Drosophila* embryo (Figures 9.19A; 9.8B); the absence of the Krüppel protein causes the embryo to lack these regions. The protein products of the gap genes interact with neighboring gap gene proteins to activate the transcription of the **pair-rule genes**. The products of these genes subdivide the broad gap gene regions into parasegments. Mutations of pair-rule genes, such as *fushi tarazu* (Figures 9.8C, 9.19B, 9.20), usually delete portions of alternate segments. Finally, the **segment polarity** genes are responsible for maintaining certain repeated structures within each segment. Mutations in these genes cause a portion of each segment to be deleted and replaced by a mirror-image structure of another portion of the segment. For instance, in *engrailed* mutants, portions of the posterior part of each segment are replaced by duplications of the anterior region of the subsequent segment (Figures 9.19C, 9.8D). Thus, the segmentation genes are transcription factors that use the gradients of the early-cleavage embryo to transform the embryo into a periodic, parasegmental structure.

After the parasegmental boundaries are set, the pair-rule and gap genes interact to regulate the homeotic selector genes, which determine the identity of each segment. By the end of the cellular blastoderm stage, each segment primordium has been given an individual identity by its unique constellation of gap, pair-rule, and homeotic gene products (Levine and Harding 1989).

The gap genes

The gap genes were originally discovered through a series of mutant embryos that lacked groups of consecutive segments (Figure 9.21; Nüsslein-Volhard and Wieschaus 1980). Deletions caused by mutations of the *hunchback*, *Krüppel*, and *knirps* genes span the entire segmented region of the *Drosophila* embryo. The *giant* gap gene overlaps with these three, and mutations of the *tailless* and *huckebein* genes delete portions of the unsegmented termini of the embryo.

The expression of the gap genes is dynamic. There is usually a low level of transcriptional activity across the entire embryo that becomes defined into discrete regions of high activity as cleavage continues (Jäckle et al. 1986). The critical element appears to be the expression of the Hunchback protein, which by the end of nuclear division cycle 12 is found at high levels across the anterior part of the embryo, and then forms a steep gradient through about 15 nuclei. The last third of the embryo has undetectable Hunchback levels. The transcription patterns of the anterior gap genes are initiated by the different concentrations of the Hunchback and Bicoid proteins. High levels of Hunchback protein induce the expression of *giant*, while the *Krüppel* transcript appears over the region where Hunchback begins to decline. High levels of Hunchback protein also prevent the transcription of the posterior gap genes (such as *knirps*) in the anterior part of the embryo (Struhl et al. 1992). It is thought that a gradient of the Caudal protein, highest at the posterior pole, is responsible for activating the abdominal gap genes *knirps* and *giant*. The *giant* gene has two methods for its activation, one for its anterior expression band and one for its posterior expression band (Rivera-Pomar 1995; Schulz and Tautz 1995).

After the establishment of these patterns by the maternal effect genes and Hunchback, the expression of each gap gene

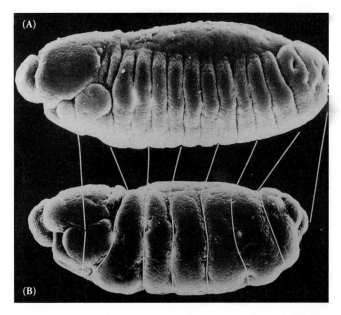

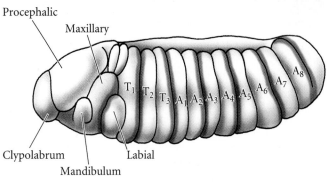

Figure 9.20
Defects seen in the *fushi tarazu* mutant. (A) Scanning electron micrograph of a wild-type embryo, seen in lateral view. (B) Same stage of a *fushi tarazu* mutant embryo. The white lines connect the homologous portions of the segmented germ band. (C) Diagram of wild-type embryonic segmentation. The shaded areas show the parasegments of the germ band that are missing in the mutant embryo. (After Kaufman et al. 1990; photographs courtesy of T. Kaufman.)

(A)   Expression of the gap genes

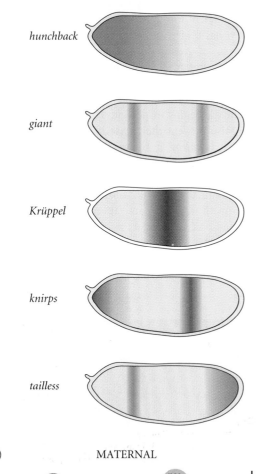

hunchback

giant

Krüppel

knirps

tailless

(B)   MATERNAL

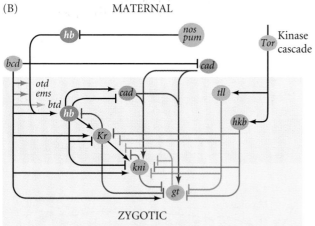

ZYGOTIC

Figure 9.21

Conversion of maternal protein gradients into zygotic gap gene expression. (A) Gap gene expression patterns. (B) The gradients of maternal transcription factors Bicoid, Caudal, and Hunchback regulate the transcription of the gap genes. Hunchback and Caudal proteins come from both maternal messages and new zygotic transcription. These gap-gene encoded proteins diffuse, and the interactions between them are critical in activating the transcription of the pair-rule genes. At the two termini of the embryo, the interaction between Torso and Torsolike activates the *tailless* (*tll*) and *huckebein* (*hkb*) gap genes. (B after Rivera-Pomar and Jäckle 1996.)

becomes stabilized and maintained by interactions between the different gap gene products themselves (Figure 9.21B).* For instance, *Krüppel* gene expression is negatively regulated on its anterior boundary by the Hunchback and Giant proteins and on its posterior boundary by the Knirps and Tailless proteins (Jäckle et al. 1986; Harding and Levine 1988; Hoch et al. 1992). If Hunchback activity is lacking, the domain of *Krüppel* expression extends anteriorly. If Knirps activity is lacking, *Krüppel* gene expression extends more posteriorly. The boundaries between the regions of gap gene transcription are probably created by mutual repression. Just as the Giant and Hunchback proteins can control the anterior boundary of *Krüppel* transcription, so Krüppel protein can determine the posterior boundaries of *giant* and *hunchback* transcription. If an embryo lacks the *Krüppel* gene, *hunchback* transcription continues into the area usually allotted to *Krüppel* (Jäckle et al. 1986; Kraut and Levine 1991). These boundary-forming inhibitions are thought to be directly mediated by the gap gene products, because all four major gap genes (*hunchback*, *giant*, *Krüppel*, and *knirps*) encode DNA-binding proteins that can activate or repress the transcription of other gap genes (Knipple et al. 1985; Gaul and Jäckle 1990; Capovilla et al. 1992).

The pair-rule genes

The first indication of segmentation in the fly embryo comes when the pair-rule genes are expressed during the thirteenth division cycle. The transcription patterns of these genes are striking in that they divide the embryo into the areas that are the precursors of the segmental body plan. As can be seen in Figure 9.22B–E and Figure 9.8C, one vertical band of nuclei (the cells are just beginning to form) expresses a pair-rule gene, then another band of nuclei does not express it, and then another band of nuclei expresses it again. The result is a "zebra stripe" pattern along the anterior-posterior axis, dividing the embryo into 15 subunits (Hafen et al. 1984). Eight genes are currently known to be capable of dividing the early embryo in this fashion; they are listed in Table 9.2. It is important to note that not all nuclei express the same pair-rule genes. In fact, within each parasegment, each row of nuclei has its own constellation of pair-rule gene expression that distinguishes it from any other row.

How are some nuclei of the *Drosophila* embryo told to transcribe a particular gene while their neighbors are told not to transcribe it? The answer appears to come from the distribution of the protein products of the gap genes. Whereas the RNA of each of the gap genes has a very discrete distribution that defines abutting or slightly overlapping regions of expression, the *protein* products of these genes extend more broadly. In fact, they overlap by at least 8–10 nuclei (which at this stage accounts for about two to three segment primordia).

*The interactions between genes and gene products are facilitated by the fact that these reactions occur within a syncytium, in which the cell membranes have not yet formed.

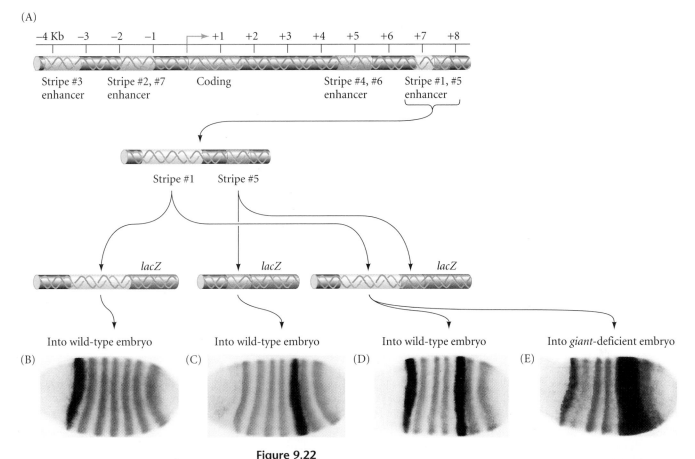

(A)

−4 Kb −3 −2 −1 → +1 +2 +3 +4 +5 +6 +7 +8

Stripe #3   Stripe #2, #7   Coding   Stripe #4, #6   Stripe #1, #5
enhancer   enhancer   enhancer   enhancer

Stripe #1   Stripe #5

lacZ   lacZ   lacZ

Into wild-type embryo   Into wild-type embryo   Into wild-type embryo   Into *giant*-deficient embryo

(B)   (C)   (D)   (E)

Figure 9.22
Specific promoter regions of the *even-skipped* gene control specific transcription bands in the embryo. (A) Partial map of the *eve* promoter, showing the regions responsible for the various stripes. (B–E) A reporter β-galactosidase gene was fused to different regions of the *even-skipped* promoter and injected into fly oocytes. The resulting embryos were stained (orange bands) for the presence of Even-skipped protein. B–D are wild-type embryos that were injected with *lacZ* transgenes containing the enhancer region specific for stripe 1 (B), stripe 5 (C), or both regions (D). In E, the enhancer region for stripes 1 and 5 was injected into an embryo deficient in Giant. Here, the posterior border of stripe 5 is missing. (After Fujioka et al. 1999 and Sackerson et al. 1999; photographs courtesy of M. Fujioka and J. B. Jaynes.)

This was demonstrated in a striking manner by Štanojević and co-workers (1989). They fixed cellularizing blastoderms (i.e., the stage when cells are beginning to form at the rim of the syncytial embryo), stained the Hunchback protein with an antibody carrying a red dye, and simultaneously stained the Krüppel protein with an antibody carrying a green dye. Cellularizing regions that contained both proteins bound both antibodies and were stained bright yellow (see Figure 9.8B). Krüppel protein overlaps with Knirps protein in a similar manner in the posterior region of the embryo (Pankratz et al. 1990).

Three genes are known to be the primary pair-rule genes. These genes—*hairy*, *even-skipped*, and *runt*—are essential for the formation of the periodic pattern, and they are directly controlled by the gap gene proteins. The enhancers of the primary pair-rule genes are recognized by gap

Table 9.2 Major genes affecting segmentation pattern in *Drosophila*

Category		Category	
Gap genes	*Krüppel (Kr)*	Pair-rule genes Secondary	*fushi tarazu (ftz)*
	knirps (kni)		*odd-paired (opa)*
	hunchback (hb)		*odd-skipped (odd)*
	giant (gt)		*sloppy-paired (slp)*
	tailless (tll)		*paired (prd)*
	huckebein (hkb)		
	buttonhead (btd)	Segment polarity genes	*engrailed (en)*
	empty spiracles (ems)		*wingless (wg)*
	orthodenticle (otd)		*cubitus interruptus[D] (ci[D])*
			hedgehog (hh)
Pair-rule genes Primary	*hairy (h)*		*fused (fu)*
	even-skipped (eve)		*armadillo (arm)*
	runt (run)		*patched (ptc)*
			gooseberry (gsb)
			pangolin (pan)

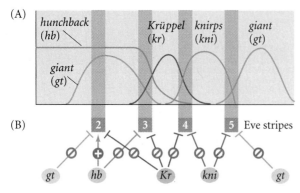

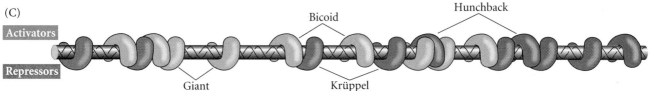

Figure 9.23

Hypothesis for the formation of the second stripe of transcription from the *even-skipped* gene. (A) The *even-skipped* gene is active where concentrations of most of the gap gene proteins are low. (B) Thus, the boundaries of *eve* transcription are determined by high concentrations of these proteins. Different enhancer elements contain binding sequences for different transcription factors. In the enhancer for the second *eve* transcription band, the binding of Hunchback protein stimulates transcription. In the enhancer for the third band, it inhibits transcription. (C) Enhancer element for stripe 2 regulation, containing binding sequences for Krüppel, Giant, Bicoid, and Hunchback proteins. Note that nearly every activator site is closely linked to a repressor site, suggesting competitive interactions at these positions. (A and B after Reinitz and Sharp 1995; C after Štanojevíc et al. 1991.)

gene proteins, and it is thought that the different concentrations of gap gene proteins determine whether a pair-rule gene is transcribed or not. The enhancers of the primary pair-rule genes are often modular: the control over each stripe is located in a discrete region of the DNA. One of the best-studied enhancers is that for the *even-skipped* gene. The structure of this enhancer is shown in Figure 9.22A. It is composed of modular units arranged in such a way that each unit regulates a separate stripe. For instance, the second *even-skipped* stripe is repressed by both Giant and Krüppel proteins and is activated by Hunchback protein and low concentrations of Bicoid (Figure 9.23; Small et al. 1991, 1992; Štanojevíc et al. 1991). DNase I footprinting (see Chapter 5) showed that the enhancer region for this stripe contains six binding sites for Krüppel protein, three for Hunchback protein, three for Giant protein, and five for Bicoid protein. Similarly, *even-skipped* stripe 5 is regulated negatively by Krüppel protein (on its anterior border) and by Giant protein (on its posterior border) (Small et al. 1996; Fujioka 1999).

The importance of these enhancers can be shown by both genetic and biochemical means. First, a mutation in a particular enhancer can delete its particular stripe and no other. Second, if a reporter gene such as *lacZ* (encoding β-galactosidase) is fused to one of these enhancer elements, the *lacZ* gene is expressed only in that particular stripe (see Figure 9.22; Fujioka et al. 1999). Third, the placement of the stripes can be altered by deleting the gap genes that regulate them. Thus, the placement of the stripes of pair-rule gene expression is a result of (1) the modular *cis*-regulatory enhancer elements of the pair-rule genes and (2) the *trans*-regulatory gap gene proteins that bind to these enhancer sites.

Once initiated by the gap gene proteins, the transcription pattern of the primary pair-rule genes becomes stabilized by their interactions among themselves (Levine and Harding

1989). The primary pair-rule genes also form the context that allows or inhibits the expression of the later-acting secondary pair-rule genes. One such secondary pair-rule gene is **fushi tarazu** (**ftz**; Japanese, "too few segments;" Figures 9.8, 9.19, 9.20). Early in cycle 14, *ftz* RNA and protein are seen throughout the segmented portion of the embryo. However, as the proteins from the primary pair-rule genes begin to interact with the *ftz* enhancer, the *ftz* gene is repressed in certain bands of nuclei to create interstripe regions. Meanwhile, the Ftz protein interacts with its own promoter to stimulate more transcription of the *ftz* gene (Figure 9.24; Edgar et al. 1986b; Karr and Kornberg 1989; Schier and Gehring 1992).

The expression of the each pair-rule gene in seven stripes divides the embryo into fourteen parasegments, with each pair-rule gene being expressed in alternate parasegments. Moreover, each row of nuclei within each parasegment expresses a particular and unique combination of pair-rule products. These products will activate the next level of segmentation genes, the segment polarity genes.

The segment polarity genes

So far, our discussion has described interactions between molecules within the syncytial embryo. But once cells form, interactions take place between the cells. These intercellular interactions are mediated by the segment polarity genes, and they accomplish two important tasks. First, they reinforce the parasegmental periodicity established by the earlier transcription factors. Second, through this cell-to-cell signaling, cell fates are established within each parasegment.

The segment polarity genes encode proteins that are constituents of the Wingless and Hedgehog signal transduction pathways (see Chapter 6). Mutations in these genes lead to defects in segmentation and in gene expression pattern across each parasegment. The development of the normal pattern re-

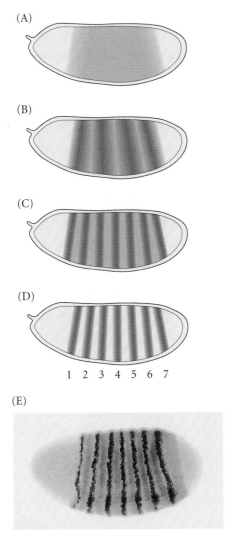

(A)

(B)

(C)

(D)

1 2 3 4 5 6 7

(E)

Figure 9.24
Transcription of the *fushi tarazu* gene in the *Drosophila* embryo.
(A–D) At the beginning of cycle 14, there is low-level transcription
of *ftz* in each of the nuclei in the segmented region of the embryo.
Within the next 30 minutes, the expression pattern alters as *ftz* transcription is enhanced in certain regions (which form stripes) and
repressed in the interstripe regions. (E) Double labeling of the *even-
skipped* (blue bands) and *fushi tarazu* (brown bands) transcripts,
showing that *ftz* is expressed between the *eve* bands at this stage.
(A–D after Karr and Kornberg 1989; E courtesy of J. B. Jaynes and
M. Fujioka.)

lies on the fact only one row of cells in each parasegment is
permitted to express the Hedgehog protein, and only one row
of cells in each parasegment is permitted to express the
Wingless protein. The key to this pattern is the activation of
the *engrailed* gene in those cells that are going to express the
Hedgehog protein. The *engrailed* gene is activated when cells
have high levels of the Even-skipped, Fushi tarazu, or Paired
transcription factors. Moreover, it is repressed in those cells
that receive high levels of Odd-skipped, Runt, or Sloppy-

paired proteins. As a result, Engrailed is expressed in fourteen
stripes across the anterior-posterior axis of the embryo (see
Figure 9.8D). (Indeed, in mutations that cause the embryo to
be deficient in Fushi tarazu, only seven bands of Engrailed are
expressed.) These stripes of *engrailed* transcription mark the
anterior boundary of each parasegment (and the posterior
border of each segment). The *wingless* gene is activated in
those bands of cells that receive little or no Even-skipped or
Fushi tarazu proteins, but which do contain the Sloppy-paired
protein. This causes *wingless* to be transcribed solely in the
row of cells directly anterior to the cells where *engrailed* is
transcribed (Figure 9.25).

Once *wingless* and *engrailed* expression is established in
adjacent cells, this pattern must be maintained to retain the
parasegmental periodicity of the body plan established by the
pair-rule genes. It should be remembered that the mRNAs and
proteins involved in initiating these patterns are short-lived,
and that the patterns must be maintained after their initiators
are no longer being synthesized. The maintenance of these
patterns is regulated by interactions between cells expressing
wingless and those expressing *engrailed*. The Wingless protein,
secreted from the *wingless*-expressing cells, diffuses to adjacent cells. The cells expressing *engrailed* can bind this protein
because they contain the *Drosophila* membrane receptor protein for Wingless, D-Frizzled-2 (see Figure 6.23; Bhanot et al.
1996). This receptor activates the Wnt signal transduction
pathway, resulting in the continued expression of *engrailed*
(Siegfried et al. 1994).

Moreover, this activation starts another portion of this
reciprocal pathway. The Engrailed protein activates the transcription of the *hedgehog* gene in the *engrailed*-expressing cells.
The Hedgehog protein can bind to the Hedgehog receptor (the
Patched protein) on neighboring cells. When it binds to the adjacent posterior cells, it stimulates the expression of the *wingless* gene. The result is a reciprocal loop wherein the Engrailed-
synthesizing cells secrete the Hedgehog protein, which
maintains the expression of the *wingless* gene in the neighboring cells, while the Wingless-secreting cells maintain the expression of the *engrailed* and *hedgehog* genes in their neighbors
in turn (Heemskerk et al. 1991; Ingham et al. 1991; Mohler and
Vani 1992). In this way, the transcription pattern of these two
types of cells is stabilized. This interaction creates a stable
boundary, as well as a signaling center from which Hedgehog
and Wingless proteins diffuse across the parasegment.

The diffusion of these proteins is thought to provide the
gradients by which the cells of the parasegment acquire their
identities. This process can be seen in the dorsal epidermis,
where the rows of larval cells produce different cuticular
structures depending on their position within the segment.
The 1° row consists of large, pigmented spikes called denticles.
Posterior to these cells, the 2° row produces a smooth epidermal cuticle. The next two cell rows have a 3° fate, making
small, thick hairs, and these are followed by several rows of
cells that adopt the 4° fate, producing fine hairs (Figure 9.26).

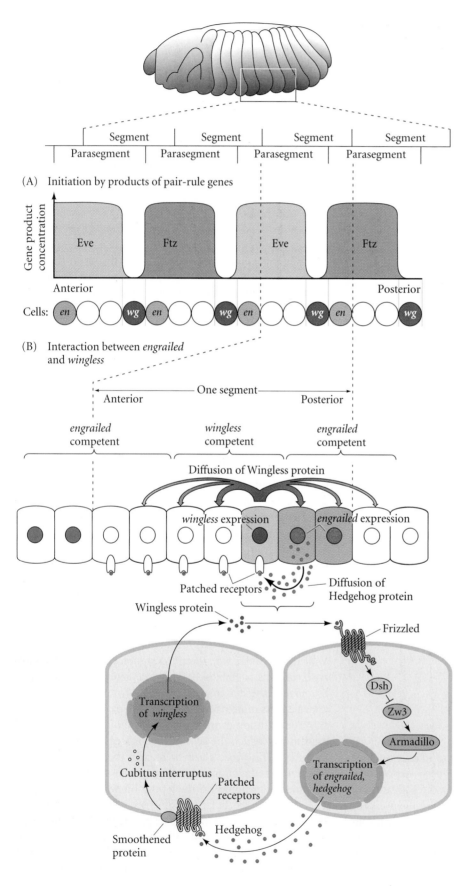

(A) Initiation by products of pair-rule genes

(B) Interaction between *engrailed* and *wingless*

Figure 9.25
Model for the transcription of the segment polarity genes *engrailed* (*en*) and *wingless* (*wg*). (A) The expression of *wg* and *en* is initiated by pair-rule genes. The *en* gene is expressed when the cells contain high concentrations of either Even-skipped or Fushi tarazu proteins. The *wg* gene is transcribed when neither *eve* or *ftz* genes are active, but a third gene (probably *odd-paired*) is expressed. (B) The continued expression of *wg* and *en* is maintained by interactions between the Engrailed- and Wingless-expressing cells. Wingless protein is secreted and diffuses to the surrounding cells. In those cells competent to express Engrailed (having Eve or Ftz proteins), Wingless protein is bound by the Frizzled receptor, which enables the activation of the *en* gene. (Armadillo is the *Drosophila* name for β-catenin.) Engrailed protein activates the transcription of the *hedgehog* gene and also activates its own (*en*) gene transcription. Hedgehog protein diffuses from these cells and binds to the Patched protein. This binding prevents the Patched protein from inhibiting signaling by the Smoothened protein. That signal enables the transcription of the *wg* gene and the subsequent secretion of the Wingless protein.

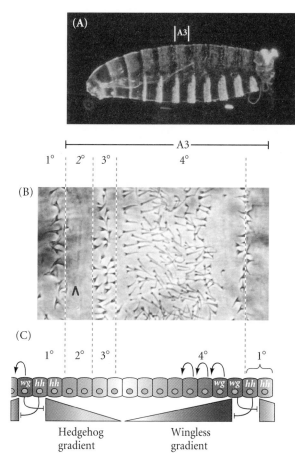

Figure 9.26
Cell specification by the Wingless/Hedgehog signaling center. (A) Bright-field photograph of wild-type *Drosophila* embryo, showing the position of the third abdominal segment. (B) Close-up of the dorsal area of the A3 segment, showing the different cuticular structures made by the 1°, 2°, 3°, and 4° rows of cells. (C) Model for the role of Wingless and Hedgehog. Each signal is responsible for about half the pattern. Either each signal acts in a graded manner (shown here as gradients decreasing with distance from their respective sources) to specify the fates of cells at a distance from these sources, or each signal acts locally on the neighboring cells to initiate a cascade of inductions (shown here as sequential arrows). (After Heemskerk and DiNardo 1994; photographs courtesy of the authors.)

WEBSITE 9.7 Asymmetrical spread of morphogens. It is unlikely that morphogens such as Wingless spread by free diffusion. The asymmetry of Wingless diffusion suggests that neighboring cells play a crucial role in moving the protein.

WEBSITE 9.8 Getting a head in the fly. The segment polarity genes may act differently in the head than in the trunk. Indeed, the formation of the *Drosophila* head may differ significantly from the way the rest of the body is formed.

The *wingless*-expressing cells lie within the region producing the fine hairs, while the *hedgehog*-expressing cells are near the 1° row of cells. The fates of the cells can be altered by experimentally increasing or decreasing the levels of Hedgehog or Wingless protein (Heemskerek and DiNardo 1994; Bokor and DiNardo 1996; Porter et al. 1996). For example, if the *hedgehog* gene is fused to a heat shock promoter and the embryos are grown at a temperature that activates the gene, more Hedgehog protein is made, and the cells normally showing 3° fates will become 2° cells. The rows of 4° cells farthest from the Wingless-secreting cells may also become 3° or 2° cells. It seems that the cells closest to the Wingless secreters cannot respond to Hedgehog, and Hedgehog cannot, by itself, specify the 1° fate (which may require the expression of certain pair-rule gene products). Thus, Hedgehog and Wingless appear necessary for elaborating the entire pattern of cell types across the parasegment. However, the mechanism by which they accomplish this specification is not clear. Either these signals act in a graded fashion, as morphogens, or they act locally to initiate a cascade of local signaling events, in which each interaction uses a different ligand and receptor (Figure 9.26). The resulting pattern of cell fates also changes the focus of patterning from parasegment to segment. There are now external markers, as the *engrailed*-expressing cells become the most posterior cells of each segment.

The Homeotic Selector Genes

Patterns of homeotic gene expression

After the segmental boundaries have been established, the characteristic structures of each segment are specified. This specification is accomplished by the **homeotic selector genes** (Lewis 1978). There are two regions of *Drosophila* chromosome 3 that contain most of these homeotic genes (Figure 9.27). One region, the **Antennapedia complex**, contains the homeotic genes *labial* (*lab*), *Antennapedia* (*Antp*), *sex combs reduced* (*scr*), *deformed* (*dfd*), and *proboscipedia* (*pb*). The *labial* and *deformed* genes specify the head segments, while *sex combs reduced* and *Antennapedia* contribute to giving the thoracic segments their identities. The *proboscipedia* gene appears to act only in adults, but in its absence, the labial palps of the mouth are transformed into legs (Wakimoto et al. 1984; Kaufman et al. 1990). The second region of homeotic genes is the **bithorax complex** (Lewis 1978). There are three protein-coding genes found in this complex: *ultrabithorax* (*ubx*), which is required for the identity of the third thoracic segment; and the *abdominal A* (*abdA*) and *Abdominal B* (*AbdB*) genes, which are responsible for the segmental identities of the abdominal segments (Sánchez-Herrero et al. 1985). The lethal phenotype of the triple-point mutant Ubx⁻, abdA⁻, AbdB⁻ is identical to that resulting from a deletion of the entire bithorax complex (Casanova et al. 1987). The chromosome region containing both the Antennapedia complex and the bithorax complex is often referred to as the **homeotic complex (Hom-C)**.

Figure 9.27
Homeotic gene expression in *Drosophila*. In the center are the genes of the Antennapedia and bithorax complexes and their functional domains. Below and above the gene map are the regions of homeotic gene expression (both mRNA and protein) in the blastoderm of the *Drosophila* embryo and the regions that form in the adult fly. Darker shaded areas are those segments or parasegments with the most product. (After Dessain et al. 1992 and Kaufman et al. 1990.)

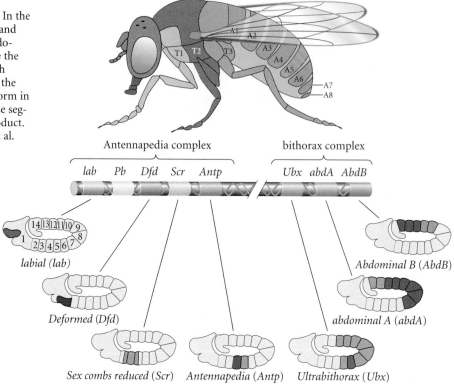

Because these genes are responsible for the specification of fly body parts, mutations in them lead to bizarre phenotypes. In 1894, William Bateson called these organisms "homeotic mutants," and they have fascinated developmental biologists for decades.* For example, the body of the normal adult fly contains three thoracic segments, all of which produce a pair of legs. The first thoracic segment does not produce any further appendages, but the second thoracic segment produces both a set of legs and a set of wings. The third thoracic segment produces a set of wings and a set of balancers known as **halteres**. In homeotic mutants, these specific segmental identities can be changed. When the *ultrabithorax* gene is deleted, the third thoracic segment (which is characterized by halteres) becomes transformed into another second thoracic segment. The result (Figure 9.28) is a fly with four wings—an embarrassing situation for a classic dipteran.† Similarly, the Antennapedia protein is usually used to specify the second

thoracic segment of the fly. But when flies have a mutation wherein the *Antennapedia* gene is expressed in the head (as well as in the thorax), legs rather than antennae grow out of the head sockets (Figure 9.29). In the recessive mutant of *Antennapedia*, the gene fails to be expressed in the second thoracic segment, and antennae sprout out of the leg positions (Struhl 1981; Frischer et al. 1986; Schneuwly et al. 1987).

These major homeotic selector genes have been cloned and their expression analyzed by in situ hybridization (Harding et al. 1985; Akam 1987). Transcripts from each gene can be detected in specific regions of the embryo and are especially prominent in the central nervous system (see Figure 9.27).

Initiating the patterns of homeotic gene expression

The initial domains of homeotic gene expression are influenced by the gap genes and pair-rule genes. For instance, the expression of the *abdA* and *AbdB* genes is repressed by the gap gene proteins Hunchback and Krüppel. This inhibition prevents these abdomen-specifying genes from being expressed in the head and thorax (Casares and Sánchez-Herrero 1995). Conversely, the *Ultrabithorax* gene is activated by certain lev-

---

Homeo means "similar." *Homeotic mutants* are mutants in which one structure is replaced by another (as where an antenna is replaced by a leg). *Homeotic genes* are genes whose mutation can cause such transformations; thus, they are genes that specify the identity of a particular body segment. The *homeobox* is a conserved DNA sequence of about 180 base pairs that is shared by many homeotic genes. This sequence encodes the 60-amino acid *homeodomain*, which recognizes specific DNA sequences. The homeodomain is an important region of the transcription factors encoded by homeotic genes (see Sidelights & Speculations). Not all genes with homeoboxes are homeotic genes, however.

†Dipterans (two-winged insects such as flies) are thought to have evolved from four-winged insects; it is possible that this change arose via alterations in the bithorax complex. Chapter 22 includes more speculation on the relationship between the homeotic complex and evolution.

Figure 9.28
This four-winged fruit fly was constructed by putting together three mutations in *cis* regulators of the *ultrabithorax* gene. These mutations effectively transform the third thoracic segment into another second thoracic segment (i.e., halteres into wings). (Photograph courtesy of E. B. Lewis.)

els of the Hunchback protein, so that it is originally transcribed in a broad band in the middle of the embryo, and the transcription of *Antennapedia* is activated by Krüppel (Harding and Levine 1988; Struhl et al. 1992). The boundaries of homeotic gene expression are soon confined to the parasegments defined by the Fushi tarazu and Even-skipped proteins (Ingham and Martinez-Arias 1986; Müller and Bienz 1992).

Maintaining the patterns of homeotic gene expression

The expression of homeotic genes is a dynamic process. The *Antennapedia* gene (*Antp*), for instance, although initially expressed in presumptive parasegment 4, soon appears in parasegment 5. As the germ band expands, *Antp* expression is seen in the presumptive neural tube as far posterior as parasegment 12. During further development, the domain of *Antp* expression contracts again, and *Antp* transcripts are localized strongly to parasegments 4 and 5. Like that of other homeotic genes, *Antp* expression is negatively regulated by all the homeotic gene products expressed posterior to it (Harding and Levine 1989; González-Reyes and Morata 1990). In other words, each of the bithorax complex genes represses the expression of *Antennapedia*. If the *Ultrabithorax* gene is deleted, *Antp* activity extends through the region that would normally have expressed *Ubx* and stops where the *Abd* region begins. (This allows the third thoracic segment to form wings like the second thoracic segment, as in Figure 9.29.) If the entire bithorax complex is deleted, *Antp* expression extends throughout the abdomen. (Such a larva does not survive, but the cuticle pattern throughout the abdomen is that of the second thoracic segment.)

As we saw above, the gap gene and pair-rule gene proteins are transient, but the identities of the segments must be stabi-

lized so that differentiation can occur. Thus, once the transcription patterns of the homeotic genes have become stabilized, they are "locked" into place by alteration of the chromatin conformation in these genes. The repression of homeotic genes appears to be maintained by the Polycomb family of proteins, while the active chromatin conformation appears to be maintained by the Trithorax proteins (Ingham and Whittle 1980; McKeon and Brock 1991; Simon et al. 1992).

Realisator genes

The search is now on for "realisator genes," those genes that are the targets of the homeotic gene proteins and which function to form the specified tissue or organ primordia. In the formation of the second thoracic segment, for example, *Antennapedia* is expressed. Casares and Mann (1998) have shown that Antennapedia protein binds to the enhancer of the

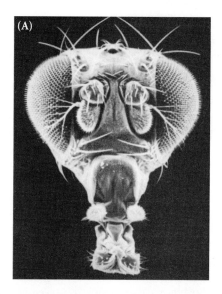

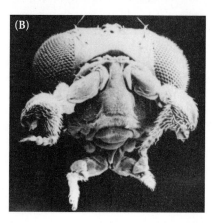

Figure 9.29
(A) Head of a wild-type fly. (B) Head of a fly containing the Antennapedia mutation that converts antennae into legs. (From Kaufman et al. 1990; photographs courtesy of T. C. Kaufman.)

(A)   (B)

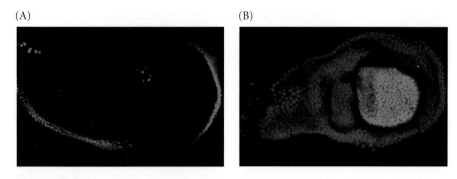

Figure 9.30
Antibody staining of the Ultrabithorax protein in (A) the wing disc and (B) the haltere disc of third instar *Drosophila* larvae. These are the cells that will give rise to a wing and a haltere, respectively. In the wing disc, Ultrabithorax staining can be seen only on the cells that form the peripodial membrane, and not on those that form the wing itself. In the haltere disc, the Ultrabithorax protein is found in those cells that will produce the major portion of the haltere. (From Weatherbee et al. 1998; photographs courtesy of S. D. Weatherbee and S. Carroll.)

homothorax gene and prevent its expression. Homothorax is necessary for producing a transcription factor critical for antenna formation. Therefore, one of Antennapedia's functions is to suppress those genes necessary for antenna development.

The Ultrabithorax protein is able to repress the expression of the *Wingless* gene in those cells that will become the halteres of the fly. One of the major differences between the appendage-forming cells of the second and the third thoracic segments is that Wingless expression occurs in the appendage-forming cells of the second thoracic segment, but not in those of the third thoracic segment. Wingless acts as a growth promoter and morphogen in these tissues. In the third thoracic segment, Ubx protein is found in these cells, and it prevents the expression of the *Wingless* gene (Figure 9.30; Weatherbee et al. 1998). Thus, one of the ways in which Ubx protein specifies the third thoracic segment is by preventing the expression of those genes that would generate the wing tissue.

Another target of the homeotic proteins, the *distal-less* gene (itself a homeobox-containing gene: see Sidelights and Speculations), is necessary for limb development and is active solely in the thorax. Distal-less expression is repressed in the abdomen, probably by a combination of Ubx and AbdA proteins that can bind to its enhancer and block transcription (Vachon et al. 1992; Castelli-Gair and Akam 1995). This presents a paradox, since parasegment 5 (entirely thoracic and leg-producing) and parasegment 6 (which includes most of the legless first abdominal segment) both expressUbx. How can these two very different segments be specified by the same gene? Castelli-Gair and Akam (1995) have shown that the mere presence of Ubx

protein in a group of cells is not sufficient for specification. Rather, the time and place of its expression within the parasegment can be critical. Before *Ubx* expression, parasegments 4–6 have similar potentials. At division cycle 10, *Ubx* expression in the anterior parts of parasegments 5 and 6 prevents those parasegments from forming structures (such as the anterior spiracle) characteristic of parasegment 4. Moreover, in the posterior compartment of parasegment 6 (but not parasegment 5), Ubx protein blocks the formation of the limb primordium by repressing the *distal-less* gene. At division cycle 11, by which time Ubx has pervaded all of parasegment 6, the *distal-less* gene has become self-regulatory and cannot be repressed by Ubx (Figure 9.31).

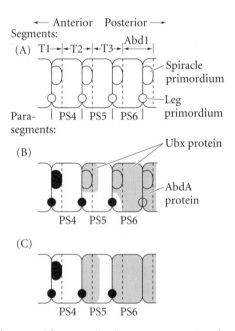

Figure 9.31
Schematic representation of the differences between *Ubx* expression in parasegments 5 and 6. (A) Before *Ubx* expression, each parasegment is competent to make both spiracles and legs. (B) At division cycle 10, early *Ubx* expression blocks the formation of the anterior spiracle in PS5 and PS6, and prevents limb formation in the posterior compartment of PS6. AbdA protein plays the same role in other abdominal segments. (C) At cycle 11, the *Ubx* expression domain extends to the limb primordia of PS5 and PS6, but it is "too late" to repress *distal-less* gene expression. (After Castelli-Gair and Akam 1995.)

The Homeodomain Proteins

HOMEODOMAIN PROTEINS are a family of transcription factors characterized by a 60-amino acid domain (the **homeodomain**) that binds to certain regions of DNA. The homeodomain was first discovered in those proteins whose absence or misregulation caused homeotic transformations of *Drosophila* segments. It is thought that homeodomain proteins activate batteries of genes that specify the particular properties of each segment. The homeodomain proteins include the products of the eight genes of the homeotic complex, as well as other proteins such as Fushi tarazu, Caudal, Distal-less and Bicoid. Homeodomain proteins are important in determining the anterior-posterior axes of both invertebrates and vertebrates. In *Drosophila*, the presence of certain homeodomain proteins is also necessary for the determination of specific neurons. Without these transcription factors, the fates of these neuronal cells are altered (Doe et al. 1988).

The homeodomain is encoded by a 180-base-pair DNA sequence known as the **homeobox**. The homeodomains appear to be the sites of these proteins that bind DNA, and they are critical in specifying cell fates. For instance, if a chimeric protein is constructed mostly of Antennapedia but with the carboxyl terminus (including the homeodomain) of Ultrabithorax, it can substitute for Ultrabithorax and specify the appropriate cells as parasegment 6 (Mann and Hogness 1990). The isolated homeodomain of Antennapedia will bind to the same promoters as the entire Antennapedia protein, indicating that the binding of this protein is dependent on its homeodomain (Müller et al. 1988).

The homeodomain folds into three α helices, the latter two folding into a helix-turn-helix conformation that is characteristic of transcription factors that bind DNA in the major groove of the double helix (Otting et al. 1990; Percival-Smith et al. 1990). The third

helix is the recognition helix, and it is here that the amino acids make contact with the bases of the DNA. A four-base motif, TAAT, is conserved in nearly all sites recognized by homeodomains; it probably distinguishes those sites to which homeodomain proteins can bind. The 5′ terminal T appears to be critical in this recognition, as mutating it destroys all homeodomain binding. The base pairs following the TAAT motif are important in distinguishing between similar recognition sites. For instance, the next base pair is recognized by amino acid 9 of the recognition helix. Mutation studies have shown that the Bicoid and Antennapedia homeodomain proteins use lysine and glutamine, respectively, at position 9 to distinguish related recognition sites. The lysine of the Bicoid homeodomain recognizes the G of CG pairs, while the glutamine of the Antennapedia homeodomain recognizes the A of AT pairs (Figure 9.32; Hanes and Brent 1991). If the lysine in

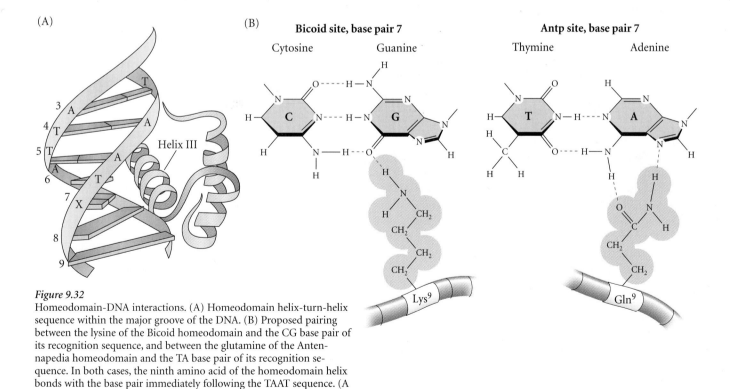

(A)

(B) **Bicoid site, base pair 7**
Cytosine   Guanine

Antp site, base pair 7
Thymine   Adenine

Figure 9.32
Homeodomain-DNA interactions. (A) Homeodomain helix-turn-helix sequence within the major groove of the DNA. (B) Proposed pairing between the lysine of the Bicoid homeodomain and the CG base pair of its recognition sequence, and between the glutamine of the Antennapedia homeodomain and the TA base pair of its recognition sequence. In both cases, the ninth amino acid of the homeodomain helix bonds with the base pair immediately following the TAAT sequence. (A after Riddihough 1992; B after Hanes and Brent 1991)

Bicoid is replaced by glutamine, the resulting protein will recognize Antennapedia-binding sites (Hanes and Brent 1989, 1991). Other homeodomain proteins show a similar pattern, in which one portion of the homeodomain recognizes the common TAAT sequence, while another portion recognizes a specific structure adjacent to it.

Cofactors for the *Hom-C* Genes

The genes of the *Drosophila* homeotic complex specify segmental fates, but they may need some help in doing it. The DNA-binding sites recognized by the homeodomains of the Hom-C proteins are very similar, and there is some overlap in their binding specificity. In 1990, Peifer and Wieschaus discovered that the product of the *Extradenticle* (*Exd*) gene interacts with several Hom-C proteins and may help specify the segmental identities. For instance, the Ubx protein is responsible for specifying the identity of the first abdominal segment (A1). Without Extradenticle protein, it will transform this segment into A3. Moreover, the Exd and Ubx proteins are both needed for the regulation of the *decapentaplegic* gene, and the structure of the *decapentaplegic* promoter suggests that the Extradenticle protein may dimerize with the Ubx protein on the enhancer of this target gene (Raskolb and Wieschaus 1994; van Dyke and Murre 1994). The Extradenticle protein includes a homeodomain, and the human protein PBX1, which resembles the Extradenticle protein, may play a similar role as a cofactor for human homeotic genes.

The product of the *teashirt* gene may also be an important cofactor. This zinc finger transcription factor is necessary for the functioning of the Sex combs reduced protein, which distinguishes between the labial and first thoracic segments. It is critical for the specification of the anterior prothoracic (parasegment 3) identity, and it may be the gene that specifies the "groundstate condition" of the homeotic complex. If the bithorax complex and the *Antennapedia* gene are removed, all the segments become anterior prothorax. The product of the *teashirt* gene appears to work with the Scr protein to distinguish thorax from head and to work throughout the trunk to prevent head structures from forming (Roder et al. 1992). ■

WEBSITE **9.9 Homeotic genes and their protein products.** The *Hom-C* genes have fascinating structures. The protein products of these genes bind to DNA in the presence of other proteins that may allow them to recognize specific sequences of DNA.

The Generation of Dorsal-Ventral Polarity

In 1936, embryologist E. E. Just criticized those geneticists who sought to explain *Drosophila* development by looking at specific mutations affecting eye color, bristle number, and wing shape. He said that he wasn't interested in the development of the bristles of a fly's back; rather, he wanted to know how the fly embryo makes the back itself. Fifty years later, embryologists and geneticists are finally answering that question.*

The Morphogenetic Agent for Dorsal-Ventral Polarity

Dorsal-ventral polarity is established by the gradient of a transcription factor called **Dorsal**. Unlike Bicoid, whose gradient is established within a syncytium, Dorsal forms a gradient over a field of cells that is established as a consequence of cell-to-cell signaling events.

The specification of the dorsal-ventral axis takes place in several steps. The critical step is the translocation of the Dorsal protein from the cytoplasm into the nuclei of the ventral cells during the fourteenth division cycle. Anderson and Nüsslein-Volhard (1984) isolated 11 maternal effect genes, each of whose absence is associated with a lack of ventral structures (Figure 9.33). The absence of another maternal effect gene, *cactus*, causes the ventralization of all cells. The proteins encoded by these maternal genes are critical for making certain that the Dorsal protein gets into only those nuclei on the ventral surface of the embryo.† After its translocation, the Dorsal protein acts on cell nuclei to specify the different regions of the embryo. Different concentrations of Dorsal protein in the nuclei appear to specify different fates in those cells.

The Translocation of Dorsal Protein

The protein that actually distinguishes dorsum (back) from ventrum (belly) is the product of the *dorsal* gene. The RNA transcript of the mother's *dorsal* gene is placed in the oocyte by her ovarian cells. However, Dorsal protein is not synthesized from this maternal message until about 90 minutes after fertilization. When this protein is translated, it is found throughout the embryo, not just on the ventral or dorsal side. How, then, can this protein act as a morphogen if it is located everywhere in the embryo?

In 1989, the surprising answer was found (Roth et al. 1989; Rushlow et al. 1989; Steward 1989). While Dorsal protein can be found throughout the syncytial blastoderm of the early *Drosophila* embryo, it is translocated into nuclei only in the ventral part of the embryo (Figure 9.34A, B).In the nucleus, Dorsal binds to certain genes to activate or suppress their transcription. If Dorsal does not enter the nucleus, the genes

---

*In a manner that Just could not have predicted, it turns out that some of the genes (such as *decapentaplegic*) that are involved in regulating bristle number or wing shape also have earlier functions regulating dorsal-ventral polarity.

†Remember that a gene in *Drosophila* is usually named after its *mutant* phenotype. Thus, the product of the *dorsal* gene is necessary for the differentiation of *ventral* cells. That is, in the absence of the *dorsal* gene, the ventral cells become dorsalized.

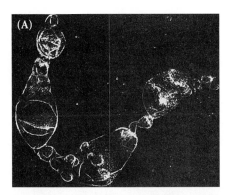

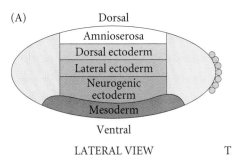

Figure 9.33

Effect of mutations affecting the distribution of the Dorsal protein. (A) Deformed larva consisting entirely of dorsal cells. Larvae like these developed from the eggs of a female homozygous for a mutation of the *snake* gene, one of the maternal effect genes involved in the signaling cascade that establishes a gradient of Dorsal in the embryo. (B) Larvae developed from snake mutant eggs that received injections of mRNA from wild-type eggs. These larvae have a wild-type appearance. (From Anderson and Nüsslein-Volhard 1984; photographs courtesy of C. Nüsslein-Volhard.)

responsible for specifying ventral cell types (*snail* and *twist*) are not transcribed, the genes responsible for specifying dorsal cell types (*decapentaplegic* and *zerknüllt*) are not repressed, and all the cells of the embryo become specified as dorsal cells.

This model of dorsal-ventral axis formation in *Drosophila* is supported by analyses of mutations that give rise to an entirely dorsalized or an entirely ventralized phenotype (see Figures 9.33A and 9.34). In those mutants in which all the cells are dorsalized (as is evident by their dorsal cuticle), Dorsal protein does not enter the nucleus in any cell. Conversely, in those mutants in which all cells have a ventral phenotype, Dorsal protein is found in every cell nucleus.

The follicular epithelium surrounding the developing oocyte is initially symmetrical, but this symmetry is broken by a signal from the oocyte nucleus. The oocyte nucleus is originally located at the posterior end of the oocyte, away from the nurse cells. It then moves to an anterior dorsal position and signals the overlying follicle cells to become the more columnar dorsal follicle cells (Montell et al. 1991; Schüpbach et al. 1991). The dorsalizing signal from the oocyte nucleus appears to be the product of the **gurken** gene (Schüpbach 1987; Forlani et al. 1993). The *gurken* message becomes localized in a crescent between the oocyte nucleus and the oocyte plasma membrane, and its protein product forms an anterior-posterior gradient along the dorsal surface of the oocyte (Neuman-Silberg and Schüpbach 1993; Figure 9.36). Since it can diffuse only a short distance, the Gurken protein reaches only those follicle cells closest to the oocyte nucleus. Mutations of the Gurken gene in the mother (and thus in the oocyte) cause the ventralization of both the embryo and its surrounding follicle cells. Mutations of this gene in the mother (and thus in the oocyte) cause the ventralization of both the embryo and its surrounding follicle cells. (If the mutation is in the follicle cells and not in the egg, the embryo is normal.)

The signal cascade

SIGNAL FROM THE OOCYTE NUCLEUS TO THE FOLLICLE CELLS. If Dorsal protein is found throughout the embryo, but gets translocated into the nuclei of only ventral cells, then something else must be providing asymmetrical cues (Figure 9.35). It appears that this signal is mediated through a complex interaction between the oocyte and its surrounding follicle cells.

Figure 9.34

Translocation of Dorsal protein into ventral, but not lateral or dorsal, nuclei. (A) Fate map of a cross section through the *Drosophila* embryo. The most ventral part becomes the mesoderm; the next higher portion becomes the neurogenic (ventral) ectoderm. The lateral and epidermal ectoderm can be distinguished in the cuticle, and the dorsalmost region becomes the amnioserosa, the extraembryonic layer that surrounds the embryo. (B–D) Cross sections of embryos stained with antibody to show the presence of Dorsal protein (dark-stained area). (B) A wild-type embryo, showing Dorsal protein in the ventralmost nuclei. (C) A dorsalized mutant, showing no localization of Dorsal protein in any nucleus. (D) A ventralized mutant, in which Dorsal protein has entered the nucleus of every cell. (A from Rushlow et al. 1989; B–D from Roth et al. 1989, photographs courtesy of the authors.)

(A)

Dorsal

Amnioserosa
Dorsal ectoderm
Lateral ectoderm
Neurogenic ectoderm
Mesoderm

Ventral

LATERAL VIEW

TRANSVERSE SECTION

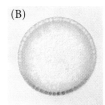

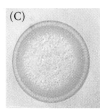

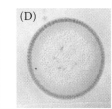

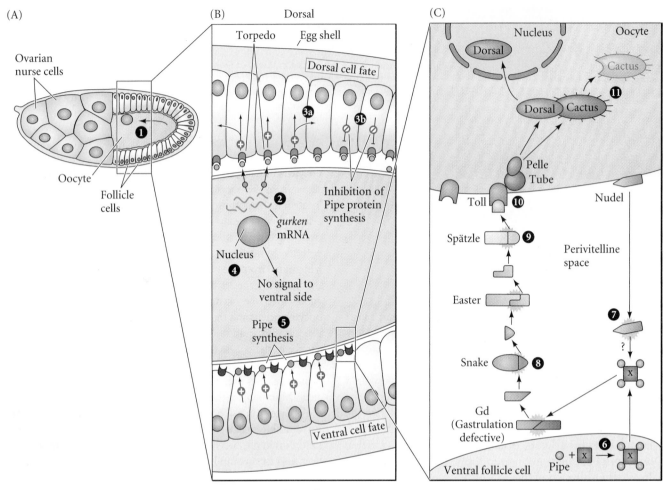

(A)

Ovarian
nurse cells

Oocyte

Follicle
cells

(B)

Dorsal

Torpedo   Egg shell

Dorsal cell fate

3a   3b

Inhibition of
Pipe protein
synthesis

2

gurken
mRNA

Nucleus

4

No signal to
ventral side

Pipe 5
synthesis

Ventral cell fate

Ventral

(C)

Nucleus   Oocyte

Dorsal   Cactus

11

Dorsal   Cactus

Pelle
Tube

Toll   10   Nudel

Spätzle   9   Perivitelline
space

Easter

7

?

Snake   8   x

Gd
(Gastrulation
defective)   6

Ventral follicle cell   Pipe   + x → x

❶ Oocyte nucleus travels to anterior
dorsal side of oocyte. It synthesizes
gurken mRNA which remains between
the nucleus and the follicle cells.

❷ *gurken* messages are translated.
The Gurken protein is received
by Torpedo proteins during
mid-oogenesis.

❸ₐ Torpedo signal causes follicle cells to
differentiate to a dorsal morphology.

❸ᵦ Synthesis of Pipe protein is inhibited
in dorsal follicle cells.

❹ Gurken protein does not diffuse
to ventral side.

❺ Ventral follicle cells synthesize Pipe
proteins.

❻ In ventral follicle cells, Pipe completes
the modification of unknown factor (x).

❼ Nudel and factor (x) interact to split
the Gastrulation-deficient (Gd) protein.

❽ The activated Gd protein splits the Snake
protein, and the activated Snake protein
cleaves the Easter protein.

❾ The activated Easter protein splits Spätzle;
activated Spätzle binds to Toll receptor protein.

❿ Toll activation activates Tube and Pelle, which
phosphorylate the Cactus protein. Cactus is
degraded, releasing it from Dorsal.

⓫ Dorsal protein enters the nucleus and
ventralizes the cell.

◀ **Figure 9.35**
Schematic representation of the generation of dorsal-ventral polarity in *Drosophila*. (A) The oocyte develops in an ovarian follicle consisting of 15 nurse cells (which supply maternal proteins and messages to the developing egg) and numerous follicle cells. (B) The nucleus of the oocyte travels to what will become the dorsal side of the embryo. The *gurken* genes of the oocyte synthesize mRNA that becomes localized between the oocyte nucleus and the cell membrane, where it is translated into Gurken protein. The Gurken signal is received by the receptor protein made by the *torpedo* gene of the follicle cells. Given the short diffusibility of the signal, only the follicle cells closest to the oocyte nucleus (i.e., the dorsal follicle cells) receive this signal. The signal from the Torpedo receptor causes the follicle cells to take on a characteristic dorsal follicle morphology and (somehow) inhibit the synthesis of Pipe protein. Therefore, this protein is made only by the ventral follicle cells. (C) Pipe modifies an unknown protein (X) and allows it to be secreted from the ventral follicle cells. Nudel protein interacts with this modified factor to split the products of the *gastrulation defective* and *snake* genes to create an active enzyme that will split the zymogen form of the Easter protein into an active Easter protease. The Easter protease splits the Spätzle protein into a form that can bind to the Toll receptor (which is found throughout the embryonic cell membrane). Thus, only the ventral cells receive the Toll signal. This signal separates the Cactus protein from the Dorsal protein, allowing Dorsal to be translocated into the nuclei and ventralize the cells. (After van Eeden and St. Johnston 1999.)

The Gurken signal is received by the follicle cells through a receptor encoded by the **torpedo** gene. Molecular analysis has now established that *gurken* encodes a homologue of the vertebrate epidermal growth factor (EGF), while *torpedo* encodes a homologue of the vertebrate EGF receptor (Price et al. 1989; Neuman-Silberberg and Schüpbach 1993). *Maternal deficiency of *torpedo* causes the ventralization of the embryo. Moreover, the *torpedo* gene is active in the ovarian follicle cells, not in the embryo. This was discovered by making germ line/somatic chimeras. Schüpbach (1987) transplanted germ cell precursors from wild-type embryos to embryos whose mothers carried the torpedo mutation. Conversely, she transplanted the germ cells of the torpedo mutant embryos to the wild-type embryos (Figure 9.37). When mated to wild-type males, the wild-type eggs produced ventralized embryos when they developed within the torpedo mutant mothers' follicles. The torpedo mutant eggs were able to produce normal embryos if they developed within a wild-type ovary. Thus, unlike the *gurken* gene product, the wild-type *torpedo* gene is needed in the follicle cells, not in the egg itself.

SIGNAL FROM THE FOLLICLE CELLS TO THE OOCYTE CYTOPLASM. The activated Torpedo receptor protein inhibits the expression of the *pipe* gene. As a result, the Pipe protein is made only in the ventral follicle cells (Sen et al. 1998). The Pipe protein (in some as yet unknown way) activates the Nudel protein, which is secreted to the cell membrane of the ventral embryonic cells. A few hours later in development, the activated Nudel protein initiates the activation of three serine proteases that are secreted by the embryo into the perivitelline fluid (see Figure 9.35C; Hong and Hashimoto 1995). These three serine proteases are the products of the *gastrulation defective* (*gd*), *snake* (*snk*), and *easter* (*ea*) genes. Like most extracellular proteases, they are secreted in an inactive

(A)

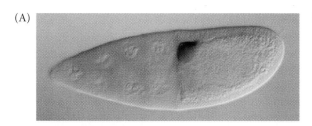

(D)

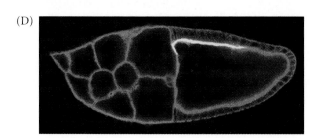

(B)

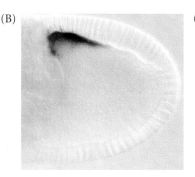

(C)

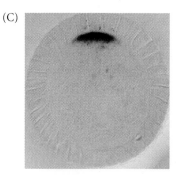

Figure 9.36
Expression of the *gurken* message and protein between the oocyte nucleus and the dorsal anterior cell membrane. (A) The mRNA for the Gurken protein is localized between the oocyte nucleus and the dorsal follicle cells of the ovary. (B) The Gurken protein is similarly located (shown here is a younger stage than A). (C) Cross section of the egg through the region of Gurken protein expression. (D) More mature oocyte, showing Gurken protein (yellow) across the dorsal region. The actin has been stained red, showing cell boundaries. As the oocyte grows, follicle cells migrate across the top of the oocyte, becoming exposed to Gurken. (A after Ray et al. 1996, courtesy of T. Schüpbach; B and C after Peri et al. 1999, courtesy of S. Roth; D courtesy of C. van Buskirk and T. Schüpbach.)

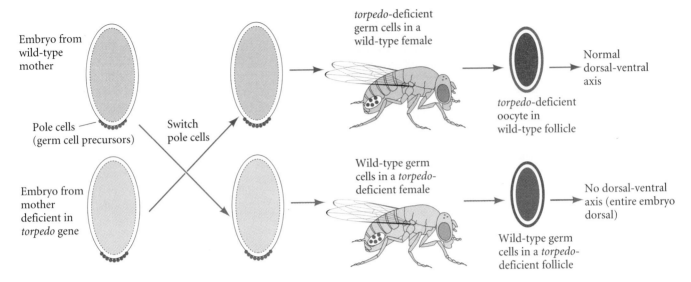

Figure 9.37
Germ line chimeras made by interchanging pole cells (germ cell precursors) between wild-type embryos and embryos from mothers homozygous for the *torpedo* gene. These transplants produce wild-type females whose eggs come from the mutant mothers, and *torpedo*-deficient females that lay wild-type eggs. The *torpedo*-deficient eggs produce normal embryos if they develop in the wild-type ovary, while the wild-type eggs produce ventralized embryos if they eggs develop in the mutant mother's ovary.

form and become activated by peptide cleavage. It is thought that the activated Nudel protein first tethers and activates the Gastrulation defective protein. This protease cleaves the Snake protein. This cleavage activates the Snake protease, which in turn cleaves the Easter protein. This cleavage activates the Easter protein, which cleaves the Spätzle protein (Chasan et al. 1992; Hong and Hashimoto 1995).

The cleaved Spätzle protein is now able to bind to its receptor in the oocyte cell membrane, the product of the *Toll* gene. Toll protein is a maternal product that is evenly distributed throughout the cell membrane of the egg (Hashimoto et al. 1988, 1991), but it becomes activated only by binding the Spätzle protein, which is produced only on the ventral side of the egg. Therefore, the Toll proteins on the ventral side of the egg are transducing a signal into the egg, while the Toll receptors on the dorsal side of the egg are not.

Establishing the dorsal protein gradient

SEPARATION OF THE DORSAL AND CACTUS PROTEINS. The crucial outcome of signaling through the Toll protein is the establishment of a gradient of Dorsal protein in the ventral cell nuclei. How is this gradient established? It appears that the Cactus protein is blocking the portion of the Dorsal protein that enables the Dorsal protein to get into nuclei. As long as this Cactus protein is bound to it, Dorsal protein remains in the cytoplasm. Thus, this entire complex signaling system is

organized to split the Cactus protein from the Dorsal protein in the ventral region of the egg. When Spätzle binds to and activates the Toll protein, the Toll protein can activate the Pelle protein kinase. (The Tube protein is probably necessary for bringing Pelle to the cell membrane, where it can be activated: Galindo et al. 1995.)

The activated Pelle protein kinase can (probably through an intermediate) phosphorylate the Cactus protein. Once phosphorylated, the Cactus protein is degraded, and the Dorsal protein can enter the nucleus (Kidd 1992; Shelton and Wasserman 1993; Whalen and Steward 1993; Reach et al. 1996). Since the signal transduction cascade creates a gradient of Spätzle protein that is highest in the most ventral region, there is a gradient of Dorsal translocation into the ventral cells of the embryo, with the highest concentrations of Dorsal protein in the most ventral cell nuclei.

The process described for the translocation of Dorsal protein into the nucleus is very similar to the process for the translocation of the NF-κB transcription factor into the nucleus of mammalian lymphocytes. In fact, there is substantial homology between NF-κB and Dorsal, between IκB and Cactus, between the Toll protein and the interleukin 1 receptor, between Pelle protein and an IL-1-associated protein kinase, and between the DNA sequences recognized by Dorsal and by NF-κB[*] (González-Crespo and Levine 1994; Cao et al. 1996). Thus, the biochemical pathway used to specify dorsal-ventral polarity in *Drosophila* appears to be homologous to that used to differentiate lymphocytes in mammals (Figure 9.38).

[*]Lemaitre and colleagues (1996) have shown that Toll and its ligand (Spätzle) are also involved in the *Drosophila* immune response against fungal infections. Moreover, several *toll*-related genes have been discovered in humans, which may be involved in both the immune response and early development (Rock et al. 1998; Qureshi et al. 1999).

(A) *Drosophila*   (B) Mammals

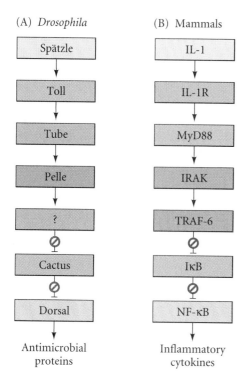

Figure 9.38

Model of a conserved pathway for regulating nuclear transport of transcription factors in *Drosophila* and mammals. (A) In *Drosophila*, the Toll protein binds the signal from the Spätzle protein and activates the kinase region of the Pelle protein. The Pelle protein phosphorylates Cactus and Dorsal, causing the two proteins to separate from each other. The Dorsal protein can then enter the nucleus and regulate the transcription of ventrally specific genes. (B) In mammalian lymphocytes, the IL-1 receptor can cause the phosphorylation of IκB (through the protein IRAK kinase). This enables the NF-κB protein to enter the nucleus and effect the transcription of several lymphocyte-specific genes involved in the inflammatory response. The particular colors indicate homologous proteins in the homologous pathway. (After Qureshi et al. 1999.)

in nuclei that have received high concentrations of Dorsal protein, since their enhancers do not bind Dorsal with a very high affinity (Thisse et al. 1988, 1991; Jiang et al. 1991; Pan et al. 1991). The Twist protein activates mesodermal genes, while the Snail protein represses particular nonmesodermal genes that might otherwise be active. The *rhomboid* gene is interesting because it is activated by Dorsal but repressed by Snail. Thus, *rhomboid* is not expressed in the most ventral cells (i.e., the mesodermal precursors), but is expressed in the cells adjacent to the mesoderm that form the presumptive neural ectoderm (Figure 9.41; Jiang and Levine 1993). Both Snail and Twist are needed for the complete mesodermal phenotype and proper

EFFECTS OF THE DORSAL PROTEIN GRADIENT. What does the Dorsal protein do once it is located in the nuclei of the ventral cells? A look at the fate map of a cross section through the *Drosophila* embryo at the fourteenth division cycle (see Figure 9.35B) makes it obvious that the 16 cells with the highest concentration of Dorsal protein are those that generate the mesoderm. The next cell up from this region generates the specialized glial and neural cells of the midline. The next two cells are those that give rise to the ventral epidermis and ventral nerve cord, while the nine cells above them produce the dorsal epidermis. The most dorsal group of six cells generates the amnioserosal covering of the embryo (Ferguson and Anderson 1991).

This fate map is generated by the gradient of Dorsal protein in the nuclei. Large amounts of Dorsal instruct the cells to become mesoderm, while lesser amounts instruct the cells to become glial or ectodermal tissue (Jiang and Levine 1993). The first morphogenetic event of *Drosophila* gastrulation is the invagination of the 16 ventralmost cells of the embryo (Figure 9.39). All of the body muscles, fat bodies, and gonads derive from these mesodermal cells (Foe 1989). The Dorsal protein specifies these cells to become mesoderm in two ways. First, Dorsal activates specific genes that create the mesodermal phenotype. Three of the target genes for Dorsal are *twist*, *snail*, and *rhomboid* (Figure 9.40). These genes are transcribed only

Figure 9.39

Gastrulation in *Drosophila*. In this cross section, the mesodermal cells at the ventral portion of the embryo buckle inward, forming a tube, which then flattens and generates the mesodermal organs. The nuclei are stained with antibody to the Twist protein. (From Leptin 1991a; photographs courtesy of M. Leptin.)

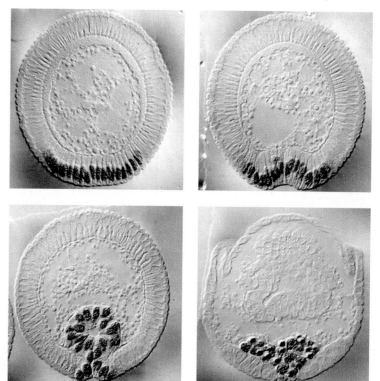

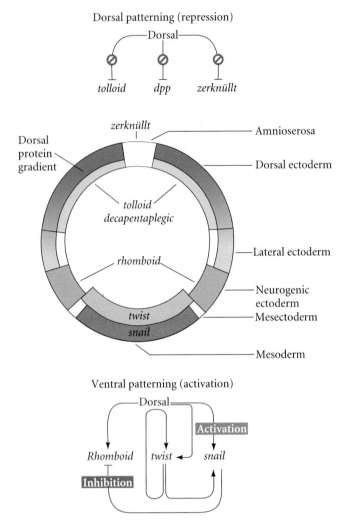

Figure 9.40
Subdivision of the dorsal-ventral axis by the gradient of Dorsal protein in the nuclei. The Dorsal protein activates the zygotic genes *rhomboid*, *twist*, and *snail*, depending on its nuclear concentration. The Snail protein, formed most ventrally, inhibits the transcription of the Rhomboid protein. The Dorsal protein also inhibits the expression of the *tolloid*, *decapentaplegic*, and *zerknüllt* genes in the ventral region. Differing concentrations of Zerknüllt protein determine the fates of the dorsal cells. (After Steward and Govind 1993.)

gastrulation (Leptin et al. 1991b). The sharp border between the mesodermal cells and those cells adjacent to them that generate glial cells (mesectoderm) is produced by the presence of Snail and Twist in the ventralmost cells, but of only Twist in the next cell up (Kosman et al. 1991). In mutants of *snail*, the ventralmost cells still have the *twist* gene activated, and they resemble the more lateral cells (Nambu et al. 1990).

The Dorsal protein also determines the mesoderm directly. In addition to activating the mesoderm-stimulating genes (*twist* and *snail*), it directly inhibits the dorsalizing genes *zerknüllt* (*zen*) and *decapentaplegic* (*dpp*). Thus, in the same cells, the Dorsal protein can act as an activator of some genes

and a repressor of others. Whether the Dorsal protein functions to activate or repress depends on the structure of the genes' enhancers. The *zen* enhancer contains a silencer region that contains a binding site for Dorsal and a second binding site for two other DNA-binding proteins. These two other proteins enable the Dorsal protein to bind a transcriptional repressor protein (Groucho) and bring it to the DNA (Valentine et al. 1998). Mutants of *dorsal* express *dpp* and *zen* genes throughout the embryo (Rushlow et al. 1987), and embryos deficient in *dpp* and *zen* fail to form dorsal structures (Irish and Gelbart 1987). Thus, in wild-type embryos, the mesodermal precursors express *twist* and *snail* (but not *zen* or *dpp*); precursors of the dorsal epidermis and amnioserosa express *zen* and *dpp* but not *twist* or *snail*. Glial (mesectoderm) precursors express only *snail*, while the lateral neural ectodermal precursors do not express any of these four genes (Kosman et al. 1991; Ray et al. 1991). Thus, as a consequence of the responses to the Dorsal protein gradient, the axis becomes subdivided into mesoderm, mesectoderm, neurogenic ectoderm, epidermis, and amnioserosa.

Axes and Organ Primordia: The Cartesian Coordinate Model

The anterior-posterior and dorsal-ventral axes of *Drosophila* embryos form a coordinate system that can be used to specify positions within the embryo. Theoretically, cells that are initially equivalent in developmental potential can respond to their position by expressing different sets of genes. This type of specification has been demonstrated in the formation of the salivary gland rudiments (Panzer et al. 1992). First, salivary glands form only in the strip of cells defined by the activity of the *sex combs reduced* (*scr*) gene along the anterior-posterior axis (parasegment 2). No salivary glands form in *scr*-deficient mutants. Moreover, if *scr* is experimentally expressed throughout the embryo, salivary gland primordia form in a ventrolateral stripe along most of the length of the embryo. The formation of salivary glands along the dorsal-ventral axis is repressed by both Decapentaplegic and Dorsal. These proteins inhibit salivary gland formation both dorsally and ventrally. Thus, the salivary glands form at the intersection of the vertical *scr* expression band (second parasegment) and the horizontal region in the middle of the embryo's circumference that has neither Decapentaplegic nor Dorsal gene products (Figure 9.41). The cells that form the salivary glands are directed to do so by the intersecting gene activities along the anterior-posterior and dorsal-ventral axes.

A similar situation is seen with tissues that are found in every segment of the fly. Neuroblasts arise from ten clusters of four to six cells each that form on each side in every segment in the strip of neural ectoderm at the midline of the embryo (Skeath and Carroll 1992). The potential to form neural cells is conferred on these cells by the expression of proneural genes from the achaete-scute gene complex: *achaete* (*ac*), *scute* (*sc*), and *lethal of scute* (*l'sc*). The cells in each cluster interact (in ways that are discussed in Chapters 8 and 12) to generate a single neural cell from the cluster. Skeath and colleagues (1993) have shown that the pattern of *achaete* and *scute* transcription is imposed by a coordinate system. Their expression is repressed by the Decapentaplegic and Snail proteins along the dorsal-ventral axis, while positive enhancement by pair-rule genes along the anterior-posterior axis causes their repetition in each half-segment. The enhancer recognized by these axis-specifying proteins lies between the *achaete* and *scute* genes and appears to regulate both of them. It is very likely, then, that the positions of organ primordia are specified throughout the fly through a two-dimensional coordinate system based on the intersection of the anterior-posterior and dorsal-ventral axes.

Coda

Genetic studies on the *Drosophila* embryo have uncovered numerous genes that are responsible for the specification of the anterior-posterior and dorsal-ventral axes. We are far from a complete understanding of *Drosophila* pattern formation, but we are much more aware of its complexity than we were five years ago. The mutations of *Drosophila* genes have given us our first glimpses of the multiple levels of pattern regulation in a complex organism and have enabled the isolation of these genes and their products. Moreover, as we will see in the forthcoming chapters, these genes provide clues to a general mechanism of pattern formation used throughout the animal kingdom.

We are beginning to learn how the genome influences the construction of the organism. The genes regulating pattern formation in *Drosophila* operate according to certain principles:

- There are morphogens—such as Bicoid and Dorsal—whose gradients determine the specification of different cell types. These morphogens can be transcription factors.
- There is a temporal order wherein different classes of

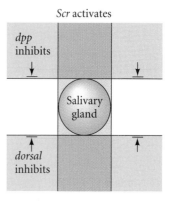

Figure 9.41
Cartesian coordinate system for the expression of genes giving rise to *Drosophila* salivary glands. These genes are activated by the protein product of the *sex combs reduced* (*scr*) homeotic gene in a narrow band along the anterior-posterior axis, and they are inhibited in the regions marked by *decapentaplegic* (*dpp*) and *dorsal* gene products along the dorsal-ventral axis. This pattern allows salivary glands to form in the midline of the embryo in the second parasegment. (After Panzer et al. 1992.)

genes are transcribed, and the products of one gene often regulate the expression of another gene.

- In *Drosophila*, boundaries of gene expression can be created by the interaction between transcription factors and their gene targets. Here, the transcription factors transcribed earlier regulate the expression of the next set of genes.
- Translational control is extremely important in the early embryo, and localized mRNAs are critical in patterning the embryo.
- Individual cell fates are not defined immediately. Rather, there is a stepwise specification wherein a given field is divided and subdivided, eventually regulating individual cell fates.

Snapshot Summary: Drosophila *Development and Axis Specification*

1. *Drosophila* cleavage is superficial. The nuclei divide 13 times before forming cells. Before cell formation, the nuclei reside in a syncytial blastoderm. Each nucleus is surrounded by an actin-filled cytoplasm.

2. When the cells form, the *Drosophila* embryo undergoes a midblastula transition, wherein the cleavages become asynchronous and new mRNA is made. The amount of chromatin determines the timing of this transition.

3. Gastrulation begins with the invagination of the most ventral region, the presumptive mesoderm. This causes the formation of a ventral furrow. The germ band expands such

that the future posterior segments curl just behind the presumptive head.

4. Maternal effect genes are responsible for the initiation of anterior-posterior polarity. *Bicoid* mRNA is sequestered by its 3′ UTR in the future anterior by the cytoskeleton; *nanos* mRNA is sequestered by its 3′ UTR in the future posterior pole. *Hunchback* and *caudal* messages are seen throughout the embryo.

5. At fertilization, *bicoid* and *nanos* messages are translated. A gradient of Bicoid protein activates more hunchback transcription in the anterior. Moreover, Bicoid inhibits the translation of

caudal mRNA. A gradient of Nanos in the posterior inhibits the translation of hunchback mRNA. Caudal protein is made in the posterior.

6. Bicoid and Hunchback proteins activate the genes responsible for the anterior portion of the fly; Caudal activates genes responsible for posterior development.

7. The unsegmented anterior and posterior are regulated by the activation of the Torso protein at the anterior and posterior poles of the egg.

8. The gap genes respond to concentrations of the maternal effect gene proteins. Their protein products interact with each other such that each gap gene protein defines specific regions of the embryo.

9. The gap gene proteins activate and repress the pair-rule genes. The pair-rule genes have modular promoters such that they become activated in the seven "stripes." Their boundaries are defined by the gap genes. These genes form seven bands of transcription along the anterior-posterior axis, each one comprising two parasegments.

10. The pair-rule gene products activate *engrailed* and *wingless* expression in adjacent cells. The *engrailed*-expressing cells form the anterior boundary of each parasegment. These cells form a signaling center that organizes the cuticle formation and segmental structure of the embryo.

11. The homeotic selector genes are found in two complexes on chromosome 3 of *Drosophila*. Together these are called Hom-C, the homeotic gene complex. The genes are arranged in the same order as their transcriptional expression. These genes specify each segment, and mutations in these genes are capable of transforming one segment into another.

12. The expression of each homeotic selector gene is regulated by the gap and pair-rule genes. Their expression is refined and maintained by interactions whereby the protein products interact with genes, preventing the transcription of neighboring Hom-C genes.

13. In Ultrabithorax mutations, the third thoracic segment becomes specifed as the second thoracic segment. This converts the halteres into wings. When Antennapedia is expressed in the head as well as in the thorax, it represses antenna formation, allowing legs to form where the antenna should be.

14. The targets of the Hom-C proteins are the realisator genes. These include *Distal-less* and *Wingless* genes (in the thoracic segments).

15. Dorsal-ventral polarity is regulated by the entry of the Dorsal protein into the nucleus. Dorsal-ventral polarity is initiated by the nucleus being positioned in the dorsal-anterior of the oocyte and transcribing the *gurken* message. This message is transported to the region above the nucleus and adjacent to the follicle cells.

16. The *gurken* mRNA is translated into the Gurken protein, which is secreted from the oocyte and binds to its receptor, Torpedo, on the follicle cells. This dorsalizes the follicle cells, preventing them from synthesizing Pipe.

17. The Pipe protein in the ventral follicle cells modifies an as yet unknown factor that modifies the Nudel protein. This allows the Nudel protein to activate a cascade of proteolysis in the space between the ventral follicle cells and the ventral cells of the embryo.

18. As a result of the cascade, the Spätzle protein is activated and binds to the Toll protein on the ventral embryonic cells.

19. The activated Toll protein activates Pelle and Tube to phosphorylate the Cactus protein, which has been bound to the Dorsal protein. Phosphorylated Cactus protein is degraded, allowing Dorsal protein to enter the nucleus.

20. Once in the nucleus, Dorsal protein activates the genes responsible for the ventral cell fates and represses those genes whose proteins would specify dorsal cell fates. Since a gradient of Dorsal protein enters the various nuclei, those at the most ventral surface become mesoderm, those more lateral become neurogenic ectoderm.

21. Organs form at the intersection of dorsal-ventral and anterior-posterior regions of gene expression.

Literature Cited

Akam, M. E. 1987. The molecular basis for metameric pattern in the *Drosophila* embryo. *Development* 101: 1–22.

Anderson, K. V. 1989. *Drosophila*: The maternal contribution. *In* D. M. Glover and B. D. Hames (eds.), *Genes and Embryos*. IRL, New York, pp. 1–37.

Anderson, K. V. and C. Nüsslein-Volhard. 1984. Information for the dorsal-ventral pattern of the *Drosophila* embryo is stored as maternal mRNA. *Nature* 311: 223–227.

Baker, J., W. E. Theurkauf and G. Schubiger. 1993. Dynamic changes in microtubule configuration correlate with nuclear migration in the preblastoderm *Drosophila* embryo. *J. Cell Biol.* 122: 113–121.

Barker, D. D., C. Wang, J. Moore, L. K. Dickinson and R. Lehmann. 1992. Pumilio is essential for function but not for distribution of the *Drosophila* abdominal determinant, Nanos. *Genes Dev.* 6: 2312–2326.

Bate, M. and A. Martinez-Arias. 1993. *The Development of Drosophila melanogaster*. Cold Spring Harbor Laboratory Press, Cold Spring Harbor, NY.

Bateson, W. 1894. *Materials for the Study of Variation.* Macmillan, London.

Baumgartner, S. and M. Noll. 1990. Networks of interaction among pair-rule genes regulating paired expression during primordial segmentation of *Drosophila. Mech. Dev.* 1: 1–18.

Berleth, T. and 7 others. 1988. The role of localization of *bicoid* RNA in organizing the anterior pattern of the *Drosophila* embryo. *EMBO J.* 7: 1749–1756.

Bhanot, P. and 8 others. 1996. A new member of the frizzled family from *Drosophila* functions as a Wingless receptor. *Nature* 382: 225–230.

Bokor, P. and S. DiNardo. 1996. The roles of Hedgehog and Wingless in patterning the dorsal epidermis in *Drosophila. Development* 122: 1083–1092.

Campos-Ortega, J. A. and V. Hartenstein. 1985. *The Embryonic Development of Drosophila melanogaster.* Springer-Verlag, New York.

Cao, Z., W. J. Henzel and X. Gao. 1996. IRAK: A kinase associated with the interleukin-1 receptor. *Science* 271: 1128–1131.

Capovilla, M., E. D. Eldon and V. Pirrotta. 1992. The *Giant* gene of *Drosophila* encodes a bZIP DNA-binding protein that regulates the expression of other segmentation gap genes. *Development* 114: 99–112.

Casanova, J. and G. Struhl. 1989. Localized surface activity of torso, a receptor tyrosine kinase, specifies body pattern in *Drosophila. Genes Dev.* 3: 2025–2038.

Casanova, J., E. Sánchez-Herrero, A. Busturia and G. Morata. 1987. Double and triple mutant combination of the bithorax complex of *Drosophila. EMBO J.* 6: 3103–3109.

Casanova, J., M. Furrioles, C. A. McCormick and G. Struhl. 1995. Similarities between *trunk* and *spätzle*, putative extracellular ligands specifying body pattern in *Drosophila. Genes Dev.* 9: 2539–2544.

Casares, F. and R. S. Mann. 1998. Control of antennal versus leg development in *Drosophila. Nature* 392: 723–726.

Casares, F. and E. Sánchez-Herrero. 1995. Regulation of the infraabdominal regions of the bithorax complex of *Drosophila* by gap genes. *Development* 121: 1855–1866.

Castelli-Gair, J. and M. Akam. 1995. How the Hox gene *Ultrabithorax* specifies two different segments: The significance of spatial and temporal regulation within metameres. *Development* 121: 2973–2982.

Chasan, R., Y. Jin and K. V. Anderson. 1992. Activation of the easter zymogen is regulated by five other genes to define dorsal-ventral polarity in the *Drosophila* embryo. *Development* 115: 607–615.

Cohen, S. M. and G. Jürgens. 1990. Mutations of *Drosophila* head development by gap-like segmentation genes. *Nature* 346: 482–485.

Degelmann, A., P. A. Hardy, N. Perrimon and A. P. Mahowald. 1986. Developmental analysis of the torsolike phenotype in *Drosophila* produced by a maternal-effect locus. *Dev. Biol.* 115: 479–489.

Dessain, S., C. T. Gross, M. A. Kuziora and W. McGinnis. 1992. *Antp*-type homeodomains have distinct DNA-binding specificities that correlate with their different regulatory functions in embryos. *EMBO J.* 11: 991–1002.

Doe, C. Q., Y. Hiromi, W. J. Gehring and C. S. Goodman. 1988. Expression and function of the segmentation gene *fushi tarazu* during *Drosophila* neurogenesis. *Science* 239: 170–175.

Driever, W. and C. Nüsslein-Volhard. 1988a. The bicoid protein determines position in the *Drosophila* embryo in a concentration-dependent manner. *Cell* 54: 95–104.

Driever, W. and C. Nüsslein-Volhard. 1988b. A gradient of bicoid protein in *Drosophila* embryos. *Cell* 54: 83–93.

Driever, W. and Nüsslein-Volhard, C. 1989. The bicoid protein is a positive regulator of *hunchback* transcription in the early *Drosophila* embryo. *Nature* 337: 138–143.

Driever, W., G. Thoma and C. Nüsslein-Volhard. 1989. Determination of spatial domains of zygotic gene expression in the *Drosophila* embryo by the affinity of binding sites for the bicoid morphogen. *Nature* 340: 363–367.

Driever, W., V. Siegel and C. Nüsslein-Volhard. 1990. Autonomous determination of anterior structures in the early *Drosophila* embryo by the bicoid morphogen. *Development* 109: 811–820.

Dubnau, J. and G. Struhl. 1996. RNA recognition and translational regulation by a homeodomain protein. *Nature* 379: 694–699.

Edgar, B. A. and P. H. O'Farrell. 1989. Genetic control of cell division patterns in the *Drosophila* embryo. *Cell* 57: 177–187.

Edgar, B. A., C. P. Kiehle and G. Schubiger. 1986a. Cell cycle control by the nucleo-cytoplasmic ratio in early *Drosophila* development. *Cell* 44: 365–372.

Edgar, B. A., M. P. Weir, G. Schubiger and T. Kornberg. 1986b. Repression and turnover pattern of *fushi tarazu* RNA in the early *Drosophila* embryo. *Cell* 47: 747–754.

Ferguson, E. L and K. V. Anderson. 1991. Dorsal-ventral pattern formation in the *Drosophila* embryo: The role of zygotically active genes. *Curr. Top. Dev. Biol.* 25: 17–43.

Ferrandon, D., I. Koch, E. Westhof and C. Nüsslein-Volhard. 1997. RNA-RNA interaction is required for the formation of specific *bicoid* mRNA 3′ UTR-Stufen ribonucleoprotein particles. *EMBO J.* 16: 1751–1758.

Finkelstein, R. and N. Perrimon. 1990. The *orthodenticle* gene is regulated by bicoid and torso and specifies *Drosophila* head development. *Nature* 346: 485–488.

Foe, V. E. 1989. Mitotic domains reveal early commitment of cells in *Drosophila* embryos. *Development* 107: 1–22.

Foe, V. E. and B. M. Alberts. 1983. Studies of nuclear and cytoplasmic behavior during the five mitotic cycles that precede gastrulation in *Drosophila* embryogenesis. *J. Cell Sci.* 61: 31–70.

Foe, V. E., G. M. Odell and B. A. Edgar. 1993. Mitosis and morphogenesis in the *Drosophila* embryo: Point and counterpoint. *In* B. M. Bate and A. Martinez-Arias (eds.), *The Development of* Drosophila melanogaster. Cold Spring Harbor Laboratory Press, Cold Spring Harbor, NY.

Forlani, S., D. Ferrandon, O. Saget and E. Mohier. 1993. A regulatory function for K10 in the establishment of dorsoventral polarity in the *Drosophila* egg and embryo. *Mech. Dev.* 41: 109–120.

Frigerio, G., M. Burri, D. Bopp, S. Baumgartner and M. Noll. 1986. Structure of the segmentation gene *paired* and the *Drosophila* PRD gene set as part of a gene network. *Cell* 47: 735–746.

Frischer, L. E., F. S. Hagen and R. L. Garber. 1986. An inversion that disrupts the *Antennapedia* gene causes abnormal structure and localization of RNAs. *Cell* 47: 1017–1023.

Fujioka, M., Y. Emi-Sarker, G. L. Yusibova, T. Goto and J. B. Jaynes. 1999. Analysis of an *even-skipped* rescue transgene reveals both composite and discrete neuronal and early blastoderm enhancers, and multi-stripe positioning by gap gene repressor gradients. *Development* 126: 2527–2538.

Fullilove, S. L. and Jacobson, A. G. 1971. Nuclear elongation and cytokinesis in *Drosophila montana. Dev. Biol.* 26: 560–578.

Furriols, M., A. Casali and J. Casanova. 1998. Dissecting the mechanism of torso receptor activation. *Mech. Dev.* 70: 111–118.

Gabay, L., R. Seger and B. Z. Shilo. 1997. MAP kinase in situ activation atlas during *Drosophila* embryogenesis. *Development* 124: 3535–3541.

Galindo, R. L., D. N. Edwards, S. K. H. Gillespie and S. A. Wasserman. 1995. Interaction of the pelle kinase with the membrane-associated protein tube is required for transduction of the dorsoventral signal in *Drosophila* embryos. *Development* 121: 2209–2218.

Gaul, U. and H. Jäckle. 1990. Role of gap genes in early *Drosophila* development. *Annu. Rev. Genet.* 27: 239–275.

Gavis, E. R. and R. Lehmann. 1992. Localization of *nanos* RNA controls embryonic polarity. *Cell* 71: 301–313.

González-Crespo, S. and M. Levine. 1994. Related target enhancers for dorsal and NF-κB signalling pathways. *Science* 264: 255–258.

González-Reyes, A. and G. Morata. 1990. The developmental effect of overexpressing a Ubx product in *Drosophila* embryos is dependent on its interactions with other homeotic products. *Cell* 61: 515–522.

Grossniklaus, U., K. M. Cadigan and W. J. Gehring. 1994. Three maternal coordinate systems cooperate in the patterning of the *Drosophila* head. *Development* 120: 3155–3171.

Hafen, E., M. Levine and W. J. Gehring. 1984. Regulation of *Antennapedia* transcript distribution by the bithorax complex in *Drosophila*. *Nature* 307: 287–289.

Hanes, S. D. and R. Brent. 1989. DNA specificity of the bicoid activator protein is determined by homeodomain recognition helix residue 9. *Cell* 57: 1275–1283.

Hanes, S. D. and R. Brent. 1991. A genetic model for interaction of the homeodomain recognition helix with DNA. *Science* 251: 426–430.

Harding, K. and M. Levine. 1988. Gap genes define the limits of *Antennapedia* and *Bithorax* gene expression during early development in *Drosophila*. *EMBO J.* 7: 205–214.

Harding, K. and M. Levine. 1989. *Drosophila*: The zygotic contribution. *In* D. M. Glover and B. D. Hames (eds.), *Genes and Embryos*. IRL, New York, pp. 38–90.

Harding, K., C. Wedeen, W. McGinnis and M. Levine. 1985. Spatially regulated expression of homeotic genes in *Drosophila*. *Science* 229: 1236–1242.

Hashimoto, C., K. L. Hudson and K. V. Anderson. 1988. The *Toll* gene of *Drosophila*, required for dorsal-ventral embryonic polarity, appears to encode a transmembrane protein. *Cell* 52: 269–279.

Hashimoto, C., S. Gerttula and K. V. Anderson. 1991. Plasma membrane localization of the Toll protein in the syncytial *Drosophila* embryo: Importance of transmembrane signalling for dorsal-ventral pattern formation. *Development* 11: 1021–1028.

Heemskerk, J. and S. DiNardo. 1994. *Drosophila* hedgehog acts as a morphogen in cellular patterning. *Cell* 76: 449–460.

Heemskerk, J., S. DiNardo, R. Kostriken and P. H. O'Farrell. 1991. Multiple modes of engrailed regulation in the progression towards cell fate determination. *Nature* 352: 404–410.

Hoch, M., N. Gerwin, H. Taubert and H. Jäckle. 1992. Competition for overlapping sites in the regulatory region of the *Drosophila* gene *Krüppel*. *Science* 256: 94–97.

Hong, C. C. and C. Hashimoto. 1995. An unusual mosaic protein with a protease domain, encoded by the *nudel* gene, is involved in defining embryonic dorsoventral polarity in *Drosophila*. *Cell* 82: 785–794.

Ingham, P. W. and A. Martinez-Arias. 1986. The correct activation of *Antennapedia* and *bithorax* complex genes requires the *fushi tarazu* gene. *Nature* 324: 592–597.

Ingham, P. W. and R. Whittle. 1980. Trithorax: A new homeotic mutation of *Drosophila* causing transformations of abdominal and thoracic imaginal segments. I. Putative role during embryogenesis. *Mol. Gen. Genet.* 179: 607–614.

Ingham, P. W., A. M. Taylor and Y. Nakano. 1991. Role of *Drosophila patched* gene in positional signalling. *Nature* 353: 184–187.

Irish, V. F. and W. M. Gelbart. 1987. The *decapentaplegic* gene is required for dorsal-ventral patterning of the *Drosophila* embryo. *Genes Dev.* 1: 868–879.

Jäckle, H., D. Tautz, R. Schuh, E. Seifert and R. Lehmann. 1986. Cross-regulatory interactions among the gap genes of *Drosophila*. *Nature* 324: 668–670.

Jiang, J. and M. Levine. 1993. Binding affinities and cooperative interactions with bHLH activators delimit threshold responses to the dorsal gradient morphogen. *Cell* 72: 741–752.

Jiang, J., D. Kosman, Y. T. Ip and M. Levine. 1991. The dorsal morphogen gradient regulates the mesoderm determinant twist in early *Drosophila* embryos. *Genes Dev.* 5: 1881–1891.

Just, E. E. 1936. Quoted in R. G. Harrison, 1937. Embryology and its relations. *Science* 85: 369–374.

Kalthoff, K. 1969. Der Einfluss vershiedener Versuchparameter auf die Häufigkeit der Missbildung "Doppelabdomen" in UV-bestrahlten Eiern von *Smittia* sp. (Diptera, Chironomidae). *Zool. Anz. Suppl.* 33: 59–65.

Kalthoff, K. and K. Sander. 1968. Der Enwicklungsgang der Missbildung "Doppelabdomen" im partiell UV-bestrahlten Ei von *Smittia parthenogenetica* (Diptera, Chironomidae). *Wilhelm Roux Arch. Entwicklungsmech. Org.* 161: 129–146.

Kandler-Singer, I. and K. Kalthoff. 1976. RNase sensitivity of an anterior morphogenetic determinant in an insect egg (*Smittia* sp., Chironomidae, Diptera). *Proc. Natl. Acad. Sci. USA* 73: 3739–3743.

Karr, T. L. and B. M. Alberts. 1986. Organization of the cytoskeleton in early *Drosophila* embryos. *J. Cell Biol.* 102: 1494–1509.

Karr, T. L. and T. B. Kornberg. 1989. *fushi tarazu* protein expression in the cellular blastoderm of *Drosophila* detected using a novel imaging technique. *Development* 105: 95–103.

Kaufman, T. C., M. A. Seeger and G. Olsen. 1990. Molecular and genetic organization of the Antennapedia gene complex of *Drosophila melanogaster*. *Adv. Genet.* 27: 309–362.

Kidd, S. 1992. Characterization of the *Drosophila cactus* locus and analysis of interactions between cactus and dorsal proteins. *Cell* 71: 623–635.

Klingler, M., M. Erdélyi, J. Szabad and C. Nüsslein-Volhard. 1988. Function of *torso* in determining the terminal anlagen of the *Drosophila* embryo. *Nature* 335: 275–277.

Knipple, D. C., E. Seifert, U. B. Rosenberg, A. Preiss and H. Jäckle. 1985. Spatial and temporal patterns of *Krüppel* gene expression in early *Drosophila* embryos. *Nature* 317: 40–44.

Kohler, R. 1994. *Lords of the Fly*. University of Chicago Press, Chicago.

Kosman, D., Y. T. Ip, M. Levine and K. Arora. 1991. Establishment of the mesoderm-neuroectoderm boundary in the *Drosophila* embryo. *Science* 254: 118–122.

Kraut, R. and M. Levine. 1991. Mutually repressive interactions between the gap genes *giant* and *Krüppel* define middle body regions of the *Drosophila* embryo. *Development* 111: 611–621.

Lehmann, R. and C. Nüsslein-Volhard. 1986. Abdominal segmentation, pole cell formation, and embryonic polarity require the localized activity of oskar, a maternal gene in *Drosophila*. *Cell* 47: 141–152

Lehmann, R. and Nüsslein-Volhard, C. 1991. The maternal gene *nanos* has a central role in posterior pattern formation of the *Drosophila* embryo. *Development* 112: 679–691.

Lemaitre, B., E. Nicolas, L. Michaut, J.-M. Reichhart and J. A. Hoffmann. 1996. The dorsoventral regulatory gene casette *spätzle/Toll/cactus* controls the potent antifungal response in *Drosophila* adults. *Cell* 86: 973–983.

Leptin, M. 1991a. Mechanics and genetics of cell shape changes during *Drosophila* ventral furrow formation. *In* R. Keller et al. (eds.), *Gastrulation: Movements, Patterns, and Molecules*. Plenum, New York, pp. 199–212.

Leptin, M. 1991b. *twist* and *snail* as positive and negative regulators during *Drosophila* mesoderm development. *Genes Dev.* 5: 1568–1576.

Levine, M. S. and K. W. Harding. 1989. *Drosophila*: The zygotic contribution. *In* D. M. Glover and B. D. Hames (eds.), *Genes and Embryos*. IRL, New York, pp. 39–94.

Lewis, E. B. 1978. A gene complex controlling segmentation in *Drosophila*. *Nature* 276: 565–570.

Macdonald, P. M. and K. Kerr. 1998. Mutational analysis of an RNA recognition element that mediates localization of *bicoid* mRNA. *Mol. Cell Biol.* 18: 3788–3795.

Macdonald, P. M. and C. A. Smibert. 1996. Translational regulation of maternal mRNAs. *Curr. Opin. Genet. Dev.* 6: 403–407.

Macdonald, P. M. and G. Struhl. 1986. A molecular gradient in early *Drosophila* embryos and its role in specifying body pattern. *Nature* 324: 537–545.

Mann, R. S. and D. S. Hogness. 1990. Functional dissection of Ultrabithorax proteins in *D. melanogaster*. *Cell* 60: 597–610.

Martin, J. R., A. Railbaud and R. Ollo. 1994. Terminal elements in *Drosophila* embryo induced by torsolike protein. *Nature* 367: 741–745.

Martinez-Arias, A. and P. A. Lawrence. 1985. Parasegments and compartments in the *Drosophila* embryo. *Nature* 313: 639–642.

McKeon, J. and H. W. Brock. 1991. Interactions of the *Polycomb* group of genes with homeotic loci of *Drosophila*. *Wilhelm Roux Arch. Dev. Biol.* 199: 387–396.

Mohler, J. and K. Vani. 1992. Molecular organization and embryonic expression of the *hedgehog* gene involved in cell-cell communication in segmental patterning in *Drosophila*. *Development* 115: 957–971.

Montell, D. J., H. Keshishian and A. C. Spradling. 1991. Laser ablation studies of the role of the *Drosophila* oocyte nucleus in pattern formation. *Science* 254: 290–293.

Müller, J. and M. Bienz. 1992. Sharp anterior boundary of homeotic gene expression conferred by the fushi tarazu protein. *EMBO J.* 11: 3653–3661.

Müller, M., M. Affolter, W. Leupin, G. Otting, K. Wüthrich and W. J. Gehring. 1988. Isolation and sequence-specific DNA binding of the Antennapedia homeodomain. *EMBO J.* 7: 4299–4304.

Nambu, J. R., R. G. Franks, S. Hong and S. Crews. 1990. The *single-minded* gene of *Drosophila* is required for the expression of genes important for the development of CNS midline cells. *Cell* 63: 63–75.

Neuman-Silberberg, F. S. and T. Schüpbach. 1993. The *Drosophila* dorsoventral patterning gene *gurken* produces a dorsally localized RNA and encodes a TGF-α-like protein. *Cell* 75: 165–174.

Newport, J. W. and M. W. Kirschner. 1982. A major developmental transition in early *Xenopus* embryos: I. Characterization and timing of cellular changes at midblastula stage. *Cell* 30: 687–696.

Nüsslein-Volhard, C. and E. Wieschaus. 1980. Mutations affecting segment number and polarity in *Drosophila*. *Nature* 287: 795–801.

Otting, G., Y. Q. Qian, M. Billeter, M. Müller, M. Affolter, W. J. Gehring and K. Wüthrich. 1990. Protein-DNA contacts in the structure of a homeodomain-DNA complex determined by nuclear magnetic resonance spectroscopy in solution. *EMBO J.* 9: 3085–3092.

Pan, D., J.-D. Huang and A. J. Courey. 1991. Functional analysis of the *Drosophila* twist promoter reveals a dorsal-binding ventral activator region. *Genes Dev.* 5: 1892–1901.

Pankratz, M. J., E. Seifert, N. Gerwin, B. Billi, U. Nauber and H. Jäckle. 1990. Gradients of *Krüppel* and *knirps* gene products direct pair-rule gene stripe patterning in the posterior region of the *Drosophila* embryo. *Cell* 61: 309–317.

Panzer, S., D. Weigel and S. K. Beckendorf. 1992. Organogenesis in *Drosophila melanogaster*: Embryonic salivary gland determination is controlled by homeotic and dorsoventral patterning genes. *Development* 114: 49–57.

Paroush, Z., S. M. Wainwright and D. Ish-Horowitz. 1997. Torso signaling mediates terminal patterning in *Drosophila* by antagonizing Groucho-mediated repression. *Development* 124: 3827–3834.

Peifer, M. and E. Wieschaus. 1990. Mutations in the *Drosophila* gene *extradenticle* affect the way specific homeodomain proteins regulate segment identity. *Genes Dev.* 4: 1209–1223.

Percival-Smith, A., M. Müller, M. Affolter and W. J. Gehring. 1990. The interaction with DNA of wild-type and mutant *fushi tarazu* homeodomains. *EMBO J.* 9: 3967–3974.

Peri, F., C. Bökel and S. Roth. 1999. Local Gurken signaling and dynamic MAPK activation during *Drosophila* oogenesis. *Mech. Dev.* 81: 75–88.

Pignoni, F., E. Steingrímsson and J. A. Lengyel. 1992. *bicoid* and the terminal system activate *tailless* expression in the early *Drosophila* embryo. *Development* 115: 239–251.

Porter, J. A. and 10 others. 1996. Hedgehog patterning activity: Role of a lipophilic modification by the carboxy-terminal autoprocessing domain. *Cell* 86: 21–34.

Price, J. V., R. J. Clifford and T. Schüpbach. 1989. The maternal ventralizing gene *torpedo* is allelic to *faint little ball*, an embryonic lethal, and encodes the *Drosophila* EGF receptor homolog. *Cell* 56: 1085–1092.

Qureshi, S. T., P. Gros and D. Malo. 1999. Host resistance to infection. *Trends Genet.* 15: 291–294.

Raskolb, C. and E. Wieschaus. 1994. Coordinate regulation of downstream genes by extradenticle and homeotic selector proteins. *EMBO J.* 15: 3561–3569.

Ray, R. P. and T. Schüpbach. 1996. Intercellular signaling and the polarization of body axes during *Drosophila* oogenesis. *Genes Dev.* 10: 1711–1723.

Ray, R. P., K. Arora, C. Nüsslein-Volhard and W. M. Gelbart. 1991. The control of cell fate along the dorsal-ventral axis of the *Drosophila* embryo. *Development* 113: 35–54.

Reach, M., R. L. Galindo, P. Towb, J. L. Allen, M. Karin and S. A. Wasserman. 1996. A gradient of Cactus protein degradation establishes dorsoventral polarity in the *Drosophila* embryo. *Dev. Biol.* 180: 353–364.

Reinitz, J. and D. H. Sharp. 1995. Mechanism of eve stripe formation. *Mech. Dev.* 49: 133–158.

Reinitz, J., E. Mjolsness and D. H. Sharp. 1995. Model for cooperative control of positional information in *Drosophila* by bicoid and maternal hunchback. *J. Exp. Zool.* 271: 47–56.

Riddihough, G. 1992. Homing in on the homeobox. *Nature* 357: 643–644.

Rivera-Pomar, R. and H. Jäckle. 1996. From gradients to stripes in *Drosophila* embryogenesis: Filling in the gaps. *Trends Genet.* 12: 478–483.

Rivera-Pomar, R., D. Niessling, U. Schmidt-Ott, W. J. Gehring and H. Jäckle. 1996. RNA binding and translational suppression by bicoid. *Nature* 379: 746–749.

Rock, F. L., G. Hardiman, J. C. Timans, R. A. Kasterlein and J. F. Bazan. 1998. A family of human receptors structurally related to *Drosophila* Toll. *Proc. Natl. Acad. Sci. USA* 95: 588–593.

Roder, L., C. Vola and S. Kerridge. 1992. The role of teashirt in trunk segmental identity in *Drosophila*. *Development* 115: 1017–1033.

Roth, S., D. Stein and C. Nüsslein-Volhard. 1989. A gradient of nuclear localization of the dorsal protein determines dorsoventral pattern in the *Drosophila* embryo. *Cell* 59: 1189–1202.

Rushlow, C., J. Frasch, H. Doyle and M. Levine. 1987. Maternal regulation of a homeobox gene controlling differentiation of dorsal tissues in *Drosophila*. *Nature* 330: 583–586.

Rushlow, C. A., K. Han, J. L. Manley and M. Levine. 1989. The graded distribution of the dorsal morphogen is initiated by selective nuclear transport in *Drosophila*. *Cell* 59: 1165–1177.

Sackerson, C. M. Fujioka and T. Goto. 1999. The *even-skipped* locus is contained in a 16-kb chromatin domain. *Dev. Biol.* 210: 39–52.

Sánchez-Herrero, E., I. Verños, R. Marco and G. Morata. 1985. Genetic organization of *Drosophila bithorax* complex. *Nature* 313: 108–113.

Sander, K. 1975. Pattern specification in the insect embryo. In *Cell Patterning*. CIBA Foundation Symposium, new series, 29. Associated Scientific Publishers, Amsterdam, New York, pp. 241–263.

Schejter, E. D. and E. Wieschaus. 1993. Bottleneck acts as a regulator of the microfilament network governing cellularization of the *Drosophila* embryo. *Cell* 75: 373–385.

Schier, A. F. and A. J. Gehring. 1992. Direct homeodomain-DNA interaction in the autoregulation of the *fushi tarazu* gene. *Nature* 356: 804–807.

Schneuwly, S., A. Kuroiwa and W. J. Gehring. 1987. Molecular analysis of the dominant homeotic *Antennapedia* phenotype. *EMBO J.* 6: 201–206.

Schüpbach, T. 1987. Germ line and soma cooperate during oogenesis to establish the dorsoventral pattern of egg shell and embryo in *Drosophila melanogaster*. *Cell* 49: 699–707.

Schüpbach, T. and E. Wieschaus. 1986. Maternal effect mutations altering the anterior-posterior pattern of the *Drosophila* embryo. *Wilhelm Roux Arch. Dev. Biol.* 195: 302–317.

Schüpbach, T., R. J. Clifford, L. J. Manseau and J. V. Price. 1991. Dorsoventral signaling processes in *Drosophila* oogenesis. *In* J. Gerhart (ed.), *Cell-Cell Interactions in Early Development*. Wiley-Liss, New York, pp. 163–174.

Schwalm, F. 1997. The insects. *In* S. F. Gilbert and A. M. Raunio (eds.), *Embryology: Constructing the Organism*. Sinauer Associates, Sunderland, MA, pp. 259–278.

Sen, J., J. S. Goltz, L. Stevens and D. Stein. 1998. Spatially restricted expression of *pipe* in the *Drosophila* egg chamber defines embryonic dorsal-ventral polarity. *Cell* 95: 471–481.

Shelton, C. A. and S. A. Wasserman. 1993. *pelle* encodes a protein kinase required to establish dorsoventral polarity in the *Drosophila* embryo. *Cell* 72: 515–525.

Siegfried, E., E. L. Wilder and N. Perrimon. 1994. Components of *wingless* signalling in *Drosophila*. *Nature* 367: 76–80.

Simon, J., A. Chiang and W. Bender. 1992. Ten different Polycomb genes are required for spatial control of the *abdA* and *AbdB* homeotic products. *Development* 114: 493–505.

Simpson-Brose, M., J. Treisman and C. Desplan. 1994. Synergy between the hunchback and bicoid morphogens is required for anterior patterning in *Drosophila*. *Cell* 78: 855–865.

Skeath, J. B. and S. B. Carroll. 1992. Regulation of proneural gene expression and cell fate during neuroblast segregation in the *Drosophila* embryo. *Development* 114: 939–946.

Skeath, J. B., G. Panganiban, J. Selegue and S. B. Carroll. 1993. Gene regulation in two dimensions: The proneural *achaete* and *scute* genes are controlled by combinations of axis-patterning genes through a common intergenic control region. *Genes Dev.* 6: 2606–2619.

Slack, J. M. W. 1983. *From Egg to Embryo: Determinative Events in Early Development.* Cambridge University Press, Cambridge.

Small, S., R. Kraut, T. Hoey, R. Warrior and M. Levine. 1991. Transcriptional regulation of a pair-rule stripe in *Drosophila*. *Genes Dev.* 5: 827–839.

Small, S., A. Blair and M. Levine. 1992. Regulation of *even-skipped* stripe 2 in the *Drosophila* embryo. *EMBO J.* 11: 4047–4057.

Small, S., A. Blair and M. Levine. 1996. Regulation of two pair-rule stripes by a single enhancer in the *Drosophila* embryo. *Dev. Biol.* 175: 314–324.

Smibert, C. A., J. E. Wilson, K. Kerr and P. M. Macdonald. 1996. Smaug protein represses translation of unlocalized *nanos* mRNA in the *Drosophila* embryo. *Genes Dev.* 10: 2600–2609.

Štanojević, D., T. Hoey and M. Levine. 1989. Sequence-specific DNA-binding activities of the gap proteins encoded by *hunchback* and *Krüppel* in *Drosophila*. *Nature* 341: 331–335.

Štanojević, D., S. Small and M. Levine. 1991. Regulators of a segmentation stripe by overlapping activators and repressors in the *Drosophila* embryo. *Science* 254: 1385–1387.

Stephanson, E. C., Y.-C. Chao and J. D. Frackenthal. 1988. Molecular analysis of the *swallow* gene of *Drosophila melanogaster*. *Genes Dev.* 2: 1655–1665.

Stevens, L. M., H. G. Frohnhöfer, M. Klingler and C. Nüsslein-Volhard. 1990. Localized requirement for *torsolike* expression in follicle cells for development of terminal anlagen of the *Drosophila* embryo. *Nature* 346: 660–662.

Steward, R. 1989. Relocalization of the dorsal protein from the cytoplasm to the nucleus correlates with its function. *Cell* 59: 1179–1188.

Steward, R. and S. Govind. 1993. Dorsal-ventral polarity in the *Drosophila* embryo. *Curr. Opin. Genet. Dev.* 3: 556–561.

St. Johnston, D. and C. Nüsslein-Volhard. 1992. The origin of pattern and polarity in the *Drosophila* embryo. *Cell* 68: 201–219.

Struhl, G. 1981. A homeotic mutation transforming leg to antenna in *Drosophila*. *Nature* 292: 635–638.

Struhl, G., K. Struhl and P. M. Macdonald. 1989. The gradient morphogen bicoid is a concentration-dependent transcriptional activator. *Cell* 57: 1259–1273.

Struhl, G., P. Johnson and P. Lawrence. 1992. Control of a *Drosophila* body pattern by the Hunchback morphogen gradient. *Cell* 69: 237–249.

Sullivan, W., P. Fogarty and W. E. Theurkauf. 1993. Mutations affecting the cytoskeletal organization of syncytial *Drosophila* embryos. *Development* 118: 1245–1254.

Tautz, D. 1988. Regulation of the *Drosophila* segmentation gene *hunchback* by two maternal morphogenetic centers. *Nature* 332: 281–284.

Thisse, B., C. Stoetzel, C. Gorostiza-Thisse and F. Perrin-Schmidt. 1988. Sequence of the *twist* gene and nuclear localization of its protein in endomesodermal cells of early *Drosophila* embryos. *EMBO J.* 7: 2175–2183.

Thisse, C., F. Perrin-Schmidt, C. Stoetzel and B. Thisse. 1991. Sequence-specific *trans* activation of the *Drosophila twist* gene by the *dorsal* gene product. *Cell* 65: 1191–1201.

Turner, F. R. and A. P. Mahowald. 1977. Scanning electron microscopy of *Drosophila melanogaster* embryogenesis. *Dev. Biol.* 57: 403–416.

Tyler, M. S. and J. W. Schetzer. 1996. *The Lives of a Fly.* (Videocassette.) ASAP Media Services, Orono, ME, and Sinauer Associates, Sunderland, MA.

Vachon, G., B. Cohen, C. Pfeifle, M. E. McGuffin, J. Botas and S. M. Cohen. 1992. Homeotic genes of the bithorax complex repress limb development in the abdomen of the *Drosophila* embryo through the target gene *Distal-less*. *Cell* 71: 437–450.

Valentine, S. A., G. Chen, T. Shandala, J. Fernandez, S. Mische, R. Saint and A. J. Courney. 1998. Dorsal-mediated repression requires the formation of a multiprotein repression complex at the *ventral* silencer. *Mol. Cell Biol.* 18: 6584–6594.

Van Dyke, M. A. and C. Murre. 1994. Extradenticle raises the DNA binding specificity of homeotic selector gene products. *Cell* 78: 617–624.

Van Eeden, F. and D. St. Johnston. 1999. The polarisation of the anterior-posterior and dorsal-ventral axes during *Drosophila* oogenesis. *Curr. Opin. Genet. Devel.* 9: 396–404.

Wakimoto, B. T., F. R. Turner and T. C. Kaufman. 1984. Defects in embryogenesis in mutants associated with the *Antennapedia* gene complex of *Drosophila melanogaster*. *Dev. Biol.* 102: 147–172.

Wang, C. and R. Lehman. 1991. Nanos is the localized posterior determinate in *Drosophila*. *Cell* 66: 637–647.

Weatherbee, S. D., G. Halder, J. Kim, A. Hudson and S. Carroll. 1998. Ultrabithorax regulates genes at several levels of the wing-patterning hierarchy to shape the development of the *Drosophila* haltere. *Genes Dev.* 12: 1474–1482.

Whalen, A. M. and R. Steward. 1993. Dissociation of the dorsal-cactus complex and phosphorylation of the dorsal protein correlate with the nuclear localization of dorsal. *J. Cell. Biol.* 123: 523–534.

Wharton, R. P. and G. Struhl. 1991. RNA regulatory elements mediate control of *Drosophila* body pattern by the posterior morphogen nanos. *Cell* 67: 955–967.

Wreden, C., A. C. Verrotti, J. A. Schisa, M. E. Lieberfarb and S. Strickland. 1997. Nanos and pumilio establish embryonic polarity in *Drosophila* by promoting posterior deadenylation of *hunchback* mRNA. *Development* 124: 3015–3023.

Wu, L. H. and J. A. Lengyel. 1998. Role of *caudal* in hindgut specification and gastrulation suggests homology between *Drosophila* amnioproctodeal invagination and vertebrate blastopore. *Development* 125: 2433–2442.

10

Early development and axis formation in amphibians

AMPHIBIAN EMBRYOS WERE THE ORGANISMS OF CHOICE for experimental embryology. With their large cells and their rapid development, salamander and frog embryos were excellently suited for transplantation experiments. However, amphibian embryos fell out of favor during the early days of developmental genetics, since frogs and salamanders undergo a long period of growth before they become fertile, and their chromosomes are often found in several copies, precluding easy mutagenesis.* However, new molecular techniques such as in situ hybridization, antisense oligonucleotides, and dominant negative proteins have allowed researchers to return to studying amphibian embryos and to integrate molecular analyses of development with earlier experimental findings. The results have been spectacular, and we are enjoying our first appreciation of how vertebrate bodies are patterned and structured.

Early Amphibian Development

Cleavage in Amphibians

Cleavage in most frog and salamander embryos is radially symmetrical and holoblastic, just like echinoderm cleavage. The amphibian egg, however, contains much more yolk. This yolk, which is concentrated in the vegetal hemisphere, is an impediment to cleavage. Thus, the first division begins at the animal pole and slowly extends down into the vegetal region (Figure 10.1; see also Figures 2.2D and 8.4). In the axolotl salamander, the cleavage furrow extends through the animal hemisphere at a rate close to 1 mm per minute. The cleavage furrow bisects the gray crescent and then slows down to a mere 0.02–0.03 mm per minute as it approaches the vegetal pole (Hara 1977).

Figure 10.2A is a scanning electron micrograph showing the first cleavage in a frog egg. One can see the difference in the furrow between the animal and the vegetal hemispheres. Figure 10.2B shows that while the first cleavage furrow is still cleaving the yolky cytoplasm of the vegetal hemisphere, the second cleavage has already started near the animal pole. This cleavage is at right angles to the first one and is also meridional. The third cleavage, as expected, is equatorial. However, because of the

---

*In the 1960s, *Xenopus laevis* replaced the *Rana* frogs and the salamanders because it could be induced to mate throughout the year. Unfortunately, *Xenopus laevis* has four copies of each chromosome rather than the more usual two, and takes 1–2 years to reach sexual maturity. These attributes make genetic studies difficult. Recently another *Xenopus* species, *X. tropicalis*, has begun to be used in the laboratory. It has all the advantages of *X. laevis*, plus it is diploid and reaches sexual maturity in a mere 6 months.

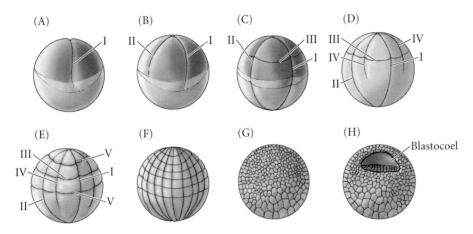

(A) (B) (C) (D)

(E) (F) (G) (H)

Blastocoel

Figure 10.1
Cleavage of a frog egg. Cleavage furrows, designated by Roman numerals, are numbered in order of appearance. (A, B) Because the vegetal yolk impedes cleavage, the second division begins in the animal region of the egg before the first division has divided the vegetal cytoplasm. (C) The third division is displaced toward the animal pole. (D–H) The vegetal hemisphere ultimately contains larger and fewer blastomeres than the animal half. H represents a cross section through a midgastrula stage embryo. (After Carlson 1981.)

vegetally placed yolk, this cleavage furrow in amphibian eggs is not actually at the equator, but is displaced toward the animal pole. It divides the frog embryo into four small animal blastomeres (micromeres) and four large blastomeres (macromeres) in the vegetal region. This unequal holoblastic cleavage establishes two major embryonic regions: a rapidly dividing region of micromeres near the animal pole and a more slowly dividing vegetal macromere area (Figure 10.2C; Figure 2.2E). As cleavage progresses, the animal region becomes packed with numerous small cells, while the vegetal region contains only a relatively small number of large, yolk-laden macromeres.

An amphibian embryo containing 16 to 64 cells is commonly called a **morula** (plural: **morulae;** from the Latin, "mulberry," whose shape it vaguely resembles). At the 128-cell stage, the blastocoel becomes apparent, and the embryo is considered a blastula. Actually, the formation of the blastocoel has been traced back to the very first cleavage furrow. Kalt (1971) demonstrated that in the frog *Xenopus laevis*, the first cleavage furrow widens in the animal hemisphere to create a

small intercellular cavity that is sealed off from the outside by tight intercellular junctions (Figure 10.3). This cavity expands during subsequent cleavages to become the blastocoel.

The blastocoel probably serves two major functions in frog embryos: (1) it permits cell migration during gastrulation, and (2) it prevents the cells beneath it from interacting prematurely with the cells above it. When Nieuwkoop (1973) took embryonic newt cells from the roof of the blastocoel, in the animal hemisphere, and placed them next to the yolky vegetal cells from the base of the blastocoel, these animal cells differentiated into mesodermal tissue instead of ectoderm. Because mesodermal tissue is normally formed from those animal cells that are adjacent to the vegetal endoderm precursors, it seems plausible that the vegetal cells influence adjacent cells to differentiate into mesodermal tissues. Thus, the blastocoel appears to prevent the contact of the vegetal cells destined to become endoderm with those cells fated to give rise to the skin and nerves.

While these cells are dividing, numerous cell adhesion molecules keep the blastomeres together. One of the most im-

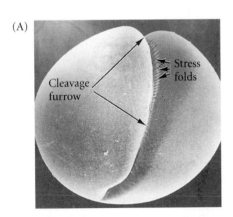

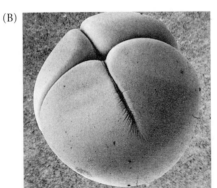

Figure 10.2
Scanning electron micrographs of the cleavage of a frog egg. (A) First cleavage. (B) Second cleavage (4 cells). (C) Fourth cleavage (16 cells), showing the size discrepancy between the animal and vegetal cells after the third division. (A from Beams and Kessel 1976, photograph courtesy of the authors; B and C courtesy of L. Biedler.)

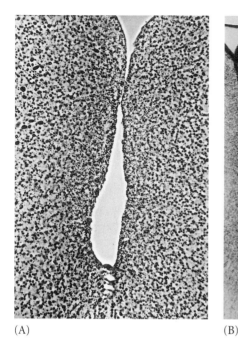
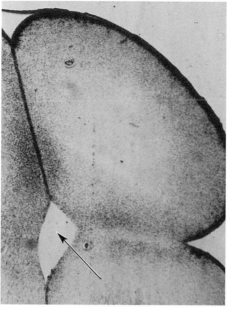

(A)   (B)

Figure 10.3
Formation of the blastocoel in a frog egg. (A) First cleavage furrow, showing a small cleft, which later develops into the blastocoel. (B) 8-cell embryo showing a small blastocoel (arrow) at the junction of the three cleavagefurrows. (From Kalt 1971; photographs courtesy of M. R. Kalt.)

way amphibians gastrulate. Different species employ different means toward the same goal (Smith and Malacinski 1983; Minsuk and Keller 1996). In recent years, the most intensive investigations have focused on the frog *Xenopus laevis*, so we will concentrate on its mode of gastrulation.

The fate map of Xenopus

Amphibian blastulae are faced with the same tasks as the invertebrate blastulae we followed in Chapters 8 and 9—name-

portant of these molecules is EP-cadherin. The mRNA for this protein is supplied in the oocyte cytoplasm. If this message is destroyed (by injecting antisense oligonucleotides complementary to this mRNA into the oocyte), the EP-cadherin is not made, and the adhesion between the blastomeres is dramatically reduced (Heasman et al. 1994a,b), resulting in the obliteration of the blastocoel (Figure 10.4).

Amphibian Gastrulation

The study of amphibian gastrulation is both one of the oldest and one of the newest areas of experimental embryology. Even though amphibian gastrulation has been extensively studied for the past century, most of our theories concerning the mechanisms of these developmental movements have been revised over the past decade. The study of amphibian gastrulation has been complicated by the fact that there is no single

ly, to bring inside the embryo those areas destined to form the endodermal organs, to surround the embryo with cells capable of forming the ectoderm, and to place the mesodermal cells in the proper positions between them. The movements whereby this is accomplished can be visualized by the technique of vital dye staining (see Chapter 1). Fate mapping by Løvtrup (1975; Landstrom and Løvtrup 1979) and by Keller (1975, 1976) has shown that cells of the *Xenopus* blastula have different fates depending on whether they are located in the deep or the superficial layers of the embryo (Figure 10.5). In *Xenopus*, the mesodermal precursors exist mostly in the deep layer of cells, while the ectoderm and endoderm arise from the superficial layer on the surface of the embryo. Most of the precursors for the notochord and other mesodermal tissues are located beneath the surface in the equatorial (marginal) region of the embryo. In urodeles (salamanders such as *Triturus* and *Ambystoma*) and in some frogs other than *Xenopus*, many

(A)   (B)

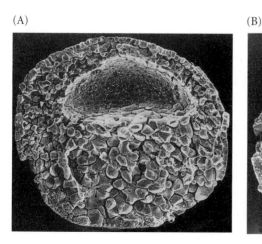
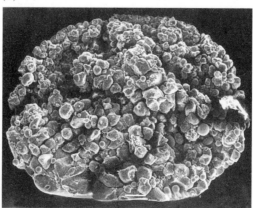

Figure 10.4
Depletion of EP-cadherin mRNA in the *Xenopus* oocyte results in the loss of adhesion between blastomeres and the obliteration of the blastocoel. (A) control embryo; (B) EP-cadherin-depleted embryo. (From Heasman et al. 1994b; photographs courtesy of J. Heasman.)

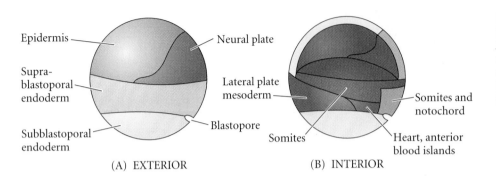

Figure 10.5
Fate maps of the blastula of the frog *Xenopus laevis*: (A) exterior; (B) interior. Most of the mesodermal derivatives are formed from the interior cells. (After Lane and Smith 1999; Newman and Krieg 1999.)

more of the notochord and mesoderm precursors are among the surface cells (Purcell and Keller 1993).

As we have seen, the unfertilized egg has a polarity along the animal-vegetal axis. Thus, the germ layers can be mapped onto the oocyte even before fertilization. The surface of the animal hemisphere will become the cells of the ectoderm (skin and nerves), the vegetal hemisphere surface will form the cells of the gut and associated organs (endoderm), and the mesodermal cells will form from the internal cytoplasm around the equator. This general fate map is thought to be imposed upon the egg by the transcription factor **VegT** and the paracrine factor **Vg1**. The mRNAs for these proteins are located in the cortex of the vegetal hemisphere of *Xenopus* oocytes, and they are apportioned to the vegetal cells during cleavage (see Figure 5.33). By using antisense oligonucleotides, Zhang and colleagues (1998) were able to deplete maternal VegT protein in early embryos. The resulting embryos lacked the normal fate map. The animal third of the embryo produced only ventral epidermis, while the marginal cells (which normally produced mesoderm) generated epidermal and neural tissue. The vegetal third (which usually produces endoderm) produced a mixture of ectoderm and mesoderm (Figure 10.6). Joseph and Melton (1998) demonstrated that embryos that lacked functional Vg1 lacked endoderm and dorsal mesoderm.

These findings tell us nothing, however, about which part of the egg will form the belly and which the back. The anterior-posterior, dorsal-ventral, and left-right axes are specified by the events of fertilization and are realized during gastrulation.

VADE MECUM **Amphibian development.** The events of cleavage and gastrulation are difficult to envision without three-dimensional models. You can see movies of such 3-D models, as well as footage of a living *Xenopus* embryo, in the segments on amphibian development.
[Click on Amphibian]

Cell movements during amphibian gastrulation

Before we look at the process of gastrulation in detail, let us first trace the movement patterns of the germ layers. Gastrulation in frog embryos is initiated on the future dorsal side of the embryo, just below the equator in the region of the gray crescent (Figure 10.7). Here, the cells invaginate to form a slitlike blastopore. These cells change their shape dramatically. The main body of each cell is displaced toward the inside of the embryo while the cell maintains contact with the outside surface by way of a slender neck (Figure 10.8). These **bottle cells** line the archenteron as it forms. Thus, as in the gastrulating sea urchin, an invagination of cells initiates archenteron formation. However, unlike gastrulation in sea urchins, gastrulation in the frog begins not at the most vegetal region, but in the **marginal zone**: the zone surrounding the equator of the blastula, where the animal and vegetal hemispheres meet. Here the endodermal cells are not as large or as yolky as the most vegetal blastomeres.

The next phase of gastrulation involves the involution of the marginal zone cells while the animal cells undergo epiboly and converge at the blastopore (Figure 10.7C, D). When the

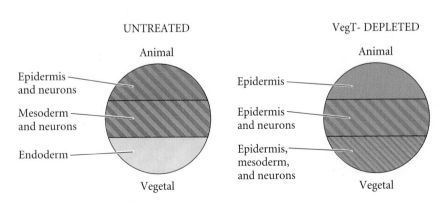

Figure 10.6
The fates of the three regions of the *Xenopus* blastula are altered by the depletion of VegT. (A) In normal embryos, the animal third forms epidermal and neural ectoderm, the equatorial third forms mesoderm, and the vegetal third contains the VegT protein and forms the endoderm. In VegT-depleted embryos, the animal cap forms only ventral epidermis, while the equatorial third produces epidermal and neural ectoderm. The vegetal third of these embryos produces ectoderm (both epidermal and neural) as well as mesoderm. No endoderm is produced. (After Zhang et al. 1998.)

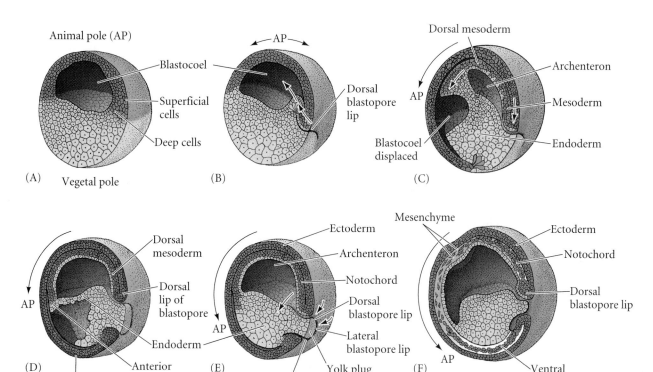

Figure 10.7
Cell movements during frog gastrulation. The meridional sections are cut through the middle of the embryo and positioned so that the vegetal pole is tilted toward the observer and slightly to the left. The major cell movements are indicated by arrows, and the superficial animal hemisphere cells are colored so that their movements can be followed. (A, B) Early gastrulation. The bottle cells of the margin move inward to form the dorsal lip of the blastopore, and the mesodermal precursors involute under the roof of the blastocoel. AP marks the position of the animal pole, which will change as gastrulation continues. (C, D) Mid-gastrulation. The archenteron forms and displaces the blastocoel, and cells migrate from the lateral and ventral lips of the blastopore into the embryo. The cells of the animal hemisphere migrate down toward the vegetal region, moving the blastopore to the region near the vegetal pole. (E, F) Toward the end of gastrulation, the blastocoel is obliterated, the embryo becomes surrounded by ectoderm, the endoderm has been internalized, and the mesodermal cells have been positioned between the ectoderm and endoderm. (After Keller 1986.)

mesoderm cells. These cells will form the **notochord**, a transient mesodermal "backbone" that plays an important role in distinguishing and patterning the nervous system.

As the new cells enter the embryo, the blastocoel is displaced to the side opposite the dorsal lip of the blastopore. Meanwhile, the blastopore lip expands laterally and ventrally as the processes of bottle cell formation and involution continue about the blastopore. The widening blastopore "crescent" develops lateral lips and finally a ventral lip over which

migrating marginal cells reach the **dorsal lip** of the blastopore, they turn inward and travel along the inner surface of the outer animal hemisphere cells. Thus, the cells constituting the lip of the blastopore are constantly changing. The first cells to compose the dorsal blastopore lip are the bottle cells that invaginated to form the leading edge of the archenteron. These cells later become the pharyngeal cells of the foregut. As these first cells pass into the interior of the embryo, the dorsal blastopore lip becomes composed of cells that involute into the embryo to become the **prechordal plate** (the precursor of the head mesoderm). The next cells involuting into the embryo through the dorsal blastopore lip are called the **chorda-**

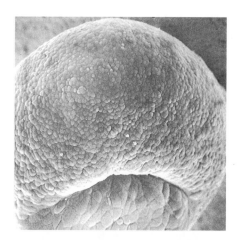

Figure 10.8
Surface view of an early dorsal blastopore lip of *Xenopus*. The size difference between the animal and vegetal blastomeres is readily apparent. (Micrograph courtesy of C. Phillips.)

additional mesodermal and endodermal precursor cells pass. With the formation of the ventral lip, the blastopore has formed a ring around the large endodermal cells that remain exposed on the vegetal surface. This remaining patch of endoderm is called the **yolk plug**; it, too, is eventually internalized (Figure 10.9). At that point, all the endodermal precursors have been brought into the interior of the embryo, the ectoderm has encircled the surface, and the mesoderm has been brought between them.

The midblastula transition: Preparing for gastrulation

Now that we have seen an overview of amphibian gastrulation, we can look more deeply into its mechanisms. The first precondition for gastrulation is the activation of the genome. In *Xenopus*, the nuclear genes are not transcribed until late in the twelfth cell cycle (Newport and Kirschner 1982a,b). At that time, different genes begin to be transcribed in different cells, and the blastomeres acquire the capacity to become motile. This dramatic change is called the midblastula transition (see Chapters 8 and 9). It is thought that different transcription factors (such as the VegT protein, mentioned above) become active in different cells at this time, giving the cells new properties. For instance, the vegetal cells (probably under the direction of the maternal VegT protein) become the endoderm and begin secreting the factors that cause the cells above them to become the mesoderm (Wylie et al. 1996).

Positioning the blastopore

The vegetal cells are critical in determining the location of the blastopore, as is the point of sperm entry. The microtubules of the sperm direct cytoplasmic movements that empower the vegetal cells opposite the point of sperm entry to induce the blastopore in the mesoderm above them. This region of cells opposite the point of sperm entry will form the blastopore and become the dorsal portion of the body.

In Chapter 7, we saw that the internal cytoplasm of the fertilized egg remains oriented with respect to gravity because of its dense yolk accumulation, while the cortical cytoplasm actively rotates 30 degrees animally ("upward"), toward the point of sperm entry (see Figure 7.33). In this way, a new state of symmetry is acquired. Whereas the unfertilized egg was radially symmetrical about the animal-vegetal axis, the fertilized egg now has a dorsal-ventral axis. It has become bilaterally symmetrical (having right and left sides). The inner cytoplasm moves as well. Fluorescence microscopy of early embryos has shown that the cytoplasm of the presumptive dorsal cells differs from that of the presumptive ventral cells (see Figure 7.35; Danilchik and Denegre 1991). These cytoplasmic movements activate the cytoplasm opposite the point of sperm entry, enabling it to initiate gastrulation. The side where the sperm enters marks the future ventral surface of the embryo; the opposite side, where gastrulation is initiated, marks the future dorsum of the embryo (Gerhart et al. 1981, 1986; Vincent et al. 1986). If cortical rotation is blocked, there is no dorsal development, and the embryo dies as a mass of ventral (primarily gut) cells (Vincent and Gerhart 1987).

Although the sperm is not needed to induce these movements in the egg cytoplasm, it is important in determining the *direction* of the rotation. If an egg is artificially activated, the cortical rotation still takes place at the correct time. However, the direction of this movement is unpredictable. The direc-

Figure 10.9

Epiboly of the ectoderm. (A) Changes in the region around the blastopore as the dorsal, lateral, and ventral lips are formed in succession. When the ventral lip completes the circle, the endoderm becomes progressively internalized. Numbers ii–v correspond to Figures 10.7 B–E, respectively. (B) Summary of epiboly of the ectoderm and involution of the mesodermal cells migrating into the blastopore and then under the surface. The endoderm beneath the blastopore lip is not mobile and is enclosed by these movements. (A from Balinsky 1975; photographs courtesy of B. I. Balinsky.)

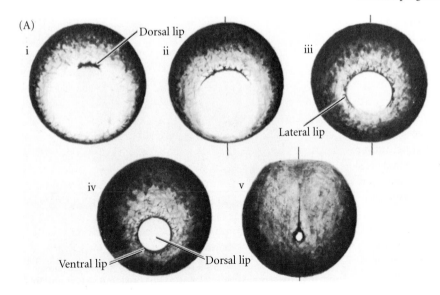

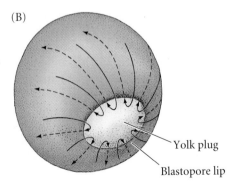

(A)

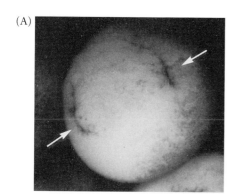

(B)

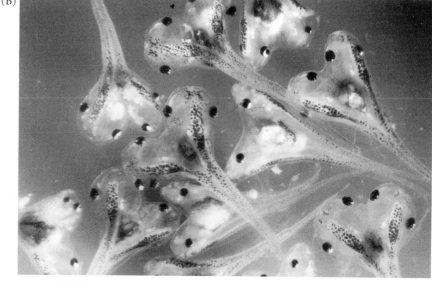

Figure 10.10

Twin blastopores produced by rotating dejellied *Xenopus* eggs ventral side (sperm entry point) up at the time of first cleavage. (A) Two blastopores are instructed to form: the original one (opposite the point of sperm entry) and the new one created by the displacement of cytoplasmic material. (B) These eggs develop two complete axes, which form twin tadpoles, joined ventrally. (Photographs courtesy of J. Gerhart.)

tional bias provided by the point of sperm entry can be overridden by mechanically redirecting the spatial relationship between the cortical and internal cytoplasms. When a *Xenopus* egg is turned 90 degrees, so that the point of sperm entry faces upward, the cytoplasm rotates such that the embryo initiates gastrulation on the *same* side as sperm entry (Gerhart et al. 1981; Cooke 1986). One can even produce two gastrulation initiation sites by combining the natural sperm-oriented rotation with an artificially induced rotation of the egg. Black and Gerhart (1985, 1986) let the initial sperm-directed rotation occur, but then immobilized eggs in gelatin and gently centrifuged them so that the internal cytoplasm would flow toward the point of sperm entry. When the centrifuged eggs were then allowed to develop in normal water, two sites of gastrulation emerged, leading to conjoined twin larvae (Figure 10.10). Black and Gerhart hypothesized that the twinning was caused by the formation of two areas of interaction: one axis formed where the normal cortical rotation caused the normal cytoplasmic interactions in the vegetal region of the cell, the other where the centrifugation-driven cytoplasm interacted with the vegetal components.

It appears that cortical rotation enables the vegetal blastomeres opposite the point of sperm entry to induce the cells above them to initiate gastrulation. Gimlich and Gerhart (1984), using transplantation experiments on 64-cell *Xenopus* embryos, showed that the three vegetal blastomeres opposite the point of sperm entry are able to induce the formation of the dorsal lip of the blastopore and of a complete dorsal axis when transplanted into UV-irradiated embryos (which otherwise would have failed to properly initiate gastrulation: Figure 10.11A). Moreover, these three blastomeres, which underlie the prospective dorsal lip region, can also induce a secondary blastopore and axis when transplanted into the ventral side of a normal, unirradiated embryo (Figure 10.11B). Holowacz and Elinson (1993) found that cortical cytoplasm from the dorsal vegetal cells of the 16-cell *Xenopus* embryo was able to

induce the formation of secondary axes when injected into ventral vegetal cells. Neither cortical cytoplasm from animal cells nor the deep cytoplasm from ventral cells could induce such axes. Later in this chapter, we will provide evidence that this dorsal signal is the transcription factor β-catenin (Wylie et al. 1996; Larabell et al. 1997).

Invagination and involution

Amphibian gastrulation is first visible when a group of marginal endoderm cells on the dorsal surface of the blastula sinks into the embryo. The outer (**apical**) surfaces of these cells contract dramatically, while their inner (**basal**) ends expand. The apical-basal length of these cells greatly increases to yield the characteristic "bottle" shape. In salamanders, these bottle cells appear to have an active role in the early movements of gastrulation. Johannes Holtfreter (1943, 1944) found that bottle cells from early salamander gastrulae could attach to glass coverslips and lead the movement of those cells attached to them. Even more convincing were Holtfreter's recombination experiments. When dorsal marginal zone cells (which would normally give rise to the dorsal lip of the blastopore) were excised and placed on inner prospective endoderm tissue, they formed bottle cells and sank below the surface of the inner endoderm (Figure 10.12). Moreover, as they sank, they created a depression reminiscent of the early blastopore. Thus, Holtfreter claimed that the ability to invaginate into the deep endoderm is an innate property of the dorsal marginal zone cells.

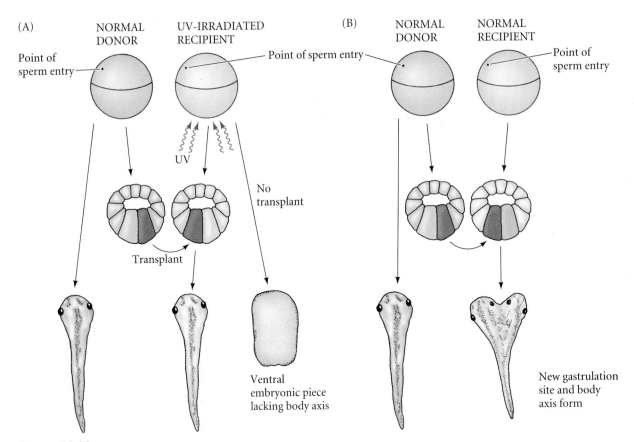

(A) NORMAL DONOR — UV-IRRADIATED RECIPIENT

Point of sperm entry

Point of sperm entry

UV

No transplant

Transplant

Ventral embryonic piece lacking body axis

(B) NORMAL DONOR — NORMAL RECIPIENT

Point of sperm entry

New gastrulation site and body axis form

Figure 10.11
Transplantation experiments on 64-cell amphibian embryos demonstrating that the vegetal cells underlying the prospective dorsal blastopore lip region are responsible for causing the initiation of gastrulation. (A) Rescue of irradiated embryos by transplanting the dorsal vegetal blastomeres of a normal embryo into a cavity made by the removal of a similar number of vegetal cells. An irradiated zygote without this transplant fails to undergo normal gastrulation. (B) Formation of a new gastrulation site and body axis by the transplantation of the most dorsal vegetal cells of one embryo into the ventralmost vegetal region of another embryo. (After Gimlich and Gerhart 1984.)

WEBSITE **10.1 Demostrating tissue affinities.** The tissue affinities that Holtfreter predicted have been demonstrated quantitatively by new studies that measure the surface tensions of different cell layers.

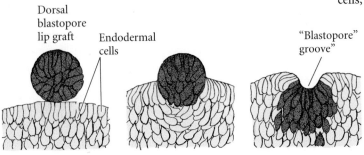

Dorsal blastopore lip graft

Endodermal cells

"Blastopore" groove"

The situation in the frog embryo is somewhat different. R. E. Keller and his students (Keller 1981; Hardin and Keller 1988) have shown that although the bottle cells of *Xenopus* may play a role in initiating the involution of the marginal zone as they become bottle-shaped, they are not essential for gastrulation to continue. The peculiar shape change of the bottle cells is needed to initiate gastrulation; it is the constriction of these cells that first forms the slit-like blastopore. However, after starting these movements, the *Xenopus* bottle cells are no longer needed for gastrulation. When bottle cells are removed after their formation, involution and blastopore formation and closure continue.

The major factor in the movement of cells into the embryo appears to be the involution of the subsurface marginal cells, rather than the superficial ones. The movements of the vegetal endoderm place the prospective pharyngeal endoderm adjacent to the roof of the blasto-

Figure 10.12
A graft of cells from the dorsal marginal zone of a salamander embryo sinks into a layer of endodermal cells and forms a blastopore-like groove. (After Holtfreter 1944.)

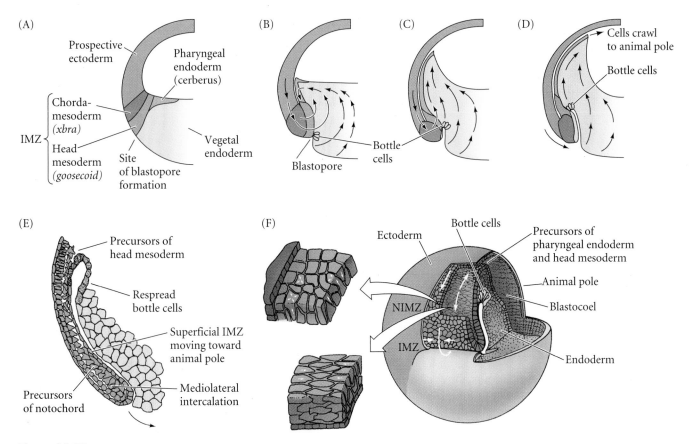

Figure 10.13
Early movements of *Xenopus* gastrulation. The yellow is vegetal endoderm. Orange represents the prospective pharyngeal endoderm (as seen by *Cerberus* expression). Dark orange represents the prospective head mesoderm (as seen by *goosecoid* expression), and the chordamesoderm (indicated by *Xbra* expression) is in red. The prospective ectoderm is blue. (A) At the beginning of gastrulation, the inner marginal zone (IMZ) is formed. (B) The vegetal rotation (white arrows) pushes the prospective pharyngeal endoderm to the side of the blastocoel. (C, D) The vegetal endoderm movements push the pharyngeal endoderm forward, driving the mesoderm passively into the embryo and toward the animal pole. The ectoderm begins epiboly. (E) As gastrulation continues, the deep marginal cells flatten, and the formerly superficial cells form the wall of the archenteron. (F) Radial intercalation, looking down at the dorsal blastopore lip from the dorsal suface. In the noninvoluting marginal zone (NIMZ) and the upper portin of the IMZ, deep (mesodermal) cells are intercalating radially to make a thin band of flattened cells. This thinning of several layers into a few causes extension toward the blastopore lip. Just above the lip, mediolateral intercalation of the cells produces stresses that pull the IMZ over the lip. After involuting over the lip, mediolateral intercalation continues, elongating and narrowing the axial mesoderm. (After Wilson and Keller 1991 and Winklbauer and Schürfeld 1999.)

coel. This places the prospective pharyngeal endoderm immediately ahead of the migrating mesoderm. The cells then migrate along the basal surface of the blastocoel roof (Figure 10.13A–D; Winklbauer and Schürfeld 1999). The superficial layer of marginal cells is pulled inward to form the endoder-

mal lining of the archenteron, merely because it is attached to the actively migrating deep cells. While experimental removal of the bottle cells does not affect the involution of the deep or superficial marginal zone cells into the embryo, the removal of the deep involuting marginal zone (IMZ) cells and their replacement with animal region cells (which do not normally undergo involution) stops archenteron formation.

The convergent extension of the dorsal mesoderm

Involution begins dorsally, led by the pharyngeal endomesoderm* and the prechordal plate. These tissues will migrate most anteriorly beneath the surface ectoderm. The next tissues to enter the dorsal blastopore lip contain notochord and somite precursors. Meanwhile, as the lip of the blastopore expands to have dorsolateral, lateral, and ventral sides, the prosepective heart mesoderm, kidney mesoderm, and ventral mesoderm enter into the embryo.

Figures 10.13D–F depict the behavior of the IMZ cells at successive stages of *Xenopus* gastrulation (Keller and Schoenwolf 1977; Keller 1980, 1981; Hardin and Keller 1988). The IMZ is originally several layers thick. Shortly before their

---

*The pharyngeal endoderm and head mesoderm cannot be separated experimentally at this stage, so they are therefore sometimes referred to collectively as the pharyngeal endomesoderm. The notochord is the basic unit of the dorsal mesoderm, but it is thought that the dorsal portion of the somites may also have similar properties.

(A)

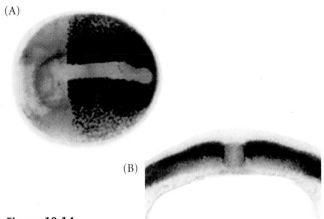

(B)

Figure 10.14
The expression of paraxial protocadherin. (A) Expression of paraxial protocadherin during late gastrulation shows the distinct downregulation in the notochord and the absence of expression in the head region. (B) Double-stained cross section through a late *Xenopus* gastrula shows the separation of notochord (brown staining for chordin) and the paraxial mesoderm (blue staining for paraxial protocadherin). (Photographs courtesy of E. M. De Robertis.)

involution through the blastopore lip, the several layers of deep IMZ cells intercalate radially to form one thin, broad layer. This intercalation further extends the IMZ vegetally. At the same time, the superficial cells spread out by dividing and flattening. When the deep cells reach the blastopore lip, they involute into the embryo and initiate a second type of intercalation. This intercalation causes a **convergent extension** along the mediolateral axis (Figure 10.13F) that integrates several mesodermal streams to form a long, narrow band. This is reminiscent of traffic on a highway when several lanes must merge to form a single lane. The anterior part of this band migrates toward the animal cap. Thus, the mesodermal stream continues to migrate toward the animal pole, and the overlying layer of superficial cells (including the bottle cells) is passively pulled toward the animal pole, thereby forming the endodermal roof of the archenteron (see Figures 10.7 and 10.13E). The radial and mediolateral intercalations of the deep layer of cells appear to be responsible for the continued movement of mesoderm into the embryo.

The adhesive changes driving convergent extension appear to be directed by two cell adhesion molecules, **paraxial protocadherin** and **axial protocadherin**. The former is initially found throughout the dorsal mesoderm, but then is turned off in the precursors of the notochord. At that time, axial protocadherin becomes expressed in the notochordal tissue (Figure 10.14). An experimental dominant negative form of paraxial protocadherin (which is secreted instead of being bound to the cell membrane) prevents convergent extension*

---

*Dominant negative proteins are mutated forms of the wild-type protein that interfere with the normal functioning of the wild-type protein. Thus, a dominant negative protein will have an effect similar to a loss-of-function mutation in the gene encoding the particular protein.

(Kim et al. 1998). Moreover, the expression domain of paraxial protocadherin separates the trunk mesodermal cells, which undergo convergent extension, from the head mesodermal cells, which do not.

Migration of the involuting mesoderm

As mesodermal movement progresses, convergent extension continues to narrow and lengthen the involuting marginal zone. The IMZ contains the prospective endodermal roof of the archenteron in its superficial layer (IMZ$_S$) and the prospective mesodermal cells, including those of the notochord, in its deep region (IMZ$_D$). During the middle third of gastrulation, the expanding sheet of mesoderm converges toward the midline of the embryo. This process is driven by the continued mediolateral intercalation of cells along the anterior-posterior axis, thereby further narrowing the band. Toward the end of gastrulation, the centrally located notochord separates from the somitic mesoderm on either side of it, and the notochord cells elongate separately (Wilson and Keller 1991). This may in part be a consequence of the different protocadherins in the axial and paraxial mesoderms (Kim et al. 1998). This convergent extension of the mesoderm appears to be autonomous, because the movements of these cells occur even if this region of the embryo is experimentally isolated from the rest of the embryo (Keller 1986).

During gastrulation, the **animal cap** and **noninvoluting marginal zone (NIMZ)** cells expand by epiboly to cover the entire embryo. The dorsal portion of the NIMZ extends more rapidly toward the blastopore than the ventral portion, thus causing the blastopore lips to move toward the ventral side. While those mesodermal cells entering through the dorsal lip of the blastopore give rise to the dorsal axial mesoderm (notochord and somites), the remainder of the body mesoderm (which forms the heart, kidneys, blood, bones, and parts of several other organs) enters through the ventral and lateral blastopore lips to create the **mesodermal mantle**. The endoderm is derived from the IMZ$_S$ cells that form the lining of the archenteron roof and from the subblastoporal vegetal cells that become the archenteron floor (Keller 1986).

Epiboly of the ectoderm

While involution is occurring at the blastopore lips, the ectodermal precursors are expanding over the entire embryo. Keller (1980) and Keller and Schoenwolf (1977) have used scanning electron microscopy to observe the changes in both the superficial cells and the deep cells of the animal and marginal regions. The major mechanism of epiboly in *Xenopus* gastrulation appears to be an increase in cell number (through division) coupled with a concurrent integration of several deep layers into one (Figure 10.15). During early gastrulation, three rounds of cell division increase the number of the deep cell layers in the animal hemisphere. At the same time, complete integration of the numerous deep cells into one layer occurs. The most superficial layer expands by cell division and

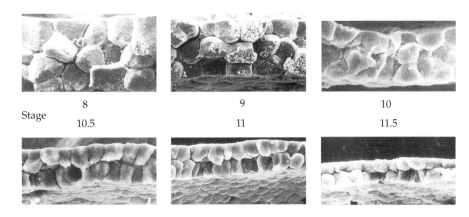

Stage 8   9   10
10.5   11   11.5

Figure 10.15
Scanning electron micrographs of the *Xenopus* blastocoel roof, showing the changes in cell shape and arrangement. Stages 8 and 9 are blastulae; stages 10–11.5 represent progressively later gastrulae. (From Keller 1980; photographs courtesy of R. E. Keller.)

flattening. The spreading of cells in the dorsal and ventral marginal zones appears to proceed by the same mechanism, although changes in cell shape appear to play a greater role than in the animal region. The result of these expansions is the epiboly of the superficial and deep cells of the animal cap and NIMZ over the surface of the embryo (Keller and Danilchik 1988). Most of the marginal zone cells, as previously mentioned, involute to join the mesodermal cell stream within the embryo. As the ectoderm epibolizes over the entire embryo, it eventually internalizes all the endoderm within it. At this point, the ectoderm covers the embryo, the endoderm is located within the embryo, and the mesoderm is positioned between them.

WEBSITE **10.2 Migration of the mesodermal mantle.** Different growth rates coupled with the intercalation of cell layers allows the mesoderm to expand in a tightly coordinated fashion.

Sidelights & Speculations

Fibronectin and the Pathways for Mesodermal Migration

HOW ARE THE INVOLUTING CELLS informed where to go once they enter the inside of the embryo? In salamanders, it appears that the involuting mesodermal precursors migrate toward the animal pole on a fibronectin lattice secreted by the cells of the blastocoel roof. Shortly before gastrulation, the presumptive ectoderm of the blastocoel roof secretes an extracellular matrix that contains fibrils of fibronectin (Figure 10.16A; Boucaut et al. 1984; Nakatsuji et al. 1985). The involuting mesoderm appears to travel along these fibronectin fibers. Confirmation of this hypothesis was obtained by chemically synthesizing a "phony" fibronectin that can compete with the genuine fibronectin of the extracellular matrix. Cells bind to a region of the fibronectin protein that contains a three-amino acid sequence (Arg-Gly-Asp; RGD). Boucaut and co-workers injected large amounts of a small peptide containing this sequence into the blastocoels of salamander embryos shortly before gastrulation began. If fibronectin were essential for cell migration, then cells binding this soluble peptide fragment instead of the real extracellular fibronectin should stop migrating. Unable to find their "road," the mesodermal cells should cease their involution. That is precisely what happened (Figure 10.16B–E). No migrating cells were seen along the underside of the ectoderm in the experimental embryos. Instead, the mesodermal precursors remained outside the embryos, forming a convoluted cell mass. Other small synthetic peptides (including other fragments of the fibronectin molecule) did not impede migration.

The mesodermal cells are thought to adhere to fibronectin through the $\alpha v \beta 1$ integrin protein (Alfandari et al. 1995). Mesodermal migration can also be arrested by the microinjection of antibodies against either fibronectin or the integrin subunit that serve as part of the fibronectin receptor (D'Arribère et al. 1988, 1990). Alfandari and colleagues (1995) have shown that the αv subunit of integrin appears on the mesodermal cells just prior to gastrulation, persists on their surfaces throughout gastrulation, and disappears when gastrulation ends. It seems, then, that the synthesis of this fibronectin receptor may signal the times for the mesoderm to begin, continue, and stop migration.

In addition to permitting the attachment of mesodermal cells, the fibronectin-containing extracellular matrix appears to provide the cues for the direction of cell migration. Shi and colleagues (1989) showed that salamander IMZ cells would migrate in the wrong direction if extra fibronectin lattices were placed in their path. In *Xenopus*, convergent extension pushes the migrating cells upward toward the animal pole. Fibronectin appears to delineate the boundaries within which this movement can occur (see Figure 6.32). The fibronectin fibrils are necessary for the head mesodermal cells to flatten and to extend broad (lamelliform) processes in the direction of migration (Winklbauer et al. 1991; Winklbauer and Keller 1996). The im-

(A)

(B)

(C)

(D)

(E)

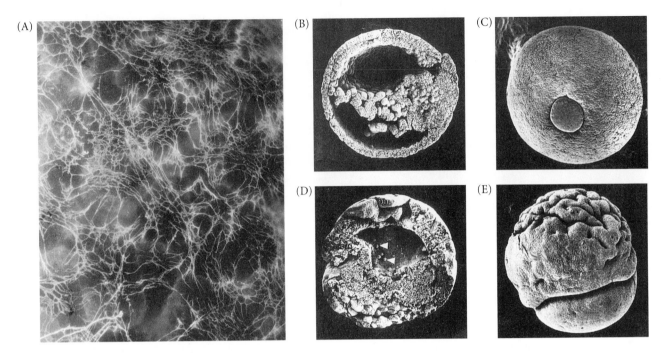

Figure 10.16

Fibronectin and amphibian gastrulation. (A) Immunofluorescence reveals a fibrillar network of fibronectin on the basal surface of the prospective ectodermal cells lining the blastocoel roof in the salamander embryo. (B–E) Scanning electron micrographs of normal (B, C) and abnormal (D, E) salamander gastrulation. The blastocoel in (D) and (E) was injected with the cell-binding fragment of fibronectin, while the normally gastrulating blastula was injected with a control solution. (B) Section during mid-gastrulation. (C) The yolk plug toward the end of gastrulation. (D, E) The finishing stages of the arrested gastrulation, wherein the mesodermal precursors, having bound the synthetic fibronectin, cannot recognize the normal fibronectin-lined migration route. The archenteron fails to form, and the noninvoluted mesodermal precursors remain on the surface. (A from Boucaut et al. 1985; B–E from Boucaut et al. 1984; photographs courtesy of J.-C. Boucaut and J.-P. Thiery.)

portance of these fibronectin fibrils is also seen in inviable interspecific hybrids. Delarue and colleagues (1985) have shown that certain inviable hybrids between two species of toads arrest during gastrulation because they do not secrete these fibronectin fibrils. It appears, then, that the extracellular matrix of the blastocoel roof, and particularly its fibronectin component, is important in the migration of the mesodermal cells during amphibian gastrulation. ■

Axis Formation in Amphibians: The Phenomenon of the Organizer

The Progressive Determination of the Amphibian Axes

Vertebrate axes do not form from localized determinants in the various blastomeres, as in *Drosophila*. Rather, they arise progressively through a sequence of interactions between neighboring cells. Amphibian axis formation is an example of regulative development. In Chapter 3, we discussed the concept of regulative development, wherein (1) an isolated blastomere has a potency greater than its normal embryonic fate, and (2) a cell's fate is determined by interactions between neighboring cells. Such interactions are called inductions (see Chapter 6). That such inductive interactions were responsible for amphibian axis determination was demonstrated by the laboratory of Hans Spemann at the University of Freiburg.

The experiments of Spemann and his students framed the questions that experimental embryologists asked for most of the twentieth century, and they resulted in a Nobel Prize for Spemann in 1935. More recently, the discoveries of the molecules associated with these inductive processes have provided some of the most exciting moments in contemporary science.

The experiment that began this research program was performed in 1903, when Spemann demonstrated that early newt blastomeres have identical nuclei, each capable of producing an entire larva. His procedure was ingenious: Shortly after fertilizing a newt egg, Spemann used a baby's hair taken from his daughter to lasso the zygote in the plane of the first cleavage. He then partially constricted the egg, causing all the nuclear divisions to remain on one side of the constriction. Eventually, often as late as the 16-cell stage, a nucleus would escape across the constriction into the non-nucleated side. Cleavage then began on this side, too, whereupon Spemann tightened the lasso until the two halves were completely sepa-

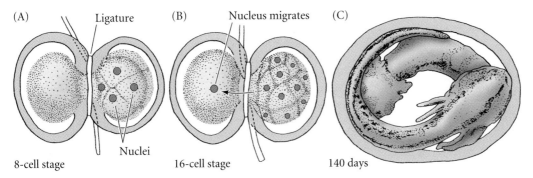

Figure 10.17
Spemann's demonstration of nuclear equivalence in newt cleavage. (A) When the fertilized egg of the newt *Triturus taeniatus* was constricted by a ligature, the nucleus was restricted to one-half of the embryo. The cleavage on that side of the embryo reached the 8-cell stage, while the other side remained undivided. (B) At the 16-cell stage, a single nucleus entered the as yet undivided half, and the ligature was constricted to complete the separation of the two halves. (C) After 140 days, each side had developed into a normal embryo. (After Spemann 1938.)

terial passes into only one of the two blastomeres. Spemann found that when these two blastomeres are separated, only the blastomere containing the gray crescent develops normally.

WEBSITE **10.3 Embryology and individuality.** One egg usually makes only one adult. However, there are exceptions to this rule, and Spemann was drawn into embryology through the paradoxes of creating more than one individual from a single egg.

rated. Twin larvae developed, one slightly older than the other (Figure 10.17). Spemann concluded from this experiment that early amphibian nuclei were genetically identical and that each cell was capable of giving rise to an entire organism.

However, when Spemann performed a similar experiment with the constriction still longitudinal, but perpendicular to the plane of the first cleavage (separating the future dorsal and ventral regions rather than the right and left sides), he obtained a different result altogether. The nuclei continued to divide on both sides of the constriction, but only one side—the future dorsal side of the embryo—gave rise to a normal larva. The other side produced an unorganized tissue mass of ventral cells, which Spemann called the *Bauchstück*—the belly piece. This tissue mass was a ball of epidermal cells (ectoderm) containing blood and mesenchyme (mesoderm) and gut cells (endoderm), but no dorsal structures such as nervous system, notochord, or somites (Figure 10.18).

Why should these two experiments give different results? One possibility was that when the egg was divided perpendicular to the first cleavage plane, some *cytoplasmic* substance was not equally distributed into the two halves. Fortunately, the salamander egg was a good place to test that hypothesis. As we have seen in Chapter 7 and above, there are dramatic movements in the cytoplasm following the fertilization of amphibian eggs, and in some amphibians these movements expose a gray, crescent-shaped area of cytoplasm in the region directly opposite the point of sperm entry. This area has been called the **gray crescent**. Moreover, the first cleavage plane normally splits the gray crescent equally into the two blastomeres. If these cells are then separated, two complete larvae develop. However, should this cleavage plane be aberrant (either in the rare natural event or in an experiment), the gray crescent ma-

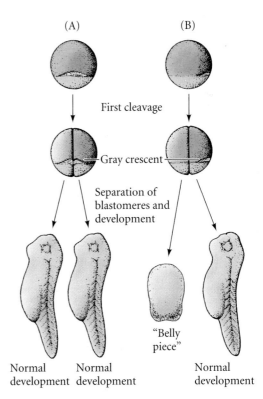

Figure 10.18
Asymmetry in the amphibian egg. (A) When the egg is divided along the plane of first cleavage into two blastomeres, each of which gets one-half of the gray crescent, each experimentally separated cell develops into a normal embryo. (B) When only one of the two blastomeres receives the entire gray crescent, it alone forms a normal embryo. The other half produces a mass of unorganized tissue lacking dorsal structures. (After Spemann 1938.)

(A) TRANSPLANTATION IN EARLY GASTRULA

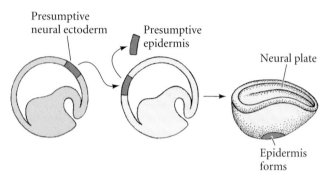

Presumptive neural ectoderm

Presumptive epidermis

Neural plate

Epidermis forms

(B) TRANSPLANTATION IN LATE GASTRULA

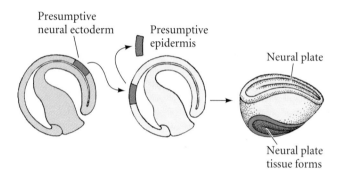

Presumptive neural ectoderm

Presumptive epidermis

Neural plate

Neural plate tissue forms

Figure 10.19
Determination of ectoderm during newt gastrulation. Presumptive neural ectoderm from one newt embryo is transplanted into a region in another embryo that normally becomes epidermis. (A) When the tissues are transferred between early gastrulas, the presumptive neural tissue develops into epidermis, and only one neural plate is seen. (B) When the same experiment is performed using late-gastrula tissues, the presumptive neural cells form neural tissue, thereby causing two neural plates to form on the host. (After Saxén and Toivonen 1962.)

It appeared, then, that something in the gray crescent region was essential for proper embryonic development. But how did it function? What role did it play in normal development? The most important clue came from the fate map of this area of the egg, for it showed that the gray crescent region gives rise to the cells that initiate gastrulation. These cells form the dorsal lip of the blastopore. The cells of the dorsal lip are committed to invaginate into the blastula, thus initiating gastrulation and the formation of the notochord. Because all fu-

ture amphibian development depends on the interaction of cells rearranged during gastrulation, Spemann speculated that the importance of the gray crescent material lies in its ability to initiate gastrulation, and that crucial developmental changes occur during gastrulation.

In 1918, Spemann demonstrated that enormous changes in cell potency do indeed take place during gastrulation. He found that the cells of the *early* gastrula were uncommitted, but that the fates of *late* gastrula cells were determined. Spemann demonstrated this by exchanging tissues between the gastrulae of two species of newts whose embryos were differently pigmented (Figure 10.19). When a region of prospective epidermal cells from an *early* gastrula was transplanted into an area in another early gastrula where the neural tissue normally formed, the transplanted cells gave rise to neural tissue. When prospective neural tissue from early gastrulae was transplanted to the region fated to become belly skin, the neural tissue became epidermal (Table 10.1). Thus, these early newt gastrula cells were not yet committed to a specific fate. Such cells are said to exhibit **conditional** (i.e., regulative or dependent) **development** because their ultimate fates depend on their location in the embryo. However, when the same interspecies transplantation experiments were performed on *late* gastrulae, Spemann obtained completely different results. Rather than differentiating in accordance with their new location, the transplanted cells exhibited **autonomous** (or independent, or mosaic) **development**. Their prospective fate was determined, and the cells developed independently of their new embryonic location. Specifically, prospective neural cells now developed into brain tissue even when placed in the region of prospective epidermis, and prospective epidermis formed skin even in the region of the prospective neural tube. Within the time separating early and late gastrulation, the potencies of these groups of cells had become restricted to their eventual paths of differentiation. Something was causing them to become determined to epidermal and neural fates. What was happening?

Table 10.1 Results of tissue transplantation during early- and late-gastrula stages in the newt

Donor region	Host region	Differentiation of donor tissue	Conclusion
EARLY GASTRULA			
Prospective neurons	Prospective epidermis	Epidermis	Dependent (conditional) development
Prospective epidermis	Prospective neurons	Neurons	Dependent (conditional) development
LATE GASTRULA			
Prospective neurons	Prospective epidermis	Neurons	Independent (autonomous) development (determined)
Prospective epidermis	Prospective neurons	Epidermis	Independent (autonomous) development (determined)

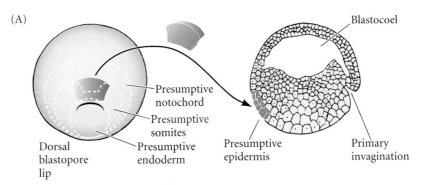

Figure 10.20
Organization of a secondary axis by dorsal blastopore lip tissue. (A) Dorsal lip tissue from an early gastrula is transplanted into another early gastrula in the region that normally becomes ventral epidermis. (B) The donor tissue invaginates and forms a second archenteron, and then a second embryonic axis. Both donor and host tissues are seen in the new neural tube, notochord, and somites. (C) Eventually, a second embryo forms that is joined to the host. (D) Structure of the dorsal blastopore lip region in an early *Xenopus* gastrula. (A–C after Hamburger 1988; D after Winklbauer and Schürfeld 1999 and Arendt and Nübler-Jung 1999.)

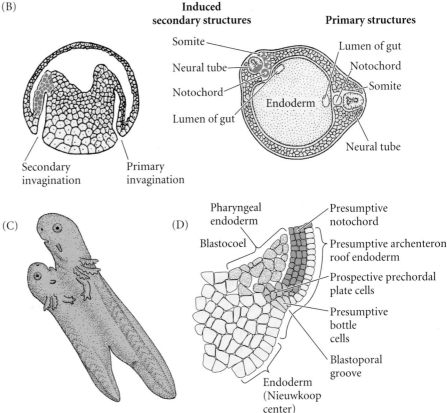

Hans Spemann and Hilde Mangold: Primary Embryonic Induction

The most spectacular transplantation experiments were published by Hans Spemann and Hilde Mangold in 1924.* They showed that, of all the tissues in the early gastrula, only one has its fate determined. This self-differentiating tissue is the dorsal lip of the blastopore, the tissue derived from the gray crescent cytoplasm. When this dorsal lip tissue was transplanted into the presumptive belly skin region of another gastrula, it not only continued to be blastopore lip, but also initiated gastrulation and embryogenesis in the surrounding tissue (Figure 10.20). Two conjoined embryos were formed instead of one!

In these experiments, Spemann and Mangold used differently pigmented embryos from two newt species: the darkly pigmented *Triturus taeniatus* and the non-pigmented *Triturus cristatus*. So when Spemann and Mangold prepared these transplants, they were able to readily identify host and donor tissues on the basis of color.[†] When the dorsal lip of an early *T. taeniatus* gastrula was removed and implant-

ed into the region of an early *T. cristatus* gastrula fated to become ventral epidermis (belly skin), the dorsal lip tissue invaginated just as it would normally have done (showing self-determination), and disappeared beneath the vegetal cells. The pigmented donor tissue then continued to self-differentiate into the chordamesoderm (notochord) and other mesodermal structures that normally form from the dorsal lip. As the new donor-derived mesodermal cells moved forward, host cells began to participate in the production of the new embryo, becoming organs that normally they never would have formed. In this secondary embryo, a somite could be seen containing both pigmented (donor) and unpigmented (host) tissue. Even more spectacularly, the dorsal lip cells were able to interact with the host tissues to form a complete neural plate from host ectoderm. Eventually, a secondary embryo formed, face

*Hilde Proescholdt Mangold died in a tragic accident in 1924, when her kitchen's gasoline heater exploded. At the time she was 26 years old, and her paper was just being published. Hers is one of the very few doctoral theses in biology that have directly resulted in the awarding of a Nobel Prize. For more information about Hilde Mangold and her times and the experiments that identified the organizer, see Hamburger 1984, 1988, and Fässler and Sander 1996.

[†]It was fortunate that Spemann's laboratory, and those of his students, usually used salamander embryos for their experiments. It turns out that frog ectoderm is much more difficult to induce than that of these urodeles.

to face with its host. These technically difficult experiments have been repeated using nuclear markers, and the results of Spemann and Mangold have been confirmed (Smith and Slack 1983; Recanzone and Harris 1985).

> WEBSITE **10.4 Spemann, Mangold, and the organizer.** Spemann did not see the importance of this work the first time they did it. This website provides a more detailed account of why Spemann and Mangold did this experiment.

Spemann (1938) referred to the dorsal lip cells and their derivatives (notochord, prechordal mesoderm) as the **organizer** because (1) they induced the host's ventral tissues to change their fates to form a neural tube and dorsal mesodermal tissue (such as somites), and (2) they organized host and donor tissues into a secondary embryo with clear anterior-posterior and dorsal-ventral axes. He proposed that during normal development, these cells organize the dorsal ectoderm into a neural tube and transform the flanking mesoderm into the anterior-posterior body axis. It is now known (thanks largely to Spemann and his students) that the interaction of the chordamesoderm and ectoderm is not sufficient to "organize" the entire embryo. Rather, it initiates a series of sequential inductive events. As discussed in Chapter 6, the process by which one embryonic region interacts with a second region to influence that second region's differentiation or behavior is called induction. Because there are numerous inductions during embryonic development, this key induction wherein the progeny of dorsal lip cells induce the dorsal axis and the neural tube is traditionally called **primary embryonic induction**.[*]

The Mechanisms of Axis Formation in Amphibians

The experiments of Spemann and Mangold showed that the dorsal lip of the blastopore, and the notochord that forms from it, constituted an "organizer" that could instruct the formation of new embryonic axes. But the mechanisms by which the organizer was constructed and through which it operated were totally unknown. Indeed, it is said that Spemann and Mangold's paper posed more questions than it answered. Among these questions were:

- How did the organizer get its properties? What caused the dorsal blastopore lip to differ from any other region of the embryo?
- What factors were being secreted from the organizer to cause the formation of the neural tube and to create the anterior-posterior, dorsal-ventral, and left-right axes?

[*] This classic term has been a source of confusion because the induction of the neural tube by the notochord is no longer considered the first inductive process in the embryo. We will soon discuss inductive events that precede this "primary" induction.

(A) Dissected blastula fragments give rise to different tissue in culture:

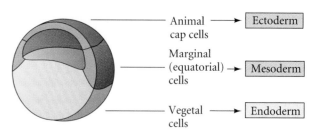

(B) Animal and vegetal fragments give mesoderm

(C)

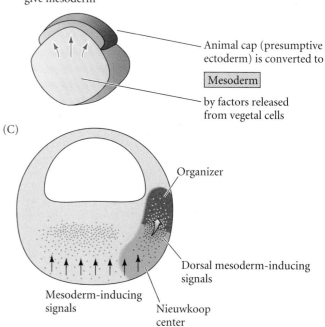

Figure 10.21
Summary of experiments by Nieuwkoop and by Nakamura and Takasaki (1970), showing mesodermal induction by vegetal endoderm. (A) Isolated animal cap cells become a mass of ciliated epidermis, isolated vegetal cells generate gutlike tissue, and isolated equatorial (marginal zone) cells become mesoderm. (B) If animal cap cells are combined with vegetal cap cells, many of the animal cells generate mesodermal tissue. (C) Model for mesoderm induction in *Xenopus*. A ventral signal (probably FGF2 or BMP4) is released throughout the vegetal region of the embryo. This induces the marginal cells to become mesoderm. On the dorsal side (away from the point of sperm entry), a signal is released by the vegetal cells of the Nieuwkoop center. This dorsal signal induces the formation of the Spemann organizer in the overlying marginal zone cells. The possible identity of this signal will be discussed later in this chapter. (C after De Robertis et al. 1992.)

- How did the different parts of the neural tube become established, with the most anterior becoming the sensory organs and forebrain, and the most posterior becoming spinal cord?

We will take up each of these questions in turn.

The origin of the Nieuwkoop center

The major clue in determining how the dorsal blastopore lip obtained its properties came from the experiments of Pieter Nieuwkoop (1969, 1973, 1977). He and his colleagues in the Netherlands demonstrated that the properties of this newly formed mesoderm were induced by the vegetal (presumptive endoderm) cells underlying them. He removed the equatorial cells (i.e., presumptive mesoderm) from a blastula and showed that neither the animal cap (presumptive ectoderm) nor the vegetal cap (presumptive endoderm) produced any mesodermal tissue. However, when the two caps were recombined, the animal cap cells were induced to form mesodermal structures such as notochord, muscles, kidney cells, and blood cells (Figure 10.21). The polarity of this induction (whether the animal cells formed dorsal mesoderm or ventral mesoderm) depended on the dorsal-ventral polarity of the endodermal (vegetal) fragment. While the ventral and lateral vegetal cells (those closer to the side of sperm entry) induced ventral (mesenchyme, blood) and intermediate (muscle, kidney) mesoderm, the dorsalmost vegetal cells specified dorsal mesoderm components (somites, notochord), including those having the properties of the organizer. The dorsalmost vegetal cells of the blastula, which are capable of inducing the organizer, have been called the **Nieuwkoop center** (Gerhart et al. 1989).

The Nieuwkoop center was demonstrated in the 32-cell *Xenopus* embryo by transplantation and recombination experiments. First, Gimlich and Gerhart (Gimlich and Gerhart 1984; Gimlich 1985, 1986) performed an experiment analogous to the Spemann and Mangold studies, except that they used blastulae rather than gastrulae. When they transplanted the dorsalmost vegetal blastomere from one blastula into the ventral vegetal side of another blastula, two embryonic axes were formed (see Figure 10.11B). Second, Dale and Slack (1987) recombined single vegetal blastomeres from a 32-cell *Xenopus* embryo with the uppermost animal tier of a fluorescently labeled embryo of the same stage. The

dorsalmost vegetal cell, as expected, induced the animal pole cells to become dorsal mesoderm. The remaining vegetal cells usually induced the animal cells to produce either intermediate or ventral mesodermal tissues (Figure 10.22). Thus, dorsal vegetal cells can induce animal cells to become dorsal mesodermal tissue.

The Nieuwkoop center is created by the cytoplasmic rotation that occurs during fertilization (see Chapter 7). When this rotation is inhibited by UV light, the resulting embryo will not form dorsal-anterior structures such as the head or neural tube (Vincent and Gerhart 1987). However, these UV-treated embryos can be rescued by transplantation of the dorsalmost vegetal blastomeres from a normal embryo at the 32-cell stage (Dale and Slack 1987; see Figure 10.11A). If eggs are rotated toward the end of the first cell cycle so that the future ventral side is upward, two Nieuwkoop centers are formed, leading to two dorsal blastopore lips and two embryonic axes (see Figure 10.10). Therefore, the specification of the dorsal-ventral axis begins at the moment of sperm entry.

The molecular biology of the Nieuwkoop center

In *Xenopus*, the endoderm is able to induce the formation of mesoderm by causing the presumptive mesodermal cells to express the *Xenopus Brachyury* (*Xbra*) gene. The mechanism of this induction is not well understood (see Harland and Gerhart 1997), but the Xbra protein is a transcription factor that activates the genes that produce mesoderm-specific proteins. While all the vegetal cells appear to be able to induce the overlying marginal cells to become mesoderm, only the dor-

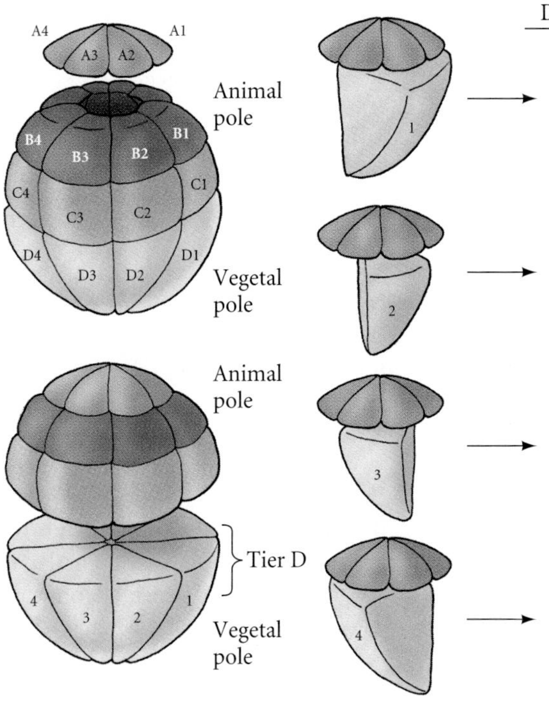

Figure 10.22
The regional specificity of mesoderm induction can be demonstrated by recombining cells of 32-cell *Xenopus* embryos. Animal pole cells were labeled with fluorescent polymers so that their descendants could be identified, then combined with individual vegetal blastomeres. The inductions resulting from these recombinations are summarized at the right. D1, the dorsalmost vegetal blastomere, was the most likely to induce the animal pole cells to form dorsal mesoderm. (After Dale and Slack 1987.)

	Percentage of total inductions		
	Dorsal	Intermediate	Ventral
1	77	23	0
2	11	61	28
3	5	45	50
4	16	42	42

(A)

(B)

(C)

(D)

(E)

(F)

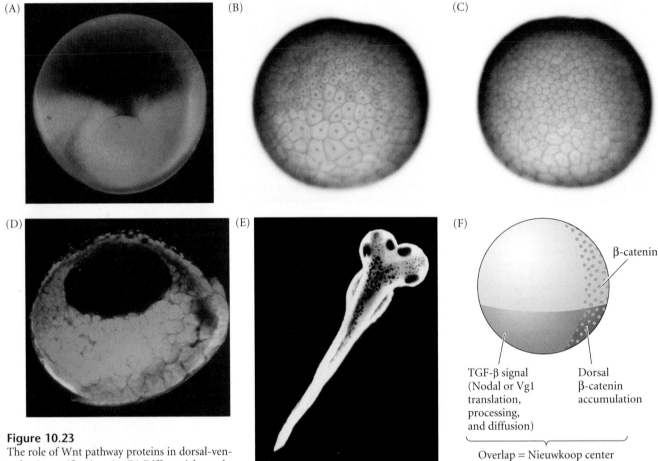

β-catenin

TGF-β signal
(Nodal or Vg1
translation,
processing,
and diffusion)

Dorsal
β-catenin
accumulation

Overlap = Nieuwkoop center

Figure 10.23
The role of Wnt pathway proteins in dorsal-ventral axis specification. (A–D) Differential translocation of β-catenin into *Xenopus* blastomere nu-
clei. (A) Early 2-cell stage of *Xenopus*, showing β-catenin (orange) predominantly at the
dorsal surface. (B) Presumptive dorsal side of a *Xenopus* blastula stained for β-catenin shows
nuclear localization. (C) Such nuclear localization is not seen on the ventral side of the same
embryo. (D) β-catenin dorsal localization persists through the gastrula stage. (E) Dorsal axis
formation caused by the injection of both blastomeres of a 2-cell *Xenopus* embryo with
dominant inactive GSK-3. Dorsal fate is actively suppressed by wild-type GSK-3. (F) Irenic
model whereby the Nieuwkoop center (characterized by *siamois* gene expression and the
ability to induce dorsal mesoderm) is created by the synergy of the activation of β-catenin
dorsally and the activation of the TGF-β signal vegetally. (A, D from R. T. Moon; B and C
from Schneider et al. 1996, photographs courtesy of P. Hausen; E from Pierce and Kimelman
1995, photograph courtesy of D. Kimelman.)

salmost vegetal cells can instruct the overlying dorsal marginal cells to become the organizer. The major candidate for the
factor that forms the Nieuwkoop center in these dorsalmost
vegetal cells is β-**catenin**.

WEBSITE **10.5 Mesoderm induction.** There are numerous theories concerning how the generic mesoderm is induced by the endoderm. Evidence points to three molecules
as possible mesoderm inducers: bFGF, Vg1, and an activin-
like protein. These proteins can activate *Xbra* as well as
other mesodermal proteins.

β-catenin is a multifunctional protein
that can act as an anchor for cell membrane cadherins (see Chapter 3) or as a
nuclear transcription factor (see Chapter
6). In *Xenopus* embryos, β-catenin begins to accumulate in the dorsal region
of the egg during the cytoplasmic movements of fertilization. β-catenin continues to accumulate preferentially at the
dorsal side throughout early cleavage,
and this accumulation is seen in the nu-
clei of the dorsal cells (Figure 10.23A–D; Schneider et al. 1996;
Larabell et al. 1997). This region of β-catenin accumulation
originally appears to cover both the Nieuwkoop center and organizer regions. During later cleavage, the cells containing β-
catenin may reside specifically in the Nieuwkoop center
(Heasman et al. 1994a; Guger and Gumbiner 1995).

β-catenin is necessary for forming the dorsal axis, since
experimental depletion of β-catenin transcripts with antisense oligonucleotides results in the lack of dorsal structures
(Heasman et al. 1994a). Moreover, the injection of exogenous
β-catenin into the ventral side of the embryo produces a secondary axis (Funayama et al. 1995; Guger and Gumbiner

1995). β-catenin is part of the Wnt signal transduction pathway and is negatively regulated by the glycogen synthase kinase 3 (GSK-3; see Chapter 6). GSK-3 also plays a critical role in axis formation by suppressing dorsal fates. Activated GSK-3 blocks axis formation when added to the egg (Pierce and Kimelman 1995; He et al. 1995; Yost et al. 1996). If endogenous GSK-3 is knocked out by a dominant negative protein in the ventral cells of the early embryo, a second axis forms (Figure 10.23E).

So how can β-catenin become localized to the future dorsal cells of the blastula? Labeling experiments (Yost et al. 1996; Larabell et al. 1997) suggest that β-catenin is initially synthesized (from maternal messages) throughout the embryo, but is degraded by GSK-3-mediated phosphorylation specifically in the ventral cells. The critical event for axis determination may be the movement of an inhibitor of GSK-3 to the cytoplasm opposite the point of sperm entry (i.e., to the future dorsal cells). One candidate for this agent is the Disheveled protein. This protein is the normal suppressor of GSK-3 in the Wnt pathway (see Figure 6.23), and it is originally found in the vegetal cortex of the unfertilized *Xenopus* egg. However, upon fertilization, Disheveled is translocated along the microtubular array to the dorsal side of the embryo (Figure 10.24; Miller et al. 1999). Thus, on the dorsal side of the embryo, β-catenin should be stable, since GSK-3 is not able to degrade it; while in the ventral portion of the embryo, GSK-3 should initiate the degradation of β-catenin.

WEBSITE **10.6 GBP.** In addition to Disheveled, a second inhibitor of GSK-3 has been identified in *Xenopus* eggs. This protein, GBP, can rescue axis formation in UV-treated eggs.

β-catenin is a transcription factor that can associate with other transcription factors to give them new properties. It is known that *Xenopus* β-catenin can combine with a ubiquitous transcription factor known as **Tcf3**, and that a mutant form of Tcf3 lacking a β-catenin binding domain results in embryos without dorsal axes (Molenaar et al. 1996). The β-catenin/Tcf3 complex appears to bind to the promoters of several genes whose activity is critical for axis formation. One of these genes is *siamois*, which is expressed in the Nieuwkoop center immediately following the midblastula transition. If this gene is ec-

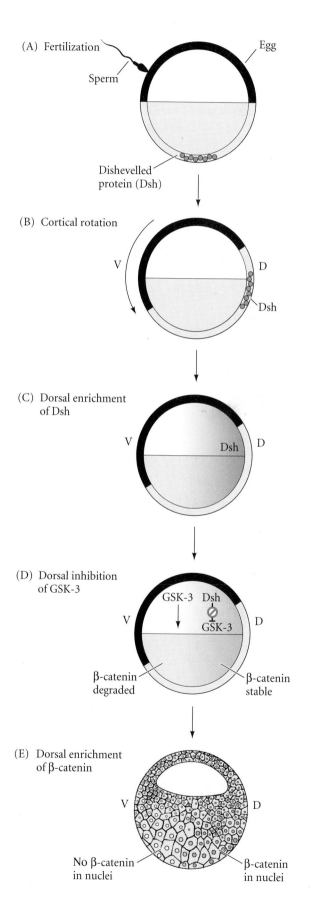

Figure 10.24
Model of the mechanism by which the Disheveled protein stabilizes β-catenin in the dorsal portion of the amphibian egg. (A) Disheveled (Dsh) associates with a particular set of proteins at the vegetal pole of the unfertilized egg. (B) Upon fertilization, these protein vesicles are translocated dorsally along subcortical microtubule tracks. (C) Disheveled is then released from its vesicles and is distributed in the future dorsal third of the 1-cell embryo. (D) Disheveled binds to and blocks the action of GSK-3, thereby preventing the degradation of β-catenin on the dorsal side of the embryo. (E) The nuclei of the blastomeres in the dorsal region of the embryo receive β-catenin, while the nuclei of those in the ventral region do not.

topically expressed in the ventral vegetal cells, a secondary axis emerges on the former ventral side of the embryo, and if cortical rotation is prevented, *siamois* expression is eliminated (Lemaire et al. 1995; Brannon and Kimelman 1996). The Tcf3 protein is thought to inhibit *siamois* transcription when it binds to that gene's promoters in the absence of β-catenin. However, when the Tcf3/β-catenin complex binds to its promoter, *siamois* is activated (Figure 10.25; Brannon et al. 1997).

The Siamois protein is critical for the expression of organizer-specific genes (Fan and Sokol 1997; Kessler 1997). Siamois protein binds to the promoter of the *goosecoid* gene and activates its expression (Laurent et al. 1997). The protein product of *goosecoid* appears to be essential for activating numerous genes in the Spemann organizer. So one could expect

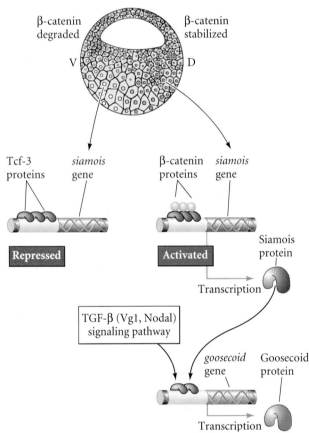

Figure 10.25
Summary of events hypothesized to bring about the induction of the organizer in the dorsal mesoderm. Cortical rotation causes the translocation of Disheveled protein to the dorsal side of the embryo. Dsh binds GSK-3, thereby allowing β-catenin to accumulate in the future dorsal portion of the embryo. During cleavage, β-catenin enters the nuclei and binds with Tcf3 to form a transcription factor that activates genes encoding proteins such as Siamois. Siamois and Lim-1, a transcription factor activated by the TGF-β pathway, function together to activate the *goosecoid* gene in the organizer. Goosecoid is a transcription factor that can activate genes whose proteins are responsible for the organizer's activities. (After Moon and Kimelman 1998).

that the dorsal side of the embryo would contain β-catenin, that β-catenin would allow this region to express Siamois, and that Siamois would initiate the formation of the organizer. However, Siamois alone is not sufficient for generating the organizer; another protein also appears to be critical in the activation of *goosecoid* and the formation of the organizer. Recent studies suggest that maximum *goosecoid* expression occurs when there is synergism between the Siamois protein and a vegetally expressed TGF-β signal (see Chapter 6) (Brannon and Kimelman 1996). While the cortical rotation may activate the β-catenins and allow the expression of *siamois* in the dorsal region of the embryo, the translation of vegetally localized messages encoding a factor of the TGF-β family may generate a protein that permits the activation of *goosecoid* best in the cells that will become the organizer. The TGF-β family protein in the Nieuwkoop center could induce the cells in the dorsal marginal zone above them to express some transcription factor that would also bind to the promoter of the *goosecoid* gene and cooperate with *siamois* to activate it (see Figure 10.25).

The candidates for this TGF-β factor include Vg1, VegT, and Nodal-related proteins. Each of these proteins is made in the endoderm (see Figure 5.33). Agius and colleagues (2000) have provided evidence that all of these proteins may act in a pathway, and that the critical proteins are the Nodal-related factors. When they repeated the Nieuwkoop animal-vegetal recombination experiments (see Figure 10.21) but included a specific inhibitor of Nodal-related proteins, the induction by the vegetal cells failed to occur. (The inhibitor did not inhibit Vg1, VegT, or activin.) Moreover, they found that during the late blastula stages, three Nodal-related proteins (Xnr1, Xnr2, and Xnr4) are expressed in a dorsal-to-ventral gradient in the endoderm. This gradient is formed by the activation of *Xenopus* Nodal-related gene expression by the synergistic action of VegT and Vg1 with β–catenin. Agius and his colleagues present a model, shown in Figure 10.26, in which the dorsally located β–catenin and the vegetally located VegT and Vg1 signals interact to create a gradient of Nodal-related proteins (Xnr1, 2, 4) across the endoderm. These Nodal-related proteins specify the mesoderm such that those regions with little or no Nodal-related protein become ventral mesoderm, those regions with some Nodal protein become lateral mesoderm, and those regions with a great deal of Nodal protein become the organizer. These Nodal-related proteins will activate the *goosecoid* gene, and the specific inhibitor of Nodal-related proteins prevents this activation.

The Functions of the Organizer

While the Nieuwkoop center cells remain endodermal, the cells of the organizer become the dorsal mesoderm and migrate underneath the dorsal ectoderm. There, the dorsal mesoderm induces the central nervous system to form. The properties of the organizer tissue can be divided into five major functions:

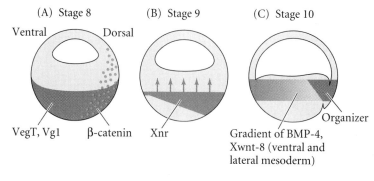

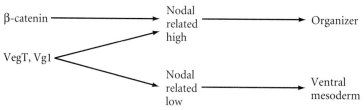

Figure 10.26
Model for mesoderm induction and organizer formation by the interaction of β-catenin and TGF-β proteins. (A) At late blastula stages, Vg1 and VegT are found in the vegetal hemisphere, while β-catenin is located in the dorsal region. (B) β-catenin acts synergistically with Veg1 and VegT to activate the *Xenopus Nodal-related (Xnr)* genes. This creates a gradient of Xnr proteins across the endoderm, highest in the dorsal region. (C) The mesoderm is specified by the gradient of Xnr proteins. Mesodermal regions with little or no Xnr proteins have high levels of BMP-4 and Xwnt-8; they become ventral mesoderm. Those having intermediate concentrations of Xnrs become lateral mesoderm. Where there is a high concentration of Xnrs, the *goosecoid* gene and other dorsal mesodermal genes are activated, and the mesodermal tissue becomes the organizer. (These results may explain the activity concentration experiments mentioned in Chapter 3.) (After Agius et al. 2000.)

1. The ability to become dorsal mesoderm (prechordal plate, chordamesoderm, etc.)
2. The ability to dorsalize the surrounding mesoderm into lateral mesoderm (when it would otherwise form ventral mesoderm)
3. The ability to dorsalize the ectoderm into neural ectoderm
4. The ability to initiate the movements of gastrulation
5. The ability to cause the neural plate (the induced neural ectoderm) to become the neural tube

In *Xenopus* (and other vertebrates), the formation of the anterior-posterior axis follows the formation of the dorsal-ventral axis. Once the dorsal portion of the embryo is established, the movement of the involuting mesoderm establishes the anterior-posterior axis. The mesoderm that migrates first through the dorsal blastopore lip gives rise to the anterior structures; the mesoderm migrating through the lateral and ventral lips forms the posterior structures.

It is now thought that the cells of the organizer ultimately contribute to four cell types—pharyngeal endoderm, head mesoderm (prechordal plate), dorsal mesoderm (primarily

the notochord), and the dorsal blastopore lip (Keller 1976; Gont et al. 1993). The pharyngeal endoderm and prechordal plate lead the migration of the organizer tissue and appear to induce the forebrain and midbrain. The dorsal mesoderm induces the hindbrain and trunk. The dorsal blastopore lip forms the dorsal mesoderm and eventually becomes the chordaneural hinge that induces the tip of the tail.

When the organizer was first described, it started one of the first truly international scientific research programs: the search for the organizer molecules. Researchers from Britain, Germany, France, the United States, Belgium, Finland, Japan, and the Soviet Union all tried to find these remarkable substances (see Gilbert and Saxén 1993). R. G. Harrison (quoted by Twitty 1966, p. 39) referred to the amphibian gastrula as the "new Yukon to which eager miners were now rushing to dig for gold around the blastopore." Unfortunately, their picks and shovels proved too blunt to uncover the molecules involved. The analysis of organizer molecules had to wait until recombinant DNA technologies enabled investigators to make cDNA clones from blastopore lip mRNA and to see which of these clones encoded factors that could dorsalize the embryo.

WEBSITE 10.7 Early attempts to locate the organizer molecules. While Spemann did not believe that molecules alone could organize the embryo, his students began a long quest for these factors.

The formation of the dorsal (organizer) mesoderm involves the activation of several genes. The secreted proteins of the Nieuwkoop center are thought to activate a set of transcription factors in the mesodermal cells above them. These transcription factors then activate the genes encoding the secreted products of the organizer. Several organizer-specific transcription factors have been found and are listed in Table 10.2.

Table 10.2 Proteins expressed solely or almost exclusively in the organizer (partial list)

Nuclear proteins	Secreted proteins
XLim1	Chordin
Xnot	Dickkopf
Otx2	ADMP
XFD1	Frzb
XANF1	Noggin
Goosecoid	Follistatin
HNF3β-related proteins (e.g., Forkhead, Pintallavis)	Sonic hedgehog
	Cerberus
	Nodal-related proteins (several)

(A) (B) (C) (D)

Figure 10.27
Ability of *goosecoid* mRNA to induce a new axis. (A) At the gastrula stage, a control embryo (either uninjected or given an injection of *goosecoid*-like mRNA but lacking the homeobox) has one dorsal blastopore lip. (B) An embryo whose ventral vegetal blastomeres were injected at the 16-cell stage with *goosecoid* message. Note the secondary dorsal lip. (C) The top two embryos, which were injected with *goosecoid* mRNA, show two axes; the bottom two control embryos do not. In the upper embryos, two dorsal axes are seen. (D) Twinned embryo produced by *goosecoid* injection. Two complete sets of head structures have been induced. (After Cho et al. 1991a; Niehrs et al. 1993; photographs courtesy of E. De Robertis.)

As mentioned earlier, one of the important targets of the Nieuwkoop center appears to be the *goosecoid* gene. The area of expression of *goosecoid* mRNA correlates with the organizer domain in both normal and experimentally treated animals. When lithium chloride treatment is used to increase the organizer mesoderm throughout the marginal zone, the expression of *goosecoid* likewise is expanded. Conversely, when eggs are treated with UV light prior to first cleavage, both dorsal-anterior induction and *goosecoid* expression are significantly inhibited. Injection of the full-length *goosecoid* message into the two ventral blastomeres of a 4-cell *Xenopus* embryo causes the progeny of those blastomeres to involute, undergo convergent extension, and form the dorsal mesoderm and head endoderm of a secondary axis (Figure 10.27; Niehrs et al. 1993). Labeling experiments (Niehrs et al. 1993) have shown that such *goosecoid*-injected cells are able to recruit neighboring host cells into the dorsal axis as well. Thus, the Nieuwkoop center activates the *goosecoid* gene in the organizer tissues, and this gene encodes a DNA-binding protein that (1) activates the migration properties (involution and convergent extension) of the dorsal blastopore lip cells, (2) autonomously determines the dorsal mesodermal fates of those cells expressing it, and (3) enables the *goosecoid*-expressing cells to recruit neighboring cells into the dorsal axis. Goosecoid also has been

found to activate *Xotx2*, a gene that is critical for brain formation, in the anterior mesoderm and in the presumptive brain ectoderm (Blitz and Cho 1995).

The diffusible proteins of the organizer I: The BMP inhibitors

Goosecoid protein works in the nucleus. It must activate (either directly or indirectly) those genes encoding the soluble proteins that function to organize the dorsal-ventral and anterior-posterior axis. Early evidence for diffusible signals from the notochord came from several sources. First, Hans Holtfreter (1933) showed that if amphibian embryos are placed in a high salt solution, the mesoderm will *e*vaginate rather than *in*vaginate, and will not underlie the ectoderm. Such ectoderm is not underlain by the notochord, and it does not form neural structures. Further evidence for soluble factors came from the transfilter studies of Finnish investigators (Saxén 1961; Toivonen et al. 1975; Toivonen and Wartiovaara 1976). Newt dorsal lip tissue was placed on one side of a filter thin enough so that no processes could fit through the pores, and competent gastrula ectoderm was placed on the other side. After several hours, neural structures were observed in the ectodermal tissue (Figure 10.28). The identities of the factors diffusing from the organizer, however, took another quarter of a century to identify.

Recent studies on induction have resulted in a remarkable and non-obvious conclusion: The ectoderm is actually in-

Figure 10.28
Neural structures induced in presumptive ectoderm by newt dorsal lip tissue, separated from the ectoderm by a Nucleopore filter with an average pore diameter of 0.05 mm. Anterior neural tissues are evident, including some induced eyes. (From Toivonen 1979; photograph courtesy of L. Saxén.)

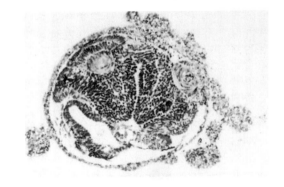

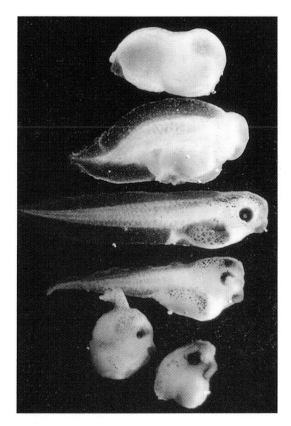

Figure 10.29
Rescue of dorsal structures by Noggin protein. When *Xenopus* eggs are exposed to ultraviolet radiation, cortical rotation fails to occur, and the embryos lack dorsal structures (top). If such an embryo is injected with *noggin* mRNA, it develops dorsal structures in a dosage-related fashion (top to bottom). If too much *noggin* message is injected, the embryo produces dorsal anterior tissue at the expense of ventral and posterior tissue, becoming little more than a head (bottom). (Photograph courtesy of R. M. Harland.)

mids whose RNAs rescued the dorsal axis in these embryos were split into smaller sets, and so on, until single-plasmid clones were isolated whose mRNAs were able to restore the dorsal axis in such embryos. One of these clones contained the *noggin* gene (Figure 10.29). Smith and Harland (1992) have shown that newly transcribed (as opposed to maternal) *noggin* mRNA is first localized in the dorsal blastopore lip region and then becomes expressed in the notochord (Figure 10.30). Injection of *noggin* mRNA into 1-cell, UV-irradiated embryos completely rescued the dorsal axis and allowed the formation of a complete embryo. The mRNA sequence for the Noggin protein suggests strongly that it is a secreted protein. In 1993, Smith and his colleagues found that Noggin could accomplish two of the major functions of the Spemann-Mangold organizer: it induced dorsal ectoderm to form neural tissue, and it dorsalized mesoderm cells that would otherwise contribute to the ventral mesoderm. Noggin binds to BMP4 and BMP2 and inhibits their binding to receptors (Zimmerman et al. 1996).

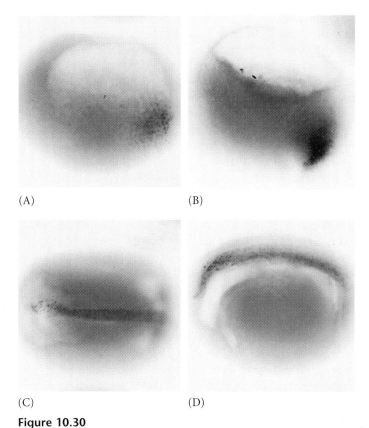

(A)   (B)

(C)   (D)

Figure 10.30
Localization of the *noggin* mRNA in the organizer tissue, shown by in situ hybridization. (A) At gastrulation, *noggin* message (dark areas) accumulates in the dorsal marginal zone. (B) When cells involute, *noggin* mRNA is seen in the dorsal blastopore lip. (C) During convergent extension, *noggin* message is expressed in the precursors of the notochord, prechordal plate, and pharyngeal endoderm, which extend (D) beneath the ectoderm in the center of the embryo. (Photographs courtesy of R. M. Harland.)

duced to become epidermal. The agents of this induction are bone morphogenetic proteins (BMPs). The nervous system forms from that region of the ectoderm that is protected from this epidermal induction (Hemmati-Brivanlou and Melton 1997). In other words, (1) the "default fate" of the ectoderm is to become neural; (2) certain parts of the embryo induce the ectoderm to become epidermal tissue, and (3) the organizer tissues act by secreting molecules that block this induction, thereby allowing the ectoderm "protected" by these factors to become neural.

NOGGIN. In 1992, the first of the soluble organizer molecules was isolated. Smith and Harland (1992) constructed a cDNA plasmid library from dorsalized (lithium chloride-treated) gastrulae. RNAs synthesized from sets of these plasmids were injected into ventralized embryos (having no neural tube) produced by irradiating early embryos with ultraviolet light. Those sets of plas-

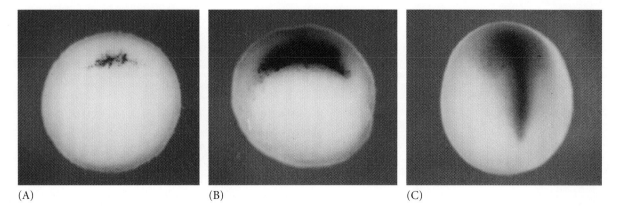

(A)   (B)   (C)

Figure 10.31

Chordin mRNA localization. (A) Whole-mount in situ hybridization shows that just prior to gastrulation, chordin message (dark area) is expressed in the region that will become the dorsal blastopore lip. (B) As gastrulation begins, Chordin is expressed at the dorsal blastopore lip. (C) In later stages of gastrulation, Chordin message is seen in the organizer tissues. (From Sasai et al. 1994; photographs courtesy of E. De Robertis.)

CHORDIN AND NODAL-RELATED 3. The second organizer protein found was **chordin**. It was isolated from clones of cDNA whose mRNAs were present in dorsalized, but not in ventralized, embryos (Sasai et al. 1994). These clones were tested by injecting them into ventral blastomeres and seeing whether they induced secondary axes. One of the clones capable of inducing a secondary neural tube contained the *chordin* gene. The chordin mRNA was found to be localized in the dorsal blastopore lip and later in the dorsal mesoderm of the notochord (Figure 10.31). Like Noggin, chordin binds directly to BMP4 and BMP2 and prevents their complexing with their receptors (Piccolo et al. 1996). In zebrafish, a loss-of-function mutation of chordin (the "chordino" mutant) has a greatly reduced neural plate and an enlarged region of ventral mesoderm (Hammerschmidt et al. 1996). Nodal-related-3 (Xnr-3) is synthesized by the superficial cells of the organizer and is also able to block BMP4 (Smith et al. 1995; Hansen et al. 1997).

FOLLISTATIN. The fourth organizer-secreted protein, **follistatin**, was found in the organizer through an unexpected result of an experiment that was looking for something else. Ali Hemmati-Brivanlou and Douglas Melton (1992, 1994) wanted to see whether the protein activin was crucial for mesoderm induction, so they constructed a dominant negative activin receptor and injected it into *Xenopus* embryos. Remarkably, the ectoderm of these embryos began to express neural-specific proteins. It appeared that the activin receptor (which also binds other structurally similar molecules such as the bone morphogenetic proteins) normally functioned to bind an inhibitor of neurulation. When its function was blocked, all the ectoderm became neural. In 1994, Hemmati-Brivanlou and Melton proposed a "default model of neurulation" whereby the organizer functioned by producing inhibitors of whatever was blocking neurulation. That is to say, the "normal" fate of an ectodermal cell was to become a neuron; it had to be induced to become an epidermal skin cell. The organizer somehow prevented the ectodermal cells from being induced. This model was supported by, and explained, some cell dissociation experiments that had also produced odd results. Three studies, by Grunz and Tacke (1989), Sato and Sargent (1989), and Godsave and Slack (1989) had shown that when whole embryos or their animal caps were dissociated, they formed neural tissue. This result would be explainable if the "default state" of the ectoderm was not epidermal, but neural, and the tissue had to be induced to have an epidermal phenotype. The organizer, then, would block this epidermalizing induction.

Since the naturally occurring protein follistatin binds to and inhibits activin (and other related proteins), it was hypothesized that it might be one of the factors secreted by the organizer. Using in situ hybridization, Hemmati-Brivanlou and Melton (1994) found the mRNA for follistatin in the dorsal blastopore lip and notochord.

So it appeared that there might be a neural default state and an actively induced epidermal fate. This hypothesis was counter to the neural induction model that had preceded it for 70 years. But what proteins were inducing the epidermis, and were they really being blocked by the molecules secreted by the organizer?

The leading candidate for the epidermal inducer is bone morphogenesis protein-4 (BMP4). It was known that there is an antagonistic relationship between BMP4 and the organizer. If the mRNA for BMP4 is injected into *Xenopus* eggs, all the mesoderm in the embryo becomes ventrolateral mesoderm, and no involution occurs at the blastopore lip (Dale et al. 1992; Jones et al. 1992). Conversely, overexpression of a dominant negative BMP4 receptor resulted in the formation of two dorsal axes (Graff et al. 1994; Suzuki et al. 1994). In 1995, Wilson and Hemmati-Brivanlou demonstrated that BMP4 induced ectodermal cells to become epidermal. By 1996, several laboratories had demonstrated that Noggin, chordin, and follistatin each was secreted by the organizer and that each prevented BMP from binding to the ectoderm and mesoderm near the organizer (Piccolo et al. 1996; Zimmerman et al. 1996; Iemura et al. 1998).

(A)

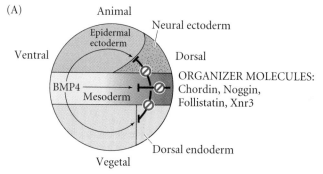

Figure 10.32
Model for the action of the organizer. (A) BMP4 (and certain other molecules) is a powerful ventralizing factor. Organizer proteins such as Chordin, Noggin, and Follistatin block the action of BMP4. The antagonistic effects of these proteins can be seen in all three germ layers. (B) BMP4 may elicit the expression of different genes in a concentration-dependent fashion; in that way, the mesoderm could be patterned. BMP4 is expressed throughout the marginal zone (prospective mesoderm) except in the dorsal domain. Noggin and Chordin are expressed in the dorsal domain. These proteins bind to BMP4 and prevent it from reaching the mesodermal cells. In the regions of noggin and chordin expression, BMP4 is totally prevented from binding, and these tissues become notochord (organizer) tissue. Slightly farther away from the organizer, *myf5*, a marker for the dorsolateral muscles, is activated. As more and more BMP4 molecules are allowed to bind to the cells, *Xvent2* (ventrolateral) and *Xvent1* (ventral) genes become expressed. (After Dosch et al. 1997.)

When BMP binds to ectodermal cells, it activates the expression of genes such as *msx1*, which induce the expression of epidermal-specific genes, while inhibiting those genes that would produce a neural phenotype (Suzuki et al. 1997b). In the mesoderm, BMP4 activates genes such as *Xvent1*, which give the mesoderm a ventral phenotype. Low doses of BMP4 appear to activate muscle formation; intermediate levels instruct cells to become kidney; and high doses activate those genes that instruct the mesoderm to become blood cells

(B)

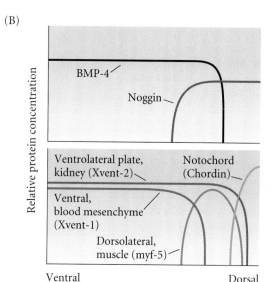

(Hemmati-Brivanlou and Thomsen 1995; Gawantka et al. 1995; Dosch et al. 1997). The varying doses are created by the interaction of BMP4 (coming from the ventral and lateral mesoderm) with the BMP antagonists coming from the organizer* (Figure 10.32).

Thus, by 1996, there was a consensus that BMP4 was the active inducer of ventral ectoderm (epidermis) and the ventralizer of the mesoderm (blood cells and connective tissue), and that Noggin, chordin, and follistatin could prevent its function. The organizer worked by secreting inhibitors of BMP4, not by directly inducing neurons.

*The details of BMP4 production and degradation are discussed in Chapter 22.

Sidelights & Speculations

BMP4 and Geoffroy's Lobster

THE HYPOTHESIS that the organizer secretes proteins that block BMP4 received further credence from an unexpected source—the emerging field of evolutionary developmental biology. De Robertis, Kimelman, and others (Holley et al. 1995; Schmidt et al. 1995; De Robertis and Sasai 1996) found that the same chordin-BMP4 interaction that instructed the formation of the neural tube in vertebrates also formed neural tissue in fruit flies. The *dorsal* neural tube of the vertebrate and the *ventral*

neural cord of the fly appear to be generated by the same set of instructions, conserved throughout evolution. This was the second paradigm shift occasioned by the newly acquired information on the molecular biology of induction.

The *Drosophila* homologue of the *bmp4* gene is *decapentaplegic* (*dpp*) (Figure 10.33). As discussed in the previous chapter, the Dpp protein is responsible for the patterning of the dorsal-ventral axis of *Drosophila*, and it is present in the dorsal portion of the embryo

and diffuses ventrally. It is opposed by a protein called Short-gastrulation (Sog). Short-gastrulation is the *Drosophila* homologue of chordin. These homologues not only appear to be similar; they can substitute for each other. When *Sog* mRNA is injected into ventral regions of *Xenopus* embryos, it induces the *Xenopus* notochord and neural tube. Injection of chordin mRNA into *Drosophila* produces ventral nervous tissue. Although the *Xenopus* chordin usually functions to dorsalize the embryo, it ventralizes the

Drosophila embryo. In *Drosophila*, Dpp is made dorsally; in *Xenopus*, BMP4 is made ventrally. In both cases, Sog/chordin makes neural tissue by blocking the effects of Dpp/BMP4. In *Drosophila*, Dpp interacts with the product of the *screw* gene to function. In *Xenopus*, the homologue of *screw*, *bmp7*, appears to be essential for the ventralizing effect of BMP4 (Hawley et al. 1995).

In 1822, the French anatomist Etienne Geoffroy Saint-Hilaire provoked one of the most heated and critical confrontations in biology when he proposed that the lobster was but the vertebrate upside down. He claimed that the ventral side of the lobster (with its nerve cord) was homologous to the dorsal side of the vertebrate (Appel 1987). It seems that he was correct on the molecular level, if not on the anatomical. De Robertis and Sasai (1996) have proposed that there was a common ancestor for all bilateral phyla—a hypo-

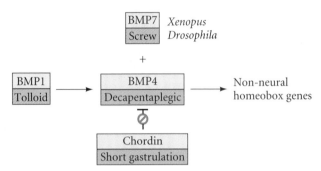

Figure 10.33
Homologous developmental pathways in the formation of the central nervous systems of a vertebrate (*Xenopus*) and an invertebrate (*Drosophila*). The *Xenopus* factor is on the top (yellow boxes), the homologous *Drosophila* protein is underneath (red boxes). (After De Robertis and Sasai 1996; Sasai et al. 1996.)

thetical creature (dubbed Urbilateria) of some 600 million years ago that was the ancestor for both the protostome and the deuterostome subkingdoms. The BMP4-(Dpp)/chordin(Sog) interaction is an example of "homologous processes," suggesting a unity of developmental principles among all animals (Gilbert et al. 1996). ∎

The diffusible proteins of the organizer II: The Wnt inhibitors

It had been thought that all the neural tissue induced by the organizer was induced to become forebrain, and that the notochord represented the most anterior portion of the organizer. However, the most anterior regions of the head and brain are underlain not by notochord, but by pharyngeal endoderm and head (prechordal) mesoderm. This "endomesoderm" constitutes the leading edge of the dorsal blastopore lip. Recent studies have shown that these cells not only induce the most anterior head structures, but that they do it by blocking the Wnt pathway as well as blocking BMP4.

CERBERUS. In 1993, Christian and Moon showed that Xwnt8, a member of the Wnt family of growth and differentiation factors, inhibited neural induction. It was found to be synthesized throughout the marginal mesoderm—except in the region forming the dorsal lip. Thus, in addition to BMP4, there was a second anti-neuralizing protein being secreted from the nonorganizer mesoderm. Were there any proteins in the organizer that countered this activity?

In 1996, Bouwmeester and colleagues showed that the induction of the most anterior head structures could be accomplished by a secreted protein called **Cerberus.*** Unlike the other proteins secreted by the organizer, Cerberus promotes the formation of the cement gland (the most anterior region of tadpole ectoderm), eyes, and olfactory (nasal) placodes.

*"Cerberus" is another name out of Greek mythology; the protein is named after the three-headed dog that guarded the entrance to Hades.

When *cerberus* mRNA was injected into a vegetal ventral *Xenopus* blastomere at the 32-cell stage, ectopic head structures were formed (Figure 10.34). These head structures were made from the injected cell as well as from neighboring cells. The *cerberus* gene is expressed in the pharyngeal endomesoderm cells that arise from the deep cells of the early dorsal lip, and the Cerberus protein can bind both BMPs and Xwnt8 (Glinka et al. 1997; Piccolo et al. 1999).

FRZB AND DICKKOPF. Shortly after the attributes of Cerebus were demonstrated, two other proteins, Frzb and Dickkopf,

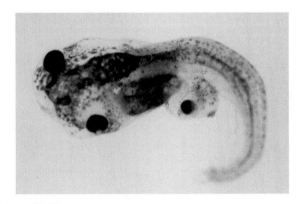

Figure 10.34
Cerberus mRNA injected into a single D4 (ventral vegetal) blastomere of a 32-cell *Xenopus* embryo induces head structures as well as a duplicated heart and liver. The secondary eye (a single cyclopic eye) and olfactory placode can be readily seen. (From Bouwmeester et al. 1996; photograph courtesy of E. M. De Robertis.)

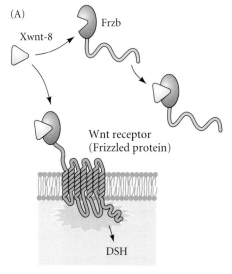

(A)

Xwnt-8

Frzb

Wnt receptor
(Frizzled protein)

DSH

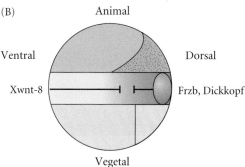

(B)

Animal

Ventral

Dorsal

Xwnt-8

Frzb, Dickkopf

Vegetal

Figure 10.35
Xwnt8 is capable of ventralizing the mesoderm and preventing anterior head formation in the ectoderm. (A) Frzb is a protein secreted by the anterior region of the organizer. It binds to Xwnt8 before that inducer can bind to its receptor. Frzb resembles the Wnt-binding domain of the Wnt receptor (Frizzled protein), but is a soluble molecule. (B) Xwnt8 is made throughout the marginal zone.

were discovered to be synthesized in the involuting endomesoderm. **Frzb** (pronounced "frisbee") is a small, soluble form of Frizzled, the Wnt receptor, that is capable of binding Wnt proteins in solution (Figure 10.35; Leyns et al. 1997; Wang et al. 1997). It is synthesized predominantly in the endomesoderm cells beneath the head (Figure 10.36A). If embryos are made to synthesize excess Frzb, Wnt signaling fails to occur, and the embryos lack ventral posterior structures, becoming solely head (Figure 10.36B). The **Dickkopf** (German; "big head," "stubborn") protein also appears to interact directly with Wnt proteins extracellularly. Injection of antibodies against Dickkopf protein into the blastocoel causes the resulting embryos to have small, deformed heads with no forebrain (Glinka et al. 1998).

Glinka and colleagues (1997) have thus proposed a new model for embryonic induction. The induction of trunk structures may be caused by the blockade of BMP signaling from the notochord. However, to produce a head, both the BMP signal and the Wnt signal must be blocked. This blockade comes from the endomesoderm, now considered the most anterior portion of the organizer.

Conversion of the ectoderm into neural plate cells

So far, we have discussed the factors that prevent the dorsal ectoderm from becoming epidermis. Obviously, once that is accomplished, other genes must transform the ectoderm into neural tissue. The key protein involved in activating the neural phenotype in the ectoderm appears to be **neurogenin** (Ma et al. 1996). The transcription factors that appear in the ectoderm in the absence of BMP are able to induce the expression of neurogenin, and the transcription factors (such as Msx1) induced in the ectoderm by BMP signals are able to repress neurogenin expression (Figure 10.37; see Sasai 1998). Neurogenin is itself a transcription factor, and it activates a series of genes whose products are responsible for the neural phenotype. One of the genes activated by neurogenin is the gene for NeuroD, a transcription factor that activates the genes producing the structural neural-specific proteins (Lee et al. 1995). In addition, Noggin or Cerberus can induce another transcription factor in the ectoderm. This protein is called Xenopus brain factor-2 (XBF-2), and it appears to repress the epidermal genes (Mariani and Harland 1998). By these pushes and pulls, the dorsal ectoderm is converted into neural plate tissue.

WEBSITE **10.8 The autonomous specification of the endoderm.** While the mesoderm is induced, and the differences between neural and epidermal ectoderm are induced, the endoderm appears to be specified autonomously. Recent studies are investigating how this is done.

Figure 10.36
Frzb expression and function.
(A) Double in situ hybridization localizing *frzb* (dark blue) and chordin (brown) messages. *frzb* mRNA can be seen to be transcribed in the head endomesoderm of the organizer, but not in the notochord (where chordin is expressed). (B) Microinjection of *frzb* mRNA into the marginal zone leads to the inhibition of trunk formation. (From Leyns et al. 1997; photographs courtesy of E. M. De Robertis.)

(A)

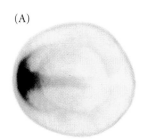

(B)

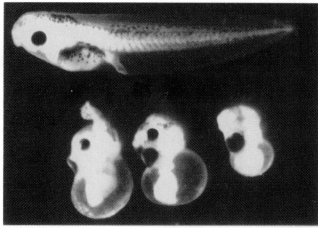

Figure 10.37
Hypothetical pathways differentiating ectoderm into epidermis or neural ectoderm. In the presence of BMP signaling, epidermalizing transcription factors are generated, leading to the activation of the pathway enabling the cell to become an epidermal keratinocyte. In the absence of BMP signaling, neuralizing transcription factors are produced. These factors activate the gene for neurogenin. Neurogenin acts as a transcription factor to activate the *NeuroD* gene, and NeuroD acts as a transcription factor to cause the differentiation of the cell into a neuron.

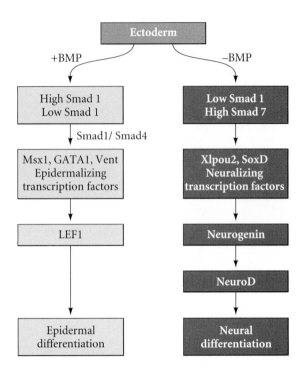

Sidelights & Speculations

Competence, Bias, and Neurulation

IN ADDITION to the signals coming from the underlying chordal plate and dorsal mesoderm, there may also be a bias in the cells of the dorsal part of the embryo toward becoming neural. Phillips and colleagues (London et al. 1988; Savage and Phillips 1989) have shown that the dorsal and ventral animal pole cells of the early-cleavage embryo differ in their expression of the Epi1 protein. Not only do the presumptive epidermal cells express this protein, which is not expressed in the presumptive neural cells, but the region of cells failing to express Epi1 increases during gastrulation. Moreover, in the ventral mesoderm, proteins encoded by *Xvex-1* and *Xvex-2* block the expression of dorsal genes (Shapira et al. 2000). Other differences between dorsal and ventral ectoderm also become apparent at this time, prior to the notochord's movement beneath the ectoderm (Otte and Moon 1992). The cues for this "bias" toward neurulation may be provided by signals from the dorsal lip traveling in a planar (horizontal) fashion through the ectoderm (Figure 10.38; Sharpe et al. 1987; Dixon and Kintner 1989; Doniach et al. 1993). The molecular basis of this planar "ectodermal" signaling system is not known. It is possible that it is more important in some species than in others. ■

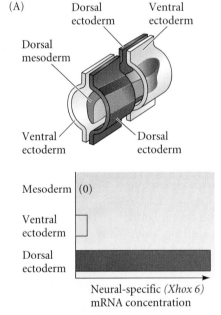

WEBSITE 10.9 Planar induction. Several studies suggest that the planar mode of signal transduction—from the dorsal lip through the ectoderm—may also help pattern the ectoderm. Other studies argue against planar induction playing any role in patterning the ectoderm.

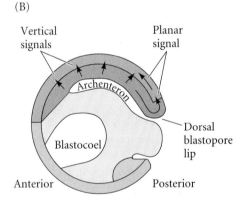

Figure 10.38
Ectodermal bias toward neurulation. (A) Neural gene expression occurs more readily in dorsal ectoderm than in ventral ectoderm. Ectoderm fragments from the dorsal and ventral thirds of a *Xenopus* embryo were wrapped around dorsal mesoderm (notochord) tissue. The fragments were then separated and tested for the induction of neural mRNAs. The mesoderm had none, the ventral ectoderm had some, and the dorsal ectoderm expressed high concentrations of neural messages. (B) Model showing the two modes of signaling from the dorsal blastopore lip: a vertical mechanism through the dorsal mesoderm, and a horizontal (planar) mechanism through the ectoderm. (A after Sharpe et al. 1987; B after Doniach 1993.)

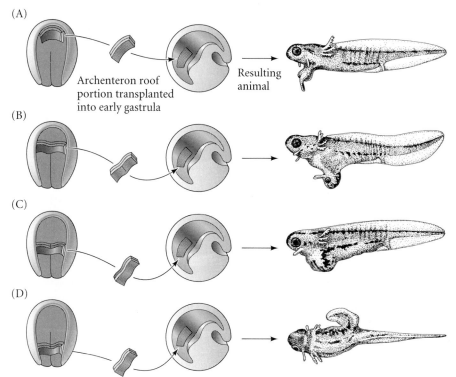

(A)

Archenteron roof portion transplanted into early gastrula

Resulting animal

(B)

(C)

(D)

Figure 10.39
Regional specificity of induction can be demonstrated by implanting different regions (color) of the archenteron roof into early *Triturus* gastrulae. The resulting embryos develop secondary dorsal structures. (A) Head with balancers. (B) Head with balancers, eyes, and forebrain. (C) Posterior part of head, deuterencephalon, and otic vesicles. (D) Trunk-tail segment. (After Mangold 1933.)

The Regional Specificity of Induction

The determination of regional differences

One of the most fascinating phenomena in neural induction is the regional specificity of the neural structures that are produced. Forebrain, hindbrain, and spinocaudal regions of the neural tube must all be properly organized in an anterior-to-posterior direction. The organizer tissue not only induces the neural tube, but also specifies the regions of the neural tube. This region-specific induction was demonstrated by Hilde Mangold's husband, Otto Mangold (1933). He transplanted four successive regions of the archenteron roof of late-gastrula newt embryos into the blastocoels of early-gastrula embryos (Figure 10.39). The most anterior portion of the archenteron roof induced balancers and portions of the oral apparatus (Figure 10.39A); the next most anterior section induced the formation of various head structures, including nose, eyes, balancers, and otic vesicles (Figure 10.39B); the third section induced the hindbrain structure (Figure 10.39C); and the most posterior section induced the formation of dorsal trunk and tail mesoderm* (Figure 10.39D). Moreover, when dorsal blastopore lips from *early* salamander gastrulae were transplanted into other early salamander gas-

trulae, they formed secondary heads. When dorsal lips from *later* gastrulas were transplanted into early salamander gastrulae, however, they induced the formation of secondary tails (Figure 10.40; Mangold 1933). These results show that the first cells of the organizer to enter the embryo induce the formation of brains and heads, while those cells that form the dorsal lip of later-stage embryos induce the cells above them to become spinal cords and tails.

Molecular correlates of neural caudalization

In the 1950s, evidence accumulated for the existence of two gradients in amphibian embryos: a dorsal gradient of "neuralizing" activity and a caudal gradient of posteriorizing ("mesodermalizing") activity (Nieuwkoop 1952; Toivonen and Saxén 1955). The neuralizing activity came from the organizer and induced the ectoderm to be neural. The posteriorizing activity originated in the posterior of the embryo and weakened anteriorly. Recent studies have extended this model and provided candidates for the posteriorizing molecules. As predicted, the two signaling systems—neuralizing and posteriorizing—work independently (Kolm and Sive 1997). Chordin, Noggin, and the other molecules discussed above constitute the neuralizing factors secreted by the organizer. The candidates for the posteriorizing factor include eFGF, retinoic acid, and Wnt3a.

WEBSITE **10.10 Regional specification.** The research into regional specification has been a fascinating endeavor involving scientists from all over the world. Before molecular biology gave us the tools to uncover morphogenetic proteins, embryologists developed ingenious ways of finding out what those proteins were doing.

*The induction of dorsal mesoderm—rather than the dorsal ectoderm of the nervous system—by the posterior end of the notochord was confirmed by Bïjtel (1931) and Spofford (1945), who showed that the posterior fifth of the neural plate gives rise to tail somites and the posterior portions of the pronephric kidney duct.

(A) Transplantation of young gastrula dorsal lip

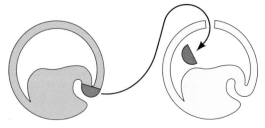

(B) Transplantation of advanced gastrula dorsal lip

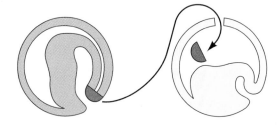

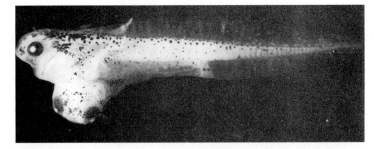

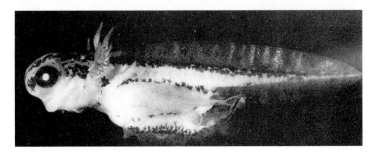

Figure 10.40
Regionally specific inducing action of the dorsal blastopore lip. (A) Young dorsal lips (which will form the anterior portion of the organizer) induce anterior dorsal structures when transplanted into early newt gastrulae. (B) Older dorsal lips transplanted into early newt gastrulae produce more posterior dorsal structures. (From Saxén and Toivonen 1962; photographs courtesy of L. Saxén.)

eFGF. Fibroblast growth factors are able to turn anterior neural tissue into posterior neural tissue. When early-gastrula ectoderm (which has not yet been underlain by dorsal mesoderm) was isolated and neuralized by Noggin, Chordin, or Follistatin, anterior-type neural markers were found in that tissue. However, when isolated early-gastrula ectoderm was incubated with a neural inducer plus FGF2, it expressed more posterior neural markers. FGF2 will also induce forebrain tissue to express hindbrain-specific genes (Cox and Hemmati-Brivanlou 1995; Lamb and Harland 1995). Furthermore, when FGF signaling is blocked in vivo by a dominant negative FGF receptor, the resulting tadpoles lack their posterior segments (Amaya et al. 1991). FGF2 is probably not the natural posteriorizing FGF in *Xenopus*, since it is not secreted by the embryo at this site, and it is not localized to any side of the embryo. However, embryonic FGF (**eFGF**, a *Xenopus* FGF similar to mammalian FGF4) is found in *Xenopus* posterior and tailbud mesoderm, and it has the same effects as FGF2 (Isaacs et al. 1992). Overexpression of eFGF up-regulates several posteriorly expressed genes, including the *Xenopus* homologue of *caudal*. These genes, in turn, encode proteins that appear to regulate the Hox genes controlling the specification of body segments along the anterior-posterior axis. This leads to the more posterior specification of the caudal nervous system (Pownall et al. 1996). Interestingly, eFGF may be induced by the posterior notochord (Taira et al. 1997).

RETINOIC ACID. **Retinoic acid** is also likely to play a role in posteriorizing the neural tube. If *Xenopus* neurulae are treated with nanomolar to micromolar concentrations of retinoic acid (RA), their forebrain and midbrain development is impaired in a concentration-dependent fashion (Figure 10.41; Papalopulu et al. 1991; Sharpe 1991). When lower concentrations are used, the actual induction of neural tissue does not appear to be inhibited, but fewer forebrain messages and structures are produced (Durston et al. 1989; Sive et al. 1990). Retinoic acid appears to affect both the mesoderm and the ectoderm. Ruiz i Altaba and Jessell (1991) found that anterior dorsal mesoderm from RA-treated gastrulae was unable to induce head structures in host embryos, and Sive and Cheng (1991) found that RA-treated ectoderm was unable to respond to the anterior-inducing mesoderm of untreated gastrulae. An RA gradient (tenfold higher in the posterior than in the anterior) has been detected in the dorsal mesoderm of early *Xenopus* neurulae (Chen et al. 1994). Like eFGF, retinoic acid has been shown to activate the expression of more posterior Hox genes (Cho et al. 1991b; Sive and Cheng 1991; Kolm and Sive 1997).

WNT3A. Another candidate for the caudalizing factor is *Xenopus* **Wnt3a** (McGrew et al. 1995). This protein is found in the neural ectoderm of the early neurula. When ectoderm is isolated from *Xenopus* gastrulae but remains connected to the dorsal blastopore lip, the ectoderm develops an anterior-posterior array of neural markers. If the embryo is first injected with *Xwnt3a* RNA (causing the overexpression of this protein), the anterior markers are lost. The basic model of neural induction, then, looks like the diagram in Figure 10.42.

Concentration of RA presented to late neurula stage

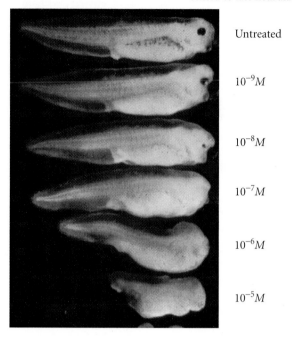

Untreated

$10^{-9}M$

$10^{-8}M$

$10^{-7}M$

$10^{-6}M$

$10^{-5}M$

Figure 10.41
Effects of RA and FGF on the anterior-posterior axis. Late neurula embryos exposed to progressively higher concentrations of retinoic acid (and allowed to grow until the controls reached tadpole stage) lose posterior regions of the embryo (as well as having eye deformities). (From Ruiz i Altaba and Jessell 1991; photograph courtesy of T. Jessell.)

The relationship between these three pathways of posteriorization has yet to be worked out. They may play different roles in different regions of the embryo, and they may work together. Retinoic acid has its major effect on hindbrain patterning, while eFGF is more important in the regionalization of the spinal cord (Blumberg et al. 1997; Kolm et al. 1997; Godsave et al. 1998). Wnt3a may suppress anterior genes and be a permissive factor allowing the other two proteins to function (McGrew et al. 1997).

Specifying the left-right axis

While the developing tadpole looks symmetrical from the outside, there are several internal organs, such as the heart and the gut tube, that are not evenly balanced on the right and left sides. In other words, in addition to its dorsal-ventral and anterior-posterior axes, the embryo has a left-right axis. Somehow, the body must be given clues as to which half is right and which half is left.

In all vertebrates studied so far, the crucial event in left-right axis formation is the expression of a **nodal** gene in the lateral plate mesoderm on the *left* side of the embryo. In *Xenopus*, this gene is *Xnr-1* (*Xenopus* nodal-related-1). If the expression of this gene is also permitted to occur on the right-hand side, the position of the heart (which is normally on the left side) and the coiling of the gut are randomized.

But what causes the expression of *Xnr-1* solely on the left-hand side? In *Xenopus*, it is possible that the first clue is given at fertilization. The microtubules in-

volved in cytoplasmic rotation appear to be crucial, since if their formation is blocked, no left-right axis appears (Yost 1998). Moreover, the Vg1 protein, which appears to be expressed throughout the vegetal hemisphere, seems to be processed into its active form predominantly on the left-hand side of the embryo. Injecting active Vg1 protein into the left vegetal blastomeres has no effect, but adding it to the right vegetal blastomeres leads to the expression of *Xnr-1* in both the right and left lateral plate and to the randomization of heart and gut positions (Hyatt et al. 1996). Moreover, when active Vg1 protein is injected into a particular right-side vegetal blastomere (the third vegetal cell to the right of the dorsal midline), the entire left-right axis is inverted. No other TGF-β family member is able to accomplish this reversal (Hyatt and Yost 1998). It is not yet known how the events at fertilization lead to the expression of *Xnr-1* in the left lateral plate mesoderm during gastrulation.

The pathway by which the Xnr-1 protein instructs the heart and gut to fold properly is also unknown, but one of the key genes activated by Xnr-1 appears to be **pitx2**. Since this gene is activated by Xnr-1, it is normally expressed only on the

(3)
Wnt, FGF, or RA gradient

(1)
BMP inhibitors:
Chordin, Noggin, Follistatin

(2)
Wnt inhibitors:
Cerberus, Frzb, Dickkopf

Dorsal mesoderm

Dorsal blastopore lip

Endoderm

Ectoderm

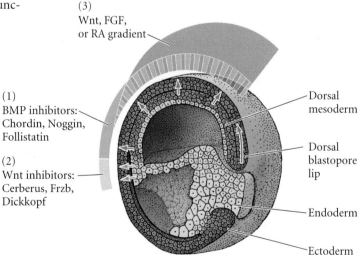

Figure 10.42
Model for the organizer function and axis specification in *Xenopus* gastrula. (1) BMP inhibitors from organizer tissue (dorsal mesoderm and pharyngeal mesendoderm) block the formation of epidermis, ventrolateral mesoderm, and ventrolateral endoderm. (2) Wnt inhibitors in the anterior of the organizer (pharyngeal mesendoderm) allow the induction of head structures. (3) A gradient of caudalizing factors (eFGF, retinoic acid, and/or Wnt3a) specify the regional expression of Hox genes.

Figure 10.43
Pitx2 determines the direction of heart looping and gut coiling. (A) Wild-type (stage 45) *Xenopus* tadpole viewed from the ventral side, showing rightward heart looping and counterclockwise gut coiling. (B) If an embryo is injected with Pitx2 so that this protein is present in the mesoderm of both the right and left sides (instead of just the left side), heart looping and gut coiling are random with respect to each other. Sometimes this results in complete reversals, as in this embryo, in which the heart loops to the left and the gut coils in a clockwise manner. (From Ryan et al. 1998; photograph courtesy of J. C. Izpisúa-Belmonte.)

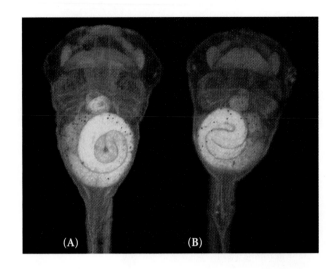

left side of the embryo. However, if the Pitx2 protein is injected into the right side, too, the placement of the heart and the coiling of the gut are randomized (Figure 10.43; Ryan et al. 1998). Pitx2 persists on the left side of the embryo as the heart and gut develop, controlling their respective positions. Pitx2 may be at the "heart of the heart" (Strauss 1998).

We are finally putting names to the "agents" and "soluble factors" of the experimental embryologists. We are finally delineating the intercellular pathways of paracrine factors and transcription factors that constitute the first steps in the processes of organogenesis. The international research program initiated by Spemann's laboratory in the 1920s is reaching toward a conclusion. But this research has found levels of complexity far deeper than Spemann would have conceived, and just as Spemann's experiments told us how much we didn't know, so today, we are faced with a whole new set of problems generated by our solutions to older ones.

Surveying the field in 1927, Spemann remarked:

We still stand in the presence of riddles, but not without hope of solving them. And riddles with the hope of solution—what more can a scientist desire?

The challenge still remains.

Snapshot Summary: Early Development and Axis Formation in Amphibians

1. Amphibian cleavage is holoblastic, but unequal due to the presence of yolk in the vegetal hemisphere.

2. Amphibian gastrulation begins with the invagination of the bottle cells, followed by the coordinated involution of the mesoderm and the epiboly of the ectoderm. Vegetal rotation plays a significant role in directing the involution.

3. The driving forces for ectodermal epiboly and the convergent extension of the mesoderm are the intercalation events in which several tissue layers merge. Fibronectin plays a critical role in enabling the mesodermal cells to migrate into the embryo.

4. The dorsal lip of the blastopore forms the organizer tissue of the amphibian gastrula. This tissue dorsalizes the ectoderm, transforming it into neural tissue, and it transforms ventral mesoderm into lateral mesoderm.

5. The organizer consists of pharyngeal endoderm, head mesoderm, notochord, and dorsal blastopore lip. The organizer functions by secreting proteins (Noggin, chordin, and follistatin) that block the BMP signal that would otherwise ventralize the mesoderm and activate the epidermal genes in the ectoderm.

6. In the head region, an addition set of proteins (Cerberus, Frzb, Dickkopf) block the Wnt signal from the ventral and lateral mesoderm.

7. The organizer is itself induced by the Nieuwkoop center, located in the dorsalmost vegetal cells. This center is formed by cortical rotation during fertilization, which translocates the Dishevelled protein to the dorsal side of the egg.

8. The Dishevelled protein stabilizes β-catenin in the dorsal cells of the embryo. Thus, the Nieuwkoop center is formed by the accumulation of β-catenin, which can complex with Tcf3 to form a transcription factor complex that can activate the transcription of the *siamois* gene.

9. The *siamois* product and a TGF-β signal (perhaps from Vg1) can activate the *goosecoid* gene in the organizer. The *goosecoid* gene can activate other genes that cause the organizer to function.

10. Other posteriorizing signals (Wnt3a, retinoic acid, eFGF) can influence the anterior-posterior specification of the neural tube.

11. The left-right axis appears to be initiated at fertilization through the Vg1 protein. In a still unknown fashion, this protein activates a Nodal protein solely on the left side of the body. As in other vertebrates, the Nodal protein activates expression of Pitx2, which is critical in distinguishing left-sidedness from right-sidedness in the heart and gut tubes.

Literature Cited

Agius, E., M. Oelgeschläager, O. Wessely, C. Kemp, and E. M. De Robertis. 2000. Endodermal Nodal-related signals and mesoderm induction in *Xenopus*. *Development* 127: 1151–1159.

Alfandari, D., C. A. Whittaker, D. W. DeSimone and T. Darribère. 1995. Integrin αv subunit is expressed on mesodermal cell surfaces during amphibian gastrulation. *Dev. Biol.* 170: 249–261.

Amaya, E., T. J. Musci and M. W. Kirschner. 1991. Expression of a dominant negative mutant of the FGF receptor disrupts mesoderm formation in *Xenopus* embryos. *Cell* 66: 257–270.

Amaya, E., M. F. Offield and R. M. Grainger. 1998. Frog genetics: *Xenopus tropicalis* jumps into the future. *Trends Genet.* 14: 253–255.

Appel, T. A. 1987. *The Cuvier-Geoffroy Debate: French Biology in the Decades before Darwin.* Oxford University Press, New York.

Arendt, D. and K. Nübler-Jung. 1999. Rearranging gastrulation in the name of yolk: Evolution of gastrulation in yolk-rich amniote eggs. *Mech. Dev.* 81: 3–22.

Balinsky, B. I. 1975. *Introduction to Embryology.* 4th Ed. Saunders, Philadelphia.

Beams, H. W. and R. G. Kessel. 1976. Cytokinesis: A comparative study of cytoplasmic division in animal cells. *Am. Sci.* 64: 279–290.

Bïjtel, J. H. 1931. Über die Entwicklung des Schwanzes bei Amphibien. *Wilhelm Roux Arch. Entwicklungsmech. Org.* 125: 448–486.

Black, S. D. and J. Gerhart. 1985. Experimental control of the site of embryonic axis formation in *Xenopus laevis* eggs centrifuged before first cleavage. *Dev. Biol.* 108: 310–324.

Black, S. D. and J. Gerhart. 1986. High frequency twinning of *Xenopus laevis* embryos from eggs centrifuged before first cleavage. *Dev. Biol.* 116: 228–240.

Blitz, I. L. and K. W. Y. Cho. 1995. Anterior neurectoderm is progressively induced during gastrulation: The role of the *Xenopus* homeobox gene *orthodenticle*. *Development* 121: 993–1004.

Blumberg, B., J. Bolado, T. Moreno, C. Kintner, R. Evans and N. Papalopulu. 1997. An essential role for signaling in anteroposterior neural patterning. *Development* 124: 373–379.

Boucaut, J.-C., T. D'Arribère, T. J. Poole, H. Aoyama, K. M. Yamada and J.-P. Thiery. 1984. Biologically active synthetic peptides as probes of embryonic development: A competitive peptide inhibition of fibronectin function inhibits gastrulation in amphibian embryos and neural crest cell migration in avian embryos. *J. Cell Biol.* 99: 1822–1830.

Boucaut, J.-C., T. D'Arribère, S. D. Li, H. Boulekbache, K. M. Yamada and J.-P. Thiery. 1985. Evidence for the role of fibronectin in amphibian gastrulation. *J. Embryol. Exp. Morphol.* 89 [Suppl.]: 211–217.

Bouwmeester, T., S.-H. Kim, Y. Sasai, B. Lu and E. M. De Robertis. 1996. Cerberus is a head-inducing secreted factor expressed in the anterior endoderm of Spemann's organizer. *Nature* 382: 595–601.

Brannon, M. and D. Kimelman. 1996. Activation of *siamois* by the Wnt pathway. *Dev. Biol.* 180: 344–347.

Brannon, M., M. Gomperts, L. Sumoy, R. T. Moon and D. Kimelman. 1997. β-catenin/XTcf-3 complex binds to the *siamois* promoter to regulate dorsal axis specification in *Xenopus*. *Genes Dev.* 11: 2359–2370.

Carlson, B. M. 1981. *Patten's Foundations of Embryology.* McGraw-Hill, New York.

Casellas, R. and A. Hemmati-Brivanlou. 1998. *Xenopus* smad7 inhibits both the activin and BMP pathways and acts as a neural inducer. *Dev. Biol.* 198: 1–12.

Chen, Y. P., L. Huang and M. Solursh. 1994. A concentration gradient of retinoids in the early *Xenopus laevis* embryo. *Dev. Biol.* 161: 70–76.

Cho, K. W. Y., B. Blumberg, H. Steinbeisser and E. De Robertis. 1991a. Molecular nature of Spemann's organizer: The role of the *Xenopus* homeobox gene *goosecoid*. *Cell* 67: 1111–1120.

Cho, K. W. Y., A. A. Morita, C. V. E. Wright and E. M. De Robertis. 1991b. Overexpression of a homeodomain protein confers axis-forming activity to uncommitted *Xenopus* embryonic cells. *Cell* 65: 55–64.

Christian, J. L. and R. T. Moon. 1993. Interactions between Xwnt8 and Spemann organizer signaling pathways generate dorsoventral pattern in the embryonic mesoderm of *Xenopus*. *Genes Dev.* 7: 13–28.

Cooke, J. 1986. Permanent distortion of positional system of *Xenopus* embryo by brief early perturbation in gravity. *Nature* 319: 60–63.

Cox, W. G. and A. Hemmati-Brivanlou. 1995. Caudalization of neural fate by tissue recombination and bFGF. *Development* 121: 4349–4358.

Cui, Y., Q. Tian and J. L. Christian. 1996. Synergistic effects of Vg1 and Wnt signals in the specification of dorsal mesoderm and endoderm. *Dev. Biol.* 180: 22–34.

Dale, L. and J. M. W. Slack. 1987. Regional specificity within the mesoderm of early embryos of *Xenopus laevis*. *Development* 100: 279–295.

Dale, L., G. Howes, B. M. J. Price and J. C. Smith. 1992. Bone morphogenetic protein 4: A ventralizing factor in early *Xenopus* development. *Development* 115: 573–585.

Danilchik, M. V. and J. M. Denegre. 1991. Deep cytoplasmic rearrangements during early development in *Xenopus laevis*. *Development* 111: 845–856.

D'Arribère, T., K. M. Yamada, K. E. Johnson and J.-C. Boucaut. 1988. The 140-kD fibronectin receptor complex is required for mesodermal cell adhesion during gastrulation in the amphibian *Pleurodeles waltii*. *Dev. Biol.* 126: 182–194.

D'Arribère, T., K. Guida, H. Larjava, K. E. Johnson, K. M. Yamada, J.-P. Thiery and J.-C. Boucaut. 1990. In vivo analysis of integrin β1 subunit function in fibronectin matrix assembly. *J. Cell Biol.* 110: 1813–1823.

Delarue, M., T. D'Arribère, C. Aimar and J.-C. Boucaut. 1985. Bufonid nucleocytoplasmic hybrids arrested at the early gastrula stage lack a fibronectin-containing fibrillar extracellular matrix. *Wilhelm Roux Arch. Dev. Biol.* 194: 275–280.

De Robertis, E. M. and Y. Sasai. 1996. A common plan for dorsoventral patterning in Bilateria. *Nature* 380: 37–40.

De Robertis, E. M., M. Blum, C. Niehrs and H. Steinbeisser. 1992. Goosecoid and the organizer. *Development* 1992 [Suppl.]: 167–171.

Dixon, J. E. and C. R.Kintner. 1989. Cellular contacts required for neural induction in *Xenopus laevis* embryos: Evidence for two signals. *Development* 106: 749–757.

Doniach, T. 1993. Planar and vertical induction of anteroposterior pattern during the development of the amphibian central nervous system. *J. Neurobiol.* 24: 1256–1275.

Dosch, R., V. Gawantka, H. Delius, C. Blumenstock and C. Niehrs. 1997. BMP-4 acts as a morphogen in dorsolateral mesoderm patterning in *Xenopus*. *Development* 124: 2325–2334.

Durston, A., A. Timmermans, W. J. Hage, H. F. J. Hendriks, N. J. de Vries, M. Heideveld and P. D. Nieuwkoop. 1989. Retinoic acid causes an anteroposterior transformation in the developing central nervous system. *Nature* 330: 140–144.

Fan, C-M. and S. Y. Sokol. 1997. A role for Siamois in Spemann organizer formation. *Development* 124: 2581–2589.

Fässler, P. E. and K. Sander. 1996. Hilde Mangold (1898–1924) and Spemann's organizer: Achievement and tragedy. *Wilhelm Roux Arch. Dev. Biol.* 205: 323–332.

Funayama, N., F. Fagotto, P. McCrea and B. M. Grumbiner. 1995. Embryonic axis induction by the armadillo repeat domain of β-catenin: Evidence for intracellular signalling. *J. Cell Biol.* 128: 959–968.

Gawantka, V., H. Delius, K. Hirschfeld, C. Blumenstock and C. Niehrs. 1995. Antagonizing the Spemann organizer: Role of the homeobox gene Xvent-1. *EMBO J.* 14: 6268–6279.

Geoffroy Saint-Hilaire, E. 1822. Considérations génerales sur la vertèbre. *Mém. Mus. Hist. Nat.* 9: 89–1110.

Gerhart, J., G. Ubbels, S. Black, K. Hara and M. Kirschner. 1981. A reinvestigation of the role of the grey crescent in axis formation in *Xenopus laevis*. *Nature* 292: 511–516.

Gerhart, J. and 7 others. 1986. Amphibian early development. *BioScience* 36: 541–549.

Gerhart, J. C., M. Danilchik, T. Doniach, S. Roberts, B. Rowning, and R. Stewart. 1989. Cortical rotation of the *Xenopus* egg: Consequences for the anteroposterior pattern of embryonic dorsal development. *Development* [Suppl.] 107: 37–51.

Gilbert, S. F. and L. Saxén. 1993. Spemann's organizer: Models and molecules. *Mech. Dev.* 41: 73–89.

Gilbert, S. F., J. Opitz and R. A. Raff. 1996. Resynthesizing evolutionary and developmental biology. *Dev. Biol.* 173: 357–372.

Gimlich, R. L. 1985. Cytoplasmic localization and chordamesoderm induction in the frog embryo. *J. Embryol. Exp. Morphol.* 89: 89–111.

Gimlich, R. L. 1986. Acquisition of developmental autonomy in the equatorial region of the *Xenopus* embryo. *Dev. Biol.* 116: 340–352.

Gimlich, R. L. and J. C. Gerhart. 1984. Early cellular interactions promote embryonic axis formation in *Xenopus laevis*. *Dev. Biol.* 104: 117–130.

Glinka, A., W. Wu, D. Onichtchouk, C. Blumenstock and C. Niehrs. 1997. Head induction by simultaneous repression of BMP and Wnt signalling in *Xenopus*. *Nature* 389: 517–519.

Glinka, A., W. Wu, A. P. Monaghan, C. Blumenstock and C. Niehrs. 1998. Dickkopf-1 is a member of a new family of secreted proteins and functions in head induction. *Nature* 391: 357–362.

Godsave, S. F. and J. M. W. Slack. 1989. Clonal analysis of mesoderm induction in *Xenopus*. *Dev. Biol.* 134: 486–490.

Godsave, S. F., C. H. Koster, A. Getahun, M. Mathu, M. Hooiveld, J. van der Wees, J Hendriks and A. J. Durston. 1998. Graded retinoid responses in the developing hindbrain. *Dev. Dynam.* 213: 39–49.

Gont, L. K., H. Steinbeisser, B. Blumberg and E. M. De Robertis. 1993. Tail formation as a continuation of gastrulation: The multiple tail populations of the *Xenopus* tailbud derive from the late blastopore lip. *Development* 119: 991–1004.

Graff, J. M., R. S. Thies, J. J. Song, A. J. Celeste and D. A. Melton. 1994. Studies with a *Xenopus* BMP receptor suggest that ventral mesoderm-inducing signals override dorsal signals in vivo. *Cell* 79: 169–179.

Grunz, H. and L. Tacke. 1989. Neural differentiation of *Xenopus laevis* ectoderm takes place after disaggregation and delayed reaggregation without induces. *Cell Diff. Dev.* 32: 117–124.

Guger, K. A. and B. M. Gumbiner. 1995. β-catenin has wnt-like activity and mimics the Nieuwkoop signaling center in *Xenopus* dorsal-ventral patterning. *Dev. Biol.* 172: 115–125.

Hamburger, V. 1984. Hilde Mangold, co-discoverer of the organizer. *J. Hist. Biol.* 17: 1–11.

Hamburger, V. 1988. *The Heritage of Experimental Embryology: Hans Spemann and the Organizer.* Oxford University Press, Oxford.

Hammerschmidt, M., G. N. Serbedzija and A. P. McMahon. 1996. Genetic analysis of dorsoventral pattern formation in the zebrafish: Requirement of a BMP-like ventralizing activity and its dorsal repressor. *Genes Dev.* 10: 2452–2461.

Hansen, C. S., C. D. Marion, K. Steele, S.. George and W. C. Smith. 1997. Direct neural induction and selective inhibition of mesoderm formation and epidermis inducers by Xnr3. *Development* 124: 483–492.

Hara, K. 1977. The cleavage pattern of the axolotl egg studied by cinematography and cell counting. *Wilhelm Roux Arch. Entwicklungsmech. Org.* 181: 73–87.

Hardin, J. D. and R. Keller. 1988. The behaviour and function of bottle cells during gastrulation of *Xenopus laevis*. *Development* 103: 211–230.

Harland, R. M. and J. Gerhart. 1997. Formation and function of Spemann's organizer. *Annu. Rev. Cell Dev. Biol.* 13: 611–667.

Hawley, S. H. B., K. Wünnenberg-Stapleton, C. Hashimoto, M. N. Laurent, T. Watabe, B. W. Blumberg and K. W. Y. Cho. 1995. Disruption of BMP signals in embryonic *Xenopus* ectoderm leads to direct neural induction. *Genes Dev.* 9: 2923–2935.

He, X., J.-P. Saint-Jeannet, J. R. Woodgett, H. E. Varmus and I. B. Dawid. 1995. Glycogen synthase kinase-3 and dorsoventral patterning in *Xenopus* embryos. *Nature* 374: 617–622.

Heasman, J. M. and 8 others. 1994a. Overexpression of cadherins and underexpression of β-catenin inhibit dorsal mesoderm induction in early *Xenopus* embryos. *Cell* 79: 791–803.

Heasman, J., D. Ginsberg, K. Goldstone, T. Pratt, C. Yoshidanaro and C. Wylie. 1994b. A functional test for maternally inherited cadherin in *Xenopus* shows its importance in cell adhesion at the blastula stage. *Development* 120: 49–57.

Hemmati-Brivanlou, A. and D. A. Melton. 1992. A truncated activin receptor inhibits mesoderm induction and formation of axial structures in *Xenopus* embryos. *Nature* 359: 609–614.

Hemmati-Brivanlou, A. and D. A. Melton. 1994. Inhibition of activin signalling promotes neuralization in *Xenopus*. *Cell* 77: 273–281.

Hemmati-Brivanlou A. and D. A. Melton. 1997. Vertebrate embryonic cells will become nerve cells unless told otherwise. *Cell* 88: 13–17.

Hemmati-Brivanlou, A. and G. H. Thomsen. 1995. Ventral mesodermal patterning in *Xenopus* embryos: Expression patterns and activities of BMP-2 and BMP-4. *Dev. Genet.* 17: 78–810.

Holley, S. A., P. D. Jackson, Y. Sasai, B. Lu, E. M. De Robertis, F. M. Hoffmann and E. L. Ferguson. 1995. A conserved system for dorsal-ventral patterning in insects and vertebrates involving sog and chordin. *Nature* 376: 249–253.

Holowacz, T. and R. P. Elinson. 1993. Cortical cytoplasm, which induces dorsal axis formation in *Xenopus*, is inactivated by UV irradiation of the oocyte. *Development* 119: 277–285.

Holtfreter, H. 1933. Die totale Exogastrulation, eine Selbststablösung des Ektoderms von Entomesoderm. Entwicklung und funktionelles Verhalten nervenloser Organe. *Arch. Entwick. Mech. Org.* 129: 669–793.

Holtfreter, J. 1943. A study of the mechanics of gastrulation, Part I. *J. Exp. Zool.* 94: 261–318.

Holtfreter, J. 1944. A study of the mechanics of gastrulation, Part II. *J. Exp. Zool.* 95: 171–212.

Hyatt, B. A. and H. J. Yost. 1998. The left/right coordinator: The role of Vg1 in organizing left-right axis formation. *Cell* 93: 37–46.

Hyatt, B. A., J. L. Lohr and H. J. Yost. 1996. Initiation of vertebrate left-right axis by maternal vg1. *Nature* 384: 62–65.

Iemura, S.-I. and 7 others. 1998. Direct binding of follistatin to a complex of bone morphogenetic protein and its receptor inhibits ventral and epidermal cell fates in early *Xenopus* embryo. *Proc. Natl. Acad. Sci. USA* 95: 9337–9342.

Isaacs, H. V., D. Tannahill and J. M. W. Slack. 1992. Expression of a novel FGF in the *Xenopus* embryo: A new candidate inducing factor for mesoderm formation and anteroposterior specification. *Development* 114: 711–720.

Jones, C. M., K. M. Lyons, P. M. Lapan, C. V. E. Wright and B. L. M. Hogan. 1992. DVR-4 (bone morphogenetic protein-4) as a posterior ventralizing factor in *Xenopus* mesoderm induction. *Development* 115: 639–647.

Joseph, E. M. and D. A. Melton. 1998. Mutant Vg₁ ligands disrupt endoderm and mesoderm formation in *Xenopus* embryos. *Development* 125: 2677–2685.

Kalt, M. R. 1971. The relationship between cleavage and blastocoel formation in *Xenopus laevis*. I. Light microscopic observations. *J. Embryol. Exp. Morphol.* 26: 37–410.

Keller, R. E. 1975. Vital dye mapping of the gastrula and neurula of *Xenopus laevis*. I. Prospective areas and morphogenetic movements of the superficial layer. *Dev. Biol.* 42: 222–241.

Keller, R. E. 1976. Vital dye mapping of the gastrula and neurula of *Xenopus laevis*. II. Prospective areas and morphogenetic movements of the deep layer. *Dev. Biol.* 51: 118–137.

Keller, R. E. 1980. The cellular basis of epiboly: An SEM study of deep cell rearrangement during gastrulation of *Xenopus laevis*. *J. Embryol. Exp. Morphol.* 60: 201–243.

Keller, R. E. 1981. An experimental analysis of the role of bottle cells and the deep marginal zone in the gastrulation of *Xenopus laevis*. *J. Exp. Zool.* 216: 81–101.

Keller, R. E. 1986. The cellular basis of amphibian gastrulation. *In* L. Browder (ed.), *Develop-*

mental Biology: A Comprehensive Synthesis, Vol. 2. Plenum, New York, pp. 241–327.

Keller, R. and M. Danilchik. 1988. Regional expression, pattern and timing of convergence and extension during gastrulation of *Xenopus laevis*. *Development* 103: 193–209.

Keller, R. E. and G. C. Schoenwolf. 1977. An SEM study of cellular morphology, contact, and arrangement as related to gastrulation in *Xenopus laevis*. *Wilhelm Roux Arch. Dev. Biol.* 182: 165–186.

Kessler, D. S. 1997. Siamois is required for formation of Spemann's organizer. *Proc. Natl. Acad. Sci. USA* 94: 13017–13022.

Kessler, D. S. and D. A. Melton. 1995. Induction of dorsal mesoderm by soluble, mature Vg1 protein. *Development* 121: 2155–2164.

Kim, S.-H., A. Yamamoto, T. Bouwmeester, E. Agius and E. M. De Robertis. 1998. The role of paraxial protocadherin in selective adhesion and cell movements of the mesoderm during *Xenopus* gastrulation. *Development* 125: 4681–4691.

Kolm, P. J. and H. L. Sive. 1997. Retinoids and posterior neural induction: A reevaluation of Nieuwkoop's two-step hypothesis. *Cold Spring Harbor Symp. Quant. Biol.* 62: 511–521.

Kolm. P. J., V. Apekin and H. Sive. 1997. *Xenopus* hindbrain patterning requires retinoic acid signaling. *Dev. Biol.* 192: 1–16.

Lamb, T. M. and R. M. Harland. 1995. Fibroblast growth factor is a direct neural inducer, which combined with noggin generates anterior-posterior pattern. *Development* 121: 3627–3636.

Landstrom, U. and Løvtrup, S.1979. Fate maps and cell differentiation in the amphibian embryo: An experimental study. *J. Embryol. Exp. Morphol.* 54: 113–130.

Lane, M. C. and W. C. Smith. 1999. The origins of primitive blood in *Xenopus*: Implications for axial patterning. *Development* 126: 423–434.

Larabell, C. A. and 7 others. 1997. Establishment of the dorsal-ventral axis in *Xenopus* embryos is presaged by early asymmetries in β-catenin which are modulated by the Wnt signaling pathway. *J. Cell Biol.* 136: 1123–1136.

Laurent, M. N., I. L. Blitz, C. Hashimoto, U. Rothbacher and K. W.-Y. Cho. 1997. The *Xenopus* homeobox gene twin mediates Wnt induction of goosecoid in establishment of Spemann's organizer. *Development* 124: 4905–4916.

Lee, J. E., S. M. Hollenberg, L. Snider, D. L. Turner, N. Lipnick and H. Weintraub. 1995. Conversion of *Xenopus* ectoderm into neurons by neuroD, a basic helix-loop-helix protein. *Science* 268: 836–844.

Lemaire, P., N. Garrett and J. B. Gurdon. 1995. Expression cloning of *Siamois*, a *Xenopus* homeobox gene expressed in dorsal-vegetal cells of blastulae and able to induce a complete secondary axis. *Cell* 81: 85–94

Leyns, L., T. Bouwmeester, S.-H. Kim, S. Piccolo and E. M. De Robertis. 1997. Frzβ-1 is a secret-

ed antagonist of Wnt signaling expressed in the Spemann organizer. *Cell* 88: 747–756.

Løvtrup, S. 1975. Fate maps and gastrulation in amphibia: A critique of current views. *Can. J. Zool.* 53: 473–479.

Lundmark, C. 1986. Roles of bilateral zones of ingressing superficial cells during gastrulation of *Ambystoma mexicanum*. *J. Embryol. Exp. Morphol.* 97: 47–62.

Ma, Q. F., C. Kintner and D. J. Anderson. 1996. Identification of neurogenin, a vertebrate neuronal determination gene. *Cell* 87: 43–52.

Mangold, O. 1933. Über die Induktionsfahigkeit der verschiedenen Bezirke der Neurula von Urodelen. *Naturwissenschaften* 21: 761–766.

Mariani, F. and R. Harland. 1998. XBF-2 is a transcriptional repressor that converts ectoderm into neural tissue. *Development* 125: 5019–5031.

McGrew, L. L., C.-J. Lai and R. T. Moon. 1995. Specification of the anteroposterior neural axis through synergistic interaction of the wnt signaling cascade with noggin and follistatin. *Dev. Biol.* 172: 337–342.

McGrew, L. L., S. Hoppler and R. T. Moon. 1997. Wnt and FGF pathways cooperatively pattern anteroposterior neural ectoderm in *Xenopus*. *Mech. Dev.* 69: 105–114.

Miller, J. R., B. A. Rowning, C. A. Larabell, J. A. Yang-Snyder, R. L. Bates and R. T. Moon. 1999. Establishment of the dorsal-ventral axis in *Xenopus* embryos coincides with the dorsal enrichment of Dishevelled that is dependent on cortical rotation. *J. Cell Biol.* 146: 427–437.

Minsuk, S. B. and R. E. Keller. 1996. Dorsal mesoderm has a dual origin and forms by a novel mechanism in *Hymenochirus*, a relative of *Xenopus*. *Dev. Biol.* 174: 92–103.

Molenaar, M. and 8 others. 1996. Xtcf-3 transcription factor mediates β-catenin-induced axis formation in *Xenopus* embryos. *Cell* 86: 391–399.

Moon, R. T. and D. Kimelman. 1998. From cortical rotation to organizer gene expression: Toward a molecular explanation of axis specification in *Xenopus*. *BioEssays* 20: 536–545.

Nakamura, O. and H. Takasaki. 1970. Further studies on the differentiation capacity of the dorsal marginal zone in the morula of *Triturus pyrrhogaster*. *Proc. Japan Acad.* 46: 700–705.

Nakatsuji, N., M. A. Smolira and C. C. Wylie. 1985. Fibronectin visualized by scanning electron microscope immunocytochemistry on the substratum for cell migration in *Xenopus laevis* gastrulae. *Dev. Biol.* 107: 264–268.

Newman, C. S. and P. A. Krieg. 1999. Specification and differentiation of the heart in amphibia. *In* S. A. Moody, *Cell Lineage and Fate Determination*. Academic Press, New York, pp. 341–351.

Newport, J. W. and M. W. Kirschner. 1982a. A major developmental transition in early *Xenopus* embryos: I. Characterization and timing of cellular changes at midblastula stage. *Cell* 30: 675–686.

Newport, J. W. and M. W. Kirschner. 1982b. A major developmental transition in early *Xenopus* embryos. II. Control of the onset of transcription. *Cell* 30: 687–696.

Niehrs, C., R. Keller, K. W. Y. Cho and E. M. De Robertis. 1993. The homeobox gene *goosecoid* controls cell migration in *Xenopus* embryos. *Cell* 72: 491–503.

Nieuwkoop, P. D. 1952. Activation and organization of the central nervous system in amphibians. III. Synthesis of a new working hypothesis. *J. Exp. Zool.* 120: 83–108.

Nieuwkoop, P. D. 1969. The formation of the mesoderm in urodele amphibians. I. Induction by the endoderm. *Wilhelm Roux Arch. Entwicklungsmech. Org.* 162: 341–373.

Nieuwkoop, P. D. 1973. The "organisation center" of the amphibian embryo: Its origin, spatial organisation and morphogenetic action. *Adv. Morphogenet.* 10: 1–310.

Nieuwkoop, P. D. 1977. Origin and establishment of embryonic polar axes in amphibian development. *Curr. Top. Dev. Biol.* 11: 115–132.

Otte, A. P. and R. T. Moon. 1992. Protein kinase C isozymes have distinct roles in neural induction and competence in *Xenopus*. *Cell* 68: 1021–1029.

Papalopulu, N., J. D. W. Clarke, L. Bradley, D. Wilkinson, R. Krumlauf and N. Holder. 1991. Retinoic acid causes abnormal development and segmental patterning of the anterior hindbrain in *Xenopus* embryos. *Development* 113: 1145–1158.

Piccolo, S., Y. Sasai, B. Lu and E. M. De Robertis. 1996. Dorsoventral patterning in *Xenopus*: Inhibition of ventral signals by direct binding of chordin to BMP-4. *Cell* 86: 589–598.

Piccolo, S., E. Agius, L. Leyns, S. Bhattacharyya, H. Grunz, T. Bouwmeester and E. M. DeRobertis. 1999. The head inducer Cerberus is a multifunctional antagonist of Nodal, BMP, and Wnt signals. *Nature* 397: 707–710.

Pierce, S. B. and D. Kimelman. 1995. Regulation of Spemann organizer formation by the intracellular kinase Xgsk-3. *Development* 121: 755–765.

Pownall, M. E., A. S. Tucker, J. M. W. Slack and H. V. Isaacs. 1996. eFGF, Xcad3 and Hox genes form a molecular pathway that establishes the anteroposterior axis in *Xenopus*. *Development* 122: 3881–3892.

Purcell, S. M. and R. Keller. 1993. A different type of amphibian mesoderm morphogenesis in *Ceratophrys ornata*. *Development* 117: 307–317.

Recanzone, G. and W. A. Harris. 1985. Demonstration of neural induction using nuclear markers in *Xenopus*. *Wilhelm Roux Arch. Dev. Biol.* 194: 344–354.

Ruiz i Altaba, A. and T. Jessell. 1991. Retinoic acid modifies mesodermal patterning in early *Xenopus* embryos. *Genes Dev.* 5: 175–187.

Ryan, A. and 14 others. 1998. Pitx2 determines left-right asymmetries in vertebrates. *Nature* 394: 54–55.

Sasai, Y. 1998. Identifying the missing links: Genes that connect neural induction and primary neurogenesis in vertebrates. *Neuron* 21: 455–458.

Sasai, Y., B. Lu, H. Steinbeisser, D. Geissert, L. K. Gont and E. M. De Robertis. 1994. *Xenopus* chordin: A novel dorsalizing factor activated by organizer-specific homeobox genes. *Cell* 79: 779–790.

Sasai, Y., B. Lu, S. Piccolo and E. M. De Robertis. 1996. Endoderm induction by the organizer-secreted factors chordin and noggin in *Xenopus* animal caps. *EMBO J.* 15: 4547–4555.

Sato, S. M. and T. D. Sargent. 1989. Development of neural inducing capacity in dissociated *Xenopus* embryos. *Dev. Biol.* 134: 263–366.

Saxén, L. 1961. Transfilter neural induction of amphibian ectoderm. *Dev. Biol.* 3: 140–152.

Saxén, L. and S. Toivonen. 1962. *Embryonic Induction.* Prentice-Hall, Englewood Cliffs, NJ.

Schmidt, J., V. Francoise, E. Bier and D. Kimelman. 1995. *Drosophila* short gastrulation induces an ectopic axis in *Xenopus*: Evidence for conserved mechanisms of dorsoventral patterning. *Development* 121: 4319–4328.

Schneider, S., H. Steinbeisser, R. M. Warga and P. Hausen. 1996. β-catenin translocation into nuclei demarcates the dorsalizing centers in frog and fish embryos. *Mech. Dev.* 57: 191–198.

Shapira, E., K. Marom, V. Levy, R. Yelin and A. Fainsod. 2000. The Xvex-1 antimorph reveals the temporal competence for organizer formation and an early role for ventral homeobox genes. *Mech. Dev.* 90: 77–87.

Sharpe, C. R. 1991. Retinoic acid can mimic endogenous signals involved in transformation of the *Xenopus* nervous system. *Neuron* 7: 239–247.

Sharpe, C. R., A. Fritz, E. M. De Robertis and J. B. Gurdon. 1987. A homeobox-containing marker of posterior neural differentiation shows importance of predetermination in neural induction. *Cell* 50: 749–758.

Shi, D.-L., T. D'Arribère, K. E. Johnson and J.-C.Boucaut. 1989. Initiation of mesodermal cell migration and spreading relative to gastrulation in the urodele amphibian *Pleurodeles walti*. *Development* 105: 351–363.

Sive, H. L. and P. F. Cheng. 1991. Retinoic acid perturbs the expression of *Xhox-lab* genes and alters mesodermal determination in *Xenopus laevis*. *Genes Dev.* 5: 1321–1332.

Sive, H. L., B. W. Draper, R. M. Harland and H. Weintraub. 1990. Identification of a retinoic acid sensitive period during primary axis formation in *Xenopus laevis*. *Genes Dev.* 4: 932–942.

Smith, J. C. and G. M. Malacinski. 1983. The origin of the mesoderm in an anuran, *Xenopus laevis*, and a urodele, *Ambystoma mexicanum*. *Dev. Biol.* 98: 250–254.

Smith, J. C. and J. M. W. Slack. 1983. Dorsalization and neural induction: Properties of the organizer in *Xenopus laevis*. *J. Embryol. Exp. Morphol.* 78: 299–317.

Smith, W. C. and R. M. Harland. 1992. Expression cloning of noggin, a new dorsalizing factor localized to the Spemann organizer in *Xenopus* embryos. *Cell* 70: 829–840.

Smith, W. C., A. K. Knecht, M. Wu and R. M. Harland. 1993. Secreted noggin mimics the Spemann organizer in dorsalizing *Xenopus* mesoderm. *Nature* 361: 547–5410.

Smith, W. C., R. McKendry, S. Ribisi and R. M. Harland. 1995. A *nodal*-related gene defines a physical and functional domain within the Spemann organizer. *Cell* 82: 37–46.

Spemann, H. 1903. Entwicklungsphysiologische Studien am Tritonei. III. *Arch. Entwicklungsmech.* 16: 551–631.

Spemann, H. 1918. Über die Determination der ersten Organanlagen des Amphibienembryo. *Wilhelm Roux Arch. Entwicklungsmech. Org.* 43: 448–555.

Spemann, H. 1927. Neue Arbieten über Organisatoren in der tierischen Entwicklung. *Naturwissenschaften* 15: 946–951.

Spemann, H. 1938. *Embryonic Development and Induction.* Yale University Press, New Haven.

Spemann, H. and H. Mangold. 1924. Induction of embryonic primordia by implantation of organizers from a different species. *In* B. H. Willier and J. M. Oppenheimer (eds.), *Foundations of Experimental Embryology.* Hafner, New York, pp. 144–184.

Spofford, W. R. 1945. Observations on the posterior part of the neural plate in *Amblystoma*. *J. Exp. Zool.* 99: 35–52.

Strauss, E. 1998. How embryos shape up. *Science* 281: 166–167.

Suzuki, A., R. S. Thies, N. Yamaji, J. J. Song, J. M. Wozney, K. Muramaki and N. Ueno. 1994. A truncated bone morphogenetic protein receptor affects dorsal-ventral patterning in early *Xenopus* embryo. *Proc. Natl. Acad. Sci. USA* 91: 10255–10259.

Suzuki, A, C. Chang, J. M. Yingling, X.-F. Wang and A. Hemmati-Brivanlou. 1997a. Smad5 induces ventral fates in *Xenopus* embryo. *Dev. Biol.* 184: 402–405.

Suzuki, A., N. Ueno and A. Hemmati-Brivanlou. 1997b. *Xenopus msx1* mediates epidermal induction and neural inhibition by BMP4. *Development* 124: 3037–3044.

Taira, M., J.-P. Saint-Jeannet and I. B. Dawid. 1997. Role of the *Xlim-1* and *Xbra* genes in anteroposterior patterning of neural tissue by the head and trunk organizer. *Proc. Natl. Acad. Sci. USA* 94: 895–900.

Toivonen, S. 1979. Transmission problem in primary induction. *Differentiation* 15: 177–181.

Toivonen, S. and L. Saxén. 1955. The simultaneous inducing action of liver and bone marrow of the guinea pig in implantation and explantation experiments with embryos of *Triturus*. *Exp. Cell Res.* [Suppl.] 3: 346–357.

Toivonen, S. and J. Wartiovaara. 1976. Mechanism of cell interaction during primary induction studied in transfilter experiments. *Differentiation* 5: 61–66.

Toivonen, S., D. Tarin, L. Saxén, P. J. Tarin and J. Wartiovaara. 1975. Transfilter studies on neural induction in the newt. *Differentiation* 4: 1–7.

Twitty, V. C. 1966. *Of Scientists and Salamanders.* Freeman, San Francisco.

Vincent, J. P. and J. C. Gerhart. 1987. Subcortical rotation in *Xenopus* eggs: An early step in embryonic axis specification. *Dev. Biol.* 123: 526–539.

Vincent, J. P., G. F. Oster and J. C. Gerhart. 1986. Kinematics of gray crescent formation in *Xenopus* eggs. Displacement of subcortical cytoplasm relative to the egg surface. *Dev. Biol.* 113: 484–500.

Wang, S., M. Krinks, K. Lin, F. P. Luyten and M. Moos Jr. 1997. Frzb, a secreted protein expressed in the Spemann organizer, binds and inhibits Wnt-8. *Cell* 88: 757–766.

Wilson, P. A. and A. Hemmati-Brivanlou. 1995. Induction of epidermis and inhibition of neural fate by BMP-4. *Nature* 376: 331–333.

Wilson, P. and R. Keller. 1991. Cell rearrangement during gastrulation of *Xenopus*: Direct observation of cultured explants. *Development* 112: 289–300.

Winklbauer, R. and R. E. Keller. 1996. Fibronectin, mesoderm migration, and gastrulation in *Xenopus*. *Dev. Biol.* 177: 413–426.

Winklbauer, R. and M. Schürfeld. 1999. Vegetal rotation, a new gastrulation movement involved in the internalization of the mesoderm and endoderm in *Xenopus*. *Development* 126: 3703–3713.

Wylie, C., M. Kofron, C. Payne, R. Anderson, M. Hosobuchi, E. Joseph and J. Heasman. 1996. Maternal β-catenin establishes a "dorsal signal" in early *Xenopus* embryos. *Development* 122: 2987–2996.

Yost, C., M. Torres, J. R. Miller, E. Huang, D. Kimelman and R. T. Moon. 1996. The axis-inducing ability, stability, and subcellular localization of β-catenin is regulated in *Xenopus* embryos by glycogen synthase kinase-3. *Genes Dev.* 10: 1443–1454.

Yost, H. J. 1998. Left-right development in *Xenopus* and zebrafish. *Semin. Cell Dev. Biol.* 9: 61–66.

Zhang, J., D. W. Houston, M. L. King, C. Payne, C. Wylie and J. Heasman. 1998. The role of maternal VegT in establishing the primary germ layers in *Xenopus* embryos. *Cell* 94: 515–524.

Zimmerman, L. B., J. M. de Jesús Escobar and R. M. Harland. 1996. The Spemann organizer signal noggin binds and inactivates bone morphogenesis protein 4. *Cell* 86: 599–606.

The early development of vertebrates: Fish, birds, and mammals

THIS FINAL CHAPTER ON THE PROCESSES OF EARLY DEVELOPMENT will extend our survey of vertebrate development to include fish, birds, and mammals. The amphibian embryos described in the previous chapter divide by means of radial holoblastic cleavage. Cleavage in bird, reptile, and fish eggs is meroblastic, with only a small portion of the cytoplasm being used to make cells. Mammals modify their holoblastic cleavage to make a placenta, which enables the embryo to develop inside another organism. Although methods of gastrulation also differ among the vertebrate classes, there are some underlying principles in common throughout the vertebrates.

Early Development in Fish

In recent years, the teleost fish *Danio rerio*, known as the zebrafish, has become a favorite organism of those who wish to study vertebrate development. Zebrafish have large broods, breed all year, are easily maintained, have transparent embryos that develop outside the mother (an important feature for microscopy), and can be raised so that mutants can be readily screened and propagated. In addition, they develop rapidly, so that at 24 hours after fertilization, the embryo has formed most of its tissues and organ primordia and displays the characteristic tadpole-like form (see Granato and Nüsslein-Volhard 1996; Langeland and Kimmel 1997). Therefore, much of the description of fish development below is based on studies of this species

WEBSITE **11.1 Gene manipulation in the zebrafish.** Zebrafish can be bred so that they express altered genes rapidly. This technology makes *Danio* an important contributor to developmental studies.

Cleavage in Fish Eggs

In fish eggs, cleavage occurs only in the **blastodisc**, a thin region of yolk-free cytoplasm at the animal cap of the egg. Most of the egg cell is full of yolk. The cell divisions do not completely divide the egg, so this type of cleavage is called **meroblastic** (Greek, *meros*, "part"). Since only the cytoplasm of the blastodisc becomes the embryo, this type of meroblastic cleavage is called **discoidal**. Scanning electron micrographs show beautifully the incomplete nature of discoidal meroblastic cleavage in fish eggs (Figure 11.1). The calcium waves initiated at fertilization stimulate the contraction of the actin cytoskeleton to squeeze non-yolky cytoplasm into the animal pole of the egg. This converts the spherical egg into a more pear-shaped structure, with an apical blastodisc (Leung et al. 1998). Early cleavage divisions follow a highly

339

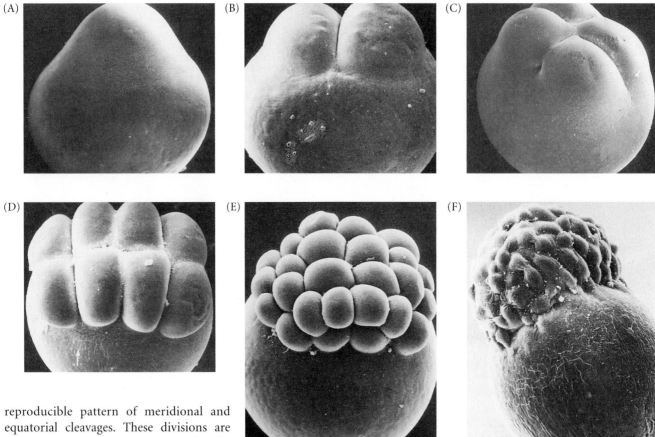

Figure 11.1
Discoidal cleavage in a zebrafish egg. (A) 1-cell embryo. The mound atop the cytoplasm is the blastodisc region. (B) 2-cell embryo. (C) 4-cell embryo. (D) 8-cell embryo, wherein two rows of four cells are formed. (E) 32-cell embryo. (F) 64-cell embryo, wherein the blastodisc can be seen atop the yolk cell. (From Beams and Kessel 1976; photographs courtesy of the authors.)

reproducible pattern of meridional and equatorial cleavages. These divisions are rapid, taking about 15 minutes each. The first 12 divisions occur synchronously, forming a mound of cells that sits at the animal pole of a large **yolk cell**. These cells constitute the **blastoderm**. Initially, all the cells maintain some open connection with one another and with the underlying yolk cell so that moderately sized (17-kDa) molecules can pass freely from one blastomere to the next (Kimmel and Law 1985).

Beginning at about the tenth cell division, the onset of the midblastula transition can be detected: zygotic gene transcription begins, cell divisions slow, and cell movement becomes evident (Kane and Kimmel 1993). At this time, three distinct cell populations can be distinguished. The first of these is the **yolk syncytial layer** (**YSL**). The YSL is formed at the ninth or tenth cell cycle, when the cells at the vegetal edge of the blastoderm fuse with the underlying yolk cell. This fusion produces a ring of nuclei within the part of the yolk cell cytoplasm that sits just beneath the blastoderm. Later, as the blastoderm expands vegetally to surround the yolk cell, some of the yolk syncytial nuclei will move under the blastoderm to form the **internal YSL**, and some of the nuclei will move vegetally, staying ahead of the blastoderm margin, to form **external YSL** (Figure 11.2A,B). The YSL will be important for directing some of the cell movements of gastrulation. The second cell population distinguished at the midblastula tran-

sition is the **enveloping layer** (**EVL**; Figure 11.2A). It is made up of the most superficial cells of the blastoderm, which form an epithelial sheet a single cell layer thick. The EVL eventually becomes the **periderm**, an extraembryonic protective covering that is sloughed off during later development.

Between the EVL and the YSL are the **deep cells**. These are the cells that give rise to the embryo proper. The fates of the early blastoderm cells are not determined, and cell lineage studies (in which a nondiffusible fluorescent dye is injected into one of the cells so that the descendants of that cell can be followed) show that there is much cell mixing during cleavage. Moreover, any one of these cells can give rise to an unpredictable variety of tissue descendants (Kimmel and Warga 1987; Helde et al. 1994). The fate of the blastoderm cells appears to be fixed shortly before gastrulation begins. At this time, cells in specific regions of the embryo give rise to certain tissues in a highly predictable manner, allowing a fate map to be made (Figure 11.2C; see also Figure 1.6; Kimmel et al 1990).

(A)

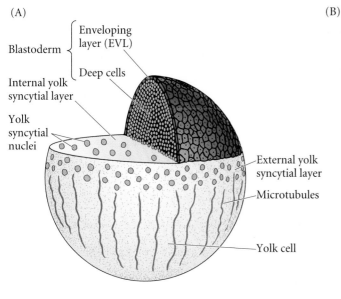

Blastoderm {
Enveloping layer (EVL)
Deep cells

Internal yolk syncytial layer

Yolk syncytial nuclei

External yolk syncytial layer

Microtubules

Yolk cell

(B)

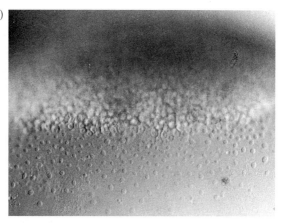

Figure 11.2
Fish blastula. (A) Prior to gastrulation, the deep cells are surrounded by the EVL. The animal surface of the yolk cell is flat and contains the nuclei of the YSL. Microtubules extend through the yolky cytoplasm and through the external region of the YSL. (B) Late-blastula stage embryo of the minnow *Fundulus*, showing the external YSL. The nuclei of these cells were derived from cells at the margin of the blastoderm, which released their nuclei into the yolky cytoplasm. (C) Fate map of the deep cells after cell mixing has stopped. The lateral view is shown, and not all organ fates are labeled (for the sake of clarity). (A and C after Langeland and Kimmel 1997; B from Trinkaus 1993, photograph courtesy of the author.)

(C)

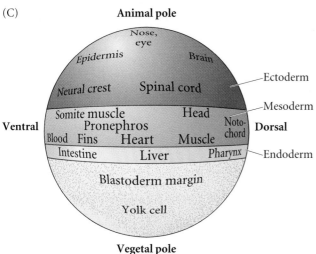

Animal pole

Nose, eye
Epidermis
Brain
Neural crest
Spinal cord — Ectoderm
Somite muscle
Head — Mesoderm
Ventral
Pronephros
Noto-chord **Dorsal**
Blood Fins Heart Muscle
Intestine Liver Pharynx — Endoderm
Blastoderm margin
Yolk cell

Vegetal pole

Gastrulation in Fish Embryos

The first cell movement of fish gastrulation is the epiboly of the blastoderm cells over the yolk. In the initial phase, the deep blastoderm cells move outwardly to intercalate with the more superficial cells (Warga and Kimmel 1990). Later, these cells move over the surface of the yolk to envelop it completely (Figure 11.3). This movement is not due to the active crawling of the blastomeres, however. Rather, the movement is provided by the autonomously expanding YSL "within" the animal pole yolk cytoplasm. The EVL is tightly joined to the YSL and is dragged along with it. The deep cells of the blastoderm then fill in the space between the YSL and the EVL as epiboly proceeds. This can be demonstrated by severing the attachments between the YSL and the EVL. When this is done, the EVL and deep cells spring back to the top of the yolk, while the YSL continues its expansion around the yolk cell (Trinkaus 1984, 1992). The expansion of the YSL depends on a network

of microtubules in the YSL, and radiation or drugs that block the polymerization of tubulin inhibit epiboly (Strahle and Jesuthasan 1993; Solnica-Krezel and Driever 1994).

During migration, one side of the blastoderm becomes noticeably thicker than the other. Cell-labeling experiments indicate that the thicker side marks the site of the future dorsal surface of the embryo (Schmidt and Campos-Ortega 1995).

The formation of germ layers

After the blastoderm cells have covered about half the zebrafish yolk cell (and earlier in fish eggs with larger yolks), a thickening occurs throughout the margin of the epibolizing blastoderm. This thickening is called the **germ ring**, and it is composed of a superficial layer, the **epiblast**, and an inner layer, the **hypoblast**. We do not understand how the hypoblast is made. Some research groups claim that the hypoblast is formed by the *involution* of superficial cells under the margin followed by their migration toward the animal pole (see Figure 11.3C). The involution begins at the future dorsal portion of the embryo, but occurs all around the margin. Other laboratories claim that these superficial cells *ingress* to form the hypoblast (see Trinkaus 1996). It is possible that both mechanisms are at work, with different modes of hypoblast formation predominating in different species. Once formed, however, the cells of both the epiblast and hypoblast intercalate on the future dorsal side of the embryo to form a localized thickening, the **embryonic shield** (Figure 11.4). As we will see, this shield is functionally equivalent to the dorsal blastopore

(A)

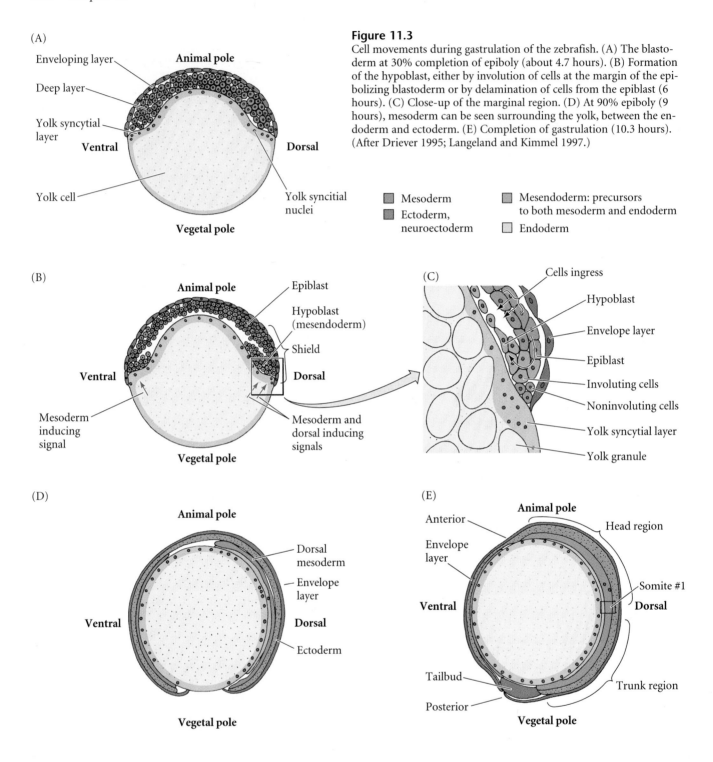

Animal pole

Enveloping layer

Deep layer

Yolk syncytial layer

Ventral

Yolk cell

Dorsal

Yolk syncitial nuclei

Vegetal pole

Figure 11.3
Cell movements during gastrulation of the zebrafish. (A) The blastoderm at 30% completion of epiboly (about 4.7 hours). (B) Formation of the hypoblast, either by involution of cells at the margin of the epibolizing blastoderm or by delamination of cells from the epiblast (6 hours). (C) Close-up of the marginal region. (D) At 90% epiboly (9 hours), mesoderm can be seen surrounding the yolk, between the endoderm and ectoderm. (E) Completion of gastrulation (10.3 hours). (After Driever 1995; Langeland and Kimmel 1997.)

■ Mesoderm

■ Ectoderm, neuroectoderm

■ Mesendoderm: precursors to both mesoderm and endoderm

□ Endoderm

(B)

Animal pole

Epiblast

Hypoblast (mesendoderm)

Shield

Ventral

Dorsal

Mesoderm inducing signal

Mesoderm and dorsal inducing signals

Vegetal pole

(C)

Cells ingress

Hypoblast

Envelope layer

Epiblast

Involuting cells

Noninvoluting cells

Yolk syncytial layer

Yolk granule

(D)

Animal pole

Dorsal mesoderm

Envelope layer

Ventral

Dorsal

Ectoderm

Vegetal pole

(E)

Animal pole

Anterior

Head region

Envelope layer

Ventral

Somite #1

Dorsal

Tailbud

Trunk region

Posterior

Vegetal pole

lip of amphibians, since it can organize a secondary embryonic axis when transplanted to a host embryo (Oppenheimer 1936; Ho 1992). Thus, as the cells undergo epiboly around the yolk, they are also involuting at the margins and converging anteriorly and dorsally toward the embryonic shield (Trinkaus 1992). The hypoblast cells of the embryonic shield converge and extend anteriorly, eventually narrowing along the dorsal midline of the hypoblast. This movement forms the **chordamesoderm**, the precursor of the **notochord** (Figure 11.4B,C). The cells adjacent to the chordamesoderm, the

paraxial mesoderm cells, are the precursors of the mesodermal somites (Figure 11.4D,E). The concomitant convergence and extension in the epiblast brings the presumptive neural cells from all over the epiblast into the dorsal midline, where

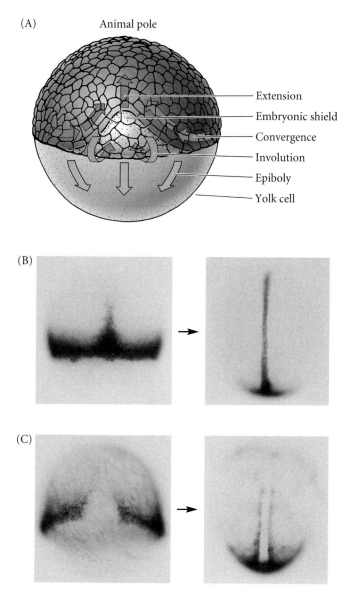

Figure 11.4
Convergence and extension in the zebrafish gastrula. (A) Dorsal view of the convergence and extension movements during zebrafish gastrulation. Epiboly spreads the blastoderm over the yolk; involution or ingression generates the hypoblast; convergence and extension bring the hypoblast and epiblast cells to the dorsal side to form the embryonic shield. Within the shield, intercalation extends the chordamesoderm toward the animal pole. (B) Convergent extension of chordamesoderm is shown by those cells expressing the gene *no tail*, a gene that is expressed by notochord cells. (C) Convergent extension of paraxial mesodermal cells (marked by their expression of the *snail* gene) to flank the notochord. (From Langeland and Kimmel 1997; photographs courtesy of the authors.)

they form the **neural keel**. The rest of the epiblast becomes the skin of the fish. The zebrafish fate map, then, is not much different from that of the frog or other vertebrates (as we will soon see). If one conceptually opens a *Xenopus* blastula at the vegetal pole and stretches the opening into a marginal ring, the resulting fate map closely resembles that of the zebrafish embryo at the stage when half of the yolk has been covered by the blastoderm (see Figure 1.9; Langeland and Kimmel 1997).

Axis Formation in Fish Embryos

Dorsal-ventral axis formation: The embryonic shield

The embryonic shield is critical in establishing the dorsal-ventral axis in fishes. It can convert lateral and ventral mesoderm (blood and connective tissue precursors) into dorsal mesoderm (notochord and somites), and it can cause the ectoderm to become neural rather than epidermal. This was shown by transplantation experiments in which the embryonic shield of one early-gastrula embryo was transplanted to the ventral side of another (Figure 11.5; Oppenheimer 1936; Koshida et al. 1998). Two axes formed, sharing a common yolk cell. Although the prechordal plate and the notochord were derived from the donor embryonic shield, the other organs of the secondary axis came from host tissues that would have formed ventral structures. The new axis has been induced by the donor cells. In the embryo that had had its embryonic shield removed, no dorsal structures formed, and the embryo lacked a nervous system. These experiments are similar to those performed on amphibian gastrulae by Spemann and Mangold (1924; see Chapter 10), and they demonstrate that the embryonic shield is the homologue of the dorsal blastopore lip, the amphibian organizer.

Like the amphibian dorsal blastopore lip, the embryonic shield forms the prechordal plate and the notochord of the developing embryo. The precursors of these two regions are responsible for inducing the ectoderm to become neural ectoderm. Moreover, the presumptive notochord and prechordal plate appear to do this in a manner very much like that of their homologous structures in amphibians.* In both fishes and amphibians, BMP proteins made in the ventral and lateral regions of the embryo would normally cause the ectoderm to become epidermis. The notochord of both fishes and amphibians secretes factors that block this induction and thereby allow the ectoderm to become neural. In fishes, the BMP that ventralizes the embryo is **BMP2B**. The protein secreted by the chordame-

*Another similarity between the amphibian and fish organizers is that they can be duplicated by rotating the egg and changing the orientation of the microtubules (Fluck et al. 1998). One difference in the axial development of these groups is that in amphibians (see Chapter 10), the prechordal plate is necessary for inducing the anterior brain to form. In *Danio*, the prechordal plate appears to be necessary for forming ventral neural structures, but the anterior regions of the brain can form in its absence (Schier et al. 1997; Schier and Talbot 1998).

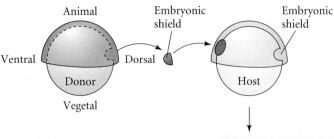

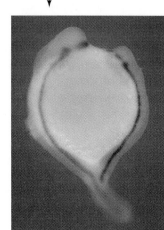

Figure 11.5
The embryonic shield as organizer in the fish embryo. A donor embryonic shield (about 100 cells from a stained embryo) is transplanted into a host embryo at the same early-gastrula stage. The result is two embryonic axes joined to the host's yolk cell. In the photograph, both axes have been stained for *sonic hedgehog* mRNA, which is expressed in the ventral midline. (The embryo to the right is the secondary axis.) (After Shinya et al. 1999; photograph courtesy of the authors.)

soderm that binds with and inactivates BMP2B is a chordin-like paracrine factor called **Chordino** (Figure 11.6B; Kishimoto et al. 1997; Schulte-Merker et al. 1997). If the *chordino* gene is mutated, the neural tube fails to form. It is hypothesized (Nguyen et al. 1998) that different concentrations of BMP2B pattern the ventral and lateral regions of the zebrafish ectoderm and mesoderm, and that the ratio between Chordino and BMP2B may specify the position along the dorsal-ventral axis. In fishes, however, the notochord may not be the only structure capable of producing the proteins that block BMP2B. If the notochord fails to form (as in the *floating head* or *no tail* mutations), the neural tube will still be produced. It is possible that the notochordal precursor cells (which are produced in these mutations) are still able to induce the neural tube, or that the dorsal portion of the somite precursors can compensate for the lack of a notochord (Halpern et al. 1993; 1995; Hammerschmidt et al. 1996).

The embryonic shield appears to acquire its organizing ability in much the same way as its amphibian counterparts. In amphibians, the endoderm cells beneath the dorsal blastopore lip (i.e., the Nieuwkoop center) accumulate β-catenin. This protein is critical in amphibians for the ability of the endoderm to induce the cells above them to become the dorsal lip (organizer) cells. In zebrafish, the nuclei in that part of the yolk syncytial layer that lies beneath the cells that will become the embryonic shield similarly accumulate β-catenin. This protein distinguishes the dorsal YSL from the lateral and ventral YSL regions (Figure 11.7; Schneider et al. 1996). Inducing β-catenin accumulation on the ventral side of the egg causes dorsalization and a second embryonic axis (Kelly et al. 1995). In addition, just prior to gastrulation, the cells of the dorsal blastopore margin synthesize and se-

Figure 11.6
Axis formation in the zebrafish embryo. (A) Prior to gastrulation, the zebrafish blastoderm is arranged with the presumptive ectoderm near the animal pole, the presumptive mesoderm beneath it, and the presumptive endoderm sitting atop the yolk cell. The yolk syncitial layer(and possibly the endoderm) sends two signals to the presumptive mesoderm. One signal (lighter arrows) induces the mesoderm, while a second signal (heavy arrow) specifically induces an area of mesoderm to become the dorsal mesoderm (embryonic shield). (B) Formation of the dorsal-ventral axis. During gastrulation, the ventral mesoderm secretes BMP2B (arrows) to induce the ventral and lateral mesodermal and epidermal differentiation. The dorsal mesoderm secretes factors (such as Chordino) that block BMP2B and dorsalize the mesoderm and ectoderm (converting the latter into neural tissue). (C) Recent studies have identified two signaling centers for anterior-posterior polarity, one (1) at the border of the neural and non-neural ectoderm, which induces anterior neural cell types, and the other (2) at the lateral margin, which generates a posteriorizing signal. (After Schier and Talbot 1998.)

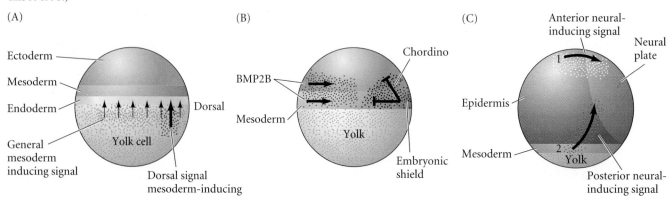

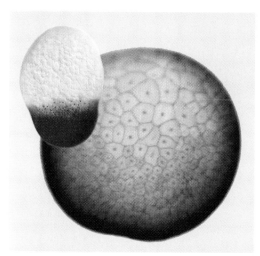

Figure 11.7
Nuclear localization of β-catenin marks the dorsal side of the *Xenopus* blastula (the larger one) and helps form its Nieuwkoop center beneath the organizer. In the zebrafish late blastula, nuclear localization of β-catenin is seen in the yolk syncytial layer nuclei beneath the future embryonic shield. (Photograph courtesy of S. Schneider.)

W E B S I T E **11.2 The fish Hox complexes.** In birds and mammals, Hox genes play important roles in the identity of different parts of the body. Studies of fishes show that the Hox genes may similarly specify somite identity along the anterior-posterior axis. These studies also show important evolutionary changes that may have led to the evolution of the vertebrates.

Left-right axis formation

Little is known about left-right axis formation in fishes. In zebrafish, a gene called ***nodal-related-2*** is expressed solely in the left side of the lateral plate mesoderm* (Rebagliati et al. 1998). Nodal-related-2 is a paracrine factor of the TGF-β family. As we will see later in this chapter, proteins of the TGF-β family are critical in the establishment of the left-right axis throughout the vertebrate classes.

crete Nodal-related proteins. These induce the precursors of the notochord and prechordal plate to activate *goosecoid* and other genes (Sampath et al. 1998; Gritsman et al. 2000.) Thus, the embryonic shield is considered equivalent to the amphibian organizer, and the dorsal part of the yolk cell can be thought of as the Nieuwkoop center of the fish embryo.

Anterior-posterior axis formation: Two signaling centers

As is evident from Figure 11.5, when a second dorsal-ventral axis is experimentally induced in zebrafish eggs, both the regular and the induced axes have the same anterior-posterior polarity. Both heads are at the former animal cap, and both tails are located vegetally. Indeed, the anterior-posterior axis is specified during oogenesis, and the animal cap marks the anterior of the embryo. This axis becomes stabilized during gastrulation through two distinct signaling centers. First, a small group of anterior neural cells at the border between the neural and surface ectoderm (a region that become the pituitary gland, nasal placode, and anterior forebrain) secrete compounds that cause anterior development. If these anterior neural cells are experimentally placed more posteriorly in the embryo, they will cause the neural cells near them to assume the characteristics of forebrain neurons. The second signaling center, in the posterior of the embryo, consists of lateral mesendoderm precursors at the margin of the gastrulating blastoderm. These cells produce caudalizing compounds , most likely Nodal-related proteins and activin (Figure 11.6C; Woo and Fraser 1997; Houart et al. 1998; Thisse et al. 2000). If transplanted adjacent to anterior neural ectoderm, this tissue will transform the presumptive forebrain tissue into hindbrain-like structures.

Early Development in Birds

Cleavage in Bird Eggs

Ever since Aristotle first followed its 3-week development, the domestic chicken has been a favorite organism for embryological studies. It is accessible all year and is easily raised. Moreover, at any particular temperature, its developmental stage can be accurately predicted. Thus, large numbers of embryos can be obtained at the same stage. The chick embryo can be surgically manipulated and, since it forms most of its organs in ways very similarly to those of mammals, it has often served as a surrogate for human embryos.

Fertilization of the chick egg occurs in the oviduct, before the albumen and the shell are secreted upon it. The egg is telolecithal (like that of the fish), with a small disc of cytoplasm sitting atop a large yolk. Like fish eggs, the yolky eggs of birds undergo discoidal meroblastic cleavage. Cleavage occurs only in the blastodisc, a small disc of cytoplasm 2–3 mm in diameter at the animal pole of the egg cell. The first cleavage furrow appears centrally in the blastodisc, and other cleavages follow to create a single-layered blastoderm (Figure 11.8). As in the fish embryo, these cleavages do not extend into the yolky cytoplasm, so the early-cleavage cells are continuous with each other and with the yolk at their bases (Figure 11.8E). Thereafter, equatorial and vertical cleavages divide the blastoderm into a tissue five to six cell layers thick. These cells become linked together by tight junctions (Bellairs et al. 1975; Eyal-Giladi 1991). Between the blastoderm and the yolk is a space called the **subgerminal cavity** (Figure 11.9A). This space is created when the blastoderm cells absorb fluid from the albumin ("egg white") and secrete it between themselves and the yolk (New 1956). At this stage, the deep cells in the

---

*The lateral plate mesoderm, discussed in Chapter 15, is the region that forms the heart and body cavities.

Figure 11.8

Discoidal meroblastic cleavage in a chick egg. (A–D) Four stages viewed from the animal pole (the future dorsal side of the embryo). (E) An early-cleavage embryo viewed from the side. (After Bellairs et al. 1978.)

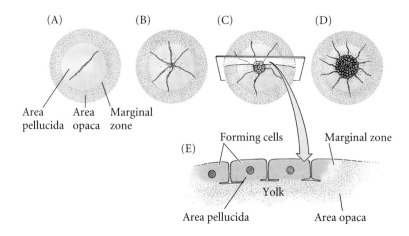

(A)   (B)   (C)   (D)

Area pellucida   Area opaca   Marginal zone

(E) Forming cells   Marginal zone

Yolk

Area pellucida   Area opaca

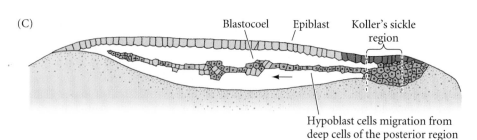

(A)

Blastoderm

Anterior   Epiblast   Posterior marginal zone

Subgerminal space

Area opaca

Yolk

(B)

Hypoblast cells delaminating from epiblast

Area pellucida

Area opaca

(C)

Blastocoel   Epiblast   Koller's sickle region

Hypoblast cells migration from deep cells of the posterior region

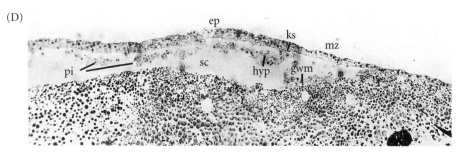

(D)

ep   ks   mz

pi   sc   hyp   gwm

Figure 11.9

Formation of the two-layered blastoderm of the chick embryo. (A, B) Primary hypoblast cells delaminate individually to form islands of cells beneath the epiblast. (C) Secondary hypoblast cells from the posterior margin (Koller's sickle and the posterior marginal cells behind it) migrate beneath the epiblast and incorporate the polyinvagination islands. As the hypoblast moves anteriorly, epiblast cells collect at the region anterior to Koller's sickle to form the primitive streak. (D) This sagittal section of an embryo near the posterior margin shows an upper layer consisting of a central epiblast that trails into the cells of Koller's sickle (ks) and the posterior marginal zone (mz). Certain cells have delaminated from the epiblast (ep) to form polyinvagination islands (pi) of 5 to 20 cells each. These cells will be joined by those hypoblast cells (hyp) migrating anteriorly from Koller's sickle to form the lower (secondary hypoblastic) layer. (sc, subgerminal cavity; gwm, germ wall margin.) (From Eyal-Giladi et al. 1992, photograph courtesy of H. Eyal-Giladi.)

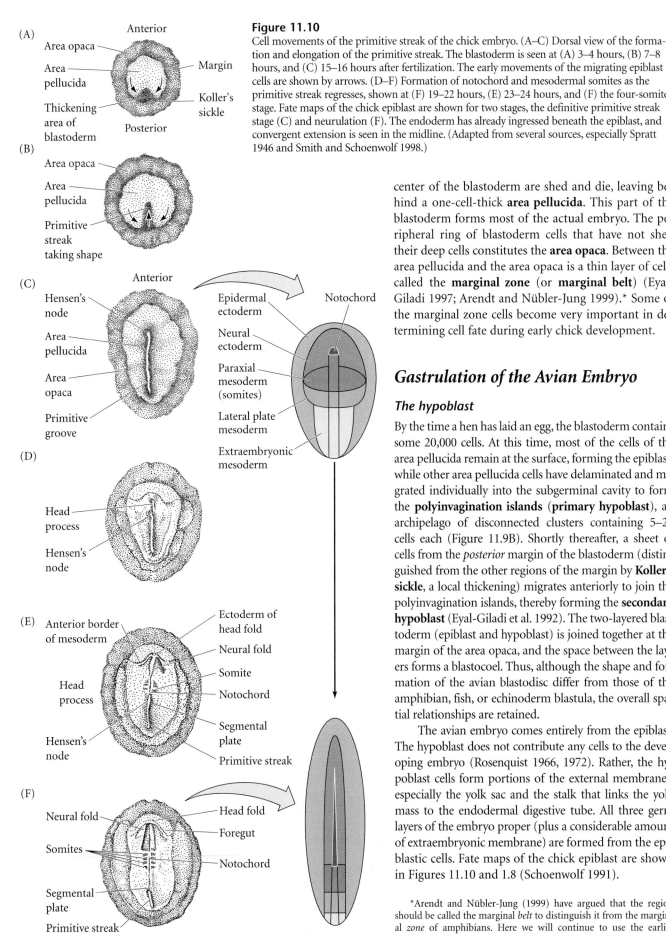

(A)
Anterior
Area opaca
Area pellucida
Margin
Koller's sickle
Thickening area of blastoderm
Posterior

(B)
Area opaca
Area pellucida
Primitive streak taking shape

(C)
Anterior
Hensen's node
Area pellucida
Area opaca
Primitive groove

Epidermal ectoderm
Neural ectoderm
Paraxial mesoderm (somites)
Lateral plate mesoderm
Extraembryonic mesoderm
Notochord

(D)
Head process
Hensen's node

(E)
Anterior border of mesoderm
Head process
Hensen's node
Ectoderm of head fold
Neural fold
Somite
Notochord
Segmental plate
Primitive streak

(F)
Neural fold
Somites
Segmental plate
Primitive streak
Head fold
Foregut
Notochord

Figure 11.10
Cell movements of the primitive streak of the chick embryo. (A–C) Dorsal view of the formation and elongation of the primitive streak. The blastoderm is seen at (A) 3–4 hours, (B) 7–8 hours, and (C) 15–16 hours after fertilization. The early movements of the migrating epiblast cells are shown by arrows. (D–F) Formation of notochord and mesodermal somites as the primitive streak regresses, shown at (F) 19–22 hours, (E) 23–24 hours, and (F) the four-somite stage. Fate maps of the chick epiblast are shown for two stages, the definitive primitive streak stage (C) and neurulation (F). The endoderm has already ingressed beneath the epiblast, and convergent extension is seen in the midline. (Adapted from several sources, especially Spratt 1946 and Smith and Schoenwolf 1998.)

center of the blastoderm are shed and die, leaving behind a one-cell-thick **area pellucida**. This part of the blastoderm forms most of the actual embryo. The peripheral ring of blastoderm cells that have not shed their deep cells constitutes the **area opaca**. Between the area pellucida and the area opaca is a thin layer of cells called the **marginal zone** (or **marginal belt**) (Eyal-Giladi 1997; Arendt and Nübler-Jung 1999).* Some of the marginal zone cells become very important in determining cell fate during early chick development.

Gastrulation of the Avian Embryo

The hypoblast

By the time a hen has laid an egg, the blastoderm contains some 20,000 cells. At this time, most of the cells of the area pellucida remain at the surface, forming the epiblast, while other area pellucida cells have delaminated and migrated individually into the subgerminal cavity to form the **polyinvagination islands** (**primary hypoblast**), an archipelago of disconnected clusters containing 5–20 cells each (Figure 11.9B). Shortly thereafter, a sheet of cells from the *posterior* margin of the blastoderm (distinguished from the other regions of the margin by **Koller's sickle**, a local thickening) migrates anteriorly to join the polyinvagination islands, thereby forming the **secondary hypoblast** (Eyal-Giladi et al. 1992). The two-layered blastoderm (epiblast and hypoblast) is joined together at the margin of the area opaca, and the space between the layers forms a blastocoel. Thus, although the shape and formation of the avian blastodisc differ from those of the amphibian, fish, or echinoderm blastula, the overall spatial relationships are retained.

The avian embryo comes entirely from the epiblast. The hypoblast does not contribute any cells to the developing embryo (Rosenquist 1966, 1972). Rather, the hypoblast cells form portions of the external membranes, especially the yolk sac and the stalk that links the yolk mass to the endodermal digestive tube. All three germ layers of the embryo proper (plus a considerable amount of extraembryonic membrane) are formed from the epiblastic cells. Fate maps of the chick epiblast are shown in Figures 11.10 and 1.8 (Schoenwolf 1991).

*Arendt and Nübler-Jung (1999) have argued that the region should be called the marginal *belt* to distinguish it from the marginal *zone* of amphibians. Here we will continue to use the earlier nomenclature.

The primitive streak

The major structural characteristic of avian, reptilian, and mammalian gastrulation is the **primitive streak**. This streak is first visible as a thickening of the epiblast at the posterior region of the embryo, just anterior to Koller's sickle (Figure 11.10A). This thickening is caused by the ingression of endodermal precursors from the epiblast into the blastocoel and by the migration of cells from the lateral region of the posterior epiblast toward the center (Figure 11.10B; Vakaet 1984; Bellairs 1986; Eyal-Giladi et al. 1992). As these cells enter the primitive streak, the streak elongates toward the future head region. At the same time, the secondary hypoblast cells continue to migrate anteriorly from the posterior margin of the blastoderm. The elongation of the primitive streak appears to be coextensive with the anterior migration of these secondary hypoblast cells. The streak eventually extends 60–75% of the length of the area pellucida.

The primitive streak defines the axes of the embryo. It extends from *posterior* to *anterior*; migrating cells enter through its *dorsal* side and move to its *ventral* side; and it separates the *left* portion of the embryo from the *right*. Those elements close to the streak will be the **medial** (central) structures, while those farther from it will be the **distal** (lateral) structures (Figure 11.10C–E).

As cells converge to form the primitive streak, a depression forms within the streak. This depression is called the **primitive groove**, and it serves as an opening through which migrating cells pass into the blastocoel. Thus, the primitive groove is analogous to the amphibian blastopore. At the anterior end of the primitive streak is a regional thickening of cells called the **primitive knot** or **Hensen's node**. The center of this node contains a funnel-shaped depression (sometimes called the **primitive pit**) through which cells can pass into the blastocoel. Hensen's node is the functional equivalent of the dorsal lip of the amphibian blastopore (i.e., the organizer) and the fish embryonic shield.

As soon as the primitive streak has formed, epiblast cells begin to migrate through it and into the blastocoel (Figure 11.11). The primitive streak has a continually changing cell population. Cells migrating through Hensen's node pass down into the blastocoel and migrate anteriorly, forming foregut, head mesoderm, and notochord; cells passing through the lateral portions of the primitive streak give rise to the majority of endodermal and mesodermal tissues (Schoenwolf et al. 1992). Unlike the *Xenopus* mesoderm, which migrates as sheets of cells into the blastocoel, cells entering the inside of the avian embryo ingress as individuals after undergoing an epithelial-to-mesenchymal transformation. At Hensen's node and throughout the primitive streak, the breakdown of the basal lamina and the release of these cells into the embryo is thought to be accomplished by **scatter factor**, a 190-kDa protein secreted by the cells as they enter the streak (Stern et al. 1990;

(A)

(B)

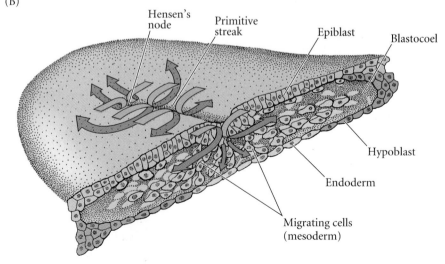

Figure 11.11
Migration of endodermal and mesodermal cells through the primitive streak. (A) Scanning electron micrograph shows epiblast cells passing into the blastocoel and extending their apical ends to become bottle cells. (B) Stereogram of a gastrulating chick embryo, showing the relationship of the primitive streak, the migrating cells, and the two original layers of the blastoderm. The lower layer becomes a mosaic of hypoblast and endodermal cells; the hypoblast cells eventually sort out to form a layer beneath the endoderm and contribute to the yolk sac. (A from Solursh and Revel 1978, courtesy of M. Solursh; B after Balinsky 1975.)

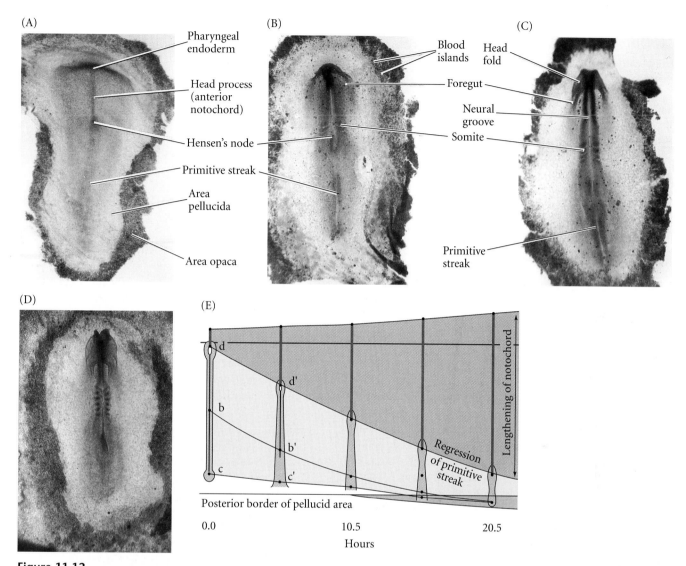

Figure 11.12
Chick gastrulation from about 24 to about 28 hours. (A) The primitive streak at full extension (24 hours). The head process (anterior notochord) can be seen extending from Hensen's node. (B) Two-somite stage (25 hours). Pharyngeal endoderm is seen anteriorly, while the anterior notochord pushes up the head process beneath it. The primitive streak is regressing. (C) Four-somite stage (27 hours). (D) At 28 hours, the primitive streak has regressed to the caudal portion of the embryo. (E) Regression of the primitive streak, leaving the notochord in its wake. Various points of the streak (represented by letters) were followed after it achieved its maximum length. Time represents hours after achieving maximum length (the reference line, about 18 hours after incubation). (Photographs courtesy of K. Linask; E after Spratt 1947.)

DeLuca et al. 1999). Scatter factor can convert epithelial sheets into mesenchymal cells in several ways, and it is probably involved both in downregulating E-cadherin expression and in preventing E-cadherin from functioning

MIGRATION THROUGH THE PRIMITIVE STREAK: FORMATION OF ENDODERM AND MESODERM. The first cells to migrate through Hensen's node are those destined to become the pharyngial endoderm of the foregut. Once inside the blastocoel, these endodermal cells migrate anteriorly and eventually displace the hypoblast cells, causing the hypoblast cells to be confined to a region in the anterior portion of the area pellucida. This region, the **germinal crescent**, does not form any embryonic structures, but it does contain the precursors of the germ cells, which later migrate through the blood vessels to the gonads (see Chapter 19). The next cells entering the blastocoel through Hensen's node also move anteriorly, but they do not move as far ventrally as the presumptive foregut endodermal cells. Rather, they remain between the endoderm and the epiblast to form the **head mesenchyme** and the **prechordal plate mesoderm** (see Psychoyos and Stern 1996). These early-ingressing cells all move anteriorly, pushing up the anterior midline region of the epiblast to form the **head process** (Figure 11.12). Thus, the head of the avian embryo forms anterior (**rostral**) to Hensen's node. The next cells migrating through Hensen's node become chordamesoderm (notochord) cells. These cells extend up to the presumptive midbrain, where they meet the prechordal plate. The hindbrain and trunk form from the chordamesoderm at the level of Hensen's node and caudal to it.

Meanwhile, cells continue migrating inwardly through the lateral portion of the primitive streak. As they enter the blastocoel, these cells separate into two layers. The deep layer joins the hypoblast along its midline and displaces the hypoblast cells to the sides. These deep-moving cells give rise to all the endodermal organs of the embryo as well as to most of the extraembryonic membranes (the hypoblast forms the rest). The second migrating layer spreads between this endoderm and the epiblast, forming a loose layer of cells. These middle layer cells generate the mesodermal portions of the embryo and extraembryonic membranes. By 22 hours of incubation, most of the presumptive endodermal cells are in the interior of the embryo, although presumptive mesodermal cells continue to migrate inward for a longer time.

REGRESSION OF THE PRIMITIVE STREAK. Now a new phase of gastrulation begins. While mesodermal ingression continues, the primitive streak starts to regress, moving Hensen's node from near the center of the area pellucida to a more posterior position (see Figure 11.12). It leaves in its wake the dorsal axis of the embryo and the notochord. As the node moves posteriorly, the notochord is laid down, starting at the level of the future midbrain. While the anterior portion of the notochord is formed by the ingression of cells through Hensen's node, the posterior notochord (after somite 17 in the chick) forms from the condensation of mesodermal tissue that has ingressed through the primitive streak (i.e., not through Hensen's node). This portion of the notochord extends posteriorly to form the tail of the embryo (Le Douarin et al. 1996). Finally, Hensen's node regresses to its most posterior position, forming the anal region. At this time, all the presumptive endodermal and mesodermal cells have entered the embryo, and the epiblast is composed entirely of presumptive ectodermal cells.

As a consequence of the sequence by which the head mesoderm and notochord are established, avian (and mammalian) embryos exhibit a distinct anterior-to-posterior gradient of developmental maturity. While cells of the posterior portions of the embryo are undergoing gastrulation, cells at the anterior end are already starting to form organs (see Darnell et al. 1999). For the next several days, the anterior end of the embryo is more advanced in its development (having had a "head start," if you will) than the posterior end.

Epiboly of the ectoderm

While the presumptive mesodermal and endodermal cells are moving inward, the ectodermal precursors proliferate. Moreover, the ectodermal cells migrate to surround the yolk by epiboly. The enclosure of the yolk by the ectoderm (again

reminiscent of the epiboly of amphibian ectoderm) is a Herculean task that takes the greater part of 4 days to complete. It involves the continuous production of new cellular material and the migration of the presumptive ectodermal cells along the underside of the vitelline envelope (New 1959; Spratt 1963). Interestingly, only the cells of the outer edge of the area opaca attach firmly to the vitelline envelope. These cells are inherently different from the other blastoderm cells, as they can extend enormous (500 µm) cytoplasmic processes onto the vitelline envelope. These elongated filopodia are believed to be the locomotor apparatus of these marginal cells, by which they pull the other ectodermal cells around the yolk (Schlesinger 1958). The filopodia appear to bind to fibronectin, a laminar protein that is a component of the chick vitelline envelope. If the contact between the marginal cells and the fibronectin is experimentally broken (by adding a soluble polypeptide similar to fibronectin), the filopodia retract, and epidermal migration ceases (Lash et al. 1990).

Thus, as avian gastrulation draws to a close, the ectoderm has surrounded the yolk, the endoderm has replaced the hypoblast, and the mesoderm has positioned itself between these two regions. We have identified many of the processes involved in avian gastrulation, but we remain ignorant as to the mechanisms by which many of these processes are carried out.

WEBSITE **11.3 Epiblast cell heterogeneity.** Although the early epiblast appears uniform, different cells have different molecules on their cell surfaces. This allows some of them to remain in the epiblast while others migrate into the embryo.

VADE MECUM **Chick development.** You can view movies of 3-D models of chick cleavage and gastrulation. **[Click on Chick-Early]**

Axis Formation in the Chick Embryo

While the formation of the body axes is accomplished during gastrulation, their specification occurs earlier, during the cleavage stage.

The role of pH in forming the dorsal-ventral axis

The dorsal-ventral (back-belly) axis is critical to the formation of the hypoblast and to the further development of the embryo. This axis is established when the cleaving cells of the blastoderm establish a barrier between the basic (pH 9.5) albumin above the blastodisc and the acidic (pH 6.5) subgerminal space below it. Water and sodium ions are transported from the albumin through the cells and into the subgerminal space, creating a membrane potential difference of 25 mV across the epiblast cell layer (positive at the ventral side of the cells). This process distinguishes two sides of the epiblast: a side facing the negative and basic albumin, which becomes the dorsal side, and a side facing the positive and acidic subgerminal space fluid, which becomes the ventral side. This axis can be reversed experimentally either by reversing the pH gra-

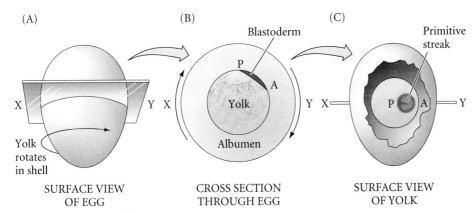

(A)   (B)   (C)

Blastoderm

P

A

Yolk

Albumen

X ——— Y   X   Y   X ——— Y

Yolk
rotates
in shell

Primitive
streak

P   A

SURFACE VIEW
OF EGG

CROSS SECTION
THROUGH EGG

SURFACE VIEW
OF YOLK

Figure 11.13
Specification of the chick anterior-posterior axis by gravity. Rotation in the shell gland (A) results in the lighter components of the yolk pushing up one side of the blastoderm (B). That more elevated region becomes the posterior of the embryo (C). (After Wolpert et al. 1998.)

dient or by inverting the potential difference across the cell layer (reviewed in Stern and Canning 1988).

The role of gravity in forming the anterior-posterior axis

The conversion of the radially symmetrical blastoderm into a bilaterally symmetrical structure is determined by gravity. As the ovum passes through the hen's reproductive tract, it is rotated for about 20 hours in the shell gland. This spinning, at a rate of 10 to 12 revolutions per hour, shifts the yolk such that its lighter components lie beneath one side of the blastoderm. This tips up that end of the blastoderm, and that end becomes the posterior portion of the embryo—the part where primitive streak formation begins (Figure 11.13; Kochav and Eyal-Giladi 1971; Eyal-Giladi and Fabian 1980).

It is not known what interactions cause this portion of the blastoderm to become the posterior margin and to initiate gastrulation. The ability to initiate a primitive streak is found throughout the marginal zone, and if the blastoderm is separated into parts, each having its own marginal zone, each part will form its own primitive streak (Figure 11.14; Spratt and Haas 1960). However, once a **posterior marginal zone (PMZ)** has formed, it controls the other regions of the margin. Not only do these PMZ cells initiate gastrulation, but they also prevent other regions of the margin from forming their own primitive streaks* (Khaner and Eyal-Giladi 1989; Eyal-Giladi et al. 1992).

*Earlier investigators (Waddington 1932) thought that the hypoblast induced the formation of the primitive streak and provided it with anterior-posterior polarity. However, Khaner (1995) rotated the epiblast with respect to the hypoblast at different stages of chick development and showed that the epiblast initiates primitive streak formation and maintains its polarity independently of the orientation of the hypoblast.

It now seems apparent that the PMZ contains cells that act as the equivalent of the amphibian Nieuwkoop center. When placed in the anterior region of the marginal zone, a graft of posterior marginal zone tissue (posterior to and not including Koller's sickle) is able to induce a primitive streak and Hensen's node without contributing cells to either structure (Bachvarova et al. 1998; Khaner 1998). Like the amphibian Nieuwkoop center, this region is thought to be the place where β-catenin localization in the nucleus and a TGF-β family signal coincide. The posterior marginal zone is the only region of the embryo secreting Vg1 (Mitrani et al. 1990; Hume and Dodd 1993; Seleiro et al. 1996).

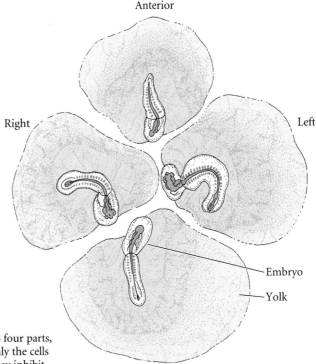

Anterior

Right   Left

Embryo

Yolk

Posterior

Figure 11.14
Regulation of the chick blastoderm. When the blastoderm is divided into four parts, each part can initiate gastrulation and give rise to an embryo. Usually, only the cells of the posterior marginal zone are able to form a primitive streak, and they inhibit other areas of the marginal zone from doing so. (After Spratt and Haas 1960.)

(A)

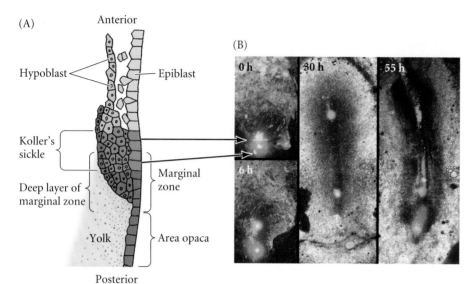

Figure 11.15
Formation of Hensen's node from Koller's sickle. (A) Diagram of the posterior end of an early (pre-streak) embryo, showing the cells labeled with fluorescent dyes in the photographs. (B) Just before gastrulation, cells in the anterior end of Koller's sickle (the epiblast and middle layer) were labeled with green dye. Cells of the posterior portion of Koller's sickle were labeled with red dye. As the cells migrate, the anterior cells formed Hensen's node and its notochord derivatives. The posterior cells formed the posterior region of the primitive streak. The time after dye injection is labeled on each photograph. (After Bachvarova et al. 1998; photographs courtesy of R. F. Bachvarova.)

The epiblast and middle layer cells in the anterior portion of Koller's sickle become Hensen's node (Bachvarova et al. 1998). The posterior portion of Koller's sickle contributes to the posterior portion of the primitive streak (Figure 11.15). Hensen's node has long been known to be the avian equivalent of the amphibian dorsal blastopore lip, since it is (1) the site where gastrulation begins, (2) the region whose cells become the chordamesoderm, and (3) the region whose cells can organize a second embryonic axis when transplanted into other locations of the gastrula (Figure 11.16; Waddington 1933; 1934; Diaz and Schoenwolf 1990). Moreover, the transplantation of Koller's sickle can also cause the formation of new axes, and the middle layer cells in Koller's sickle express *goosecoid*, just like the cells of Spemann's organizer (Izpisua-Belmonte et al. 1993; Callebaut and van Neuten 1994).

As is the case in all vertebrates, the dorsal mesoderm is able to induce the formation of the central nervous system in the ectoderm overlying it. The cells of Hensen's node and its derivatives act like the amphibian organizer, and they secrete Chordin, Noggin, and Nodal proteins. These antagonize BMPs and dorsalize the ectoderm and mesoderm. The mech-

anisms for neural induction in the chick, however, probably differ from those of the frog, and the antagonism of the BMP signal does not appear to be sufficient for neural induction. In chick embryos, BMPs do not inhibit neural induction, nor does ectopic expression of chordin in the non-neural epiblast cause neural induction (see Streit and Stern 1999). Rather, fibroblast growth factors (FGFs) appear to generate neuronal phenotypes in the epiblast cells. Fibroblast growth factors are produced in Hensen's node and the primitive streak, and beads containing certain FGFs can induce trunk and hindbrain neuronal expression in the epiblast cells* (Alvarez et al. 1998; Storey et al. 1998). The factor or factors regulating anterior neuron production remain unknown, but recent evidence (Diaz and Schoenwolf 1990; Darnell et al. 1999) suggests that the anterior visceral endoderm is providing these signals.

---

*A precondition for the BMP-chordin system to neuralize ectoderm is that dissociated (uninduced) presumptive ectoderm cells will become neural. In chick embryos, however, the dissociated epiblast cells develop into muscles (George-Weinstein et al. 1996).

Figure 11.16
Induction of a new embryo by transplantation of Hensen's node. (A) A Hensen's node from a duck embryo is transplanted into the epiblast of a chick embryo. (B) A secondary embryo is induced (as is evident by the neural tube) from host tissues at the graft site. (After Waddington 1933.)

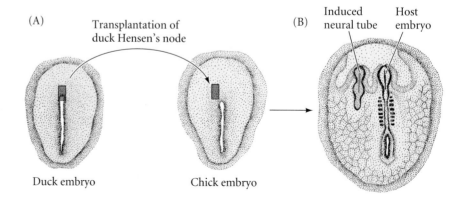

Left-right axis formation

As we have seen, the vertebrate body is not symmetrical. Rather, it has distinct right and left sides. The heart and spleen, for instance, are generally on the left side of the body, while the liver is usually on the right side. The distinction between the right and left sides of vertebrates is regulated by two major proteins: the paracrine factor **Nodal** and the transcription factor **Pitx2**. However, the mechanism by which *nodal* gene expression is activated in the left side of the body differs among vertebrate classes. The ease with which chick embryos can be manipulated has allowed scientists to elucidate the pathways of left-right discrimination in birds more readily than in other vertebrates.

As the primitive streak reaches its maximum length, transcription of the *sonic hedgehog* gene ceases on the right side of the embryo, due to the expression on this side of activin and its receptor (Figure 11.17). Activin signaling blocks the expression of *sonic hedgehog* and activates the expression of *fgf8*. FGF8 prevents the transcription of the **caronte** gene. In the absence of Caronte, bone morphogenetic proteins (BMPs) are able to block the expression of *nodal* and *lefty-2*. This activates the *snail* gene (*cSnR*) that is characteristic of the right side of avian embryonic organs. On the left side of the body, the Lefty-1 protein blocks the expression of *fgf8*, while Sonic hedgehog activates *caronte* (Figure 11.18). Caronte is a paracrine factor that prevents BMPs from repressing the *nodal* and *lefty-2* genes, and also inhibits BMPs from blocking the expression of *lefty-1* on the ventral midline structures (Rodriguez-Esteban et al. 1999; Yokouchi et al. 1999). Nodal and Lefty-2 activate *pitx2* and repress *snail* (*cSnR*). Lefty-1 in the ventral midline prevents the Caronte signal from passing to the right side of the embryo. As in *Xenopus*, *pitx2* is crucial in directing the asymmetry of the embryonic structures.

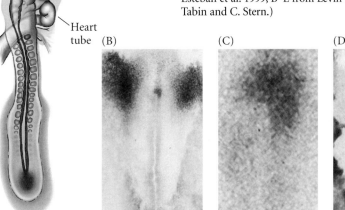

Figure 11.17

Pathway for left-right asymmetry in the chick embryo. (A) Left side: On the left side of Hensen's node, *sonic hedgehog* (*shh*) and *lefty-1* are expressed. Lefty-1 blocks *fgf8* expression, while Sonic hedgehog activates the expression of *caronte* (*car*). Meanwhile, retinoic acid permits the expression of Lefty-1 along the ventral midline. Caronte expression blocks bone morphogenetic proteins (BMPs) on the left side, which would otherwise block the expression of *nodal* and *lefty-2*. In the presence of Nodal and Lefty-2, the *pitx2* gene is activated and the *snail* gene (*cSnR*) is repressed. Pitx2 is active in the various organ primordia and specifies the side to be left. In the right side of the embryo, activin is expressed, along with activin receptor IIa. This activates FGF8, a protein that blocks the expression of *caronte*. In the absence of Caronte, BMP represses the activation of *nodal* and *lefty-2*. This allows the *snail* gene to be active while the *pitx2* gene remains repressed. (B, C) Dorsal and close-up views of in situ hybridization of the *sonic hedgehog* mRNA. (D, E) Dorsal and close-up views of the activin receptor IIa message. (A after Rodriguez-Esteban et al. 1999; B–E from Levin et al. 1995, photographs courtesy of C. Tabin and C. Stern.)

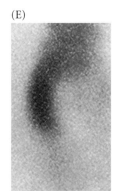

(A) 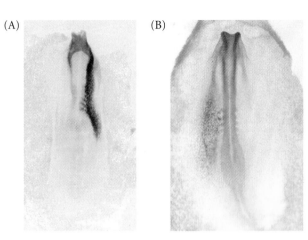 (B) (C)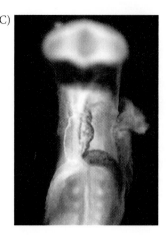

Figure 11.18
Transmission of the left-side signal from Hensen's node to the lateral plate mesoderm: the asymmetric expression of *caronte*, *nodal*, and *pitx2* genes in the left side of the chick embryo. (A) Whole mount in situ hybridization of *caronte* mRNA. The view is from the ventral surface, "from below," so the expression seems to be on the right. Dorsally, the expression pattern would be on the left. (B) Whole mount in situ hybridization using probes for the chick *nodal* message (stained purple) shows its expression in the lateral plate mesoderm only on the left side of the embryo. This view is from the dorsal side (looking "down" at the embryo). (C) A similar in situ hybridization, using the probe for *pitx2* at a later stage of development. At this stage, the heart is forming, and *pitx2* expression can be seen on the left side of the heart tube (as well as symmetrically in more anterior tissues). The embryo is seen from its ventral surface. (A from Rodriguez-Esteban et al. 1999, photograph courtesy of Dr. J. Izpisúa-Belmonte; B courtesy of C. Stern; C from Logan et al. 1998, photograph courtesy of C. Tabin.)

Experimentally induced expression of either *nodal* or *pitx2* on the right side of the chick is able to reverse the asymmetry or cause randomization of the asymmetry on the right or left sides* (Logan et al. 1998; Ryan et al. 1998).

The course to either left- or right-sidedness can be interfered with at any point along the pathway. If activin is blocked by the experimental addition of follistatin to the embryo, the asymmetry of *sonic hedgehog* expression vanishes, and the heart has an equal chance of looping either way (Levin et al. 1995; 1997). Conversely, when activin-soaked beads are placed to the left side of Hensen's node, the *sonic hedgehog* gene (usually expressed only on the left side) is repressed. This, in turn, suppresses the transcription of *nodal*. In this situation, too, the heart tube forms randomly, having an equal probability of being left or right. A similar condition can be produced by implanting cells secreting Sonic hedgehog into the right side of the node. In these cases, *nodal* is induced symmetrically in the lateral plate mesoderm, and the heart has a 50% chance of having a left-handed tube (Figure 11.19).

*In humans, homozygous loss of *PITX2* causes Rieger syndrome, a condition characterized by asymmetry anomalies. A similar condition is caused by knocking out this gene in mice (Fu et al. 1999; Lin et al. 1999).

Early Mammalian Development

Cleavage in Mammals

It is not surprising that mammalian cleavage has been the most difficult to study. Mammalian eggs are among the smallest in the animal kingdom, making them hard to manipulate experimentally. The human zygote, for instance, is only 100 μm in diameter—barely visible to the eye and less than one-thousandth the volume of a *Xenopus* egg. Also, mammalian zygotes are not produced in numbers comparable to sea urchin or frog zygotes, so it is difficult to obtain enough material for biochemical studies. Usually, fewer than ten eggs are ovulated by a female at a given time. As a final hurdle, the development of mammalian embryos is accomplished within another organism, rather than in the external environment. Only recently has it been possible to duplicate some of these internal conditions and observe development in vitro.

The unique nature of mammalian cleavage

With all these difficulties, knowledge of mammalian cleavage was worth waiting for, as mammalian cleavage turned out to be strikingly different from most other patterns of embryonic cell division. The mammalian oocyte is released from the ovary and swept by the fimbriae into the oviduct (Figure 11.20). Fertilization occurs in the ampulla of the oviduct, a region close to the ovary. Meiosis is completed at this time, and first cleavage begins about a day later (see Figure 7.32). Cleavages in mammalian eggs are among the slowest in the animal kingdom—about 12–24 hours apart. Meanwhile, the cilia in the oviduct push the embryo toward the uterus; the first cleavages occur along this journey.

In addition to the slowness of cell division, there are several other features of mammalian cleavage that distinguish it from other cleavage types. The second of these differences is the unique orientation of mammalian blastomeres with relation to one another. The first cleavage is a normal meridional division; however, in the second cleavage, one of the two blastomeres divides meridionally and the other divides equatorially (Figure 11.21). This type of cleavage is called **rotational cleavage** (Gulyas 1975).

(A)

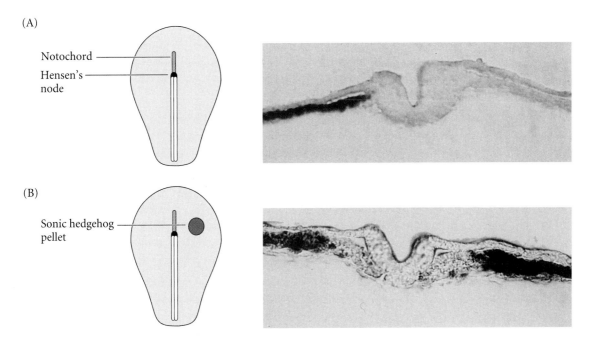

(B)

Figure 11.19
Ectopic expression of *sonic hedgehog* leads to symmetrical *nodal* expression and randomization of heart looping. (A) Wild-type expression of the chick *nodal* gene, showing expression on the left side (dark area). Nearly all hearts develop right-sided loops. This pattern is also seen when pellets containing control substances are implanted into the right side of Hensen's node, or when a Sonic hedgehog-containing pellet is implanted on the left side (where *sonic hedgehog* is usually expressed). (B) When a Sonic hedgehog pellet is implanted on the right side of the node, the expression of *nodal* becomes bilaterally symmetrical. (From Levin et al. 1995; photographs courtesy of the authors.)

Figure 11.20
Development of a human embryo from fertilization to implantation. Compaction of the human embryo occurs on day 4 when it is at the 10-cell stage. The embryo "hatches" from the zona pellucida upon reaching the uterus. During its migration to the uterus, the zona prevents the embryo from prematurely adhering to the oviduct rather than traveling to the uterus. (After Tuchmann-Duplessis et al. 1972.)

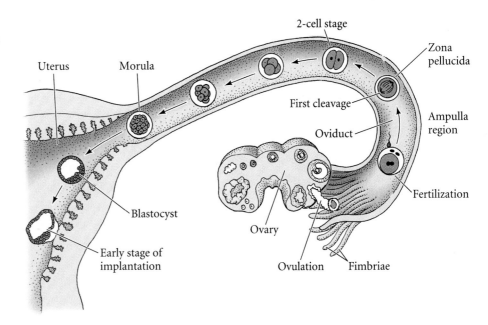

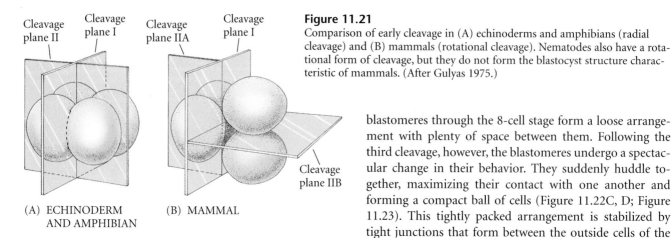

Cleavage plane II   Cleavage plane I   Cleavage plane IIA   Cleavage plane I

Cleavage plane IIB

(A) ECHINODERM AND AMPHIBIAN

(B) MAMMAL

Figure 11.21
Comparison of early cleavage in (A) echinoderms and amphibians (radial cleavage) and (B) mammals (rotational cleavage). Nematodes also have a rotational form of cleavage, but they do not form the blastocyst structure characteristic of mammals. (After Gulyas 1975.)

The third major difference between mammalian cleavage and that of most other embryos is the marked asynchrony of early cell division. Mammalian blastomeres do not all divide at the same time. Thus, mammalian embryos do not increase exponentially from 2- to 4- to 8-cell stages, but frequently contain odd numbers of cells. Fourth, unlike almost all other animal genomes, the mammalian genome is activated during early cleavage, and produces the proteins necessary for cleavage to occur. In the mouse and goat, the switch from maternal to zygotic control occurs at the 2-cell stage (Piko and Clegg 1982; Prather 1989).

Most research on mammalian development has focused on the mouse embryo, since mice are relatively easy to breed throughout the year, have large litters, and can be housed easily. Thus, most of the studies discussed here will concern **murine** (mouse) development.

Compaction

The fifth, and perhaps the most crucial, difference between mammalian cleavage and all other types involves the phenomenon of **compaction**. As seen in Figure 11.22, mouse blastomeres through the 8-cell stage form a loose arrangement with plenty of space between them. Following the third cleavage, however, the blastomeres undergo a spectacular change in their behavior. They suddenly huddle together, maximizing their contact with one another and forming a compact ball of cells (Figure 11.22C, D; Figure 11.23). This tightly packed arrangement is stabilized by tight junctions that form between the outside cells of the ball, sealing off the inside of the sphere. The cells within the sphere form gap junctions, thereby enabling small molecules and ions to pass between them.

The cells of the compacted 8-cell embryo divide to produce a 16-cell **morula** (Figure 11.22E). The morula consists of a small group of internal cells surrounded by a larger group of external cells (Barlow et al. 1972). Most of the descendants of the external cells become the **trophoblast (trophectoderm)** cells. This group of cells produces no embryonic structures. Rather, it forms the tissue of the **chorion**, the embryonic portion of the **placenta**. The chorion enables the fetus to get oxygen and nourishment from the mother. It also secretes hormones that cause the mother's uterus to retain the fetus, and produces regulators of the immune response so that the mother will not reject the embryo as she would an organ graft.

The mouse embryo proper is derived from the descendants of the inner cells of the 16-cell stage, supplemented by cells dividing from the trophoblast during the transition to the 32-cell stage (Pedersen et al. 1986; Fleming 1987). These cells generate the **inner cell mass** (**ICM**), which will give rise to the embryo and its associated yolk sac, allantois, and amnion. By the 64-cell stage, the inner cell mass (approximately 13 cells) and the trophoblast cells have become separate cell layers, neither contributing cells to the other group (Dyce et al. 1987; Fleming 1987). Thus, the distinction between trophoblast and inner cell mass blastomeres represents the first differentiation event in mammalian development. This differentiation is required for the early mammalian embryo to adhere to the uterus (Figure 11.24). The development of the embryo proper can wait until after that

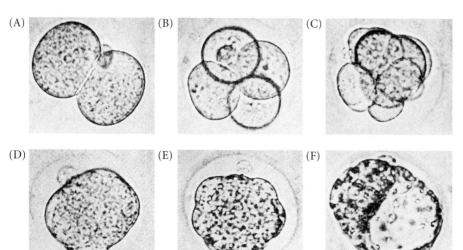

(A)   (B)   (C)

(D)   (E)   (F)

Figure 11.22
The cleavage of a single mouse embryo in vitro. (A) 2-cell stage. (B) 4-cell stage. (C) Early 8-cell stage. (D) Compacted 8-cell stage. (E) Morula. (F) Blastocyst. (From Mulnard 1967; photographs courtesy of J. G. Mulnard.)

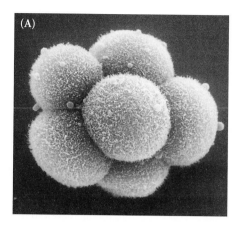

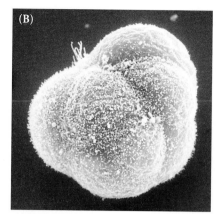

Figure 11.23
Scanning electron micrographs of (A) un-compacted and (B) compacted 8-cell mouse embryos. (Photographs courtesy of C. Ziomek.)

attachment occurs. The inner cell mass actively supports the trophoblast, secreting proteins (such as FGF4) that cause the trophoblast cells to divide (Tanaka et al. 1998).

Initially, the morula does not have an internal cavity. However, during a process called **cavitation**, the trophoblast cells secrete fluid into the morula to create a blastocoel. The inner cell mass is positioned on one side of the ring of trophoblast cells (see Figures 11.23 and 11.25). The resulting structure, called the **blastocyst**, is another hallmark of mammalian cleavage.

> WEBSITE **11.4 Mechanisms of compaction and the formation of the inner cell mass.** What determines whether a cell is to become a trophoblast cell or a member of the inner cell mass? It may just be a matter of chance.

> WEBSITE **11.5 Human cleavage and compaction.** There is a slight growth advantage to XY blastomeres that may have had profound effects for in vitro fertility operations.

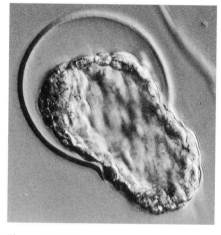

Figure 11.25
Mouse blastocyst hatching from the zona pellucida. (Photograph from Mark et al. 1985, courtesy of E. Lacy.)

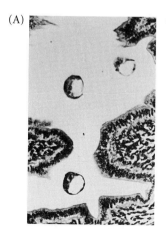

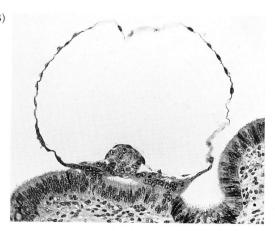

Figure 11.24
Implantation of the mammalian blastocyst into the uterus. (A) Mouse blastocysts entering the uterus. (B) Initial implantation of the blastocyst in a rhesus monkey. (A from Rugh 1967; B courtesy of the Carnegie Institution of Washington, Chester Reather, photographer.)

Escape from the Zona Pellucida

While the embryo is moving through the oviduct en route to the uterus, the blastocyst expands within the **zona pellucida** (the extracellular matrix of the egg that was essential for sperm binding during fertilization; see Chapter 7). The plasma membranes of the trophoblast cells contain a sodium pump (a Na^+/K^+-ATPase) facing the blastocoel, and these proteins pump sodium ions into the central cavity. This accumulation of sodium ions draws in water osmotically, thus enlarging the blastocoel (Borland 1977; Wiley 1984). During this time, the zona pellucida prevents the blastocyst from adhering to the oviduct walls. When such adherence does take place in humans, it is called an ectopic or **tubal pregnancy**. This is a dangerous condition because the implantation of the embryo into the oviduct can cause a life-threatening hemorrhage. When the embryo reaches the uterus, however, it must "hatch" from the zona so that it can adhere to the uterine wall.

The mouse blastocyst hatches from the zona by lysing a small hole in it and squeezing through that hole as the blastocyst expands (Figure 11.25). A trypsin-like protease, **strypsin**, is located on the trophoblast cell membranes and lyses a hole in the fibrillar matrix of the zona (Perona and Wassarman 1986; Yamazaki and Kato 1989). Once out, the blastocyst can make direct contact with the uterus. The uterine epithelium (**endometrium**) "catches" the blastocyst on an extracellular matrix containing collagen, laminin, fibronectin, hyaluronic acid, and heparan sulfate receptors. The trophoblast cells contain integrins that will bind to the uterine collagen, fibronectin, and laminin, and they synthesize heparan sulfate proteoglycan precisely prior to implantation (see Carson et al. 1993). Once in contact with the endometrium, the trophoblast secretes another set of proteases, including collagenase, stromelysin, and plasminogen activator. These protein-digesting enzymes digest the extracellular matrix of the uterine tissue, enabling the blastocyst to bury itself within the uterine wall (Strickland et al. 1976; Brenner et al. 1989).

WEBSITE **11.6 The mechanisms of implantation**. The molecular mechanisms by which the blastocyst adheres to and enters into the uterine wall constitute a fascinating story of cell adhesion and reciprocal interactions between two organisms, the mother and the embryo.

Gastrulation in Mammals

Birds and mammals are both descendants of reptilian species. Therefore, it is not surprising that mammalian development parallels that of reptiles and birds. What is surprising is that the gastrulation movements of reptilian and avian embryos, which evolved as an adaptation to yolky eggs, are retained even in the absence of large amounts of yolk in the mammalian embryo. The mammalian inner cell mass can be envisioned as sitting atop an imaginary ball of yolk, following instructions that seem more appropriate to its reptilian ancestors.

Modifications for development within another organism

The mammalian embryo obtains nutrients directly from its mother and does not rely on stored yolk. This adaptation has entailed a dramatic restructuring of the maternal anatomy (such as expansion of the oviduct to form the uterus) as well as the development of a fetal organ capable of absorbing ma-

Figure 11.26
Schematic diagram showing the derivation of tissues in human and rhesus monkey embryos. (After Luckett 1978; Bianchi et al. 1993.)

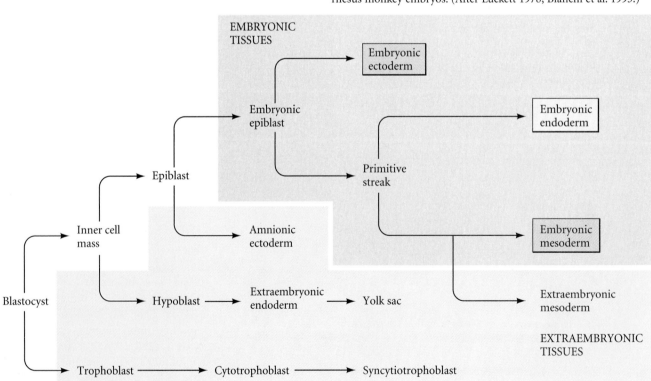

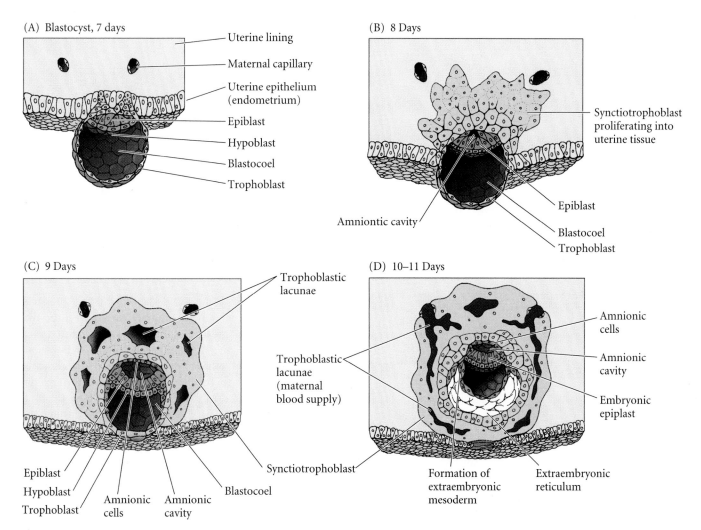

(A) Blastocyst, 7 days

— Uterine lining
— Maternal capillary
— Uterine epithelium (endometrium)
— Epiblast
— Hypoblast
— Blastocoel
— Trophoblast

(B) 8 Days

Synctiotrophoblast proliferating into uterine tissue

Amniontic cavity

Epiblast
Blastocoel
Trophoblast

(C) 9 Days

Trophoblastic lacunae

Trophoblastic lacunae (maternal blood supply)

Epiblast
Hypoblast
Trophoblast
Amnionic cells
Amnionic cavity
Blastocoel
Synctiotrophoblast

(D) 10–11 Days

Amnionic cells
Amnionic cavity
Embryonic epiblast

Formation of extraembryonic mesoderm
Extraembryonic reticulum

Figure 11.27
Tissue formation in the human embryo between days 7 and 11. (A, B) Human blastocyst immediately prior to gastrulation. The inner cell mass delaminates hypoblast cells that line the blastocoel, forming the extraembryonic endoderm of the primitive yolk sac and a two-layered (epiblast and hypoblast) blastodisc similar to that seen in avian embryos. The trophoblast in some mammals can be divided into the polar trophoblast, which covers the inner cell mass, and the mural trophoblast, which does not. The trophoblast divides into the cytotrophoblast, which will form the villi, and the syncytiotrophoblast, which will ingress into the uterine tissue. (C) Meanwhile, the epiblast splits into the amnionic ectoderm (which encircles the amnionic cavity) and the embryonic epiblast. The adult mammal forms from the cells of the embryonic epiblast. (D) The extraembryonic endoderm forms the yolk sac. (After Gilbert 1989; Larsen 1993.)

ternal nutrients. This fetal organ—the chorion—is derived primarily from embryonic trophoblast cells, supplemented with mesodermal cells derived from the inner cell mass. The chorion forms the fetal portion of the placenta. It will induce the uterine cells to form the maternal portion of the placenta, the **decidua**. The decidua becomes rich in the blood vessels that will provide oxygen and nutrients to the embryo.

The origins of early mammalian tissues are summarized in Figure 11.26. The first segregation of cells within the inner cell mass results in the formation of the hypoblast (sometimes called the **primitive endoderm**) layer (Figure 11.27A). The hypoblast cells delaminate from the inner cell mass to line the blastocoel cavity, where they give rise to the **extraembryonic endoderm**, which forms the yolk sac. As in avian embryos, these cells do not produce any part of the newborn organism. The remaining inner cell mass tissue above the hypoblast is now referred to as the epiblast. The epiblast cell layer is split by small clefts that eventually coalesce to separate the **embryonic epiblast** from the other epiblast cells, which form the **amnionic cavity** (Figures 11.27 B, C). Once the lining of the amnion is completed, it fills with a secretion called **amnionic (amniotic) fluid**, which serves as a shock absorber for the developing embryo while preventing its desiccation. The embryonic epiblast is believed to contain all the cells that will generate the actual embryo, and it is similar in many ways to the avian epiblast.

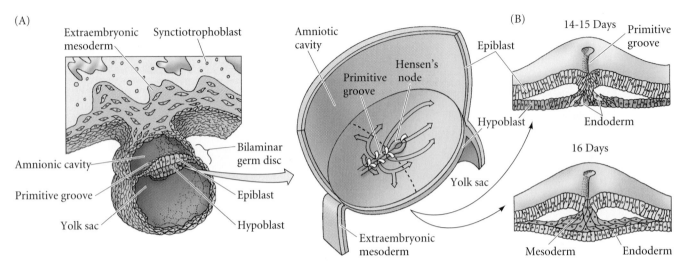

Figure 11.28
Amnion structure and cell movements during human gastrulation. (A) Human embryo and uterine connections at day 15 of gestation. In the upper view, the embryo is cut sagittally through the midline; the lower view looks down upon the dorsal surface of the embryo. (B) The movements of the epiblast cells through the primitive streak and Hensen's node and underneath the epiblast are superimposed on the dorsal surface view. At days 14 and 15, the ingressing epiblast cells are thought to replace the hypoblast cells (which contribute to the yolk sac lining), while at day 16, the ingressing cells fan out to form the mesodermal layer. (After Larsen 1993.)

By labeling individual cells of the epiblast with horseradish peroxidase, Kirstie Lawson and her colleagues (1991) were able to construct a detailed fate map of the mouse epiblast (Figure 1.6). Gastrulation begins at the posterior end of the embryo, and this is where the **node** forms* (Figure 11.28). Like the chick epiblast cells, the mammalian mesoderm and

endoderm migrate through a primitive streak, and like their avian counterparts, the migrating cells of the mammalian epiblast lose E-cadherin, detach from their neighbors, and migrate through the streak as individual cells (Burdsall et al. 1993). Those cells migrating through the node give rise to the notochord. However, in contrast to notochord formation in the chick, the cells that form the mouse notochord are thought to become integrated into the endoderm of the primitive gut (Jurand 1974; Sulik et al. 1994). These cells can be seen as a band of small, ciliated cells extending rostrally from the node (Figure 11.29). They form the notochord by converging medially and folding off in a dorsal direction from the roof of the gut.

The ectodermal precursors end up anterior to the fully extended primitive streak, as in the chick epiblast; but whereas the mesoderm of the chick forms from cells posterior to the farthest extent of the streak, the mouse mesoderm forms from cells anterior to the

---

*In mammalian development, Hensen's node is usually just called "the node," despite the fact that Hensen discovered this structure in rabbit embryos.

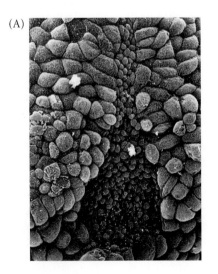

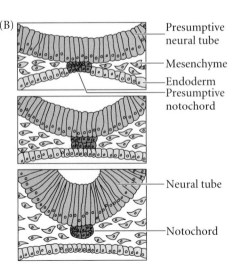

Presumptive neural tube
Mesenchyme
Endoderm
Presumptive notochord

Neural tube

Notochord

Figure 11.29
Formation of the notochord in the mouse. (A) The ventral surface a the 7.5-day mouse embryo, seen by scanning electron microscopy. The presumptive notochord cells are the small, ciliated cells in the midline that are flanked by the larger endodermal cells of the primitive gut. The node (with its ciliated cells) is seen at the bottom. (B) The formation of the notochord by the dorsal infolding of the small, ciliated cells. (From Sulik et al. 1994; photograph courtesy of K. Sulik and G. C. Schoenwolf.)

primitive streak. In some instances, a single cell gives rise to descendants in more than one germ layer, or to both embryonic and extrabryonic derivatives. Thus, at the epiblast stage, these lineages have not become separate from one another. As in avian embryos, the cells migrating in between the hypoblast and epiblast layers are coated with hyaluronic acid, which they synthesize as they leave the primitive streak. This acts to keep them separate while they migrate (Solursh and Morriss 1977). It is thought (Larsen 1993) that the replacement of human hypoblast cells by endoderm precursors occurs on days 14–15 of gestation, while the migration of cells forming the mesoderm does not start until day 16 (Figure 11.28B).

Formation of extraembryonic membranes

While the embryonic epiblast is undergoing cell movements reminiscent of those seen in reptilian or avian gastrulation, the extraembryonic cells are making the distinctly mammalian tissues that enable the fetus to survive within the maternal uterus. Although the initial trophoblast cells of mice and humans divide like most other cells of the body, they give rise to a population of cells wherein nuclear division occurs in the absence of cytokinesis. The original type of trophoblast cells constitute a layer called the **cytotrophoblast**, whereas the multinucleated type of cell forms the **syncytiotrophoblast**. The cytotrophoblast initially adheres to the endometrium through a series of adhesion molecules. Moreover, these cells also contain proteolytic enzymes that enable them to enter the uterine wall and remodel the uterine blood vessels so that the maternal blood bathes fetal blood vessels. The syncytiotro-

phoblast tissue is thought to further the progression of the embryo into the uterine wall by digesting uterine tissue (Fisher et al. 1989). The uterus, in turn, sends blood vessels into this area, where they eventually contact the syncytiotrophoblast. Shortly thereafter, mesodermal tissue extends outward from the gastrulating embryo (see Figure 11.27D). Studies of human and rhesus monkey embryos have suggested that the yolk sac (and hence the hypoblast) is the source of this extraembryonic mesoderm (Bianchi et al. 1993). The extraembryonic mesoderm joins the trophoblastic extensions and gives rise to the blood vessels that carry nutrients from the mother to the embryo. The narrow connecting stalk of extraembryonic mesoderm that links the embryo to the trophoblast eventually forms the vessels of the **umbilical cord**. The fully developed extraembryonic organ, consisting of trophoblast tissue and blood vessel-containing mesoderm, is called the chorion, and it fuses with the uterine wall to create the placenta. Thus, the placenta has both a maternal portion (the uterine endometrium, which is modified during pregnancy) and a fetal component (the chorion). The chorion may be very closely apposed to maternal tissues while still being readily separable from them (as in the contact placenta of the pig), or it may be so intimately integrated with maternal tissues that the two cannot be separated without damage to both the mother and the developing fetus (as in the deciduous placenta of most mammals, including humans).*

Figure 11.30 shows the relationships between the embryonic and extraembryonic tissues of a 6-week human embryo. The embryo is seen encased in the amnion and is further shielded by the chorion. The blood vessels extending to and from the chorion are readily observable, as are the villi that project from the outer surface of the chorion. These villi contain the blood vessels and allow the chorion to have a large area exposed to the maternal blood. Although fetal and maternal circulatory systems normally never merge, diffusion of soluble substances can occur through the villi (Figure 11.31). In this manner, the mother provides the fetus with nutrients and oxygen, and the fetus sends its waste products (mainly carbon dioxide and urea) into the maternal circulation. The maternal and fetal blood cells, however, usually do not mix.

*There are numerous types of placentas, and the extraembryonic membranes form differently in different orders of mammals (see Cruz and Pedersen 1991). Although mice and humans gastrulate and implant in a similar fashion, their extraembryonic structures are distinctive. It is very risky to extrapolate developmental phenomena from one group of mammals to another. Even Leonardo da Vinci got caught (Renfree 1982). His remarkable drawing of the human fetus inside the placenta is stunning art, but poor science: the placenta is that of a cow.

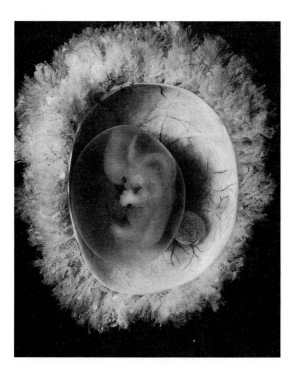

Figure 11.30
Human embryo and placenta after 40 days of gestation. The embryo lies within the amnion, and its blood vessels can be seen extending into the chorionic villi. The small sphere to the right of the embryo is the yolk sac. (The Carnegie Institution of Washington, courtesy of C. F. Reather.)

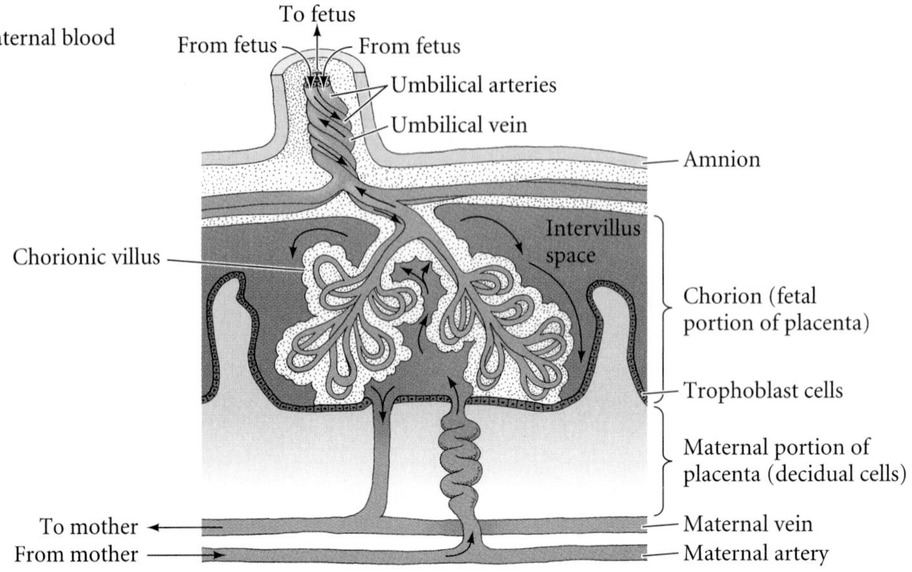

Figure 11.31
Relationship of the chorionic villi to the maternal blood in the uterus.

WEBSITE **11.7 Placental functions**. Placentas are nutritional, endocrine, and immunological organs. They provide hormones that enable the uterus to retain the pregnancy and also accelerate mammary gland development. Placentas also block the potential immune response of the mother against the developing fetus. Recent studies suggest that the placenta uses several mechanisms to block the mother's immune response.

Sidelights & Speculations

Twins

THE EARLY CELLS OF THE EMBRYO can replace each other and compensate for a missing cell. This was first shown in 1952, when Seidel destroyed one cell of a 2-cell rabbit embryo, and the remaining cell produced an entire embryo. Once the inner cell mass has become separate from the trophoblast, the ICM cells constitute an **equivalence group**. In other words, each ICM cell has the same potency (in this case, each cell can give rise to all the cell types of the embryo, but not to the trophoblast), and their fates will be determined by interactions among their descendants. Gardiner and Rossant (1976) also showed that if cells of the ICM (but not trophoblast cells) are injected into blastocysts, they contribute to the new embryo. Since its blastomeres can generate any cell type in the body, the cells of the blastocyst are called **totipotent** (see Chapter 4).

This regulative capacity of the ICM blastomeres is also seen in humans. Human twins are classified into two major groups: **monozygotic** (one-egg, or identical) twins and **dizygotic** (two-egg, or fraternal) twins. Fraternal twins are the result of two separate fertilization events, whereas identical twins are formed from a single embryo whose cells

somehow dissociated from one another. Identical twins may be produced by the separation of early blastomeres, or even by the separation of the inner cell mass into two regions within the same blastocyst.

Identical twins occur in roughly 0.25% of human births. About 33% of identical twins have two complete and separate chorions, indicating that separation occurred before the formation of the trophoblast tissue at day 5 (Figure 11.32A). The remaining identical twins share a common chorion, suggesting that the split occurred within the inner cell mass after the trophoblast formed. By day 9, the human embryo has completed the construction of another extraembryonic layer, the lining of the amnion. This tissue forms the amnionic sac (or water sac), which surrounds the embryo with amnionic fluid and protects it from desiccation and abrupt movement. If the separation of the embryo were to come after the formation of the chorion on day 5 but before the formation of the amnion on day 9, then the resulting embryos should have one chorion and two amnions (Figure 11.32B). This happens in about two-thirds of human identical twins. A small percentage of identical twins are born within

a single chorion and amnion (Figure 11.32C). This means that the division of the embryo came after day 9. Such newborns are at risk of being conjoined ("Siamese") twins.

The ability to produce an entire embryo from cells that normally would have contributed to only a portion of the embryo is called regulation, and is discussed in Chapter 3. Regulation is also seen in the ability of two or more early embryos to form one chimeric individual rather than twins, triplets, or a multiheaded individual. Chimeric mice can be produced by artificially aggregating two or more early-cleavage (usually 4- or 8-cell) embryos to form a composite embryo. As shown in Figure 11.33A, the zonae pellucidae of two genetically different embryos can be artificially removed and the embryos brought together to form a common blastocyst. These blastocysts are then implanted into the uterus of a foster mother. When they are born, the chimeric offspring have some cells from each embryo. This is readily seen when the aggregated blastomeres come from mouse strains that differ in their coat colors. When blastomeres from white and black strains are aggregated, the result is commonly a mouse with black and white bands. There is even ev-

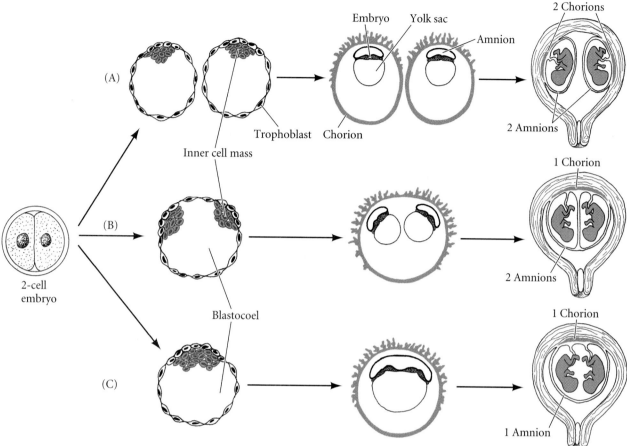

Figure 11.32
Diagram showing the timing of human monozygotic twinning with relation to extraembryonic membranes. (A) Splitting occurs before the formation of the trophoblast, so each twin has its own chorion and amnion. (B) Splitting occurs after trophoblast formation but before amnion formation, resulting in twins having individual amnionic sacs but sharing one chorion. (C) Splitting after amnion formation leads to twins in one amnionic sac and a single chorion. (After Langman 1981).

idence (de la Chappelle et al. 1974; Mayr et al. 1979) that human embryos can form chimeras. Some individuals have two genetically different cell types (XX and XY) within the same body, each with its own set of genetically defined characteristics. The simplest explanation for such a phenomenon is that these individuals resulted from the aggregation of two embryos, one male and one female, that were developing at the same time. If this explanation is correct, then two fraternal twins have fused to create a single composite individual.

Markert and Petters (1978) have shown that three early 8-cell embryos can unite to form a common compacted morula and that the resulting mouse can have the coat colors of the three different strains (Figure 11.33B). Moreover, they showed that each of the three embryos gave rise to precursors of

the gametes. When a chimeric (black/brown/white) female mouse was mated to a white-furred (recessive) male, offspring of each of the three colors were produced.

According to our observations of twin formation and chimeric mice, each blastomere of the inner cell mass should be able to produce any cell of the body. This hypothesis has been confirmed, and it has very important consequences for the study of mammalian development. When ICM cells are isolated and grown under certain conditions, they remain undifferentiated and continue to divide in culture (Evans and Kaufman 1981; Martin 1981). Such cells are called **embryonic stem cells** (**ES cells**). As shown in Chapter 4, cloned genes can be inserted into the nuclei of these cells, or the existing genes can be mutated. When these manipulated ES cells are injected into a mouse blastocyst, they can in-

tegrate into the host inner cell mass. The resulting embryo has cells coming from both the host and the donor tissue. This technique has become extremely important in determining the function of genes during mammalian development. ■

WEBSITE **11.8 Non-identical monozygotic twins.** Although monozygotic twins have the same genome, random developmental factors or the uterine environment may give them dramatically different phenotypes.

WEBSITE **11.9 Conjoined twins.** There are rare events in which more than one set of axes is induced in the same embryo. This can produce conjoined twins, twins that share some parts of their bodies. The medical and social issues raised by conjoined twins provide a fascinating look at what has constituted "individuality" throughout history.

(A)

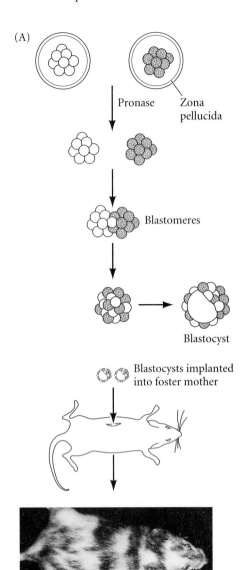

(B)

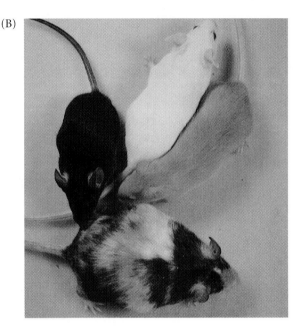

Figure 11.33
Production of chimeric mice. (A) The experimental procedures used to produce chimeric mice. Early 8-cell embryos of genetically distinct mice (here, with coat color differences) are isolated from mouse oviducts and brought together after their zonae are removed by proteolytic enzymes. The cells form a composite blastocyst, which is implanted into the uterus of a foster mother. The photograph shows one of the actual chimeric mice produced in this manner. (B) An adult female chimeric mouse (bottom) produced from the fusion of three 4-cell embryos: one from two white-furred parents, one from two black-furred parents, and one from two brown-furred parents. The resulting mouse has coat colors from all three embryos. Moreover, each embryo contributed germ line cells, as is evidenced by the three colors of offspring (above) produced when this chimeric female was mated with recessive (white-furred) males. (A, photograph courtesy of B. Mintz; B from Markert and Petters 1978, photograph courtesy of C. Markert.)

Mammalian Anterior-Posterior Axis Formation

Two signaling centers

The mammalian embryo appears to have two signaling centers: one in the node ("the organizer") and one in the **anterior visceral endoderm** (Figure 11.34A; Beddington et al. 1994). The node appears to be responsible for the creation of all of the body, and these two signaling centers work together to form the forebrain (Bachiller et al. 2000). Both the mouse node and the anterior visceral endoderm express many of the genes known to be expressed in the chick and frog organizer tissue. The node produces Chordin and Noggin (which the anterior

visceral endoderm does not), while the anterior visceral endoderm expresses several genes that are necessary for head formation. These include the genes for transcription factors Hesx-1, Lim-1, and Otx-2, as well as the gene for the paracrine factor Cerberus. The anterior visceral endoderm is established before the node, and the primitive streak always forms on the side of the epiblast *opposite* this anterior site. Homozygosity for mutant alleles of any of the above-mentioned head-specific organizing genes produces mice lacking forebrains (Figure 11.35; Thomas and Beddington 1996; Beddington and Robertson 1999). While knockouts of either *chordin* or *noggin* do not affect development, mice missing both sets of genes develop a body lacking forebrain, nose, and facial structures.

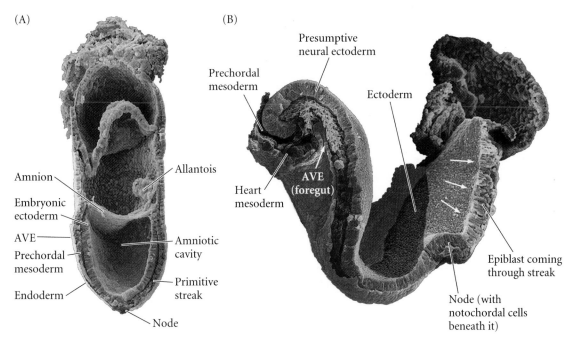

(A)

Amnion
Embryonic ectoderm
AVE
Prechordal mesoderm
Endoderm
Allantois
Amniotic cavity
Primitive streak
Node

(B)

Presumptive neural ectoderm
Prechordal mesoderm
Ectoderm
Heart mesoderm
AVE (foregut)
Epiblast coming through streak
Node (with notochordal cells beneath it)

Figure 11.34
The two signaling centers of the mammalian embryo. (A) In the day 7 mouse embryo, the dorsal surface of the epiblast (embryonic ectoderm) is in contact with the amnionic cavity. The ventral surface of the epiblast contacts the newly formed mesoderm. In this cuplike arrangement, the endoderm covers the surface of the embryo. The node is at the bottom of the cup, and it has generated chordamesoderm. The two signaling centers, the node and the anterior visceral endoderm, are located on opposite sides of the cup. Eventually, the notochord will link them. The caudal side of the embryo is marked by the presence of the allantois. (B) By embryonic day 8, the anterior visceral endoderm lines the foregut, and the prechordal mesoderm is now in contact with the forebrain ectoderm. The node is now farther caudal, due largely to the rapid growth of the anterior portion of the embryo. The cells in the midline of the epiblast migrate through the primitive streak (white arrows). (Photographs courtesy of K. Sulik.)

(Indeed, in other insects, such as the flour beetle *Tribolium*, it is a single unit.) The Hom-C genes are arranged in the same general order as their expression pattern along the anterior-posterior axis, the most 3′ gene (*labial*) being required for producing the most anterior structures, and the most 5′ gene (*AbdB*) specifying the development of the posterior abdomen. Mouse and human genomes contain four copies of the Hox complex per haploid set, located on four different chromosomes (*Hoxa* through *Hoxd* in the mouse, *HOXA* through *HOXD* in humans: Boncinelli et al. 1988; McGinnis and Krumlauf 1992; Scott 1992). Not only are the same general types of homeotic genes found in both flies and mammals, but the order of these genes on their respective chromosomes is

WEBSITE 11.10 Gastrulation in the mouse. The mouse gastrula is shaped like a cup and has a more complicated structure than the human gastrula. The extraembryonic tissues of the mouse appear to be critical in establishing the position of the node and anterior visceral endoderm signaling centers.

Patterning the anterior-posterior axis: The Hox code hypothesis

Once gastrulation begins, anterior-posterior polarity in all vertebrates becomes specified by the expression of **Hox genes**. These genes are homologous to the homeotic gene complex (Hom-C) of the fruit fly (see Chapter 9). The *Drosophila* homeotic gene complex on chromosome 3 contains the Antennapedia and bithorax clusters of homeotic genes (see Figure 9.28), and can be seen as a single functional unit.

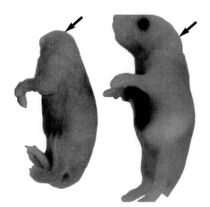

Figure 11.35
Headless phenotype of the *Lim-1* knockout mouse. A *Lim-1*-deficient mouse is shown next to a wild-type littermate. The ear pinnae (arrows) are the most anterior structures seen in these mutants. (From Shawlot and Behringer 1995, photograph courtesy of the authors.)

(A)

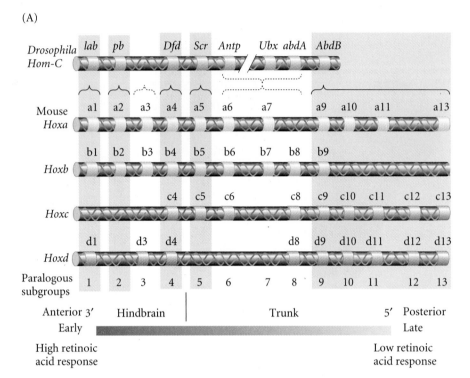

(B)

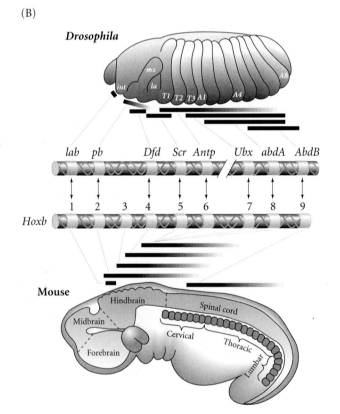

Figure 11.36
Evolutionary conservation of homeotic gene organization and transcriptional expression in fruit flies and mice. (A) Similarity between the *Hom-C* cluster on *Drosophila* chromosome 3 and the four Hox gene clusters in the mouse genome. The shaded regions show particularly strong structural similarities between species, and one can see that the order of the genes on the chromosomes has been conserved. Those genes at the 5′ end (since all mouse Hox genes are transcribed in the same direction) are those that are expressed more posteriorly, are expressed later, and can be induced only by high doses of retinoic acid. Genes having similar structures, the same relative positions on each of the four chromosomes, and similar expression patterns belong to the same paralogous group. (B) Comparison of the transcription patterns of the *Hom-C* and *Hoxb* genes of *Drosophila* (at 10 hours) and mice (at 12 days), respectively. The homologous human genes are called (capitalized) HOX genes. (A after Krumlauf 1993; B after McGinnis and Krumlauf 1992.)

remarkably similar. In addition, the expression of these genes follows the same pattern: those mammalian genes homologous to the *Drosophila labial, proboscipedia,* and *deformed* genes are expressed anteriorly, while those genes that are homologous to the *Drosophila Abdominal-B* gene are expressed posteriorly. Another set of genes that controls the formation of the fly head (*orthodenticle* and *empty spiracles*) has homologues in the mouse that show expression in the midbrain and forebrain.

While Hox genes appear to specify the anterior-posterior axis throughout the vertebrates, we shall discuss mammals here, since the experimental evidence is particularly strong for this class. The mammalian Hox/HOX genes are numbered from 1 to 13, starting from that end of each complex that is expressed most anteriorly. Figure 11.36 shows the relationships between the *Drosophila* and mouse homeotic gene sets. The equivalent genes in each mouse complex (such as *Hoxa-1, Hoxb-1,* and *Hoxd-1*) are called a **paralogous group**. It is thought that the four mammalian Hox complexes were formed from chromosome duplications. Because there is not a one-to-one correspondence between the *Drosophila* Hom-C genes and the mouse Hox genes, it is likely that independent gene duplications have occurred since these two animal branches diverged (Hunt and Krumlauf 1992; see Chapter 22).

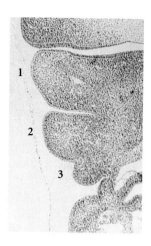
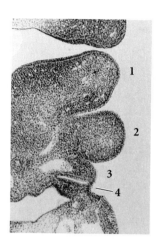

Figure 11.37
Deficient development of neural crest-derived pharyngeal arch and pouch structures in *Hoxa-3*-deficient mice. The arches are numbered. (Right) A 10.5-day embryo of a heterozygous *Hoxa-3* mouse (wild-type), showing normal development of pouch 3 (thymus), pouch 4 (parathyroid), and other structures. (Left) A homozygous mutant *Hoxa-3*-deficient mouse lacks the proper development of these structures. (From Chisaka and Capecchi 1991.)

Expression of Hox genes along the dorsal axis

Hox gene expression can be seen along the dorsal axis (in the neural tube, neural crest, paraxial mesoderm, and surface ectoderm) from the anterior boundary of the hindbrain through the tail. The different regions of the body from the midbrain through the tail are characterized by different constellations of Hox gene expression, and the pattern of Hox gene expression is thought to specify the different regions. In general, the genes of paralogous group 1 are expressed from the tip of the tail to the most anterior border of the hindbrain. Paralogue 2 genes are expressed throughout the spinal cord, but the anterior limit of expression stops two segments more caudally than that of the paralogue 1 genes (see Figure 11.36; Wilkinson et al. 1989; Keynes and Lumsden 1990). The higher-numbered Hox paralogues are expressed solely in the posterior regions of the neural tube, where they also form a "nested" set.

Experimental analysis of the Hox code

The expression patterns of the murine Hox genes suggest a code whereby certain combinations of Hox genes specify a particular region of the anterior-posterior axis (Hunt and Krumlauf 1991). Particular sets of paralogous genes provide segmental identity along the anterior-posterior axis of the body. Evidence for such a code comes from three sources:

- Gene targeting or "knockout" experiments (see Chapter 4), in which mice are constructed that lack both copies of one or more Hox genes
- Retinoic acid teratogenesis, in which mouse embryos exposed to retinoic acid show a different pattern of Hox gene expression along the anterior-posterior axis and abnormal differentiation of their axial structures
- Comparative anatomy, in which the types of vertebrae in different vertebrate species are correlated with the constellation of Hox gene expression

GENE TARGETING. When Chisaka and Capecchi (1991) knocked out the *Hoxa-3* gene from inbred mice, these homozygous mutants died soon after birth. Autopsies of these mice revealed that their neck cartilage was abnormally short and thick and that they had severely deficient or absent thymuses, thyroids, and parathyroid glands (Figure 11.37). The heart and major blood vessels were also malformed. Further analysis showed that the *number* and *migration* of the neural crest cells that normally form these structures were not affected. Rather, it appears that the *Hoxa-3* genes are responsible for specifying cranial neural crest cell fate and for enabling these cells to differentiate and proliferate into neck cartilage and the glands that form the fourth and sixth pharyngeal pouches (Manley and Capecchi 1995).

Knockout of the *Hoxa-2* gene also produces mice whose neural crest cells have been respecified, but the defects in these mice are anterior to those in the *Hoxa-3* knockouts. Cranial elements normally formed by the neural crest cells of the second pharyngeal arch (stapes, styloid bones) are missing and are replaced by duplicates of the structures of the first pharyngeal arch (incus, malleus, etc.) (Gendron-Maguire et al. 1993; Rijli et al. 1993). Thus, without certain Hox genes, some regionally specific organs along the anterior-posterior axis fail to form, or become respecified as other regions. Similarly, when the *Hoxc-8* gene is knocked out (Le Mouellic et al. 1992), several axial skeletal segments resemble more anterior segments, much like what is seen in *Drosophila* loss-of-function homeotic mutations. As can be seen in Figure 11.38, the first lumbar vertebra of the mouse has formed a rib—something characteristic of the thoracic vertebrae anterior to it.

One can get severe axial transformations by knocking out two or more genes of a paralogous group. Mice homozygous for the *Hoxd-3* deletion have mild abnormalities of the first cervical vertebra (the **atlas**), while mice homozygous for the *Hoxa-3* deletion have no abnormality of this bone, though they have other malformations (see the discussion of this mutant above). When both sets of mutations a bred into the same mouse, both sets of problems become more severe. Mice with neither *Hoxa-3* nor *Hoxd-3* have no atlas bone at all, and the hyoid and thyroid cartilage is so reduced in size that there are holes in the skeleton (Condie and Capecchi 1994; Greer et al. 2000). It appears that there are interactions occurring between the products of the Hox genes, and that in some functions, one of the paralogues can replace the other.

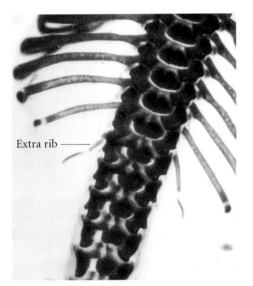

Extra rib ———

Figure 11.38
Partial transformation of the first lumbar vertebra into a thoracic vertebra by the knockout of the *Hoxc-8* gene. The first lumbar vertebra of this mouse has formed a rib—a structure normally formed only by the thoracic vertebrae anterior to it. (Photograph from Le Mouellic et al. 1992; courtesy of the authors.)

retinoic acid (RA). Exogenous retinoic acid given to mouse embryos in utero can cause certain Hox genes to become expressed in groups of cells that usually do not express them (Conlon and Rossant 1992; Kessel 1992). Moreover, the craniofacial abnormalities of mouse embryos exposed to teratogenic doses of RA (Figure 11.39) can be mimicked by causing the expression of *Hoxa-7* throughout the embryo (Balling et al. 1989). If high doses of RA can activate Hox genes in inappropriate locations along the anterior-posterior axis, and if the constellation of active Hox genes specifies the region of the anterior-posterior axis, then mice given RA in utero should show homeotic transformations manifested as rostralizing malformations occurring along that axis.

Kessel and Gruss (1991) found this to be the case. Wild-type mice have 7 cervical (neck) vertebrae, 13 thoracic (ribbed) vertebrae, and 6 lumbar (abdominal) vertebrae, in addition to the sacral and caudal (tail) vertebrae. In embryos exposed to RA on day 8 of gestation (during gastrulation), the first one or two lumbar vertebrae were transformed into thoracic (ribbed) vertebrae, while the first sacral vertebra often became a lumbar vertebra. In some cases, the entire posterior region of the mouse embryo failed to form (Figure 11.39E). These changes in structure were correlated with changes in the

Thus, the evidence from gene knockouts supports the hypotheses that (1) different sets of Hox genes are necessary for the specification of any region of the anterior-posterior axis, (2) that the members of a paralogous group of Hox genes may be responsible for different subsets of organs within these regions, and (3) that the defects caused by knocking out particular Hox genes occur in the most anterior region of that gene's expression.

RETINOIC ACID TERATOGENESIS. Homeotic changes are also seen when mouse embryos are exposed to teratogenic doses of

(A) 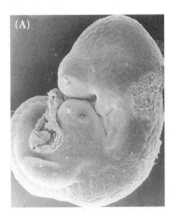   (B) 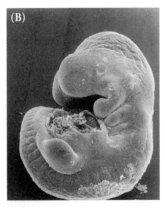   (C)    (D)

(E)

Figure 11.39
Mouse embryos cultured at day 8 in control medium (A, C) or in medium containing retinoic acid (B, D). At day 10 (A, B), the first pharyngeal arch of the treated embryos has a shortened and flattened appearance and has apparently fused with the second pharyngeal arch. At day 17 (C, D), craniofacial malformations can be seen in the neural crest-derived cartilage of the treated embryos. Meckel's cartilage has been completely displaced from the mandibular (lower jaw) to the maxillary (upper mouth) region, and the malleus and incus cartilages have not formed. (E) In some cases, RA exposure causes the loss of the lumbar, sacral, and caudal vertebrae. (A and B from Goulding and Pratt 1986; C and D from Morriss-Kay 1993; E from Kessel 1992, photographs courtesy of the authors.)

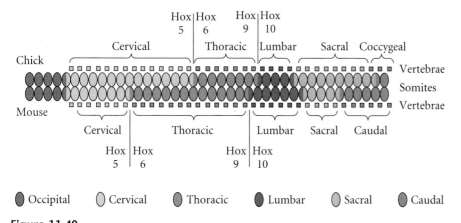

Figure 11.40
Schematic representation of the mouse and chick vertebral pattern along the anterior-posterior axis. The boundaries of expression of certain Hox gene paralogous groups have been mapped onto these domains. (After Burke et al. 1995.)

constellation of Hox genes expressed in these tissues. For example, when RA was given to embryos on day 8, *Hoxa-10* expression was shifted posteriorly, and an additional set of ribs formed on what had been the first lumbar vertebra. When posterior Hox genes were not expressed at all, the caudal part of the embryo failed to form.*

Retinoic acid probably plays a role in axis specification during normal development, and the source of retinoic acid is probably the node (Hogan et al. 1992; Maden et al. 1996). It is possible that the specification of mesoderm cells depends on the amount of time spent within the high retinoic acid concentrations of the node: the more time spent in the node, the more posterior the specification. This pattern can be demonstrated in culture, as embryonal carcinoma cells express more "posterior" Hox genes the longer they are exposed to retinoic acid (Simeone et al. 1990). Giving RA exogenously would mimic the RA concentrations normally encountered only by the posterior cells. The evidence points to a Hox code wherein different constellations of Hox genes, activated by different retinoic acid concentrations, specify the regional characteristics along the anterior-posterior axis.

COMPARATIVE ANATOMY. A new type of comparative embryology is now emerging, and it is based on the comparison of gene expression patterns. Gaunt (1994) and Burke and her collaborators (1995) have compared the vertebrae of the mouse and the chick. Although the mouse and the chick have a similar number of vertebrae, they apportion them differently. Mice (like all mammals, be they giraffes or whales) have only 7 cervical vertebrae. These are followed by 13 thoracic vertebrae, 6

lumbar vertebrae, 4 sacral vertebrae, and a variable (20+) number of caudal vertebrae (Figure 11.40). The chick, on the other hand, has 14 cervical vertebrae, 7 thoracic vertebrae, 12 or 13 (depending on the strain) lumbosacral vertebrae, and 5 coccygeal (fused tail) vertebrae. The researchers asked, Does the constellation of Hox gene expression correlate with the type of vertebra formed (e.g., cervical or thoracic) or with the relative position of the vertebrae (e.g., number 8 or 9)?

The answer is that the constellation of Hox gene expression predicts the type of vertebra formed. In the mouse, the transition between cervical and thoracic vertebrae is between vertebrae 7 and 8; in the chick it is between vertebrae 14 and 15. In both cases, the *Hox-5* paralogues are expressed in the last cervical vertebrae, while the anterior boundary of the *Hox-6* paralogues extends to the first thoracic vertebra. Similarly, in both animals, the thoracic-lumbar transition is seen at the boundary between the *Hox-9* and *Hox-10* paralogous groups. It appears there is a code of differing Hox gene expression along the anterior-posterior axis, and this code determines the type of vertebra formed.

WEBSITE **11.11 Why do mammals have only seven cervical vertebrae?** Recent speculation predicts that the mammalian Hox genes function simultaneously in several processes. To alter a Hox gene's expression so as to change vertebral type might lead to lethal changes in the other processes.

The Dorsal-Ventral and Left-Right Axes in Mammals

The dorsal-ventral axis

Very little is known about the mechanisms of dorsal-ventral axis formation in mammals. In mice and humans, the hypoblast forms on the side of the inner cell mass that is exposed to the blastocyst fluid, while the dorsal axis forms from those ICM cells that are in contact with the trophoblast. Thus, the dorsal-ventral axis of the embryo is, in part, defined by the embryonic-abembryonic axis of the blastocyst. This axis (wherein the **embryonic region** contains the ICM while the **abembryonic region** is that part of the blastocyst opposite the ICM) may be determined within the oocyte, as it develops perpendicularly to the animal-vegetal axis of the newly fertilized egg (Figure 11.41; Gardner 1997). As development proceeds, the notochord maintains dorsal-ventral polarity by inducing specific dorsal-ventral patterns of gene expression in the overlying ectoderm (Goulding et al. 1993).

---

*Hoxa-10 is also important in specifying the axial pattern of the genital ducts. Knockouts of *Hoxa-10* create mice wherein the upper region of the uterus is transformed into tissue resembling the oviduct. This region coincides with the anterior limit of *Hoxa-10* expression in the wild-type Müllerian duct (Benson et al. 1996).

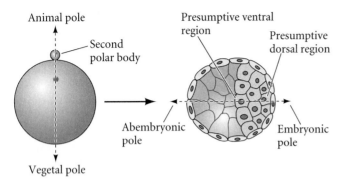

Figure 11.41
Relationship between the animal-vegetal axis of the egg and the embryonic-abembryonic axis of the blastocyst. The polar body marks the animal pole of the embryo. The dorsal-ventral axis of the embryo appears to form at right angles to the animal-vegetal axis of the egg.

The left-right axis

The mammalian body is not symmetrical. Although the heart begins its formation at the midline of the embryo, it moves to the left side of the chest cavity and loops to the right (Figure 11.42). The spleen is found solely on the left side of the abdomen, the major lobe of the liver forms on the right side of the abdomen, the large intestine loops right to left as it traverses the abdominal cavity, and the right lung has one more lobe than the left lung.

Mutations in mice have shown that there are two levels of regulating the left-right axis: a global level and an organ-specific level. Some genes, such as *situs inversus viscerum* (*iv*), randomizes the left-right axis for each asymmetrical organ (Hummel and Chapman 1959; Layton 1976). This means that the heart may loop to the left in one homozygous animal, but loop to the right in another (Figure 11.43). Moreover, the di-

Figure 11.42
Left-right asymmetry in the developing human. (A) Abdominal cross sections show that the originally symmetrical organ rudiments acquire asymmetric positions by week 11. The liver moves to the right and the spleen moves to the left. (B) Not only does the heart move to the left side of the body, but the originally symmetrical veins of the heart regress differentially to form the superior and inferior venae cavae, which connect only to the right side of the heart. (C) The right lung branches into three lobes, while the left lung (near the heart) forms only two lobes. In human males, the scrotum also forms asymmetrically. (After Kosaki and Casey 1998.)

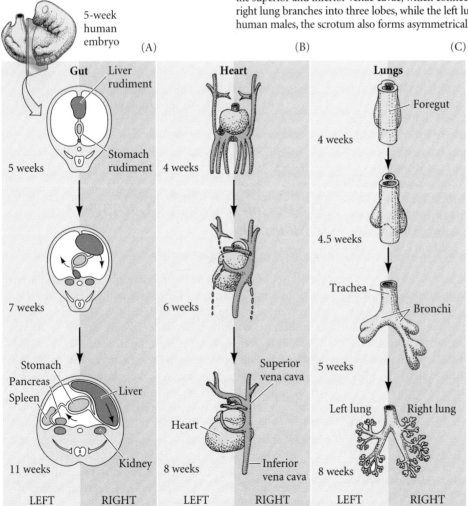

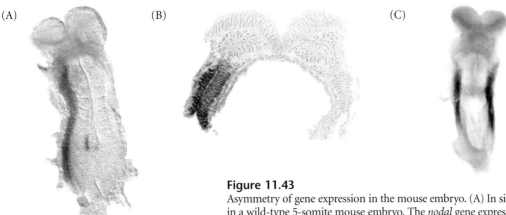

Figure 11.43
Asymmetry of gene expression in the mouse embryo. (A) In situ hybridization for *nodal* mRNA in a wild-type 5-somite mouse embryo. The *nodal* gene expression is confined to the lateral plate mesoderm on the left side of the embryo. (B) Cross section through the embryo at the same stage as (A). (C) In mice with the *situs inversus viscerum* (*iv*) mutation, *nodal* expression is seen in the lateral plate mesoderm on both sides of the embryo. The heart has an equal chance of looping to either side. (After Lowe et al. 1996; photographs courtesy of M. R. Kuehn.)

rection of heart looping is not coordinated with the placement of the spleen or the stomach. This can cause serious problems, even death. The second gene, *inversion of embryonic turning* (*inv*), causes a more global phenotype. Mice homozygous for an insertion mutation at this locus were found to have all their asymmetrical organs on the wrong side of the body (Yokoyama et al. 1993).* Since all the organs are reversed, this asymmetry does not have dire consequences for the mice.

*This gene was discovered accidentally when Yokoyama and colleagues (1993) made transgenic mice wherein the transgene (for the tyrosinase enzyme) was inserted randomly into the genome. In one instance, this gene inserted itself into a region of chromosome 4, knocking out the existing *inv* gene. The resulting homozygous mice had laterality defects.

Recently, several additional asymmetrically expressed genes have been discovered, and their influence on one another has enabled scientists to put them into a possible pathway. The end of this pathway—the activation of Nodal proteins and the Pitx2 transcription factor on the left side of the lateral plate mesoderm—appears to be the same as in frog and chick embryos, although the path leading to this point differs between the species (Figure 11.44; see Figure 11.17; Collignon et al. 1996; Lowe et al. 1996; Meno et al. 1996).

In frogs, the pathway begins with the placement of Vg1; in chicks it begins with the suppression of *sonic hedgehog* expression. In mammals, the distinction between left and right sides begins in the ciliary cells of the node (Figure 11.44B). The cilia cause the flow of fluid in the yolk sac cavity from right to left.

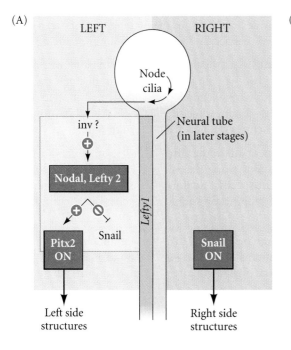

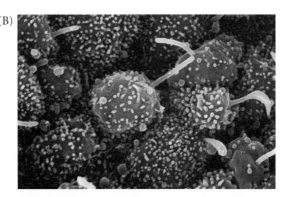

Figure 11.44
Situs formation in mammals. (A) Proposed pathway for left-right axis formation in the mouse. The leftward movement of cilia in the node activates some as yet unidentified factor (possibly the product of the *inv* gene). This product activates the *nodal* and *lefty2* genes. The diffusion of Nodal and Lefty2 proteins to the right-hand side is restricted by the product of the *Lefty1* gene which coats the bottom of the neural tube on the left side. Nodal activates *Pitx2*, the gene whose product activates left-sided properties in the various organs containing it. Either Nodal or Lefty2 (perhaps both) repress the *Snail* gene whose product is needed to instruct right-sidedness. (B) Ciliated cells of the mammalian node. This photograph is a close-up of the node seen in Figure 11.29A. (Photograph courtesy of K. Sulik and G. C. Schoenwolf.)

When Nonaka and colleagues (1998) knocked out a mouse gene encoding the ciliary motor protein dynein (see Chapter 7), the nodal cilia did not move, and the situs (lateral position) of each asymmetrical organ was randomized. This finding correlated extremely well with other data. First, it had long been known that humans having a dynein deficiency had immotile cilia and a random chance of having their hearts on the left or right side of the body (Afzelius 1976). Second, when the gene for the *iv* mutant mice described above was cloned, it was found to encode the ciliary dynein protein (Supp et al. 1997).

In some way (perhaps through the product of the *inv* gene), this leftward motion of the cilia activates the genes for two paracrine factors, Nodal and Lefty-2, in the lateral plate mesoderm on the left side of the embryo (Figure 11.44A). The proteins produced by these factors spread throughout the left side of the embryo, and appear to be restrained to that side by the Lefty-1 protein, which is secreted by the bottom left side of the neural tube (Meno et al. 1998). Lefty-2 appears to be able to block the Snail protein (which becomes specific for the right side of the body), while Nodal activates *pitx2* gene expression (Pedra et al. 1998).

The development of mammals has enormous importance for understanding the bases of numerous human diseases. In the next chapters, we will discuss later aspects of vertebrate development and the relationship between genetics and development during organ formation.

Snapshot Summary: The Early Development of Vertebrates

1. Fishes, reptiles, and birds undergo discoidal meroblastic cleavage, wherein the early cell divisions do not cut through the yolk of the egg. These cells form a blastoderm.

2. In fishes, the deep cells form between the yolk syncytial layer and the enveloping layer. These cells migrate over the top of the yolk, forming the hypoblast and epiblast layers. On the future dorsal side, these layers intercalate to form the embryonic shield, a structure homologous to the amphibian organizer. Transplantation of the embryonic shield into the ventral side of another embryo will cause the formation of a second embryonic axis.

3. There appear to be two signaling centers supplying anterior-posterior information in fishes, one located at the border between the neural and surface ectoderm, the other in the lateral mesoderm.

4. In chick embryos, early cleavage forms an area opaca and an area pellucida. The region between them is the marginal zone. Gastrulation begins at the posterior marginal zone, as the hypoblast and primitive streak both start there.

5. The primitive streak is derived from anterior epiblast cells and the central cells of the posterior marginal zone. As the primitive streak extends rostrally, Hensen's node is formed. Cells migrating through Hensen's node become chordamesoderm (notochord) cells. These extend up to the presumptive midbrain, where they meet the prechordal plate.

6. The prechordal plate induces the formation of the forebrain; the chordamesoderm induces the formation of the midbrain, hindbrain, and spinal cord. The first cells migrating laterally through the primitive streak become endoderm, displacing the hypoblast. The mesoderm cells then migrate through. Meanwhile, the surface ectoderm undergoes epiboly around the entire yolk.

7. In birds, gravity is critical in determining the anterior-posterior axis, while pH differences appear crucial for distinguishing dorsal from ventral. The left-right axis is formed by the expression of *nodal* on the left side of the embryo, which signals *pitx2* expression on the left side of developing organs.

8. Mammals undergo holoblastic rotational cleavage, characterized by a slow rate of division, a unique cleavage orientation, lack of divisional synchrony, and the formation of a blastocyst.

9. The blastocyst forms after the blastomeres undergo compaction. It contains outer cells—the trophoblast cells—that become the chorion, and an inner cell mass that becomes the amnion and the embryo.

10. The chorion forms the fetal portion of the placenta, which functions to provide oxygen and nutrition to the embryo, to provide hormones for the maintenance of pregnancy, and to provide barriers to the mother's immune system.

11. Mammalian gastrulation is not unlike that of birds. There appear to be two signaling centers—one in the node and one in the anterior visceral endoderm. The latter is critical for generating the forebrain, while the former is critical in inducing the axial structures caudally from the midbrain.

12. Hox genes pattern the anterior-posterior axis and help to specify positions along that axis. If Hox genes are knocked out, segment-specific malformations can arise. Similarly, causing the ectopic expression of Hox genes can alter the body axis.

13. The homology of gene structure and the similarity of expression patterns between *Drosophila* and mammalian Hox genes suggests that this patterning mechanism is extremely ancient.

14. The mammalian left-right axis is specified similarly to that of the chick.

Literature Cited

Afzelius, B. A. 1976. A human syndrome caused by immotile cilia. *Science* 193: 317–319.

Alvarez, I. S., M. Araujo and M. A. Nieto. 1998. Neural induction in whole embryo cultures by FGF. *Dev. Biol.* 199: 42–54.

Arendt, D. and K. Nübler-Jung. 1999. Rearranging gastrulation in the name of yolk: Evolution of gastrulation in yolk-rich amniote eggs. *Mech. Dev.* 81: 3–22.

Bachiller, D. and 10 others. 2000. The organizer factors Chordin and Noggin are required for mouse forebrain development. *Nature* 403: 658–661.

Bachvarova, R. F., I. Skromme and C. D. Stern. 1998. Induction of primitive streak and Hensen's node by the posterior marginal zone in the early chick embryo. *Development* 125: 3521–3534.

Balinsky, B. I. 1975. *Introduction to Embryology*, 4th Ed. Saunders, Philadelphia.

Ballard, W. B. 1981. Morphogenetic movements and fate maps of vertebrates. *Am. Zool.* 21: 391–399.

Balling, R., G. Mutter, P. Gruss and M. Kessel. 1989. Craniofacial abnormalities induced by ectopic expression of the homeobox gene *Hox-1.1* in transgenic mice. *Cell* 58: 337–347.

Barlow, P., D. A. J. Owen and C. Graham. 1972. DNA synthesis in the preimplantation mouse embryo. *J. Embryol. Exp. Morphol.* 27: 432–445.

Beams, H. W. and R. G. Kessel. 1975. Cytokineses: A comparative study of cytoplasmic division in animal cells. *Am. Sci.* 64: 279–290.

Beddington, R. S. P. 1994. Induction of a second neural axis by the mouse node. *Development* 120: 613–620.

Beddington, R. S. P. and E. J. Robertson. 1999. Axis development and early asymmetry in mammals. *Cell* 96: 195–209.

Bellairs, R. 1986. The primitive streak. *Anat. Embryol.* 174: 1–14.

Bellairs, R., A. S. Breathnach and M. Gross. 1975. Freeze–fracture replication of junctional complexes in unincubated and incubated chick embryos. *Cell Tissue Res.* 162: 235–252.

Bellairs, R., F. W. Lorenz and T. Dunlap. 1978. Cleavage in the chick embryo. *J. Embryol. Exp. Morphol.* 43: 55–69.

Benson, G. V., H. Lim, B. C. Paria, I. Satokata, S. K. Dey and R. L. Maas. 1996. Mechanisms of reduced fertility in Hoxa-10 mutant mice: Uterine homeosis and loss of maternal Hoxa-10 expression. *Development* 122: 2687–2696.

Bianchi, D. W., L. E. Wilkins-Haug, A. C. Enders and E. D. Hay . 1993. Origin of extraembryonic mesoderm in experimental animals: Relevance to chorionic mosaicism in humans. *Am. J. Med. Genet.* 46: 542–550.

Boncinelli, E., R. Somma, D. Acampora, M. Pannese, M. D'Esposito, A. Faiella and A. Simeone. 1988. Organization of human homeobox genes. *Hum. Reprod.* 3: 880–886.

Borland, R. M. 1977. Transport processes in the mammalian blastocyst. *Dev. Mammals* 1: 31–67.

Brenner, C. A., R. R. Adler, D. A. Rappolee, R. A. Pedersen and Z. Werb. 1989. Genes for extracellular matrix-degrading metalloproteases and their inhibitor, TIMP, are expressed during early mammalian development. *Genes Dev.* 3: 848–859.

Burdsall, C. A., C. H. Damsky and R. A. Pedersen. 1993. The role of E-cadherin and integrins in mesoderm differentiation and migration at the mammalian primitive streak. *Development* 118: 829–844.

Burke, A. C., A. C. Nelson, B. A. Morgan and C. Tabin. 1995. Hox genes and the evolution of vertebrate axial morphology. *Development* 121: 333–346.

Callebaut, M and E. van Neuten. 1994. Rauber's (Koller's) sickle: The early gastrulation organizer of the avian blastoderm. *Eur. J. Morphol.* 34: 347–361.

Carson, D. D., J.-P. Tang and J. Julian. 1993. Heparan sulfate proteoglycan (perlecan) expression by mouse embryos during acquisition of attachment competence. *Dev. Biol.* 155: 97–106.

Chisaka, O. and M. Capecchi. 1991. Regionally restricted developmental defects resulting from targeted disruption of the mouse homeobox gene *Hox-1.5*. *Nature* 350: 473–479.

Collignon, J., I. Varlet and E. J. Robertson. 1996. Relationship between asymmetric nodal expression and the direction of embryonic turning. *Nature* 381: 155–158.

Condie, B. G. and M. R. Capecchi. 1994. Mice with targeted disruptions in the paralogous genes *hoxa-3* and *hoxd-3* reveal synergistic interactions. *Nature* 370: 304–307.

Conlon, R. A. and J. Rossant. 1992. Exogenous retinoic acid rapidly induces anterior ectopic expression of murine *Hox-2* genes in vivo. *Development* 116: 357–368.

Cruz, Y. P. and R. A. Pedersen. 1991. Origin of embryonic and extraembryonic cell lineages in mammalian embryos. *In Animal Applications of Research in Mammalian Development*. Cold Spring Harbor Press, Cold Spring Harbor, NY, pp. 147–204.

Darnell, D. K., M. R. Stark and G. C. Schoenwolf. 1999. Timing and cell interactions underlying neural induction in the chick embryo. *Development* 126:2505–2514

de la Chappelle, A., J. Schroder, P. Rantanen, B. Thomasson, M. Niemi, A. Tilikainen, R. Sanger and E. E. Robson. 1974. Early fusion of two human embryos? *Ann. Hum. Genet.* 38: 63–75.

DeLuca, S. M. and 7 others. 1999. Hepatocyte growth factor/scatter factor promotes a switch from E- to N-cadherin in chick embryo epiblast cells. *Exp. Cell Res.* 251: 3–15.

Diaz, M. S. and G. C. Schoenwolf. 1990. Formation of ectopic neuroepithelium in chick blastoderms: Age-related capacities for induction and self-differentiation following transplantation of quail Hensen's nodes. *Anat. Rec.* 229: 437–448.

Driever, W. 1995. Axis formation in zebrafish. *Curr. Opin. Genet. Dev.* 5: 610–618.

Dyce, J., M. George, H. Goodall and T. P. Fleming. 1987. Do trophectoderm and inner cell mass cells in the mouse blastocyst maintain discrete lineages? *Development* 100: 685–698.

Elinson, R. P. 1987. Changes in developmental patterns: Embryos of amphibians with large eggs. *In* R. A. Raff and E. C. Raff (eds.), *Development as an Evolutionary Process*. Alan R. Liss, New York, pp. 1–21.

Evans, M. J. and M. H. Kaufman. 1981. Establishment in culture of pluripotent cells from mouse embryos. *Nature* 292: 154–156.

Eyal-Giladi, H. 1991. The early embryonic development of the chick, an epigenetic process. *Crit. Rev. Poultry Biol.* 3: 143–166.

Eyal-Giladi, H. 1997. Establishment of the axis in chordates: Facts and speculations. *Development* 124: 2285–2296.

Eyal-Giladi, H. and B. Fabian. 1980. Axis determination in uterine chick blastocysts under changing spatial positions during the sensitive period of polarity. *Dev. Biol.* 77: 228–232.

Eyal-Giladi, H., A. Debby and N. Harel. 1992. The posterior section of the chick's area pellucida and its involvement in hypoblast and primitive streak formation. *Development* 116: 819–830.

Fisher, S. J., T.-Y. Cui, L. Zhang, K. Grahl, Z. Guo-Yang, J. Tarpey and C. H. Damsky. 1989. Adhesive and degradative properties of the human placental cytotrophoblast cells in vitro. *J. Cell Biol.* 109: 891–902.

Fleming, T. P. 1987. Quantitative analysis of cell allocation to trophectoderm and inner cell mass in the mouse embryo. *Dev. Biol.* 119: 520–531.

Fluck, R. A., K. L. Krok, B. A. Bast, S. E. Michaud and C. E. Kim. 1998. Gravity influences the position of the dorsoventral axis in medaka fish embryos (*Oryzias latipes*). *Dev. Growth Diff.* 40: 509–518.

Fu, M.-F., C. Pressman, R. Dyer, R. L. Johnson and J. F. Martin. 1999. Function of Rieger syndrome gene in left-right asymmetry and craniofacial development. *Nature* 401: 276–278.

Gardiner, R. C. and J. Rossant. 1976. Determination during embryogenesis in mammals. *Ciba Found. Symp.* 40: 5–18.

Gardner, R. L. 1997. The early blastocyst is bilaterally symmetrical and its axis of symmetry is aligned with the animal-vegetal axis of the zygote in the mouse. *Development* 124: 289–301.

Gaunt, S. J. 1994. Conservation in the Hox code during morphological evolution. *Int. J. Dev. Biol.* 38: 549–552.

Gendron-Maguire, M., M. Mallo, M. Zhang and T. Gridley. 1993. Hoxa-2 mutant mice exhibit homeotic transformation of skeletal elements derived from cranial neural crest. *Cell* 75: 1317–1331.

Gilbert, S. G. 1989. *Pictorial Human Embryology.* University of Washington Press, Seattle.

Goulding, E. H. and R. M. Pratt. 1986. Isotretinoin teratogenicity in mouse whole embryo culture. *J. Craniofac. Genet. Dev. Biol.* 6: 99–112.

Goulding, M. D., A. Lumsden and P. Gruss. 1993. Signals from the notochord and floor plate regulate the region-specific expression of two *Pax* genes in the developing spinal cord. *Development* 117: 1001–1016.

Granato, M. and C. Nüsslein-Volhard. 1996. Fishing for genes controlling development. *Curr. Opin. Gen. Dev.* 6: 461–468.

Greer, J. M., J. Puetz, K. Thomas and M. R. Capecchi. 2000. Maintenance of functional equivalence during paralogous Hox gene evolution. *Nature* 403: 661–664.

Gritsman, K., W. S. Talbot and A. F. Schier. 2000. Nodal signaling patterns the organizer. *Development* 127: 921-932.

Gulyas, B. J. 1975. A reexamination of the cleavage patterns in eutherian mammalian eggs: Rotation of the blastomere pairs during second cleavage in the rabbit. *J. Exp. Zool.* 193: 235–248.

Halpern, M., R. Ho, C. Walker and C. Kimmel. 1993. Induction of somitic muscle pioneers and floor plate is distinguished by the zebrafish *no tail* mutation. *Cell* 75: 99–111.

Halpern, M. E., C. Thisse, R. K. Ho, B. Thise, B. Riggeleman, E. S. Trevarrow and J. H. Weinberg. 1995. Cell-autonomous shift from axial to paraxial mesoderm development in zebrafish *floating head* mutants. *Development* 121: 4257–4264.

Hammerschmidt, M. and 14 others. 1996. *dino* and *mercedes*, two genes regulating dorsal development in the zebrafish embryo. *Development* 123: 95–102.

Helde, K. A., E. T. Wilson, C. J. Cretekos and D. J. Grunwald. 1994. Contribution of early cells to the fate map of the zebrafish gastrula. *Science* 265: 517–520.

Hogan, B. L. M., C. Thaller and G. Eichele. 1992. Evidence that Hensen's node is a site of retinoic acid synthesis. *Nature* 359: 237–241.

Houart, C., M. Westerfield and S. W. Wilson. 1998. A small population of anterior cells patterns the forebrain during zebrafish gastrulation. *Nature* 391: 788–792.

Hume, C. R. and J. Dodd. 1993. *Cwnt-8C*: A novel Wnt gene with a potential role in primitive streak formation and hindbrain organization. *Development* 119: 1147–1160.

Hummel, K. P. and D. B. Chapman. 1959. Visceral inversion and associated anomalies in the mouse. *J. Hered.* 50: 9–13.

Hunt, P. and R. Krumlauf. 1991. Deciphering the Hox code: Clues to patterning branchial regions of the head. *Cell* 66: 1075–1078.

Hunt, P. and R. Krumlauf. 1992. Hox codes and positional specification in vertebrate embryonic axes. *Annu. Rev. Cell Biol.* 8: 227–256.

Hunt, P. and 7 others. 1991. A distinct *Hox* code for the branchial region of the head. *Nature* 353: 861–864.

Izpisua-Belmonte, J. C., E. M. De Robertis, K. G. Storey and C. D. Stern. 1993. The homeobox gene *goosecoid* and the origin of organizer cells in the early chick blastoderm. *Cell* 74: 645–659.

Jurand, A. 1974. Some aspects of the development of the notochord in mouse embryos. *J. Embryol. Exp. Morphol.* 32: 1–33.

Kane, D. A. and C. B. Kimmel. 1993. The zebrafish midblastula transition. *Development* 119: 447–456.

Kelly, G. M., D. F. Erezyilmaz and R. T. Moon. 1995. Induction of a secondary embryonic axis in zebrafish occurs following the overexpression of beta-catenin. *Mech. Dev.* 53: 261–273.

Kessel, M. 1992. Respecification of vertebral identities by retinoic acid. *Development* 115: 487–501.

Kessel, M. and P. Gruss. 1991. Homeotic transformations of murine vertebrae and concomitant alteration of Hox codes induced by retinoic acid. *Cell* 67: 89–104.

Keynes, R. and A. Lumsden. 1990. Segmentation and the origin of regional diversity in the vertebrate central nervous system. *Neuron* 2: 1–9.

Khaner, O. 1995. The rotated hypoblast of the chicken embryo does not initiate an ectopic axis in the epiblast. *Proc. Natl. Acad. Sci. USA* 92: 10733–10737.

Khaner, O. 1998. The ability to initiate an axis in the avian blastula is concentrated mainly at a posterior site. *Dev. Biol.* 194: 257–266.

Khaner, O. and Eyal-Giladi, H. 1989. The chick's marginal zone and primitive streak formation. I. Coordinative effect of induction and inhibition. *Dev. Biol.* 134: 206–214.

Kimmel, C. B. and R. D. Law. 1985. Cell lineage of zebrafish blastomeres. II. Formation of the yolk syncytial layer. *Dev. Biol.* 108: 86–93.

Kimmel, C. B. and R. M. Warga. 1987. Indeterminate cell lineage of the zebrafish embryo. *Dev. Biol.* 124: 269–280.

Kimmel, C. B., R. M. Warga and T. F. Schilling. 1990. Origin and organization of the zebrafish fate map. *Development* 108: 581–594.

Kishimoto, Y., K. H. Lee, L. Zon, M. Hammerschmidt and S. Schulte-Merker. 1997. The molecular nature of zebrafish *swirl*: BMP2 function is essential during early dorsoventral patterning. *Development* 124: 4457–4466.

Kochav, S. M. and H. Eyal-Giladi. 1971. Bilateral symmetry in the chick embryo determination by gravity. *Science* 171: 1027–1029.

Kosaki, K. and B. Casey. 1998. Genetics of human left-right axis malformations. *Semin. Cell Dev.* 9: 89–99.

Koshida, S., M. Shinya, T. Mizuno, A. Kuroiwa and H. Takeda. 1998. Initial anteroposterior pattern of zebrafish central nervous system is determined by differential competence of the epiblast. *Development* 125: 1957–1966.

Krumlauf, R. 1993. Hox genes and pattern formation in the branchial region of the vertebrate head. *Trends. Genet.* 9: 106–112.

Langeland, J. and C. B. Kimmel. 1997. The embryology of fish. *In* S. F. Gilbert and A. M. Raunio (eds.), *Embryology: Constructing the Organism.* Sinauer Associates, Sunderland, MA, pp. 383–407.

Langman, J. 1981. *Medical Embryology*, 4th Ed. Williams & Wilkins, Baltimore.

Larsen, W. J. 1993. *Human Embryology.* Churchill Livingston, New York.

Lash, J. W., E. Gosfield III, D. Ostrovsky and R. Bellairs. 1990. Migration of chick blastoderm under the vitelline membrane: The role of fibronectin. *Dev. Biol.* 139: 407–416.

Lawson, K. A., J. J. Meneses and R. A. Pedersen. 1991. Clonal analysis of epiblast fate during germ layer formation in the mouse embryo. *Development* 113: 891–911.

Layton, W. M., Jr. 1976. Random determination of a developmental process. *J. Hered.* 67: 336–338.

Le Douarin, N., A. Grapin-Botton and M. Catala. 1996. Patterning of the neural primordium in the avian embryo. *Semin. Dev. Biol.* 7: 157–167.

Le Mouellic, H., Y. Lallemand and P. Brûlet. 1992. Homeosis in the mouse induced by a null mutation in the *Hox-3.1* gene. *Cell* 69: 251–264.

Leung, C., S. E. Webb, and A. Miller. 1998. Calcium transients accompany ooplasmic segregation in zebrafish embryos. *Dev. Growth Differ.* 40: 313–326.

Levin, M., R. L. Johnson, C. Stern, M. Kuehn and C. Tabin. 1995. A molecular pathway determining left-right asymmetry in chick embryogenesis. *Cell* 82: 803–814.

Levin, M., Pagan, S., Roberts, D., Cooke, J., Kuehn, M. R. and Tabin, C. J. 1997. Left/Right patterning signals and the independent regulation of different aspects of *situs* in the chick embryo. *Dev. Biol.* 189: 57–67.

Lin, C. R. and 7 others. 1999. Pitx2 regulates lung asymmetry, cardiac positioning, and pituitary and tooth morphogenesis. *Nature* 401: 279–282.

Logan, M., S. M. Pagán-Westphal, D. M. Smith, L. Paganessi and C. J. Tabin. 1998. The transcription factor Pitx2 mediates situs-specific morphogenesis in response to left-right asymmetric signals. *Cell* 94: 307–317.

Lowe, L. A. and 8 others. 1996. Conserved left-right asymmetry of nodal expression and alterations in murine situs inversus. *Nature* 381: 158–161.

Luckett, W. P. 1978. Origin and differentiation of the yolk sac and extraembryonic mesoderm in presomite human and rhesus monkey embryos. *Am. J. Anat.* 152: 59–98.

Maden, M., E. Gale, I. Kostetski and M. Zile. 1996. Vitamin A-deficient quail embryos have half a hindbrain and other neural defects. *Curr. Biol.* 6: 417–426.

Manley, N. R. and M. R. Capecchi. 1995. The role of *Hoxa-3* in mouse thymus and thyroid development. *Development* 121: 1989–2003.

Mark, W. H., K. Signorelli and E. Lacy. 1985. An inserted mutation in a transgenic mouse line results in developmental arrest at day 5 of gestation. *Cold Spring Harbor Symp. Quant. Biol.* 50: 453–463.

Markert, C. L. and R. M. Petters. 1978. Manufactured hexaparental mice show that adults are derived from three embryonic cells. *Science* 202: 56–58.

Martin, G. R. 1981. Isolation of a pluripotent cell line from early mouse embryos cultured in medium conditioned by teratocarcinoma stem cells. *Proc. Natl. Acad. Sci. USA* 78: 7634–7638.

Mayr, W. R., V. Pausch and W. Schnedl. 1979. Human chimaera detectable only by investigation of her progeny. *Nature* 277: 210–211.

McGinnis, W. and R. Krumlauf. 1992. Homeobox genes and axial patterning. *Cell* 68: 283–302.

Meno, C. and 7 others. 1996. Left-right asymmetric expression of the TGFβ-family member *lefty* in mouse embryos. *Nature* 381: 151–155.

Meno, C. and 8 others. 1998. *lefty-1* is required for left-right determination as a regulator of *lefty-2* and *nodal*. *Cell* 94: 287–297.

Mitrani, E., T. Ziv, G. Thomsen, Y. Shimoni, D. A. Melton and A. Bril. 1990. Activin can induce the formation of axial structures and is expressed in the hypoblast of the chick. *Cell* 63: 495–501.

Morriss-Kay, G. 1993. Retinoic acid and craniofacial development: Molecules and morphogenesis. *BioEssays* 15: 9–15.

Mulnard, J. G. 1967. Analyse microcinematographique du developpement de l'oeuf de souris du stade II au blastocyste. *Arch. Biol.* (Liege) 78: 107–138.

New, D. A. T. 1956. The formation of sub-blastodermic fluid in hens' eggs. *J. Embryol. Exp. Morphol.* 43: 221–227.

New, D. A. T. 1959. Adhesive properties and expansion of the chick blastoderm. *J. Embryol. Exp. Morphol.* 7: 146–164.

Nguyen, V. H., B. Schmid, J. Trout, S. A. Connors, M. Ekker and M. C. Mullins. 1998. Ventral and lateral regions of the zebrafish gastrula, including the neural crest progenitors, are established by a *bmp2b/swirl* pathway of genes. *Dev. Biol.* 199: 93–110.

Nonaka, S. and 7 others. 1998. Randomization of left-right asymmetry due to loss of nodal cilia generating leftward flow of extraembryonic fluid in mice lacking KIF3B motor protein. *Cell* 95: 829–837.

Oppenheimer, J. M. 1936. Transplantation experiments on developing teleosts (*Fundulus* and *Perca*). *J. Exp. Zool.* 72: 409–437.

Pedersen, R. A., K. Wu and H. Batakier. 1986. Origin of the inner cell mass in mouse embryos: Cell lineage analysis by microinjection. *Dev. Biol.* 117: 581–595.

Pedra, M. E., J. M. Icardo, M. Albajar, J. C. Rodriguez-Rey and M. A. Ros. 1998. *Pitx2* participates in the late phase of the pathway controlling left-right asymmetry. *Cell* 94: 319–324.

Perona, R. M. and P. M. Wassarman. 1986. Mouse blastocysts hatch in vitro by using a trypsin-like proteinase associated with cells of mural trophectoderm. *Dev. Biol.* 114: 42–52.

Piko, L. and K. B. Clegg. 1982. Quantitative changes in total RNA, total poly(A), and ribosomes in early mouse embryos. *Dev. Biol.* 89: 362–378.

Prather, R. S. 1989. Nuclear transfer in mammals and amphibia. *In* H. Schatten and G. Schatten (eds.), *The Molecular Biology of Fertilization.* Academic Press, New York, pp. 323–340.

Psychoyos, D. and C. D. Stern. 1996. Fates and migratory routes of primitive streak cells in the chick embryo. *Development* 122: 1523–1534.

Rebagliati, M. R., R. Toyama, C. Fricke, P. Haffter and I. B. Sawid. 1998. Zebrafish *nodal*-related genes are implicated in axial patterning and establishing left-right asymmetry. *Dev. Biol.* 199: 261–272.

Renfree, M. B. 1982. Implantation and placentation. *In* C. R. Austin and R. V. Short (eds.), *Embryonic and Fetal Development.* Cambridge University Press, Cambridge, pp. 26–69.

Richardson, M. K. 1995. Heterochrony and the phylotypic period. *Dev. Biol.* 172: 412–421.

Richardson, M. K., J. Hanken, M. J. Gooneratne, C. Pieau, A. Raynaud, L. Selwood and G. M. Wright. 1997. There is no highly conserved embryonic stage in the vertebrates: Implications for current theories of evolution and development. *Anat. Embryol.* 196: 91–106.

Rijli, F. M., M. Mark, S. Lakkaraju, A. Dierich, P. Dollé and P. Chambon. 1993. A homeotic transformation is generated in the rostral branchial region of the head by disruption on *Hoxa-2*, which acts as a selector gene. *Cell* 75: 1333–1349.

Rodriguez-Esteban, C., J. Capdevilla, A. N. Economides, J. Pascual, Á. Ortiz and J. C. Izpisúa-Belmonte. 1999. The novel Cer-like protein Caronte mediates the establishment of embryonic left-right asymmetry. *Nature* 401: 243–251.

Rosenquist, G. C. 1966. A radioautographic study of labeled grafts in the chick blastoderm: Development from primitive-streak stages to stage 12. *Carnegie Inst. Wash. Contrib. Embryol.* 38: 31–110.

Rosenquist, G. C. 1972. Endoderm movements in the chick embryo between the short streak and head process stages. *J. Exp. Zool.* 180: 95–104.

Rugh, R. 1967. *The Mouse.* Burgess, Minneapolis.

Ryan, A. K. and 14 others. 1998. Pitx2 determines left-right asymmetry of internal organs in vertebrates. *Nature* 394: 545–551.

Sampath, K. and 8 others. 1998. Induction of the zebrafish ventral brain and floorplate requires cyclops/nodal signalling. *Nature* 375: 185–189.

Schier, A. F. and Talbot, W. S. 1998. The zebrafish organizer. *Curr. Opin. Genet. Dev.* 8: 464–471.

Schier, A. F., S. C. Neuhauss, K. A. Held, W. S. Talbot and W. Driever. 1997 The *one eyed pinhead* gene functions in mesoderm and endoderm formation in zebrafish and interacts with *no tail*. *Development* 124: 327–342.

Schlesinger, A. B. 1958. The structural significance of the avian yolk in embryogenesis. *J. Exp. Zool.* 138: 223–258.

Schmidt, B. and J. Campos-Ortega. 1994. Dorsoventral polarity of the zebrafish embryo is distinguishable prior to the onset of gastrulation. *Wilhelm Roux Arch. Dev. Biol.* 203: 374–380.

Schneider, S., H. Steinbeisser, R. M. Warga and P. Hausen. 1996. Beta-catenin translocation into nuclei demarcates the dorsalizing centers in frog and fish embryos. *Mech. Dev.* 57: 191–198.

Schoenwolf, G. C. 1991. Cell movements in the epiblast during gastrulation and neurulation in avian embryos. *In* R. Keller, W. H. Clark, Jr. and F. Griffin (eds.), *Gastrulation: Movements, Patterns, and Molecules.* Plenum, New York, pp. 1–28.

Schoenwolf, G. C., V. Garcia-Martinez and M. S. Diaz. 1992. Mesoderm movement and fate during amphibian gastrulation and neurulation. *Dev. Dynam.* 193: 235–248.

Schulte-Merker, S., K. J. Lee, A. P. McMahon and M. Hammerschmidt. 1997. The zebrafish organizer requires *chordino*. *Nature* 387: 862–863.

Scott, M. 1992. Vertebrate homeobox gene nomenclature. *Cell* 71: 551–553.

Seidel, F. 1952. Die Entwicklungspotenzen einer isolierten Blastomere des Zweizellenstadiums im Säugetiere. *Naturwissenschaften* 39: 355–356.

Seleiro, E. A. P., D. J. Connoly and J. Cooke. 1996. Early developmental expression and experimental axis determination by the chick *Vg1* gene. *Curr. Biol.* 6: 1476–1486.

Shawlot, W. and R. R. Behringer. 1995. Requirement for *Lim1* in head-organizer function. *Nature* 374:425–430.

Shinya, M., M. Furutani-Seiki, A. Kuroiwa and H. Takeda. 1999. Mosaic analysis with oep mutant reveals a repressive interaction between floor-plate and non-floor-plate mutant cells in the zebrafish neural tube. *Dev. Growth Diff.* 41: 135–142.

Simeone, A., D. Acampora, M. Arcioni, E. Boncinelli and F. Mavilio. 1990. Sequential activation of *Hox 2* genes by retinoic acid in human embryonal carcinoma cells. *Nature* 34: 763–766.

Smith, J. L. and G. C. Schoenwolf. 1998. Getting organized: New insights into the organizer of higher vertebrates. *Curr. Top. Dev. Biol.* 40: 79–110.

Solnica-Krezel, L. and W. Driever. 1994. Microtubule arrays of the zebrafish yolk cell: Organization and function during epiboly. *Development* 120: 2443–2455.

Solursh, M. and G. M. Morriss. 1977. Glycosaminoglycan synthesis in rat embryos during the formation of the primary mesenchyme and neural folds. *Dev. Biol.* 57: 75–86.

Solursh, M. and J. P. Revel. 1978. A scanning electron microscope study of cell shape and cell appendages in the primitive streak region of the rat and chick embryo. *Differentiation* 11: 185–190.

Spemann, H. and H. Mangold. 1924. Induction of embryonic primordia by implantation of organizers from a different species. *In* B. H. Willier and J. M. Oppenheimer (eds.), *Foundations of Experimental Embryology.* Hafner, New York, pp. 144–184.

Spratt, N. T., Jr. 1946. Formation of the primitive streak in the explanted chick blastoderm marked with carbon particles. *J. Exp. Zool.* 103: 259–304.

Spratt, N. T., Jr. 1947. Regression and shortening of the primitive streak in the explanted chick blastoderm. *J. Exp. Zool.* 104: 69–100.

Spratt, N. T., Jr. 1963. Role of the substratum, supracellular continuity, and differential growth in morphogenetic cell movements. *Dev. Biol.* 7: 51–63.

Spratt, N. T., Jr. and H. Haas. 1960. Integrative mechanisms in development of early chick blastoderm. I. Regulated potentiality of separate parts. *J. Exp. Zool.* 145: 97–138.

Stern, C. D. and D. R. Canning. 1988. Gastrulation in birds: A model system for the study of animal morphogenesis. *Experientia* 44: 651–657.

Stern, C. D., G. W. Ireland, S. E. Herrick, E. Gherardi, J. Gray, M. Perryman and M. Stoker. 1990. Epithelial scatter factor and development of the chick embryonic axis. *Development* 110: 1271–1284.

Storey, K. G., A. Goriely, C. M. Sargent, J. M. Brown, H. D. Burns, H. M. Abud and J. K. Heath. 1998. Early posterior neural tissue is induced by FGF in the chick embryo. *Development* 125: 473–484.

Strahle, U. and S. Jesuthasan. 1993. Ultraviolet irradiation impairs epiboly in zebrafish embryos: Evidence for a microtubule-dependent mechanism of epiboly. *Development* 119: 451–453.

Streit, A. and C. D. Stern. 1999. More to neural induction than inhibition of BMPs. *In* S. Mooney (ed.), *Cell Lineage and Fate Determination.* Academic Press, New York, pp. 437–449.

Strickland, S., E. Reich and M. I. Sherman. 1976. Plasminogen activator in early embryogenesis: Enzyme production by trophoblast and parietal endoderm. *Cell* 9: 231–240.

Sulik, K., D. B. Dehart, J. L. Carson, T. Vrablic, K. Gesteland and G. C. Schoenwolf. 1994. Morphogenesis of the murine node and notochordal plate. *Dev. Dynam.* 201: 260–278.

Supp, D. M., D. P. Witte, S. S. Potter and M. Brueckner. 1997. Mutation of an axonal dynein affects left-right asymmetry in *inversus viscerum* mice. *Nature* 389: 963–966.

Tanaka, S., T. Kunath, A.-K. Hadjantonakis, A. Nagy and J. Rossant. 1998. Promotion of trophoblast stem cell proliferation by FGF4. *Science* 282: 2072–2075.

Thisse, B., C. V. E. Wright and C. Thisse. 2000. Activin and Nodal-related factors control anterior-posterior patterning of the zebrafish embryo. *Nature* 403: 425–427.

Thomas, P. Q. and R. S. P. Beddington. 1996. Anterior primitive endoderm may be responsible for patterning the anterior neural plate in the mouse embryo. *Curr. Biol.* 6: 1487–1496.

Trinkaus, J. P. 1984. Mechanisms of *Fundulus* epiboly—a current view. *Am. Zool.* 24: 673–688.

Trinkaus, J. P. 1992. The midblastula transition, the YSL transition, and the onset of gastrulation in *Fundulus. Development* [Suppl.] 1992: 75–80.

Trinkaus, J. P. 1993. The yolk syncitial layer of *Fundulus*: Its orign and history and its significance for early embryogenesis. *J. Exp. Zool.* 265: 258–284.

Trinkaus, J. P. 1996. Ingression during early gastrulation of *Fundulus. Dev. Biol.* 177: 356–370.

Tuchmann-Duplessis, H., G. David and P. Haegel. 1972. *Illustrated Human Embryology,* Vol. 1. Springer-Verlag, New York.

Vakaet, L. 1984. The initiation of gastrula ingression in the chick blastoderm. *Am. Zool.* 24: 555–562.

von Baer, K. E. 1828. *Entwicklungsgeschichte der Thirer: Beobachtung und Reflexion.* Bornträger, Konigsberg.

von Baer. K. E. 1886. *Autobiography of Karl Ernst von Baer* (trans. H. Schneider). Science History Publications, Canton, MA, pp. 261–262.

Waddington, C. H. 1932. Experiments in the development of chick and duck embryos cultivated in vitro. *Phil. Trans. R. Soc. Lond.* [B] 13: 221.

Waddington, C. H. 1933. Induction by the primitive streak and its derivatives in the chick. *J. Exp. Zool.* 10: 38–46.

Waddington, C. H. 1934. Experiments on embryonic induction. *J. Exp. Zool.* 11: 211–227.

Warga, R. M. and C. B. Kimmel. 1990. Cell movements during epiboly and gastrulation in zebrafish. *Development* 108: 569–580.

Wiley, L. M. 1984. Cavitation in the mouse preimplantation embryo: Na/K ATPase and the origin of nascent blastocoel fluid. *Dev. Biol.* 105: 330–342.

Wilkinson, D. G., S. Bhatt, M. Cook, E. Boncinelli and R. Kruflauf. 1989. Segmental expression of *Hox-2* homeobox-containing genes in the developing mouse hindbrain. *Nature* 341: 405–409.

Wolpert, L., R. Beddington, J. Brockes, T. Jessell, P. Lawrence and E. Meyerowitz. 1998. *Principles of Development.* Current Biology, Ltd., London.

Woo, K. and S. E. Fraser. 1997. Specification of the zebrafish nervous system by nonaxial signals. *Science* 277: 254–257.

Yamazaki, K. and Y. Kato. 1989. Sites of zona pellucida shedding by mouse embryo other than mural trophectoderm. *J. Exp. Zool.* 249: 347–349.

Yokouchi, Y., K. J. Vogan, R. V. Pearse II and C. J. Tabin. 1999. Antagonistic signaling by *Caronte,* a novel *Cerberus*-related gene, establishes left-right asymmetric gene expression. *Cell* 98: 573–583.

Yokoyama, T., N . G. Copeland, N. A. Jenkins, C. A. Montgomery, F. F. B. Elder and P. A. Overbeek. 1993. Reversal of left-right symmetry: A *situs inversus* mutation. *Science* 260: 679–682.

c h a p t e r 12   *The central nervous system and the epidermis*

For the real amazement, if you wish to be amazed, is this process. You start out as a single cell derived from the coupling of a sperm and an egg; this divides in two, then four, then eight, and so on, and at a certain stage there emerges a single cell which has as all its progeny the human brain. The mere existence of such a cell should be one of the great astonishments of the earth. People ought to be walking around all day, all through their waking hours calling to each other in endless wonderment, talking of nothing except that cell.

LEWIS THOMAS (1979)

Even more appealing than virgin forest was the jungle lying before me at that moment: the central nervous system.

RITA LEVI-MONTALCINI (1988)

"WHAT IS PERHAPS THE MOST INTRIGUING QUESTION of all is whether the brain is powerful enough to solve the problem of its own creation." So Gregor Eichele (1992) ended a review of research on mammalian brain development. The construction of an organ that perceives, thinks, loves, hates, remembers, changes, fools itself, and coordinates our conscious and unconscious bodily processes is undoubtedly the most challenging of all developmental enigmas. A combination of genetic, cellular, and organismal approaches is giving us a preliminary understanding of how the basic anatomy of the brain becomes ordered.

The fates of the vertebrate ectoderm are shown in Figure 12.1. In the past two chapters, we have seen how the ectoderm is instructed to form the nervous system and the epidermis. A portion of the dorsal ectoderm is specified to become neural ectoderm, and its cells become distinguishable by their columnar appearance. This region of the embryo is called the **neural plate**. The process by which this tissue forms a **neural tube**, the rudiment of the central nervous system, is called **neurulation**, and an embryo undergoing such changes is called a **neurula** (Figure 12.2). The neural tube will form the brain anteriorly and the spinal cord. This chapter will look at the processes by which the neural tube and the epidermis arise and acquire their distinctive patterns.

Formation of the Neural Tube

There are two major ways of forming a neural tube. In **primary neurulation**, the cells surrounding the neural plate direct the neural plate cells to proliferate, invaginate, and pinch off from the surface to form a hollow tube. In **secondary neurulation**, the neural tube arises from a solid cord of cells that sinks into the embryo and subsequently hollows out (cavitates) to form a hollow tube. The extent to which these modes of construction are used varies among vertebrate classes. Neurulation in fishes is exclusively secondary. In birds, the anterior portions of the neural tube are constructed by primary neurulation, while the neural tube caudal to the twenty-seventh somite pair (i.e., everything posterior to the hindlimbs) is made by secondary neurulation (Pasteels 1937; Catala et al. 1996). In amphibians, such as *Xenopus*, most of the tadpole neural tube is made by primary neurulation, but the tail neural tube is derived from secondary neurulation (Gont et al. 1993). In mice (and probably humans, too), secondary neurulation begins at or around the level of somite 35 (Schoenwolf 1984; Nievelstein et al. 1993).

379

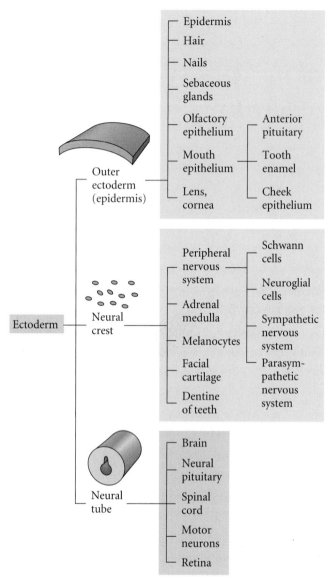

Figure 12.1
Major derivatives of the ectoderm germ layer. The ectoderm is divided into three major domains—the surface ectoderm (primarily epidermis), the neural tube (brain and spinal cord), and the neural crest (peripheral neurons, pigment, facial cartilage).

Primary neurulation

The events of primary neurulation in the chick and the frog are illustrated in Figures 12.3 and 12.4, respectively. During primary neurulation, the original ectoderm is divided into three sets of cells: (1) the internally positioned neural tube, which will form the brain and spinal cord, (2) the externally positioned epidermis of the skin, and (3) the neural crest cells. The neural crest cells form in the region that connects the neural tube and epidermis, but then migrate elsewhere; they will generate the peripheral neurons and glia, the pigment cells of the skin, and several other cell types.

VADE MECUM **Chick Neurulation.** By 33 hours of incubation, neurulation in the chick embryo is well underway. Both whole mounts and a complete set of serial cross sections through a 33-hour chick embryo are included in this segment so that you can see this amazing event. The serial sections can be displayed either as a continuum in movie format or individually, along with labels and color-coding that designates germ layers.
[Click on Chick-Mid]

The process of primary neurulation appears to be similar in amphibians, reptiles, birds, and mammals (Gallera 1971). Shortly after the neural plate has formed, its edges thicken and move upward to form the **neural folds**, while a U-shaped **neural groove** appears in the center of the plate, dividing the future right and left sides of the embryo (see Figures 12.2C and 12.3). The neural folds migrate toward the midline of the

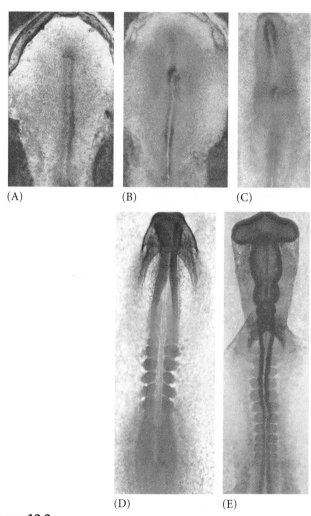

Figure 12.2
Neurulation in a chick embryo (dorsal view). (A) Flat neural plate. (B) Flat neural plate with underlying notochord (head process). (C) Neural groove. (D) Incipient neural tube. (E) Neural tube, showing the three brain regions and the spinal cord. (Photographs courtesy of G. C. Schoenwolf.)

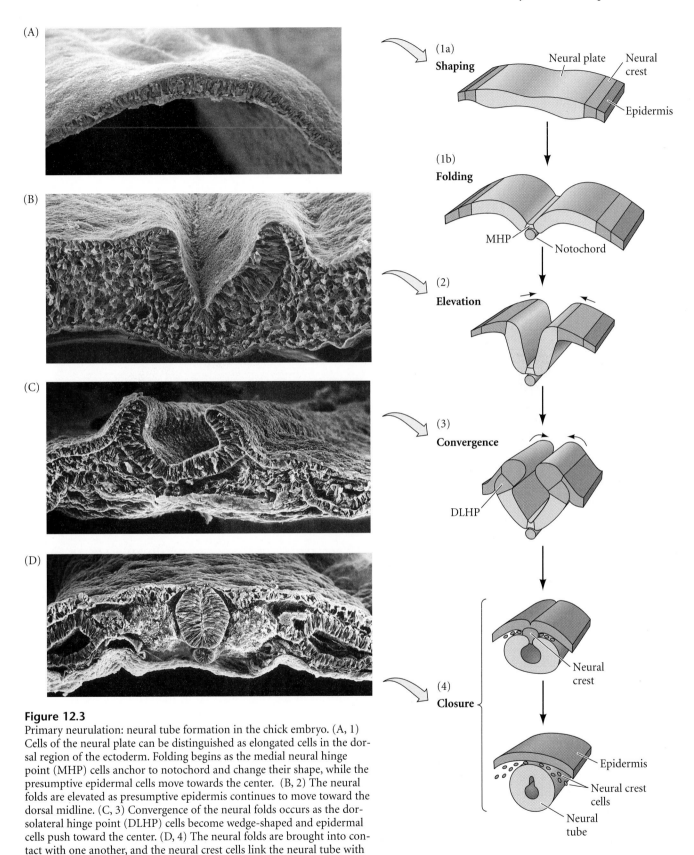

Figure 12.3
Primary neurulation: neural tube formation in the chick embryo. (A, 1) Cells of the neural plate can be distinguished as elongated cells in the dorsal region of the ectoderm. Folding begins as the medial neural hinge point (MHP) cells anchor to notochord and change their shape, while the presumptive epidermal cells move towards the center. (B, 2) The neural folds are elevated as presumptive epidermis continues to move toward the dorsal midline. (C, 3) Convergence of the neural folds occurs as the dorsolateral hinge point (DLHP) cells become wedge-shaped and epidermal cells push toward the center. (D, 4) The neural folds are brought into contact with one another, and the neural crest cells link the neural tube with the epidermis. The neural crest cells then disperse, leaving the neural tube separate from the epidermis. (Photographs courtesy of K. Tosney and G. Schoenwolf; drawings after Smith and Schoenwolf 1997.)

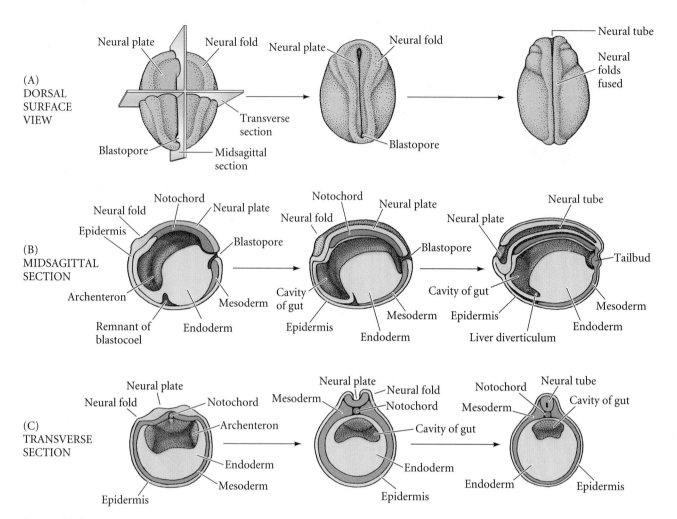

Figure 12.4
Three views of neurulation in an amphibian embryo, showing early (left), middle (center), and late (right) neurulae in each case. (A) Looking down on the dorsal surface of the whole embryo. (B) Sagittal section through the medial plane of the embryo. (C) Transverse section through the center of the embryo. (After Balinsky 1975.)

embryo, eventually fusing to form the neural tube beneath the overlying ectoderm. The cells at the dorsalmost portion of the neural tube become the **neural crest** cells.

Neurulation occurs in somewhat different ways in different regions of the body. The head, trunk, and tail each form their region of the neural tube in ways that reflect the inductive relationship of the pharyngeal endoderm, prechordal plate, and notochord to its overlying ectoderm (Chapters 10 and 11). The head and trunk regions both undergo variants of primary neurulation, and this process can be divided into four distinct but spatially and temporally overlapping stages: (1) formation of the neural plate; (2) shaping of the neural plate; (3) bending of the neural plate to form the neural groove; and (4) closure of the neural groove to form the neural tube (Smith and Schoenwolf 1997; see Figure 12.2).

FORMATION AND SHAPING OF THE NEURAL PLATE. The process of neurulation begins when the underlying dorsal mesoderm (and pharyngeal endoderm in the head region) signals the ectodermal cells above it to elongate into columnar neural plate cells (Smith and Schoenwolf 1989; Keller et al. 1992). Their elongated shape distinguishes the cells of the prospective neural plate from the flatter pre-epidermal cells surrounding them. As much as 50% of the ectoderm is included in the neural plate. The neural plate is shaped by the intrinsic movements of the epidermal and neural plate regions. The neural plate lengthens along the anterior-posterior axis, narrowing itself so that subsequent bending will form a tube (instead of a spherical capsule).

In both amphibians and amniotes, the neural plate lengthens and narrows by convergent extension, intercalating several layers of cells into a few layers. In addition, the cell divisions of the neural plate cells are preferentially in the **rostral-caudal** (beak-tail; anterior-posterior) direction (Jacobson and Sater 1988; Schoenwolf and Alvarez 1989; Sausedo et al. 1997; see Figures 12.2 and 12.3). These events will occur even if the tissues involved are isolated. If the neural plate is isolated, its cells converge and extend to make a thinner plate, but fail to roll up into a neural tube. However, if the "border region" containing both presumptive epidermis and neural plate tissue is isolated, it will form small neural folds in culture (Jacobson and Moury 1995; Moury and Schoenwolf 1995).

> WEBSITE **12.1 Formation of the floor plate cells.** One of the major controversies in developmental neurobiology concerns the origin of the cells that form the ventral floor of the neural tube. It is possible that these cells are derived directly from the notochord and do not arise from the surface ectoderm.

BENDING OF THE NEURAL PLATE. The bending of the neural plate involves the formation of **hinge regions** where the neural tube contacts surrounding tissues. In these regions, the presumptive epidermal cells adhere to the lateral edges of the neural plate and move them toward the midline (see Figure 12.3B). In birds and mammals, the cells at the midline of the neural plate are called the **medial hinge point** (**MHP**) **cells**. They are derived from the portion of the neural plate just anterior to Hensen's node and from the anterior midline of Hensen's node (Schoenwolf 1991a,b; Catala et al. 1996). The MHP cells become anchored to the notochord beneath them and form a hinge, which forms a furrow at the dorsal midline. The notochord induces the MHP cells to decrease their height and to become wedge-shaped (van Straaten et al. 1988; Smith and Schoenwolf 1989). The cells lateral to the MHP do not undergo such a change (Figures 12.3B, C). Shortly thereafter, two other hinge regions form furrows near the connection of the neural plate with the remainder of the ectoderm. These regions are called the **dorsolateral hinge points** (**DLHPs**), and they are anchored to the surface ectoderm of the neural folds. These cells, too, increase their height and become wedge-shaped.

Cell wedging is intimately linked to changes in cell shape. In the DLHPs, microtubules and microfilaments are both involved in these changes. Colchicine, an inhibitor of microtubule polymerization, inhibits the elongation of these cells, while cytochalasin B, an inhibitor of microfilament formation, prevents the apical constriction of these cells, thereby inhibiting wedge formation (Burnside 1973; Karfunkel 1972; Nagele and Lee 1987). After the initial furrowing of the neural plate, the plate bends around these hinge regions. Each hinge acts as a pivot that directs the rotation of the cells around it (Smith and Schoenwolf 1991).

Meanwhile, extrinsic forces are also at work. The surface ectoderm of the chick embryo pushes toward the midline of the embryo, providing another motive force for the bending of the neural plate (see Figure 12.3C; Alvarez and Schoenwolf 1992). This movement of the presumptive epidermis and the anchoring of the neural plate to the underlying mesoderm may also be important for ensuring that the neural tube invaginates into the embryo and not outward. If small pieces of neural plate are isolated from the rest of the embryo (includ-

ing the mesoderm), they tend to roll inside out (Schoenwolf 1991a). The pushing of the presumptive epidermis toward the center and the furrowing of the neural tube creates the neural folds.

CLOSURE OF THE NEURAL TUBE. The neural tube closes as the paired neural folds are brought together at the dorsal midline. The folds adhere to each other, and the cells from the two folds merge. In some species, the cells at this junction form the neural crest cells. In birds, the neural crest cells do not migrate from the dorsal region until after the neural tube has been closed at that site. In mammals, however, the cranial neural crest cells (which form facial and neck structures) migrate while the neural folds are elevating (i.e., prior to neural tube closure), whereas in the spinal cord region, the crest cells wait until closure has occurred (Nichols 1981; Erickson and Weston 1983).

The closure of the neural tube does not occur simultaneously throughout the ectoderm. This is best seen in those vertebrates (such as birds and mammals) whose body axis is elongated prior to neurulation. Figure 12.5 depicts neurulation in a 24-hour chick embryo. Neurulation in the **cephalic** (head) region is well advanced, while the **caudal** (tail) region of the embryo is still undergoing gastrulation. Regionalization of the neural tube also occurs as a result of changes in the shape of the tube. In the cephalic end (where the brain will form), the wall of the tube is broad and thick. Here, a series of swellings and constrictions define the various brain compartments. Caudal to the head region, however, the neural tube remains a simple tube that tapers off toward the tail. The two open ends of the neural tube are called the **anterior neuropore** and the **posterior neuropore**.

Unlike neurulation in chicks (in which neural tube closure is initiated at the level of the future midbrain and "zips up" in both directions), neural tube closure in mammals is initiated at several places along the anterior-posterior axis (Golden and Chernoff 1993; Van Allen et al. 1993). Different **neural tube defects** are caused when various parts of the neural tube fail to close (Figure 12.6). Failure to close the human *posterior* neural tube regions at day 27 (or the subsequent rupture of the posterior neuropore shortly thereafter) results in a condition called **spina bifida**, the severity of which depends on how much of the spinal cord remains exposed. Failure to close the *anterior* neural tube regions results in a lethal condition, **anencephaly**. Here, the forebrain remains in contact with the amniotic fluid and subsequently degenerates. Fetal forebrain development ceases, and the vault of the skull fails to form. The failure of the entire neural tube to close over the entire body axis is called **craniorachischisis**. Collectively, neural tube defects are not rare in humans, as they are seen in about 1 in every 500 live births. Neural tube closure defects can often be detected during pregnancy by various physical and chemical tests.

Human neural tube closure requires a complex interplay between genetic and environmental factors. Certain genes,

Figure 12.5
Stereogram of a 24-hour chick embryo. Cephalic portions are finishing neurulation while the caudal portions are still undergoing gastrulation. (From Patten 1971; after Huettner 1949.)

Figure 12.6
Neurulation in the human embryo. (A) Dorsal and transverse sections of a 22-day human embryo initiating neurulation. Both anterior and posterior neuropores are open to the amniotic fluid. (B) Dorsal view of a neurulating human embryo a day later. The anterior neuropore region is closing while the posterior neuropore remains open. (C) Regions of neural tube closure postulated by genetic evidence (superimposed on newborn body). (D) Anencephaly is caused by the failure of neural plate fusion in region 2. (E) Spina bifida is caused by the failure of region 5 to fuse (or of the posterior neuropore to close). (C–E after Van Allen et al. 1993.)

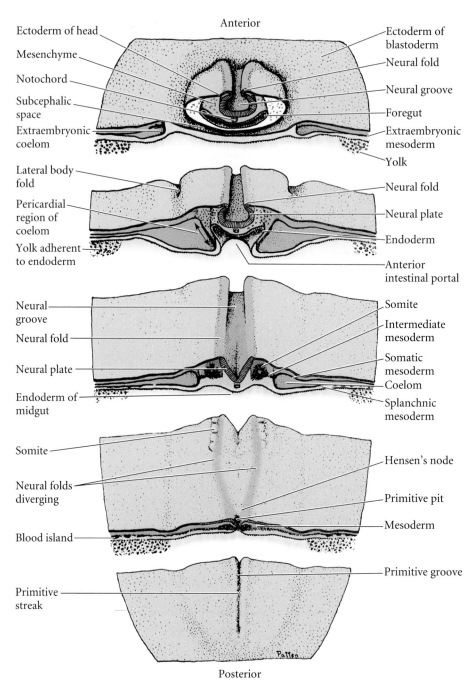

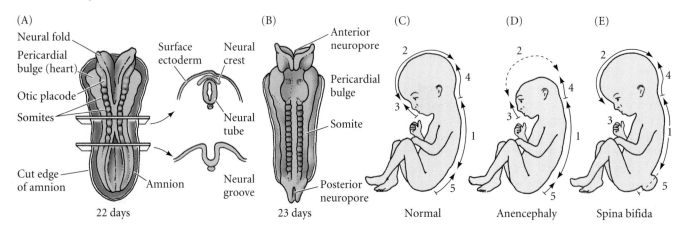

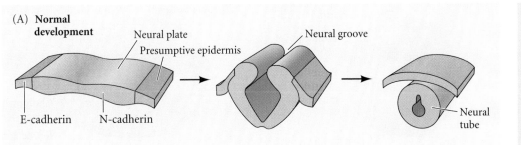

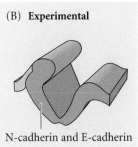

(A) **Normal development**

Neural plate
Presumptive epidermis
Neural groove

E-cadherin   N-cadherin

Neural tube

(B) **Experimental**

N-cadherin and E-cadherin

Figure 12.7
Expression of N-cadherin and E-cadherin adhesion proteins during neurulation in *Xenopus*. (A) Normal development. In the neural plate stage, N-cadherin is seen in the neural plate, while E-cadherin is seen on the presumptive epidermis. Eventually, the N-cadherin-bearing neural cells separate from the E-cadherin-containing epidermal cells. (The neural crest cells have neither cadherin, and they disperse.) (B) No separation of the neural tube occurs when one side of the frog embryo is injected with N-cadherin mRNA, so that N-cadherin is expressed in the epidermal cells as well as in the presumptive neural tube.

such as *Pax3*, *sonic hedgehog*, and *openbrain*, are essential for the formation of the mammalian neural tube, but dietary factors, such as cholesterol and folic acid, also appear to be critical. It has been estimated that 50% of human neural tube defects could be prevented by a pregnant woman's taking supplemental folic acid (vitamin B_{12}), and the U.S. Public Health Service recommends that all women of childbearing age take 0.4 mg of folate daily to reduce the risk of neural tube defects during pregnancy (Milunsky et al. 1989; Czeizel and Dudas 1992; Centers for Disease Control 1992).

The neural tube eventually forms a closed cylinder that separates from the surface ectoderm. This separation is thought to be mediated by the expression of different cell adhesion molecules. Although the cells that will become the neural tube originally express E-cadherin, they stop producing this protein as the neural tube forms, and instead synthe-

size N-cadherin and N-CAM (Figure 12.7). As a result, the surface ectoderm and neural tube tissues no longer adhere to each other. If the surface ectoderm is experimentally made to express N-cadherin (by injecting N-cadherin mRNA into one cell of a 2-cell *Xenopus* embryo), the separation of the neural tube from the presumptive epidermis is dramatically impeded (Detrick et al. 1990; Fujimori et al. 1990).

WEBSITE **12.2 Neural tube closure.** The closing of the neural tube is a complex event that can be influenced by both genes and environment. The interactions between genetic and environmental factors are now being untangled.

Secondary neurulation

Secondary neurulation involves the making of a **medullary cord** and its subsequent hollowing into a neural tube (Figure 12.8). Knowledge of the mechanisms of secondary neurulation may be important in medicine, given the prevalence of human posterior spinal cord malformations.

In frogs and chicks, secondary neurulation is usually seen in the neural tube of the lumbar (abdominal) and tail vertebrae. In both cases, it can be seen as a continuation of gastrulation. In the frog, instead of involuting into the embryo, the cells of the dorsal blastopore lip keep growing ventrally (Figure 12.9A, B). The growing region at the tip of the lip is called the **chordoneural hinge** (Pasteels 1937), and it contains precur-

(A)   Medullary cord

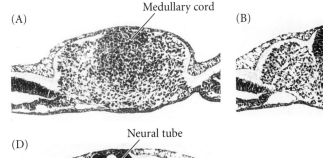

(B)

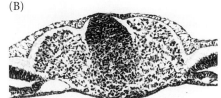

(C)

Notochord

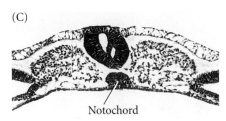

(D)   Neural tube

Figure 12.8
Secondary neurulation in the caudal region of a 25-somite chick embryo. (A) The medullary cord forming at the most caudal end of the chick tailbud. (B) The medullary cord at a slightly more anterior position in the tailbud. (C) The neural tube is cavitating and the notochord forming. (D) The lumens coalesce to form the central canal of the neural tube. (From Catala et al. 1995; photographs courtesy of N. M. Le Douarin.)

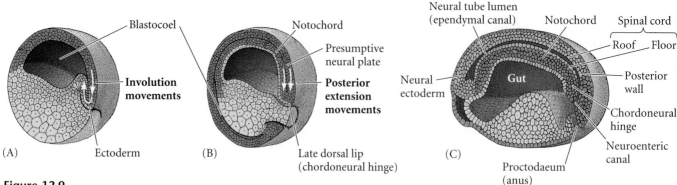

Figure 12.9
Movements of cells during secondary neurulation in *Xenopus*.
(A) Involution of the mesoderm at the mid-gastrula stage.
(B) Movements of the dorsal blastopore lip at the late gastrula/early neurula stage. Involution has ceased, and both the ectoderm and the mesoderm of the late blastopore lip move posteriorly. (C) Early tadpole stage, wherein the cells lining the blastopore form the neurenteric canal, part of which becomes the lumen of the secondary neural tube. (From Gont et al. 1993.)

sors for both the posteriormost portion of the neural plate and the posterior portion of the notochord. The growth of this region converts the roughly spherical gastrula, 1.2 mm in diameter, into a linear tadpole some 9 mm long. The tip of the tail is the direct descendant of the dorsal blastopore lip, and the cells lining the blastopore form the **neurenteric canal**. The proximal part of the neurenteric canal fuses with the anus, while the distal portion becomes the **ependymal canal** (i.e., the lumen of the neural tube) (Figure 12.9C; Gont et al. 1993).

Differentiation of the Neural Tube

The differentiation of the neural tube into the various regions of the central nervous system occurs simultaneously in three different ways. On the gross anatomical level, the neural tube and its lumen bulge and constrict to form the chambers of the brain and the spinal cord. At the tissue level, the cell populations within the wall of the neural tube rearrange themselves to form the different functional regions of the brain and the spinal cord. Finally, on the cellular level, the neuroepithelial cells themselves differentiate into the numerous types of nerve cells (**neurons**) and supportive cells (**glia**) present in the body.

The early development of most vertebrate brains is similar, but because the human brain may be the most organized piece of matter in the solar system and is arguably the most interesting organ in the animal kingdom, we will concentrate on the development that is supposed to make *Homo* sapient.

Figure 12.10
Early human brain development. The three primary brain vesicles are subdivided as development continues. At the right is a list of the adult derivatives formed by the walls and cavities of the brain. (After Moore and Persaud 1993.)

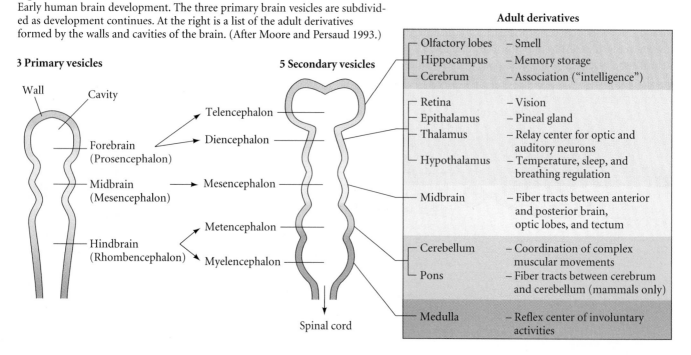

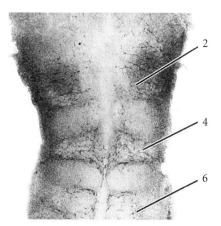

Figure 12.11
A 2-day embryonic chick hindbrain, splayed to show the lateral walls. Neurons were visualized with an antibody staining neurofilament proteins. Rhombomeres 2, 4, and 6 are distinguished by the high density of axons at this early developmental stage. (From Lumsden and Keynes 1989; photograph courtesy of A. Keynes.)

The anterior-posterior axis

The early mammalian neural tube is a straight structure. However, even before the posterior portion of the tube has formed, the most anterior portion of the tube is undergoing drastic changes. In this region, the neural tube balloons into three primary vesicles (Figure 12.10): forebrain (**prosencephalon**), midbrain (**mesencephalon**), and hindbrain (**rhombencephalon**). By the time the posterior end of the neural tube closes, secondary bulges—the **optic vesicles**—have extended laterally from each side of the developing forebrain.

The prosencephalon becomes subdivided into the anterior **telencephalon** and the more caudal **diencephalon**. The telencephalon will eventually form the cerebral hemispheres, and the diencephalon will form the thalamic and hypothalamic brain regions that receive neural input from the retina. Indeed, the retina itself is a derivative of the diencephalon. The mesencephalon does not become subdivided, and its lumen eventually becomes the cerebral aqueduct. The rhombencephalon becomes subdivided into a posterior **myelencephalon** and a more anterior **metencephalon**. The myelencephalon eventually becomes the medulla oblongata, whose neurons generate the nerves that regulate respiratory, gastrointestinal, and cardiovascular movements. The metencephalon gives rise to the cerebellum, the part of the brain responsible for coordinating movements, posture, and balance. The rhombencephalon develops a segmental pattern that specifies the places where certain nerves originate. Periodic swellings called **rhombomeres** divide the rhombencephalon into smaller compartments. These rhombomeres represent separate developmental "territories" in that the cells within each rhombomere can mix freely within it, but not with cells from adjacent rhombomeres (Guthrie and Lumsden 1991). Moreover, each rhombomere has

a different developmental fate. Each rhombomere will form **ganglia**—clusters of neuronal cell bodies whose axons form a nerve. The generation of the cranial nerves from the rhombomeres has been most extensively studied in the chick, in which the first neurons appear in the even-numbered rhombomeres, r2, r4, and r6 (Figure 12.11; Lumsden and Keynes 1989). Neurons originating from r2 ganglia form the fifth (trigeminal) cranial nerve; those from r4 form the seventh (facial) and eighth (vestibuloacoustic) cranial nerves; and the ninth (glossopharyngeal) cranial nerve exits from r6.

The ballooning of the early embryonic brain is remarkable in its rate, in its extent, and in its being the result primarily of an increase in cavity size, not tissue growth. In the chick embryo, brain volume expands thirty-fold between days 3 and 5 of development. This rapid expansion is thought to be caused by positive fluid pressure exerted against the walls of the neural tube by the fluid within it. It might be expected that this fluid pressure would be dissipated by the spinal cord, but this does not appear to happen. Rather, as the neural folds close in the region between the presumptive brain and the presumptive spinal cord, the surrounding dorsal tissues push in to constrict the neural tube at the base of the brain (Figure 12.12; Schoenwolf and Desmond 1984; Desmond and

Figure 12.12
Occlusion of the neural tube allows expansion of the future brain region. (A) Dye injected into the anterior portion of a 3-day chick neural tube fills the brain region, but does not pass into the spinal region. (B, C) Section of the chick neural tube at the base of the brain (B) before occlusion and (C) during occlusion. (D) Reopening of the occlusion after initial brain enlargement allows dye to pass from the brain region into the spinal cord region. (Photographs courtesy of M. Desmond.)

(A)   (B)   (D)

(C)

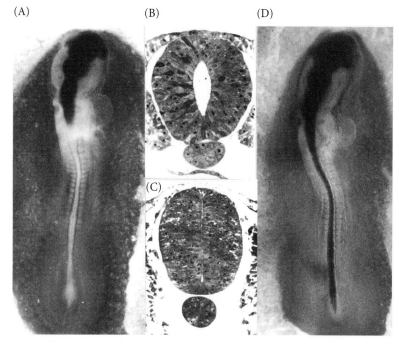

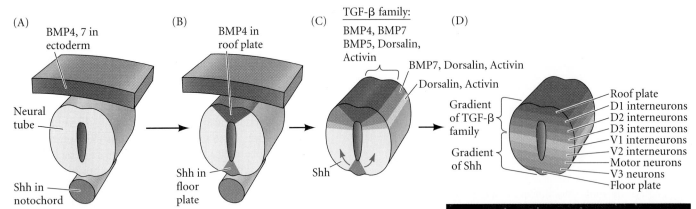

(A) BMP4, 7 in ectoderm

Neural tube

Shh in notochord

(B) BMP4 in roof plate

Shh in floor plate

(C) TGF-β family: BMP4, BMP7 BMP5, Dorsalin, Activin

BMP7, Dorsalin, Activin

Dorsalin, Activin

Shh

(D) Gradient of TGF-β family

Gradient of Shh

Roof plate
D1 interneurons
D2 interneurons
D3 interneurons
V1 interneurons
V2 interneurons
Motor neurons
V3 neurons
Floor plate

Figure 12.13
Dorsal-ventral specification of the neural tube. (A) The newly formed neural tube is influenced by two signaling centers. The roof of the neural tube is exposed to BMP4 and BMP7 from the epidermis, and the floor of the neural tube is exposed to Sonic hedgehog protein from the notochord. (B) Secondary signaling centers are established within the neural tube. BMP4 is expressed and secreted from the roof plate cells; Sonic hedgehog is expressed and secreted from the floor plate cells. (C) BMP4 establishes a nested cascade of TGF-β-related factors, spreading ventrally into the neural tube from the roof plate. Sonic hedgehog diffuses dorsally as a gradient from the floor plate cells. (D) The neurons of the spinal cord are given their identities by their exposure to these gradients of paracrine factors. The amount and type of paracrine factors present cause different transcription factors to be activated in the nuclei of these cells, depending on their position in the neural tube. (E) Chick neural tube, showing areas of Sonic hedgehog (green) and Dorsalin expression (blue). Motor neurons induced by a particular concentration of Sonic hedgehog are stained orange/yellow. (Photograph courtesy of T. M. Jessell.)

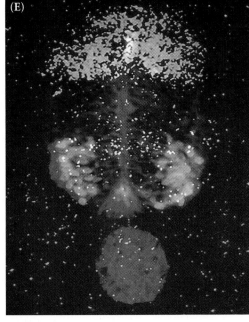

(E)

Schoenwolf 1986; Desmond and Field 1992). This occlusion (which also occurs in the human embryo) effectively separates the presumptive brain region from the future spinal cord (Desmond 1982). If the fluid pressure in the anterior portion of such an occluded neural tube is experimentally removed, the chick brain enlarges at a much slower rate and contains many fewer cells than are found in normal embryos. The occluded region of the neural tube reopens after the initial rapid enlargement of the brain ventricles.

> WEBSITE **12.3 Mapping the mesencephalon.** The chick cerebellum has a dual origin, with most of it coming from the metencephalon, but some portions coming from the mesencephalon. Moreover, genetic abnormalities can be traced to specific brain regions by transplantations between wild-type and mutant brain regions.

> WEBSITE **12.4 Specifying the brain boundaries.** The Pax transcription factors and the paracrine factor FGF8 are critical in establishing the boundaries of the forebrain, midbrain, and hindbrain.

The dorsal-ventral axis

The neural tube is polarized along its dorsal-ventral axis. In the spinal cord, for instance, the *dorsal* region is the place where the spinal neurons receive input from sensory neurons, while the *ventral* region is where the motor neurons reside. In

the middle are numerous interneurons that relay information between them. The polarity of the neural tube is induced by signals coming from its immediate environment. The dorsal pattern is imposed by the epidermis, while the ventral pattern is induced by the notochord (Figure 12.13).

VENTRAL PATTERNING OF THE NEURAL TUBE. The specification of the ventral neural tube appears to be mediated by external tissues. One agent of ventral specification is the **Sonic hedgehog** protein, probably originating from the notochord; another agent specifying the ventral neural cell types is retinoic acid, which probably comes from the adjacent somites (Pierani et al. 1999). Sonic hedgehog establishes a gradient, and different levels of this protein cause the formation of different cell types. Sonic hedgehog is initially synthesized in the notochord. Here it is processed by cholesterol-mediated cleavage, and the active form of the protein (the amino-terminal portion) is secreted from the notochord. The secreted Sonic hedgehog induces the medial hinge cells to become the **floor plate** of the neural tube. These floor plate cells also secrete Sonic hedgehog, which

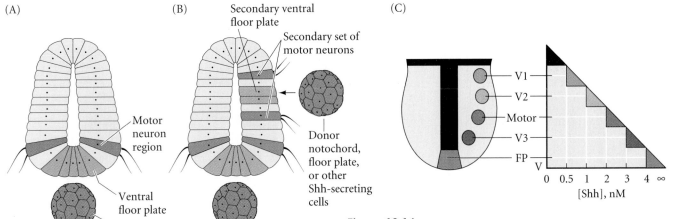

Figure 12.14
Cascade of inductions initiated by the notochord in the ventral neural tube. (A) Two cell types in the newly formed neural tube. Those closest to the notochord become the ventral floor plate neurons; motor neurons emerge on the ventrolateral sides. (B) If a second notochord, floor plate, or any other Sonic hedgehog-secreting cell is placed adjacent to the neural tube, it induces a second set of ventral floor plate neurons, as well as two other sets of motor neurons. (C) Relationship between Sonic hedgehog concentrations, the generation of particular neuronal types in vitro, and distance from notochord. (A, B after Placzek et al. 1990; C after Briscoe et al. 1999.)

forms a gradient highest at the most ventral portion of the neural tube (Roelink et al. 1995; Briscoe et al. 1999). Those cells adjacent to the floor plate that receive high concentrations of Sonic hedgehog become the ventral (V3) neurons, while the next group of cells, exposed to slightly less Sonic hedgehog, become motor neurons (Figure 12.14). The next two groups of cells, receiving progressively less of this protein, become the V2 and V1 interneurons. The different concentrations of Sonic hedgehog function by causing the different groups of neurons to express different types of transcription factors. These transcription factors, in turn, activate the genes whose protein products give the cell its identity. Sonic hedgehog may also work by repressing the expression of genes encoding dorsal neural tube transcription factors. These genes would otherwise be expressed throughout the neural tube.

The importance of Sonic hedgehog in inducing and patterning the ventral portion of the neural tube can be shown experimentally. If notochord fragments are taken from one embryo and transplanted to the lateral side of a host neural tube, the host neural tube will form another set of floor plate cells at its sides (Figure 12.14B). The floor plate cells, once induced, induce the formation of motor neurons on either side of them. The same results can be obtained if the notochord fragments are replaced by pellets of cultured cells secreting Sonic hedgehog (Echelard et al. 1993; Roelink et al. 1994). Moreover, if a piece of notochord is removed from an embryo, the neural tube adjacent to the deleted region will have no floor plate cells (Placzek et al. 1990; Yamada et al. 1991, 1993).

DORSAL PATTERNING OF THE NEURAL TUBE. The dorsal fates of the neural tube are established by proteins of the TGF-β superfamily, especially the bone morphogenetic proteins 4 and 7, dorsalin, and activin (Liem et al. 1995, 1997). Initially, BMP4 and BMP7 are found in the epidermis. Just as the notochord establishes a secondary signaling center—the floor plate cells— on the ventral side of the neural tube, the epidermis establishes a secondary signaling center by inducing

BMP4 expression in the **roof plate** cells of the neural tube. The BMP4 protein from the roof plate induces a cascade of TGF-β superfamily proteins in adjacent cells (Figure 12.13C). Different sets of cells are thus exposed to different concentrations of TGF-β superfamily proteins at different times (the most dorsal being exposed to more factors at higher concentrations and at earlier times). The temporal and concentration gradients of the TGF-β superfamily proteins induce different types of transcription factors in cells at different distances from the roof plate, thereby giving them different identities.

Tissue Architecture of the Central Nervous System

The neurons of the brain are organized into layers (**cortices**) and clusters (**nuclei**), each having different functions and connections. The original neural tube is composed of a **germinal neuroepithelium** that is one cell layer thick. This is a layer of rapidly dividing neural stem cells. Sauer (1935) and others have shown that all the cells of the germinal epithelium are continuous from the luminal surface of the neural tube to the outside surface, but that the nuclei of these cells are at different heights, thereby giving the superficial impression that the neural tube has numerous cell layers. The nuclei move within their cells as they go through the cell cycle. DNA synthesis (S phase) occurs while the nucleus is at the outside edge of the

(A)

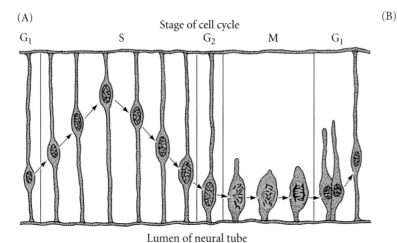

Lumen of neural tube

Figure 12.15
(A) Schematic section of a chick embryo neural tube, showing the position of the nucleus in a neuroepithelial cell as a function of the cell cycle. Mitotic cells are found near the center of the neural tube, adjacent to the lumen. (B) Scanning electron micrograph of a newly formed chick neural tube, showing cells at different stages of their cell cycles. (A after Sauer 1935; B courtesy of K. Tosney.)

(B)

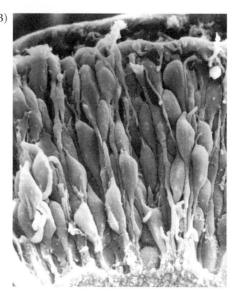

neural tube, and the nucleus migrates luminally as the cell cycle proceeds (Figure 12.15). Mitosis occurs on the luminal side of the cell layer. If mammalian neural tube cells are labeled with radioactive thymidine during early development, 100% of them will incorporate this base into their DNA (Fujita 1964). Shortly thereafter, certain cells stop incorporating these DNA precursors, thereby indicating that they are no longer participating in DNA synthesis and mitosis. These neuronal and glial cells then migrate and differentiate outside the neural tube (Fujita 1966; Jacobson 1968).

If dividing cells in the germinal neuroepithelium are labeled with radioactive thymidine at a single point in their development, and their progeny are found in the outer cortex in the adult brain, then those neurons must have migrated to their cortical positions from the germinal neuroepithelium. This happens because a neuroepithelial stem cell divides "vertically" instead of "horizontally." Thus, the daughter cell adjacent to the lumen remains connected to the ventricular surface (and usually remains a stem cell), while the other daughter cell migrates away (Chenn and McConnell 1995). The time of this vertical division is the last time the latter cell will divide, and is called that neuron's **birthday**. Different types of neurons and glial cells have their birthdays at different times. Labeling at different times during development shows that cells with the earliest birthdays migrate the shortest distances. The cells with later birthdays migrate through these layers to form the more superficial regions of the cortex. Subsequent differentiation depends on the positions these neurons occupy once outside the germinal neuroepithelium (Letourneau 1977; Jacobson 1991).

Spinal chord and medulla organization

As the cells adjacent to the lumen continue to divide, the migrating cells form a second layer around the original neural tube. This layer becomes progressively thicker as more cells are added to it from the germinal neuroepithelium. This new layer is called the **mantle** (or **intermediate**) **zone**, and the germinal epithelium is now called the **ventricular zone** (and, later, the **ependyma**) (Figure 12.16). The mantle zone cells differentiate into both neurons and glia. The neurons make connections among themselves and send forth axons away from the lumen, thereby creating a cell-poor **marginal zone**. Eventually, glial cells cover many of the axons in the marginal zone in myelin sheaths, giving them a whitish appearance. Hence, the mantle zone, containing the neuronal cell bodies, is often referred to as the **gray matter**; the axonal, marginal layer is often called the **white matter**.

In the spinal cord and medulla, this basic three-zone pattern of ependymal, mantle, and marginal layers is retained throughout development. The gray matter (mantle) gradually becomes a butterfly-shaped structure surrounded by white matter; and both become encased in connective tissue. As the neural tube matures, a longitudinal groove—the **sulcus limitans**— divides it into dorsal and ventral halves. The dorsal portion receives input from sensory neurons, whereas the ventral portion is involved in effecting various motor functions (Figure 12.17).

Cerebellar organization

In the brain, cell migration, differential neuronal proliferation, and selective cell death produce modifications of the three-zone pattern (Figure 12.16). In the cerebellum, some neuronal precursors enter the marginal zone to form clusters of neurons called **nuclei**. Each nucleus works as a functional unit, serving as a relay station between the outer layers of the

At the top of figure (A): Stage of cell cycle — G_1, S, G_2, M, G_1

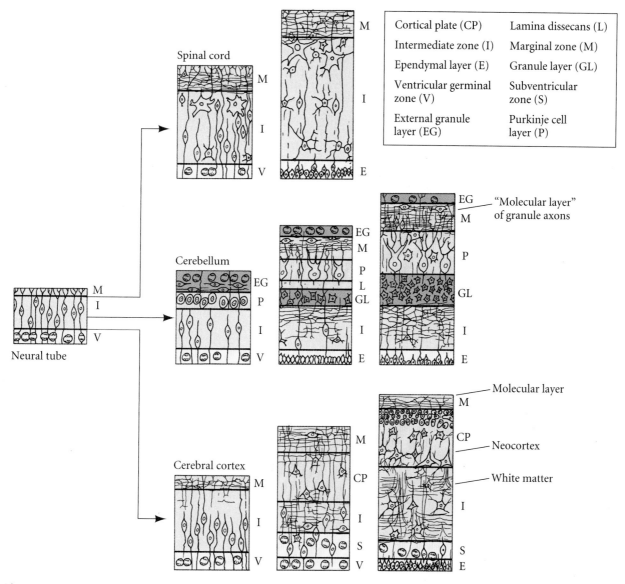

Cortical plate (CP)   Lamina dissecans (L)

Intermediate zone (I)   Marginal zone (M)

Ependymal layer (E)   Granule layer (GL)

Ventricular germinal zone (V)   Subventricular zone (S)

External granule layer (EG)   Purkinje cell layer (P)

Figure 12.16
Differentiation of the walls of the neural tube. A section of a 5-week human neural tube (left) reveals three zones: ventricular (ependymal), intermediate (mantle), and marginal. In the spinal cord and medulla (top row), the ependymal zone remains the sole source of neurons and glial cells. In the cerebellum (middle row), a second mitotic layer, the external granule layer, forms at the region farthest removed from the ependyma. Neuroblasts from this layer migrate back into the intermediate zone to form the granule cells. In the cerebral cortex (bottom row), the migrating neuroblasts and glioblasts form a cortical plate containing six layers. (After Jacobson 1991.)

cerebellum and other parts of the brain. In the cerebellum, some neuronal precursors can also migrate away from the germinal epithelium. These precursor cells, called **neuroblasts**, migrate to the outer surface of the developing cerebellum and form a new germinal zone, the **external granule layer**, near the outer boundary of the neural tube. At the outer boundary of the external granule layer (which is one to two cells thick), neuroblasts proliferate. The inner compartment of the external granule layer contains postmitotic neuroblasts that are the precursors of the major neurons of the cerebellar cortex, the **granule neurons**. These granule neurons migrate back into the developing cerebellar white matter to produce a region called the **internal granule layer**. Meanwhile, the original ependymal layer of the cerebellum generates a wide variety of neurons and glial cells, including the distinctive and large **Purkinje neurons**. Purkinje neurons are not only critical in the electrical pathway of the cerebellum, they also support the granule neurons. The Purkinje cell secretes Sonic hedgehog, which sustains the division of granule neuron precursors in the external granule layer (Wallace 1999). Each Purkinje neuron has an enormous **dendritic arbor**, which spreads like a tree above a bulblike cell body. A typical Purkinje neuron may

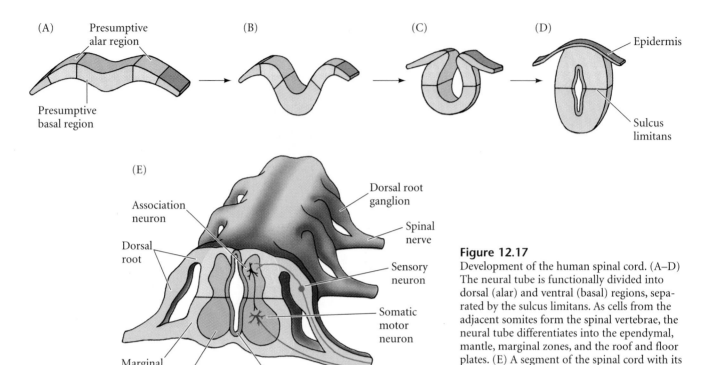

Figure 12.17
Development of the human spinal cord. (A–D) The neural tube is functionally divided into dorsal (alar) and ventral (basal) regions, separated by the sulcus limitans. As cells from the adjacent somites form the spinal vertebrae, the neural tube differentiates into the ependymal, mantle, marginal zones, and the roof and floor plates. (E) A segment of the spinal cord with its sensory (alar) and motor (basal) roots. (After Larsen 1993.)

form as many as 100,000 connections (synapses) with other neurons, more than any other neuron studied. Each Purkinje neuron also emits a slender axon, which connects to neurons in the deep cerebellar nuclei.

The development of spatial organization is critical for the proper functioning of the cerebellum. All electrical impulses eventually regulate the activity of the Purkinje cells, which are the only output neurons of the cerebellar cortex. For this to happen, the proper cells must differentiate at the appropriate place and time. How is this accomplished?

One mechanism thought to be important for positioning young neurons within the developing mammalian brain is **glial guidance** (Rakic 1972; Hatten 1990). Throughout the cortex, neurons are seen to ride "the glial monorail" to their respective destinations. In the cerebellum, the granule cell precursors travel on the long processes of the **Bergmann glia** (Figure 12.18; Rakic and Sidman 1973; Rakic 1975). This neural-glial interaction is a complex and fascinating series of events, involving reciprocal recognition between glia and neuroblasts (Hatten 1990; Komuro and Rakic 1992). The neuron maintains its adhesion to the glial cell through a number of proteins, one of them an adhesion protein called **astrotactin**. If the astrotactin on a neuron is masked by antibodies to that protein, the neuron will fail to adhere to the glial processes (Edmondson et al. 1988; Fishell and Hatten 1991).

Much insight into the mechanisms of spatial ordering in the brain has come from the analysis of neurological mutations in mice. Over 30 mutations are known to affect the arrangement of cerebellar neurons. Many of these cerebellar mutants have been found because the phenotype of such mutants—namely, the inability to keep balance while walking—can be easily recognized. For obvious reasons, these mutations are given names such as *weaver*, *reeler*, *staggerer*, and *waltzer*.

WEBSITE **12.5 Cerebellar mutations of the mouse.** The mouse mutations affecting cerebellar function have given us remarkable insights into the ways in which the cerebellum is constructed. The *reeler* mutation, in particular, has been extremely important in our knowledge of how cerebellar neurons migrate.

Cerebral organization

The three-zone arrangement of the neural tube is also modified in the cerebrum. The cerebrum is organized in two distinct ways. First, like the cerebellum, it is organized vertically into layers that interact with one another (see Figure 12.16). Certain neuroblasts from the mantle zone migrate on glial processes through the white matter to generate a second zone of neurons at the outer surface of the brain. This new layer of gray matter is called the **neocortex**. The neocortex eventually stratifies into six layers of neuronal cell bodies; the adult forms of these layers are not completed until the middle of childhood. Each layer of the neocortex differs from the others in its functional properties, the types of neurons found there, and the sets of connections that they make. For instance, neurons in layer 4 receive their major input from the thalamus (a region that forms from the diencephalon), while neurons in layer 6 send their major output back to the thalamus.

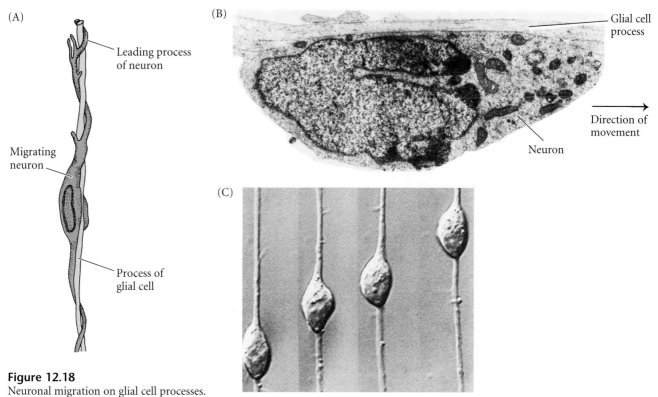

Figure 12.18
Neuronal migration on glial cell processes.
(A) Diagram of a cortical neuron migrating
on a glial cell process. (B) Electron micrograph of the region where
the neuronal cell body adheres to the glial process. (C) Sequential
photographs of a neuron migrating on a cerebellar glial process.
The leading process has several filopodial extensions. The neuron
reaches speeds around 40 μm/hr as it progresses on the glial
process. (A after Rakic 1975; B from Gregory et al. 1988; C from
Hatten 1990; photographs courtesy of M. Hatten.)

Second, the cerebral cortex is organized horizontally into
over 40 regions that regulate anatomically and functionally
distinct processes. For instance, neurons in cortical layer 6 of
the "visual cortex" project axons to the lateral geniculate nucleus of the thalamus (see Chapter 13), while layer 6 neurons
of the auditory cortex (located more anteriorly than the visual cortex) project axons to the medial geniculate nucleus of the
thalamus (for hearing).

Neither the vertical nor the horizontal organization of the
cerebral cortex is clonally specified. Rather, the developing
cortex forms from the mixing of cells derived from numerous
stem cells. After their final mitosis, most of the neuronal precursors generated in the ventricular (ependymal) zone migrate outward along glial processes to form the **cortical plate**
at the outer surface of the brain. As in the rest of the brain,
those neuronal precursors with the earliest "birthdays" form
the layer closest to the ventricle. Subsequent neurons travel
greater distances to form the more superficial layers of the
cortex. This process forms an "inside-out" gradient of development (Figure 12.19; Rakic 1974). A single stem cell in the
ventricular layer can give rise to neurons (and glial cells) in

any of the cortical layers (Walsh and Cepko 1988). But how do
the cells know which layer to enter?

McConnell and Kaznowski (1991) have shown that the
determination of laminar identity (i.e., which layer a cell migrates to) is made during the final cell division. Newly generated neuronal precursors transplanted after this last division
from young brains (where they would form layer 6) into older
brains whose migratory neurons are forming layer 2 are committed to their fate, and migrate only to layer 6. However, if
these cells are transplanted prior to their final division (during mid-S phase), they are uncommitted, and can migrate to
layer 2 (Figure 12.20). The fates of neuronal precursors from
older brains are more fixed. While the neuronal precursor cells
formed early in development have the potential to become
any neuron (at layers 2 or 6, for instance), later precursor cells
give rise only to upper-level (layer 2) neurons (Frantz and
McConnell 1996). We still do not know the nature of the information given to the cell as it becomes committed.

Not all neurons, however, migrate radially. O'Rourke and
her colleagues (1992) labeled young ferret neurons with fluorescent dye and followed their migration through the brain.
While a great majority of the young neurons migrated radially on glial processes from the ventricular zone into the cortical plate, about 12% of them migrated laterally from one
functional region of the cerebral cortex into another. These
observations meshed well with those of Walsh and Cepko
(1992), who infected ventricular stem cells with a retrovirus
and were able to stain these cells and their progeny after birth.

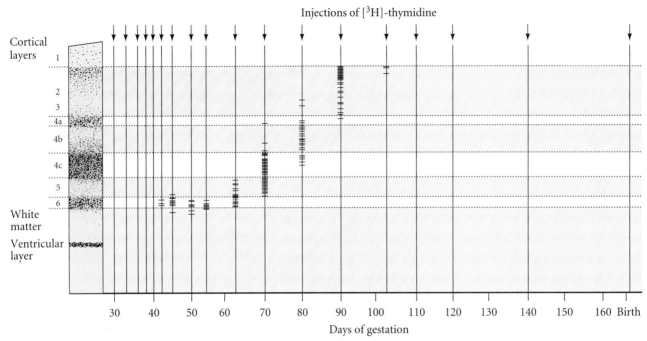

Figure 12.19
"Inside-out" gradient of cerebral cortex formation in the rhesus monkey. The "birthdays" of the cortical neurons were determined by injecting pregnant animals intravenously with [³H]-thymidine at certain times of gestation. Those fetal cells undergoing the S phase of their final cell division cycle at the time of the injections were heavily labeled. When the neurons were "born," these cells migrated to various regions and were detected by autoradiography of microscopic slices. Cells labeled at different times were found to have migrated to various cortical regions. The figure represents the positions of those neurons in the layers of the visual cortex. Full gestation in rhesus monkeys is 165 days. The youngest neurons are found at the periphery of the neural tube. (After Rakic 1974.)

They found that the neural descendants of a single ventricular stem cell were dispersed across the functional regions of the cortex. Thus, the specification of the cortical areas into specific functional domains occurs after neurogenesis. Once the cells arrive at their final destination, it is thought that they produce particular adhesion molecules that organize them together as brain nuclei (Matsunami and Takeichi 1995).

The cerebrum is quite plastic. The development of the human neocortex is particularly striking in this regard. The human brain continues to develop at fetal rates even after birth (Holt et al. 1975). Based on morphological and behavioral criteria and on comparisons with other primates, Portmann (1941, 1945) suggested that human gestation should really last 21 months instead of 9. However, no woman could deliver a 21-month-old fetus because the head would not pass through the birth canal; thus, humans give birth at the end of 9 months. Montagu (1962) and Gould (1977) have suggested that during our first year of life, we are essentially extrauterine fetuses, and they speculate that much of human intelligence comes from the stimulation of the nervous system as it is forming during that first year.*

WEBSITE **12.6 Constructing the cerebral cortex.** Three genes have recently been shown to be necessary for the proper lamination of the mammalian brain. They appear to be important for cortical neural migration, and when mutated in humans can produce profound mental retardation.

WEBSITE **12.7 What makes us human.** Human brain size is off the scale when we compare ourselves with other apes. Our retention of fetal neural stem cell division rates for a year after birth may be the critical step in our becoming human.

Adult neural stem cells

Until recently, it had been generally thought that once the nervous system was mature, no new neurons were "born." The neurons we formed in utero and during the first few years of life were all we could ever expect to have. However, the good news from recent studies is that environmental stimulation can increase the number of new neurons in the mammalian

*Contrary to the claims of a widely circulated anti-abortion film, the human cerebral cortex has no neuronal connections at 12 weeks' gestation (and therefore cannot move in response to a thought, nor experience consciousness or fear). Measurable electrical activity characteristic of neural cells (the electroencephalogram, or EEG, pattern) is first seen at 7 months' gestation. Morowitz and Trefil (1992) put forth the provocative opinion that since society in the United States has defined death as the loss of the EEG pattern, perhaps it should accept the acquisition of the EEG pattern as the start of human life.

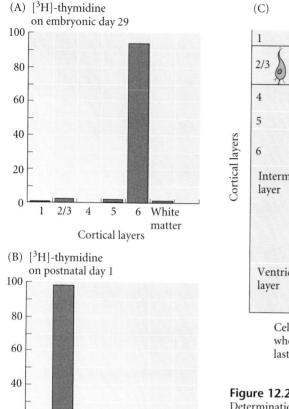

(A) [³H]-thymidine on embryonic day 29

(B) [³H]-thymidine on postnatal day 1

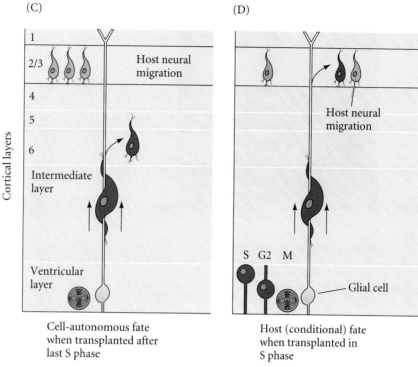

Cell-autonomous fate when transplanted after last S phase

Host (conditional) fate when transplanted in S phase

Figure 12.20

Determination of laminar identity in the ferret cerebrum. (A) "Early" neuronal precursors (birthdays on embryonic day 29) migrate to layer 6. (B) "Late" neuronal precursors (birthdays on postnatal day 1) migrate farther, into layers 2 and 3. (C) When early precursors (dark blue) are transplanted into older ventricular zones after their last mitotic S phase, the neurons they form migrate into layer 6. (D) If these precursors are transplanted before or during their last S phase, however, the neurons migrate (with the host neurons) to layer 2. (After McConnell and Kaznowski 1991.)

brain (Kemperman et al. 1997a,b; Gould et al. 1999a,b; Praag et al. 1999). To do these experiments, researchers injected adult mice, rats, or marmosets with bromodeoxyuridine (BrdU), a nucleoside that resembles thymidine and which will be incorporated into a cell's DNA only if the cell is undergoing DNA replication. Thus, any cell labeled with BrdU must have been undergoing mitosis during the time when it was exposed to BrdU. This technique showed that thousands of new neurons were being made each day in adult mammals. Injecting humans with BrdU is usually unethical, since large doses of BrdU are often lethal. However, in certain cancer patients, the progress of chemotherapy is monitored by transfusing the patient with a small amount of BrdU. Gage and colleagues (Erikkson et al. 1998) took postmortem samples from the brains of five such patients who had died between 16 and 781 days after the BrdU infusion. In all five subjects, they saw new neurons in the granular cell layer of the hippocampal dentate gyrus (a part of the brain where memories may be formed). The BrdU-labeled cells also stained for neuron-specific markers (Figure 12.21). It appears that the stem cells pro-

ducing these neurons are located in the ependyma, the former ventricular layer in which the embryonic neural stem cells once resided (Doetsch et al. 1999; Johansson et al. 1999). These results are surprising, since the ependyma consists of differentiated glial cells whose ciliated surface keeps the cerebral spinal fluid flowing. Indeed, one author (Barres 1999) described them as "the most boring of all glial subtypes." It appears that these glial cells (or perhaps only some of them) can dedifferentiate and become neural stem cells. Thus, although the rate of new neuron formation in adulthood may be relatively small, the human brain is not an anatomical *fait accompli* at birth, or even after childhood.*

Neuronal Types

The human brain consists of over 10^{11} neurons associated with over 10^{12} glial cells. Those cells that remain integral com-

*The use of cultured neural stem cells to regenerate or repair parts of the brain will be considered in the next chapter.

(A)

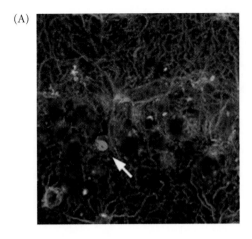

(B)

Figure 12.21
Evidence of adult neural stem cells. (A) Newly generated neurons in the adult human brain. This cell is located in the dentate gyrus of the hippocampus. The green staining, which indicates newly divided cells, is from a fluorescent antibody against BrdU (a thymidine analogue that is only taken up during S phase). The red fluorescence is from an antibody that only stains neural cells. Glial cells are stained purple. The overlap of green and red (arrow) shows a cell that is a newly formed neural cell. (B) Model for the production of neuronal stem cells in the adult mouse brain. (A from Eriksson et al. 1998, photograph courtesy of F. H. Gage; B after Doetsch et al. 1999.)

ponents of the neural tube lining become **ependymal cells**. These cells can give rise to the precursors of neurons and glial cells. As we have seen above, it is thought that the differentiation of these precursor cells is largely determined by the environment that they enter (Rakic and Goldman 1982) and that, at least in some cases, a given ependymal cell can form both neurons and glia (Turner and Cepko 1987).

The brain contains a wide variety of neuronal and glial types (as is evident from a comparison of the relatively small granule cell with the enormous Purkinje neuron). The fine extensions of the neuron that are used to pick up electrical impulses from other cells are called **dendrites** (Figure 12.22). Some neurons develop only a few dendrites, whereas other cells (such as the Purkinje neurons) develop extensive dendritic trees. Very few dendrites can be found on cortical neurons at birth, but one of the amazing things about the first year of human life is the increase in the number of these receptive regions. During this year, each cortical neuron develops enough dendritic surface to accommodate as many as 100,000 synapses with other neurons. The average cortical neuron connects with 10,000 other neural cells. This pattern of synapses enables the human cortex to function as the center for learning, reasoning, and memory, to develop the capacity for symbolic expression, and to produce voluntary responses to interpreted stimuli.

Another important feature of a developing neuron is its **axon** (sometimes called a **neurite**). Whereas dendrites are often numerous and do not extend far from the neuronal cell body, or **soma**, axons may extend for several feet. The pain receptors on your big toe, for example, must transmit their messages all the way to the spinal cord. One of the fundamental concepts of neurobiology is that the axon is a continuous extension of the nerve cell body. At the turn of the twentieth century,

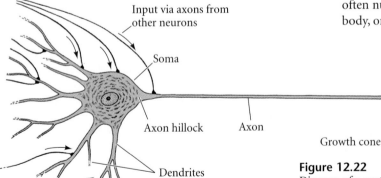

Figure 12.22
Diagram of a motor neuron. Electrical impulses are received by the dendrites. and the stimulated neuron transmits impulses through the axon (which may be 2–3 feet long) to its target tissue. The axon is the cellular process through which the neuron sends its signals. The growth cone of the axon is both a locomotor and a sensory apparatus. It actively explores the environment and picks up directional cues as to where to go. Eventually it will form a synapse with its target tissue.

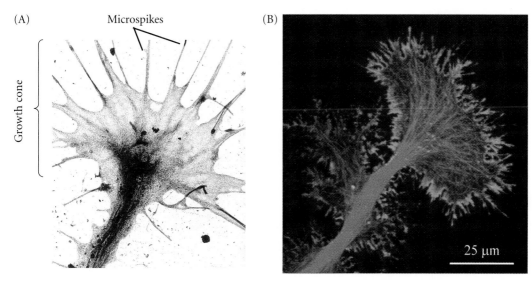

Figure 12.23
Axon growth cones. (A) Actin microspikes in an axon growth cones, seen by transmission electron microscopy. (B) Growth cone of the hawkmoth *Manduca sexta* during axon extension and pathfinding. The actin in the filopodia is stained green with fluorescent phalloidin, while the microtubules are stained red with a fluorescent antibody to tubulin. (A from Letourneau 1979; B courtesy of R. B. Levin and R. Luedemanan.)

there were many competing theories of axon formation. Theodor Schwann, one of the founders of the cell theory, believed that numerous neural cells linked themselves together in a chain to form an axon. Viktor Hensen, the discoverer of the embryonic node, thought that the axon formed around preexisting cytoplasmic threads between the cells. Wilhelm His (1886) and Santiago Ramón y Cajal (1890) postulated that the axon was an outgrowth (albeit an extremely large one) of the neuronal soma. In 1907, Ross Harrison demonstrated the validity of the outgrowth theory in an elegant experiment that founded both the science of developmental neurobiology and the technique of tissue culture. Harrison isolated a portion of the neural tube from a 3-mm frog tadpole. (At this stage, shortly after the closure of the neural tube, there is no visible differentiation of axons.) He placed this neuroblast-containing tissue in a drop of frog lymph on a coverslip and inverted the coverslip over a depression slide so he could watch what was happening within this "hanging drop." What Harrison saw was the emergence of the axons as outgrowths from the neuroblasts, elongating at about 56 μm/hr.

Such nerve outgrowth is led by the tip of the axon, called the **growth cone** (see Figure 12.23). The growth cone does not proceed in a straight line, but rather "feels" its way along the substrate. The growth cone moves by the elongation and contraction of pointed filopodia called **microspikes**. These microspikes contain microfilaments, which are oriented parallel to the long axis of the axon. (This mechanism is similar to the one seen in the filopodial microfilaments of secondary mes-

enchyme cells in echinoderms; see Chapter 8) Treating neurons with cytochalasin B destroys the actin microspikes, inhibiting their further advance (Yamada et al. 1971; Forscher and Smith 1988). Within the axon itself, structural support is provided by microtubules, and the axon will retract if the neuron is placed in a solution of colchicine. Thus, the developing neuron retains the same mechanisms that we have already noted in the dorsolateral hinge points of the neural tube; namely, elongation by microtubules and apical shape changes by microfilaments. As in most migrating cells, the exploratory filopodia of the growth cone attach to the substrate and exert a force that pulls the rest of the cell forward. Axons will not grow if the growth cone fails to advance (Lamoureux et al. 1989). In addition to their structural role in axonal migration, the microspikes also have a sensory function. Fanning out in front of the growth cone, each microspike samples the microenvironment and sends signals back to the cell body (Davenport et al. 1993). As we will see in Chapter 13, the microspikes are the fundamental organelles involved in neuronal pathfinding.

Neurons transmit electrical impulses from one region of the body to another. These impulses usually go from the dendrites into the soma, where they are focused into the axon. To prevent dispersion of the electrical signal and to facilitate its conduction, that part of the axon within the central nervous system is insulated at intervals by processes that originate from a type of glial cell called an **oligodendrocyte**. The oligodendrocyte wraps itself around the developing axon. It then produces a specialized cell membrane called a **myelin sheath**. In the peripheral nervous system, a glial cell type called the **Schwann cell** accomplishes this myelination (Figure 12.24). The myelin sheath is essential for proper neural function, and demyelination of nerve fibers is associated with convulsions, paralysis, and several severely debilitating or lethal afflictions. In the *trembler* mouse mutant, the Schwann cells are unable to

(A)

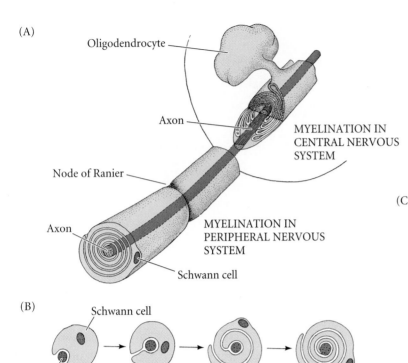

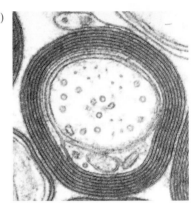

Figure 12.24
Myelination in the central and peripheral nervous systems. (A) In the peripheral nervous system, Schwann cells wrap themselves around the axon; in the central nervous system, myelination is accomplished by the processes of oligodendrocytes. (B) The mechanism of this wrapping entails the production of an enormous membrane complex. (C) Micrograph of an axon enveloped by the myelin membrane of a Schwann cell. (Photograph courtesy of C. S. Raine.)

produce a particular protein component of the myelin sheath, so that myelination is deficient in the peripheral nervous system, but normal in the central nervous system. Conversely, in another mouse mutant, called *jimpy*, the central nervous system is deficient in myelin, while the peripheral nerves are unaffected (Sidman et al. 1964; Henry and Sidman 1988).

The axon must also be specialized for secreting a specific neurotransmitter across the small gaps (synaptic clefts) that separate the axon of one cell from the surface of its target cell (the soma, dendrites, or axon of a receiving neuron or a receptor site on a peripheral organ). Some neurons develop the

ability to synthesize and secrete acetylcholine, while other neurons develop the enzymatic pathways for making and secreting epinephrine, norepinephrine, octopamine, serotonin, γ-aminobutyric acid (GABA), dopamine, or some other neurotransmitter. Each neuron must activate those genes responsible for making the enzymes that can synthesize its neurotransmitter. Thus, neuronal development involves both structural and molecular differentiation.*

*The regeneration of neurons and their axons will be discussed in Chapter 18. The glial cells are probably very important in permitting or forbidding the axons to regenerate.

Sidelights & Speculations

Homologous Specification of Neural Tissue between Vertebrates and Arthropods

THE INSECT NERVOUS SYSTEM and the vertebrate nervous system appear to have little in common. However, the molecular *instructions* for forming the neural tube and specifying its regions appear to be homologous. This homology can be seen at many stages of neural development.

Specification of neural ectoderm: The BMP/Dpp and chordin/Sog pathway.
First, as we have seen in previous chapters, both the insect and the vertebrate form their respective neural tubes by inhibiting a BMP signal to the ectoderm. The pathways that process these proteins in flies and vertebrates

also are homologous (see Chapters 10 and 11). During *Drosophila* gastrulation, the vegetalmost cells, the precursors of the mesoderm, invaginate into the yolky blastocoel, causing the neurogenic ectoderm to be in the ventral region of the embryo (Figure 12.25). The neurogenic ectoderm delaminates about

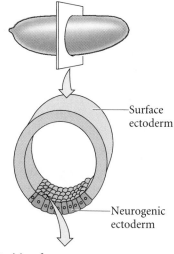

(A) Positional cues:
Segmentation genes (A/P),
dorsal/ventral genes

(B) Neuroblast specification:
Neuroblast identity genes

(C) Neuroblast formation:
Proneural genes *(achaete, scute)*

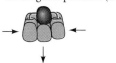

(D) Lateral inhibition:
Neurogenic proteins (Delta, Notch)

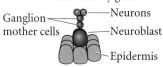

(E) Neuroblast cell lineage:
Ganglion mother cells and
neural identity genes

Ganglion — Neurons
mother cells — Neuroblast
— Epidermis

(F)

Figure 12.25
Sequential specification of the neuroblast lineage in *Drosophila*. (A) The neurogenic ectoderm is specified by positional cues along the dorsal-ventral and anterior-posterior axes. (B, C) Clusters of potential neuroblasts are specified by the proneural genes, which encode transcription factors such as Achaete (shown in F). (D) Interaction between the potential neuroblasts selects one cell from the cluster to be the neuroblast, and this cell inhibits the other cells of the cluster from becoming neuroblasts. (These interactions, mediated by Notch and Delta, were discussed in Chapter 6.)(E) The neuroblast buds off ganglion mother cells (in a manner that will be discussed in Chapter 13), each of which forms two neurons. (F) *Drosophila* embryo stained for the *achaete* transcript. The neurogenic clusters express this gene. The bracket indicates a domain of neurogenic activity. (After Goodman and Doe 1993; F from Skeath and Carroll 1992, photograph courtesy of J. Skeath.)

60 cells (30 per side) into the embryo, and these (along with the cells in the ventral midline) are the neuroblasts, the precursors of the neurons.

Neural competence: The *achaete-scute* genes.

In *Drosophila*, the genes activated in the neural ectoderm that enable a cell to become a neuroblast are called the **proneural genes**. These genes encode the transcription factors Achaete and Scute, among others. In mammals, the *MASH1* gene (mammalian achaete-scute homologue) is expressed in subsets of neurons, and may influence neuronal differentiation in olfactory (smell) receptor cells as well as other cells of the central nervous system (Johnson et al. 1990; Skeath and Carroll 1992; Cau et al. 1997; Shou et al. 1999).

Singling out neuronal precursors: The Notch-Delta interaction.

Not every ectodermal cell expressing the proneural genes becomes a neuron. Many more become glial or skin cells. As we saw in Chapter 6, the Delta protein (a ligand) and the Notch protein (a receptor) interact such that Delta activates signal transduction in the adjacent cell through that cell's Notch protein. The

cell with slightly more Delta protein on its cell surface will inhibit its neighboring cells from becoming neurons. In flies, frogs, and chicks, Delta is found in those cells that will become neurons, while Notch is elevated in those cells that become the glial cells (Figure 12.26; Chitnis et al. 1995). The down-regulation of Notch in new neurons (to sustain neural differentiation) also appears to be the same in chicks and flies (Wakamatsu et al. 1999).

Differentiation of specific neurons.

In mice, the *Nkx2.2* gene is activated by the Sonic hedgehog signal and encodes a transcription factor that specifies the identities of ventral nerve cells. Its *Drosophila* homologue, *vnd*, controls dorsal-ventral neuronal identities in the fly central nervous system in a similar manner (Skeath et al. 1994; Briscoe et al. 1999).

Therefore, while the nervous system of the adult fly and the adult vertebrate are remarkably different, the genes and interactions that instruct the formation of these nervous systems are remarkably similar. As we shall see in the example of eye development below, this similarity in instructions unifies many aspects of the central nervous system across phyla. ■

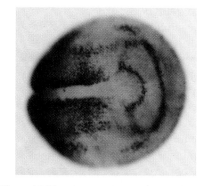

Figure 12.26
Expression of the *Delta* gene during *Xenopus* neurulation is found in those cells that are the precursors of the motor, intermediate, and sensory neurons. The expression of *Delta* distinguishes these neural cells from the presumptive epidermal and glial cells. (From Chitnis et al. 1995; photograph courtesy of C. Kintner.)

Development of the Vertebrate Eye

An individual gains knowledge of its environment through its sensory organs. The major sensory organs of the head develop from the interactions of the neural tube with a series of epidermal thickenings called the **cranial ectodermal placodes**. The most anterior of these are the two **olfactory pla-**

(A) 4-mm embryo

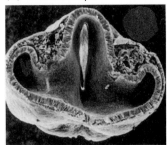

(B) 4.5-mm embryo

Lens placode

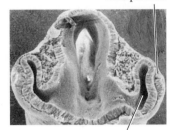

Optic vesicle

(C) 5-mm embryo

Lens vesicle

Optic cup

(D) 7-mm embryo

Retina   Lens

Cornea

Figure 12.27
Development of the vertebrate eye (see also Figure 6.5). (A) The optic vesicle evaginates from the brain and contacts the overlying ectoderm, inducing a lens placode. (B, C) The overlying ectoderm differentiates into lens cells as the optic vesicle folds in on itself, and the lens placode becomes the lens vesicle. (C) The optic vesicle becomes the neural and pigmented retina as the lens is internalized. (D) The lens vesicle induces the overlying ectoderm to become the cornea (A–C from Hilfer and Yang 1980, photographs courtesy of S. R. Hilfer; D courtesy of K. Tosney.)

codes that form the ganglia for the olfactory nerves, which are responsible for the sense of smell. The **auditory placodes** similarly invaginate to form the inner ear labyrinth, whose neurons form the acoustic ganglion, which enables us to hear. In this section, we will focus on the eye.

The dynamics of optic development

The induction of the eye was discussed in Chapter 6, and will only be summarized here (Figure 12.27). At gastrulation, the involuting endoderm and mesoderm interact with the adjacent prospective head ectoderm to give the head ectoderm a lens-forming bias (Saha et al. 1989). But not all parts of the head ectoderm eventually form lenses, and the lens must have a precise spatial relationship with the retina. The activation of the head ectoderm's latent lens-forming ability and the positioning of the lens in relation to the retina is accomplished by the optic vesicle. It extends from the diencephalon, and when it meets the head ectoderm, it induces the formation of a **lens placode**, which then invaginates to form the lens. The optic vesicle becomes the two-walled **optic cup**, whose two layers differentiate in different directions. The cells of the outer layer produce melanin pigment (being one of the few tissues other than the neural crest cells that can form this pigment) and ultimately become the **pigmented retina**. The cells of the inner layer proliferate rapidly and generate a variety of glia, ganglion cells, interneurons, and light-sensitive photoreceptor neurons. Collectively, these cells constitute the **neural retina**. The retinal **ganglion cells** are neurons whose axons send electrical impulses to the brain. Their axons meet at the base of the eye and travel down the optic stalk. This stalk is then called the **optic nerve**.

But how is it that a specific region of neural ectoderm is informed that it will become the optic vesicle? It appears that a group of transcription factors—Six3, Pax6, and Rx1—are expressed together in the most anterior tip of the neural plate. This single domain will later split into the bilateral regions that will form the optic vesicles. Again, we see the similarities between the *Drosophila* and the vertebrate nervous system, for these three proteins are also necessary for the formation of the *Drosophila* eye. As discussed in Chapters 4 and 5, the Pax6 protein appears to be especially important in the development of the lens and retina. Indeed, it appears to be a common denominator for photoreceptive cells in all phyla. If the mouse *Pax6* gene is inserted into the *Drosophila* genome and activated randomly, *Drosophila* eyes form in those cells where the mouse *Pax6* is being expressed (see Chapter 22; Halder et al. 1995)! While *Pax6* is also expressed in the murine forebrain, hindbrain, and nasal placodes, the eyes seem to be most sensitive to its absence. In humans and mice, *Pax6* heterozygotes have small eyes, while homozygotic mice and humans (and *Drosophila*) lack eyes altogether (Jordan et al. 1992; Glaser et al. 1994; Quiring et al. 1994).

The separation of the single eye field into two bilateral fields depends upon the secretion of Sonic hedgehog. If the *sonic hedgehog* gene is mutated, or if the processing of this protein is inhibited, the single median eye field will not split. The result is **cyclopia**—a single eye in the center of the face (and usually below the nose) (Figure 12.28; see also Figure 6.25; Chiang et al. 1996; Kelley et al. 1996; Roessler et al. 1996; Li et al. 1997). It is thought that Sonic hedgehog protein from the prechordal plate suppresses *Pax6* expression in the center of the embryo, dividing the field in two.

WEBSITE **12.8 Human cyclopia.** Mutations in *sonic hedgehog* have been implicated in causing cyclopia in humans. Moreover, mutations of genes involved in cholesterol synthesis have also been implicated.

(A) Wild-type

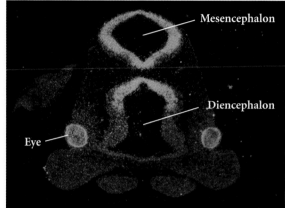

Mesencephalon

Diencephalon

Eye

(B) *Shh⁻/Shh⁻*

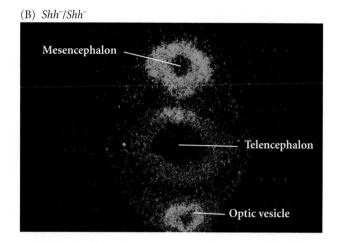

Mesencephalon

Telencephalon

Optic vesicle

(C)

Telencephalon

Diencephalon

Mesencephalon

Metencephalon

Prechordal plate

Myelencephalon

Hypothalamus

Notochord

Figure 12.28
Brain defects in embryos lacking *sonic hedgehog*. (A) A wild-type mouse and (B) a 12.5-day embryo lacking *sonic hedgehog*. The expression of the *otx-2* gene is seen in red to highlight certain regions. In the mutant, no midline forms, and there is a single, continuous optic vesicle in the ventral region. The nose will form above it. (C) Drawing showing the location of the prechordal plate in the 12-day mouse embryo. (Photographs courtesy of P. A. Beachy and C. Chiang.)

Neural retina differentiation

Like the cerebral and cerebellar cortices, the neural retina develops into a layered array of different neuronal types (Figure 12.29). These layers include the light- and color-sensitive photoreceptor (rod and cone) cells, the cell bodies of the ganglion cells, and the bipolar interneurons that transmit electrical stimuli from the rods and cones to the ganglion cells. In addition, there are numerous Müller glial cells, which maintain the integrity of the retina, as well as amacrine neurons (which lack large axons) and horizontal neurons that transmit electrical impulses in the plane of the retina.

In the early stages of retinal development, cell division in a germinal layer and the migration and differential death of the resulting cells form the striated, laminar pattern of the neural retina. The formation of this highly structured tissue is one of the most intensely studied problems of developmental neurobiology. It has been shown (Turner and Cepko 1987) that a single neuroblast precursor cell from the retinal germinal layer can give rise to at least three types of neurons or to two types of neurons and a glial cell. This analysis was performed using an ingenious technique to label the cells generated by one particular neuroblast precursor cell. Newborn rats

(whose retinas are still developing) were injected in the back of their eyes with a virus that can integrate into their DNA. This virus contained a β-galactosidase gene (not present in rat retina) that would be expressed only in the infected cells. A month after the rats' eyes were infected, the retinas were removed and stained for the presence of β-galactosidase. Only the progeny of the infected cells should have stained blue. Figure 12.30 shows one of the strips of cells derived from an infected precursor cell. The stain can be seen in five rods, a bipolar neuron, and a retinal (Müller) glial cell.

Lens and cornea differentiation

During its continued development into a lens, the lens placode rounds up and contacts the new overlying ectoderm. The lens vesicle then induces the ectoderm to form the transparent cornea. Here, physical parameters play an important role in the development of the eye. Intraocular fluid pressure is necessary for the correct curvature of the cornea so that light can be focused on the retina. The importance of this pressure can be demonstrated experimentally: the cornea will not develop its characteristic curve when a small glass tube is inserted through the wall of a developing chick eye to drain away the intraocular fluid (Coulombre 1956, 1965). Intraocular pressure is sustained by a ring of scleral bones (probably derived from the neural crest), which acts as an inelastic restraint.

The differentiation of the lens tissue into a transparent membrane capable of directing light onto the retina involves

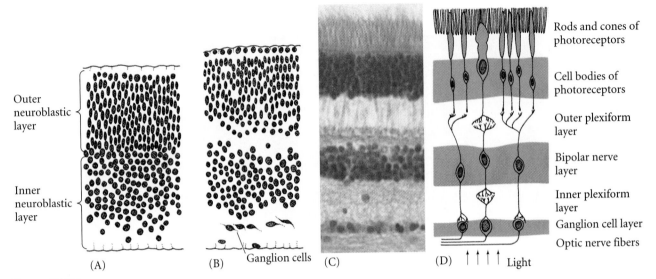

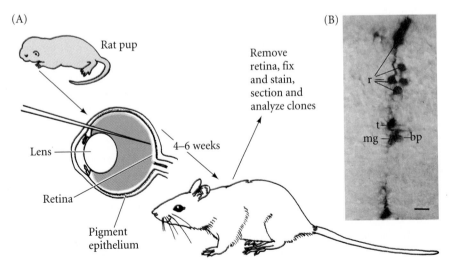

Figure 12.29

Development of the human retina. Retinal neurons sort out into functional layers during development. (A, B) Initial separation of neuroblasts within the retina. (C) The three layers of neurons in the adult retina and the synaptic layers between them. (D) A functional depiction of the major neuronal pathway in the retina. Light traverses the layers until it is received by the photoreceptors. The axons of the photoreceptors synapse with bipolar neurons, which transmit electrical signals to the ganglion cells. The axons of the ganglion cells join to form the optic nerve, which enters the brain. (A and B after Mann 1964; C, photograph courtesy of G. Grunwald.)

changes in cell structure and shape as well as the synthesis of transparent, lens-specific proteins called **crystallins** (Figure 12.31). The cells at the inner portion of the lens vesicle elongate and, under the influence of the neural retina, become the lens fibers (Piatigorsky 1981). As these fibers continue to grow, they synthesize crystallins, which eventually fill up the cell and cause the extrusion of the nucleus. The crystallin-synthesizing fibers continue to grow and eventually fill the space between the two layers of the lens vesicle. The anterior cells of

the lens vesicle constitute a germinal epithelium, which keeps dividing. These dividing cells move toward the equator of the vesicle, and as they pass through the equatorial region, they, too, begin to elongate (Figure 12.31D). Thus, the lens contains three regions: an anterior zone of dividing epithelial cells, an equatorial zone of cellular elongation, and a posterior and central zone of crystallin-containing fiber cells. This arrangement persists throughout the lifetime of the animal as fibers are continuously being laid down. In the adult chicken, the process of differentiation from an epithelial cell to a lens fiber takes 2 years (Papaconstantinou 1967).

Directly in front of the lens is a pigmented and muscular tissue called the **iris**. The iris muscles control the size of the pupil (and give an individual his or her characteristic eye

Figure 12.30

Determination of the lineage of a neuroblast in the rat retina. (A) Technique whereby a virus containing a functional β-galactosidase gene is injected into the back of the eye of a newborn rat to infect some of the retinal precursor cells. After a month to 6 weeks, the eye is removed and the retina is stained for the presence of β-galactosidase. (B) Stained cells forming a strip across the neural retina, including five rods (r), a bipolar neuron (bp), a rod terminal (t), and a Müller glial cell (mg). The identities of these cells were confirmed by Nomarski-phase contrast microscopy. (Scale bar, 20 μm.) (From Turner and Cepko 1987; photograph courtesy of D. Turner.)

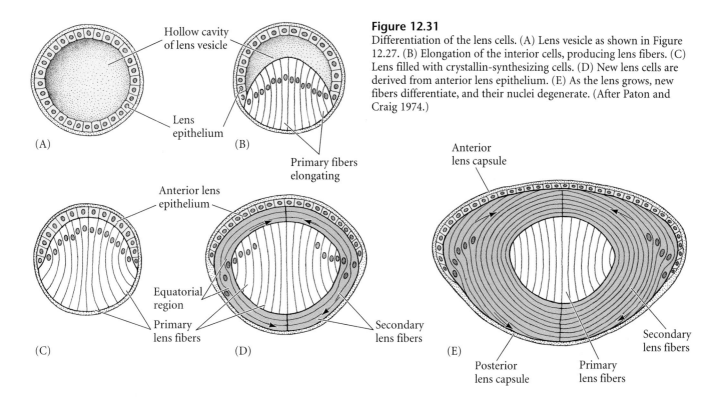

Figure 12.31
Differentiation of the lens cells. (A) Lens vesicle as shown in Figure 12.27. (B) Elongation of the interior cells, producing lens fibers. (C) Lens filled with crystallin-synthesizing cells. (D) New lens cells are derived from anterior lens epithelium. (E) As the lens grows, new fibers differentiate, and their nuclei degenerate. (After Paton and Craig 1974.)

color). Unlike the other muscles of the body (which are derived from the mesoderm), part of the iris is derived from the ectodermal layer. Specifically, this region of the iris develops from a portion of the optic cup that is continuous with the neural retina, but does not make photoreceptors.

> WEBSITE **12.9 Why babies don't see well.** The retinal photoreceptors are not fully developed at birth. As one gets older, the density of photoreceptors increases, allowing far better discrimination and nearly 350 times the light-absorbing capacity.

The Epidermis and the Origin of Cutaneous Structures

The origin of epidermal cells

The cells covering the embryo after neurulation form the presumptive epidermis. Originally, this tissue is one cell layer thick, but in most vertebrates it shortly becomes a two-layered structure. The outer layer gives rise to the **periderm**, a temporary covering that is shed once the inner layer differentiates to form a true epidermis. The inner layer, called the **basal layer** (or **stratum germinativum**), is a germinal epithelium that gives rise to all the cells of the epidermis (Figure 12.32). The basal layer divides to give rise to another, outer population of cells that constitutes the **spinous layer**. These two epidermal layers together are referred to as the **Malpighian layer**. The cells of the Malpighian layer divide to produce the **granular layer** of the epidermis, so called because its cells are character-

ized by granules of the protein keratin. Unlike the cells remaining in the Malpighian layer, the cells of the granular layer do not divide, but begin to differentiate into epidermal skin cells, the **keratinocytes**. The keratin granules become more prominent as the keratinocytes of the granular layer age and migrate outward to form the **cornified layer** (**stratum corneum**) These cells become flattened sacs of keratin protein, and their nuclei are pushed to one edge of the cell.. The depth of the cornified layer varies from site to site, but it is usually 10 to 30 cells thick. Shortly after birth, the outer cells of the cornified layer are shed and are replaced by new cells coming up from the granular layer. Throughout life, the dead keratinized cells of the cornified layer are shed (humans lose about 1.5 grams of these cells each day*) and are replaced by new cells, the source of which is the mitotic cells of the Malpighian layer. Pigment cells (melanocytes) from the neural crest also reside in the Malpighian layer, where they transfer their pigment sacs (melanosomes) to the developing keratinocytes.

The epidermal stem cells of the Malpighian layer are bound to the basal lamina by their integrin proteins. However, as these cells become committed to differentiate, they down-regulate their integrins and eventually lose them as they migrate into the spinous layer (Jones and Watt 1993).

Several growth factors stimulate the development of the epidermis. One of these is **transforming growth factor-α** (**TGF-α**). TGF-α is made by the basal cells and stimulates

---

*Most of this skin becomes "house dust" on furniture and floors. If you doubt this, burn some dust; it will smell like singed skin.

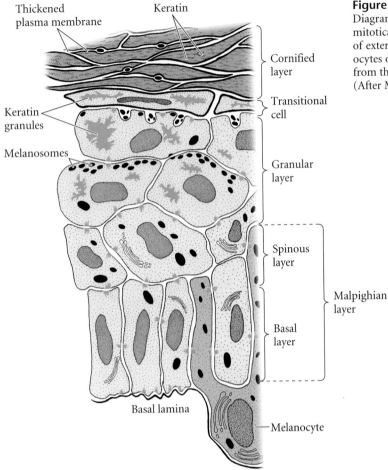

Thickened plasma membrane

Keratin

Keratin granules

Melanosomes

Cornified layer

Transitional cell

Granular layer

Spinous layer

Malpighian layer

Basal layer

Basal lamina

Melanocyte

Figure 12.32
Diagram of the layers of the human epidermis. The basal cells are mitotically active, whereas the fully keratinized cells characteristic of external skin are dead and are continually shed. The keratinocytes obtain their pigment through the transfer of melanosomes from the processes of melanocytes that reside in the basal layer. (After Montagna and Parakkal 1974.)

KGF is received by the basal cells of the epidermis and is thought to regulate their proliferation. If the gene encoding KGF is fused with the keratin 14 promoter, the KGF becomes autocrine in the resulting transgenic mice (Figure 12.33C). These mice have a thickened epidermis, baggy skin, far too many basal cells, and no hair follicles, not even whisker follicles (Guo et al. 1993). The basal cells are "forced" into the epidermal pathway of differentiation. The alternative pathway for basal cells leads to the generation of hair follicles.

Cutaneous appendages

The epidermis and dermis also interact at specific sites to create the sweat glands and the **cutaneous appendages**: hairs, scales, or feathers (depending on the species). In mammals, the first indication that a hair follicle primordium, or **hair germ**, will form at a particular place is an aggregation of cells in the basal layer of the epidermis. This aggregation is directed by the underlying dermal fibroblast cells and occurs at different times and different places in the embryo. It is probable that the dermal signals cause the stabilization of β-catenin in the ectoderm (Gat et al. 1998). The basal cells elongate, divide, and sink into the dermis. The dermal fibroblasts respond to this ingression of epidermal cells by forming a small node (the **dermal papilla**) beneath the hair germ. The dermal papilla then pushes up on the basal stem cells and stimulates them to divide more rapidly. The basal cells respond by producing postmitotic cells that will differentiate into the keratinized hair shaft (see Hardy 1992; Miller et al. 1993). Melanoblasts, which were present among the epidermal cells as they ingressed, differentiate into melanocytes and transfer their pigment to the shaft (Figure 12.34). As this is occurring, two epithelial swellings begin to grow on the side of the hair germ. The cells of the lower swelling may retain a population of stem cells that will regenerate the hair shaft periodically when it is shed (Pinkus and Mehregan 1981; Cotsarelis et al. 1990). The cells of the upper bulge form the **sebaceous glands**, which produce an oily secretion, **sebum**. In many mammals, including humans, the sebum mixes with the shed peridermal cells to form the whitish vernix caseosa, which surrounds the fetus at birth. Just as there is a pluripotent neural stem cell whose offspring become neural and glial cells, so

their own division. When a growth factor is made by the same cell that receives it, that factor is called an **autocrine growth factor**. Such factors have to be carefully regulated because if their levels are elevated, more cells are rapidly produced. In adult skin, a cell born in the Malpighian layer takes roughly 8 weeks to reach the cornified layer, and remains there for about 2 weeks. In individuals with psoriasis, a disease characterized by the exfoliation of enormous amounts of epidermal cells, a cell's time in the cornified layer is only 2 days (Weinstein and van Scott 1965; Halprin 1972). This condition has been linked to the overexpression of TGF-α (which occurs secondarily to an inflammation) (Elder et al. 1989). Similarly, if the TGF-α gene is linked to a promoter for keratin 14 (one of the major skin proteins) and inserted into the mouse pronucleus, the resulting transgenic mice activate the TGF-α gene in their skin cells and cannot down-regulate it. The result is a mouse with scaly skin, stunted hair growth, and an enormous surplus of keratinized epidermis over its single layer of basal cells (Figure 12.33B; Vassar and Fuchs 1991).

Another growth factor needed for epidermal production is **keratinocyte growth factor** (**KGF**; also known as fibroblast growth factor 7), a paracrine factor that is produced by the fibroblasts of the underlying (mesodermally derived) dermis.

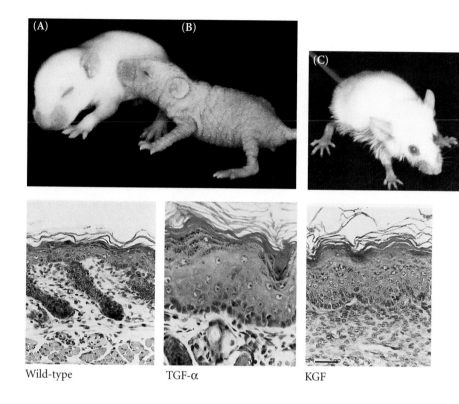

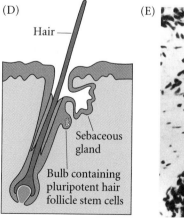

Wild-type   TGF-α   KGF

Figure 12.33
Growth factors and epidermal proliferation. (A) A wild-type mouse pup. (B) A littermate of (A) that is expressing high levels of TGF-α in its keratinocytes. It has scaly skin and very little hair. Below each mouse is a cross section through its skin. The mouse overexpressing TGF-α has very extensive layers of keratinized epithelia, which it sheds. (C) A transgenic mouse expressing low levels of KGF in its keratinocytes. Note the sparsity of hair around the legs, eyes, and nose. The skin section shows an absence of hair follicles and an increased number of basal epidermal cells. Beneath each mouse is a section of its skin. (From Vassar and Fuchs 1991; Guo et al. 1993; photographs courtesy of E. Fuchs.)

there appears to be a pluripotent epidermal stem cell whose progeny can become epidermis, sebaceous gland, or hair shaft.

The first hairs in the human embryo are of a thin, closely spaced type called **lanugo**. This type of hair is usually shed before birth and is replaced (at least in part, by new follicles) by the short and silky **vellus**. Vellus remains on many parts of the human body usually considered hairless, such as the forehead and eyelids. In other areas of the body, vellus gives way to "terminal" hair. During a person's life, some of the follicles that produced vellus can later form terminal hair and still later revert to vellus production. The armpits of infants, for instance, have follicles that produce vellus until adolescence, then begin producing terminal shafts. Conversely, in normal masculine

Figure 12.34
Development of the hair follicles in fetal human skin. (A) Basal epidermal cells become columnar and bulge slightly into the dermis. (B) Epidermal cells continue to proliferate, and dermal mesenchyme cells collect at the base of the primary hair germ to form a dermal papilla. (C) Hair shaft differentiation begins in the elongated hair germ. (D) The keratinized hair shaft extends from the hair follicle, the secondary bud forms the sebaceous gland, and beneath it is a region that may contain the hair stem cells for the next cycle of hair production. (E) Photograph of elongated hair germ. (After Hardy 1992 and Miller et al. 1993; photograph courtesy of W. Montagna.)

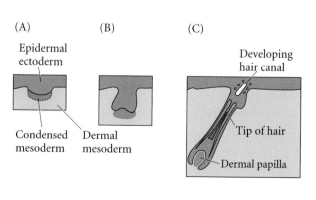

(A) Epidermal ectoderm / Condensed mesoderm / Dermal mesoderm
(B)
(C) Developing hair canal / Tip of hair / Dermal papilla

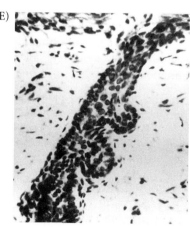

(D) Hair / Sebaceous gland / Bulb containing pluripotent hair follicle stem cells
(E)

pattern baldness, the scalp follicles revert to producing unpigmented and very fine vellus hair (Montagna and Parakkal 1974).

> WEBSITE **12.10 Normal variation in human hair production.** The human hair has a complex life cycle. Moreover, some hairs grow short (such as those of our eyelashes) while other hairs (such as those of our scalp) grow long. The pattern of hair size and of baldness is determined by paracrine and endocrine factors.

> WEBSITE **12.11 Mutations of human hair production.** In addition to normal variation, there are also inherited mutations that interfere with normal hair development. Some people are born without the ability to grow hair, while others develop hair over their entire bodies. These genetic conditions give us insights into the mechanisms of normal hair growth.

Patterning of cutaneous appendages

It is obvious that cutaneous appendages such as hair, feathers, or scales do not grow randomly over the body. Rather, there are spaces between them, and these spaces (for instance, on the scalp) are very similar from region to region (Widelitz and Chuong 1998). Recent research suggests that a reaction-diffusionprocess may be responsible for this pattern (see Chapter 1). The activator is Sonic hedgehog, a paracrine factor that acts locally without much diffusion (Nohno et al. 1995). The inhibitor is believed to be BMP4 or BMP2, both of which are paracrine factors with a greater range of diffusion (Jung et al. 1998; Noramly and Morgan 1998). BMPs may prevent the dermal fibroblasts from aggregating, while Sonic hedgehog may support the formation and retention of the dermal papilla (see Figure 6.6).

Snapshot Summary: Central Nervous System and Epidermis

1. The neural tube forms from the shaping and folding of the neural plate. In primary neurulation, the surface ectoderm folds into a tube that separates from the surface. In secondary neurulation, the ectoderm forms a cord and then forms a cavity within it.

2. Primary neurulation is regulated by both intrinsic and extrinsic forces. Intrinsic wedging occurs within cells of the hinge regions to bend the neural plate. Extrinsic forces include the migration of the surface ectoderm towards the center of the embryo.

3. Neural tube closure is also a mixture of extrinsic and intrinsic forces. In humans, if the neural tube fails to close various diseases can result.

4. The neural crest cells arise at the lateral borders of the neural tube and surface ectoderm. They become located between the neural tube and surface ectoderm, and they migrate away from this region to become peripheral neural, glial, and pigment cells.

5. There is a gradient of maturity in many embryos, especially those of amniotes. The anterior develops earlier than the posterior.

6. The dorsal-ventral patterning of the neural tube is accomplished by proteins of the TGF-β family secreted from the surface ectoderm and roof of the neural tube, and from Sonic hedgehog protein secreted by the notochord and floor plate cells. Both types of protein appear to work through gradients.

7. The brain forms three primary vesicles: prosencephalon (forebrain), mesencephalon (midbrain), and rhombencephalon (hindbrain). The prosencephalon and rhombencephalon will become subdivided.

8. The brain expands through fluid secretion putting positive pressure on the vesicles.

9. The neurons of the brain are organized into cortices (layers) and nuclei (clusters).

10. New neurons are formed by mitosis in the neural tube. The neural precursors can migrate away from the neural tube and form a new layer. Neurons forming later have to migrate through the existing layers. This forms the cortical layers. The germinal zone at the lumen of the neural tube is called the ventricular zone. The new layer is called the mantle zone (gray matter).

11. In the cerebellum, a second germinal zone—the external granule layer—is formed. Other neurons migrate out of the ventricular zone on the processes of glial cells.

12. The cerebral cortex in humans has six layers, and the mantle zone is called the neocortex. Cell fates are often fixed as they undergo their last division. Neurons derived from the same stem cell may end up in different functional regions of the brain.

13. Neural stem cells have been observed in the adult human brain. We now believe humans can continue making neurons throughout life, although at nowhere near the fetal rate.

14. Dendrites receive signals from other neurons, while axons transmit them. The place where the signaling takes place (through the release of neurotransmitters) is called a synapse.

15. Axons grow from the nerve cell body, or soma. They are led by the growth cone.

16. The chordate and arthropod systems, though structurally very different, appear to be specified through the same set of genetic instructions.

17. The retina forms from the optic vesicle that extends from the brain. Pax6 plays a major role in eye formation, and the downregulation of Pax6 by Sonic hedgehog in the center of the brain splits the eye-forming region of the brain in half. If Sonic hedeghog is not expressed there, a single medial eye results.

18. The photoreceptor cells gather the light and transmit the impulse through interneurons to the retinal ganglion cells. The axons of the retinal ganglion cells form the optic nerve.

19. The lens and cornea form from the surface ectoderm. Both must become transparent.

20. The basal layer of the surface ectoderm becomes the stratum germinativum, or germinal layer of the skin. These cells divide to produce a stem cell and a cell committed to become an epidermal cell (keratinocyte). Stem cells appear to be able to make hair.

21. Paracrine factors such as TGF-α and FGF7 are important in normal skin development.

22. Cutaneous appendages—hair, feathers, and scales—are formed by epithelial-mesenchymal interactions between the epidermis and the dermal mesoderm.

Literature Cited

Alvarez, I. S. and G. C. Schoenwolf. 1992. Expansion of surface epithelium provides the major extrinsic force for bending of the neural plate. *J. Exp. Zool.* 261: 340–348.

Balinsky, B. I. 1975. *Introduction to Embryology,* 4th Ed. Saunders, Philadelphia.

Briscoe, J., L. Sussel, D. Hartigan-O'Connor, T. M. Jessell, J. L. R. Rubenstein and J. Ericson. 1999. Homeobox gene *Nkx2.2* and specification of neuronal identity by graded Sonic hedgehog signaling. *Nature* 398: 622–627.

Burnside, B. 1973. Microtubules and microfilaments in amphibian neurulation. *Am. Zool.* 13: 989–1006.

Catala, M., M.-A. Teillet and N. M. Le Douarin. 1995. Organization and development of the tail bud analyzed with the quail-chick chimaera system. *Mech. Dev.* 51: 51–65.

Catala, M., M.-A. Teillet, E. M. De Robertis and N. M. Le Douarin. 1996. A spinal cord fate map in the avian embryo: While regressing, Hensen's node lays down the notochord and floor plate thus joining the spinal cord lateral walls. *Development* 122: 2599–2610.

Cau, E., G. Gradwohl, C. Fode and F. Guillemot. 1997. Mash1 activates a cascade of bHLH regulators in olfactory neuron progenitors. *Development* 124: 1611–1621.

Centers for Disease Control. 1992. Recommendations for the use of folic acid to reduce the number of cases of spina bifida and other neural tube defects. *Morb. Mortal. Wkly. Rep.* 41: 1–7.

Chenn, A. and S. K. McConnell. 1995. Cleavage orientation and the asymmetric inheritance of Notch1 immunoreactivity in mammalian neurogenesis. *Cell* 82: 631–641.

Chiang, C., Y. Litingung, E. Lee, K. K. Young, J. E. Corden, H. Westphal and P. A. Beachy. 1996. Cyclopia and defective axial patterning in mice lacking *sonic hedgehog* gene function. *Nature* 383: 407–413.

Chitnis, A., D. Henrique, J. Lewis, D. Ish-Horowicz and C. Kintner. 1995. Primary neurogenesis in *Xenopus* embryos regulated by a homolog of the *Drosophila* neurogenic gene *Delta. Nature* 375: 761–766.

Cotsarelis, G., T.-T. Sun and R. M. Lavker. 1990. Label-retaining cells reside in the bulge area of pilosebaceous unit: Implications for follicular stem cells, hair cycle and skin carcinogenesis. *Cell* 61: 1329–1337.

Coulombre, A. J. 1956. The role of intraocular pressure in the development of the chick eye. I. Control of eye size. *J. Exp. Zool.* 133: 211–225.

Coulombre, A. J. 1965. The eye. *In* R. DeHaan and H. Ursprung (eds.), *Organogenesis.* Holt, Rinehart & Winston, New York, pp. 217–251.

Czeizel, A. and I. Dudas. 1992. Prevention of first occurrence of neural tube defects by periconceptional vitamin supplementation. *N. Engl. J. Med.* 327: 1832–1835.

Davenport, R. W., P. Dou, V. Rehder and S. B. Kater. 1993. A sensory role for neuronal growth cone filopodia. *Nature* 361: 721–724.

Desmond, M. E. 1982. A description of the occlusion of the lumen of the spinal cord in early human embryos. *Anat. Rec.* 204: 89–93.

Desmond, M. E. and M. C. Field. 1992. Evaluation of neural fold fusion and coincident initiation of spinal cord occlusion in the chick embryo. *J. Comp. Neurol.* 319: 246–260.

Desmond, M. E. and G. C. Schoenwolf. 1986. Evaluation of the roles of intrinsic and extrinsic factors in occlusion of the spinal neurocoel during rapid brain enlargement in the chick embryo. *J. Embryol. Exp. Morphol.* 97: 25–46.

Detrick, R. J., D. Dickey and C. R. Kintner. 1990. The effects of N-cadherin misexpression on morphogenesis in *Xenopus* embryos. *Neuron* 4: 493–506.

Doetsch, F., I. Caillé, D. A. Lim, J. M. García-Verdugo and A. Alvarez-Buylla. 1999. Subventricular zone astrocytes are neural stem cells in the adult mammalian brain. *Cell* 97: 703–716.

Echelard, Y., D. J. Epstein, B. St-Jacques, L. Shen, J. Mohler, J. A. McMahon and A. McMahon. 1993. Sonic hedgehog, a member of a family of putative signaling molecules, is implicated in the regulation of CNS polarity. *Cell* 75: 1417–1430.

Edmondson, J. C., R. K. H. Liem, J. C. Kuster and M. E. Hatten. 1988. Astrotactin: A novel neuronal cell surface antigen that mediates neuronal-astroglial interactions in cerebellar microcultures. *J. Cell Biol.* 106: 505–517.

Eichele, G. 1992. Budding thoughts. *The Sciences* (Jan. 1992): 30–36.

Elder, J. T. and 8 others. 1989. Overexpression of transforming growth factor in psoriatic epidermis. *Science* 243: 811–814.

Erickson, C. A. and J. A. Weston. 1983. An SEM analysis of neural crest migration in the mouse. *J. Embryol.. Exp. Morphol.* 74: 97–118.

Erikksson, P. S., E. Perfiliea, T. Björn-Erikksson, A.-M. Alborn, C. Nordberg, D. A. Peterson and F. H. Gage. 1998. Neurogenesis in the adult human hippocampus. *Nature Med.* 4: 1313–1317.

Fishell, G. and M. E. Hatten. 1991. Astrotactin provides a receptor system for glia-guided neuronal migration. *Development* 113: 755–765.

Forscher, P. and S. J. Smith. 1988. Actions of cytochalasins on the organization of actin filaments and microtubules in a neural growth cone. *J. Cell Biol.* 107: 1505–1516.

Frantz, G. D. and S. K. McConnell. 1996. Restriction of late cerebral cortical progenitors to an upper-layer fate. *Neuron* 17: 55–61.

Fujimori, T., S. Miyatani and M. Takeichi. 1990. Ectopic expression of N-cadherin perturbs histogenesis in *Xenopus* embryos. *Development* 110: 97–104.

Fujita, S. 1964. Analysis of neuron differentiation in the central nervous system by tritiated thymidine autoradiography. *J. Comp. Neurol.* 122: 311–328.

Fujita, S. 1966. Application of light and electron microscopy to the study of the cytogenesis of the forebrain. *In* R. Hassler and H. Stephen (eds.), *Evolution of the Forebrain.* Plenum, New York, pp. 180–196.

Gallera, J. 1971. Primary induction in birds. *Adv. Morphogenet.* 9: 149–180.

Gat, U., R. Das Gupta, L. Degenstein and E. Fuchs. 1998. De-novo hair follicle morphogenesis and hair tumors in mice expressing a truncated β-catenin in skin. *Cell* 95: 605–614.

Glaser, T., L. Jepeal, J. G. Edwards, S. R. Young, J. Favor and R. L. Maas. 1994. *PAX6* gene dosage effect in a family with congenital cataracts, aniridia, anophthalmia and central nervous system defects. *Nature Genet.* 7: 463–471.

Golden, J. A. and Chernoff, G. F. 1993. Intermittent pattern of neural tube closure in two strains of mice. *Teratology* 47: 73–80.

Gont, L. K., H. Steinbeisser, B. Blumberg and E. M. De Robertis. 1993. Tail formation as a continuation of gastrulation: The multiple cell populations of the *Xenopus* tailbud derive from the late blastopore lip. *Development* 119: 991–1004.

Goodman, C. S. and C. Q. Doe. 1993. Embryonic development of the *Drosophila* central nervous system. *In* M. Bate and A. Martinez Arias (eds.), *The Development of* Drosophila melanogaster. Cold Spring Harbor Press, Cold Spring Harbor, NY, pp. 1131–1206.

Gould, E., A. J. Reeves, M. S. A. Graziano and C. Gross. 1999a. Neurogenesis in the adult cortex of adult primates. *Science* 286: 548–552.

Gould, E., A. Beylin, P. Tanapat, A. Reeves and T. J. Shors. 1999b. Learning enhances adult neurogenesis in the hippocampal formation. *Nature Neurosci.* 2: 260–265.

Gould, S. J. 1977. *Ontogeny and Phylogeny.* Harvard University Press, Cambridge, MA.

Guo, L., Q.-C. Yu and E. Fuchs. 1993. Targeting expression of keratinocyte growth factor to keratinocytes elicits striking changes in epithelial differentiation in transgenic mice. *EMBO J.* 12: 973–986.

Guthrie, S. and A. Lumsden. 1991. Formation and regeneration of rhombomere boundaries in the developing chick hindbrain. *Development* 112: 221–229.

Gregory, W. A., J. C. Edmondson, M. E. Hatten and C. A. Mason. 1988. Cytology and neural-glial apposition of migrating cerebellar granule cells in vitro. *J. Neurosci.* 8: 1728–1738.

Halder, G., P. Callaerts and W. J. Gehring. 1995. Induction of ectopic eyes by targeted expression of the *eyeless* gene in *Drosophila. Science* 267: 1788–1792.

Halprin, K. M. 1972. Epidermal "turnover time"—a reexamination. *J. Invest. Dermatol.* 86: 14–19.

Hardy, M. H. 1992. The secret life of the hair follicle. *Trends Genet.* 8: 55–61.

Harrison, R. G. 1907. Observations on the living developing nerve fiber. *Anat. Rec.* 1: 116–118.

Hatten, M. E. 1990. Riding the glial monorail: A common mechanism for glial-guided neuronal migration in different regions of the mammalian brain. *Trends Neurosci.* 13: 179–184.

Henry, E. W. and R. L. Sidman. 1988. Long lives for homozygous trembler mutant mice despite virtual absence of peripheral nerve myelin. *Science* 241: 344–346.

Hilfer, S. R. and J.-J. W. Yang. 1980. Accumulation of CPC-precipitable material at apical cell surfaces during formation of the optic cup. *Anat. Rec.* 197: 423–433.

His, W. 1886. Zur Geschichte des mensch-lichen Rückenmarks und der Nervenwurzeln. *Ges. Wiss.* 13, S. 477.

Hogan, B. L. M. 1999. Morphogenesis. *Cell* 96: 225–233.

Holt, A. B., D. B. Cheek, E. D. Mellitz and D. E. Hill. 1975. Brain size and the relation of the primate to the non-primate. *In* D. B. Cheek (ed.), *Fetal and Postnatal Cellular Growth: Hormones and Nutrition.* Wiley, New York, pp. 23–44.

Huettner, A. F. 1949. *Fundamentals of Comparative Embryology of the Vertebrates*, 2nd Ed. Macmillan, New York.

Jacobson, A. G. and J. G. Moury. 1995. Tissue boundaries and cell behavior during neurulation. *Dev. Biol.* 171: 98–110.

Jacobson, A. G. and A. K. Sater. 1988. Features of embryonic induction. *Development* 104: 341–359.

Jacobson, M. 1968. Cessation of DNA synthesis in retinal ganglion cells correlated with the time of specification of their central connections. *Dev. Biol.* 17: 219–232.

Jacobson, M. 1991. *Developmental Neurobiology*, 2nd Ed. Plenum, New York.

Johansson, C. B., S. Momma, D. L. Clarke, M. Risling, U. Lendahl and J. Frisén. 1999. Identification of a neural stem cell in the adult mammalian central nervous system. *Cell* 96: 25–34.

Johnson, J. E., S. J. Birren and D. J. Anderson. 1990. Two rat homologs of *Drosophila achaete-scute* specifically expressed in neuronal precursors. *Nature* 346: 858–861.

Jones, P. H. and F. M. Watt. 1993. Separation of human epidermal stem cells from transit amplifying cells on the basis of differences in integrin function and expression. *Cell* 73: 713–724.

Jordan, T. and 7 others. 1992. The human *PAX6* gene is mutated in two patients with aniridia. *Nature Genet.* 1: 328–332.

Jung, H.-S. and 7 others. 1998. Local inhibitory action of BMPs and their relationships with activators of feather formation: Implications for periodic patterning. *Dev. Biol.* 196: 11–23.

Karfunkel, P. 1972. The activity of microtubules and microfilaments in neurulation in the chick. *J. Exp. Zool.* 181: 289–302.

Keller, R., J. Shih, A. K. Sater and C. Moreno. 1992. Planar induction of convergence and extension of the neural plate by the organizer of *Xenopus. Dev. Dynam.* 193: 218–234.

Kelley, R. I. and 7 others. 1996. Holoprosencephaly in RSH/Smith-Lemli-Opitz syndrome: Does abnormal cholesterol metabolism affect the function of Sonic hedgehog? *Am. J. Med. Genet.* 66: 478–484.

Kempermann, G., H. G. Kuhn and F. H. Gage. 1997a. Genetic influence on neurogenesis in the dentate gyrus of adult mice. *Proc. Natl. Acad. Sci. USA* 94: 10409–10414.

Kempermann, G., H. G. Kuhn and F. H. Gage. 1997b. More hippocampal neurons in adult mice living in an enriched environment. *Nature* 386: 493–495.

Komuro, H. and P. Rakic. 1992. Selective role of N-type calcium channels in neuronal migration. *Science* 157: 806–809.

Lamoureux, P., R. E. Buxbaum and S. R. Heidemann. 1989. Direct evidence that growth cones pull. *Nature* 340: 159–162.

Larsen, W. J. 1993. *Human Embryology.* Churchill Livingstone, New York.

Letourneau, P. C. 1977. Regulation of neuronal morphogenesis by cell-substratum adhesion. *Soc. Neurosci. Symp.* 2: 67–81.

Letourneau, P. C. 1979. Cell substratum adhesion of neurite growth cones and its role in neurite elongation. *Exp. Cell Res.* 124: 127–138.

Liem, K. F., Jr., G. Tremmi, H. Roelink and T. M. Jessell. 1995. Dorsal differentiation of neural plate cells induced by BMP-mediated signals from epidermal ectoderm. *Cell* 82: 969–979.

Liem, K. F., Jr., G. Tremmi and T. M. Jessell. 1997. A role for the roof plate and its resident TGF-β-related proteins in neuronal patterning in the dorsal spinal cord. *Cell* 91: 127–138.

Lumsden, A. and R. Keynes. 1989. Segmental patterns of neuronal development in the chick hindbrain. *Nature* 337: 424–428.

Mann, I. 1964. *The Development of the Human Eye.* Grune and Stratton, New York.

Matsunami, H. and M. Takeichi. 1995. Fetal brain subdivisions defined by T- and E- cadherins expressions: Evidence for the role of cadherin activity in region-specific, cell-cell adhesion. *Dev. Biol.* 172: 466–478.

McConnell, S. K. and C. E. Kaznowski. 1991. Cell cycle dependence of laminar determination in developing cerebral cortex. *Science* 254: 282–285.

Miller, S. J., R. M. Lavker and T.-T. Sun. 1993. Keratinocyte stem cells of corneal, skin, and hair follicle. *Semin. Dev. Biol.* 4: 217–240.

Milunsky, A., H. Jick, S. S. Jick, C. L. Bruell, D. S. Maclaughlen, K. J. Rothman and W. Willett. 1989. Multivitamin folic acid supplementation in early pregnancy reduces the prevalence of neural tube defects. *JAMA* 262: 2847–2852.

Montagna, W. and P. F. Parakkal. 1974. The piliary apparatus. *In* W. Montagna (ed.), *The Structure and Formation of Skin*. Academic Press, New York, pp. 172–258.

Montagu, M. F. A. 1962. Time, morphology, and neoteny in the evolution of man. *In* M. F. A. Montagu (ed.), *Culture and Evolution of Man*. Oxford University Press, New York.

Moore, K. L. and T. V. N. Persaud. 1993. *Before We Are Born: Essentials of Embryology and Birth Defects*. W. B. Saunders, Philadelphia.

Morowitz, H. J. and J. S. Trefil. 1992. *The Facts of Life: Science and the Abortion Controversy*. Oxford University Press, New York.

Moury, J. D. and G. C. Schoenwolf. 1995. Cooperative model of epithelial shaping and bending during avian neurulation: Autonomous movements of the neural plate, autonomous movements of the epidermis, and interactions in the neural plate/epidermis transition zone. *Dev. Dynam.* 204: 323–337.

Nagele, R. G. and H. Y. Lee. 1987. Studies in the mechanism of neurulation in the chick. Morphometric analysis of the relationship between regional variations in cell shape and sites of motive force generation. *J. Exp. Biol.* 24: 197–205.

Nichols, D. H. 1981. Neural crest formation in the head of the mouse embryo as observed using a new histological technique. *J. Embryol. Exp. Morphol.* 64: 105–120.

Nievelstein, R. A. J., N. G. Hartwig, C. Vermeij-Keers and J. Valk. 1993. Embryonic development of the mammalian caudal neural tube. *Teratology* 48: 21–31.

Nohno, T. W., Y. Kawakami, H. Ohuchi, A. Fujiwara, H. Yoshioka and S. Noji. 1995. Involvement of the *sonic hedgehog* gene in chick feather formation. *Biochem. Biophys. Res. Comm.* 206: 33–39.

Noramly, S. and B. A. Morgan. 1998. BMPs mediate lateral inhibition at successive stages of feather tract development. *Development* 125: 3775–3787.

O'Rourke, N. A., M. E. Dailey, S. J. Smith and S. K. McConnell. 1992. Diverse migratory pathways in the developing cerebral cortex. *Science* 258: 299–302.

Papaconstantinou, J. 1967. Molecular aspects of lens cell differentiation. *Science* 156: 338–346.

Pasteels, J. 1937. Etudes sur la gastrulation des vertébrés méroblastiques. III. Oiseaux. IV. Conclusions générales. *Arch. Biol.* 48: 381–488.

Paton, D. and J. A. Craig. 1974. Cataracts: Development, diagnosis, and management. *CIBA Clin. Symp.* 26(3): 2–32.

Patten, B. M. 1971. *Early Embryology of the Chick*, 5th Ed. McGraw-Hill, New York.

Piatigorsky, J. 1981. Lens differentiation in vertebrates: A review of cellular and molecular features. *Differentiation* 19: 134–153.

Pierani, A., S. Brenner-Morton, C. Chiang and T. M. Jessell. 1999. A Sonic hedgehog-independent, retinoid-activated pathway of neurogenesis in the ventral spinal cord. *Cell* 97: 903–915.

Pinkus, H. and A. H. H. Mehregan. 1981. *A Guide to Dermohistopathology*. Appleton Century Crofts, New York.

Placzek, M., M. Tessier-Lavigne, T. Yamada, T. Jessell and J. Dodd. 1990. Mesodermal control of neural cell identity: Floor plate induction by the notochord. *Science* 250: 985–988.

Portmann, A. 1941. Die Tragzeiten der Primaten und die Dauer der Schwangerschaft beim Menschen: Ein Problem der vergleichen Biologie. *Rev. Suisse Zool.* 48: 511–518.

Portmann, A. 1945. Die Ontogenese des Menschen als Problem der Evolutionsforschung. *Verh. Schweiz. Naturf. Ges.* 125: 44–53.

Praag, H. van, G. Kempermann and F. H. Gage. 1999. Running increases cell proliferation and neurogenesis in the adult mouse gentate gyrus. *Nature Neurosci.* 2: 266–270.

Quiring, R., U. Walldorf, U. Kloter and W. J. Gehring. 1994. Homology of the *eyeless* gene of *Drosophila* to the *Small eye* gene in mice and *Aniridia* in humans. *Science* 265: 785–789.

Rakic, P. 1972. Mode of cell migration to superficial layers of fetal monkey neocortex. *J. Comp. Neurol.* 145: 61–84.

Rakic, P. 1974. Neurons in rhesus visual cortex: Systematic relation between time of origin and eventual disposition. *Science* 183: 425–427.

Rakic, P. 1975. Cell migration and neuronal ectopias in the brain. *In* D. Bergsma (ed.), *Morphogenesis and Malformations of Face and Brain*. Birth Defects Original Article Series, vol. 11, no.7. Alan R. Liss, New York, pp. 95–129.

Rakic P. and P. S. Goldman. 1982. Development and modifiability of the cerebral cortex. *Neurosci. Rev.* 20: 429–612.

Rakic, P. and R. L. Sidman. 1973. Organization of cerebellar cortex secondary to deficit of granule cells in weaver mutant mice. *J. Comp. Neurol.* 152: 133–162.

Ramón y Cajal, S. 1890. Sur l'origene et les ramifications des fibres neuveuses de la moelle embryonnaire. *Anat. Anz.* 5: 111–119.

Roelink, H. and 10 others. 1994. Floor plate and motoneuron induction by *vhh-1*, a vertebrate homolog of *hedgehog* expressed by the notochord. *Cell* 76: 761–775.

Roelink, H., J. A. Porter, C. Chiang, Y. Tanabe, D. T. Chang, P. A. Beachy and T. M. Jessell. 1995. Floor plate and motor neuron induction by different concentrations of amino terminal cleavage product of Sonic hedgehog autoproteolysis. *Cell* 81: 445–455.

Roessler, E. and 7 others. 1996. Mutations in the human *sonic hedgehog* gene cause holoprosencephaly. *Nature Genet.* 14: 357–360.

Saha, M., C. L. Spann and R. M. Grainger. 1989. Embryonic lens induction: More than meets the optic vesicle. *Cell Diff. Dev.* 28: 153–172.

Sauer, F. C. 1935. Mitosis in the neural tube. *J. Comp. Neurol.* 62: 377–405.

Sausedo, R. A., J. Smith and G. C. Schoenwolf. 1997. Role of oriented cell division in shaping and bending of the neural plate. *J. Comp. Neurol.* 381: 473–488.

Schoenwolf, G. C. 1984. Histological and ultrastructural studies of secondary neurulation in mouse embryos. *Am. J. Anat.* 169: 361–374.

Schoenwolf, G. C. 1991a. Cell movements driving neurulation in avian embryos. *Development* 2 [Suppl.]: 157–168.

Schoenwolf, G. C. 1991b. Cell movements in the epiblast during gastrulation and neurulation in avian embryos. *In* R. Keller et al. (eds.), *Gastrulation*. Plenum, New York, pp. 1–28.

Schoenwolf, G. C. and I. S. Alvarez. 1989. Roles of neuroepithelial cell rearrangement and division in shaping of the avian neural plate. *Development* 106: 427–439.

Schoenwolf, G. C. and N. E. Desmond. 1984. Descriptive studies of the occlusion and reopening of the spinal canal of the early chick embryo. *Anat. Rec.* 209: 251–263.

Shou, J., P. C. Rim and A. L. Calof. 1999. BMPs inhibit neurogenesis by a mechanism involving degradation of a transcription factor. *Nature Neurosci.* 2: 339–345.

Skeath, J. B. and S. B. Carroll. 1992. Regulation of proneural gene expression and cell fate during neuroblast segregation in the *Drosophila* embryo. *Development* 114: 939–46.

Skeath, J. B., G. F. Panganiban and S. B. Carroll. 1994. The ventral nervous system defective gene controls proneural gene expression at two distinct steps during neuroblast formation in *Drosophila. Development* 120: 1517–24

Sidman, R. L., M. M. Dickie and S. H. Appel. 1964. Mutant mice (*quaking* and *jimpy*) with deficient myelination in the central nervous system. *Science* 144: 309–312.

Smith, J. L. and G. C. Schoenwolf. 1989. Notochordal induction of cell wedging in the chick neural plate and its role in neural tube formation. *J. Exp. Zool.* 250: 49–62.

Smith, J. L. and G. C. Schoenwolf. 1991. Further evidence of extrinsic forces in bending of the neural plate. *J. Comp. Neurol.* 307: 225–236.

Smith, J. L. and G. C. Schoenwolf. 1997. Neurulation: Coming to closure. *Trends Neurosci.* 11: 510–517.

Spemann, H. 1938. *Embryonic Development and Induction.* Yale University Press, New Haven.

Turner, D. L. and C. L. Cepko. 1987. A common progenitor for neurons and glia persists in rat retina late in development. *Nature* 328: 131–136.

Van Allen, M. I. and 15 others. 1993. Evidence for multi-site closure of the neural tube in humans. *Am. J. Med. Genet.* 47: 723–743.

van Straaten, H. W. M., J. W. M. Hekking, E. J. L. M. Wiertz-Hoessels, F. Thors and J. Drukker. 1988. Effect of the notochord on the differentiation of a floor plate area in the neural tube of the chick embryo. *Anat. Embryol.* 177: 317–324.

Vassar, R. and E. Fuchs. 1991. Transgenic mice provide new insights into the role of TGF-α during epidermal development and differentiation. *Genes Dev.* 5: 714–727.

Wakamatsu, Y., T. M. Maynard, S. U. Jones and J. A. Weston. 1999. Numb localizes in the basal cortex of mitotic avian neuroepithelial cells and modulates neuronal differentiation by binding to Notch-1. *Neuron* 23: 71–81.

Wallace, V. A. 1999. Purkinje cell-derived Sonic hedgehog regulates granule neuron precursor cell proliferation in the developing mouse cerebellum. *Curr. Biol.* 9: 445–448.

Walsh, C. and C. L. Cepko. 1988. Clonally related cortical cells show several migration patterns. *Science* 241: 1342–1345.

Walsh, C. and C. L. Cepko. 1992. Widespread dispersion of neuronal clones across functional regions of the cerebral cortex. *Science* 255: 434–440.

Weinstein, G. D. and E. J. van Scott. 1965. Turnover times of normal and psoriatic epidermis. *J. Invest. Dermatol.* 45: 257–262.

Widelitz, R. B. and C.-M. Chuong. 1998. Early molecular events in feather morphogenesis: Induction and dermal condensation. *In* Chung, C.-M. (editor) *Molecular Basis of Epithelial Appendage Morphogenesis*, R. G. Landes, Austin. Pp. 243–263.

Yamada, K. M., B. S. Spooner and N. K. Wessells. 1971. Ultrastructure and function of growth cones and axons of cultured nerve cells. *J. Cell Biol.* 49: 614–635.

Yamada, T., M. Placzek, H. Tanaka, J. Dodd and T. M. Jessell. 1991. Control of cell pattern in the developing nervous system: Polarizing activity of floor plate and notochord. *Cell* 64: 635–647.

Yamada, T., S. L. Pfaff, T. Edlund and T. M. Jessell. 1993. Control of cell pattern in the neural tube: Motor neuron induction by diffusible factors from notochord and floor plate. *Cell* 73: 673–686.

13 *Neural crest cells and axonal specificity*

IN THIS CHAPTER WE CONTINUE OUR DISCUSSION of ectodermal development. Here we will focus on neural crest cells and axonal guidance. Neural crest cells and axon growth cones share the property of having to migrate far from their source of origin to specific places in the embryo. They both have to recognize cues to begin this migration, and they both have to respond to signals that guide them along specific routes to their final destination. Recent research has discovered that many of the signals recognized by the neural crest cells and by the axon growth cones are the same.

The Neural Crest

Although derived from the ectoderm, the neural crest has sometimes been called the fourth germ layer because of its importance. It has even been said, perhaps hyperbolically, that "the only interesting thing about vertebrates is the neural crest" (quoted in Thorogood 1989). The neural crest cells originate at the dorsalmost region of the neural tube. Transplantation experiments wherein a quail neural plate is grafted into a chick non-neural ectoderm have shown that juxtaposing these tissues induces the formation of neural crest cells, and that both the prospective neural plate and the prospective epidermis contribute to the neural crest (Selleck and Bronner-Fraser 1995; see also Mancilla and Mayor 1996).

The neural crest cells migrate extensively to generate a prodigious number of differentiated cell types. These cell types include (1) the neurons and glial cells of the sensory, sympathetic, and parasympathetic nervous systems, (2) the epinephrine-producing (medulla) cells of the adrenal gland, (3) the pigment-containing cells of the epidermis, and (4) many of the skeletal and connective tissue components of the head. The fate of the neural crest cells depends, to a large degree, on where they migrate to and settle. Table 13.1 is a summary of some of the cell types derived from the neural crest.

The neural crest can be divided into four main functional (but overlapping) domains (Figure 13.1):

• The **cranial (cephalic) neural crest**, whose cells migrate dorsolaterally to produce the craniofacial mesenchyme that differentiates into the cartilage, bone, cranial neurons, glia, and connective tissues of the face. These cells enter the pharyngeal arches and pouches to give rise to thymic cells, odontoblasts of the tooth primordia, and the bones of middle ear and jaw.

• The **trunk neural crest**, whose cells take one of two major pathways. Neural crest cells that become the pigment-synthesizing **melanocytes** migrate dorsolaterally

411

Table 13.1 Some derivatives of the neural crest

Derivative	Cell type or structure derived
Peripheral nervous system (PNS)	Neurons, including sensory ganglia, sympathetic and parasympathetic ganglia, and plexuses Neuroglial cells Schwann cells
Endocrine and paraendocrine derivatives	Adrenal medulla Calcitonin-secreting cells Carotid body type I cells
Pigment cells	Epidermal pigment cells
Facial cartilage and bone	Facial and anterior ventral skull cartilage and bones
Connective tissue	Corneal endothelium and stroma Tooth papillae Dermis, smooth muscle, and adipose tissue of skin of head and neck Connective tissue of salivary, lachrymal, thymus, thyroid, and pituitary glands Connective tissue and smooth muscle in arteries of aortic arch origin

Source: After Jacobson 1991, based on multiple sources.

thus to the absence of peristaltic movement in the bowels.

- The **cardiac neural crest** is located between the cranial and trunk neural crests. In chick embryos, this neural crest region extends from the first to the third somites, overlapping the anterior portion of the vagal neural crest. (Kirby 1987; Kirby and Waldo 1990). The cardiac neural crest cells can develop into melanocytes, neurons, cartilage, and connective tissue (of the third, fourth, and sixth pharyngeal arches). In addition, this region of the neural crest produces the entire musculoconnective tissue wall of the large arteries as they arise from the heart, as well as contributing to the septum that separates the pulmonary circulation from the aorta (Le Lièvre and Le Douarin 1975).

into the ectoderm and continue on their way toward the ventral midline of the belly. The second migratory pathway takes the trunk neural crest cells ventrolaterally through the anterior half of each sclerotome. (**Sclerotomes** are blocks of mesodermal cells, derived from somites, that will differentiate into the vertebral cartilage of the spine.) Those trunk neural crest cells that remain in the sclerotome form the **dorsal root ganglia** containing the sensory neurons. Those cells that continue more ventrally form the sympathetic ganglia, the adrenal medulla, and the nerve clusters surrounding the aorta.

- The **vagal** and **sacral neural crest**, whose cells generate the **parasympathetic** (**enteric**) **ganglia** of the gut (Le Douarin and Teillet 1973; Pomeranz et al. 1991). The vagal (neck) neural crest lies opposite chick somites 1–7, while the sacral neural crest lies posterior to somite 28. Failure of neural crest cell migration from these regions to the colon results in the absence of enteric ganglia and

Figure 13.1

Regions of the neural crest. The cranial neural crest migrates into the branchial arches and the face to form the bones and cartilage of the face and neck. It also produces pigment and cranial nerves. The vagal neural crest (near somites 1–7) and the sacral neural crest (posterior to somite 28) form the parasympathetic nerves of the gut. The cardiac neural crest cells arise from the neural crest by somites 1–3; they are critical in making the division beween the aorta and the pulmonary artery. Neural crest cells of the trunk (about somite 6 through the tail) make the sympathetic neurons, and a subset of these (at the level of somites 18–24) form the medulla portion of the adrenal gland. (After Le Douarin 1982.)

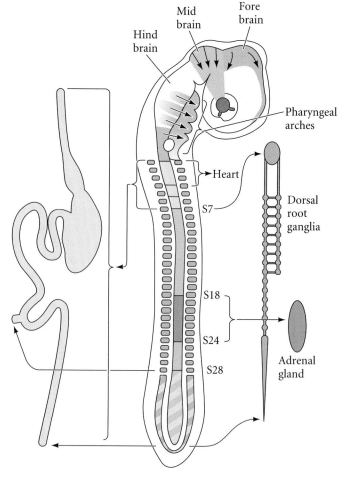

The Trunk Neural Crest

Migration pathways of trunk neural crest cells

The trunk neural crest is a transient structure, its cells dispersing soon after the neural tube closes. There are two major pathways taken by the migrating trunk neural crest cells (Figure 13.2A). Those cells migrating along the **dorsolateral pathway** become melanocytes, the melanin-forming pigment cells. They travel through the dermis, entering the ectoderm through minute holes in the basal lamina (which they may make). Here they colonize the skin and hair follicles (Mayer 1973; Erickson et al. 1992). This pathway was demonstrated in a series of classic experiments by Mary Rawles and others (1948), who transplanted the neural tube and crest from a pigmented strain of chickens into the neural tube of an albino chick embryo (see Figure 1.11).

Fate mapping of the neural crest cells has also shown that there is a **ventral pathway** wherein trunk neural crest

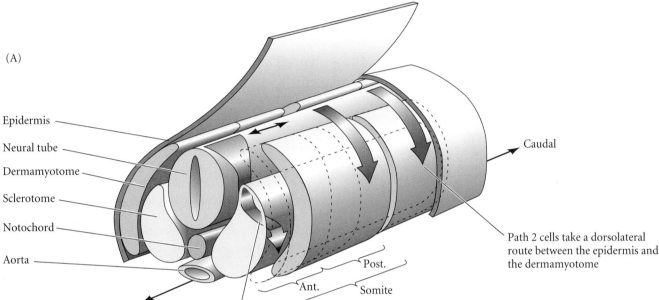

(A)

Epidermis

Neural tube

Dermamyotome

Sclerotome

Notochord

Aorta

Rostral

Caudal

Path 2 cells take a dorsolateral route between the epidermis and the dermamyotome

Post.

Ant.

Somite

Path 1 cells travel ventrally through the anterior sclerotome

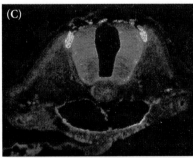

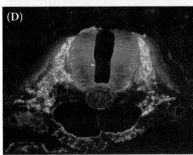

Anterior sclerotome of somite

Neural tube

(B)

Posterior sclerotome of somite

(C)

(D)

Figure 13.2

Neural crest cell migration in the trunk of the chick embryo. (A) Schematic diagram of neural crest cell migration. In Path 1 (the ventral pathway), cells travel ventrally through the anterior of the sclerotome (that portion of the somite that generates vertebral cartilage). Those cells initially opposite the posterior portions of the sclerotomes migrate along the neural tube until they come to an opposite anterior region. These cells contribute to the sympathetic and parasympathetic ganglia as well as to the adrenal medullary cells and dorsal root ganglia. Other trunk neural crest cells enter Path 2 (the dorsolateral pathway) somewhat later. These cells travel along a dorsolateral route beneath the ectoderm, and become pigment-producing melanocytes. (Migration pathways are shown on only one side of the embryo.) (B) These fluorescence photomicrographs of longitudinal sections of a 2-day chick embryo are stained red with antibody HNK-1, which selectively recognizes neural crest cells. Extensive staining is seen in the anterior, but not in the posterior, half of each sclerotome. (C, D) Cross sections through these areas, showing extensive migration through the anterior portion of the sclerotome, but no migration through the posterior portion. (B from Wang and Anderson 1997; C–D from Bronner-Fraser 1986; photographs courtesy of the authors.)

cells become sensory (dorsal root) and sympathetic neurons, adrenomedullary cells, and Schwann cells (Weston 1963; Le Douarin and Teillet 1974). In birds and mammals (but not fishes and frogs), these cells migrate ventrally through the *anterior* but not through the *posterior* section of the sclerotomes (Figure 13.2B, C; Rickmann et al. 1985; Bronner-Fraser 1986; Loring and Erickson 1987; Teillet et al. 1987). By transplanting quail neural tubes into chick embryos, Teillet and co-workers were able to mark neural crest cells both genetically and immunologically. The antibody marker recognized and labeled neural crest cells of both species; the genetic marker enabled the investigators to distinguish between quail and chick cells. These studies showed that neural crest cells initially located opposite the posterior regions of the somites migrate anteriorly or posteriorly along the neural tube and then enter the *anterior* region of their own or adjacent somites. These neural crest cells join with the neural crest cells that were initially opposite the anterior portion of the somite, and they form the same structures. Thus, each dorsal root ganglion is composed of three neural crest populations: one from the neural crest opposite the anterior portion of the somite and one from each of the adjacent neural crest regions opposite the posterior portions of the somites.

The mechanisms of trunk neural crest migration

EMIGRATION FROM THE NEURAL TUBE. Any analysis of migration (be it of birds, butterflies, or neural crest cells) has to ask four questions:

1. How is migration initiated?
2. How do the migratory agents know the route on which to travel?
3. What signals indicate that the destination has been reached and that migration should end?
4. When does the migrating agent become competent to respond to these signals?

Neural crest cells originate from the neural folds through interactions of the neural plate with the presumptive epidermis. In cultures of embryonic chick ectoderm, presumptive epidermis can induce neural crest formation in the neural plate to which it is connected (Dickinson et al. 1995). These changes can be mimicked by culturing neural plate cells with **bone morphogenetic proteins 4** and **7**, two proteins that are known to be secreted by the presumptive epidermis (Liem et al. 1997; see Chapter 12). BMP4 and BMP7 induce the expression of the **Slug protein** and the **RhoB protein** in the cells destined to become neural crest (Figure 13.3; Nieto et al. 1994; Mancilla and Mayor 1996; Liu and Jessell 1998). If either of these proteins is inactivated or inhibited from forming, the neural crest cells fail to emigrate from the neural tube.*

For cells to leave the neural crest, there must be pushes as well as pulls. The RhoB protein may be involved in establish-

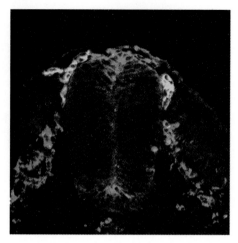

Figure 13.3
All migrating neural crest cells express HNK-1 (red stain), as in Figure 13.2. The RhoB protein (green stain) is expressed in cells as they leave the neural crest. Cells expressing both HNK-1 and RhoB appear yellow. (After Liu and Jessell 1998; photograph courtesy of T. M. Jessell.)

ing the cytoskeletal conditions that promote migration (Hall 1998). However, the cells cannot leave the neural tube as long as they are tightly connected to one another. One of the functions of the Slug protein is to activate the factors that dissociate the tight junctions between the cells (Savagne et al. 1997). Another factor in the initiation of neural crest cell migration is the loss of the N-cadherin that had linked them together. Originally found on the surface of the neural crest cells, this cell adhesion protein is downregulated at the time of cell migration. Migrating trunk neural crest cells have no N-cadherin on their surfaces, but they begin to express it again as they aggregate to form the dorsal root and sympathetic ganglia (Takeichi 1988; Akitaya and Bronner-Fraser 1992).

RECOGNITION OF SURROUNDING EXTRACELLULAR MATRICES. The path taken by the migrating trunk neural crest cells is controlled by the extracellular matrices surrounding the neural tube (Newgreen and Gooday 1985; Newgreen et al. 1986). But what are the extracellular matrix molecules that enable or forbid migration? One set of proteins promotes migration. These proteins include fibronectin, laminin, tenascin, various collagen molecules, and proteoglycans, and they are seen throughout the matrix encountered by the neural crest cells. Another set of proteins impedes migration and provides the specificity for cellular movements. The main proteins involved in this restriction of neural crest cell migration are the **ephrin proteins**. These proteins are expressed in the posterior section of each sclerotome, and wherever they are, neural crest

*The BMP signal may be amplified by other signals coming from the presumptive ectoderm. Fibroblast growth factors and Wnt proteins may be essential for maintaining the BMP-initiated events (see Mayor et al. 1997; LaBonne and Bronner-Fraser 1998).

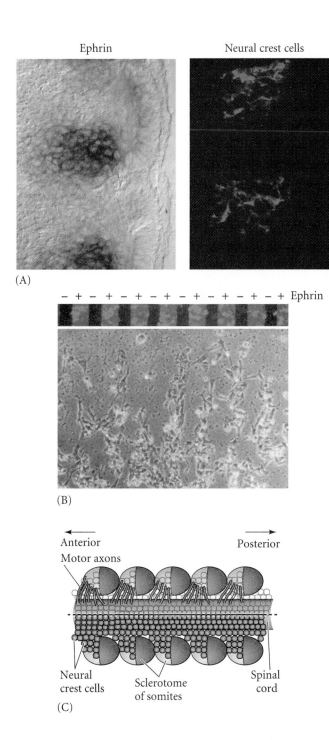

Ephrin   Neural crest cells

(A)

– + – + – + – + – + – + – + Ephrin

(B)

Anterior ← → Posterior
Motor axons

Neural crest cells   Sclerotome of somites   Spinal cord
(C)

Figure 13.4
Segmental restriction of neural crest cells and motor neurons by the ephrin proteins of the sclerotome. (A) Negative correlation between regions of ephrin in the sclerotome (dark blue stain, left) and neural crest cell migration (green HNK-1 stain, right). (B) When neural crest cells are plated on fibronectin-containing matrices with alternating stripes of ephrin, they bind to those regions lacking ephrin. (C) Composite scheme showing migration of spinal cord neural crest cells and motor neurons through the ephrin-deficient anterior regions of the sclerotomes. (For clarity, the neural crest cells and motor neurons are each depicted on only one side of the spinal cord.) (A and B from Krull et al. 1997; C after O'Leary and Wilkinson 1999.)

the ephrins activates the tyrosine kinase domains of the Eph receptors in the neural crest cells, and these kinases probably phosphorylate proteins that interfere with the actin cytoskeleton that is critical for cell migration. In addition to ephrins, there are other proteins in the posterior portion of each sclerotome that also appear to contribute to the inhospitable nature of these regions (Krull et al. 1995). This patterning of neural crest cell migration generates the overall segmental character of the peripheral nervous system, reflected in the positioning of the dorsal root ganglia and other neural crest-derived structures.

Chemotactic and maintenance factors are also important in neural crest cell migration. Stem cell factor (mentioned in Chapter 6) is critical in allowing the continued proliferation of those neural crest cells that enter the skin, and it may also serve as an anti-apoptosis factor and a chemotactic factor. If stem cell factor is secreted from cells that do not usually synthesize this protein (such as the cheek epithelium or footpads), neural crest cells will enter those regions and become melanocytes (Kunisada et al. 1998). Therefore, the migration of neural crest cells appears to be regulated both by the extracellular matrix and by soluble factors secreted by potential destinations. As we will see later in this chapter, the movement of axonal growth cones can also be regulated by similar (and sometimes identical) cues.

WEBSITE **13.1 The specificity of the extracellular matrix.** The importance of the extracellular matrix for neural crest cell migration was first shown in a series of creative experiments using mutant salamanders.

WEBSITE **13.2 Mouse neural crest cell mutants.** Some of the most important insights into neural crest cell development and migration have come from studies of mutant mice. These mice can be recognized by their altered pigmentation, resulting from abnormalities of proliferation, migration, or differentiation.

Trunk neural crest cell differentiation

THE PLURIPOTENCY OF TRUNK NEURAL CREST CELLS. One of the most exciting features of neural crest cells is their pluripotency. A single neural crest cell can differentiate into any of several different cell types, depending on its location within the embryo.

cells do not go (Figure 13.4). If neural crest cells are plated into a culture dish that contains alternate rows of extracellular matrix with or without ephrins, these cells will leave the ephrin-containing matrix and move along the matrix stripes that lack ephrin (Figure 13.4B; Krull et al. 1997; Wang and Anderson 1997). The neural crest cells recognize the ephrin proteins through their cell surface **Eph receptors**. Thus, the neural crest cells contain an Eph receptor in their plasma membranes, while the posterior portions of the trunk sclerotomes contain an Eph ligand in their membranes. Binding to

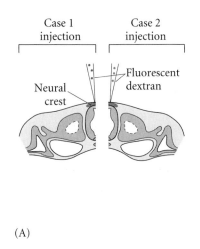

(A)

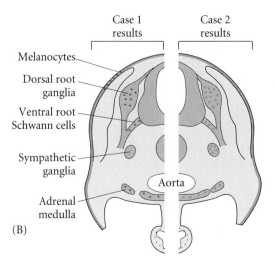

(B)

Figure 13.5
Pluripotency of trunk neural crest cells. (A) A single neural crest cell is injected with highly fluorescent dextran shortly before migration of the neural crest cells is initiated. The progeny of this cell will each receive some of these fluorescent molecules. (B) Two days later, neural crest-derived tissues contain dextran-labeled cells descended from the injected precursor. The figure summarizes data from two different experiments (case 1 and case 2). (After Lumsden 1988a.)

For example, the parasympathetic neurons formed by the vagal (neck) neural crest cells produce acetylcholine as their neurotransmitter; they are therefore cholinergic neurons. The sympathetic neurons formed by the thoracic (chest) neural crest cells produce norepinephrine; they are adrenergic neurons. But when chick vagal and thoracic neural crests are reciprocally transplanted, the former thoracic crest produces the *cholinergic* neurons of the parasympathetic ganglia, and the former vagal crest forms *adrenergic* neurons in the sympathetic ganglia (Le Douarin et al. 1975). Kahn and co-workers (1980) found that premigratory neural crest cells from both the thoracic and the vagal regions have the enzymes for synthesizing both acetylcholine and norepinephrine. Thus, the thoracic neural crest cells are capable of developing into cholinergic neurons when they are placed into the neck, and the vagal neural crest cells are capable of becoming adrenergic neurons when they are placed in the trunk.

The pluripotency of some neural crest cells is such that even regions of the neural crest that never produce nerves in normal embryos can be made to do so under certain conditions. Cranial neural crest cells from the midbrain region normally migrate into the eye and interact with the pigmented retina to become scleral cartilage cells (Noden 1978). However, if this region of the neural crest is transplanted into the trunk region, it can form sensory ganglion neurons, adrenomedullary cells, glia, and Schwann cells (Schweizer et al. 1983).

The research we have just described studied the potential of *populations* of cells. It is still uncertain whether most of the individual cells that leave the neural crest are pluripotent or whether most have already become restricted to certain fates. Bronner-Fraser and Fraser (1988, 1989) provided evidence that some, if not most, of the individual neural crest cells are pluripotent as they leave the crest. They injected fluorescent dextran molecules into individual neural crest cells while the cells were still above the neural tube, and then looked to see what types of cells their descendants became after migration. The progeny of a single neural crest cell could become sensory neurons, pigment cells, adrenomedullary cells, and glia (Figure 13.5). In mammals, the neural crest cell is similarly seen as a stem cell that can generate further multipotent neural crest cells.

However, D. J. Anderson's laboratory (Lo et al. 1997; Ma et al. 1998; Perez et al. 1999) found evidence that some populations of neural crest cells are committed very soon after leaving the neural tube. They have shown that the sensory neurons from the neural crest are specified by the transcription factor **neurogenin**, whereas the sympathetic and parasympathetic neurons from the neural crest are specified by the related transcription factor **Mash-1**. The expression of neurogenin (which would prevent a neural crest cell from becoming anything but a sensory neuron) is seen almost immediately after the neural crest cells emigrate from the neural tube. It is not yet known how committed these cells are to retaining these original biases. It appears, then, that some neural crest cells retain the ability to differentiate into a large number of different cell types, while other neural crest cell populations are specified early in development.

WEBSITE **13.3 Committed neural crest cells**. Not all neural crest cells are pluripotent. The neural crest cells leaving the dorsal neural tube are a population of heterogeneous cells, some of which have a restricted potency.

FINAL DIFFERENTIATION OF THE TRUNK NEURAL CREST CELLS. The final differentiation of neural crest cells is determined in large part by the environment to which they migrate. It does not involve the selective death of those cells already committed to secreting a type of neurotransmitter other than the one called for (Coulombe and Bronner-Fraser 1987). Heart cells, for example, secrete a protein, leukemia inhibition factor (LIF), that can convert adrenergic sympathetic neurons into cholinergic neurons without affecting their survival or growth (Chun and Patterson 1977; Fukada 1980; Yamamori et al.

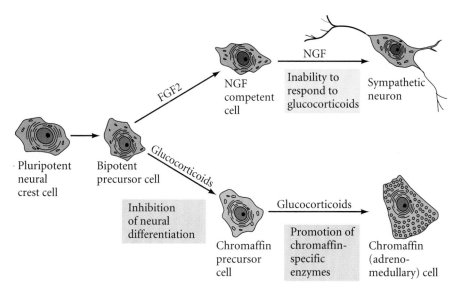

Figure 13.6
Final differentiation of a trunk neural crest cell committed to become either an adrenomedullary (chromaffin) cell or a sympathetic neuron. Glucocorticoids appear to act at two places in this pathway: first, they inhibit the actions of those factors that promote neuronal differentiation, and second, they induce those enzymes characteristic of the adrenomedullary cells. Those cells exposed sequentially to basic fibroblast growth factor (FGF2) and nerve growth factor (NGF) differentiate into the sympathetic neurons.

1989). Similarly, bone morphogenetic protein 2 (BMP2), a protein secreted by the heart, lung, and dorsal aorta, influences rat neural crest cells to differentiate into cholinergic neurons. Such neurons form the sympathetic ganglia in the region of these organs (Shah et al. 1996). While BMP2 may induce neural crest cells to become neurons, glial growth factor (GGF; neuregulin) suppresses neuronal differentiation and directs development toward glial fates (Shah et al. 1994). Another paracrine factor, endothelin-3, appears to stimulate neural crest cells to become melanocytes in the skin and adrenergic neurons in the gut (Baynash et al. 1994; Lahav et al. 1996). To distinguish between these two fates, the neural crest cells entering the skin also encounter Wnt proteins that inhibit neural development and promote melanocyte differentiation (Dorsky et al. 1998). Similarly, the chick trunk neural crest cells that migrate into the region destined to become the adrenal medulla can differentiate in two directions. The presence of certain paracrine factors induces these cells to become sympathetic neurons (Varley et al. 1995), while those cells that also encounter glucocorticoids like those made by the cortical cells of the adrenal gland differentiate into adrenomedullary cells (Figure 13.6; Anderson and Axel 1986; Vogel and Weston 1990). Thus, the fate of a neural crest cell can be directed by the milieu of the tissue environment in which it settles.

The Cranial Neural Crest

Cranial neural crest cells have a different repertoire of fates from the trunk neural crest cells. While both types of neural crest cells can form melanocytes, neurons, and glia, only the cells of the cranial neural crest are able to produce cartilage and bone. Moreover, if transplanted into the trunk region, the cranial neural crest participates in forming trunk cartilage that normally does not arise from neural crest components.

The "face" is largely the product of the cranial neural crest, and the evolution of the jaws, teeth, and facial cartilage occurs through changes in the placement of these cells (see Chapter 22). As mentioned in Chapter 12, the hindbrain is segmented along the anterior-posterior axis into compartments called rhombomeres. The chick cranial neural crest cells migrate from those regions anterior to rhombomere 6, taking one of three major pathways (Figure 13.7; Lumsden and Guthrie 1991). First, cells from rhombomeres 1 and 2 migrate to the first pharyngeal (mandibular) arch, forming the jawbones as well as the incus and malleus bones of the ear. They are also pulled by the expanding epidermis to form the **frontonasal process**. The neural crest cells of the frontonasal process generate the bones of the face. Second, cells from rhombomere 4 populate the second pharyngeal arch (forming the hyoid cartilage of the neck). Third, cells from rhombomere 6 migrate into the third and fourth pharyngeal arches and pouches* to form the thymus, parathyroid, and thyroid glands. If the neural crest is removed from those regions including rhombomere 6, the thymus, parathyroid glands, and thyroid fail to form (Bockman and Kirby 1984).

Neural crest cells from rhombomeres 3 and 5 do not migrate through the mesoderm surrounding them, but enter into the migrating streams of neural crest cells on either side of them. Those that do not enter these streams of cells will die (Graham et al. 1993, 1994; Sechrist et al. 1993). There is evidence in frog embryos that the separate streams are kept apart by ephrins. Blocking the activity of the Eph receptors causes the cells from the different streams to mix together (Smith et al. 1997; Helbling et al. 1998).

In mammalian embryos, cranial neural crest cells migrate before the neural tube is closed (Tan and Morriss-Kay 1985) and give rise to the facial mesenchyme (Johnston et al. 1985). The neural crest cells originating in the forebrain and mid-

*The pharyngeal, or branchial, arches are outpocketings of the head and neck region into which neural crest cells migrate (see Figure 13.1). The pharyngeal pouches form between these arches and become the thyroid, parathyroid, and thymus.

(A)

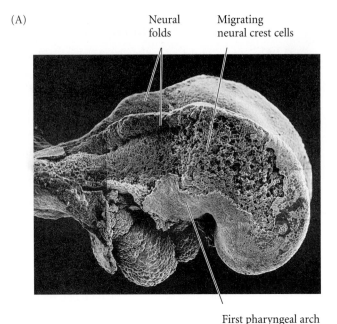

Neural folds

Migrating neural crest cells

First pharyngeal arch

(B)

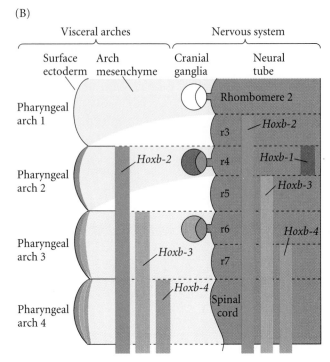

Figure 13.7
Cranial neural crest cell migration in the mammalian head. (A) Scanning electron micrograph of a rat embryo with part of the lateral ectoderm removed from the surface. Neural crest cell migration can be seen over the midbrain, and the column of neural crest cells migrating into the future first pharyngeal arch is evident. (B) Schematic drawing of cranial neural crest cell migration into the pharyngeal arches, showing Hox expression patterns. Note that the pattern is staggered between the neural tube and the pharyngeal arches, such that crest cells from the fourth rhombomere enter the second pharyngeal arch. The Hox gene expression boundaries coincide with rhombomere borders. (C) Structures formed in the human face by the neural crest-derived mesenchymal cells of the neural crest. The cartilaginous elements of the pharyngeal arches are indicated by colors, and the stippled region indicates the facial skeleton produced by anterior regions of the cranial neural crest. (A from Tan and Morriss-Kay 1985, courtesy of S.-S. Tan; B after McGinnis and Krumlauf 1992; C after Carlson 1999.)

(C)

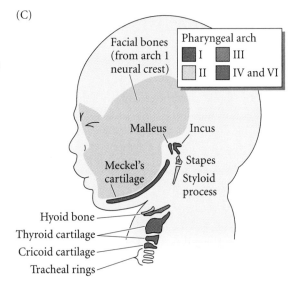

brain contribute to the frontonasal process, palate, and mesenchyme of the first pharyngeal arch. This structure becomes part of the gill apparatus in fishes; in humans, it gives rise to the jawbones and to the incus and malleus bones of the middle ear. The neural crest cells originating in the anterior hindbrain region generate the mesenchyme of the second pharyngeal arch, which generates the human stapes bone as well as much of the facial cartilage (see Figure 13.7C; Table 13.2). The cranial neural crest cells also give rise to the mesenchyme of the third, fourth, and sixth pharyngeal arches (the fifth degenerates in humans), which produce the neck bones and muscles.

In at least some cases, these cranial neural crest cells are instructed quite early as to what tissues they can form. Noden (1983) removed regions of chick neural crest that would normally seed the second pharyngeal arch and replaced them

with cells that would migrate into the first pharyngeal arch. The host embryos developed two sets of lower jaw structures, since the graft-derived cells also produced a mandible.

It appears that the combinations of Hox genes expressed in the various regions of neural crest cells specify their fates. When *Hoxa-2* is knocked out from mouse embryos, the neural crest cells of the second pharyngeal arch are transformed into those structures of the first pharyngeal arch (Gendron-Maguire et al. 1993; Rijli et al. 1993). As we discussed in

Table 13.2 Some derivatives of the pharyngeal arches

Pharyngeal arch	Skeletal elements (neural crest plus mesoderm)	Arches, arteries (mesoderm)	Muscles (mesoderm)	Cranial nerves (neural tube)
1	Incus and malleus (from neural crest); mandible, maxilla, and temporal bone regions (from crest dermal mesenchyme)	Maxillary branch of the carotid artery (to the ear, nose, and jaw)	Jaw muscles; floor of mouth; muscles of the ear and soft palate	Maxillary and mandibular divisions of trigeminal nerve (V)
2	Stapes bone of the middle ear; styloid process of temporal bone; part of hyoid bone of neck (all from neural crest cartilage)	Arteries to the ear region: cortico-tympanic artery (adult); stapedial artery (embryo)	Muscles of facial expression; jaw and upper neck muscles	Facial nerve (VII)
3	Lower rim and greater horns of hyoid bone (from neural crest)	Common carotid artery; root of internal carotid	Stylopharyngeus (to elevate the pharynx)	Glossopharyngeal nerve (IX)
4	Laryngeal cartilages (from lateral plate mesoderm)	Arch of aorta; right subclavian artery; original spouts of pulmonary arteries	Constrictors of pharynx and vocal cords	Superior laryngeal branch of vagus nerve (X)
6	Laryngeal cartilages (from lateral plate mesoderm)	Ductus arteriosus; roots of definitive pulmonary arteries	Intrinsic muscles of larynx	Recurrent laryngeal branch of vagus nerve (X)

Source: Based on Larsen 1992.

Chapter 11, Chisaka and Capecchi (1991) knocked out the *Hoxa-3* gene from inbred mice and found that these mutant mice had severely deficient or absent thymuses, thyroids, and parathyroid glands, shortened neck vertebrae, and malformed major heart vessels (see Figure 11.38). It appears that the *Hoxa-3* genes are responsible for specifying the cranial neural crest cells that give rise to the neck cartilage and pharyngeal arch derivatives. *Hoxa-1* and *Hoxb-1* are both required for the migration of rhombomere 4 neural crest cells into the second pharyngeal pouch. If both genes are knocked out from mice, no migration occurs from r4, and the second pouch-derived middle ear structures are absent (Gavalas et al. 1998; Studer et al. 1998). Moreover, retinoic acid induces the anterior expression of Hox genes that are usually expressed only more posteriorly, and they cause rhombomeres 2 and 3 to assume the identity of rhombomeres 4 and 5 (Figure 13.8; Marshall et al.

1992; Kessel 1993). In these circumstances, the trigeminal nerve (which arises from r2) is transformed into another facial nerve (characteristic of r4), and abnormalities of the first pharyngeal arch indicate that the neural crest cells of the second and third rhombomeres have been transformed into more posterior phenotypes.

Once in the pharyngeal arches and pouches, the neural crest cells have to continue proliferating and then differentiate. Mice that are deficient in the gene for endothelin-1 have specific abnormalities of pharyngeal arches 3 and 4 (as well as the heart); the neural crest cells enter the arches but are not stimulated to divide (Thomas et al. 1998). As a result, these mice have a spectrum of defects that resemble a human condition called **CATCH-22**—an acronym for *c*ardiac defects, *ab*normal face, *t*hymic hypoplasia, *c*left palate, *h*ypocalcemia, and deletions of chromosome *22*.

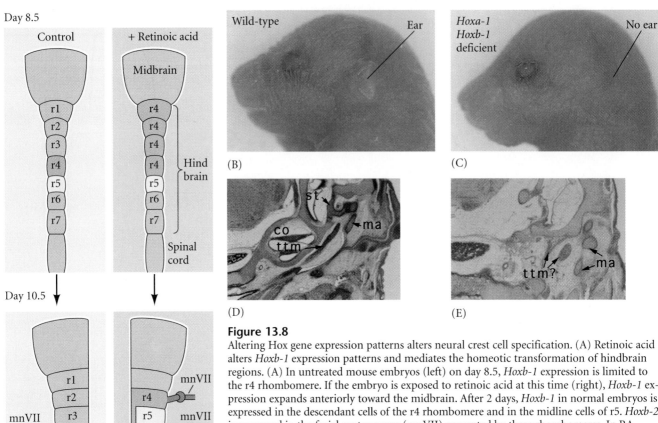

Figure 13.8
Altering Hox gene expression patterns alters neural crest cell specification. (A) Retinoic acid alters *Hoxb-1* expression patterns and mediates the homeotic transformation of hindbrain regions. (A) In untreated mouse embryos (left) on day 8.5, *Hoxb-1* expression is limited to the r4 rhombomere. If the embryo is exposed to retinoic acid at this time (right), *Hoxb-1* expression expands anteriorly toward the midbrain. After 2 days, *Hoxb-1* in normal embryos is expressed in the descendant cells of the r4 rhombomere and in the midline cells of r5. *Hoxb-2* is expressed in the facial motor nerve (mnVII) generated by these rhombomeres. In RA-treated embryos, the normal pattern of r4/5 has been duplicated in r2/3. The neural crest expression pattern of *Hoxb-2* is also duplicated, and a second facial motor nerve is formed. These results suggest that retinoic acid mediates the homeotic transformation of r2/3 to r4/5. (B–E) In mutants of *Hoxa-1* and *Hoxb-1* that fail to express these genes in r4, the mouse fails to develop ears. (B) Wild-type mouse, showing external ear. (C) Double homozygote *Hoxa-1*, *Hoxb-1*-deficient mouse without ear. (D–E) Transverse sections through wild-type and double mutant skulls, showing the absence of middle and inner ear structures in the mutant mouse. st, stapes; ma, malleus; co, cochlea; ttm, tympanic membrane. (A after Krumlauf 1993; B–E from Gavalas et al. 1998, photographs courtesy of P. Chambon.)

Sidelights & Speculations

Tooth Development

URING THE MORPHOGENESIS of any organ, numerous dialogues are occurring between the interacting tissues. In epithelial-mesenchymal interactions, the mesenchyme influences the epithelium; the epithelial tissue, once changed by the mesenchyme, can secrete factors that change the mesenchyme. Such interactions continue until an organ is formed with organ-specific mesenchyme cells and organ-specific epithelia. Some of the most extensively studied interactions are those that form the mammalian tooth. Here, the neural crest-derived mesenchyme cells become the dentin-secreting **odontoblasts**, while the jaw epithelium differentiates into the enamel-secreting **ameloblasts**. A summary of recent research that correlates mesenchyme induction and differentiation in the mammalian tooth is shown in Figure 13.9.

Tooth development begins when the mandibular (jaw) epithelium causes neural crest-derived **ectomesenchyme** (i.e., mesenchyme produced from the ectoderm) to aggregate at specific sites. The polarity of the mandibular epithelium is determined by interactions between BMP4, which is located distally, and FGF8, which is located proximally (closest to the skull). Those teeth formed in the FGF8 regions will become molars, while those teeth that develop in the BMP4 regions will become incisors (Tucker et al. 1998). Soon afterward, the expression pattern of BMP4 and

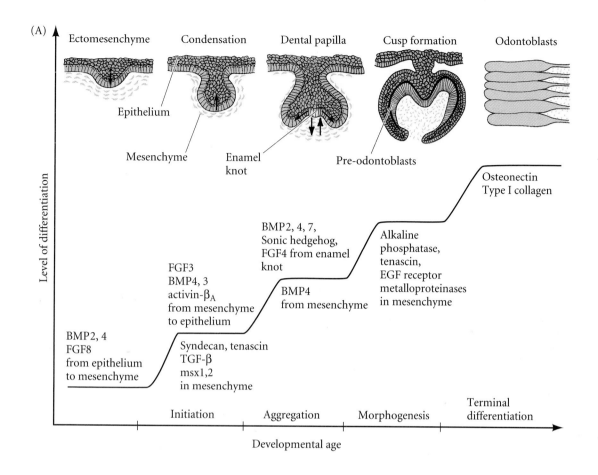

(A)

Ectomesenchyme   Condensation   Dental papilla   Cusp formation   Odontoblasts

Epithelium

Mesenchyme   Enamel
knot

Pre-odontoblasts

Level of differentiation

Osteonectin
Type I collagen

BMP2, 4, 7,
Sonic hedgehog,
FGF4 from enamel
knot

Alkaline
phosphatase,
tenascin,
EGF receptor
metalloproteinases
in mesenchyme

FGF3
BMP4, 3
activin-β_A
from mesenchyme
to epithelium

BMP4
from mesenchyme

BMP2, 4
FGF8
from epithelium
to mesenchyme

Syndecan, tenascin
TGF-β
msx1,2
in mesenchyme

Initiation   Aggregation   Morphogenesis   Terminal
differentiation

Developmental age

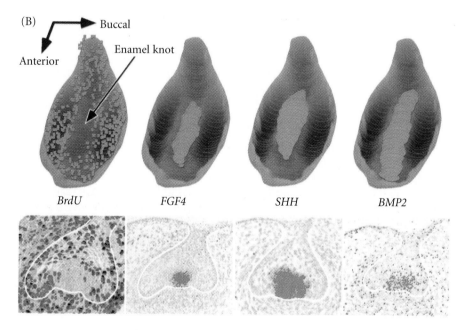

(B)

Buccal

Anterior

Enamel knot

BrdU   FGF4   SHH   BMP2

Figure 13.9
Coordinated differentiation and morphogenesis in the mammalian tooth. (A) As development progresses, the neural crest-derived mesenchyme of the jaw undergoes stepwise differentiation as it interacts with the jaw epithelium. (B) Concentration of paracrine growth and differentiation factors in the region where morphogenesis and differentiation are occurring in the 14-day embryonic mouse lower molar. The boundary of the tooth epithelium is shown in white. The paracrine factors are being secreted by the enamel knot, a mass of nondividing epithelial cells. (The panel on the left shows that the cells of the enamel knot are not replicating DNA.) Above each in situ hybridization photograph is a serial reconstruction of the area of gene expression. (A after Thesleff et al. 1990 and Thesleff and Sahlberg 1996; B from Jernvall 1995, photographs courtesy of A. Vaahtokari, J. Jernvall, and I. Thesleff.)

FGF8 changes, and the sites of the tooth primordia are determined by the interactions between these same molecules in the epithelium. FGF8 induces *Pax9* expression in the underlying ectomesenchyme, while BMP4 inhibits *Pax9* expression. Pax9 is a transcription factor whose expression in the ectomesenchyme is critical for the initiation of tooth morphogenesis, and in Pax9-deficient mice, tooth development ceases early. The only places where ectomesenchyme condense and teeth develop are where FGF8 is present and BMPs are absent (Vainio et al. 1993; Neubüser et al. 1997). Thus, spaces develop between the teeth.

At this time, the epithelium possesses the potential to generate tooth structures out of several types of mesenchyme cells (Mina and Kollar 1987; Lumsden 1988b). However, this tooth-forming potential soon becomes transferred to the ectomesenchyme that has aggregated beneath it. These ectomesenchymal cells form the **dental papilla** and are now able to induce tooth morphogenesis in other epithelia (Kollar and Baird 1970). At this stage, the jaw epithelium has lost its ability to instruct tooth formation in other mesenchymes. Thus, the "odontogenic potential" has shifted from the epithelium to the mesenchyme. This shift in the odontogenic potential coincides with a shift in the synthesis of BMP4 from the epithelium to the ectomesenchyme.

As the dental mesenchyme cells condense, they are induced to synthesize the membrane protein syndecan and the extracellular matrix protein tenascin. These proteins (which can bind each other) appear at the time the epithelium induces mesenchymal aggregation, and Thesleff and her colleagues (1990) have proposed that these two molecules may interact to bring about this condensation. Moreover, after the ectomesenchyme has ag-

gregated, it begins to secrete BMP4 as well as other growth and differentiation factors (FGF3, BMP3, HGF, and activin) (Wilkinson et al. 1989; Thesleff and Sahlberg 1996). These proteins from the ectomesenchyme induce a critical structure in the epithelium. This structure is called the **enamel knot**, and it functions as the major signaling center for tooth development (Jernvall et al. 1994). This group of cells appears as a nondividing population of cells in the center of the growing cusps. Moreover, in situ hybridization has demonstrated that the enamel knot is the source of Sonic hedgehog, FGF4, BMP7, BMP4, and BMP2 secretion (Figure 13.9B; Koyama et al. 1996; Vaahtokari et al. 1996a). As a nondividing population secreting growth factors capable of being received by both the epithelium and the ectomesenchyme, the enamel knot is thought to direct the cusp morphogenesis of the tooth and

to be critical in directing the evolutionary changes of tooth structure in mammals (Jernvall 1995).

The mesenchyme cells begin to differentiate into odontoblasts, and tenascin expression is induced at much higher levels and at the same sites as alkaline phosphatase expression. Both of these proteins have been associated with bone and cartilage differentiation, and they may promote the mineralization of the extracellular matrix (Mackie et al. 1987). Finally, as the odontoblast phenotype emerges, osteonectin and type I collagen are secreted as components of the extracellular matrix. The enamel knot disappears through apoptosis, responding to its own BMP4 (Vaahtokari et al. 1996b; Jernvall et al. 1998). By this steplike process, the cranial neural crest cells of the jaw are transformed into the dentin-secreting odontoblasts. ■

The Cardiac Neural Crest

As we will see in Chapter 15, the heart originally forms in the neck region, directly beneath the pharyngeal arches, so it should not be surprising that it acquires cells from the neural crest. However, the contributions of the neural crest to the heart have only recently been appreciated. The caudal region of the cranial neural crest is sometimes called the cardiac neural crest, since its cells (and only these particular neural crest cells) can generate the endothelium of the aortic arch arteries and the septum between the aorta and the pulmonary artery (Waldo et al. 1998; see Figure 13.10). In the chick, the cardiac neural crest lies above the neural tube region from rhombomere 7 through the portion of the spinal cord apposing the third somite, and its cells migrate into pharyngeal arches 3, 4, and 6. The cardiac neural crest is unique in that if it is removed and replaced by anterior cranial or trunk neural crest, cardiac abnormalities (notably the failure of the truncus arteriosus to separate into the aortic and pulmonary arteries) occur. Thus, the cardiac neural crest is already determined to generate cardiac cells, and the other regions of the neural crest cannot substitute for it (Kirby 1989; Kuratani and Kirby 1991).

In mice, the cardiac neural crest cells are peculiar in that they express the transcription factor Pax3. Mutations of Pax3 result in persistent truncus arteriosus (the failure of the aorta and pulmonary artery to separate), as well as defects in the thymus, thyroid, and parathyroid glands (Conway et al. 1997). Congenital heart defects in humans and mice often occur with defects in the parathyroid, thyroid, or thymus glands. It would not be surprising if all of these were linked to defects in the migration of cells from the neural crest.

WEBSITE 13.4 Communication between migrating neural crest cells. Recent research has shown that neural crest cells might cooperate with one another as they migrate. There may be subtle communication between these cells through their gap junctional complexes, and this communication may be important for heart development.

Neuronal Specification and Axonal Specificity

Not only do neuronal precursor cells and neural crest cells migrate to their place of function, but so do the axons extending from the cell bodies of neurons. Unlike most cells, whose parts all stay in the same place, the neuron is able to produce axons that may extend for meters. As we saw in Chapter 12, the axon has its own locomotory apparatus, which resides in the growth cone. The growth cone can respond to the same types of signals that migrating cells can sense. The cues for axonal migration, moreover, may be even more specific than those used to guide certain cell types to particular areas. Each of the 10^{11} neurons in the human brain has the potential to interact specifically with thousands of other neurons, and a large neuron (such as a Purkinje cell or motor neuron) can receive input from more than 10^5 other cells (Figure 13.11; Gershon et al. 1985). Understanding the generation of this stunningly ordered complexity is one of the greatest challenges to modern science.

Goodman and Doe (1993) list eight stages of neurogenesis:

1. The induction and patterning of a neuron-forming (neurogenic) region
2. The birth and migration of neurons and glia

(A)

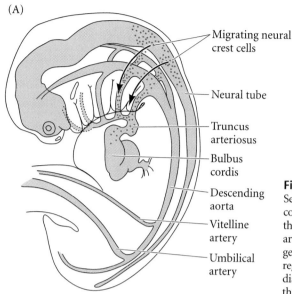

Migrating neural crest cells

Neural tube

Truncus arteriosus

Bulbus cordis

Descending aorta

Vitelline artery

Umbilical artery

(B)

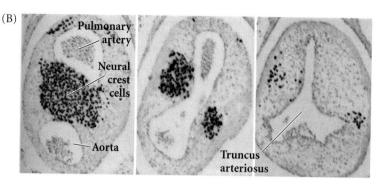

Pulmonary artery

Neural crest cells

Aorta

Truncus arteriosus

Figure 13.10

Separation of the truncus arteriosus into the pulmonary artery and aorta. The truncoconal septa (between the aorta and the pulmonary trunk) forms from the cells of the cardiac neural crest. (A) Human cardiac neural crest cells migrate to pharyngeal arches 4 and 6 during the fifth week of gestation and enter the truncus arteriosus to generate the septa. (B) Quail cardiac crest cells were transplanted into the analogous region of a chick embryo, and the embryos were allowed to develop. The quail cardiac neural crest cells can be recognized by a quail-specific antibody, which stains them darkly. In the heart, these cells can be seen separating the truncus arteriosus (right) into the pulmonary artery and the aorta (left). (A after Kirby and Waldo 1990; B from Waldo et al. 1998, photographs courtesy of K. Waldo and M. L. Kirby.)

3. The specification of cell fates
4. The guidance of axonal growth cones to specific targets
5. The formation of synaptic connections
6. The binding of trophic factors for survival and differentiation
7. The competitive rearrangement of functional synapses
8. Continued synaptic plasticity during the organism's lifetime

The first two of these processes were the topics of the previous chapter. Here, we continue our investigation of the processes of neural development.

WEBSITE **13.5 The evolution of developmental neurobiology.** Santiago Ramón y Cajal, Viktor Hamburger, and Rita Levi-Montalcini helped bring order to the study of neuronal development by identifying some of the important questions that still occupy us today.

The Generation of Neuronal Diversity

Neurons are specified in a hierarchical manner. The first decision is whether a given cell is to become a neuron or epidermis. If the cell is to become a neuron, the next decision is what type of neuron it will be: whether it is to become a motor neuron, a sensory neuron, a commissural neuron, or some other type. After this fate is determined, still another decision gives the neuron a specific target. To illustrate this process of progressive specification, we will focus on the motor neurons of vertebrates.

Vertebrates form a dorsal neural tube by blocking a BMP signal, and the specification of neural (as opposed to glial or epidermal) fate is accomplished through the Notch-Delta

pathway (see Chapter 12). The specification of the type of neuron appears to be controlled by the position of the neuronal precursor within the neural tube and by its birthday. As described in Chapter 12, neurons at the ventrolateral margin of the vertebrate neural tube become the motor neurons, while different interneurons are derived from cells in the dorsal region of the tube. Since the grafting of floor plate or notochord cells (which secrete Sonic hedgehog protein) to lateral areas can respecify dorsolateral cells as motor neurons, the decision

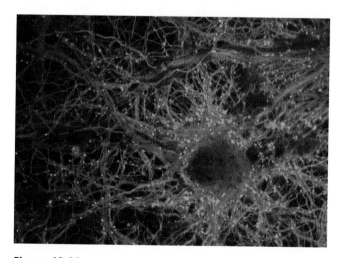

Figure 13.11

Connections of axons to a cultured rat hippocampal neuron. The rat neuron has been stained red with fluorescent antibodies to tubulin. The neuron (red) appears to be outlined by the synaptic protein synapsin (stained green), which is present in the terminals of axons that contact it. (Photograph courtesy of R. Fitzsimmons and PerkinElmer Life Sciences.)

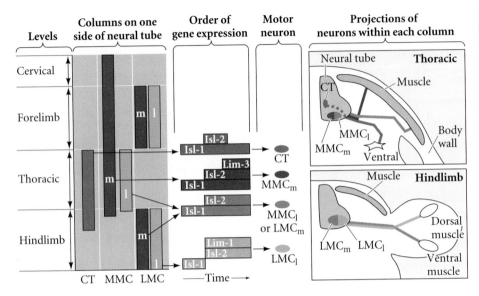

Figure 13.12
Motor neuron organization and LIM specification. At the left is half the spinal cord, showing the division of motor neurons into three columns. Neurons in the different columns display specific sets of LIM genes, and neurons within each column make similar pathfinding decisions. CT motor neurons project ventrally to the sympathetic ganglia. Neurons of the MMC project to the axial muscles, and neurons of the LMC send axons to the limb musculature. Where these columns are subdivided, medial (m) subdivisions project to ventral positions and lateral (l) subdivisions send axons to dorsal regions of the target tissues. (After Tsushida et al. 1994; Tosney et al. 1995.)

as to type of neuron is probably a function of the cell's position relative to the floor plate. Ericson and colleagues (1996) have shown that two periods of Sonic hedgehog signaling are needed to specify the motor neurons: an early period wherein the cells of the ventrolateral margin are instructed to become ventral neurons, and a later period (which includes the S phase of its last cell division) that instructs a ventral neuron to become a motor neuron (rather than an interneuron). The first decision is probably regulated by the secretion of Sonic hedgehog from the notochord, while the latter decision is more likely regulated by Sonic hedgehog from the floor plate cells. Sonic hedgehog appears to specify motor neurons by inducing certain transcription factors at different concentrations (Ericson et al. 1992; Tanabe et al. 1998; see Figure12.14).

The next decision involves target specificity. If a cell is to become a neuron and, specifically, a motor neuron, will that motor neuron be one that innervates the thigh, the forelimb, or the tongue? The anterior-posterior specification of the neural tube is regulated primarily by the Hox genes from the hindbrain through the spinal cord, and by specific head genes (such as *Otx*) in the brain. Within a region of the body, motor neuron specificity is regulated by the cell's age when it last divides. As discussed in Chapter 12, a neuron's birthday determines which layer of the cortex it will enter. As the younger cells migrate to the periphery, they must pass through neurons that differentiated earlier in development. As younger motor neurons migrate through the region of older motor neurons in the intermediate zone, they express new transcription factors as a result of a retinoic acid (or other retinoid) signal secreted by the early-born motor neurons (Sockanathan and Jessell 1998). These transcription factors are encoded by the **Lim genes** and are structurally related to those encoded by the Hox genes.

As a result of their differing birthdays and migration patterns, motor neurons form three major groupings (Landmesser 1978; Hollyday 1980). The cell bodies of the motor neurons projecting to a single muscle are clustered in a longitudinal column called a "pool." These pools are grouped together into three larger columns according to their targets. Motor neurons in the column of Terni (CT) project ventrally into the sympathetic ganglia. Motor neuron pools of the lateral motor column (LMC) extend to the limb musculature, while those of the medial motor column (MMC) project to the axial muscles. The lateral and medial motor columns are subdivided along the mediolateral axis in a manner that correlates with the dorsal-ventral positions of their respective targets (Figure 13.12; Tosney et al. 1995). This arrangement of motor neurons is constant throughout the vertebrates.

WEBSITE **13.6 Horseradish peroxidase staining**. Many of the fundamental discoveries of axon specificity used a technique wherein the plant enzyme horseradish peroxidase was injected into nerves.

The targets of these motor neurons are specified before their axons extend into the periphery. This was shown by Lance-Jones and Landmesser (1980), who reversed segments of the chick spinal cord so that the motor neurons were in new locations. The axons went to their original targets, not to the ones expected from their new positions (Figure 13.13). The molecular basis for this specificity resides in the members of the LIM family of proteins that are induced during neuronal migration (see Figure 13.12; Tsushida et al. 1994). For instance, all motor neurons express Islet-1 and (slightly later) Islet-2. If no other LIM protein is expressed, the neurons project to the ventral body wall muscles. Those neurons in the medial portion of the MMC also express Lim-3, which distinguishes them from the other motor neurons. The lateral pools of the LMC are distinguished by their short expression of Lim-1, while the CT motor neurons cease to express Islet-2. Thus, each group of neurons is characterized by a particular constellation of LIM transcription factors.

(A) Stages 15–16

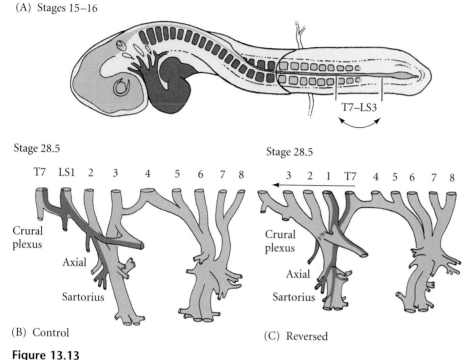

Figure 13.13
Compensation for small dislocations of axonal initiation position in the chick embryo. (A) A length of spinal cord comprising segments T7–LS3 (seventh thoracic to third lumbosacral segments) is reversed in the 2.5-day embryo. (B) Normal pattern of axon projection to the muscles at 6 days. (C) Projection of axons in the reversed segment. The ectopically placed neurons eventually found their proper neural pathways and innervated the appropriate muscles. (From Lance-Jones and Landmesser 1980.)

Pattern Generation in the Nervous System

The functioning of the vertebrate brain depends not only on the differentiation and positioning of the neurons, but also on the specific connections these cells make among themselves and their peripheral targets. In some manner, nerves from a sensory organ such as the eye must connect to specific neurons in the brain that can interpret visual stimuli, and axons from the nervous system must cross large expanses of tissue before innervating their target tissue. How does the neuronal axon "know" to traverse numerous potential target cells to make its specific connection?

Ross G. Harrison (see Figure 4.3) suggested that the specificity of axonal growth is due to pioneer nerve fibers, which go ahead of other axons and serve as guides for them* (Harrison 1910). This observation simplified, but did not solve, the problem of how neurons form appropriate patterns of interconnection. Harrison also noted, however, that axons must grow on a solid substrate, and he speculated that differ-

ences among embryonic surfaces might allow axons to travel in certain specified directions. The final connections would occur by complementary interactions on the target cell surface:

That it must be a sort of a surface reaction between each kind of nerve fiber and the particular structure to be innervated seems clear from the fact that sensory and motor fibers, though running close together in the same bundle, nevertheless form proper peripheral connections, the one with the epidermis and the other with the muscle.... The foregoing facts suggest that there may be a certain analogy here with the union of egg and sperm cell.

Research on the specificity of neuronal connections has focused on two major systems: motor neurons, whose axons travel from the nerve to a specific muscle, and the optic system, wherein axons originating in the retina find their way back into the brain. In both cases, the specificity of axonal connections is seen to unfold in three steps (Goodman and Shatz 1993):

- **Pathway selection**, wherein the axons travel along a route that leads them to a particular region of the embryo.
- **Target selection**, wherein the axons, once they reach the correct area, recognize and bind to a set of cells with which they may form stable connections.
- **Address selection**, wherein the initial patterns are refined such that each axon binds to a small subset (sometimes only one) of its possible targets.

The first two processes are independent of neuronal activity. The third process involves interactions between several active neurons and converts the overlapping projections into a fine-tuned pattern of connections.

It has been known since the 1930s that motor axons can find their appropriate muscles even if the neuronal activity of the axons is blocked. Twitty (who had been Harrison's student) and his colleagues found that embryos of the newt *Taricha torosa* secreted a toxin, tetrodotoxin, that blocked neural transmission in other species. By grafting pieces of *T. torosa* embryos onto embryos of other salamander species, they were able to paralyze the host embryos for days while development occurred. Normal neuronal connections were made, even though no neuronal activity could occur. At about the time the tadpoles were ready to feed, the toxin wore off, and the young salamanders swam and fed normally (Twitty

---

*The growth cones of pioneer neurons migrate to their target tissue while embryonic distances are still short and the intervening embryonic tissue is still relatively uncomplicated. Later in development, other neurons bind to the pioneer neuron and thereby enter the target tissue. Klose and Bentley (1989) have shown that in some cases, the pioneer neurons die after the other neurons reach their destination. Yet, if those pioneer neurons were prevented from differentiating, the other axons would not have reached their target tissue.

and Johnson 1934; Twitty 1937). More recent experiments using zebrafish mutants with nonfunctional neurotransmitter receptors similarly demonstrated that motor neurons can establish their normal patterns of innervation in the absence of neuronal activity (Westerfield et al. 1990).

But the question remains: How are the axons instructed where to go?

Cell adhesion and contact guidance: Attractive and permissive molecules

The initial pathway a growth cone follows is determined by the environment the growth cone experiences. The extracellular environment can provide substrates upon which to migrate, and these substrates can provide navigational information to the growth cone. Some of the substrates the growth cone encounters will permit it to adhere to them, and thus promote axon migration. Other substrates will cause the growth cone to retract, and will not allow its axon to grow in that direction. Some of the substrates can give extremely specific cues, recognized by only a small set of neuronal growth cones, while other cues are recognized by large sets of neurons. The migrational cues of the substrate can come from anatomical structures, the extracellular matrix, or from adjacent cell surfaces.

Anatomical cues can provide nonspecific barriers or pathways. For instance, cellular channels have been detected in the mouse retina (Silver and Sidman 1980), and these appear to guide the retinal ganglion cell growth cones into the optic stalk as they develop. Extracellular matrix cues often involve gradients of adhesivity. Growth cones prefer to migrate on surfaces that are more adhesive than their surroundings, and a gradient of increasingly adhesive molecules can direct them to their respective targets. This phenomenon of migrating to a target on adhesive gradients is called **haptotaxis**. The presence of adhesive extracellular matrix molecules delineates microscopic roads through the embryo on which axons travel (Akers et al. 1981; Gundersen 1987), and many of these roads appear to be paved with the glycoprotein **laminin** (Figure 13.14). Letourneau and co-workers (1988) have shown that the axons of certain spinal neurons travel through the neuroepithelium over a transient laminin-coated surface that precisely marks their path. Similarly, there is a very good correlation between the elongation of retinal axons and the presence of laminin on the neuroepithelial cells and astrocytes in the embryonic mouse brain (Cohen et al. 1987; Liesi and Silver 1988). Punctate laminin deposits are seen on the glial cell surfaces along the pathway leading from the retinal ganglion cells to the optic tectum in the brain, whereas adjacent areas (where the optic nerve axons fail to grow) lack these laminin deposits. After the retinal axons have reached the tectum, the glial cells differentiate and lose their laminin. At this point, the retinal ganglion neurons that have formed the optic nerve lose their integrin receptor for laminin.

Guidance can also be provided by cell adhesion molecules, such as N-CAM, L1, or NrCAM. Mutations in the

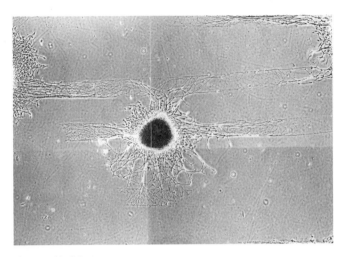

Figure 13.14
Outgrowth of sensory neurons placed on a patterned substrate consisting of parallel stripes of laminin applied to a background of type IV collagen. (From Gundersen 1987; photograph courtesy of R. W. Gundersen.)

human L1 protein appear to cause several defects in brain development that are associated with the failure of neuronal migration. L1-deficient mice have a reduced number of neurons that have been able to cross the midline in the posterior hindbrain (see Kenwrick and Doherty 1998).

Guidance by axon-specific migratory cues: The labeled pathways hypothesis

Because extracellular matrix molecules such as laminin and L1 are found in several places throughout the embryo, they can usually provide only general cues for growth cone movement. It would be difficult for such general molecules to direct the growth cones of several different types of neurons in several different directions. Yet in *Drosophila*, grasshoppers, and *Caenorhabditis* (and probably most invertebrates), the patterning of axonal migration is an astoundingly precise process, and adjacent axons are given different migratory instructions. For example, within each segment of the grasshopper, 61 neuroblasts emerge (30 on each side and 1 in the center). One of them, the 7-4 neuroblast, is a stem cell that gives rise to a family of six neurons, termed C, G, Q1, Q2, Q5, and Q6. This family of neurons is shown in Figure 13.15. The axonal growth cones of these neurons reach their targets by following specific pathways formed by other, earlier neurons. Q1 and Q2 follow a straight path together, traversing numerous other cells, until they meet the axon from the dorsal midline precursor neuron (dMP2), which they then follow posteriorly. The other four neurons of the 7-4 family migrate across the dMP2 axon as if it did not exist. Axons from the C and G neurons progress a long way together, but ultimately, C follows nerves X1 and X2 toward the posterior of the segment, while G adheres to axons P1 and P2 (which go posteriorly) and

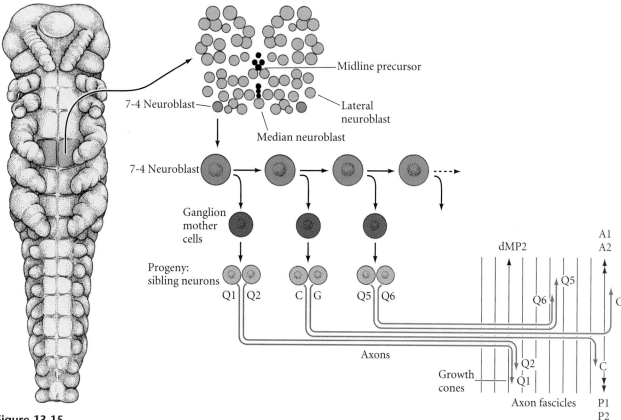

Figure 13.15
Each of the 17 segments of the early grasshopper embryo has the same pattern of neuroblast development. There are 30 lateral neuroblasts on each side, 1 median neuroblast, and 7 midline precursors. The midline precursors divide once, while the lateral neuroblasts are stem cells that divide repeatedly to form "ganglion mother cells." Each of these cells divides once to yield two sibling neurons. Lateral neuroblast 7-4 has nearly 100 neuronal progeny, of which the first six are shown here. These neurons send out axons that specifically recognize the cell surfaces of different axons and migrate with them. (After Goodman and Bastiani 1984.)

moves anteriorly on their surfaces (Goodman et al. 1984; Taghert et al. 1984). The G growth cone will have encountered over a hundred different surfaces to which it could have adhered, but it adheres only to the P neurons. If the P neurons are ablated by laser, the G growth cone acts abnormally, its filopodia searching randomly for its proper migratory surface. If any of the other hundred or so axons are destroyed, the G growth cone behaves normally.

This model of axonal pathfinding in insects has been called the **labeled pathways hypothesis** because it implies that a given neuron can specifically recognize the surface of another neuron that has grown out before it. Evidence for this specificity comes from studies using antibodies that recognize cell surface molecules produced by these neurons (Bastiani et al. 1987). Neurons aCC and pCC are sibling neurons in the grasshopper (both are derived from neuroblast 1-1) that have

very different fates. Moreover, different sets of axons adhere to each of them, creating independent bundles of axons, called **fascicles**. The specificity of this fasciculation depends on the presence of the protein **fasciclin I**. Fasciclin I is found on the two aCC neurons of each segment in the 10-hour embryo. It is not present on the pCC neurons. By hour 11, however, other neurons (but not pCC) express this cell surface molecule. These neurons are precisely those (RP1, RP2, U1, U2) whose axons fasciculate with aCC. There are at least four fasciclin molecules expressed on different subsets of neurons, and each of these molecules allows the growth cones of certain neurons to recognize specifically those axons with which they will fasciculate (Harrelson and Goodman 1988; Zinn et al. 1988).

WEBSITE 13.7 **Kallmann syndrome.** Some infertile men have no sense of smell. The relationship between sense of smell and male fertility was elusive until the gene for Kallmann syndrome was identified as producing a protein that was necessary for the proper migration of both olfactory axons and hormone-secreting nerve cells.

Contact guidance by specific growth cone repulsion

In addition to specific adhesion, there is also the possibility of specific repulsion by extracellular substrates. Just as neural crest cells are inhibited from migrating across the posterior

(A) Top

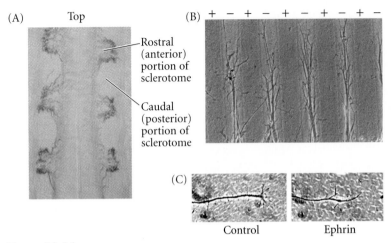

Rostral (anterior) portion of sclerotome

Caudal (posterior) portion of sclerotome

(B) + − + − + − + − + −

(C)

Control   Ephrin

Figure 13.16
Repulsion of dorsal root ganglion growth cones. (A) Motor axons migrating through the rostral (anterior), but not the caudal (posterior), compartments of each sclerotome. (B) In vitro assay, wherein ephrin stripes were placed on a background surface of laminin. Motor axons placed on the dish grew only where the ephrin was absent. (C) Inhibition of growth cones by ephrin after 10 minutes of incubation. The left side photograph shows a control axon subjected to a similar (but not inhibitory) compound; the right axon was exposed to an ephrin found in the posterior sclerotome. (From Wang and Anderson 1997; photographs courtesy of the authors.)

portion of the sclerotome, axons from the dorsal root ganglion and motor neurons pass only through the anterior portion of each sclerotome area and avoid migrating through the posterior regions of the sclerotome (Figure 13.16; see Figure 13.4). Davies and colleagues (1990) showed that membranes isolated from the posterior portion of the somite cause the collapse of growth cones of these neurons (Figure 13.16B,C). Recent studies (Wang et al. 1997; Krull et al. 1999) have shown that these growth cones contain Eph receptors that are responsive to the ephrin proteins in the posterior sclerotome cells. Thus the same signals that pattern the neural crest cells also pattern the spinal neuronal outgrowths.

In addition to the ephrins, the **semaphorin proteins** also guide growth cones by selective repulsion. These membrane proteins are seen throughout the animal kingdom and are responsible for steering many axons to their targets. They are especially important in making "turns" where an axon does not grow in a straight line, but must change direction. Semaphorin I (also known as fasciclin IV) is a transmembrane protein that is expressed in a band of epithelial cells in the developing insect limb. This protein appears to inhibit the growth cones of the Ti1 sensory neurons from moving forward, causing them to turn (Figure 13.17; Kolodkin et al. 1992, 1993). In *Drosophila*, semaphorin II is secreted by a single large thoracic muscle. In this way, the thoracic muscle prevents itself from being innervated by inappropriate axons (Matthes et al. 1995). Semaphorin III, found in mammals and birds, is also known as **collapsin** (Luo et al. 1993). This secreted protein was found to collapse the growth cones of axons originating in the dorsal root ganglia. There are several types of neurons in the dorsal

root ganglia whose axons enter the dorsal spinal cord. Most of these axons are prevented from traveling farther and entering the ventral spinal cord. However, a subset of these axons do travel ventrally through the other neural cells (Figure 13.18). These particular axons are not inhibited by semaphorin III, while those of the other neurons are (Messersmith et al. 1995). This finding suggests that semaphorin/collapsin patterns sensory projections from the dorsal root ganglia by selectively repelling certain axons so that they terminate dorsally.

WEBSITE **13.8 The pathways of motor neurons.** To innervate the limb musculature, a motor axon extends over hundreds of cells in a complex and changing environment. Recent research has discovered several paths and several barriers that help guide these axons to their appropriate destinations.

Guidance by diffusible molecules

NETRINS AND THEIR RECEPTORS. The idea that chemotactic cues guide axons in the developing nervous system was first proposed by Santiago Ramón y Cajal (1892). He suggested that the commissural neurons of the spinal cord might be told by diffusible factors to send axons from their dorsal positions to the ventral floor plate. Commissural neurons are interneurons that cross the ventral midline to coordinate right and left motor activities. Thus, they somehow must migrate to (and through) the ventral midline. The axons of these neurons begin growing ventrally down the side of the neural tube. However, about two-thirds of the way down, their direction changes, and they project through the ventrolateral (motor)

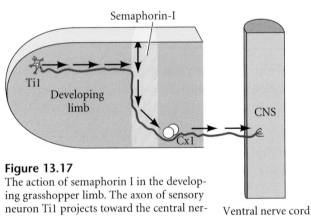

Semaphorin-I

Ti1

Developing limb

Cx1

CNS

Ventral nerve cord

Figure 13.17
The action of semaphorin I in the developing grasshopper limb. The axon of sensory neuron Ti1 projects toward the central nervous system. (The dark arrows represent sequential steps en route.) When it reaches a band of semaphorin-I-expressing epithelial cells, it reorients its growth cone and extends ventrally along the distal boundary of the semaphorin-I-expressing cells. When its filopodia connect to the Cx1 pair of cells, it crosses the boundary and projects into the CNS. When semaphorin I is blocked by antibodies, the growth cone searches randomly for the Cx1 cells. (After Kolodkin et al. 1993.)

(A)

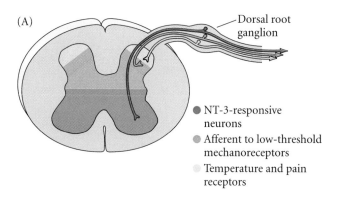

Dorsal root ganglion

● NT-3-responsive neurons
● Afferent to low-threshold mechanoreceptors
○ Temperature and pain receptors

Figure 13.18
Semaphorin III as a selective inhibitor of axonal projections into the ventral spinal cord. (A) Trajectory of axons in relation to semaphorin III expression in the spinal cord of the 14-day embryonic rat. The neurotrophin-3-responsive neurons can travel to the ventral region of the spinal cord, but the afferent axons for the mechanoreceptors and for temperature and pain receptor neurons terminate dorsally. (B) Transgenic chick fibroblast cells that secrete semaphorin III inhibit the outgrowth of mechanoreceptor axons. These axons are growing in medium treated with NGF, which stimulates their growth, but are still inhibited from growing toward the source of semaphorin III. (C) Neurons that are responsive to NT-3 for growth are not inhibited from extending toward the source of semaphorin III. Dotted circles mark the explant of NT-3 responsive neurons. (A after Marx 1995; B and C from Messersmith et al. 1995, photographs courtesy of A. Kolodkin.)

(B)

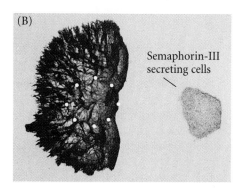

Semaphorin-III secreting cells

(C)

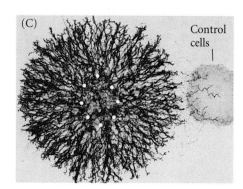

Control cells

neuron area of the neural tube toward the floor plate cells (Figure 13.19).

In 1994, Serafini and colleagues developed an assay that would allow them to screen for such a diffusible molecule. When dorsal spinal cord explants from chick embryos were plated onto collagen gels, the presence of floor plate cells near them promoted the outgrowth of commissural axons. Serafini and his coworkers took fractions of embryonic chick brain homogenate and tested them to see if any of the proteins therein mimicked this activity. This process resulted in the identification of two proteins, **netrin-1** and **netrin-2**. Netrin-1 is made by and secret-

ed from the floor plate cells, whereas netrin-2 is synthesized in the lower region of the spinal cord, but not in the floor plate (Figure 13.19B). The chemotactic effects of these netrins were demonstrated by transforming chick fibroblast cells (which usually do not make these proteins) into netrin-producing cells using a vector containing an active netrin gene (Kennedy et al. 1994). Aggregates of these netrin-secreting fibroblast cells elicited commissural axon outgrowth from dorsal rat spinal cord explants, while control cells that were given the vector without the active netrin gene did not elicit such activity (Figure 13.20). It is possible that the commissural neurons first encounter a gradient of netrin-2, which brings

Figure 13.19
Trajectory of the commissural axons in the rat spinal cord. (A) Schematic drawing of a model wherein commissural neurons first experience a gradient of netrin-2 and then a steeper gradient of netrin-1. The commissural axons are thus chemotactically guided ventrally down the lateral margin of the spinal cord toward the floor plate. Upon reaching the floor plate, the commissural axons change their direction owing to contact guidance from the floor plate cells. (B) Autoradiographic localization of netrin-1 mRNA by in situ hybridization to the hindbrain of a young rat embryo using antisense RNA. Hybridization gives an intense signal from the floor plate neurons. (B from Kennedy et al. 1994; photograph courtesy of M. Tessier-Lavigne.)

(A)

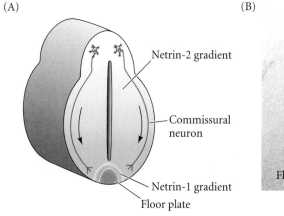

Netrin-2 gradient

Commissural neuron

Netrin-1 gradient

Floor plate

(B)

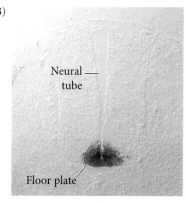

Neural tube

Floor plate

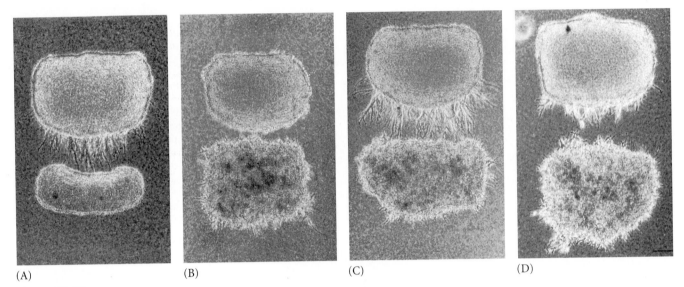

(A)   (B)   (C)   (D)

Figure 13.20
Transformed COS cells secreting netrins elicit commissural axon outgrowth from 11-day embryonic rat dorsal spinal cord explants. (A) Outgrowth of commissural neurons is seen when a dorsal spinal cord explant (upper tissue) encounters a floor plate explant. (B) No outgrowth is seen when the dorsal spinal cord explant is exposed to aggregated chick fibroblast cells that have been transfected with the cloning vector alone (no netrin gene). (C, D) Commissural neuron outgrowth in response to aggregated chick fibroblast cells that expressed the gene for netrin-1 (C) and netrin-2 (D). The identity of the outgrowths was confirmed by immunohistology showing commissural-specific antigens on these axons. (Scale bar, 100 μm.) (From Kennedy et al. 1994; photographs courtesy of M. Tessier-Lavigne.)

them into the domain of the steeper netrin-1 gradient (see Figure 13.19A).

In these experiments, both netrins become associated with the extracellular matrix.* This association can play important roles. In vertebrates, netrin-1 may serve as both an attractive and a repulsive signal. The growth cones of *Xenopus* retinal neurons are attracted to netrin-1 and are guided to the optic nerve head by this diffusible factor. Once there, however, the combination of netrin-1 and laminin prevents the axons from departing from the optic nerve. It appears that the laminin of the extracellular matrix surrounding the optic nerve converts the netrin from being an attractive molecule to being a repulsive one (Höpker et al. 1999).

The netrins have numerous regions of homology with UNC-6, a protein implicated in directing the circumferential migration of axons around the body of *Caenorhabditis elegans*. In the wild-type nematode, UNC-6 induces axons from certain centrally located positions to move ventrally, and it induces some ventrally placed cell bodies to extend axons dorsally (Figure 13.21). In loss-of-function mutations of *unc-6*,

*The binding of a soluble factor to the extracellular matrix makes for an interesting ambiguity between chemotaxis, haptotaxis, and labeled pathways. Nature doesn't necessarily conform to the categories we create.

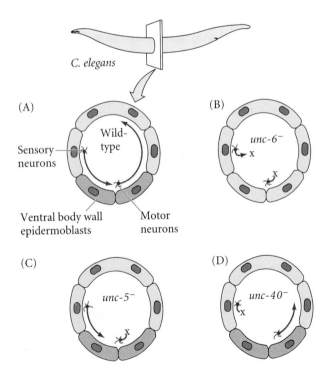

Figure 13.21
UNC expression and function in axonal guidance. (A) In the body of the wild-type *C. elegans* embryo, sensory neurons project ventrally and motor neurons project dorsally. The ventral body wall epidermoblasts expressing *unc-6* are darkly shaded. (B) In the unc-6 mutant embryo, neither of these migrations occurs. (C) The *unc-5* loss-of-function mutation affects only the dorsal movements of the motor neurons. (D) The *unc-40* loss-of-function mutation affects only the ventral migration of the sensory growth cones. (After Goodman 1994.)

neither of these axonal movements occurs (Hedgecock et al. 1990; Ishii et al. 1992; Hamelin et al. 1993). Mutations of the *unc-40* gene disrupt the ventral (but not the dorsal) axonal migration, while mutations of the *unc-5* gene prevent only the dorsal migration. Leonardo and colleagues (1997) have proposed that UNC-5 and UNC-40 are portions of the UNC-6 receptor complex, and that UNC-5 might convert an attraction (mediated by UNC-40 alone) into a repulsion.

If UNC-6 is attractive to some neurons and repulsive to others, one might expect that this dual role might also be ascribed to netrins. Colamarino and Tessier-Lavigne (1995) have shown this to be the case by looking at the trajectory of the trochlear (fourth cranial) nerve. On their way to innervate an eye muscle, the axons of the trochlear nerve originate near the floor plate of the brain stem and migrate dorsally away from the floor plate region. This pathway is maintained when the brain stem region is explanted into collagen gel. The dorsal outgrowth of the trochlear neurons can be prevented by placing floor plate cells or transgenic, netrin-1-secreting chick fibroblast cells within 450 μm of the dorsal portion of the explant. This dorsal outgrowth is not prevented by dorsal explants of the neural tube or by chick fibroblast cells that do not contain the active netrin-1 gene (Figure 13.22). Therefore, netrins and UNC-6 appear to be chemotactic to some neurons and chemorepulsive to others.

There is some reciprocity in science, and just as research on vertebrate netrin genes led to the discovery of their *C. elegans* homologues, research on the nematode UNC-5 gene led to the discovery of the gene encoding the human netrin receptor. This turns out to be a gene whose mutation in mice causes a disease called rostral cerebellar malformation (Ackerman et al. 1997; Leonardo et al. 1997).

WEBSITE **13.9 Genetic control of neuroblast migration in *C. elegans*.** The homeotic gene *mab-5* controls the direction in which certain neurons migrate in the nematode. The expression of this gene can alter which way a neuron travels.

SLIT PROTEINS. Diffusible proteins can also provide guidance by repulsion. One very important chemorepulsive molecule is the **Slit protein**. In *Drosophila*, the Slit protein is secreted by the neural cells in the midline, and it acts to prevent most neurons from crossing the midline from either side. The growth cones of *Drosophila* neurons contain the **Roundabout** (**Robo**) protein, which is the receptor for the Slit protein. In this way, most *Drosophila* neurons are prevented from migrating across the midline. However, the commissural neurons that traverse the embryo from side to side find a way to avoid this repulsion. They accomplish this task by downregulating the Robo protein as they approach the midline. Once they have traveled across the middle of the embryo, they re-express this protein and become sensitive again to the midline inhibitory actions of Slit (Figure 13.23; Brose et al. 1999; Kidd et al. 1999). Slit also functions in the vertebrate nervous system to guide neuronal growth cones. There are three Slit proteins in vertebrates, and they are used to turn away motor neurons and olfactory (smell) neurons (Brose et al. 1999; Li et al. 1999). This repulsive activity may be extremely important for the guidance of the olfactory neurons, since they are guided to specific regions of the telencephalon by repulsive cues emanating from the midline of the forebrain (Pini 1993).

WEBSITE **13.10 The early evidence for chemotaxis.** Before molecular techniques, investigators using transplantation experiments and ingenuity found evidence that chemotactic molecules were being released by target tissues.

Target selection

Once a neuron reaches a group of cells wherein lie its potential targets, it is responsive to various proteins that are produced by the target cells. In addition to the proteins already mentioned, some target cells produce a set of chemotactic factors collectively called **neurotrophins**. These proteins include nerve

(A)

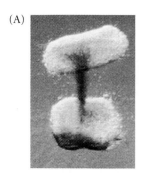

(B)

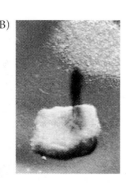

(C)

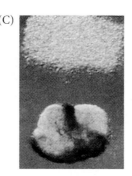

(D)

Figure 13.22
Netrin inhibits the outgrowth of trochlear axons from the dorsal spinal cord. Trochlear axons, stained for trochlear-axon-specific antigen, emerge dorsally and are not inhibited by a dorsal spinal cord explant (A) or by unaltered chick fibroblast cells (B). They are inhibited by transgenic fibroblast cells secreting netrin-1 (C) or by the floor plate of the spinal cord (D). (After Colamarino and Tessier-Lavigne 1995; photographs courtesy of M. Tessier-Lavigne.)

(A)

Robo protein distribution

Slit

Midline

Comm

Robo downregulated

Robo upregulated

Commissural axons cross

No down-regulation of Robo

Non-commissural axons do not cross

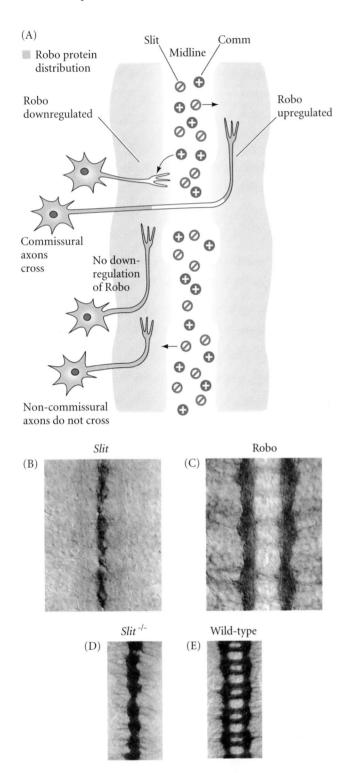

(B) *Slit*

(C) Robo

(D) *Slit*$^{-/-}$

(E) Wild-type

Figure 13.23
Model for chemotactic factors directing commissural axons to cross the midline while keeping other axons on one side of the midline. (A) The midline secretes Comm protein, which is stimulatory to commissural axons, and Slit protein, which is inhibitory to non-commissural axons. When they reach the midline, commissural axons have little or no Robo protein, the receptor of Slit. Stimulated by Comm, these axons cross the midline. Once across the midline, they re-express Robo, and therefore cannot return. Non-commissural neurons express Robo and therefore are inhibited from crossing the midline. (B) Expression of the Slit protein (as shown by antibody staining) in the midline neurons of *Drosophila*. (C) Expression of the Robo protein (as shown by antibody staining) along the neurons of the longitudinal tracts of the CNS scaffold. (D) Staining of the CNS axon scaffold (stained with antibodies to all CNS neurons) in a Slit loss-of-function mutant shows axons entering but failing to leave the midline (instead of running alongside it). (E) Wild-type CNS axon scaffold shows the ladder-like arrangement of neurons crossing the midline. (A after Thomas 1998; B–E after Kidd et al. 1999, photographs courtesy of C. S. Goodman.)

rons from the rat dorsal root ganglia, and these same neurons can be induced to turn toward a source of NT-3 (Figure 13.24).

Although genetic and biochemical techniques enable us to look at the effect of one type of molecule at a time, we must remember that any growth cone is sensing a wide range of chemotactic and chemorepulsive molecules, both in solution and on the substratum upon which it migrates. Growth cones do not rely on a single type of molecule to recognize their target. Rather, they integrate the simultaneously presented attractive and repulsive cues and select their targets based on the

Figure 13.24
Embryonic axon from a rat dorsal root ganglion turning in response to a source of NT-3. The photographs document the turning over a 10-minute period. The same growth cone was insensitive to other neurotrophins. (After Paves and Saarma 1997; photographs courtesy of M. Saarma.)

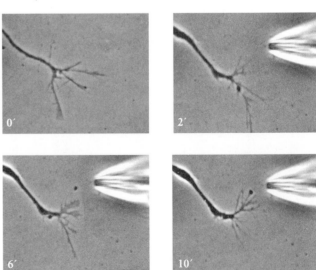

growth factor (NGF), brain-derived neurotrophic factor (BDNF), neurotrophin 3 (NT-3), and NT-4/5. These proteins are released from potential target tissues and work at short ranges as either chemotactic factors or chemorepulsive factors (Paves and Saarma 1997). Each neurotrophin can promote the growth of some axons to its source while inhibiting other axons. For instance, BDNF can stop the growth of certain sensory neu-

(A) Growth cone contacts myotube

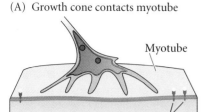

Myotube

ACh receptors

(B) Neural agrin induces ACh receptors to cluster

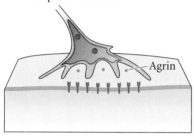

Agrin

(C) Synaptic basal lamina forms

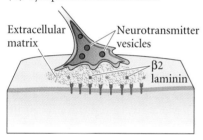

Extracellular matrix

Neurotransmitter vesicles

β2 laminin

(D)

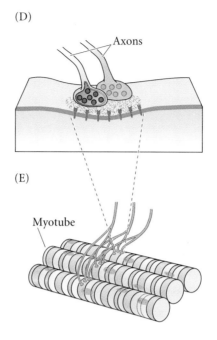

Axons

(E)

Myotube

(F)

Sheathing by Schwann cells

(G) In maturity

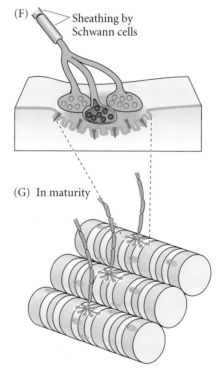

Figure 13.25
Differentiation of motor neuron synapse with muscle. Parts (E) and (G) are depicted at a lower magnification than the others to give an overview of the region where axon meets muscle. (A) A growth cone approaches a developing muscle cell. (B) The axon stops and forms an unspecialized contact on the muscle surface. Agrin, released by the neural tube, causes the clustering of acetylcholine receptors near the axon. (C) Neurotransmitter vesicles enter the axon terminal, and an extracellular matrix connects the axon terminal to the muscle cell as the synapse widens. This matrix contains a nerve-specific laminin. (D) Other axons converge on the same synaptic site. (E) Overview of muscle innervation by several axons (seen in mammals at birth). (F) All axons but one are eliminated. The remaining axon can branch to form a complex junction with the muscle. Each axon terminal is sheathed by a Schwann cell process, and folds form in the muscle cell membrane. (G) Overview of the muscle innervation several weeks after birth. (After Hall and Sanes 1993; Purves 1994; Hall 1995.)

combined input of these multiple cues (Winberg et al. 1998).

WEBSITE **13.11 The neurotrophin receptors**. Neurotrophins can bind to high-affinity receptors or to low-affinity receptors, and the pattern of binding can determine whether the signal is stimulatory or inhibitory.

Address selection: Activity-dependent development

When an axon contacts its target (usually either a muscle or another neuron), it forms a specialized junction called a **synapse**. Neurotransmitters from the axon terminal are released at these synapses to depolarize or hyperpolarize the membrane of the target cell across the synaptic cleft.

The construction of the synapse involves several steps (Figure 13.25; Burden 1998). When motor neurons in the spinal cord extend axons to muscles, growth cones that contact newly formed muscle cells migrate over their surfaces. When the growth cone first adheres to the muscle cell membrane, no specializations can be seen in either membrane. However, the axon terminals soon begin to accumulate neurotransmitter-containing synaptic vesicles, the membranes of both cells thicken at the region of contact, and the synaptic cleft between the cells fills with extracellular matrix that includes a specific form of laminin. This muscle-derived laminin specifically binds the growth cones of motor neurons and may act as a "stop signal" for axonal growth (Martin et al. 1995; Noakes et al. 1995). In at least some neuron-to-neuron synapses, the synapse is stabilized by N-cadherin. The activity of the synapse releases N-cadherin from storage vesicles in the growth cone (Tanaka et al. 2000).

After the first contact is made, growth cones from other axons converge at the site to form additional synapses. During development, all mammalian muscles studied are innervated by at least two axons. However, this polyneuronal innervation is transient. During early postnatal life, all but one of these axon branches are retracted. This rearrangement is based on "competition" between the axons (Purves and Lichtman 1980; Thompson 1983; Colman et al. 1997). When one of the motor neurons is active, it suppresses the synapses of the other neurons, possibly through a nitric oxide-dependent mechanism (Dan and Poo 1992; Wang et al. 1995). Eventually, the less active synapses are eliminated. The remaining axon terminal expands and is ensheathed by a Schwann cell.

Differential survival after innervation: Neurotrophic factors

One of the most puzzling phenomena in the development of the nervous system is neuronal cell death. In many parts of the vertebrate central and peripheral nervous systems, over half the neurons die during the normal course of development (see Chapter 6, especially Figure 6.28). Moreover, there do not seem to be great similarities across species. For instance, in the cat retina, about 80% of the retinal ganglion cells die, while in the chick retina, this figure is only 40%. In the retinas of fishes and amphibians, no retinal ganglion cells appear to die (Patterson 1992).

The apoptotic death of a neuron is not caused by any obvious defect. Indeed, these neurons have differentiated and successfully extended axons to their targets. Rather, it appears that the target tissue regulates the number of axons innervating it by limiting the supply of a neurotrophin. In addition to their roles as chemotrophic factors (see above), neurotrophins have been shown to regulate the survival of different subsets of neurons (Figure 13.26). NGF is necessary for the survival of sympathetic and sensory neurons. Treating mouse embryos with anti-NGF antibodies reduces the number of trigeminal sympathetic and dorsal root ganglion neurons to 20% of their control numbers (Levi-Montalcini and Booker 1960; Pearson et al. 1983). Removal of these neurons' target tissues causes the death of the neurons that would have innervated them, and there is a good correlation between the amount of NGF secreted and the survival of neurons that innervate these tissues (Korsching and Thoenen 1983; Harper and Davies 1990). BDNF does not affect sympathetic or sensory neurons, but it can rescue fetal motor neurons in vivo from normally occurring cell death and from induced cell death following the removal of their target tissue. The results of these in vitro studies have been corroborated by gene knockout experiments, wherein the deletion of particular neurotrophic factors causes the loss of only particular subsets of neurons (Crowley et al. 1994; Jones et al. 1994). Another neurotrophin, glial cell line-derived neurotrophic factor (GDNF), enhances the survival of another group of neurons: the midbrain dopaminergic neurons whose destruction characterizes Parkinson disease (Lin et al. 1993). This neurotrophic factor can prevent the death of these neurons in adult brains (see Lindsay 1995).

The actual survival of any given neuron in the embryo may depend on a combination of agents. Schmidt and Kater (1993) have shown that neurotrophic factors, depolarization (activation), and interactions with the substrate all combine synergistically to determine neuronal survival. For instance, the survival of chick ciliary ganglion neurons in culture was promoted by FGF, laminin, or depolarization. However, FGF did not promote survival when laminin was absent, and the combined effects of laminin, FGF, and depolarization were greater than the summed effects of each of them (Figure 13.27). The neurotrophic factors and the other environmental agents appear to function by suppressing an apoptotic "suicide program" that would be constitutively expressed unless repressed by these factors (Raff et al. 1993). The survival of retinal ganglion cells in culture is dependent on neurotrophic factors, but these cells can respond to these factors only if they have been depolarized (Meyer-Franke et al. 1995). Moreover, since neuronal activity stimulates the production of neurotrophins by the active nerves, it is likely that neurons receiving a signal produce more neurotrophins (Thoenen 1995). These factors could have an effect on nearby synapses that are active (i.e., capable of responding to the neurotrophins), thereby stabilizing a set of active synapses to the exclusion of inactive ones.

The discovery and purification of neurotrophic proteins and the analysis of their interactions with substrate and electrical conditions may make new therapies for neurodegenerative

(A) Sympathetic   (B) Dorsal root   (C) Nodose

NGF

BDNF

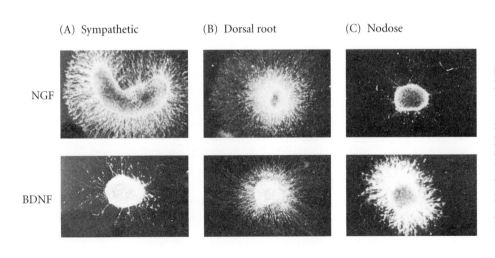

Figure 13.26
Effects of NGF (top row) and BDNF (bottom) on axonal outgrowths from (A) sympathetic ganglia, (B) dorsal root ganglia, and (C) nodose (taste perception) ganglia. While both NGF and BDNF had a mild stimulatory effect on dorsal root ganglion axonal outgrowth, the sympathetic ganglia responded dramatically to NGF and hardly at all to BDNF, while the converse was true with the nodose ganglia. (From Ibáñez et al. 1991.)

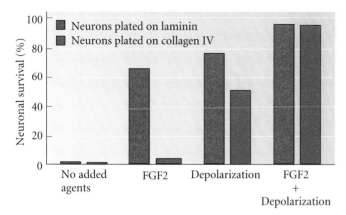

Figure 13.27
Interactions between substrate, depolarization, and the neurotrophin basic FGF (FGF2) in the survival of ciliary ganglion neurons. Neurons were plated either on laminin (a survival-enhancing substrate) or collagen IV (which does not enhance neuron survival) and observed after 24 hours of culture in the presence or absence of depolarization or FGF2. When cells were depolarized and grown in the presence of FGF2, it did not matter on what substrate they grew. However, when FGF2 was present without depolarization, the substrate made a large difference. (From Schmidt and Kater 1993.)

diseases possible. Numerous pharmaceutical companies are starting clinical trials of neurotrophic factors for the possible alleviation of spinal cord injuries (NGF), Parkinson disease (GDNF), amyotrophic lateral sclerosis (BDNF, CNTF), peripheral neuropathies (NGF, NT-3), and Alzheimer disease (NGF, GDNF).

Paths to glory: Migration of the retinal ganglion axons

Nearly all the mechanisms for neural specification and selectivity mentioned in this chapter can be seen in the ways in which individual retinal neurons send axons to the appropriate areas of the brain. Even when retinal neurons are transplanted far away from the eye, they are able to find these areas (Harris 1986). This ability to guide the axons of translocated neurons to their appropriate target sites implies that the guidance cues are not distributed solely along the normal pathway, but exist throughout the embryonic brain. Guiding an axon from a nerve cell body to its destination across the embryo is a complex process, and several different types of cues may be used simultaneously to ensure that the correct connections get established.

GROWTH OF THE RETINAL GANGLION AXON TO THE OPTIC NERVE. The first steps in getting the retinal ganglion axons to their specific regions of the optic tectum take place within the retina (Figure 13.28A). As the retinal ganglion cells differentiate, their position in the inner margin of the retina is determined by cadherin molecules (N-cadherin as well as retina-specific R-cadherin) on their cell membranes (Matsunaga et al. 1988; Inuzuka et al. 1991). The axons from these cells grow along the inner surface of the retina toward the optic disc, the head of the optic nerve (Figure 13.28B). The mature human optic nerve will contain over a million retinal ganglion axons. The adhesion and growth of the retinal cell axons along the inner surface of the retina may be governed by the laminin-containing basal lamina. However, the attachment to laminin cannot explain the directionality of this growth. N-CAM appears to be especially important here, since the directional migration of the retinal ganglion growth cones depends on the N-CAM-expressing glial endfeet at the inner retinal surface (Stier and Schlosshauer 1995). Also, the secretion of netrin-1 by the cells of the optic disc (where the axons are assembled from the optic nerve) probably plays a role in this migration. Mice lacking *netrin-1* genes (or the genes for the netrin receptor found in the retinal ganglion axons) have poorly formed optic nerves, as many of the axons fail to leave the eye and grow randomly around the disc (Deiner et al. 1997). Condroitin sulfate proteoglycan, a repulsive factor for retinal neurons, may provide pushes toward the disc (Hynes and Lander 1992).

Figure 13.28
Multiple guidance cues direct the movement of retinal ganglion axons to the optic tectum. (After Hynes and Lander 1992.)

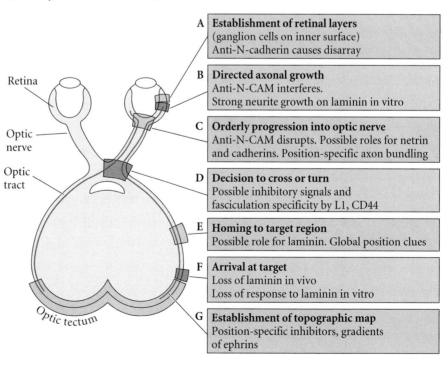

GROWTH OF THE RETINAL GANGLION AXON THROUGH THE OPTIC CHIASM. When the axons enter the optic nerve, they grow on glial cells toward the midbrain. In non-mammalian vertebrates, the axons will go to a portion of the brain called the optic tectum. (Mammalian axons go to the lateral geniculate nuclei; this pathway will be discussed further in Chapter 21 .) In vitro studies suggest that numerous cell adhesion molecules—N-CAM, cadherins, and integrins—play roles in orienting the axon toward the optic tectum (Neugebauer et al. 1988). Upon their arrival at the optic nerve, the axons fasciculate with axons that are already present there. N-CAM is critical to this fasciculation, and antibodies against N-CAM (or removal of its polysialic acid component) cause the axons to enter the optic nerve in a disorderly fashion, which in turn causes them to emerge at the wrong positions in the tectum (Figure 13.28C; Thanos et al. 1984; Yin et al. 1995).

Upon entering the brain, mammalian retinal ganglion axons reach the optic chiasm, where they have to "decide" if they are to continue straight or if they are to turn 90° and enter the other side of the brain (Figure 13.28D). It appears that those axons that are not destined to cross to the other side of the brain are repulsed from doing so when they enter the optic chiasm (Godement et al. 1990), but the molecular basis of this repulsion is not known. However, two guidance molecules, the L1 adhesion molecule (which promotes the crossing of the chiasm) and the CD44 protein (which inhibits crossing), are expressed on the brain neurons in the region of the chiasm. On their way to the optic tectum, the axons travel on a pathway (the optic tract) over glial cells whose surfaces are coated with laminin (Figure 13.28E). Very few areas of the brain contain laminin, and the laminin in this pathway exists only when the optic nerve fibers are growing on it (Cohen et al. 1987). At this point, the retinal ganglion axons have reached the optic region of the brain (Figure 13.26F), and target selection begins.

TARGET SELECTION. When the axons come to the end of the laminin-lined optic tract, they spread out and find their specific targets in the optic tectum. Studies on frogs and fishes (in which retinal neurons from each eye project to the opposite side of the brain) have indicated that each retinal ganglion axon sends its impulse to one specific site (a cell or small group of cells) within the tectum (Sperry 1951). As shown in Figure 13.29, there are two optic tecta in the frog brain. The axons from the right eye enter the left optic tectum, while those from the left eye form synapses with the cells of the right optic tectum. The growth of axons in the *Xenopus* optic tract appears to be mediated by fibroblast growth factors secreted by the cells lining the tract. The retinal ganglion axons express FGF receptors in their growth cones. However, as the axons reach the tectum, the amount of FGF rapidly diminishes, perhaps slowing down the axons and allowing them to find their targets (McFarlane et al. 1995).

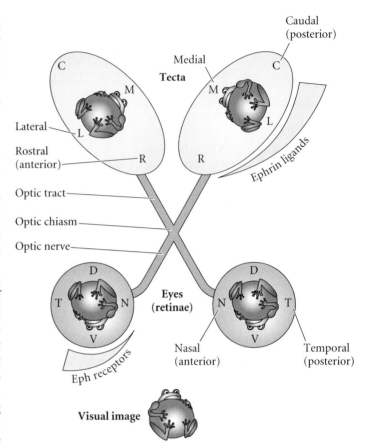

Figure 13.29
Map of the normal retinotectal projection in the adult *Xenopus*. The right eye innervates the left tectum, and the left eye innervates the right tectum. The dorsal portion of the retina (D) innervates the lateral (L) regions of the tectum. The nasal (anterior) region of the retina projects to the caudal (C) region of the tectum. (After Holt, in press; courtesy of C. Holt.)

The map of retinal connections to the frog optic tectum (the retinotectal projection) was detailed by Marcus Jacobson (1967). Jacobson defined this map by shining a narrow beam of light on a small, limited region of the retina and noting, by means of a recording electrode in the tectum, which tectal cells were being stimulated. The retinotectal projection of *Xenopus laevis* is shown in Figure 13.29. Light illuminating the ventral part of the retina stimulates cells on the lateral surface of the tectum. Similarly, light focused on the posterior part of the retina stimulates cells in the caudal portion of the tectum. These studies demonstrated a point-for-point correspondence between the cells of the retina and the cells of the tectum. When a group of retinal cells is activated, a very small and specific group of tectal cells is stimulated. We also can observe that the points form a continuum; in other words, adjacent points on the retina project onto adjacent points on the tectum. This arrangement enables the frog to see an unbroken image. This intricate specificity caused Sperry (1965) to put forward the **chemoaffinity hypothesis**:

> *The complicated nerve fiber circuits of the brain grow, assemble, and organize themselves through the use of intricate chemical codes under genetic control. Early in development, the nerve cells, numbering in the millions, acquire and retain thereafter, individual identification tags, chemical in nature, by which they can be distinguished and recognized from one another.*

Current theories do not propose a point-for-point specificity between each axon and the nerve that it contacts. Rather, evidence now demonstrates that gradients of adhesivity (especially those involving repulsion) play a role in defining the territories that the axons enter, and that activity-driven competition between these neurons determines the final connection of each axon.*

ADHESIVE SPECIFICITIES IN DIFFERENT REGIONS OF THE TECTUM. There is good evidence that retinal ganglion cells can distinguish between regions of the tectum. Cells taken from the ventral half of the chick neural retina preferentially adhere to dorsal halves of the tectum, and vice versa (Roth and Marchase 1976; Gottlieb et al. 1976; Halfter et al. 1981).

Retinal ganglion cells are specified along the dorsal-ventral axis by a gradient of transcription factors. Dorsal retina are characterized by high concentrations of the Tbx5 transcription factor, while ventral retinal cells have high levels of Pax2. These transcription factors are induced by paracrine factors (BMP4 and retinoic acid, respectively) from nearby tissues (Koshiba-Takeuchi et al. 2000). Misexpression of Tbx in the early chick retina results in marked abnormalities of the retinal-tectal projection. Therefore, the retinal ganglial cells are specified according to this location.

One gradient that has been identified functionally is a gradient of repulsion that is highest in the posterior tectum and weakest in the anterior tectum. Bonhoeffer and colleagues (Walter et al. 1987; Baier and Bonhoeffer 1992) prepared a "carpet" of tectal membranes having alternating

stripes derived from the posterior and the anterior tecta. They then let cells from the nasal (anterior) or temporal (posterior) regions of the retina extend axons into this carpet. The ganglion cells from the nasal portion of the retina extended axons equally well on both the anterior and posterior tectal membranes. The neurons from the temporal side of the retina, however, extended axons only on the anterior tectal membranes (Figure 13.30). When the growth cone of a temporal retinal ganglion axon contacted a posterior tectal cell membrane, the filopodia of the growth cone withdrew, and the growth cone collapsed and retracted (Cox et al. 1990).

The basis for this specificity appears to be gradients of ephrin proteins and their receptors. In the optic tectum, ephrin proteins (especially ephrins A2 and A5) are found in gradients that are highest at the posterior (caudal) of the tectum and decline anteriorly (rostrally) (Figure 13.29A). Moreover, cloned ephrin proteins have the ability to repulse axons (Figure 13.29B), and ectopically expressed ephrin will prohibit axons from the temporal (but not from the nasal) regions of the retina from projecting to where it is expressed (Drescher et al. 1995; Nakamoto et al. 1996). The complementary Eph receptors have been found on chick retinal ganglion cells, and they are expressed in a temporal-to-nasal gradient along the retinal ganglion axons (Cheng et al. 1995). There appear to be several Eph receptors and ligands in the tectum and retina, and they may play push-and-pull roles in guiding the temporal retinal ganglion axons to the anterior tectum and allowing the nasal retinal ganglion axons to project to the posterior portion of the tectum.

Activity-dependent synapse formation also appears to be involved in the final stages of retinal projection to the brain. In frog, bird, and rodent embryos treated with tetrodotoxin,

---

*In recent years, researchers have discovered dozens of mutations in zebrafish that affect the migration of the retinal ganglion axons to the tectum or the specificity of the retinotectal connections. These mutants are only now being analyzed, but they promise to provide major insights into the mechanisms by which our sensory input enters the brain. The December 1996 (Volume 123) issue of *Development* contains several articles mapping the genes involved in the migration of the axon from the retina to the optic cortex. Over 30 mutant genes have been found that affect either the ability of zebrafish retinal ganglion axons to find the optic tectum or the ability of the axons to find their appropriate connections within the tectum (Karlstrom et al. 1997).

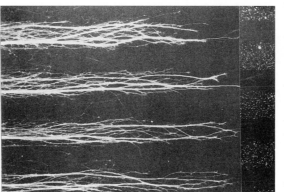

Tectal membranes

Anterior

Posterior

Anterior

Posterior

Anterior

Posterior

Anterior

Figure 13.30
Differential repulsion of temporal retinal ganglion axons on tectal membranes. Alternating stripes of anterior and posterior tectal membranes were absorbed onto filter paper. When axons from temporal (posterior) retinal ganglion cells were grown on such alternating carpets, they preferentially extended axons on the anterior tectal membranes. (From Walter et al. 1987.)

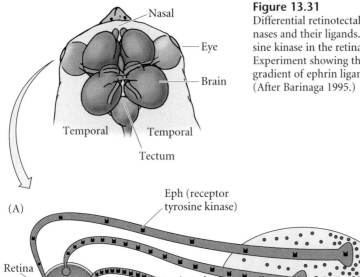

Figure 13.31
Differential retinotectal adhesion is guided by gradients of Eph receptor tyrosine kinases and their ligands. (A) Representation of the dual gradients of Eph receptor tyrosine kinase in the retina and its ligand (ephrinA2 and ephrinA5) in the tectum. (B) Experiment showing that temporal, but not nasal, retinal ganglion axons respond to a gradient of ephrin ligand in tectal membranes by turning away or slowing down. (After Barinaga 1995.)

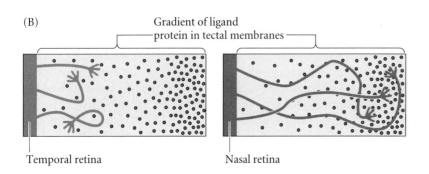

axons will grow normally to their respective territories and will make synapses with the tectal neurons. However, the retinotectal map is a coarse one, lacking fine resolution. Just as in the final specification of motor neuron synapses, neural activity is needed for the point-for-point retinal projection onto the tectal neurons (Harris 1984; Fawcett and O'Leary 1985; Kobayashi et al. 1990). This elimination of transient retinal contacts by the tectum may also involve nitric oxide expression by the target tectal cells (Wu et al. 1994).

Activity-dependent synapse formation is extremely important during the development of the mammalian visual system, and this process will be detailed in Chapter 21.

Sidelights & Speculations

Fetal Neurons in Adult Hosts

IN 1976, LUND AND HAUSCHKA implanted fetal rat brain tissue into the brain of a newborn rat. The fetal neurons made the appropriate connections within the host brain. This study offered the possibility that transplants of fetal neurons might be able to repair damaged regions in human brains.

There are many neural degenerative diseases, and Parkinson disease is one of the most prevalent, afflicting about a million Americans. In Parkinson disease, dopamine-producing neurons of the substantia nigra (a cluster of cells in the brain stem) are destroyed, and their axon terminals in the caudate nucleus and putamen (two brain nuclei) degenerate. This leads to muscle tremors, difficulty in initiating voluntary movements, and problems in cognition. The injection of L-dopa (which the body metabolizes into dopamine) relieves the symptoms temporarily, but L-dopa loses its effect with prolonged use, and it sometimes has adverse side effects.

In 1990, Lindvall and colleagues implanted human neural cells from the substantia nigra of 8- to 9-week fetuses into a patient with Parkinson disease. The donor and recipient did not need to be related, since the brain is separated from the immune system by the blood-brain barrier, which shelters tissue implanted there from rejection by the immune system. Within 5 months, the transplant had restored much of the dopamine normally made by the substantia nigra, as well as the patient's capacity for voluntary movement. Other laboratories have reported similar restoration of function following the transplantation of fetal neurons into Parkinson patients (Freed et al. 1992; Spencer et al. 1992). According to Björkland (1987), the optimal donor tissue is that containing presumptive dopamine-secreting neurons that have undergone their last cell division, but have not yet formed extensive synaptic connections. In 1992, Widner and colleagues showed that grafts from fetal mesencephalons were able to restore motor functions in two patients who had destroyed their substantiae nigrae by injecting themselves with synthetic heroin contaminated with the by-product MPTP. This compound had created a condition that resembled severe Parkinson disease.

Two recent studies have shown that fetal human cell transplants are not the only way to restore the functional anatomy of the substantia nigra in Parkinson patients. First, studies by Isacson and colleagues (1995) have suggested that the donor embryonic cells need not be human. Embryonic pig mesencephalon cells reconstructed the normal neuronal connections when injected into the striata of adult rats with a Parkinson-like disease. Second, when Gash and colleagues (1996) injected GDNF into the cerebra of monkeys in which Parkinson-like syndromes had been induced with MPTP, the monkeys showed a functional recovery from their symptoms. Moreover, dopamine concentrations and numbers of dopamine-producing neurons in their brains increased substantial-

ly. Since Parkinson disease is progressive, it is not known whether the grafted or newly divided neurons will fall victim to the same disease process that destroyed the endogenous neurons (Kordower et al. 1997; Hauser et al. 1999). However, it appears likely that fetal grafts and new neurons are able to reestablish the synaptic connections that the destroyed neurons once made.

In recent years (see Chapter 12), it has become evident that even the adult central nervous system contains pluripotent neural stem cells. Research is underway to culture these cells in the hopes of transplanting a patient's own neural stem cells into the damaged region of the brain (Ourednik et al. 1999). ■

The Development of Behaviors: Constancy and Plasticity

One of the most fascinating aspects of developmental neurobiology is the correlation of certain neuronal connections with certain behaviors. There are two remarkable aspects of this phenomenon. First, there are cases in which complex behavioral patterns are inherently present in the "circuitry" of the brain at birth. The heartbeat of a 19-day chick embryo quickens when it hears a distress call, and no other call will evoke this response (Gottlieb 1965). Furthermore, a newly hatched chick will immediately seek shelter if presented with the shadow of a hawk. An actual hawk is not needed; the shadow cast by a paper silhouette will suffice, and the shadow of no other bird will cause this response (Tinbergen 1951). There appear, then, to be certain neuronal connections that lead to inherent behaviors in vertebrates.

Equally remarkable are those instances in which the nervous system is so plastic that new experiences can modify the original set of neuronal connections, causing the creation of new neurons or the formation of new synapses between exist-

ing neurons. We will discuss neural plasticity at greater length in Chapter 21, but suffice it to say here that the brain does not stop developing at birth. The Nobel Prize-winning research of Hubel and Wiesel (1962, 1963) demonstrated that there is competition between the retinal neurons of each eye for targets in the cortex, and that their connections must be strengthened by experience. As mentioned in the preceding chapter, new experiences lead to the generation of new neurons in adult birds and mammals. Thus, the nervous system continues to develop in adult life, and the pattern of neuronal connections is a product of both inherited patterning and patterning produced by experience.

As one investigator (Purves 1994) recently concluded his analysis of brain development:

Although a vast majority of this construction must arise from developmental programs laid down during the evolution of each species, neural activity can modulate and instruct this process, thus storing the wealth of idiosyncratic information that each of us acquires through individual experience and practice.

Snapshot Summary: Neural Crest Cells and Axonal Specificity

1. The neural crest is a transitory structure. Its cells migrate to become numerous different cell types.

2. Trunk neural crest cells can migrate dorsolaterally into the ectoderm, where they become melanocytes. They can also migrate ventrally, to become sympathetic and parasympathetic neurons and adrenal medulla cells.

3. A portion of the anterior trunk neural crest enters the heart and forms the separation between the pulmonary artery and aorta.

4. The cranial neural crest cells enter the pharyngeal arches to become the cartilage of the jaw and the bones of the middle ear. They also form the bones of the frontonasal process, the papillae of the teeth, and the cranial nerves.

5. The formation of the neural crest depends on interactions between the prospective epidermis and the neural plate. Paracrine factors from these regions induce the formation of transcription factors that enable neural crest cells to emigrate.

6. The path a neural crest cell takes depends on the extracellular matrix it meets.

7. Trunk neural crest cells will migrate through the anterior portion of each somite, but not through the posterior portion of a somite. Ephrin proteins are expressed in the posterior portion of each somite and appear to prevent neural crest cell migration.

8. Some neural crest cells appear to be capable of forming large repertoire of cell types. Other neural crest cells may be committed to a fate even before migrating. The final destination of the neural crest cell can sometimes change the specification of the neural crest cell.

9. The fates of the cranial neural crest cells are to a great extent controlled by the Hox genes.

10. Teeth develop through an elaborate dialogue between the neural crest-derived mesenchyme and the jaw epithelium. The mesenchyme becomes the odontoblasts, while the epithelium generates the ameloblasts.

11. The major signaling center of the tooth is the enamel knot. It secretes several paracrine factors that regulate cell proliferation and differentiation in both the mesencyme and epithelium.

12. The specification of the motor neurons is done according to their place in the neural tube. The LIM family of transcription factors plays an important role in this specification.

13. Targets of the motor neurons are specified before the motor neurons extend into the periphery.

14. The growth cone is the locomotor organelle of the neuron, and it senses the environmental cues. (It has been called a "neural crest cell on a leash" because the growth cone and the neural crest cell both are migratory and sense the environment.)

15. Axons can find their targets without neuronal activity.

16. Some proteins are generally permissive to neuron adhesion and provide substrates on which axons can migrate. Other substances prohibit migration.

17. Some growth cones recognize molecules which are present in very specific areas and therefore will be guided by these molecules to their respective targets.

18. Some neurons are "kept in line" by repulsive molecules. If they wander off the path to their target, these molecules bring them back. Some molecules, such as the semaphorins, are selectively repulsive to a particular set of neurons.

19. Some neurons sense gradients of a protein and are brought to their target by following these gradients. The netrins may work in this fashion.

20. Target selection can be brought about by neurotrophins, proteins that are made by the target tissue that stimulate the particular set of axons that can innervate it. In some cases, the target makes only enough of these factors to support a single axon.

21. Address selection is activity-dependent. An active neuron can suppress synapse formation by other neurons on the same target.

22. Retinal ganglial axons in frogs and chick send axons that bind to specific regions of the optic tectum. This process is mediated by numerous interactions, and the target selection appears to be mediated through ephrins.

23. In some instances, fetal neurons can integrate into adult brains and re-establish damaged synapses.

24. Some behaviors appear to be innate ("hard-wired") while others are learned. Experience can strengthen certain neural connections.

Literature Cited

Ackerman, S. L., L. P. Kozak, S. A. Przyborski, L. A. Rund, B. B., Boyer and B. B. Knowles. 1997. The rostral malformation gene encodes an UNC-5 protein. *Nature* 386: 838–842.

Akers, R. M., D. F. Mosher and J. E. Lilien. 1981. Promotion of retinal neurite outgrowth by substratum-bound fibronectin. *Dev. Biol.* 86: 179–188.

Akitaya, T. and M. Bronner-Fraser. 1992. Expression of cell adhesion molecules during initiation and cessation of neural crest cell migration. *Dev. Dyn.* 194: 12–20.

Anderson, D. J. and R. Axel. 1986. A bipotential neuroendocrine precursor whose choice of cell fate is determined by NGF and glucocorticoids. *Cell* 47: 1079–1090.

Baier, H. and F. Bonhoeffer. 1992. Axon guidance by gradients of a target-derived component. *Science* 255: 472–475.

Barinaga, M. 1995. Receptors find work as guides. *Science* 269: 1668–1670.

Bastiani, M. J., A. L. Harrelson, P. M. Snow and C. S. Goodman. 1987. Expression of fasciclin I and II glycoproteins on subsets of axon pathways during neuronal development in the grasshopper. *Cell* 48: 745–755.

Baynash, A. G., K. Hosoda, A. Giaid, J. A. Richardson, N. Emoto, R. E. Hammer and M. Yanagisawa. 1994. Interaction of endothelin-3 with endothelin-B receptor is essential for development of epidermal melanocytes and enteric neurons. *Cell* 79: 1277–1285.

Björkland, A. 1987. Brain implants, transplants. *In* G. Adelman (ed.), *Encyclopedia of Neuroscience*, Vol. 1. Birkhauser, Boston, pp. 165–167.

Bockman, D. E. and M. L. Kirby. 1984. Dependence of thymus development on derivatives of the neural crest. *Science* 223: 498–500.

Bronner-Fraser, M. 1986. Analysis of the early stages of trunk neural crest migration in avian embryos using monoclonal antibody HNK-1. *Dev. Biol.* 115: 44–55.

Bronner-Fraser, M. and S. E. Fraser. 1988. Cell lineage analysis reveals multipotency of some avian neural crest cells. *Nature* 335: 161–164.

Bronner-Fraser, M. and S. E. Fraser. 1989. Developmental potential of avian trunk neural crest cells in situ. *Neuron* 3: 755–766.

Brose, K. and 7 others. 1999. Slit proteins bind robo receptors and have an evolutionary conserved role in axon guidance. *Cell* 96: 795–806.

Burden, S. J. 1998. The formation of neuromuscular synapses. *Genes Dev.* 12: 133–148.

Cheng, H.-J., M. Nakamoto, A. D. Bergemann and J. G. Flanagan. 1995. Complementary gradients in expression and binding of ELF-1 and Mek4 in development of the topographic retinotectal projection map. *Cell* 82: 371–381.

Chisaka, O. and M. Capecchi. 1991. Regionally restricted developmental defects resulting from targeted disruption of the mouse homeobox gene *Hox1.5*. *Nature* 350: 473–479.

Chun, L. L. Y. and P. H. Patterson. 1977. Role of nerve growth factor in development of rat sympathetic neurons in vitro: Survival, growth, and differentiation of catecholamine production. *J. Cell Biol.* 75: 694–704.

Cohen, J., J. F. Burne, C. McKinlay and J. Winter. 1987. The role of laminin and the laminin/fibronectin receptor complex in the outgrowth of retinal ganglial cell axons. *Dev. Biol.* 122: 407–418.

Colamarino, S. A. and M. Tessier-Lavigne. 1995. The axonal chemoattractant netrin-1 is also a chemorepellent for trochlear motor axons. *Cell* 81: 621–629.

Colman, H., J., Nabekura and J. W. Lichtman. 1997. Alterations in synaptic strength precede axon withdrawal. *Science* 275: 356–361.

Conway, S. J., D. J., Henderson and A. J. Copp. 1997. Pax3 is required for cardiac neural crest migration in the mouse: Evidence from the *splotch* (*Sp²ᴴ*) mutant. *Development* 124: 505–514.

Coulombe, J. N. and M. Bronner-Fraser. 1987. Cholinergic neurones acquire adrenergic neurotransmitters when transplanted into an embryo. *Nature* 324: 569–572.

Cox, E. C., B. Müller and F. Bonhoeffer. 1990. Axonal guidance in chick visual system: Posterior tectal membranes induce collapse of growth cones from temporal retina. *Neuron* 2: 31–37.

Crowley, C. and 10 others. 1994. Mice lacking nerve growth factor display perinatal loss of sensory and sympathetic neurons yet develop basal forebrain cholinergic neurons. *Cell* 76: 1001–1011.

Dan, Y. and M.-M. Poo. 1992. Hebbian depression of isolated neuromuscular synapses in vitro. *Science* 256: 1570–1573.

Davies, J. A., G. W. M. Cook, C. D. Stern and R. J. Keynes. 1990. Isolation from chick somites of a glycoprotein fraction that causes collapse of dorsal root ganglion growth cones. *Neuron* 2: 11–20.

Deiner, M. S., T. E. Kennedy, A. Fazeli, T. Serafini, M. Tessier-Lavigne and D. W. Sretavan. 1997. Netrin-1 and DCC mediate axon guidance locally at the optic disk: Loss of function leads to optic nerve hypoplasia. *Neuron* 19: 575–589.

Dickinson, M. E., M. A. Selleck, A. P. McMahon and F. M. Bronner. 1995. Dorsalization of the neural tube by the non-neural ectoderm. *Development* 121: 2099–2106.

Dorsky, R. I., R. T., Moon and D. W. Raible. 1998. Control of neural crest cell fate by the Wnt signalling pathway. *Nature* 396: 370–372.

Drescher, U., C. Kremoser, C. Handwerker, J. Löschinger, M. Noda and F. Bonhoeffer. 1995. In vitro guidance of retinal ganglion cell axons by RAGS, a 25 kDa protein related to ligands for Eph receptor tyrosine kinases. *Cell* 82: 359–370.

Erickson, C. A., T. D. Duong and K. W. Tosney. 1992. Descriptive and experimental analysis of the dispersion of neural crest cells along the dorsolateral pathway and their entry into ectoderm in the chick embryo. *Dev. Biol.* 151: 251–272.

Ericson, J., S. Thor, T. Edlund, T. J. Jessell and T. Yamada. 1992. Early stages of motor neuron differentiation revealed by expression of homeobox gene *islet-1*. *Science* 256: 1555–1560.

Ericson, J., S. Morton, A. Kawakami, H. Roelink and T. M. Jessell. 1996. Two critical periods of Sonic hedgehog signaling required for the specification of motor neuron identity. *Cell* 87: 661–673.

Fawcett, J. W. and D. D. M. O'Leary. 1985. The role of electrical activity in the formation of topographic maps in the nervous system. *Trends Neurosci.* 8: 201–206.

Freed, C. R. and 18 others. 1992. Survival of implanted dopamine cells and neurological improvement 12 to 46 months after transplantation for Parkinson's disease. *N. Engl. J. Med.* 327: 1549–1555.

Fukada, K. 1980. Hormonal control of neurotransmitter choice in sympathetic neuron cultures. *Nature* 287: 553–555.

Gash, D. M. and 11 others. 1996. Functional recovery in parkinsonian monkeys treated with GDNF. *Nature* 380: 252–255.

Gavalas, A., M. Studer, A. Lumsden, F. Rijli, R. Krumlauf and P. Chambon. 1998. *Hoxa1* and *Hoxb1* synergize in patterning the hindbrain, cranial nerves and second pharyngeal arch. *Development* 125: 1123–1136.

Gendron-Maguire, M., M. Mallo, M. Zhang and T. Gridley. 1993. *Hoxa2* mutant mice exhibit homeotic transformation of skeletal elements derived from cranial neural crest. *Cell* 75: 1317–1331.

Gershon, M. D., J. H. Schwartz and E. R. Kandel. 1985. Morphology of chemical synapses and pattern of interconnections. *In* E. R. Kandel and J. H. Schwartz (eds.), *Principles of Neural Science*, 2nd Ed. Elsevier, New York, pp. 132–147.

Godement, P., J. Salaun and C. A. Mason. 1990. Retinal axon pathfinding in the optic chiasm: Divergence of crossed and uncrossed fibers. *Neuron* 5: 173–186.

Goodman, C. S. 1994. The likeness of being: Phylogenetically conserved molecular mechanisms of growth cone guidance. *Cell* 78: 353–356.

Goodman, C. S. and M. J. Bastiani. 1984. How embryonic nerve cells recognize one another. *Sci. Am.* 251(6): 58–66.

Goodman, C. S. and C. Q. Doe. 1993. Embryonic development of the *Drosophila* central nervous system. *In* M. Bate and A. Martinez Arias, *The Development of* Drosophila melanogaster. Cold Spring Harbor Press, Cold Spring Harbor, NY, pp. 1131–1206.

Goodman, C. S. and C. J. Shatz. 1993. Developmental mechanisms that generate precise patterns of neuronal connectivity. *Neuron* 10 [Suppl.]: 77–98.

Goodman, C. S., M. J. Bastiani, C. Q. Doe, S. du Lac, S. L. Helfand, J. Y. Kuwada and J. B. Thomas. 1984. Cell recognition during neuronal development. *Science* 225: 1271–1287.

Gottlieb, D. I., K. Rock and L. Glaser. 1976. A gradient of adhesive specificity in developing avian retina. *Proc. Natl. Acad. Sci. USA* 73: 410–414.

Gottlieb, G. 1965. Prenatal auditory sensitivity in chickens and ducks. *Science* 147: 1596–1598.

Graham, A., I. Heyman and A. Lumsden. 1993. Even-numbered rhombomeres control the apoptotic elimination of neural crest cells from odd-numbered rhombomeres of the chick hindbrain. *Development* 119: 233–245.

Graham, A., P. Francis-West, P. Brickell and A. Lumsden. 1994. The signalling molecule BMP-4 mediates apoptosis in the rhombencephalic neural crest. *Nature* 372: 684–686.

Gundersen, R. W. 1987. Response of sensory neurites and growth cones to patterned substrata of laminin and fibronectin in vitro. *Dev. Biol.* 121: 423–431.

Halfter, W., M. Claviez and U. Schwarz. 1981. Preferential adhesion of tectal membranes to anterior embryonic chick retina neurites. *Nature* 292: 67–70.

Hall, A. 1998. The Rho GTPases and the actin cytoskeleton. *Science* 279: 509–514.

Hall, Z. W. 1995. Laminin β2 (S-laminin): A new player at the synapse. *Science* 269: 362–363.

Hall, Z. W. and J. R. Sanes. 1993. Synaptic structure and development: The neuromuscular junction. *Neuron* 10 [Suppl.]: 99–121.

Hamelin, M., Y. Zhou, M.-W. Su, I. M. Scott and J. G. Culotti. 1993. Expression of the unc-5 guidance receptor in the touch neurons of *C. elegans* steers their axons dorsally. *Nature* 364: 327–330.

Harper, S. and A. M. Davies. 1990. NGF mRNA expression in developing cutaneous epithelium related to innervation density. *Development* 110: 515–519.

Harrelson, A. L. and C. S. Goodman. 1988. Growth cone guidance in insects: Fasciclin II is a member of the immunoglobulin superfamily. *Science* 242: 700–708.

Harris, W. A. 1984. Axonal pathfinding in the absence of normal pathways and impulse activity. *J. Neurosci.* 4: 1153–1162.

Harris, W. A. 1986. Homing behavior of axons in the embryonic vertebrate brain. *Nature* 320: 266–269.

Harrison, R. G. 1910. The outgrowth of the nerve fiber as a mode of protoplasmic movement. *J. Exp. Zool.* 9: 787–848.

Hauser, R. A., T. B. Freeman, B. J. Snow, M. Nauert, L. Gauger, J. H. Kordower and C. W. Olanow. 1999. Long-term evaluation of bilateral fetal nigral transplantation in Parkinson's disease. *Arch. Neurol.* 56: 179–187.

Hedgecock, E. M., J. G. Culotti and D. H. Hall. 1990. The *unc-5*, *unc-6*, and *unc-40* genes guide circumferential migrations of pioneer axons and mesodermal cells on the epidermis in *C. elegans*. *Neuron* 2: 61–85.

Helbling, P. M., C. T. Tran and A. W. Brändli. 1998. Requirement for EphA receptor signaling in the segregation of *Xenopus* third and fourth arch neural crest cells. *Mech. Dev.* 78: 63–79.

Hollyday, M. 1980. Organization of motor pools in the chick lumbar lateral motor column. *J. Comp. Neurol.* 194: 143–170.

Holt, C. In press. Retinal-tectal projection. *Macmillan Encyclopedia of Life Sciences.* Macmillan, London.

Höpker, V. H., D. Shewan, M. Tessier-Lavigne, M.-M. Poo and C. Holt. 1999. Growth-cone attraction to netrin-1 is converted to repulsion by laminin-1. *Nature* 401: 69–73.

Hubel, D. H. and T. N. Wiesel. 1962. Receptive fields, binocular interaction and functional architecture in the cat's visual cortex. *J. Physiol.* 160: 106–154.

Hubel, D. H. and T. N. Wiesel. 1963. Receptive fields of cells in striate cortex of very young, visually inexperienced kittens. *J. Neurophysiol.* 26: 944–1002.

Hynes, R. O. and A. D. Lander. 1992. Contact and adhesive specificities in the associations, migrations, and targeting of cells and axons. *Cell* 68: 303–322.

Ibáñez, C. F., T. Ebendal and H. Persson. 1991. Chimeric molecules with multiple neurotrophic activities reveal structural elements determining the specificities of NGF and BDNF. *EMBO J.* 10: 2105–2110.

Inuzuka, H., S. Miyatani and M. Takeichi. 1991. R-cadherin: A novel Ca^{2+}-dependent cell-cell adhesion molecule expressed in the retina. *Neuron* 7: 69–79.

Isacson, O., T. Deacon, P. Pakzaban, W. R. Galpern, J. Dinsmore and L. H. Burns. 1995. Transplanted xenogenic neural cells in neurodegenerative disease models exhibit remarkable axonal target specificity and distinct growth patterns of glial and axonal fibres. *Nature Med.* 1: 1189–1194.

Ishii, N., W. G. Wadsworth, B. D. Stern, J. G. Culotti and E. M. Hedgecock. 1992. UNC-6, a laminin-related protein, guides pioneer axon migrations in *C. elegans*. *Neuron* 9: 873–881.

Jacobson, M. 1967. Retinal ganglion cells: Specification of central connections in larval *Xenopus laevis*. *Science* 155: 1106–1108.

Jacobson, M. 1991. *Developmental Neurobiology*, 2nd Ed. Plenum, New York.

Jernvall, J. 1995. Mammalian molar cusp patterns: Developmental mechanisms of diversity. *Acta Zool. Fennica* 198: 1–61.

Jernvall, J., P. Kettunen, I. Karavanova, L. B. Martin and I. Theseleff. 1994. Evidence for the role of the enamel knot as a control center in mammalian tooth cusp formation: Non-dividing cells express growth stimulating Fgf-4 gene. *Int. J. Dev. Biol.* 38: 463–469.

Jernvall, J., T. Aberg, P. Kettunen, S. Keränen and I. Thesleff. 1998. The life history of an embryonic signaling center: BMP4 induces *p21* and is associated with apoptosis in the mouse tooth enamel knot. *Development* 125: 161–169.

Johnston, M. C., K. K. Sulik, W. S. Webster and B. L. Jarvis. 1985. Isotretinoin embryopathy in a mouse model: Cranial neural crest involvement. *Teratology* 31: 26A.

Jones, K. R., I. Farinas, C. Backus and L. F. Reichart. 1994. Targeted disruption of the BDNF gene perturbs brain and sensory neuron development, but not motor neuron development. *Cell* 76: 989–999.

Kahn, C. R., J. T. Coyle and A. M. Cohen. 1980. Head and trunk neural crest in vitro: Autonomic neuron differentiation. *Dev. Biol.* 77: 340–348.

Karlstrom, R. O., T. Trowe and F. Bonhoeffer. 1997. Genetic analysis of axon guidance and mapping in the zebrafish. *Trends Neurosci.* 20: 3–8.

Kennedy, T. E., T. Serafini, J. R. de la Torre and M. Tessier-Lavigne. 1994. Netrins are diffusible chemotropic factors for commissural axons in the embryonic spinal cord. *Cell* 78: 425–435.

Kenwrick, S. and P. Doherty. 1998. Neural cell adhesion molecule L1: Relating disease to function. *BioEssays* 20: 668–676.

Kessel, M. 1993. Reversal of axonal pathways from rhombomere 3 correlates with extra Hox expression domains. *Neuron* 10: 379–393.

Kidd, T., K. S. Bland and C. S. Goodman. 1999. Slit is the midline repellent for the Robo receptor in *Drosophila*. *Cell* 96: 785–794.

Kirby, M. L. 1987. Cardiac morphogenesis: Recent research advances. *Pediatr. Res.* 21: 219–224.

Kirby, M. L. 1989. Plasticity and predetermination of mesencephalic and trunk neural crest transplanted into the region of the cardiac neural crest. *Dev. Biol.* 134: 401–412.

Kirby, M. L. and K. L. Waldo. 1990. Role of neural crest in congenital heart disease. *Circulation* 82: 332–340.

Klose, M. and D. Bentley. 1989. Transient pioneer neurons are essential for forming an embryonic peripheral nerve. *Science* 245: 982–984.

Kobayashi, T., H. Nakamura and M. Yasuda. 1990. Disturbance of refinement of retinotectal projection in chick embryos by tetrodotoxin and grayanotoxin. *Dev. Brain Res.* 57: 29–35.

Kollar, E. J. and G. Baird. 1970. Tissue interaction in developing mouse tooth germs. II. The inductive role of the dental papilla. *J. Embryol. Exp. Morphol.* 24: 173–186.

Kolodkin, A. L., D. J. Matthes, T. P. O'Connor, N. H. Patel, D. Bentley and C. S. Goodman. 1992. Fasciclin IV: Sequence, expression, and function during growth cone guidance in the grasshopper embryo. *Neuron* 9: 831–845.

Kolodkin, A. L., D. J. Matthes and C. S. Goodman. 1993. The semaphorin genes encode a family of transmembrane and secreted growth cone guidance molecules. *Cell* 75: 1389–1399.

Kordower, J. H., T. B. Freeman and C. W. Olanow. 1998. Neuropathology of fetal nigral grafts in patients with Parkinson's disease. *Mov. Dis.* 13: 88–95.

Korsching, S. and H. Thoenen. 1983. Nerve growth factor in sympathetic ganglia and corresponding target organs of the rat: Correlation with density of sympathetic innervation. *Proc. Natl. Acad. Sci. USA* 80: 3513–3516.

Koshiba-Takeuchi, K. and 10 others. 2000. Tbx5 and retinotectal projection. *Science* 287: 134–137.

Koyama, E. and 10 others. 1996. Polarizing activity, Sonic hedgehog, and tooth development in embryonic and postnatal mouse. *Dev. Dyn.* 206: 59–72.

Krull, C. E, A. Collazo, S. E. Fraser and M. Bronner-Fraser. 1995. Segmental migration of trunk neural crest: Time lapse analysis reveals a role for PNA binding molecules. *Development* 121: 3733–3743.

Krull, C. and and 7 others. 1997. Interactions between Eph-related receptors and ligands confer rostrocaudal pattern to trunk neural crest migration. *Curr. Biol.* 7: 571–580.

Krull, C. E., J. Eberhart, R. McLennan, S. A. Koblar, D. P. Cerretti and E. B. Pasquale. 1999. Segmental patterning of the peripheral nervous system. *Dev. Biol.* 210: 203.

Krumlauf, R. 1993. Hox genes and pattern formation in the branchial region of the vertebrate head. *Trends Genet.* 9: 106–112.

Kunisada, T. and 8 others. 1998. Transgene expression of steel factor in the basal layer of epidermis promotes survival, proliferation, differentiation, and migration of melanocyte precursors. *Development* 125: 2915–2923.

Kuratani, S. C. and M. L. Kirby. 1991. Initial migration and distribution of the cardiac neural crest in the avian embryo: An introduction to the concept of the circumpharyngeal crest. *Am. J. Anat.* 191: 215–227.

LaBonne, C. and M. Bronner-Fraser. 1998. Neural crest induction in *Xenopus*: Evidence for a two-signal model. *Development* 125: 2403–2414.

Lahav, R., C. Ziller, E. Dupin and N. M. Le Douarin. 1996. Endothelin 3 promotes neural crest cell proliferation and mediates a vast increase in melanocyte number in culture. *Proc. Natl. Acad. Sci. USA* 93: 3892–3897.

Lance-Jones, C. and L. Landmesser. 1980. Motor neuron projection patterns in chick hindlimb following partial reversals of the spinal cord. *J. Physiol.* 302: 581–602.

Landmesser, L. 1978. The development of motor projection patterns in the chick hindlimb. *J. Physiol.* 284: 391–414.

Le Douarin, N. M. 1982. *The Neural Crest*. Cambridge University Press, New York.

Le Douarin, N. M. and M.-A. Teillet. 1973. The migration of neural crest cells to the wall of the digestive tract in avian embryo. *J. Embryol. Exp. Morphol.* 30: 31–48.

Le Douarin, N. M. and M.-A. Teillet. 1974. Experimental analysis of the migration and differentiation of neuroblasts of the autonomic nervous system and of neuroectodermal mesenchyme derivatives, using a biological cell marking technique. *Dev. Biol.* 41: 162–184.

Le Douarin, N. M., D. Renaud, M.-A. Teillet and G. H. Le Douarin. 1975. Cholinergic differentiation of presumptive adrenergic neuroblasts in interspecific chimeras after heterotopic transplantation. *Proc. Natl. Acad. Sci. USA* 72: 728–732.

Le Lièvre, C. S. and N. M. Le Douarin. 1975. Mesenchymal derivatives of the neural crest: Analysis of chimaeric quail and chick embryos. *J. Embryol. Exp. Morphol.* 34: 125–154.

Leonardo, E. D., L. Hinck, M. Masu, K. Keino-Masu, S. L. Ackerman and M. Tessier-Lavigne. 1997. Vertebrate homologues of *C. elegans* UNC-5 are candidate netrin receptors. *Nature* 386: 833–838.

Letourneau, P., A. M. Madsen, S. M. Palm and L. T. Furcht. 1988. Immunoreactivity for laminin in the developing ventral longitudinal pathway of the brain. *Dev. Biol.* 125: 135–144.

Levi-Montalcini, R. and B. Booker. 1960. Destruction of the sympathetic ganglia in mammals by an antiserum to the nerve growth factor protein. *Proc. Natl. Acad. Sci. USA* 46: 384–390.

Li, H.-S. and 12 others. 1999. Vertebrate slit, a secreted ligand for transmembrane protein Roundabout, is a repellent for olfactory bulb axons. *Cell* 96: 807–818.

Liem, K. F., Jr., G. Tremml and T. M. Jessell. 1997. A role for the roof plate and its resident TGFβ-related proteins in neuronal patterning of the dorsal spinal cord. *Cell* 91: 127–138.

Liesi, P. and J. Silver. 1988. Is astrocyte laminin involved in axon guidance in the mammalian CNS? *Dev. Biol.* 130: 774–785.

Lin, L.-F. H., D. H. Doherty, J. D. Lile, S. Bektesh and F. Collins. 1993. GDNF: A glial cell-line derived neurotrophic factor for midbrain dopaminergic neurons. *Science* 260: 1130–1132.

Lindsay, R. M. 1995. Neuron saving schemes. *Nature* 373: 289–290.

Lindvall, O., P. Brundin and H. Widner. 1990. Grafts of fetal dopamine neurons survive and improve motor functions in Parkinson's disease. *Science* 247: 574–577.

Liu, J.-P. and T. M. Jessell. 1998. A role for rhoB in the delamination of neural crest cells from the dorsal neural tube. *Development* 125: 5055–5067.

Lo, L., L. Sommer and D. J. Anderson. 1997. MASH1 maintains competence for BMP2-induced neuronal differentiation in post-migratory neural crest cells. *Curr. Biol.* 7: 440–450.

Loring, J. F. and C. A. Erickson. 1987. Neural crest cell migratory pathways in the trunk of the chick embryo. *Dev. Biol.* 121: 220–236.

Lumsden, A. G. S. 1988a. Multipotent cells in the avian neural crest. *Trends Neurosci.* 12: 81–83.

Lumsden, A. G. S. 1988b. Spatial organization of the epithelium and the role of neural crest cells in the initiation of the mammalian tooth germ. *Development* 103 [Suppl.]: 155–169.

Lumsden, A. G. S. and S. Guthrie. 1991. Alternating patterns of cell surface properties and neural crest cell migration during segmentation of the chick embryo. *Development* [Suppl.] 2: 9–15.

Lund, R. D. and S. D. Hauschka. 1976. Transplanted neural tissue develops connection with host rat brain. *Science* 193: 582–584.

Luo, Y., D. Raibile and J. A. Raper. 1993. Collapsin: A protein in brain that induces the collapse and paralysis of neuronal growth cones. *Cell* 75: 217–227.

Ma, Q., Z. Chen, I. del Barco Barrantes, J. L. de la Pompa and D. J. Anderson. 1998. Neurogenin-1 is essential for the determination of neuronal precursors for proximal cranial sensory ganglia. *Neuron* 20: 469–482.

Mackie, E. J., I. Thesleff and R. Chiquet-Ehrismann. 1987. Tenascin is associated with chondrogenic and osteogenic differentiation in vivo and promotes chondrogenesis in vivo. *J. Cell Biol.* 105: 2569–2579.

Mancilla, A. and R. Mayor. 1996. Neural crest formation in *Xenopus laevis*: Mechanisms of Xslug induction. *Dev. Biol.* 177: 580–589.

Marshall, H., S. Nonchev, M. H. Sham, I. Muchamore, A. Lumsden and R. Krumlauf. 1992. Retinoic acid alters hindbrain Hox code and induces transformation of rhombomeres 2/3 into a 4/5 identity. *Nature* 360: 737–741.

McGinnis, W. and R. Krumlauf. 1992. Homeobox genes and axial patterning. *Cell* 68; 283-302.

Martin, P. T., A. J. Ettinger and J. R. Sanes. 1995. Synaptic localization domain in the synaptic cleft protein laminin β2 (s-laminin). *Science* 269: 413–416.

Marx, J. 1995. Helping neurons find their way. *Science* 268: 971–973.

Matsunaga, M., K. Hatta and M. Takeichi. 1988. Role of N-cadherin cell adhesion molecules in the histogenesis of neural retina. *Neuron* 1: 289–295.

Matthes, D. J., H. Sink, A. L. Kolodkin and C. S. Goodman. 1995. Semaphorin II can function as a selective inhibitor of specific synaptic arborizations. *Cell* 81: 631–639.

Mayer, T. C. 1973. The migratory pathway of neural crest cells into the skin of mouse embryos. *Dev. Biol.* 34: 39–46.

Mayor, R., N. Guerrero and C. Martinez. 1997. Role of FGF and Noggin in neural crest cell induction. *Dev. Biol.* 189: 1–12.

McFarlane, S., L. McNeill and C. E. Holt. 1995. FGF signaling and target recognition in the developing *Xenopus* visual system. *Neuron* 15: 1017–1028.

Messersmith, E. K., E. D. Leonardo, C. J. Shatz, M. Tessier-Lavigne, C. S. Goodman and A. Kolodkin. 1995. Semaphorin III can function as a selective chemorepellent to pattern sensory projections in the spinal cord. *Neuron* 14: 949–959.

Meyer-Franke, A., M. R. Kaplan, F. W. Pfrieger and B. A. Barres. 1995. Characterization of the signaling interactions that promote the survival and growth of developing retinal ganglion cells in culture. *Neuron* 15: 805–819.

Mina, M. and E. J. Kollar. 1987. The induction of odontogenesis in non-dental mesenchyme combined with early murine mandibular arch epithelium. *Arch. Oral Biol.* 32: 123–127.

Nakamoto, M. and 7 others. 1996. Topographically specific effects of ELF-1 on retinal axon guidance in vitro and retinal axon mapping in vivo. *Cell* 86: 755–766.

Neubüser, A., H. Peters, R. Ballig and G. R. Martin. 1997. Antagonistic interactions between FGF and BMP signaling pathways: A mechanism for positioning the sites of tooth formation. *Cell* 90: 247–255.

Neugebauer, K. M., K. J. Tomaselli, J. Lilien and L. F. Reichardt. 1988. N-cadherin, N-CAM, and integrins promote retinal neurite outgrowth on astrocytes in vitro. *J. Cell Biol.* 107: 1177–1187.

Newgreen, D. F. and D. Gooday. 1985. Control of onset of migration of neural crest cells in avian embryos: Role of Ca^{++}-dependent cell adhesions. *Cell Tissue Res.* 239: 329–336.

Newgreen, D. F., M. Scheel and V. Kaster. 1986. Morphogenesis of sclerotome and neural crest cells in avian embryos: In vivo and in vitro studies on the role of notochordal extracellular material. *Cell Tissue Res.* 244: 299–313.

Nieto, M. A., M. G. Sargent, D. G. Wilkinson and J. Cooke. 1994. Control of cell behavior during vertebrate development by *slug*, a zinc finger gene. *Science* 264: 835–839.

Noakes, P. G., M. Gautam, J. Mudd, J. R. Sanes and J. P. Merlie. 1995. Aberrant differentiations of neuromuscular junctions in mice lacking s-laminin/laminin β2. *Nature* 374: 258–262.

Noden, D. M. 1978. The control of avian cephalic neural crest cytodifferentiation. I. Skeletal and connective tissue. *Dev. Biol.* 69: 296–312.

Noden, D. M. 1983. The role of the neural crest in patterning of avian cranial skeletal, connective, and muscle tissues. *Dev. Biol.* 96: 144–165.

O'Leary, D. D. M. and D. G. Wilkinson. 1999. Eph receptors and ephrins in neural development. *Curr. Opin. Neurobiol.* 9: 65–73.

Ourednik, V., J. Ourednik, K. I. Park, and E. Y. Snyder. 1999. Neural stem cells: A versatile tool for cell replacement and gene therapy in the central nervous system. *Clin. Genet.* 56: 267–278.

Patterson, P. H. 1992. Neuron-target interactions. *In* Z. Hall (ed.), *An Introduction to Molecular Neurobiology*. Sinauer Associates, Sunderland, MA, pp. 428–459.

Paves, H. and M. Saarma. 1997. Neurotrophins as in vitro growth cone guidance molecules for embryonic sensory neurons. *Cell Tissue Res.* 290: 285–297.

Pearson, J., E. M. Johnson, Jr. and L. Brandeis. 1983. Effects of antibodies to nerve growth factor on intrauterine development of derivatives of cranial neural crest and placode in the guinea pig. *Dev. Biol.* 96: 32–36.

Perez, S. E., S. Rebelo and D. J. Anderson. 1999. Early specification of sensory neuron fate revealed by expression and function of neurogenins in the chick embryo. *Development* 126: 1715–1728.

Pini, A. 1993. Chemorepulsion of axons in the developing mammalian central nervous system. *Science* 261: 95–98.

Pomeranz, H. D., T. P. Rothman and M. D. Gershon. 1991. Colonization of the post-umbilical bowel by cells derived from the sacral neural crest: Direct tracing of cell migration using an intercalating probe and a replication-deficient retrovirus. *Development* 111: 647–655.

Purves, D. 1994. *Neural Activity and the Growth of the Brain*. Cambridge University Press, New York.

Purves, D. and J. W. Lichtman. 1980. Elimination of synapses in the developing nervous system. *Science* 210: 153–157.

Raff, M. C., B. A. Barres, J. F. Burne, H. S. Coles, Y. Ishizaki and M. D. Jacobson. 1993. Programmed cell death and the control of cell survival: Lessons from the nervous system. *Science* 262: 695–700.

Ramón y Cajal, S. 1892. Le Rètine de Vertèbrès. *La Cellule* 9: 119–258.

Rawles, M. E. 1948. Origin of melanophores and their role in development of color patterns in vertebrates. *Physiol. Rev.* 28: 383–408.

Rickmann, M., J. W. Fawcett and R. J. Keynes. 1985. The migration of neural crest cells and the growth of motor neurons through the rostral half of the chick somite. *J. Embryol. Exp. Morphol.* 90: 437–455.

Rijli, F. M., M. Mark, S. Lakkaraju, A. Dierich, P. Dollé and P. Chambon. 1993. A homeotic transformation is generated in the rostral branchial region of the head by a disruption of *Hoxa-2*, which acts as a selector gene. *Cell* 75: 1333–1349.

Roth, S. and R. B. Marchase. 1976. An in vitro assay for retinotectal specificity. *In* S. H. Barondes (ed.), *Neuronal Recognition*. Plenum, New York, pp. 227–248.

Savagne, R. P., K. M. Yamada and J. P. Thiery. 1997. The zinc finger protein slug causes desmosome dissociation, an initial and necessary step for growth factor-induced epithelial-mesenchymal transition. *J. Cell Biol.* 137: 1403–1419.

Schmidt, M. and S. B. Kater. 1993. Fibroblast growth factors, depolarization, and substrate interact in a combinatorial way to promote neuronal survival. *Dev. Biol.* 158: 228–237.

Schweizer, G., C. Ayer-LeLièvre and N. M. Le Douarin. 1983. Restrictions in developmental capacities in the dorsal root ganglia during the course of development. *Cell Diff.* 13: 191–200.

Sechrist, J., G. N. Serbedzija, T. Scherson, S. E. Fraser and M. Bronner-Fraser. 1993. Segmental migration of the hindbrain neural crest does not arise from its segmental origin. *Development* 118: 691–703.

Selleck, M. A. and M. Bronner-Fraser. 1995. Origins of the avian neural crest: The role of neural plate-epidermal interactions. *Development* 121: 525–538.

Serafini, T., T. E. Kennedy, M. J. Galko, C. Mirayan, T. M. Jessell and M. Tessier-Lavigne. 1994. The netrins define a family of axon outgrowth-promoting proteins homologous to *C. elegans* UNC-6. *Cell* 78: 409–424.

Shah, N. M., M. A. Marchionni, I. Isaacs, P. Stroobant and D. J. Anderson. 1994. Glial growth factor restricts mammalian neural crest stem cells to a glial fate. *Cell* 77: 349–360.

Shah, N. M., A. K. Groves and D. J. Anderson. 1996. Alternative neural crest cell fates are instructively promoted by TGF-β superfamily members. *Cell* 85: 331–343.

Silver, J. and R. L. Sidman. 1980. A mechanism for guidance and topologic patterning of retinal ganglion cell axons. *J. Comp. Neurol.* 189: 101–111.

Smith, A., V. Robinson, K. Patel and D. G. Wilkinson. 1997. The EphA4 and EphB1 receptor tyrosine kinases and ephrin-B2 ligand regulate targeted migration of branchial neural crest cells. *Curr. Biol.* 7: 561–570.

Sockanathan, S. and T. M. Jessell. 1998. Motor neuron-derived retinoid signal specifies subtype identity of spinal motor neurons. *Cell* 94: 503–514.

Spencer, D. D. and 15 others. 1992. Unilateral transplantation of human fetal mesencephalic tissue into the caudate nucleus of patients with Parkinson disease. *N. Engl. J. Med.* 327: 1541–1548.

Sperry, R. W. 1951. Mechanisms of neural maturation. *In* S. S. Stevens (ed.), *Handbook of Experimental Psychology*. Wiley, New York, pp. 236–280.

Sperry, R. W. 1965. Embryogenesis of behavioral nerve nets. *In* R. L. DeHaan and H. Ursprung (eds.), *Organogenesis*. Holt, Rinehart & Winston, New York, pp. 161–186.

Stier, H. and B. Schlosshauer. 1995. Axonal guidance in the chicken retina. *Development* 121: 1443–1454.

Studer, M., A. Gavalas, H. Marshall, L. Ariza-McNaughton, F. Rijli, P. Chambon and R. Krumlauf. 1998. Genetic interactions between *Hoxa1* and *Hoxb1* reveal new roles in regulation of early hindbrain patterning. *Development* 125: 1025–1036.

Taghert, P. H., C. Q. Doe and C. S. Goodman. 1984. Cell determination and regulation during development of neuroblasts and neurones in grasshopper embryo. *Nature* 307: 163–165.

Takeichi, M. 1988. The cadherins: Cell-cell adhesion molecules controlling animal morphogenesis. *Development* 102: 639–656.

Tan, S.-S and G. Morriss-Kay. 1985. The development and distribution of the cranial neural crest in the rat embryo. *Cell Tissue Res.* 240: 403–416.

Tanabe, Y., C. William and T. M. Jessell. 1998. Specification of motor neuron identity by the MNR2 homeodomain protein. *Cell* 95: 67–80.

Tanaka, H. and 8 others, 2000. Molecular modification of N-cadherin in response to synaptic activity. *Neuron* 25; 93–107.

Teillet, M.-A., C. Kalcheim and N. M. Le Douarin. 1987. Formation of the dorsal root ganglia in the avian embryo: Segmental origin and migratory behavior of neural crest progenitor cells. *Dev. Biol.* 120: 329–347.

Thanos, S., F. Bonhoeffer and U. Rutishauser. 1984. Fiber-fiber interaction and tectal cues influence the development of the chicken retinotectal projection. *Proc. Natl. Acad. Sci. USA* 81: 1906–1910.

Thesleff, I. and C. Sahlberg. 1996. Growth factors as inductive signals regulating tooth morphogenesis. *Semin. Cell Dev. Biol.* 7: 185–193.

Thesleff, I., A. Vaahtokari and S. Vainio. 1990. Molecular changes during determination and differentiation of the dental mesenchyme cell lineage. *J. Biol. Bucalle* 18: 179–188.

Thoenen, H. 1995. Neurotrophins and neuronal plasticity. *Science* 270: 593–598.

Thomas, J. B. 1998. Axon guidance: Crossing the midline. *Curr. Biol.* 8: R102–R104.

Thomas, L. 1992. *The Fragile Species.* Macmillan, New York.

Thomas, T., H. Kurihara, H. Yamagishi, Y. Kurihara, Y. Yazaki, E. N. Olson and D. Srivastava. 1998. A signaling cascade involving endothelin-1, dHAND, and Msx1 regulates development of neural crest-derived branchial arch mesenchyme. *Development* 125: 3005–3014.

Thompson, W. J. 1983. Synapse elimination in neonatal rat muscle is sensitive to pattern of muscle use. *Nature* 302: 614–616.

Thorogood, P. 1989. Review of developmental and evolutionary aspects of the neural crest. *Trends Neurosci.* 12: 38–39.

Tinbergen, N. 1951. *The Study of Instinct.* Clarendon Press, Oxford.

Tosney, K. W., K. B. Hotary and C. Lance-Jones. 1995. Specificity of motoneurons. *BioEssays* 17: 379–382.

Tsushida, T., M. Ensini, S. B. Morton, M. Baldassare, T. Edlund, T. M. Jessell and S. L. Pfaff. 1994. Topographic organization of embryonic motor neurons defined by expression of LIM homeobox genes. *Cell* 79: 957–970.

Tucker, A. S., K. L. Matthews and P. T. Sharpe. 1998. Transformation of tooth type induced by inhibition of BMP signaling. *Science* 282: 1136–1138.

Twitty, V. C. 1937. Experiments on the phenomenon of paralysis produced by a toxin occurring in *Triturus* embryos. *J. Exp. Zool.* 76: 67–104.

Twitty, V. C. and H. H. Johnson. 1934. Motor inhibition in *Amblystoma* produced by *Triturus* transplants. *Science* 80: 78–79.

Vaahtokari, A., T. Aberg, J. Jernvall, S. Keränen and I. Thesleff. 1996a. The enamel knot as a signalling center in the developing mouse tooth. *Mech. Dev.* 54: 39–43.

Vaahtokari, A., T. Aberg and I. Thesleff. 1996b. Apoptosis in the developing tooth: association with an embryonic signaling center and suppression by EGF and FGF-4. *Development* 122: 121–129.

Vainio, S., I. Karavanova, A. Jowett and I. Thesleff. 1993. Identification of BMP-4 as a signal mediating secondary induction between epithelial and mesenchymal tissues during early tooth development. *Cell* 75: 45–58.

Varley, J. E., R. G. Wehby, D. C. Rueger and G. D. Maxwell. 1995. Number of adrenergic and islet-1 immunoreactive cells is increased in avian trunk neural crest cultures in the presence of human recombinant osteogenic protein-1. *Dev. Dyn.* 203: 434–447.

Vogel, K. S. and J. A. Weston. 1990. The sympathoadrenal lineage in avian embryos. II. Effects of glucocorticoids on cultured neural crest cells. *Dev. Biol.* 139: 13–23.

Waldo, K., S. Miyagawa-Tomita, D. Kumiski and M. L. Kirby. 1998. Cardiac neural crest cells provide new insight into septation of the cardiac outflow tract: Aortic sac to ventricular septal closure. *Dev. Biol.* 196: 129–144.

Walter, J., S. Henke-Fahle and F. Bonhoeffer. 1987. Avoidance of posterior tectal membranes by temporal retinal axons. *Development* 101: 909–913.

Wang, H. U. and D. J. Anderson. 1997. Eph family transmembrane ligands can mediate repulsive guidance of trunk neural crest migration and motor axon outgrowth. *Neuron* 18: 383–396.

Wang, T., Z. Xie and B. Lu. 1995. Nitric oxide mediates activity-dependent synaptic suppression at developing neuromuscular synapses. *Nature* 374: 262–266.

Westerfield, M., D. W. Liu, C. B. Kimmel and C. Walker. 1990. Pathfinding and synapse formation in a zebrafish mutant lacking functional acetylcholine receptors. *Neuron* 4: 867–874.

Weston, J. 1963. A radiographic analysis of the migration and localization of trunk neural crest cells in the chick. *Dev. Biol.* 6: 274–310.

Widner, H. and 8 others. 1992. Bilateral fetal mesencephalic grafts in two patients with parkinsonism induced by 1-methyl-4-phenyl-1,2,3,6 tetrahydropryridine (MPTP). *N. Engl. J. Med.* 327: 1556–1563.

Wilkinson, D. G., S. Bhatt and A. P. McMahon. 1989. Expression of the FGF-related proto-oncogene *int-2* suggests multiple roles in fetal development. *Development* 105: 131–136.

Winberg, M. L., K. J. Mitchell and C. S. Goodman. 1998. Genetic analysis of the mechanisms controlling target selection: Complementary and combinatorial functions of netrins, semaphorins, and IgCAMs. *Cell* 93: 581–591.

Wu, H. H., C. V. Williams and S. C. McLoon. 1994. Involvement of nitric oxide in the elimination of a transient retinotectal projection in development. *Science* 265: 1593–1596.

Yamamori, T., K. Fukada, R. Aebersold, S. Korsching, M.-J. Fann and P. H. Patterson. 1989. The cholinergic neuronal differentiation factor from heart cells is identical to leukemia inhibitory factor. *Science* 246: 1412–1416.

Yin, X., M. Watanabe and U. Rutishauser. 1995. Effects of polysialic acid on the behavior of retinal ganglion cell axons during growth into the optic tract and tectum. *Development* 121: 3439–3446.

Zinn, K., L. McAllister and C. S. Goodman. 1988. Sequence analysis and neuronal expression of fasciclin I in grasshopper and *Drosophila*. *Cell* 53: 577–587.

chapter *14* *Paraxial and intermediate mesoderm*

I N CHAPTERS 12 AND 13, we followed the various tissues formed by the vertebrate ectoderm. In this chapter and the next, we will follow the development of the mesodermal and endodermal germ layers. We will see that the endoderm forms the lining of the digestive and respiratory tubes, with their associated organs. The mesoderm will be seen to generate all the organs between the ectodermal wall and the endodermal tissues.

The mesoderm of a neurula-stage embryo can be divided into five regions (Figure 14.1). The first region is the **chordamesoderm**. This tissue forms the notochord, a transient organ whose major functions include inducing the formation of the neural tube and establishing the anterior-posterior body axis. (The formation of the notochord on the future dorsal side of the embryo was discussed in Chapters 10 and 11.) The second region is the **paraxial mesoderm** (or **somitic dorsal mesoderm**). The term *dorsal* refers to the observation that the tissues developing from this region will be in the back of the embryo, along the spine. The cells in this region will form **somites**, blocks of mesodermal cells on both sides of the neural tube that will produce many of the connective tissues of the back (bone, muscle, cartilage, and dermis). The third region, the **intermediate mesoderm**, forms the urogenital system. Farther away from the notochord, the **lateral plate mesoderm** gives rise to the heart, blood vessels, and blood cells of the circulatory system, as well as to the lining of the body cavities and to all the mesodermal components of the limbs except the muscles. It will also form a series of extraembryonic membranes that are important for transporting nutrients to the embryo. Finally, the **head mesenchyme** contributes to the connective tissues and musculature of the face.

Paraxial Mesoderm: The Somites and Their Derivatives

One of the major tasks of gastrulation is to create a mesodermal layer between the endoderm and the ectoderm. As shown in Figure 14.2, the formation of mesodermal and endodermal organs is not subsequent to neural tube formation, but occurs synchronously. The notochord extends beneath the neural tube from the base of the head into the tail. On either side of the neural tube lie thick bands of mesodermal cells. These bands of paraxial mesoderm are referred to as the **segmental plate** (in birds) and the **unsegmented mesoderm** (in mammals). As the primitive streak regresses and the neural folds begin to gather at the center of the embryo, the paraxial mesoderm separates into blocks of cells called somites. Although somites are transient structures, they are extremely important in organizing the segmental pattern of vertebrate embryos. As we saw in the preceding chapter, the somites determine the

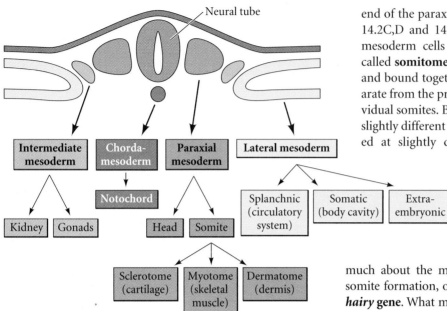

Figure 14.1
The major lineages of the mesoderm.

migration paths of neural crest cells and spinal nerve axons. Somites give rise to the cells that form the vertebrae and ribs, the dermis of the dorsal skin, the skeletal muscles of the back, and the skeletal muscles of the body wall and limbs.

> **VADE MECUM** **Mesoderm in the vertebrate embryo.** The organization of the mesoderm in the neurula stage is similar for all vertebrates. You can see this organization by viewing serial sections of the chick embryo.
> **[Click on Chick-Mid]**

The initiation of somite formation

PERIODICITY. The important components of somitogenesis (somite formation) are periodicity, epithelialization, specification, and differentiation. The first somites appear in the anterior portion of the trunk, and new somites "bud off" from the rostral

end of the paraxial mesoderm at regular intervals (Figures 14.2C,D and 14.3). Somite formation begins as paraxial mesoderm cells become organized into whorls of cells called **somitomeres**. The somitomeres become compacted and bound together by an epithelium, and eventually separate from the presomitic paraxial mesoderm to form individual somites. Because individual embryos can develop at slightly different rates (as when chick embryos are incubated at slightly different temperatures), the number of somites present is usually the best indicator of how far development has proceeded. The total number of somites formed is characteristic of a species (50 in chicks, 65 in mice, and about 500 in some snakes).

Although we do not know much about the mechanisms controlling the periodicity of somite formation, one of the key agents in this process is the *hairy* gene. What makes the *hairy* gene so interesting is that it is expressed in a dynamic pattern. First, it is expressed in the caudal portion of each somite, and it persists in these caudal regions for at least 15 hours. Second, *hairy* is expressed in the presomitic segmental plate or unsegmented mesoderm in a cyclic wavelike manner, cresting every 90 minutes. Its expression is detected first in the caudalmost region of the presomitic mesoderm, and this region of expression moves anteriorly as each somite forms. Eventually, like a wave leaving shells upon a beach, its expression recedes caudally, while the most anterior expression region remains. The caudalmost region of this an-

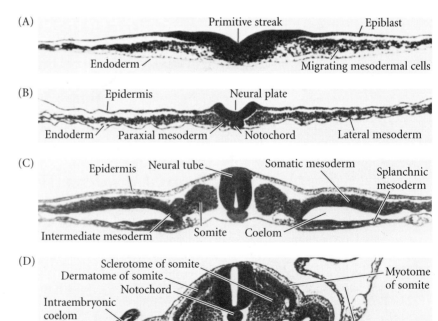

Figure 14.2
Gastrulation and neurulation in the chick embryo, focusing on the mesodermal component. (A) Primitive streak region, showing migrating mesodermal and endodermal precursors. (B) Formation of the notochord and paraxial mesoderm. (C, D) Differentiation of the somites, coelom, and the two aortae (which will eventually fuse). A–C, 24-hour embryos; D, 48-hour embryo.

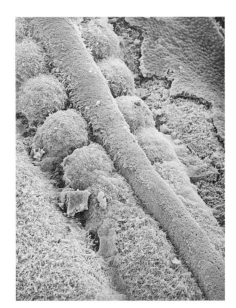

Figure 14.3
Figure 14.3
Neural tube and somites. Scanning electron micrograph showing well-formed somites and paraxial mesoderm (bottom right) that has not yet separated into distinct somites. A rounding of the paraxial mesoderm into a somitomere can be seen at the lower left, and neural crest cells can be seen migrating ventrally from the roof of the neural tube. (Photograph courtesy of K. W. Tosney.)

critical for separating the somites. In the zebrafish, the boundary between the most recently separated somite and the presomitic mesoderm forms between ephrinB2 in the posterior of the somite and EphA4 in the most anterior portion of the presomitic mesoderm (Figure 14.5A; Durbin et al. 1998). As somites form, this pattern of gene expression is reiterated caudally. Interfering with this signaling (by injecting embryos with RNA encoding dominant negative Ephs) leads to abnormal somite boundary formation. Eph signaling is thought to mediate cell shape changes, and these could be responsible for separating the presomitic mesoderm at the EphA4-ephrinB2 border. Thus, ephrin-Eph signaling may be responsible for converting the prepattern established by the Hairy protein in the presomitic mesoderm into actual somites.

EPITHELIALIZATION. Several studies in chicks have shown that the conversion from mesenchymal tissue into an epithelial

terior expression band correlates with the posterior terminus of the next somite to be formed. This process can be seen in Figure 14.4.

Thus, the expression pattern of the *hairy* gene correlates with the positioning of the place where a somite will separate from the unsegmented mesoderm. Labeling of cells with diI shows that this wave of *hairy* expression is not caused by the migration of particular cells that express the *hairy* gene. Rather, the wavelike pattern of *hairy* expression is an autonomous property of the presomitic mesoderm. Even if the presomitic mesoderm is separated from the caudalmost mesoderm, the ectoderm, neural tube, notochord, Hensen's node, and lateral plate regions, the dynamic expression of the *hairy* gene remains (Palmeirim et al. 1997; Jouve et al. 2000).

SEPARATION. The *hairy* gene encodes a transcription factor (indeed, one that is also used for forming the segmental units of *Drosophila*—see Chapter 9), but it is not known what its targets are. One set of possible targets (directly or indirectly) are the genes for ephrin and its receptor. The Eph receptor proteins and their ephrin ligands were discussed in Chapter 13 as being able to cause cell-cell repulsion between the posterior somite and migrating neural crest cells. Ephrin and Eph also may be

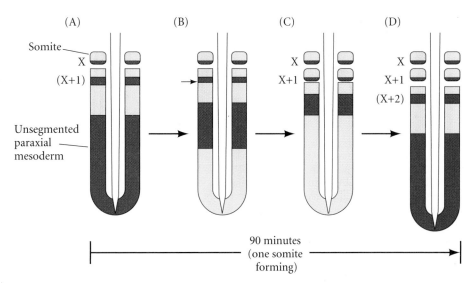

Figure 14.4
Somite formation correlates with the expression of the *hairy* gene. (A) Schematic representation of the posterior portion of an X-somite chick embryo. Somite X is marked. The expression of the *hairy* gene (purple) is seen in the caudal half of this somite, as well as in the posterior portion of the presomitic mesoderm and in a thin band that will form the caudal half of the next somite. (B) A caudal fissure (small arrow) begins to separate the somite from the presomitic mesoderm. The posterior region of *hairy* expression extends anteriorly. (C) The newly formed X+1 somite retains the expression of *hairy* in its caudal half, as the posterior domain of *hairy* expression moves farther anteriorly and shortens. (D) The formation of somite X+1 is completed, and the anterior region of what had been the posterior *hairy* expression pattern is now the anterior expression pattern. It will become the caudal domain of somite X+2. The entire process takes 90 minutes. (After Palmeirim et al. 1997.)

(A)

Somites:
Anterior
Posterior

EphA4

EphrinB2

Unsegmented
paraxial mesoderm

(B)

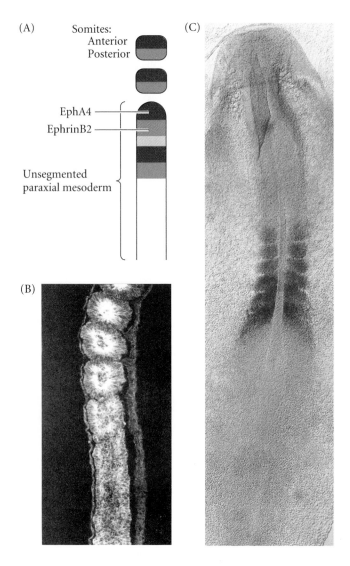

(C)

Figure 14.5
Transition from somitomere to somite. (A) Expression pattern of receptor tyrosine kinase EphA4 (blue) and its ligand, ephrinB2 (red) as somites develop. The somite boundary forms at the junction between the region of ephrin expression on the posterior of the last formed somite and the region of Eph expression on the anterior of the next somite to form. In the presomitic mesoderm, the pattern is created anew as each somite buds off. The posteriormost region of the next somite to form does not express ephrin until that somite is ready to separate. (B) N-cadherin expression (white area) correlates with the conversion of loose mesenchyme cells into the epithelial somite. (C) In situ hybridization showing the expression of *Paraxis* mRNA (red) in the 6-somite chick embryo. It is seen both in the somites and in the region of the unsegmented mesoderm that will give rise to the next somite. (A after Durbin et al. 1998; B from Hatta et al. 1987, courtesy of M. Takeichi; C from Barnes et al. 1997, courtesy of R. Tuan.)

block occurs even before each somite splits off. As seen in Figure 14.3, the cells of the somitomere are randomly organized as a mesenchymal mass, but the synthesis of two extracellular matrix proteins, fibronectin and N-cadherin, links them into arrays that will form tight junctions and generate their own basal laminae (Figure 14.5B; Ostrovsky et al. 1984; Lash and Yamada 1986; Hatta et al. 1987). These extracellular matrix proteins, in turn, may be regulated by the expression of the *Paraxis* gene. This gene encodes a transcription factor that is also expressed at the rostral (anterior) end of the unsegmented mesoderm of mouse embryos, and which is seen in precisely that region that will form the somite (Figure 14.5C). Injection of antisense oligonucleotides complementary to the *Paraxis* message produces defects of the paraxial mesoderm, and in the somites of *Paraxis*-deficient mice, no epithelial structures are formed. These defective somites have segregated from the segmental plate and their cells have differentiated, but they are completely disorganized (Burgess et al. 1995; Barnes et al. 1997). The Paraxis protein is therefore an essential part of the conversion from mesenchyme to epithelium (Burgess et al. 1996; Barnes et al. 1997; Tajbakhsh and Spörle 1998).

Figure 14.6
The segmental plate mesoderm is determined as to its position along the anterior-posterior axis before somitogenesis. When segmental plate mesoderm that would ordinarily form thoracic somites is transplanted into a region in a younger embryo (caudal to the first somite) that would ordinarily give rise to cervical (neck) somites, the grafted mesoderm differentiates according to its original position and forms ribs in the neck. Note that only one side is affected. (After Kieny et al. 1972.)

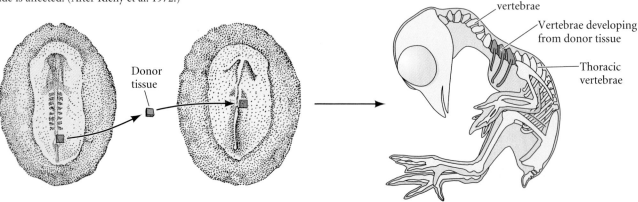

Donor
tissue

Cervical
vertebrae

Vertebrae developing
from donor tissue

Thoracic
vertebrae

Specification and commitment of somitic cell types

AXIAL SPECIFICATION. Although all the somites look identical, they will form different structures at different positions along the anterior-posterior axis. For instance, the ribs are derived from somites. The somites that form the cervical vertebrae of the neck and the lumbar vertebrae of the abdomen are not capable of forming ribs; ribs are generated only by the somites forming the thoracic vertebrae. Moreover, the specification of the thoracic vertebrae occurs very early in development. If one isolates the region of chick segmental plate that will give rise to a thoracic somite, and transplants this mesoderm into the cervical (neck) region of a younger embryo, the host embryo will develop ribs in its neck. Those ribs will form only on the side where the thoracic mesoderm has been transplanted (Figure 14.6; Kieny et al. 1972; Nowicki and Burke 1999). As discussed in Chapter 11 (see Figure 11.41), the somites are specified in this manner according to the Hox genes they express. Mice that are homozygous for a loss-of-function mutation of *Hoxc-8* will convert a lumbar vertebra into an extra ribbed thoracic vertebra (see Figure 11.39).

DIFFERENTIATION WITHIN THE SOMITE. Somites form (1) the cartilage of the vertebrae and ribs, (2) the muscles of the rib cage, limbs, and back, and (3) the dermis of the dorsal skin. Unlike the early commitment of the mesoderm along the anterior-posterior axis, the commitment of the cells within a somite to their respective fates occurs relatively late, after the somite has already formed. When the somite is first separated from the presomitic mesoderm, any of its cells can become any of the somite-derived structures. However, as the somite matures, its various regions become committed to forming only certain cell types. The ventral medial cells of the somite (those cells located farthest from the back but closest to the neural tube) undergo mitosis, lose their round epithelial characteristics, and become mesenchymal cells again. The portion of the somite that gives rise to these cells is called the **sclerotome**, and these mesenchymal cells ultimately become the cartilage cells (chondrocytes) of the vertebrae and part (if not all) of each rib (Figures 14.2 and 14.7).

Fate mapping with chick-quail chimeras (Ordahl and Le Douarin 1992; Brand-Saberi et al. 1996; Kato and Aoyama 1998) has revealed that the remaining epithelial portion of the somite is arranged into three regions (Figure 14.7). The cells in

Figure 14.7
Diagram of a transverse section through the trunk of a chick embryo on days 2–4. (A) The 2-day somite can be divided into sclerotome cells and dermamyotome cells. (B) On day 3, the sclerotome cells lose their adhesion to one another and migrate toward the neural tube. (C) On day 4, the remaining dermamyotome cells divide. The medial cells form an epaxial myotome beneath the dermamyotome, while the lateral cells form a hypaxial myotome. (D) A layer of muscle cell precursors (the myotome) forms beneath the epithelial dermamyotome. (A, B after Langman 1981; C, D after Ordahl 1993.)

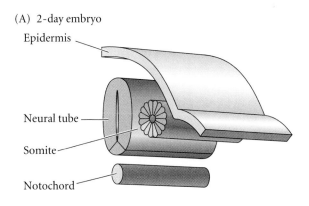

(A) 2-day embryo

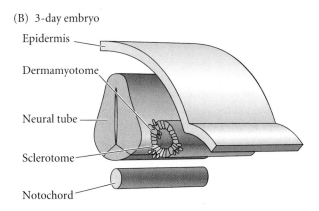

(B) 3-day embryo

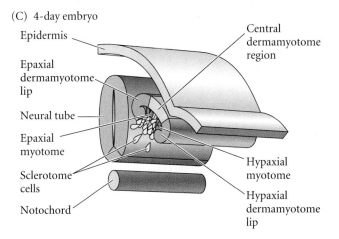

(C) 4-day embryo

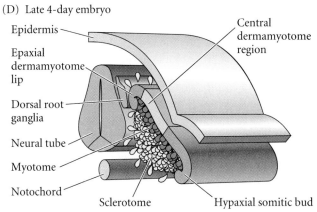

(D) Late 4-day embryo

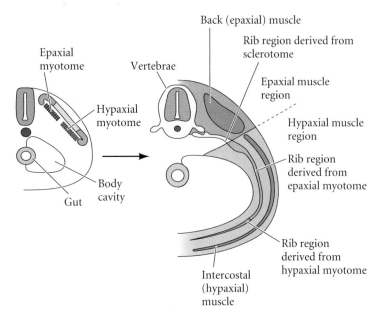

Figure 14.8
Myotome derivatives of the mouse embryo. The epaxial muscles form from the region of the dermamyotome closest to the neural tube. The hypaxial muscles form from the region of dermamyotome furthest from the neural tube. The epaxial myotome will form the back muscles and the central portion of the rib. The hypaxial myotome will form the intercostal muscles and distal portion of the rib. The proximal (ventral) proteion of the rib is formed by sclerotome cells.* (After Kato and Aoyama 1998; Ordahl and Williams 1998.)

the two lateral portions of the epithelium (those regions closest to and farthest from the neural tube) are muscle-forming cells. They divide to produce a lower layer of muscle precursor cells, the **myoblasts**. The resulting double-layered structure is called the **dermamyotome**, and the lower layer is called the **myotome**. Those myoblasts formed from the region closest to the neural tube form the **epaxial muscles** (the deep muscles of the back), while those myoblasts formed in the region farthest from the neural tube produce the **hypaxial muscles** of the body wall, limbs, and tongue (Figures 14.7 and 14.8; see Christ and Ordahl 1995; Venters et al. 1999). The central region of the dorsal layer of the dermamyotome is called the **dermatome**, and it generates the mesenchymal connective tissue of the back skin: the **dermis**. (The dermis of other areas of the body forms from other mesenchymal cells, not from the somites.) The dermamyotome may also produce the distal cartilage of the ribs, its lateral edge producing the most ventral portion of the rib (Figure 14.8; Kato and Aoyama 1998).

Determining somitic cell fates

Like the proverbial piece of real estate, the destiny of a somitic region depends on three things: location, location, and location.

*This model has recently been challenged by Huang et al. (2000), whose transplantation experiments confirm the traditional view that the entire rib originates from the sclerotome.

DETERMINATION OF THE SCLEROTOME AND DERMATOME. The specification of the somite is accomplished by the interaction of several tissues. The ventral-medial portion of the somite is induced to become the sclerotome by paracrine factors, especially Sonic hedgehog, secreted from the notochord and the neural tube floor plate (Fan and Tessier-Lavigne 1994; Johnson et al. 1994). If portions of the notochord (or another source of Sonic hedgehog) are transplanted next to other regions of the somite, those regions, too, will become sclerotome cells. Sclerotome cells express a new transcription factor, Pax1, that is required for their differentiation into cartilage and whose presence is necessary for the formation of the vertebrae (Figure 14.9; Smith and Tuan 1996). They also express I-mf, an inhibitor of the myogenic bHLH family of transcription factors that initiate muscle formation (Chen et al. 1996).

The dermatome differentiates in response to another factor secreted by the neural tube, neurotrophin 3 (NT-3). Antibodies that block the activities of NT-3 prevent the conversion of the epithelial dermatome into the loose dermal mesenchyme that migrates beneath the epidermis (Brill et al. 1995).

DETERMINATION OF THE MYOTOME. In similar ways, the myotome is induced by at least two distinct signals. Studies involving transplantation and knockout mice indicate that the epaxial muscle cells coming from the medial portion of the somite are induced by factors from the neural tube, probably Wnt1 and Wnt3a from the dorsal region and low levels of Sonic hedgehog from the ventral region (Münsterberg et al. 1995; Stern et al. 1995; Ikeya and Takada 1998). The hypaxial muscles coming from the lateral edge of the somite are probably induced by a combination of Wnt proteins from the epidermis and bone morphogenetic protein 4 (BMP4) from the lateral plate mesoderm (Cossu et al. 1996a; Pourquié et al. 1996; Dietrich et al. 1998). These factors cause the myotome cells to express particular transcription factors that activate the muscle-specific genes.

In addition to these positive signals, there are inhibitory signals that prevent a signal from affecting an inappropriate group of cells. For example, Sonic hedgehog not only activates sclerotome and myotome development; it also inhibits BMP4 signal from the lateral plate mesoderm from extending medially and ventrally (thus preventing the conversion of sclerotome into muscle) (Watanabe et al. 1998). Similarly, Noggin is produced by the most medial portion of the dermamyotome and prevents BMP4 from giving these cells the migratory characteristics of hypaxial muscle (Marcelle et al. 1997).

And what happens to the notochord, that central mesodermal structure? After it has provided the axial integrity of the early embryo and has induced the formation of the dorsal neural tube, most of it degenerates. Wherever the sclerotome cells have formed a vertebral body, the notochordal cells die.

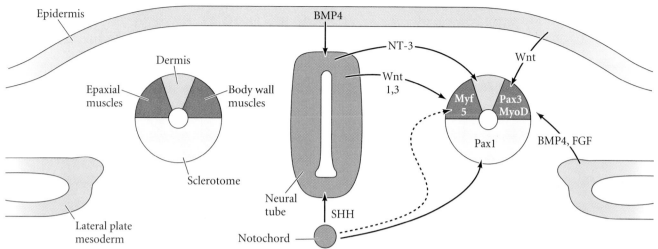

Figure 14.9
Model of major postulated interactions in the patterning of the somite. A combination of Wnts (probably Wnt1 and Wnt3a) are induced by BMP4 in the dorsal neural tube. These Wnt proteins, in combination with low concentrations of Sonic hedgehog from the notochord and floor plate, induce the epaxial myotome, which synthesizes the myogenic transcription factor Myf5. High concentrations of Sonic hedgehog induce Pax1 expression in those cells fated to become the sclerotome. Certain concentrations of neurotrophin-3 (NT-3) from the dorsal neural tube appear to specify the dermatome, while Wnt proteins from the epidermis, in conjunction with BMP4 and FGF5 from the lateral plate mesoderm, are thought to induce the hypaxial myotome. (After Cossu et al. 1996b.)

However, in between the vertebrae, the notochordal cells form the tissue of the intervertebral discs, the nuclei pulposi. These are the discs that "slip" in certain back injuries.

> WEBSITE **14.1 Calling the competence of the somite into question.** When the *tbx6* gene was knocked out from mice, the resulting embryos had three neural tubes in the posterior of theirbodies. Without the *tbx6* gene, the somitic tissue responded to the notochord and epidermal signals as if it were neural ectoderm.

> WEBSITE **14.2 Cranial paraxial mesoderm.** Most of the head musculature does not come from somites. Rather, it comes from the cranial paraxial mesoderm. These cells originate adjacent to the sides of the brain, and they migrate to their respective destinations.

Myogenesis: The Development of Muscle

Specification and differentiation by the myogenic bHLH proteins

As we have seen, muscle cells come from two cell lineages in the somite. In both instances, paracrine factors instruct the myotome cells to become muscles by inducing them to synthesize the MyoD protein (Maroto et al. 1997; Tajbakhsh et al. 1997). In the lateral portion of the somite, which forms the hypaxial muscles, factors from the surrounding environment induce the Pax3 transcription factor. In the absence of other inhibitory transcription factors (such as those found in the sclerotome cells), Pax3 then activates the genes encoding two muscle-spe-

cific transcription factors, **Myf5** and **MyoD**. In the medial region of the somite, which forms the epaxial muscles, MyoD is induced through a slightly different pathway. MyoD and Myf5 belong to a family of transcription factors called the **myogenic bHLH** (basic helix-loop-helix) **proteins** (sometimes also referred to as the MyoD family). The proteins of this family all bind to similar sites on the DNA and activate muscle-specific genes. For instance, the MyoD protein appears to directly activate the muscle-specific creatine phosphokinase gene by binding to the DNA immediately upstream from it (Lassar et al. 1989), and there are two MyoD-binding sites on the DNA adjacent to the genes encoding a subunit of the chicken muscle acetylcholine receptor (Piette et al. 1990). MyoD also directly activates its own gene. Therefore, once the *myoD* gene is activated, its protein product binds to the DNA immediately upstream of the *myoD* gene and keeps this gene active.

While Pax3 is found in several other cell types, the myogenic bHLH proteins are specific for muscle cells. Any cell making a myogenic bHLH transcription factor such as MyoD or Myf5 is committed to becoming a muscle cell. Transfection of genes encoding any of these myogenic proteins into a wide range of cultured cells converts those cells into muscles (Thayer et al. 1989; Weintraub et al. 1989).

> WEBSITE **14.3 Myogenic bHLH proteins and their regulators.** Since the MyoD protein and its relatives are so powerful that they can turn nearly any cell into a muscle cell, the synthesis of this protein has to be inhibited at numerous steps. Numerous inhibitors of MyoD family gene expression and protein function have been found.

Muscle cell fusion

The myotome cells producing the myogenic bHLH proteins are the myoblasts—committed muscle cell precursors. Experiments with chimeric mice and cultured myoblasts showed

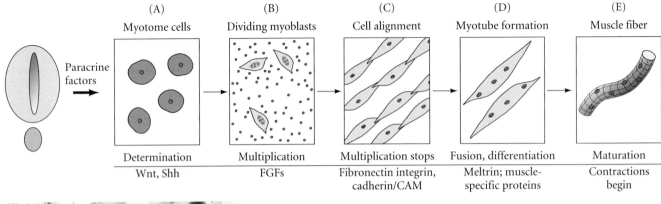

(A)	(B)	(C)	(D)	(E)
Myotome cells	Dividing myoblasts	Cell alignment	Myotube formation	Muscle fiber
Determination	Multiplication	Multiplication stops	Fusion, differentiation	Maturation
Wnt, Shh	FGFs	Fibronectin integrin, cadherin/CAM	Meltrin; muscle-specific proteins	Contractions begin

Paracrine factors

(F)

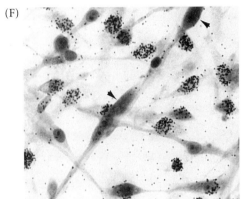

Figure 14.10
Conversion of myoblasts into muscles in culture. (A) Determination of myotome cells by paracrine factors. (B) Committed myoblasts divide in the presence of growth factors (primarily FGFs), but show no obvious muscle-specific proteins. (C–D) When the growth factors are used up, the myoblasts cease dividing, align, and fuse into myotubes. (E) The myotubes become organized into muscle fibers that spontaneously contract. (F) Autoradiograph showing DNA synthesis in myoblasts and the exit of fusing cells from the cell cycle. Phospholipase C can "freeze" myoblasts after they have aligned with other myoblasts, but before their membranes fuse. Cultured myoblasts were treated with phospholipase C and then exposed to radioactive thymidine. Unattached myoblasts still divide and incorporate the radioactive thymidine into their DNA. Aligned (but not yet fused) cells (arrows) do not incorporate the label. (Drawings after Wolpert 1998; photograph from Nameroff and Munar 1976, courtesy of M. Nameroff.)

conclusively that these cells align together and fuse to form the multinucleated **myotubes** characteristic of muscle tissue. Thus, the multinucleated myotube cells are the product of several myoblasts joining together and dissolving the cell membranes between themselves (Konigsberg 1963; Mintz and Baker 1967).

Muscle cell fusion begins when the myoblasts leave the cell cycle. As long as particular growth factors (particularly fibroblast growth factors) are present, the myoblasts will proliferate without differentiating. When these factors are depleted, the myoblasts stop dividing, secrete fibronectin onto their extracellular matrix, and bind to it through α5β1 integrin, their major fibronectin receptor (Menko and Boettiger 1987; Boettiger et al. 1995). If this adhesion is experimentally blocked, no further muscle development ensues, so it appears that the signal from the integrin-fibronectin attachment is critical for instructing the myoblasts to differentiate into muscle cells (Figure 14.10).

The second step is the alignment of the myoblasts together into chains. This step is mediated by cell membrane glycoproteins, including several cadherins and CAMs (Knudsen 1985: Knudsen et al. 1990). Recognition and alignment between cells takes place only if the two cells are myoblasts. Fusion can occur even between chick and rat myoblasts in culture (Yaffe and Feldman 1965); the identity of the species is not critical.

The third step is the cell fusion event itself. As in most membrane fusions, calcium ions are critical, and fusion can be activated by calcium ionophores, such as A23187, that carry calcium ions across cell membranes (Shainberg et al. 1969; David et al. 1981). Fusion appears to be mediated by a set of metalloproteinases called **meltrins**. These proteins were discovered during a search for myoblast proteins that would be homologous to fertilin, a protein implicated in sperm-egg membrane fusion. Yagami-Hiromasa and colleagues (1995) found that one of these meltrins (meltrin-α) is expressed in myoblasts at about the same time that fusion begins, and that antisense RNA to the meltrin-α message inhibited fusion when added to myoblasts.

WEBSITE **14.4 Myotube formation**. Multinucleated myotubes could arise either as (1) a fusion event between several mononuclear myoblasts, or (2) as a string of mitoses within a single myoblast. The former is thought to be the mechanism for myotube formation in skeletal muscle; the latter is thought to be the way heart myotubes are formed.

Osteogenesis: The Development of Bones

Some of the most obvious structures derived from the paraxial mesoderm are bones. We can only begin to outline the mechanisms of bone formation here; students wishing further details are invited to consult histology textbooks that devote entire chapters to this topic.

There are three distinct lineages that generate the skeleton. The somites generate the axial skeleton, the lateral plate mesoderm generates the limb skeleton, and the cranial neural crest gives rise to the branchial arch and craniofacial bones and cartilage.* There are two major modes of bone formation, or **osteogenesis**, and both involve the transformation of a preexisting mesenchymal tissue into bone tissue. The direct conversion of mesenchymal tissue into bone is called **intramembranous ossification**. This process occurs primarily in the bones of the skull. In other cases, the mesenchymal cells differentiate into cartilage, and this cartilage is later replaced by bone. The process by which a cartilage intermediate is formed and replaced by bone cells is called **endochondral ossification**.

Intramembranous ossification

Intramembranous ossification is the characteristic way in which the flat bones of the skull and the turtle shell are formed. During intramembranous ossification in the skull, neural crest-derived mesenchymal cells proliferate and condense into compact nodules. (Thus, intramembranous ossification is not occurring from sclerotome-derived cells.) Some of these cells develop into capillaries; others change their shape to become **osteoblasts**, committed bone precursor cells (Figure 14.11A). The osteoblasts secrete a collagen-proteoglycan matrix that is able to bind calcium salts. Through this binding, the prebone (osteoid) matrix becomes calcified. In most cases, osteoblasts are separated from the region of calcification by a layer of the osteoid matrix they secrete. Occasionally, though, osteoblasts become trapped in the calcified matrix and become **osteocytes**—bone cells. As calcification proceeds, bony spicules radiate out from the region where ossification began (Figure 14.11B). Furthermore, the entire region of calcified spicules becomes surrounded by compact mesenchymal cells that form the **periosteum** (a membrane that surrounds the bone). The cells on the inner surface of the periosteum also become osteoblasts and deposit osteoid matrix parallel to that of the existing spicules. In this manner, many layers of bone are formed.

*Craniofacial cartilage development was discussed in Chapter 13 and will be revisited in Chapter 22; the development of the limbs will be detailed in Chapter 16.

The mechanism of intramembranous ossification involves bone morphogenetic proteins and the activation of a transcription factor called **CBFA1**. Bone morphogenetic proteins (probably BMP2, BMP4, and BMP7) from the head epidermis are thought to instruct the neural crest-derived mesenchymal cells to become bone cells directly (Hall 1988). The BMPs activate the *Cbfa1* gene in the mesenchymal cells. Just as the myogenic bHLH family of transcription factors is competent to transform primitive mesenchyme cells (or just about any other cell) into muscle-forming myoblasts, the CBFA1 transcription factor appears to be able to transform mesenchyme cells into osteoblasts. Ducy and her colleagues (1997) found that the mRNA for mouse CBFA1 is severely restricted to the mesenchymal condensations that form bone, and is limited to the osteoblast lineage. The protein appears to activate the genes for osteocalcin, osteopontin, and other bone-specific extracellular matrix proteins.

Confirmation and extension of this conclusion was obtained from gene targeting experiments wherein the mouse *Cbfa1* gene was knocked out (Komori et al. 1997; Otto et al. 1997). Mice homozygous for this deletion died shortly after birth without taking a breath, and their skeletons completely lacked bone. The mutants had only the cartilaginous skeletal model (Figure 14.12). In these mice, both endochondral and intramembranous ossification had been eliminated. The osteoblasts were in an arrested state of development, expressing neither osteocalcin nor osteopontin.

Mice that were heterozygous for *Cbfa1* showed skeletal defects similar to those of a human syndrome called cleidocranial dysplasia (CCD). In this syndrome, the skull sutures fail to close (adults retain the fontanel associated with young

Figure 14.11
Schematic diagram of intramembranous ossification. (A) Mesenchymal cells condense to produce osteoblasts, which deposit osteoid matrix. These osteoblasts become arrayed along the calcified region of the matrix. Osteoblasts that are trapped within the bone matrix become osteocytes. (B) Intramembranous ossification in the plastron (ventral shell) of the red-ear slider turtle *Trachemys scripta*. The plastron of a one-month-old hatchling was stained with alcian blue (for cartilage) and alizarin red (for bone). No cartilage was seen to precede the formation of bone. (Photograph courtesy of G. Loredo, A. Brukman, and S. F. Gilbert.)

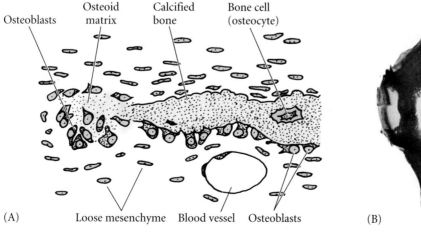

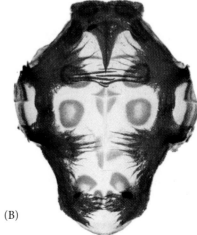

(A) Osteoblasts — Osteoid matrix — Calcified bone — Bone cell (osteocyte) — Loose mesenchyme — Blood vessel — Osteoblasts

(B)

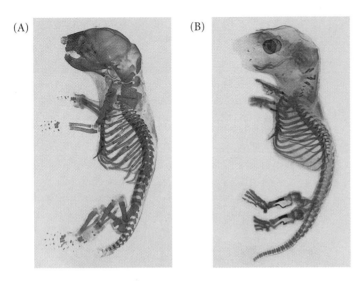

(A)　(B)

Figure 14.12
Gene targeting of *Cbfa1* in mice causes lack of bone formation. Newborn mice (wild-type and homozygotes for *Cbfa1*) were stained with alcian blue (for cartilage) and alizarin red (for bone). Cartilage development in both mice was normal. (A) Wild-type littermate. (B) Homozygous mutant showing cartilage, but an absence of ossification throughout the entire body. (From Otto et al. 1997; photographs courtesy of *Cell* and MIT Press.)

infants), growth is stunted, and the clavicle (collarbone) is often absent or deformed.* When DNA from patients with CCD was analyzed, each patient had either deletions or point mutations in the *CBFA1* gene. Control individuals did not have such mutations. Therefore, it appears that cleidocranial dysplasia is caused by heterozygosity of the *CBFA1* gene (Mundlos et al. 1997).

*CCD may have been responsible for the phenotype of Thersites, the Greek soldier described in the *Iliad* as having "both shoulders humped together, curving over his caved-in chest, and bobbing above them his skull warped to a point...." (Dickman 1997).

WEBSITE 14.5 **Human dermal ossification syndromes**. The human dermis appears to remain responsive to BMPs throughout life. The misexpression of BMP genes in children and adults can retrigger the embryonic bone growth pathway and lead to progressive and debilitating ossification of the skin.

Endochondral ossification

Endochondral ossification involves the formation of **cartilage** tissue from aggregated mesenchymal cells, and the subsequent replacement of cartilage tissue by bone (Horton 1990). The process of endochondral ossification can be divided into five stages (Figure 14.13). First, the mesenchymal cells are commited to become cartilage cells. This committment is caused by paracrine factors that induce the nearby mesodermal cells to express two transcription factors, Pax1 and **Scleraxis**. These transcription factors are thought to activate cartilage-specific genes (Cserjesi et al. 1995; Sosic et al. 1997). Thus, Scleraxis is expressed in the mesenchyme from the sclerotome, in the facial mesenchyme that forms cartilaginous precursors to bone, and in the limb mesenchyme (Figure 14.14).

Figure 14.13
Schematic diagram of endochondral ossification. (A, B) Mesenchymal cells condense and differentiate into chondrocytes to form the cartilaginous model of the bone. (C) Chondrocytes in the center of the shaft undergo hypertrophy and apoptosis while they change and mineralize their extracellular matrix. Their deaths allow blood vessels to enter. (D, E) Blood vessels bring in osteoblasts, which bind to the degenerating cartilaginous matrix and deposit bone matrix. (F–H) Bone formation and growth consist of ordered arrays of proliferating, hypertrophic, and mineralizing chondrocytes. Secondary ossification centers also form as blood vessels enter near the tips of the bone. (After Horton 1990.)

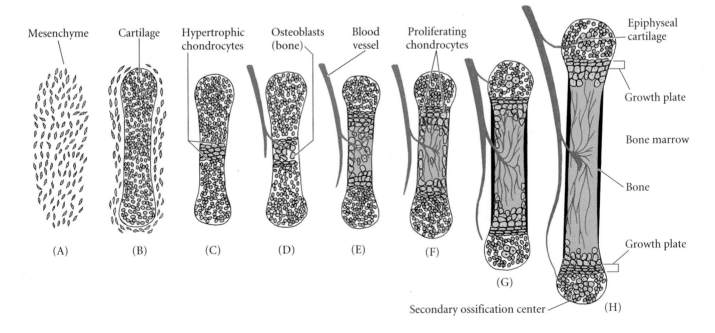

Mesenchyme　Cartilage　Hypertrophic chondrocytes　Osteoblasts (bone)　Blood vessel　Proliferating chondrocytes　Epiphyseal cartilage

Growth plate

Bone marrow

Bone

Growth plate

(A)　(B)　(C)　(D)　(E)　(F)　(G)

Secondary ossification center　(H)

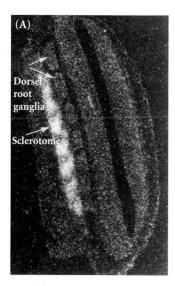

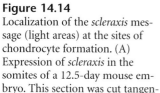

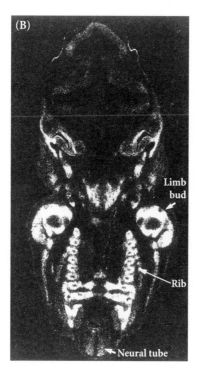

Figure 14.14
Localization of the *scleraxis* message (light areas) at the sites of chondrocyte formation. (A) Expression of *scleraxis* in the somites of a 12.5-day mouse embryo. This section was cut tangentially, and the neural tube runs along the anterior-posterior axis. (B) Section through a 11.5-day mouse embryo, in which *scleraxis* transcripts are seen in the condensing cartilage of the nose and face and in the precursors of the limbs and ribs. (After Cserjesi et al. 1995; photographs courtesy of E. Olson.)

During the second phase of endochondral ossification, the committed mesenchyme cells condense into compact nodules and differentiate into **chondrocytes**, the cartilage cells. N-cadherin appears to be important in the initiation of these condensations, and N-CAM seems to be critical for maintaining them (Oberlender and Tuan 1994; Hall and Miyake 1995). In humans, the *SOX9* gene, which encodes a DNA-binding protein, is expressed in the precartilaginous condensations. Mutations of the *SOX9* gene cause camptomelic dysplasia, a rare disorder of skeletal development that results in deformities of most of the bones of the body. Most affected babies die from respiratory failure due to poorly formed tracheal and rib cartilage (Wright et al. 1995).

During the third phase of endochondral ossification, the chondrocytes proliferate rapidly to form the model for the bone. As they divide, the chondrocytes secrete a cartilage-specific extracellular matrix. In the fourth phase, the chondrocytes stop dividing and increase their volume dramatically, becoming **hypertrophic chondrocytes**. These large chondrocytes alter the matrix they produce (by adding collagen X and more fibronectin) to enable it to become mineralized by calcium carbonate. The fifth phase involves the invasion of the cartilage model by blood vessels. The hypertrophic chondrocytes die by apoptosis. This space will become bone marrow. As the cartilage cells die, a group of cells that have surrounded the cartilage model differentiate into osteoblasts. The osto-

blasts begin forming bone matrix on the partially degraded cartilage (Bruder and Caplan 1989; Hatori et al. 1995). Eventually, all the cartilage is replaced by bone. Thus, the cartilage tissue serves as a model for the bone that follows. The skeletal components of the vertebral column, the pelvis, and the limbs are first formed of cartilage and later become bone.

The replacement of chondrocytes by bone cells is dependent on the mineralization of the extracellular matrix. This is clearly illustrated in the developing skeleton of the chick embryo, which utilizes the calcium carbonate of the eggshell as its calcium source. During development, the circulatory system of the chick embryo translocates about 120 mg of calcium from the shell to the skeleton (Tuan 1987). When chick embryos are removed from their shells at day 3 and grown in shell-less cultures (in plastic wrap) for the duration of their development, much of the cartilaginous skeleton fails to mature into bony tissue (Figure 14.15; Tuan and Lynch 1983). A number of events lead to the hypertrophy and mineralization of the chondrocytes, including an initial switch from aerobic to anaerobic respiration, which alters their cell metabolism and mitochondrial energy potential (Shapiro et al. 1992). Hypertrophic chondrocytes secrete numerous small membrane-bound vesicles into the extracellular matrix. These vesicles contain enzymes that are active in the generation of calcium and phosphate ions and initiate the mineralization process within the cartilaginous matrix (Wu et al. 1997). The hypertrophic chondrocytes, their metabolism and mitochondrial membranes altered, then die by apoptosis (Hatori et al. 1995; Rajpurohit et al. 1999).

Figure 14.15
Skeletal mineralization in 19-day chick embryos that developed (A) in shell-less culture and (B) inside the egg during normal incubation. The embryos were fixed and stained with alizarin red to show the calcified bone matrix. (From Tuan and Lynch 1983; photographs courtesy of R. Tuan.)

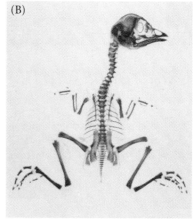

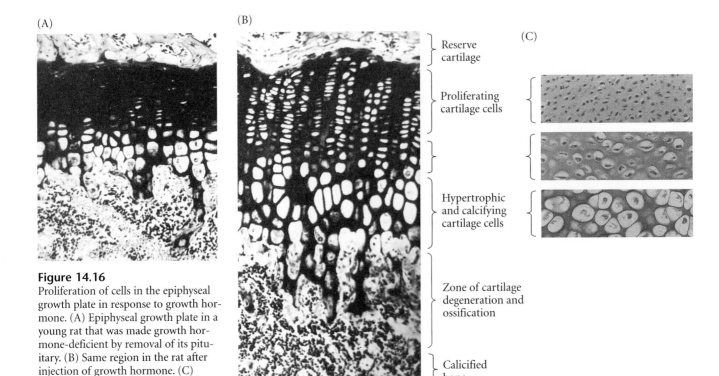

Figure 14.16
Proliferation of cells in the epiphyseal growth plate in response to growth hormone. (A) Epiphyseal growth plate in a young rat that was made growth hormone-deficient by removal of its pituitary. (B) Same region in the rat after injection of growth hormone. (C) Stained cartilage from specific regions of the epiphyseal growth plate. (A, B, I. Gersh's photographs from Bloom and Fawcett 1975; C from Chen et al. 1995, photograph courtesy of P. Goetinck.)

Reserve cartilage

Proliferating cartilage cells

Hypertrophic and calcifying cartilage cells

Zone of cartilage degeneration and ossification

Caliciﬁed bone

In the long bones of many mammals (including humans), endochondral ossification spreads outward in both directions from the center of the bone (see Figure 14.13). If all of our cartilage were turned into bone before birth, we would not grow any larger, and our bones would be only as large as the original cartilaginous model. However, as the ossification front nears the ends of the cartilage model, the chondrocytes near the ossification front proliferate prior to undergoing hypertrophy, pushing out the cartilaginous ends of the bone. These cartilaginous areas at the ends of the long bones are called **epiphyseal growth plates**. These plates contain three regions: a region of chondrocyte proliferation, a region of mature chondrocytes, and a region of hypertrophic chondrocytes (Figure 14.16; Chen et al. 1995). As the inner cartilage hypertrophies and the ossification front extends farther outward, the remaining cartilage in the epiphyseal growth plate proliferates. As long as the epiphyseal growth plates are able to produce chondrocytes, the bone continues to grow.

WEBSITE 14.6 **Paracrine factors, their receptors, and human bone growth.** Mutations in the genes encoding paracrine factors and their receptors cause numerous skeletal anomalies in humans and mice. The FGF and Hedgehog pathways are especially important.

Control of Cartilage Maturation at the Growth Plate

Recent discoveries of human and murine mutations resulting in abnormal skeletal development have provided remarkable insights into how the differentiation, proliferation, and patterning of chondrocytes are regulated.

Fibroblast Growth Factor Receptors
The proliferation of the epiphyseal growth plate cells and facial cartilage can be halted by the presence of fibroblast growth factors (Deng et al. 1996; Webster and Donoghue 1996). These factors appear to instruct the cartilage precursors to differentiate rather than to divide. In humans, mutations of the receptors for fibroblast growth factors can cause these receptors to become activated prematurely. Such mutations give rise to the major types of human dwarfism. Achondro-

plasia is a dominant condition caused by mutations in the transmembrane region of fibroblast growth factor receptor 3 (FGFR3). Roughly 95% of achondroplastic dwarfs have the same mutation of FGFR3, a base pair substitution that converts glycine to arginine at position 380 in the transmembrane region of the protein. In addition, mutations in the extracellular portion of the FGFR3 protein or in the tyrosine kinase intracellular domain may result in thanatophoric dysplasia, a lethal form of dwarfism that resembles homozygous achondroplasia (see Figure 6.22; Bellus et al. 1995; Tavormina et al. 1995). Mutations in FGFR1 can cause Pfeiffer syndrome, characterized by limb defects and premature fusion of the cranial sutures (craniosynostosis), resulting in abnormal skull and facial shape. Different mutations in FGFR2 can give rise to various abnormalities of the limbs and face (Park et al. 1995; Wilkie et al. 1995).

Insulin-like Growth Factors

The epiphyseal growth plate cells are very responsive to hormones, and their proliferation is stimulated by growth hormone and insulin-like growth factors. Nilsson and colleagues (1986) showed that growth hormone stimulates the production of insulin-like growth factor I (IGF-I) in the epiphyseal chondrocytes, and that these chondrocytes respond to it by proliferating. When they added growth hormone to the tibial growth plates of young mice who could not manufacture their own growth hormone (because their pituitaries had been removed), it stimulated the formation of IGF-I in the chondrocytes of the proliferative zone (see Figure 14.16). The combination of growth hormone and IGF-I appears to provide an extremely strong mitotic signal. It appears that IGF-I is essential for the normal growth spurt at puberty. The pygmies of the Ituri Forest of Zaire have normal growth hormone and IGF-I levels until puberty. However, at puberty, their IGF-I levels fall to about one-third that of other adolescents (Merimee et al. 1987).

Estrogen Receptors

The pubertal growth spurt and the subsequent cessation of growth are induced by sex hormones (Kaplan and Grumbach 1990). At the end of puberty, high levels of estrogen or testosterone cause the remaining epiphyseal plate cartilage to undergo hypertrophy. These cartilage cells grow, die, and are replaced by bone. Without any further cartilage formation, growth of these bones ceases, a process known as **growth plate closure**.

In conditions of precocious puberty, there is an initial growth spurt (making the individual taller than his or her peers), followed by the cessation of epiphyseal cell division (allowing that person's peers to catch up and surpass his or her height). In males, it was not thought that estrogen played any role in these events. However, in 1994, Smith and colleagues published the case history of a man whose growth was still linear despite his undergoing a normal puberty. His epiphyseal plates had not matured, and he still had proliferating chondrocytes at 28 years of age. His "bone age"—the amount of ephiphyseal cartilage he retained—was roughly half his chronological age. This person was found to lack any functional estrogen receptor. At present, at least three human males have been reported who either cannot make estrogens or who lack the estrogen receptor. All three are close to 7 feet tall and are still growing (Sharpe 1997). Therefore, estrogen plays a role in epiphyseal maturation in males as well as in females.

Thyroid hormone and parathyroid-related hormone are also important in regulating the maturation and hypertrophy program of the epiphyseal growth plate (Ballock and Reddi 1994). Thus, children with hypothyroidism are prone to developing growth plate disorders.

Extracellular Matrix Proteins

The extracellular matrix of the cartilage is also critical for the proper differentiation and organization of growth plate chondrocytes. Mutations that affect type XI collagen or the sulfation of cartilage proteoglycans can cause severe skeletal abnormalities. Mice with deficiencies of type XI collagen die at birth with abnormalities of limb, mandible, rib, and tracheal cartilage (Li et al. 1995). Failure to add sulfate groups to cartilage proteoglycans causes diastrophic dysplasia, a human dwarfism characterized by severe curvature of the spine, clubfoot, and deformed earlobes (Hästbacka et al. 1994). ∎

Osteoclasts

As new bone material is added peripherally from the internal surface of the periosteum, there is a hollowing out of the internal region to form the bone marrow cavity. This destruction of bone tissue is due to **osteoclasts**, multinucleated cells that enter the bone through the blood vessels (Kahn and Simmons 1975; Manolagas and Jilka 1995). Osteoclasts are probably derived from the same precursors as macrophage blood cells, and they dissolve both the inorganic and the protein portions of the bone matrix (Ash et al. 1980; Blair et al. 1986). Each osteoclast extends numerous cellular processes into the matrix and pumps out hydrogen ions onto the surrounding material, thereby acidifying and solubilizing it* (Figure 14.17; Baron et al. 1985, 1986). The blood vessels also import the blood-forming cells that will reside in the marrow for the duration of the organism's life. The number and activity of osteoclasts must be tightly regulated. If there are too many active osteoclasts, too much bone will be dissolved, and **osteoporosis** will result. Conversely, if not enough osteoblasts are produced, the bones are not hollowed out for the marrow, and **osteopetrosis** results (Tondravi et al. 1997).

> W E B S I T E **14.7 Osteoclast differentiation.** Hormones regulate the production of osteoclasts, and the hormonal changes of aging may cause osteoporosis by increasing the number of osteoclasts. The conversion of a macrophage stem cell into a osteoclast is regulated by osteoprotegerin and its ligand. It is thought that signals on osteoblasts instruct the progenitor cell to become an osteoclast.

*Given the physiology of the osteoclast, we can now appreciate H. L. Mencken's (1919) prescient intuition: "Life is a struggle, not against sin, not against the Money Power, not against malicious animal magnetism, but against hydrogen ions."

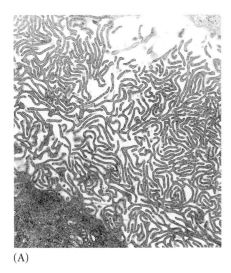
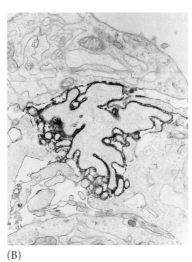
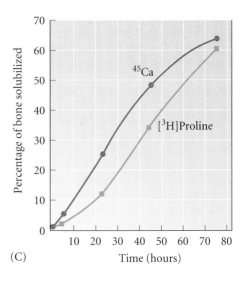

(A)   (B)   (C)

Figure 14.17
Osteoclast activity on the bone matrix. (A) Electron micrograph of the ruffled membrane of a chick osteoclast cultured on reconstituted bone matrix. (B) Section of ruffled membrane stained for the presence of an ATPase capable of transporting hydrogen ions from the cell. The ATPase is restricted to the membrane of the cell process. (C) Solubilization of inorganic and collagenous matrix components (as measured by the release of [^{45}Ca] and [^{3}H] proline, respectively) by 10,000 osteoclasts incubated on labeled bone fragments. (A and C from Blair et al. 1986; B from Baron et al. 1986, photograph courtesy of the authors.)

Intermediate Mesoderm

The intermediate mesoderm generates the urogenital system—the kidneys, the gonads, and their respective duct systems. Saving the gonads for our discussion of sex determination in Chapter 17, we will concentrate here on the development of the mammalian kidney.

Progression of kidney types

The importance of the kidney cannot be overestimated. As Homer Smith noted (1953), "our kidneys constitute the major foundation of our philosophical freedom. Only because they work the way they do has it become possible for us to have bone, muscles, glands, and brains."

While this statement may smack of hyperbole, the kidney is an incredibly intricate organ. Its functional unit, the **nephron**, contains over 10,000 cells and at least 12 different cell types, with each cell type located in a particular place in relation to the others along the length of the nephron.

The development of the mammalian kidney progresses through three major stages. The first two stages are transient; only the third and last persists as a functional kidney. Early in development (day 22 in humans; day 8 in mice), the **pronephric duct** arises in the intermediate mesoderm just ventral to the anterior somites. The cells of this duct migrate caudally, and the anterior region of the duct induces the adjacent mesenchyme to form the tubules of the initial kidney, the **pronephros** (Figure 14.18A). While the pronephric tubules

form functioning kidneys in fish and in amphibian larvae, they are not thought to be active in mammalian amniotes. In mammals, the pronephric tubules and the anterior portion of the pronephric duct degenerate, but the more caudal portions of the pronephric duct persist and serve as the central component of the excretory system throughout its development (Toivonen 1945; Saxén 1987). This remaining duct is often referred to as the **nephric** or **Wolffian duct**.

As the pronephric tubules degenerate, the middle portion of the nephric duct induces a new set of kidney tubules in the adjacent mesenchyme. This set of tubules constitutes the **mesonephros**, or mesonephric kidney (Figure 14.18B; Sainio and Raatikainen-Ahokas 1999). In some mammalian species, the mesonephros functions briefly in urine filtration, but in humans and rodents, it does not function as a working kidney. In humans, about 30 mesonephric tubules form, beginning around day 25. As more tubules are induced caudally, the anterior mesonephric tubules begin to regress through apoptosis (although in mice, the anterior tubules remain while the posterior ones regress: Figure 14.18C, D). During its brief existence, however, the mesonephros provides important developmental functions. First, as will be discussed in Chapter 15, it is the source of hematopoietic stem cells necessary for blood cell development (Medvinsky and Dzierzak 1996; Wintour et al. 1996). Second, in male mammals, some of the mesonephric tubules persist to become the sperm-carrying tubes (the vas deferens and efferent ducts) of the testes (see Chapter 17).

The permanent kidney of amniotes, the **metanephros**, is generated by some of the same components as the earlier, transient kidney types (Figure 14.18C). It is thought to originate through a complex set of interactions between epithelial and mesenchymal components of the intermediate mesoderm. In the first steps, the **metanephrogenic mesenchyme**

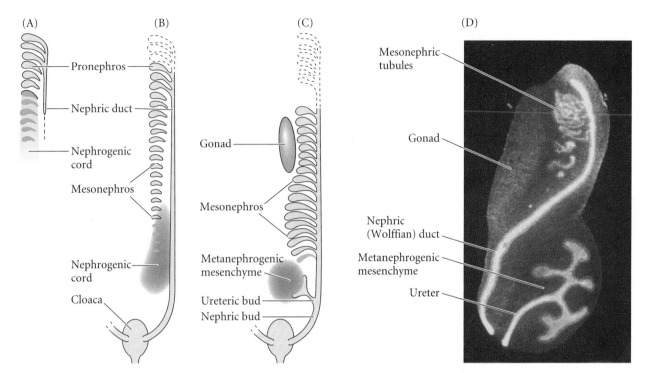

Figure 14.18
General scheme of development in the vertebrate kidney. (A) The original tubules, constituting the pronephros, are induced from the nephrogenic mesenchyme by the pronephric duct as it migrates caudally. (B) As the pronephros degenerates, the mesonephric tubules form. (C) The final mammalian kidney, the metanephros, is induced by the ureteric bud, which branches from the nephric duct. (D) The intermediate mesoderm of a 13-day mouse embryo showing the initiation of the metanephric kidney (bottom) while the mesonephros is still apparent. The duct tissue is stained with a fluorescent antibody to a cytokeratin found in the pronephric duct and its derivatives. (A–C after Saxén 1987; D courtesy of S. Vainio.)

forms in posteriorly located regions of the intermediate mesoderm, and it induces the formation of a branch from each of the paired nephric ducts. These epithelial branches are called the **ureteric buds**. These buds eventually separate from the nephric duct to become the ureters that take the urine to the bladder. When the ureteric buds emerge from the nephric duct, they enter the metanephrogenic mesenchyme. The ureteric buds induce this mesenchymal tissue to condense around them and differentiate into the nephrons of the mammalian kidney. As this mesenchyme differentiates, it tells the ureter bud to branch and grow.

Reciprocal interaction of kidney tissues

Thus, the two intermediate mesodermal tissues—the ureteric bud and the metanephrogenic mesenchyme—interact and reciprocally induce each other to form the kidney (Figure 14.19). The metanephrogenic mesenchyme causes the ureteric bud to elongate and branch. The tips of these branches induce the loose mesenchyme cells to form epithelial aggregates. Each aggregated nodule of about 20 cells will proliferate and differentiate into the intricate structure of the renal nephron. Each

nodule first elongates into a "comma" shape and then forms a characteristic S-shaped tube. Soon afterward, the cells of this epithelial structure begin to differentiate into regionally specific cell types, including the capsule cells, the podocytes, and the distal and proximal tubule cells.

While this is happening, the epithelializing nodules break down the basal lamina of the ureter bud ducts and fuse with them. This creates a connection between the ureteric bud and the newly formed tube, thereby enabling material to pass from one into the other (Bard et al. 2000). The tubes derived from the mesenchyme form the secretory nephrons of the functioning kidney, and the branched ureteric bud gives rise to the renal collecting ducts and to the ureter, which drains the urine from the kidney.

Clifford Grobstein (1955, 1956) documented this reciprocal induction in vitro. He separated the ureteric bud from the metanephrogenic mesenchyme and cultured them either individually or together. In the absence of mesenchyme, the ureteric bud does not branch. In the absence of the ureteric bud, the mesenchyme soon dies. When they are placed together, however, the ureteric bud grows and branches, and nephrons form throughout the mesenchyme (Figure 14.20).

The mechanisms of reciprocal induction

The induction of the metanephros can be viewed as a dialogue between the ureteric bud and the metanephrogenic mesenchyme. As the dialogue continues, both tissues are altered.

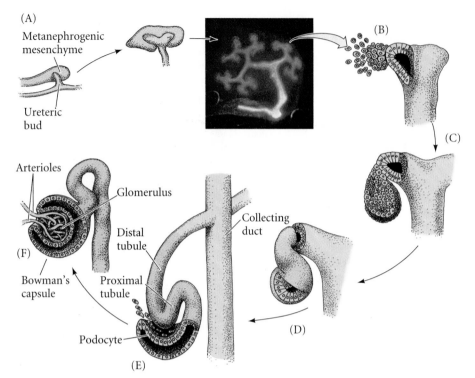

(A)

Metanephrogenic mesenchyme

Ureteric bud

(B)

(C)

Arterioles

Glomerulus

Distal tubule

Collecting duct

(F)

Bowman's capsule

Proximal tubule

Podocyte

(E)

(D)

Figure 14.19
Reciprocal induction in the development of the mammalian kidney. (A) As the ureteric bud enters the metanephrogenic mesenchyme, the mesenchyme induces the bud to branch. (B–F) At the tips of the branches, the epithelium induces the mesenchyme to aggregate and cavitate to form the renal tubules and glomeruli (where the blood from the arteriole is filtered). When the mesenchyme has condensed into an epithelium, it digests the basement membrane of the ureteric bud cells that induced it and connects to the ureteric bud epithelium. The mesenchyme becomes the nephron (renal tubules and Bowman's capsule), while the ureteric bud becomes the collecting duct for the urine. (After Saxén 1987.)

There appear to be at least eight sets of signals operating in the reciprocal induction of the metanephros.

STEP 1: FORMATION OF THE METANEPHROGENIC MESENCHYME.
Only the metanephrogenic mesenchyme has the competence to respond to the ureteric bud to form kidney tubules, and if induced by other tissues (such as embryonic salivary gland or neural tube tissue), the metanephrogenic mesenchyme will respond by forming kidney tubules and no other structures (Saxén 1970; Sariola et al. 1982). Thus, the metanephrogenic mesenchyme cannot become any tissue other than nephrons. Its competence to respond to ureteric bud inducers is thought to be regulated by a transcription factor called WT1, and if the metanephrogenic mesenchyme lacks this factor, the uninduced cells die (Kriedberg et al. 1993). In situ hybridization shows that WT1 is normally first expressed in the intermediate mesoderm prior to kidney formation and is then expressed in the developing kidney, gonad, and mesothelium (Pritchard-Jones et al. 1990; van Heyningen et al. 1990; Armstrong et al. 1992). Although the metanephrogenic mesenchyme appears homogeneous, it may contain both mesodermally derived tissue and some cells of neural crest origin (Le Douarin and Tiellet 1974; Sariola et al. 1989; Sainio et al. 1994).

Figure 14.20
Kidney induction observed in vitro. (A) An 11-day mouse metanephric rudiment includes both ureteric bud and metanephrogenic mesenchyme. (B) After the first day of culture, nephrons can be seen at the tips of the branching ureters. (C) The branching collecting ducts formed by the ureteric bud and the nephrons formed by the mesenchymal condensations at the tips of these buds can be clearly seen after 8 days of culture. (A and B from Saxén and Sariola 1987; C from Grobstein 1955; all photographs courtesy of the authors.)

(A)   (B)

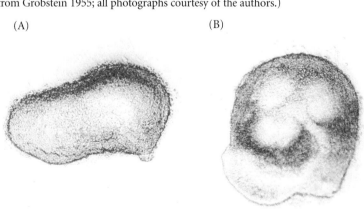

(C)

Renal tubules   Collecting ducts

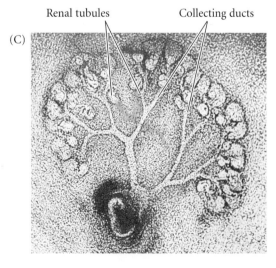

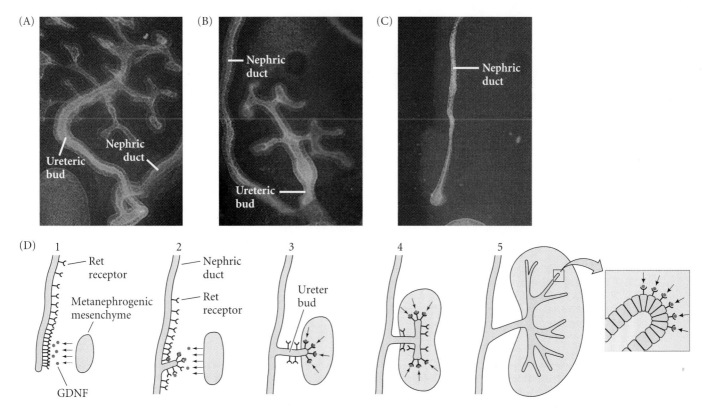

Figure 14.21

Ureteric bud growth is dependent on GDNF and its receptor. (A) The ureteric bud from a 11.5-day wild-type mouse embryonic kidney cultured for 72 hours has a characteristic branching pattern. (B) In embryonic mice heterozygous for the genes encoding GDNF, the size of the ureteric bud and the number and length of its branches are reduced. (C) In mouse embryos missing both copies of the *gdnf* gene, the ureteric bud does not form. (Scale bars = 100 μm.) (D) The receptors for GDNF are concentrated in the posterior portion of the nephric duct. GDNF secreted by the metanephrogenic mesenchyme stimulates the growth of the ureteric bud from this duct. At later stages, the GDNF receptor is found exclusively at the tips of the ureteric buds. (A–C from Pichel et al. 1996, photographs courtesy of J. G. Pichel and H. Sariola; D after Schuchardt et al. 1996.)

STEP 2: THE METANEPHROGENIC MESENCHYME SECRETES GDNF AND HGF TO INDUCE AND DIRECT THE URETERIC BUD. The second signal in kidney development is a set of diffusible molecules that cause the two ureteric buds to grow out from the nephric ducts. Recent research has shown that glial-derived neurotrophic factor (GDNF) is a critical component of this signal. GDNF is synthesized in the metanephrogenic mesenchyme, and mice whose *gdnf* genes were knocked out died soon after birth from their lack of kidneys (Moore et al. 1996; Pichel et al. 1996; Sánchez et al. 1996). The GDNF receptor (the c-Ret protein) is synthesized in the nephric ducts and later becomes concentrated in the growing ureteric buds (Figure 14.21; Schuchardt et al. 1996). Mice lacking the GDNF receptor also die of renal agenesis. Another protein synthesized by the metanephrogenic mesenchyme is hepatocyte growth factor (HGF; also known as scatter factor), and the receptor for HGF (the c-met protein) is made by the ureteric buds. Antibodies to HGF will block ureteric bud outgrowth in cultured kidney rudiments (Santos et al. 1994; Woolf et al. 1995). The synthesis of GDNF and HGF by the mes-

enchyme is thought to be regulated by the WT1 protein.

In another mouse mutation, the Danforth short-tail mutant, the ureteric bud is initiated but does not enter the metanephrogenic mesenchyme (Gluecksohn-Schoenheimer 1943). Here, too, the kidney does not form. This failure of the ureteric bud to grow has been correlated with the absence of Wnt11 expression in the tips of the ureteric bud. Wnt11 expression is maintained by proteoglycans made by the mesenchyme. It appears that once the ureteric bud has entered the mesenchyme, these mesenchymal proteoglycans stimulate its continued growth by maintaining the expression and secretion of Wnt11 (Davies et al. 1995; Kispert et al. 1996).

STEP 3: THE URETERIC BUD SECRETES FGF2 AND BMP7 TO PREVENT MESENCHYMAL APOPTOSIS. The third signal in kidney development is sent from the ureteric bud to the metanephrogenic mesenchyme, and it alters the fate of the mesenchyme cells. If left uninduced by the ureteric bud, the mesenchyme cells undergo apoptosis (Grobstein 1955; Koseki et al. 1992). However, if induced by the ureteric bud, the mesenchyme cells are rescued from the precipice of death and are converted into proliferating stem cells (Bard and Ross 1991; Bard et al. 1996). The factors secreted from the ureteric bud include fibroblast growth factor 2 (FGF2) and bone morphogenetic protein 7

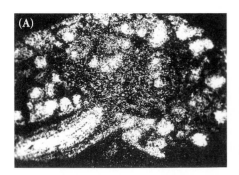

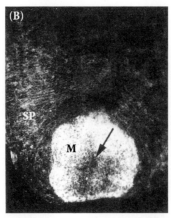

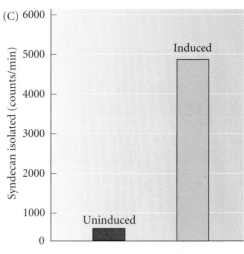

Figure 14.22

Syndecan expression in induced and uninduced kidney mesenchymes. (A) In situ hybridization localizing syndecan mRNA in the mesenchymal aggregates of a 15-day embryonic mouse kidney. Visualization of the autoradiograph is by dark-field illumination (so the RNA appears white). (B) Isolated kidney mesenchyme, M, induced by spinal cord tissue, SPC, shows intense syndecan expression when stained with fluorescent antibodies to syndecan. Uninduced mesenchyme does not. (C) The amount of syndecan (labeled with radioactive sulfur) isolated from induced kidney mesenchyme is tenfold greater than that isolated from a similar amount of uninduced mesenchyme. (After Vainio et al. 1992; photographs courtesy of S. Vainio.)

(BMP7). FGF2 has three modes of action in that it inhibits apoptosis, promotes the condensation of mesenchyme cells, and maintains the synthesis of WT1 (Perantoni et al. 1995). BMP7 has similar effects, and in the absence of BMP7, the mesenchyme of the kidney undergoes apoptosis (see Figure 4.21; Dudley et al. 1995; Luo et al. 1995).

STEP 4: LIF FROM THE URETERIC BUD INDUCES THE MESENCHYME CELLS TO AGGREGATE AND TO SECRETE WNT4. The ureteric bud also causes dramatic changes in the behavior of the metanephrogenic mesenchyme cells, converting them into an epithelium. The newly induced mesenchyme synthesizes two adhesive proteins, E-cadherin and syndecan (Figure 14.22; Vainio et al. 1989, 1992), which clump the mesenchyme together. These aggregated nodes of mesenchyme now synthesize an epithelial basal lamina containing type IV collagen and laminin. At the same time, the mesenchyme cells synthesize receptors for laminin, and this allows the aggregated cells to become an epithelium (Ekblom et al. 1994; Müller et al. 1997). The cytoskeleton also changes from one

*The mesenchymal to epithelial transition appears to be mediated by the expression of Pax2 in the newly induced mesenchyme cells. When antisense RNA to Pax2 prevents the translation of the Pax2 mRNA that is transcribed as a response to induction, the mesenchyme cells of cultured kidney rudiments fail to condense (Rothenpieler and Dressler 1993).

characteristic of mesenchyme cells to one typical of epithelia* (Ekblom et al. 1983; Lehtonen et al. 1985).

The transition from mesenchymal to epithelial organization may be mediated by several molecules. FGF2 is needed to induce the aggregation of the mesenchyme cells, but it is not capable of turning these aggregates into epithelial cells (Karavanova et al. 1996). Leukemia inhibitory factor (LIF) is able to convert these mesenchymal aggregates into kidney tubule epithelium (but only if they have been exposed to FGF2) (Figure 14.23; Barasch et al. 1999). The ureteric bud secretes FGFs and LIF, and the mesenchyme has receptors for these proteins.

Once induced, and after it has started to condense, the mesenchyme begins to secrete Wnt4, which acts in an autocrine fashion to complete the transition from mesenchymal

Figure 14.23

LIF induces kidney tubule formation. (A) Metanephrogenic mesenchyme treated with FGF2 or TGFα will form clumps but will not form epithelia. (B) Tubular epithelium is induced when FGF-treated mesenchyme is exposed to ureter bud secretions or LIF. (Photographs courtesy of J. Barasch.)

(A)

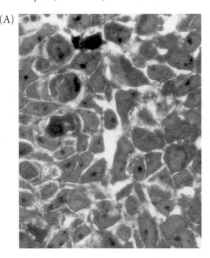

(B)

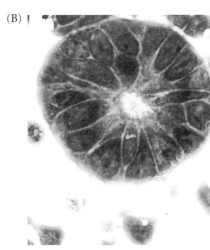

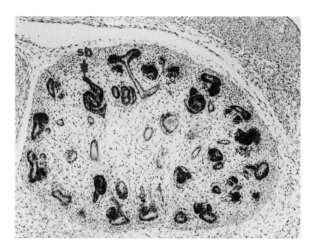

Figure 14.24
Lim-1 expression (dark stain) in the 19-day embryonic mouse kidney. In situ hybridization shows high levels of expression in the newly epithelialized comma-shaped and S-shaped bodies. (From Karavanov et al. 1998; photograph courtesy of A. A. Karavanov.)

mass to epithelium (Stark et al. 1994; Kispert et al. 1998). Wnt4 expression is found in the condensing mesenchyme cells, in the S-shaped tubes, and in the region where the newly epithelialized cells fuse with the ureteric bud tips. In mice lacking the *Wnt4* gene, the mesenchyme becomes condensed but does not form epithelia. Therefore, the ureteric bud induces the changes in the metanephrogenic mesenchyme by secreting FGFs and LIF, but these changes are mediated by the effects of the mesenchyme's secretion of Wnt4 on itself.

STEP 5: CONVERSION OF THE AGGREGATED CELLS INTO A NEPHRON. Not much is known about the transition from pre-tubular aggregate to nephron. One molecule that may be involved in this transition is the Lim-1 homeodomain transcription factor (Karavanov et al. 1998). This protein is found in the mesenchyme cells after they have condensed around the ureteric bud, and its expression persists in the developing nephron (Figure 14.24). Two other proteins that may be critical for the conversion of the aggregated cells into a nephron are polycystins 1 and 2. These proteins are the products of the genes whose loss-of-function alleles give rise to human polycystic kidney disease. Mice deficient in these genes have abnormal, swollen nephrons (Ward et al. 1996; van Adelsberg et al. 1997).

STEP 6: SIGNALS FROM THE MESENCHYME INDUCE THE BRANCHING OF THE URETERIC BUD. We still do not know for certain the identities of those molecules secreted by the metanephrogenic mesenchyme that are responsible for inducing the epithelial branching patterns of the ureteric bud. Recent evidence has implicated several paracrine factors in these events, and these factors probably work as pushes and pulls. Some factors may preserve the extracellular matrix surrounding the epithelium, thereby preventing branching from taking place. Conversely, other factors may cause the digestion of this extracellular matrix, permitting branching to occur. The first candidate for regulating kidney branching is GDNF (Sainio et al. 1997). Not only can GDNF from the mesenchyme induce the initial ureteric bud from the nephric duct; it can also induce secondary buds from the ureteric bud once the bud has entered the mesenchyme (Figure 14.25). The second candidate molecule is **transforming growth factor β1 (TGF-β1)**. When exogenous TGF-β1 is added to cultured kidneys, it prevents the epithelium from branching (Figure 14.26B; Ritvos et al. 1995). TGF-β1 is known to promote the synthesis of extracellular matrix proteins and to inhibit the

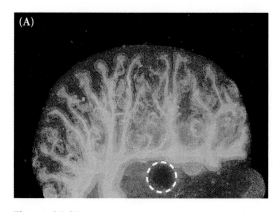
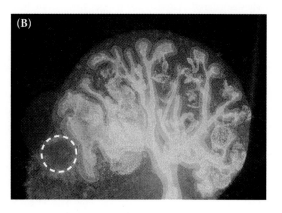

Figure 14.25
The effect of GDNF on the branching of the ureteric epithelium. The ureteric bud and its branches are stained orange (with antibodies to cytokeratin 18), while the nephrons are stained green (with antibodies to nephron brush border antigens). (A) 13-day embryonic mouse kidney cultured 2 days with a control bead (circle) has a normal branching pattern. (B) A similar kidney cultured 2 days with a GDNF-soaked bead shows a distorted pattern, as new branches are induced in the vicinity of the bead. (From Sainio et al. 1997; photographs courtesy of K. Sainio.)

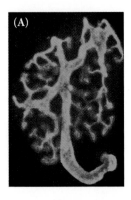

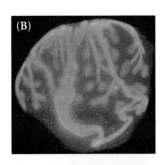

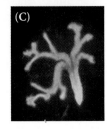

Figure 14.26
The effect of TGF-β1 and activin on the morphogenesis of kidney epithelium. (A) An 11-day mouse kidney cultured for 4 days in control medium has a normal branching pattern. (B) An 11-day mouse kidney cultured in TGF-β1 shows no branching until reaching the periphery of the mesenchyme, and the branches formed are elongated. (C) An 11-day mouse kidney cultured in activin shows a distortion of branching, suggesting a failure to stabilize branches after they have been made. (From Ritvos et al. 1995.)

metalloproteinases that can digest these matrices (Penttinen et al. 1988; Nakamura et al. 1990). It is possible that TGF-β1 stabilizes branches once they form. A third molecule that may be important in epithelial branching is **activin**. When activin is added exogenously to embryonic mouse kidney rudiments, it severely distorts the normal branching pattern (Figure 14.26C; Ritvos et al. 1995). The epithelial cells do not die and are still capable of inducing the mesenchyme cells to form nephrons, but the branches are grossly disordered. Activin may trigger the

digestion of the extracellular matrix at the site of a new branch, and adding it exogenously may cause the breakdown of the extracellular matrix throughout the epithelium.

STEPS 7 AND 8. INDUCTION OF MESENCHYMAL STEM CELLS BY THE URETERIC BUD, AND THE DIFFERENTIATION OF THE NEPHRON AND GROWTH OF THE URETERIC BUD. After the initial interactions create the first pretubular aggregates, the metanephrogenic mesenchyme cells near the kidney border begin to proliferate to form stem cells. These stem cells can interact with the ureteric bud branches to form new nephrons, or they can produce stromal cells. The stromal cells migrate to the central portion of the kidney and produce factors (as yet unknown) that (1) enable the continued growth of the ureteric bud and (2) stimulate the differentiation of the nephron into the convoluted renal tubules, Henle's loop, the glomerulus, and the juxtaglomerular apparatus.

The transcription factor BF-2 is synthesized in these stromal cells. When it is knocked out in mouse embryos, the resulting kidney lacks a branched ureteric tree (it branches only three or four times instead of the normal seven or eight, resulting in an 8- to 16-fold reduction in the number of branches), and the aggregates do not differentiate into nephrons (Hatini et al. 1996). So it appears that the factors necessary for ureteric epithelial growth and nephron differentiation are synthesized by the stromal cells and are regulated by transcription factor BF-2.

The stromal cells have been found to secrete FGF7, a growth factor whose receptor is found on the ureteric bud. FGF7 is seen to be critical for maintaining ureteric epithelial growth, and consequently, ensuring an appropriate number of nephrons in the kidney (Qiao et al. 1999).

In the developing kidney, we see an epitome of the reciprocal interactions needed to form an organ. We also see that we have only begun to understand how organs form.

Snapshot Summary: Paraxial and Intermediate Mesoderm

1. The paraxial mesoderm forms blocks of tissue called somites. Somites give rise to three major divisions: the dermatome, the myotome, and the sclerotome.

2. The dermatome of the somite forms the back dermis. The sclerotome of the somite forms the vertebral cartilage. In thoracic vertebrae, the sclerotome cells also form the proximal portions of the ribs.

3. The epaxial myotome forms the back musculature. The hypaxial myotome forms the muscles of the body wall, limb, and tongue.

4. Somites are formed from the segmental plate (unsegmented mesoderm) by a combination of proteins. The Hairy protein, a transcription factor, appears to specify the somite

boundaries. Notch and Eph receptor systems may be involved in the separation of the somites from the unsegmented paraxial mesoderm. N-cadherin and fibronectin appear to be important in causing these cells to become epithelial.

5. The somite regions are specified by paracrine factors secreted by neighboring tissue. The sclerotome is specified to a large degree by the Sonic hedgehog protein, secreted by the notochord and floor plate cells. The dermatome is specified by neurotrophin-3, secreted by the roof plate cells of the neural tube.

6. The two myotome regions are specified by different factors. The epaxial myotome is specified by Wnt proteins from the dorsal neural tube. The hypaxial myotome is specified by

BMP4 (and perhaps other proteins) secreted by the lateral plate mesoderm. In both instances, myogenic bHLH transcription factors are induced in the cells that will become muscles—the hypaxial and epaxial myoblasts.

7. Muscle formation involves the myoblasts ceasing to divide, aligning themselves, and fusing.

8. The major lineages that form the skeleton are the somites (axial skeleton), lateral plate mesoderm (appendages), and neural crest (skull and face).

9. There are two major types of ossification. Intramembranous ossification occurs in the skull bones and in turtle shells. Here, mesenchyme is directly converted into bone. In endochondral ossification, the mesenchyme cells become cartilage. These cartilagenous models are later become replaced by bone cells.

10. The replacement of cartilage by bone during endochondral ossification depends upon the mineralization of the cartilage matrix.

11. In the long bones of humans and other mammals, the ends contain cartilagenous regions called epiphyseal growth plates. The cartilage in these regions proliferate to make the bone bigger. Eventually, the cartilage is replaced by bone and growth stops.

12. The hollowing out of bone for the bone marrow is accomplished by osteoclasts. Osteoclasts continually remodel bone throughout a person's lifetime.

13. The intermediate mesoderm generates the kidneys and gonads.

14. The metanephric kidney of mammals is formed by the reciprocal interactions of the metanephrogenic mesenchyme and a branch of the nephric duct called the ureter bud.

15. The metanephrogenic mesenchyme becomes competent to become kidney tubules by expressing WT1. WT1 is also thought to enable that mesenchyme to secrete GDNF and HGF. These two factors are secreted by the mesoderm and induce the formation of the ureteric bud.

16. The ureteric bud secretes FGF2 and BMP7 to prevent apoptosis in the kidney mesenchyme. Without these factors, the kidney mesenchyme dies. FGF2 also makes the mesenchyme competent to respond to LIF.

17. The ureter bud also secretes LIF, and this protein induces the competent kidney mesenchyme to become epithelial tubules. As they form the kidney tubules, the cells secrete Wnt4, which promotes and maintains their epithelialization.

18. The condensing mesenchyme secretes paracrine factors (most of them members of the TGF-β superfamily) that mediate the branching of the ureter bud. The branching depends upon the extracellular matrix of the epithelium.

Literature Cited

Armstrong, J. F., K. Pritchard-Jones, W. A. Bickmore, N. D. Hastie and J. B. L. Bard. 1992. The expression of the Wilms' tumor gene, *WT-1*, in the developing mammalian embryo. *Mech. Dev.* 40: 85–97.

Ash, P. J., J. F. Loutit and K. M. S. Townsend. 1980. Osteoclasts derived from haematopoietic stem cells. *Nature* 283: 669–670.

Ballock, R. T. and A. H. Reddi. 1994. Thyroxine is the serum factor that regulates morphogenesis of columnar cartilage from isolated chondrocytes in chemically defined medium. *J. Cell Biol.* 126: 1311–1318.

Barasch, J. and 10 others. 1999. Mesenchymal to epithelial conversion in rat metanephros is induced by LIF. *Cell* 99: 377–386.

Bard, J. B. L. and A. S. A. Ross. 1991. LIF, the ES cell inhibition factor, reversibly blocks nephrogenesis in cultured mouse kidney rudiments. *Development* 113: 193–198.

Bard, J. B. L., J. A. Davies, I. Karavanova, E. Lehtonen, H. Sariola and S. Vainio. 1996. Kidney development: the inductive interactions. *Semin. Cell Dev. Biol.* 7: 195–202.

Bard, J. B. L., A. Gordon, L. Sharp and W. Sellers. 2000. The early stages of nephron formation in the developing mouse kidney. In press.

Barnes, G. L., C. W. Hsu, B. D. Mariani and R. S. Tuan. 1997. Cloning and characterization of chicken *paraxis*: A regulator of paraxial mesoderm development and somite formation. *Dev. Biol.* 189: 95–111.

Baron, R., L. Neff, D. Louvard and P. J. Courtoy. 1985. Cell mediated extracellular acidification and bone resorption: Evidence for a low pH in resorbing lacuna and localization of a 100-kD lysosomal membrane protein at the osteoclast ruffled border. *J. Cell Biol.* 101: 2210–2222.

Baron, R., L. Neff, C. Roy, A. Boisvert and M. Caplan. 1986. Evidence for a high and specific concentration of (Na⁺,K⁺) ATPase in the plasma membrane of the osteoclast. *Cell* 46: 311–320.

Bellus, G. A. and 8 others. 1995. A recurrent mutation in the tyrosine kinase domain of fibroblast growth factor receptor 3 causes hypochondroplasia. *Nature Genet.* 10: 357–359.

Bishop-Calame, S. 1966. Étude experimentale de l'organogenese du systéme urogénital de l'embryon de poulet. *Arch. Anat. Microsc. Morphol. Exp.* 55: 215–309.

Blair, H. C., A. J. Kahn, E. C. Crouch, J. J. Jeffrey and S. L. Teitelbaum. 1986. Isolated osteoclasts resorb the organic and inorganic components of bone. *J. Cell Biol.* 102: 1164–1172.

Bloom, W. and D. W. Fawcett. 1975. *Textbook of Histology*, 10th Ed. Saunders, Philadelphia.

Boettiger, D., M. Enomoto-Iwamoto, H. Y. Yoon, U. Hofer, A. S. Menko and R. Chiquet-Ehrismann. 1995. Regulation of integrin a5b1 affinity during myogenic differentiation. *Dev. Biol.* 169: 261–272.

Brand-Saberi, B., J. Wilting, C., Ebensperger and B. Christ. 1996. The formation of somite compartments in the avian embryo. *Int. J. Dev. Biol.* 40: 411–420.

Brill, G., N. Kahane, C. Carmeli, D. von Schack, Y.-A. Barde and C. Kalcheim. 1995. Epithelial-mesenchymal conversion of dermatome progenitors requires neural tube-derived signals: Characterization of the role of neurotrophin-3. *Development* 121: 2583–2594.

Bruder, S. P. and A. I. Caplan. 1989. Cellular and molecular events during embryonic bone development. *Connect. Tiss. Res.* 20: 65–71.

Burgess, R., P. Cserjesi, K. L. Ligon and E. N. Olson. 1995. *Paraxis*: A basic helix-loop-helix protein expressed in paraxial mesoderm and developing somites. *Dev. Biol.* 168: 296–306.

Burgess, R., Rawls., D. Brown, A. Bradley and E. N. Olson. 1996. Requirement of the *paraxis* gene for somite formation and musculoskeletal patterning. *Nature* 384: 570–573.

Chen, C.-M., N. Kraut, M. Groudine and H. Weintraub. 1996. I-mf, a novel myogenic repressor, interacts with members of the MyoD family. *Cell* 86: 731–741.

Chen, Q., D. M. Johnson, D. R. Haudenschild and P. F. Goetinck. 1995. Progression and recapitulation of the chondrocyte differentiation program: Cartilage matrix protein is a marker for cartilage maturation. *Dev. Biol.* 172: 293–306.

Christ, B. and C. P. Ohrdahl. 1995. Early stages of chick somite development. *Anat. Embryol.* 191: 381–396.

Cossu, G., R. Kelly, S. Tajbakhsh, S. Di Donna, E. Vivarelli and M. Buckingham. 1996a. Activation of different myogenic pathways: *myf-5* is induced by the neural tube and MyoD by the dorsal ectoderm in mouse paraxial mesoderm. *Development* 122: 429–437.

Cossu, G., S. Tajbakhsh and M. Buckingham. 1996b. How is myogenesis initiated in the embryo? *Trends Genet.* 12: 218–223.

Cserjesi, P. and 7 others. 1995. A basic helix-loop-helix protein that prefigures skeletal formation during mouse embryogenesis. *Development* 121: 1099–1110.

David, J. D., W. M. See and C. A. Higginbotham. 1981. Fusion of chick embryo skeletal myoblasts: Role of calcium influx preceding membrane union. *Dev. Biol.* 82: 297–307.

Davies, J. A., M. Lyon, J. Gallagher and D. R. Garrod. 1995. Sulphated proteoglycan is required for collecting duct growth and branching but not nephron formation during kidney development. *Development* 121: 1507–1517

Deng, C., A. Wynshaw-Boris, F. Zhou, A. Kuo and P. Leder. 1996. Fibroblast growth factor receptor-3 is a negative regulator of bone growth. *Cell* 84: 911–921.

Dickman, S. 1997. No bones about a genetic switch for bone development. *Science* 276: 1502.

Dietrich, S., F. R. Schubert, C. Healy, P. T., Sharpe and A. Lumsden. 1998. Specification of hypaxial musculature. *Development* 125: 2235–2249.

Ducy, P., R. Zhang, V. Geoffroy, A. L., Ridall and G. Karsenty. 1997. Osf2/Cba1: A transcriptional activator of osteoblast differentiation. *Cell* 89: 747–754.

Dudley, A. T., K. M. Lyons and E. J. Robertson. 1995. A requirement for bone morphogenesis protein-7 during development of the mammalian kidney and eye. *Genes Dev.* 9: 2795–2807.

Durbin, L. and 8 others. 1998. Eph signaling is required for segmentation and differentiation of the somites. *Genes Dev.* 12: 3096–3109.

Ekblom, P., I. Thesleff, L. Saxén, A. Miettinen and R. Timpl. 1983. Transferrin as a fetal growth factor: Acquisition of responsiveness related to embryonic induction. *Proc. Natl. Acad. Sci. USA* 80: 2651–2655.

Ekblom, P. and 8 others. 1994. Role of mesenchymal nidogen for epithelial morphogenesis in vitro. *Development* 120: 2003–2014.

Fan, C. M. and M. Tessier-Lavigne. 1994. Patterning of mammalian somites by surface ectoderm and notochord: Evidence for sclerotome induction by a hedgehog homolog. *Cell* 79: 1175–1186.

Gluecksohn-Schoenheimer, S. 1943. The morphological manifestations of a dominant mutation in mice affecting tail and urogenital system. *Genetics* 28: 341–348.

Grobstein, C. 1955. Induction interaction in the development of the mouse metanephros. *J. Exp. Zool.* 130: 319–340.

Grobstein, C. 1956. Trans-filter induction of tubules in mouse metanephrogenic mesenchyme. *Exp. Cell Res.* 10: 424–440.

Hall, B. K. 1988. The embryonic development of bone. *Am. Sci.* 76: 174–181.

Hall, B. K. and T. Miyake. 1995. Divide, accumulate, differentiate: Cell condensations in skeletal development revisited. *Int. J. Dev. Biol.* 39: 881–893.

Hästbacka, J., A. de la Chapelle and M. M. Mahtani. 1994. The diastrophic dysplasia gene encodes a novel sulfate transporter: Positional cloning by fine-structure linkage disequilibrium mapping. *Cell* 78: 1073–1087.

Hatini, V., S. O. Huh, D. Herzlinger, V. C. Soares and E. Lai. 1996. Essential role of stromal morphogenesis in kidney morphogenesis revealed by targeted disruption of winged helix transcription factor, BF-2. *Genes Dev.* 10: 1467–1478.

Hatori, M., K. J. Klatte, C. C., Teixeira and I. M. Shapiro. 1995. End labeling studies of fragmented DNA in avian growth plate: Evidence for apoptosis in terminally differentiated chondrocytes. *J. Bone Miner. Res.* 10: 1960–1968.

Hatta, K., S. Takagi, H. Fujisawa and M. Takeichi. 1987. Spatial and temporal expression of N-cadherin cell adhesion molecule correlated with morphogenetic processes of chicken embryo. *Dev. Biol.* 120: 218–227.

Horton, W. A. 1990. The biology of bone growth. *Growth Genet. Horm.* 6(2): 1–3.

Huang, R., Q. Zhi, C. Schmidt, J. Wilting, B. Brand-Saberi and B. Christ. 2000. Sclerotomal origin of the ribs. *Development* 127: 527–532.

Ikeya, M. and S. Takada. 1998. Wnt signaling from the dorsal neural tube is required for the formation of the medial dermomyotome. *Development* 125: 4969–4976.

Johnson, R. L., E. Laufer, R. D. Riddle and C. Tabin. 1994. Ectopic expression of *Sonic hedgehog* alters dorsal-ventral patterning of somites. *Cell* 79: 1165–1173.

Jouve, C., I. Palmeirim, D. Henrique, J. Beckers, A. Gossler, D. Ish-Horowicz, and O. Pourquié. 2000. Notch signalling is required for cyclic expression of the hairy-like gene HES1 in the presomitic mesoderm. *Development* 127: 1421–1429.

Kahn, A. J. and D. J. Simmons. 1975. Investigation of cell lineage in bone using a chimaera of chick and quail embryonic tissue. *Nature* 258: 325–327.

Kaplan, S. L. and M. M. Grumbach. 1990. Pathophysiology and treatment of sexual precocity. *J. Clin. Endocrinol. Metab.* 71: 785–789.

Karavanov, A. A., I. Karavanova, A., Perantoni and I. B. Dawid. 1998. Expression pattern of the rat *Lim-1* homeobox gene suggests a dual role during kidney development. *Int. J. Dev. Biol.* 42: 61–66.

Karavanova, I. D., L. F. Dove, J. H. Resau and A. O. Perantoni. 1996. Conditioned medium from a rat ureteric bud cell line in combination with bFGF induces complete differentiation of isolated metanephric mesenchyme. *Development* 122: 4159–4167.

Kato, N. and H. Aoyama. 1998. Dermamyomal origin of the ribs as revealed by extirpation and transplantation experiments in chick and quail embryos. *Development* 125: 3437–3443.

Kieny, M., A. Mauger and P. Segel. 1972. Early regionalization of somitic mesoderm as studied by the development of the axial skeleton of the chick embryo. *Dev. Biol.* 28: 142–161.

Kispert, A., S. Vainio, L. Shen, D. R. Rowitch and A. P. McMahon. 1996. Proteoglycans are required for maintenance of Wnt-11 expression in the ureter tips. *Development* 122: 3627–3637.

Kispert, A., S. Vainio and A. P. McMahon. 1998. Wnt-4 is a mesenchymal signal for epithelial transformation of metanephric mesenchyme in the developing kidney. *Development* 125: 4225–4234.

Knudsen, K. A. 1985. The calcium-dependent myoblast adhesion that precedes cell fusion is mediated by glycoproteins. *J. Cell Biol.* 101: 891–897.

Knudsen, K. A., S. A. McElwee and L. Myers. 1990. A role for the neural cell adhesion molecule, N-CAM, in myoblast interaction during myogenesis. *Dev. Biol.* 138: 159–168.

Komori, T. and 14 others. 1997. Targeted disruption of *Cba1* results in a complete lack of bone formation owing to maturational arrest of osteoblasts. *Cell* 89: 755–764.

Konigsberg, I. R. 1963. Clonal analysis of myogenesis. *Science* 140: 1273–1284.

Koseki, C., D. Herzlinger and Q. Al-Auqati. 1992. Apoptosis in metanephric development. *J. Cell Biol.* 119: 1327–1333.

Kreidberg, J. A., H. Sariola, J. M. Loring, M. Maeda, J. Pelletier, D. Housman and R. Jaenisch. 1993. WT-1 is required for early kidney development. *Cell* 74: 679–691.

Langman, J. 1981. *Medical Embryology*, 4th Ed. Williams & Wilkins, Baltimore.

Lash, J. W. and K. M. Yamada. 1986. The adhesion recognition signal of fibronectin: A possible trigger mechanism for compaction during somitogenesis. *In* R. Bellairs, D. H. Ede and J. W. Lash (eds.), *Somites in Developing Embryos*. Plenum, New York, pp. 201–208.

Lassar, A. B., J. N. Buskin, D. Lockshon, R. L. Davis, S. Apone, S. D. Hauschka and H. Weintraub. 1989. MyoD is a sequence-specific DNA binding protein requiring a region of *myc* homology to bind to the muscle creatine kinase enhancer. *Cell* 58: 823–831.

Le Douarin, N. and M.-A. Tiellet. 1974. Experimental analysis of the migration and differentiation of neuroblasts of the autonomic nervous system and of neuroectodermal derivatives, using a biological cell marking technique. *Dev. Biol.* 41: 162–184.

Lehtonen, E., I. Virtanen and L. Saxén. 1985. Reorganization of the intermediate cytoskeleton in induced mesenchyme cells is independent of tubule morphogenesis. *Dev. Biol.* 108: 481–490.

Li, Y. and 16 others. 1995. A fibrillar collagen gene, *Col1a1*, is essential for skeletal morphogenesis. *Cell* 80: 423–430.

Luo, G., C. Hofmann, A. L. J. J. Bronckers, M. Sohocki, A. Bradley and G. Karsenty. 1995. BMP-7 is an inducer of nephrogenesis and is also required for eye development and skeletal patterning. *Genes Dev.* 9: 2808–2820.

Manolagas, S. and R. L. Jilka. 1995. Bone marrow, cytokines, and bone remodeling. *N. Engl. J. Med.* 332: 305–310.

Marcelle, C., M. R. Stark and M. Bronner-Fraser. 1997. Coordinate actions of BMPs, Wnts, Shh, and Noggin mediate patterning of the dorsal somite. *Development* 124: 3955–3963.

Maroto, M., R. Reshef, A. E. Münsterberg, S. Koester, M. Goulding and A. B. Lassar. 1997. Ectopic Pax-3 activates *MyoD* and *Myf-5* expression in embryonic mesoderm and neural tissue. *Cell* 89: 139–148.

Medvinsky, A. and E. Dzierzak. 1996. Definitive hematopoiesis is autonomously initiated by the AGM region. *Cell* 86: 897–906.

Mencken, H. L. 1919. Exeunt omnes. *Smart Set* 60: 138–145.

Menko, A. S. and D. Boettiger. 1987. Occupation of the extracellular matrix integrin is a control point for myogenic differentiation. *Cell* 51: 51–57.

Merimee, T. J., J. Zapf, B. Hewlett and L. L. Cavalli-Sforza. 1987. Insulin-like growth factors in pygmies: The role of puberty in determining final stature. *N. Engl. J. Med.* 316: 906–911.

Mintz, B. and W. W. Baker. 1967. Normal mammalian muscle differentiation and gene control of isocitrate dehydrogenase synthesis. *Proc. Natl. Acad. Sci. USA* 58: 592–598.

Moore, M. W. and 9 others. 1996. Renal and neuronal abnormalities in mice lacking GDNF. *Nature* 382: 76–79.

Müller, U., D. Wang, S. Denda, J. Menesis, R. Pedersen and L. Reichardt. 1997. Integrin alpha-8/beta-1 is critically important for epithelial-mesenchymal interactions during kidney morphogenesis. *Cell* 88: 603–613.

Mundlos, S. and 13 others. 1997. Mutations involving the transcription factor CBFA1 cause cleiocranial dysplasia. *Cell* 89: 773–779.

Münsterberg, A. E., J. Kitajewski, D. A. Bumcroft, A. P. McMahon and A. B. Lassar. 1995. Combinatorial signaling by sonic hedgehog and Wnt family members induce myogenic bHLH gene expression in the somite. *Genes Dev.* 9: 2911–2922.

Nakamura, T., S. Okuda, D. Miller, E. Ruoslahti and W. Border. 1990. Transforming factor-β (TGF-β) regulates production of extracellular matrix (ECM) components by glomerular epithelial cells. *Kidney Int.* 37, 221.

Nameroff, M. and E. Munar. 1976. Inhibition of cellular differentiation by phospholipase C. II. Separation of fusion and recognition among myogenic cells. *Dev. Biol.* 49: 288–293.

Nilsson, A., J. Isgaard, A. Lindahl, A. Dahlström, A. Skottner and O. G. P. Isaksson. 1986. Regulation by growth hormone of number of chondrocytes containing IGF-I in rat growth plate. *Science* 233: 571–574.

Nowicki, J. L. and A. C. Burke. 1999. Testing *Hox* genes by surgical manipulation. *Dev. Biol.* 210: 228.

Oberlender, S. A. and R. S. Tuan. 1994. Expression and functional involvement of N-cadherin in embryonic limb chondrogenesis. *Development* 120: 177–187.

Ordahl, C. P. 1993. Myogenic lineages within the developing somite. *In* M. Bernfield (ed.), *Molecular Basis of Morphogenesis*. Wiley-Liss, New York, pp. 165–170.

Ordahl, C. P. and N. Le Douarin. 1992. Two myogenic lineages within the developing somite. *Development* 114: 339–353.

Ordahl, C. P. and B. A. Williams. 1998. Knowing chops from chuck: Roasting MyoD redundancy. *BioEssays* 20: 357–362.

Ostrovsky, D., C. M. Cheney, A. W. Seitz and J. W. Lash. 1984. Fibronectin distribution during somitogenesis in the chick embryo. *Cell Diff.* 13: 217–223.

Otto, F. and 11 others. 1997. *Cba1*, a candidate gene for cleidocranial dysplasia syndrome, is essential for osteoblast differentiation and bone development. *Cell* 89: 765–771.

Palmeirim, I., D. Henrique, D. Ish-Horowicz and O. Pourquié. 1997. Avian *hairy* gene expression identifies a molecular clock linked to vertebrate segmentation and somitogenesis. *Cell* 91: 639–648.

Park, W.-J., G. Bellus and E. W. Jabs. 1995. Mutations in fibroblast growth factor receptors:

Phenotypic consequences during eukaryotic development. *Am. J. Hum. Genet.* 57: 748–754.

Penttinen, R. P., S. Kobayashi and P. Bornstein. 1988. Transforming growth factor-β increases mRNA for matrix proteins in the presence and in the absence of the changes in mRNA stability. *Proc. Natl. Acad. Sci. USA* 85: 1105–1108.

Perantoni, A. O., L. F. Dove and I. Karavanova. 1995. Basic fibroblast growth factor can mediate the early inductive events in renal development. *Proc. Natl. Acad. Sci USA* 92: 4696–4700.

Pichel, J. G. and 11 others. 1996. Defects in enteric innervation and kidney development in mice lacking GDNF. *Nature* 382: 73–76.

Piette, J., J.-L. Bessereau, M. Huchet and J.-P. Changeaux. 1990. Two adjacent MyoD1-binding sites regulate expression of the acetylcholine receptor α-subunit gene. *Nature* 345:353–355.

Pourquié, O. and 9 others. 1996. Lateral and axial signals involved in somite patterning: A role for BMP4. *Cell* 84: 461–471.

Pritchard-Jones, K. and 11 others. 1990. The candidate Wilms' tumour gene is involved in genitourinary development. *Nature* 346: 194–197.

Qiao, J., R. Uzzo, T. Obara-Ishihara, L. Degenstein, E. Fuchs and D. Herzlinger. 1999. FGF-7 modulates ureteric bud growth and nephron number in the developing kidney. *Development* 126: 547–554.

Rajpurohit, R., K. Mansfield, K. Ohyama, D. Ewert and I. M. Shapiro. 1999. Chondrocyte death is linked to development of a mitochondrial membrane permeability transition in the growth plate. *J. Cell. Physiol.* 179: 287–296.

Ritvos, O., T. Tuuri, M. Erämaa, K. Sainio, K. Hilden, L. Saxén and S. F. Gilbert. 1995. Activin disrupts epithelial branching morphogenesis in developing murine kidney, pancreas, and salivary gland. *Mech. Dev.* 50: 229–245.

Rothenpieler, U. W. and G. R. Dressler. 1993. Pax-2 is required for mesenchyme-to-epithelium conversion during kidney development. *Development* 119: 711–720.

Sainio, K. and A. Raatikainen-Ahokas. 1999. Mesonephric kidney: A stem cell factory. *Int. J. Dev. Biol.* 43: 435–439.

Sainio, K., D. Nonclercq, M. Saarma, J. Palgi, L. Saxén and H. Sariola. 1994. Neuronal characteristics of embryonic renal stroma. *Int. J. Dev. Biol.* 38: 77–84.

Sainio, K. and 10 others. 1997. Glial cell derived neurotrophic factor is required for bud initiation from ureteric epithelium. *Development* 124: 4077–4087.

Sánchez, M. P., I. Silos-Santiago, J. Frisén, B. He, S. A. Lira and M. Barbacid. 1996. Renal agenesis and the absence of enteric neurons in mice lacking GDNF. *Nature* 382: 70–73.

Santos, O. F. P., E. J. G. Baras, X.-M. Yang, K. Matsumoto, T. Nakamura, M. Park and S. K. Nigam. 1994. Involvement of hepatocyte growth factor in kidney development. *Dev. Biol.* 163: 525–529.

Sariola, H., P. Ekblom and L. Saxén. 1982. Restricted developmental options of the metanephric mesenchyme. *In* M. Burger and R. Weber (eds.), *Embryonic Development*, Part B: *Cellular Aspects*. Alan R. Liss, New York, pp. 425–431.

Sariola, H., K. Holm-Sainio and S. Henke-Fahle. 1989. The effect of neuronal cells on kidney differentiation. *Int. J. Dev. Biol.* 33: 149–155.

Saxén, L. 1970. Failure to demonstrate tubule induction in heterologous mesenchyme. *Dev. Biol.* 23: 511–523.

Saxén, L. 1987. *Organogenesis of the Kidney.* Cambridge University Press, Cambridge.

Saxén, L. and H. Sariola. 1987. Early organogenesis of the kidney. *Pediatr. Nephrol.* 1: 385–392.

Schuchardt, A., V. D-Agati, V. Pachnis and F. Constantini. 1996. Renal agenesis and hypodysplasia in *ret-k⁻* mutant mice result from defects in ureteric bud development. *Development* 122: 1919–1929.

Shainberg, A., G. Yagil and D. Yaffe. 1969. Control of myogenesis in vitro by Ca^{2+} concentration in nutritional medium. *Exp. Cell Res.* 58: 163–167.

Shapiro, I., K. DeBolt, V. Funanage, S. Smith and R. Tuan. 1992. Developmental regulation of creatine kinase activity in cells of the epiphyseal growth plate. *J. Bone Miner. Res.* 7: 493–500.

Sharpe, R. M. 1997. Do males rely on female hormones? *Nature* 390: 447–448.

Smith, C. A. and R. S. Tuan. 1996. Functional involvement of Pax-1 in somite development: Somite dysmorphogenesis in chick embryos treated with *Pax-1* paired-box antisense oligonucleotide. *Teratology* 52: 333–345.

Smith, E. P. and 8 others. 1994. Estrogen resistance caused by a mutation in the estrogen-receptor gene in a man. *N. Engl. J. Med.* 331: 1056–1061.

Smith, H. 1953. *From Fish to Philosopher.* Little, Brown, Boston.

Sosic, D., B. Brand-Saberi, C. Schmidt, B. Christ and E. Olson. 1997. Regulation of *paraxis* expression and somite formation by ectoderm- and neural tube-derived signals. *Dev. Biol.* 185: 229–243.

Stark, K., S. Vainio, G. Vassileva and A. P. McMahon. 1994. Epithelial transformation of metanephric mesenchyme in the developing kidney regulated by Wnt-4. *Nature* 372: 679–683.

Stern, H. M., A. M. C. Brown and S. D. Hauschka. 1995. Myogenesis in paraxial mesoderm: Preferential induction by dorsal neural tube and by cells expressing Wnt-1. *Development* 121: 3675–3686.

Tajbakhsh, S. and R. Spörle. 1998. Somite development: Constructing the vertebrate body. *Cell* 92: 9–16.

Tajbakhsh, S., D. Rocancourt, G. Cossu and M. Buckingham. 1997. Redefining the genetic hierarchies controlling skeletal myogenesis: Pax-3 and Myf-5 act upstream of MyoD. *Cell* 89: 127–138.

Tavormina, P. L. and 9 others. 1995. Thanatophoric dysplasia (types I and II) caused by distinct mutations in fibroblast growth factor receptor 3. *Nature Genet.* 9: 321–328.

Thayer, M. J., S. J. Tapscott, R. L. Davis, W. E. Wright, A. B. Lassar and H. Weintraub. 1989. Positive autoregulation of the myogenic determination gene *MyoD1*. *Cell* 58: 241–248.

Toivonen, S. 1945. Uber die Entwicklung der Vor- und Uriniere beim Kaninchen. *Ann. Acad. Sci. Fenn.* Ser. A. 8: 1–27.

Tondravi, M. M., S. R. McKercher, K. Anderson, J. M. Erdmann, M. Quiroz, R. Maki and S. L. Teitelbaum. 1997. Osteopetrosis in mice lacking haematopoietic transcription factor PU.1. *Nature* 386: 81–84.

Tuan, R. 1987. Mechanisms and regulation of calcium transport by the chick embryonic chorioallantoic membrane. *J. Exp. Zool.* [Suppl.] 1: 1–13.

Tuan, R. S. and M. H. Lynch. 1983. Effect of experimentally induced calcium deficiency on the developmental expression of collagen types in chick embryonic skeleton. *Dev. Biol.* 100: 374–386.

Vainio, S., E. Lehtonen, M. Jalkanen, M. Bernfield and L. Saxén. 1989. Epithelial-mesenchymal interactions regulate the stage-specific expression of a cell surface proteoglycan, syndecan, in the developing kidney. *Dev. Biol.* 134: 382–391.

Vainio, S., M. Jalkanen, M. Bernfield and L. Saxén. 1992. Transient expression of syndecan in mesenchymal cell aggregates of the embryonic kidney. *Dev. Biol.* 152: 221–232.

van Adelsberg, J., S. Chamberlain and V. D'Agati. 1997. *Polycystin* expression is temporally and spatially regulated during renal development. *Am. J. Physiol.* 272: F602–F609.

van Heyningen, V. and 11 others. 1990. Role for Wilms tumor gene in genital development? *Proc. Natl. Acad. Sci. USA* 87: 5383–5386.

Venters, S. J., S. Thorsteindöttir and M. J. Duxton. 1999. Early development of the myotome in the mouse. *Devel. Dynam.* 216: 219–232.

Ward, C. J. and 8 others. 1996. Polycystin, the polycystic kidney disease protein, is expressed by epithelial cells in fetal, adult, and polycystic kidney. *Proc. Nat. Acad. Sci. USA* 93: 1524–1528.

Watanabe, Y., D. Duprez, A.-H. Monsoro-Burq, C. Vincent and N. Le Douarin. 1998. Two domains in vertebral development: Antagonistic regulation by SHH and BMP4 proteins. *Development* 125: 2631–2639.

Webster, M. K. and D. J. Donoghue. 1996. Constitutive activation of fibroblast growth factor receptor 3 by the transmembrane domain point mutation found in achondroplasia. *EMBO J.* 15: 520–527.

Weintraub, H., S. J. Tapscott, R. L. Davis, M. J. Thayer, M. A. Adam, A. B. Lassar and D. Miller. 1989. Activation of muscle-specific genes in pigment, nerve, fat, liver, and fibroblast cell lines by forced expression of MyoD. *Proc. Natl. Acad. Sci. USA* 86:5434–5438.

Wilkie, A. O. M., G. M. Morriss-Kay, E. Y. Jones and J. K. Heath. 1995. Functions of fibroblast growth factors and their receptors. *Curr. Biol.* 5: 500–507.

Wintour, E. M., A. Butkus, L. Earnest and S. Pompolo. 1996. The erythropoietin gene is expressed strongly in the mammalian mesonephric kidney. *Blood* 88: 3349–3353.

Woolf, A. S. and 8 others. 1995. Roles of hepatocyte growth factor/scatter factor and the Met receptor in the early development of the metanephros. *J. Cell Biol.* 128: 171–184.

Wright, E. and 8 others. 1995. The Sry-related gene *Sox9* is expressed during chondrogenesis in mouse embryos. *Nature Genet.* 9: 15–20.

Wu, L. N., B. R. Genge, D. G. Dunkelberger, R. Z. LeGeros, B. Concannon and R. E. Wuthier. 1997. Physiochemical characterization of the nucleational core of matrix vesicles. *J. Biol. Chem.* 272: 4404–4411.

Yaffe, D. and M. Feldman. 1965. The formation of hybrid multinucleated muscle fibres from myoblasts of different genetic origin. *Dev. Biol.* 11: 300–317.

Yagami-Hiromasa, T., T. Sato, T. Kurisaki, K. Kamijo, Y.-I. Nabeshima and A. Fujisawa-Sehara. 1995. A metalloprotease-disintegrin participating in myoblast fusion. *Nature* 377: 652–656.

chapter 15 Lateral plate mesoderm and endoderm

The Heart of Creatures is the Foundation of Life, the Prince of All, the Sun of the Microcosm, on which all Vegetation doth depend, from whence all Vigor and Strength doth flow.
WILLIAM HARVEY (1628)

Blut is ein ganz besodrer Saft. (Blood is a very special juice.)
WOLFGANG GOETHE (1805)

IN THE CHAOS OF THE ENGLISH CIVIL WARS, William Harvey, physician to the King and discoverer of the blastoderm, was comforted by viewing the heart as the undisputed ruler of the body, through whose divinely ordained powers the lawful growth of the organism was assured. Later embryologists looked at the heart as more of a servant than a ruler, the chamberlain of the household who assured that the nutrients reached the centrally located brain and peripherally located muscles. In either metaphor, the heart, its circulation, and the digestive system were seen as being absolutely critical during development. As Harvey (1651) persuasively argued, the chick embryo must form its own blood without any help from the hen, and this blood is crucial in embryonic growth. How this happened was a mystery. "What artificer," he wrote, could create blood "when there is yet no liver in being?" The nutrition provided by the egg was also paramount to Harvey. His conclusion about the nutritive value of the yolk and albumin was that "The egge is, as it were, an exposed womb; wherein there is a substance concluded as the Representative and Substitute, or Vicar of the breasts."

This chapter will outline the mechanisms by which the circulatory system, the respiratory system, and the digestive system emerge in the amniote embryo.

Lateral Plate Mesoderm

On either side of the intermediate mesoderm resides the lateral plate mesoderm. Each plate splits horizontally into the dorsal **somatic (parietal) mesoderm**, which underlies the ectoderm, and the ventral **splanchnic (visceral) mesoderm**, which overlies the endoderm. The space between these layers becomes the body cavity—the **coelom**—which stretches from the future neck region to the posterior of the body. During later development, the right- and left-side coeloms fuse, and folds of tissue extend from the somatic mesoderm, dividing the coelom into separate cavities. In mammals, the coelom is subdivided into the **pleural, pericardial**, and **peritoneal** cavities, enveloping the thorax, heart, and abdomen, respectively. The mechanism for creating the linings of these body cavities from the lateral plate mesoderm has changed little throughout vertebrate evolution, and the development of the chick mesoderm can be compared with similar stages of frog embryos (Figure 15.1).

The Heart

The circulatory system is one of the great achievements of the lateral plate mesoderm. Consisting of a heart, blood cells, and an intricate system of blood vessels, the circu-

471

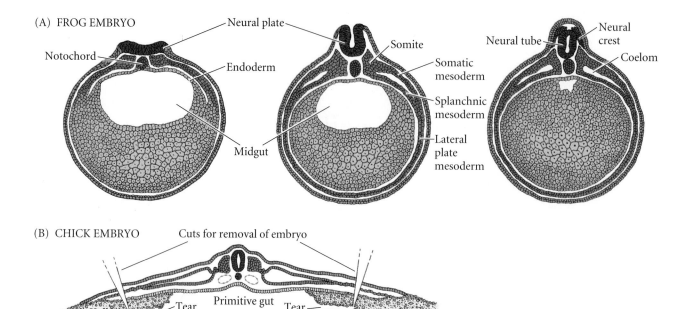

(A) FROG EMBRYO

Neural plate

Notochord

Endoderm

Midgut

Somite

Somatic mesoderm

Splanchnic mesoderm

Lateral plate mesoderm

Neural tube

Neural crest

Coelom

(B) CHICK EMBRYO

Cuts for removal of embryo

Yolk

Tear

Primitive gut

Tear

(C) CHICK "MADE INTO" FROG

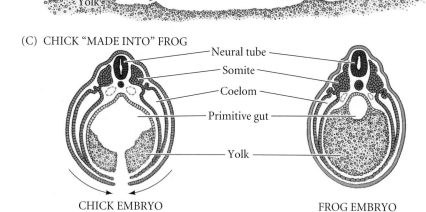

Neural tube

Somite

Coelom

Primitive gut

Yolk

CHICK EMBRYO
(removed from yolk, edges pulled together)

FROG EMBRYO

Figure 15.1
Comparison of mesodermal development in frog and chick embryos. (A) Neurula-stage frog embryos, showing progressive development of the mesoderm and coelom. (B) Transverse section of a chick embryo. (C) When the chick embryo is separated from its enormous yolk mass, it resembles the amphibian neurula at a similar stage. (A after Rugh 1951; B and C after Patten 1951.)

latory system provides nourishment to the developing vertebrate embryo. The circulatory system is the first functional unit in the developing embryo, and the heart is the first functional organ. The vertebrate heart arises from two regions of splanchnic mesoderm—one on each side of the body—that interact with adjacent tissue to become specified for heart development.

> VADE MECUM **Early heart development.** The vertebrate heart begins to function early in its development. You can see this in movies of the living chick embryo at early stages when the heart is little more than a looped tube. **[Click on Chick-Late]**

Specification of heart tissue and fusion of heart rudiments

In amniote vertebrates, the embryo is a flattened disc, and the lateral plate mesoderm does not completely encircle the yolk sac. The presumptive heart cells originate in the early primitive streak, just posterior to Hensen's node and extending about halfway down its length. These cells migrate through the streak and form two groups of mesodermal cells lateral to (and at the same level as) Hensen's node (Figure 15.2; Garcia-Martinez and Schoenwolf 1993). These groups of cells are called the **cardiogenic mesoderm**. The cells forming the atrial and ventricular musculature, the cushion cells of the valves, the Purkinje conducting fibers, and the endothelial lining of the heart are all generated from these two clusters (Mikawa 1999).

When the chick embryo is only 18–20 hours old, the presumptive heart cells move anteriorly between the ectoderm and endoderm toward the middle of the embryo, remaining in close contact with the endodermal surface (Linask and Lash 1986). When these cells reach the lateral walls of the anterior gut tube, migration ceases. The directionality for this migration appears to be provided by the foregut endoderm. If the cardiac region endoderm is rotated with respect to the rest of the embryo, migration of the cardiogenic mesoderm cells is reversed. It is thought that the endodermal component responsible for this movement is an anterior-to-posterior concentration gradient of fibronectin. Antibodies against fibronectin stop the migration, while antibodies against other extracellular matrix components do not (Linask and Lash 1988a,b).

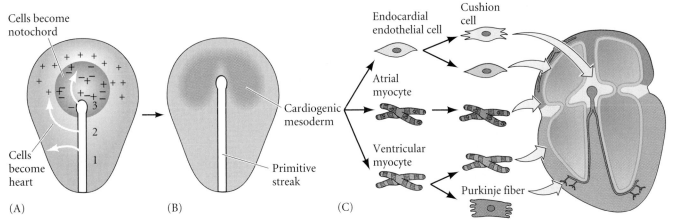

Figure 15.2

Heart-forming cells of the chick embryo. (A) Model for the specification of cardiogenic mesoderm. The routes of mesodermal migration at various regions of the primitive streak are represented by arrows. Signals that induce cardiac myogenesis are represented by plus
signs (+), inhibitors of cardiac induction by minus signs (−).
Migrating mesoderm in region 1 encounters neither inducers nor
repressors. Cells migrating from region 3 encounter both. Only
those cells migrating from region 2 encounter the inducer without
the inhibitor. (B) This process produces a horseshoe-shaped region
of cardiogenic mesoderm. (C) The cardiogenic mesoderm contains
the precursors of the three cell types of the endocardium and myocardium. The endocardium provides the endothelial lining of the
heart as well as the cushion cells that form the valves. The atrial myocyte of the myocardium and the ventricular myocyte (some of
which becomes the conducting Purkinje fibers) are also generated
by the cardiogenic mesoderm. The neural crest cells and endothelium of the cardiac vessels are specified separately at different locations. (A after Garcia-Martinez and Schoenwolf 1993; B after
Schultheiss et al. 1995; C after Mikawa 1999.)

The endoderm and primitive streak also specify the some
of the cardiogenic cells to become heart muscles. Cerberus and
an unknown factor, possibly BMP2 in the anterior endoderm,
induce the synthesis of the **Nkx2-5** transcription factor in the
migrating mesodermal cells that will become the heart*
(Komuro and Izumo 1993; Lints et al. 1993; Sugi and Lough
1994; Schultheiss et al. 1995; Andrée et al. 1998). Nkx2-5 is a
critical protein in instructing the mesoderm to become heart
tissue, and it activates the synthesis of other transcription factors (especially members of the GATA and MEF2 families).
Working together, these transcription factors activate the expression of genes encoding cardiac muscle-specific proteins
(such as cardiac actin, atrial naturetic factor, and the alpha
myosin heavy chains). Specification of the heart cells occurs

gradually, with the ventricular cells becoming specified prior to
the atrial cells (Markwald et al. 1998).

Cell differentiation occurs independently in the two heart-
forming primordia that are migrating toward each other
(Figure 15.3). As they migrate, the cells begin to express N-cadherin on their apices and join into an epithelium. A small population of these cells then downregulates N-cadherin and delaminates from the epithelium to form the **endocardium**, the
lining of the heart that is continuous with the blood vessels.[†]
The epithelial cells form the **myocardium** (Manasek 1968;
Linask and Lash 1993; Linask et al. 1997). The myocardium
will form the heart muscles that will pump for the lifetime of
the organism. The endocardial cells produce many of the heart
valves, secrete the proteins that regulate myocardial growth,
and regulate the placement of nervous tissue in the heart.

As neurulation proceeds, the foregut is formed by the inward folding of the splanchnic mesoderm (Figures 15.3 and
15.4). This movement brings the two cardiac tubes together,
eventually uniting the myocardium into a single tube. The bilateral origin of the heart can be demonstrated by surgically
preventing the merger of the lateral plate mesoderm (Gräper
1907; DeHaan 1959). This results in a condition called **cardia
bifida**, in which a separate heart forms on each side of the body
(Figure 15.4E). The two endocardia lie within this common
tube for a short while, but these will also fuse. At this time, the
originally paired coelomic chambers unite to form the body
cavity in which the heart resides.

This fusion occurs at about 29 hours in chick development and at 3 weeks in human gestation (see Figure 15.3C,D).
The unfused posterior portions of the endocardium become
the openings of the **vitelline veins** into the heart. These veins
will carry nutrients from the yolk sac into the **sinus venosus**.

---

*The Nkx2-5 homeodomain transcription factor is homologous to the
Tinman transcription factor that is active in specifying the heart tube of
Drosophila. Moreover, neither Tinman nor Nkx2-5 alone is sufficient to complete heart development in their respective organisms. Mice lacking *Nkx2-5*
start forming their heart tubes, but the tube fails to thicken or to loop (Lyons
et al. 1995), and humans with a mutation in one of their NKX2-5 genes have
congenital heart malformations (Schott et al. 1998).

[†]The endocardial population is distinct from the myocardial population
even before gastrulation. Cell lineage studies using retroviral markers show
that clones of myocardial cells during the pre-streak stages have no endocardial cells, and no clone of endocardial cells has any myocardial cells. Thus, the
cardiogenic mesoderm already has two committed populations of cells
(Cohen-Gould and Mikawa 1996).

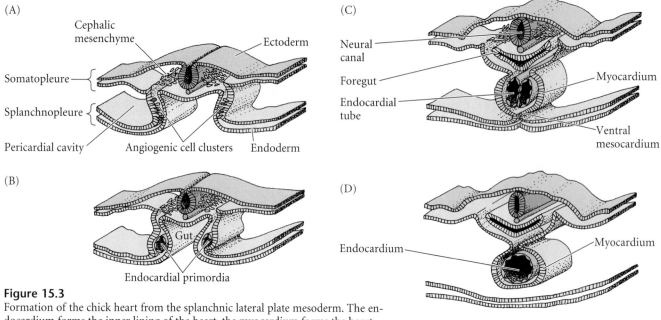

Figure 15.3
Formation of the chick heart from the splanchnic lateral plate mesoderm. The endocardium forms the inner lining of the heart, the myocardium forms the heart muscles, and the epicardium will eventually cover the heart. Transverse sections through the heart-forming region of the chick embryo are shown at (A) 25 hours, (B) 26 hours, (C) 28 hours, and (D) 29 hours. (After Carlson 1981.)

Figure 15.4
Fusion of the right and left heart rudiments to form a single cardiac tube. The cells fated to form the heart myocardium are shown by staining for the *Xin* message, whose protein product will be essential for the looping of the heart tube. (A) Stage 9 chick neurula, in which expression of *Xin* (purple) is seen in the two symmetrical heart-forming fields (arrowheads). (B) Cross section of the same stage neurula, showing the heart-forming cells (purple) in the splanchnic mesoderm of the lateral plate. (C) Stage 10 chick embryo, showing the fusion of the two heart-forming regions prior to looping. (D) Cross section through a similar stage 10 chick embryo, showing *Xin* expression in the myocardium. (E) Cardia bifida in chick embryo, caused by surgically cutting the ventral midline, thereby preventing the two heart primordia from fusing. (A–D from Wang et al. 1999, photographs courtesy of J. J.-C. Lin; E, photograph courtesy of R. L. DeHaan.)

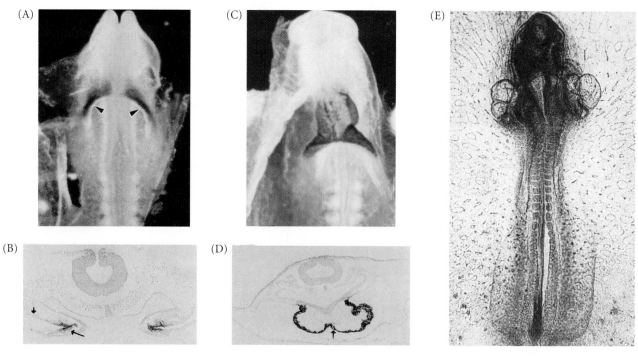

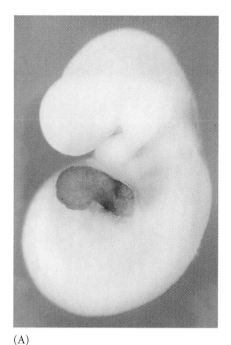

(A)

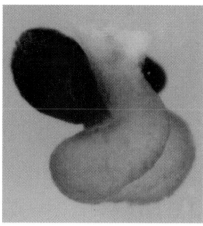

(B)

Figure 15.5
Specification of the atrium and ventricles occurs even before heart looping. The atrium and ventricles of the mouse embryo have separate types of myosin proteins, which allows them to be differentially stained. In these photographs, the atrium is stained blue, while the ventricular myosin is stained orange. (A) In the tubular heart (prior to looping), the two myosins (and their respective stains) overlap at the atrioventricular channel joining the future regions of the heart. (B) After looping, the dark blue stain is seen in the definitive atria and inflow tract, while the orange stain is seen in the ventricles. The unstained region above the ventricles is the truncus arteriosus, which becomes separated into the aorta and pulmonary arteries (see Figure 13.10). (After Xavier-Neto et al. 1999; photographs courtesy of N. Rosenthal.)

The blood then passes through a valvelike flap into the atrial region of the heart. Contractions of the **truncus arteriosus** speed the blood into the aorta.

Pulsations of the heart begin while the paired primordia are still fusing. The pacemaker of this contraction is the sinus venosus. Contractions begin here, and a wave of muscle contraction is propagated up the tubular heart. In this way, the heart can pump blood even before its intricate system of valves has been completed. Heart muscle cells have their own inherent ability to contract, and isolated heart cells from 7-day rat or chick embryos will continue to beat in petri dishes (Harary and Farley 1963; DeHaan 1967). In the embryo, these contractions become regulated by electrical stimuli from the medulla oblongata via the vagus nerve, and by 4 days, the electrocardiogram of a chick embryo approximates that of an adult.

Looping and formation of heart chambers

In 3-day chick embryos and 5-week human embryos, the heart is a two-chambered tube, with one atrium and one ventricle (Figure 15.5). In the chick embryo, the unaided eye can see the remarkable cycle of blood entering the lower chamber and being pumped out through the aorta. The looping of the heart converts the original anterior-posterior polarity of the heart tube into the right-left polarity seen in the adult (Figures 15.5 and 15.6). Thus, the portion of the heart tube destined to become the right ventricle lies anterior to the portion that will become the left ventricle. This looping is dependent upon the left-right patterning proteins (Nodal, Lefty-2) discussed in Chapter 11. Within the heart primordium, Nkx2–5 regulates the Hand1 and Hand2 transcription factors. Although the Hand proteins appear to be synthesized throughout the early

heart tube, Hand1 becomes restricted to the future left ventricle, and Hand2 to the right, as looping commences. Without these proteins, looping fails to occur normally and the ventricles fail to form properly (Srivastava et al. 1995; Biben and Harvey 1997). The Pitx-2 transcription factor, activated solely in the left side of the lateral plate mesoderm, is also critical for proper heart looping, and it may regulate the expression of proteins such as the extracellular matrix protein flectin to regulate the physical tension of the heart tissues on the different sides (Figure 15.7; Tsuda et al. 1996). Transcription factors Nkx2-5 and MEF2C also activate the *Xin* gene, whose protein product, Xin (Chinese for "heart"), may mediate the cytoskeletal changes essential for heart looping (Wang et al. 1999).

> WEBSITE **15.1 Transcription factors and heart formation.** The formation of the heart and blood cells is presided over by several transcription factors whose combinations enable the different parts of the heart to form and allow different types of blood cells to develop.

The separation of atrium from ventricle is specified by the several transcription factors that become restricted to either the anterior or the posterior portion of the heart tube (see Figure 15.6; Bao et al. 1999; Bruneau et al. 1999; Wang et al. 1999). The partitioning of this tube into a distinctive atrium and ventricle is accomplished when cells from the myocardium produce a factor (probably transforming growth factor β3) that causes cells from the adjacent endocardium to detach and enter the hyaluronate-rich "cardiac jelly" between the two layers (Markwald et al. 1977; Potts et al. 1991). In humans, these cells cause the formation of an **endocardial cushion** that divides the tube into right and left atrioventricular channels (Figure 15.8). Meanwhile, the primitive atrium is partitioned by two **septa** that grow ventrally toward the endocardial cushion. The septa, however, have holes in them, so blood can still cross from one side into the other. This crossing of blood is needed for the survival of the fetus before circulation to func-

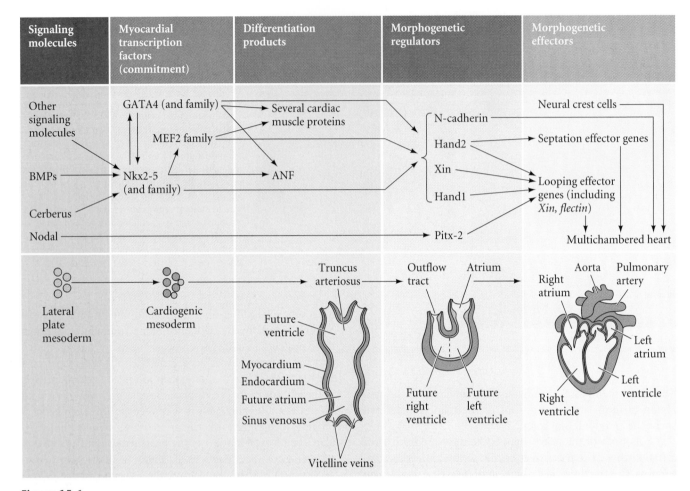

Signaling molecules	Myocardial transcription factors (commitment)	Differentiation products	Morphogenetic regulators	Morphogenetic effectors

Figure 15.6
Cascade of heart development. A correlation is made between the morphological stage and the transcription factors present in the nucleus of the heart precursor cells. The cardioblasts are the committed heart precursor cells containing Nkx2-5 and GATA family proteins. These proteins convert the cardioblast into cardiomyocytes (heart muscles), which make the cardiac muscle-specific proteins. These cardiomyocytes join together to form the cardiac tube. Under the influence of the Hand proteins, Xin, and Pitx-2, the heart loops and the formation of the chambers begins.

tional lungs has been established. Upon the first breath, however, these holes close, and the left and right circulatory loops become established (see Sidelights & Speculations). With the formation of the septa (which usually occurs in the seventh week of human development), the heart is a four-chambered structure with the pulmonary artery connected to the right ventricle and the aorta connected to the left.

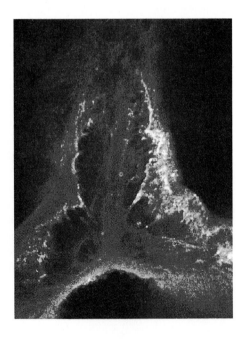

Figure 15.7
Asymmetric expression of the flectin protein in the developing chick heart. This extracellular protein accumulates predominantly on the left side of the heart in the stage 10 chick embryo. This photograph is taken from the ventral side (looking "up" at the embryo, so the flectin accumulation appears on the "right." (Scanning laser confocal photograph courtesy of K. Linask.)

(A) 33 DAYS

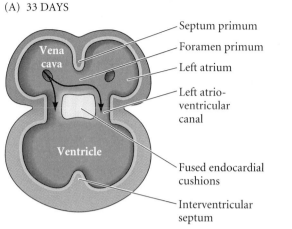

Septum primum

Foramen primum

Left atrium

Left atrio-
ventricular
canal

Fused endocardial
cushions

Interventricular
septum

Vena
cava

Ventricle

(B) THIRD MONTH

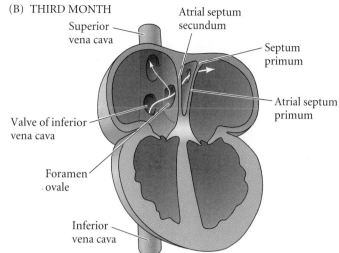

Superior
vena cava

Atrial septum
secundum

Septum
primum

Atrial septum
primum

Valve of inferior
vena cava

Foramen
ovale

Inferior
vena cava

Figure 15.8
Formation of the chambers of the heart. (A) Diagrammatic cross section of the human heart at 4.5 weeks. The atrial and ventricular septa
are growing toward the endocardial cushion. (B) Cross section of the
human heart during the third month of gestation. Blood can cross
from the right side of the heart to the left side through the openings
in the primary and secondary atrial septa. (After Larsen 1993.)

Sidelights & Speculations

Redirecting Blood Flow in the Newborn Mammal

ALTHOUGH THE DEVELOPING MAMMALIAN
fetus shares with the adult the need
to get oxygen and nutrients to its tissues, the physiology of the fetus differs drastically from that of the adult. Chief among
these differences is the lack of functional
lungs and intestines. All oxygen and nutrients
must come from the placenta. This raises two
questions. First, how does the fetus obtain
oxygen from maternal blood? And second,
how is blood circulation redirected to the
lungs once the umbilical cord is cut and
breathing becomes necessary?

Human Embryonic Circulation
The human embryonic circulatory system is
a modification of that used in other amniotes, such as birds and reptiles. The circulatory system to and from the chick embryo
and yolk sac is shown in Figure 15.9. Blood
pumped through the dorsal aorta passes over
the aortic arches and down into the embryo.

Some of this blood leaves the embryo
through the vitelline arteries and enters the
yolk sac. Nutrients and oxygen are absorbed
from the yolk, and the blood returns through
the vitelline veins back into the heart through
the sinus venosus. In mammalian embryos,
food and oxygen are obtained from the placenta. Thus, although the mammalian embryo has vessels analogous to the vitelline
veins, the main supply of food and oxygen
comes from the umbilical vein, which unites
the embryo with the placenta (Figure 15.10).
This vein, which takes the oxygenated and
food-laden blood back into the embryo, is
derived from what would be the right
vitelline vein in birds. The umbilical artery,
carrying wastes to the placenta, is derived
from what would have become the allantoic
artery of the chick. It extends from the caudal
portion of the aorta and proceeds along the
allantois and then out to the placenta.

Fetal Hemoglobin
The solution to the fetus's problem of getting
oxygen from its mother's blood involves the
development of a fetal hemoglobin. The hemoglobin in fetal red blood cells differs
slightly from that in adult corpuscles. Two of
the four peptides of the fetal and adult hemoglobin chains are identical—the alpha (α)
chains—but adult hemoglobin has two beta
(β) chains, while the fetus has two gamma (γ)
chains (Figure 15.11). Normal β-chains bind
the natural regulator diphosphoglycerate,
which assists in the unloading of oxygen. The
γ-chain isoforms do not bind diphosphoglycerate as well and therefore have a higher affinity for oxygen. in the low-oxygen environment of the placenta, oxygen is released from
adult hemoglobin. in this same environment,
fetal hemoglobin does not give away oxygen,
but binds it. This small difference in oxygen
affinity mediates the transfer of oxygen from
the mother to the fetus. Within the fetus, the

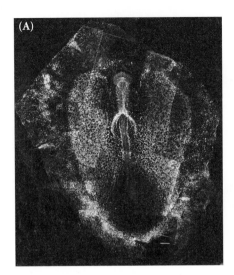

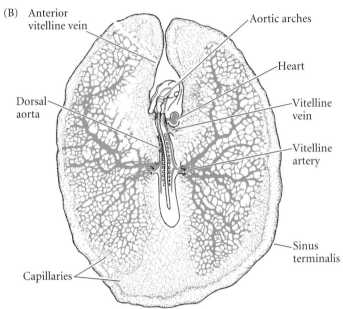

Figure 15.9
Circulatory system of the early avian embryo. (A) Construction of the vasculature in a 7-somite quail embryo, stained with a fluorescent antibody that recognizes endothelial cells. Blood islands, which will form the blood vessels and blood cells, can be seen at the edges. (B) Circulatory system of a 44-hour chick embryo. This view shows arteries in color; the veins are stippled. The sinus terminalis is the outer limit of the circulatory system and the site of blood cell generation. (A, photographic montage from Pardanaud et al. 1987, courtesy of F. Dieterlen-Lièvre; B after Carlson 1981.)

myoglobin of the fetal muscles has an even higher affinity for oxygen, so oxygen molecules pass from fetal hemoglobin for storage and use in the fetal muscles. Fetal hemoglobin is not deleterious to the newborn, and in humans, the replacement of fetal hemoglobin-containing blood cells with adult hemoglobin-containing blood cells is not complete until about 6 months after birth. (The molecular basis for this switch in globins was discussed in Chapter 5.)

From Fetal to Newborn Circulation

But once the fetus is not getting its oxygen from the mother, how does it restructure its circulation to get oxygen from its own lungs? During fetal development, an opening—the **ductus arteriosus**—diverts blood from the pulmonary artery into the aorta (and thus to the placenta). Because blood does not return from the pulmonary vein in the fetus, the de-

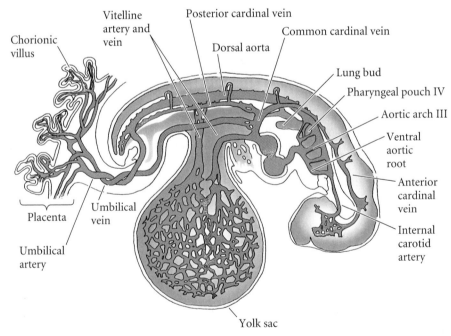

Figure 15.10
Circulatory system of a 4-week human embryo. Although at this stage all the major blood vessels are paired left and right, only the right vessels are shown. Arteries are shown in color. (From Carlson 1981.)

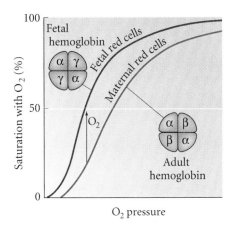

Figure 15.11
Transfer of oxygen from the mother to the fetus in human embryos. Adult and fetal hemoglobin molecules differ in their protein subunits. The fetal γ-chain binds diphosphoglycerate less avidly than does the adult β-chain. Consequently, fetal hemoglobin can bind oxygen more efficiently than can adult hemoglobin. In the placenta, there is a net flow (arrow) of oxygen from the mother's blood (which gives up oxygen to the tissues at the lower oxygen pressure) to the fetal blood, which is still picking it up.

veloping mammal has to have some other way of getting blood into the left ventricle to be pumped. This is accomplished by the foramen ovale, a hole in the septum separating the right and left atria. Blood can enter the right atrium, pass through the foramen to the left atrium, and then enter the left ventricle (Figure 15.12). When the first breath is drawn, blood pressure in the left side of the heart increases. It causes the septa over the foramen ovale to close, thereby separating the pulmonary and systemic circulation. Moreover, the decreases in prostaglandins experienced by the newborn cause the muscles surrounding the ductus arteriosus to close that opening (Nguyen et al. 1997).* Thus, when breathing begins, the respiratory circulation is shunted from the placenta to the lungs. ■

*In some infants, the septa fail to close, and the foramen ovale is left open. Usually the opening is so small that such children have no physical symptoms, and the foramen eventually closes. If it does not close completely, however, and the secondary septum fails to form, the atrial septal opening may cause enlargement of the right side of the heart, which can lead to heart failure during early adulthood.

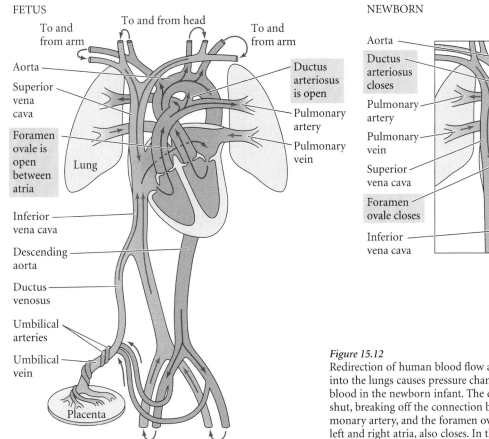

Figure 15.12
Redirection of human blood flow at birth. The expansion of air into the lungs causes pressure changes that redirect the flow of blood in the newborn infant. The ductus arteriosus squeezes shut, breaking off the connection between the aorta and the pulmonary artery, and the foramen ovale, a passageway between the left and right atria, also closes. In this way, pulmonary circulation is separated from systemic circulation.

(A) 29 DAYS

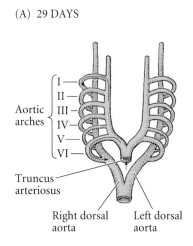

(B) 49 DAYS

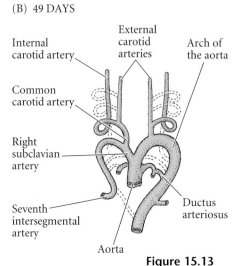

(C) 56 DAYS

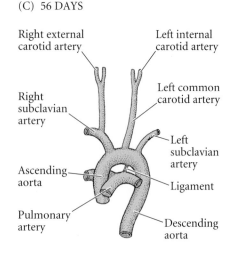

Figure 15.13

The aortic arches of the human embryo. (A) Originally, the truncus arteriosus pumps blood into the aorta, which branches on either side of the foregut. The six aortic arches take blood from the truncus arteriosus and allow it to flow into the dorsal aorta. (B) The arches begin to disintegrate or become modified (the dotted lines indicate degenerating structures). (C) Eventually, the remaining arches are modified and the adult arterial system is formed. (After Langman 1981.)

Formation of Blood Vessels

Although the heart is the first functional organ of the body, it does not even begin to pump until the vascular system of the embryo has established its first circulatory loops. Rather than sprouting from the heart, the blood vessels form independently, linking up to the heart soon afterward. Everyone's circulatory system is different, since the genome cannot encode the intricate series of connections between the arteries and veins. Indeed, chance plays a major role in establishing the microanatomy of the circulatory system. However, all circulatory systems in a given species look very much alike, because the development of the circulatory system is severely constrained by physiological, physical, and evolutionary parameters.

Constraints on how blood vessels may be constructed

The first constraint on vascular development is *physiological*. Unlike new machines, which do not need to function until they have left the assembly line, new organisms have to function even as they develop. The embryonic cells must obtain nourishment before there is an intestine, use oxygen before there are lungs, and excrete wastes before there are kidneys. All these functions are mediated through the embryonic circulatory system. Therefore, the circulatory physiology of the developing embryo must differ from that of the adult organism. Food is absorbed not through the intestine, but from either the yolk or the placenta, and respiration is conducted not through the gills or lungs, but through the chorionic or allantoic membranes. The major embryonic blood vessels must be constructed to serve these extraembryonic structures.

The second constraint is *evolutionary*. The mammalian embryo extends blood vessels to the yolk sac even though there is no yolk inside. Moreover, the blood leaving the heart via the truncus arteriosus passes through vessels that loop over the foregut to reach the dorsal aorta. Six pairs of aortic arches loop over the pharynx (Figure 15.13). In primitive fishes, these arches persist and enable the gills to oxygenate the blood. In the adult bird or mammal, in which lungs oxygenate the blood,

such a system makes little sense, but all six pairs of aortic arches are formed in mammalian and avian embryos before the system eventually becomes simplified into a single aortic arch. Thus, even though our physiology does not require such a structure, our embryonic condition reflects our evolutionary history.

The third set of constraints is *physical*. According to the laws of fluid movement, the most effective transport of fluids is performed by large tubes. As the radius of a blood vessel gets smaller, resistance to flow increases as r^{-4} (Poiseuille's law). A blood vessel that is half as wide as another has a resistance to flow 16 times greater. However, diffusion of nutrients can take place only when blood flows slowly and has access to cell membranes. So here is a paradox: the constraints of diffusion mandate that vessels be small, while the laws of hydraulics mandate that vessels be large. Living organisms have solved this paradox by evolving circulatory systems with a hierarchy of vessel sizes (LaBarbera 1990). This hierarchy is formed very early in development (and is already well established in the 3-day chick embryo). In dogs, blood in the large vessels (aorta and vena cava) flows over 100 times faster than it does in the capillaries. With a system of large vessels specialized for transport and small vessels specialized for diffusion (where the blood spends most of its time), nutrients and oxygen can reach the individual cells of the growing organism.

But this is not the entire story. If fluid under constant pressure moves directly from a large-diameter tube into a small-diameter tube (as in a hose nozzle), the fluid velocity increases. The evolutionary solution to this problem was the emergence of many smaller vessels branching out from a larg-

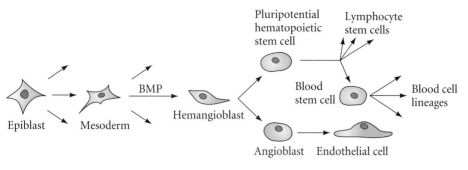

Figure 15.14
Origin and fate of the hemangioblast cells. Hemangioblasts are derived from mesodermal cells that are exposed to relatively high concentrations of certain bone morphogenetic proteins during early development. They give rise to the angioblasts—the precursors of the vascular endothelial cells—and to the pluripotential hematopoietic stem cells that generate the blood cells and lymphocytes. (After Liao and Zon 1999.)

er one, making the collective cross-sectional area of all the smaller vessels greater than that of the larger vessel. Circulatory systems show a relationship (known as Murray's law) in which the cube of the radius of the parent vessel approximates the sum of the cubes of the radii of the smaller vessels. The construction of any circulation system must negotiate among all of these physical, physiological, and evolutionary constraints.

Vasculogenesis: Formation of blood vessels from blood islands

Blood vessel formation is intimately connected to blood cell formation. Indeed, blood vessels and blood cells are believed to share a common precursor, the **hemangioblast*** (Figure

*The prefix hem- (or hemato-) refers to blood (as in hemoglobin). Similarly, the prefix angio- refers to blood vessels. The suffix -blast denotes a rapidly dividing cell, usually a stem cell. The suffix -poesis refers to generation or formation, and is also the root of the word poetry. The adjectival form of poesis is poietic. So hematopoietic stem cells are those cells that generate the different types of blood cells. The Latin suffix –genesis (as in angiogenesis) means the same as the Greek -poiesis.

15.14). Not only do blood vessels and blood cells share common sites of origin, but mutations of certain transcription factors in mice and zebrafish will delete both blood cells and blood vessels. In addition, the earliest blood cells and the earliest capillary cells share many of the same rare proteins on their cell surfaces (Wood et al. 1997; Choi et al. 1998; Liao and Zon 1999).

Blood vessels are constructed by two processes: vasculogenesis and angiogenesis (Figure 15.15). During **vasculogenesis**,

Figure 15.15
Vasculogenesis and angiogenesis. Vasculogenesis involves the formation of hemangioblastic blood islands and the construction of capillary networks from them. Angiogenesis involves the formation of new blood vessels by remodeling and building on older ones. Angiogenesis finishes the circulatory connections begun by vasculogenesis and builds arteries and veins from the capillaries. In this diagram, the major paracrine factors involved in each step are shown boxed, and their receptors (on the vessel-forming cells) are shown beneath them. (After Hanahan 1997; Risau 1997.)

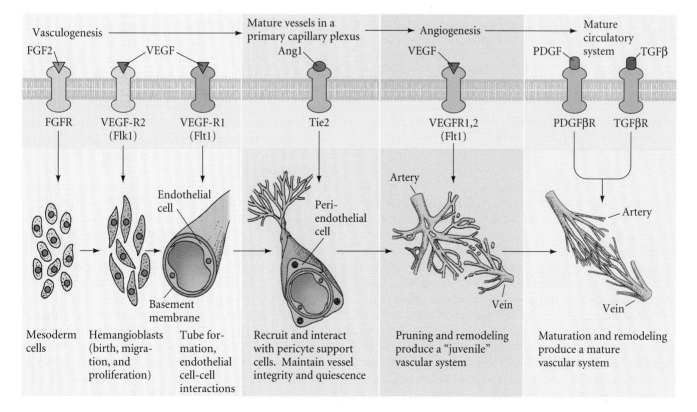

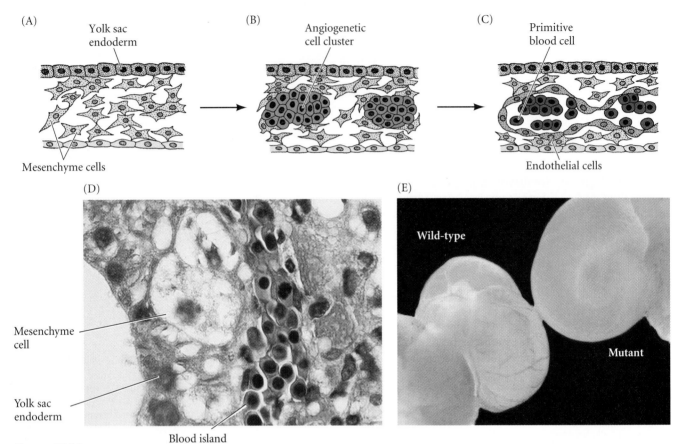

(A) Yolk sac endoderm

Mesenchyme cells

(B) Angiogenetic cell cluster

(C) Primitive blood cell

Endothelial cells

(D)

Mesenchyme cell

Yolk sac endoderm

Blood island

(E)

Wild-type

Mutant

Figure 15.16
Vasculogenesis. Blood vessel formation is first seen in the wall of the yolk sac, where (A) undifferentiated mesenchyme condenses to form (B) angiogenetic cell clusters. (C) The centers of these clusters form the blood cells, and the outsides of the clusters develop into blood vessel endothelial cells. (D) Photograph of a human blood island in the mesoderm surrounding the yolk sac. (The photograph is from a tubal pregnancy—an embryo that had to be evacuated since it had implanted into an oviduct rather than into the uterus.) (E) Yolk sacs of a wild-type mouse and a littermate heterozygous for a loss-of-function mutation of *VEGF*. The heterozygote embryo lacks blood vessels in its yolk sac and dies. (A–C after Langman 1981; D from Katayama and Kayano 1999; E from Ferrara and Alitalo 1999. Photographs courtesy of the authors.)

blood vessels are created de novo from the lateral plate mesoderm. In the first phase of vasculogenesis, groups of splanchnic mesoderm cells are specified to become hemangioblasts, the precursors of both the blood cells and the blood vessels (Shalaby et al. 1997). These cells condense into aggregations that are often called **blood islands**. The inner cells of these blood islands become **hematopoietic stem cells** (the precursors of all the blood cells), while the outer cells become **angioblasts**, the precursors of the blood vessels. In the second phase of vasculogenesis, the angioblasts multiply and differentiate into **endothelial cells**, which form the lining of the blood vessels. In the third phase, the endothelial cells form tubes and connect to form the **primary capillary plexus**, a network of

capillaries. In the second process, **angiogenesis**, this primary network will be remodeled and pruned into a distinct capillary bed, arteries, and veins (Risau 1997; Hanahan 1997). It is important to realize that the capillary networks of each organ arise within the organ itself, and are not extensions from larger vessels (Auerbach et al. 1985; Pardanaud et al. 1989).

The aggregation of hemangioblasts in extraembryonic regions is a critical step in amniote development, for the blood islands that line the yolk sac produce the vitelline (**omphalomesenteric**) veins that bring nutrients to the embryo and transport gases to and from the sites of respiratory exchange (Figure 15.16). These cells are first seen in the area opaca of chick embryogenesis, when the primitive streak is at its fullest extent (Pardanaud et al. 1987). These cords of hemangioblasts soon become hollow. The outer cells become the flat endothelial cell lining of the vessel. The central cells of the blood islands differentiate into the embryonic blood cells. As the blood islands grow, they eventually merge to form the capillary network draining into the two vitelline veins, which bring food and blood cells to the newly formed heart.

Three growth factors may be responsible for initiating vasculogenesis (see Figure 15.15). One of these, **basic fibroblast growth factor** (FGF2) is required for the generation of hemangioblasts from the splanchnic mesoderm. When cells from quail blastodiscs are dissociated in culture, they do not form blood islands or endothelial cells. However, when these cells

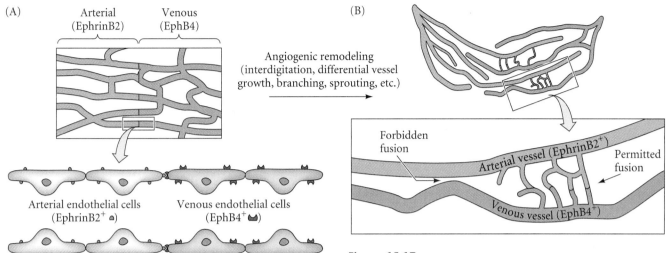

(A)

Arterial (EphrinB2)   Venous (EphB4)

Arterial endothelial cells (EphrinB2⁺ ◭)   Venous endothelial cells (EphB4⁺ ◡)

Site of EphrinB2/EphB4 interaction

Angiogenic remodeling (interdigitation, differential vessel growth, branching, sprouting, etc.)

(B)

Forbidden fusion

Arterial vessel (EphrinB2⁺)

Permitted fusion

Venous vessel (EphB4⁺)

Figure 15.17
Model of the roles of ephrin and Eph receptors during angiogenesis. (A) Primary capillary plexus produced by vasculogenesis. The arterial and venous endothelial cells have sorted themselves out by the presence of EphrinB2 or EphB4 in their respective cell membranes. (B) A maturing vascular network wherein the ephrin-Eph interaction mediates the joining of small branches (future capillaries) and may prevent fusion laterally. (After Yancopopoulos et al. 1998.)

are cultured in FGF2, blood islands emerge and form endothelial cells (Flamme and Risau 1992). FGF2 is synthesized in the chick embryonic chorioallantoic membrane and is responsible for the vascularization of this tissue (Ribatti et al. 1995). The second protein involved in vasculogenesis is **vascular endothelial growth factor** (**VEGF**). VEGF appears to enable the differentiation of the angioblasts and their multiplication to form endothelial tubes. VEGF is secreted by the mesenchymal cells near the blood islands, and the hemangioblasts and angioblasts have receptors for VEGF (Millauer et al. 1993). If mouse embryos lack the genes encoding either VEGF or the major receptor for VEGF (the Flk1 receptor tyrosine kinase), yolk sac blood islands fail to appear, and vasculogenesis fails to take place (Figure 15.16E; Ferrara et al. 1996). Mice lacking genes for the second receptor for VEGF (the Flt1 receptor tyrosine kinase) have differentiated endothelial cells and blood islands, but these cells are not organized into blood vessels (Fong et al. 1995; Shalaby et al. 1995). A third protein, **angiopoietin-1** (**Ang1**), mediates the interaction between the endothelial cells and the **pericytes**—smooth musclelike cells they recruit to cover them. Mutations of either angiopoietin-1 or its receptor lead to malformed blood vessels, deficient in the smooth muscles that usually surround them (Davis et al. 1996; Suri et al. 1996; Vikkula et al. 1996).

Angiogenesis: Sprouting of blood vessels and remodeling of vascular beds

After an initial phase of vasculogenesis, the primary capillary networks are remodeled. At this time, veins and arteries are made. This process is called angiogenesis (see Figure 15.15). First, VEGF acting alone on the newly formed capillaries causes a loosening of cell contacts and a degradation of the extracellular matrix at certain points. The exposed endothelial cells proliferate and sprout from these regions, eventually forming a

new vessel. New vessels can also be formed in the primary capillary bed by splitting an existing vessel in two. The loosening of the cell-cell contacts may also allow the fusion of capillaries to form wider vessels—the arteries and veins. Eventually, the mature capillary network forms and is stabilized by TGF-β (which strengthens the extracellular matrix) and **platelet-derived growth factor** (**PDGF**, which is necessary for the recruitment of the pericyte cells that contribute to the mechanical flexibility of the capillary wall) (Lindahl et al. 1997).

A key to our understanding of the mechanism by which veins and arteries form was the discovery that the primary capillary plexus in mice actually contains two types of endothelial cells. The precursors of the arteries contain EphrinB2 in their cell membranes. The precursors of the veins contain one of the receptors for this molecule, EphB4 tyrosine kinase, in their cell membranes (Figure 15.17; Wang et al. 1998). If EphrinB2 is knocked out in mice, vasculogenesis occurs, but angiogenesis does not. It is thought that EphB4 interacts with its ligand, EphrinB2, during angiogenesis in two ways. First, at the borders of the venous and arterial capillaries, it ensures that arterial capillaries connect only to venous ones. Second, in non-border areas, it might ensure that the fusion of capillaries to make larger vessels occurs only between the same type of vessel.*

In some instances, angiogenesis may be the major way of making blood vessels. In the forelimb bud, for instance, the

*Interestingly, the same Eph-ephrin system is implicated in the patterning of the paraxial mesoderm into somites (see Chapter 14).

(A)

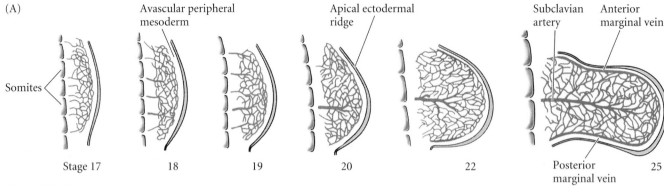

Somites

Avascular peripheral mesoderm

Apical ectodermal ridge

Subclavian artery

Anterior marginal vein

Posterior marginal vein

Stage 17   18   19   20   22   25

Figure 15.18
Vascularization of the chick forelimb. (A) Development of the vascular system during the early development of the chick wing bud. The periphery of the bud is avascular, and more avascular regions will form at the regions where chondrocytes condense to form the cartilaginous precursors to the bones. (B) Dorsal view of a stage 22 wing bud injected with India ink. (A after Feinberg 1991; B from Feinberg and Cafasso 1995, photograph courtesy of R. N. Feinberg.)

(B)

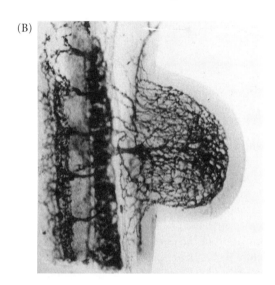

capillary network is probably derived by the sprouting of cells from the aorta (Evans 1909; Feinberg 1991). Within this capillary network, a central artery (which becomes the subclavian artery) forms as the major feeding vessel. Blood returns to the body through marginal veins that form from the anterior and posterior capillaries (Figure 15.18). The organ-forming regions of the body are thought to secrete angiogenesis factors that promote sprouting by enabling the mitosis and migration of endothelial cells into those areas. VEGF (mentioned earlier as a vasculogenesis factor) also promotes the migration of endothelial cells into the organs from preexisting blood vessels on the organs' surfaces.

WEBSITE **15.2 Angioblast migration in the chick.** After their initial formation in the yolk sac mesoderm, angioblast stem cells are seen in the somites and splanchnic mesoderm. These stem cells populate different regions of the embryo.

Several organs make their own angiogenesis factors. The placenta is one organ whose function depends on redirecting existing blood vessels into it. When the placenta is first being formed, it induces angiogenesis by secreting **proliferin (PLF)**, a factor that resembles growth hormone. When the placental blood vessels have become established (after day 12 in the mouse), the placenta secretes **proliferin-related protein (PRP)**, a peptide that acts as an inhibitor of angiogenesis (Jackson et al. 1994). Ovarian follicle cells and placental cells also secrete leptin, a hormone that is involved in appetite suppression in the adult. However, it can also act locally to induce angiogenesis and cause endothelial cells to organize into tubes (Figure 15.19; Antczak et al. 1997; Sierra-Honigmann et al. 1998).

The developing bone is another organ that redirects blood vessels into it while it is forming. As mentioned in Chapter 14, cartilage is usually an avascular tissue, except when capillaries invade the growth plate to convert the cartilage into bone. Hypertrophic cartilage (but not mature or dividing cartilage) secretes a 120-kDa angiogenesis factor (Alini et al. 1996). It is interesting that this factor is made only when the early hypertrophic chondrocytes have been exposed to vitamin D. This finding suggests an explanation for the bone deformities seen in patients with rickets.

Angiogenesis is critical in the growth of any tissue, including tumors. Tumors are "successful" only when they are able to direct blood vessels into them. Therefore, tumors secrete angiogenesis factors. The ability to inhibit such factors may have important medical applications as a way to prevent tumor growth and metastasis (Fidler and Ellis 1994).

WEBSITE **15.3 Angiogenesis in diabetes and tumor formation.** Angiogenesis is a critical part of tumor formation and diabetes. Some newly discovered proteins, such as angiostatin, endostatin, and squalamine, can inhibit angiogenesis and may provide cures for cancers.

(A)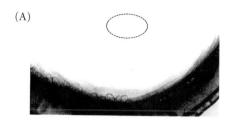

(B)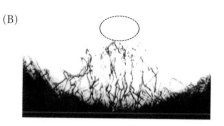

(C)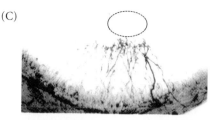

(D)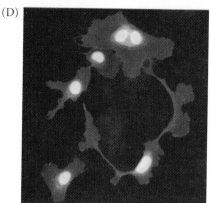

Figure 15.19

Leptin-induced angiogenesis. (A–C) Corneal response assay. Pellets (indicated by dashed circles) containing either saline (A), leptin (B), or VEGF (C) were implanted into rat corneas. The cornea normally lacks blood vessels, but is close to a highly vascularized region (the limbus) of the eye. If the pellet contains a substance with angiogenetic activity, it will induce the proliferation, migration, and orientation of the endothelial cells into new capillaries directed toward the pellet. VEGF and leptin both had angiogenetic activity. (D) Cultured endothelial cells treated with leptin crawl together to form tubes. (From Sierra-Honigmann et al. 1998; photographs courtesy of M. R. Sierra-Honigmann.)

The Development of Blood Cells

The stem cell concept

Many adult tissues are formed such that once the cells are created, they are not expected to be replaced. Most neurons and bones, for instance, cannot be replaced if they are damaged or lost. There are several populations of cells, however, that are constantly dying and being replaced. Each day, we lose and replace about 1.5 grams of skin cells and about 10^{11} red blood cells. The skin cells are sloughed off, and the red blood cells are killed in the spleen. Where are their replacements coming from? They come from populations of stem cells. A **stem cell** is a cell that is capable of extensive proliferation, creating more stem cells (self-renewal) as well as more differentiated cellular progeny (Figure 15.20). Stem cells are, in effect, a population of embryonic cells, continuously producing cells that can undergo further development within an adult organism. Thus, the adult vertebrate body retains populations of stem cells, and these stem cell populations can produce both more stem cells

and a population of cells that can undergo further development (Potten and Loeffler 1990). Our blood cells, intestinal crypt cells, epidermis, and (in males) spermatocytes are populations in a steady-state equilibrium in which cell production balances cell loss (Hay 1966). In most cases, stem cells can produce either more stem cells or more differentiated cells when body equilibrium is stressed by injury or environment. (This is seen by the production of enormous numbers of red blood cells when the body suffers from anoxia.)

Stem cells have been identified in all the tissues mentioned above, but they are most readily studied in blood cell development. Blood cell formation in the adult occurs in the bone marrow and spleen; it also occurs in the fetal liver. The path of development that a stem cell descendant enters depends on the molecular milieu in which it finds itself. This became apparent when experimental evidence showed that red blood cells (erythrocytes), white blood cells (granulocytes, neutrophils, and platelets), and lymphocytes shared a common precursor—the **pluripotential hematopoietic stem cell**.

Pluripotential stem cells and hematopoietic microenvironments

THE PLURIPOTENTIAL HEMATOPOIETIC STEM CELL. The hemangioblast cells of the lateral plate mesoderm can give rise to both the angioblasts of the vascular system and the pluripotential hematopoietic stem cells of the blood and lymphoid systems. The pluripotential hematopoietic stem cell is one of our body's most impressive cells. From it will emerge erythrocytes, neutrophils, basophils, eosinophils, platelets, mast cells, monocytes, tissue macrophages, osteoclasts, and the T and B lymphocytes (Figure 15.21). The pluripotential hematopoietic

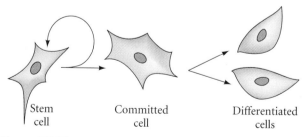

Figure 15.20

The concept of stem cells. A stem cell divides to produce another stem cell and a committed cell. The committed cell can divide again, but its progeny will form a very restricted set of differentiated cell types. A stem cell can also produce two stem cells or two committed cells. If a significant number of stem cells divide to produce two committed cells, the longevity of that lineage becomes limited.

Stem cell   Committed cell   Differentiated cells

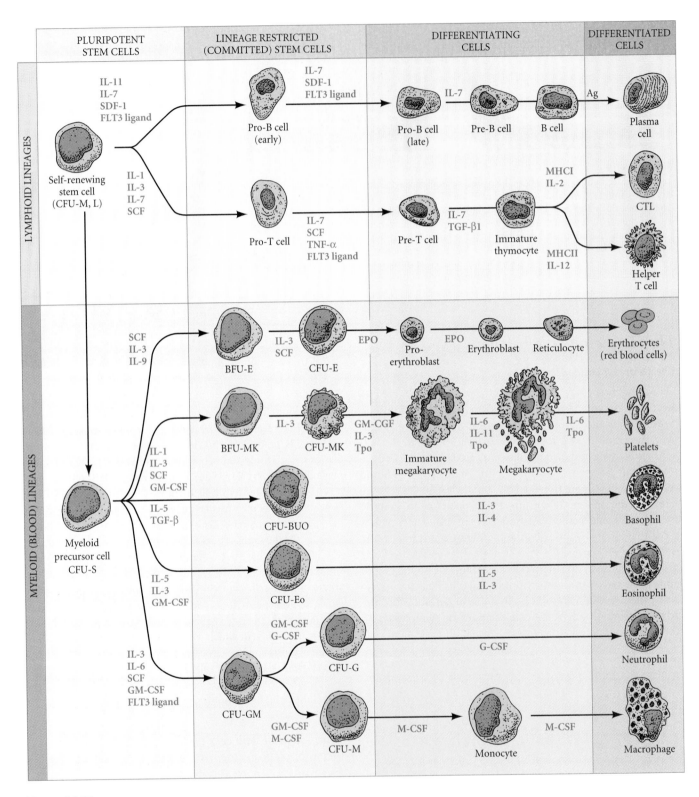

Figure 15.21

A model for the origin of mammalian blood and lymphoid cells. (Other models are consistent with the data, and this one summarizes features from several models.) Ag, antigen; EPO, erythropoietin; G-CSF, granulocyte colony-stimulating factor; GM-CSF, granulocyte-macrophage colony-stimulating factor; IL, interleukin; LIF, leukemia inhibiting factor; M-CSF, macrophage colony-stimulating factor; MHCI, major histocampatability complex type I protein; MHCII, major histocampatability complex type II protein; SCF, stem cell factor; SDF-1, stromal-derived factor-1; TNF, tumor necrosis factor; Tpo, thrombopoietin. (After Nakauchi and Gachelin 1993; R&D Systems 1997.)

stem cell (also known as the **colony-forming unit of the myeloid and lymphoid cells,** or **CFU-M,L,** and the **CD34 cell**) was discovered when irradiated bone marrow (which contains blood-forming cells) was injected into mice that had a hereditary deficiency of blood-forming cells. Abramson and her colleagues (1977) showed that the same radiation-induced chromosomal abnormalities were present in both the nucleated blood cells and the circulating lymphocytes of these mice. This finding was confirmed by studies in which marrow cells were injected with certain viruses that become incorporated into cellular DNA at various random places. The same virally derived genes were seen in the same region of the genome in both lymphocytes and blood cells (Keller et al. 1985; Lemischka et al. 1986). In 1995, Berardi and colleagues isolated a fraction of cells that may be the human CFU-M,L. By killing all the cells that divided when exposed to proteins that would activate more committed stem cells, they were left with about one nucleated cell for every 10,000 originally found in the bone marrow. These cells could generate both the blood and lymphoid lineages.

THE CFU-S. If the pluripotential hematopoietic stem cell can give rise to both the lymphocytes and the blood cells, the next stem cell in the sequence is somewhat more committed, being able to give rise to all the blood cells, but not to the lymphocytes. The existence of a such a multipotent hematopoietic stem cell was demonstrated by Till and McCulloch (1961), who injected bone marrow cells into lethally irradiated mice of the same genetic strain as the marrow donors. (Irradiation kills the hematopoietic cells of the host, so that any new blood cells formed must come from the donor mouse.) Some of these donor cells produced discrete nodules, or colonies, on the spleens of the host animals (Figure 15.22). Microscopic studies showed these colonies to be composed of erythrocyte, granulocyte, and platelet precursors. Thus, a single cell from the bone marrow was capable of forming many different blood cell types. The cell type responsible was called the **CFU-S,** the *colony-forming unit of the spleen.* This type of cell has also been called the **CFU-GEMM,** the *colony-forming unit for granulocytes, erythrocytes, macrophages,* and *megakaryocytes.* Further studies used chromosomal markers to prove that the different types of cells within a single colony were formed from the same CFU-S. In these studies, bone marrow cells were irradiated so that very few survived. Many of those that did survive developed chromosomal abnormalities that could be detected microscopically. When such irradiated CFU-S cells were injected into a mouse whose own blood-forming stem cells had been destroyed, each cell of a spleen colony, be it granulocyte or erythrocyte precursor, had the same chromosomal anomaly (Becker et al. 1963).

An important part of the stem cell concept is the requirement that the stem cell be able to form more stem cells in addition to its differentiated cell types. This has indeed been found to be the case for hematopoietic stem cells. When

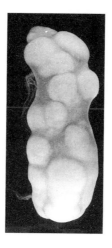

Figure 15.22
Isolated blood-forming colonies. When bone marrow containing hematopoietic stem cells is injected into an irradiated mouse, discrete colonies of blood cells derived from a single precursor cell are seen on the surface of the spleen of that mouse. The colonies contain several types of blood cells. (From Till 1981; photograph courtesy of J. E. Till.)

spleen colonies derived from a single CFU-S are resuspended and injected into other mice, many spleen colonies emerge (Jurśyśková and Tkadleÿcek 1965; Humphries et al. 1979). Thus, we see that a single marrow cell can form numerous different cell types and can also undergo self-renewal; in other words, the CFU-S is a multipotent hematopoietic stem cell.

The CFU-M,L and the CFU-S are both supported by **stem cell factor** (**SCF**), a paracrine protein that promotes cell division in numerous stem cell populations. As mentioned in Chapter 6, SCF also promotes the division of melanoblasts (which will form pigment cells) and germ stem cells (which will form the gametes). Thus, mice lacking SCF or its receptor (the c-Kit protein) are sterile (no germ cells), white (no pigment cells), anemic (no red blood cells), and immunodeficient (no lymphocytes).

BLOOD AND LYMPHOCYTE LINEAGES. Figure 15.21 summarizes the development of the blood and lymph cells and the paracrine factors involved in this process. The first pluripotential hematopoietic stem cell is the CFU-M,L. The development of this cell type appears to be dependent on the transcription factor SCL. Mice lacking this protein die from the absence of all blood and lymphocyte lineages. SCL may specify the ventral mesoderm to a blood cell fate, or it may enable the formation or maintenance of the CFU-M,L cells (Porcher et al. 1996; Robb et al. 1996). The CFU-M,L give rise to the CFU-S (blood cells) and the several lymphocytic stem cell types. The CFU-S is also pluripotent because its progeny, too, can differentiate into numerous cell types. The immediate progeny of the CFU-S, however, are **lineage-restricted stem cells.** Each can produce only one type of cell in addition to renewing itself. The **BFU-E** (*burst-forming unit, erythroid*), for instance, is a lineage-restricted stem cell formed from the CFU-S, and it can form only one cell type in addition to itself. That cell type is the **CFU-E** (*colony-forming unit, erythroid*), which is capable of responding to the hormone **erythropoietin** to produce the first recognizable differentiated member of the erythrocyte lineage, the **proerythroblast,** a red blood cell precursor. Erythropoietin is a glycoprotein that rapidly induces the synthesis of the mRNA for globin (Krantz and Goldwasser 1965). It is produced predominantly in the kidney, and its synthesis is responsive to environmental conditions. If the level of blood oxygen falls, erythropoietin production is increased, leading to the production of more red blood cells. As the proerythroblast matures, it becomes an **erythroblast**, synthesizing enormous amounts of hemoglobin. Eventually, the mammalian erythroblast expels its nucleus, becoming a **reticulocyte**. Reticulocytes can no longer synthesize globin mRNA, but they can still translate existing messages into globin. The final stage of differentiation is the **erythrocyte**, or mature red blood cell. Here, no division, RNA synthesis, or protein synthesis takes place. The DNA of the erythrocyte condenses and makes no further messages. Amphibians, fish, and birds retain the functionless nucleus; mammals extrude it from the cell. The cell leaves the bone marrow and delivers oxygen to the body tissues.* Similarly, there are lineage-restricted stem cells that give rise to platelets, to granulocytes (neutrophils, basophils, and eosinophils), and to macrophages.

Some hematopoietic growth factors (such as IL-3) stimulate the division and maturation of the early stem cells, thus increasing the numbers of all blood cell types. Other factors (such as erythropoietin) are specific for certain cell lineages only. A cell's ability to respond to these factors is dependent upon the presence of receptors for the factors on its surface. The number of these receptors is quite low. There are only about 700 receptors for erythropoietin on a CFU-E, and most other progenitor cells have similar low numbers of growth factor receptors. The exception is the receptor for macrophage colony-stimulating factor—M-CSF, also known as CSF-1—which can number up to 73,000 per cell on certain progenitor cells.

HEMATOPOIETIC INDUCTIVE MICROENVIRONMENTS. Different paracrine factors are important in causing hematopoietic stem cells to differentiate along particular pathways (see Figure 15.21). The paracrine factors involved in blood cell and lymphocyte formation are called **cytokines**. Cytokines can be made by several cell types, but they are collected and concentrated by the extracellular matrix of the stromal (mesenchymal) cells at the sites of hematopoiesis (Hunt et al. 1987; Whitlock et al. 1987). For instance, granulocyte-macrophage colony-stimulating factor (GM-CSF) and the multilineage growth factor IL-3 both bind to the heparan sulfate glycosaminoglycan of the bone marrow stroma (Gordon et al. 1987; Roberts et al. 1988). The extracellular matrix is then able to present these factors to the stem cells in concentrations high enough to bind to their receptors.

The developmental path taken by the descendant of a pluripotential stem cell depends on which growth factors it meets, and is therefore determined by the stromal cells. Wolf and Trentin (1968) demonstrated that short-range interactions between stromal cells and stem cells determine the developmental fates of the stem cells' progeny. These investigators placed plugs of bone marrow in a spleen and then injected stem cells into it. Those CFU-S cells that came to reside in the spleen formed colonies that were predominantly erythroid, whereas those colonies that formed in the marrow were predominantly granulocytic. In fact, colonies that straddled the borders of the two tissue types were predominantly erythroid in the spleen and granulocytic in the marrow. Such regions of determination are referred to as **hematopoietic inductive microenvironments** (**HIMs**). In early blood cell formation (in the mesoderm surrounding the mammalian yolk sac), the endothelial cells of the blood islands appear to be heterogeneous HIMs, inducing the stem cells to form different blood and lymphocyte lineages (Lu et al. 1996; Auerbach et al. 1997).

Sites of hematopoiesis

Vertebrate blood development occurs in two phases: a transient **embryonic** ("primitive") **phase** of hematopoiesis is followed by the **definitive** ("adult") **phase**. These phases differ in their sites of blood cell production, the timing of hematopoiesis, the morphology of the cells produced, and even the type of globin genes used in the red blood cells. The embryonic phase of hematopoiesis is probably used to initiate the circulation that provides the embryo with its initial blood cells and with its capillary network to the yolk. The definitive phase of hematopoiesis is used to generate more cell types and to provide the stem cells that will last for the lifetime of the individual.

Embryonic hematopoiesis is associated with the blood islands in the ventral mesoderm near the yolk. In the mouse, embryonic erythropoiesis is seen in the blood islands in the mesoderm surrounding the yolk sac. In chick embryos, the first blood cells are seen in those blood islands forming in the posterior marginal zone near the site of hypoblast initiation (Wilt 1974; Azar and Eyal-Giladi 1979). In *Xenopus*, the ventral mesoderm forms a large blood island that is the first site of hematopoiesis. Zebrafish are the exceptions to this pattern, as their first blood cells form in the paraxial mesoderm. However, this region of paraxial mesoderm contains the same hematopoietic transcription factors as the ventral mesoderm in *Xenopus*, mice, and fish (Detrich et al. 1995). Since vertebrate ventral mesoderm is associated with high concentrations of BMPs, it is not surprising that ectopic BMP 2 and 4 can induce blood and blood vessel formation in *Xenopus*, and that interference with BMP signaling prevents blood formation (Maeno et al. 1994; Hemmati-Brivanlou and Thomsen 1995). Moreover, in the zebrafish, the *swirl* mutation, which prevents

---

*In 1846, the young Joseph Leidy (then a struggling coroner, later the most famous biologist in America) was the first to use a microscope to solve a murder mystery. A man accused of killing a Philadelphia farmer had blood on his clothes and hatchet. The suspect claimed the blood was from chickens he had been slaughtering. Leidy examined the blood and found no nuclei in the erythrocytes; thus, it could not have been chicken blood. The suspect subsequently confessed (Warren 1998).

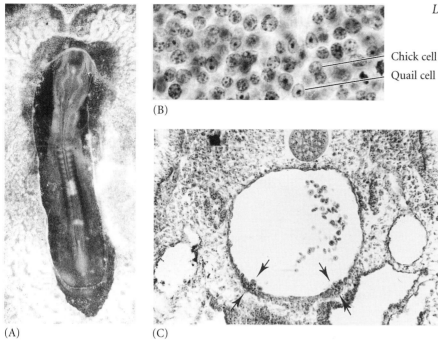

Chick cell
Quail cell

(B)

(A)

(C)

Figure 15.23
Blood cell mapping using chick-quail chimeras. (A) Photograph of a "yolk sac chimera," created by transplanting the blastoderm of a quail onto the yolk sac of a chick. (B) Photograph of chick and quail cells in the thymus of a chimeric animal, showing the difference in nuclear staining. The lymphoid cells are all chick, whereas the structural cells of the thymus are of quail origin. (C) Section through the aorta of a 3-day chick embryo, showing the cells (arrows) that give rise to the hematopoietic stem cells. If cells from this region are taken from quail embryos and placed into chick embryos, the chick embryos will have quail blood. (From Martin et al. 1978, Dieterlen-Lièvre and Martin 1981; photographs courtesy of F. Dieterlen-Lièvre.)

BMP2 signaling, also abolishes ventral mesoderm and blood cell production (Mullins et al. 1996).

The embryonic hematopoietic cell population, however, is transitory. The hematopoietic stem cells that last the lifetime of the organism are derived from the mesodermal area surrounding the aorta. This was shown in the chick by a series of elegant experiments by Dieterlen-Lièvre, who grafted the blastoderm of chickens onto the yolk of Japanese quail (Figure 15.23). Chick cells are readily distinguishable from quail cells because the quail cell nucleus stains much more darkly (owing to its dense nucleoli), thus providing a permanent marker for distinguishing the two cell types (see Figure 1.10). Using these "yolk sac chimeras," Dieterlen-Lièvre and Martin (1981) showed that the yolk sac stem cells do not contribute cells to the adult animal. Instead, the definitive stem cells are formed within nodes of mesoderm that line the mesentery and the major blood vessels. In the 4-day chick embryo, the aortic wall appears to be the most important source of new blood cells, and it has been found to contain numerous hematopoietic stem cells (Cormier and Dieterlen-Lièvre 1988).

Similarly, studies in fishes, mammals, and frogs indicate that the definitive hematopoietic cells are formed near the aorta in a domain called the **aorta-gonad-mesonephros (AGM) region**. The first blood islands in the mouse embryo appear in the mesoderm around the yolk sac, but by day 11, pluripotential hematopoietic stem cells and CFU-S cells can be found in the AGM (Kubai and Auerbach 1983; Godlin et al. 1993; Medvinsky et al. 1993). These are the hematopoietic stem cells that will colonize the liver and constitute the fetal and adult circulatory system (Medvinsky and Dzierzak 1996). Müller and colleagues (1994) have proposed that two waves of cells colonize the fetal liver. According to this hypothesis, the first population of these cells comes from the yolk sac and comprises predominantly pluripotential stem cells. The majority population, however, comes from the AGM, and comprises both CFU-S and CFU-M,L cells (Figure 15.24). This hypothesis was supported by the finding that mice deficient in the tran-

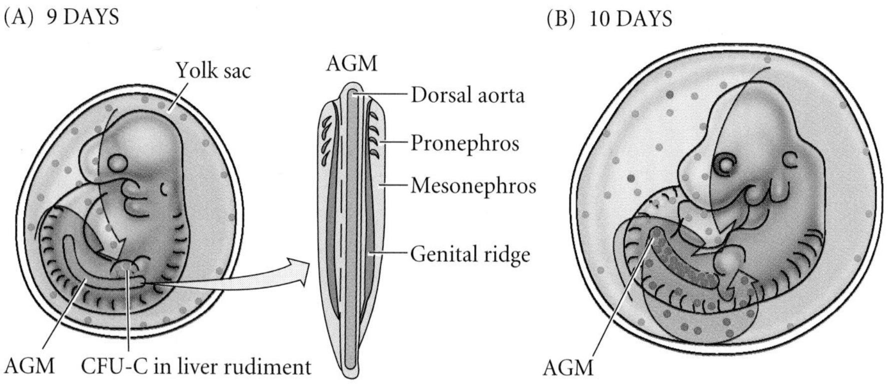

(A) 9 DAYS

Yolk sac

AGM
— Dorsal aorta
— Pronephros
— Mesonephros
— Genital ridge

AGM   CFU-C in liver rudiment

(B) 10 DAYS

AGM

Figure 15.24
Colonization of the mouse liver by two waves of hematopoietic stem cells. The two main sources of hematopoietic progenitor cells are the yolk sac and the AGM region. (A) At day 9, the yolk sac contributes an early line of CFU-C cells that probably does not last long after birth, and which makes a population of predominantly red blood cells. This cell population is thought to be the major source of the first wave of hematopoiesis in the liver. (B) At day 10, the AGM-derived cells provide CFU-S cells and pluripotential hematopoietic stem cells. These constitute the major cells of the second wave. (After Dzierzak and Medvinsky 1995.)

Wild-type   *spadetail*   *cloche*

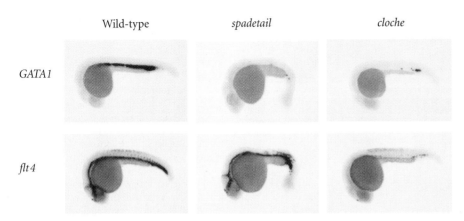

GATA1

flt4

Figure 15.25
Expression of the *flt4* and *GATA1* genes (dark areas) in wild-type and mutant zebrafish embryos. GATA1 is a transcription factor essential for blood cell development, and it is absent in both the *cloche* and *spadetail* mutants. The *flt4* gene is essential for vascular development. It is expressed in the *spadetail* mutant, but not in the *cloche* mutatnt. This indicates that both *spadetail* and *cloche* have defects in blood cell development, but only *cloche* is deficient for vascular development as well. (After Thompson et al. 1998; photographs courtesy of J. Zon.)

scription factor AML1 undergo normal yolk sac hematopoiesis, but no definitive (AGM) hematopoiesis (Okuda et al. 1996). These mutant mice die at embryonic day 12.5. Their livers contain a small number of primitive nucleated red blood cells, whereas control livers are full of blood cells derived from the AGM. The AML protein is critical for activating the genes involved in definitive hematopoiesis. At around the time of birth, stem cells from the liver populate the bone marrow, which then becomes the major site of blood formation throughout adult life.

The pathways of hematopoiesis are difficult to unravel, but recently a new approach has gained momentum. Since zebrafish can be easily screened for mutations of developmentally important genes, over 26 different mutations of hematopoiesis have been found in this species (Liao and Zon 1999). Some of these, such as the swirl mutation, inhibit the production of ventral mesoderm. Mutations of the *moonshine* and *cloche* genes prevent hemangioblast development; mutations of *frascati* and *thunderbird* act at the level of the pluripotential hematopoietic stem cells, and several other mutations affect the pathways leading to the various differentiated blood cell types. By studying gene expression patterns in these mutants, the stage at which the mutant gene operates can be discerned (Figure 15.25; Thompson et al. 1998). For instance, the cloche and spadetail mutants have defects in both primitive and definitive hematopoiesis. The cloche mutant, however, also has defects in vascularization, indicating that *cloche* works at the level of the hemangioblast, while the *spadetail* gene works later, probably at the level of the pluripotential hematopoietic stem cell.

Endoderm

The Pharynx

The function of the embryonic endoderm is to construct the linings of two tubes within the body. The first tube, extending throughout the length of the body, is the **digestive tube**. Buds from this tube form the liver, gallbladder, and pancreas. The second tube, the **respiratory tube**, forms as an outgrowth of the digestive tube, and it eventually bifurcates into two lungs. The digestive and respiratory tubes share a common chamber in the anterior region of the embryo; this region is called the **pharynx**. Epithelial outpockets of the pharynx give rise to the tonsils and the thyroid, thymus, and parathyroid glands.

The respiratory and digestive tubes are both derived from the primitive gut (Figure 15.26). As the endoderm pinches in toward the center of the embryo, the foregut and hindgut regions are formed. At first, the oral end is blocked by a region of ectoderm called the **oral plate**, or **stomodeum**. Eventually (at about 22 days in human embryos), the stomodeum breaks, thereby creating the oral opening of the digestive tube. The opening itself is lined by ectodermal cells. This arrangement creates an interesting situation, because the oral plate ectoderm is in contact with the brain ectoderm, which has curved around toward the ventral portion of the embryo. These two ectodermal regions interact with each other. The roof of the oral region forms Rathke's pouch and becomes the *glandular* part of the pituitary gland. The neural tissue on the floor of the diencephalon gives rise to the infundibulum, which becomes the *neural* portion of the pituitary. Thus, the pituitary gland has a dual origin, which is reflected in its adult functions.

The anterior endodermal portion of the digestive and respiratory tubes begins in the pharynx. Here, the mammalian embryo produces four pairs of **pharyngeal pouches** (Figure 15.27). Between these pouches are the **pharyngeal arches**. The first pair of pharyngeal pouches becomes the auditory cavities of the middle ear and the associated eustachian tubes. The second pair of pouches gives rise to the walls of the tonsils. The thymus is derived from the third pair of pharyngeal pouches; it will direct the differentiation of T lymphocytes during later stages of development. One pair of parathyroid glands is also derived from the third pair of pharyngeal pouches, and the other pair is derived from the fourth. In addition to these paired pouches, a small, central diverticulum is formed between the second pharyngeal pouches on the floor of the phar-

(A) 16 DAYS

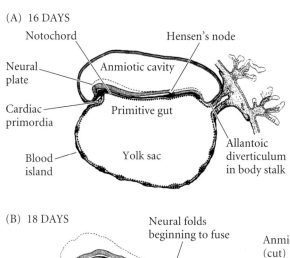

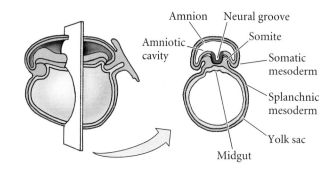

(B) 18 DAYS

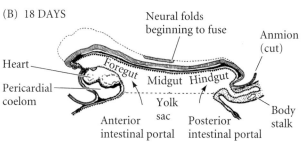

(C) 22 DAYS

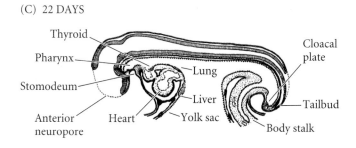

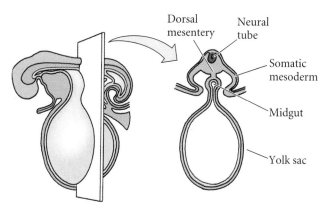

(D) 28 DAYS

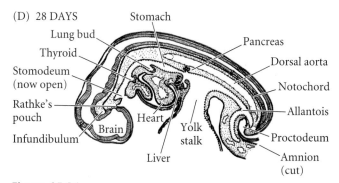

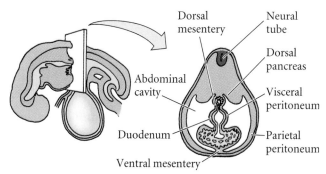

Figure 15.26
Formation of the human digestive system, depicted at about (A) 16 days, (B) 18 days, (C) 22 days, and (D) 28 days. (After Crelin 1961.)

ynx. This pocket of endoderm and mesenchyme will bud off from the pharynx and migrate down the neck to become the thyroid gland. The respiratory diverticulum sprouts from the pharyngeal floor, between the fourth pair of pharyngeal pouches, to form the lungs, as we will see below.

The Digestive Tube and Its Derivatives

Posterior to the pharynx, the digestive tube constricts to form the esophagus, which is followed in sequence by the stomach, small intestine, and large intestine. The endodermal cells generate only the lining of the digestive tube and its glands; mesodermal mesenchyme cells will surround this tube to provide the muscles for peristalsis.

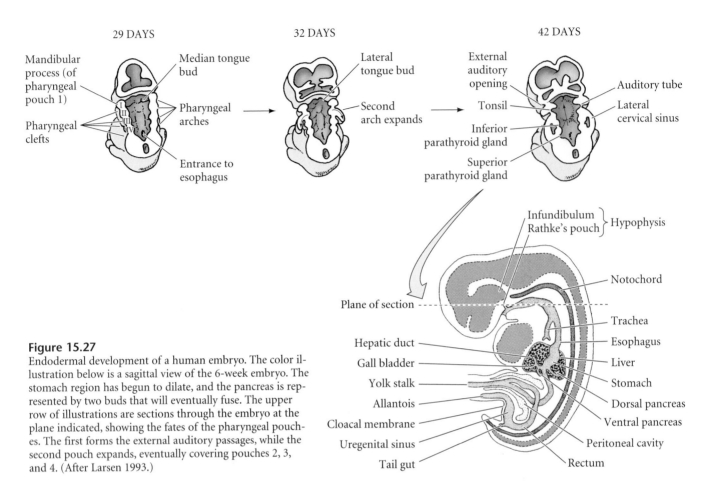

29 DAYS

Mandibular process (of pharyngeal pouch 1)

Median tongue bud

Pharyngeal arches

Pharyngeal clefts

Entrance to esophagus

32 DAYS

Lateral tongue bud

Second arch expands

42 DAYS

External auditory opening

Auditory tube

Tonsil

Lateral cervical sinus

Inferior parathyroid gland

Superior parathyroid gland

Infundibulum
Rathke's pouch } Hypophysis

Notochord

Plane of section

Hepatic duct

Gall bladder

Yolk stalk

Allantois

Cloacal membrane

Uregenital sinus

Tail gut

Trachea

Esophagus

Liver

Stomach

Dorsal pancreas

Ventral pancreas

Peritoneal cavity

Rectum

Figure 15.27
Endodermal development of a human embryo. The color illustration below is a sagittal view of the 6-week embryo. The stomach region has begun to dilate, and the pancreas is represented by two buds that will eventually fuse. The upper row of illustrations are sections through the embryo at the plane indicated, showing the fates of the pharyngeal pouches. The first forms the external auditory passages, while the second pouch expands, eventually covering pouches 2, 3, and 4. (After Larsen 1993.)

As Figure 15.27 shows, the stomach develops as a dilated region close to the pharynx. More caudally, the intestines develop, and the connection between the intestine and yolk sac is eventually severed. At the caudal end of the intestine, a depression forms where the endoderm meets the overlying ectoderm. Here, a thin **cloacal membrane** separates the two tissues. It eventually ruptures, forming the opening that will become the anus.

Specification of the gut tissue

As the endodermal tubes form, the endodermal epithelium is able to respond differently to different regionally specific mesodermal mesenchymes. This enables the digestive tube and respiratory tube to develop different structures in different regions. Thus, as the mammalian digestive tube meets new mesenchymes, it differentiates into esophagus, stomach, small intestine, and colon (Gumpel-Pinot et al. 1978; Fukumachi and Takayama 1980; Kedinger et al. 1990).

The specificity of the mesoderm is thought to be controlled by its interactions with the endodermal tube during earlier stages of development. As the gut tube begins to form from the anterior and posterior ends, it induces the splanchnic mesoderm to become regionally specific. Roberts and colleagues (1995, 1997) have implicated Sonic hedgehog in this

specification. Early in development, the expression of Shh is limited to the posterior endoderm of the hindgut and the pharynx. As the tubes extend toward the center of the embryo, the domains of Shh expression increase, eventually extending throughout the gut endoderm. Shh is secreted in different concentrations at different sites, and its target appears to be the mesoderm surrounding the gut tube. (The splanchnic mesoderm cells around the gut contain Patched protein, the receptor for the Hedgehog protein family, in their cell membranes.) In the hindgut, the secretion of Shh by the endoderm induces in the mesoderm a nested set of "posterior" Hox gene expression. As in the vertebrae (see Chapter 11), the anterior borders of the expression pattern delineate the morphological boundaries of the regions that will form the cloaca, large intestine, cecum, mid-cecum (at the midgut/hindgut border), and the posterior portion of the midgut (Figure 15.28; Roberts et al. 1995; Yokouchi et al. 1995). When Hox-containing viruses cause the misexpression of these Hox genes in the mesoderm, the mesodermal cells alter the differentiation of the adjacent endoderm (Roberts et al. 1998). Thus, the endodermal expression of Shh in the hindgut seems to induce a nested expression of Hox genes in the mesoderm. These Hox genes probably specify the mesoderm so that it can interact with the endodermal tube and specify its regions.

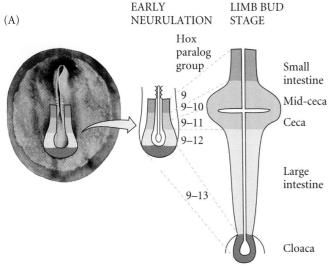

EARLY NEURULATION   LIMB BUD STAGE

Hox paralog group

9
9–10
9–11
9–12

9–13

Small intestine
Mid-ceca
Ceca

Large intestine

Cloaca

Figure 15.28
Regional specification of the visceral mesoderm through interactions with the posterior gut endoderm. (A) The expression and secretion of Sonic hedgehog in the endoderm generates a nested set of Hox gene expression in the adjacent mesoderm. After the mesoderm is specified, it can act on the endodermal tube to induce specific morphological regions. (B) Possible course of interactions between the endoderm and the visceral mesoderm. (A after Roberts et al. 1995.)

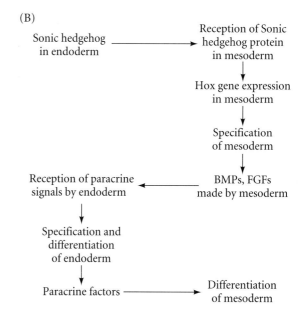

(B)

Sonic hedgehog in endoderm → Reception of Sonic hedgehog protein in mesoderm

↓

Hox gene expression in mesoderm

↓

Specification of mesoderm

↓

Reception of paracrine signals by endoderm ← BMPs, FGFs made by mesoderm

↓

Specification and differentiation of endoderm

↓

Paracrine factors → Differentiation of mesoderm

Liver, pancreas, and gallbladder

The endoderm also forms the lining of three accessory organs that develop immediately caudal to the stomach. The **hepatic diverticulum** is the tube of endoderm that extends out from the foregut into the surrounding mesenchyme. The mesenchyme induces this endoderm to proliferate, to branch, and to form the glandular epithelium of the liver. A portion of the hepatic diverticulum (that region closest to the digestive tube) continues to function as the drainage duct of the liver, and a branch from this duct produces the gallbladder (Figure 15.29).

The pancreas develops from the fusion of distinct dorsal and ventral diverticula. Both of these primordia arise from the endoderm immediately caudal to the stomach, and as they grow, they come closer together and eventually fuse. In humans, only the ventral duct survives to carry digestive enzymes into the intestine. In other species (such as the dog), both the dorsal and ventral ducts empty into the intestine.

Figure 15.29
Pancreatic development in humans. (A) At 30 days, the ventral pancreatic bud is close to the liver primordium. (B) By 35 days it begins migrating posteriorly, and (C) comes into contact with the dorsal pancreatic bud during the sixth week of development. (D) In most individuals, the dorsal pancreatic bud loses its duct into the duodenum; however, in about 10% of the population, the dual duct system persists. (After Langman 1981.)

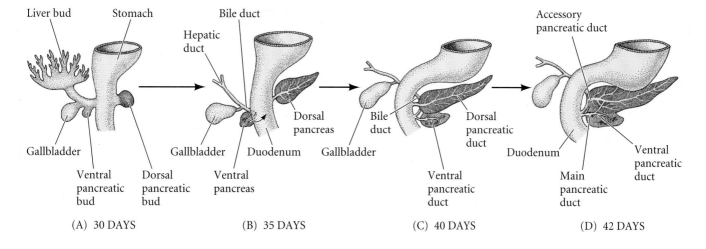

(A) 30 DAYS   (B) 35 DAYS   (C) 40 DAYS   (D) 42 DAYS

The Specification of Liver and Pancreas

THERE IS AN INTIMATE RELATIONSHIP between the lateral plate splanchnic mesoderm and the foregut endoderm. Just as the foregut endoderm is critical in specifying the cardiogenic mesoderm, so the cardiogenic mesoderm induces the endodermal tube to produce the liver primordium. This induction is probably due to FGFs secreted by the developing heart cells (Le Douarin 1975; Gualdi et al. 1996; Jung et al. 1999).

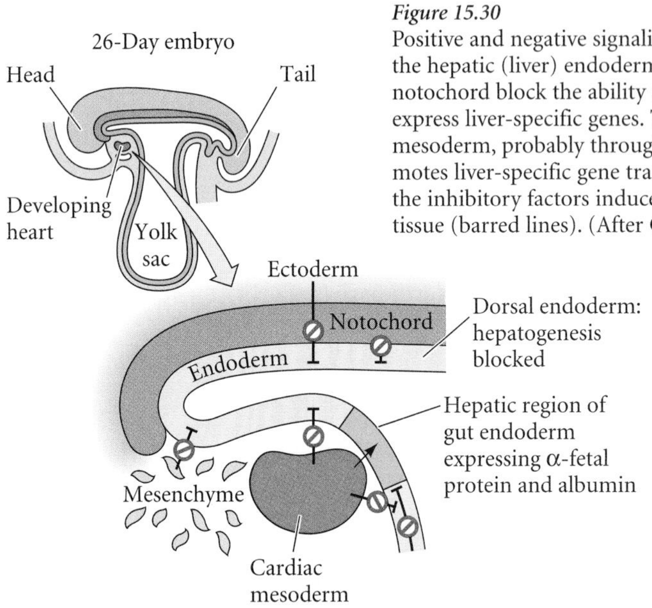

Liver Formation

Liver-specific gene expression (such as the genes for α-fetoprotein and albumin) can occur anywhere in the gut tube, if that tube is exposed to cardiogenic mesoderm. However, this induction can occur only if the notochord is removed. If the notochord is placed by the portion of the endoderm normally induced by the cardiac tissue to become liver, the endoderm will not form liver tissue. Therefore, the developing heart appears to induce the liver to form, while the presence of the notochord inhibits liver formation (Figure 15.30). As mentioned in Chapter 5, the heart (and therefore the FGFs) appears to act by blocking an inhibitor of liver-specific gene transcription.

Pancreas Formation

The formation of the pancreas may be the flip side of liver formation. While the heart cells promote and the notochord prevents liver formation, the notochord may actively promote pancreas formation, and the heart may block the pancreas from forming. It has been hypothesized (Gannon and Wright 1999) that the region of *pdx1* expression in the endoderm endows a particular region of the digestive tube (including future portions of the stomach, duodenum, liver, and pancreas rudiments) with the ability to become either pancreas or liver. One set of conditions (presence of heart, absence of notochord) induces the liver, while another set of conditions (presence of notochord, absence of heart) causes the pancreas to form.

The notochord activates pancreas development by repressing Sonic hedgehog expression in the endoderm (Apelqvist et al. 1997; Hebrok et al. 1998). (This was a sur-

Figure 15.30
Positive and negative signaling in the formation of the hepatic (liver) endoderm. The ectoderm and notochord block the ability of the endoderm to express liver-specific genes. The cardiogenic mesoderm, probably through FGF1 or FGF2, promotes liver-specific gene transcription by blocking the inhibitory factors induced by the surrounding tissue (barred lines). (After Gualdi et al. 1996.)

prising finding, since we saw in Chapter 13 that the notochord is a source of Sonic hedgehog and an inducer of further *shh* gene expression in ectodermal tissues.) Sonic hedgehog is expressed throughout the gut endoderm, except in the region that will form the pancreas. The notochord in this region of the embryo secretes FGF2 and activin, which are able to down-regulate *shh* expression in the endoderm. Once the endodermal expression pattern of *shh* is established, the Pdx1 transcription factor becomes expressed in the pancreatic epithelium. If *shh* is experimentally expressed in this region, the tissue reverts to being intestinal. The *pdx1* gene appears to give the pancreatic epithelium the ability to respond to its mesenchyme. Mice without this gene lack a pancreas, although their epithelium does differentiate into pre-islet cells that synthesize small amounts of glucagon and insulin (Jonnson et al. 1994; Ahlgren et al. 1996; Offield et al. 1996).

Once the pancreatic rudiments are initiated, they begin to form both exocrine and endocrine tissue. The exocrine tissue produces amylase and α-fetoprotein, while the endocrine tissue makes insulin, glucagon, and

somatostatin. The ratio of exocrine and endocrine cells is regulated by follistatin, a protein secreted by the pancreatic mesenchyme. Follistatin (which inhibits certain BMPs and activin) promotes the development of exocrine cells and represses the formation of endocrine cells (Miralles et al. 1998). The initial endocrine cells are thought to express the Pax6 and Pax4 transcription factors. The Pax6 transcription factor (in conjunction with Pdx1) may regulate both the expression of the pancreatic hormone genes (see Chapter 5) and the expression of adhesion proteins that link the endocrine cells together into the islets of Langerhans. Pax6-deficent mice have deficient pancreatic hormone production and malformed islets. Those cells that retain both Pax6 and Pax4 become the β-cells of the islets of Langerhans, and they produce insulin. Those islet cells that down-regulate Pax4 and synthesize only Pax6 become the α-cells that secrete glucagon (Sosa-Pineda et al. 1997; St-Onge et al. 1997). These transcription factors are important both for the morphogenesis of particular pancreatic cell types and for the expression of the hormone genes characteristic of the mature cell. ■

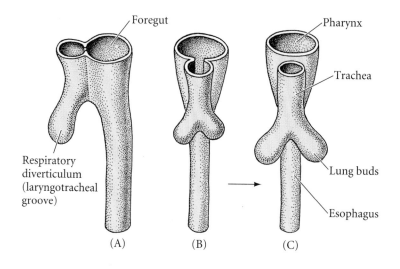

Foregut

Pharynx

Trachea

Respiratory
diverticulum
(laryngotracheal
groove)

Lung buds

Esophagus

(A)   (B)   (C)

Figure 15.31
Partitioning of the foregut into the esophagus and respiratory diverticulum during the third and fourth weeks of human gestation. (A) Lateral view, end of week 3. (B, C) Ventral views, week 4. (After Langman 1981.)

counters (Shannon et al. 1998). Thus, the respiratory epithelium is extremely malleable and can differentiate according to its mesenchymal instructions.

WEBSITE 15.4 **Induction of the lung.** The induction of the lung also involves the interplay between FGFs and Shh. However, it appears to be different from the induction of either the pancreas or the liver.

The Respiratory Tube

The lungs are a derivative of the digestive tube, even though they serve no role in digestion. In the center of the pharyngeal floor, between the fourth pair of pharyngeal pouches, the **laryngotracheal groove** extends ventrally (Figure 15.31). This groove then bifurcates into the two branches that form the paired bronchi and lungs. The laryngotracheal endoderm becomes the lining of the trachea, the two bronchi, and the air sacs (alveoli) of the lungs.

The lungs are an evolutionary novelty, and they are among the last of the mammalian organs to fully differentiate. The lungs must be able to draw in oxygen at the newborn's first breath. To accomplish this, the alveolar cells secrete a surfactant into the fluid bathing the lungs. This surfactant, consisting of phospholipids such as sphingomyelin and lecithin, is secreted very late in gestation, and it usually reaches physiologically useful levels at about week 34 of human gestation. The surfactant enables the alveolar cells to touch one another without sticking together. Thus, infants born prematurely often have difficulty breathing and have to be placed on respirators until their surfactant-producing cells mature.

As in the digestive tube, the regional specificity of the mesenchyme determines the differentiation of the developing respiratory tube. In the developing mammal, the respiratory epithelium responds in two distinct fashions. In the region of the neck, it grows straight, forming the trachea. After entering the thorax, it branches, forming the two bronchi and then the lungs. The respiratory epithelium can be isolated soon after it has split into two bronchi, and the two sides can be treated differently. Figure 15.32 shows the result of such an experiment. The right bronchial epithelium retained its lung mesenchyme, whereas the left bronchus was surrounded with tracheal mesenchyme (Wessells 1970). The right bronchus proliferated and branched under the influence of the lung mesenchyme, whereas the left bronchus continued to grow in an unbranched manner. Moreover, the differentiation of the respiratory epithelia into trachea or lung cells depends on the mesenchyme it en-

The Extraembryonic Membranes

In reptiles, birds, and mammals, embryonic development has taken a new evolutionary direction. Reptiles evolved a mechanism for laying eggs on dry land, thus freeing them to explore niches that were not close to water. To accomplish this, the embryo produces four sets of extraembryonic membranes to mediate between it and the environment (see Chapter 11). Even though most mammals have evolved placentas instead of shells, the basic pattern of extraembryonic membranes remains the same. In developing reptiles, birds, and mammals, there initially is no distinction between embryonic and extraembryonic domains. However, as the body of the embryo takes shape, the epithelia at the border between the embryo and the extraembryonic domain divide unequally to create body folds that isolate the embryo from the yolk and delineate which areas are to be embryonic and which extraem-

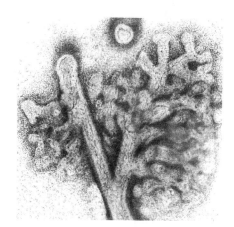

Figure 15.32
Ability of presumptive lung epithelium to differentiate with respect to the source of the inducing mesenchyme. After embryonic mouse lung epithelium had branched into two bronchi, the entire rudiment was excised and cultured. The right bronchus was left untouched, while the tip of the left bronchus was covered with tracheal mesenchyme. The tip of the right bronchus has formed the branches characteristic of the lung, whereas hardly any branching has occurred in the tip of the left bronchus. (From Wessells 1970; photograph courtesy of N. Wessells.)

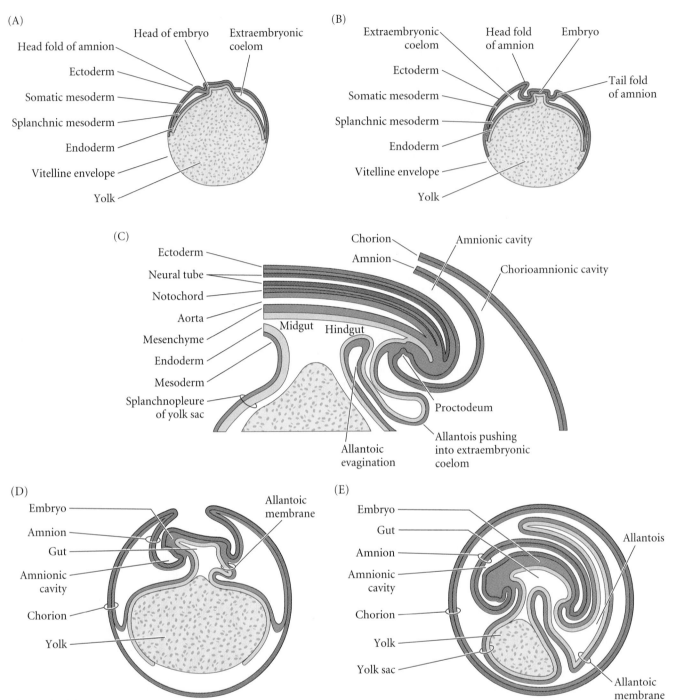

Figure 15.33
Schematic drawings of the extraembryonic membranes of the chick.
The embryo is cut longitudinally, and the albumin and shell coatings are not shown. (A) A 2-day embryo. (B) A 3-day embryo. (C)
Detailed schematic diagram of the caudal (hind) region of the chick
embryo, showing the formation of the allantois. (D) A 5-day embryo. (E) A 9-day embryo. (After Carlson 1981.)

bryonic (Miller et al. 1994, 1999). These membranous folds are
formed by the extension of ectodermal and endodermal epithelium underlain with lateral plate mesoderm. The combination
of ectoderm and mesoderm, often referred to as the **somatopleure**, forms the **amnion** and **chorion**; the combination of
endoderm and mesoderm—the **splanchnopleure**—forms the
yolk sac and **allantois**. The endodermal or ectodermal tissue

supplies functioning epithelial cells, and the mesoderm generates the essential blood supply to and from this epithelium. The
formation of these folds can be followed in Figure 15.33.

The amnion and chorion

The first problem of a land-dwelling egg is desiccation.
Embryonic cells would quickly dry out if they were not in an
aqueous environment. This environment is supplied by the
amnion. The cells of this membrane secrete **amnionic fluid**;
thus, embryogenesis still occurs in water. This evolutionary
adaptation is so significant and characteristic that reptiles,
birds, and mammals are grouped together as the **amniote vertebrates**, or amniotes.

The second problem of a land-dwelling egg is gas exchange. This exchange is provided for by the chorion, the outermost extraembryonic membrane. In birds and reptiles, this membrane adheres to the shell, allowing the exchange of gases between the egg and the environment. In mammals, as we have seen, the chorion has developed into the **placenta**, which has evolved endocrine, immune, and nutritive functions in addition to those of respiration.

The allantois and yolk sac

The third problem for a land-dwelling egg is waste disposal. The allantois stores urinary wastes and also helps mediate gas exchange. In reptiles and birds, the allantois becomes a large sac, as there is no other way to keep the toxic by-products of metabolism away from the developing embryo. In some amniote species, such as chickens, the mesodermal layer of the allantoic membrane reaches and fuses with the mesodermal layer of the chorion to create the **chorioallantoic membrane**. This extremely vascular envelope is crucial for chick development and is responsible for transporting calcium from the eggshell into the embryo for bone production (Tuan 1987). In mammals, the size of the allantois depends on how well nitrogenous wastes can be removed by the chorionic placenta. In humans (in which nitrogenous wastes can be efficiently removed through the maternal circulation), the allantois is a vestigial sac. In pigs, however, the allantois is a large and important organ.

The fourth problem that a land-dwelling egg has to solve is nutrition. The yolk sac is the first extraembryonic membrane to be formed, as it mediates nutrition in developing birds and reptiles. It is derived from splanchnopleural cells that grow over the yolk to enclose it. The yolk sac is connected to the midgut by an open tube, the **yolk duct**, so that the walls of the yolk sac and the walls of the gut are continuous. The blood vessels within the mesoderm of the splanchnopleure transport nutrients from the yolk into the body, for yolk is not taken directly into the body through the yolk duct. Rather, endodermal cells digest the protein into soluble amino acids that can then be passed on to the blood vessels within the yolk sac. Other nutrients, including vitamins, ions, and fatty acids, are stored in the yolk sac and transported by the yolk sac blood vessels into the embryonic circulation. In these ways, the four extraembryonic membranes enable the amniote embryo to develop on land.

Snapshot Summary: Lateral Mesoderm and Endoderm

1. The lateral plate mesoderm splits into two layers. The dorsal layer is the somatic (parietal) mesoderm, which underlies the ectoderm and forms the somatopleure. The ventral layer is the splanchnic (visceral) mesoderm, which overlies the endoderm and forms the splanchnopleure.

2. The space between these two layers is the body cavity, the coelom.

3. The heart arise from splanchnic mesoderm on both sides of the body. This region of cells is called the cardiogenic mesoderm.

4. The Nkx2-5 transcription factor is important in specifying cells to become cardiogenic mesoderm. These cells migrate from the sides to the midline of the embryo, in the neck region.

5. Cardiogenic mesoderm forms the endocardium (which is continuous with the blood vessels) and the myocardium (the muscular component of the heart).

6. The endocardial tubes form separately and then fuse. The looping of the heart transforms the original anterior-posterior polarity of the heart tube into a right-left polarity.

7. In mammals, fetal circulation differs dramatically from adult circulation. When the infant takes its first breath, changes in air pressure close the foramen ovale through which blood had passed from the right to the left atrium. At that time, the lungs, rather than the placenta, become the source of oxygen.

8. Blood vessel formation is constrained by physiological, evolutionary, and physical parameters. The subdividing of a large vessel into numerous smaller ones allows rapid transport of the blood to regions of gas and nutrient diffusion.

9. Blood vessels are constructed by two processes, vaculogenesis and angiogenesis. Vasculogenesis involves the condensing of visceral mesoderm cells to form blood islands. The outer cells of these islands become endothelial (blood vessel) cells. Angiogenesis involves remodeling existing blood vessels.

10. Numerous paracrine factors are essential in blood vessel formation. FGF2 is needed for specifying the angioblasts. VEGF is essential for the differentiation of angioblasts. Angiopoietin-1 allows the smooth muscle cells (and smooth muscle-like pericytes) to cover the vessels. Ephrin ligands and Eph receptor tyrosine kinases are critical for capillary bed formation.

11. The pluripotential hematopoietic stem cell generates other pluripotential stem cells, as well as lineage-restricted stem cells. It gives rise to both blood cells and lymphocytes.

12. The CFU-S is a blood stem cell that can generate the more committed stem cells for the different blood lineages. Hematopoietic inductive microenvironments determine the direction of the blood cell differentiation.

13. In mammals, embryonic blood stem cells are provided by the blood islands near the yolk. The definitive adult blood stem cells come from the aorta-gonad-mesonephros region within the embryo.

14. The endoderm constructs the digestive tube and the respiratory tube.

15. Four pairs of pharyngeal pouches become the endodermal lining of the eustacian tube, tonsils, thymus, and parathyroid glands. The thyroid also forms in this region of endoderm.

16. The gut tissue forms by reciprocal interactions between the endoderm and the mesoderm. Sonic hedgehog from the endoderm appears to play a role in inducing a nested pattern of Hox gene expression in the mesoderm surrounding the gut. The regionalized mesoderm then instructs the endodermal tube to become the different organs of the digestive tube.

17. The pancreas forms in a region of endoderm that lacks Sonic hedgehog expression. The Pdx1 transcription factor is expressed in this region.

18. The respiratory tube is derived as an outpocketing of the digestive tube. The regional specificity of the mesenchyme it meets determines whether the tube remains straight (as in the tracheae) or branches (as in the alveoli).

19. The yolk sac and allantois are derived from the splanchnopleure. The yolk sac (in birds and reptiles) allows yolk nutrients to pass into the blood. The allantois collects nitrogenous wastes.

20. The chorion and amnion are made by the somatopleure. In birds and reptiles, the chorion abuts the shell and allows for gas exchange. The amnion in birds, reptiles, and mammals bathes the embryo in amnionic fluid.

Literature Cited

Abramson, S., R. G. Miller and R. A. Phillips. 1977. The identification in adult bone marrow of pluripotent and restricted stem cells of the myeloid and lymphoid systems. *J. Exp. Med.* 145: 1567–1579.

Ahlgren, U., J. Jonnson and H. Edlund. 1996. The morphogenesis of the pancreatic mesenchyme is uncoupled from that of the pancreatic epithelium in IPF/PDX1-deficient mice. *Development* 122: 1409–1416.

Alini, M., A. Marriott, T. Chen, S. Abe and A. R. Poole. 1996. A novel angiogenic molecule produced at the time of chondrocyte hypertrophy during endochondral bone formation. *Dev. Biol.* 176: 124–132.

Andrée, B., D. Duprez, B. Vorbusch, H.-H., Arnold and T. Brand. 1998. BMP-2 induces ectopic expression of cardiac lineage markers and interferes with somite formation in chicken embryos. *Mech. Dev.* 70: 119–131.

Antczak, A. M., J. van Blerkom and A. Clark. 1997. A novel mechanism of VEGF, leptin, and TGFβ2 sequestration in human ovarian follicle cells. *Hum. Reprod.* 12: 2226–2234.

Apelqvist, A, U. Ahlgren and H. Edlund. 1997. Sonic hedgehog directs specialised mesoderm differentiation in the intestine and pancreas. *Curr. Biol.* 7: 801–804.

Auerbach, R., L. Alby, L. Morrissey, M. Tu and J. Joseph. 1985. Expression of organ-specific antigens on capillary endothelial cells. *Microvasc. Res.* 29: 401–411.

Auerbach, R., B. Gilligan, L.-S. Lu and S.-J. Wang. 1997. Cell interactions in the mouse yolk sac: Vasculogenesis and hematopoiesis. *J. Cell. Physiol.* 173: 202–205.

Azar, Y. and H. Eyal-Giladi. 1979. Marginal zone cells—the primitive streak-inducing component of the primary hypoblast in the chick. *J. Embryol. Exp. Morphol.* 52: 79–88.

Bao, Z.-Z., B. G. Bruneau, J. G. Seidman, C. E. Seidman and C. L. Cepko. 1999. Regulation of chamber-specific gene expression in the developing heart by *Irx4*. *Science* 283: 1161–1164.

Becker, A. J., E. A. McCulloch and J. E. Till. 1963. Cytological demonstration of the clonal nature of spleen cells derived from transplanted mouse marrow cells. *Nature* 197: 452–454.

Berardi, A. C., A. Wang, J. D. Levine, P. Lopez and D. T. Scadden. 1995. Functional characterization of human hematopoietic stem cells. *Science* 267: 104–108.

Biben, C. and R. P. Harvey. 1997. Homeodomain factor Nkx2-5 controls left-right asymmetric expression of bHLH gene *eHand* during murine heart development. *Genes Dev.* 11: 1357–1369.

Bruneau, B. G., M. Logan, N. Davis, T. Levi, C. J. Tabin, J. G. Seidman and C. E. Seidman. 1999. Chamber-specific cardiac expression of *Tbx5* and heart defects in Holt-Oram syndrome. *Dev. Biol.* 211: 100–108.

Carlson, B. M. 1981. *Patten's Foundations of Embryology.* McGraw-Hill, New York.

Choi, H., M. Kennedy, A. Kazarov, J. C. Padaimitriou and G. Keller. 1998. A common precursor for hematopoietic and endothelial cells. *Development* 125: 725–732.

Cohen-Gould, L. and T. Mikawa. 1996. The fate diversity of mesodermal cells within the heart field during chicken early embryogenesis. *Dev. Biol.* 177: 265–273.

Cormier, F. and F. Dieterlen-Lièvre. 1988. The wall of the chick aorta harbours M-CFC, G-CFC, GM-CFC and BFU-E. *Development* 102: 279–285.

Crelin, E. S. 1961. Development of the gastrointestinal tract. *Clin. Symp.* 13: 68–82.

Davis, S. and 10 others. 1996. Isolation of angiopoietin-1, a ligand for the TIE2 receptor, by secretion trap expression cloning. *Cell* 87: 1161–1169.

DeHaan, R. L. 1959. Cardia bifida and the development of pacemaker function in the early chicken heart. *Dev. Biol.* 1: 586–602.

DeHaan, R. L. 1967. Regulation of spontaneous activity and growth of embryonic chick heart cells in tissue culture. *Dev. Biol.* 16: 216–249.

Detrich, H. W. III and 8 others. 1995. Intraembryonic hematopoietic cell migration during vertebrate development. *Proc. Natl. Acad. Sci. USA* 92: 10713–10717.

Dieterlen-Lièvre, F. and C. Martin. 1981. Diffuse intraembryonic hemopoiesis in normal and chimeric avian development. *Dev. Biol.* 88: 180–191.

Dzierzak, E. and A. Medvinsky. 1995. Mouse embryonic hematopoiesis. *Trends Genet.* 11: 359–366.

Evans, H. M. 1909. On the earliest blood vessels in the anterior limbs of birds and their relation to the primary subclavian artery. *Am. J. Anat.* 9: 281–319.

Feinberg, R. N. 1991. Vascular development in the embryonic limb bud. In R. N. Feinberg, G. K. Sherer and R. Auerbach (eds.), *The Development of the Vascular System.* Issues in Biomedicine 14. Karger, Basel, pp. 136–148.

Feinberg, R. N. and E. Cafasso. 1995. Macromolecular permeability of chick wing microvessels: An intravital study. *Anat. Embryol.* 191: 337–342.

Ferrara, N. and K. Alitalo. 1999. Clinical applications angiogenic growth factors and their inhibitors. *Nature Med.* 5: 1359–1364.

Ferrara, N. and 8 others. Heterozygous embryonic lethality induced by targeted inactivation of the VEGF gene. *Nature* 380: 439–442.

Fidler, I. J. and L. M. Ellis. 1994. The implications of angiogenesis for the biology and therapy of cancer metastasis. *Cell* 70: 185–188.

Flamme, I. and W. Risau. 1992. Induction of vasculogenesis and hematogenesis in vitro. *Development* 116: 435–439.

Fong, G.-H., J. Rossant, M. Gertenstein and M. L. Breitman. 1995. Role of the Flt-1 receptor tyrosine kinase in regulating the assembly of vascular endothelium. *Nature* 376: 66–70.

Fukumachi, H. and S. Takayama. 1980. Epithelial-mesenchymal interaction in differentiation of duodenal epithelium of fetal rats in organ culture. *Experientia* 36: 335–336.

Gannon, M. and C. Wright. 1999. Endodermal patterning and organogenesis. *In* S. A. Moody (ed.), *Cell Lineage and Fate Determination.* Academic Press, San Diego. pp. 583–615.

Garcia-Martinez, V. and G. C. Schoenwolf. 1993. Primitive-streak origin of the cardiovascular system in avian embryos. *Dev. Biol.* 159: 706–719.

Godlin, I. E., J. A. Garcia-Porrero, A. Coutinho, F. Dieterlen-Lièvre and M. A. R. Marcos. 1993. Para-aortic splanchnopleura from early mouse embryos contain B1a cell progenitors. *Nature* 364: 67–70.

Gordon, M. Y., G. P. Riley, S. M. Watt and M. F. Greaves. 1987. Compartmentalization of a haematopoietic growth factor (GM-CSF) by glycosaminoglycans in the bone marrow microenvironment. *Nature* 326: 403–405.

Gräper, L. 1907. Untersuchungen über die Herzbildung der Vögel. *Wilhelm Roux Arch. Entwicklungsmech. Org.* 24: 375–410.

Gualdi, R., P. Bossard, M. Zheng, Y. Hamada, J. R. Coleman and K. S. Zaret. 1996. Hepatic specification of the gut endoderm in vitro: Cell signaling and transcriptional control. *Genes Dev.* 10: 1670–1682.

Gumpel-Pinot, M., S. Yasugi and T. Mizuno. 1978. Différenciation d'épithéliums endodermiques associaés au mésoderme splanchnique. *Comp. Rend. Acad. Sci.* (Paris) 286: 117–120.

Hanahan, D. 1997. Signaling vascular morphogenesis and maintenance. *Science* 277: 48–50.

Harary, I. and B. Farley. 1963. In vitro studies on single beating rat heart cells. II. Intercellular communication. *Exp. Cell Res.* 29: 466–474.

Harvey, William 1651. *Exercitationes de Generatione Animalium.* Jansson, Amsterdam. Quoted in J. Needham, *A History of Embryology.* Abelard-Schuman, New York. pp. 133–153.

Hay, E. 1966. *Regeneration.* Holt, Rinehart & Winston, New York.

Hebrok, M., S. Kim and D. A. Melton. 1998. Notochord repression of endodermal sonic hedgehog permits pancreas development. *Genes Dev.* 12: 1705–1713.

Hemmati-Brivanlou, A. and G. H. Thomsen. 1995. Ventral mesodermal patterning in *Xenopus* embryos: Expression patterns and activities of BMP2 and BMP4. *Dev. Genet.* 17: 78–89.

Humphries, R. K., P. B. Jacky, F. J. Dill, A. C. Eaves and C. J. Eaves. 1979. CFUs in individual erythroid colonies derived in vitro from adult mature mouse marrow. *Nature* 279: 718–720.

Hunt, P., D. Robertson, D. Weiss, D. Rennick, F. Lee and O. N. Witte. 1987. A single bone marrow stromal cell type supports the in vitro growth of early lymphoid and myeloid cells. *Cell* 48: 997–1007.

Jackson, D., O. V. Volpert, N. Bouk and D. I. H. Linzer. 1994. Stimulation and inhibition of angiogenesis by placental proliferin and proliferin-related protein. *Science* 266: 1581–1584.

Jonnson, J., L. Carlsson, T. Edlund and H. Edlund. 1994. Insulin-promote-factor 1 is required for pancreas development in mice. *Nature* 371: 606–608.

Jung, J., M. Goldfarb and K. S. Zaret. 1999. Initiation of mammalian liver development from endoderm by fibroblast growth factors. *Science* 284: 1998–2003.

Jurssková V. and L. Tkadlcek. 1965. Character of primary and secondary colonies of haematopoiesis in the spleen of irradiated mice. *Nature* 206: 951–952.

Katayama, I. and H. Kayano. 1999. Yolk sac with blood island. *New Engl. J. Med.* 340: 617.

Kedinger, M., P. M. Simon-Assman, F. Bouziges, C. Arnold, E. Alexandre and K. Haffen. 1990. Smooth muscle actin expression during rat gut development and induction in fetal skin fibroblastic cells associated with intestinal embryonic epithelium. *Differentiation* 43: 87–97.

Keller, G., C. Paige, E. Gilboa and E. F. Wagner. 1985. Expression of a foreign gene in myeloid and lymphoid cells derived from multipotent hematopoietic precursors. *Nature* 318: 149–154.

Komuro, I. and S. Izumo. 1993. Csx: A murine homeobox-containing gene specifically expressed in the developing heart. *Proc. Natl. Acad. Sci. USA* 90: 8145–8149.

Krantz, S. B. and E. Goldwasser. 1965. On the mechanism of erythropoietin induced differentiation. II. The effect on RNA synthesis. *Biochim. Biophys. Acta* 103: 325–332.

Kubai, L. and R. Auerbach. 1983. A new source of embryonic lymphocytes in the mouse. *Nature* 301: 154–156.

LaBarbera, M. 1990. Principles of design of fluid transport systems in zoology. *Science* 249: 992–1000.

Langman, J. 1981. *Medical Embryology,* 4th Ed. Williams & Wilkins, Baltimore.

Larsen, W. J. 1993. *Human Embryology.* Churchill-Livingstone, New York.

Le Douarin, N. 1975. An experimental analysis of liver development. *Med. Biol.* 53: 427–455.

Lemischka, I. R., D. H. Raulet and R. C. Mulligan. 1986. Developmental potential and dynamic behavior of hematopoietic stem cells. *Cell* 45: 917–927.

Liao, E. C. and L./ I. Zon. 1999. Conservation of themes in vertebrate blood development. *In* S. A. Moody (ed.), *Cell Lineage and Fate Determination.* Academic Press, San Diego. pp. 569–582.

Linask, K. K. and J. W. Lash. 1986. Precardiac cell migration: Fibronectin localization at mesoderm-endoderm interface during directional movement. *Dev. Biol.* 114: 87–101.

Linask, K. K. and J. W. Lash. 1988a. A role for fibronectin in the migration of avian precardiac cells. I. Dose-dependent effects of fibronectin antibody. *Dev. Biol.* 129: 315–323.

Linask, K. K. and J. W. Lash. 1988b. A role for fibronectin in the migration of avian precardiac cells. II. Rotation of the heart-forming region during different stages and their effects. *Dev. Biol.* 129: 324–329.

Linask, K. K. and J. W. Lash. 1993. Early heart development: Dynamics of endocardial cell sorting suggests a common origin with cardiomyocytes. *Dev. Dynam.* 195: 62–66.

Linask, K. K., K. A. Knudsen and Y.-H. Gui. 1997. N-cadherin-catenin interaction: Necessary component of cardiac cell compartmentalization during early vertebrate development. *Dev. Biol.* 185: 148–164.

Lindahl, P., B. R. Johansson, P. Levéen and C. Betscholtz. 1997. Pericyte loss and microaneurysm formation in PDGF-B-deficient mice. *Science* 277: 242–245.

Lints, T. J., L. M. Parsons, L. Hartley, I. Lyons and R. P. Harvey. 1993. Nkx2.5: A novel murine homeobox gene expressed in early heart progenitor cells and their myogenic descendants. *Development* 119: 419–431.

Lu, L.-S., S.-J Wang and R. Auerbach. 1996. In vitro and in vivo differentiation into B cells, T cells, and myeloid cells of primitive yolk sac hematopoietic precursor cells expanded >100 fold by coculture with a clonal yolk sac endothelial line. *Proc. Natl. Acad. Sci. USA* 93: 14782–14787.

Lyons, I., L. M. Parsons, L. Hartley, R. Li, J. E. Andrews, L. Robb and R. P. Harvey. 1995. Myogenic and morphogenetic defects in the heart tubes of murine embryos lacking the homeobox gene Nkx2.5. *Genes Dev.* 9: 1654–1666.

Maeno, M., R. C. Ong, A. Suzuki, N. Ueno and H. F. Kung. 1994. A truncated BMP4 receptor alters the fate of ventral mesoderm to dorsal mesoderm: Roles of animal pole tissue in the development of ventral mesoderm. *Proc. Natl. Acad. Sci. USA* 91: 10260–10264.

Manasek, F. J. 1968. Embryonic development of the heart: A light and electron microscopic study of myocardial development in the early chick embryo. *J. Morphol.* 125: 329–366.

Markwald, R. R., T. P. Fitzharris and J. J. Manasek. 1977. Structural development of endocardial cushions. *Am. J. Anat.* 148: 85–120.

Markwald, R. R., T. Trusk and R. Moreno-Rodriguez. 1998. Formation and septation of the tubular heart: Integrating the dynamics of morphology with emerging molecular concepts. *In* M. de la Cruz and R. R. Markwald (ed.), *Living Morphogenesis of the Heart*. Birkhauser Press, Boston.

Martin, C., D. Beaupain and F. Dieterlen-Lièvre. 1978. Developmental relationships between vitelline and intraembryonic haemopoiesis studied in avian yolk sac chimeras. *Cell Diff.* 7: 115–130.

Medvinsky, A. and E. Dzierzak. 1996. Definitive hematopoiesis is autonomously initiated by the AGM region. *Cell* 86: 897–906.

Medvinsky, A. L., N. L. Samoylina, A. M. Müller and E. A. Dzierzak. 1993. An early pre-liver intraembryonic source of CFU-S in the developing mouse. *Nature* 364: 64–67.

Mikawa, T. 1999. Determination of heart cell lineages. *In* S. A. Moody (ed.), *Cell Lineage and Fate Determination*. Academic Press, San Diego. pp. 451–462.

Millauer, B., S. Wizigmann-Voos, H. Schnürch, R. Martinez, N. P. H. Müller, W. Risau and A. Ullrich. 1993. High-affinity VEGF binding and developmental expression suggest *flk-1* as a major regulator of vasculogenesis and angiogenesis. *Cell* 72: 835–846.

Miller, S. A., K. L. Bresee, C. L. Michaelson and D. A. Tyrell. 1994. Domains of differential cell proliferation and formation of amnion folds in chick embryo ectoderm. *Anat. Rec.* 238: 225–236.

Miller, S. A. and 10 others. Domains of differential cell proliferation suggest hinged folding in avian gut endoderm. *Dev. Dynam.* 216: 398–410.

Miralles, F., P. Czernichow and R. Scharfman. 1998. Follistatin regulates the relative proportions of endocrine versus exocrine tissue during pancreatic development. *Development* 125: 1017–1024.

Müller, A. M., A. Medvinsky, J. Strouboulis, F. Grosveld and E. Dzierzak. 1994. Development of hematopoietic stem cell activity in the mouse embryo. *Immunity* 1: 291–301.

Mullins, M. C. and 12 others. 1996. Genes establishing dorsoventral pattern formation in the zebrafish embryo: the ventral specifying genes. *Development* 123: 81–93.

Nakauchi, H. and G. Gachelin. 1993. Les cellules souches. *La Recherche* 254: 537–541.

Nguyen, M.-T. and 7 others. 1997. The prostaglandin receptor EP$_4$ triggers remodelling of the cardiovascular system at birth. *Nature* 390: 78–81.

Offield, M. F. and 7 others. 1996. PDX-1 is required for pancreatic outgrowth and differentiation of the rostral duodenum. *Development* 122: 983–995.

Okuda, T., J. van Deursen, S. W. Hiebert, G. Grosveld and J. R. Downing. 1996. *AML*, the target of multiple chromosomal translocations in human leukemia, is essential for normal fetal liver hematopoiesis. *Cell* 84: 321–330.

Pardanaud, L., C. Altmann, P. Kitos, F. Dieterlen-Lièvre and C. Buck. 1987. Vasculogenesis in the early quail blastodisc as studied with a monoclonal antibody recognizing endothelial cells. *Development* 100: 339–349.

Pardanaud, L., F. Yassine and F. Dieterlen-Lièvre. 1989. Relationship between vasculogenesis, angiogenesis, and hemopoiesis during avian ontogeny. *Development* 105: 473–485.

Patten, B. M. 1951. *Early Embryology of the Chick*, 4th Ed. McGraw-Hill, New York.

Porcher, C., W. Swat, K. Rockwell, Y. Fujiwara, F. W. Alt and S. H. Orkin. 1996. The T cell leukemia oncoprotein SCL/tal-1 is essential for development of all hematopoietic lineages. *Cell* 86: 47–57.

Potten, C. S. and M. Loeffler. 1990. Stem cells: Attributes, spirals, pitfalls, and uncertainties. Lessons for and from the crypt. *Development* 110: 1001–1020.

Potts, J. D., J. M. Dagle, J. A. Walder, D. L. Weeks and R. B. Runyon. 1991. Epithelial-mesenchymal transformation of embryonic cardiac endothelial cells is inhibited by a modified antisense oligodeoxynucleotide to transforming growth factor β3. *Proc. Natl. Acad. Sci. USA* 88: 1516–1520.

R&D Systems. 1997. Human ELISAs for the cytokines of hematopoiesis. *Science* 278 (5340): back cover.

Ribatti, D., C. Urbinati, B. Nico, M. Rusnati, L. Roncali and M. Presta. 1995. Endogenous basic fibroblast growth factor is implicated in the vascularization of the chick embryo chorioallantoic membrane. *Dev. Biol.* 170: 39–49.

Risau, W. 1997. Mechanisms of angiogenesis. *Nature* 386: 671–674.

Robb, L., N. J. Elwood, A. G. Elefanty, F. Köntgen, R. Li, L. D. Barnett and C. G. Begley. 1996. The *scl* gene product is required for the generation of all hematopoietic lineages in the adult mouse. *EMBO J.* 15: 4123–4129.

Roberts, D. J., R. L. Johnson, A. C. Burke, C. E. Nelson, B. A. Morgan and C. Tabin. 1995. Sonic hedgehog is an endodermal signal inducing Bmp-4 and Hox genes during induction and regionalization of the chick hindgut. *Development* 121: 3163–3174.

Roberts, D. J., D. M. Smith, D. J. Goff and C. J. Tabin. 1998. Epithelial-mesenchymal signaling during the regionalization of the chick gut. *Development* 125: 2791–2801.

Roberts, R., J. Gallagher, E. Spooncer, T. D. Allen, F. Bloomfield and T. M. Dexter. 1988. Heparan sulphate-bound growth factors: A mechanism for stromal cell mediated haemopoiesis. *Nature* 332: 376–378.

Rugh, R. 1951. *The Frog: Its Reproduction and Development*. Blakiston, Philadelphia.

Schott, J.-J. and 8 others. 1998. Congenital heart disease caused by mutations in the transcription factor NKX2-5. *Science* 281: 108–111.

Schultheiss, T. M., S. Xydas and A. B. Lassar. 1995. Induction of avian cardiac myogenesis by anterior endoderm. *Development* 121: 4203–4214.

Shalaby, F., J. Rossant, T. P. Yamaguchi, M. Gertenstein, X.-F. Wu, M. L. Breitman and A. C. Schuh. 1995. Failure of blood-island formation and vasculogenesis in flk-1-deficient mice. *Nature* 376: 62–66

Shalaby, F. and 7 others. 1997. A requirement for Flk1 in primitive and definitive hematopoiesis and vasculogenesis. *Cell* 89: 981–990.

Shannon, J. M., L. D. Nielsen, S. A. Gebb and S. H. Randell. 1998. Mesenchyme specifies epithelial differentiation in reciprocal recombinants of embryonic lung and trachea. *Devel. Dynam.* 212: 482–494.

Sierra-Honigmann, M. R. and 10 others. 1998. Biological action of leptin as an angiogenic factor. *Science* 281: 1683–1686.

Sosa-Pineda, B., K. Chowdhury, M. Torres, G. Oliver and P. Gruss. 1997. The *Pax4* gene is essential for differentiation of insulin-producing β cells in the mammalian pancreas. *Nature* 386: 399–402.

Srivastava, D., P. Cserjesi and E. N. Olson. 1995. A subclass of bHLH proteins required for cardiac morphogenesis. *Science* 270: 1995–1999.

St-Onge, L., B. Sosa-Pineda, K. Chowdhury, A. Mansouri and P. Gruss. 1997. *Pax6* is required for differentiation of glucagon-producing α-cells in mouse pancreas. *Nature* 387: 406–409.

Sugi, Y. and J. Lough. 1994. Anterior endoderm is a specific effector of terminal cardiac myocyte differentiation of cells from the embryonic heart forming region. *Dev. Dynam.* 200: 155–162.

Suri, C. and 7 others. 1996. Requisite role of angiopoietin-1, a ligand for the TIE2 receptor, during embryonic angiogenesis. *Cell* 87: 1171–1180.

Thompson, M. A. and 12 others. 1998. The *cloche* and *spadetail* genes differentially affect hematopoiesis and vasculogenesis. *Dev. Biol.* 197: 248–269.

Till, J. E. 1981. Cellular diversity in the blood-forming system. *Am. Sci.* 69: 522–527.

Till, J. E. and E. A. McCulloch. 1961. A direct measurement of the radiation sensitivity of normal mouse bone marrow cells. *Radiat. Res.* 14: 213–222.

Tsuda, T., N. Philp, M. H. Zile and K. K. Linask. 1996. Left-right asymmetric localization of flectin in the extracellular matrix during heart looping. *Dev. Biol.* 173: 39–50.

Tuan, R. 1987. Mechanisms and regulation of calcium transport by the chick embryonic chorioallantoic membrane. *J. Exp. Zool.* [Suppl.] 1: 1–13.

Vikkula, M. and 11 others. 1996. Vascular dysmorphogenesis caused by an activating mutation in the receptor tyrosine kinase TIE2. *Cell* 87: 1181–1190.

Wang, D.-Z. and 8 others. 1999. Requirement of a novel gene, *Xin*, in cardiac morphogenesis. *Development* 126: 1281–1294.

Wang, H. U., Z.-F. Chen and D. J. Anderson. 1998. Molecular distinction and angiogenic interaction between embryonic arteries and veins revealed by ephrin-B2 and its receptor Eph-B4. *Cell* 93: 741–753.

Warren, L. 1998. *Joseph Leidy: The Last Man Who Knew Everything.* Yale University Press, New Haven.

Wessells, N. K. 1970. Mammalian lung development: Interactions in formulation and morphogenesis of tracheal buds. *J. Exp. Zool.* 175: 455–466.

Whitlock, C. A., G. F. Tidmarsh, C. Muller-Sieburg and I. L. Weissman. 1987. Bone marrow stromal cell lines with lymphopoietic activity express high levels of a pre-B neoplasia-associated molecule. *Cell* 48: 1009–1021.

Wilt, F. H. 1974. The beginnings of erythropoiesis in the yolk sac of the chick embryo. *Ann. NY Acad. Sci.* 241: 99–112.

Wolf, N. S. and J. J. Trentin. 1968. Hemopoietic colony studies. V. Effect of hemopoietic organ stroma on differentiation of pluripotent stem cells. *J. Exp. Med.* 127: 205–214.

Wood, H. B., G. May, L. Healy, T. Enver and G. M. Morriss-Kay. 1997. CD34 expression patterns during early mouse development are related to modes of blood vessel formation and reveal additional sites of hematopoiesis. *Blood* 90: 2300–2311.

Xavier-Neto, J. and 7 others. 1999. A retinoic acid-inducible transgenic marker of sino-atrial development in the mouse heart. *Development* 126: 2667–2687.

Yancopopoulos, G. D., M. Klagsbrun and J. Folkman. 1998. Vasculogenesis, angiogenesis, and growth factors: Ephrins enter the fray at the border. *Cell* 93: 661–664.

Yokouchi, Y., J. Sakiyama and A. Kuroiwa. 1995. Coordinate expression of Abd-B subfamily genes of the HoxA cluster in developing digestive tract of the chick embryo. *Dev. Biol.* 169: 76–89.

chapter *16* Development of the tetrapod limb

My arms are longer than my legs...I am my own sculptor: I am shaping myself from within with living, wet, malleable materials: what other artist has ever had available to him as perfect a design as the one possessed by my hammers and chisels: the cells move to the exact spot for building an arm: it's the first time they've ever done it, never before and never again, do your mercies benz understand what I'm saying? I will never be repeated.

CARLOS FUENTES (1989)

What can be more curious than that the hand of a man, formed for grasping, that of a mole for digging, the leg of a horse, the paddle of the porpoise, and the wing of the bat should all be constructed on the same pattern and should include similar bones, and in the same relative positions?

CHARLES DARWIN (1859)

PATTERN FORMATION IS THE PROCESS by which embryonic cells form ordered spatial arrangements of differentiated tissues. The ability to carry out this process is one of the most dramatic properties of developing organisms, and one that has provoked a sense of awe in scientists and laypeople alike. How is it that the embryo is able not only to generate all the different cell types of the body, but also to produce them in a way that forms functional tissues and organs? It is one thing to differentiate the chondrocytes and osteocytes that synthesize the cartilage and bone matrices, respectively; it is another thing to produce those cells in a temporal-spatial orientation that generates a functional bone. It is still another thing to make that bone a humerus and not a pelvis or a femur. The ability of limb cells to sense their relative positions and to differentiate with regard to those positions has been the subject of intense debate and experimentation. How are the cells that differentiate into the embryonic bone specified so as to form an appendage with digits at one end and a shoulder at the other? (It would be quite a useless appendage if the order were reversed.) Here the cell types are the same, but the patterns they form are different.

The vertebrate limb is an extremely complex organ with an asymmetrical arrangement of parts. There are three major axes to consider, one of which is the proximal (close) to distal (far) axis. The bones of the limb, be it wing, foot, hand, or flipper, consist of a proximal **stylopod** (humerus/femur) adjacent to the body wall, a **zeugopod** (radius-ulna/tibia-fibula) in the middle region, and a distal **autopod** (carpals-fingers/tarsals-toes) (Figure 16.1). Originally, these structures are cartilaginous, but eventually, most of the cartilage is replaced by bone. The positions of each of the bones and muscles in the limb are precisely organized. The second axis is the anterior (front) to posterior (back) axis. Our little fingers, for instance, mark the posterior side, while our thumbs are in the anterior. In humans, it is obvious that each hand develops as a mirror image of the other. One can imagine other arrangements to exist—such as the thumb developing on the left side of both hands—but these do not occur. The third axis is the dorsal-ventral axis. The palm (ventral) is readily distinguishable from the knuckles (dorsal). In some manner, the three-dimensional pattern of the forelimb is routinely produced. The fundamental problem of morphogenesis—how specific structures arise in particular places—is exemplified in limb development. How is it that one part of the lateral plate mesoderm develops limb-forming capacities? How is it that the fingers form at one end of the limb and nowhere else? How is it that the little finger develops at one edge of the limb and the thumb at the other?

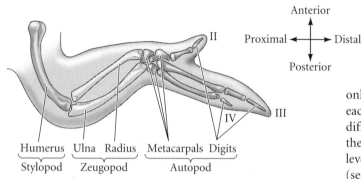

Anterior

Proximal ←——→ Distal

Posterior

Humerus | Ulna | Radius | Metacarpals | Digits
Stylopod | Zeugopod | | Autopod

Figure 16.1
Skeletal pattern of the chick wing. According to convention, the digits are numbered II, III, and IV. (Digits I and V are not found in chick wings.)

The basic "morphogenetic rules" for forming a limb appear to be the same in all tetrapods (see Hinchliffe 1991). Fallon and Crosby (1977) showed that grafted pieces of reptile or mammalian limb buds can direct the formation of chick limbs, and Sessions and co-workers (1989) found that regions of frog limb buds can direct the patterning of salamander limbs, and vice versa. Moreover, as will be detailed in Chapter 18, the regeneration of salamander limbs appears to follow the same rules as developing limbs (Muneoka and Bryant 1982). But what are these morphogenetic rules?

The positional information needed to construct a limb has to function in a three-dimensional coordinate system.* During the past decade, particular proteins have been identified that play a role in the formation of each of these limb axes. The proximal-distal (shoulder-finger; hip-toe) axis appears to be regulated by the fibroblast growth factor (FGF) family of proteins. The anterior-posterior (thumb-pinky) axis seems to be regulated by the Sonic hedgehog protein, and the dorsal-ventral (knuckle-palm) axis is regulated, at least in part, by Wnt7a. The interactions of these proteins determine the differentiation of the cell types and also mutually support one another.

WEBSITE **16.1 The mathematical modeling of limb development.** The specification of the limb axes and the patterns in which limb outgrowth might occur were predicted mathematically before the actual molecular interactions were discovered.

Formation of the Limb Bud

Specification of the limb fields: Hox genes and retinoic acid

Limbs will not form just anywhere along the body axis. Rather, there are discrete positions where limb fields are generated. Using the techniques described in Chapter 3, researchers have precisely localized the limb fields of many vertebrate species. Interestingly, in all land vertebrates, there are

only four limb buds per embryo, and they are always opposite each other with respect to the midline. Although the limbs of different vertebrates differ with respect to which somite level they arise from, their position is constant with respect to the level of Hox gene expression along the anterior-posterior axis (see Chapter 11). For instance, in fishes (in which the pectoral and pelvic fins correspond to the anterior and posterior limbs, respectively), amphibians, birds, and mammals, the forelimb buds are found at the most anterior expression region of *Hoxc-6*, the position of the first thoracic vertebra[†] (Oliver et al. 1988; Molven et al. 1990; Burke et al. 1995). The lateral plate mesoderm in the limb field is also special in that it will induce myoblasts to migrate out from the somites and enter the limb bud. No other region of the lateral plate mesoderm will do that (Hayashi and Ozawa 1995).

Retinoic acid appears to be critical for the initiation of limb bud outgrowth, since blocking the synthesis of retinoic acid with certain drugs prevents limb bud initiation (Stratford et al. 1996). Bryant and Gardiner (1992) suggested that a gradient of retinoic acid along the anterior-posterior axis might activate certain homeotic genes in particular cells and thereby specify them to become included in the limb field. The source of this retinoic acid is probably Hensen's node (Hogan et al. 1992). The specification of limb fields by retinoic acid-activated Hox genes might explain a bizarre observation made by Mohanty-Hejmadi and colleagues (1992) and repeated by Maden (1993). When the tails of tadpoles were amputated and the stumps exposed to retinoic acid during the first days of regeneration, the tadpoles regenerated several legs from the tail stump (Figure 16.2). It appears that the retinoic acid caused a homeotic transformation in the regenerating tail by respecifying the tail tissue as a limb-forming pelvic region (Müller et al. 1996).

Induction of the early limb bud: Fibroblast growth factors

Limb development begins when mesenchyme cells proliferate from the somatic layer of the limb field lateral plate mesoderm (limb *skeletal* precursors) and from the somites (limb *muscle* precursors; Figure 16.3) These cells accumulate under the epidermal tissue to create a circular bulge called a **limb bud**. Recent studies on the earliest stages of limb formation have shown that the signal for limb bud formation comes from the lateral plate mesoderm cells that will become the prospective limb mesenchyme. These cells secrete the paracrine factor FGF10. FGF10 is capable of initiating the limb-forming inter-

---

*Actually, it is a *four*-dimensional system, in which time is the fourth axis. Developmental biologists get used to seeing nature in four dimensions.

[†]Interestingly, the Hox gene expression pattern in at least some snakes (such as *Python*) creates a pattern in which each somite is specified to become a thoracic (ribbed) vertebra. The patterns of Hox gene expression associated with limb-forming regions are not seen (Cohn and Tickle 1999; see Chapter 22).

(A)

(B)

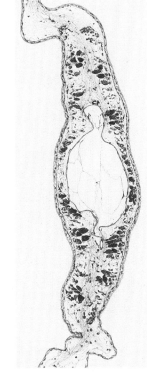

(C)

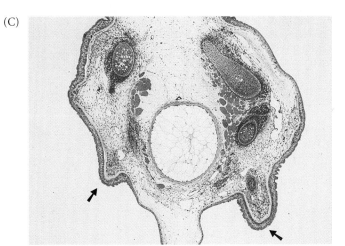

Figure 16.2

Legs regenerating from retinoic acid-treated tadpole tail. (A) The tail stump of a balloon frog tadpole treated with retinoic acid after amputation will form limbs from the amputation site. (B) Normal tail regeneration in a *Rana temporaria* tadpole 4 weeks after amputation. A small neural tube can be seen above a large notochord, and the muscles are arranged in packets. No cartilage or bone is present. (C) A retinoic acid-treated tadpole tail makes limb buds (arrows) as well as pelvic cartilage and bone. The cartilaginous rudiment of the femur can be seen in the right limb bud. (A from Mohanty-Hejmadi et al. 1992, courtesy of P. Mohanty-Hejmadi; B and C from Müller et al. 1996, courtesy of G. Müller.)

actions between the ectoderm and the mesoderm. If beads containing FGF10 are placed ectopically beneath the flank ectoderm, extra limbs emerge (Figure 16.4) (Ohuchi et al. 1997; Sekine et al. 1999).

Specification of forelimb or hindlimb: Tbx4 and Tbx5

The limb buds have to be specified as being those of either the forelimb or the hindlimb. How are these distinguished? In 1996, Gibson-Brown and colleagues made a tantalizing correlation: The gene encoding the **Tbx5** transcription factor is

Figure 16.3

Limb bud formation. (A) Proliferation of mesodermal cells from the somatic region of the lateral plate mesoderm causes the limb bud in the amphibian embryo to bulge outward. These cells generate the skeletal elements of the limb. Contributions of cells from the myotome provide the source of the limb's musculature. (B) Entry of myotome cells (purple) into the limb bud. This computer reconstruction was made from sections from an in situ hybridization to the *myf5* mRNA found in developing muscle cells. If you can cross your eyes, the three-dimensionality of the stereogram will become apparent. (B courtesy of J. Streicher and G. Müller.)

(A)

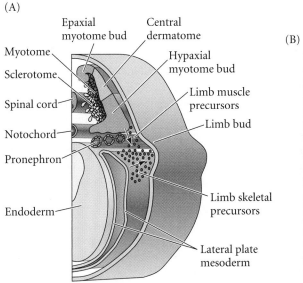

Epaxial myotome bud

Central dermatome

Myotome

Sclerotome

Spinal cord

Notochord

Pronephron

Endoderm

Hypaxial myotome bud

Limb muscle precursors

Limb bud

Limb skeletal precursors

Lateral plate mesoderm

(B)

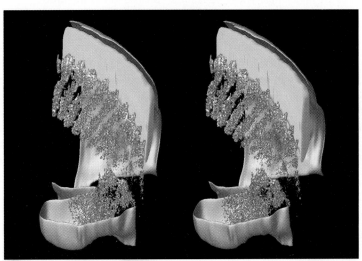

Figure 16.4
FGF10 expression and action in the developing chick limb. (A) FGF10 becomes expressed in the lateral plate mesoderm in precisely those positions where limbs normally form. (B) When cells genetically constructed to secrete FGF10 are placed into the flanks of chick embryos, the FGF10 can cause the formation of an ectopic limb (arrow). (From Ohuchi et al. 1997; courtesy of S. Noji.)

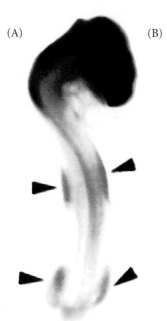

(A)

(B)

transcribed in mouse *forelimbs*, while the gene encoding the closely related transcription factor **Tbx4** is expressed in *hindlimbs*.* Could these two transcription factors be involved in directing forelimb versus hindlimb specificity? The loss-of-function data were equivocal: humans heterozygous for the *TBX5* gene have Holt-Oram syndrome, characterized by abnormalities of the heart and upper limbs (Basson et al. 1996; Li et al. 1996). The legs are not affected, but neither are the arms transformed into a pair of legs.

In 1998 and 1999, however, several laboratories (Ohuchi et al. 1998; Logan et al. 1998; Takeuchi et al. 1999; Rodriguez-Esteban et al. 1999, among others) provided gain-of-function evidence that Tbx4 and Tbx5 specify hindlimbs and forelimbs, respectively. First, if FGF-secreting beads were used to induce an ectopic limb between the chick hindlimb and forelimb buds, the type of limb produced was determined by the Tbx protein expressed. Those buds induced by placing FGF beads close to the hindlimb (opposite somite 25) expressed *Tbx4* and became hindlimbs. Those buds induced close to the forelimb (opposite somite 17) expressed *Tbx5* and developed as forelimbs (wings). Those buds induced in the center of the flank tissue expressed *Tbx5* in the anterior portion of the limb and *Tbx4* in the posterior portion of the limb. These limbs developed as chimeric structures, with the anterior resembling a forelimb and the posterior resembling a hindlimb (Figure 16.5). Moreover, when a chick embryo was made to express *Tbx4* throughout the flank tissue (by infecting the tissue with a virus that expressed Tbx4), limbs induced in the anterior region of the flank often became legs instead of wings (Figure 16.6). Thus, Tbx4 and Tbx5 appear to be critical in instructing the limbs to become hindlimbs and forelimbs, respectively.

*Tbx stands for T-box. The *T* (*Brachyury*) gene and its relatives have a sequence that encodes this specific DNA-binding domain.

WEBSITE 16.2 Specifying forelimbs and hindlimbs. While Tbx4 and Tbx5 are central to limb type specification, we still need to know how these two transcription factors become expressed in their respective limb buds, and what they do to make the limbs different.

Induction of the apical ectodermal ridge

As mesenchyme cells enter the limb region, they secrete factors that induce the overlying ectoderm to form a structure called the **apical ectodermal ridge** (**AER**) (Figure 16.7; Saunders 1948; Kieny 1960; Saunders and Reuss 1974). This ridge runs along the distal margin of the limb bud and will become a major signaling center for the developing limb. Its roles include (1) maintaining the mesenchyme beneath it in a plastic, proliferating phase that enables the linear (proximal-distal) growth of the limb; (2) maintaining the expression of those molecules that generate the anterior-posterior (thumb-pinky) axis; and (3) interacting with the proteins specifying the anterior-posterior and dorsal-ventral axes so that each cell is given instructions on how to differentiate.

The factor secreted by the mesenchyme cells to induce the AER is probably FGF10 (Xu et al. 1998; Yonei-Tamura et al. 1999). (Other FGFs, such as FGF2, FGF4, and FGF8, will also induce an AER to form; but FGF10 appears to be the FGF synthesized at the appropriate time and in the appropriate places.) FGF10 is capable of inducing the AER in the competent ectoderm between the dorsal and ventral sides of the embryo. This junction is important. In mutants in which the limb bud is dorsalized and there is no dorsal-ventral junction (as in the chick mutant *limbless*), the AER fails to form, and limb development ceases (Carrington and Fallon 1988; Laufer et al. 1997; Rodriguez-Esteban et al. 1997; Tanaka et al. 1997).

WEBSITE 16.3 Induction of the AER. The induction of the AER is a complex event involving the interaction between the dorsal and ventral compartments of the ectoderm. The Notch pathway may be critical in this process. Misexpression of these genes can cause absence or duplication of limbs.

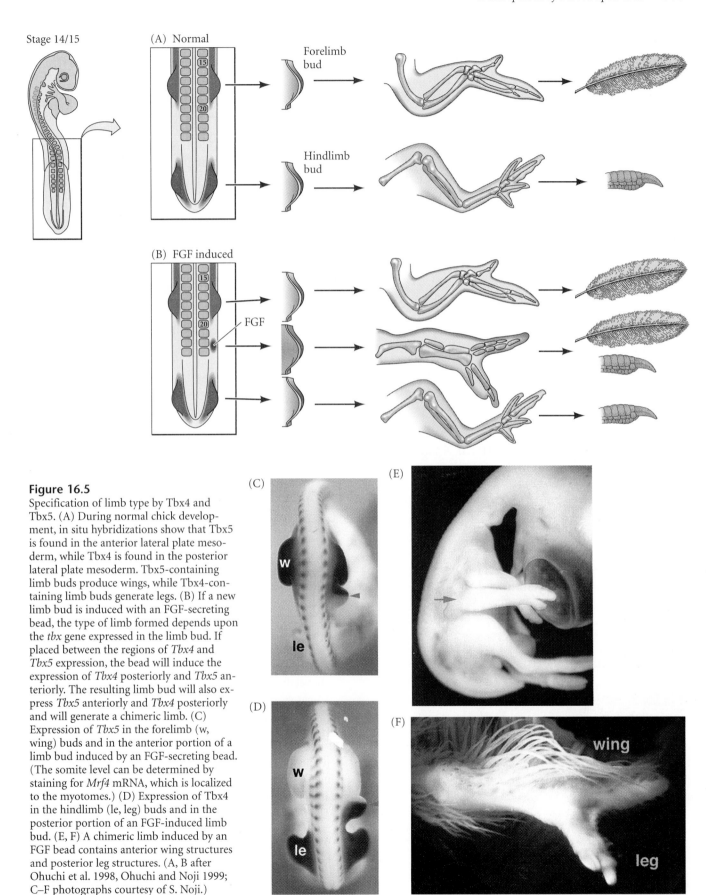

Figure 16.5
Specification of limb type by Tbx4 and Tbx5. (A) During normal chick development, in situ hybridizations show that Tbx5 is found in the anterior lateral plate mesoderm, while Tbx4 is found in the posterior lateral plate mesoderm. Tbx5-containing limb buds produce wings, while Tbx4-containing limb buds generate legs. (B) If a new limb bud is induced with an FGF-secreting bead, the type of limb formed depends upon the *tbx* gene expressed in the limb bud. If placed between the regions of *Tbx4* and *Tbx5* expression, the bead will induce the expression of *Tbx4* posteriorly and *Tbx5* anteriorly. The resulting limb bud will also express *Tbx5* anteriorly and *Tbx4* posteriorly and will generate a chimeric limb. (C) Expression of *Tbx5* in the forelimb (w, wing) buds and in the anterior portion of a limb bud induced by an FGF-secreting bead. (The somite level can be determined by staining for *Mrf4* mRNA, which is localized to the myotomes.) (D) Expression of Tbx4 in the hindlimb (le, leg) buds and in the posterior portion of an FGF-induced limb bud. (E, F) A chimeric limb induced by an FGF bead contains anterior wing structures and posterior leg structures. (A, B after Ohuchi et al. 1998, Ohuchi and Noji 1999; C–F photographs courtesy of S. Noji.)

Figure 16.6
Respecification of forelimb into hindlimb by ectopic expression of Tbx4. (A) An FGF-secreting bead opposite somite 21 usually induces a Tbx5-expressing limb bud that forms a new wing. (B) If the entire flank is experimentally made to express Tbx4 (by infecting it with a Tbx4-expressing virus), the FGF-induced limb bud expresses Tbx4, and often becomes a leg. (After Rodriguez-Esteban et al. 1999; photographs courtesy of J. C. Izpisúa-Belmonte.)

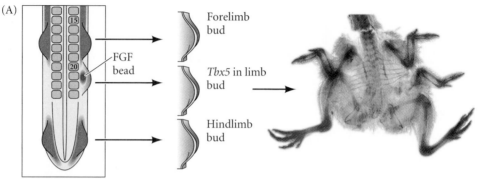

(A)

Forelimb bud

FGF bead

Tbx5 in limb bud

Hindlimb bud

(B)

Forelimb bud

FGF bead

Tbx4 limb bud

Ectopic *Tbx4* expression

Hindlimb bud

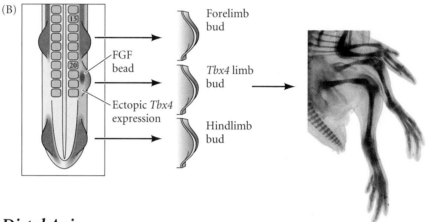

Generating the Proximal-Distal Axis of the Limb

The apical ectodermal ridge: The ectodermal component

The proximal-distal growth and differentiation of the limb bud is made possible by a series of interactions between the limb bud mesenchyme and the AER (Figure 16.8; Harrison 1918; Saunders 1948). These interactions were demonstrated by the results of several experiments on chick embryos:

1. If the AER is removed at any time during limb development, further development of distal limb skeletal elements ceases.
2. If an extra AER is grafted onto an existing limb bud, supernumerary structures are formed, usually toward the distal end of the limb.
3. If leg mesenchyme is placed directly beneath the wing AER, distal hindlimb structures (toes) develop at the end of the limb. (However, if this mesenchyme is placed farther from the AER, the hindlimb mesenchyme becomes integrated into wing structures.)
4. If limb mesenchyme is replaced by nonlimb mesenchyme beneath the AER, the AER regresses and limb development ceases.

Thus, although the mesenchyme cells induce and sustain the AER and determine the type of limb to be formed, the AER is responsible for the sustained outgrowth and development of the limb (Zwilling 1955; Saunders et al. 1957; Saunders 1972; Krabbenhoft and Fallon 1989). The AER keeps the mesenchyme cells directly beneath it in a state of mitotic proliferation and prevents them from forming cartilage. Hurle and co-workers

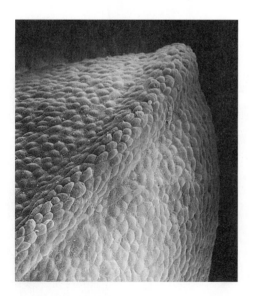

Figure 16.7
Scanning electron micrograph of an early chick forelimb bud, with its apical ectodermal ridge in the foreground. (Courtesy of K. W. Tosney.)

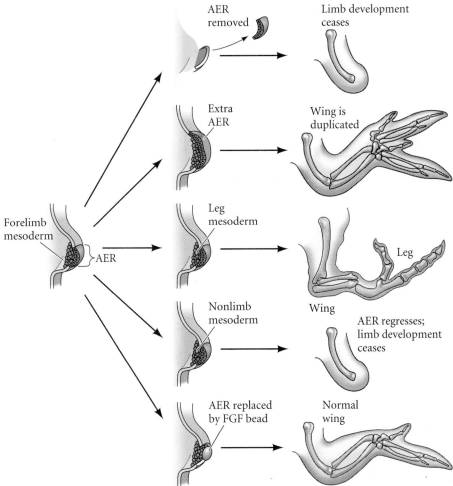

Figure 16.8
Summary of experiments demonstrating the effect of the apical ectodermal ridge (AER) on the underlying mesenchyme. (Modified from Wessells 1977.)

(1989) found that if they cut away a small portion of the AER in a region that would normally fall between the digits of the chick leg, an extra digit emerged at that place* (Figure 16.9).

moved from an early limb bud, only the most proximal parts of the stylopod are made. However, if an FGF-containing bead is placed in the hole left by the removal of the AER, a normal limb will form (see Figure 16.8; Niswander et al. 1993; Fallon et al. 1994; Crossley et al. 1996).

The progress zone: The mesodermal component

The proximal-distal axis is defined only after the induction of the apical ectodermal ridge by the underlying mesoderm. The limb bud elongates by means of the proliferation of the mesenchyme cells underneath the AER. This region of cell division is called the **progress zone**, and it extends about 200 μm in from the AER. Molecules from the AER are thought to keep the progress zone mesenchyme cells dividing, and it is now thought that FGFs are the molecules responsible. When the AER is re-

*When referring to the hand, one has an orderly set of names to specify each digit (digitus pollicis, d. indicis, d. medius, d. annularis, and d. minimus, respectively, from thumb to little finger). No such nomenclature exists for the pedal digits, but the plan proposed by Phillips (1991) has much merit. The pedal digits, from hallux to small toe, would be named porcellus fori, p. domi, p. carnivorus, p. non voratus, and p. plorans domi, respectively.

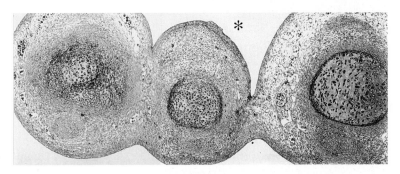

Figure 16.9
Cross section through the distal region of a chick limb 3 days after a wedge of the AER was removed from an area that would normally form interdigital tissue. Instead of degenerating, the remaining interdigital tissue formed an extra digit (marked by an asterisk). (From Hurle et al. 1989; photograph courtesy of the authors.)

Figure 16.10
Dorsal view of chick skeletal pattern after removal of the entire AER from the right wing bud of chick embryos at various stages. The last photo (E) is of a normal wing skeleton. (From Iten 1982; photographs courtesy of L. Iten.)

(A) (B) (C)

(D) (E)

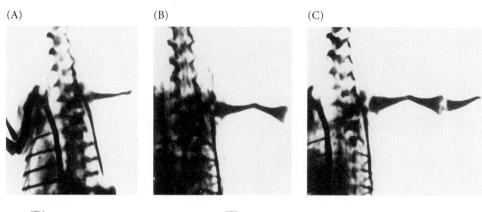

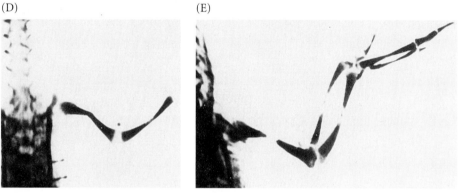

When the mesenchyme cells leave the progress zone, they differentiate in a regionally specific manner. The first cells leaving the progress zone form proximal (stylopod) structures; those cells that have undergone numerous divisions in the progress zone become the more distal structures (Saunders 1948; Summerbell 1974). Therefore, if the AER is removed from an early-stage wing bud, the cells of the progress zone stop dividing, and only a humerus forms. If the AER is removed slightly later, humerus, radius, and ulna form (Figure 16.10; Iten, 1982; Rowe et al. 1982).

Proximal-distal polarity resides in the mesodermal compartment of the limb. If the AER provided the positional information—somehow instructing the undifferentiated mesoderm beneath it as to what structures to make—then older AERs combined with younger mesoderm should produce limbs with deletions in the middle, while younger AERs combined with older mesoderm should produce duplications of structures. This was not found to be the case, however (Rubin and Saunders 1972). Rather, normal limbs form in both experiments. But when the entire progress zone, including both the mesoderm and AER, from an early embryo is placed on the limb bud of a later-stage embryo, new proximal structures are produced beyond those already present. Conversely, when old progress zones are added to young limb buds, distal structures immediately develop, so that digits are seen to emerge from the humerus without the intervening ulna and radius (Figure 16.11; Summerbell and Lewis 1975).

The mitotic state of the progress zone is maintained by interactions between the FGF proteins of the progress zone

(A)

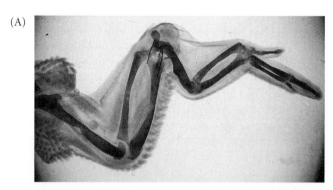

(B)

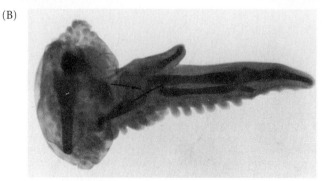

Figure 16.11
Control of proximal-distal specification by the cells of the progress zone. (A) Extra set of ulna and radius formed when an early-bud progress zone was transplanted to a late wing bud that had already formed ulna and radius. (B) Lack of intermediate structures seen when a late-bud progress zone was transplanted to an early limb bud. The hinges indicate the locations of the grafts. (From Summerbell and Lewis 1975; photographs courtesy of D. Summerbell.)

(A)
(B)
(C)

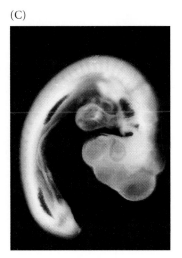

Figure 16.12
FGF8 in the AER. (A) In situ hybridization showing expression of *Fgf8* message in the ectoderm as the limb bud begins to form. (B) Expression of *Fgf8* RNA in the apical ectodermal ridge, the source of mitotic signals to the underlying mesoderm. (C) In normal 3-day chick embryos, FGF8 is expressed in the apical ectodermal ridge of both the forelimb and hindlimb buds. It is also expressed in several other places in the embryo including the pharyngeal arches. (A and B courtesy of J. C. Izpisúa-Belmonte; C courtesy of A. López-Martínez and J. F. Fallon.)

and of the AER. FGF10 secretion by the mesenchyme cells induces the AER, and it also induces the AER to express FGF8 (Figure 16.12). The FGF8 secreted by the AER reciprocates by maintaining the mitotic activity of the progress zone mesenchyme cells (Figure 16.13; Mahmood et al. 1995; Crossley et al. 1996; Vogel et al. 1996; Ohuchi et al. 1997).

Hox genes and the specification of the proximal-distal axis

The type of structure formed along the proximal-distal axis is specified by the Hox genes. The products of the Hox genes have already played a role in specifying the place where the limbs will form. Now they will play a second role in specifying whether a particular mesenchymal cell will become stylopod, zeugopod,

or autopod. The 5′ (AbdB-like) portions (paralogues 9–13) of the *HoxA* and *HoxD* complexes appear to be active in the forelimb buds of mice. Based on the expression patterns of these genes, and on naturally occurring and gene knockout mutations, Davis and colleagues (1995) proposed a model wherein these Hox genes specify the identity of a limb region (Figure 16.14). For instance, when they knocked out all four loci for the paralogous genes *Hoxa-11* and *Hoxd-11*, the resulting mice lacked the ulna and radius of their forelimbs (Figure 16.14A,B,). Similarly, knocking out all four *Hoxa-13* and *Hoxd-13* loci resulted in loss of the autopod (Fromental-Ramain et al. 1996). Humans homozygous for a *HOXD13* mutation show abnormalities of the hands and feet wherein the digits fuse, and human patients with homozygous mutant alleles of *HOXA13* also have deformities of their autopods (Figure 16.14C; Muragaki et al. 1996; Mortlock and Innis 1997). In both mice and humans, the autopod (the most distal portion of the limb) is affected by the loss of function of the most 5′ Hox genes.

The mechanism by which Hox genes could specify the proximal-distal axis is not yet understood, but one clue comes from the analysis of chicken *Hoxa-13*. Ectopic expression of this gene (which is usually expressed in the distal ends of developing chick limbs) appears to make the cells expressing it stickier. This, in turn, would cause the cartilaginous nodules to condense in specific ways (Yokouchi et al. 1995; Newman 1996).

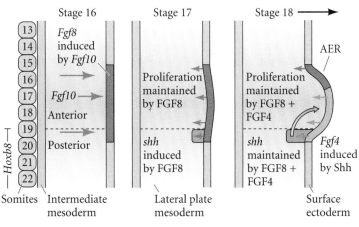

Figure 16.13
A molecular model for the initiation of the limb bud. FGF10 secreted by the lateral plate mesoderm induces FGF8 expression in the competent ectoderm at the dorsal-ventral boundary. The anterior-posterior boundary is present at stage 16 (and perhaps earlier). The FGF8 secretion by the ectoderm induces the proliferation of the mesenchyme cells and induces Sonic hedgehog expression in the posterior region of the limb bud. Sonic hedgehog induces FGF4 expression in the posterior portion of the limb bud ectoderm. FGF2 is also made by the AER ectoderm, although it is not yet clear whether it is induced by the FGF10.

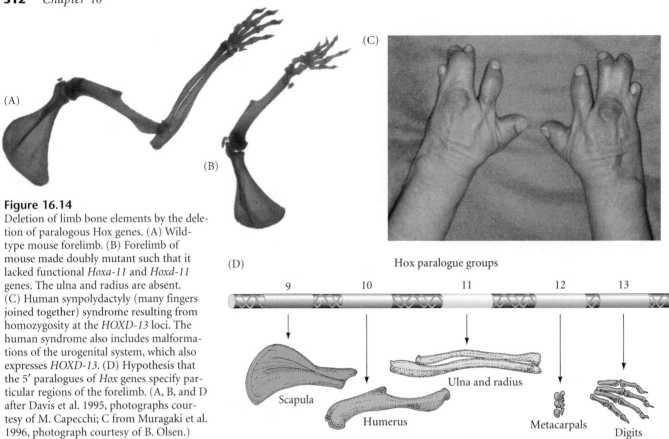

Figure 16.14
Deletion of limb bone elements by the deletion of paralogous Hox genes. (A) Wild-type mouse forelimb. (B) Forelimb of mouse made doubly mutant such that it lacked functional *Hoxa-11* and *Hoxd-11* genes. The ulna and radius are absent. (C) Human synpolydactyly (many fingers joined together) syndrome resulting from homozygosity at the *HOXD-13* loci. The human syndrome also includes malformations of the urogenital system, which also expresses *HOXD-13*. (D) Hypothesis that the 5′ paralogues of *Hox* genes specify particular regions of the forelimb. (A, B, and D after Davis et al. 1995, photographs courtesy of M. Capecchi; C from Muragaki et al. 1996, photograph courtesy of B. Olsen.)

As the limb grows outward, the pattern of Hox gene expression changes. When the stylopod is forming, *Hoxd-9* and *Hoxd-10* are expressed in the progress zone mesenchyme (Figure 16.15; Nelson et al. 1996). When the zeugopod bones are being formed, the pattern shifts remarkably, displaying a nested sequence of *Hoxd* gene expression. The posterior region expresses all the *Hoxd* genes from *Hoxd-9* to *Hoxd-13*, while only *Hoxd-9* is expressed anteriorly. In the third phase of limb development, when the autopod is forming, there is a further redeployment of Hox gene products. *Hoxd-9* is no longer expressed. Rather, *Hoxa-13* is expressed in the anterior tip of the limb bud and in a band marking the boundary of the autopod. *Hoxd-13* products join those of *Hoxa-13* in the anterior region of the limb bud, while *Hoxa-12*, *Hoxa-11*, and *Hoxd-10–12* are expressed throughout the posterior two-thirds of the limb bud.

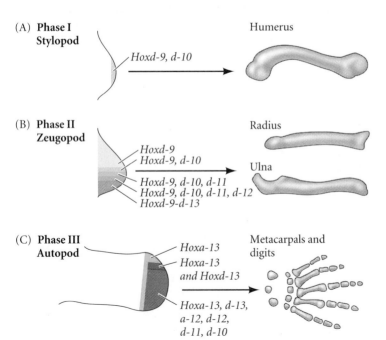

Figure 16.15
Hox gene expression changes during the formation of the tetrapod limb. (A) During the formation of the stylopod (phase I), *Hoxd-9* and *Hoxd-10* are expressed in the newly formed limb bud. (B) During zeugopod formation (phase 2), there is a nested expression of *Hoxd* genes such that *Hoxd-9* through *Hoxd-13* are expressed in the posterior of the limb bud, while only *Hoxd-9* is expressed both anteriorly and posteriorly. (C) Inversion of Hox gene expression during autopod formation. *Hoxd-13* and *Hoxa-13* are expressed anteriorly and posteriorly, while *Hoxd-10* through *Hoxd-12* and *Hoxa-12* are expressed posteriorly. (After Shubin et al. 1997.)

Hox Genes and the Evolution of the Tetrapod Limb

MACROEVOLUTION, the generation of morphological novelties in the evolution of new species and higher taxa, results from alterations of development. One of the most obvious macroevolutionary changes is that from the fish fin to the amphibian leg. As Richard Owen (1849) pointed out, there is considerable homology between the bones of the fish fin and the tetrapod limb, the pectoral and pelvic fins of the fish being homologous to the tetrapod forelimb and hindlimb, respectively. While specific homologies were able to be made between the proximal elements of the fin and the limb, the homologies proposed between the autopod of the limb (the hand or foot at the distal end) and the rays of the fins "did not hold water." This was true even when one compared the tetrapod limb with the fins of the crossopterygian (lobe-finned) fishes thought to have been closely related to the ancestors of the amphibians (see Coates 1994; Hinchliffe 1994). While there seems to be homology for the proximal and central elements of the limb, the autopod seems to be something new—what evolutionary biologists call a neomorphic structure.

Recent studies have strongly suggested that the expression of the 5′ genes of the Hoxd group may be crucial in the change from fin to limb. Tetrapods and fishes share the first two phases of the Hox expression pattern in their appendages. Thus, both groups form stylopods and zeugopods. However, the phase III pattern of Hox gene expression is unique to tetrapods and is not found in fishes. Moreover, this change in Hox gene expression is mediated by a single enhancer element that is not found in fishes (Gerard et al. 1993; van der Hoeven et al. 1996). This phase III change represents an inversion of gene expression, placing the most 5′ Hox gene products in the anterior of the limb bud. Instead of being restricted to the posterior of the limb bud, the expression of the 5′ Hox genes sweeps across the distal mesenchyme, just beneath the AER. This band of expression is coincident with the "digital arch" from which the digits form (Figure 16.16; Morgan and Tabin 1994; Sordino et al. 1995; Nelson et al. 1996).

Thus, while the Hox gene expression pattern is homologous between fish and tetrapod limbs in the proximal regions, the expression pattern in the late-bud distal mesenchyme is new. These studies also confirm the paleontological interpretations of Shubin and Alberch (1986; Shubin et al. 1997), who proposed that the path of digit formation was not (as previously believed) through the fourth digit (making the fin rays homologous to the other digits), but through an arch of distal wrist condensations (metapterygia) that begins posteriorly and turns anteriorly across the distal mesenchyme (Figure 16.16). Thus, the border of 5′ *HoxD* gene expression follows the metapterygial axis that Shubin and Alberch hypothesized as being the origin of digits. The foot and hand, then, appear to be new structures in evolution, and they appear to have been formed by the repositioning of *HoxD* gene expression during fin development. The use of the same enhancer to generate both fingers and toes also helps solve the problem of how these structures were evolved at the same time. ■

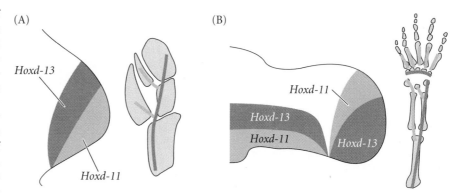

Figure 16.16
Differences in *Hoxd-11* and *Hoxd-13* expression in fish and tetrapod embryonic appendages. (A) Fin of a fish, wherein *Hoxd-11* expression is distal to *Hoxd-13* expression. The fin axis extends distally. (B) In tetrapods, *Hoxd-13* expression becomes distal to *Hoxd-11* expression, and the limb axis shifts anteriorly from its original proximal-distal orientation. The digits originate from the posterior side of the axis.

WEBSITE **16.4 Dinosaurs and the origin of birds.** Whether birds are descendants of certain dinosaurs is a controversial subject that concerns hip structure, and whether the autopod of the dinosaur is made of the same digits as the bird autopod.

Specification of the Anterior-Posterior Limb Axis

The zone of polarizing activity

The specification of the anterior-posterior axis of the limb is the earliest change from the pluripotent condition. In chicks, this axis is specified shortly before a limb bud is recognizable. Hamburger (1938) showed that as early as the 16-somite stage, prospective wing mesoderm transplanted to the flank area develops into a limb with the anterior-posterior and dorsal-ventral polarities of the donor graft, not those of the host tissue.

Although the differentiation of the proximal-distal structures is thought to depend on how many divisions a cell un-

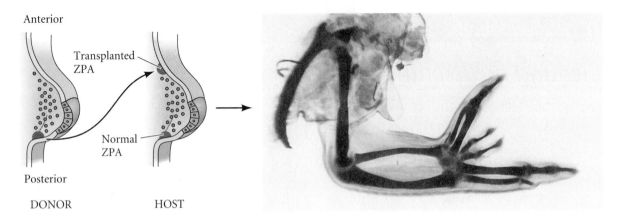

Anterior

Transplanted ZPA

Normal ZPA

Posterior

DONOR   HOST

Figure 16.17
When a ZPA is grafted to anterior limb bud mesoderm, duplicated digits emerge as a mirror image of the normal digits. (From Honig and Summerbell 1985; photograph courtesy of D. Summerbell.)

dergoes while in the progress zone, information instructing a cell as to its position on the anterior-posterior and dorsal-ventral axes must come from other sources. Several experiments (Saunders and Gasseling 1968; Tickle et al. 1975) suggest that the anterior-posterior axis is specified by a small block of mesodermal tissue near the posterior junction of the young limb bud and the body wall. This region of the mesoderm has been called the **zone of polarizing activity** (**ZPA**). When this tissue is taken from a young limb bud and transplanted into a position on the anterior side of another limb bud (Figure

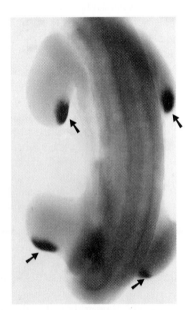

Figure 16.18
In situ hybridization showing the sites of Sonic hedgehog expression (arrows) in the posterior mesoderm of the chick limb buds. These are precisely the regions that transplantation experiments defined as the ZPA. (Photograph courtesy of R. D. Riddle.)

16.17), the number of digits of the resulting wing is doubled. Moreover, the structures of the extra set of digits are mirror images of the normally produced structures. The polarity has been maintained, but the information is now coming from both an anterior and a posterior direction.

Sonic hedgehog defines the ZPA

The search for the molecule(s) conferring this polarizing activity on the ZPA became one of the most intensive quests in developmental biology. In 1993, Riddle and colleagues showed by in situ hybridization that *sonic hedgehog* (*shh*), a vertebrate homologue of the *Drosophila hedgehog* gene, was expressed specifically in that region of the limb bud known to be the ZPA (Figure 16.18).

As evidence that this association between the ZPA and *sonic hedgehog* was more than just a correlation, Riddle and co-workers (1993) demonstrated that the secretion of Sonic hedgehog protein is sufficient for ZPA activity. They transfected embryonic chick fibroblasts (which normally would never synthesize this protein) with a viral vector containing the *shh* gene (Figure 16.19). The gene became expressed and translated in these fibroblasts, which were then inserted under the anterior ectoderm of an early chick limb bud. Mirror-image digit duplications like those induced by ZPA transplants were the result. More recently, beads containing Sonic hedgehog protein were shown to cause the same duplications (López-Martínez et al. 1995). Thus, Sonic hedgehog appears to be the active agent of the ZPA.

SPECIFICATION OF THE POSTERIOR LIMB BUD TO EXPRESS SONIC HEDGEHOG. Two new questions emerged from this discovery: First, how does Sonic hedgehog become expressed only in the posterior region of the limb bud? And second, what does it do once it is expressed? We do not yet know what causes the activation of the *sonic hedgehog* genes specifically in the cells of the posterior limb bud and not in the cells located more anteriorly. The *sonic hedgehog* gene appears to be activated by an FGF protein coming from the newly formed apical ectodermal ridge (see Figure 16.13). FGF8 is secreted from

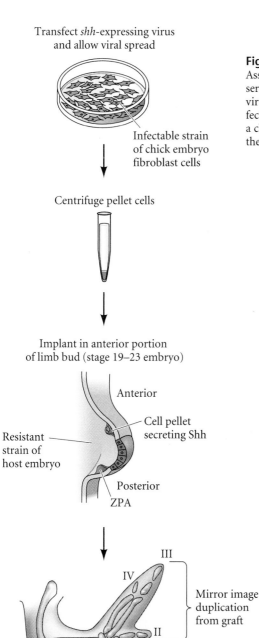

Transfect *shh*-expressing virus
and allow viral spread

Infectable strain
of chick embryo
fibroblast cells

Centrifuge pellet cells

Implant in anterior portion
of limb bud (stage 19–23 embryo)

Anterior

Cell pellet
secreting Shh

Resistant
strain of
host embryo

Posterior
ZPA

III

IV

Mirror image
duplication
from graft

II

II

Normal
development

IV

III

Figure 16.19
Assay for polarizing activity of Sonic hedgehog. The *sonic hedgehog* gene was inserted adjacent to an active promoter of a chicken virus, and the recombinant virus was placed into cultured chick embryo fibroblast cells (CEF). The virally infected cells were pelleted and implanted into the anterior margin of a limb bud of a chick embryo. The resulting limbs produced mirror-image digits, showing that the secreted protein had polarizing activity. (After Riddle et al. 1993.)

constructed transgenic mice in which the *Hoxb-8* gene was placed under the control of a new promoter that would cause its expression throughout the forelimb buds. This resulted in the expression of *sonic hedgehog* in the anterior portion of the limb buds, the creation of a new ZPA, and mirror-image forelimb duplications. This evidence suggests that the Hoxb-8 protein is involved in specifying the domain of *sonic hedgehog* expression and thus establishing the ZPA.

THE ACTION OF SONIC HEDGEHOG. When Sonic hedgehog was first shown to define the ZPA, it was thought to act as a morphogen. In other words, it was thought to diffuse from the ZPA where it was being synthesized and to form a concentration gradient from the posterior to the anterior of the limb bud. However, recent research has provided evidence that Sonic hedgehog protein (or its active amino terminal region) does not diffuse outside the ZPA (Yang et al. 1997). It is now thought that Sonic hedgehog works by initiating and sustaining a cascade of other proteins, such as BMP2 and BMP7 (Laufer et al. 1994; Kawakami et al. 1996; Drossopoulou et al. 2000). A gradient of BMPs may emanate from the ZPA and specify the digits.

However it works, Sonic hedgehog (directly or with help from the BMP cascade) regulates the expression of the 5′ *HoxD* genes. The transition from phase I to phase II Hox expression patterns (see Figure 16.15) is coincident with Sonic hedgehog expression in the ZPA. Moreover, transplantation of either the ZPA or other Sonic hedgehog-secreting cells to the anterior margin of the limb bud at this stage leads to the formation of mirror-image patterns of *HoxD* gene expression and results in mirror-image digit patterns (Izpisúa-Belmonte et al. 1991; Nohno et al. 1991; Riddle et al. 1993).

The Generation of the Dorsal-Ventral Axis

The third axis of the limb distinguishes the dorsal limb (knuckles, nails) from the ventral limb (pads, soles). In 1974, MacCabe and co-workers demonstrated that the dorsal-ventral polarity of the limb bud is determined by the ectoderm encasing it. If the ectoderm is rotated 180° with respect to the limb bud mesenchyme, the dorsal-ventral axis is partially reversed; the distal elements (digits) are "upside down." This suggested that the late specification of the dorsal-ventral axis of the limb is regulated by its ectodermal component. One molecule that appears to be particularly important in specifying the dorsal-ventral polarity is Wnt7a. The *Wnt7a* gene is

the AER, and is capable of activating *sonic hedgehog* (Heikinheimo et al. 1994; Crossley and Martin 1995). But why doesn't FGF8 activate all the mesenchyme cells beneath the AER? The answer may reside in the differential competence of certain mesenchyme cells to respond to the FGF signal. Charité and colleagues (1994) have suggested that the *Hoxb-8* protein may be critical in providing this restricted competence. They observed that the *Hoxb-8* gene is usually expressed in the posterior half of the mouse forelimb bud. They then

(A)

(B)

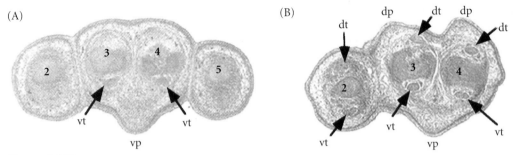

Figure 16.20
Dorsal-to-ventral transformations of limb regions in mice deficient for both *Wnt7a* genes. (A) Histological section (stained with hemotoxylin and eosin) of wild-type 15.5-day embryonic mouse forelimb paw. The ventral tendons and ventral footpads are readily seen. (B) Same section through a mutant embryo deficient in *Wnt7a*. Tendons and footpads are duplicated on what would normally be the dorsal surface of the paw. dt, dorsal tendons; dp, dorsal footpad; vp, ventral footpad; vt, ventral tendon. Numbers indicate digit identity. (From Parr and McMahon 1995; photographs courtesy of the authors.)

expressed in the dorsal (but not the ventral) ectoderm of the chick and mouse limb buds (Dealy 1993; Parr et al. 1993). In 1995, Parr and McMahon genetically deleted *Wnt7a* from mouse embryos. The resulting embryos had sole pads on both surfaces of their paws, showing that Wnt7a is needed for the dorsal patterning of the limb (Figure 16.20).

Wnt7a induces activation of the *Lmx1* gene in the dorsal mesenchyme, and this gene encodes a transcription factor that appears to be essential for specifying dorsal cell fates in the limb. If this factor is expressed in the ventral mesenchyme cells, they develop a dorsal phenotype (Riddle et al. 1995; Vogel et al. 1995). Mutants of *Lmx1* in humans and mice also show its importance for specifying dorsal limb fates. Knockouts of this gene in mice produce a syndrome in which the dorsal limb phenotype is lacking, and loss-of-function mutations in humans produce the nail-patella syndrome, a condition in which the dorsal sides of the limbs have been ventralized (Chen et al. 1998; Dreyer et al. 1998).

Coordination among the Three Axes

The three axes of the tetrapod limb are all interrelated and coordinated. Some of the principal interactions among the mechanisms specifying the axes are shown in Figure 16.21. Indeed, the molecules that define one of these axes are often used to maintain another axis. For instance, Sonic hedgehog in the ZPA activates the expression of the *Fgf4* gene in the posterior region of the AER (see figure 16.13). *Fgf4* expression is important in recruiting mesenchyme cells into the progress zone, and it is also critical in maintaining the expression of Sonic hedgehog in the ZPA (Li and Muneoka 1999). Therefore, the AER and the ZPA mutually support each other through the positive loop of Sonic hedgehog and FGF4 (Todt and Fallon 1987; Laufer et al. 1994; Niswander et al. 1994).

The *Wnt7a*-deficient mice mentioned above not only lacked dorsal limb structures; they also lacked posterior digits,

suggesting that Wnt7a is also needed for the anterior-posterior axis. Yang and Niswander (1995) made a similar set of observations in chick embryos. These investigators removed the dorsal ectoderm from the developing limb and found that such an operation resulted in the loss of posterior skeletal elements from the limbs. The reason that these limbs lacked posterior digits was that Sonic hedgehog expression was missing. Viral-induced expression of Wnt7a was able to replace the dorsal ectoderm and restore Sonic hedgehog expression and posterior phenotypes. These findings showed that the synthesis of Sonic hedgehog is stimulated by the combination of FGF4 and Wnt7a proteins.

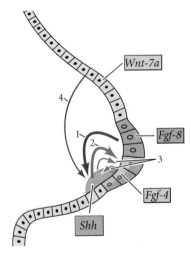

Figure 16.21
Some of the molecular interactions by which limb bud formation is initiated and maintained. Some of the major loops include (1) the establishment of the ZPA (Sonic hedgehog) by the AER (FGF8); (2) the induction of FGF4 (in the AER) by Sonic hedgehog (in the ZPA); (3) the mutual maintenance of FGF4 and Sonic hedgehog; and (4) the maintenance of Sonic hedgehog by Wnt7a (in the dorsal ectoderm). The broad arrows indicate induction; the finer arrows indicate maintenance. (After Pearse and Tabin 1998.)

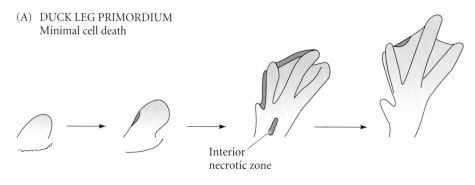

(A) DUCK LEG PRIMORDIUM
Minimal cell death

Interior
necrotic zone

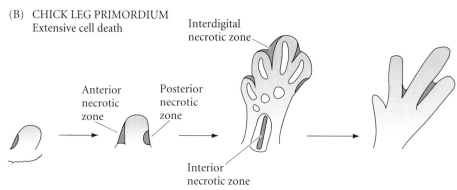

(B) CHICK LEG PRIMORDIUM
Extensive cell death

Interdigital
necrotic zone

Anterior
necrotic
zone

Posterior
necrotic
zone

Interior
necrotic zone

Figure 16.22
Patterns of cell death in leg primordia of (A) duck and (B) chick embryos. Shading indicates areas of cell death. In the duck, the regions of cell death are very small, whereas there are extensive regions of cell death in the interdigital tissue of the chicken leg. (After Saunders and Fallon 1966.)

Cell Death and the Formation of Digits and Joints

Sculpting the autopod

Cell death plays a role in sculpting the limb. Indeed, it is essential if our joints are to form and if our fingers are to become separate (Zaleske 1985). The death (or lack of death) of specific cells in the vertebrate limb is genetically programmed and has been selected for during evolution. One such case involves the webbing or nonwebbing of feet. The difference between a chicken's foot and that of a duck is the presence or absence of cell death between the digits (Figure 16.22A,B). Saunders and co-workers (1962; Saunders and Fallon 1966) have shown that after a certain stage, chick cells between the digit cartilage are destined to die, and will do so even if transplanted to another region of the embryo or placed into culture. Before that time, however, transplantation to a duck limb will save them. Between the time when the cell's death is determined and when death actually takes place, levels of DNA, RNA, and protein synthesis in the cell decrease dramatically (Pollak and Fallon 1976).

In addition to the **interdigital necrotic zone**, there are three other regions that are "sculpted" by cell death. The ulna and radius are separated from each other by an **interior necrotic zone**, and two other regions, the **anterior** and **posterior necrotic zones**, further shape the end of the limb (Figure 16.22B; Saunders and Fallon 1966). Although these zones are said to be "necrotic," this term is a holdover from the days when

no distinction was made between necrotic cell death and apoptotic cell death (see chapter 6). These cells die by apoptosis, and the death of the interdigital tissue is associated with the fragmentation of their DNA (Mori et al. 1995).

The signal for apoptosis in the autopod is probably provided by the BMP proteins. BMP2, BMP4, and BMP7 are each expressed in the interdigital mesenchyme, and blocking BMP signaling (by infecting progress zone cells with retroviruses carrying dominant negative BMP receptors) prevents interdigital apoptosis (Figure 16.23A; Zou and Niswander 1996; Yokouchi et al. 1996). Since these BMPs are expressed throughout the progress zone mesenchyme, it is thought that cell death would be the "default" state unless there were active suppression of the BMPs. This suppression may come from the Noggin protein, which is made in the developing digits and in the perichondrial cells surrounding them (Figure 16.23B; Capdevila and Johnson 1998; Merino et al. 1998). If *noggin* is expressed throughout the limb bud, no apoptosis is seen.

Forming the joints

The function originally ascribed to BMPs was the formation, not the prevention, of bone and cartilage tissue. In the developing limb, BMPs induce the mesenchymal cells either to undergo apoptosis or to become cartilage-producing chondrocytes—depending on the stage of development. The same BMPs can induce death or differentiation, depending on the age of the target cell. This "context dependency" of signal action is a critical concept in developmental biology. It is also critical for the formation of joints. Macias and colleagues (1997) have shown that during early limb bud stages (before cartilage condensation), beads secreting BMP2 or BMP7 cause

Figure 16.23
Blocking the BMP receptor can prevent apoptosis in the chick autopod. (A) The left limb bud was untreated, while the right limb bud was infected with a virus that transcribed a dominant negative BMP receptor. This mutant receptor blocked BMP signaling in the right limb bud and prevented apoptosis and digit growth. (B) *noggin* expression in developing leg digits. This protein is expressed in the regions of cartilage condensation and is absent from the regions of interdigital apoptosis and joint formation at this stage. (A from Zou and Niswander 1996, photograph courtesy of L. Niswander; B from Merino et al. 1998, photograph courtesy of J. M. Hurle.)

(A)
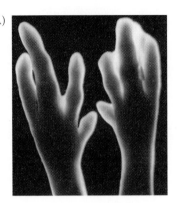

(B)
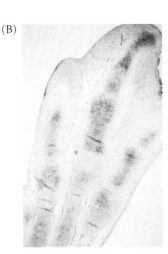

apoptosis. Two days later, the same beads cause the limb bud cells to form cartilage.

In the normally developing limb, BMPs use both these properties to form joints. BMP7 is made in the perichondrial cells surrounding the condensing chondrocytes and promotes cartilage formation (Figure 16.24A; Macias et al. 1997) . Two other BMP proteins, BMP2 and GDF5, are expressed at the regions between the bones, where joints will form (Figure 16.24B; Macias et al. 1997; Brunet et al. 1998). Mouse mutations have suggested that the function of these proteins in joint formation is critical. Mutations of *Gdf5* produce brachy-

podism, a condition characterized by a lack of limb joints (Storm and Kingsley 1999). In mice homozygous for loss-of-function alleles of *noggin*, no joints form, either. It appears that the BMP7 in these *noggin*-defective embryos is able to recruit nearly all the surrounding mesenchyme into the digits (Figure 16.24C,D; Brunet et al. 1998). The roles of BMP2 and GDF5 are more controversial. They may either be destroying mesenchymal cells to form the joint or inducing them to rapidly differentiate and join one or the other cartilaginous nodule. In either way, a space is made between the nodules, and a joint can form.

Limb development is an exciting meeting place for developmental biology, evolutionary biology, and medicine. Within the next decade, we can expect to know the bases for numerous congenital diseases of limb formation, and perhaps we will understand how limbs are modified into flippers, wings, hands, and legs.

Figure 16.24
Possible involvement of BMPs in stabilizing cartilage and forming joints. (A, B) Expression pattern of (A) BMP7 and (B) BMP2 in two consecutive microscopic sections of chick digits III and IV during a late stage of limb formation. The asterisks in (A) and the arrows in (B) denote places where joints will form. (C, D) The effects of Noggin. (C) 16.5-day autopod from a wild-type mouse, showing GDF5 expression (dark blue) at the joints. (D) 16.5-day *noggin*-deficient mutant mouse autopod, showing neither joints nor GDF5 expression. Presumably, in the absence of Noggin, BMP7 was able to convert nearly all the mesenchyme into cartilage. (A, B from Macias et al. 1997; C, D from Brunet et al. 1998. Photographs courtesy of J. M. Hurle and A. P. McMahon.)

(A)
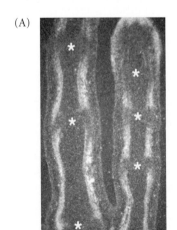

(B)

(C)
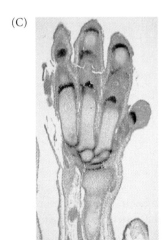

(D)
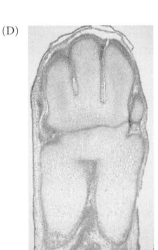

Snapshot Summary: The Tetrapod Limb

1. The places where limbs emerge from the body axis depend upon Hox gene expression.

2. The specification of the limb field into a hindlimb or forelimb bud is determined by *Tbx4* and *Tbx5* expression.

3. The proximal-distal axis of the developing limb is determined by the induction of the ectoderm at the dorsal-ventral boundary to form the apical ectodermal ridge (AER). This induction is caused by an FGF, probably FGF10. The AER secretes FGF8, which keeps the underlying mesenchyme proliferative and undifferentiated. This mesenchyme is called the progress zone.

4. As the limb grows outward, the stylopod forms first, then the zeugopod, and the autopod is formed last. Each of these phases involves the expression of Hox genes, and the formation of the autopod involves a reversal of Hox gene expression that distinguishes fish fins from tetrapod limbs.

5. The anterior-posterior axis is defined by the expression of Sonic hedgehog in the posterior mesoderm of the limb bud. This region is called the zone of polarizing activity (ZPA). If the ZPA or Sonic hedgehog-secreting cells or beads are placed in the anterior margin, they establish a second, mirror-image pattern of Hox gene expression and a corresponding mirror-image duplication of the digits.

6. The ZPA is established by the interaction of FGF8 from the AER and mesenchyme made competent to express Sonic hedgehog by its expression of particular Hox genes. Sonic hedgehog acts, probably in an indirect manner, to change the expression of the Hox genes in the limb bud.

7. The dorsal-ventral axis is formed, in part, by the expression of Wnt7a in the dorsal portion of the limb ectoderm. Wnt7a also maintains the expression of Sonic hedgehog in the ZPA and FGF4 in the posterior AER. FGF4 and Sonic hedgehog reciprocally maintain each other's expression.

8. Cell death in the limb is necessary for the formation of digits and joints. It is mediated by BMPs. The effects of BMPs can be regulated by the Noggin protein, and the BMPs can be involved both in inducing apoptosis and in differentiating the mesenchymal cells into cartilage.

Literature Cited

Basson, C. T. and 13 others. 1996. Mutations in human TBX5 cause limb and cardiac malformation in Holt-Oram syndrome. *Nature Genet.* 15: 30–35.

Brunet, L. J., J. A. McMahon, A. P. McMahon and R. M. Harland. 1998. Noggin, cartilage morphogenesis and joint formation in the mammalian skeleton. *Science* 280: 1455–1457.

Bryant, S. V. and D. M. Gardiner. 1992. Retinoic acid, local cell-cell interactions, and pattern formation in vertebrate limbs. *Dev. Biol.* 152: 1–25.

Burke, A. C., C. E. Nelson, B. A. Morgan and C. Tabin. 1995. Hox genes and the evolution of vertebrate axial morphology. *Development* 121: 333–346.

Capdevila, J. and R. L. Johnson. 1998. Endogenous and ectopic expression of *noggin* suggests a conserved mechanism for regulation of BMP function during limb and somite patterning. *Dev. Biol.* 197: 205–217.

Carrington, J. L. and J. F. Fallon. 1988. Initial limb budding is independent of apical ectodermal ridge activity: Evidence from a *limbless* mutant. *Development* 104: 361–367.

Charité, J., W. de Graaff, S. Shen and J. Duchamps. 1994. Ectopic expression of *Hoxb-8* causes duplications of the ZPA in the forelimb and homeotic transformation of axial structures. *Cell* 78: 559–601.

Chen, H. and 8 others. 1998. Limb and kidney defects in *Lmx1b* mutant mice suggest an involvement of *LMX1B* in human nail patella syndrome. *Nature Genet.* 19: 51–55.

Coates, M. I. 1994. The origin of vertebrate limbs. *Development* [Suppl.]: 169–180.

Cohn, M. J. and C. Tickle. 1999. Developmental basis of limblessness and axial patterning in snakes. *Nature* 399: 474–479.

Crossley, P. H. and G. R. Martin. 1995. The mouse *Fgf8* gene encodes a family of polypeptides and is expressed in regions that direct outgrowth and patterning in the developing embryo. *Development* 121: 439–451.

Crossley, P. H., G. Monowada, C. A. MacArthur and G. R. Martin. 1996. Roles for FGF8 in the induction, initiation, and maintenance of chick development of the tetrapod limb. *Cell* 84: 127–136.

Davis, A. P., D. P. Witte, H. M. Hsieh-Li, S. Potter and M. R. Capecchi. 1995. Absence of radius and ulna in mice lacking *hoxa-11* and *hoxd-11*. *Nature* 375: 791–795.

Dealy, C. N., A. Roth, D. Ferrari, A. M. C. Brown and R. A. Kosher. 1993. *Wnt-5a* and *Wnt-7a* are expressed in the developing chick limb bud in a manner that suggests roles in pattern formation along the proximodistal and dorsoventral axes. *Mech. Dev.* 43: 175–186.

De Robertis, E. M., E. A. Morita and K. W. Y. Cho. 1991. Gradient fields and homeobox genes. *Development* 112: 669–678.

Dreyer, S. D. and 7 others. 1998. Mutations in *LMX1B* cause abnormal skeletal patterning and renal dysplasia in nail patella syndrome. *Nature Genet.* 19: 47–50.

Drossopoulou, G., K. E. Lewis, J. J. Sanz-Ezguerro, N. Nikbaklt, A. P. McMahon, C. Hofman and C. Tickle. 2000. A model for anterior-posterior patterning of the vertebrate limb based on sequential long-and short-range *Shh* signalling and BMP signalling. *Development* 127: 1337–1348.

Fallon, J. F. and G. M. Crosby. 1977. Polarizing zone activity in limb buds of amniotes. In D. A. Ede, J. R. Hinchliffe and M. Balls (eds.), *Vertebrate Limb and Somite Morphogenesis.* Cambridge University Press, Cambridge, pp. 55–59.

Fallon, J. F., A. Lopez, M. A. Ros, M. P. Savage, B. B. Olwin and B. K. Simandl. 1994. FGF-2: Apical ectodermal ridge growth signal for chick development of the tetrapod limb. *Science* 264: 104–107.

Fromental-Ramain, C., X. Warot, N. Messadecq, M. LeMeur, P. Dollé and P. Chambon. 1996. *Hoxa-13* and *Hoxd-13* play a crucial role in the patterning of the limb autopod. *Development* 122: 2997–3011.

Gerard, M., D. Duboule and J. C. Zajkany. 1993. Cooperation of regulatory elements in the activation of the *Hoxd11* gene. *Compt. R. Acad. Sci. III* 316: 985–994.

Gibson-Brown, J. J., S. I. Agulnik, D. L. Chapman, M. Alexiou, N. Garvey, L. M. Silver and V. E. Papaioannou. 1996. Evidence of a role for T-box genes in the evolution of limb morphogenesis and the specification of forelimb/hindlimb identity. *Mech. Dev.* 56: 93–101.

Hamburger, V. 1938. Morphogenetic and axial self-differentiation of transplanted limb primordia of 2-day chick embryos. *J. Exp. Zool.* 77: 379–400.

Harrison, R. G. 1918. Experiments on the development of the forelimb of *Ambystoma*, a self-differentiating equipotential system. *J. Exp. Zool.* 25: 413–461.

Hayashi, K. and E. Ozawa. 1995. Myogenic cell migration from somites is induced by tissue contact with medial region of the presumptive limb mesoderm in chick embryos. *Development* 121: 661–669.

Heikinheimo, M., A. Lawshe, G. M. Shackleford, D. B. Wilson and C. A. MacArthur. 1994. Fgf-8 expression in the postgastrula mouse suggests roles in the development of the face, limbs, and central nervous system. *Mech. Dev.* 48: 129–138.

Hinchliffe, J. R. 1991. Developmental approaches to the problem of transformation of limb structure in evolution. *In* J. R. Hinchliffe (ed.), *Developmental Patterning of the Vertebrate Limb.* Plenum, New York, pp. 313–323.

Hinchliffe, J. R. 1994. Evolutionary biology of the tetrapod limb. *Development* [Suppl.] 163–168.

Hogan, B. L., M. Thaller, C. Thaller and G. Eichele. 1992. Evidence that Hensen's node is a source of retinoic acid synthesis. *Nature* 359: 237–241.

Honig, L. S. and D. Summerbell. 1985. Maps of strength of positional signaling activity in the developing chick wing bud. *J. Embryol. Exp. Morphol.* 87: 163–174.

Hurle, J. M., Y. Gañan and D. Macias. 1989. Experimental analysis of the in vivo chondrogenic potential of the interdigital mesenchyme of the chick limb bud subjected to local ectodermal removal. *Dev. Biol.* 132: 368–374.

Huxley, J. S. and G. R. De Beer. 1934. *The Elements of Experimental Embryology.* Cambridge University Press, Cambridge.

Iten, L. E. 1982. Pattern specification and pattern regulation in the embryonic chick limb bud. *Am. Zool.* 22: 117–129.

Izpisúa-Belmonte, J.-C., C. Tickle, P. Dollé, L. Wolpert and D. Duboule. 1991. Expression of the homeobox *Hox-4* genes and the specification of position in chick wing development. *Nature* 350: 585–589.

Kawakami, Y. and 9 others. 1996. BMP signaling during bone pattern determination in the developing limb. *Development* 122: 3557–3566.

Kieny, M. 1960. Rôle inducteur du mésoderme dans la différenciation précoce du bourgeon de membre chez l'embryon de poulet. *J. Embryol. Exp. Morphol.* 8: 457–467.

Krabbenhoft, K. M. and J. F. Fallon. 1989. The formation of leg or wing specific structures by leg bud cells grafted to the wing bud is influenced by proximity to the apical ridge. *Dev. Biol.* 131: 373–382.

Laufer, E., C. E. Nelson, R. L. Johnson, B. A. Morgan and C. Tabin. 1994. *Sonic hedgehog* and *Fgf-4* act through a signaling cascade and feedback loop to integrate growth and patterning of the developing limb bud. *Cell* 79: 993–1003.

Laufer, E. and 7 others. 1997. The *Radical fringe* expression boundary in the limb bud ectoderm regulates AER formation. *Nature* 386: 366–367.

Li, Q.Y. and 16 others. 1996. Holt-Oram syndrome is caused by mutations in *TBX5*, a member of the Brachyury (T) gene family. *Nature Genet.* 15: 21–29.

Li, S. and K. Muneoka. 1999. Cell migration and chick limb development: Chemotactic action of FGF-4 and the AER. *Dev. Biol.* 211: 335–347.

Logan, M., H.-H. Simon and C. Tabin. 1998. Differential regulation of T-box and homeobox transcription factors suggests roles in controlling chick limb-type identity. *Development* 125: 2825–2835.

López-Martínez, A. and 7 others. 1995. Limb-patterning activity and restricted posterior localization of the amino-terminal product of sonic hedgehog cleavage. *Curr. Biol.* 5: 791–796.

MacCabe, J. A., J. Errick and J. W. Saunders, Jr. 1974. Ectodermal control of dorso-ventral axis in leg bud of chick embryo. *Dev. Biol.* 39: 69–82.

Macias, D., Y. Gañon, T. K. Sampath, M. E. Piedra, M. A. Ros and J. M. Hurle. 1997. Role of BMP2 and OP-1 (BMP7) in programmed cell death and skeletogenesis during chick limb development. *Development* 124: 1109–1117.

Maden, M. 1993. The homeotic transformation of tails into limbs in *Rana temporaria* by retinoids. *Dev. Biol.* 159: 379–391.

Mahmood, R. and 9 others. 1995. A role for FGF-8 in the initiation and maintenance of vertebrate limb outgrowth. *Curr. Biol.* 5: 797–806.

Merino, R., Y. Gañan, D. Macias, A. N. Economides, K. T. Sampath and J. M. Hurle. 1998. Morphogenesis of digits in the avian limb is controlled by FGFs, TGFβs, and noggin through BMP signaling. *Dev. Biol.* 200: 35–45.

Mohanty-Hejmadi, P., S. K. Dutta and P. Mahapatra. 1992. Limbs generated at the site of tail amputation in marbled balloon frog after vitamin A treatment. *Nature* 355: 352–353.

Molven, A., C. V. E. Wright, R. Bremiller, E. M. De Robertis and C. B. Kimmel. 1990. Expression of a homeobox gene product in normal and mutant zebrafish embryos: Evolution of the tetrapod body plan. *Development* 109: 279–288.

Morgan, B. A. and C. Tabin. 1994. Hox genes and the evolution of vertebrate axial morphology. *Development* [Suppl.]: 181–186.

Mori, C., N. Nakamura, S. Kimura, H. Irie, T. Takigawa and K. Shiota. 1995. Programmed cell death in the interdigital tissue of the fetal mouse limb is apoptosis with DNA fragmentation. *Anat. Rec.* 242: 103–110.

Mortlock, D. P. and J. W. Innis. 1997. Mutation of *HOXA13* in hand-foot-genital syndrome. *Nature Genet.* 15: 179–181.

Müller, G., J. Streicher and R. Müller. 1996. Homeotic duplicate of the pelvic body segment in regenerating tadpole tails induced by retinoic acid. *Dev. Genes Evol.* 206: 344–348.

Muneoka, K. and S. V. Bryant. 1982. Evidence that patterning mechanisms in developing and regenerating limbs are the same. *Nature* 298: 369–371.

Muragaki, Y., S. Mundlos, J. Upton and B. Olsen. 1996. Altered growth and branching patterns in synpolydactyly caused by mutations in *HOXD13*. *Science* 272: 548–551.

Nelson, C. E. and 9 others. 1996. Analysis of Hox gene expression in the chick limb bud. *Development* 122: 1449–1466.

Newman, S. A. 1996. Sticky fingers: Hox genes and cell adhesion in vertebrate development of the tetrapod limb. *BioEssays* 18: 171–174.

Niswander, L., C. Tickle, A. Vogel, I. Booth and G. R. Martin. 1993. FGF-4 replaces the apical ectodermal ridge and directs outgrowth and patterning of the limb. *Cell* 75: 579–587.

Niswander, L., S. Jeffrey, G. R. Martin and C. Tickle. 1994. A positive feedback loop coordinates growth and patterning in the vertebrate limb. *Nature* 371: 609–612.

Nohno, T. and 7 others. 1991. Involvement of the *Chox-4* chicken homeobox genes in determination of anteroposterior axial polarity during development of the tetrapod limb. *Cell* 64: 1197–1205.

Ohuchi, H. and Noji, S. 1999. Fibroblast growth factor-induced additional limbs in the study of initiation of limb formation, limb identity, myogenesis, and innervation. *Cell Tissue Res.* 296: 45–56.

Ohuchi, H. and 11 others. 1997. The mesenchymal factor, FGF10, initiates and maintains the outgrowth of the chick limb bud through interaction with FGF8, and apical ectodermal factor. *Development* 124: 2235–2244.

Ohuchi, H. and 7 others. 1998. Correlation of wing-leg identity in ectopic FGF-induced chimeric limbs with the differential expression of chick *Tbx5* and *Tbx4*. *Development* 125: 51–60.

Oliver, G., C. V. E. Wright, J. Hardwicke and E. M. De Robertis. 1988. A gradient of homeodomain protein in developing forelimbs of *Xenopus* and mouse embryos. *Cell* 55: 1017–1024.

Opitz, J. M. 1985. The developmental field concept. *Am. J. Med. Genet.* 21: 1–11.

Owen, R. 1849. *On the Nature of Limbs.* J. Van Voor, London.

Parr, B. A. and A. P. McMahon. 1995. Dorsalizing signal *wnt-7a* required for normal polarity of D-V and A-P axes of the mouse limb. *Nature* 374: 350–353.

Parr, B. A., M. J. Shea, G. Vassileva and A. P. McMahon. 1993. Mouse *Wnt* genes exhibit discrete domains of expression in early embryonic CNS and limb buds. *Development* 119: 247–261.

Pearse, R. V. II and C. J. Tabin. 1998. The molecular ZPA. *J. Exp. Zool.* 282: 677–690.

Phillips, J. 1991. Higgledy, piggledy. *N. Engl. J. Med.* 324: 497.

Pollak, R. D. and J. F. Fallon. 1976. Autoradiographic analysis of macromolecular synthesis in prospectively necrotic cells of the chick limb bud. II. Nucleic acids. *Exp. Cell Res.* 100: 15–22.

Riddle, R. D., R. L. Johnson, E. Laufer and C. Tabin. 1993. *Sonic hedgehog* mediates the polarizing activity of the ZPA. *Cell* 75: 1401–1416.

Riddle, R. D., M. Ensini, C. Nelson, T. Tsuchida, T. M. Jessell and C. Tabin. 1995. Induction of the LIM homeobox gene *Lmx1* by WNT7a establishes dorsoventral pattern in the vertebrate limb. *Cell* 83: 631–640.

Rodriguez-Esteban, C., J. W. R. Schwabe, J. De La Peña, B. Foys, B. Eshelman and J. C. Izpisúa-Belmonte. 1997. Radical fringe positions the apical ectodermal ridge at the dorsoventral boundary of the vertebrate limb. *Nature* 386: 360–366.

Rodriguez-Esteban, C., T. Tsukui, S. Yonei, J. Magallon, K. Tamura and J. C. Izpisúa-Belmonte. 1999. T-box genes *Tbx4* and *Tbx5* regulate limb outgrowth and identity. *Nature* 398: 814–818.

Rowe, D. A., J. M. Cairnes and J. F. Fallon. 1982. Spatial and temporal patterns of cell death in limb bud mesoderm after apical ectodermal ridge removal. *Dev. Biol.* 93: 83–91.

Rubin, L. and J. W. Saunders, Jr. 1972. Ectodermal–mesodermal interactions in the growth of limbs in the chick embryo: Constancy and temporal limits of the ectodermal induction. *Dev. Biol.* 28: 94–112.

Saunders, J. W., Jr. 1948. The proximal-distal sequence of origin of the parts of the chick wing and the role of the ectoderm. *J. Exp. Zool.* 108: 363–404.

Saunders, J. W., Jr. 1972. Developmental control of three-dimensional polarity in the avian limb. *Ann. NY Acad. Sci. USA* 193: 29–42.

Saunders, J. W., Jr. and J. F. Fallon. 1966. Cell death in morphogenesis. *In* M. Locke (ed.), *Major Problems of Developmental Biology.* Academic Press, New York, pp. 289–314.

Saunders, J. W., Jr. and M. T. Gasseling. 1968. Ectodermal-mesodermal interactions in the origin of limb symmetry. *In* R. Fleischmajer and R. E. Billingham (eds.), *Epithelial-Mesenchymal Interactions.* Williams & Wilkins, Baltimore, pp. 78–97.

Saunders, J. W., Jr. and C. Reuss. 1974. Inductive and axial properties of prospective wing-bud mesoderm in the chick embryo. *Dev. Biol.* 38: 41–50.

Saunders, J. W., Jr., J. M. Cairns and M. T. Gasseling. 1957. The role of the apical ridge of ectoderm in the differentiation of the morphological structure and inductive specificity of limb parts of the chick. *J. Morphol.* 101: 57–88.

Saunders, J. W., Jr., M. T. Gasseling and L.C. Saunders. 1962. Cellular death in morphogenesis of the avian wing. *Dev. Biol.* 5: 147–178.

Sekine, K. and 10 others. 1999. Fgf10 is essential for limb and lung formation. *Nature Genet.* 21: 138–141.

Sessions, S. K., D. M. Gardiner and S. V. Bryant. 1989. Compatible limb patterning mechanisms in urodeles and anurans. *Dev. Biol.* 131: 294–301.

Shubin, N. H. and P. Alberch. 1986. A morphogenetic approach to the origin and basic organization of the tetrapod limb. *Evol. Biol.* 20: 319–387.

Shubin, N., C. Tabin and S. Carroll. 1997. Fossils, genes, and the evolution of animal limbs. *Nature* 388: 639–648.

Sordino, P., F. van der Hoeven and D. Duboule. 1995. Hox gene expression in teleost fins and the origin of the vertebrate digits. *Nature* 375: 678–681.

Storm, E. E. and D. M. Kingsley. 1999. GDF5 coordinates bone and joint formation during digit development. *Dev. Biol.* 209: 11–27.

Stratford, T., C. Horton and M. Maden. 1996. Retinoic acid is required for the initiation of outgrowth in the chick limb bud. *Curr Biol.* 6: 1124–1133.

Summerbell, D. 1974. A quantitative analysis of the effect of excision of the AER from the chick limb bud. *J. Embryol. Exp. Morphol.* 32: 651–660.

Summerbell, D. and Lewis, J. H. 1975. Time, place and positional value in the chick limb bud. *J. Embryol. Exp. Morphol.* 33: 621–643.

Takeuchi, J. K. and 8 others. 1999. *Tbx5* and *Tbx4* genes determine the wing/leg identity of limb buds. *Nature* 398: 810–814.

Tanaka, M., K. Tamura, S. Noji, T. Nohno and H. Ide. 1997. Induction of additional limb at the dorsal-ventral boundary of a chick embryo. *Dev. Biol.* 182: 191–203.

Tickle, C., D. Summerbell and L. Wolpert. 1975. Positional signaling and specification of digits in chick limb morphogenesis. *Nature* 254: 199–202.

Todt, W. L. and J. F. Fallon. 1987. Posterior apical ectodermal ridge removal in chick wing bud triggers a series of events resulting in defective anterior pattern. *Development* 101: 505–515.

van der Hoeven, F., J. Zakay and D. Duboule. 1996. Gene transpositions in the HoxD complex reveal a hierarchy of regulatory controls. *Cell* 85: 1025–1035.

Vogel, A., C. Rodriguez, W. Warnken and J.-C. Izpisúa-Belmonte. 1995. Dorsal cell fate specified by chick *Lmx1* during vertebrate development of the tetrapod limb. *Nature* 378: 716–720.

Vogel, A., C. Rodriguez and J.-C. Izpisúa-Belmonte. 1996. Involvement of FGF-8 in initiation, outgrowth, and patterning of the vertebrate limb. *Development* 122: 1737–1750.

Weiss, P. 1939. *Principles of Development.* Holt, Rinehart & Winston, New York.

Wessells, N. K. 1977. *Tissue Interaction and Development.* Benjamin Cummings, Menlo Park, CA.

Wolpert, L. 1977. *The Development of Pattern and Form in Animals.* Carolina Biological, Burlington, NC.

Xu, X. L. and 7 others. 1998. Fibroblast growth factor receptor Z (FGFR2)-mediated reciprocal regulatory loop between FGF8 and FGF10 is essential for limb induction. *Development* 125: 753–765.

Yang, Y. Z. and L. Niswander. 1995. Interaction between signaling molecules Wnt7a and Shh during vertebrate limb development: Dorsal signals regulate anteroposterior patterning. *Cell* 80: 939–947.

Yang, Y. Z. and 10 others.1997. Relationship between dose, distance, and time in Sonic hedgehog-mediated regulation of anteroposterior polarity in the chick limb. *Development* 124: 4393–4404.

Yonei-Tamura, S., T. Endo, H. Yajima, H. Ohuichi, H. Ida and K. Tamura. 1999. FGF7 and FGF10 directly induce the apical ectodermal ridge in chick embryos. *Dev. Biol.* 211: 133–143.

Yokouchi, Y., S. Nakazato, M. Yamamoto, Y. Goto, T. Kameda, H. Iba and A. Kuroiwa. 1995. Misexpression of *Hoxa-13* induces cartilage homeotic transformation and changes in adhesiveness in chick limb buds. *Genes Dev.* 9: 2509–2522.

Yokouchi, Y., J. Sakiyama, T. Kameda, H. Iba, A. Suzuki, N. Ueno and A. Kuroiwa. 1996. BMP-2/-4 mediate programmed cell death in chicken limb buds. *Development* 122: 3725–3734.

Zaleske, D. J. 1985. Development of the upper limb. *Hand Clin.* 1985(3): 383–390.

Zou, H. and L. Niswander. 1996. Requirement for BMP signaling in interdigital apoptosis and scale formation. *Science* 272: 738–741.

Zwilling, E. 1955. Ectoderm-mesoderm relationship in the development of the chick embryo limb bud. *J. Exp. Zool.* 128: 423–441.

c h a p t e r 17 *Sex determination*

Sexual reproduction is … the master-piece of nature.

ERASMUS DARWIN (1791)

It is quaint to notice that the number of speculations connected with the nature of sex have well-nigh doubled since Drelincourt, in the eighteenth century, brought together two hundred and sixty-two "groundless hypotheses," and since Blumenbach caustically remarked that nothing was more certain than that Drelincourt's own theory formed the two hundred and sixty-third.

J. A. THOMSON (1926)

HOW AN INDIVIDUAL'S SEX IS DETERMINED has been one of the great questions of embryology since antiquity. Aristotle, who collected and dissected embryos, claimed that sex was determined by the heat of the male partner during intercourse. The more heated the passion, the greater the probability of male offspring. (Aristotle counseled elderly men to conceive in the summer if they wished to have male heirs.) Aristotle (ca. 335 B.C.E.) promulgated a very straightforward hypothesis of sex determination: women were men whose development was arrested too early. The female was "a mutilated male" whose development had stopped because the coldness of the mother's womb overcame the heat of the father's semen. Women were therefore colder and more passive than men, and female sexual organs had not matured to the point at which they could provide active seeds. This view was accepted by the Christian Church and by Galen (whose anatomy texts were to be the standard for over a thousand years). Around the year 200 C.E., Galen wrote:

> *Just as mankind is the most perfect of all animals, so within mankind, the man is more perfect than the woman, and the reason for this perfection is his excess heat, for heat is Nature's primary instrument … the woman is less perfect than the man in respect to the generative parts. For the parts were formed within her when she was still a fetus, but could not because of the defect in heat emerge and project on the outside.*

The view that women were but poorly developed men and that their genitalia were like men's, only turned inside out, was a very popular one for over a thousand years. As late as 1543, Andreas Vesalius, the Paduan anatomist who overturned much of Galen's anatomy (and who risked censure by the church for arguing that men and women have the same number of ribs), held this view. The illustrations from his two major works, *De Humani Corporis Fabrica* and *Tabulae Sex*, show that he saw the female genitalia as internal representations of the male genitalia (Figure 17.1). Nevertheless, Vesalius' books sparked a revolution in anatomy, and by the end of the 1500s, anatomists had dismissed Galen's representation of female anatomy. During the 1600s and 1700s, females were seen as producing eggs that could transmit parental traits, and the physiology of the sex organs began to be studied. Still, there was no consensus about how the sexes became determined (see Horowitz 1976; Tuana 1988; Schiebinger 1989).

WEBSITE **17.1 Social critique of sex determination research.** In numerous cultures, women have been seen as the "default state" of men. Historians and biologists have shown that until recently such biases characterized the scientific study of human sex determination.

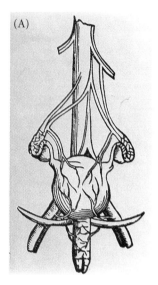 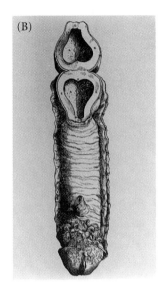

Figure 17.1
Vesalius' representations (1538, 1543) of the female reproductive organs. (A) Vesalius' rendering of Galen's conception of the female tract from the vagina to the uterus. (B) Rendering of the female reproductive system. (Reprinted in Schiebinger 1989.)

Until the twentieth century, the environment—temperature and nutrition, in particular—was believed to be important in determining sex. In 1890, Geddes and Thomson summarized all available data on sex determination and came to the conclusion that the "constitution, age, nutrition, and environment of the parents must be especially considered" in any such analysis. They argued that factors favoring the storage of energy and nutrients predisposed one to have female offspring, whereas factors favoring the utilization of energy and nutrients influenced one to have male offspring.

This environmental view of sex determination remained the only major scientific theory until the rediscovery of Mendel's work in 1900 and the rediscovery of the sex chromosome by McClung in 1902. It was not until 1905, however, that the correlation (in insects) of the female sex with XX sex chromosomes and the male sex with XY or XO chromosomes was established (Stevens 1905; Wilson 1905). This finding suggested strongly that a specific nuclear component was responsible for directing the development of the sexual phenotype. Thus, evidence accumulated that sex determination occurs by nuclear inheritance rather than by environmental happenstance.

Today we know that both environmental and internal mechanisms of sex determination can operate in different species. We will first discuss the chromosomal mechanisms of sex determination and then consider the ways by which the environment regulates the sexual phenotype.

Chromosomal Sex Determination in Mammals

Primary and secondary sex determination

Primary sex determination is the determination of the gonads. In mammals, primary sex determination is strictly chromosomal and is not usually influenced by the environment. In most cases, the female is XX and the male is XY. Every individual must have at least one X chromosome. Since the female is XX, each of her eggs has a single X chromosome. The male, being XY, can generate two types of sperm: half bear the X chromosome, half the Y. If the egg receives another X chromosome from the sperm, the resulting individual is XX, forms ovaries, and is female; if the egg receives a Y chromosome from the sperm, the individual is XY, forms testes, and is male. The Y chromosome carries a gene that encodes a testis-determining factor. This factor organizes the gonad into a testis rather than an ovary. Unlike the situation in *Drosophila* (discussed below), the mammalian Y chromosome is a crucial factor for determining sex in mammals. A person with five X chromosomes and one Y chromosome (XXXXXY) would be male. Furthermore, an individual with only a single X chromosome and no second X or Y (i.e., XO) develops as a female and begins making ovaries, although the ovarian follicles cannot be maintained. For a complete ovary, a second X chromosome is needed.

In mammalian primary sex determination, there is no "default state." The formation of ovaries and testes are both active, gene-directed processes. Moreover, as we shall see, both diverge from a common precursor, the bipotential gonad.

Secondary sex determination affects the bodily phenotype outside the gonads. A male mammal has a penis, seminal vesicles, and prostate gland. A female mammal has a vagina, cervix, uterus, oviducts, and mammary glands. In many species, each sex has a sex-specific size, vocal cartilage, and musculature. These secondary sex characters are usually determined by hormones secreted from the gonads. However, in the absence of gonads, the female phenotype is generated. When Jost (1953) removed fetal rabbit gonads before they had differentiated, the resulting rabbits had a female phenotype, regardless of whether they were XX or XY. They each had oviducts, a uterus, and a vagina, and each lacked a penis and male accessory structures.

The general scheme of mammalian sex determination is shown in Figure 17.2. If the Y chromosome is absent, the gonadal primordia develop into ovaries. The ovaries produce **estrogen**, a hormone that enables the development of the **Müllerian duct** into the uterus, oviducts, and upper end of the vagina. If the Y chromosome is present, testes form and secrete two major hormones. The first hormone—**anti-Müllerian duct hormone** (**AMH**; also referred to as Müllerian-inhibiting substance, MIS)—destroys the Müllerian duct. The

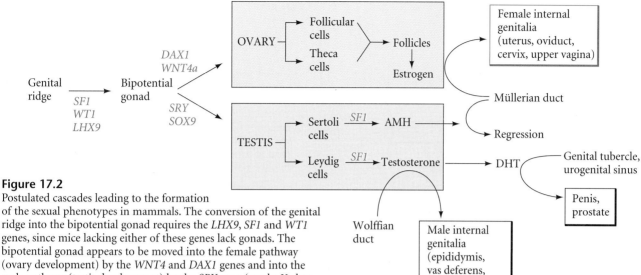

Figure 17.2
Postulated cascades leading to the formation of the sexual phenotypes in mammals. The conversion of the genital ridge into the bipotential gonad requires the *LHX9*, *SF1* and *WT1* genes, since mice lacking either of these genes lack gonads. The bipotential gonad appears to be moved into the female pathway (ovary development) by the *WNT4* and *DAX1* genes and into the male pathway (testis development) by the *SRY* gene (on the Y chromosome) in conjunction with autosomal genes such as *SOX9*. The ovary makes thecal cells and granulosa cells, which together are capable of synthesizing estrogen. Under the influence of estrogen (first from the mother, then from the fetal gonads), the Müllerian duct differentiates into the female genitalia, and the offspring develops the secondary sex characteristics of a female. The testis makes two major hormones. The first, anti-Müllerian duct factor (AMH), causes the Müllerian duct to regress. The second, testosterone, causes the differentiation of the Wolffian duct into the male internal genitalia. In the urogenital region, testosterone is converted into dihydrotestosterone (DHT), and this hormone causes the morphogenesis of the penis and prostate gland. (After Marx 1995 and Birk et al. 2000.)

second hormone—**testosterone**—masculinizes the fetus, stimulating the formation of the penis, scrotum, and other portions of the male anatomy, as well as inhibiting the development of the breast primordia. Thus, the body has the female phenotype unless it is changed by the two hormones secreted by the fetal testes. We will now take a more detailed look at these events.

The developing gonads

The gonads embody a unique embryological situation. All other organ rudiments can normally differentiate into only one type of organ. A lung rudiment can become only a lung, and a liver rudiment can develop only into a liver. The gonadal rudiment, however, has two normal options. When it differentiates, it can develop into either an ovary or a testis. The path of differentiation taken by this rudiment determines the future sexual development of the organism. But, before this decision is made, the mammalian gonad first develops through a **bipotential (indifferent) stage**, during which time it has neither female nor male characteristics.

In humans, the gonadal rudiments appear in the intermediate mesoderm during week 4 and remains sexually indif-

ferent until week 7. The gonadal rudiments are paired regions of the intermediate mesoderm; they form adjacent to the developing kidneys. The ventral portions of the gonadal rudiments are composed of the genital ridge epithelium. During the indifferent stage, the genital ridge epithelium proliferates into the loose connective mesenchymal tissue above it (Figure 17.3A,B). These epithelial layers form the **sex cords**. The germ cells migrate into the gonad during week 6, and are surrounded by the sex cords. In both XY and XX gonads, the sex cords remain connected to the surface epithelium.

If the fetus is XY, the sex cords continue to proliferate through the eighth week, extending deeply into the connective tissue. These cords fuse, forming a network of internal (medullary) sex cords and, at its most distal end, the thinner **rete testis** (Figure 17.3C, D). Eventually, the sex cords—now called **testis cords**—lose contact with the surface epithelium and become separated from it by a thick extracellular matrix, the **tunica albuginea**. Thus, the germ cells are found in the cords within the testes. During fetal life and childhood, the testis cords remain solid. At puberty, however, the cords will hollow out to form the **seminiferous tubules**, and the germ cells will begin to differentiate into sperm.

The cells of the seminiferous tubule are called **Sertoli cells**. The Sertoli cells of the testis cords nurture the sperm and secrete anti-Müllerian duct hormone. The sperm are transported from the inside of the testis through the rete testis, which joins the **efferent ducts**. These efferent tubules are the remnants of the mesonephric kidney, and they link the testis to the Wolffian duct, which used to be the collecting tube of the mesonephric kidney (see Chapter 15). In males, the Wolffian duct differentiates to become the **epididymis** (adjacent to the

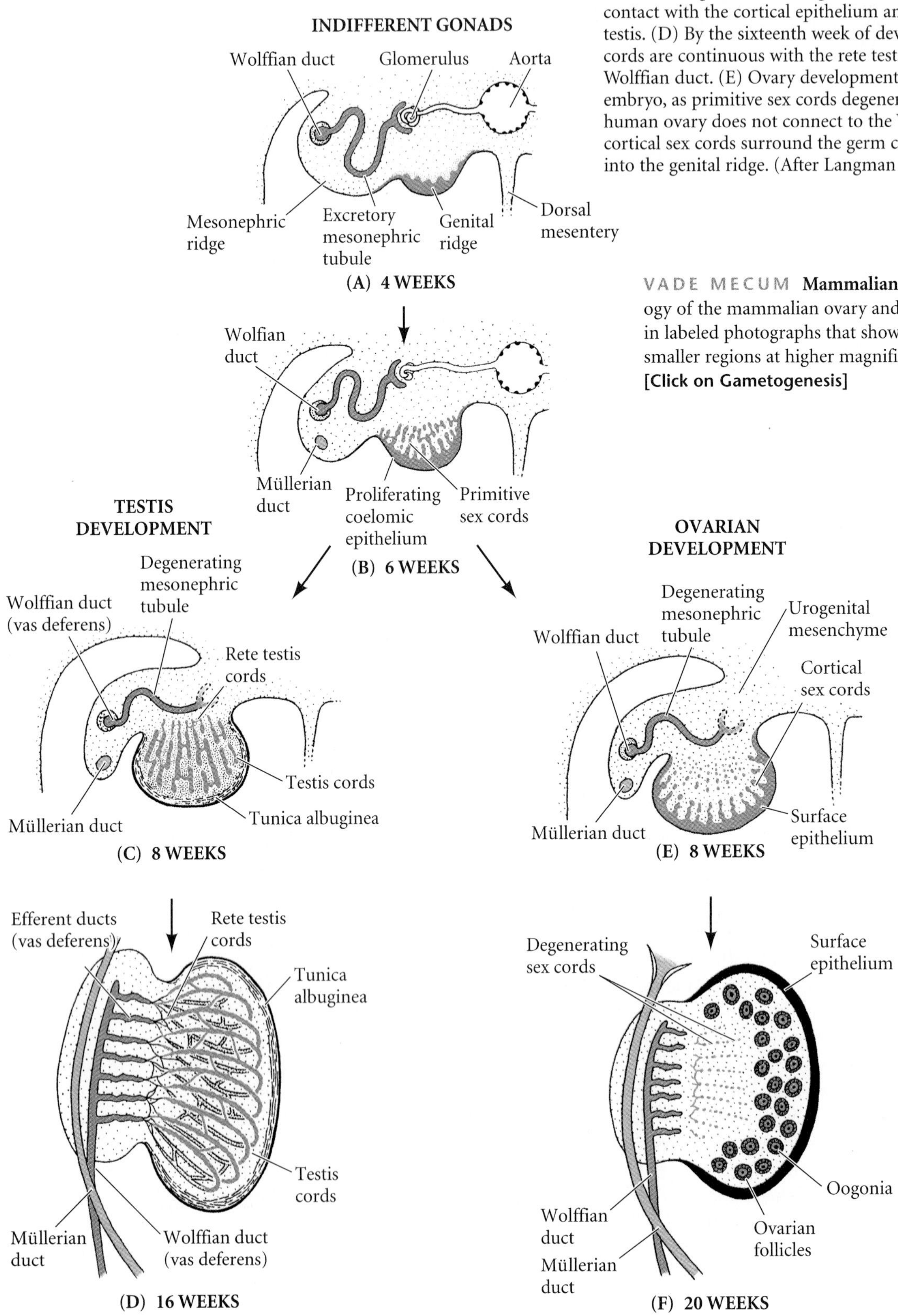

INDIFFERENT GONADS

(A) 4 WEEKS

Figure 17.3
Differentiation of human gonads shown in transverse section. (A) Genital ridge of a 4-week embryo. (B) Genital ridge of a 6-week indifferent gonad showing primitive sex cords. (C) Testis development in the eighth week. The sex cords lose contact with the cortical epithelium and develop the rete testis. (D) By the sixteenth week of development, the testis cords are continuous with the rete testis and connect with the Wolffian duct. (E) Ovary development in an 8-week human embryo, as primitive sex cords degenerate. (F) The 20-week human ovary does not connect to the Wolffian duct, and new cortical sex cords surround the germ cells that have migrated into the genital ridge. (After Langman 1981.)

VADE MECUM **Mammalian gonads.** The histology of the mammalian ovary and testis can be seen in labeled photographs that show progressively smaller regions at higher magnifications. **[Click on Gametogenesis]**

(B) 6 WEEKS

TESTIS DEVELOPMENT

(C) 8 WEEKS

(D) 16 WEEKS

OVARIAN DEVELOPMENT

(E) 8 WEEKS

(F) 20 WEEKS

Figure 17.4
Summary of the development of the gonads and their ducts in mammals. Note that both the Wolffian and Müllerian ducts are present at the indifferent gonad stage.

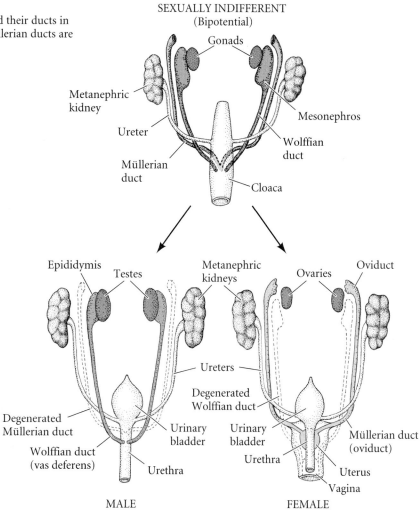

testis) and the **vas deferens**, the tube through which the sperm pass into the urethra and out of the body. Meanwhile, during fetal development, the interstitial mesenchyme cells of the testes differentiate into **Leydig cells**, which make testosterone.

In females, the germ cells will reside near the outer surface of the gonad. Unlike the sex cords in males, which continue their proliferation, the initial sex cords of XX gonads degenerate. However, the epithelium soon produces a new set of sex cords, which do not penetrate deeply into the mesenchyme, but stay near the outer surface (cortex) of the organ. Thus, they are called **cortical sex cords**. These cords are split into clusters, with each cluster surrounding a germ cell (Figure 17.3E, F). The germ cells will become the ova, and the surrounding cortical sex cords will differentiate into the **granulosa cells**. The mesenchyme cells of the ovary differentiate into the **thecal cells**. Together, the thecal and granulosa cells will form the **follicles** that envelop the germ cells and secrete steroid hormones. Each follicle will contain a single germ cell. In females, the Müllerian duct remains intact, and it differentiates into the oviducts, uterus, cervix, and upper vagina. The Wolffian duct, deprived of testosterone, degenerates. A summary of the development of mammalian reproductive systems is shown in Figure 17.4.

GONADS		
Gonadal type	Testis	Ovary
Sex cords	Medullary (internal)	Cortical (external)
DUCTS		
Remaining duct for germ cells	Wolffian	Müllerian
Duct differentiation	Vas deferens, epididymis, seminal vessicle	Oviduct, uterus, cervix, upper portion of vagina

The mechanisms of mammalian primary sex determination

Several genes have been found whose function is necessary for normal sexual differentiation. Unlike those that act in other developing organs, the genes involved in sex determination differ extensively between phyla, so one cannot look at *Drosophila* sex-determining genes and expect to see their homologues directing mammalian sex determination. However, since the phenotype of mutations in sex-determining genes is often sterility, clinical studies have been used to identify those genes that are active in determining whether humans become male or female. Experimental manipulations to confirm the functions of these genes can be done in mice.

SRY: THE Y CHROMOSOME SEX DETERMINANT. In humans, the major gene for the testis-determining factor resides on the short arm of the Y chromosome. Individuals who are born with the short arm but not the long arm of the Y chromosome are male, while individuals born with the long arm of the Y

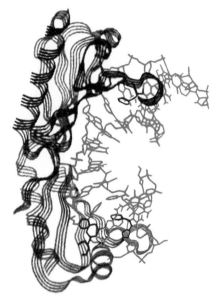

Figure 17.5
Association of DNA with the SRY protein can cause the DNA to bend 70–80 degrees. The black structure represents the HMG box of the SRY protein. The red coil is the double helix of DNA specifically bound by SRY. (After Haqq et al. 1994 and Werner et al. 1995.)

chromosome but not the short arm are female. By analyzing the DNA of rare XX men and XY women, the position of the testis-determining gene has been narrowed down to a 35,000-base-pair region of the Y chromosome located near the tip of the short arm. In this region, Sinclair and colleagues (1990) found a male-specific DNA sequence that could encode a peptide of 223 amino acids. This peptide is probably a transcription factor, since it contains a DNA-binding domain called the **HMG** (*h*igh-*m*obility *g*roup) **box**. This domain is found in several transcription factors and nonhistone chromatin proteins, and it induces bending in the region of DNA to which it binds (Figure 17.5; Giese et al. 1992). This gene is called **SRY** (*s*ex-determining *r*egion of the *Y* chromosome), and there is extensive evidence that it is indeed the gene that encodes the human testis-determining factor. *SRY* is found in normal XY males and in the rare XX males, and it is absent from normal XX females and from many XY females. Another group of XY females was found to have point or frameshift mutations in the *SRY* gene; these mutations prevent the SRY protein from binding to or bending DNA (Pontiggia et al. 1994; Werner et al. 1995). It is thought that several testis-specific genes contain SRY-binding sites in their promoters or enhancers, and that the binding of SRY to these sites begins the developmental pathway to testis formation (Cohen et al. 1994).

If *SRY* actually does encode the major testis-determining factor, one would expect that it would act in the genital ridge immediately before or during testis differentiation. This prediction has been met in studies of the homologous gene found in mice. The mouse gene (*Sry*) also correlates with the presence of testes; it is present in XX males and absent in XY females (Gubbay et al. 1990; Koopman et al. 1990). The *Sry* gene is expressed in the somatic cells of the bipotential mouse gonad immediately before or during its differentiating into a testis; its expression then disappears (Hacker et al. 1995).

The most impressive evidence for *Sry* being the gene for testis-determining factor comes from transgenic mice. If *Sry* induces testis formation, then inserting *Sry* DNA into the genome of a normal XX mouse zygote should cause that XX mouse to form testes. Koopman and colleagues (1991) took the 14-kilobase region of DNA that includes the *Sry* gene (and presumably its regulatory elements) and microinjected this sequence into the pronuclei of newly fertilized mouse zygotes. In several instances, the XX embryos injected with this sequence developed testes, male accessory organs, and penises (Figure 17.6). (Functional sperm were not formed, but they were not expected, either, because the presence of two X chromosomes prevents sperm formation in XXY mice and men, and the transgenic mice lacked the rest of the Y chromosome, which contains genes needed for spermatogenesis.) Therefore, there are good reasons to think that *Sry/SRY* is the major gene on the Y chromosome for testis determination in mammals.

Sry/SRY is necessary, but not sufficient, for the development of the mammalian testis. Studies on mice (Eicher and Washburn 1983; Washburn and Eicher 1989; Eicher et al. 1996) have shown that the *Sry* gene of some strains of mice failed to produce testes when placed into a different strain of mouse. When the Sry protein binds to its sites on DNA, it

Figure 17.6
An XX mouse transgenic for *Sry* is male. (A) Polymerase chain reaction followed by electrophoresis shows the presence of the *Sry* gene in normal XY males and in a transgenic XX *Sry* mouse. The gene is absent in a female XX littermate. (B) The external genitalia of the transgenic mouse are male (right) and essentially the same as in an XY male (left). (From Koopman et al. 1991; photographs courtesy of the authors.)

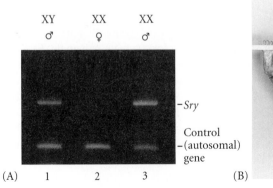

(A)   1   2   3

— *Sry*

Control
—(autosomal)
gene

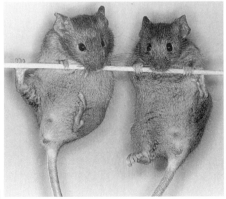

(B)

probably creates large conformational changes. It unwinds the double helix in its vicinity and bends the DNA as much as 80 degrees (Pontiggia et al. 1994; Werner et al. 1995). This bending may bring distantly bound proteins of the transcription apparatus into close contact, enabling them to interact and influence transcription. The identities of these proteins are not yet known, but they, too, are needed for testis determination.

SRY may have more than one mode of action in converting the bipotential gonads into testes. It had been assumed for the past decade that SRY worked directly in the genital ridge to convert the epithelium into male-specific Sertoli cells. Recent studies (Capel et al. 1999), however, have suggested that SRY works via an indirect mechanism: SRY in the genital ridge cells induces the cells to secrete a chemotactic factor that permits the migration of mesonephric cells into the XY gonad. These mesonephric cells induce the gonadal epithelium to become Sertoli cells with male-specific gene expression patterns. The researchers found that when they cultured XX gonads with either XX or XY mesonephrons, the mesonephric cells did not enter the gonads. However, when they cultured XX or XY mesonephrons with XY gonads, or with gonads from XX mice containing the *Sry* transgene, the mesonephric cells did enter the gonads (Figure 17.7). There was a strict correlation between the presence of Sry in the gonadal cells, mesonephric

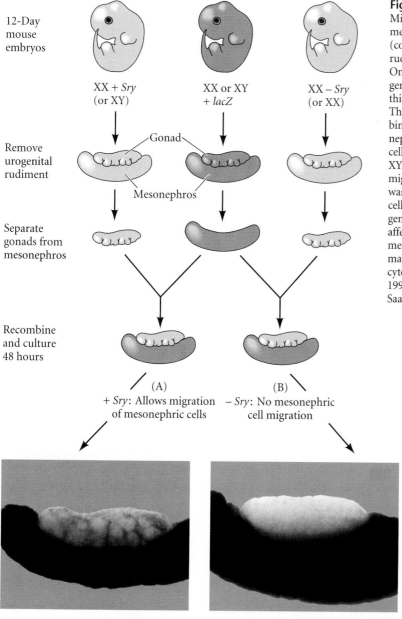

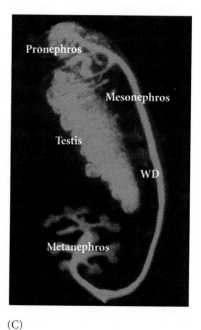

Figure 17.7

Migration of the mesonephric cells into *Sry*⁺ gonadal rudiments. In the experiment diagrammed, urogenital ridges (containing both the mesonephric kidneys and gonadal rudiments) were collected from 12-day embryonic mice. One of the mice was marked with a β-galactosidase transgene (*lacZ*) that is active in every cell. Thus every cell of this mouse turns blue when stained for β-galactosidase. The gonad and mesonephros were separated and recombined, using gonadal tissue from unlabeled mice and mesonephros from labeled mice. (A) Migration of mesonephric cells into the gonad was seen when the gonadal cells were XY or when they were XX with an *Sry* transgene. (B) No migration of mesonephric tissue into the bipotential gonad was seen when the gonad contained either XX cells or XXY cells in which the Y chromosome had a deletion in the *Sry* gene. The sex chromosomes of the mesonephros did not affect the migration. (C) Intimate relation between the mesonephric ducts and the developing gonad in the 16-day male mouse embryo. The duct tissue has been stained for cytokeratin-8. WD, Wolfian duct. (A, B after Capel et al. 1999, photographs courtesy of B. Capel; C from Sariola and Saarma 1999, photograph courtesy of H. Sariola.)

cell migration, and the formation of testis cords. Tilmann and Capel (1999) showed that mesonephric cells are critical for testis cord formation and that the migrating mesonephric cells can induce XX gonadal cells to form testis cords. It appears, then, that Sry may function indirectly to create testes by inducing mesonephric cell migration into the gonad.

> WEBSITE 17.2 **Finding the male-determining genes.** The mapping of the testis-determining factor to the *SRY* region took scientists more than 50 years to accomplish. Moreover, other testis-forming genes have been found on the autosomes.

SOX9: AUTOSOMAL SEX REVERSAL. One of the autosomal genes involved in sex determination is *SOX9*, which encodes a putative transcription factor that also contains an HMG box. XX humans who have an extra copy of *SOX9* develop as males, even though they have no *SRY* gene (Huang et al. 1999). Individuals having only one functional copy of this gene have a syndrome called campomelic dysplasia, a disease involving numerous skeletal and organ systems. About 75% of XY patients with this syndrome develop as phenotypic females or hermaphrodites (Foster et al. 1994; Wagner et al. 1994; Mansour et al. 1995). It appears that *SOX9* is essential for testis formation. The mouse homologue of this gene, *Sox9*, is expressed only in male (XY) but not in female (XX) genital ridges. Moreover, *Sox9* expression is seen in the same genital ridge cells as *Sry*, and it is expressed just slightly after *Sry* expression (Wright et al. 1995; Kent et al. 1996). The Sox9 protein binds to a promoter site on the *Amh* gene, providing a critical link in the pathway toward a male phenotype (Figure 17.8; Arango et al. 1999).

While *Sry* is found specifically in mammals, *Sox9* is found throughout the vertebrates. *Sox9* may be the older and more central sex determination gene, although in mammals it became activated by its relative, *Sry*.

SF1: THE LINK BETWEEN SRY AND THE MALE DEVELOPMENTAL PATHWAYS. Another protein that may be directly or indirectly activated by SRY is the transcription factor **SF1** (*steroidogenic factor 1*). *Sf1* is necessary to make the bipotential gonad; but while Sf1 levels decline in the genital ridge of XX mouse embryos, the *Sf1* gene stays on in the developing testis. Sf1 appears to be active in masculinizing both the Leydig and the Sertoli cells. In the Sertoli cells, Sf1, working in collaboration with Sox9, is needed to elevate the levels of AMH transcription (see Figure 17.8; Shen et al. 1994; Arango et al. 1999). In the Leydig cells, Sf1 activates the genes encoding the enzymes that make testosterone. The importance of SF1 for testis development and AMH regulation in humans is demonstrated by an XY patient who is heterozygous for *SF1*. Although the genes for SRY and SOX9 are normal, this individual has malformed fibrous gonads and retains fully developed Müllerian duct structures (Achermann et al. 1999). It is thought that SRY (directly or indirectly) activates the *SF1* gene, and the SF1 protein then activates both components of the male sexual differentiation pathway (Sertoli AMH and Leydig testosterone).

DAX1: A POTENTIAL OVARY-DETERMINING GENE ON THE X CHROMOSOME. In 1980, Bernstein and her colleagues reported two sisters who were genetically XY. Their Y chromosomes were normal, but they had a duplication of a small portion of the short arm of the X chromosome. Subsequent cases were found, and it was concluded that if there were two copies of this region on the active X chromosome, the SRY signal would be reversed (Figure 17.9). Bardoni and her colleagues (1994) proposed that this region contains a gene for a protein that competes with the SRY factor and that is important in directing the development of the ovary. In testicular development, this gene would be suppressed, but having two active copies of the gene would override this suppression. This gene, *DAX1*, has been cloned and shown to encode a member of the nuclear hormone receptor family (Muscatelli et al. 1994; Zanaria 1994). *Dax1* is expressed in the genital ridges of the mouse embryo, shortly after *Sry* expression. Indeed, in XY mice, *Sry* and *Dax1* are expressed in the same cells. DAX1 appears to antagonize the function of SRY, and it down-regulates SF1 expression (Nachtigal et al. 1998; Swain et al. 1998). Thus, *DAX1* is probably a gene that is involved in ovary determination.

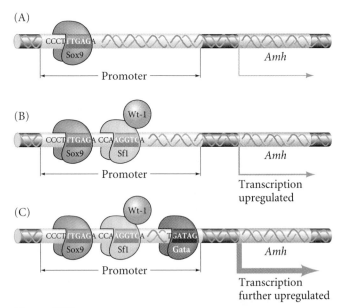

Figure 17.8
Synergism of Sox9 and Sf1 to activate the expression of the *Amh* gene. (A) The binding of Sox9 to the *Amh* promoter initiates transcription of the *Amh* gene in the Sertoli cells. (B) After Sox9 binds, the expression of AMH is upregulated by the binding of SF1 and Wt-1. AMH is believed to position SF1 on its DNA-binding site, and Wt-1 is joined to the Sf1 protein. (C) Gata (a transcription factor common to many cell types) upregulates *Amh* expression further. Neither Sf1 nor Gata can function if Sox9 is absent. (After Arango et al. 1999.)

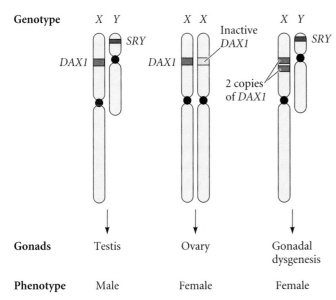

Genotype

Gonads Testis   Ovary   Gonadal dysgenesis

Phenotype Male   Female   Female

Figure 17.9
Phenotypic sex reversal in humans having two copies of the *DAX1* locus. *DAX1* (on the X chromosome) plus *SRY* (on the Y chromosome) produces testes. *DAX1* without *SRY* (since the other *DAX1* locus is on the inactive X chromosome) produces ovaries. Two active copies of *DAX1* (on the active X chromosome) plus *SRY* (on the Y chromosome) lead to a poorly formed gonad. Since the gonad makes neither AMH nor testosterone, the phenotype is female. (After Genetics Review Group 1995.)

WNT4: A POTENTIAL OVARY-DETERMINING GENE ON AN AUTO-SOME. The *WNT4* gene is another gene that may be critical in ovary determination. This gene is expressed in the mouse genital ridge while it is still in its bipotential stage. *Wnt4* expression then becomes undetectable in XY gonads (which become testes), whereas it is maintained in XX gonads as they begin to form ovaries. In transgenic XX mice that lack the *Wnt4* genes, the ovary fails to form properly, and its cells express testis-specific markers, including AMH- and testosterone-producing enzymes (Vainio et al. 1999). Sry may form testes by repressing *Wnt4* expression in the genital ridge, as well as by promoting *Sf1*. One possible model is shown in Figure 17.10.

It should be realized that both testis and ovary development are active processes. In mammalian primary sex determination, neither is a "default state" (Eicher and Washburn 1986). Although remarkable progress has been made in recent years, we still do not know what the testis- or ovary-determining genes are doing, and the problem of primary sex determination remains (as it has since prehistory) one of the great unsolved problems of biology.

Secondary sex determination: Hormonal regulation of the sexual phenotype

Primary sex determination involves the formation of either an ovary or a testis from the bipotential gonad. This, however, does not give the complete sexual phenotype. Secondary sex determination in mammals involves the development of the female and male phenotypes in response to hormones secreted by the ovaries and testes. Both female and male secondary sex determination have two major temporal phases. The first occurs within the embryo during organogenesis; the second occurs during adolescence.

As mentioned earlier, if the bipotential gonads are removed from an embryonic mammal, the female phenotype is realized: the Müllerian ducts develop while the Wolffian duct degenerates. This pattern also is seen in certain humans who are born without functional gonads. Individuals whose cells have only one X chromosome (and no Y chromosome) originally develop ovaries, but these ovaries atrophy before birth, and the germ cells die before puberty. However, under the influence of estrogen, derived first from the ovary but then from the mother and placenta, these infants are born with a female genital tract (Langman and Wilson 1982).

The formation of the male phenotype involves the secretion of two testicular hormones. The first of these hormones is AMH, the hormone made by the Sertoli cells that causes the degeneration of the Müllerian duct. The second is the steroid testosterone, which is secreted from the fetal Leydig cells. This hormone causes the Wolffian duct to differentiate into the epididymis, vas deferens, and seminal vesicles, and it causes the urogenital swellings to develop into the scrotum and penis.

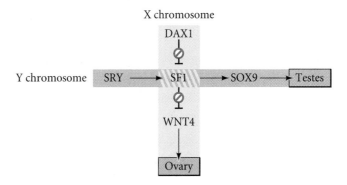

Figure 17.10
Possible mechanism for primary sex determination in mammals. While we do not know the specific interactions involved, this model attempts to organize the data into a coherent sequence. Other models are possible. In this model, SRY competes with the DAX1 protein to activate or repress the *SF1* gene. If a single X and Y chromosome are present, the SRY will be favored, and the activation of *SF1* will occur. If two copies of *DAX1* are present on the X chromosome (or if there is no Y chromosome), the *SF1* gene will not be activated. The SF1 protein is thought to activate the *SOX9* gene, which instructs the sex cords to develop into the Sertoli cells of the testes, and may also repress *WNT4*. WNT4 would otherwise cause the differentiation of the gonad into an ovary. Most of the genes activated by WNT4 and SOX9 have not been identified, and the mechanisms by which SRY and DAX1 function are not yet known.

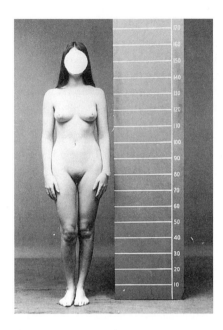

Figure 17.11
An XY individual with androgen insensitivity syndrome. Despite the XY karyotype and the presence of testes, such individuals develop female secondary sex characteristics. Internally, however, these women lack the Müllerian duct derivatives and have undescended testes. (Photograph courtesy of C. B. Hammond.)

The existence of these two independent systems of masculinization is demonstrated by people having **androgen insensitivity syndrome**. These XY individuals have the *SRY* gene, and thus have testes that make testosterone and AMH. However, they lack the testosterone receptor protein, and therefore cannot *respond* to the testosterone made by their testes (Meyer et al. 1975). Because they are able to respond to estrogen made in their adrenal glands, they develop the female phenotype (Figure 17.11). However, despite their distinctly female appearance, these individuals do have testes, and even though they cannot respond to testosterone, they produce and respond to AMH. Thus, their Müllerian ducts degenerate. These people develop as normal but sterile women,* lacking a uterus and oviducts and having testes in the abdomen.

TESTOSTERONE AND DIHYDROTESTOSTERONE. Although testosterone is one of the two primary masculinizing hormones, there is evidence that it might not be the active masculinizing hormone in certain tissues. Testosterone appears to be responsible for promoting the formation of the male reproductive structures (the epididymis, seminal vesicles, and vas deferens) that develop from the Wolffian duct primordium. However, it does not directly masculinize the male urethra, prostate, penis, or scrotum. These latter functions are controlled by **5α-dihydrotestosterone** (Figure 17.12). Siiteri and Wilson (1974) showed that testosterone is converted to 5α-dihydrotestosterone in the urogenital sinus and swellings, but not in the Wolffian duct. 5α-dihydrotestosterone appears to be a more potent hormone than testosterone.

The importance of 5α-dihydrotestosterone was demonstrated by Imperato-McGinley and her colleagues (1974). They found a small community in the Dominican Republic in which several inhabitants had a genetic deficiency of the enzyme 5α-ketosteroid reductase 2, the enzyme that converts testosterone to dihydrotestosterone. These individuals lack a functional gene for this enzyme (Andersson et al. 1991; Thigpen et al. 1992). Although XY children with this syndrome have functioning testes, they have a blind vaginal pouch and an enlarged clitoris. They appear to be girls and are raised as such. Their internal anatomy, however, is male: they have testes, Wolffian duct development, and Müllerian duct degeneration. Thus, it appears that the formation of the external genitalia is under the control of dihydrotestosterone, whereas Wolffian duct differentiation is controlled by testosterone itself. Interestingly, when the testes of these children produce more testosterone at puberty, the external genitalia are able to respond to the higher levels of the hormone, and they differentiate. The penis enlarges, the scrotum descends, and the person originally thought to be a girl is shown to be a young man.

---

*Androgen insensitivity syndrome is one of several conditions called *pseudohermaphroditism*. In a pseudohermaphrodite, there is only one type of gonad, but the secondary sex characteristics differ from what would be expected from the gonadal sex. In humans, male pseudohermaphroditism can be caused by mutations in the androgen receptor or by mutations affecting testosterone synthesis (Geissler et al. 1994). Female pseudohermaphroditism can be caused by an overproduction of testosterone.

True hermaphrodites (rare in humans, but the norm in some invertebrates such as nematodes and earthworms) contain both male and female gonadal tissue. Mammalian true hermaphrodites result from abnormalities of primary sex determination. Such abnormalities can occur when the Y chromosome is translocated to the X chromosome. In those tissues where the translocated X chromosome is inactivated during dosage compensation, the *SRY* gene will be turned off. However, in those tissues where the translocated X chromosome is not inactivated, the *SRY* gene will be on (Berkovitz et al. 1992; Margarit et al. 2000).

WEBSITE **17.3 Dihydrotestosterone in adult men.** The drug finasteride, which inhibits the conversion of testosterone to dihydrotestosterone, is being used to treat prostate growth and male pattern baldness.

WEBSITE **17.4 Insulin-like hormone 3.** In addition to testosterone, the Leydig cells secrete another hormone, insulin-like hormone 3 (Insl3). This hormone is required for the descent of the gonads into the scrotum. Males lacking this hormone are infertile because the testes do not descend. In females, lack of this hormone deregulates the menstrual cycle.

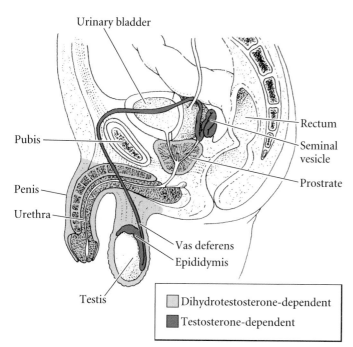

Urinary bladder

Pubis

Penis

Urethra

Testis

Rectum

Seminal vesicle

Prostrate

Vas deferens

Epididymis

Dihydrotestosterone-dependent

Testosterone-dependent

Figure 17.12
Testosterone- and dihydrotestosterone-dependent regions of the human male genital system. (After Imperato-McGinley et al. 1974.)

ANTI-MÜLLERIAN DUCT HORMONE. Anti-Müllerian hormone (AMH), the hormone that causes the degeneration of the Müllerian duct, is a 560-amino acid glycoprotein secreted from the Sertoli cells (Tran et al. 1977; Cate et al. 1986). When fragments of fetal testes or isolated Sertoli cells are placed adjacent to cultured tissue segments containing portions of the Wolffian and Müllerian ducts, the Müllerian duct atrophies even though no change occurs in the Wolffian duct (Figure 17.13). AMH is thought to bind to the mesenchyme cells surrounding the Müllerian duct and to cause these cells to secrete a paracrine factor that induces apoptosis in the Müllerian duct epithelium (Trelstad et al. 1982; Roberts et al. 1999).

WEBSITE **17.5 Roles of AMH.** AMH may have other roles in sex determination besides causing the breakdown of the Müllerian ducts. It may cause sex reversal in some mammals, and may become useful as an anti-tumor drug.

ESTROGEN. Estrogen is needed for the complete development of both the Müllerian and the Wolffian ducts. In

females, estrogen secreted from the fetal ovaries appears sufficient to induce the differentiation of the Müllerian duct into its various components: the uterus, oviducts, and cervix. The extreme sensitivity of the Müllerian duct to estrogenic compounds is demonstrated by the teratogenic effects of **diethylstilbesterol** (**DES**), a powerful synthetic estrogen that can cause infertility by changing the patterning of the Müllerian duct (see Mittendorf 1995). In mice, DES can cause the oviduct epithelium to take on the appearance of the uterus, and the uterine epithelium to resemble that of the cervix (Ma et al. 1998).

In males, estrogen is actually needed for fertility. One of the functions of the efferent duct (vas efferens) cells is to absorb about 90% of the water from the lumen of the rete testis. This concentrates the sperm, giving them a longer lifespan and providing more sperm per ejaculate. This absorption of water is regulated by estrogen. If estrogen or its receptor is absent in mice, this water is not absorbed, and the mouse is sterile (Hess et al. 1997). While blood concentrations of estrogen are higher in females than in males, the concentration of estrogen in the rete testis is even higher than that in female blood.

WEBSITE **17.6 The mechanisms of breast development.** Breast tissue has a sexually dimorphic mode of development. Testosterone inhibits breast development, while estrogen promotes it. Most breast development is accomplished after birth, and different hormones act during puberty and pregnancy to cause breast enlargement and differentiation.

WEBSITE **17.7 The action of DES.** Diethylstilbesterol was a drug given to women to ease their pregnancies. Unfortunately, it was later found to alter the reproductive tract of female fetuses. The mechanism of DES action is thought to involve the repression of the Hox genes that instruct the regional specificity of the Müllerian duct.

Figure 17.13
Assay for AMH activity in the anterior segment of a 14.5-day fetal rat reproductive tract. (A) At the start of the experiment, both the Müllerian duct (arrow at left) and Wolffian duct (arrow at right) are open. (B) After 3 days in culture with AMH-secreting tissue, the Wolffian duct (arrow) is open, but the Müllerian duct has degenerated and closed. (Photograph courtesy of N. Josso.)

(A)   (B)

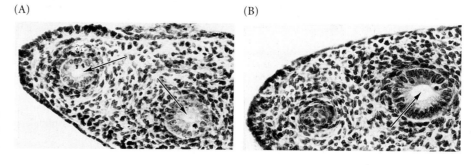

Sex Determination and Behaviors

Organization/Activation Hypothesis

Does prenatal (or neonatal) exposure to particular steroid hormones impose permanent sex-specific changes on the central nervous system? Such sex-specific neural changes have been shown in regions of the brain that regulate "involuntary" sexual physiology. The cyclic secretion of luteinizing hormone by the adult female rat pituitary, for example, is dependent on the lack of testosterone during the first week of the animal's life. The luteinizing hormone secretion of female rats can be made noncyclic by giving them testosterone 4 days after birth; conversely, the luteinizing hormone secretion of males can be made cyclical by removing their testes within a day of birth (Barraclough and Gorski 1962). It is thought that sex hormones may act during the fetal or neonatal stage of a mammal's life to organize the nervous system in a sex-specific manner, and that during adult life, the same hormones may have transitory, activational effects. This idea is called the **organization/activation hypothesis**.

Interestingly, the hormone chiefly responsible for determining the male brain pattern is **estradiol**, a type of estrogen.* Testosterone in fetal or neonatal blood can be converted into estradiol by the enzyme P450 aromatase, and this conversion occurs in the hypothalamus and limbic system—two areas of the brain known to regulate hormone secretion and reproductive behavior (Reddy et al. 1974; McEwen et al. 1977). Thus, testosterone exerts its effects on the nervous system by being converted into estradiol. But the fetal environment is rich in estrogens from the gonads and placenta. What stops these estrogens from masculinizing the nervous system of a female fetus? Fetal estrogen (in both males

---

*The terms *estrogen* and *estradiol* are often used interchangeably. However, estrogen refers to a class of steroid hormones responsible for establishing and maintaining specific female characteristics. Estradiol is one of these hormones, and in most mammals (including humans), it is the most potent of the estrogens.

and females) is bound by **α-fetoprotein**. This protein is made in the fetal liver and becomes a major component of the fetal blood and cerebrospinal fluid. It will bind and inactivate estrogen, but not testosterone.

Attempts to extend the organization/activation hypothesis to "voluntary" sexual behaviors are more controversial because there is no truly sex-specific behavior that distinguishes the two sexes of many mammals, and because hormonal treatment has multiple effects on the developing mammal. For instance, injecting testosterone into a week-old female rat will increase pelvic thrusting behavior and diminish lordosis—a posture that stimulates mounting behavior in the male—when she reaches adulthood (Phoenix et al. 1959; Kandel et al. 1995). These behavioral changes can be ascribed to testosterone-mediated changes in the central nervous system, but they could also be due to hormonal effects on other tissues. Testosterone enables the growth of the muscles that allow pelvic thrusting. And since testosterone causes females to grow larger and to close their vaginal orifices, one cannot conclude that the lack of lordosis is due solely to testosterone-mediated changes in the neural circuitry (Harris and Levine 1965; De Jonge et al. 1988; Moore 1990; Moore et al. 1992; Fausto-Sterling 1995).

In addition, the effects of sex steroids on the brain are very complicated, and the steroids may be metabolized differently in different regions of the brain. Male mice lacking the testosterone receptor still retain a male-specific preoptic morphology in the brain, and male mice lacking the aromatase enzyme are capable of breeding (Breedlove 1992; Fisher et al. 1998). These studies show that there is more to sex-specific morphology and behavior than steroid hormones. Despite best-selling books that pretend to know the answers, we have much more to learn regarding the relationship between development, steroids, and behavior. Moreover, extrapolating from rats to humans is a very risky business, as no sex-specific behavior has yet been identified in humans, and what is "masculine"

in one culture may be considered "feminine" in another (see Jacklin 1981; Bleier 1984; Fausto-Sterling 1992). As one review (Kandel et al. 1995) concludes:

> There is ample evidence that the neural organization of reproductive behaviors, while importantly influenced by hormonal events during a critical prenatal period, does not exert an immutable influence over adult sexual behavior or even over an individual's sexual orientation. Within the life of an individual, religious, social, or psychological motives can prompt biologically similar persons to diverge widely in their sexual activities.

Male Homosexuality

Certain behaviors are often said to be part of the "complete" male or female phenotype. The brain of a mature man is said to be formed such that it causes him to desire mating with a mature woman, and the brain of a mature woman causes her to desire to mate with a mature man. However, as important as desires are in our lives, they cannot be detected by in situ hybridization or isolated by monoclonal antibodies. We do not yet know if sexual desires are primarily instilled in us by our social education or are fundamentally "hardwired" into our brains by genes or hormones during our intrauterine development or by other means.

In 1991, Simon LeVay proposed that part of the anterior hypothalamus of homosexual men has the anatomical form typical of women rather than of heterosexual men. The hypothalamus is thought to be the source of our sexual urges, and rats have a sexually dimorphic area in their anterior hypothalamus that appears to regulate their sexual behavior. Thus, this study generated a great deal of publicity and discussion. The major results are shown in Figure 17.14. The interstitial nuclei (neuron clusters) of the anterior hypothalamus (INAH) were divided into four regions. Three of them showed no signs of sexual dimorphism. However, one of them, INAH3, showed a statistically significant dif-

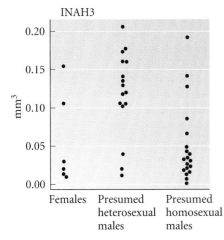

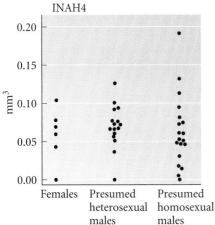

Figure 17.14
A portion of the data claiming a biological basis for homosexuality. INAH4 and INAH3 are two groups of hypothalamic neurons. INAH4 shows no sexual dimorphism in volume, while INAH3 shows a statistically significant clustering, although the range is similar. The INAH3 from autopsies of "homosexual" male brains cluster toward the female distribution. However, no cause-and-effect relationship can be posited. (After LeVay 1991.)

ference in volume between males and females; it was claimed that the male INAH3 is, on average, more than twice as large as the female INAH3. Moreover, LeVay's data suggested that the INAH3 of homosexual men was similar in volume to that of women and less than half the size of heterosexual men's INAH3. This finding, LeVay claimed, "suggests that sexual orientation has a biological substrate."

There have been several criticisms of LeVay's interpretation of the data. First, the data are from populations, not individuals. One can also say that there is a statistical range and that men and women have the same general range. Indeed, one of the INAH3 from a homosexual male was larger than all but one of those from the 16 "heterosexual males" in the study. Second, the "heterosexual men" were not necessarily hetero-

sexual, nor were the "homosexual men" necessarily homosexual; the brains came from corpses of people whose sexual preferences were not known. This brings up another issue: homosexuality has many forms, and is probably not a single phenotype. Third, the brains of the "homosexual men" were taken from patients who had died of AIDS. AIDS affects the brain, and its effect on the hypothalamic neurons is not known.

Fourth, because the study was done on the brains of dead subjects, one cannot infer cause and effect. Such data show only correlations, not causation. It is as likely that behaviors can affect regional neuronal density as it is that regional neuronal density can affect behaviors. If one interprets the data as indicating that the INAH3 of male homosexuals is smaller than that of male heterosexuals, one still does not know whether that is a cause of homosexuality or a result of it. Indeed, Breedlove (1997) has shown that the density and size of certain neurons in rat spinal ganglia depend on the frequency of sexual intercourse. In this case, the behavior was affecting the neurons. Fifth, even if a difference in INAH3 does exist, there is no evidence that the difference has anything to do with sexuality. Sixth, these studies do not indicate when such differences (if they exist) emerge. The question of whether differences among the heterosexual male, female, and homosexual male INAH3 occur during embryonic development, shortly after birth, during the first few years of life, during adolescence, or at some other time was not addressed.

In 1993, a correlation was made between a particular DNA sequence on the X chromosome and a particular subgroup of male homosexuals: homosexual men who had a homosexual brother. Out of 40 pairs of homosexual brothers wherein one brother had inherited a particular region of the X chromosome from his mother, the other brother had also inherited this region in 33 cases (Hamer et al. 1993). One would have

expected both brothers to have done so in only 20 cases, on average. Again, this is only a statistical concordance, and one that could be coincidental. Moreover, the control (the incidence of the same marker in the "nonhomosexual" males of these families) was not reported, and the statistical bias of the observations has been called into question, especially since other laboratories have not been able to repeat the result (Risch et al. 1993; Marshall 1995). More recent studies (Hu et al. 1995; Rice et al. 1999) found little or no increase in the incidence of this DNA sequence when homosexual men were compared to their nonhomosexual brothers. Hu and colleagues concluded that this sequence is "neither necessary nor sufficient for a homosexual orientation." Thus, despite the reports of these studies in the public media, no "gay gene" has been found.

Genes encode RNAs and proteins, not behaviors. While genes may bias behavioral outcomes, we have no evidence for their "controlling" them. The observance of people with schizophrenia, or people whose personalities change radically after a religious conversion or a traumatic experience, indicates that a single genotype can support a wide range of personalities. This is certainly a problem with any definition of a "homosexual phenotype," since people can alternate between homosexual and heterosexual behavior, and the definition of what is homosexual behavior differs between cultures (see Carroll and Wolpe 1996). Thus, whether homosexual desires are formed by genes within the nucleus, by sex hormones during fetal development, or by experiences after birth is still an open question. ∎

WEBSITE **17.8 Sex and the central nervous system.** There is ample evidence that estrogens and testosterone can cause changes in the central nervous system. Birds appear especially susceptible to hormonally induced changes in their behaviors.

Table 17.1 Ratios of X chromosomes to autosomes in different sexual phenotypes in *Drosophila melanogaster*

X chromosomes	Autosome sets	(A)X:A ratio	Sex
3	2	1.50	Metafemale
4	3	1.33	Metafemale
4	4	1.00	Normal female
3	3	1.00	Normal female
2	2	1.00	Normal female
2	3	0.66	Intersex
1	2	0.50	Normal male
1	3	0.33	Metamale

Source: After Strickberger 1968.

Chromosomal Sex Determination in Drosophila

The sexual development pathway

Although both mammals and fruit flies produce XX females and XY males, their chromosomes achieve these ends using very different means. The sex-determining mechanisms in mammals and in insects such as *Drosophila* are very different. In mammals, the Y chromosome plays a pivotal role in determining the male sex. Thus, XO mammals are females, with ovaries, a uterus, and oviducts (but usually very few, if any, ova). In *Drosophila*, sex determination is achieved by a balance of female determinants on the X chromosome and male determinants on the autosomes. Normally, flies have either one or two X chromosomes and two sets of autosomes. If there is but one X chromosome in a diploid cell (1X:2A), the fly is male. If there are two X chromosomes in a diploid cell (2X:2A), the fly is female (Bridges 1921, 1925). Thus, XO *Drosophila* are sterile males. In flies, the Y chromosome is not involved in deter-

mining sex. Rather, it contains genes active in forming sperm in adults. Table 17.1 shows the different X-to-autosome ratios and the resulting sex.

In *Drosophila*, and in insects in general, one can observe **gynandromorphs**—animals in which certain regions of the body are male and other regions are female (Figure 17.15). This can happen when an X chromosome is lost from one embryonic nucleus. The cells descended from that cell, instead of being XX (female), are XO (male). Because there are no sex hormones in insects to modulate such events, each cell makes its own sexual "decision." The XO cells display male characteristics, whereas the XX cells display female traits. This situation provides a beautiful example of the association between insect X chromosomes and sex.

Any theory of *Drosophila* sex determination must explain how the X-to-autosome (X:A) ratio is read and how this information is transmitted to the genes controlling the male or female phenotypes. Although we do not yet know the intimate mechanisms by which the X:A ratio is made known to the cells, research in the past two decades has revolutionized our view of *Drosophila* sex determination. Much of this research has focused on the identification and analysis of the genes that are necessary for sexual differentiation and the placement of those genes in a developmental sequence. Several genes with roles in sex determination have been found. Loss-of-function

Figure 17.15
Gynandromorphs. (A) Gynandromorph of *D. melanogaster* in which the left side is female (XX) and the right side is male (XO). The male side has lost an X chromosome bearing the wild-type alleles of eye color and wing shape, thereby allowing the expression of the recessive alleles *eosin eye* and *miniature wing* on the remaining X chromosome. (B) Photograph of a gynandromorphic *Io* moth, divided bilaterally into a rose-brown female half and a smaller, yellow male half. (A from Morgan and Bridges 1919, drawn by Edith Wallace. B; photograph by T. R. Manley, courtesy of *The Journal of Heredity*.)

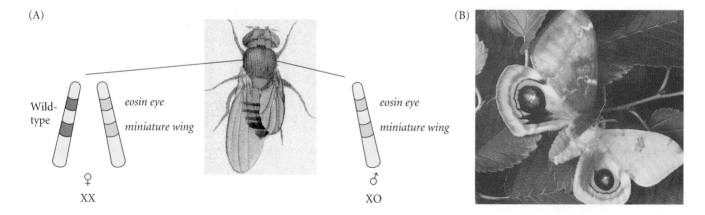

(A)

Wild-type — *eosin eye* — *miniature wing* — ♀ XX

eosin eye — *miniature wing* — ♂ XO

(B)

mutations in most of these genes—*Sex-lethal (Sxl)*, *transformer (tra)*, and *transformer-2 (tra2)*—transform XX individuals into males. Such mutations have no effect on sex determination in XY males. Homozygosity of the *intersex (ix)* gene causes XX flies to develop an intersex phenotype having portions of male and female tissue in the same organ. The *doublesex (dsx)* gene is important for the sexual differentiation of both sexes. If *dsx* is absent, both XX and XY flies turn into intersexes (Baker and Ridge 1980; Belote et al. 1985a). The positioning of these genes in a developmental pathway is based on (1) the interpretation of genetic crosses resulting in flies bearing two or more of these mutations and (2) the determination of what happens when there is a complete absence of the products of one of these genes. Such studies have generated the model of the regulatory cascade seen in Figure 17.16.

The sex-lethal gene as the pivot for sex determination

INTERPRETING THE X:A RATIO. The first phase of *Drosophila* sex determination involves reading the X:A ratio. What elements on the X chromosome are "counted," and how is this information used? It appears that high values of the X:A ratio are responsible for activating the feminizing switch gene ***Sex-lethal (Sxl)***. In XY cells, *Sxl* remains inactive during the early stages of development (Cline 1983; Salz et al. 1987). In XX *Drosophila*, *Sxl* is activated during the first 2 hours after fertilization, and this gene transcribes a particular embryonic type of *Sxl* mRNA that is found for only about 2 hours more (Salz

Figure 17.16
Proposed regulation cascade for *Drosophila* somatic sex determination. Arrows represent activation, while a block at the end of a line indicates suppression. The *msl* loci, under the control of the *Sxl* gene, regulate the dosage compensatory transcription of the male X chromosome. (After Baker et al. 1987.)

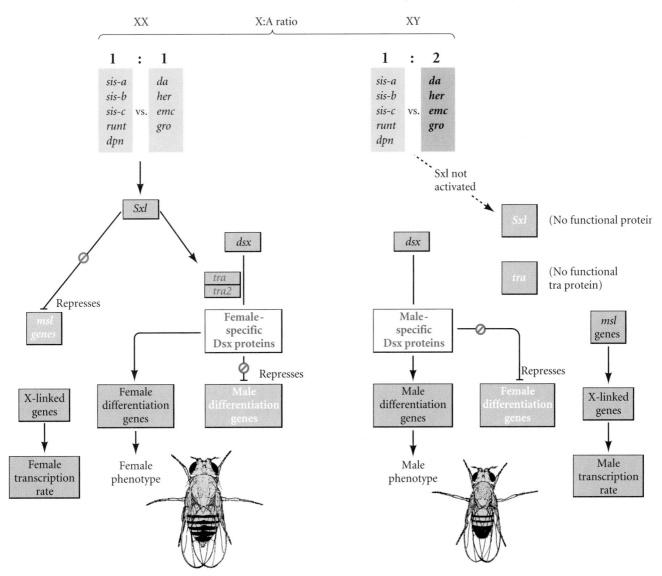

et al. 1989). Once activated, the *Sxl* gene remains active because its protein product is able to bind to and activate its own promoter (Sánchez and Nöthiger 1983).

This female-specific activation of *Sxl* is thought to be stimulated by **"numerator proteins"** encoded by the X chromosome. These constitute the X part of the X:A ratio. Cline (1988) has demonstrated that these numerator proteins include Sisterless-a and Sisterless-b. These proteins bind to the "early" promoter of the *Sxl* gene to promote its transcription shortly after fertilization.

The **"denominator proteins"** are autosomally encoded proteins such as Deadpan and Extramacrochaetae. These proteins block the binding or activity of the numerator proteins (Van Doren et al. 1991; Younger-Shepherd et al. 1992). The denominator proteins may actually be able to form inactive heterodimers with the numerator proteins (Figure 17.17). It appears, then, that the X:A ratio is measured by competition between X-encoded activators and autosomally encoded repressors of the promoter of the *Sxl* gene.

Figure 17.17
The differential activation of the *sxl* gene in females and males. (A) In wild-type *Drosophila* with two X chromosomes and two sets of autosomes (2X:2A), the numerator proteins encoded on the X chromosomes (sis-a, sis-b, etc.) are not all bound by inhibitory denominator proteins derived from genes (such as *deadpan*) on the autosomes. The numerator proteins activate the early promoter of the *Sxl* gene. Eventually, in both males and females, constitutive transcription of *sxl* starts from the late promoter. If Sxl is already available (i.e., from early transcription), the *Sxl* pre-mRNA is spliced to form the functional female-specific message. (B) In wild-type *Drosophila* with one X chromosome and two sets of autosomes (1X:2A), the numerator proteins are bound by the denominator proteins and cannot activate the early promoter. When the *Sxl* gene is transcribed from the late promoter, RNA splicing does not exclude the male-specific exon in the mRNA. The resulting message encodes a truncated and nonfunctional peptide, since the male-specific exon contains a translation termination codon. (After Keyes et al. 1992.)

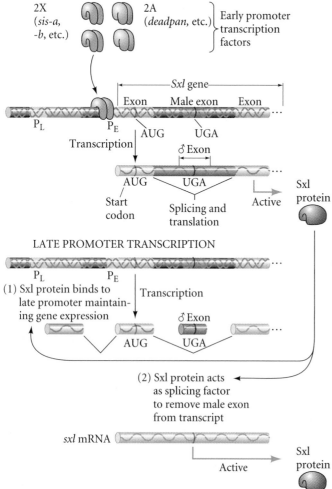

(A) Female   2X:2A

EARLY PROMOTER TRANSCRIPTION

sxl mRNA

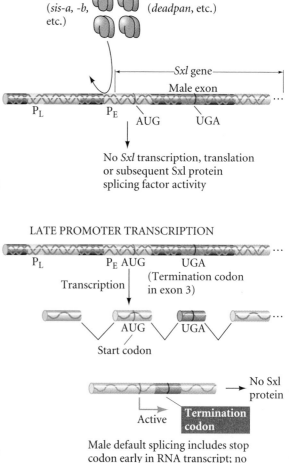

(B) Male   1X:2A

Heterodimer transcription factors do not initiate *Sxl* transcription

No *Sxl* transcription, translation or subsequent Sxl protein splicing factor activity

LATE PROMOTER TRANSCRIPTION

Male default splicing includes stop codon early in RNA transcript; no protein translated

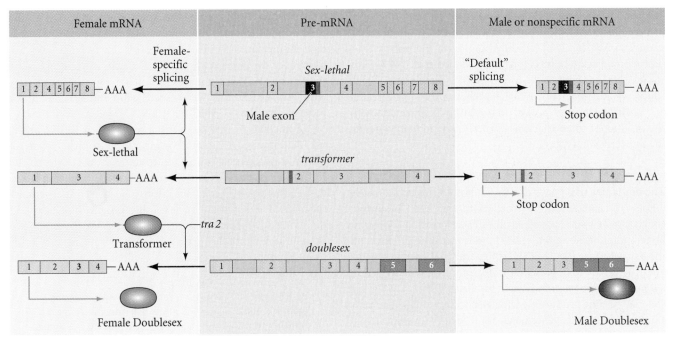

Figure 17.18
The pattern of sex-specific RNA splicing in three major *Drosophila* sex-determining genes. The pre-mRNAs are located in the center of the diagram and are identical in both male and female nuclei. In each case, the female-specific transcript is shown at the left, while the default transcript (whether male or nonspecific) is shown to the right. Exons are numbered, and the positions of the termination codons and poly(A) sites are marked. (After Baker 1989.)

MAINTENANCE OF *Sxl* FUNCTION. Shortly after *Sxl* transcription has taken place, a second, "late" promoter on the *Sex-lethal* gene is activated, and the gene is now transcribed in both males and females. However, analysis of the cDNA from *Sxl* mRNA shows that the *Sxl* mRNA of males differs from *sxl* mRNA of females (Bell et al. 1988). This difference is the result of differential RNA processing. Moreover, the Sxl protein appears to bind to its own mRNA precursor to splice it in the female manner. Since males do not have any available Sxl protein when the late promoter is activated, their new *Sxl* transcripts are processed in the male manner (Keyes et al. 1992). The male *Sxl* mRNA is nonfunctional. While the female-specific *Sxl* message encodes a protein of 354 amino acids, the male-specific *Sxl* transcript contains a translation termination codon (UGA) after amino acid 48. The differential RNA processing that puts this termination codon into the male-specific mRNA is shown in Figures 17.17B and 17.18. In males, the nuclear transcript is spliced in a manner that yields eight exons, and the termination codon is within exon 3. In females, RNA processing yields only seven exons, and the male-specific exon 3 is now spliced out as a large intron. Thus, the female-specific mRNA lacks the termination codon.

The protein made by the female-specific *Sxl* transcript contains two regions that are important for binding to RNA. These regions are similar to regions found in nuclear RNA-binding proteins. Bell and colleagues (1988) have shown that there are two targets for the female-specific Sxl protein. One of these targets is the pre-mRNA of *Sxl* itself. The second is the pre-mRNA of the next gene on the pathway, *transformer*.

WEBSITE **17.9 Other sex determination proteins in *Drosophila*.** Sex-lethal does not work alone, but in concert with several other proteins whose presence is essential for its function. Many of these proteins have other roles during development.

The transformer genes

The *Sxl* gene regulates somatic sex determination by controlling the processing of the **transformer** (**tra**) gene transcript. The *tra* message is alternatively spliced to create a female-specific mRNA as well as a nonspecific mRNA that is found in both females and males. Like the male *sxl* message, the nonspecific *tra* mRNA contains a termination codon early in the message, making the protein nonfunctional (Boggs et al. 1987). In *tra*, the second exon of the nonspecific mRNA has the termination codon. This exon is not utilized in the female-specific message (see Figure 17.18). How is it that the females make a different transcript than the males? The female-specific protein from the *Sxl* gene activates a female-specific 3' splice site in the transformer pre-mRNA, causing it to be processed in a way that splices out the second exon. To do this, the *Sxl* protein blocks the binding of splicing factor U2AF to the nonspecific splice site by specifically binding to the polypyrimidine tract adjacent to it (Figure 17.19; Handa et al. 1999). This causes U2AF to bind to the lower-affinity (female-specific) 3′ splice site and generate a female-specific mRNA (Valcárcel et al. 1993). The female-specific *tra* product acts in concert with the product of the *transformer-2* (*tra2*) gene to help generate the female phenotype.

Doublesex: *The switch gene of sex determination*

The **doublesex** (**dsx**) gene is active in both males and females, but its primary transcript is processed in a sex-specific manner (Baker et al. 1987). This alternative RNA processing appears to be the result of the action of the *transformer* gene products on the *dsx* gene (see Figure 5.31). If the Tra2 and female-specific Tra proteins are both present, the *dsx* transcript is processed in a female-specific manner (Ryner and Baker 1991). The female splicing pattern produces a female-specific protein that activates female-specific genes (such as those of the yolk proteins) and inhibits male development. As discussed in Chapter 5, if functional Tra is not produced, a male-specific transcript of *dsx* is made. This transcript encodes an active protein that inhibits female traits and promotes male traits.

The functions of the Doublesex proteins can be seen in the formation of the *Drosophila* genitalia. Male and female genitalia in *Drosophila* are derived from separate cell populations. In male (XY) flies, the female primordium is repressed, and the male primordium differentiates into the adult genital structures. In female (XX) flies, the male primordium is repressed, and the female primordium differentiates. If the *dsx*

gene is absent (and thus neither transcript is made), both the male and the female primordia develop, and intersexual genitalia are produced. Similarly, in the fat body of *Drosophila*, activation of the genes for egg yolk production is stimulated by the female Dsx protein and is inhibited by the male Dsx protein (Schüpbach et al. 1978; Coschigano and Wensink 1993; Jursnich and Burtis 1993).

According to this model (Baker 1989), the result of the sex determination cascade comes down to what type of mRNA is going to be processed from the *dsx* transcript. If the X:A ratio is 1, then *Sxl* makes a female-specific splicing factor that causes the *tra* gene transcript to be spliced in a female-specific manner. This female-specific protein interacts with the Tra2 splicing factor to cause the *doublesex* pre-mRNA to be spliced in a female-specific manner. If the *doublesex* transcript is not acted on in this way, it will be processed in a "default" manner to make the male-specific message.

> **WEBSITE 17.10 Conservation of sex-determining genes.** While the pathways of sex determination appear to differ between humans and flies, the discovery of a human gene similar to *doublesex* suggests that there may be a common end point to the two pathways.

> **WEBSITE 17.11 Hermaphrodites.** In *C. elegans* and many other invertebrates, hermaphroditism is the general rule. These animals are born with both ovaries and testes. In some fishes, sequential hermaphroditism is seen, with an individual fish being female some seasons and male in others. In humans, hermaphrodes are rare and usually sterile.

Environmental Sex Determination

Temperature-dependent sex determination in reptiles

While the sex of most snakes and most lizards is determined by sex chromosomes at the time of fertilization, the sex of most turtles and all species of crocodilians is determined by the environment after fertilization. In these reptiles, the temperature of the eggs during a certain period of development is the deciding factor in determining sex, and small changes in temperature can cause dramatic changes in the sex ratio (Bull 1980). Often, eggs incubated at low temperatures (22–27°C) produce one sex, whereas eggs incubated at higher temperatures (30°C and above) produce the other. There is only a small range of temperatures that permits both males and females to hatch from the same brood of eggs. Figure 17.20 shows the abrupt temperature-induced change in sex ratios for the red-eared slider turtle. If eggs are incubated below 28°C, all the turtles hatching from them will be male. Above 31°C, every egg gives rise to a female. At temperatures in between, the broods will give rise to individuals of both sexes. Variations on this theme also exist. The eggs of the snapping turtle *Macroclemys*, for instance, become female at either cool (22°C or lower) or hot (28°C or above) temperatures. Between these extremes, males predominate.

Figure 17.19
Stereogram showing binding of *tra* pre-mRNA by the cleft of the Sxl protein. The bound 12-nucleotide RNA (GUUGUUUUUUUU) is shown in yellow. The strongly positive regions are shown in blue, while the scattered negative regions are in red. It is worth crossing your eyes to get the three-dimensional effect. (From Handa et al. 1999; stereogram courtesy of S. Yokoyama.)

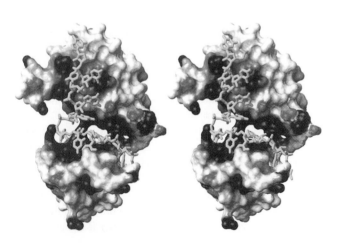

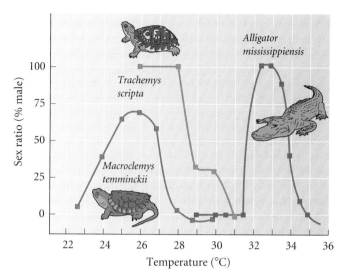

Figure 17.20
Temperature-dependent sex determination in three reptile species: the American alligator (*Alligator mississippiensis*), the red-eared slider turtle (*Trachemys scripta elegans*), and the alligator snapping turtle (*Macroclemys temminckii*). (After Crain and Guillette 1998.)

One of the best-studied reptiles is the European pond turtle, *Emys obicularis*. In laboratory studies, incubating *Emys* eggs at temperatures above 30°C produces all females, while temperatures below 25°C produce all-male broods. The threshold temperature (at which the sex ratio is even) is 28.5°C (Pieau et al. 1994). The developmental period during which sex determination occurs can be discovered by incubating eggs at the male-producing temperature for a certain amount of time and then shifting the eggs to an incubator at the female-producing temperature (and vice versa). In *Emys*, the last third of development appears to be the most critical for sex determination. It is not thought that turtles can reverse their sex after this period.

The pathways toward maleness and femaleness in reptiles are just being delineated. Unlike the situation in mammals, sex determination in reptiles (and birds) is hormone-dependent. In birds and reptiles, estrogen is essential for ovarian development. Estrogen can override temperature and induce ovarian differentiation even at masculinizing temperatures. Similarly, injecting eggs with inhibitors of estrogen synthesis will produce male offspring, even if the eggs are incubated at temperatures that usually produce females (Dorizzi et al. 1994; Rhen and Lang 1994). Moreover, the sensitive time for the effects of estrogens and their inhibitors coincides with the time when sex determination usually occurs (Bull et al. 1988; Gutzke and Chymiy 1988).

It appears that the enzyme **aromatase** (which can convert testosterone into estrogen) is important in temperature-dependent sex determination. The estrogen synthesis inhibitors used in the experiments mentioned above worked by blocking the aromatase enzyme, showing that experimentally low aromatase conditions yield male offspring. This correlation is seen to hold under natural conditions as well. The aromatase activity of *Emys* is very low at the male-promoting temperature of 25°C. At the female-promoting temperature of 30°C, aromatase activity increases dramatically during the critical period for sex determination (Desvages et al. 1993; Pieau et al. 1994). Temperature-dependent aromatase activity is also seen in diamondback terrapins, and its inhibition masculinizes their gonads (Jeyasuria et al. 1994). One remarkable finding is that the injection of an aromatase inhibitor into the eggs of an all-female parthenogenetic species of lizards causes the formation of males (Wibbels and Crews 1994).

It is not known whether the temperature sensitivity resides in the aromatase gene or protein itself or in other proteins that regulate it. One hypothesis is that the temperature is sensed by neurons in the central nervous system and transduced to the bipotential gonad by nerve fibers (see Lance 1997). Another hypothesis is that aromatase activity may be regulated by *Sox9*. This sex-determining gene is seen throughout the vertebrates, where its expression in gonads correlates extremely well with the production of testes. When two species of turtles were raised at female-promoting temperatures, *Sox9* expression was down-regulated during the critical time for sex determination. However, in the bipotential gonads of those turtles raised at male-promoting temperatures, *Sox9* expression was retained in the medullary sex cords destined to become Sertoli cells (Spotila et al. 1998; Moreno-Mendoza et al. 1999).

The evolutionary advantages and disadvantages of temperature-dependent sex determination are discussed in Chapter 21. Recent studies (Bergeron et al. 1994, 1999) have shown that polychlorinated biphenyl compounds (PCBs), a class of widespread pollutants that can act like estrogens, are able to reverse the sex of turtles raised at "male" temperatures. This knowledge may have important consequences in environmental conservation efforts to protect endangered turtle species.

Location-dependent sex determination in Bonellia *and* Crepidula

As mentioned in Chapter 3, the sex of the echiuroid worm *Bonellia* depends on where a larva settles. If a *Bonellia* larva lands on the ocean floor, it develops into a 10-cm-long female. If the larva is attracted to a female's proboscis, it travels along the tube until it enters the female's body. Therein it differentiates into a minute (1–3-mm-long) male that is essentially a sperm-producing symbiont of the female (see Figure 3.1).

Another example in which sex determination is affected by the location of the organism is the case of the slipper snail *Crepidula fornicata*. In this species, individuals pile up on top of one another to form a mound (Figure 17.21). Young individuals are always male. This phase is followed by the degeneration of the male reproductive system and a period of lability. The next phase can be either male or female, depending on

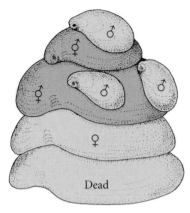

Figure 17.21
Cluster of *Crepidula* snails. Two individuals are changing from male to female. After these molluscs become female, they will be fertilized by the male above them. (After Coe 1936.)

the animal's position in the mound. If the snail is attached to a female, it will become male. If such a snail is removed from its attachment, it will become female. Similarly, the presence of large numbers of males will cause some of the males to become females. However, once an individual becomes female, it will not revert to being male (Coe 1936). More examples of context-dependent sex determination will be studied in Chapter 21.

Nature has provided many variations on her masterpiece. In some species, including most mammals and insects, sex is determined solely by chromosomes; in other species, sex is a matter of environmental conditions. We are finally beginning to understand the mechanisms by which this masterpiece is created.

Snapshot Summary: Sex Determination

1. In mammals, primary sex determination (the determination of gonadal sex) is a function of the sex chromosomes. XX individuals are females, XY individuals are males.

2. The Y chromosome plays a key role in male sex determination. XY and XX mammals both have a bipotential gonad that makes the primary sex cords. In XY animals, these cords continue to be formed within the gonad, and eventually differentiate into the Sertoli cells of the testes. The interstitial mesenchyme becomes the Leydig cells.

3. In XX individuals, the internal sex cords degenerate, and a second set of cortical sex cords emerges. These remain on the periphery of the gonad. Germ cells enter the sex cords, but will not be released from the gonad until puberty. The epithelium of the sex cords becomes the granulosa cells; the mesenchyme becomes the thecal cells.

4. In humans, the *SRY* gene is the testis-determining factor on the Y chromosome. It synthesizes a DNA-binding protein that is thought to compete with the *DAX1* protein. It is thought that if SRY is produced at a high enough level, it activates (either directly or indirectly) the *SF1* gene and inhibits the *WNT4* gene.

5. The *SF1* product is believed to activate the *SOX9* gene, as well as several other genes involved in synthesizing steroid hormones and anti-Müllerian duct hormone (AMH). *SOX9* may organize the genital ridge epithelium to form testes, but the corresponding ovary-forming genes have not yet been found, although the *WNT-4* gene may be important in this regard.

6. Secondary sex determination in mammals involves the hormones produced by the developing gonads. Under estrogenic stimulation, the Müllerian duct differentiates into the oviducts, uterus, cervix, and upper portion of the vagina. In male mammals, the Müllerian duct is destroyed by the AMH

produced by the Sertoli cells, while the testosterone produced by the Leydig cells enables the Wolffian duct to differentiate into the vas deferens and seminal vesicle. In female mammals, the Wolffian duct degenerates because of the lack of testosterone.

7. The conversion of testosterone to dihydrotestosterone in the genital rudiment and prostate gland precursor enables the differentiation of the penis, scrotum, and prostate gland.

8. Individuals with mutations of these hormones or their receptors may have a distinction between their primary and secondary sex characteristics.

9. In *Drosophila*, sex is determined by the ratio of X chromosomes to autosomes, and the Y chromosome does not play a role in sex determination. There are no sex hormones, so each cell makes a sex determination decision.

10. The *Drosophila Sxl* gene is activated in females (by proteins encoded on the X chromosomes) and is repressed in males (by factors encoded on the autosomes). Sxl protein acts as an RNA splicing factor to splice an inhibitory exon from the *tra* transcript. Therefore, female flies have an active Tra protein, while males do not.

11. The Tra protein also acts as an RNA splicing factor to splice exons of the *doublesex* transcript. The *doublesex* gene is transcribed in both XX and XY cells, but its pre-mRNA is processed to form different mRNAs, depending on whether Tra is present. The proteins translated from both messages are active, and they activate or inhibit transcription of a set of genes involved in producing the sexually dimorphic traits of the fly.

12. In turtles and alligators, sex is often determined by the temperature during the time of gonad determination. Since estrogen is necessary for ovary development, it is possible that differing levels of aromatase (the enzyme that can convert testosterone into estrogen) distinguish male from female patterns of gonadal differentiation.

13. In some species, such as *Bonellia* and *Crepidula*, sex determination is brought about by the position of the individual with regard to other individuals of the same species.

Literature Cited

Achermann, J. C., M. Ito, P. Hindmarsh and J. Jameson. 1999. A mutation in the gene encoding steroidogenic factor-1 causes XY sex reversal and adrenal failure in humans. *Nature Genet.* 22: 125–126.

Andersson, S., D. M. Berman, E. P. Jenkins and D. W. Russell. 1991. Deletion of steroid 5α-reductase 2 gene in male pseudohermaphroditism. *Nature* 354: 159–161.

Arango, N. A., R. Lovell-Badge and R. E. Behringer. 1999. Targeted mutagenesis of the endogenous mouse *Mis* gene promoter: In vivo definition of genetic pathways of vertebrate sexual development. *Cell* 99: 409–419.

Aristotle. ca. 335 B.C.E. The generation of animals. Translated by A. Platt. *In* J. Barnes (ed.), *The Complete Works of Aristotle*, Vol. 8. Princeton University Press, Princeton, NJ, 1984.

Baker, B. S. 1989. Sex in flies: The splice of life. *Nature* 340: 521–524.

Baker, B. S. and K. A. Ridge. 1980. Sex and the single cell. I. On the action of major loci affecting sex determination in *Drosophila melanogaster*. *Genetics* 94: 383–423.

Baker, B. S., R. N. Nagoshi and K. C. Burtis. 1987. Molecular genetic aspects of sex determination in *Drosophila*. *BioEssays* 6: 66–70.

Bardoni, B. and 11 others. 1994. A dosage sensitive locus at chromosome Xp21 is involved in male to female sex reversal. *Nature Genet.* 7: 497–501.

Barraclough, C. A. and R. A. Gorski. 1962. Studies on mating behavior in the androgen-sterilized female rat in relation to the hypothalamic regulation of sexual behavior. *J. Endocrinol.* 25: 175–182.

Bell, L. R., E. M. Maine, P. Schedl and T. W. Cline. 1988. *Sex-lethal*, a *Drosophila* sex determination switch gene, exhibits sex-specific RNA splicing and sequence similarity to RNA binding proteins. *Cell* 55: 1037–1046.

Belote, J. M. and B. S. Baker. 1987. Sexual behavior: Its genetic control during development and adulthood in *Drosophila melanogaster*. *Proc. Natl. Acad. Sci. USA* 84: 8026–8030.

Belote, J. M., M. B. McKeown, D. J. Andrew, T. N. Scott, M. F. Wolfner and B. S. Baker. 1985a. Control of sexual differentiation in *Drosophila melanogaster*. *Cold Spring Harbor Symp. Quant. Biol.* 50: 605–614.

Belote, J. M., A. M. Handler, M. F. Wolfner, K. L. Livak and B. S. Baker. 1985b. Sex-specific regulation of yolk protein gene expression in *Drosophila*. *Cell* 40: 339–348.

Bergeron, J. M., D. Crews and J. A. MacLachlan. 1994. PCBs as environmental estrogens: Turtle sex determination as a biomarker of environmental contamination. *Environ. Health Perspect.* 102: 780–781.

Bergeron, J. M., E. Willingham, C. T. Osborn III, T. Rhen and D. Crews. 1999. Developmental synergism of steroidal estrogens in sex determination. *Environ. Health Perspect.* 107: 93–97.

Berkovitz, G. D. and 7 others. 1992. The role of the sex-determining region of the Y chromosome (SRY) in the etiology of 46,XX true hermaphrodites. *Hum. Genet.* 88: 411–416.

Bernstein, R., T. Jenkins, B. Dawson, J. Wagner, G. Devald, G. C. Koo and S. S. Wachtel. 1980. Female phenotype and multiple abnormalities in sibs with a Y chromosome and partial X-chromosome duplication: H-Y antigen and Xg blood group findings. *J. Med. Genet.* 17: 291–300.

Birk, O. S. and 11 others. 2000. The LIM homeobox gene *LHX9* is essential for mouse gonad formation. *Nature* 403: 909–913.

Bleier, R. 1984. *Science and Gender*. Pergamon, New York, pp. 80–114.

Boggs, R. T., P. Gregor, S. Idriss, J. M. Belote and M. McKeown. 1987. Regulation of sexual differentiation in *D. melanogaster* via alternative splicing of RNA from the *transformer* gene. *Cell* 50: 739–747.

Breedlove, S. M. 1992. Sexual dimorphism in the vertebrate nervous system. *J. Neurosci.* 12: 4133–4142.

Breedlove, S. M. 1997. Sex on the brain. *Nature* 389: 801.

Bridges, C. B. 1921. Triploid intersexes in *Drosophila melanogaster*. *Science* 54: 252–254.

Bridges, C. B. 1925. Sex in relation to chromosomes and genes. *Am. Nat.* 59: 127–137.

Buehr, M., S. Gu and A. McLaren. 1993. Mesonephric contribution to testis differentiation in the fetal mouse. *Development* 117: 273–281.

Bull, J. J. 1980. Sex determination in reptiles. *Q. Rev. Biol.* 55: 3–21.

Bull, J. J., W. H. N. Gutzke and D. Crews. 1988. Sex reversal by estradiol in three reptilian orders. *Gen. Comp. Endocrinol.* 70: 425–428.

Capel, B., K. H. Albrecht, L. L. Washburn and E. M. Eicher. 1999. Migration of mesonephric cells into the mammalian gonad depends on *Sry*. *Mech. Dev.* 84: 127–131.

Carroll, J. and P. R. Wolpe. 1996. *Sexuality and Gender in Society*. HarperCollins, New York.

Cate, R. L. and 18 others. 1986. Isolation of the bovine and human genes for Müllerian inhibiting substance and expression of the gene in animal cells. *Cell* 45: 685–698.

Cline, T. W. 1983. The interaction between daughterless and *Sex-lethal* in triploids: A novel sex-transforming maternal effect linking sex determination and dosage compensation in *Drosophila melanogaster*. *Dev. Biol.* 95: 260–274.

Cline, T. W. 1988. Evidence that *sisterless-a* and *sisterless-b* are two of several discrete "numerator elements" of the X/A sex determination signal in *Drosophila* that switch *Sxl* between two alternative stable expression states. *Genetics* 119: 829–862.

Coe, W. R. 1936. Sexual phases in *Crepidula*. *J. Exp. Zool.* 72: 455–477.

Cohen, D. R., A. H. Sinclair and J. D. McGovern. 1994. SRY protein enhances transcription of Fos-related antigen 1 promoter constructs. *Proc. Natl. Acad. Sci. USA* 91: 4372–4376.

Coschigano, K. T. and P. C. Wensink. 1993. Sex-specific transcriptional regulation of the male and female doublesex proteins of *Drosophila*. *Genes Dev.* 7: 42–54.

De Jonge, F. H., J.-W. Muntjewerff, A. L. Louwerse and N. E. Van de Poll. 1988. Sexual behavior and sexual orientation of the female rat after hormonal treatment during various stages of development. *Horm. Behav.* 22: 100–115.

Desvages, G., M. Girondot and C. Pieau. 1993. Sensitive stages for the effects of temperature on gonadal aromatase activity in embryos of the marine turtle *Dermochelys coriacea*. *Gen. Comp. Endocrinol.* 92: 54–61.

Dorizzi, M., G. Richard-Mercier, G. Desvages and C. Pieau. 1994. Masculinization of gonads by aromatase inhibitors in a turtle with temperature-dependent sex determination. *Differentiation* 58: 1–8.

Eicher, E. M. and L. L. Washburn. 1983. Inherited sex reversal in mice: Identification of a new sex-determining gene. *J. Exp. Zool.* 228: 297–304.

Eicher, E. M. and L. L. Washburn. 1986. Genetic control of primary sex determination in mice. *Annu. Rev. Genet.* 20: 327–360.

Eicher, E. M. and 8 others. 1996. Sex-determining genes on mouse autosomes identified by linkage analysis of C57BL/6J-Y-POS sex reversal. *Nature Genet.* 14: 206–209.

Fausto-Sterling, A. 1992. *Myths of Gender*. Basic Books, New York.

Fausto-Sterling, A. 1995. Animal models for the development of human sexuality: A critical evaluation. *J. Homosexuality* 28: 217–236.

Fisher, C. R., K. H. Graves, A. F. Parlow and E. R. Simpson. 1998. Characterization of mice deficient in aromatase (ArKO) because of a targeted disruption of the *cyp19* gene. *Proc. Natl. Acad. Sci. USA* 95: 6965–6970.

Foster, J. W. and 11 others. 1994. Campomelic dysplasia and autosomal sex reversal caused by mutations in an *SRY*-related gene. *Nature* 372: 525–530.

Galen, C. ca. 200 C.E. *On the Usefulness of the Parts of the Body*. Translated by M. May. Cornell University Press, Ithaca, NY, 1968.

Geddes, P. and J. A. Thomson. 1890. *The Evolution of Sex*. Walter Scott, London.

Genetics Review Group. 1995. One for a boy, two for a girl? *Curr. Biol.* 5: 37–39.

Geissler, W. M. and 9 others. 1994. Male pseudohermaphroditism caused by mutations of testicular 17b-hydroxysteroid dehydrogenase 3. *Nature Genet.* 7: 34–39.

Giese, K., J. Cox and R. Grosschedl. 1992. The HMG domain of lymphoid enhancer factor 1 bends DNA and facilitates the assembly of functional nucleoprotein structures. *Cell* 69: 185–195.

Gubbay, J. and 8 others. 1990. A gene mapping to the sex-determining region of the mouse Y chromosome is a member of a novel family of embryonically expressed genes. *Nature* 346: 245–250.

Gutzke, W. H. N. and D. B. Chymiy. 1988. Sensitive periods during embryology for hormonally induced sex determination in turtles. *Gen. Comp. Endocrinol.* 71: 265–267.

Hacker, A., B. Capel, P. Goodfellow and R. Lovell-Badge. 1995. Expression of *Sry*, the mouse sex determining gene. *Development* 121: 1603–1614.

Hamer, D. H., S. Hu, V. L. Magnuson, N. Hu and A. M. L. Pattatucci. 1993. A linkage between DNA markers on the X chromosome and male sexual orientation. *Science* 261: 321–327.

Handa, N. and 7 others. 1999. Structural basis for recognition of the *tra* mRNA precursor by the Sex-lethal protein. *Nature* 398: 579–585.

Haqq, C., C.-Y. King, E. Ukiyama, S. Falsafi, N. Haqq, P. K. Donahoe and M. A. Weiss. 1994. Molecular basis of mammalian sexual determination: Activation of Müllerian inhibiting substance gene expression by SRY. *Science* 266: 1494–1500.

Harris, G. W. and S. Levine. 1965. Sexual physiology of the brain and its experimental control. *J. Physiol.* 181: 379–400.

Hess, R. A., D. Bunick, K.-H. Lee, J. Bahr, J. A. Taylor, K. S. Korach and D. B. Lubahn. 1997. A role for oestrogen in the male reproductive system. *Nature* 390: 509–511.

Horowitz, M. C. 1976. Aristotle and women. *J. Hist. Biol.* 9: 183–213.

Hu, S. and 7 others. 1995. Linkage between sexual orientation and chromosome Xq28 in males but not in females. *Nature Genet.* 11: 248–256.

Huang, B., S. Wang, Y. Ning, A. N. Lamb and J. Bartley. 1999. Autosomal XX sex reversal caused by duplication of SOX9. *Am. J. Med. Genet.* 87: 349–353.

Imperato-McGinley, J., L. Guerrero, T. Gautier and R. E. Peterson. 1974. Steroid 5α-reductase deficiency in man: An inherited form of male pseudohermaphroditism. *Science* 186: 1213–1215.

Jacklin, D. 1981. Methodological issues in the study of sex-related differences. *Dev. Rev.* 1: 266–273.

Jeyasuria, P., W. M. Roosenburg and A. R. Place. 1994. Role of P-450 aromatase in sex determination of the diamondback terrapin, *Malaclemys terrapin*. *J. Exp. Zool.* 270: 95–111.

Jost, A. 1953. Problems of fetal endocrinology: The gonadal and hypophyseal hormones. *Recent Prog. Horm. Res.* 8: 379–418.

Jursnich, V. A. and K. C. Burtis. 1993. A positive role in differentiation for the male doublesex protein of *Drosophila*. *Dev. Biol.* 155: 235–249.

Kandel, E. R. and J. H. Schwartz. 1985. *Principles of Neural Science*. Elsevier, New York.

Kandel, E. R., J. H. Schwartz and T. M. Jessell. 1995. *Essentials of Neural Science and Behavior*. Appleton and Lange, Norwalk, CT.

Kent, J., S. C. Wheatley, J. E. Andrews, A. H. Sinclair and P. Koopman. 1996. A male-specific role for SOX9 in vertebrate sex determination. *Development* 122: 2813–2822.

Keyes, L. N., T. W. Cline and P. Schedl. 1992. The primary sex determination signal of *Drosophila* acts at the level of transcription. *Cell* 68: 933–943.

Koopman, P., A. Münsterberg, B. Capel, N. Vivian and A. Lovell-Badge. 1990. Expression of a candidate sex-determining gene during mouse testis differentiation. *Nature* 348: 450–452.

Koopman, P., J. Gubbay, N. Vivian, P. Goodfellow and R. Lovell-Badge. 1991. Male development of chromosomally female mice transgenic for *Sry*. *Nature* 351: 117–121.

Lance, A. A. 1997. Sex determination in reptiles: An update. *Am. Zool.* 37: 504–513.

Langman, J. 1981. *Medical Embryology*, 4th Ed. Williams & Wilkins, Baltimore.

Langman, J. and D. B. Wilson. 1982. Embryology and congenital malformations of the female genital tract. *In* A. Blaustein (ed.), *Pathology of the Female Genital Tract*, 2nd Ed. Springer-Verlag, New York, pp. 1–20.

LeVay, S. 1991. A difference in hypothalamic structure between heterosexual and homosexual men. *Science* 253: 1034–1037.

Ma, L., G. V. Benson, H. Lim, S. K. Dey and R. L. Maas. 1998. *Abdominal B* (*AbdB*) *Hoxa* genes:

Regulation in adult uterus by estrogen and progesterone and repression in Müllerian duct by synthetic estrogen diethylstilbesterol (DES). *Dev. Biol.* 197: 141–154.

Mansour, S., C. M. Hall, M. E. Pembrey and I. D. Young. 1995. A clinical and genetic study of campomelic dysplasia. *J. Med. Genet.* 32: 415–420.

Margarit, E., M. D. Coll, R. Olivo, D. Gómez, A. Soler and F. Ballesta. 2000. *SRY* gene transferred to the long arm of the X chromosome in a Y-positive XX true hermaphrodite. *Amer. J. Med. Genet.* 90: 25–28.

Marshall, E. 1995. NIH's "gay gene" study questioned. *Science* 268: 1841.

Marx, J. 1995. Mammalian sex determination: Snaring the genes that divide sexes for mammals. *Science* 269: 1824–1825.

McClung, C. E. 1902. The accessory chromosome sex determinant? *Biol. Bull.* 3: 72–77.

McEwen, B. S., I. Leiberburg, C. Chaptal and L. C. Krey. 1977. Aromatization: Important for sexual differentiation of the neonatal rat brain. *Horm. Behav.* 9: 249–263.

Meyer, W. J., B. R. Migeon and C. J. Migeon. 1975. Locus on human X chromosome for dihydrotestosterone receptor and androgen insensitivity. *Proc. Natl. Acad. Sci. USA* 72: 1469–1472.

Mittendorf, R. 1995. Teratogen update: Carcinogenesis and teratogenesis associated with exposure to diethylstilbesterol (DES) in utero. *Teratology* 51: 435–445.

Moore, C. L. 1990. Comparative development of vertebrate sexual behavior: Levels, cascades, and webs. *In* D. A. Dewsbury (ed.), *Contemporary Issues in Comparative Psychology*. Sinauer Associates, Sunderland, MA, pp. 278–299.

Moore, C. L., H. Dou and J. M. Juraska. 1992. Maternal stimulation affects the number of motor neurons in a sexually dimorphic nucleus of the lumbar spinal cord. *Brain Res.* 572: 52–56.

Moreno-Mendoza, N., V. R. Harley and H. Merchant-Larios. 1999. Differential expression of SOX9 in gonads of the sea turtle *Lepidochelys olivacea* at male- or female-promoting temperatures. *J. Exp. Zool.* 284: 705–710.

Morgan, T. H. and C. B. Bridges. 1919. The origin of gynandromorphs. In *Contributions to the Study of Drosophila*. Carnegie Institution of Washington Publication no. 278. Carnegie Institution of Washington, Washington, DC, pp. 1–122.

Muscatelli, F. and 14 others. 1994. Mutations in the DAX-1 gene give rise to both X-linked adrenal hypoplasia congenita and hypogonadotropic hypogonadism. *Nature* 372: 672–634.

Nachtigal, M. W., Y. Hirokawa, D. L. Enyeart-van Houten, J. N. Flanagan, G. D. Hammer and H. A. Ingraham. 1998. Wilms' tumor 1 and Dax-1 modulate the orphan nuclear receptor SF-1 in sex-specific gene expression. *Cell* 93: 445–454.

Phoenix, C. H., R. W. Goy, A. A. Gerall and W. C. Young. 1959. Organizing action of prenatally administered testosterone proprionate on the tissues mediating mating behavior in the female guinea pig. *Endocrinology* 65: 369–382.

Pieau, C., N. Girondot, G. Richard-Mercier, M. Desvages, P Dorizzi and P. Zaborski. 1994. Temperature sensitivity of sexual differentiation of gonads in the European pond turtle. *J. Exp. Zool.* 270: 86–93.

Pontiggia, A., R. Rimini, P. N. Goodfellow, R. Lovell-Badge and M. E. Bianchi. 1994. Sex-reversing mutations affect the architecture of Sry/DNA complexes. *EMBO J.* 13: 6115–6124.

Reddy, V. R., F. Naftolin and K. J. Ryan. 1974. Conversion of androstenedione to estrone by neural tissues from fetal and neonatal rats. *Endocrinology* 94: 117–121.

Rhen, T. and J. W. Lang, 1994. Temperature-dependent sex determination in the snapping turtle: Manipulation of the embryonic sex steroid environment. *Gen. Comp. Endocrinol.* 96: 243–254.

Rice, G., C. Anderson, N. Risch and G. Ebers. 1999. Male homosexuality: Absence of linkage to microsatellite markers at Xq28. *Science* 284: 665–667.

Risch, N., E. Squires-Wheeler and B. J. B. Keats. 1993. Male sexual orientation and genetic evidence. *Science* 262: 2063–2065.

Roberts, L. M., Y. Hirokawa, M. W. Nachtigal and H. A. Ingraham. 1999. Paracrine-mediated apoptosis in reproductive tract development. *Development* 208: 110–122.

Ryner, L. C. and B. S. Baker. 1991. Regulation of *doublesex* pre-mRNA processing occurs by 3′-splice site activation. *Genes Dev.* 5: 2071–2085.

Salz, H. K., T. W. Cline and P. Schedl. 1987. Functional changes associated with structural alterations induced by mobilization of a P element inserted into the *Sex-lethal* gene of *Drosophila. Genetics* 117: 221–231.

Salz, H. K., E. M. Maine, L. N. Keyes, M. E. Samuels, T. W. Cline and P. Schedl. 1989. The *Drosophila* female-specific sex-determination gene, *Sex-lethal,* has stage-, tissue-, and sex-specific RNAs suggesting multiple modes of regulation. *Genes Dev.* 3: 708–719.

Sánchez, L. and R. Nöthiger. 1983. Sex determination and dosage compensation in *Drosophila melanogaster:* Production of male clones in XX females. *EMBO J.* 1: 485–491.

Sariola, H. and M. Saarma. 1999. GDNF and its receptors in the regulation of ureter branching. *Int. J. Dev. Biol.* 43: 413–418.

Schiebinger, L. 1989. *The Mind Has No Sex?* Harvard University Press, Cambridge, MA.

Schüpbach, T., E. Wieschaus and R. Nöthiger. 1978. The embryonic organization of the genital disc studied in genetic mosaics of *Drosophila melanogaster. Wilhelm Roux Arch. Dev. Biol.* 185: 249–270.

Shen, W.-H., C. C. D. Moore, Y. Ikeda, K. L. Parker and H. A. Ingraham. 1994. Nuclear receptor steroidogenic factor 1 regulates the Müllerian-inhibiting substance gene: A link to the sex determination cascade. *Cell* 77: 651–661.

Siiteri, P. K. and J. D. Wilson. 1974. Testosterone formation and metabolism during male sexual differentiation in the human embryo. *J. Clin. Endocrinol. Metab.* 38: 113–125.

Sinclair, A. H. and 9 others. 1990. A gene from the human sex-determining region encodes a protein with homology to a conserved DNA-binding motif. *Nature* 346: 240–244.

Spotila, L. D., J. R. Spotila and S. E. Hall. 1998. Sequence and expression analysis of *WT1* and *SOX9* in the red-eared slider turtle, *Trachemys scripta. J. Exp. Zool.* 281: 417–427.

Stevens, N. M. 1905. Studies in spermatogenesis with especial reference to the "accessory chromosome."Carnegie Institution of Washington publication no. 36. Washington, DC: Carnegie Institution of Washington.

Swain, A., V. Narvaez, P. Burgoyne, G. Camerino and R. Lovell-Badge. 1998. *Dax1* antagonizes *Sry* action in mammalian sex determination. *Nature* 391: 761–767.

Thigpen, A. E., D. L. Davis, J. Imperato-McGinley and D. W Russell. 1992. The molecular basis of steroid 5α-reductase deficiency in a large Dominican kindred. *N. Engl. J. Med.* 327: 1216–1219.

Tilmann, C. and B. Capel. 1999. Mesonephric cell migration induces testis cord formation and Sertoli cell differentiation in the mammalian gonad. *Development* 126: 2883–2890.

Tran, D., N. Meusy-Dessolle and N. Josso. 1977. Anti-Müllerian hormone is a functional marker of foetal Sertoli cells. *Nature* 269: 411–412.

Trelstad, R. L., A. Hayashi, K. Hayashi and P. K. Donahoe. 1982. The epithelial-mesenchymal

interface of the male rate Müllerian duct: Loss of basement membrane integrity and ductal regression. *Dev. Biol.* 92: 27–40.

Tuana, N. 1988. The weaker seed. *Hypatia* 3: 35–59.

Vainio, S., M. Heikkilä, A. Kispert, A. Chin and A. P. McMahon. 1999. Female development in mammals is regulated by Wnt4 signalling. *Nature* 397: 405–409.

Van Doren, H. M. Ellis and J. W. Posakony. 1991. The *Drosophila* extramacrochaetae protein antagonizes sequence-specific DNA binding by daughterless/achaete-scute protein complexes. *Development* 113: 245–255.

Wagner, T. and 13 others. 1994. Autosomal sex reversal and campomelic dysplasia are caused by mutations in and around the *SRY*-related gene *SOX9. Cell* 79: 1111–1120.

Washburn. L. L. and E. M. Eicher. 1989. Normal testis determination in the mouse depends on genetic interaction of a locus on chromosome 17 and the Y chromosome. *Genetics* 123: 173–179.

Werner, M. H., J. R. Huth, A. M. Groneborn and G. M. Clore. 1995. Molecular basis of human 46X,Y sex reversal revealed from the three-dimensional solution structure of the human *SRY*–DNA complex. *Cell* 81: 705–714.

Wibbels, T. and, D. Crews. 1994. Putative aromatase inhibitor induces male sex determination in a female unisexual lizard and in a turtle with temperature-dependent sex determination. *J. Endocrinol.* 141: 295–299.

Wilson, E. B. 1905. The chromosomes in relation to the determination of sex in insects. *Science* 22: 500–502.

Wright, E. and 8 others. 1995. The *Sry*-related gene *Sox9* is expressed during chondrogenesis in mouse embryos. *Nature Genet.* 9: 15–20.

Younger-Shepherd, S., H. Vaessin, E. Bier, L. Y. Jan and Y. N. Jan. 1992. *deadpan,* an essential pan-neural gene encoding an HLH protein, acts as a denominator in *Drosophila* sex determination. *Cell* 70: 911–922.

Zanaria, E. and 13 others. 1994. An unusual member of the nuclear hormone receptor superfamily responsible for X-linked adrenal hypoplasia congenita. *Nature* 372: 635–641.

Metamorphosis, regeneration, and aging

The old order changeth, yielding place to the new.

ALFRED LORD TENNYSON (1886)

The earth-bound early stages built enormous digestive tracts and hauled them around on caterpillar treads. Later in the life-history these assets could be liquidated and reinvested in the construction of an entirely new organism—a flying-machine devoted to sex.

CARROLL M. WILLIAMS (1958)

I'd give my right arm to know the secret of regeneration.

OSCAR E. SCHOTTE (1950s)

DEVELOPMENT NEVER CEASES. Throughout life, we continuously generate new blood cells, lymphocytes, keratinocytes, and digestive tract epithelium from stem cells. In addition to these continuous daily changes, there are instances in which development during adult life is obvious—sometimes even startling. One of these instances is **metamorphosis**, the transition from a larval stage to an adult stage. In many instances of metamorphosis, a large proportion of the animal's structure changes, and the larva and the adult are unrecognizable as being the same individual (see Figure 2.4). Another startling type of development in the adult is **regeneration**, the creation of a new organ after the original one has been removed. Some adult salamanders, for instance, can regrow limbs after these appendages have been amputated. The third category of developmental change in the adult is a more controversial area. It encompasses those alterations of form associated with **aging**. Some scientists believe that the processes of degeneration are not properly part of the study of developmental biology. Other investigators point to the genetically determined, species-specific patterns of aging and claim that **gerontology**, the science of aging, studies an important part of the life cycle and is therefore rightly included in developmental biology. Whatever their relationship to normative development, metamorphosis, regeneration, and aging are poised to be critical topics for the biology of the twenty-first century.

Metamorphosis: The Hormonal Reactivation of Development

In most species of animals, embryonic development leads to a larval stage with characteristics very different from those of the adult organism. Very often, larval forms are specialized for some function, such as growth or dispersal. The pluteus larva of the sea urchin, for instance, can travel on ocean currents, whereas the adult urchin leads a sedentary existence. The caterpillar larvae of butterflies and moths are specialized for feeding, whereas their adult forms are specialized for flight and reproduction, often lacking the mouthparts necessary for eating. The division of functions between larva and adult is often remarkably distinct (Wald 1981). Cecropia moths, for example, hatch from eggs and develop as wingless juveniles (caterpillars) for several months. All this development enables them to spend a day or so as fully developed winged insects, mating quickly before they die. The adults never eat, and in fact have no mouthparts during this short reproductive phase of the life cycle. As might be expected, the juvenile and adult forms often live in different environments.

During metamorphosis, developmental processes are reactivated by specific hormones, and the entire organism changes to prepare itself for its new mode of existence. These changes are not solely ones of form. In amphibian tadpoles, metamorphosis causes the developmental maturation of liver enzymes, hemoglobin, and eye pigments, as well as the remodeling of the nervous, digestive, and reproductive systems. Thus, metamorphosis is often a time of dramatic developmental change affecting the entire organism.

Amphibian Metamorphosis

Morphological changes associated with metamorphosis

In amphibians, metamorphosis is generally associated with the changes that prepare an aquatic organism for a primarily terrestrial existence. In **urodeles** (salamanders), these changes include the resorption of the tail fin, the destruction of the external gills, and a change in skin structure. In **anurans** (frogs and toads), the metamorphic changes are more dramatic, and almost every organ is subject to modification (see Figure 2.4; Table 18.1). Regressive changes include the loss of the tadpole's horny teeth and internal gills, as well as the destruction of the tail. At the same time, constructive processes such as limb development and dermoid gland morphogenesis are also evident. The means of locomotion changes as the paddle tail recedes while the hindlimbs and forelimbs develop. The tadpole's cartilaginous skull is replaced by the predominantly bony skull of the frog. The horny teeth used for tearing pond plants disappear as the mouth and jaw take a new shape, and the tongue muscle develops. Meanwhile, the large intestine characteristic of herbivores shortens to suit the more carnivorous diet of the adult frog. The gills regress, and the gill arches degenerate. The lungs enlarge, and muscles and cartilage develop for pumping air in and out of the lungs. The sensory apparatus changes, too, as the lateral line system of the tadpole degenerates, and the eye and ear undergo further differentiation (see Fritzsch et al. 1988). The middle ear develops, as does the tympanic membrane characteristic of frog and toad outer ears. In the eye, both nictitating membranes and eyelids emerge.

When an animal changes its habitat and mode of nutrition, one would expect the nervous system to undergo dramatic changes, and it certainly does. One readily observed consequence of anuran metamorphosis is the movement of the eyes forward from their originally lateral position (Figure 18.1).* The lateral eyes of the tadpole are typical of preyed-upon herbivores, whereas the frontally located eyes of the frog befit its more predatory lifestyle. To catch its prey, the frog needs to see in three dimensions. That is, it has to acquire a binocular field of vision wherein input from both eyes converges in the brain (see Chapter 13). In the tadpole, the right eye innervates the left side of the brain, and vice versa. There are no ipsilateral (same-side) projections of the retinal neurons. During metamorphosis, however, these additional ipsi-

*One of the most spectacular movements of the eyes during metamorphosis occurs in flatfish such as flounder. Originally, the eyes are on opposite sides of the face. However, during metamorphosis, one of the eyes migrates across the head to meet the other eye on the same side of the fish. This allows the fish to dwell on the bottom, looking upward.

Table 18.1 Summary of some metamorphic changes in anurans

System	Larva	Adult
Locomotory	Aquatic; tail fins	Terrestrial; tailless tetrapod
Respiratory	Gills, skin, lungs; larval hemoglobins	Skin, lungs; adult hemoglobins
Circulatory	Aortic arches; aorta; anterior, posterior, and common jugular veins	Carotid arch; systemic arch; cardinal veins
Nutritional	Herbivorous: long spiral gut; intestinal symbionts; small mouth, horny jaws, labial teeth	Carnivorous: Short gut; proteases; large mouth with long tongue
Nervous	Lack of nictitating membrane; porphyropsin, lateral line system, Mauthner's neurons	Development of ocular muscles, nictitating membrane, rhodopsin; loss of lateral line system, degeneration of Mauthner's neurons; tympanic membrane
Excretory	Largely ammonia, some urea (ammonotelic)	Largely urea; high activity of enzymes of ornithine-urea cycle (ureotelic)
Integumental	Thin, bilayered epidermis with thin dermis; no mucous glands or granular glands	Stratified squamous epidermis with adult keratins; well-developed dermis contains mucous glands and granular glands secreting antimicrobial peptides

Source: Data from Turner and Bagnara 1976 and Reilly et al. 1994.

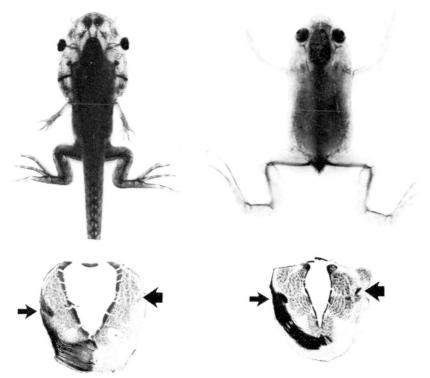

Figure 18.1
Eye migration and associated neuronal changes during metamorphosis of the *Xenopus laevis* tadpole. The eyes of the tadpole are laterally placed, so there is relatively little binocular field of vision. The eyes migrate dorsally and rostrally during metamorphosis, creating a large binocular field for the adult frog. Below each stage is a representation of the optic region of the brain. When horseradish peroxidase is injected (at the site of the broad arrows) into the retina of the tadpole, the optic neurons translocate it to the contralateral (opposite) side of the brain (small arrows), but not to the ipsilateral (same) side. As metamorphosis continues, the ipsilateral projections involved in binocular vision begin to be seen. (From Hoskins and Grobstein 1984; photographs courtesy of P. Grobstein.)

tadpoles (as in freshwater fishes), the major retinal photopigment is porphyropsin. During metamorphosis, the pigment changes to rhodopsin, the characteristic photopigment of terrestrial and marine vertebrates (Wald 1945, 1981; Smith-Gill and Carver 1981; Hanken and Hall 1988). Tadpole hemoglobin is changed into an adult hemoglobin that binds oxygen more slowly and releases it more rapidly than does tadpole hemoglobin (McCutcheon 1936; Riggs 1951). The liver enzymes change also, reflecting the change in habitat. Tadpoles, like most freshwater fishes, are **ammonotelic**; that is, they excrete ammonia. Many adult frogs (such as the genus *Rana*, but not the more aquatic *Xenopus*) are **ureotelic**, excreting urea, like most terrestrial vertebrates, which requires less water than excreting ammonia. During metamorphosis, the liver begins to synthesize the urea cycle enzymes necessary to create urea from carbon dioxide and ammonia (Figure 18.2).

Hormonal control of amphibian metamorphosis

The control of metamorphosis by thyroid hormones was demonstrated by Gudernatsch (1912), who discovered that tadpoles metamorphosed prematurely when fed powdered sheep thyroid gland. In a complementary study, Allen (1916) found that when he removed or destroyed the thyroid rudiment from early tadpoles (thus performing a thyroidectomy), the larvae never metamorphosed, instead becoming giant tadpoles.

The metamorphic changes of frog development are all brought about by the secretion of the hormones **thyroxine** (T_4) and **triiodothyronine** (T_3) from the thyroid during metamorphosis (Figure 18.3). It is thought that T_3 is the more important hormone, as it will cause metamorphic changes in thyroidectomized tadpoles in much lower concentrations than will T_4 (Kistler et al. 1977; Robinson et al. 1977).

REGIONALLY SPECIFIC CHANGES. The various organs of the body respond differently to hormonal stimulation. The same stimulus causes some tissues to degenerate while causing others to develop and differentiate. For instance, tail degeneration is clearly associated with increasing levels of thyroid hormones. The degeneration of tail structures is relatively rapid, as the bony skeleton does not extend to the tail, which is supported only by the notochord (Wassersug 1989). The regression of the tail is brought about by apoptosis, and it occurs in four stages.

lateral pathways emerge, enabling input from both eyes to reach the same area of the brain (Currie and Cowan 1974; Hoskins and Grobstein 1985a). In *Xenopus*, these new neuronal pathways result not from the remodeling of existing neurons, but from the formation of new neurons that differentiate in response to thyroid hormones (Hoskins and Grobstein 1985a,b). Some larval neurons, such as certain motor neurons in the tadpole jaw, switch their allegiances from larval muscle to the newly formed adult muscle (Alley and Barnes 1983). Still other neurons, such as those innervating the tongue (a newly formed muscle not present in the larva), have lain dormant during the tadpole stage and first form synapses during metamorphosis (Grobstein 1987). Thus, the anuran nervous system undergoes enormous restructuring during metamorphosis. Some neurons die, others are born, and others change their specificity.

Biochemical changes associated with metamorphosis

In addition to the obvious morphological changes, important biochemical transformations occur during metamorphosis. In

(A)

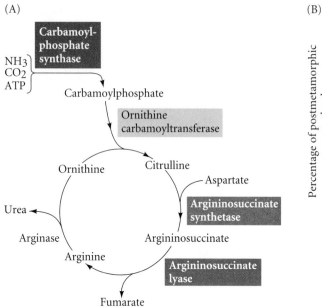

(B)

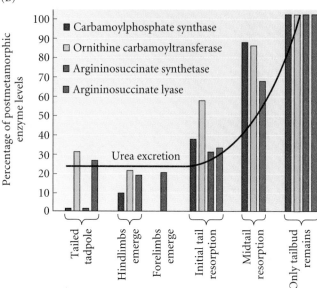

Figure 18.2
Development of the urea cycle during anuran metamorphosis. (A) The major features of the urea cycle, by which nitrogenous wastes can be detoxified and excreted with minimal water loss. (B) Emergence of urea cycle enzyme activities correlated with metamorphic changes in the frog *Rana catesbeiana*. (After Cohen 1970.)

First, protein synthesis decreases in the striated muscle cells of the tail (Little et al. 1973). Next, there is an increase in concentrations of digestive enzymes within the cells. Concentrations of lysosomal proteases, RNase, DNase, collagenase, phosphatase, and glycosidases all rise in the epidermis, notochord, and nerve cord cells (Fox 1973). Cell death is probably caused by the release of these enzymes into the cytoplasm. After cell

Figure 18.3
Formulae of thyroxine (T_4) and triiodothyronine (T_3).

HO—⬡—O—⬡—CH_2—CH—COOH
                             |
                            NH_2

Thyroxine (T_4)

HO—⬡—O—⬡—CH_2—CH—COOH
                             |
                            NH_2

Triiodothyronine (T_3)

death occurs, macrophages collect in the tail region, digesting the debris with their own proteolytic enzymes (Kaltenbach et al. 1979). The result is that the tail becomes a large sac of proteolytic enzymes (Figure 18.4). The major proteolytic enzymes involved appear to be collagenases and other metalloproteinases whose synthesis depends on thyroid hormones. If a metalloproteinase inhibitor (TIMP) is added to the tail, it prevents tail regression (Oofusa and Yoshizato 1991; Patterson et al. 1995).

The response to thyroid hormones is specific to the region of the body. Tadpole head and body epidermis differentiate a new set of glands when exposed to T_3. In the tail, however, T_3 causes the death of the epidermal cells and a tail-specific suppression of stem cell divisions that could give rise to more epidermal cells. The result is the death of the tail epidermal cells, while the head and body epidermis continues to function (Nishikawa et al. 1989). These regional epidermal responses appear to be controlled by the regional specificity of the dermal mesoderm. If tail dermatome cells (mesodermal cells that generate the tail dermis) are transplanted into the trunk, the epidermis they contact will degenerate upon metamorphosis. Conversely, when trunk dermatome is transplanted into the tail, those regions of skin persist. Changing the ectoderm does not alter the regional response to thyroid hormones (Kinoshita et al. 1989).

Organ-specific response to thyroid hormones is dramatically demonstrated by transplanting a tail tip to the trunk region or by placing an eye cup in the tail (Schwind 1933; Geigy 1941). The tail tip placed in the trunk is not protected from degeneration, but the eye retains its integrity despite the fact that it lies within the degenerating tail (Figure 18.5). Thus, the degeneration of the tail represents an organ-specific programmed cell death. Only specific tissues die when a signal is

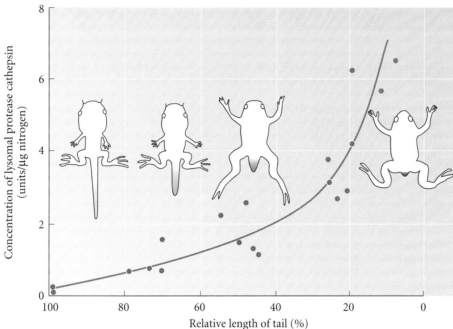

Figure 18.4
Increase in lysosomal protease activity during tail regression in *Xenopus laevis*. The lysosomal enzymes are thought to be responsible for digesting the tail cells. (After Karp and Berrill 1981.)

given. Such programmed cell deaths are important in molding the body. The degeneration of the human tail during week 4 of development resembles the regression of the tadpole tail (Fallon and Simandl 1978).

COORDINATION OF DEVELOPMENTAL CHANGES. One of the major problems of metamorphosis is the coordination of developmental events. For instance, the tail should not degenerate until some other means of locomotion—the limbs—has developed, and the gills should not regress until the animal can utilize its newly developed lung muscles. The means of coordinating metamorphic events appears to be a difference among tissues and organs in their responsiveness to different amounts of hormone (Saxén et al. 1957; Kollros 1961). This model is

called the **threshold concept**. As the concentration of thyroid hormones gradually builds up, different events occur at different concentrations of the hormones. If tadpoles are deprived of their thyroids and are placed in a dilute solution of thyroid hormones, the only morphological effects are the shortening of the intestines and accelerated hindlimb growth. However, at higher concentrations of thyroid hormones, tail regression is seen before the hindlimbs are formed. These experiments suggest that as thyroid hormone levels gradually rise, the hindlimbs develop first and then the tail regresses. Similarly, when T_3 is given to tadpoles, it induces the earliest-forming bones at the lowest dosages and the last bones at higher dosages, mimicking the natural situation (Hanken and Hall 1988). Thus, the timing of metamorphosis appears to be regulated by the sensitivity of different tissues to thyroid hormones.

To ensure that this timing system works, two of the organs most sensitive to thyroxine are the thyroid itself and the pituitary gland, which regulates thyroid hormone production. Thyroid hormones initially create positive feedback to the pituitary gland, causing the anterior pituitary to induce the thyroid to produce more T_3 and T_4 (Saxén et al. 1957; White and Nicoll 1981). Later, as an effect of metamorphosis, the thyroid partially degenerates, and inhibitors of thyroid hormone functions are made (Goos 1978).

VADE MECUM **Amphibian metamorphosis and frog calls.** For photographs of amphibian metamorphosis (and for the sounds of the adult frogs), check out the metamorphosis and frog call sections of the CD-ROM. **[Click on Amphibian]**

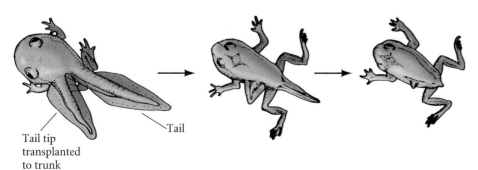

Tail tip transplanted to trunk

Tail

Figure 18.5
Regional specificity during frog metamorphosis. Tail tips regress even when transplanted to the trunk. Eye cups, however, remain intact even when transplanted into the regressing tail. (After Schwind 1933.)

Raising Tadpoles

OST TEMPERATE ZONE FROGS do not invest time or energy in providing for their tadpoles. However, among tropical frogs, there are numerous species in which adult frogs take painstaking care of their tadpoles. The poison arrow frog, *Dendrobates*, for example, is found in the rain forests of Central America. Most of the time, these highly toxic frogs live in the leaf litter of the forest floor. After laying eggs in a damp leaf, a parent (sometimes male, sometimes female) stands guard over the eggs. If the ground gets too dry, the frog will urinate on the eggs to keep them moist. When the eggs mature into tadpoles, the guarding frog allows them to wriggle onto its back (see Figure 18.6A). The frog then climbs into the canopy until it finds a bromeliad with a small pool of

water in its leaf base. Here it deposits one of its tadpoles, then goes back for another, and so on, until the brood has been placed in numerous small pools. Then each day the female returns to these pools and deposits a small number of unfertilized eggs into them, replenishing the dwindling food supply for the tadpoles until they finish metamorphosis (Mitchell 1988; van Wijngaarden and Bolanos 1992; Brust 1993). It is not known how the female frog remembers—or is informed about—where the tadpoles have been deposited.

Marsupial frogs carry their developing eggs in depressions in their skin, and they will often brood their tadpoles in their mouths. When the tadpoles undergo metamorphosis, the frogs spit out their progeny. Even more

impressive, the gastric-brooding frogs of Australia, *Rheobatrachus silus* and *Rheobatrachus vitellinus* eat their eggs. The eggs develop into larvae, and the larvae undergo metamorphosis in the mother's stomach. About 8 weeks after being swallowed alive, about two dozen small frogs emerge from the female's mouth (Figure 18.6B; Corben et al. 1974; Tyler 1983). What stops the eggs from being digested or excreted? It appears that the eggs secrete prostaglandin hormones that stop acid secretion and prevent peristaltic contractions in the stomach (Tyler et al. 1983). During this time, the stomach is fundamentally a uterus, and the frog does not eat. After the oral birth, the parent's stomach morphology and function return to normal. Unfortunately, both of these remarkable frog species are now feared extinct. No member of either species has been seen since the mid-1980s. ■

(A)

(B)

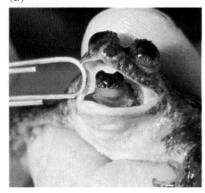

Figure 18.6
Parental care of tadpoles. (A) Tadpoles of the reticulate poison dart frog *Dendrobates* are carried on their parent's back to small pools of water in the Peruvian rain forest canopy. (B) The female *Rheobatrachus* of Australia brooded over a dozen tadpoles in its stomach. They emerged after completing metamorphosis. The last time anyone saw a *Rheobatrachus* frog alive was in 1985. (A courtesy of M. Fogden, DRK Photo; B courtesy of M. Tyler.)

Molecular responses to thyroid hormones during metamorphosis

Thyroid hormones appear to work largely at the level of transcription, activating the transcription of some genes and repressing the transcription of others (Lyman and White 1987; Mathison and Miller 1987). The transcription of the genes for albumin, carbamoylphosphate synthase, adult globin, adult skin keratin, and the *Xenopus* homologue of *sonic hedgehog* is activated by thyroid hormones. The transcription of the *sonic hedgehog* gene in the intestine is particularly interesting, since it suggests that the regional patterning of the organs formed during metamorphosis might be generated by the reappearance of some of the same molecules that structured the embryo (Stolow and Shi 1995; Stolow et al. 1997).

But these are relatively late responses to thyroid hormones. The earliest response to T_3 is the transcriptional activation of the **thyroid hormone receptor** (TR) genes (Yaoita and Brown 1990; Kawahara et al. 1991). Thyroid hormone receptors are members of the steroid hormone receptor superfamily of transcription factors. There are two major types of T_3 receptors, TRα and TRβ. Interestingly, the mRNAs and proteins of both TRs are present at relatively low levels in the premetamorphosis tadpole and then increase before thyroid hormone is released or metamorphosis begins (Table 18.2; Kawahara et al. 1991; Baker and Tata 1992). The thyroid hormone receptors may bind to their specific sites on the chromatin even before thyroid hormones are present, and they are thought to repress gene transcription. When T_3 or T_4 enters

Table 18.2 Relative accumulation of TRα and TRβ mRNA in *Xenopus* tadpoles following treatment with T₃ and prolactin

Treatment	Relative units	
	TRα	TRβ
None	505	24
T₃	1290	368
Prolactin + T₃	799	<10
Prolactin	405	43

Source: After Baker and Tata 1992.

the cell and binds to the chromatin-bound receptors, the hormone-receptor complex is converted from a repressor to a strong transcriptional activator (Wolffe and Shi 1999). At this time, the synthesis of TRs accelerates dramatically, coinciding with the onset of metamorphosis.

The injection of exogenous T₃ causes a twofold to fivefold increase in TRα message and a 20- to 50-fold increase in the mRNA for TRβ. Thus, T₃ binds to its TR and transcribes the *TR* gene. This "autoinduction" of T₃ receptor message by T₃ may play a significant role in the acceleration of metamorphosis (Figure 18.7). The more T₃ receptors a tissue has, the more competent it should be to respond to small amounts of

T₃. Thus, **metamorphic climax**, that time when the visible changes of metamorphosis occur rapidly, may be brought about by the enhanced production and induction of more T₃ receptors. The TR does not work alone, however, but forms a dimer with the retinoid receptor, RXR. This dimer binds thyroid hormones and can enter the nucleus to effect transcription (Wong and Shi 1995; Wolffe and Shi 1999).

The hormone prolactin has been found to inhibit the up-regulation of TRα and TRβ mRNAs. Moreover, if the up-regulation of the TR is experimentally blocked by prolactin, the tail is not resorbed, and the adult-specific keratin gene is not activated (Tata et al. 1991; Baker and Tata 1992). Injections of

Figure 18.7
Hypothetical model for the acceleration of metamorphosis in *Xenopus* by the autoinduction of T₃ receptors by T₃. (A) The premetamorphosis tadpole is characterized by low levels of thyrotropin (thyroid hormone-releasing factor), thyroid hormones, and T₃ receptors. (B) At the onset of metamorphosis, the levels of thyrotropin increase (probably due to the developmental maturation of the pituitary gland). This increases the amount of T₃. The T₃ binds to the small amount of T₃ receptor present to stimulate the transcription of more T₃ receptor mRNA. Some other T₃-induced proteins are also needed for the transcription of more T₃ message. (C) At metamorphic climax, the large concentrations of T₃ induce the synthesis of still more T₃ receptors, which cause a more rapid response to T₃.

(A) PREMETAMORPHOSIS

Low thyrotropin levels

Low T₃, T₄ concentration — Low T₃ receptor concentration

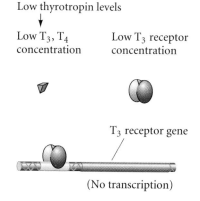

T₃ receptor gene

(No transcription)

(B) EARLY METAMORPHOSIS (PROMETAMORPHOSIS)

Thyrotropin levels rise

T₃, T₄ concentration increases — T₃ receptor concentration increases

Transcription

T₃ receptor binding stimulates production of more T₃ receptors

T₃

Binding of T₃ receptor stimulates transcription of other genes

Transcription

(C) METAMORPHIC CLIMAX

High thyrotropin levels

High T₃, T₄ concentration — High T₃ receptor concentration

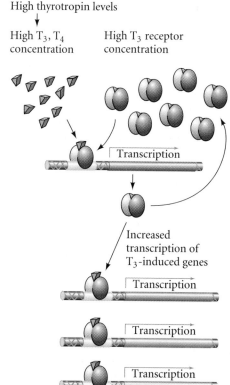

Transcription

Increased transcription of T₃-induced genes

Transcription

Transcription

Transcription

prolactin stimulate larval growth and inhibit metamorphosis (Bern et al. 1967; Etkin and Gona 1967), but there is dispute as to whether this finding reflects the natural role of prolactin (Takahashi et al. 1990; Buckbinder and Brown 1993). We still do not know the mechanisms by which levels of thyroid hormone are regulated in the tadpole, nor do we know how the reception of thyroid hormone elicits different responses (proliferation, differentiation, cell death) in different tissues.

Sidelights & Speculations

Heterochrony

MOST SPECIES OF ANIMALS develop through a larval phase. However, some species have modified their life cycles by either greatly extending or shortening their larval period. The phenomenon wherein animals change the relative time of appearance and rate of development of characters present in their ancestors is called **heterochrony**. Here we will discuss three extreme types of heterochrony: neoteny, progenesis, and direct development. **Neoteny** refers to the retention of the juvenile form owing to the retardation of body development relative to that of the germ cells and gonads, which achieve maturity at the normal time. **Progenesis** also involves the retention of the juvenile form, but in this case, the gonads and germ cells develop at a faster rate than normal, and become sexually mature while the rest of the body is still in a juvenile phase. In **direct development**, the embryo abandons the stages of larval development entirely and proceeds to construct a small adult.

Neoteny

In certain salamanders, sexual maturity occurs in what is usually considered a larval state. The reproductive system and germ cells mature, while the rest of the body retains its juvenile form throughout life. In most instances, metamorphosis fails to occur and sexual maturity takes place in a "larval" body. The Mexican axolotl, *Ambystoma mexicanum*, does not undergo metamorphosis in nature because its pituitary gland does not release a thyrotropin (thyroid-stimulating hormone) to activate T_3 synthesis in its thyroid glands (Prahlad and DeLanney 1965; Norris et al. 1973; Taurog et al. 1974). Thus, when investigators gave *A. mexicanum* either thyroid hormones or thyrotropin, they found that the salamander metamorphosed into an adult form not seen in nature (Figure 18.8; Huxley 1920). Other species, such as *A. tigrinum*, metamorphose only if given cues from the environment. Otherwise, they become neotenic, successfully mating as larvae. In part of its range, *A. tigrinum* is a neotenic salamander, paddling its way through the cold ponds of the Rocky Mountains. Its gonads and germ cells mature while the rest of the body retains its larval form. However, in the warmer region of its range, the larval form of *A. tigrinum* is transitory, leading to the land-dwelling tiger salamander. Neotenic populations from the Rockies can be induced to undergo metamorphosis simply by placing them in water at higher temperatures. It appears that the hypothalamus of this species cannot produce thyrotropin-releasing factor at low temperatures.

Some salamanders are permanently neotenic, even in the laboratory. Whereas T_3 is able to produce the long-lost adult form of *A. mexicanum*, the neotenic species of *Necturus* and *Siren* remain unresponsive to thyroid hormones (Frieden 1981); their neoteny is permanent. Yaoita and Brown (1990) noted that the mRNA for TRβ is absent in *Necturus*, so that this species cannot respond to T_3. The genetic changes thought to be responsible for neoteny in several salamander species are shown in Figure 18.9.

De Beer (1940) and Gould (1977) have speculated that neoteny is a major factor in the evolution of more complex taxa. By retarding the development of somatic tissues, neoteny may give natural selection a flexible substrate. According to Gould (1977, p. 283), neoteny would "provide an escape from specialization. Animals can relinquish their highly specialized adult forms, return to the lability of youth, and prepare themselves for new evolutionary directions."

Figure 18.8
Metamorphosis induced in the axolotl.
(A) Normal adult form of the axolotl.
(B) Specimen treated with thyroxine to induce metamorphosis. (Photographs courtesy of G. Malacinski.)

(A)

(B)

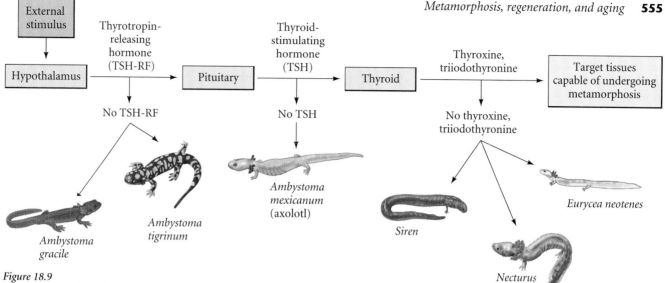

Figure 18.9
Stages along the hypothalamus-pituitary-thyroid axis of salamanders at which various species are thought to have blocked metamorphosis. Normally, thyrotropin-releasing factor from the hypothalamus causes the release of thyrotropin from the pituitary. Thyrotropin causes the thyroid gland to synthesize and release T_3 and T_4. These thyroid hormones bind to their receptors in competent tissues. *Ambystoma tigrinum* and *A. gracile* have defects in thyrotropin-releasing hormone, while *A. mexicanum* has a defect in thyrotropin production. *Eurycea, Necturus,* and *Siren* appear to have a receptor defect in the thyroid hormone-responsive tissues. *Eurycea* will metamorphose when exposed to extremely high concentrations of thyroxine, while *Necturus* and *Siren* do not respond to any dose. (After Frieden 1981.)

Progenesis

In progenesis, gonadal maturation is accelerated while the rest of the body develops normally to a certain stage. Progenesis has enabled some salamander species to find new ecological niches. *Bolitoglossa occidentalis* is a tropical salamander that, unlike other members of its genus, lives in trees. This salamander's webbed feet and small body size suit it for arboreal existence, the webbed feet producing suction for climbing and the small body making such traction efficient. Alberch and Alberch (1981)

have shown that *B. occidentalis* resembles juveniles of the related species *B. subpalmata* and *B. rostrata* (whose young are small, with digits that have not yet grown past their webbing; see Figure 22.17). *B. occidentalis* undergoes metamorphosis at a much smaller size than its relatives. This appears to have given it a phenotype that made tree-dwelling possible.

Direct Development

While some animals have extended their larval period of life, others have "accelerated"

their development by abandoning their "normal" larval forms. This latter phenomenon, called direct development, is typified by frog species that lack tadpoles and by sea urchins that lack pluteus larvae. Elinson and his colleagues (del Pino and Elinson 1983; Elinson 1987) have studied *Eleutherodactylus coqui*, a small frog that is one of the most populous vertebrates on the island of Puerto Rico. Unlike the eggs of *Rana* and *Xenopus*, the eggs of *E. coqui* are fertilized while they are still within the female's body. Each egg is about 3.5 mm in diameter (roughly 20 times the volume of *Xenopus* eggs). After the eggs are laid, the male gently sits on the developing embryos, protecting them from predators and desiccation (Taigen et al. 1984). Early development is like that of most frogs. Cleavage is holoblastic, gastrulation is initiated at a subequatorial position, and the neural folds become elevated from the surface. However, shortly after the neural tube closes, limb buds appear on the surface (Figure 18.10A,B). This early emergence of

Figure 18.10
Direct development of the frog *Eleutherodactylus coqui*. (A) Limb buds are seen as the frog develops on the yolk. (B) As the yolk is used up, the limb buds are easily seen. (C) Three weeks after fertilization, tiny froglets hatch. They are seen here in a petri dish and on a Canadian dime. (Photographs courtesy of R. P. Elinson.)

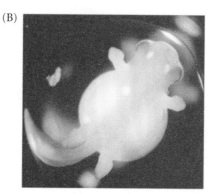

limb buds is the first indication that development is direct and will not pass through a limbless tadpole stage. Moreover, the emergence of the limbs does not depend on thyroid hormones (Lynn and Peadon 1955). What emerges from the egg jelly 3 weeks after fertilization is not a tadpole, but a little frog (Figure 18.10C). The froglet has a transient tail that is used for respiration rather than locomotion. Such direct-developing frogs do not need ponds for their larval stages and can therefore colonize new regions inaccessible to other frogs. Direct development also occurs in other phyla, in which it is also correlated with a large egg. It seems that if nutrition can be provided in the egg, the life cycle need not have a food-gathering larval stage. ■

Metamorphosis in Insects

Types of insect metamorphosis

Whereas amphibian metamorphosis is characterized by the remodeling of existing tissues, insect metamorphosis often involves the destruction of larval tissues and their replacement by an entirely different population of cells. Insects grow by molting—shedding their cuticle—and growing new cuticle as their size increases. There are three major patterns of insect development. A few insects, such as springtails and mayflies, have no larval stage and undergo direct development. These are called the **ametabolous** insects (Figure 18.11A). These insects have a **pronymph** stage immediately after hatching, bearing the structures that have enabled it to get out of the egg. But after this transitory stage, the insect begins to look like a small adult; after each molt, they are bigger, but unchanged in form (Truman and Riddiford 1999). Other insects, notably grasshoppers and bugs, undergo a gradual, **hemimetabolous** metamorphosis (Figure 18.11B). After spending a very brief period of time as a pronymph (whose cuticle is often shed as the insect hatches), the insect looks like an immature adult. This immature stage is called a **nymph**. The rudiments of the wings, genital organs, and other adult structures are present, and these structures become more mature with each molt. At the last molt, the emerging insect is a winged and sexually mature adult.

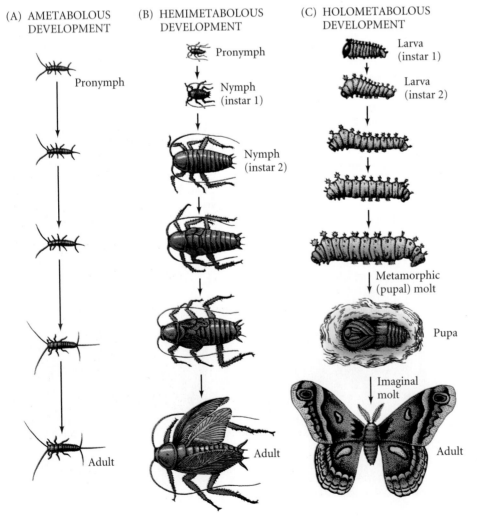

(A) AMETABOLOUS DEVELOPMENT

Pronymph

Adult

(B) HEMIMETABOLOUS DEVELOPMENT

Pronymph

Nymph (instar 1)

Nymph (instar 2)

Adult

(C) HOLOMETABOLOUS DEVELOPMENT

Larva (instar 1)

Larva (instar 2)

Metamorphic (pupal) molt

Pupa

Imaginal molt

Adult

Figure 18.11
Modes of insect development. Molts are represented as arrows. (A) Ametabolous (direct) development in a silverfish. After a brief pronymph stage, the insect looks like a small adult. (B) Hemimetabolous (gradual) metamorphosis in a cockroach. After a very brief pronymph phase, the insect becomes a nymph. After each molt, the next nymphal instar looks more like an adult, gradually growing wings and genital organs. (C) Holometabolous (complete) metamorphosis in a moth. After hatching as a larva, the insect undergoes successive larval molts until a metamorphic molt causes it to enter the pupal stage. Then an imaginal molt turns it into an adult.

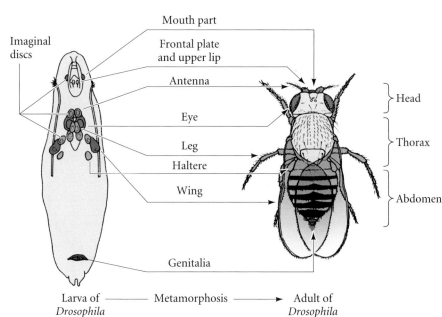

Discs for:

Mouth part

Frontal plate and upper lip

Imaginal discs

Antenna

Eye

Leg

Haltere

Wing

Head

Thorax

Abdomen

Genitalia

Larva of *Drosophila* —— Metamorphosis —→ Adult of *Drosophila*

Figure 18.12
The locations and developmental fates of the imaginal discs in *Drosophila melanogaster*. (After Fristrom et al. 1969.)

VADE MECUM **Development of a holometabolous insect.** To watch the stages of fruit fly development, see the QuickTime movies of the growing larva. This segment also shows methods for "peeling" a pupa. Photographs of each of the imaginal discs are also shown, along with instructions for dissecting out these discs.
[Click on Fruit Fly]

In the **holometabolous** insects (Figure 18.11C: flies, beetles, moths, and butterflies), there is no pronymph stage. The juvenile form that hatches from the egg is called a **larva**. The larva (caterpillar, grub, maggot) undergoes a series of molts as it becomes larger. The stages between these larval molts are called **instars**. The number of molts before becoming an adult is characteristic for the species, although environmental factors can increase or decrease the number. The instar stages grow in a stepwise fashion, each being qualitatively larger than the previous one. Finally, there is a dramatic and sudden transformation between the larval and adult stages. After the last instar stage, the larva undergoes a **metamorphic molt** to become a **pupa**. The pupa does not feed, and its energy must come from those foods it ingested while a larva. During pupation, the adult structures are formed and replace the larval structures. Eventually, an **imaginal molt** enables the adult ("imago") to shed the pupal case and emerge. While the larva is said to *hatch* from an egg, adults are said to **eclose** from the pupa.

Eversion and differentiation of the imaginal discs

In holometabolous insects, the transformation from juvenile into adult occurs within the pupal cuticle. Most of the old body of the larva is systematically destroyed by apoptosis, while new adult organs develop from undifferentiated nests of cells, the **imaginal discs** . Thus, within any larva, there are two distinct populations of cells: the larval cells, which are used for the functions of the juvenile insect, and the thousands of imaginal cells, which lie within the larva in clusters, awaiting the signal to differentiate.

In *Drosophila*, there are ten major pairs of imaginal discs, which construct many of the adult organs, and an unpaired genital disc, which forms the reproductive structures (Figure 18.12). The abdominal epidermis forms from a small group of imaginal cells called **histoblasts**, which lie in the region of the larval gut.

Other nests of histoblasts located throughout the larva form the internal organs of the adult. The imaginal discs can be seen in the newly hatched larva as local thickenings of the epidermis. Whereas most of the larval cells have a very limited mitotic capacity, the imaginal discs divide rapidly at specific characteristic times. As the cells proliferate, they form a tubular epithelium that folds in upon itself in a compact spiral (Figure 18.13A). The largest disc, that of the wing, contains some 60,000 cells,

(A)

(B)

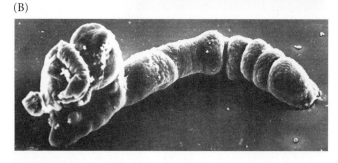

Figure 18.13
Imaginal disc elongation. Scanning electron micrograph of *Drosophila* last-instar leg disc (A) before and (B) after elongation. (From Fristrom et al. 1977; photograph courtesy of D. Fristrom.)

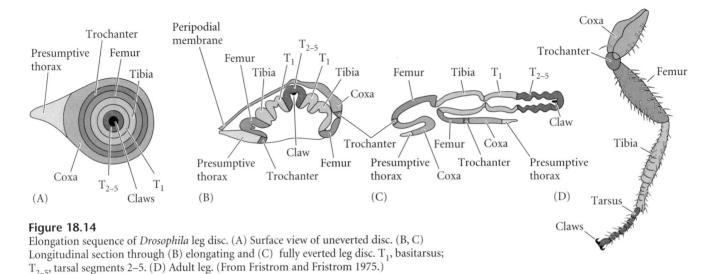

Figure 18.14
Elongation sequence of *Drosophila* leg disc. (A) Surface view of uneverted disc. (B, C) Longitudinal section through (B) elongating and (C) fully everted leg disc. T_1, basitarsus; T_{2-5}, tarsal segments 2–5. (D) Adult leg. (From Fristrom and Fristrom 1975.)

whereas the leg and haltere discs contain around 10,000 (Fristrom 1972). At metamorphosis, these cells proliferate, differentiate, and elongate (Figure 18.13B).

The fate map and elongation sequence of the leg disc are shown in Figure 18.14. At the end of the third instar, just before pupation, the leg disc is an epithelial sac connected by a thin stalk to the larval epidermis. On one side of the sac, the epithelium is coiled into a series of concentric folds "reminiscent of a Danish pastry" (Kalm et al. 1995). As pupation begins, the cells at the center of the disc telescope out to become the most distal portions of the leg—the claws and the tarsus. The outer cells become the proximal structures—the coxa and the adjoining epidermis (Schubiger 1968). After differentiating, the cells of the appendages and epidermis secrete a cuticle appropriate for the specific region. Although the disc is composed primarily of epidermal cells, a small number of **adepithelial cells** migrate into the disc early in development. During the pupal period, these cells give rise to the muscles and nerves that serve that structure.

Studies by Condic and her colleagues (1990) have demonstrated that the elongation of imaginal discs is due primarily to cell shape change within the disc epithelium. Using fluorescently labeled phalloidin to stain the periph-

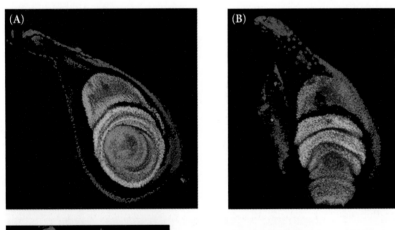

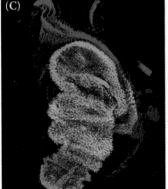

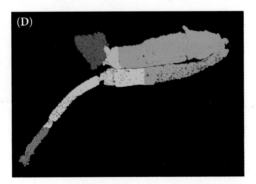

Figure 18.15
The fates of the imaginal disc cells are directed by transcription factors found in different regions of the imaginal disc. At the periphery, the *homothorax* gene (purple) establishes the boundary for the coxa. The expression of the *dachshund* gene (green) locates the femur and proximal tibia. The most distal structures, the claw and lower tarsal segments, arise from the expression of *distal-less* (red) in the center of the imaginal disc. A, B, and C are taken at successively later stages of pupal development. D localizes the expression domains of the genes onto a leg immediately prior to eclosion. The areas where there is overlap are shown in yellow, aqua, and orange. (From Abu-Shaar and Mann 1998; photographs courtesy of R. S. Mann.)

eral microfilaments of leg disc cells, they showed that the cells of early third-instar discs are tightly compressed along the proximal-distal axis. This compression is maintained through several rounds of cell division. Then, when the tissue begins elongating, the compression is removed, and the cells "spring" into their rounder state. This conversion of an epithelium of compressed cells into a longer epithelium of noncompressed cells represents a novel mechanism for the extension of an organ during development.

The type of leg structure generated is determined by the interactions between several genes in the imaginal disc. Figure 18.15 shows the expression of three genes involved in determining the proximal-distal axis of the fly leg. In the third-instar leg disc, the center of the disc secretes the highest concentration of two morphogens, Wingless (Wg) and Decapen-

taplegic (Dpp). High concentrations of these paracrine factors cause the expression of the *Distal-less* gene. Moderate concentrations cause the expression of the *dachshund* gene, and lower concentrations cause the expression of the *homothorax* gene. Those cells expressing *Distal-less* telescope out to become the most distal structures of the leg—the claw and distal tarsal segments. Those expressing *homothorax* become the most proximal structure, the coxa. Cells expressing *dachshund* become the femur and proximal tibia. Areas of overlap produce the trochanter and distal tibia (Abu-Shaar and Mann 1998). These regions of gene expression are stabilized by inhibitory interactions between the protein products of these genes and of the neighboring genes. In this manner, the gradient of Wg and Dpp proteins is converted into discrete domains of gene expression that specify the different regions of the *Drosophila* leg.

Sidelights & Speculations

Determination of the Wing Imaginal Discs

Determination of Discs from Ectoderm: Distal-less Protein

The molecular biology of insect metamorphosis begins with the specification of certain epidermal cells to become imaginal disc precursors. As we discussed in Chapter 9, the organ rudiments in *Drosophila* are specified on an orthogonal grid by intersecting anterior-posterior and dorsal-ventral signals. In most segments, Hox gene products prevent *Distal-less* gene expression and the establishment of limb primordia; but in those segments that are specified to be thoracic, limb formation is permitted. Cohen and his colleagues (1993) have demonstrated that the leg and wing originate from the same set of imaginal precursors, specified at the intersections between the anterior-posterior stripes of Wingless (Wg) protein expression and the

horizontal band of cells expressing the Decapentaplegic (Dpp) protein. Both proteins are soluble and have a limited range of diffusion. In the early *Drosophila* embryo (at germ band extension about 4.5 hours after fertilization), a single group of cells at these intersections forms the imaginal disc precursors in each thoracic segment. These cells (and only these cells) express the Distal-less protein. As the cells expressing Dpp are moved dorsally, some of these Distal-less-expressing cells move with them to establish a

secondary cluster of imaginal cells (derived from the original ventral cluster). The initial clusters form the leg imaginal disc, while the secondary clusters form the wing or haltere disc. Thus, the leg and wing discs have a common origin (Figure 18.16).

Determination of Disc Identity: Vestigial Protein

Despite their common origin, it is obvious that the leg and wing discs are determined to become different structures. The determina-

Figure 18.16
Schematic model for the initiation and separation of the leg-wing disc in the *Drosophila* thorax. The embryo is divided into an orthogonal grid by vertical stripes of Wingless (Wg) and a horizontal band of Decapentaplegic (Dpp) synthesis and secretion. The initial disc forms at the intersection of these secretory domains. The Dpp-secreting cells migrate dorsally, bringing with them some of the imaginal disc cells. These dorsal disc cells generate the wing disc, while the cells remaining form the leg disc. (After Cohen et al. 1993.)

◯ Cell expressing Distal-less

⬢ Range of Wg signal

⬡ Range of Dpp signal

○ Cell expressing Dpp

● Cell expressing Wg

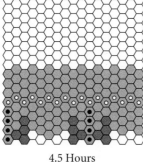

4.5 Hours

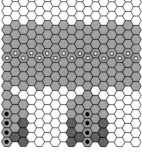

10 Hours

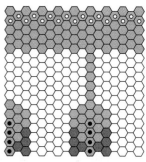

Mature embryo

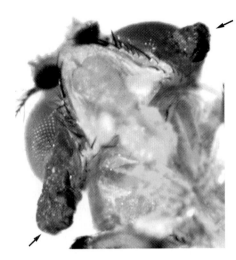

Figure 18.17
The *vestigial* gene determines wing disc identity. If the *vestigial* gene is expressed in some of the cells of an eye imaginal disc, those cells will be determined as wing cells, and that part of the eye will form wing tissue (arrows). (After Kim et al. 1996; photograph courtesy of S. Carroll and S. Paddock.)

tion of the wing disc appears to be regulated by the *vestigial* gene. Using a targeted gene expression system, Kim and colleagues (1996) have caused the *vestigial* gene to be expressed in eye, antenna, and leg discs (Figure 18.17). When this happens, regions of the normal structure are converted into wing tissue.

Determination of the Anterior-Posterior Axis: Engrailed and Decapentaplegic Proteins

The axes of the wing are specified by interactions at their compartmental boundaries (Meinhardt 1980; Causo et al. 1993; Tabata et al. 1995). After these initial interactions, a polar coordinate system may subdivide the wing regions more finely (Held 1995).

During the first larval instar, the leg and wing imaginal discs acquire their anterior-posterior axis. The discs become split into two compartments representing the future anterior and posterior regions of the appendage (i.e., the front of the wing and the rear of the wing). Based on the position of its cells in the segment, the posterior compartment of the wing disc expresses the *engrailed* gene (Figure 18.18; Garcia-Bellido et al. 1973; Lawrence and Morata 1976). If *engrailed* function is absent, all the disc cells become anteriorized. The boundary between the posterior and anterior compartments is strictly observed. Cells from one side cannot produce descendants that cross over the boundary to the other.

The Engrailed protein is a transcription factor, and it activates the *hedgehog* gene in the cells of the posterior compartment. The posterior wing disc cells express the Hedgehog protein, which acts as a short-range signal to induce the expression of Dpp in adjacent *anterior* cells, while the expression of *engrailed* in the posterior cells renders them nonresponsive to the Hedgehog they secrete (preventing them from expressing Dpp). Dpp is a TGF-β family paracrine factor that acts as a long-range signal to establish the anterior-posterior axis of the wing (Guillen et al. 1995; Tabata et al. 1995; Nellen et al. 1996). Cells in both compartments close to the area of Dpp expression are exposed to relatively high concentrations of this protein and activate the *spalt* and *oculomotor-blind* (*omb*) genes. Those cells farther away receive lower concentrations of Dpp and activate only the *omb* gene. These two genes encode transcription factors that specify the region of the wing from the center (where Dpp is expressed) to the periphery.

Dorsal-Ventral Axis: Wingless and Apterous Proteins

During the second larval instar, the dorsal-ventral axis of the wing disc is determined. The dorsal-ventral boundary lies at the future margin of the wing blade, separating the upper surface of the wing from the lower (Bryant 1970; Garcia-Bellido et al. 1973). The gene involved in this compartmentalization event is *apterous*. Cells expressing *apterous* become the dorsal cells (Figure 18.19A; Blair 1993; Diaz-Benjumea and Cohen 1993). When *apterous* is deleted, all cells in the disc acquire ventral fates. The Apterous protein is a transcription factor that activates the genes for the Serrate and Fringe proteins. Serrate is a ligand for the Notch receptor, and Fringe is involved in regulating Notch ligand binding (Irvine and Wieschaus 1994; Williams et al.

Figure 18.18
Compartmentalization and anterior-posterior patterning in the wing disc. (A) In the first-instar larva, the anterior-posterior axis has been formed and is manifested by the expression of the *engrailed* gene in the posterior compartment. The Engrailed protein functions as a transcription factor to activate the *hedgehog* gene. Hedgehog acts as a short-range paracrine factor to activate the *decapentaplegic* gene in the anterior cells adjacent to the posterior compartment. The Decapentaplegic protein (Dpp) acts as a long-range transcription factor spreading in both directions from its source. (B) Dpp forms a concentration gradient, wherein high concentrations of Dpp near the source activate the *spalt* and *omb* genes. Lower concentrations (near the periphery) activate *omb* without activating *spalt*.

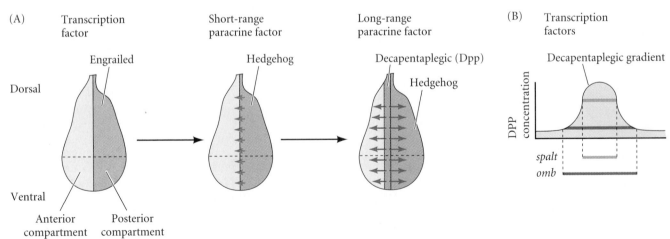

Figure 18.19
Determining the dorsal-ventral axis. (A) The prospective ventral surface of the wing is stained by antibodies to the Vestigial protein (green), while the prospective dorsal side is stained by antibodies to the Apterous protein (red). The region of yellow shows where they overlap. (B) The Wingless protein (purple) synthesized at this juncture organizes the wing disc along the dorsal-ventral boundary. The expression of the Vestigial protein (green) is seen in those cells close to those expressing Wingless. (C) Anterior-posterior and dorsal-ventral boundaries. Apterous protein (red) is seen in the dorsal compartment; Vestigial protein (green) is seen in the ventral compartment. The yellow shows the overlap. Cubitus interruptus (blue) (induced by Dpp expression) marks the anterior compartment as distinct from the posterior one. (D) The dorsal and ventral portions of the wing disc telescope out to form the two-layered wing. Gene expression patterns are indicated on the double-layered wing. (Photographs courtesy of S. Carroll and S. Paddock.)

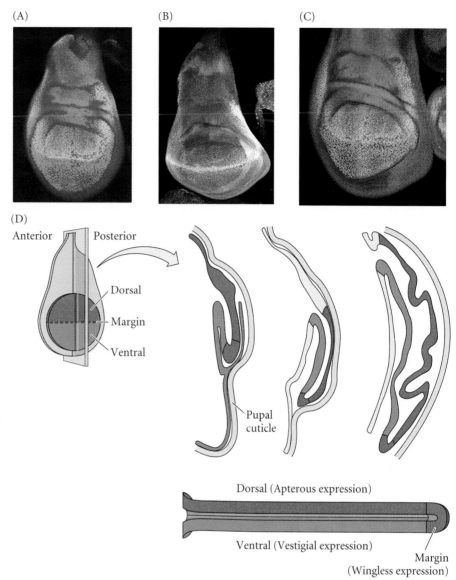

1994; Kim et al. 1995). The Notch receptor is found in the ventral cells, so the binding of Notch on the ventral side with Serrate on the dorsal side stabilizes the wing margin.

The future wing margin also becomes a signaling center, analogous to the Dpp-secreting band that forms the anterior-posterior axis (Zecca et al. 1996; Neumann and Cohen 1997). The Fringe and Serrate proteins act to promote the transcription of the *Wingless* gene at the border of the dorsal and ventral compartments (Figure 18.19B–D). Wingless is secreted by the cells adjacent to the border in both compartments, and it acts as a long-range morphogen. Those cells that are close to this border and receive high concentrations of Wingless protein express the *achaete-scute* genes. The next group of cells receives moderate concentrations of Wingless and expresses the *distal-less* gene. Those wing cells receiving relatively low amounts of Wingless express the *vestigial* gene. However, only those cells expressing the Apterous protein (i.e., the dorsal cells) are able to respond to these signals. Thus, Wingless protein acts as a morphogen on the dorsal surface only.

One of the problems with this model of diffusible gradients is that although the influence of Wingless extends to the edges of the wing disc, neither Hedgehog nor Wingless protein travels well in the extracellular environment. The transport of these proteins does not seem to be due to simple diffusion. One possibility is that the cells actively transport Wingless in a particular direction (Pfeiffer and Vincent 1999). Alternatively, Wingless might be transported locally by diffusion and more widely by cytoplasmic filaments that extend from the peripheral cells into the central region producing the Wingless protein. Evidence for this latter model comes from Ramírez-Weber and Kornberg (1999), who identified extremely thin processes extending from the peripheral cells, across the wing disc, to the sites of Wingless synthesis (Figure 18.20). These actin-based extensions are called **cytonemes**, and they are similar to the thin filopodia of sea urchin mesenchymal cells. Evidence has yet to show that they are active in transporting the Wingless protein to the peripheral cells. However, this is an exciting and unexpected mechanism for some types of long-range cell-cell communication.

Proximal-Distal Axis: Wingless Protein

In addition to acting as a morphogen defining the dorsal-ventral axis, the Wingless protein also acts to promote cell division and the extension of the wing (Neumann and Cohen 1996). The interaction between the dorsal-ventral and anterior-posterior axes at their boundaries is critical for the outgrowth along the proximal-distal axis. During metamorphosis, the "distalization" of the proximal-distal axis from the base of the thorax outward to the tip of the wing or leg is

Figure 18.20
Cytonemes in the third-instar wing disc of *Drosophila*. (A) The GFP reporter gene was expressed in cells of the anterior and posterior flanks of the wing disc, but not among the central cells that are synthesizing Wingless. (B) Enlargement of a peripheral region of the disc, showing thin processes, called cytonemes, extending from the peripheral cells (bright regions) into the area producing Wingless. (C) Single cells in culture from the peripheral regions extend cytonemes. (After Ramírez-Weber and Kornberg 1999; photographs courtesy of T. B. Kornberg.)

(A)

(B)

(C)

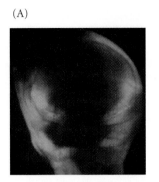
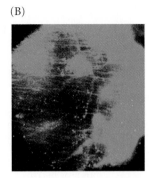

accomplished by cell interactions at the boundaries between the other two axes.* ■

*Yes, this process is complex, and likely to get more so as we learn more about it. This situation is not without its humor. Sydney Brenner (1996) recalls Nobel laureate Francis Crick being frustrated by this complexity and saying, "God knows how these imaginal discs work." Brenner fantasized a meeting wherein Crick asked the Deity how He constructed these entities, only to have God bewildered by their complexity as well. Eventually, all God could do was to reassure Crick that "we've been building flies here for 200 million years and we have had no complaints."

WEBSITE **18.1 Creation of the dorsal-ventral wing surfaces**. The juxtaposition of those cells expressing Apterous with those that do not initiates a cascade of gene expression that results in markedly different cell types. These events were predicted by theoretical biologists years before the molecules were discovered.

WEBSITE **18.2 Homologous specification**. If a group of cells in one imaginal disc are mutated such that they give rise to a structure characteristic of another imaginal disc (for instance, cells from a leg disc giving rise to antennal structures), the regional specification of those structures will be in accordance with their position in the original disc.

Hormonal control of insect metamorphosis

Although the detailed mechanisms of insect metamorphosis differ among species, the general pattern of hormone action is very similar. Like amphibian metamorphosis, the metamorphosis of insects appears to be regulated by effector hormones, which are controlled by neurohormones in the brain (for reviews, see Gilbert and Goodman 1981; Riddiford 1996). Insect molting and metamorphosis are controlled by two effector hormones: the steroid **20-hydroxyecdysone** and the lipid **juvenile hormone** (**JH**) (Figure 18.21). 20-hydroxyecdysone initiates and coordinates each molt and regulates the changes in gene expression that occur during metamorphosis. Juvenile hormone prevents the ecdysone-induced changes in gene expression that are necessary for metamorphosis. Thus, its presence during a molt ensures that the result of that molt produces another instar, not a pupa or an adult.

The molting process is initiated in the brain, where neurosecretory cells release **prothoracicotropic hormone** (**PTTH**) in response to neural, hormonal, or environmental signals. PTTH is a peptide hormone with a molecular weight of approximately 40,000, and it stimulates the production of ecdysone by the **prothoracic gland**. This ecdysone is modified in peripheral tissues to become the active molting hormone 20-hydroxyecdysone.* Each molt is initiated by one or more

*Since its discovery in 1954, when Butenandt and Karlson isolated 25 mg of ecdysone from 500 kg of silkworm moth pupae, 20-hydroxyecdysone has gone under several names, including β-ecdysone, ecdysterone, and crustecdysone.

pulses of 20-hydroxyecdysone. For a larval molt, the first pulse produces a small rise in the hydroxyecdysone concentration in the larval hemolymph (blood) and elicits a change in cellular commitment. A second, large pulse of hydroxyecdysone initiates the differentiation events associated with molting. The hydroxyecdysone produced by these pulses commits and stimulates the epidermal cells to synthesize enzymes that digest and recycle the components of the cuticle.

Juvenile hormone is secreted by the **corpora allata**. The secretory cells of the corpora allata are active during larval molts but inactive during the metamorphic molt. As long as JH is present, the hydroxyecdysone-stimulated molts result in a new larval instar. In the last larval instar, however, the medial nerve from the brain to the corpora allata inhibits the gland from producing JH, and there is a simultaneous increase in the body's ability to degrade existing JH (Safranek and Williams 1989). Both these mechanisms cause JH levels to drop below a critical threshold value. This triggers the release of PTTH from the brain (Nijhout and Williams 1974; Rountree and Bollenbacher 1986). PTTH, in turn, stimulates the prothoracic glands to secrete a small amount of ecdysone. The resulting hydroxyecdysone, in the absence of high levels of JH, commits the cells to pupal development. Larva-specific mRNAs are not replaced, and new mRNAs are synthesized whose protein products inhibit the transcription of the larval messages. After the second ecdysone pulse, new pupa-specific gene products are synthesized (Riddiford 1982), and the subsequent molt shifts the organism from larva to pupa. It appears, then, that

(A)

(B)

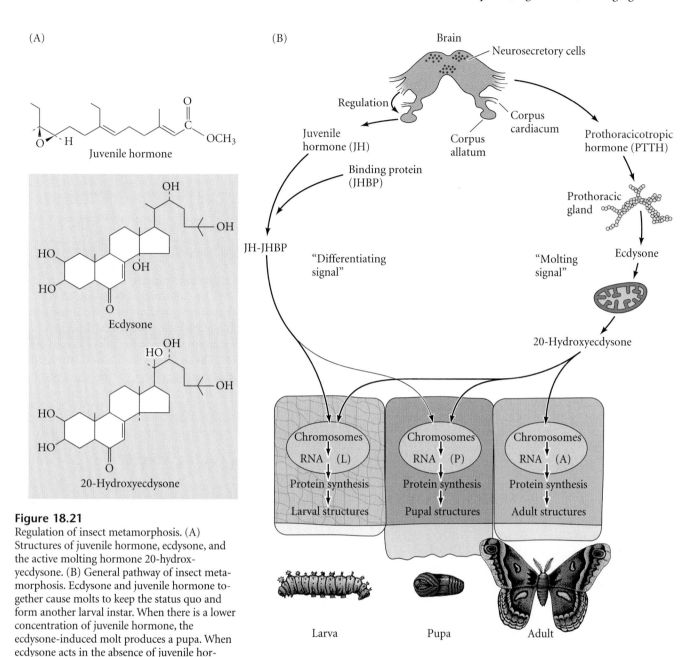

Juvenile hormone

Ecdysone

20-Hydroxyecdysone

Figure 18.21
Regulation of insect metamorphosis. (A) Structures of juvenile hormone, ecdysone, and the active molting hormone 20-hydroxyecdysone. (B) General pathway of insect metamorphosis. Ecdysone and juvenile hormone together cause molts to keep the status quo and form another larval instar. When there is a lower concentration of juvenile hormone, the ecdysone-induced molt produces a pupa. When ecdysone acts in the absence of juvenile hormone, the imaginal discs differentiate, and the molt gives rise to the adult. (After Gilbert and Goodman 1981.)

the first ecdysone pulse during the last larval instar triggers the processes that inactivate the larva-specific genes and prepare the pupa-specific genes to be transcribed. The second ecdysone pulse transcribes the pupa-specific genes and initiates the molt (Nijhout 1994). At the imaginal molt, when ecdysone acts in the absence of juvenile hormone, the imaginal discs differentiate, and the molt gives rise to the adult.

WEBSITE **18.3 Insect metamorphosis.** Four websites discuss (1) the experiments of Wigglesworth and others who identified the hormones of metamorphosis and the glands producing them; (2) the variations that *Drosophila*

and other insects play on the general theme of metamorphosis; (3) the remodeling of the insect nervous system during metamorphosis, and (4) a microarray analysis of *Drosophila* metamorphosis wherein several thousand genes were simultaneously screened.

The molecular biology of hydroxyecdysone activity

ECDYSONE RECEPTORS. 20-hydroxyecdysone cannot bind to DNA by itself. Like amphibian thyroid hormones, 20-hydroxyecdysone first binds to receptors. These receptors are almost identical in structure to the thyroid hormone receptors. The receptors specifically binding 20-hydroxyecdysone are called the **ecdysone receptors** (**EcR**). An EcR protein forms an active

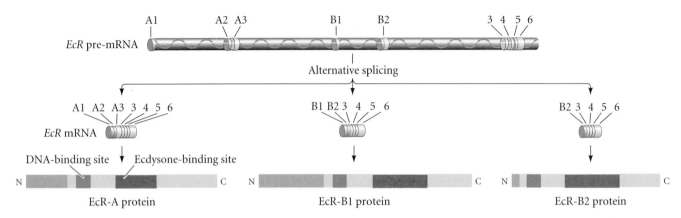

Figure 18.22
Formation of the ecdysone receptors. Alternative mRNA splicing of the ecdysone receptor (*EcR*) transcript creates three types of *EcR* mRNAs. These generate proteins having the same DNA-binding site (blue) and hydroxyecdysone-binding site (red), but with very different amino termini. (After Talbot et al. 1993.)

molecule by pairing with an Ultraspiracle (Usp) protein, the homologue of the amphibian RXR that helps form the thyroid hormone receptor (Koelle et al. 1991; Yao et al. 1992; Thomas et al. 1993). Although there is only one type of gene for Usp in *Drosophila*, and only one type of gene for EcR, the EcR gene transcript can be spliced in at least three different ways to form three distinct proteins. All three EcR proteins have the same domains for 20-hydroxyecdysone and DNA binding but they differ in their N-terminal domains (Figure 18.22). The type of EcR in a cell may inform the cell how to act when it receives a hormonal signal (Talbot et al. 1993; Truman et al. 1994). All cells appear to have some of each type, but the strictly larval tissues and neurons that die when exposed to 20-hydroxyecdysone are characterized by their great abundance of the EcR-B1 form of the ecdysone receptor. Imaginal discs and differentiating neurons, on the other hand, show a preponderance of the EcR-A isoform. It is therefore possible that the different receptors activate different sets of genes when they bind 20-hydroxyecdysone.

BINDING OF 20-HYDROXYECDYSONE TO DNA. During molting and metamorphosis, certain regions of the polytene chromosomes of *Drosophila* puff out in the cells of certain organs at certain times (see Figure 4.13; Clever 1966; Ashburner 1972; Ashburner and Berondes 1978). These chromosome puffs represent areas where DNA is being actively transcribed.

Figure 18.23
Hydroxyecdysone-induced puffs in cultured salivary gland cells of *D. melanogaster*. The chromosome region here is the same as in Figure 4.13. (A) Uninduced control. (B–E) Hydroxyecdysone-stimulated chromosomes at (B) 25 minutes, (C) 1 hour, (D) 2 hours, and (E) 4 hours. (Photographs courtesy of M. Ashburner.)

Moreover, these organ-specific patterns of chromosome puffing can be reproduced by culturing larval tissue and adding hormones to the medium or by adding hydroxyecdysone to an earlier-stage larva. When 20-hydroxyecdysone is added to larval salivary glands, certain puffs are produced and others regress (Figure 18.23). The puffing is mediated by the binding of hydroxyecdysone at specific places on the chromosomes;

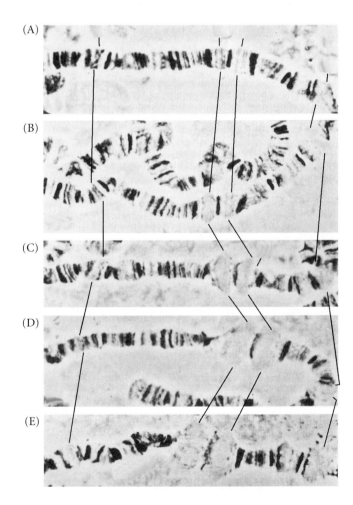

Figure 18.24
The Ashburner model of hydroxyecdysone regulation of transcription. Hydroxyecdysone binds to its receptor, and this compound binds to an early puff gene and a late puff gene. The early puff gene is activated, and its protein product (1) represses the transcription of its own gene and (2) activates the late puff gene, perhaps by displacing the ecdysone receptor. (After Richards 1992.)

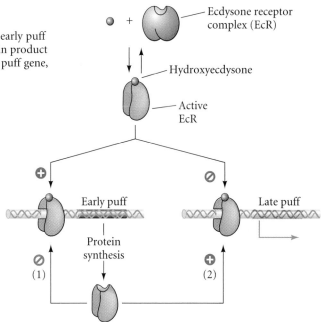

fluorescent antibodies against hydroxyecdysone find this hormone localized to the regions of the genome that are sensitive to it (Gronemeyer and Pongs 1980).

Hydroxyecdysone-regulated chromosome puffs occurring during the late stages of the third-instar larva (as it prepares to form the pupa) can be divided into three categories: "early" puffs that hydroxyecdysone causes to regress; "early" puffs that hydroxyecdysone induces rapidly; and "late" puffs that are first seen several hours after hydroxyecdysone stimulation. For example, in the larval salivary gland, about six puffs emerge within a few minutes of hydroxyecdysone treatment. No new protein has to be made in order for these puffs to be induced. A much larger set of genes are induced later in development, and these genes do need protein synthesis to become transcribed. Ashburner (1974, 1990) hypothesized that the "early" genes make a protein product that is essential for the activation of the "late" genes and that, moreover, this protein itself turns off the transcription of the early genes (Figure 18.24). These insights have been confirmed by molecular analyses. The three early puffs include the genes for EcR and two other transcription factors, BR-C and E74B.

The *broad-complex* (*BR-C*) gene is particularly interesting. Like the ecdysone receptor gene, the *BR-C* gene can generate several different transcription factor proteins through differentially initiated and spliced messages. It appears that the variants of the ecdysone receptor may signal particular variants of the BR-C protein to be synthesized. Organs such as the larval salivary gland that are destined for death during metamorphosis express the Z1 isoform; imaginal discs destined for cell differentiation express the Z2 isoform; and the central nervous system (which undergoes marked remodeling during metamorphosis) expresses all isoforms, with Z3 predominating (Emery et al. 1994; Crossgrove et al. 1996).

In addition to the restricted activities of the different isoforms of BR-C, there appear to be common processes that all of the isoforms accomplish. Restifo and Wilson (1998) provided evidence that these common functions are prevented by juvenile hormone. Deletions of the *BR-C* gene lead to faulty muscle development, retention of larval structures that would normally degenerate, abnormal nervous system morphology, and eventually the death of the larva (Restifo and White 1991). This syndrome is very similar to that induced by adding excess JH to *Cecropia* silkworm larvae (Riddiford 1972) or by adding juvenile hormone analogues to *Drosophila*. Thus, it appears that juvenile hormone prevents ecdysone-inducible gene expression by interfering with the BR-C proteins.

The BR-C proteins are themselves transcription factors, and their targets remain to be identified. However, we are beginning to get a glimpse at the molecular level of one of the most basic areas of all developmental biology—the transformation of a larva into a fly, butterfly, or moth.

WEBSITE **18.4 Precocenes and synthetic JH.** Given the voracity of insect larvae, it's amazing that any plant exists. However, many plants get revenge on their predators by making compounds that alter their metamorphoses and prevent the animals from developing or reproducing.

Regeneration

Regeneration—the reactivation of development in later life to restore missing tissues—is so "unhuman" that it has been a source of fascination to humans since the beginnings of biological science. It is difficult to behold the phenomenon of limb regeneration in newts or starfish without wondering why we cannot grow back our own arms and legs. What gives salamanders this ability we so sorely lack? Experimental biology was born in the efforts of eighteenth-century naturalists to document regeneration and to answer this question. The regeneration experiments of Tremblay (hydras), Réaumur (crustaceans), and Spallanzani (salamanders) set the standard for experimental research and for the intelligent discussion of one's data (see Dinsmore 1991). Réaumur, for instance, noted that crayfish had the ability to regenerate their limbs because their limbs broke easily at the joints. Human limbs, he wrote, were not so vulnerable, so Nature provided us not with regenerable limbs, but with "a beautiful opportunity to admire her foresight." Tremblay's advice to researchers who would enter this new field is pertinent to read even today. He tells us to go directly to nature and to avoid the prejudices that our education has given us.* Moreover, "one should not become disheartened by want of success, but should try anew whatever has failed. It is even good to repeat successful experiments a number of times. All that is possible to see is not discovered, and often cannot be discovered, the first time."

More than two centuries later, we are beginning to find answers to the great problem of regeneration, and we may soon be able to alter the human body so as to permit our own limbs, nerves, and organs to regenerate. This would mean that severed limbs could be restored, that diseased organs could be removed and regrown, and that nerve cells altered by age, disease, or trauma could once again function normally. To bring these benefits to humanity, we first have to understand how regeneration occurs in those species that have this ability. Our new knowledge of the roles of paracrine factors in organ formation, and our ability to clone the genes that produce those factors, has propelled what Susan Bryant (1999) has called "a regeneration renaissance." Since "renaissance" literally means "rebirth," and since regeneration can be seen as a return to the embryonic state, the term is apt in many ways.

There are three major ways by which regeneration can occur. The first mechanism involves the dedifferentiation of adult structures to form an undifferentiated mass of cells that then becomes respecified. This type of regeneration is called **epimorphosis** and is characteristic of regenerating limbs. The second mechanism is called **morphallaxis**. Here, regeneration occurs through the repatterning of existing tissues, and there is little new growth. Such regeneration is seen in hydras. A third type of regeneration is an intermediate type, and can be thought of as **compensatory regeneration**. Here, the cells divide, but maintain their differentiated functions. They produce cells similar to themselves and do not form a mass of undifferentiated tissue. This type of regeneration is characteristic of the mammalian liver. We discussed regeneration of flatworms (Chapter 3) and of the amphibian eye (Chapter 4) earlier in the book. Here we will concentrate on salamander limb, hydras, and mammalian liver regeneration.

Epimorphic Regeneration of Salamander Limbs

When an adult salamander limb is amputated, the remaining cells are able to reconstruct a complete limb, with all its differentiated cells arranged in the proper order. In other words, the new cells construct only the missing structures and no more. For example, when a wrist is amputated, the salaman-

---

*The tradition of leaving one's books and classrooms and going directly to nature is very strong in developmental biology. There is a sign at Woods Hole Marine Biological Laboratory, the scene of some of the most important embryological research in America. The sign, attributed to Louis Agassiz, reads, "Study Nature, Not Books." It hangs at the entrance to the library.

der forms a new wrist and not a new elbow (Figure 18.25). In some way, the salamander limb "knows" where the proximal-distal axis has been severed and is able to regenerate from that point on.

Formation of the apical ectodermal cap and regeneration blastema

Salamanders accomplish this feat by dedifferentiation and respecification. Upon limb amputation, a plasma clot forms, and within 6 to 12 hours, epidermal cells from the remaining stump migrate to cover the wound surface, forming the **wound epidermis**. This single-layered structure is required for the regeneration of the limb, and it proliferates to form the **apical ectodermal cap**. Thus, in contrast to wound healing in mammals, no scar forms, and the dermis does not move with the epidermis to cover the site of amputation. The nerves innervating the limb degenerate for a short distance proximal to the plane of amputation (see Chernoff and Stocum 1995). During the next 4 days, the cells beneath the developing cap undergo a dramatic dedifferentiation: bone cells, cartilage cells, fibroblasts, myocytes, and neural cells lose their differentiated characteristics and become detached from one another. Genes that are expressed in differentiated tissues (such as the *MRF4* and *myf5* genes expressed in the muscle cells) are downregulated, while there is a dramatic increase in the ex-

Figure 18.25
Regeneration of a salamander forelimb. The amputation shown on the left was made below the elbow; the amputation shown on the right cut through the humerus. In both cases, the correct positional information is respecified. (From Goss 1969; photograph courtesy of R. J. Goss.)

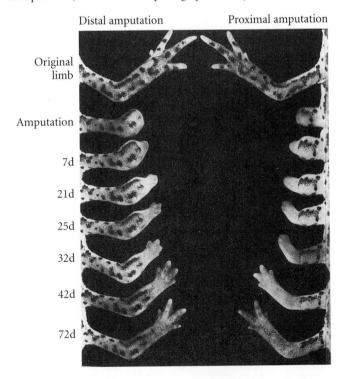

Distal amputation   Proximal amputation

Original limb

Amputation

7d

21d

25d

32d

42d

72d

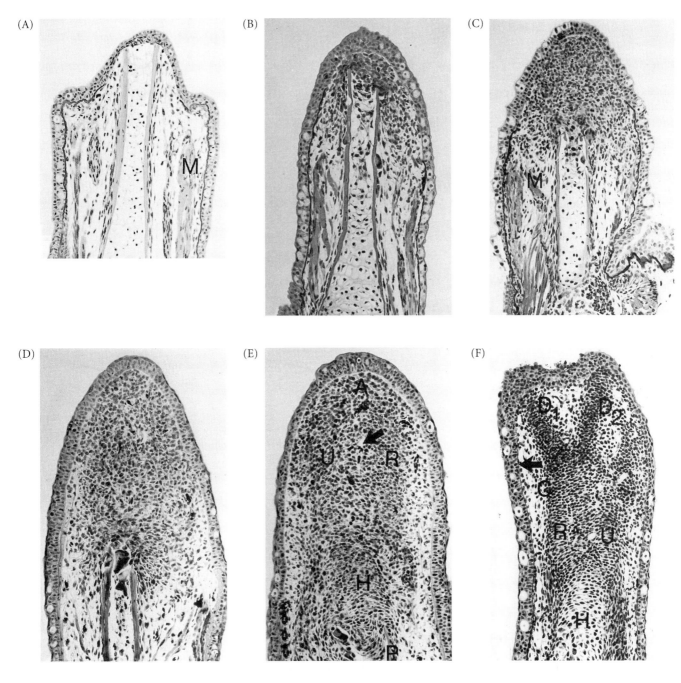

Figure 18.26
Regeneration in the larval forelimb of the spotted salamander *Ambystoma maculatum*. (A) Longitudinal section of the upper arm, 2 days after amputation. The skin and muscle have retracted from the tip of the humerus. (B) At 5 days after amputation, a thin accumulation of blastema cells is seen beneath a thickened epidermis. (C) At 7 days, a large population of mitotically active blastema cells lies distal to the humerus. (D) At 8 days, the blastema elongates by mitotic activity; much dedifferetiation has occured. (E) At 9 days, early redifferentiation can be seen. Chondrogenesis has begun in the proximal part of the regenerating humerus, H. The letter A marks the apical mesenchyme of the blastema, and U and R are the precartilaginous condensations that will form the ulna and radius, respectively. P represents the stump where the amputation was made. (F) At 10 days after amputation, the precartilaginous condensations for the carpal bones (ankle, C), and the first two digits (D_1, D_2) can also be seen. (From Stocum 1979; photographs courtesy of D. L. Stocum.)

pression of genes, such as *msx1*, that are associated with the proliferating progress zone mesenchyme of the embryonic limb (Simon et al. 1995). The formerly well-structured limb region at the cut edge of the stump thus forms a proliferating mass of indistinguishable, dedifferentiated cells just beneath the apical ectodermal cap. This dedifferentiated cell mass is called the **regeneration blastema**. These cells will continue to proliferate, and will eventually redifferentiate to form the new structures of the limb (Figure 18.26; Butler 1935).

The creation of the blastema depends upon the formation of single, mononucleated cells. It is probable that the macrophages that are released into the wound site secrete metalloproteinases that digest the extracellular matrices holding epithelial cells together (Yang et al. 1999). But many of these cells are differentiated and have left the cell cycle. How do they regain the ability to divide? Microscopy and tracer dye studies have shown that when multinucleated myotubes (whose nuclei are removed from the cell cycle: see Chapter 14) are introduced into a blastema, they give rise to labeled mononucleated cells that proliferate and can differentiate into many tissues of the regenerated limb (Hay 1959; Lo et al. 1993). It appears that myotube nuclei are forced to enter the cell cycle by a serum factor created by thrombin, the same protease that is involved in forming clots. Thrombin is released when the amputation is made, and when serum is exposed to thrombin, it forms a factor capable of inducing cultured newt myotubes to enter the cell cycle. Mouse myotubes, however, do not respond to this chemical. This difference in responsiveness may relate directly to the difference in regenerative ability between salamanders and mammals* (Tanaka et al. 1999).

Proliferation of the blastema cells: The requirement for nerves

The proliferation of the salamander limb regeneration blastema is dependent on the presence of nerves. Singer (1954) demonstrated that a minimum number of nerve fibers must be present for regeneration to take place. It is thought that the neurons release mitosis-stimulating factors that increase the proliferation of the blastema cells (Singer and Caston 1972; Mescher and Tassava 1975). There are several candidates for these neural-derived mitotic factors, and each may be important. **Glial growth factor** (**GGF**) is known to be produced by newt neural cells, is present in the blastema, and is lost upon denervation. When this peptide is added to a denervated blastema, the mitotically arrested cells are able to divide again (Brockes and Kinter 1986). A fibroblast growth factor may also be involved. FGFs infused into denervated blastemas are able to restore mitosis (Mescher and Gospodarowicz 1979). Another important neural agent necessary for cell cycling is **transferrin**, an iron-transport protein that is necessary for mitosis in all dividing cells (since ribonucleotide reductase, the rate-limiting enzyme of DNA synthesis, re-

quires a ferric ion in its active site). When a hindlimb is severed, the sciatic nerve transports transferrin along the axon and releases large quantities of this protein into the blastema (Munaim and Mescher 1986; Mescher 1992). Neural extracts and transferrin are both able to stimulate cell division in denervated limbs, and chelation of ferric ions from a neural extract abolishes its mitotic activity (Munaim and Mescher 1986; Albert and Boilly 1988).

Pattern formation in the regeneration blastema

The regeneration blastema resembles in many ways the progress zone of the developing limb. The dorsal-ventral and anterior-posterior axes between the stump and the regenerating tissue are conserved, and cellular and molecular studies have confirmed that the patterning mechanisms of developing and regenerating limbs are very similar. By transplanting regenerating limb blastemas onto developing limb buds, Muneoka and Bryant (1982) showed that the blastema cells could respond to limb bud signals and contribute to the developing limb. At the molecular level, just as Sonic hedgehog is seen in the posterior region of the developing limb progress zone mesenchyme, it is seen in the early posterior regeneration blastema (Imokawa and Yoshizato 1997; Torok et al. 1999). The initial pattern of Hox gene expression in regenerating limbs is not the same as that in developing limbs. However, the nested pattern of *Hoxa* and *Hoxd* gene expression characteristic of limb development is established as the limb regenerates (Torok et al. 1998).

Retinoic acid appears to play an important role both in the dedifferentiation of the cells to form the regeneration blastema and in the respecification processes as the cells redifferentiate. If regenerating animals are treated with sufficient concentrations of retinoic acid (or other retinoids), their regenerated limbs will have duplications along the proximal-distal axis (Figure 18.27; Niazi and Saxena 1978; Maden 1982). This response is dose-dependent and at maximal dosage can result in a complete new limb regenerating (starting at the most proximal bone), regardless of the original level of amputation. Dosages higher than this result in inhibition of regeneration. It appears that the retinoic acid causes the cells to be respecified to a more proximal position (Figure 18.28; Crawford and Stocum 1988b; Pecorino et al. 1996).

Retinoic acid is synthesized in the regenerating limb wound epidermis and is seen to form a gradient along the proximal-distal axis of the blastema (Brockes 1992; Scadding and Maden 1994; Viviano et al. 1995). This gradient of retinoic acid is thought to activate genes differentially across the blastema, resulting in the specification of pattern in the regenerating limb. One of these retinoic acid-responsive genes is the *msx1* gene that is associated with mesenchyme proliferation (Shen et al. 1994). Another set of genes that may be respecified by retinoic acid is the *Hoxa* genes. Gardiner and colleagues (1995) have shown that the expression pattern of

---

*This thrombin-produced factor has not yet been isolated. But it is not the only difference between urodele and mammalian limbs. Another difference is that salamanders retain Hox gene expression in their appendages even when they are adults. It should be noted that, although most of the regeneration blastema comes from the dedifferentiation of the tissue at the edge of the stump, another source of cells is also possible. While multinucleated skeletal muscle tissue will dedifferentiate and supply uncommitted mononucleated cells to the blastema, the limb also contains muscle satellite cells—mononucleated cells committed to the muscle lineage—that may be used during regeneration. Thus, the regenerated limb's muscles may come from both the blastema and from these reserve cells.

(A)

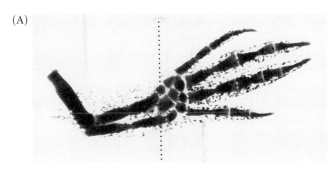

Figure 18.27
Effects of vitamin A (a retinoid) on regenerating salamander limbs. (A) Normal regenerated axolotl limb (9×) with humerus, paired radius and ulna, carpals, and digits. Dotted line shows plane of amputation through the carpal area. (B) Regeneration after amputation through the carpal area, but after the regenerating animal had been placed in retinol palmitate (vitamin A) for 15 days. A new humerus, ulna, radius, carpal set, and digit set have emerged (5×). (From Maden et al. 1982; photographs courtesy of M. Maden.)

(B)

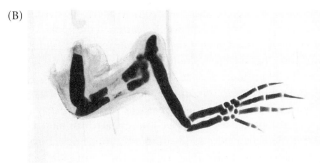

certain *Hoxa* genes in the distal cells of the regeneration blastema is changed by exogenous retinoic acid into an expression pattern characteristic of more proximal cells. It is probable that during normal regeneration, the wound epidermis/apical ectodermal cap secretes retinoic acid, which activates the genes needed for cell proliferation, downregulates the genes that are specific for differentiated cells, and activates a set of Hox genes that tells the cells where they are in the limb and how much they need to grow. The mechanism by which the Hox genes do this is not known, but changes in cell-cell adhesion and other surface qualities of the cells have been observed (Nardi and Stocum 1983; Stocum and Crawford 1987; Bryant and Gardiner 1992). Thus, in salamander limb regeneration, adult cells can go "back to the future," returning to an "embryonic" condition to begin the formation of the limb anew.

WEBSITE **18.5 The polar coordinate model**. The phenomena of epimorphic regeneration can be seen formally as events that reestablish continuity among tissues that the amputation has severed. The polar coordinate model attempts to explain the numerous phenomena of limb regeneration.

(A)

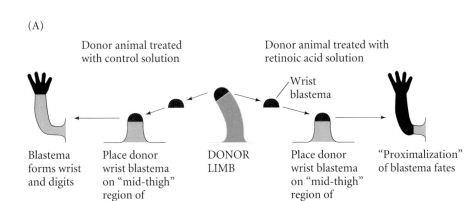

Figure 18.28
Proximalization of blastema respecification by retinoic acid. (A) When a wrist blastema from a recently cut axolotl forelimb is placed onto a host hindlimb cut at the mid-thigh level, it will generate only the wrist. The host (whose own leg was removed) will fill the gap and regenerate up to the wrist. However, if the donor animal is treated with retinoic acid, the wrist blastema will regenerate a complete limb and, when grafted, will fail to cause the host to fill the gap. (B) Wrist blastema from a darkly pigmented axolotl was treated with retinoic acid and placed onto the amputated mid-thigh region of a golden axolotl. The treated blastema regenerated a complete limb. (Data from Crawford and Stocum 1998a,b; photograph courtesy of K. Crawford.)

(B)

WEBSITE **18.6 Regeneration in annelid worms.** An easy laboratory exercise can discover the rules by which worms regenerate their segments. This website details some of those experiments.

Compensatory Regeneration in the Mammalian Liver

The liver's ability to regenerate seems to have been known since ancient times. According to Greek mythology, Prometheus's punishment (for giving civilization to humans) was to be chained to a rock and to have an eagle eat a portion of his liver each day. His liver recovered from this **partial hepatectomy** each night, thereby continually supplying food for the eagle and eternal punishment for Prometheus. The standard assay for liver regeneration is to remove (after anesthesia) specific lobes of the liver, leaving the others intact. The removed lobe does not grow back, but the remaining lobes enlarge to compensate for the loss of the missing liver tissue (Higgins and Anderson 1931). Even in humans, the amount of liver regenerated is equivalent to the amount of liver removed.

The liver regenerates by the proliferation of the existing tissues. Surprisingly, the regenerating liver cells do not fully dedifferentiate when they reenter the cell cycle. No blastema is formed. Rather, the five types of liver cells—hepatocytes, duct cells, fat-storing (Ito) cells, endothelial cells, and Kupffer macrophages—each begin dividing to produce more of themselves (Figure 18.29). Each type of cell retains its cellular identity, and the liver retains its ability to synthesize the liver-specific

Figure 18.29
Kinetics of DNA synthesis in the four major cell types of the mammalian liver. It is possible that, since the hepatocytes respond fastest, they are secreting paracrine factors that induce DNA replication in the other cells. (After Michalopoulos and DeFrances 1997.)

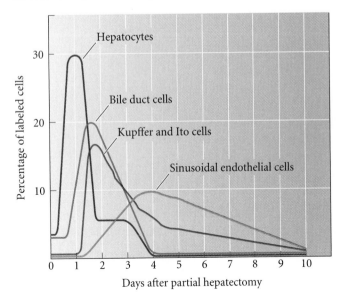

enzymes necessary for glucose regulation, toxin degradation, bile synthesis, albumin production, and other hepatic functions (Michalopoulos and DeFrances 1997).

As in the regenerating salamander limb, there is a return to an embryonic condition in the regenerating liver. Fetal transcription factors and products are made, as are the cyclins that control cell division. But the return to the embryonic state is not as complete as in the amphibian limb. Although other paracrine and endocrine factors are necessary for liver regeneration, one of the most important proteins for returning liver cells to the cell cycle is **hepatocyte growth factor** (**HGF**). This protein, also known and mentioned earlier in the book as scatter factor, induces many of the embryonic proteins. Within an hour after partial hepatectomy, the blood level of HGF has risen 20-fold (Lindroos et al. 1991). However, hepatocytes that are still connected to one another in an epithelium cannot respond to HGF. The trauma of partial hepatectomy may activate metalloproteinases that digest the extracellular matrix and permit the hepatocytes to separate and proliferate. These enzymes also may cleave HGF to its active form (Mars et al. 1995). The mechanisms by which the factors interact and by which the liver is told to stop regenerating after the appropriate size is reached remain to be discovered.

Morphallactic Regeneration in Hydras

*Hydra** is a genus of freshwater cnidarians. Most hydras are about 0.5 cm long. A hydra has a tubular body, with a "head" at its distal end and a "foot" at its proximal end. The "foot," or **basal disc**, enables the hydra to stick to rocks. The "head" consists of a conical **hypostome** region (containing the mouth) surrounded by a ring of tentacles (which catch its food). Hydras have only two epithelial cell layers, lacking a true mesoderm. They can reproduce sexually, but do so only under adverse conditions, such as severe crowding. They usually multiply by budding off a new individual (Martin 1997; Figure 18.30). The buds form about two-thirds of the way down the body axis.

When a hydra is cut in half, the half containing the head will regenerate a new basal disc, and the half containing the basal disc will regenerate a new head. Moreover, if a hydra is cut into several portions, the middle portions will regenerate both heads and basal discs at their appropriate ends. No cell division is required for this to happen, and the result is a small hydra. This regeneration is morphallactic.

THE HEAD ACTIVATOR GRADIENT. The above experiments show that every portion of the hydra along the apical-basal axis is potentially able to form a basal disc, a head, or even an

---

*The *Hydra* is another character from Greek mythology. Whenever one of this serpent's many heads was chopped off, it regenerated two new ones. Hercules finally defeated the Hydra by cauterizing the stumps of its heads with fire. Hercules had a longstanding interest in regeneration, for he was also the hero who finally freed the bound Prometheus, thus stopping his daily hepatectomies.

The head activator gradient can be measured by implanting rings of tissue from various levels of a donor hydra into a particular region of the host trunk (MacWilliams 1983b). The higher the level of head activator in the donor tissue, the greater the percentage of implants that will induce the formation of new hypostomes. The head activating factor is found to be concentrated in the head and to decrease linearly toward the basal disc.

THE HEAD INHIBITOR GRADIENT. In 1926, Rand showed that the normal regeneration of the hypostome is inhibited when an intact hypostome is grafted adjacent to the amputation site. This finding suggested that one hypostome can inhibit the formation of another. Extra heads do not form in the hydra because

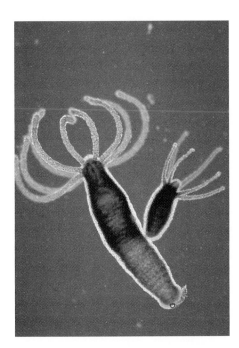

Figure 18.30
Budding in *Hydra*. A new individual buds from the right side of an adult *Hydra*. (Photograph © Biophoto/Photo Researchers Inc.)

entire hydra. However, the polarity of the hydra is coordinated by a series of morphogenetic gradients that permit the head to form only at one place and the disc to form only at another. Evidence for such gradients in hydras was first obtained by grafting experiments begun by Ethel Browne in the early 1900s. When hypostome tissue from one hydra is transplanted into the middle of another hydra, it forms a new apical-basal axis, with the hypostome extending outward (Figure 18.31A). When a basal disc is grafted to the middle of a host hydra, a new axis also forms, but with the opposite polarity, extending a basal disc (Figure 18.31B). When cells from both ends are transplanted simultaneously into the middle of a host, no new axis is formed, or the new axis has little polarity (Figure 18.31C; Browne 1909; Newman 1974). These experiments have been interpreted to indicate the existence of a head activator gradient (highest at the hypostome) and a basal activator gradient (highest at the basal disc).

Figure 18.31
Grafting experiments showing different morphogenetic capabilities in different regions of the *Hydra* axis. (A) Hypostome tissue grafted into host trunk induces a secondary axis with an extended hypostome. (B) Basal disc tissue grafted into host trunk induces a secondary axis with an extended basal disc. (C) If both hypostome and basal disc tissues are transplanted together, only weak inductions, if any, are seen. (After Newman 1974.)

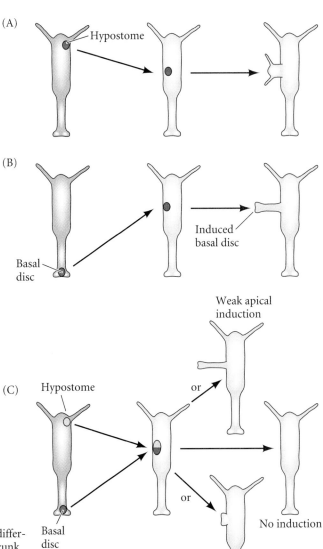

Figure 18.32
Grafting experiments providing evidence for a gradient of head inhibitor. (A) Subhypostomal tissue does not generate a new head when placed close to an existing head. (B) Subhypostomal tissue forms a head if the existing one is removed. A head will also form at the site where the host's head was amputated. (C) Subhypostomal tissue induces a new head when placed far away from the existing head. (After Newman 1974.)

(A) Intact host: No secondary axis induced

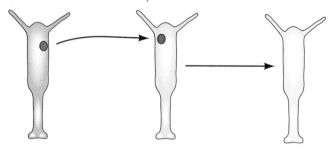

(B) Host's head removed: Secondary axis induced

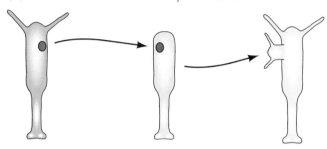

(C) Intact host: Graft away from head region induces secondary axis

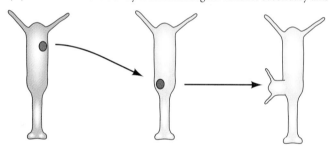

the presence of the hypostome prevents the formation of any other hypostome. The head inhibitor gradient can be measured by inserting a subhypostomal region (a region just below the hypostome, having a relatively high concentration of head activator) into various regions along the trunks of host hydras. This region will not produce a head when implanted into the apical area of an intact host hydra (Figure 18.32). However, it will form a head if the host's head has been removed. Moreover, this subhypostomal region will induce a head if placed lower on the host. Thus, there appears to be a gradient of head inhibitor as well as head activator (Wilby and Webster 1970; MacWilliams 1983a).

BASAL DISC ACTIVATOR AND INHIBITOR GRADIENTS. The basal disc also has properties suggesting that it is the source of a foot inhibitor gradient and a foot activator gradient (Hicklin and Wolpert 1973; Schmidt and Schaller 1976; Meinhardt 1993; Grens et al. 1999). The inhibitor gradients for the head and the foot may be important for determining where and when a bud can form. In young adult hydras, the gradients of head and foot inhibitors appear to block bud formation. However, as the hydra grows, the sources of these labile substances grow farther apart, creating a region of tissue, about two-thirds down the trunk, where both inhibitor gradients are minimal. Here is where the bud forms (Figure 18.33; Shostak 1974; Bode and Bode 1984; Schiliro et al. 1999). Certain mutants of *Hydra* have defects in their ability to form buds, and these defects can be explained by alterations of the morphogen gradi-

ents. The L4 mutant of *Hydra magnipapillata*, for instance, forms buds very slowly, and only after reaching a size about twice as long as wild-type individuals. The amount of head inhibitory substance in these mutants was found to be much greater than in wild-type *Hydra* (Takano and Sugiyama 1983).

The inhibitor and activator gradients also inform the hydra "which end is up" and specify positional values along the apical-basal axis. When the head is removed, the head inhibitor no longer is made, and this causes the head activator to induce a new head. The region with the most head activator will form the head. Once the head is made, it makes the head inhibitor, and the equilibrium is restored.

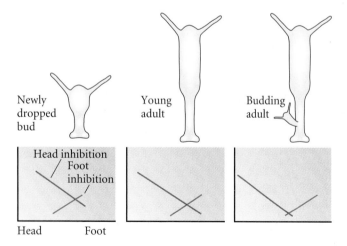

Newly dropped bud

Young adult

Budding adult

Head inhibition / Foot inhibition /

Head   Foot

Figure 18.33
Head inhibitor (blue line) and foot inhibitor (red line) gradients in newly dropped hydra buds, young adults, and budding adults. (After Bode and Bode 1984.)

Sidelights & Speculations

Medical Advances in Regeneration

HUMANS DO NOT REGENERATE their organs. Children can regenerate their fingertips, but even this ability is lost in adults. The ability to regenerate damaged human organs would constitute a medical revolution. Researchers are attempting to find ways of activating in the adult developmental programs that were used during organogenesis. One avenue of this research program is a search for stem cells that are still relatively undifferentiated, but which can form particular cell types if given the appropriate environment (see Chapters 4 and 12). The second avenue is a search for the environments that will allow such cells to initiate cell and organ formation. We will look at two of these medical efforts.

Bone Regeneration

While fractured bones can heal, bone cells in adults usually do not regrow to bridge wide gaps. The finding (Vortkamp et al. 1998) that the same paracrine and endocrine factors involved in endochondral ossification are also involved in fracture repair raises the possibility that new bone could grow if the proper paracrine and extracellular environment were provided. The problem is how to deliver these compounds to a particular location over a long period of time. One solution to the problem of delivery was devised by Bonadio and his colleagues (1999). They developed a collagen gel containing plasmids carrying the human parathyroid hormone gene. The plasmid-impregnated gel was placed between the gaps of a broken dog tibia or femur. As cells migrated into the collagen matrix, they incorporated the plasmid and made parathyroid hormone. A dose-dependent increase in new bone formation was seen in about a month (Figure 18.34). This type of treatment has potential to help people with large bone fractures as well as those with osteoporosis.

Neural Regeneration

While the central nervous system is characterized by its ability to change and make new connections, it has very little regenerative capacity. However, the peripheral nervous system, especially the motor neurons, has signif-

icant regenerative powers, even in adult mammals. The regeneration of motor neurons involves regrowing a severed axon, not replacing a missing or diseased cell body. If the cell body of a motor neuron is destroyed, it cannot be replaced. The myelin sheath that covers the axon is necessary for its regeneration. This

sheath is made by the Schwann cells, a type of glial cell in the peripheral nervous system (see Chapter 12). When an axon is severed, the Schwann cells divide to form a pathway along which the axon can grow from the proximal stump. This proliferation of the Schwann cells is critical for directing the regenerating axon

(A) Treated fracture

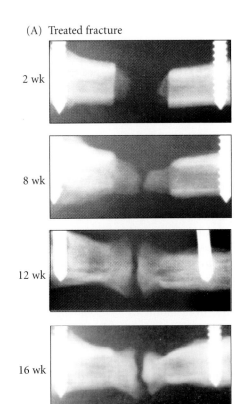

2 wk

8 wk

12 wk

16 wk

18 wk

(B) Untreated fracture
24 wk

(C) Whole bone
53 wk

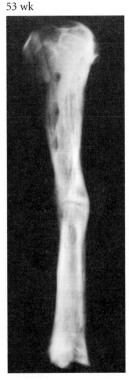

Figure 18.34
Bone formation from collagen matrix containing plasmids bearing the human parathyroid hormone. (A) A 1.6 cm gap was made in a dog femur and stabilized with screws. Plasmid-containing gel was placed on the edges of the break. X-rays of the area at 2, 8, 12, 16, and 18 weeks after the surgery show the formation of bone bridging the gap at 18 weeks. (B) Control fracture (no plasmid in the gel) at 24 weeks. (C) Whole bone a year after surgery, showing repaired region. (After Bonadio et al. 1999; photographs courtesy of J. Bonadio.)

to the original Schwann cell basement membrane. If the regrowing axon can find that basement membrane, it can be guided to its target and restore the original connection. The regenerating neuron secretes mitogens that allow the Schwann cells to divide. Some of these mitogens are specific to the developing or regenerating nervous system (Livesey et al. 1997).

The neurons of the central nervous system cannot regenerate their axons under normal conditions. Thus, spinal cord injuries can cause permanent paralysis. As mentioned in Chapter 12, one strategy to get around this block is to find ways of enlarging the population of neural stem cells and to direct their development in ways that circumvent the lesions caused by disease or trauma. The neural stem cells found in adult mammals may be very similar to embryonic neural stem cells and can respond to the same growth factors (Johe et al. 1996; Johansson et al. 1999). Another strategy for CNS neural regeneration is to create environments that encourage axonal growth. Unlike the Schwann cell of the peripheral nervous system, the myelinating cell of the central nervous system, the oligodendrocyte, produces substances that inhibit axon regeneration (Schwab and Caroni 1988). Schwann cells transplanted from the peripheral nervous system into a CNS lesion are able to encourage the growth of CNS axons to their targets (Keirstad et al. 1999; Weidner et al. 1999).

Two substances that are inhibitory to neuron outgrowth have been isolated from oligodendrocyte myelin. The first is myelin-associated glycoprotein; the second is Nogo-1 (Mukhopadyay et al. 1994; Chen et al. 2000; GrandPré et al 2000). Antibodies against Nogo-1 allow partial regeneration after spinal cord injury (Schnell and Schwab 1990).

Research into CNS axon regeneration may become one of the most important contributions of developmental biology to medicine. ■

Aging: The Biology of Senescence

Entropy always wins. Each multicellular organism, using energy from the sun, is able to develop and maintain its identity for only so long. Then deterioration prevails over synthesis, and the organism ages. **Aging** can be defined as the time-related deterioration of the physiological functions necessary for survival and fertility. The characteristics of aging—as distinguished from diseases of aging (such as cancer and heart disease)—affect all the individuals of a species.

Many evolutionary biologists (Medawar 1952; Kirkwood 1977) would deny that aging is part of the genetic repertoire of an animal. Rather, they would consider aging to be the default state occurring after the animal has fulfilled the requirements of natural selection. After its offspring are born and raised, the animal can die. Indeed, in many organisms, from moths to salmon, this is exactly what happens. As soon as the eggs are fertilized and laid, the adults die. However, recent studies have indicated that there are genetic components to senescence, and that the genetically determined life span characteristic of a species can be modulated by altering genes or diet.

Maximum Life Span and Life Expectancy

The **maximum life span** is a characteristic of the species. It is the maximum number of years a member of that species has been known to survive. The maximum human life span is estimated to be 121 years (Arking 1998). The life spans of tortoises and lake trout are both unknown, but are estimated to be more than 150 years. The maximum life span of a domestic dog is about 20 years, and that of a laboratory mouse is 4.5 years. If a *Drosophila* fruit fly survives to eclose (in the wild, over 90% die as larvae), it has a maximum life span of 3 months.

However, a person cannot expect to live 121 years, and most mice in the wild do not live to celebrate their first birthday. The **life expectancy**, the amount of time a member of a species can expect to live, is not characteristic of species, but of populations. It is usually defined as the age at which half the population still survives. A baby born in England in the 1780s could expect to live to be 35 years old. In Massachusetts during that same time, the life expectancy was 28 years. This was the normal range of human life expectancy for most of the human race in most times. Even today, the life expectancy in some areas of the world (Cambodia, Togo, Afghanistan, and several other countries) is less than 40 years. In the United States, a child born in 1986 can expect to live 71 years if male and 78 years if female.*

Given that in most times and places, humans did not live much past 40 years, our awareness of human aging is relatively new. A 65-year-old person was rare in colonial America, but is a common sight today. Some survival curves for female *Homo sapiens* in the United States are plotted in Figure 18.35. In 1900, 50% of American women were dead by age 58. In 1980, 50% of American women were dead by age 81. Thus, the phenomena of senescence and the diseases of aging are much more common today than they were a century ago. In 1900, people did not have the "luxury" of dying from heart attacks or cancers. These diseases generally occur in people over the age of 50 years. Rather, people died (as they are still dying in many parts of the world) from infectious diseases and parasites (Arking 1998). Similarly, until recently, relatively few people exhibited the more general human sensescent phenotype: graying hair, sagging and wrinkling skin, joint stiffness, osteoporosis (loss of bone calcium), loss of muscle fibers and muscular strength, memory loss, eyesight deterioration, and

---

*You can see why the funding of Social Security is problematic in the United States. When it was created in 1935, the average working citizen died before age 65. Thus, he (and it usually was a he) was not expected to get back what he had paid into the system. Similarly, marriage "until death do us part" was an easier feat when death occurred in the third or fourth decade of life. The death rate of young women due to infections associated with childbirth was high throughout the world before antibiotics.

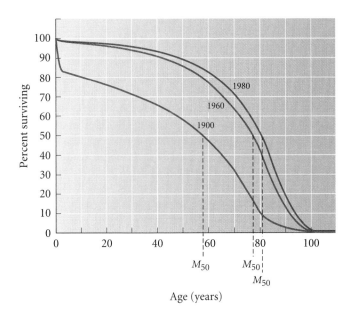

Figure 18.35
Survival curves for U.S. females in 1900, 1960, and 1980. M_{50} represents the age at which 50% of the individuals of each population survived. (After Arking 1998.)

the slowing of sexual responsiveness. As Shakespeare noted in *As You Like It*, those who did survive to senescence left the world "sans teeth, sans eyes, sans taste, sans everything."

Causes of Aging

The general senescent phenotype is characteristic of each species. But what causes it? This question can be asked at many levels. We will be looking primarily at the cellular level of organization. Even here, there is evidence for many different theories, and there is not yet a consensus on what causes aging.

Oxidative damage

One major theory sees our metabolism as the cause of our aging. According to this theory, aging is a by-product of normal metabolism; no mutations are required. About 2–3% of the oxygen atoms taken up by the mitochondria are reduced insufficiently to **reactive oxygen species** (**ROS**). These ROS include the superoxide ion, the hydroxyl radical, and hydrogen peroxide. ROS can oxidize and damage cell membranes, proteins, and nucleic acids. Evidence for this theory includes the observation that *Drosophila* that overexpress enzymes that destroy ROS (catalase, which degrades peroxide, and superoxide dismutase) live 30–40% longer than do controls (Orr and Sohal 1994; Parkes et al. 1998). Moreover, flies with mutations in the *methuselah* gene (named after the Biblical fellow said to have lived 969 years) live 35% longer than wild-type flies. The *methusaleh* mutants have enhanced resistance to paraquat, a poison that works by generating ROS within cells (Lin et al. 1998). These findings not only suggest that aging is under ge-

netic control, but also provide evidence for the role of ROS in the aging process. In *C. elegans*, too, individuals with mutations that increase the synthesis of ROS-degrading enzymes live much longer than wild-type nematodes (Larsen 1993; Vanfleteren and De Vreese 1996).

The evidence for ROS involvement in mammalian aging is not as clear. Mutations in mice that result in the lack of certain ROS-degrading enzymes do not cause premature aging (Ho et al. 1997; Melov et al. 1998). However, there may be more genetic redundancy in mammals than in invertebrates, and other genes may be up-regulated to produce related ROS-degrading enzymes. Migliaccio and colleagues (1999) have observed mutant mice that live one-third longer than their wild-type littermates. These mice lack a particular protein, p66[shc]. They develop normally, but the lack of p66[shc] apparently gives them cellular resistance to ROS, and thus higher resistance to oxygen-induced stress on membranes and proteins. The p66[shc] protein may be a component of a signal transduction pathway that leads to apoptosis upon oxygen stress, and it may be involved in mediating the life spans of mammals.

Another type of evidence does suggest that ROS may be important in mammalian aging: aging in mammals can be slowed by caloric restriction (Lee et al. 1999). However, caloric restriction can also have other effects, so it is not certain if it works by preventing ROS synthesis. Also, vitamins E and C are both ROS inhibitors, and vitamin E increases the longevity of flies and nematodes when it is added to their diet (Balin et al. 1983; Kakkar et al. 1996). However, results in mammals are not as easy to interpret, and there is no clear evidence that ROS inhibitors work as well as in invertebrates (Arking 1998).

General wear-and-tear and genetic instability

"Wear-and-tear" theories of aging are among the oldest hypotheses proposed to account for the general scenescent phenotype (Weismann 1891; Szilard 1959). As one gets older, small traumas to the body build up. Point mutations increase in number, and the efficiencies of the enzymes encoded by our genes decrease. Moreover, if a mutation occured in a part of the protein synthetic apparatus, the cell would make a large percentage of faulty proteins (Orgel 1963). If mutations arose in the DNA-synthesizing enzymes, the rate of mutations would be expected to increase markedly, and Murray and Holliday (1981) have documented such faulty DNA polymerases in senescent cells. Likewise, DNA repair may be important in preventing senescence, and species whose members' cells have more efficient DNA repair enzymes live longer (Figure 18.36; Hart and Setlow 1974). Moreover, genetic defects in DNA repair enzymes can produce premature aging syndromes in humans (Yu et al. 1996; Sun et al. 1998).

MITOCHONDRIAL GENOME DAMAGE. The mutation rate in mitochondria is 10–20 times faster than the nuclear DNA mutation rate (Johnson et al. 1999). It is thought that mutations in mitochondria could (1) lead to defects in energy produc-

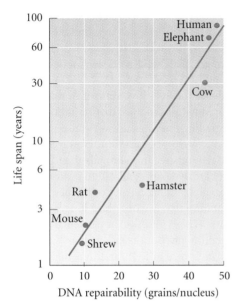

Figure 18.36
Correlation between life span and the ability of fibroblasts from various mammalian species to repair DNA. Capacity for repair is represented in autoradiography by the number of grains from radioactive thymidine per cell nucleus. Note that the *y* axis (life span) is a logarithmic scale. (After Hart and Setlow 1974.)

tion, (2) lead to the production of ROS by faulty electron transport, and/or (3) induce apoptosis. Age-dependent declines in mitochondrial function are seen in many animals, including humans (Boffoli et al. 1994). A recent report (Michikawa et al. 1999) shows that there are "hot spots" for age-related mutations in the mitochondrial genome, and that mitochondria with these mutations have a higher replication frequency than wild-type mitochondria. Thus, the mutants are able to outcompete the wild-type mitochondria and eventually dominate the cell and its progeny. Moreover, the mutations may not only allow more ROS to be made, but may make the mitochondrial DNA more susceptible to ROS-mediated damage.

TELOMERE SHORTENING. Telomeres are repeated DNA sequences at the ends of chromosomes. They are not replicated by DNA polymerase, and they will shorten at each cell division unless maintained by **telomerase**. Telomerase adds the telomere onto the chromosome at each cell division. Most mammalian somatic tissues lack telomerase, so it has been proposed (Salk 1982; Harley et al. 1990) that telomere shortening could be a "clock" that eventually prohibits the cells from dividing any more. When human fibroblasts are cultured, they can divide only a certain number of times, and their telomeres shorten. If these cells are made to express telomerase, they can continue dividing (Bodnar et al. 1998; Vaziri and Benchimol 1998).

However, there is no correlation between telomere length and the life span of an animal (humans have much shorter telomeres than mice), nor is there a correlation between human telomere length and a person's age (Cristofalo et al. 1998). Telomerase-deficient mice do not show profound aging defects, which we would expect if telomerase were the major factor in determining the rate of aging (Rudolph et al. 1999). It has been suggested that telomere-dependent inhibition of cell division might serve primarily as a defense against cancer rather than as a kind of "aging clock."

GENETIC AGING PROGRAMS. Several genes have been shown to affect aging. In humans, Hutchinson-Gilford progeria syndrome causes children to age rapidly and to die (usually of heart failure) as early as 12 years (Figure 18.37). It is caused by a dominant mutant gene, and its symptoms include thin skin with age spots, resorbed bone mass, hair loss, and arteriosclerosis. A similar syndrome is caused by mutations of the *klotho* gene in mice (Kuro-o et al. 1997). The functions of the products of these genes are not known, but they are thought to be involved in suppressing the aging phenotypes. These proteins may be extremely important in determining the timing of senescence.

In *C. elegans*, there appear to be at least two genetic pathways that affect aging. The first pathway involves the decision to remain a larva or to continue growth. After hatching, the *C. elegans* larva proceeds through four instar stages, after which it can become an adult or (if the nematodes are overcrowded or if there is insufficient food) can enter a nonfeeding, metabolically dormant **dauer stage**. It can remain a dauer larva for

Figure 18.37
Children with progeria. Although less than 8 years old, the child on the right has a phenotype similar to that of an aged person. The hair loss, fat distribution, and transparency of the skin are characteristic of the normal human aging pattern seen in elderly adults. (Photograph © Associated Press.)

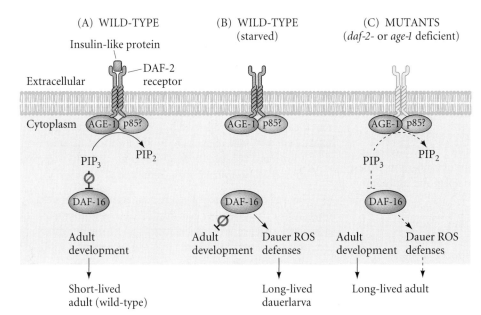

Figure 18.38
Proposed mechanism for extending the life span of *C. elegans* through the dauer larva pathway. (A) Wild-type animal in a favorable environment makes a ligand that activates a pathway that inhibits the DAF-16 protein. This allows metamorphosis to the adult stage. (B) If a wild-type larva is starved, the ligand is not made, and DAF-16 creates a long-lived dauer larva by repressing the pathway to adult development and by increasing the ROS-degrading enzymes. (C) In mutants deficient in *age-1* (or *daf-2*), DAF-16 is only partially inhibited. Adult development is permitted, and the ROS-degrading enzymes are also made. This creates a long-lived adult. (After Johnson et al. 1999.)

up to 6 months, rather than becoming an adult that lives only a few weeks. When it comes out of the dauer stage, it will live as long as if it had never been a dauer larva. In the dauer stage, adult development is suppressed, and extra defenses against ROS are synthesized. If some of the genes involved in this pathway are mutated, adult development is allowed, but the ROS defenses are still made. The resulting adults live twice to four times as long as wild-type adults (Figure 18.38; Friedman and Johnson 1988). The second pathway involves the gonads.

Germ cells appear to inhibit longevity, while the somatic cells of the gonads act to prolong the life of the nematode (Hsin and Kenyon 1999).

As human life expectancy increases due to our increased ability to prevent and cure disease, we are still left with a general aging syndrome that is characteristic of our species. Unless attention is paid to the general aging syndrome, we risk ending up like Tithonios, the miserable wretch of Greek mythology to whom the gods awarded eternal life, but not eternal youth.

Snapshot Summary: Metamorphosis, Regeneration, and Aging

1. Amphibian metamorphosis includes both morphological and biochemical changes. Some structures are remodeled, some are replaced, and some new structures are formed.

2. Many changes during amphibian metamorphosis are regionally specific. The tail epidermis dies, the head epidermis does not. An eye will persist even if transplanted into a degenerating tail.

3. The hormones responsible for amphibian metamorphosis are the thyroid hormones thyroxine (T_4) and triiodothyronine (T_3). The coordination of metamorphic changes appears to be due to early changes that occur at low concentrations of the thyroid hormones. This is called the threshold concept. The molecular basis for the autoinduction of thyroid hormones may be the ability of thyroid hormones to induce production of more thyroid hormone receptor protein. Thyroid hormones act predominantly at the transcriptional level.

4. Heterochrony involves changing the relative rate of development in different parts of the animal. In animals with direct development, the tadpole stage has been lost. Some frogs, for instance, form limbs while in the egg.

5. In neoteny, the juvenile (larval) form is slowed down, while the gonads and germ cells mature at their normal rate. In progenesis, the gonads and germ cells mature rapidly, while the rest of the body matures normally. In both instances, the animal can mate while in its larval form.

6. In ametabolous insects, there is direct development. In hemimetabolous insects, there is a nymph stage wherein the immature organism is usually a smaller version of the adult. In holometabolous insects, there is a dramatic metamorphosis from larva to pupa to sexually mature adult.

7. In the period between larval molts, the larva is called an instar. After the last instar stage, the larva undergoes a metamorphic molt to become a pupa. The pupa will undergo an instar molt to become an adult.

8. During the pupal stage, the imaginal discs and histoblasts grow and differentiate to produce the structures of the adult body.

9. The anterior-posterior, dorsal-ventral, and proximal-distal axes are sequentially specified and involve interactions between different compartments in the imaginal discs.

10. Molting is caused by the hormone hydroxyecdysone. In the presence of high titres of juvenile hormone, the molt is an instar molt. In low concentrations of juvenile hormone, the molt produces a pupa; and if no juvenile hormone is present, the molt is an imaginal molt.

11. The ecdysone receptor gene can produce nRNA that can form at least three different proteins. The types of ecdysone receptors in a cell may influence the response of that cell to hydroxyecdysone. The ecdysone receptors bind to DNA to activate or repress transcription.

12. There are three major types of regeneration. In epimorphosis (such as regenerating limbs), tissues dedifferentiate into a blastema, divide, and re-differentiate into the new structure. In morphallaxis (characteristic of hydra), there is a repatterning of existing tissue with little or no growth. In compensatory regeneration (such as in the liver), cells divide but retain their differentiated state.

13. In the regenerating salamander limb, the epidermis forms an apical ectodermal cap. The cells beneath it dedifferentiate to form a blastema. The differentiated cells lose their adhesions and re-enter the cell cycle. This does not happen in mammals.

14. In hydras, there appear to be head activation gradients, head inhibition gradients, foot activation gradients, and foot inhibition gradients. Hydra budding occurs where these gradients are minimal.

15. In mammals, medical researchers are testing whether paracrine factors may permit local regeneration. Bone and neural cells are being returned to embryonic conditions in the hopes that they will regrow. Natural inhibitors of neural regeneration have recently been discovered, and their circumvention may allow spinal cord regeneration.

16. The maximum life span of a species is how long its longest observed member has lived. It is largely characteristic of a given species. Life expectancy is the time at which approximately 50 percent of the members of a given population of a species still survive.

17. There are several levels at which we can study aging, including cellular, biochemical, and genetic studies. Reactive oxygen species (ROS) can damage cell membranes, inactivate proteins, and mutate DNA. Mutations that alter the ability to make or degrade ROS can change the life span of the mutants.

18. Mitochondria may be a target for proteins that regulate aging.

19. Aging is the time-related deterioration of the physiological functions necessary for survival and reproduction. The phenotypic changes of senescence (which affect all members of the species) are not to be confused with diseases of senescence, such as cancer and heart disease (which affect individuals).

Literature Cited

Abu-Shaar, M. and R. S. Mann. 1998. Generation of multiple antagonistic domains along the proximodistal axis during *Drosophila* leg development. *Development* 125: 3821–3830.

Alberch, P. and J. Alberch. 1981. Heterochronic mechanisms of morphological diversification and evolutionary change in the neotropical salamander *Bolitoglossa occidentalis* (Amphibia: Plethodontidae). *J. Morphol.* 167: 249–264.

Albert, P. and B. Boilly. 1988. Effect of transferrin on amphibian limb regeneration: A blastema cell culture study. *Roux Arch. Dev. Biol.* 197: 193–196.

Allen, B. M. 1916. Extirpation experiments in *Rana pipiens* larva. *Science* 44: 755–757.

Alley, K. E. and M. D. Barnes. 1983. Birthdates of trigeminal motor neurons and metamorphic reorganization of the jaw myoneural system in frogs. *J. Comp. Neurol.* 218: 395–405.

Arking, R. 1998. *The Biology of Aging*, 2nd Ed. Sinauer Associates, Sunderland, MA.

Ashburner, M. 1972. Patterns of puffing activity in the salivary glands of *Drosophila*. VI. Induction by ecdysone in salivary glands of *D. melanogaster* cultured in vitro. *Chromosoma* 38: 255–281.

Ashburner, M. 1974. Sequential gene activation by ecdysone in polytene chromosomes of *Drosophila melanogaster*. II. Effects of inhibitors of protein synthesis. *Dev. Biol.* 39: 141–157.

Ashburner, M. 1990. Puffs, genes, and hormones revisited. *Cell* 61: 1–3.

Ashburner, M. and H. D. Berondes. 1978. Puffing of polytene chromosomes. *In The Genetics and Biology of* Drosophila, Vol. 2B. Academic Press, New York, pp. 316–395.

Baker, B. S. and J. R. Tata. 1992. Prolactin prevents the autoinduction of thyroid hormone receptor mRNAs during amphibian metamorphosis. *Dev. Biol.* 149: 463–467.

Balin, A. K. 1993. Testing the free radical theory of aging. *In* R. C. Adelman and G. C. Roth (eds.) *Testing the Theories of Aging*. Van Nostrand Reinhold, New York, pp. 137–182.

Bern, H. A., C. S. Nicoll and R. C. Strohman. 1967. Prolactin and tadpole growth. *Proc. Soc. Exp. Biol. Med.* 126: 518–521.

Blair, S. S. 1993. Mechanisms of compartment formation: Evidence that non-proliferating cells do not play a role in defining the D/V lineage restriction in the developing wing of *Drosophila*. *Development* 119: 339–351.

Bode, P. M. and H. R. Bode. 1984. Patterning in *Hydra*. *In* G. M. Malacinski and S. V. Bryant (eds.), *Pattern Formation*. Macmillan, New York, pp. 213–241.

Bodnar, A. G. and 9 others. 1998. Extension of lifespan by introduction of telomerase into normal human cells. *Science* 279: 349–352.

Boffoli, D., S. C. Sacco, R. Vergari, G. Solarino, G. Santacroce and S. Papa. 1994. Decline with age of the respiratory chain activity in human skeletal muscle. *Biochim. Biophys. Acta* 1226: 73–86.

Bonadio, J., E. Smiley, P. Patil and S. Goldstein. 1999. Localized, direct plasmid gene delivery in vivo: Prolonged therapy results in reproducible tissue regeneration. *Nature Med.* 5: 753–759.

Brenner, S. 1996. Francisco Crick in Paradiso. *Curr. Biol.* 6: 1202.

Brockes, J. P. 1992. Introduction of a retinoid reporter gene into the urodele limb blastema. *Proc. Natl. Acad. Sci. USA* 89: 11386–11390.

Brockes, J. P. and C. R. Kinter. 1986. Glial growth factor and nerve-dependent proliferation in the regeneration blastema of urodele amphibians. *Cell* 45: 301–306.

Browne, E. N. 1909. The production of small hydranths in hydra by insertion of small grafts. *J. Exp. Zool.* 7: 1–23.

Brust, D. G. 1993. Maternal brood care by *Dendrobates pumilio*: A frog that feeds its young. *J. Herpetol.* 26: 102–105.

Bryant, P. J. 1970. Cell lineage relationships in the imaginal wing disc of *Drosophila melanogaster*. *Dev. Biol.* 22: 389–411.

Bryant, S. 1999. A regeneration renaissance. *Semin. Cell Dev. Biol.* 10: 313.

Bryant, S. V. and D. M. Gardiner. 1992. Retinoic acid, local cell-cell interactions, and pattern formation in vertebrate limbs. *Dev. Biol.* 152: 1–25.

Buckbinder, L. and D. D. Brown. 1993. Expression of the *Xenopus* prolactin and thyrotropin genes during metamorphosis. *Proc. Natl. Acad. Sci. USA* 90: 3820–3824.

Butenandt, A. and P. Karlson. 1954. Über die Isolierung eines Metamorphosen-Hormons der Insekten in kristallisierter Form. *Z. Naturforsch.* B 9: 389–391.

Butler, E. G. 1935. Studies on limb regeneration in X-rayed *Ambystoma* larvae. *Anat. Rec.* 62: 295–307.

Causo, J. P., M. Bate and A. Martinez-Arias. 1993. A *wingless*-dependent polar coordinate system in *Drosophila* imaginal discs. *Science* 259: 484–489.

Chen, M. S. and 7 others. 2000. Nogo-A is a myelin-associated neurite outgrowth inhibitor and an antigen for monoclonal antibody IN-1. *Nature* 403: 434–439.

Chernoff, E. A. G. and D. Stocum. 1995. Developmental aspects of spinal cord and limb regeneration. *Dev. Growth Diff.* 37: 133–147.

Clever, U. 1966. Induction and repression of a puff in *Chironomus tentans*. *Dev. Biol.* 14: 421–438.

Cohen, B., A. A. Simcox and S. M. Cohen. 1993. Allocation of the thoracic imaginal primordia in the *Drosophila* embryo. *Development* 117: 597–608.

Cohen, P. P. 1970. Biochemical differentiation during amphibian metamorphosis. *Science* 168: 533–543.

Condic, M. L., D. Fristrom and J. W. Fristrom. 1990. Apical cell shape changes during *Drosophila* imaginal leg disc elongation: A novel morphogenetic mechanism. *Development* 111: 23–33.

Corben, C. J., M. J. Ingram and M. J. Tyler. 1974. Gastric brooding: Unique form of parental care in an Australian frog. *Science* 186: 946–947.

Crawford, K. and D. L. Stocum. 1988a. Retinoic acid coordinately proximalizes regenerate pattern and blastema differential affinity in axolotl limbs. *Development* 102: 687–698.

Crawford, K. and D. L. Stocum. 1988b. Retinoic acid proximalizes level-specific properties responsible for intercalary regeneration in axolotl limbs. *Development* 104: 703–712.

Cristofalo, V. J., R. G. Allen, R. J. Pignolo, B. G. Martin and J. C. Beck. 1998. Relationship between donor age and relicative life span of human cells in culture: A reevaluation. *Proc. Natl. Acad. Sci. USA* 95: 10614–10619.

Crossgrove, K., C. A. Bayer, J. W. Fristrom and G. M. Guild. 1996. The *Drosophila* Broad Complex early gene directly regulates late gene transcription during the ecdysone-induced puffing cascade. *Dev. Biol.* 180: 745–758.

Currie, J. and W. M. Cowan. 1974. Evidence for the late development of the uncrossed retinothalamic projections in the frog *Rana pipiens*. *Brain Res.* 71: 133–139.

De Beer, G. 1940. *Embryos and Ancestors*. Clarendon Press, Oxford.

del Pino, E. M. and R. P. Elinson. 1983. A novel development pattern for frogs: Gastrulation produces an embryonic disk. *Nature* 306: 589–591.

Diaz-Benjumea, F. J. and S. M. Cohen. 1993. Interaction between dorsal and ventral cells in the imaginal disc directs wing development in *Drosophila*. *Cell* 75: 741–752.

Diaz-Benjumea, F. J., B. Cohen and S. M. Cohen. 1994. Cell interaction between compartments establishes the proximal-distal axis of *Drosophila* wings. *Nature* 372: 175–179.

Dinsmore, C. E. (ed.) 1991. *A History of Regeneration Research: Milestones in the Evolution of a Science*. Cambridge University Press, New York.

Elinson, R. P. 1987. Change in developmental patterns: Embryos of amphibians with large eggs. *In* R. A. Raff and E. C. Raff (eds.), *Development as an Evolutionary Process*. Alan R. Liss, New York, pp. 1–21.

Emery, I. F., V. Bedian and G. M. Guild. 1994. Differential expression of Broad-Complex transcription factors may forecast tissue-specific developmental fates during *Drosophila* metamorphosis. *Development* 120: 3275–3287.

Etkin, W. and A. G. Gona. 1967. Antagonism between prolactin and thyroid hormone in amphibian development. *J. Exp. Zool.* 165: 249–258.

Fallon, J. F. and B. K. Simandl. 1978. Evidence of a role for cell death in the disappearance of the embryonic human tail. *Am. J. Anat.* 152: 111–130.

Fox, H. 1973. Ultrastructure of tail degeneration in *Rana temporaria* larva. *Folia Morphol.* 21: 103–112.

Frieden, E. 1981. The dual role of thyroid hormones in vertebrate development and calorigenesis. *In* L. I. Gilbert and E. Frieden (eds.), *Metamorphosis: A Problem in Developmental Biology*. Plenum, New York, pp. 545–564.

Friedman, D. B. and T. E. Johnson. 1988. A mutation in the *age-1* gene in *Caenorhabditis elegans* lengthens life and reduces hermaphrodite fertility. *Genetics* 118: 75–86.

Fristrom, D. and J. W. Fristrom. 1975. The mechanisms of evagination of imaginal disks of *Drosophila melanogaster*. I. General considerations. *Dev. Biol.* 43: 1–23.

Fristrom, J. W. 1972. The biochemistry of imaginal disc development. *In* H. Ursprung and R. Nothiger (eds.), *The Biology of Imaginal Discs*. Springer-Verlag, Berlin, pp. 109–154.

Fristrom, J. W., R. Raikow, W. Petri and D. Stewart. 1969. In vitro evagination and RNA synthesis in imaginal discs of *Drosophila melanogaster*. *In* E. W. Hanley, *Problems in Biology: RNA in Development*. University of Utah Press, Salt Lake City, pp. 381–401.

Fristrom, J. W., D. Fristrom, E. Fekete and A. H. Kuniyuki. 1977. The mechanism of evagination of imaginal discs of *Drosophila melanogaster*. *Am. Zool.* 17: 671–684.

Fritzsch, B., U. Wahnschaffe and U. Bartsch. 1988. Metamorphic changes in the octavo-lateralis system of amphibians. *In* B. Fritzsch, M. Ryan, W. Wilczynski, T. Hetherington and W. Walkowiak (eds.), *The Evolution of the Amphibian Auditory System*. Wiley, Chichester, pp. 561–586.

Garcia-Bellido, A., P. Ripoll and G. Morata. 1973. Developmental compartmentalization of the wing disc of *Drosophila*. *Nature New Biol.* 245: 251–253.

Gardiner, D. M., B. Blumberg, Y. Konine and S. V. Bryant. 1995. Regulation of HoxA expression in developing and regenerating axolotl limbs. *Development* 121: 1731–1741.

Geigy, R. 1941. Die metamorphose als Folge gewebsspezifischer determination. *Rev. Suisse Zool.* 48: 483–494.

Gilbert, L. I. and W. Goodman. 1981. Chemistry, metabolism, and transport of hormones controlling insect metamorphosis. *In* L. I. Gilbert and E. Frieden (eds.), *Metamorphosis: A Problem in Developmental Biology*. Plenum, New York, pp. 139–176.

Goos, H. J. T. 1978. Hypophysiotropic centers in the brain of amphibians and fish. *Am. Zool.* 18: 401–410.

Goss, R. J. 1956. Regenerative inhibition following limb amputation and immediate insertion into the body cavity. *Anat. Rec.* 126: 15–28.

Goss, R. J. 1969. *Principles of Regeneration*. Academic Press, New York.

Gould, S. J. 1977. *Ontogeny and Phylogeny*. Harvard University Press, Cambridge, MA.

GrandPré, T., F. Nakamura, T. Vartanian and S. M. Strittmatter. 2000. Identification of the Nogo inhibitor of axon regeneration as a reticulon protein. *Nature* 403: 439–442.

Grens, A., H. Shimizu, S. A. Hoffmeister, H. R. Bode and T. Fijisawa. 1999. The novel signal peptides, pedibin and Hym-346, lower positional value thereby enhancing foot formation in hydra. *Development* 126: 517–524.

Grobstein, P. 1987. On beyond neuronal specificity: Problems in going from cells to networks and from networks to behavior. *In* P. Shinkman (ed.), *Advances in Neural and Behavioral Development*, Vol. 3. Ablex, Norwood, NJ, pp. 1–58.

Gronemeyer, H. and O. Pongs. 1980. Localization of ecdysterone on polytene chromosomes of *Drosophila melanogaster*. *Proc. Natl. Acad. Sci. USA* 77: 2108–2112.

Gudernatsch, J. F. 1912. Feeding experiments on tadpoles. I. The influence of specific organs given as food on growth and differentiation. A contribution to the knowledge of organs with internal secretion. *Wilhelm Roux Arch. Entwicklungsmech. Org.* 35: 457–483.

Guillen, I., J. L. Mullor, J. Capdevila, E. Sanchez-Herrero, G. Morata and I. Guerrero. 1995. The function of *engrailed* and the specification of *Drosophila* wing pattern. *Development* 121: 3447–3456.

Hanken, J. and B. K. Hall. 1988. Skull development during anuran metamorphosis. II. Role of thyroid hormones in osteogenesis. *Anat. Embryol.* 178: 219–227.

Harley, C. B., A. B. Futcher and C. W. Greider. 1990. Telomeres shorten during ageing of human fibroblasts. *Nature* 345: 458–460.

Hart, R. and R. B. Setlow. 1974. Correlation between deoxyribonucleic acid excision repair and life-span in a number of mammalian species. *Proc. Natl. Acad. Sci. USA* 71: 2169–2173.

Hay, E. D. 1959. Electron microscopic observations of muscle dedifferentiation in regenerating *Ambystoma* limbs. *Dev. Biol.* 1: 555–585.

Held, L. I., Jr. 1995. Axes, boundaries and coordinates: The ABCs of fly leg development. *BioEssays* 18: 721–732.

Hicklin, J. and L. Wolpert. 1973. Positional information and pattern regulation in hydra: Formation of the foot end. *J. Embryol. Exp. Morphol.* 30: 727–740.

Higgins, G. M. and R. M. Anderson. 1931. Experimental pathology of the liver. I. Restoration of the liver of the white rat following partial surgical removal. *Arch. Pathol.* 12: 186–202.

Ho, Y. S., J. L. Magnenat, R. T. Bronson, J. Cao, M. Gargano, M. Sugawara and C. D. Funk. 1997. Mice deficient in cellular glutathione peroxidase develop normally and show no increased sensitivity to hyperoxia. *J. Biol. Chem.* 272: 16644–16651.

Hoskins, S. G. and P. Grobstein. 1984. Thyroxine induces the ipsilateral retinothalamic projection in *Xenopus laevis*. *Nature* 307: 730–733.

Hoskins, S. G. and P. Grobstein. 1985a. Development of the ipsilateral retinothalamic projection in the frog *Xenopus laevis*. II. Ingrowth of optic nerve fibers and production of ipsilaterally projecting cells. *J. Neurosci.* 5: 920–929.

Hoskins, S. G. and P. Grobstein. 1985b. Development of the ipsilateral retinothalamic projection in the frog *Xenopus laevis*. III. The role of thyroxine. *J. Neurosci.* 5: 930–940.

Hsin, H. and C. Kenyon. 1999. Signals from the reproductive system regulate the lifespan of *C. elegans*. *Nature* 399: 362–365.

Huxley, J. 1920. Metamorphosis of axolotl caused by thyroid feeding. *Nature* 104: 436.

Imokawa, Y. and K. Yoshizato. 1997. Expression of *sonic hedgehog* gene in regenerating newt limb blastema recapitulates that in developing limb buds. *Proc. Natl. Acad. Sci. USA* 94: 9159–9164.

Irvine, K. D. and E. Wieschaus. 1994. *fringe*, a boundary-specific signaling molecule, mediates interactions between dorsal and ventral cells during *Drosophila* wing development. *Cell* 79: 595–606.

Johansson, C. B., S. Mamma, D. L. Clarke, M. Rislig, U. Lendahl and J. Frisén. 1999. Identification of a neural stem cell in the adult mammalian central nervous system. *Cell* 96: 25–34.

Johe, K. K., T. G. Hazel, T. Muller, M. M. Dugich-Djordjevic and R. D. G. McKay. 1996. Single factors direct the differentiation of stem cells from the fetal and adult central nervous system. *Genes Dev.* 10: 917–927.

Johnson, F. B., D. A. Sinclair and L. Guarente. 1999. Molecular biology of aging. *Cell* 96: 291–302.

Kakkar, R., J. S. Bains and S. P. Sharma. 1996. Effect of vitamin E on lifespan, malondialdehyde content, and antioxidant enzymes in aging *Zaprionus paravittiger*. *Gerontology* 42: 312–321.

Kalm, L. von, D. Fristrom and J. Fristrom. 1995. The making of a fly leg: A model for epithelial morphogenesis. *BioEssays* 17: 693–702.

Kaltenbach, J. C., A. E. Fry and V. K. Leius. 1979. Histochemical patterns in the tadpole tail during normal and thyroxine-induced metamorphosis. II. Succinic dehydrogenase, Mg- and Ca-adenosine triphosphatases, thiamine pyrophosphatase, and 5'-nucleotidase. *Gen. Comp. Endocrinol.* 38: 111–126.

Karp, G. and N. J. Berrill. 1981. *Development.* McGraw-Hill, New York.

Kawahara, A., B. S. Baker and J. R. Tata. 1991. Developmental and regional expression of thyroid hormone receptor genes during *Xenopus* metamorphosis. *Development* 112: 933–943.

Keirstead, H. S., S. V. Morgan, M. J. Wilby and J. W. Fawcett. 1999. Enhanced axonal regeneration following combined demyelination plus Schwann cell transplantation therapy in the injured adult spinal cord. *Exp. Neurol.* 159: 225–236.

Kim, J., K. D. Irvine and S. B. Carroll. 1995. Cell recognition, signal induction, and symmetrical gene activation at the dorsal-ventral boundary of the developing *Drosophila* wing. *Cell* 82: 795–802.

Kim, J., A. Sebring, J. J. Esch, M. E. Kraus, K. Vorwerk, J. Magee and S. B. Carroll. 1996. Integration of positional information and identity by *Drosophila vestigial* gene. *Nature* 382: 133–138.

Kinoshita, T., H. Takahama, F. Sasaki and K. Watanabe. 1989. Determination of cell death in the developmental process of anuran larval skin. *J. Exp. Zool.* 251: 37–46.

Kirkwood, T. B. L. 1977. The evolution of aging. *Nature* 270: 301–304.

Kistler, A., K. Yoshizato and E. Frieden. 1977. Preferential binding of tri-substituted thyronine analogs by bullfrog tadpole tail fin cytosol. *Endocrinology* 100: 134–137.

Koelle, M. R., W. S. Talbot, W. A. Segraves, M. T. Bender, P. Cherbas and D. S. Hogness. 1991. The *Drosophila EcR* gene encodes an ecdysone receptor, a new member of the steroid receptor superfamily. *Cell* 67: 59–77.

Kollros, J. J. 1961. Mechanisms of amphibian metamorphosis: Hormones. *Am. Zool.* 1: 107–114.

Kuro-o, M. and 15 others. 1997. Mutation of the mouse *klotho* gene leads to a syndrome resembling ageing. *Nature* 390: 45–51.

Larsen, P. L. 1993. Aging and resistance to oxidative damage in *C. elegans*. *Proc. Acad. Sci. USA* 90: 8905–8909.

Lawrence, P. A. and G. Morata. 1976. Compartments of the wing of *Drosophila*: A study of the *engrailed* gene. *Dev. Biol.* 50: 321–337.

Lee, C.-K., R. G. Klopp, R. Weindruch and T. A. Prolla. 1999. Gene expression profile of aging and its retardation by caloric restriction. *Science* 285: 1390–1393.

Lin, Y.-J., L. Seroude and S. Benzer. 1998. Extended life-span and stress resistance in the *Drosophila* mutant methuselah. *Science* 282: 943–946.

Lindroos, P. M., R. Zarnegar and G. K. Michalopoulos. 1991. Hepatocyte growth factor (hepatopoietin A) rapidly increases in plasma before DNA synthesis and liver regeneration stimulated by partial hepatectomy and carbon tetrachloride administration. *Hepatology* 13: 743–750.

Little, G., B. G. Atkinson and E. Frieden. 1973. Changes in the rates of protein synthesis and degradation in the tail of *Rana catesbeiana* tadpoles during normal metamorphosis. *Dev. Biol.* 30: 366–373.

Livesey, F. J., J. A. O'Brien, M. Li, A. G. Smith, L. J. Murphy and S. P. Hunt. 1997. A Schwann cell mitogen accompanying regeneration of motor neurons. *Nature* 390: 614–618.

Lo, D. C., F. Allen and J. P. Brockes. 1993. Reversal of muscle differentiation during urodele limb regeneration. *Proc. Natl. Acad. Sci. USA* 90: 7230–7234.

Lyman, D. F. and B. A. White. 1987. Molecular cloning of hepatic mRNAs in *Rana catesbeiana* response to thyroid hormone during induced and spontaneous metamorphosis. *J. Biol. Chem.* 262: 5233–5237.

Lynn, W. G. and A. M. Peadon. 1955. The role of the thyroid gland in direct development of the anuran *Eleutherodactylus martinicenis*. *Growth* 19: 263–286.

MacWilliams, H. K. 1983a. Hydra transplantation phenomena and the mechanism of hydra head regeneration. I. Properties of head inhibition. *Dev. Biol.* 96: 217–238.

MacWilliams, H. K. 1983b. Hydra transplantation phenomena and the mechanism of hydra head regeneration. II. Properties of head activation. *Dev. Biol.* 96: 239–272.

Maden, M. 1982. Vitamin A and pattern formation of the regenerating limb. *Nature* 295: 672–675.

Mars, W. M., M. L. Liu, R. P. Kitson, R. H. Goldfarb, M. K. Gabauer and G. K. Michalopoulos. 1995. Immediate early detection of urokinase receptor after partial hepatectomy and its implications for initiation of liver regeneration *Hepatology* 21: 1695–1701.

Martin, V. J. 1997. Cnidrians: The jellyfish and hydra. In Gilbert, S. F. and Raunio, A. M. (eds.) *Embryology: Constructing the Organism.* Sinauer Associates, Sunderland, MA. pp. 57–86.

Mathison, P. M. and L. Miller. 1987. Thyroid hormone induction of keratin genes: A two-step activation of gene expression during development. *Genes Dev.* 1: 1107–1117.

McCutcheon, F. H. 1936. Hemoglobin function during the life history of the bullfrog. *J. Cell. Comp. Physiol.* 8: 63–81.

Medawar, P. B. 1952. *An Unsolved Problem in Biology.* H. K. Lewis, London.

Meinhardt, H. 1980. Cooperation of compartments for the generation of positional information. *Z. Naturforsch. C* 35: 1086–1091.

Meinhardt, H. 1993. A model for pattern formation of hypostome, tentacles, and foot in hydra: How to form structures close to each other, how to form them at a distance. *Dev. Biol.* 157: 321–333.

Melov, S., J. A. Schnieder, B. J. Day, D. Hinerfeld, S. S. Coskunra, J. D. Crapo and D. C. Wallace. 1998. A novel neurological phenotype in mice lacking mitochondrial superoxide dismutase. *Nature Genet.* 18: 59–63.

Mescher, A. L. 1992. Trophic activity of regenerating peripheral nerves. *Comments Dev. Neurobiol.* 1: 373–390.

Mescher, A. L. and D. Gospodarowicz. 1979. Mitogenic effects of a growth factor derived from myelin on denervated regenerates of newt forelimbs. *J. Exp. Zool.* 207: 497–510.

Mescher, A. L. and R. A. Tassava. 1975. Denervation effects on DNA replication and

mitosis during the initiation of limb regeneration in adult newts. *Dev. Biol.* 44: 187–197.

Michalopoulos, G. K. and M. C. DeFrances. 1997. Liver regeneration. *Science* 276: 60–66.

Michikawa, Y., F. Mazzucchelli, N. Bresolin, G. Scarlato and G. Attardi. 1999. Aging-dependent large accumulation of point mutations in the human mtDNA control region for replication. *Science* 286: 774–779.

Migliaccio, E. and 7 others. 1999. The p66[shc] adaptor protein controls oxidative stress response and life span in mammals. *Nature* 402: 309–313.

Mitchell, A. W. 1988. *The Enchanted Canopy.* Macmillan, New York.

Mukhopadyay, G., P. Doherty, F. Walsh, P. R. Crocker and M. Filbin. 1994. A novel role for myelin-associated glycoproteins as an inhibitor of axonal regeneration. *Neuron* 13: 757–767.

Munaim, S. I. and A. L. Mescher. 1986. Transferrin and the trophic effect of neural tissue on amphibian limb regeneration blastemas. *Dev. Biol.* 116: 138–142.

Muneoka, K. and S. V. Bryant. 1982. Evidence that patterning mechanisms in developing and regenerating limbs are the same. *Nature* 298: 369–371.

Murray, V. and R. Holliday. 1981. Increased error frequency of DNA polymerases from senescent fibroblasts. *J. Mol. Biol.* 146: 55–76.

Nardi, J. B. and D. L. Stocum. 1983. Surface properties of regenerating limb cell: Evidence for gradation along the proximodistal axis. *Differentiation* 25: 27–31.

Nellen, D., R. Burke, G. Struhl and K. Basler. 1996. Direct and long-range action of a Dpp morphogen gradient. *Cell* 85: 357–368.

Neumann, C. J. and S. M. Cohen. 1996. Distinct mitogenic and cell fate specification functions of *wingless* in different regions of the wing. *Development* 122: 1781–1789.

Neumann, C. J. and S. M. Cohen. 1997. Long-range action of Wingless organizes the dorsal-ventral axis of the *Drosophila* wing. *Development* 121: 589–599.

Newman, S. A. 1974. The interaction of the organizing regions of hydra and its possible relation to the role of the cut end of regeneration. *J. Embryol. Exp. Morphol.* 31: 541–555.

Niazi, I. A. and S. Saxena. 1978. Abnormal hindlimb regeneration in tadpoles of the toad *Bufo andersonii* exposed to excess vitamin A. *Folia Biol. (Krakow)* 26: 3–8.

Nijhout, H. F. 1994. *Insect Hormones.* Princeton University Press, Princeton, NJ.

Nijhout, H. F. and C. M. Williams. 1974. Control of moulting and metamorphosis in the tobacco hornworm, *Manduca sexta*: Cessation of juvenile hormone secretion as a trigger for pupation. *J. Exp. Biol.* 61: 493–501.

Nishikawa, A., M. Kaiho and K. Yoshizato. 1989. Cell death in the anuran tadpole tail: Thyroid hormone induces keratinization and tail-specif-

ic growth inhibition of epidermal cells. *Dev. Biol.* 131: 337–344.

Norris, D. O., R. E. Jones and B. B. Criley. 1973. Pituitary prolactin levels in larval, neotenic, and metamorphosed salamanders (*Ambystoma tigrinum*). *Gen. Comp. Endocrinol.* 20: 437–442.

Oofusa, K. and K. Yoshizato. 1991. Biochemical and immunological characterization of collagenase in tissues of metamorphosing bullfrog tadpoles. *Dev. Growth Diff.* 33: 329–339.

Orgel, L. E. 1963. The maintenance of the accuracy of protein synthesis and its relevance to aging. *Proc. Natl. Acad. Sci. USA* 49: 517–521.

Orr, W. S. and R. S. Sohal. 1994. Extension of lifespan by overexpression of superoxide dismutase and catalase in *Drosophila melanogaster*. *Science* 263: 1128–1130.

Parkes, T. L., A. J. Elia, D. Dickinson, A. J. Hilliker, J. P. Phillips and G. L. Boulianne. 1998. Extension of *Drosophila* lifespan by overexpression of human *SOD1* in motoneurons. *Nature Genet.* 19: 171–174.

Patterson, D., W. P. Hayes and Y. B. Shi. 1995. Transcriptional activation of the metalloproteinase gene stromelysin-3 coincides with thyroid hormone-induced cell death during frog metamorphosis. *Dev. Biol.* 167: 252–262.

Pecorino, L. T., A. Entwistle and J. P. Brockes. 1996. Activation of a single retinoic acid receptor isoform mediates proximo-distal respecification. *Curr. Biol.* 6: 563–569.

Pfeiffer, S. and J. P. Vincent. 1999. Signalling at a distance: transport of Wingless in the epidermis of *Drosophila*. *Semin. Cell Dev. Biol.* 10: 303–309.

Prahlad, K. V. and L. E. DeLanney. 1965. A study of induced metamorphosis in the axolotl. *J. Exp. Zool.* 160: 137–146.

Ramírez-Weber, F.-A. and T. B. Kornberg. 1999. Cytonemes: Cellular processes that project to the principle signaling center in *Drosophila* imaginal discs. *Cell* 97: 599–607.

Rand, H. W., J. F. Board and D. E. Minnich. 1926. Localization of formative agencies in *Hydra*. *Proc. Natl. Acad. Sci. USA* 12: 565–570.

Reilly, D. S., N. Tomassini and M. Zasloff. 1994. Expression of magainin antimicrobial peptide genes in the developing granular glands of *Xenopus* skin and induction by the thyroid hormone. *Dev. Biol.* 162: 123–133.

Restifo, L. L. and K. White. 1991. Mutations in a steroid hormone-regulated gene disrupt the metamorphosis of the central nervous system in *Drosophila*. *Dev. Biol.* 148: 174–194.

Restifo, L. L. and T. G. Wilson. 1998. A juvenile hormone agonist reveals distinct developmental pathways mediated by ecdysone-inducible Broad Complex transcription factors. *Dev. Genet.* 22: 141–159.

Richards, G. 1992. Switching partners? *Curr. Biol.* 2: 657–658.

Riddiford, L. M. 1972. Juvenile hormone in relation to the larval-pupal transformation of the *Cecropia* silkworm. *Biol. Bull.* 142: 310–325.

Riddiford, L. M. 1982. Changes in translatable mRNAs during the larval-pupal transformation of the epidermis of the tobacco hornworm. *Dev. Biol.* 92: 330–342.

Riddiford, L. M. 1996. Molecular aspects of juvenile hormone action in insect metamorphosis. *In* L. I. Gilbert, J. R. Tata and B. G. Atkinson (eds.), *Metamorphosis: Postembryonic Reprogramming of Gene Expression in Amphibian and Insect Cells.* Academic Press, San Diego, pp. 223–251.

Riggs, A. F. 1951. The metamorphosis of hemoglobin in the bullfrog. *J. Gen. Physiol.* 35: 23–40.

Robinson, H., S. Chaffee and V. A. Galton. 1977. Sensitivity of *Xenopus laevis* tadpole tail tissue to the action of thyroid hormones. *Gen. Comp. Endocrinol.* 32: 179–186.

Rountree, D. B. and W. E. Bollenbacher. 1986. The release of the prothoracicotropic hormone in the tobacco hornworm, *Manduca sexta*, is controlled intrinsically by juvenile hormone. *J. Exp. Biol.* 120: 41–58.

Rudolph, K. L., S. Chang, H.-W. Lee, M. Blasco, G. J. Gottlieb, C. Greider and R. A. DePinho. 1999. Longevity, stress response, and cancer in aging telomerase-deficient mice. *Cell* 96: 701–712.

Safranek, L. and C. M. Williams. 1989. Inactivation of the corpora allata in the final instar of the tobacco hornworm, *Manduca sexta*, requires integrity of certain neural pathways from the brain. *Biol. Bull.* 177: 396–400.

Salk, D. 1982. Can we learn about aging from the study of Werner's syndrome? *J. Am. Geriatr. Soc.* 30: 334–339.

Saxén, L., E. Saxén, S. Toivonen and K. Salimäki. 1957. The anterior pituitary and the thyroid function during normal and abnormal development of the frog. *Ann. Zool. Soc. Fennicae Vanamo* 18 (4): 1–44.

Scadding, S. R. and M. Maden. 1994. Retinoic acid gradients during limb regeneration. *Dev. Biol.* 162: 608–617.

Schiliro, D. M., B. J. Forman and L. C. Javois. 1999. Interactions between the foot and bud patterning systems in *Hydra vulgaris. Dev. Biol.* 209: 399–408.

Schmidt, T. and H. C. Schaller. 1976. Evidence for a foot-inhibiting substance in hydra. *Cell Diff.* 5: 151–159.

Schnell, L. and M. E. Schwab. 1990. Axonal regeneration in the rat spinal cord produced by an antibody against myelin-associated neurite growth inhibitor. *Nature* 343: 269–272.

Schubiger, G. 1968. Anlageplan, Determinationszustand, und Transdeterminationsleistungen der männlichen Vorderbeinschiebe von *Drosophila melanogaster. Wilhelm Roux Arch. Entwicklungsmech. Org.* 160: 9–40.

Schwab, M. E. and P. Caroni. 1988. Oligodendrocyte and CNS myelin are non-permissive substrates for neuron growth and fibroblast spreading in vitro. *J. Neurosci.* 8: 2381–2393.

Schwind, J. L. 1933. Tissue specificity at the time of metamorphosis in frog larvae. *J. Exp. Zool.* 66: 1–14.

Shen, R. Q., Y. P. Chen, L. Huang, E. Vitale and M. Solursh. 1994. Characterization of the human *MSX1* promoter and an enhancer responsible for retinoic acid induction. *Cell. Mol. Biol. Res.* 40: 297–312.

Shostak, S. 1974. Bipolar inhibitory gradients' influence on the budding region of *Hydra viridis. Am. Zool.* 14: 619–632.

Simon, H. G., C. Nelson, D. Goff, E. Laufer, B. A. Morgan and C. Tabin. 1995. The differential expression of myogenic regulatory genes and *msx-1* during dedifferentiation and redifferentiation of regenerating amphibian limbs. *Dev. Dynam.* 202: 1–12.

Singer, M. 1954. Induction of regeneration of the forelimb of the postmetamorphic frog by augmentation of the nerve supply. *J. Exp. Zool.* 126: 419–472.

Singer, M. and J. D. Caston. 1972. Neurotrophic dependence of macromolecular synthesis in the early limb regenerate of the newt, *Triturus. J. Embryol. Exp. Morphol.* 28: 1–11.

Smith-Gill, S. J. and V. Carver. 1981. Biochemical characterization of organ differentiation and maturation. *In* L. I. Gilbert and E. Frieden (eds.), *Metamorphosis: A Problem in Developmental Biology.* Plenum, New York, pp. 491–544.

Stocum, D. L. 1979. Stages of forelimb regeneration in *Ambystoma maculatum. J. Exp. Zool.* 209: 395–416.

Stocum, D. L. and K. Crawford. 1987. Use of retinoids to analyse the cellular basis of memory in regenerating amphibian limbs. *Biochem. Cell Biol.* 65: 750–761.

Stolow, M. A. and Y. B. Shi. 1995. *Xenopus* sonic hedgehog as a potential morphogen during embryogenesis and thyroid hormone dependent metamorphosis. *Nucleic Acids Res.* 23: 2555–2562.

Stolow, M. A., A. Ishizuya-oka, Y. Su and Y.-B. Shi. 1997. Gene regulation by thyroid hormone during amphibian metamorphosis: Implications on the role of cell-cell interactions and cell-extracellular matrix interactions. *Am. Zool.* 37: 195–207.

Sun, H., J. K. Jarow, I. D. Hickson and N. Mazels. 1998. The Blooms syndrome helicase unwinds G4 DNA. *J. Biol. Chem.* 273: 27587–27592.

Szilard, L. 1959. On the nature of the aging process. *Proc. Natl. Acad. Sci. USA* 45: 30–45.

Tabata, T., E. Schwartz, E. Gustavson, Z. Ali and T. B. Kornberg. 1995. Creating a *Drosophila* wing de novo, the role of *engrailed*, and the compartment border hypothesis. *Development* 121: 3359–3369.

Taigen, T. L., F. H. Plough and M. M. Stewart. 1984. Water balance of terrestrial anuran (*Eleutherodactylus coqui*) eggs: Importance of paternal care. *Ecology* 65: 248–255.

Takahashi. N., K. Yoshihama, S. Kikuyama, K. Yamamoto, K. Wakabayashi and Y. Kato. 1990. Molecular cloning and nucleotide sequence of complementary DNA for bullfrog prolactin. *J. Mol. Endocrinol.* 5: 281–287.

Takano, J. and T. Sugiyama. 1983. Genetic analysis of developmental mechanisms in hydra. VIII. Head activation and head inhibition potentials of a slow-budding strain (L4). *J. Embryol. Exp. Morphol.* 78: 141–168.

Talbot, W. S., E. A. Swyryd and D. S. Hogness. 1993. *Drosophila* tissues with different metamorphic responses to ecdysone express different ecdysone receptor isoforms. *Cell* 73: 1323–1337.

Tanaka, E. M., D. Drechsel and J. P. Brockes. 1999. Thrombin regulates S-phase re-entry by cultured newt myotubes. *Curr. Biol.* 9: 792–799.

Tata, J. R., A. Kawahara and B. S. Baker. 1991. Prolactin inhibits both thyroid hormone-induced morphogenesis and cell death in cultured amphibian larval tissues. *Dev. Biol.* 146: 72–80.

Taurog, A., C. Oliver, R. L. Porter, J. C. McKenzie and J. M. McKenzie. 1974. The role of TRH in the neoteny of the Mexican axolotl (*Ambystoma mexicanum*). *Gen. Comp. Endocrinol.* 24: 267–279.

Thomas, H. E., H. G. Stunnenberg and A. F. Stewart. 1993. Heterodimerization of the *Drosophila* ecdysone receptor with retinoid X receptor and *ultraspiracle. Nature* 362: 471–475.

Torok, M. A., D. M. Gardiner, N. H. Shubin and S. V. Bryant. 1998. Expression of *HoxD* genes in developing and regenerating axolotl limbs. *Dev. Biol.* 200: 225–233.

Torok, M. A., D. M. Gardiner, J.-C. Izpisúa-Belmonte and S. V. Bryant. 1999. *Sonic hedgehog* (*Shh*) expression in developing and regenerating axolotl limbs. *J. Exp. Zool.* 284: 197–206.

Truman, J. W. and L. M. Riddiford. 1999. The origins of insect metamorphosis. *Nature* 401: 447–452.

Truman, J. W., W. S. Talbot, S. E. Fahrbach and D. S. Hogness. 1994. Ecdysone receptor expression in the CNS correlates with stage-specific responses to ecdysteroids during *Drosophila* and *Manduca* development. *Development* 120: 219–234.

Turner, C. D. and J. T. Bagnara. 1976. *General Endocrinology*, 6th Ed. Saunders, Philadelphia.

Tyler, M. J. 1983. *The Gastric Brooding Frog.* Croom Helm, London.

Tyler, M. J., D. J. Shearman, R. Franco, P. O'Brien, R. F. Seamark and R. Kelly. 1983. Inhibition of gastric acid secretion in the gastric brooding frog, *Rheobatrachus silus. Science* 220: 607–610.

Vanfleteren, J. R. and A. De Vreese. 1996. Rate of aerobic metabolism and superoxide production rate potential in the nematode *Caenorhabditis elegans. J. Exp. Zool.* 274: 93–100.

van Wijngaarden, R. and F. Bolanos. 1992. Parental care in *Dendrobates granuliferus* (Anura, Dendrobatidae) with a description of the tadpole. *J. Herpetol.* 26: 102–105.

Vaziri, H. and S. Benchimol. 1998. Reconstitution of telomerase activity in normal human cells leads to elongation of telomeres and extended replicative lifespan. *Curr. Biol.* 8: 279–282.

Viviano, C. M., C. E. Horton, M. Maden and J. P. Brockes. 1995. Synthesis and release of 9-*cis*-retinoic acid by the urodele wound epidermis. *Development* 121: 3753–3762.

Vortkamp, A., S. Pathi, G. M. Peretti, E. M. Caruso, D. J. Zaleske and C. J. Tabin. 1998. Recapitulation of signals regulating embryonic bone formation during postnatal growth and in fracture repair. *Mech. Dev.* 71: 65–76.

Wald, G. 1945. The chemical evolution of vision. *Harvey Lect.* 41: 117–160.

Wald, G. 1981. Metamorphosis: An overview. *In* L. I. Gilbert and E. Frieden (eds.), *Metamorphosis: A Problem in Developmental Biology.* Plenum, New York, pp. 1–39.

Wassersug, R. J. 1989. Locomotion in amphibian larvae (or why aren't tadpoles built like fish). *Am. Zool.* 29: 65–84.

Weidner, N., A. Blesch, R. J. Grill and M. H. Tuszynski. 1999. Nerve growth factor-hypersecreting Schwann cell grafts augment and guide spinal cord axonal growth and remyelinate central nervous system axons in a phenotypically appropriate manner that correlates with expression of L1. *J. Comp. Neurol.* 413: 495–506.

Weismann, A. 1891. The duration of life. *In* E. B. Poulton, S. Schonland and A. E. Shipley, (eds.), *Essays on Heredity and Kindred Subjects,* 2nd Ed. Oxford University Press, Oxford, pp. 163–256.

White, B. H. and C. S. Nicoll. 1981. Hormonal control of amphibian development. *In* L. I. Gilbert and E. Frieden (eds.), *Metamorphosis: A Problem in Developmental Biology.* Plenum, New York, pp. 363–396.

Wilby, O. K. and G. Webster. 1970. Experimental studies on axial polarity in hydra. *J. Embryol. Exp. Morphol.* 24: 595–613.

Williams, J. A., S. W. Paddock, K. Vorwek and S. B. Carroll. 1994. Organization of wing formation and induction of a wing-patterning gene at the dorsal/ventral compartment boundary. *Nature* 368: 299–305.

Wolffe, A. P. and Y.-B. Shi. 1999. A hypothesis for the transcriptional control of amphibian metamorphosis by the thyroid hormone receptor. *Am. Zool.* 39: 807–817.

Wong, J. M. and Y. B. Shi. 1995. Coordinated regulation and transcriptional activation of *Xenopus* thyroid hormone and retinoid-X receptors. *J. Biol. Chem.* 270: 18479–18483.

Yang, E. V., D. M. Gardiner, M. R Carlson, C. A. Nugas and S. V. Bryant. 1999. Expression of *Mmp-9* and related matrix metalloproteinase genes during axolotl limb regeneration. *Dev. Dynam.* 216: 2–9.

Yao, T.-P., W. A. Segraves, A. E. Oro, M. McKeown and R. M. Evans. 1992. *Drosophila ultraspiracles* modulates ecdysone receptor function via heterodimer formation. *Cell* 71: 63–72.

Yaoita, Y. and D. D. Brown. 1990. A correlation of thyroid hormone receptor gene expression with amphibian metamorphosis. *Genes Dev.* 4: 1917–1924.

Yu, C. E. and 12 others. 1996. Positional cloning of the Werner's syndrome gene. *Science* 272: 258–262.

Zecca, M., K. Basler and G. Struhl. 1996. Direct and long-range action of a Wingless morphogen gradient. *Cell* 87: 833–844.

19 *The saga of the germ line*

WE BEGAN OUR ANALYSIS of animal development by discussing fertilization, and we will finish our studies of individual development by investigating **gametogenesis,** the processes by which the sperm and the egg are formed. Germ cells provide the continuity of life between generations, and the mitotic ancestors of our own germ cells once resided in the gonads of reptiles, amphibians, fishes, and invertebrates.

In many animals, such as insects, roundworms, and vertebrates, there is a clear and early separation of germ cells from somatic cell types. In several other animal phyla (and throughout the entire plant kingdom), this division is not as well established. In these species (which include cnidarians, flatworms, and tunicates), somatic cells can readily become germ cells even in adult organisms. The zooids, buds, and polyps of many invertebrate phyla testify to the ability of somatic cells to give rise to new individuals (Liu and Berrill 1948; Buss 1987).

In those organisms in which there is an established **germ line** that separates from the somatic cells early in development, the germ cells do not arise within the gonad itself. Rather, their precursors—the **primordial germ cells** (**PGCs**)—arise elsewhere and migrate into the developing gonads. The first step in gametogenesis, then, involves forming the PGCs and getting them into the genital ridge as the gonad is forming. Our discussion of the germ line will include

1. The formation of the germ plasm and the determination of the PGCs
2. The migration of the PGCs into the developing gonads
3. The process of meiosis and the modifications of meiosis for forming sperm and eggs
4. The differentiation of the sperm and egg
5. The hormonal control of gamete maturation and ovulation

Germ Plasm and the Determination of the Primordial Germ Cells

All sexually reproducing organisms arise from the fusion of gametes—sperm and eggs. All gametes arise from the primordial germ cells. In most animal species, the determination of the primordial germ cells is brought about by the cytoplasmic localization of specific proteins and mRNAs in certain cells of the early embryo (mammals being a major exception to this general rule). These cytoplasmic components are referred to as the **germ plasm.**

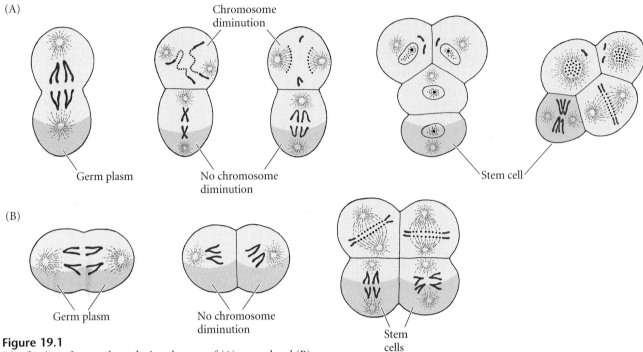

(A)

Chromosome diminution

Germ plasm

No chromosome diminution

Stem cell

(B)

Germ plasm

No chromosome diminution

Stem cells

Figure 19.1
Distribution of germ plasm during cleavage of (A) normal and (B) centrifuged zygotes of *Parascaris*. (A) The germ plasm is normally conserved in the most vegetal blastomere, as shown by the lack of chromosomal diminution in that particular cell. Thus, at the 4-cell stage, the embryo has one stem cell for its gametes. (B) When the first cleavage is displaced 90 degrees by centrifugation, both resulting cells have vegetal germ plasm, and neither cell undergoes chromosome diminution. After the second cleavage, these two cells give rise to germinal stem cells. (After Waddington 1966.)

Germ cell determination in nematodes

Theodor Boveri (Figure 4.2; 1862–1915) was the first person to observe an organism's chromosomes throughout its development. In so doing, he discovered a fascinating feature in the development of the roundworm *Parascaris aequorum* (formerly *Ascaris megalocephala*). This nematode has only two chromosomes per haploid cell, allowing for detailed observations of the individual chromosomes. The cleavage plane of the first embryonic division is unusual in that it is equatorial, separating the animal half from the vegetal half of the zygote (Figure 19.1A). More bizarre, however, is the behavior of the chromosomes in the subsequent division of these first two blastomeres. The ends of the chromosomes in the animal-derived blastomere fragment into dozens of pieces just before this cell divides. This phenomenon is called **chromosome diminution**, because only a portion of the original chromosome survives. Numerous genes are lost in these cells when the chromosomes fragment, and these genes are not included in the newly formed nuclei (Tobler et al. 1972; Müller et al. 1996). Meanwhile, in the vegetal blastomere, the chromosomes remain normal. During second cleavage, the animal cell

splits meridionally while the vegetal cell again divides equatorially. Both vegetally derived cells have normal chromosomes. However, the chromosomes of the more animally located of these two vegetal blastomeres fragment before the third cleavage. Thus, at the 4-cell stage, only one cell—the most vegetal—contains a full set of genes. At successive cleavages, nuclei with diminished chromosomes are given off from this vegetalmost line, until the 16-cell stage, when there are only two cells with undiminished chromosomes. One of these two blastomeres gives rise to the germ cells; the other eventually undergoes chromosome diminution and forms more somatic cells. The chromosomes are kept intact only in those cells destined to form the germ line. If this were not the case, the genetic information would degenerate from one generation to the next. The cells that have undergone chromosome diminution generate the somatic cells.

Boveri has been called the last of the great "observers" of embryology and the first of the great experimenters. Not content with observing the retention of the full chromosome complement by the germ cell precursors, he set out to test whether a specific region of cytoplasm protects the nuclei within it from diminution. If so, any nucleus happening to reside in this region should be protected. Boveri (1910) tested this hypothesis by centrifuging *Parascaris* eggs shortly before their first cleavage. This treatment shifted the orientation of the mitotic spindle. When the spindle forms perpendicular to its normal orientation, both resulting blastomeres should contain some of the vegetal cytoplasm (see Figure 19.1B). Indeed, Boveri found that after the first division, neither nucleus underwent chromosomal diminution. However, the next division was equatorial along the animal-vegetal axis. Here the resulting animal blastomeres both underwent diminution, whereas the two vegetal cells did not. Boveri concluded that the vegetal cytoplasm contains a factor (or factors) that protects nuclei from chromosomal diminution and determines germ cells.

In the nematode *C. elegans*, the germ line precursor cell is the P4 blastomere. The P-granules enter this cell, and they appear critical for instructing it to become the germ line precursor (see Figure 8.45). The components of the P-granules remain largely uncharacterized, but they appear to contain several transcriptional inhibitors and RNA-binding proteins, including homologues of the *Drosophila* Vasa and Nanos proteins, whose function we will see below (Kawasaki et al. 1998; Seydoux and Strome 1999; Subramanian and Seydoux 1999).

WEBSITE 19.1 **Mechanisms of chromosome diminution.** The somatic cells do not lose DNA randomly. Rather, specific regions of DNA are lost during chromosome diminution.

Germ cell determination in insects

In *Drosophila*, PGCs form as a group of cells (**pole cells**) at the posterior pole of the cellularizing blastoderm. These nuclei migrate into the posterior region at the ninth nuclear division, and they become surrounded by the **pole plasm**, a complex collection of mitochondria, fibrils, and **polar granules** (Figure 19.2; Mahowald 1971a,b; Schubiger and Wood 1977). If the pole cell nuclei are prevented from reaching the pole plasm, no germ cells will be made (Mahowald et al. 1979).

Nature has provided confirmation of the importance of both pole plasm and its polar granules. One of the components of the pole plasm is the mRNA of the ***germ cell-less* (*gcl*)** gene. This gene was discovered by Jongens and his colleagues (1992) when they mutated *Drosophila* and screened for those females who did not have "grandoffspring." They assumed that if a female did not place functional germ plasm in her eggs, she could still have offspring, but those offspring would be sterile (since they would lack germ cells). The wild-type *gcl* gene is transcribed in the nurse cells of the fly's ovary, and its

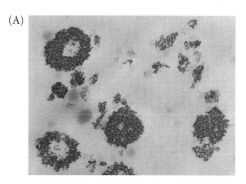

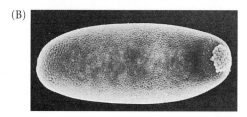

Figure 19.2
The pole plasm of *Drosophila*. (A) Electron micrograph of polar granules from particulate fraction of *Drosophila* pole cells. (B) Scanning electron micrograph of a *Drosophila* embryo just prior to completion of cleavage. The pole cells can be seen at the right of this picture. (Photographs courtesy of A. P. Mahowald.)

mRNA is transported into the egg. Once inside the egg, it is transported to the posteriormost portion and resides within what will become the pole plasm (Figure 19.3A,B). This message gets translated into protein during the early stages of cleavage (Figure 19.3C,D). The *gcl*-encoded protein appears to enter the nucleus, and it is essential for pole cell production. Flies with mutations of this gene lack germ cells.

VADE MECUM **Germ cells in the *Drosophila* embryo.** This segment follows the primordial germ cells of the living *Drosophila* embryo from their formation as pole cells through gastrulation as they move from the posterior end of the embryo into the region of the developing gonad. **[Click on Fruit Fly]**

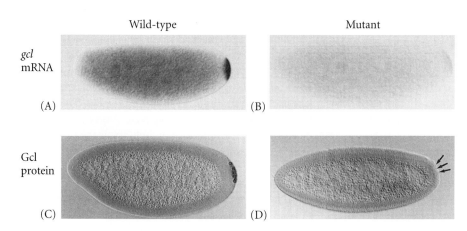

Wild-type Mutant

gcl mRNA

(A) (B)

Gcl protein

(C) (D)

Figure 19.3
Localization of Germ cell-less gene products in the posterior of the egg and embryo. The *gcl* mRNA can be seen in the posterior pole of early-cleavage embryos produced by wild-type females (A), but not in embryos produced by *gcl*-deficient mutant females (B). Antibodies against the protein encoded by the *gcl* gene can be detected in the germ cells at the cellular blastoderm stage of embryos produced by wild-type females (C), but not in embryos from mutant females (D). (From Jongens et al. 1992; photographs courtesy of T. A. Jongens.)

A second set of pole plasm components are the posterior determinants mentioned in Chapter 9. **Oskar** appears to be the critical protein of this group, since injection of *oskar* mRNA into ectopic sites of the embryo will cause the nuclei in those areas to form germ cells. The genes that restrict Oskar to the posterior pole are also necessary for germ cell formation (Ephrussi and Lehmann 1992; Newmark et al. 1997). Moreover, Oskar appears to be the limiting step of germ cell formation, since adding more *oskar* message to the oocyte causes the formation of more germ cells (Ephrussi and Lehmann 1992). Oskar functions by causing the localization of the proteins and RNAs necessary for germ cell formation. One of these RNAs is the *nanos* message, whose product is essential for posterior segment formation. **Nanos** is also essential for germ cell formation. Pole cells lacking Nanos do not migrate into the gonads and fail to become gametes. Nanos appears to be important in preventing mitosis and transcription during germ cell development (Kobayashi et al. 1996; Deshpande et al. 1999). Another one of these RNAs encodes **Vasa**, an RNA-binding protein. The mRNAs for this protein are seen in the germ plasm of many species.

A third germ plasm component was a big surprise: **mitochondrial ribosomal RNA** (**mtrRNA**). Kobayashi and Okada (1989) showed that the injection of mtrRNA into embryos formed from ultraviolet-irradiated eggs restores the ability of these embryos to form pole cells. Moreover, in normal fly eggs, the small and large mitochondrial rRNAs are located outside the mitochondria solely in the pole plasm of cleavage-stage embryos. Here they appear as components of the polar granules (Kobayashi et al. 1993; Amikura et al. 1996; Kashikawa et al. 1999). While mitochondrial RNA is involved in directing the formation of the pole cells, it does not enter them.

A fourth component of *Drosophila* pole plasm (and one that becomes localized in the polar granules) is a nontranslatable RNA called **polar granule component** (**Pgc**). While its exact function remains unknown, the pole cells of transgenic female flies making antisense RNA against Pgc fail to migrate to the gonads (Nakamura et al. 1996).

WEBSITE **19.2 The insect germ plasm.** The insect germinal cytoplasm was discovered as early as 1911, when Hegner found that removing the posterior pole cytoplasm of beetle eggs caused sterility in the resulting adults.

Germ cell determination in amphibians

Cytoplasmic localization of germ cell determinants has also been observed in vertebrate embryos. Bounoure (1934) showed that the vegetal region of fertilized frog eggs contains material with staining properties similar to those of *Drosophila* pole plasm (Figure 19.4). He was able to trace this cortical cytoplasm into the few cells in the presumptive endoderm that would normally migrate into the genital ridge. By transplanting genetically marked cells from one embryo into

Yolk platelets

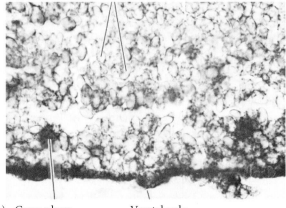

(A)   Germ plasm   Vegetal pole
                                   of zygote

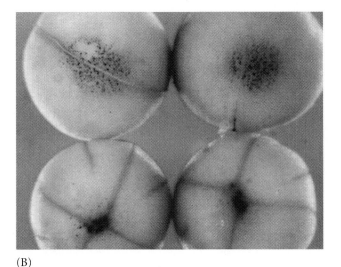

(B)

Figure 19.4
Germ plasm at the vegetal pole of frog embryos. (A) Germ plasm (dark regions) near the vegetal pole of a newly fertilized zygote. (B) In situ hybridization localizing the mRNA for Xcat2 (the *Xenopus* homologue of Nanos) in the vegetal cortex of first-cleavage (upper) and fourth-cleavage (lower) embryos. (A courtesy of A. Blackler; B after Kloc et al. 1998, photograph courtesy of L. Etkin.)

another of a differently marked strain, Blackler (1962) showed that these cells are the primordial germ cell precursors.

The early movements of the germ plasm in amphibians have been analyzed in detail by Savage and Danilchik (1993), who labeled the germ plasm with fluorescent dye. They found that the germ plasm of unfertilized eggs consists of tiny "islands" that appear to be tethered to the yolk mass near the vegetal cortex. These germ plasm islands move with the vegetal yolk mass during the cortical rotation of fertilization. After the rotation, the islands are released from the yolk mass and begin fusing together and migrating to the vegetal pole. Their aggregation depends on microtubules, and the movement of these clusters to the vegetal pole is dependent on a kinesin-like protein that may act as the motor for germ plasm movement (Robb et al. 1996). Later, periodic contractions of the vegetal cell surface also appear to push this germ plasm along the cleavage furrows of the newly formed blastomeres, enabling it to enter the embryo.

When ultraviolet light is applied to the vegetal surface (but nowhere else) of frog embryos, the resulting frogs are normal but lack germ cells in their gonads (Bounoure 1939; Smith 1966). Very few primordial germ cells reach the gonads; those few that do have about one-tenth the volume of normal PGCs and have aberrantly shaped nuclei (Züst and Dixon 1977). Savage and Danilchik (1993) found that UV light prevents vegetal surface contractions and inhibits the migration of germ plasm to the vegetal pole. The *Xenopus* homologues of *nanos* and *vasa* are specifically localized to this region (Forristall et al. 1995; Ikenishi et al. 1996; Zhou and King 1996). So, like the *Drosophila* pole plasm, the cytoplasm from the vegetal region of the frog zygote contains the determinants for germ cell formation. Moreover, several of the components are the same.

The components of the germ plasm have not all been catalogued. Indeed, in the birds and mammals, such a list has hardly even been started. Moreover, we still do not know the functions of the proteins (such as Vasa and Nanos) and non-translated RNAs found in the germ plasm. One hypothesis (Nieuwkoop and Sutasurya 1981; Wylie 1999) is that the components of the germ plasm inhibit both transcription and translation, thereby preventing the cells containing it from differentiating into anything else. According this hypothesis, the cells become germ cells because they are forbidden to become any other type of cell.

Germ Cell Migration

Germ cell migration in amphibians

The germ plasm of anuran amphibians (frogs and toads) collects around the vegetal pole in the zygote. During cleavage, this material is brought upward through the yolky cytoplasm, and eventually becomes associated with the endodermal cells lining the floor of the blastocoel (Figure 19.5A–E; Bounoure 1934; Ressom and Dixon 1988; Kloc et al. 1993). The PGCs become concentrated in the posterior region of the larval gut, and as the abdominal cavity forms, they migrate along the dorsal side of the gut, first along the dorsal mesentery (which connects the gut to the region where the mesodermal organs are forming) and then along the abdominal wall and into the genital ridges. They

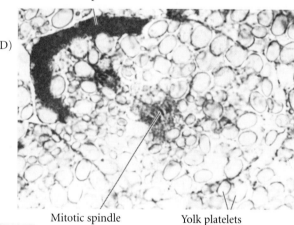

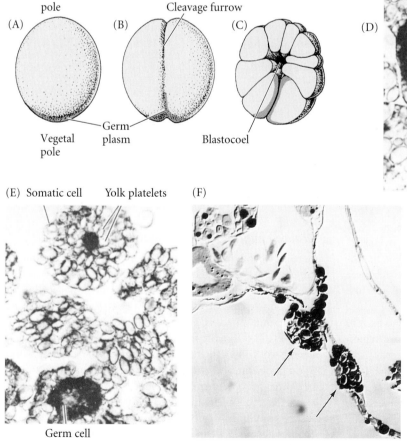

Figure 19.5
Migration of *Xenopus* germ plasm. (A–C) Changes in the position of the germ plasm (color) in an early frog embryo. Originally located near the vegetal pole of the uncleaved egg (A), the germ plasm advances along the cleavage furrows (B) until it becomes localized at the floor of the blastocoel (C). (D) Germ plasm-containing cell in the endodermal region of a blastula in mitotic anaphase. Note the germ plasm entering into only one of the two yolk-laden daughter cells. (E) Primordial germ cell and somatic cells near the floor of the blastocoel in early gastrula. (F) Migration of two primordial germ cells (arrows) along the dorsal mesentery connecting the gut region to the gonadal mesoderm (A–C after Bounoure 1934; D, E courtesy of A. Blackler; F from Heasman et al. 1977, courtesy of the authors.)

Figure 19.6
Pathway for the migration of mammalian primordial germ cells. (A) PGCs seen in the yolk sac near the junction of the hindgut and allantois. (B) The PGCs migrate through the gut and, dorsally, up the dorsal mesentery and into the genital ridges. (C) Four large PGCs in the hindgut of a mouse embryo (near the allantois and yolk sac) stain positively for high levels of alkaline phosphatase. (D) Such alkaline phosphatase-staining cells can be seen migrating up the dorsal mesentery and entering the genital ridges. (A and B from Langman 1981; C from Heath 1978; D from Mintz 1957; photographs courtesy of the authors.)

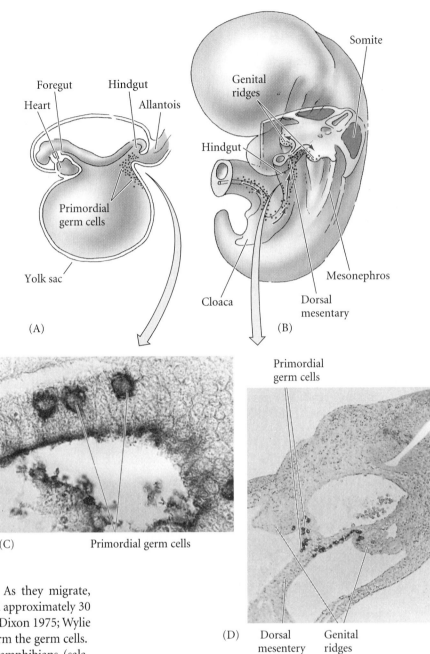

migrate up this tissue until they reach the developing gonads (Figure 19.5F). *Xenopus* PGCs move by extruding a single filopodium and then streaming their yolky cytoplasm into that filopodium while retracting their "tail." Contact guidance in this migration seems likely, as both the PGCs and the extracellular matrix over which they migrate are oriented in the direction of migration (Wylie et al. 1979). Furthermore, PGC adhesion and migration can be inhibited if the mesentery is treated with antibodies against *Xenopus* fibronectin (Heasman et al. 1981). Thus, the pathway for germ cell migration in these frogs appears to be composed of an oriented fibronectin-containing extracellular matrix. The fibrils over which the PGCs travel lose this polarity soon after migration has ended.* As they migrate, *Xenopus* PGCs divide about three times, and approximately 30 PGCs colonize the gonads (Whitington and Dixon 1975; Wylie and Heasman 1993). These will divide to form the germ cells.

The primordial germ cells of urodele amphibians (salamanders) have an apparently different origin, which has been traced by reciprocal transplantation experiments to the regions of the mesoderm that involute through the ventrolateral lips of the blastopore. Moreover, there does not seem to be any particular localized "germ plasm" in salamander eggs. Rather, the interaction of the dorsal endoderm cells and animal hemisphere cells creates the conditions needed to form germ cells in the areas that involute through the ventrolateral lips (Sutasurya and Nieuwkoop 1974; Wakahara 1996). So in salamanders, the PGCs are formed by induction within the mesodermal region and presumably follow a different path into the gonads.

---

*This does not necessarily hold true for all anurans. In the frog *Rana pipiens*, the germ cells follow a similar route, but may be passive travelers along the mesentery rather than actively motile cells (Subtelny and Penkala 1984). The migration of fish PGCs follows a similar route, too, and there may be species differences as to whether the PGCs are active or passive travellers (Braat et al. 2000).

Germ cell migration in mammals

There is no obvious germ plasm in mammals, and mammalian germ cells are not morphologically distinct during early development. However, by using monoclonal antibodies that recognize cell surface differences between the PGCs and their surrounding cells, Hahnel and Eddy (1986) showed that mouse PGCs originally reside in the epiblast of the gastrulating embryo. Ginsburg and her colleagues (1990) localized this region to the area that becomes extraembryonic mesoderm just posterior to the primitive streak of the 7-day mouse embryo. Here, about eight large, alkaline phosphatase-staining cells are seen. If this region is removed, the remaining embryo becomes devoid of germ cells, while the isolated region develops a large number of PGCs.

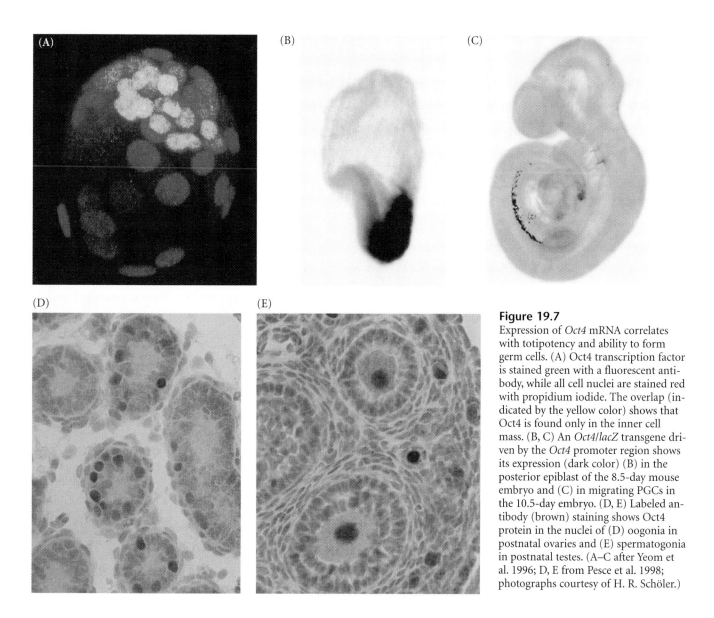

Figure 19.7
Expression of *Oct4* mRNA correlates with totipotency and ability to form germ cells. (A) Oct4 transcription factor is stained green with a fluorescent antibody, while all cell nuclei are stained red with propidium iodide. The overlap (indicated by the yellow color) shows that Oct4 is found only in the inner cell mass. (B, C) An *Oct4/lacZ* transgene driven by the *Oct4* promoter region shows its expression (dark color) (B) in the posterior epiblast of the 8.5-day mouse embryo and (C) in migrating PGCs in the 10.5-day embryo. (D, E) Labeled antibody (brown) staining shows Oct4 protein in the nuclei of (D) oogonia in postnatal ovaries and (E) spermatogonia in postnatal testes. (A–C after Yeom et al. 1996; D, E from Pesce et al. 1998; photographs courtesy of H. R. Schöler.)

In normal mouse embryos, the germ cell precursors migrate from the extraembryonic mesoderm back into the embryo, by way of the allantois. The route of mammalian PGC migration from the allantois (Figure 19.6) resembles that of anuran PGC migration. After collecting at the allantois (by day 7.5 in the mouse: Chiquoine 1954; Mintz 1957), the PGCs migrate to the adjacent yolk sac (Figure 19.6A,C). By this time, they have already split into two populations that will migrate to either the right or the left genital ridge. The PGCs then move caudally from the yolk sac through the newly formed hindgut and up the dorsal mesentery into the genital ridge (Figure 19.6B,D). Most of the PGCs have reached the developing mouse gonad by the eleventh day after fertilization. During this trek, they have proliferated from an initial population of 10–100 cells to the 2500–5000 PGCs present in the gonads by day 12.

Like the PGCs of *Xenopus*, mammalian PGCs appear to be closely associated with the cells over which they migrate, and they move by extending filopodia over the underlying cell surfaces. These cells are also capable of penetrating cell monolayers and migrating through the cell sheets (Stott and Wylie 1986). The mechanism by which the PGCs know the route of this journey is still unknown. Fibronectin is likely to be an important substrate for PGC migration (ffrench-Constant et al. 1991), and germ cells that lack the integrin receptor for such extracellular matrix proteins cannot migrate to the gonads (Anderson et al. 1999). Directionality may be provided by a gradient of soluble protein. In vitro evidence suggests that the genital ridges of 10.5-day mouse embryos secrete a diffusible TGF-β1-like protein that is capable of attracting mouse PGCs (Godin et al. 1990; Godin and Wylie 1991). Whether the genital ridge is able to provide such cues in vivo still must be tested.

Although no germ plasm has been found in mammals, the retention of totipotency has been correlated with the expression of a nuclear transcription factor, **Oct4**. This factor is expressed in all of the early-cleavage blastomere nuclei, but its expression becomes restricted to the inner cell mass. During gastrulation, it becomes expressed solely in those posterior epiblast cells thought to give rise to the primordial germ cells.

After that, this protein is seen only in the primordial germ cells and oocytes (Figure 19.7; Yeom et al. 1996; Pesce et al. 1998). (Oct4 is not seen in the developing sperm after the germ cells reach the testes and become committed to sperm production.)

The proliferation of the PGCs appears to be promoted by stem cell factor, the same growth factor needed for the proliferation of neural crest-derived melanoblasts and hematopoietic stem cells (see Chapter 6). Stem cell factor is produced by the cells along the migration pathway and remains bound to their cell membranes. It appears that the presentation of this protein

on cell membranes is important for its activity. Mice homozygous for the White mutation are deficient in germ cells (as well as melanocytes and blood cells) because their stem cells lack the receptor for stem cell factor. Mice homozygous for the Steel mutation have a similar phenotype, as they lack the ability to make this growth factor (Dolci et al. 1991; Matsui et al. 1991). The addition of stem cell factor to PGCs taken from 11-day mouse embryos will stimulate their proliferation for about 24 hours and appears to prevent programmed cell death that would otherwise occur (Pesce et al. 1993).

Sidelights & Speculations

EG Cells, ES Cells, and Teratocarcinomas

Embryonic germ (EG) cells

Stem cell factor increases the proliferation of migrating mouse primordial germ cells in culture, and this proliferation can be further increased by adding another growth factor, leukemia inhibition factor (LIF). However, the life span of these PGCs is short, and the cells soon die. But if an additional mitotic regulator—basic fibroblast growth factor (FGF2)—is added, a remarkable change takes place. The cells continue to proliferate, producing pluripotent embryonic stem cells with characteristics resembling the cells of the inner cell mass (Matsui et al. 1992; Resnick et al. 1992; Rohwedel et al. 1996). These PGC-derived cells are called **embryonic germ (EG) cells**, and they have the potential to differentiate into all the cell types of the body.

In 1998, John Gearhart's laboratory (Shamblott et al. 1998) cultured human EG cells. These cells were able to generate differentiated cells from all three primary germ layers, and they are presumably totipotent. These cells could be used medically to create neural or hematopoietic stem cells, which might be used to regenerate damaged neural or blood tissues (see Chapter 4). EG cells are often considered as embryonic stem (ES) cells (see below), and the distinction of their origin is ignored.

Embryonic stem (ES) cells

Embryonic stem (ES) cells were described in Chapter 4; these are the cells that are derived from the inner cell mass. ES cells and EG cells can be transfected with recombinant genes and inserted into the blastocyst to create

transgenic mice. Such a mammalian germ cell or stem cell contains within it all the information needed for subsequent development. What would happen if such a cell became malignant? In one type of tumor, the germ cells become embryonic stem cells, like the FGF2-treated PGCs in the experiment above. This type of tumor is called a **teratocarcinoma**. Whether spontaneous or experimentally produced, a teratocarcinoma contains an undifferentiated stem cell population that has biochemical and developmental properties remarkably similar to those of the inner cell mass (Graham 1977). Moreover, these stem cells not only divide, but can also differentiate into a wide variety of tissues, including gut and respiratory epithelia, muscle, nerve, carti-

lage, and bone (Figure 19.8). Once differentiated, these cells no longer divide, and are therefore no longer malignant. Such tumors can give rise to most of the tissue types in the body. Thus, the teratocarcinoma stem cells mimic early mammalian development, but the tumor they form is characterized by random, haphazard development.

In 1981, Stewart and Mintz formed a mouse from cells derived in part from a teratocarcinoma stem cell. Stem cells that had arisen in a teratocarcinoma of an agouti (yellow-tipped) strain of mice were cultured for several cell generations, and were seen to maintain the characteristic chromosome complement of the parental mouse. Individual stem cells descended from the tumor

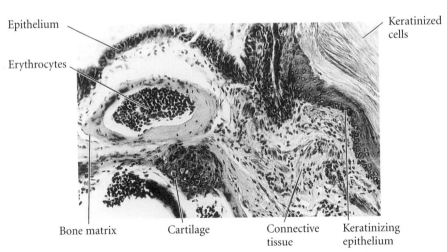

Epithelium

Erythrocytes

Keratinized cells

Bone matrix   Cartilage   Connective tissue   Keratinizing epithelium

Figure 19.8
Photomicrograph of a section through a teratocarcinoma, showing numerous differentiated cell types. (From Gardner 1982; photograph from C. Graham, courtesy of R. L. Gardner.)

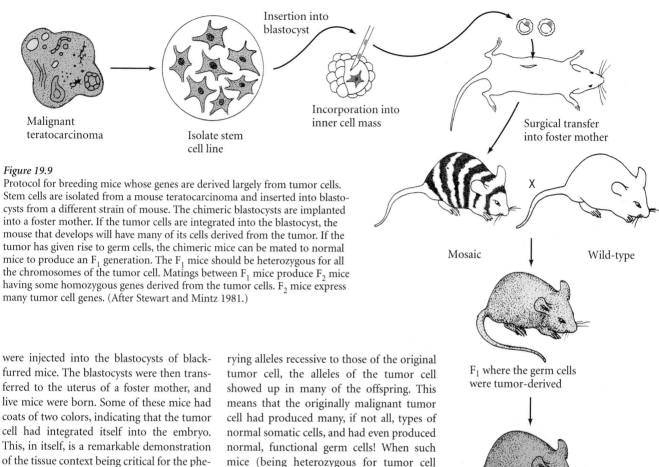

Figure 19.9
Protocol for breeding mice whose genes are derived largely from tumor cells. Stem cells are isolated from a mouse teratocarcinoma and inserted into blastocysts from a different strain of mouse. The chimeric blastocysts are implanted into a foster mother. If the tumor cells are integrated into the blastocyst, the mouse that develops will have many of its cells derived from the tumor. If the tumor has given rise to germ cells, the chimeric mice can be mated to normal mice to produce an F_1 generation. The F_1 mice should be heterozygous for all the chromosomes of the tumor cell. Matings between F_1 mice produce F_2 mice having some homozygous genes derived from the tumor cells. F_2 mice express many tumor cell genes. (After Stewart and Mintz 1981.)

were injected into the blastocysts of black-furred mice. The blastocysts were then transferred to the uterus of a foster mother, and live mice were born. Some of these mice had coats of two colors, indicating that the tumor cell had integrated itself into the embryo. This, in itself, is a remarkable demonstration of the tissue context being critical for the phenotype of a cell—a malignant cell was made nonmalignant.

But the story does not end here. When these chimeric mice were mated to mice carrying alleles recessive to those of the original tumor cell, the alleles of the tumor cell showed up in many of the offspring. This means that the originally malignant tumor cell had produced many, if not all, types of normal somatic cells, and had even produced normal, functional germ cells! When such mice (being heterozygous for tumor cell genes) were mated with each other, the resultant litter contained mice that were homozygous for a large number of genes from the tumor cell (Figure 19.9). ▪

Germ cell migration in birds and reptiles

In birds and reptiles, the primordial germ cells are derived from epiblast cells that migrate from the central region of the area pellucida to a crescent-shaped zone in the hypoblast at the anterior border of the area pellucida (Figure 19.10; Eyal-Giladi et al. 1981; Ginsburg and Eyal-Giladi 1987). This extraembryonic region is called the **germinal crescent**, and the PGCs multiply there.

Unlike those of amphibians and mammals, the PGCs of birds and reptiles migrate to the gonads primarily by means of the bloodstream (Figure 19.11). When blood vessels form in the germinal crescent, the PGCs enter those vessels and are carried by the circulation to the region where the hindgut is forming. Here, they exit from the circulation, become associated with the mesentery, and migrate into the genital ridges (Swift 1914; Mayer 1964; Kuwana 1993; Tsunekawa et al. 2000). The PGCs of the germinal crescent appear to enter the blood vessels by **diapedesis**, a type of movement common to lymphocytes and macrophages that enables cells to squeeze between the endothelial cells of small blood vessels . In some as yet undiscovered way, the PGCs are instructed to exit the blood vessels and enter the gonads (Pasteels 1953; Dubois 1969). Evidence for chemotaxis comes from studies (Kuwana et al. 1986) in which circulating chick PGCs were isolated from the blood and cultured between gonadal rudiments and other embryonic tissues. The PGCs migrated into the gonadal rudiments during a 3-hour incubation.

Germ cell migration in *Drosophila*

During *Drosophila* embryogenesis, the primordial germ cells move from the posterior pole to the gonads (Figure 19.12). The first step in this migration is a passive one, wherein the 30–40 pole cells are displaced into the posterior midgut by the

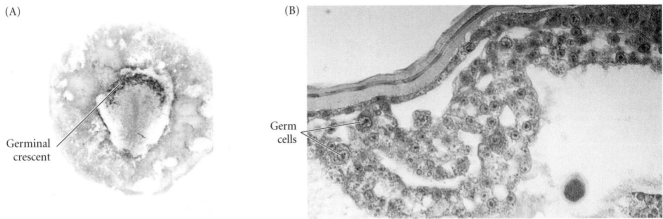

Figure 19.10
The germinal crescent of the chick embryo. (A) Germ cells of a definitive primitive streak-stage (stage 4, roughly 18-hour) chick embryo, stained (purple) for the chick Vasa homologue protein. The stained cells are confined to the germinal crescent. (B) Higher magnification of the stage 4 germinal crescent region, showing germ cells (stained brown) in the thickened epiblast. (Anterior is to the right.) (From Tsunekawa et al. 2000; photographs courtesy of N. Tsunekawa.)

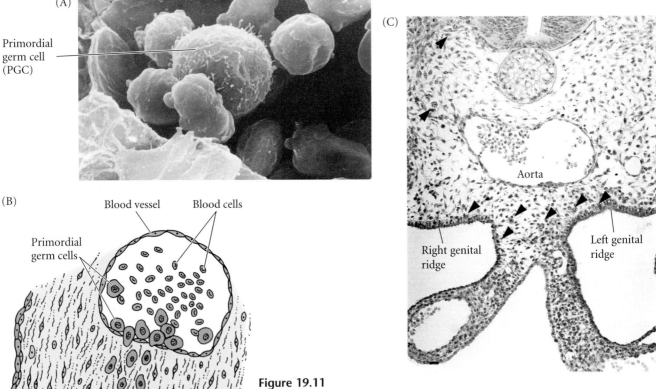

Figure 19.11
Migration of primordial germ cells in the chick embryo. (A) Scanning electron micrograph of a chick PGC in a capillary of a gastrulating embryo. The PGC can be identified by its large size and by the microvilli on its surface. (B) Diagram of transverse section near the prospective gonadal region of a chick embryo. Several PGCs within a blood vessel cluster next to the gonadal epithelium. One PGC is crossing through the blood vessel endothelium, and another PGC is already located within the gonadal epithelium. (C) Having passed through the endothelium of the dorsal aorta, chick germ cells (arrows) migrate toward the genital ridges of the embryo. (A from Kuwana 1993, courtesy of T. Kuwana; B after Romanoff 1960; C from Tsunekawa et al. 2000, courtesy of N. Tsunekawa.)

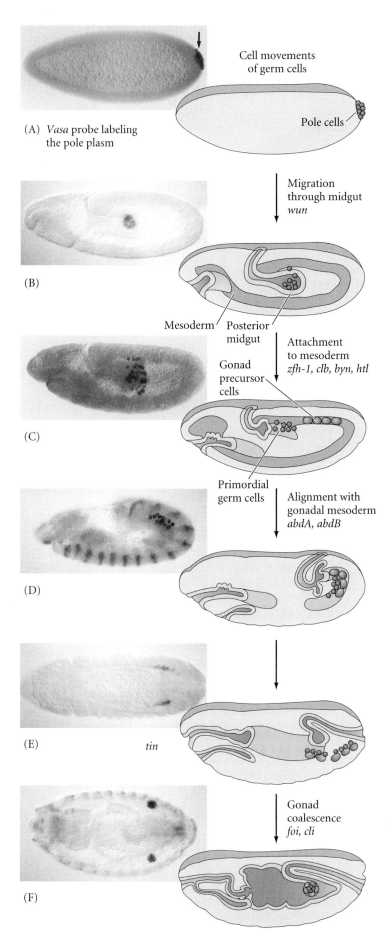

(A) *Vasa* probe labeling the pole plasm

Cell movements of germ cells

Pole cells

Migration through midgut
wun

(B)

Mesoderm   Posterior midgut

Gonad precursor cells

Attachment to mesoderm
zfh-1, clb, byn, htl

(C)

Primordial germ cells

Alignment with gonadal mesoderm
abdA, abdB

(D)

(E)   *tin*

Gonad coalescence
foi, cli

(F)

Figure 19.12
Migration of germ cells in *Drosophila* embryos. The left column shows the germ plasm as stained by antibodies against Vasa, a protein component of the germ plasm (D has been counterstained with antibodies to Engrailed protein to show the segmentation, and E and F are dorsal views.) The right column diagrams the movements of the germ cells. (A) The germ cells originate from the pole plasm at the posterior end of the egg. (B) Passive movements carry the PGCs into the posterior midgut. (C) The PGCs move through the endoderm and into the caudal visceral mesoderm by diapedesis. The *wunen* gene product expressed in the endoderm expels the PGCs, while the product of the *columbus* gene expressed in the caudal mesoderm attracts them. (D–F) The movements of the mesoderm bring the PGCs into the region of the tenth through twelfth segments, where the mesoderm coalesces around them to form the gonads. (Photographs from Warrior et al. 1994, courtesy of R. Warrior; diagrams after Howard 1998.)

movements of gastrulation. In the second step, the gut endoderm triggers active amoeboid movement in the PGCs, which travel through the blind end of the posterior midgut, migrating into the visceral mesoderm. In the third step, the PGCs split into two groups, each of which will becomes associated with a developing gonad primordium.

In the fourth step, the PGCs migrate to the gonads, which are derived from the lateral mesoderm of parasegments 10–12 (Warrior 1994; Jaglarz and Howard 1995; Broihier et al. 1998). This step involves both attraction and repulsion. The product of the *wunen* gene appears to be responsible for directing the migration of the PGCs from the endoderm into the mesoderm. This protein is expressed in the endoderm immediately before PGC migration, and it repels the PGCs. In loss-of-function mutants of this gene, the PGCs wander randomly (Zhang et al. 1997). Another gene needed for proper migration of the *Drosophila* PGCs is the product of the *columbus* gene (Moore et al. 1998; Van Doren et al. 1998). This protein is made in the mesodermal cells of the gonad, and it is necessary for the gonad to attract the PGCs. In loss-of-function mutants, the PGCs wander randomly from the endoderm, and if *columbus* is expressed in other tissues (such as the nerve cord), those tissues will attract the PGCs. In the last stage, the gonad coalesces around the germ cells, allowing the germ cells to divide and mature into gametes.

Meiosis

Once in the gonad, the primordial germ cells continue to divide mitotically, producing millions of potential gamete precursors. The PGCs of both male and female gonads are then faced with the necessity of reducing their chromosomes from the diploid to the haploid

(A)

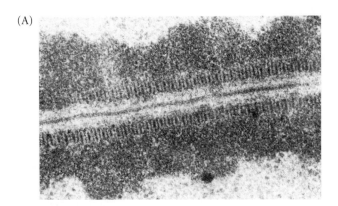

Figure 19.13
The synaptonemal complex. (A) Homologous chromosomes held together at the first meiotic prophase in a *Neottiella* oocyte. (B) Interpretive diagram of the synaptonemal complex structure. (A from von Wettstein 1971, courtesy of D. von Wettstein; B after Schmekel and Daneholt 1995.)

(B)

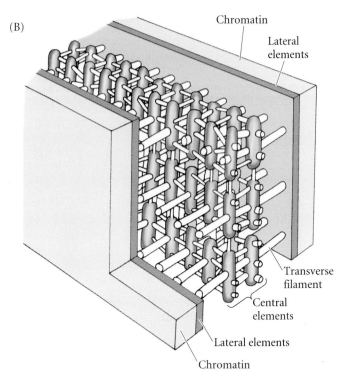

Chromatin

Lateral elements

Transverse filament

Central elements

Lateral elements

Chromatin

condition. In the haploid condition, each chromosome is represented by only one copy, whereas diploid cells have two copies of each chromosome. To accomplish this reduction, the germ cells undergo meiosis.

After the germ cell's last mitotic division, a period of DNA synthesis occurs, so that the cell initiating meiosis doubles the amount of DNA in its nucleus. In this state, each chromosome consists of two sister **chromatids** attached at a common kinetochore (centromere). (In other words, the diploid nucleus contains four copies of each chromosome, but each chromosome is seen as two chromatids bound together.) Meiosis (shown in Figure 2.9) entails two cell divisions. In the first division, homologous chromosomes (e.g., the chromosome 3 pair in the diploid cell) come together and are then separated into different cells. Hence, the first meiotic division separates homologous chromosomes into two daughter cells such that each cell has only one copy of each chromosome. But each of the chromosomes has already replicated (i.e., each has two chromatids). The second meiotic division then separates the two sister chromatids from each other. Consequently, each of the four cells produced by meiosis has a single (haploid) copy of each chromosome.

The first meiotic division begins with a long prophase, which is subdivided into five stages. During the **leptotene** (Greek, "thin thread") stage, the chromatin of the chromatids is stretched out very thinly, and it is not possible to identify individual chromosomes. DNA replication has already occurred, however, and each chromosome consists of two parallel chromatids. At the **zygotene** (Greek, "yoked threads") stage, homologous chromosomes pair side by side. This pairing is called **synapsis**, and it is characteristic of meiosis. Such pairing does not occur during mitotic divisions. Although the mechanism whereby each chromosome recognizes its homologue is not known, synapsis seems to require the presence of

the nuclear membrane and the formation of a proteinaceous ribbon called the **synaptonemal complex**. This complex is a ladderlike structure with a central element and two lateral bars (von Wettstein 1984; Schmekel and Daneholt 1995). The chromatin becomes associated with the two lateral bars, and the chromosomes are thus joined together (Figure 19.13). Examinations of meiotic cell nuclei with the electron microscope (Moses 1968; Moens 1969) suggest that paired chromosomes are bound to the nuclear membrane, and Comings (1968) has suggested that the nuclear envelope helps bring together the homologous chromosomes. The configuration formed by the four chromatids and the synaptonemal complex is referred to as a **tetrad** or a **bivalent**.

During the next stage of meiotic prophase, the chromatids thicken and shorten. This stage has therefore been called the **pachytene** (Greek, "thick thread") stage. Individual chromatids can now be distinguished under the light microscope, and crossing-over may occur. Crossing-over represents exchanges of genetic material whereby genes from one chromatid are exchanged with homologous genes from another chromatid. Crossing-over may continue into the next stage, the **diplotene** (Greek, "double threads") stage. Here, the synaptonemal complex breaks down, and the two homologous chromosomes start to separate. Usually, however, they remain attached at various places called **chiasmata**, which are thought to represent regions where crossing-over is occurring (Figure 19.14). The diplotene stage is characterized by a high level of gene transcription. In some species, the chromosomes of both male and female germ cells take on the "lampbrush" appearance characteristic of chromosomes that are actively

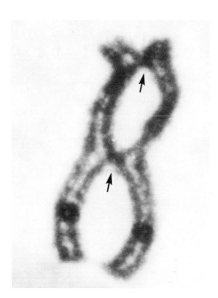

Figure 19.14
Chiasmata in diplotene bivalent chromosomes of salamander oocytes. Kinetochores are visible as darkly stained circles; arrows point to the two chiasmata. (Photograph courtesy of J. Kezer.)

making RNA (see below). During the next stage, **diakinesis** (Greek, "moving apart"), the centromeres move away from each other, and the chromosomes remain joined only at the tips of the chromatids. This last stage of meiotic prophase ends with the breakdown of the nuclear membrane and the migration of the chromosomes to the **metaphase plate**.

During anaphase I, homologous chromosomes are separated from each other in an independent fashion. This stage leads to telophase I, during which two daughter cells are formed, each cell containing one partner of the homologous chromosome pair. After a brief **interkinesis**, the second division of meiosis takes place. During this division, the centromere of each chromosome divides during anaphase so that each of the new cells gets one of the two chromatids, the final result being the creation of four haploid cells. Note that meiosis has also reassorted the chromosomes into new groupings. First, each of the four haploid cells has a different assortment of chromosomes. In humans, in which there are 23 different chromosome pairs, there can be 2^{23} (nearly 10 million) different types of haploid cells formed from the genome of a single person. In addition, the crossing-over that occurs during the pachytene and diplotene stages of prophase I further increases genetic diversity and makes the number of different gametes incalculable.

The events of meiosis appear to be coordinated through cytoplasmic connections between the dividing cells. Whereas the daughter cells formed by mitosis routinely separate from each other, the products of the meiotic cell divisions remain coupled to each other by **cytoplasmic bridges**. These bridges are seen during the formation of sperm and eggs throughout the animal kingdom (Pepling and Spradling 1998).

WEBSITE **19.3 Proteins involved in meiosis.** The phenomenon of homologous pairing and crossing-over is being analyzed in several organisms and may involve DNA repair enzymes.

WEBSITE **19.4 Human meiosis.** Nondisjunction, the failure of chromosomes to sort properly during meiosis, is not uncommon in humans. Its frequency increases with maternal age.

Sidelights & Speculations

Big Decisions: Mitosis or Meiosis? Sperm or Egg?

IN MANY SPECIES, the germ cells migrating into the gonad are bipotential and can differentiate into either sperm or ova, depending on their gonadal environment. When the ovaries of salamanders are experimentally transformed into testes, the resident germ cells cease their oogenic differentiation and begin developing as sperm (Burns 1930; Humphrey 1931). Similarly, in the housefly and mouse, the gonad is able to direct the differentiation of the germ cell (McLaren 1983; Inoue and Hiroyoshi 1986). Thus, in most organisms, the sex of the gonad and of its germ cells is the same.

But what about hermaphroditic animals, in which the change from sperm production to egg production is a naturally occurring physiological event? How is the same animal capable of producing sperm during one part of its life and oocytes during another part? Using *Caenorhabditis elegans*, Kimble and her colleagues identified two "decisions" that presumptive germ cells have to make. The first is whether to enter meiosis or to remain a mitotically dividing stem cell. The second is whether to become an egg or a sperm.

Recent evidence shows that these decisions are intimately linked. The mitotic/meiotic decision is controlled by a single nondividing cell at the end of each gonad, the **distal tip cell**. The germ cell precursors near this cell divide mitotically, forming the pool of germ cells; but as these cells get farther away from the distal tip cell, they enter meiosis. If the distal tip cell is destroyed by a focused laser beam, all the germ cells enter meiosis, and if the distal tip cell is placed in a different location in the gonad, germ line stem cells are generated near its new position (Figure 19.15; Kimble 1981; Kimble and White 1981). The distal tip cell extends long filaments that touch the distal germ cells. The

(A) Intact gonad

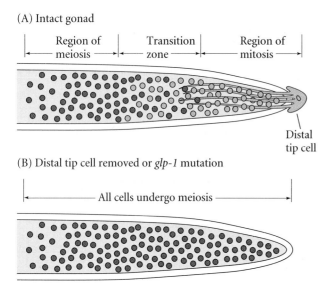

(B) Distal tip cell removed or *glp-1* mutation

Figure 19.15
Regulation of the mitosis-or-meiosis decision by the distal tip cell of the *C. elegans* ovotestis. (A) Intact gonad early in development with regions of mitosis (light-colored cells) and meiosis. The plasma membranes of the distal tip cell's extensions contain the *C. elegans* homologue of Delta, while the PGCs contain the *C. elegans* homologue of Notch. (B) Gonad after laser ablation of the distal tip cell. All germ cells enter meiosis.

extensions contain in their cell membranes the Lag-2 protein, a *C. elegans* homologue of *Delta* (Henderson et al. 1994; Tax et al. 1994; Hall et al. 1999). The Lag-2 protein maintains these germ cells in mitosis and inhibits their meiotic differentiation.

Austin and Kimble (1987) isolated a mutation that mimics the phenotype obtained when the distal tip cells are removed. It is not surprising that this mutation encodes Glp-1, the *C. elegans* homologue of Notch—the receptor for Delta. All the germ cell precursors of nematodes homozygous for the recessive mutation *glp-1* initiate meiosis, leaving no mitotic population. Instead of the 1500 germ cells usually found in the fourth larval stage of hermaphroditic development, these mutants produce only 5 to 8 sperm cells. When genetic chimeras are made in which wild-type germ cell precursors are found within a mutant larva, the wild-type cells are able to respond to the distal tip cells and undergo mitosis. However, when mutant germ cell precursors are found within wild-type larvae, they all enter meiosis. Thus, the *glp-1* gene appears to be responsible for enabling the germ cells to respond to the distal tip cell's signal.*

*The *glp-1* gene appears to be involved in a number of inductive interactions in *C. elegans*. You will no doubt recall that *glp-1* is also needed by the AB blastomere to receive inductive signals from the EMS blastomere to form pharyngeal muscles (see Chapter 8).

After the germ cells begin their meiotic divisions, they still must become either sperm or ova. Generally, in each ovotestis, the most proximal germ cells produce sperm, while the most distal (near the tip) become eggs (Hirsh et al. 1976). This means that the germ cells entering meiosis early become sperm, while those entering meiosis later become eggs. The genetics of this switch are currently being analyzed. The laboratories of Hodgkin (1985) and Kimble (Kimble et al. 1986) have isolated several genes needed for germ cell pathway selection, but the switch appears to involve the activity or inactivity of *fem-3* mRNA. Figure 19.16 presents a scheme for how these genes might function. During early development, the *fem* genes, especially *fem-3*, are critical for the specification of sperm cells. Loss-of-function mutations of these genes convert XX nematodes into females (i.e., spermless hermaphrodites). As long as the FEM proteins are made in the germ cells, sperm are produced. The active *fem* genes are thought to activate the *fog* genes (whose loss-of-function mutations cause the feminization of the germ line and eliminate spermatogenesis). The *fog* gene products activate the genes involved in transforming the germ cell into sperm and also inhibit those genes that would otherwise direct the germ cells to initiate oogenesis.

Oogenesis can begin only when *fem* activity is suppressed. This suppression appears to

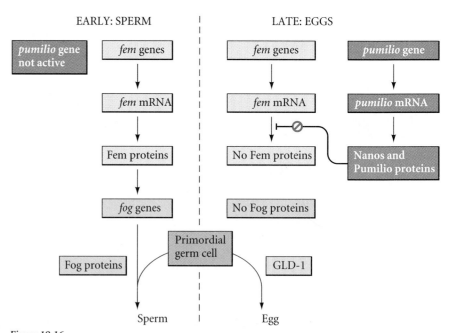

Figure 19.16
Model of sex determination switch in the germ line of *C. elegans* hermaphrodites. Sex determination in somatic tissues, showing a hierarchy of negative regulation. In the early larva, Pumilio is not synthesized, and the *fem* mRNA is able to be translated. The FEM proteins activate the *fog* genes, whose proteins cause the germ cells to undergo spermatogenesis. Later in development, *pumilio* is activated, and combines with Nanos to make a repressor of *fem* translation. Without FEM, the GLD-1 protein can function to make certain the germ cell undergoes oogenesis.

act at the level of RNA translation. The 3′ untranslated region (3′ UTR) of the *fem-3* mRNA contains a sequence that binds a repressor protein during normal development. If this region is mutated such that the repressor cannot bind, the *fem-3* mRNA remains translatable, and oogenesis never occurs. The result is a hermaphrodite body that produces only sperm (Ahringer and Kimble 1991; Ahringer et al. 1992). The *trans*-acting repressor of the *fem-3* message is a combination of the Nanos and Pumilio proteins (the same combination that represses *hunchback* message translation in *Drosophila*). The up-regulation of Pumilio expression may be critical in regulating the germ line switch from spermatogenesis to oogenesis, since Nanos is made constitutively. Nanos appears to be necessary in *C. elegans* (as it is in *Drosophila*) for the survival of all germ line cells (Kraemer et al. 1999). ■

W E B S I T E **19.5 Germ line sex determination in *C. elegans*.** The establishment of whether a germ cell is to become a sperm or an egg involves multiple levels of inhibition. Translational regulation is seen in several of these steps.

Spermatogenesis

While the reductive divisions of meiosis are conserved in every eukaryotic kingdom of life, the regulation of meiosis in mammals differs dramatically between males and females. The differences between **oogenesis**, the production of eggs, and **spermatogenesis**, the production of sperm, are outlined in Table 19.1.

Spermatogenesis is the production of sperm from the primordial germ cells. Once the vertebrate PGCs arrive at the genital ridge of a male embryo, they become incorporated into the sex cords. They remain there until maturity, at which time the sex cords hollow out to form the seminiferous tubules, and the epithelium of the tubules differentiates into the Sertoli cells. The initiation of spermatogenesis during puberty is probably regulated by the synthesis of BMP8B by the spermatogenic germ cells, the **spermatogonia**. When BMP8B reaches a critical concentration, the germ cells begin to differentiate. The differentiating cells produce high levels of BMP8B, which can then further stimulate their differentiation. Mice lacking BMP8B do not initiate spermatogenesis at puberty (Zhao et al. 1996).

The spermatogenic germ cells are bound to the Sertoli cells by N-cadherin molecules on both cell surfaces and by galactosyltransferase molecules on the spermatogenic cells that bind a carbohydrate receptor on the Sertoli cells (Newton et al. 1993; Pratt et al. 1993). The Sertoli cells nourish and protect the developing sperm cells, and spermatogenesis—the developmental pathway from germ cell to mature sperm—occurs in the recesses of the Sertoli cells (Figure 19.17). The processes by which the PGCs generate sperm have been studied in detail in several organisms, but we will focus here on spermatogenesis in mammals.

After reaching the gonad, the PGCs divide to form type A_1 spermatogonia. These cells are smaller than the PGCs and are characterized by an ovoid nucleus that contains chromatin associated with the nuclear membrane. The A_1 spermatogonia are found adjacent to the outer basement membrane of the sex cords. They are stem cells, and at maturity, they are thought to divide so as to make another type A_1 spermatogonium as well as a second, paler type of cell, the type A_2 spermatogonium. Thus, each type A_1 spermatogonium is a stem cell capable of regenerating itself as well as producing a new cell type. The A_2 spermatogonia divide to produce the A_3 spermatogonia, which then beget the type A_4 spermatogonia. It is possible that each of type A spermatogonia are stem cells, capable of self-renewal. The A_4 spermatogonium has three options: it can form another A_4 spermatogonium (self-renewal); it can undergo cell death (apoptosis); or it can differentiate into the first committed stem cell type, the **intermediate spermatogonium**. Intermediate spermatogonia are committed to becoming spermatozoa, and they divide mitotically once to form the type B spermatogonia. These cells are the precursors of the spermatocytes and are the last cells of the line that undergo mitosis. They divide once to generate the **primary spermatocytes**—the cells that enter meiosis. It is not known what causes the spermatogonia to take the path toward differentiation rather than self-renewal; nor is it known what stimulates the cells to enter meiotic rather than mitotic division (Dym 1994).

Looking at Figures 19.17 and 19.18, we find that during the spermatogonial

Table 19.1 Sexual dimorphism in mammalian meioses

Female oogenesis	Male spermatogenesis
Meiosis initiated once in a finite population of cells	Meiosis initiated continuously in a mitotically dividing stem cell population
One gamete produced per meiosis	Four gametes produced per meiosis
Completion of meiosis delayed for months or years	Meiosis completed in days or weeks
Meiosis arrested at first meiotic prophase and reinitiated in a smaller population of cells	Meiosis and differentiation proceed continuously without cell cycle arrest
Differentiation of gamete occurs while diploid, in first meiotic prophase	Differentiation of gamete occurs while haploid, after meiosis ends
All chromosomes exhibit equivalent transcription and recombination during meiotic prophase	Sex chromosomes excluded from recombination and transcription during first meiotic prophase

Source: Handel and Eppig 1998.

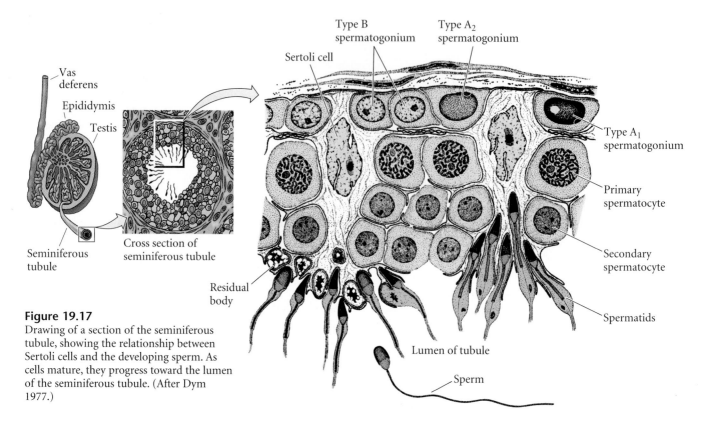

Type B spermatogonium

Type A$_2$ spermatogonium

Sertoli cell

Type A$_1$ spermatogonium

Primary spermatocyte

Secondary spermatocyte

Spermatids

Vas deferens

Epididymis

Testis

Seminiferous tubule

Cross section of seminiferous tubule

Residual body

Lumen of tubule

Sperm

Figure 19.17
Drawing of a section of the seminiferous tubule, showing the relationship between Sertoli cells and the developing sperm. As cells mature, they progress toward the lumen of the seminiferous tubule. (After Dym 1977.)

divisions, cytokinesis is not complete. Rather, the cells form a syncytium whereby each cell communicates with the others via cytoplasmic bridges about 1 μm in diameter (Dym and Fawcett 1971). The successive divisions produce clones of interconnected cells, and because ions and molecules readily pass through these intercellular bridges, each cohort matures synchronously. During this time, the spermatocyte nucleus often transcribes genes whose products will be used later to form the axoneme and acrosome.

Each primary spermatocyte undergoes the first meiotic division to yield a pair of **secondary spermatocytes**, which complete the second division of meiosis. The haploid cells thus formed are called **spermatids**, and they are still connected to one another through their cytoplasmic bridges. The spermatids that are connected in this manner have haploid nuclei, but are functionally diploid, since a gene product made in one cell can readily diffuse into the cytoplasm of its neighbors (Braun et al. 1989). During the divisions from type A$_1$ spermatogonium to spermatid, the cells move farther and farther away from the basement membrane of the seminiferous tubule and closer to its lumen (see Figure 19.17). Thus, each type of cell can be found in a particular layer of the tubule. The spermatids are located at the border of the lumen, and here they lose their cytoplasmic connections and differentiate into sperm cells. In humans, the progression from spermatogonial stem cell to mature sperm takes 65 days (Dym 1994).

WEBSITE **19.6 Gonial syncytia: Bridges to the future.** The products of meiotic divisions are connected by cytoplasmic connections. The functions of these connections may differ between those cells producing sperm and those producing eggs.

SPERMIOGENESIS. The mammalian haploid spermatid is a round, unflagellated cell that looks nothing like the mature vertebrate sperm. The next step in sperm maturation, then, is **spermiogenesis** (or **spermateliosis**), the differentiation of the sperm cell. For fertilization to occur, the sperm has to meet and bind with the egg, and spermiogenesis prepares the sperm for these functions of motility and interaction. The processes of mammalian sperm differentiation is shown in Figure 7.2. The first steps involve the construction of the acrosomal vesicle from the Golgi apparatus. The acrosome forms a cap that covers the sperm nucleus. As the acrosomal cap is formed, the nucleus rotates so that the cap will be facing the basal membrane of the seminiferous tubule. This rotation is necessary because the flagellum is beginning to form from the centriole on the other side of the nucleus, and this flagellum will extend into the lumen. During the last stage of spermiogenesis, the nucleus flattens and condenses, the remaining cytoplasm (the "cytoplasmic droplet") is jettisoned, and the mitochondria form a ring around the base of the flagellum.

One of the major changes in the nucleus is the replacement of the histones by protamines. Transcription of the gene for protamine is seen in the early haploid cells (spermatids), although translation is delayed for several days (Peschon et al. 1987). Protamines are relatively small proteins that are over 60% arginine. During spermiogenesis, the nucleosomes dissociate, and the histones of the haploid nucleus are eventually replaced by protamines. This causes the complete shutdown of transcription in the nucleus and facilitates its assuming an almost crystalline structure. The resulting sperm then enter the lumen of the tubule.

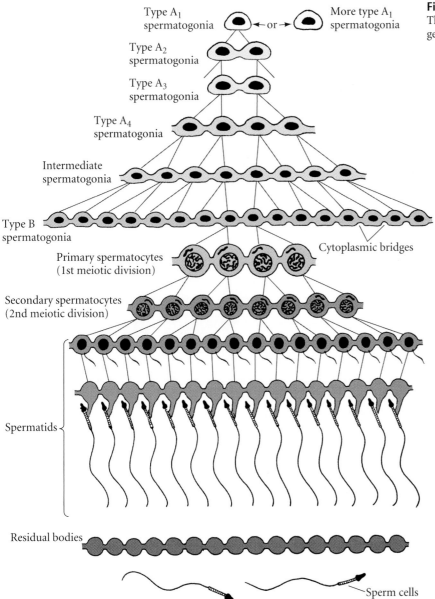

Type A$_1$ spermatogonia ← or → More type A$_1$ spermatogonia

Type A$_2$ spermatogonia

Type A$_3$ spermatogonia

Type A$_4$ spermatogonia

Intermediate spermatogonia

Type B spermatogonia

Primary spermatocytes (1st meiotic division)

Cytoplasmic bridges

Secondary spermatocytes (2nd meiotic division)

Spermatids

Residual bodies

Sperm cells

Figure 19.18
The formation of syncytial clones of human male germ cells. (After Bloom and Fawcett 1975.)

In the mouse, the entire development process from stem cell to spermatozoon takes 34.5 days. The spermatogonial stages last 8 days, meiosis lasts 13 days, and spermiogenesis takes up another 13.5 days. In humans, spermatic development takes nearly twice as long to complete. Because the type A$_1$ spermatogonia are stem cells, spermatogenesis can occur continuously. Each day, some 100 million sperm are made in each human testicle, and each ejaculation releases 200 million sperm. Unused sperm are either resorbed or passed out of the body in urine. During his lifetime, a human male can produce 10^{12} to 10^{13} sperm (Reijo et al. 1995).

VADE MECUM **Spermatogenesis in mammals.** The development of sperm is visualized with color-coded histological sections through a mammalian testis. Each stage is shown, from spermatogonium to flagellated spermatid.
[Click on Gametogenesis]

WEBSITE **19.7 Gene expression during spermatogenesis.** Transcription occurs both from the diploid spermatocyte nucleus and from the haploid spermatid nuclei. Posttranscriptional control is also important in regulating sperm gene expression.

WEBSITE **19.8 The Nebenkern.** Sperm mitochondria are often highly modified to fit the streamlined cell. The mitochondria of flies fuse together to form a structure called the Nebenkern, and this fusion is controlled by the *fuzzy onions* gene.

Oogenesis

Oogenic meiosis

Oogenesis—the differentiation of the ovum—differs from spermatogenesis in several ways. Whereas the gamete formed by spermatogenesis is essentially a motile nucleus, the gamete formed by oogenesis contains all the materials needed to initiate and maintain metabolism and development. Therefore, in addition to forming a haploid nucleus, oogenesis also builds up a store of cytoplasmic enzymes, mRNAs, organelles, and metabolic substrates. While the sperm becomes differentiated for motility, the egg develops a remarkably complex cytoplasm.

The mechanisms of oogenesis vary among species more than those of spermatogenesis. This difference should not be surprising, since patterns of reproduction vary so greatly among species. In some species, such as sea urchins and frogs, the female routinely produces hundreds or thousands of eggs at a time, whereas in other species, such as humans and most mammals, only a few eggs are produced during the lifetime of an individual. In those species that produce thousands of ova, the **oogonia** are self-renewing stem cells that endure for the lifetime of the organism. In those species that produce fewer eggs, the oogonia divide to form a limited number of egg precursor cells. In the human embryo, the thousand or so oogonia divide rapidly from the second to the seventh month of gestation to form roughly 7 million germ cells (Figure 19.19). After the seventh month of embryonic development, however, the number of germ cells drops precipitously. Most oogonia die during this period, while the remaining oogonia enter the first meiotic division (Pinkerton et al. 1961). These latter cells, called the **primary oocytes**, progress through the first

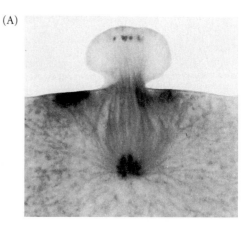

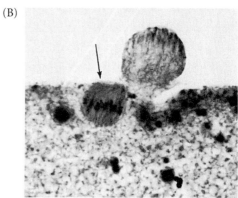

Figure 19.20
Polar body formation in the oocyte of the whitefish *Coregonus*. (A) Anaphase of first meiotic division, showing the first polar body pinching off with its chromosomes. (B) Metaphase (within the oocyte, arrow) of the second meiotic division, with the first polar body still in place. The first polar body may or may not divide again. (From Swanson et al. 1981; photographs courtesy of C. P. Swanson.)

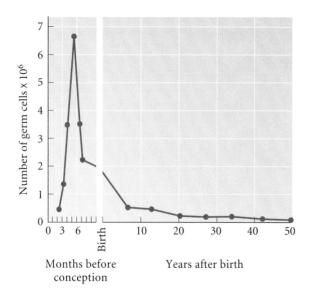

Figure 19.19
Changes in the number of germ cells in the human ovary over the life span. (After Baker 1970.)

meiotic prophase until the diplotene stage, at which point they are maintained until puberty. With the onset of adolescence, groups of oocytes periodically resume meiosis. Thus, in the human female, the first part of meiosis begins in the embryo, and the signal to resume meiosis is not given until roughly 12 years later. In fact, some oocytes are maintained in meiotic prophase for nearly 50 years. As Figure 19.19 indicates, primary oocytes continue to die even after birth. Of the millions of primary oocytes present at birth, only about 400 mature during a woman's lifetime.

Oogenic meiosis also differs from spermatogenic meiosis in its placement of the metaphase plate. When the primary oocyte divides, its nucleus, called the **germinal vesicle**, breaks down, and the metaphase spindle migrates to the periphery of the cell. At telophase, one of the two daughter cells contains hardly any cytoplasm, whereas the other cell has nearly the entire volume of cellular constituents (Figure 19.20). The smaller cell is called the **first polar body**, and the larger cell is referred to as the **secondary oocyte**. During the second division of meiosis, a similar unequal cytokinesis takes place. Most of the cytoplasm is retained by the mature egg (ovum), and a second polar body receives little more than a haploid nucleus. Thus, oogenic meiosis conserves the volume of oocyte cytoplasm in a single cell rather than splitting it equally among four progeny.

In a few species of animals, meiosis is severely modified such that the resulting gamete is diploid and need not be fertilized to develop. Such animals are said to be **parthenogenetic** (Greek, "virgin birth"). In the fly *Drosophila mangabeirai*, one of the polar bodies acts as a sperm and "fertilizes" the oocyte after the second meiotic division. In other insects (such as *Moraba virgo*) and in the lizard *Cnemidophorus uniparens*, the oogonia double their chromosome number before meiosis, so that the halving of the chromosomes restores the diploid number. The germ cells of the grasshopper *Pycnoscelus surinamensis* dispense with meiosis altogether, forming diploid ova by two mitotic divisions (Swanson et al. 1981). All of these species consist entirely of females. In other species, haploid parthenogenesis is widely used not only as a means of reproduction, but also as a mechanism of sex determination. In the Hymenoptera (bees, wasps, and ants), unfertilized eggs develop into males, whereas fertilized eggs, being diploid, develop into females. The haploid males are able to produce sperm by abandoning the first meiotic division, thereby forming two sperm cells through second meiosis.

Maturation of the oocyte in amphibians

The egg is responsible for initiating and directing development, and in some species (as seen above), fertilization is not even necessary. The accumulated material in the oocyte cytoplasm includes energy sources and energy-producing organelles (the yolk and mitochondria); the enzymes and precursors for DNA, RNA, and protein syntheses; stored messenger RNAs; structural proteins; and morphogenetic regulatory factors that control early embryogenesis. A partial catalogue of the materials stored in the oocyte cytoplasm is shown in Table 19.2, while a partial list of stored mRNAs is shown in Table 5.3. Most of this accumulation takes place during meiotic prophase I, and this stage is often subdivided into two phases, **previtellogenesis** (Greek, "before yolk formation") and **vitellogenesis**.

The eggs of fishes and amphibians are derived from an oogonial stem cell population that can generate a new cohort of oocytes each year. In the frog *Rana pipiens*, oogenesis takes 3 years. During the first 2 years, the oocyte increases its size very gradually. During the third year, however, the rapid accumulation of yolk in the oocyte causes the egg to swell to its characteristically large size (Figure 19.21). Eggs mature in yearly batches, with the first cohort maturing shortly after metamorphosis; the next group matures a year later.

Table 19.2 Cellular components stored in the mature oocyte of *Xenopus laevis*

Component	Approximate excess over amount in larval cells
Mitochondria	100,000
RNA polymerases	60,000–100,000
DNA polymerases	100,000
Ribosomes	200,000
tRNA	10,000
Histones	15,000
Deoxyribonucleoside triphosphates	2,500

Source: After Laskey 1979.

Vitellogenesis occurs when the oocyte reaches the diplotene stage of meiotic prophase. Yolk is not a single substance, but a mixture of materials used for embryonic nutrition. The major yolk component in frog eggs is a 470-kDa protein called **vitellogenin**. It is not made in the frog oocyte (as are the major yolk proteins of organisms such as annelids and crayfishes), but is synthesized in the liver and carried by the bloodstream to the ovary (Flickinger and Rounds 1956). This large protein passes between the follicle cells of the ovary, and is incorporated into the oocyte by **micropinocytosis**, the

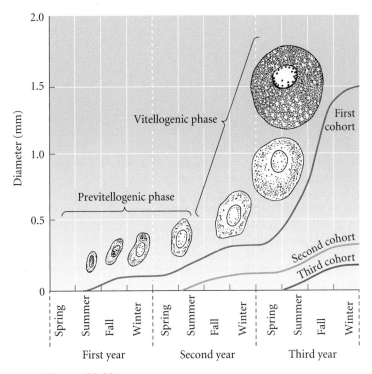

Figure 19.21
Growth of oocytes in the frog. During the first 3 years of life, three cohorts of oocytes are produced. The drawings follow the growth of the first-generation oocytes. (After Grant 1953.)

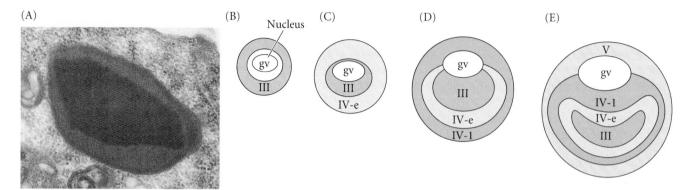

Figure 19.22

Yolk distribution in *Xenopus*. (A) An amphibian yolk platelet. (B–E) Establishment of animal-vegetal distribution of yolk platelets in the *Xenopus* oocyte. (B) In the late stage III (600 μm) oocyte, yolk platelets enter the cell equally at all points of the surface. (C, D) As the oocyte grows, the platelets at the future animal pole are displaced toward the vegetal pole, while those at the vegetal pole remain there. More yolk still enters the egg on all sides. (E) By the end of vitellogenesis, the earliest platelets (III) are all in the vegetal hemisphere, which now contains roughly 75% of the oocyte yolk. Timing of entry of yolk into oocyte platelets is indicated by shading and Roman numerals: III, stage III platelets; IV-e, early stage IV platelets; IV-l, late stage IV platelets; V, stage V platelets; gv, germinal vesicle. (After Danilchik and Gerhart 1987; photograph courtesy of L. K. Opresko.)

pinching off of membrane-bounded vesicles at the bases of microvilli (Dumont 1978). In the mature oocyte, vitellogenin is split into two smaller proteins: the heavily phosphorylated-**phosvitin** and the lipoprotein **lipovitellin**. These two proteins are packaged together into membrane-bounded **yolk platelets** (Figure 19.22A). Glycogen granules and lipochondrial inclusions store the carbohydrate and lipid components of the yolk, respectively.

Most eggs are highly asymmetrical, and it is during oogenesis that the animal-vegetal axis of the egg is specified. Danilchik and Gerhart (1987) have shown that although the concentration of yolk increases nearly tenfold as one moves from the animal to the vegetal poles of the mature *Xenopus* egg, vitellogenin uptake is uniform around the surface of the oocyte. What varies is its movement within the oocyte, and this depends on where the yolk proteins enter. When yolk platelets are formed in the future animal hemisphere, they move inward toward the center of the cell. Vegetal yolk platelets, however, do not actively move, but remain at the periphery of the cell for long periods of time, enlarging as they stay there. They are slowly displaced from the cortex as new yolk platelets come in from the surface. As a result of this differential intracellular transport, the amount of yolk steadily increases in the vegetal hemisphere, until the vegetal half of a mature *Xenopus* oocyte contains nearly 75% of the yolk (Figure 19.22B–E). The mechanism of this translocation remains unknown.

As the yolk is being deposited, the organelles also become arranged asymmetrically. The cortical granules begin to form from the Golgi apparatus; they are originally scattered randomly through the oocyte cytoplasm, but later migrate to the periphery of the cell. The mitochondria replicate at this time, dividing to form millions of mitochondria that will be apportioned to the different blastomeres during cleavage. (In *Xenopus*, new mitochondria will not be formed until after gastrulation is initiated.) As vitellogenesis nears an end, the oocyte cytoplasm becomes stratified. The cortical granules, mitochondria, and pigment granules are found at the periphery of the cell, within the actin-rich oocyte cortex. Within the inner cytoplasm, distinct gradients emerge. While the yolk platelets become more heavily concentrated at the vegetal pole of the oocyte, the glycogen granules, ribosomes, lipid vesicles, and endoplasmic reticulum are found toward the animal pole. Even specific mRNAs stored in the cytoplasm become localized to certain regions of the oocyte.

While the precise mechanisms for establishing these gradients remain unknown, studies using inhibitors have shown that the cytoskeleton is critically important in localizing specific RNAs and morphogenetic factors. There seem to be two pathways for getting mRNAs into the vegetal cortex (Forristall et al. 1995; Kloc and Etkin 1995, Klok et al. 1998). The first pathway moves messages such as those encoding the Vg1 protein, which are initially present throughout the oocyte, into the vegetal cortex in a two-step process (Yisraeli et al. 1990). In the first phase, microtubules are needed to bring *Vg1* mRNA into the vegetal hemisphere. In the second phase, microfilaments are responsible for anchoring the *Vg1* message to the cortex. The portion of the *Vg1* mRNA that binds to these cytoskeletal elements resides in its 3′ untranslated region. When a specific 340-base sequence from the Vg1 3′ UTR is placed onto a β-globin message, that β-globin mRNA is similarly localized to the vegetal cortex (see Chapter 5; Mowry and Melton 1992). Other mRNAs, including germ plasm mRNAS such as *Xlsirt* and *Xcat2*, leave the germinal vesicle and join the mitochondrial "cloud" located at the vegetal pole of the nucleus. These messages are compartmentalized into clusters associated with the germ plasm and transported to the vegetal cortex in a manner that appears to be independent of the cytoskeleton (Figure 19.23; Kloc et al. 1996). This mechanism is known as the Metro (message transport organizer) pathway.

WEBSITE **19.9 Hormonal control of yolk production.** Vitellogenesis in amphibians is mediated primarily by estrogen. Estrogen instructs the liver to express and secrete vitellogenin, and this protein is absorbed from the blood by the young oocyte.

WEBSITE **19.10 Transporting the *Vg1* mRNA.** The Vera protein specifically binds to the 3′ UTR of the *Vg1* message. Vera may link *Vg1* mRNA to a set of endoplasmic reticulum vesicles that are translocated to the vegetal cortex.

WEBSITE **19.11 Establishment of egg polarity.** In several species, the developing oocyte is a flagellated cell whose flagellum marks the future animal pole of the egg. This flagellum is lost during oogenesis.

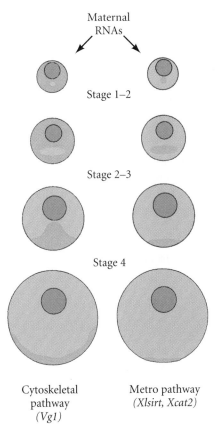

Figure 19.23
Schematic representations of the two pathways for localizing mRNAs to the vegetal region of the *Xenopus* oocyte. The respective mRNAs are shown in yellow. In the cytoskeletal (Vg1) pathway, messages are first seen throughout the egg, but they are translocated through a microtubule-driven system to the microfilaments of the vegetal cortex. The Metro (message transport organizer) pathway accumulates messages in the mitochondrial cloud, and islands of them are transported into the vegetal cortex. (After Kloc and Etkin 1995.)

Completion of amphibian meiosis: Progesterone and fertilization

Amphibian oocytes can remain for years in the diplotene stage of meiotic prophase. This state resembles the G_2 phase of the cell division cycle (see Chapter 8). Resumption of meiosis in the amphibian primary oocyte requires progesterone. This hormone is secreted by the follicle cells in response to gonadotropic hormones secreted by the pituitary gland. Within 6 hours of progesterone stimulation, **germinal vesicle breakdown** (**GVBD**) occurs, the microvilli retract, the nucleoli disintegrate, and the chromosomes contract and migrate to the animal pole to begin division. Soon afterward, the first meiotic division occurs, and the mature ovum is released from the ovary by a process called **ovulation**. The ovulated egg is in second meiotic metaphase when it is released (Figure 19.24).

How does progesterone enable the egg to break its dormancy and resume meiosis? To understand the mechanisms by which this activation is accomplished, it is necessary to briefly review the model for early blastomere division (see Chapter 8). Entry into the mitotic (M) phase of the cell cycle (in both meiosis and mitosis) is regulated by **mitosis-promoting factor**, or **MPF** (originally called "maturation-promoting factor" after its meiotic function). MPF contains two subunits, **cyclin B** and the **p34^cdc2** protein. The p34 protein is a cyclin-dependent-kinase—its activity is dependent upon the presence of cyclin. Since all the components of MPF are present in the amphibian oocyte, it is generally thought that progesterone somehow converts a pre-MPF complex into active MPF.

The mediator of the progesterone signal is the **c-mos** protein. Progesterone reinitiates meiosis by causing the egg to polyadenylate the maternal *c-mos* mRNA that has been stored in its cytoplasm (Sagata et al. 1988, 1989; Sheets et al. 1995). This message is translated into a 39-kDa phosphoprotein, known as c-mos. This protein is detectable only during oocyte maturation and is destroyed quickly upon fertilization. Yet during its brief lifetime, it plays a major role in releasing the egg from its dormancy. If the translation of c-mos is inhibited (by injecting c-mos antisense mRNA into the oocyte), germinal vesicle breakdown and the resumption of oocyte maturation do not occur. The c-mos protein activates a phosphorylation cascade that phosphorylates and activates the p34 subunit of MPF (Ferrell and Machleder 1998; Ferrell 1999). The active MPF allows the germinal vesicle to break down and the chromosomes to divide.

However, the chromosomes then encounter a second block. MPF can take the chromosomes through only the first meiotic division and the prophase of the second meiotic division. The oocyte is arrested again in the metaphase of the second meiotic division. This metaphase block is caused by the combined actions of c-mos and another protein, cyclin-dependent kinase 2 (cdk2; Gabrielli et al. 1993). These two proteins are subunits of **cytostatic factor** (**CSF**), which is found in mature frog eggs, and which can block cell cycles in

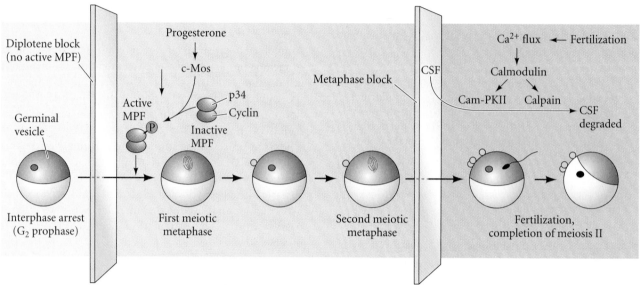

Figure 19.24
Schematic representation of *Xenopus* oocyte maturation, showing the regulation of meiotic cell division by progesterone and fertilization. Oocyte maturation is blocked in the diplotene stage of first meiotic prophase by the lack of active MPF. Progesterone activates the production of c-mos, and the c-mos protein initiates a cascade of phosphorylation that eventually phosphorylates the p34 component of MPF. This allows the MPF to become active. The MPF drives the cell cycle through the first meiotic division, but further division is blocked by CSF, a compound containing c-mos and cdk2. Upon fertilization, calcium ions released into the cytoplasm are bound by calmodulin and are used to activate two enzymes, calmodulin-dependent protein kinase II and calpain II, which inactivate and degrade CSF. Second meiosis is completed, and the two haploid pronuclei can fuse. At this time, cyclin B is resynthesized, allowing first cell cycle of cleavage to begin.

metaphase (Matsui 1974). It is thought that CSF prevents the degradation of cyclin (Figure 19.24).

The metaphase block is broken by fertilization. Evidence suggests that the calcium ion flux attending fertilization enables the calcium-binding protein **calmodulin** to become active. Calmodulin, in turn, can activate two enzymes that inactivate CSF: calmodulin-dependent protein kinase II, which inactivates p34, and calpain II, a calcium-dependent protease that degrades c-mos (Watanabe et al. 1989; Lorca et al. 1993). Without CSF, cyclin can be degraded, and the meiotic division can be completed.

Gene transcription in oocytes

In most animals (insects being a major exception), the growing oocyte is active in transcribing genes whose products are (1) necessary for cell metabolism, (2) necessary for oocyte-specific processes, or (3) needed for early development before the zygote-derived nuclei begins to function. In mice, for instance, the growing diplotene oocyte is actively transcribing

the genes for zona pellucida proteins ZP1, ZP2, and ZP3. Moreover, these genes are transcribed only in the oocyte and not in any other cell (Figure 19.25; Roller et al. 1989; Lira et al. 1990; Epifano et al. 1995).

The amphibian oocyte has certain periods of very active RNA synthesis. During the diplotene stage, certain chromosomes stretch out large loops of DNA, causing the chromosome to resemble a lampbrush (a handy instrument for cleaning test tubes in the days before microfuges). These **lampbrush chromosomes** (Figure 19.26) can be revealed as the sites of RNA synthesis by in situ hybridization. Oocyte chromosomes can be incubated with a radioactive RNA probe, and autoradiography used to visualize the precise location where the gene is being transcribed. Figure 19.26B shows diplotene chromosome I of the newt *Triturus cristatus* after incubation with radioactive histone mRNA. It is obvious that a histone gene (or set of histone genes) is located on one of these loops of the lampbrush chromosome (Old et al. 1977). Electron micrographs of gene transcripts from lampbrush chromosomes also enable one to see chains of mRNA coming off each gene as it is transcribed (Hill and MacGregor 1980).

In addition to mRNA synthesis, the patterns of ribosomal RNA and transfer RNA transcription are also regulated during oogenesis. Figure 19.27A shows the course of rRNA and tRNA synthesis during *Xenopus* oogenesis. Transcription appears to begin in early (stage I, 25–40 μm) oocytes, during the diplotene stage of meiosis. At this time, all the rRNAs and tRNAs needed for protein synthesis until the mid-blastula stage are made, and all the maternal mRNAs needed for early development are transcribed. This stage lasts for months in *Xenopus*. The rate of ribosomal RNA production is prodigious. The *Xenopus* oocyte genome has over 1800 genes encoding 18S and 28S rRNA (the two large RNAs that form the ribosomes), and these genes are selectively amplified such that

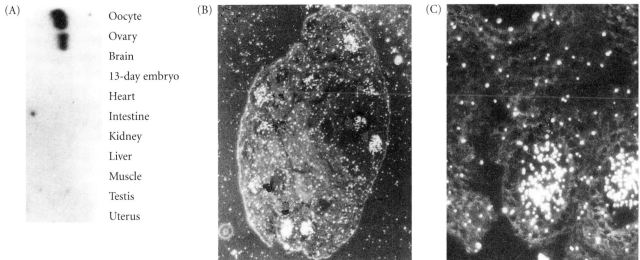

Oocyte

Ovary

Brain

13-day embryo

Heart

Intestine

Kidney

Liver

Muscle

Testis

Uterus

Figure 19.25
Expression of the *ZP3* gene in the developing mouse oocyte. (A) Northern blot of *ZP3* mRNA accumulation in embryonic mouse tissues. A radioactive probe to the ZP3 message found it expressed only in the ovary, and specifically in the oocytes. (B–C) When the *luciferinase* reporter gene is placed onto the *ZP3* promoter and inserted into the mouse genome, the *luciferinase* message is seen only in the developing oocytes of the ovary. C is a higher magnification of a section of B, showing two of the ovarian follicles containing maturing oocytes. (A from Roller et al. 1989; B–C from Lira et al. 1990; all photographs courtesy of P. Wassarman.)

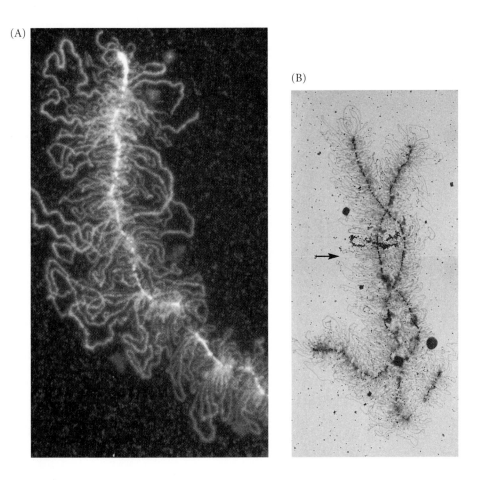

Figure 19.26
Amphibian lampbrush chromosomes are active in the diplotene germinal vesicle during first meiotic prophase. (A) A lampbrush chromosome of the salamander *Notophthalmus viridescens*. Extended DNA (white) loops out and is transcribed into RNA (red). (B) Localization (arrow) of histone genes on a lampbrush chromosome in an amphibian oocyte. The genes have been visualized by in situ hybridization and autoradiography. (A courtesy of M. B. Roth and J. Gall; B from Old et al. 1977, courtesy of H. G. Callan.)

there are over 500,000 genes making these ribosomal RNAs (Figure 19.27B; Brown and Dawid 1968). After reaching a certain size, the chromosomes of the mature (stage VI) oocyte condense, and the genes are no longer transcribed. This "mature oocyte" condition can also last for months. Upon hormonal stimulation, the oocyte completes its first meiotic division and is ovulated. The mRNAs stored by the oocyte now join with the ribosomes to initiate protein synthesis. Within hours, the second meiotic division has begun, and the secondary oocyte has been fertilized. The embryo's genes do not begin active transcription until the mid-blastula transition (Davidson 1986).

As we have seen in Chapter 5, the oocytes of several species make two classes of mRNAs—those for immediate use in the oocyte and those that are stored for use during early development. In sea urchins, the translation of stored maternal messages is initiated by fertilization, while in frogs, the signal for such translation is initiated by progesterone as the egg is about to be ovulated. One of the results of the MPF activity induced by progesterone may be the phosphorylation of proteins on the 3′ UTR of stored oocyte mRNAs. The phosphorylation of these factors is associated with the lengthening of the poly(A) tails in the stored messages and the translation of the stored mRNAs (Paris et al. 1991).

W E B S I T E   **19.12 Synthesizing oocyte ribosomes**. Ribosomes are almost a "differentiated product" of the oocyte, and the *Xenopus* oocyte contains 20,000 times as many ribosomes as somatic cells do. Gene repetition and gene amplification are both used to transcribe these enormous amounts of rRNA.

Meroistic oogenesis in insects

There are several types of oogenesis in insects, but most studies have focused on those insects, such as *Drosophila* and moths, that undergo **meroistic oogenesis**, in which cytoplasmic connections remain between the cells produced by the oogonium. In *Drosophila*, each oogonium divides four times to produce a clone of 16 cells connected to each other by **ring canals**. The production of these interconnected cells, called **cystocytes**, involves a highly ordered array of cell divisions (Figure 19.28). Only those two cells having four interconnections are capable of developing into oocytes, and of those two, only one becomes the egg. The other begins meiosis but does not complete it. Thus, only one of the 16 cystocytes can become an ovum. All the other cells become **nurse cells**. As it turns out, the cell destined to become the oocyte is that cell residing at the most posterior tip of the egg chamber, or **ovariole**, that encloses the 16-cell clone. However, since the nurse cells are connected to the oocyte by cytoplasmic bridges, the entire complex can be seen as one egg-producing unit.

THE STRUCTURE OF THE MEROISTIC OVARY. The meroistic ovary confronts us with some interesting problems. If all the cystocytes are connected so that proteins and RNAs shuttle freely among them, why should they have different developmental fates? Why should one cell become the oocyte while the others become "RNA-synthesizing factories," sending mRNAs, ribosomes, and even centrioles into the oocyte? Why is the flow of protein and RNA in one direction only? As the

(A)

Figure 19.27
Ribosomal RNA production in *Xenopus* oocytes. (A) Relative rates of DNA, tRNA, and rRNA synthesis in amphibian oogenesis during the last 3 months before ovulation. (B) The transcription of the RNA precursor of the 28S, 18S, and 5.8S ribosomal RNAs. These units are tandemly linked together, some 450 per haploid genome. (A after Gurdon 1976; B courtesy of O. L. Miller, Jr.)

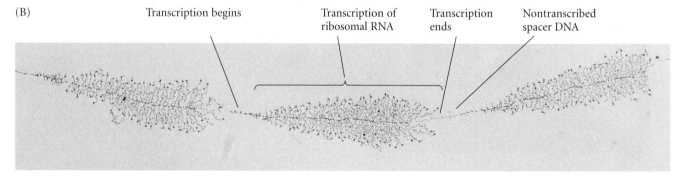

(B)   Transcription begins   Transcription of ribosomal RNA   Transcription ends   Nontranscribed spacer DNA

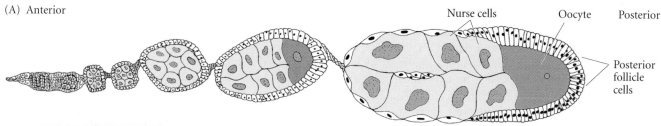

(A) Anterior — Nurse cells — Oocyte — Posterior — Posterior follicle cells

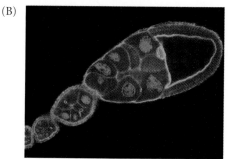

(B)

Figure 19.28
The formation of 16 interconnected cystocytes in *Drosophila*. (A) Diagram of adult ovariole, showing sequence of oogenesis as younger germinal cysts mature within the ovariole. (B) Section through center of an ovariole, showing the maturation of the oocyte. Actin is stained green, and the nuclear DNA is stained red. (C) Division of the cystocyte-forming cells (cystoblasts). The cells are represented schematically as dividing in a single plane. The stem cell divides to produce another stem cell plus an oogonium that is committed to forming the cystocytes. When the oogonium divides, the centriole of daughter cystocyte 1 retains the fusome (red), which grows through the ring canal toward its mitotic sister. The arrow shows the polarity, pointing to the cell from which the fusome grew. After three more mitotic divisions, the 16-cell cyst is formed. If intracellular transport is coordinated by the fusome, the transport of mRNAs and proteins would be toward cystocyte 1, which would thus become the oocyte. (A after Ruohola et al. 1991; B courtesy of B. M. Mechler; C after Lin and Spradling 1995.)

(C)

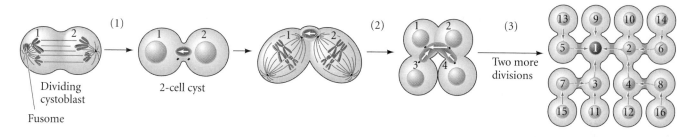

Dividing cystoblast — Fusome — (1) 2-cell cyst — (2) — (3) Two more divisions

cystocytes divide, a large, spectrin-rich structure called the **fusome** forms and spans the ring canals between the cells (Figure 19.28C). It is constructed asymmetrically, as it always grows from the spindle pole that remained in one of the cells after the first division (Lin and Spradling 1995; de Cuevas and Spradling 1998). The cell that retained the greater part of the fusome during the first division becomes the oocyte. It is not yet known if the fusome contains oogenic determinants, or if it directs the traffic of materials into this particular cell.

Once the patterns of transport are established, the cytoskeleton becomes actively involved in transporting mRNAs from the nurse cells into the oocyte cytoplasm (Cooley and Theurkauf 1994). The microtubular array is critical for oocyte determination. If this lattice is disrupted (either chemically or by mutations such as *bicaudal-D* or *egalitarian*), gene products are transmitted in all directions, and all 16 cells differentiate into nurse cells (Gutzeit 1986; Theurkauf et al. 1992, 1993; Spradling 1993). It is possible that some compounds transported from the nurse cells into the oocyte become associated with transport proteins, such as kinesin, that would enable them to travel along the tracks of microtubules extending through the ring canals (Theurkauf et al. 1992; Sun and Wyman 1993). Actin may become important for maintaining this polarity during later stages of oogenesis. Mutations that prevent actin mi-

crofilaments from lining the ring canals prevent the transport of mRNAs from nurse cells to oocyte, and disruption of the actin microfilaments randomizes the distribution of mRNA (Cooley et al. 1992; Watson et al. 1993). Thus, the cytoskeleton appears to control the movement of organelles and RNAs between nurse cells and oocyte such that developmental cues are exchanged only in the appropriate direction.

TRANSPORT OF RNA FROM NURSE CELLS TO OOCYTE. The oocytes of meroistic insects do not pass through a transcriptionally active stage, nor do they have lampbrush chromosomes. Rather, RNA synthesis is largely confined to the nurse cells, and the RNA made by those cells is actively transported into the oocyte cytoplasm. This can be seen in Figure 9.36A. Oogenesis takes place in only 12 days, so the nurse cells are very metabolically active during this time. They are aided in their transcriptional efficiency by becoming polytene. Instead of having two copies of each chromosome, they replicate their chromosomes until they have produced 512 copies. The 15 nurse cells pass ribosomal and messenger RNAs as well as proteins into the oocyte cytoplasm, and entire ribosomes may be transported as well. The mRNAs do not associate with polysomes, which suggests that they are not immediately active in protein synthesis (Paglia et al. 1976; Telfer et al. 1981).

Oogenesis in mammals

Ovulation of the mammalian egg follows one of two basic patterns, depending on the species. One type of ovulation is stimulated by the act of copulation. Physical stimulation of the cervix triggers the release of gonadotropins from the pituitary. These gonadotropins signal the egg to resume meiosis and initiate the events that will expel it from the ovary. This mechanism ensures that most copulations will result in fertilized ova, and animals that utilize this method of ovulation—rabbits and minks—have a reputation for procreative success.

Most mammals, however, have a periodic ovulation pattern, in which the female ovulates only at specific times of the year. This ovulatory time is called **estrus** (or its English equivalent, "heat"). In these cases, environmental cues, most notably the amount and type of light during the day, stimulate the hypothalamus to release gonadotropin-releasing factor. This factor stimulates the pituitary to release its gonadotropins—follicle-stimulating hormone (FSH) and luteinizing hormone (LH)—which cause the follicle cells to proliferate and secrete estrogen. The estrogen enters certain neurons and evokes the pattern of mating behavior characteristic of the species. The gonadotropins also stimulate follicular growth and initiate ovulation. Thus, estrus and ovulation occur close together.

Humans have a variation on the theme of periodic ovulation. Although human females have cyclical ovulation (averaging about once every 29.5 days) and no definitive yearly estrus, most of human reproductive physiology is shared with other primates. The characteristic primate periodicity in maturing and releasing ova is called the **menstrual cycle** because it entails the periodic shedding of blood and endothelial tissue from the uterus at monthly intervals.* The menstrual cycle represents the integration of three very different cycles: (1) the ovarian cycle, the function of which is to mature and release an oocyte, (2) the uterine cycle, the function of which is to provide the appropriate environment for the developing blastocyst, and (3) the cervical cycle, the function of which is to allow sperm to enter the female reproductive tract only at the appropriate time. These three functions are integrated through the hormones of the pituitary, hypothalamus, and ovary.

VADE MECUM **Oogenesis in mammals.** The development of the mammalian ovum and its remarkable growth during its primary oocyte stage is the subject of photographs and QuickTime movies of histological sections through a mammalian ovary.
[Click on Gametogenesis]

*The periodic shedding of the uterine lining is a controversial topic. Some scientists speculate that menstruation is an active process, with adaptive significance in evolution. Profet (1993) proposed that menstruation is a crucial immunological adaptation, protecting the uterus against infections from semen or other environmental agents. Strassmann (1996) suggests that the cyclicity of the endometrium is an energy-saving adaptation that is important in times of poor nutrition. Vaginal bleeding would be a side effect of this adaptive process. Finn (1998) claims that menstruation has no adaptive value and is necessitated by the immunological crises that are a consequence of bringing two genetically dissimilar organisms together in the uterus. Martin (1992) points out that it might even be wrong to think of there being a single function of menstruation, and that its roles might change during a woman's life cycle.

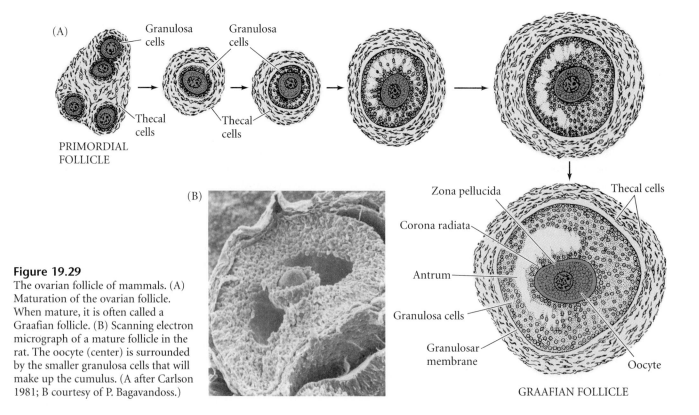

Figure 19.29
The ovarian follicle of mammals. (A) Maturation of the ovarian follicle. When mature, it is often called a Graafian follicle. (B) Scanning electron micrograph of a mature follicle in the rat. The oocyte (center) is surrounded by the smaller granulosa cells that will make up the cumulus. (A after Carlson 1981; B courtesy of P. Bagavandoss.)

Figure 19.30
The human menstrual cycle. The coordination of (B) ovarian and (D) uterine cycles is controlled by (A) the pituitary and (C) the ovarian hormones. During the follicular phase, the egg matures within the follicle, and the uterine lining is prepared to receive the blastocyst. The mature egg is released around day 14. If a blastocyst does not implant into the uterus, the uterine wall begins to break down, leading to menstruation.

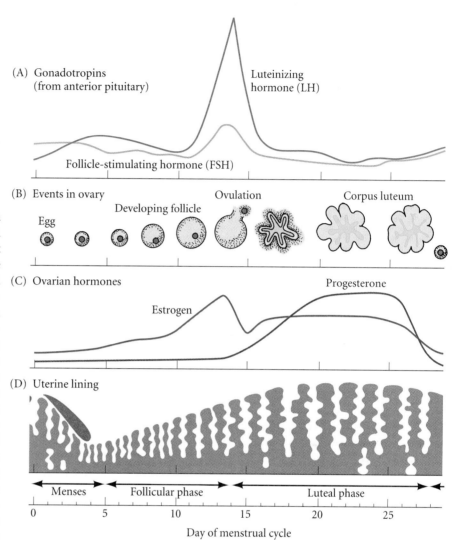

The majority of the oocytes within the adult human ovary are maintained in the prolonged diplotene stage of the first meiotic prophase (often referred to as the **dictyate state**). Each oocyte is enveloped by a primordial follicle consisting of a single layer of epithelial granulosa cells and a less organized layer of mesenchymal thecal cells (Figure 19.29). Periodically, a group of primordial follicles enters a stage of follicular growth. During this time, the oocyte undergoes a 500-fold increase in volume (corresponding to an increase in oocyte diameter from 10 μm in a primordial follicle to 80 μm in a fully developed follicle). Concomitant with oocyte growth is an increase in the number of follicular granulosa cells, which form concentric layers around the oocyte. This proliferation of granulosa cells is mediated by a paracrine factor, GDF9, a member of the TGF-β family (Dong et al. 1996). Throughout this growth period, the oocyte remains in the dictyate stage. The fully grown follicle thus contains a large oocyte surrounded by several layers of granulosa cells. The innermost of these cells will stay with the ovulated egg, forming the **cumulus**, which surrounds the egg in the oviduct. In addition, during the growth of the follicle, an **antrum** (cavity) forms, which becomes filled with a complex mixture of proteins, hormones, and other molecules. Just as the maturing oocyte synthesizes paracrine factors that allow the follicle cells to proliferate, the follicle cells secrete growth and differentiation factors (TGF-β2, VEGF, leptin, FGF2) that allow the oocyte to grow and which bring blood vessels into the follicular region (Antczak et al. 1997).

At any given time, a small group of follicles is maturing. However, after progressing to a certain stage, most oocytes and their follicles die. To survive, the follicle must be exposed to gonadotropic hormones and, "catching the wave" at the right time, must ride it until it peaks. Thus, for oocyte maturation to occur, the follicle needs to be at a certain stage of development when the waves of gonadotropin arise.

The first day of vaginal bleeding is considered to be day 1 of the menstrual cycle (Figure 19.30). This bleeding represents the sloughing off of endometrial tissue and blood vessels that would have aided the implantation of the blastocyst. In the first part of the cycle (called the **proliferative** or **follicular phase**), the pituitary starts secreting increasingly large amounts of FSH. Any maturing follicles that have reached a certain stage of development respond to this hormone with further growth and cellular proliferation. FSH also induces the formation of LH receptors on the granulosa cells. Shortly after this period of initial follicle growth, the pituitary begins secreting LH. In response to LH, the dictyate meiotic block is broken. The nuclear membranes of competent oocytes break down, and the chromosomes assemble to undergo the first meiotic division. One set of chromosomes is kept inside the oocyte, and the other ends up in the small polar body. Both are encased by the zona pellucida, which has been synthesized by the growing oocyte. It is at this stage that the egg will be ovulated.

The two gonadotropins, acting together, cause the follicle cells to produce increasing amounts of estrogen, which has at least five major activities in regulating the further progression of the menstrual cycle:

1. It causes the uterine endometrium to begin its proliferation and to become enriched with blood vessels.
2. It causes the cervical mucus to thin, thereby permitting sperm to enter the inner portions of the reproductive tract.
3. It causes an increase in the number of FSH receptors on the granulosa cells of the mature follicles (Kammerman and Ross 1975) while causing the pituitary to lower its FSH production. It also stimulates the granulosa cells to secrete the peptide hormone inhibin, which also suppresses pituitary FSH secretion (Rivier et al. 1986; Woodruff et al. 1988).
4. At low concentrations, it inhibits LH production, but at high concentrations, it stimulates it.
5. At very high concentrations and over long durations, estrogen interacts with the hypothalamus, causing it to secrete gonadotropin-releasing factor.

As estrogen levels increase as a result of follicular production, FSH levels decline. LH levels, however, continue to rise as more estrogen is secreted. As estrogens continue to be made (days 7–10), the granulosa cells continue to grow. Starting on day 10, estrogen secretion rises sharply. This rise is followed at midcycle by an enormous surge of LH and a smaller burst of FSH. Experiments with female monkeys have shown that exposure of the hypothalamus to greater than 200 pg of estrogen per milliliter of blood for more than 50 hours results in the hypothalamic secretion of gonadotropin-releasing factor. This factor causes the subsequent release of FSH and LH from the pituitary. Within 10 to 12 hours after the gonadotropin peak, the egg is ovulated (Figure 19.31; Garcia et al. 1981).

Although the detailed mechanism of ovulation is not yet known, the physical expulsion of the mature oocyte from the follicle appears to be due to an LH-induced increase in collagenase, plasminogen activator, and prostaglandin within the follicle (Lemaire et al. 1973). The mRNA for plasminogen activator has been dormant in the oocyte cytoplasm. LH causes this message to be polyadenylated and translated into this powerful protease (Huarte et al. 1987). Prostaglandins may cause localized contractions in the smooth muscles in the ovary and may also increase the flow of water from the ovarian capillaries, increasing fluid pressure in the antrum (Diaz-Infante et al. 1974; Koos and Clark 1982). If ovarian prostaglandin synthesis is inhibited, ovulation does not take place. In addition, collagenase and the plasminogen activator protease loosen and digest the extracellular matrix of the follicle (Beers et al. 1975; Downs and Longo 1983). The result of

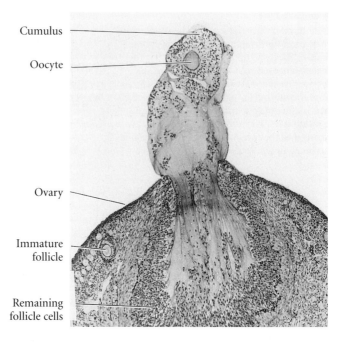

Cumulus

Oocyte

Ovary

Immature follicle

Remaining follicle cells

Figure 19.31
Ovulation in the rabbit. The ovary of a living, anesthetized rabbit was exposed and observed. When the follicle started to ovulate, the ovary was removed, fixed, and stained. (Photograph courtesy of R. J. Blandau.)

LH, then, is increased follicular pressure coupled with the degradation of the follicle wall. A hole is digested through which the ovum can burst.

Following ovulation, the **luteal phase** of the menstrual cycle begins. The remaining cells of the ruptured follicle, under the continued influence of LH, become the **corpus luteum**. (They are able to respond to this LH because the surge in FSH stimulates them to develop even more LH receptors.) The corpus luteum secretes some estrogen, but its predominant secretion is **progesterone**. This steroid hormone circulates to the uterus, where it completes the job of preparing the uterine tissue for blastocyst implantation, stimulating the growth of the uterine wall and its blood vessels. Blocking the progesterone receptor with the synthetic steroid **mifepristone** (RU486) stops the uterine wall from thickening and prevents the implantation of a blastocyst* (Couzinet et al. 1986; Greb et al.

*RU486 is thought to compete for the progesterone receptor inside the nucleus. RU486 can bind to the progesterone site in the receptor, and the receptor-RU486 complex appears to form heterodimers with the normal progesterone-carrying progesterone receptor. When this RU486-progesterone complex binds to the progesterone-responsive enhancer elements on the DNA, transcription from this site is inhibited (Vegeto et al. 1992; Spitz and Bardin 1993). In Europe, RU486 has become a widely used alternative to surgical abortion (Palka 1989; Maurice 1991).

1999). Progesterone also inhibits the production of FSH, thereby preventing the maturation of any more follicles and ova. (For this reason, a combination of estrogen and progesterone has been used in birth control pills. The growth and maturation of new ova are prevented as long as FSH is inhibited.)

If the ovum is not fertilized, the corpus luteum degenerates, progesterone secretion ceases, and the uterine wall is sloughed off. With the decline in serum progesterone levels, the pituitary secretes FSH again, and the cycle is renewed. However, if fertilization occurs, the trophoblast secretes a new hormone, **luteotropin**, which causes the corpus luteum to remain active and serum progesterone levels to remain high. Thus, the menstrual cycle enables the periodic maturation and ovulation of human ova and allows the uterus to periodically develop into an organ capable of nurturing a developing organism for 9 months.

WEBSITE **19.13 The reinitiation of mammalian meiosis.** The hormone-mediated disruption of communication between the oocyte and its surrounding follicle cells may critical in the resumption of meiosis in female mammals.

The egg and the sperm will both die if they do not meet. We are now back where we began: the stage is set for fertilization to take place. As F. R. Lillie recognized in 1919, "The elements that unite are single cells, each on the point of death; but by their union a rejuvenated individual is formed, which constitutes a link in the eternal process of Life."

Snapshot Summary: The Germ Line

1. The precursors of the gametes—the sperm and eggs—are the primordial germ cells. They form outside the gonads and migrate into the gonads during development.

2. In many species, a distinctive germ plasm exists. It often contains the Oskar, Vasa, and Nanos proteins or the mRNAs encoding them.

3. In *Drosophila*, the germ plasm becomes localized in the posterior of the embryo and forms pole cells, the precursors of the gametes. In frogs, the germ plasm originates in the vegetal portion of the oocyte.

4. In amphibians, the germ cells migrate on fibronectin matrices from the posterior larval gut to the gonads. In mammals, a similar migration is seen, and fibronectin pathways may also be used. Stem cell factor is critical in this migration, and the germ cells proliferate as they travel.

5. In birds, the germ plasm is first seen in the germinal crescent. The germ cells migrate through the blood, then leave the blood vessels and migrate into the genital ridges.

6. Germ cell migration in *Drosophila* occurs in several steps involving passive translocation, repulsion from the endoderm, and attraction to the gonads.

7. Before meiosis, the DNA is replicated and remains bound at the kinetochore. Homologous chromosomes are connected through the synaptonemal complex. The configuration of the four chromatids is called a tetrad.

8. The first division of meiosis separates the homologous chromosomes. The second division of meiosis splits the kinetochore and separates the chromatids.

9. The meiosis/mitosis decision in nematodes is regulated by the Delta proteins, which bind to the Notch proteins on the PGCs. The decision for a germ cell to become either a sperm or an egg is regulated at the level of translation of the *fem-3* message.

10. Spermatogenic meiosis in mammals is characterized by the production of four gametes per meiosis and by the absence of meiotic arrest. Oogenic meiosis is characterized by the production of one gamete per meiosis and by an arrest at first meiotic prophase to allow the egg to grow.

11. In some species, meiosis is modified such that a diploid egg is formed. These species can produce a new generation parthenogenetically, without fertilization.

12. The egg not only synthesizes numerous compounds, but also absorbs material produced by other cells. Moreover, it localizes many proteins and messages to specific regions of the cytoplasm, often tethering them to the cytoskeleton.

13. The *Xenopus* oocyte transcribes actively from lampbrush chromosomes during the first meiotic prophase.

14. In *Drosophila*, the oocyte is relatively dormant in terms of transcription. Rather, nurse cells make mRNAs that enter the developing oocyte. Which of the cells derived from the primordial germ cell becomes the oocyte and which become nurse cells is determined by the fusome and the pattern of divisions.

15. In mammals, the hormones of the menstrual cycle integrate the ovarian cycle, the uterine cycle, and the cervical cycle. This integration allows the uterus to be ready to receive an embryo shortly after ovulation occurs.

Literature Cited

Ahringer, J. and J. Kimble. 1991. Control of the sperm-oocyte switch in *Caenorhabditis elegans* hermaphrodites by the *fem-3* 3′ untranslated region. *Nature* 349: 346–348.

Ahringer, J., T. A. Rosenquist, D. N. Lawson, and J. Kimble. 1992. The *C. elegans* sex determining gene, *fem-3*, is regulated post-transcriptionally. *EMBO J.* 11: 2303–2310.

Amikura, R., S. Kobayashi, H. Saito, and M. Okada. 1996. Changes in subcellular localization of mtlrRNA outside mitochondria in oogenesis and early embryogenesis of *Drosophila melanogaster. Dev. Growth Diff.* 38: 489–498.

Anderson, R. and 8 others. 1999. Mouse primordial germ cells lacking b1 integrins enter the germline but fail to migrate normally to the gonads. *Development* 126: 1655–1664.

Antczak, M., J. van Blerkom, and A. Clark. 1997. A novel mechanism of vascular endothelial growth factor, leptin, and transforming growth factor-b2 sequestration in a population of human ovarian follicle cells. *Hum. Reprod.* 12: 2226–2234.

Austin, J. and J. Kimble. 1987. *glp-1* is required in the germ line for regulation of the decision between mitosis and meiosis in *C. elegans. Cell* 51: 589–599.

Baker, T. G. 1970. Primordial germ cells. *In* C. R. Austin and R. V. Short (eds.), *Reproduction in Mammals,* Vol. 1: *Germ Cells and Fertilization.* Cambridge University Press, Cambridge, pp. 1–13.

Beers, W. H., S. Strickland, and E. Reich. 1975. Ovarian plasminogen activator: Relationship to ovulation and hormonal regulation. *Cell* 6: 387–394.

Blackler, A. W. 1962. Transfer of primordial germ cells between two subspecies of *Xenopus laevis. J. Embryol. Exp. Morphol.* 10: 641–651.

Bloom, W. and D. W. Fawcett. 1975. *Textbook of Histology,* 10th Ed. Saunders, Philadelphia.

Bounoure, L. 1934. Recherches sur lignée germinale chez la grenouille rousse aux premiers stades au développement. *Ann. Sci. Zool.* Ser. 17, 10: 67–248.

Bounoure, L. 1939. *L'origine des Cellules Reproductries et le Probleme de la Lignée Germinale.* Gauthier-Villars, Paris.

Boveri, T. 1910. Über die Teilung centrifugierter Eier von *Ascaris megalocephala. Wilhelm Roux Arch. Entwicklungsmech. Org.* 30: 101–125.

Braat, A. K., J. E. Speksnijder and D. Zivkovic. 2000. Germ line development in fishes. *Int. J. Dev. Biol.* 43: 745–760.

Braun, R. E., R. R. Behringer, J. J. Peschon, R. L. Brinster, and R. D. Palmiter. 1989. Genetically haploid spermatids are phenotypically diploid. *Nature* 337: 373–376.

Broihier, H. T., L. A. Moore, M. Van Doren, S. Newman, and R. Lehmann. 1998. *zfh-1* is required for germ cell migration and gonadal mesoderm development in *Drosophila. Development* 125: 655–666.

Brown, D. D. and I. B. Dawid. 1968. Specific gene amplification in oocytes. *Science* 160: 272–280.

Burns, R. K., Jr. 1930. The process of sex-transformation in parabiotic *Amblystoma.* I. Transformation from female to male. *J. Exp. Zool.* 55: 123–169.

Buss, L. W. 1987. *The Evolution of Individuality.* Princeton University Press, Princeton, NJ.

Carlson, B. M. 1981. *Patten's Foundations of Embryology.* McGraw-Hill, New York.

Chiquoine, A. D. 1954. The identification, origin, and migration of the primordial germ cells in the mouse embryo. *Anat. Rec.* 118: 135–146.

Comings, D. E. 1968. The rationale for an ordered arrangement of chromatin in the interphase nucleus. *Am. J. Hum. Genet.* 20: 440–460.

Cooley, L. and W. E. Theurkauf. 1994. Cytoskeletal functions during *Drosophila* oogenesis. *Science* 266: 590–596.

Cooley, L., E. Verheyen, and K. Ayers. 1992. *chickadee* encodes a profilin required for intercellular cytoplasm transport during *Drosophila* oogenesis. *Cell* 69: 173–184.

Couzinet, B., N. Le Strat, A. Ulmann, E. E. Baulieu, and G. Schaison. 1986. Termination of early pregnancy by the progesterone antagonist RU486 (Mifepristone). *N. Engl. J. Med.* 315: 1565–1570.

Danilchik, M. V. and J. C. Gerhart. 1987. Differentiation of the animal-vegetal axis in *Xenopus laevis* oocytes: Polarized intracellular translocation of platelets establishes the yolk gradient. *Dev. Biol.* 122: 101–112.

Davidson, E. 1986. *Gene Activity in Early Development,* 3rd Ed. Academic Press, Orlando, FL.

de Cuevas, M. and A. C. Spradling. 1998. Morphogenesis of the *Drosophila* fusome and its implications for oocyte specification. *Development* 125: 2781–2789.

Deshpande, G., G. Calhoun, J. Yanowitz, and P. D. Schedl. 1999. Novel functions of *nanos* in downregulating mitosis and transcription during the development of the *Drosophila* germline. *Cell* 99: 271–281.

Diaz-Infante, A., K. H. Wright, and E. E. Wallach. 1974. Effects of indomethacin and PGF2a on ovulation and ovarian contraction in the rabbit. *Prostaglandins* 5: 567–581.

Dolci, S. and 8 others. 1991. Requirement for mast cell growth factor for primordial germ cell survival in culture. *Nature* 352: 809–811.

Dong, J., D. F. Albertini, K. Nishimori, T. R. Kumar, N. Lu, and M. Matzuk. 1996. Growth differentiation factor-9 is required during early ovarian folliculogenesis. *Nature* 383: 531–534.

Downs, S. and F. J. Longo. 1983. Prostaglandins and preovulatory follicular maturation in mice. *J. Exp. Zool.* 228: 99–108.

Dubois, R. 1969. Le mécanisme d'entrée des cellules germinales primordiales dans le réseau vasculaire, chez l'embryon de poulet. *J. Embryol. Exp. Morphol.* 21: 255– 270.

Dumont, J. N. 1978. Oogenesis in *Xenopus laevis.* VI. Route of injected tracer transport in follicle and developing oocyte. *J. Exp. Zool.* 204: 193–200.

Dym, M. 1977. The male reproductive system. *In* L. Weiss and R. O. Greep (eds.), *Histology,* 4th Ed. McGraw-Hill, New York, pp. 979–1038.

Dym, M. 1994. Spermatogonial stem cells of the testis. *Proc. Natl. Acad. Sci. USA* 91: 11287–11289.

Dym, M. and D. W. Fawcett. 1971. Further observations on the number of spermatogonia, spermatocytes, and spermatids connected by intercellular bridges in the mammalian testis. *Biol. Reprod.* 4: 195–215.

Ephrussi, A. and R. Lehmann. 1992. Induction of germ cell formation by *oskar. Nature* 358: 387–392.

Epifano, O., L.-F. Liang, M. Familari, M. C. Moos, Jr. and J. Dean. 1995. Coordinate expression of the three zona pellucida genes during mouse oogenesis. *Development* 121: 1947–1956.

Eyal-Giladi, H., M. Ginsburg, and A. Farbarou. 1981. Avian primordial germ cells are of epiblastic origin. *J. Embryol. Exp. Morphol.* 65: 139–147.

Ferrell, J. E., Jr. 1999. *Xenopus* oocyte maturation: New lessons from a good egg. *BioEssays* 21: 833–842.

Ferrell, J. E., Jr. and E. M. Machleder. 1998. The biochemical basis of an all-or-none cell fate switch in *Xenopus* oocytes. *Science* 280: 895–899.

ffrench-Constant, C., A. Hollingsworth, J. Heasman, and C. Wylie. 1991. Response to fibronectin of mouse primordial germ cells before, during, and after migration. *Development* 113: 1365–1373.

Finn, C. A. 1998. Menstruation: A nonadaptive consequence of uterine evolution. *Q. Rev. Biol.* 73: 163–173.

Flickinger, R. A. and D. E. Rounds. 1956. The maternal synthesis of egg yolk proteins as demonstrated by isotopic and serological means. *Biochem. Biophys. Acta* 22: 38–72.

Forristall, C., M. Pondel, L. Chen, and M. L. King. 1995. Patterns of localization and cytoskeletal association of two vegetally localized RNAs, Vg1 and Xcat-2. *Development* 121: 201–208.

Gabrielli, B., L. M. Roy, and J. L. Maller. 1993. Requirement for cdk2 in cytostatic factor-mediated metaphase II arrest. *Science* 259: 1766–1769.

Garcia, J. E., G. S. Jones, and G. L. Wright. 1981. Prediction of the time of ovulation. *Fertil. Steril.* 36: 308–315.

Gardner, R. L. 1982. Manipulation of development. *In* C. R. Austin and R. V. Short (eds.), *Embryonic and Fetal Development.* Cambridge University Press, Cambridge, pp. 159–180.

Ginsburg, M. and H. Eyal-Giladi. 1987. Primordial germ cells of the young chick blastoderm originate from the central zone of the area pellucida irrespective of the embryo-forming process. *Development* 101: 209–219.

Ginsburg, M., M. H. L. Snow, and A. McLaren. 1990. Primordial germ cells in the mouse embryo during gastrulation. *Development* 110: 521–528.

Godin, I. and C. C. Wylie. 1991. TGFβ1 inhibits proliferation and has a chemotactic effect on mouse primordial germ cells in culture. *Development* 113: 1451–1457.

Godin, I., C. Wylie, and J. Heasman. 1990. Genital ridges exert long-range effects on primordial germ cell numbers and direction of migration in culture. *Development* 108: 357–363.

Graham, C. E. 1977. Teratocarcinoma cells and normal mouse embryogenesis. *In* M. I. Sherman (ed.), *Concepts of Mammalian Embryogenesis.* M.I.T. Press, Cambridge, MA, pp. 315–394.

Grant, P. 1953. Phosphate metabolism during oogenesis in *Rana temporaria. J. Exp. Zool.* 124: 513–543.

Greb, R. R., L. Kiesel, A. K. Selbmann, M. Wehrmann, G. D. Hodgen, A. L. Goodman, and D. Wallwiener. 1999. Disparate actions of mifepristone (RU 486) on glands and stroma in the primate endometrium. *Hum. Reprod.* 14: 198–206.

Gurdon, J. B. 1976. *The Control of Gene Expression in Animal Development.* Harvard University Press, Cambridge, MA.

Gutzeit, H. O. 1986. The role of microfilaments in cytoplasmic streaming in *Drosophila* follicles. *J. Cell Sci.* 80: 159–169.

Hahnel, A. C. and E. M. Eddy. 1986. Cell surface markers of mouse primordial germ cells defined by two monoclonal antibodies. *Gamete Res.* 15: 25–34.

Hall, D. H. and 7 others. 1999. Ultrastructural features of the adult hermaphrodite gonad of *Caenorhabditis elegans:* Relationship between the germ line and soma. *Dev. Biol.* 212: 101–123.

Heasman, J., T. Mohun, and C. C. Wylie. 1977. Studies on the locomotion of primordial germ cells from *Xenopus laevis* in vitro. *J. Embryol. Exp. Morphol.* 42: 149–162.

Heasman, J., R. D. Hynes, A. P. Swan, V. Thomas, and C. C. Wylie. 1981. Primordial germ cells of *Xenopus* embryos: The role of fibronectin in their adhesion during migration. *Cell* 27: 437–447.

Heath, J. K. 1978. Mammalian primordial germ cells. *Dev. Mammals* 3: 272–298.

Henderson, S. T., D. Gao, E. J. Lambie and J. Kimble. 1994. Lag-2 may encode a signaling ligand of Glp-1 and Lin-12 receptors of *C. elegans. Development* 120: 2913–2924.

Hill, R. S. and H. C. MacGregor. 1980. The development of lampbrush chromosome-type transcription in the early diplotene oocytes of *Xenopus laevis:* An electron microscope analysis. *J. Cell Sci.* 44: 87–101.

Hirsh, D., D. Oppenheim, and M. Klass. 1976. Development of the reproductive system of *Caenorhabditis elegans. Dev. Biol.* 49: 200–219.

Hodgkin, J., T. Doniach, and M. Shen. 1985. The sex determination pathway in the nematode *Caenorhabditis elegans:* Variations on a theme. *Cold Spring Harbor Symp. Quant. Biol.* 50: 585–593.

Howard, K. 1998. Organogenesis: *Drosophila* goes gonadal. *Curr. Biol.* 8: R415–R417.

Huarte, J., D. Belin, A. Vassalli, S. Strickland, and J.-D. Vassalli. 1987. Meiotic maturation of mouse oocytes triggers the translation and polyadenylation of dormant tissue-type plasminogen activator mRNA. *Genes Dev.* 1: 1201–1211.

Humphrey, R. R. 1931. Studies of sex reversal in *Ambystoma.* III. Transformation of the ovary of *A. tigrinum* into a functional testis through the influence of a testis resident in the same animal. *J. Exp. Zool.* 58: 333–365.

Ikenishi, K., T. S. Tanaka, and T. Komiya. 1996. Spatio-temporal distribution of the protein of the *Xenopus* vasa homologue (*Xenopus* vasa-like gene-1, XVLG1) in embryos. *Dev. Growth Diff.* 38: 527–535.

Inoue, H. and T. Hiroyoshi. 1986. A maternal-effect sex-transformation mutant of the housefly, *Musca domestica* L. *Genetics* 112: 469–481.

Jaglarz, M. K. and K. R. Howard. 1995. The active migration of *Drosophila* primordial germ cells. *Development* 121: 3495–3503.

Jongens, T. A., B. Hay, L. Y. Jan, and Y. N. Jan. 1992. The *germ cell-less* gene product: A posteriorly localized component necessary for germ cell development in *Drosophila. Cell* 70: 569–584.

Kammerman, S. and J. Ross. 1975. Increase in numbers of gonadotropin receptors on granulosa cells during follicle maturation. *J. Clin. Endocrinol.* 41: 546–550.

Kashikawa, M. R. Amikura, A. Nakamura and S. Kobashi. 1999. Mitochondrial small ribosomal RNA is present on polar granules in early cleavage embryos of *Drosophila melanogaster. Dev. Growth Diff.* 41: 495–502.

Kawasaki, I., Y.-H. Shim, J. Kirschner, J. Kaminker, W. B. Wood, and S. Strome. 1998. PGL-1, a predicted RNA-binding component of germ granules, is essential for fertility in *C. elegans. Cell* 94: 635–645.

Kimble, J. E. 1981. Strategies for control of pattern formation in *Caenorhabditis elegans. Phil. Trans. R. Soc. Lond.* [B] 295: 539–551.

Kimble, J. E. and J. G. White. 1981. Control of germ cell development in *Caenorhabditis elegans. Dev. Biol.* 81: 208–219.

Kimble, J., M. K. Barton, T. B. Schedl, T. A. Rosenquist, and J. Austin. 1986. Controls of postembryonic germ line development in *Caenorhabditis elegans. In* J. Gall (ed.), *Gametogenesis and the Early Embryo.* Alan R. Liss, New York, pp. 97–110.

Kloc, M. and L. Etkin. 1995. Two distinct pathways for the localization of RNAs at the vegetal cortex in *Xenopus* oocytes. *Development* 121: 287–297.

Kloc, M., G. Spohr, and L. Etkin. 1993. Translocation of repetitive RNA sequences with the germ plasm in *Xenopus* oocytes. *Science* 262: 1712–1714.

Kloc, M., C. Larabell and L. Etkin. 1996. Elaboration of the messenger transport organizer pathway for localization of RNA to the vegetal cortex of *Xenopus* oocytes. *Dev. Biol.* 180: 119–130.

Kloc, M., C. Larabell, A. P. Y. Chan, and L. D. Etkin. 1998. Contributions of METRO pathway localized molecules to the organization of the germ cell lineage. *Mech. Dev.* 75: 81–93.

Kobayashi, S. and M. Okada. 1989. Restoration of pole-cell forming ability to UV-irradiated *Drosophila* embryos by injection of mitochondrial lrRNA. *Development* 107: 733–742.

Kobayashi, S., R. Amikura, and M. Okada. 1993. Presence of mitochondrial large ribosomal RNA outside mitochondria in germ plasm of *Drosophila melanogaster. Science* 260: 1521–1524.

Kobayashi, S., M. Yamada, M. Asaoka, and T. Kitamura. 1996. Essential role of the posterior morphogen nanos for germline development in *Drosophila. Nature* 380: 708–711.

Koos, R. D. and M. R. Clark. 1982. Production of 6-keto-prostaglandin F_{1a} by rat granulosa cells in vitro. *Endocrinology* 111: 1513–1518.

Kraemer, B., S. Crittenden, M. Gallegos, G. Moulder, R. Barstead, J. Kimble, and M. Wickens. 1999. NANOS-3 and FBF proteins physically interact to control the sperm-oocyte switch in *Caenorhabditis elegans. Curr. Biol.* 9: 1009–1018.

Kuwana, T. 1993. Migration of avian primordial germ cells toward the gonadal anlage. *Dev. Growth Diff.* 35: 237–243.

Kuwana, T., H. Maeda-Suga, and T. Fujimoto. 1986. Attraction of chick primordial germ cells by gonadal anlage in vitro. *Anat. Rec.* 215: 403–406.

Langman, J. 1981. *Medical Embryology,* 4th Ed. Williams & Wilkins, Baltimore.

Lemaire, W. J., N. S. T. Yang, H. H. Behram, and J. M. Marsh. 1973. Preovulatory changes in concentration of prostaglandin in rabbit graafian follicles. *Prostaglandins* 3: 367–376.

Lillie, F. R. 1919. *Problems of Fertilization.* University of Chicago Press, Chicago.

Lin, H. and A. C. Spradling. 1995. Fusome asymmetry and oocyte determination in *Drosophila*. *Dev. Genet.* 16: 6–12.

Lira, A. A., R. A. Kinloch, S. Mortillo, and P. A. Wassarman. 1990. An upstream region of the mouse ZP3 gene directs expression of firefly luciferinase specifically to growing oocytes in transgenic mice. *Proc. Natl. Acad. Sci. USA* 87: 7215–7219.

Liu, C. K. and N. J. Berrill. 1948. Gonophore formation and germ cell origin in *Tubularia*. *J. Morphol.* 83: 39–60.

Lorca, T., F. H. Cruzalegui, D. Fesquet, J.-C. Cavadore, J. Méry, A. Means, and M. Dorée. 1993. Calmodulin-dependent protein kinase II mediates inactivation of MPF and CSF upon fertilization of *Xenopus* eggs. *Nature* 366: 270–273.

Mahowald, A. P. 1971a. Polar granules of *Drosophila*. III. The continuity of polar granules during the life cycle of *Drosophila*. *J. Exp. Zool.* 176: 329–343.

Mahowald, A. P. 1971b. Polar granules of *Drosophila*. IV. Cytochemical studies showing loss of RNA from polar granules during early stages of embryogenesis. *J. Exp. Zool.* 176: 329–343.

Mahowald, A. P., J. H. Caulton, and W. J. Gehring. 1979. Ultrastructural studies of oocytes and embryos derived from female flies carrying the *grandchildless* mutation in *Drosophila subobscura*. *Dev. Biol.* 69: 118–132.

Martin, E. 1992. *The Woman in the Body: A Cultural Analysis of Reproduction*. Beacon Press, Boston.

Matsui, Y. 1974. A cytostatic factor in amphibian: Its extraction and partial characterization. *J. Exp. Zool.* 187: 141–147.

Matsui, Y., D. Toksoz, S. Nishikawa, S.-I. Nishikawa, D. Williams, K. Zsebo, and B. L. M. Hogan. 1991. Effect of Steel factor and leukemia inhibitory factor on murine primordial germ cells in culture. *Nature* 353: 750–752.

Matsui, Y., K. Zsebo, and B. L. M. Hogan. 1992. Derivation of pluripotential embryonic stem cells from murine primordial germ cells in culture. *Cell* 70: 841–847.

Maurice, J. 1991. Improvements seen for RU-486 abortions. *Science* 254: 198–200.

Mayer, D. B. 1964. The migration of primordial germ cells in the chick embryo. *Dev. Biol.* 10: 154–190.

McLaren, A. 1983. Does the chromosomal sex of a mouse cell affect its development? *Symp. Brit. Soc. Dev. Biol.* 7: 225–227.

Mintz, B. 1957. Embryological development of primordial germ cells in the mouse: Influence of a new mutation. *J. Embryol. Exp. Morphol.* 5: 396–403.

Moens, P. B. 1969. The fine structure of meiotic chromosome polarization and pairing in *Locusta migratoria*. *Chromosoma* 28: 1–25.

Moore, L. A., H. T. Broihier, M. Van Doren, L. B. Lunsford, and R. Lehmann. 1998. Identification of genes controlling germ cell migration and embryonic gonad formation in *Drosophila*. *Development* 125: 667–678.

Moses, M. J. 1968. Synaptonemal complex. *Annu. Rev. Genet.* 2: 363–412.

Mowry, K. L. and D. A. Melton. 1992. Vegetal messenger RNA localization directed by a 340-nt RNA sequence element in *Xenopus* oocytes. *Science* 255: 991–994.

Müller, F., V. Bernard, and H. Tobler. 1996. Chromatin diminution in nematodes. *BioEssays* 18: 133–138.

Nakamura, A., R. Amikura, M. Mukai, S. Kobayashi, and P. F. Lasko. 1996. Requirement for a noncoding RNA in *Drosophila* polar granules for germ cell establishment. *Science* 274: 2075–2079.

Newmark, P. A., S. E. Mohr, L. Gong, and R. E. Boswell. 1997. *mago nashi* mediates the posterior follicle cell-to-oocyte signal to organize axis formation in *Drosophila*. *Development* 124: 3194–3207.

Newton, S. C., O. W. Blaschuk, and C. F. Millette. 1993. N-cadherin mediates Sertoli cell-spermatogenic cell adhesion. *Dev. Dyn.* 197: 1–13.

Nieuwkoop, P. and L. Sutasurya. 1981. *Primordial Germ Cells of Chordates*. Cambridge University Press, Cambridge. P. 186–187.

Old, R. W., H. G. Callan, and K. W. Gross. 1977. Localization of histone gene transcripts in newt lampbrush chromosomes by in situ hybridization. *J. Cell Sci.* 27: 57–80.

Paglia, L. M., J. Berry, and W. H. Kastern. 1976. Messenger RNA synthesis, transport, and storage in silkmoth ovarian follicles. *Dev. Biol.* 51: 173–181.

Palka, J. 1989. The pill of choice? *Science* 245: 1319–1323.

Paris, J., K. Swenson, H. Piwnice-Worms, and J. D. Richter. 1991. Maturation-specific polyadenylation: In vitro activation by p34^{cdc2} and phosphorylation of a 58-kD CPE-binding protein. *Genes Dev.* 5: 1697–1708.

Pasteels, J. 1953. Contributions à l'étude du developpement des reptiles. I. Origine et migration des gonocytes chez deux Lacertiens. *Arch. Biol.* 64: 227–245.

Paules, R. S., R. Buccione, R. C. Moscel, G. F. Vande Woude, and J. J. Eppig. 1989. Mouse *mos* protoncogene product is present and functions during oogenesis. *Proc. Natl. Acad. Sci. USA* 86: 5395–5399.

Pepling, M. E. and A. C. Spradling. 1998. Female mouse germ cells form synchronously dividing cysts. *Development* 125: 3323–3328.

Pesce, M., M. G. Farrace, M. Piacentini, S. Dolci, and M. De Felici. 1993. Stem cell factor and leukemia inhibitory factor promote primordial germ cell survival by suppressing programmed cell death (apoptosis). *Development* 118: 1089–1094.

Pesce, M, X. Wang, D. J. Wolgemuth, and H. Schöler. 1998. Differential expression of the Oct-4 transcription factor during mouse germ cell differentiation. *Mech. Dev.* 71: 89–98.

Peschon, J. J., R. R. Behringer, R. L. Brinster, and R. D. Palmiter. 1987. Spermatid-specific expression of protamine-1 in transgenic mice. *Proc. Natl. Acad. Sci. USA* 84: 5316–5319.

Pinkerton, J. H. M., D. G. McKay, E. C. Adams, and A. T. Hertig. 1961. Development of the human ovary: A study using histochemical techniques. *Obstet. Gynecol.* 18: 152–181.

Pratt, S. A., N. F. Scully, and B. D. Shur. 1993. Cell surface β-1,4-galactosyltransferase on primary spermatocytes facilitates their initial adhesion to Sertoli cells in vitro. *Biol. Reprod.* 49: 470–482.

Profet, M. 1993. Menstruation as a defense against pathogens transported by sperm. *Q. Rev. Biol.* 68: 335–385.

Reijo, R. and 12 others. 1995. Diverse spermatogenic defects in humans caused by Y chromosome deletions encompassing a novel RNA-binding protein gene. *Nature Genet.* 10: 383–393.

Resnick, J. L., L. S. Bixler, L. Cheng, and P. J. Donovan. 1992. Long-term proliferation of mouse primordial germ cells in culture. *Nature* 359: 550–551.

Ressom, R. E. and K. E. Dixon. 1988. Relocation and reorganization of germ plasm in *Xenopus* embryos after fertilization. *Development* 103: 507–518.

Rivier, C., J. Rivier, and W. Vale. 1986. Inhibin-mediated feedback control of follicle-stimulating hormone secretion in the female rat. *Science* 234: 205–208.

Robb, D. L., J. Heasman, J. Raats, and C. Wylie. 1996. A kinesin-like protein is required for germ plasm aggregation in *Xenopus*. *Cell.* 87: 823–831.

Rohwedel, J., U. Sehlmeyer, J. Shan, A. Meister, and A. M. Wobus. 1996. Primordial germ cell-derived mouse embryonic germ (EG) cells in vitro resemble undifferentiated stem cells with respect to differentiation capacity and cell cycle distribution. *Cell Biol. Int.* 20: 579–587.

Roller, R. J., R. A. Kinloch, B. Y. Hiraoka, S. S.-L. Li, and P. M. Wassarman. 1989. Gene expression during mammalian oogenesis and early embryogenesis: Quantification of three messenger RNAs abundant in full grown mouse oocytes. *Development* 106: 251–261.

Romanoff, A. L. 1960. *The Avian Embryo*. Macmillan, New York.

Ruohola, H., K. A. Bremer, D. Baker, J. R. Swedlow, L. Y. Jan, and Y. N. Jan. 1991. Role of neurogenic genes in establishment of follicle cell fate and oocyte polarity during oogenesis in *Drosophila*. *Cell* 66: 433–449.

Sagata, N., M. Oskarsson, T. Copeland, J. Brumbaugh, and G. F. Vande Woude. 1988. Function of c-mos proto-oncogene product in meiotic maturation in *Xenopus* oocytes. *Nature* 335: 519–525.

Sagata, N., N. Watanabe, G. F. Vande Woude, and Y. Ikawa. 1989. The c-mos proto-oncogene product is a cytostatic factor responsible for meiotic arrest in vertebrate eggs. *Nature* 342: 512–518.

Savage, R. M. and M. V. Danilchik. 1993. Dynamics of germ plasm localization and its inhibition by ultraviolet irradiation in early cleavage *Xenopus* eggs. *Dev. Biol.* 157: 371–382.

Schmekel, K. and B. Daneholt. 1995. The central region of the synaptonemal complex revealed in three dimensions. *Trends Cell Biol.* 5: 239–242.

Schubiger, G. and W. J. Wood. 1977. Determination during early embryogenesis in *Drosophila melanogaster*. *Am. Zool.* 17: 565–576.

Seydoux, G. and S. Strome. 1999. Launching the germline in *Caenorhabditis elegans*: Regulation of gene expression in early germ cells. *Development* 126: 3275–3283.

Shamblott, M. J. and 8 others. 1998. Derivation of pluripotent stem cells from cultured human primordial germ cells. *Proc. Natl. Acad. Sci. USA* 95: 13726–13731.

Sheets, M. D., M. Wu, and M. Wickens. 1995. Polyadenylation of *c-mos* mRNA as a control point in *Xenopus* meiotic maturation. *Nature* 374: 511–516.

Smith, L. D. 1966. The role of a "germinal plasm" in the formation of primordial germ cells in *Rana pipiens*. *Dev. Biol.* 14: 330–347.

Spitz, I. M. and C. W. Bardin. 1993. Mifepristone (RU486): A modulator of progestin and glucocorticoid action. *N. Engl. J. Med.* 329: 404–412.

Spradling, A. C. 1993. Germline cysts: Communes that work. *Cell* 72: 649–651.

Stewart, T. A. and B. Mintz. 1981. Successful generations of mice produced from an established culture line of euploid teratocarcinoma cells. *Proc. Natl. Acad. Sci. USA* 78: 6314–6318.

Stott, D. and C. C. Wylie. 1986. Invasive behaviour of mouse primordial germ cells in vitro. *J. Cell Sci.* 86: 133–144.

Strassmann, B. I. 1996. The evolution of the endometrial cycles and menstruation. *Q. Rev. Biol.* 71: 181–220.

Subramanian, K. and G. Seydoux. 1999. *nos-1* and *nos-2*, two genes related to *Drosophila nanos*, regulate primordial germ cells development and survival in *Caenorhabditis elegans*. *Development* 126: 4861–4871.

Subtelny, S. and J. E. Penkala. 1984. Experimental evidence for a morphogenetic role in the emergence of primordial germ cells from the endoderm of *Rana pipiens*. *Differentiation* 26: 211–219.

Sun, Y.-A. and R. J. Wyman. 1993. Reevaluation of electrophoresis in the *Drosophila* egg chamber. *Dev. Biol.* 155: 206–215.

Sutasurya, L. A. and P. D. Nieuwkoop. 1974. The induction of primordial germ cells in the urodeles. *Wilhelm Roux Arch. Entwicklungsmech. Org.* 175: 199–220.

Swanson, C. P., T. Merz, and W. J. Young. 1981. *Cytogenetics: The Chromosome in Division, Inheritance and Evolution*. Prentice-Hall, Englewood Cliffs, NJ.

Swift, C. H. 1914. Origin and early history of the primordial germ-cells in the chick. *Am. J. Anat.* 15: 483–516.

Tax, F. E., J. J. Yeargers and J. H. Thomas. 1994. Sequence of *C. elegans* Lag-2 reveals a cell-signaling domain shared with Delta and Serrate of *Drosophila*. *Nature* 368: 150–154.

Telfer, W. H., R. I. Woodruff, and E. Huebner. 1981. Electrical polarity and cellular differentiation in meroistic ovaries. *Am. Zool.* 21: 675–686.

Theurkauf, W. E., S. Smiley, M. L. Wong, and B. M. Alberts. 1992. Reorganization of the cytoskeleton during *Drosophila* oogenesis: Implications for axis specification and intercellular transport. *Development* 115: 923–936.

Theurkauf, W. E., B. M. Alberts, Y. N. Jan, and T. A. Jongens. 1993. A control code for microtubules in the differentiation of *Drosophila* oocytes. *Development* 118: 1169–1180.

Tobler, H., K. D. Smith, and H. Ursprung. 1972. Molecular aspects of chromatin elimination in *Ascaris lumbricoides*. *Dev. Biol.* 27: 190–203.

Tsunekawa, N., M. Naito, T. Nidhida and T. Noce. 2000. Isolation of chicken *vasa* homolog gene and tracing the origin of primordial germ cells. In press.

Van Doren, M., H. T. Broihier, L. A. Moore, and R. Lehmann. 1998. HMG-CoA reductase guides migrating primordial germ cells. *Nature* 396: 466–469.

Vegeto, E., G. F. Allan, W. T. Schrader, M.-J. Tsai, D. P. McDonnell, and B. W. O'Malley. 1992. The mechanism of RU486 antagonism is dependent on the conformation of the carboxy-terminal tail of the human progesterone receptor. *Cell* 69: 703–713.

von Wettstein, D. 1971. The synaptonemal complex and four-strand crossing over. *Proc. Natl. Acad. Sci. USA* 68: 851–855.

von Wettstein, D. 1984. The synaptonemal complex and genetic segregation. In C. W. Evans and H. G. Dickinson (eds.), *Controlling Events in Meiosis*. Cambridge University Press, Cambridge, pp. 195–231.

Waddington, C. H. *Principles of Development and Differentiation*. Macmillan, New York.

Wakahara, M. 1996. Primordial germ cell development: Is the urodele pattern closer to mammals than to anurans? *Int. J. Dev. Biol.* 40: 653–659.

Warrior, R. 1994. Primordial germ cell migration and the assembly of the *Drosophila* embryonic gonad. *Dev. Biol.* 166: 180–194.

Watanabe, N., G. F. Vande Woude, Y. Ikawa, and N. Sagata. 1989. Specific proteolysis of the *c-mos* proto-oncogene product by calpain on fertilization of *Xenopus* eggs. *Nature* 342: 505–517.

Watson, C. A., I. Sauman, and S. J. Berry. 1993. Actin is a major structural and functional element of the egg cortex of giant silkmoths during oogenesis. *Dev. Biol.* 155: 315–323.

Whitington, P. M. and K. E. Dixon. 1975. Quantitative studies of germ plasm and germ cells during early embryogenesis of *Xenopus laevis*. *J. Embryol. Exp. Morphol.* 33: 57–74.

Woodruff, T. K., J. D'Agostino, N. B. Schwartz, and K. E. Mayo. 1988. Dynamic changes in inhibin messenger RNAs in rat ovarian follicles during the reproductive cycle. *Science* 239: 1296–1299.

Wylie, C. 1999. Germ cells. *Cell* 96: 165–174.

Wylie, C. C. and J. Heasman. 1993. Migration, proliferation, and potency of primordial germ cells. *Semin. Dev. Biol.* 4: 161–170.

Wylie, C. C., J. Heasman, A. P. Swan, and B. H. Anderton. 1979. Evidence for substrate guidance of primordial germ cells. *Exp. Cell Res.* 121: 315–324.

Yeom, Y. I. and 7 others. 1996. Germline regulatory element of Oct-4 specific for the totipotent cycle of embryonal cells. *Development* 122: 881–894.

Yisraeli, J. K., S. Sokol, and D. A. Melton. 1990. A two-step model for the localization of a maternal mRNA in *Xenopus* oocytes: Involvement of microtubules and microfilaments in translocation and anchoring of *Vg1* mRNA. *Development* 108: 289–298.

Zhang, N., J. Zhang, K. J. Purcell, Y. Cheng, and K. Howard. 1997. The *Drosophila* protein Wunen repels migratory germ cells. *Nature* 385: 64–67.

Zhao, G.-Q., K. Deng, P. A. Labosky, L. Liaw, and B. L. M. Hogan. 1996. The gene encoding bone morphogenetic protein 8B is required for the initiation and maintenance of spermatogenesis in the mouse. *Genes Dev.* 10: 1657–1669.

Zhou, Y. and M. L. King. 1996. Localization of Xcat-2 RNA, a putative germ plasm component, to the mitochondrial cloud in *Xenopus* stage I oocytes. *Development* 122: 2947–2953.

Züst, B. and K. E. Dixon. 1977. Events in the germ cell lineage after entry of the primordial germ cells into the genital ridges in normal and UV-irradiated *Xenopus laevis*. *J. Embryol. Exp. Morphol.* 41: 33–46.

20 *An overview of plant development*

THE FUNDAMENTAL QUESTIONS IN DEVELOPMENTAL BIOLOGY are similar for plants* and animals. Their developmental strategies, which have evolved over millions of years, have many commonalities; however, some of the challenges and solutions found in plants are sufficiently unique to warrant a separate discussion in this chapter.

Land plants have their origins in the freshwater green algae, and the transition to land correlates with the evolution of an increasingly protected embryo. Mosses, ferns, gymnosperms (conifers, cycads, and ginkgos), and angiosperms (flowering plants) all develop from protected embryos. Two examples of embryo protection are the seed coat that first appeared in the gymnosperms and the fruit that characterizes the angiosperms. As we have seen, embryo protection is also a theme in animal development. What are the differences?

1. **Plants do not gastrulate.** Plant cells are trapped within rigid cellulose walls that generally prevent cell and tissue migration. Plants, like animals, develop three basic tissue systems (dermal, ground, and vascular), but do not rely on gastrulation to establish this layered system of tissues. Plant development is highly regulated by the environment, a strategy that is adaptive for a stationary organism.

2. **Plants have sporic meiosis rather than gametic meiosis.** That is, spores, not gametes, are produced by meiosis. Gametes are produced by mitotic divisions following meiosis.

3. **The life cycle of land plants (as well as many other plants) includes both diploid and haploid multicellular stages.** This type of life cycle is referred to as alternation of generations. The evolutionary trend has been toward a reduction in the size of the haploid generation.

4. **Germ cells are not set aside early in development.** This is also the case in several animal phyla, but it is true for all plants.

5. **Plants undergo extended morphogenesis.** Clusters of actively dividing cells called **meristems**, which are similar to stem cells in animals, persist long after maturity. Meristems allow for reiterative development and the formation of new structures throughout the life of the plant.

*The term *plant* loosely encompasses many organisms, from algae to flowering plants (angiosperms). Recent phylogenetic studies show a common lineage for all green plants, distinct from the red and brown plants. This new phylogenetic tree differs from older classification schemes in which the kingdom Plantae consisted of multicellular, photosynthetic plants that develop from embryos protected by tissues of the parent plant. While this chapter focuses primarily on the flowering plants, their developmental strategies are best understood in an evolutionary context.

6. **Plants have tremendous developmental plasticity.** Many plant cells are highly plastic. While cloning in animals also illustrates plasticity, plants depend more heavily on this developmental strategy. For example, if a shoot is grazed by herbivores, meristems in the leaf often grow out to replace the lost part. (This strategy has similarities to the regeneration seen in some animals.) Whole plants can even be regenerated from some single cells. In addition, a plant's form (including branching, height, and relative amounts of vegetative and reproductive structures) is greatly influenced by environmental factors such as light and temperature, and a wide range of morphologies can result from the same genotype. This amazing level of plasticity may help compensate for the plant's lack of mobility.

7. **Plants may tolerate higher genetic loads than animals.** Plant genomes can carry a much greater load of mutations than animals before the phenotype is affected. For example, half of the maize (corn) genome appears to be made up of foreign DNA (SanMiguel et al. 1996). Most of it is in the form of retroelements that resemble retroviruses. The maize plant appears to function quite well with all of this "hitchhiking" DNA. Animals also have a significant amount of foreign DNA, but aneuploidy and polyploidy can be developmentally harmful to them. When plants are aneuploid or polyploid, the consequences can be adaptive. Many flowers found in the florist shop and the wheat used for bread flour are examples of successful polyploids.

Despite these major differences among many plants and animals, developmental genetic studies are revealing some commonalities between them in the regulation of basic molecular mechanisms of patterning, along with evolutionarily distinct solutions to the problem of creating three-dimensional form from a single cell.

Plant Life Cycles

The plant life cycle alternates between haploid and diploid generations. Embryonic development is seen only in the diploid generation. The embryo, however, is produced by the fusion of gametes, which are formed only by the haploid generation. So understanding the relationship between the two generations is important in the study of plant development.

Unlike animals (see chapter 2), plants have multicellular haploid and multicellular diploid stages in their life cycle. Gametes develop in the multicellular haploid **gametophyte** (from the Greek *phyton*, "plant"). Fertilization gives rise to a multicellular diploid **sporophyte**, which produces haploid spores via meiosis. This type of life cycle is called a **haplodiplontic** life cycle (Figure 20.1). It differs from our own **diplontic** life cycle, in which only the gametes are in the haploid state. In haplodiplontic life cycles, gametes are not the di-

rect result of a meiotic division. Diploid sporophyte cells undergo meiosis to produce haploid **spores**. Each spore goes through mitotic divisions to yield a multicellular, haploid gametophyte. Mitotic divisions within the gametophyte are required to produce the gametes. The diploid sporophyte results from the fusion of two gametes. Among the Plantae, the gametophytes and sporophytes of a species have distinct morphologies (in some algae they look alike). How a single genome can be used to create two unique morphologies is an intriguing puzzle.

All plants alternate generations. There is an evolutionary trend from sporophytes that are nutritionally dependent on autotrophic (self-feeding) gametophytes to the opposite–gametophytes that are dependent on autotrophic sporophytes. This trend is exemplified by comparing the life cycles of a moss, a fern, and an angiosperm (see Figures 20.2–20.4). (Gymnosperm life cycles bear many similarities to those of angiosperms; the distinctions will be explored in the context of angiosperm development.)

The "leafy" moss you walk on in the woods is the gametophyte generation of that plant (Figure 20.2). Mosses are **heterosporous**, which means they make two distinct types of spores; these develop into male and female gametophytes. Male gametophytes develop reproductive structures called **antheridia** (singular, antheridium) that produce sperm by mitosis. Female gametophytes develop **archegonia** (singular, archegonium) that produce eggs by mitosis. Sperm travel to a neighboring plant via a water droplet, are chemically attracted to the entrance of the archegonium, and fertilization re-

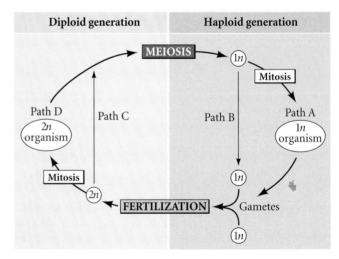

Figure 20.1
Plants have haplodiplontic life cycles that involve mitotic divisions (resulting in multicellularity) in both the haploid and diploid generations (paths A and D). Most animals are diplontic and undergo mitosis only in the diploid generation (paths B and D). Multicellular organisms with haplontic life cycles follow paths A and C.

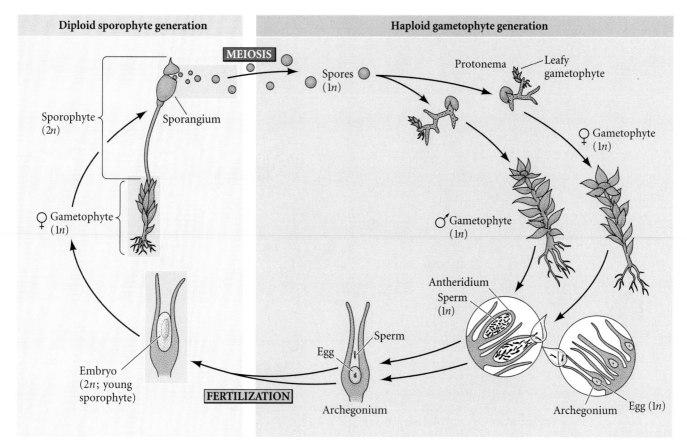

Diploid sporophyte generation

Haploid gametophyte generation

MEIOSIS

Spores (1*n*)

Protonema — Leafy gametophyte

Sporophyte (2*n*)

Sporangium

♀ Gametophyte (1*n*)

♀ Gametophyte (1*n*)

♂ Gametophyte (1*n*)

Antheridium

Sperm (1*n*)

Embryo (2*n*; young sporophyte)

FERTILIZATION

Egg

Sperm

Archegonium

Archegonium

Egg (1*n*)

Figure 20.2
Life cycle of a moss (genus *Polytrichum*). The sporophyte generation is dependent on the photosynthetic gametophyte for nutrition. Cells within the sporangium of the sporophyte undergo meiosis to produce male and female spores, respectively. These spores divide mitotically to produce multicellular male and female gametophytes. Differentiation of the growing tip of the gametophyte produces antheridia in males and archegonia in females. The sperm and eggs are produced mitotically in the antheridia and archegonia, respectively. Sperm are carried to the archegonia in water droplets. After fertilization, the sporophyte generation develops in the archegonium and remains attached to the gametophyte.

sults.* The embryonic sporophyte develops within the archegonium, and the mature sporophyte stays attached to the gametophyte. The sporophyte is not photosynthetic. Thus both the embryo and the mature sporophyte are nourished by the gametophyte. Meiosis within the capsule of the sporophyte yields haploid spores that are released and eventually germinate to form a male or female gametophyte.

Ferns follow a pattern of development similar to that of mosses, although most (but not all) ferns are **homosporous**. That is, the sporophyte produces only one type of spore within a structure called the **sporangium** (Figure 20.3). One gametophyte can produce both male and female sex organs. The greatest con-

trast between the mosses and the ferns is that both the gametophyte and the sporophyte of the fern photosynthesize and are thus autotrophic; the shift to a dominant sporophyte generation is taking place.[†]

At first glance, angiosperms may appear to have a diplontic life cycle because the gametophyte generation has been reduced to just a few cells (Figure 20.4). However, mitotic division still follows meiosis in the sporophyte, resulting in a multicellular gametophyte, which produces eggs or sperm. All of this takes place in the the organ that characterizes the angiosperms: the flower. Male and female gametophytes have distinct morphologies (i.e., angiosperms are heterosporous), but the gametes they produce no longer rely on water for fertilization. Rather, wind or members of the animal kingdom deliver the male gametophyte—**pollen**—to the female gametophyte. Another evolutionary innovation is the production of a seed coat, which adds an extra layer of protection around the embryo. The seed coat is also found in the gymnosperms. A further protective layer,

*Have you ever wondered why there are no moss trees? Aside from the fact that the gametophytes of mosses (and other plants) do not have the necessary structural support and transport systems to attain tree height, it would be very difficult for a sperm to swim up a tree!

[†]It *is* possible to have tree ferns, for two reasons. First, the gametophyte develops on the ground, where water can facilitate fertilization. Secondly, unlike mosses, the fern sporophyte has vascular tissue, which provides the support and transport system necessary to achieve substantial height.

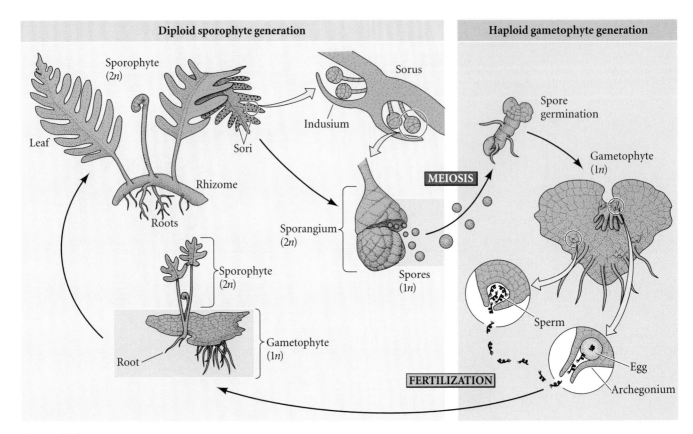

Figure 20.3
Life cycle of a fern (genus *Polypodium*). The sporophyte generation is photosynthetic and is independent of the gametophyte. The sporangia are protected by a layer of cells called the indusium. This entire structure is called a sorus. Meiosis within the sporangia yields a haploid spore. Each spore divides mitotically to produce a heart-shaped gametophyte, which differentiates both archegonia and antheridia on one individual. The gametophyte is photosynthetic and independent, although it is smaller than the sporophyte. Fertilization takes place when water is available for sperm to swim to the archegonia and fertilize the eggs. The sporophyte has vascular tissue and roots; the gametophyte does not.

the fruit, is unique to the angiosperms and aids in the dispersal of the enclosed embryos by wind or animals.

The remainder of this chapter provides a detailed exploration of angiosperm development from fertilization to senescence. Keep in mind that the basic haplodiplontic life cycle seen in the mosses and ferns is also found in the angiosperms, continuing the trend toward increased nourishment and protection of the embryo.

Gamete Production in Angiosperms

Like those of mosses and ferns, angiosperm gametes are produced by the gametophyte generation. Angiosperm gametophytes are associated with flowers. The gametes they produce join to form the sporophyte. The study of embryonic development in plants is therefore the study of early sporophyte development. In angiosperms, the sporophyte is what is com-

monly seen as the plant body. The shoot meristem of the sporophyte produces a series of vegetative structures. At a certain point in development, internal and external signals trigger a shift from vegetative to reproductive (flower-producing) development (see reviews by McDaniel et al. 1992 and Levy and Dean 1998). Once the meristem becomes floral, it initiates the development of floral parts sequentially in whorls of organs modified from leaves (Figure 20.5). The first and second whorls become **sepals** and **petals**, respectively; these organs are sterile. The pollen-producing **stamens** are initiated in the third whorl of the flower. The **carpel** in the fourth whorl contains the female gametophyte. The stamens contain four groups of cells, called the **microsporangia** (pollen sacs), within an **anther**. The microsporangia undergo meiosis to produce **microspores**. Unlike most ferns, angiosperms are heterosporous, so the prefix *micro* is used to identify the spores that mitotically yield the male gametophytes—pollen grains. The inner wall of the pollen sac, the **tapetum**, provides nourishment for the developing pollen.

Pollen

The pollen grain is an extremely simple multicellular structure. The outer wall of the pollen grain, the **exine**, is composed

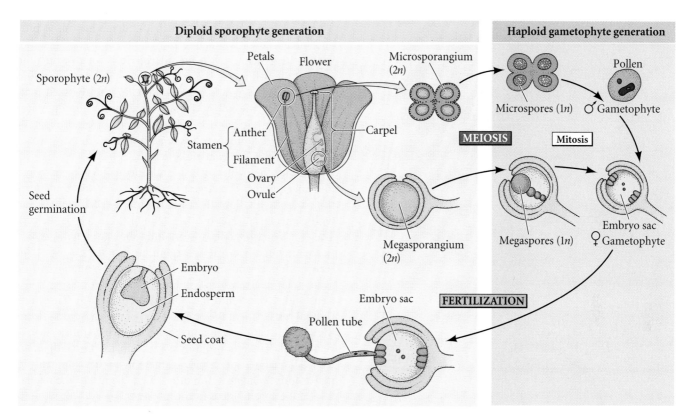

Figure 20.4
Life cycle of an angiosperm, represented here by a pea plant (genus *Pisum*). The sporophyte is the dominant generation, but multicellular male and female gametophytes are produced within the flowers of the sporophyte. Cells of the microsporangium within the anther undergo meiosis to produce microspores. Subsequent mitotic divisions are limited, but the end result is a multicellular pollen grain. The megasporangium is protected by two layers of integuments and the ovary wall. Within the megasporangium, meiosis yields four megaspores—three small and one large. Only the large megaspore survives to produce the embryo sac. Fertilization occurs when the pollen germinates and the pollen tube grows toward the embryo sac. The sporophyte generation may be maintained in a dormant state, protected by the seed coat.

of resistant material provided by both the tapetum (sporophyte generation) and the microspore (gametophyte generation). The inner wall, the **intine**, is produced by the microspore. A mature pollen grain consists of two cells, one within the other (Figure 20.6). The **tube cell** contains a **generative cell** within it. The generative cell divides to produce two sperm. The tube cell nucleus guides pollen germination and the growth of the pollen tube after the pollen lands on the stigma of a female gametophyte. One of the two sperm will fuse with the egg cell to produce the next sporophyte generation. The second sperm will participate in the formation of the endosperm, a structure that provides nourishment for the embryo.

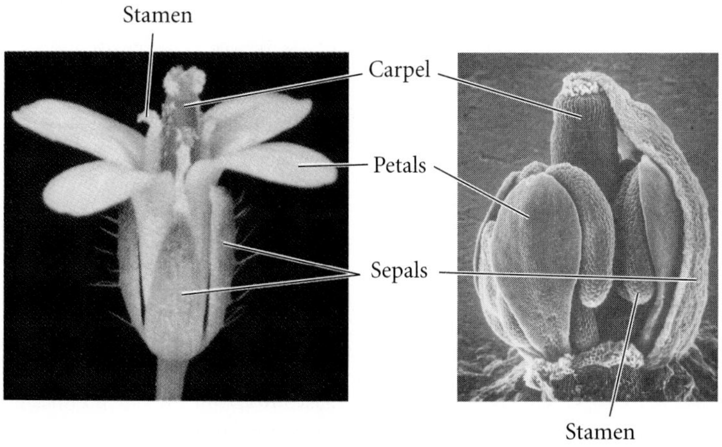

Figure 20.5
An *Arabidopsis* flower, illustrating the major organs (petals, sepals, stamen, and carpel) of the typical angiosperm flower. The micrograph on the right shows the dissected flower. Tremendous morphological variation is possible in all four organs; this variation appears to be related to reproductive strategies. (Photographs courtesy of J. Bowman.)

(A)

(B)

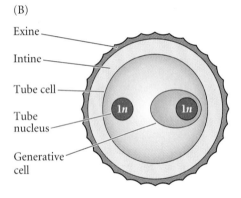

Exine

Intine

Tube cell

Tube nucleus

Generative cell

Figure 20.6
(A) Pollen grains have intricate surface patterns, as seen in this scanning electron micrograph of aster pollen. (B) A pollen grain consists of a cell within a cell. The generative cell will undergo division to produce two sperm cells. One will fertilize the egg, and the other will join with the polar nuclei, yielding the endosperm.

The ovary

The fourth whorl of organs within the flower forms the **carpel**, which gives rise to the female gametophyte (Figure 20.7). The carpel consists of the **stigma** (where the pollen lands), the **style**, and the **ovary**. Following fertilization, the ovary wall will develop into the **fruit**. This unique angiosperm structure provides further protection for the developing embryo and also enhances seed dispersal by frugivores (fruit-eating animals). Within the ovary are one or more **ovules** attached by a **placenta** to the ovary wall. Fully developed ovules are called **seeds**. The ovule has one or two outer layers of cells called the **integuments**. These enclose the **megasporangium**,

which contains sporophyte cells that undergo meiosis to produce **megaspores** (see Figure 20.4). There is a small opening in the integuments, called the **micropyle**, through which the pollen tube will grow. The integuments—an innovation first appearing in the gymnosperms—develop into the **seed coat**, which protects the embryo by providing a waterproof physical barrier. When the mature embryo disperses from the parent plant, diploid sporophyte tissue accompanies the embryo in the form of the seed coat and the fruit.

Within the ovule, meiosis and unequal cytokinesis yield four megaspores. The largest of these megaspores undergoes three mitotic divisions to produce a seven-celled **embryo sac** with eight nuclei (Figure 20.8). One of these cells is the egg. The two **synergid cells** surrounding the egg may be evolutionary remnants of the archegonium (the female sex organ seen in mosses and ferns). The **central cell** contains two or more polar nuclei, which will fuse with the second sperm nucleus and develop into the polyploid endosperm. Three **antipodal cells** form at the opposite end of the embryo sac from the synergids and degenerate before or during embryonic development. There is no known function for the antipodals. Genetic analyses of female gametophyte development in maize and *Arabidopsis** are providing insight into the regulation of the specific steps in this process (Drews et al. 1998).

*A small weed in the mustard family, *Arabidopsis* is used as a model system because of its very small genome.

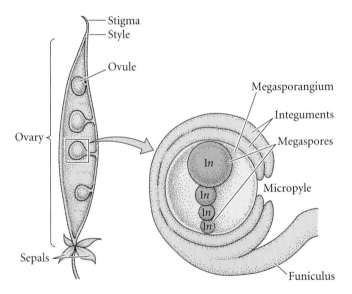

Stigma
Style
Ovule
Megasporangium
Integuments
Megaspores
Ovary
Micropyle
Sepals
Funiculus

Figure 20.7
The carpel contains one or more ovules; these contains megasporangia protected by two layers of integument cells. The megasporangia divide meiotically to produce haploid megaspores, All of the carpel is diploid except for the megaspores, which divide mitotically to produce the embryo sac (the female gametophyte).

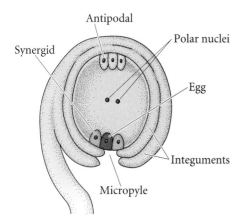

Figure 20.8
The embryo sac is the product of three mitotic divisions of the haploid megaspore. Two of the nuclei are contained within the central cell; the other six cells contain one haploid nucleus each.

Pollination

Pollination refers to the landing and subsequent germination of the pollen on the stigma. Hence it involves an interaction between the gametophytic generation of the male (the pollen) and the sporophytic generation of the female (the stigmatic surface of the carpel). Pollination can occur within a single flower (self-fertilization), or pollen can land on a different flower on the same or a different plant. About 96% of flowering plant species produce male and female gametophytes on the same plant. However, about 25% of these produce two different types of flowers on the same plant, rather than **perfect flowers** containing both male and female gametophytes. **Staminate** flowers lack carpels, while **carpellate** flowers lack stamens. Maize plants, for example, have staminate (tassel) and carpellate (ear) flowers on the same plant. Such species are considered to be **monoecious** (Greek *mono*, "one"; *oecos*, "house"). The remaining 4% of species (e.g., willows) produce staminate and carpellate flowers on separate plants . These

species are considered to be **dioecious** ("two houses"). Only a few plant species have true sex chromosomes. The terms "male" and "female" are most correctly applied only to the gametophyte generation of heterosporous plants, not to the sporophyte (Cruden and Lloyd 1995).

The arrival of a viable pollen grain on a receptive stigma does not guarantee fertilization. **Interspecific incompatibility** refers to the failure of pollen from one species to germinate and/or grow on the stigma of another species (for a review, see Taylor 1996). **Intraspecific incompatibility** is incompatibility that occurs within a species. **Self-incompatibility**—incompatibility between the pollen and the stigmas of the same individual—is an example of intraspecific incompatibility. Self-incompatibility blocks fertilization between two genetically similar gametes, increasing the probability of new gene combinations by promoting outcrossing (pollination by a different individual of the same species). Groups of closely related plants can contain a mix of self-compatible and self-incompatible species.

Several different self-incompatibility systems have evolved (Figure 20.9). Recognition of self depends on the multiallelic self-incompatibility (*S*) locus (Nasrallah et al. 1994; Dodds et al. 1996; Gaude and McCormick 1999). Gametophytic self-incompatibility occurs when the *S* allele of the pollen grain matches either of the *S* alleles of the stigma (remember that the stigma is part of the diploid sporophyte generation, which has two *S* alleles). In this case, the pollen tube begins developing, but stops before reaching the micropyle. Sporophytic self-incompatibility occurs when one of the two *S* alleles of the pollen-producing sporophyte (not the gametophyte) matches one of the *S* alleles of the stigma. Most likely, sporophyte contributions to the exine are responsible.

The *S* locus consists of several physically linked genes that regulate recognition and rejection of pollen. An *S* gene has been cloned that codes for an RNase called S RNase, which is sufficient, in the gametophytically self-incompatible petunia pistil, to recognize and reject self-pollen (Lee et al. 1994). The

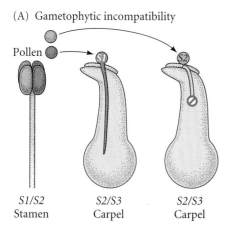

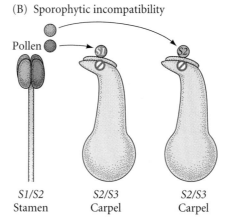

Figure 20.9
Self-incompatibility. *S1*, *S2*, and *S3* are different alleles of the self-incompatibility (*S*) locus. (A) Plants with gametophytic self-incompatibility reject pollen only when the genotype of the pollen matches one of the carpel's two alleles. (B) In sporophytic self-incompatibility, the genotype of the pollen parent, not just of the haploid pollen grain itself, can trigger an incompatibility response.

(A)

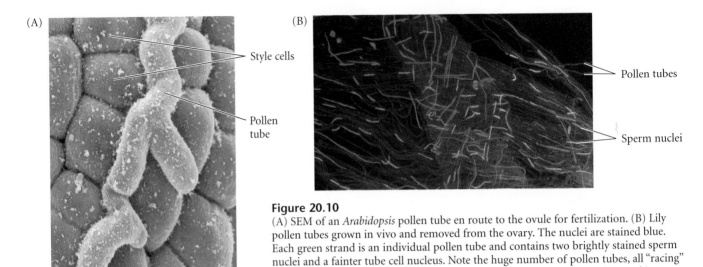

Style cells

Pollen tube

(B)

Pollen tubes

Sperm nuclei

Figure 20.10
(A) SEM of an *Arabidopsis* pollen tube en route to the ovule for fertilization. (B) Lily pollen tubes grown in vivo and removed from the ovary. The nuclei are stained blue. Each green strand is an individual pollen tube and contains two brightly stained sperm nuclei and a fainter tube cell nucleus. Note the huge number of pollen tubes, all "racing" to fertilize a single egg (and the polar nuclei). (Photographs courtesy of E. Lord.)

pollen component recognized is most likely a different gene in the *S* locus, but has not yet been identified in either gametophytically or sporophytically self-incompatible plants. In sporophytic self-incompatibility, a ligand on the pollen is thought to bind to a membrane-bound kinase receptor in the stigma that starts a signaling process leading to pollen rejection. The mechanism of pollen degradation is unclear, but appears to be highly specific.

If the pollen and the stigma are compatible, the pollen takes up water (hydrates) and the pollen tube emerges. The pollen tube grows down the style of the carpel toward the micropyle (Figure 20.10). The tube nucleus and the sperm cells are kept at the growing tip by bands of callose (a complex carbohydrate). It is possible that this may be an exception to the "plant cells do not move" rule, as the generative cell(s) appear to move ahead via adhesive molecules (Lord et al. 1996). Pollen tube growth is quite slow in gymnosperms (up to a year), while in some angiosperms the tube can grow as rapidly as 1 cm/hour.

Calcium has long been known to play an essential role in pollen tube growth (Brewbaker and Kwack 1963). Calcium accumulates in the tip of the pollen tubes, where open calcium channels are concentrated (Jaffe et al. 1975; Trewavas and Malho 1998). There is direct evidence that pollen tube growth in the field poppy is regulated by a slow-moving calcium wave controlled by the phosphoinositide signaling pathway (Figure 20.11; Franklin-Tong et al. 1996). Cytoskeletal investigations show that organelle positioning during pollen tube growth depends on interactions with cytoskeletal components. This

must link to signaling, but the specifics are still unknown (Cai and Cresti 1999).

Genetic approaches have been useful in investigating how the growing pollen tube is guided toward unfertilized ovules. In *Arabidopsis*, the pollen tube appears to be guided by a long-distance signal from the ovule (Hulskamp et al. 1995; Wilhelmi and Preuss 1999). Analysis of pollen tube growth in ovule mutants of *Arabidopsis* indicates that the haploid embryo sac is particularly important in the long-range guidance

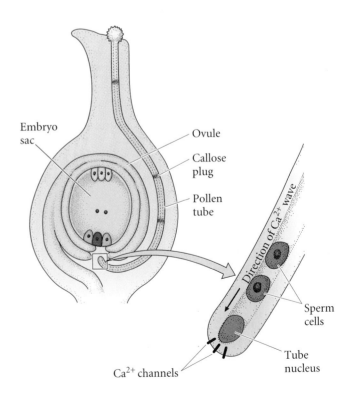

Embryo sac

Ovule

Callose plug

Pollen tube

Direction of Ca²⁺ wave

Sperm cells

Tube nucleus

Ca²⁺ channels

Figure 20.11
Calcium and pollen tube tip growth. After compatible pollen germinates, the pollen tube grows toward the micropyle. Calcium plays a key role in the growth of the tube. (After Franklin-Tong et al. 1996.)

of pollen tube growth. Mutants with defective sporophyte tissue in the ovule but a normal haploid embryo sac appear to stimulate normal pollen tube development.

While the evidence points primarily to the role of the gametophyte generation in pollen tube guidance, diploid cells may make some contribution. Two *Arabidopsis* genes, *POP2* and *POP3*, have been identified that specifically guide pollen tubes to the ovule with no other apparent effect on the plant (Wilhelmi and Preuss 1996, 1999). These genes function in both the pollen and the pistil, thus implicating the sporophyte generation in the guidance system.

Fertilization

The growing pollen tube enters the embryo sac through the micropyle and grows through one of the synergids. The two sperm cells are released, and a **double fertilization** event occurs (see review by Southworth 1996). One sperm cell fuses with the egg, producing the zygote that will develop into the sporophyte. The second sperm cell fuses with the bi- or multi-inucleate central cell, giving rise to the **endosperm**, which nourishes the developing embryo. This second event is not true fertilization in the sense of male and female gametes undergoing syngamy (fusion). That is, it does not result in a zygote, but in nutritionally supportive tissue. (When you eat popcorn, you are eating "popped" endosperm.) The other accessory cells in the embryo sac degenerate after fertilization.

The zygote of the angiosperm produces only a single embryo; the zygote of the gymnosperm, on the other hand, produces two or more embryos after cell division begins, by a process known as cleavage embryogenesis. Double fertilization, first identified a century ago, is generally restricted to the angiosperms, but it has also been found in the gymnosperms *Ephedra* and *Gnetum*, although no endosperm forms. Fried-

man (1998) has suggested that endosperm may have evolved from a second zygote "sacrificed" as a food supply in a gymnosperm with double fertilization. Investigations of the most closely related extant relative of the basal angiosperm, *Amborella*, should provide information on the evolutionary origin of the endosperm (Figure 20.12; Brown 1999).

Fertilization is not an absolute prerequisite for angiosperm embryonic development (Mogie 1992). Embryos can form within embryo sacs from haploid eggs and from cells that did not divide meiotically. This phenomenon is called **apomixis** (Greek, "without mixing"), and results in viable seeds. The viability of the resulting haploid sporophytes indicates that ploidy alone does not account for the morphological distinctions between the gametophyte and the sporophyte. Embryos can also develop from cultured sporophytic tissue. These embryos develop with no associated endosperm, and they lack a seed coat.

Embryonic Development

Experimental studies

The angiosperm zygote is embedded within the ovule and ovary and thus is not readily accessible for experimental manipulation. The following approaches, however, can yield information on the formation of the plant embryo:

- **Histological studies** of embryos at different stages show how carefully regulated cell division results in the construction of an organism, even without the ability to move cells and tissues to shape the embryo.
- **Culture experiments** using embryos isolated from ovules and embryos developing de novo from cultured sporophytic tissue provide information on the interactions between the embryo and surrounding sporophytic and endosperm tissue.
- **In vitro fertilization experiments** provide information on gamete interactions.
- **Biochemical analyses** of embryos at different stages of development provides information on such things as the stage-specific gene products necessary for patterning and establishing food reserves.
- **Genetic and molecular analyses of developmental mutants** characterized using the above approaches have greatly enhanced our understanding of embryonic development.
- **Clonal analysis** involves marking individual cells and following their fate in development (see Poethig 1987 for details on the methodology). For example, seeds heterozygous for a pigmentation gene may be irradiated so that a certain cell loses the ability to produce pigment. Its descendants will form a colorless sector that can be identified and related to the overall body pattern.

Figure 20.12
Amborella trichopoda. This plant is more closely related to the first angiosperm than any other extant species. (Photograph courtesy of Sandy Floyd.)

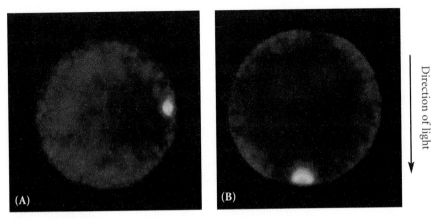

Direction of light

Figure 20.13

Axis formation in the brown alga *Pelvetia compressa*. (A) An F-actin patch (orange) is first formed at the point of sperm entry (the blue spot marks the sperm pronucleus). (B) Later, light was shone in the direction of the arrow. The sperm-induced axis was overridden, and an F-actin patch formed on the dark side, where the rhizoid will later form. (Photographs courtesy of W. Hables.)

Embryogenesis

In plants, the term **embryogenesis** covers development from the time of fertilization until dormancy occurs. The basic body plan of the sporophyte is established during embryogenesis; however, this plan is reiterated and elaborated after dormancy is broken. The major challenges of embryogenesis are

1. To establish the basic body plan. **Radial patterning** produces three tissue systems, and **axial patterning** establishes the apical-basal (shoot-root) axis.
2. To set aside meristematic tissue for postembryonic elaboration of the body structure (leaves, roots, flowers, etc.).
3. To establish an accessible food reserve for the germinating embryo until it becomes autotrophic.

Embryogenesis is similar in all angiosperms in terms of the establishment of the basic body plan (Steeves and Sussex 1989) (see Figure 20.15). There are differences in pattern elaboration, however, including differences in the precision of cell division patterns, the extent of endosperm development, cotyledon development, and the extent of shoot meristem development (Esau 1977; Johri et al. 1992).

Polarity is established in the first cell division following fertilization. The establishment of polarity has been investigated using brown algae as a model system (Belanger and Quatrano 2000). The zygotes of these plants are independent of other tissues and amenable to manipulation. The initial cell division results in one smaller cell, which will form the rhizoid (root homologue) and anchor the rest of the plant, and one larger cell, which gives rise to the thallus (main body of the sporophyte). The point of sperm entry fixes the position of the rhizoid end of the apical-basal axis. This axis is perpendicular to the plane of the first cell division. F-actin accumulates at the rhizoid pole (Kropf et al. 1999). However, light or gravity can override this fixing of the axis and establish a new position for cell division (Figure 20.13; Alessa and Kropf 1999). Once the

Figure 20.14

Asymmetrical cell division in brown algae. Time course from 8 to 25 hours after fertilization, showing algal cells stained with a vital membrane dye to visualize secretory vesicles, which appear first, and the cell plate, which begins to appear about halfway through this sequence. (Photographs courtesy of K. Belanger.)

8 hours after fertilization

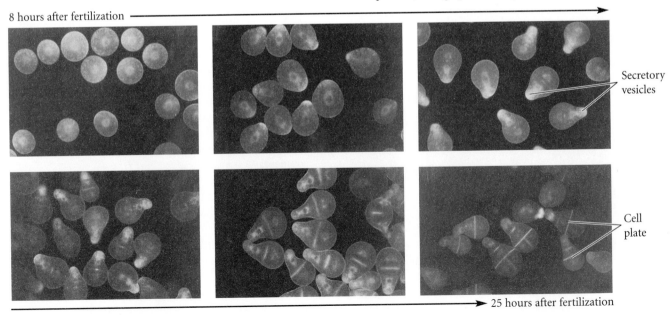

Secretory vesicles

Cell plate

25 hours after fertilization

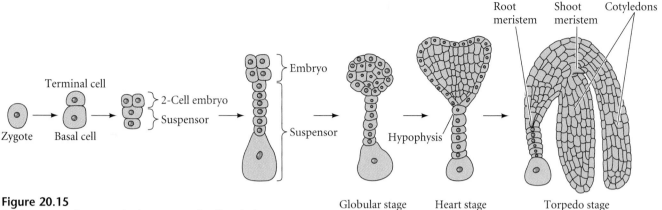

Figure 20.15

Angiosperm embryogenesis. A representative dicot is shown; a monocot would develop only a single cotyledon. While there are basic patterns of embryogenesis in angiosperms, there is tremendous morphological variation among species.

apical-basal axis is established, secretory vesicles are targeted to the rhizoid pole of the zygote (Figure 20.14). These vesicles contain material for rhizoid outgrowth, with a cell wall of distinct macromolecular composition. Targeted secretion may also help orient the first plane of cell division. Maintenance of rhizoid versus thallus fate early in development depends on information in the cell walls (Brownlee and Berger 1995). Cell wall information also appears to be important in angiosperms (reviewed in Scheres and Benfey 1999).

The basic body plan of the angiosperm laid down during embryogenesis also begins with an asymmetrical* cell division, giving rise to a **terminal cell** and a **basal cell** (Figure 20.15). The terminal cell gives rise to the **embryo proper**. The basal cell forms closest to the micropyle and gives rise to the **suspensor**. The **hypophysis** is found at the interface between the suspensor and the embryo proper. In many species it gives rise to some of the root cells. (The suspensor cells divide to form a filamentous or spherical organ that degenerates later in embryogenesis.) In both gymnosperms and angiosperms, the suspensor orients the absorptive surface of the embryo toward its food source; in angiosperms, it also appears to serve as a nutrient conduit for the developing embryo. Culturing isolated embryos of scarlet runner beans with and without the suspensor has demonstrated the need for a suspensor through the heart stage in dicots (Figure 12.16; Yeung and Sussex 1979). Embryos cultured with a suspensor are twice as likely to survive as embryos cultured without an attached suspensor at this stage. The suspensor may be a source of hormones. In scarlet runner beans, younger embryos without a suspensor can survive in culture if they are supplemented with the growth hormone gibberellic acid (Cionini et al. 1976).

As the establishment of apical-basal polarity is one of the key achievements of embryogenesis, it is useful to consider why

the suspensor and embryo proper develop unique morphologies. Here the study of embryo mutants in maize and *Arabidopsis* has been particularly helpful. Investigations of suspensor mutants (*sus1*, *sus2*, and *raspberry1*) of *Arabidopsis* have provided genetic evidence that the suspensor has the capacity

Embryo region cultured		Developed plantlets (%)
		42
		88
		100
		100

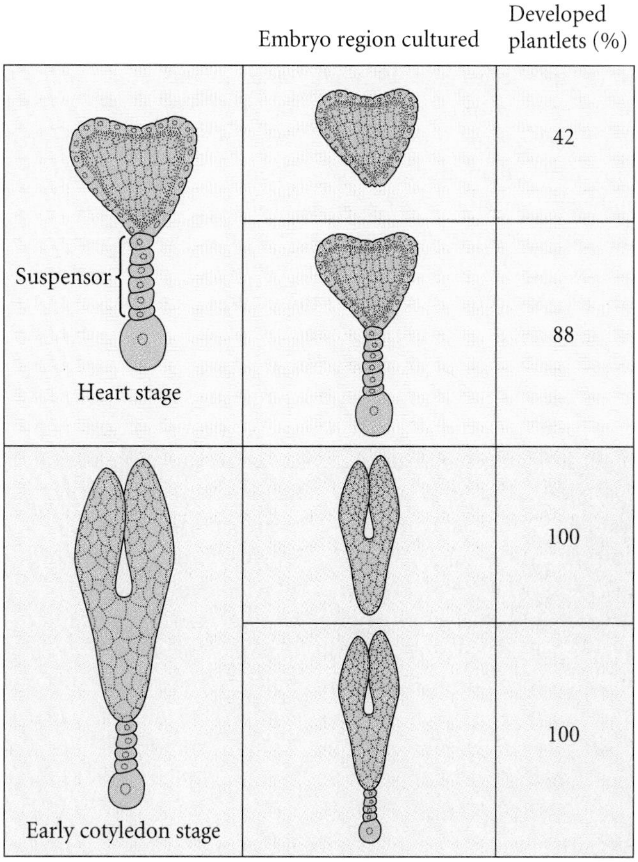

Figure 20.16

Role of the suspensor in dicot embryogenesis. Culturing scarlet runner bean embryos with and without their suspensors has demonstrated that the suspensor is essential at the heart-shaped stage, but not later. (After Yeung and Sussex 1979.)

*Asymmetrical cell division is also important in later angiosperm development, including the formation of guard cells of leaf stomata and of different cell types in the ground and vascular tissue systems.

(A)

(B)

(C)

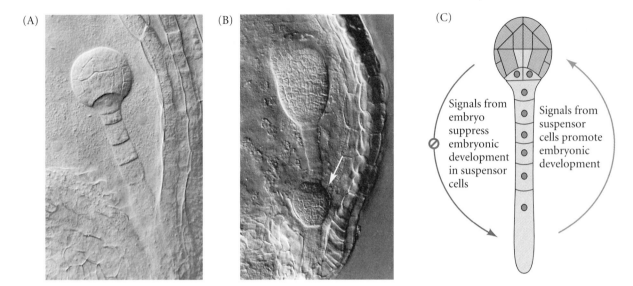

Signals from embryo suppress embryonic development in suspensor cells

Signals from suspensor cells promote embryonic development

Figure 20.17
The *SUS* gene suppresses embryonic development in the suspensor. (A) Wild-type embryo and suspensor. (B) *sus* mutant with suspensor developing like an embryo (arrow). (C) Model showing how the embryo proper suppresses embryonic development in the suspensor and the suspensor provides feedback information to the embryo. (Photographs courtesy of D. Meinke.)

to develop embryo-like structures (Figure 20.17; Schwartz et al. 1994; Yadegari et al. 1994). In these mutants, abnormalities in the embryo proper appear prior to suspensor abnormalities.* Earlier experiments in which the embryo proper was removed also demonstrated that suspensors could develop like embryos (Haccius 1963). A signal from the embryo proper to the suspensor may be important in maintaining suspensor identity and blocking the development of the suspensor as an embryo. Molecular analyses of these and other genes are providing insight into the mechanisms of communication between the suspensor and the embryo proper.

Maternal effect genes play a key role in establishing embryonic pattern in animals (see Chapter 9). The role of extrazygotic genes in plant embryogenesis is less clear, and the question is complicated by at least three potential sources of influence: sporophytic tissue, gametophytic tissue, and the polyploid endosperm. All of these tissues are in close association with the egg/zygote (Ray 1998). Endosperm development could also be affected by maternal genes. Sporophytic and gametophytic maternal effect genes have been identified in *Arabidopsis*, and it is probable that the endosperm genome influences the zygote as well. The first maternal effect gene identified, *SHORT INTEGUMENTS 1* (*SIN1*), must be expressed in the sporophyte for normal embryonic development (Ray et al. 1996). Two transcription factors (FBP7 and FBP11) are needed in the petunia sporophyte for normal endosperm development (Columbo et al. 1997). A fe-

male gametophytic maternal effect gene, *MEDEA* (after Euripides' Medea, who killed her own children), has protein domains similar to those of a *Drosophila* maternal effect gene (Grossniklaus et al. 1998). Curiously, *MEDEA* is in the Polycomb gene group (see Chapter 9), whose products alter chromatin, directly or indirectly, and affect transcription. *MEDEA* affects an imprinted gene (see Chapter 5) that is expressed by the female gametophyte and by maternally inherited alleles in the zygote, but not by paternally inherited alleles (Vielle-Calzada et al. 1999). How significant maternal effect genes are in establishing the sporophyte body plan is still an unanswered question.

Radial and axial patterns develop as cell division and differentiation continue (Figure 20.18; see also Bowman 1994 for detailed light micrographs of *Arabidopsis* embryogenesis). The cells of the embryo proper divide in transverse and longitudinal planes to form a **globular stage** embryo with several tiers of cells. Superficially, this stage bears some resemblance to cleavage in animals, but the nuclear/cytoplasmic ratio does not necessarily increase. The emerging shape of the embryo depends on regulation of the planes of cell division and expansion, since the cells are not able to move and reshape the embryo. Cell division planes in the outer layer of cells become restricted, and this layer, called the **protoderm**, becomes distinct. Radial patterning emerges at the globular stage as the three tissue systems (dermal, ground, and vascular) of the plant are initiated. The **dermal tissue** (epidermis) will form from the protoderm and contribute to the outer protective layers of the plant. **Ground tissue** (cortex and pith) forms from the ground meristem, which lies beneath the protoderm. The **procambium**, which forms at the core of the embryo will give rise to the **vascular tissue** (xylem and phloem), which will function in support and transport. The differentiation of each tissue system is at least partially independent. For example, in the *keule* mutant of *Arabidopsis*, the

---

*Another intriguing characteristic of these mutants is that cell differentiation occurs in the absence of morphogenesis. Thus, cell differentiation and morphogenesis can be uncoupled in plant development.

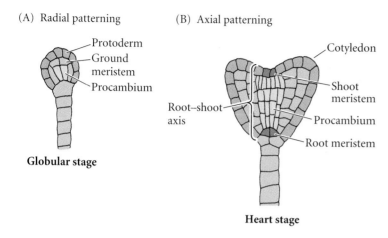

(A) Radial patterning

Protoderm
Ground meristem
Procambium

Globular stage

(B) Axial patterning

Cotyledon

Shoot meristem

Root–shoot axis

Procambium

Root meristem

Heart stage

Figure 20.18
Radial and axial patterning. (A) Radial patterning in angiosperms begins in the globular stage and results in the establishment of three tissue systems. (B) The axial pattern (shoot-root axis) is established by the heart stage.

dermal system is defective while the inner tissue systems develop normally (Mayer et al. 1991).

The globular shape of the embryo is lost as **cotyledons** ("first leaves") begin to form. Dicots have two cotyledons, which give the embryo a heart-shaped appearance as they form. The axial body plan is evident by this **heart stage** of development. Hormones (specifically, auxins) may mediate the transition from radial to bilateral symmetry (Liu et al. 1993). In monocots, such as maize, only a single cotyledon emerges.

In many plants, the cotyledons aid in nourishing the plant by becoming photosynthetic after germination (although those of some species never emerge from the ground). In some cases—peas, for example—the food reserve in the endosperm is used up before germination, and the cotyledons serve as the nutrient source for the germinating seedling.* Even in the presence of a persistent endosperm (as in maize), the cotyledons store food reserves such as starch, lipids, and proteins. In many monocots, the cotyledon grows into a large organ pressed against the endosperm and aids in nutrient transfer to the seedling. Upright cotyledons can give the embryo a torpedo shape. In some plants, the cotyledons grow sufficiently long that they must bend to fit within the confines of the seed coat. The embryo then looks like a walking stick. By this point, the suspensor is degenerating.

The **shoot apical meristem** and **root apical meristem** are clusters of stem cells that will persist in the postembryonic plant and give rise to most of the sporophyte body. The root meristem is partially derived from the hypophysis in some species. All other parts of the sporophyte body are derived from the embryo proper. Genetic evidence indicates that the

formation of the shoot and root meristems is regulated independently. This independence is demonstrated by the *dek23* maize mutant and the *shootmeristemless* (*STM*) mutant of *Arabidopsis*, both of which form a root meristem but fail to initiate a shoot meristem (Clark and Sheridan 1986; Barton and Poethig 1993). The *STM* gene, which has been cloned, is expressed in the late globular stage, before cotyledons form. Genes have also been identified that specifically affect the development of the root axis during embryogenesis. Mutations of the *HOBBIT* gene in *Arabidopsis* (Willemson et al. 1998), for example, affect the hypophysis derivatives and eliminate root meristem function.

The shoot apical meristem will initiate leaves after germination and, ultimately, the transition to reproductive development. In *Arabidopsis*, the cotyledons are produced from general embryonic tissue, not from the shoot meristem (Barton and Poethig 1993). In many angiosperms, a few leaves are initiated during embryogenesis. In the case of *Arabidopsis*, clonal analysis points to the presence of leaves in the mature embryo, even though they are not morphologically well developed (Irish and Sussex 1992). Clonal analysis has demonstrated that the cotyledons and the first two true leaves of cotton are derived from embryonic tissue rather than an organized meristem (Christianson 1986).

Clonal analysis experiments provide information on cell fates, but do not necessarily indicate whether or not cells are determined for a particular fate. Cells, tissues, and organs are shown to be determined when they have the same fate in situ, in isolation, and at a new position in the organism (see McDaniel et al. 1992 for more information on developmental states in plants). Clonal analysis has demonstrated that cells that divide in the wrong plane and "move" to a different tissue layer often differentiate according to their new position. Position, rather than clonal origin, appears to be the critical factor in embryo pattern formation, suggesting some type of cell-cell communication (Laux and Jurgens 1994). Microsurgery experiments on somatic carrot embryos demonstrate that isolated pieces of embryo can often replace the missing complement of parts (Schiavone and Racusen 1990; Scheres and Heidstra 1999). A cotyledon removed from the shoot apex will be replaced. Isolated embryonic shoots can regenerate a new root; isolated root tissue regenerates cotyledons, but is less likely to regenerate the shoot axis. Although most embryonic cells are pluripotent and can generate organs such as cotyledons and leaves, only meristems retain this capacity in the postembryonic plant body.

Dormancy

From the earliest stages of embryogenesis, there is a high level of zygotic gene expression. As the embryo reaches maturity,

---

*Mendel's famous wrinkled-seed mutant (the *rugosus* or *r* allele) has a defect in a starch branching enzyme that affects starch, lipid, and protein biosynthesis in the seed and leads to defective cotyledons (Bhattacharyya et al. 1990).

there is a shift from constructing the basic body plan to creating a food reserve by accumulating storage carbohydrates, proteins, and lipids. Genes coding for seed storage proteins were among the first to be characterized by plant molecular biologists because of the high levels of specific storage protein mRNAs that are present at different times in embryonic development. The high level of metabolic activity in the developing embryo is fueled by continuous input from the parent plant into the ovule. Eventually metabolism slows, and the connection of the seed to the ovary is severed by the degeneration of the adjacent supporting sporophyte cells. The seed dessicates (loses water), and the integuments harden to form a tough seed coat. The seed has entered **dormancy**, officially ending embryogenesis. The embryo can persist in a dormant state for weeks or years, a fact that affords tremendous survival value. There have even been examples of seeds found stored in ancient archaeological sites that germinated after thousands of years of dormancy!

Maturation leading to dormancy is the result of a precisely regulated program. The *viviparous* mutation in maize, for example, produces genetic lesions that block dormancy (Steeves and Sussex 1989). The apical meristems of *viviparous* mutants behave like those of ferns, with no pause before producing postembryonic structures. The embryo continues to develop, and seedlings emerge from the kernels on the ear attached to the parent plant. Recently a group of plant genes have been identified that belong to the Polycomb group, which regulates early development in mammals, nematodes, and insects (Preuss 1999). These genes encode chromatin silencing factors, which may play an important role in seed formation.

Plant hormones are critical in dormancy, and linking them to genetic mechanisms is an active area of research. The hormone **abscisic acid** is important in maintaining dormancy in many species. **Gibberellins**, another class of hormones, are important in breaking dormancy.

Germination

The postembryonic phase of plant development begins with **germination**. Some dormant seeds require a period of **after-ripening** during which low-level metabolic activities continue to prepare the embryo for germination. Highly evolved interactions between the seed and its environment increase the odds that the germinating seedling will survive to produce another generation. Temperature, water, light, and oxygen are all key in determining the success of germination. **Stratification** is the requirement for chilling (5°C) to break dormancy in some seeds. In temperate climates, this adaptation ensures germination only after the winter months have passed. In addition, seeds have maximum germination rates at moderate temperatures of 25°–30°C and often will not germinate at extreme temperatures. Seeds such as lettuce require light (specifically, the red wavelengths) for germination; thus seeds will not germinate so far below ground that they use up their food reserves before photosynthesis is possible.

Desiccated seeds may be only 5–20% water. **Imbibition** is the process by which the seed rehydrates, soaking up large volumes of water and swelling to many times its original size. The **radicle** (primary embryonic root) emerges from the seed first to enhance water uptake; it is protected by a root cap produced by the root apical meristem. Water is essential for metabolic activity, but so is oxygen. A seed sitting in a glass of water will not survive. Some species have such hard protective seed coats that they must be **scarified** (scratched or etched) before water and oxygen can cross the barrier. Scarification can occur by the seed being exposed to the weather and other natural elements over time, or by its exposure to acid as the seed passes through the gut of a frugivore. The frugivore thus prepares the

Figure 20.19
Meristems have delicate cells that are susceptible to damage during germination. A variety of strategies have evolved for protecting the shoot meristem during germination. (A) Some cotyledons protect the meristem as the shoot emerges from the soil. (B) Bent epicotyls (and sometimes hypocotyls) push through the soil. (C) Monocots utilize the coleoptile, a leaflike structure that sheathes the young shoot tip.

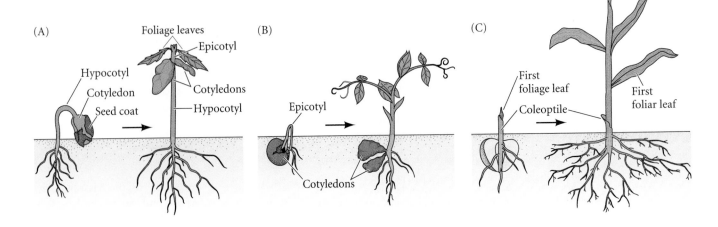

seed for germination, as well as dispersing it to a site where germination can take place.

During germination, the plant draws on the nutrient reserves in the endosperm or cotyledons. Interactions between the embryo and endosperm in monocots use gibberellin as a signal to trigger the breakdown of starch into sugar. As the shoot reaches the surface, the differentiation of chloroplasts is triggered by light. Seedlings that germinate in the dark have long, spindly stems and do not produce chlorophyll. This environmental response allows plants to use their limited resources to reach the soil surface, where photosynthesis will be productive.

The delicate shoot tip must be protected as the shoot pushes through the soil. Three strategies for protecting the shoot tip have evolved (Figure 20.19):

1. Cotyledons protect the shoot tip.
2. The **epicotyl** (the stem above the cotyledons) bends so that stem tissue, rather than the shoot tip, pushes through the soil.
3. In monocots, a special leaflike structure, the **coleoptile**, forms a protective sheath around the shoot tip.

Vegetative Growth

When the shoot emerges from the soil, most of the sporophyte body plan remains to be elaborated. Figure 20.20 shows the basic parts of the mature sporophyte plant, which will emerge from meristems.

Meristems

As has been mentioned, **meristems** are clusters of cells that allow the basic body pattern established during embryogenesis to be reiterated and extended after germination. Meristematic cells are similar to stem cells in animals.* They divide to give rise to one daughter cell that continues to be meristematic and another that differentiates. Meristems fall into three categories: apical, lateral, and intercalary.

Apical meristems occur at the growing shoot and root tips (Figure 20.21). Root apical meristems produce the root cap, which consists of lubricated cells that are sloughed off as the meristem is pushed through the soil by cell division and elongation in more proximal cells. The root apical meristem also gives rise to daughter cells that produce the three tissue systems of the root. New root apical meristems are initiated from tissue within the core of the root and emerge through the ground tissue and dermal tissue. Root meristems can also be derived secondarily from the stem of the plant; in the case of maize, this is the major source of root mass.

The shoot apical meristem produces stems, leaves, and reproductive structures. In addition to the shoot apical meris-

tem initiated during embryogenesis, axillary shoot apical meristems (axillary buds; see Figure 20.20) derived from the original one form in the axils (the angles between leaf and stem). Unlike new root meristems, these arise from the surface layers of the meristem.

Angiosperm apical meristems are composed of up to three layers of cells (labeled L1, L2, and L3) on the plant surface (Figure 20.22). One way of investigating the contributions of different layers to plant structure is by constructing chimeras. Plant chimeras are composed of layers having distinct genotypes with discernible markers. When L2, for example, has a different genotype than L1 or L3, all pollen will have the L2 genotype, indicating that pollen is derived from L2. Chimeras have also been used to demonstrate classical induction in plants, in which, as in animal development, one layer influences the developmental pathway of an adjacent layer.

The size of the shoot apical meristem is precisely controlled by intercellular signals, most likely between layers of the meristem (reviewed by Doerner 1999). Mutations in the *Arabidopsis CLAVATA* genes, for example, lead to increased meristem size and the production of extra organs.[†] *STM* has

[†]This phenomenon, called *fasciation*, is found in many species, including peas and tomatoes.

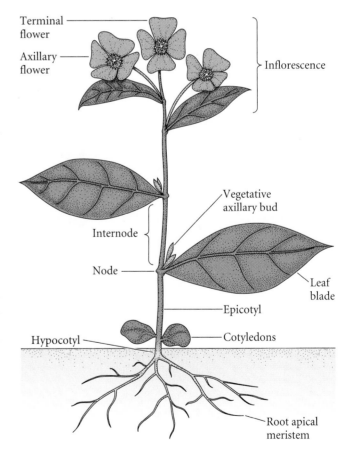

Figure 20.20
Morphology of a generalized angiosperm sporophyte.

---

*The similarities between plant meristem cells and animal stem cells may extend to the molecular level, indicating that stem cells existed before plants and animals pursued separate phylogenetic pathways. Homology has been found between genes required for plant meristems to persist and genes expressed in *Drosophila* germ line stem cells (Cox et al. 1998).

SHOOT MERISTEMS

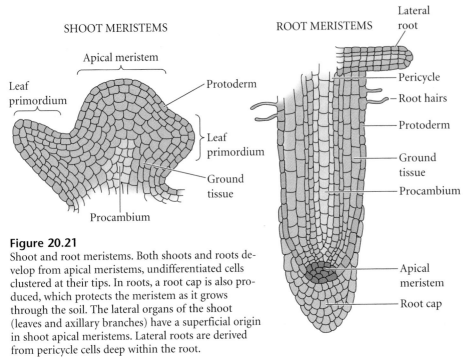

Figure 20.21

Shoot and root meristems. Both shoots and roots develop from apical meristems, undifferentiated cells clustered at their tips. In roots, a root cap is also produced, which protects the meristem as it grows through the soil. The lateral organs of the shoot (leaves and axillary branches) have a superficial origin in shoot apical meristems. Lateral roots are derived from pericycle cells deep within the root.

(which decreases meristem size) (Meyerowitz 1997).

Lateral meristems are cylindrical meristems found in shoots and roots that result in secondary growth (an increase in stem and root girth by the production of vascular tissues). Monocot stems do not have lateral meristems, but often have **intercalary meristems** inserted in the stems between mature tissues. The popping sound you can hear in a cornfield on a summer night is actually caused by the rapid increase in stem length due to intercalary meristems.

Root development

Radial and axial patterning in roots begins during embryogenesis and continues throughout development as the primary root grows and lateral roots emerge from the pericycle cells deep within the root. Laser ablation experiments eliminating single cells and clonal analyses have demonstrated that cells are plastic and that position is the primary determinant of fate in early root development. Analyses of root radial organization mutants have revealed genes with layer-specific activity (Scheres et al. 1995; Scheres and Heidstra 1999). We will illustrate these findings by looking at two *Arabidopsis* genes that regulate ground tissue fate.

In wild-type *Arabidopsis*, there are two layers of root ground tissue. The outer layer becomes the cortex, and the inner layer becomes the endodermis, which forms a tube around the vascular tissue core. The *SCARECROW* (*SCR*) and *SHORT-ROOT* (*SHR*) genes have mutant phenotypes with one, instead of two, layers of root ground tissue (Benfey et al. 1993). The *SCR* gene is necessary for an asymmetrical cell division in the initial layer of cells, yielding a smaller endodermal cell and a larger cortex cell (Figure 20.23). The *scr* mutant expresses markers for both cortex and endodermal cells, indicating that differentiation progresses in the absence of cell division (Di Laurenzio 1996). *SHR* is responsible for endodermal cell specification. Cells in the *shr* mutant do not develop endodermal features.

the opposite effect, and double mutant phenotypes are consistent with the hypothesis that the two work together to maintain meristem size (Clark et al. 1996). Perhaps they balance the rate of cell division (which enlarges the meristem) and the rate of cell differentiation in the periphery of the meristem

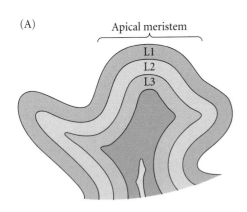

(A)

(B)

Figure 20.22

Organization of the shoot apical meristem. (A) Angiosperm meristems have two or three outer layers of cells that are histologically distinct (here labeled L1, L2, and L3). While cells in certain layers tend to have certain fates, they are not necessarily committed to those fates. If a cell is shifted to a new layer, it generally develops like the other cells in that layer. (B) The fates of the cell layers can be seen in a chimeric tobacco plant. One portion of the meristem contains three layers of wild-type cells, while the other portion has an L2 that lacks chlorophyll. This section of the meristem has given rise to the variegated leaves. In wild-type plants, the L1 layer always lacks chlorophyll (except in guard cells), but in this plant the L2, too, is genetically unable to produce chlorophyll; the L3 remains green. The L3 does not contribute to the outer edges of leaves, which is why they appear white in this plant. (Photograph courtesy of M. Marcotrigiano.)

(A)

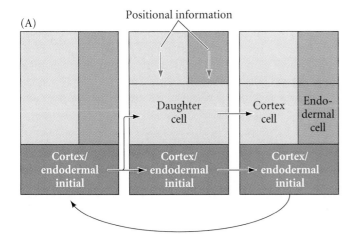

Positional information

Daughter cell → Cortex cell | Endo-dermal cell

Cortex/ endodermal initial | Cortex/ endodermal initial | Cortex/ endodermal initial

(B) Root

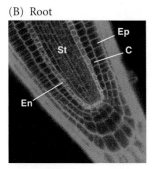

Ep
St
C
En

(C) Shoot

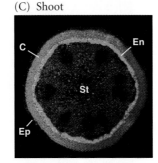

C
En
St
Ep

(D) Wild-type

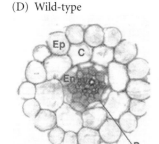

Ep
C
En
P

(E) *scr* mutant

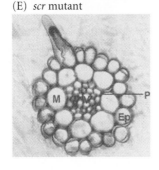

M
P
Ep

(F) *shr* mutant

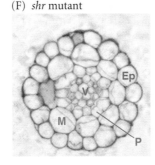

Ep
V
M
P

Figure 20.23
SCR and *SHR* regulate endodermal differentiation in root radial development. (A) Diagram of normal cell division yielding cortical and endodermal cells. *SCR* regulates this asymmetrical cell division. (B, C) *SCR* expression in root and shoot. The *SCR* promoter is linked to the GFP (green fluorescent protein). (D–F) Cross sections of primary roots of (D) wild-type *Arabidopsis*, (E) *scr* mutant, and (F) *shr* mutant. Ep, epidermis; C, cortex; En, endodermis; M, mutant layer; P, pericycle; V, vascular tissue; St, stele. (A after Scheres and Heidstra 1999; photographs courtesy of P. Benfey.

Axial patterning in roots may be morphogen-dependent, paralleling some aspects of animal development. A variety of experiments have established that the distribution of the plant hormone **auxin** organizes the axial pattern. A peak in auxin concentration at the root tip must be perceived for normal axial patterning (Sabatini et al. 1999).

As discussed earlier, distinct genes specifying root and shoot meristem formation have been identified; however, root and shoot development may share common groups of genes that regulate cell fate and patterning (Benfey 1999). This appears to be the case for the *SCR* and *SHR* genes. In the shoot, these genes are necessary for the normal gravitropic response, which is dependent on normal endodermis formation (a defect in mutants of both genes; see figure 20.23C). It's important to keep in mind that there are a number of steps between establishment of the basic pattern and elaboration of that pattern into anatomical and morphological structure. Uncovering the underlying control mechanisms is likely to be the most productive strategy in understanding how roots and shoots develop.

Shoot development

The unique aboveground architectures of different plant species have their origins in shoot meristems. Shoot architecture is affected by the amount of axillary bud outgrowth. Branching patterns are regulated by the shoot tip—a phenomenon called apical dominance—and plant hormones appear to be the factors responsible. Auxin is produced by young leaves and transported toward the base of the leaf. It can suppress the outgrowth of axillary buds. Grazing and flowering often release buds from apical dominance, at which time branching occurs. **Cytokinins** can also release buds from apical dominance. Axillary buds can initiate their own axillary buds, so branching patterns can get quite complex. Branching patterns can be regulated by environmental signals so that an expansive canopy in an open area maximizes light capture. Asymmetrical tree crowns form when two trees grow very close to each other. In addition to its environmental plasticity, shoot architecture is genetically regulated. In several species, genes have now been identified that regulate branching patterns.

Leaf primordia (clusters of cells that will form leaves) are initiated at the periphery of the shoot meristem (see Figure 20.21). The union of a leaf and the stem is called a **node**, and stem tissue between nodes is called an **internode** (see Figure 20.20). In a simplistic sense, the mature sporophyte is created by stacking node/internode units together. **Phyllotaxy**, the positioning of leaves on the stem, involves communication among existing and newly forming leaf primordia. Leaves may be arranged in various patterns, including a spiral, 180-degree alternation of single leaves, pairs, and whorls of three or more leaves at a node (Jean and Barabé 1998). Experimentation has revealed a number of mechanisms for maintaining geometrically

SIMPLE LEAF

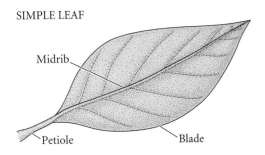

Midrib

Petiole   Blade

COMPOUND LEAF

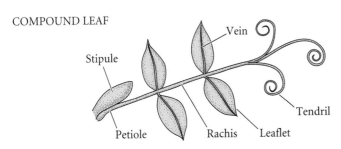

Vein

Stipule

Tendril

Petiole   Rachis   Leaflet

Figure 20.24
Simple and compound leaves.

regular spacing of leaves on a plant, including chemical and physical interactions of new leaf primordia with the shoot apex and with existing primordia (Steeves and Sussex 1989).

It is not clear how a specific pattern of phyllotaxy gets started. Descriptive mathematical models can replicate the observed patterns, but reveal nothing about the mechanism. Biophysical models (e.g., of the effects of stress/strain on deposition of cell wall material, which affects cell division and elongation) attempt to bridge this gap. Developmental genetics approaches are promising, but few phyllotactic mutants have been identified. One candidate is the *terminal ear* mutant in maize, which has irregular phyllotaxy. The wild-type gene is expressed in a horseshoe-shaped region, with a gap where the leaf will be initiated (Veit et al. 1998). The plane of the horseshoe is perpendicular to the axis of the stem.

Leaf development

Leaf development includes commitment to become a leaf, establishment of leaf axes, and morphogenesis, giving rise to a tremendous diversity of leaf shapes. Culture experiments have assessed when leaf primordia become determined for leaf development. Research on ferns and angiosperms indicates that the youngest visible leaf primordia are not determined to make a leaf; rather, these young primordia can develop as shoots in culture (Steeves 1966; Smith 1984). The programming for leaf development occurs later. The radial symmetry of the leaf primordium becomes dorsal-ventral, or flattened, in all leaves. Two other axes, the proximal-distal and lateral, are also established. The unique shapes of leaves result from

regulation of cell division and cell expansion as the leaf blade develops. There are some cases in which selective cell death (apoptosis) is involved in the shaping of a leaf, but differential cell growth appears to be a more common mechanism (Gifford and Foster 1989).

Leaves fall into two categories, simple and compound

Figure 20.25
Overexpression of Class 1 *KNOX* genes in tomato. The photograph shows the single leaves of (A) a wild-type plant, (B) a *mouse ears* mutant, with increased leaf complexity, and (C) a transgenic plant that uses a viral promoter to overexpress the tomato homologue (*LeT6*) of the *KN1* gene from maize. (Photographs courtesy of N. Sinha.)

Figure 20.26
Leaf morphology mutants in peas. (A) Wild-type pea plant. (B) The *tl* mutant, in which tendrils are converted to leaflets. (C) The *af* mutant, in which leaflets are converted to tendrils. (D) An *af tl* double mutant, which results in a "parsley leaf" phenotype. (Photographs courtesy of S. Singer.)

(Figure 20.24; see review by Sinha 1999). There is much variety in simple leaf shape, from smooth-edged leaves to deeply lobed oak leaves. Compound leaves are composed of individual leaflets (and sometimes tendrils) rather than a single leaf blade. Whether simple and compound leaves develop by the same mechanism is an open question. One perspective is that compound leaves are highly lobed simple leaves. An alternative perspective is that compound leaves are modified shoots. The ancestral state for seed plants is believed to be compound, but for angiosperms it is simple. Compound leaves have arisen multiple times in the angiosperms, and it is not clear if these are reversions to the ancestral state.

Developmental genetic approaches are being applied to leaf morphogenesis. The Class I *KNOX* genes are homeobox genes that include *STM* and the *KNOTTED 1* (*KN1*) gene in maize. Gain-of-function mutations of *KN1* cause meristem-like bumps to form on maize leaves. In wild-type plants, this gene is expressed in meristems. When *KN1*, or the tomato homologue *LeT6*, has its promoter replaced with a promoter from cauliflower mosaic virus and is inserted into the genome of tomato, the gene is expressed at high levels throughout the plant, and the leaves become "super compound" (Figure 20.25; Hareven et al. 1996; Janssen et al. 1998). Simple leaves become more lobed (but not compound) in response to overexpression of *KN1*, consistent with the hypothesis that compound leaves may be an extreme case of lobing in simple leaves (Jackson 1996). The role of *KN1* in shoot meristem and leaf development, however, is consistent with the hypothesis that compound leaves are modified shoots.

A second gene, *LEAFY*, that is essential for the transition from vegetative to reproductive development also appears to play a role in compound leaf development. It was identified in *Arabidopsis* and snapdragon (in which it is called *FLORICAULA*), and has homologues in other angiosperms. The pea homologue (*UNIFOLIATA*) has a mutant phenotype in which compound leaves are reduced to simple leaves (Hofer and Ellis 1998). This finding is also indicative of a regulatory relationship between shoots and compound leaves.

In some compound leaves, developmental decisions about leaf versus tendril formation are also made. Mutations of two leaf-shape genes can individually and in sum dramatically alter the morphology of the compound pea leaf (Figure 20.26). The *acacia* mutant (*tl*) converts tendrils to leaflets; *afilia* (*af*) converts leaflet to tendrils (Marx 1987). The *af tl* double mutant has a complex architecture and resembles a parsley leaf.

At a more microscopic level, the patterning of stomata (openings for gas and water exchange) and trichomes (hairs) across the leaf is also being investigated. In monocots, the stomata form in parallel files, while in dicots the distribution appears more random. In both cases, the patterns appear to maximize the evenness of stomata distribution . Genetic analysis is providing insight into the mechanisms regulating this distribution. A common gene group appears to be working in both shoots and roots, affecting the distribution pattern of both trichomes and root hairs (Benfey 1999).

The Vegetative-to-Reproductive Transition

Unlike some animal systems in which the germ line is set aside during early embryogenesis, the germ line in plants is established only after the transition from vegetative to reproductive development—that is, flowering. The vegetative and reproductive structures of the shoot are all derived from the shoot meristem formed during embryogenesis. Clonal analysis indicates that no cells are set aside in the shoot meristem of the embryo to be used solely in the creation of reproductive structures (McDaniel and Poethig 1988). In maize, irradiating seeds causes changes in the pigmentation of some cells. These

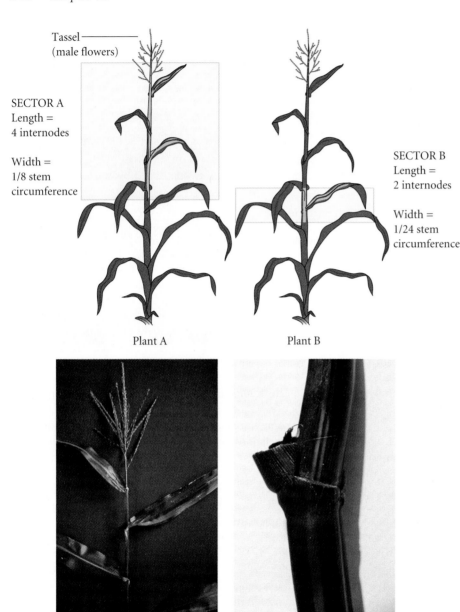

Tassel—
(male flowers)

SECTOR A
Length =
4 internodes

Width =
1/8 stem
circumference

SECTOR B
Length =
2 internodes

Width =
1/24 stem
circumference

Plant A   Plant B

Figure 20.27
Clonal analysis can be used to construct a fate map of a shoot apical meristem in maize. Seeds that are heterozygous for certain pigment genes (anthocyanins) are irradiated so that the dominant allele is lost in a few cells (a chance occurrence). All cells derived from the somatic mutant will be visually distinct from the nonmutant cells. Plants A and B have mutant sectors that reveal the fate of cells in the shoot meristem of the seed. The mutant sector in A includes both vegetative and reproductive (tassel) internodes. Thus there is no distinct developmental compartment that forms the tassel. The mutant sector in A is longer and wider than the mutant sector in B. This indicates that more cells were set aside to contribute to the lower than to the upper internodes in the shoot meristem in the seed. The actual number of cells can be calculated by taking the reciprocal of the fraction of the stem circumference the sector occupies. Sector A contributes to 1/8 of the circumference of the stem; thus 8 cells were fated to contribute to these internodes in the seed meristem. Sector B is only 1/24 of the stem circumference; thus 24 cells were fated to contribute to these internodes. In this example, only cells derived from the L1 are being analyzed. It is also important to consider the possible contributions of the L2 and L3 cell layers of the shoot meristem. (Data from McDaniel and Poethig 1988; photographs courtesy of C. McDaniel.)

seeds give rise to plants that have visually distinguishable sectors descended from the mutant cells. Such sectors may extend from the vegetative portion of the plant into the reproductive regions (Figure 20.27), indicating that maize embryos do not have distinct reproductive compartments.

Maximal reproductive success depends on the timing of flowering—and on balancing the number of seeds produced with resources allocated to individual seeds. As in animals, different strategies work best for different organisms in different environments. There is a great diversity of flowering patterns among the over 300,000 angiosperm species, yet there appears to be an underlying evolutionary conservation of flowering genes and common patterns of flowering regulation.

A simplistic explanation of the flowering process is that a signal from the leaves moves to the shoot apex and induces flowering. In some species, this flowering signal is a response

to environmental conditions. The developmental pathways leading to flowering are regulated at numerous control points in different plant organs (roots, cotyledons, leaves, and shoot apices) in various species, resulting in a diversity of flowering times and reproductive architectures. The nature of the flowering signal, however, remains unknown.

Some plants, especially woody perennials, go through a **juvenile phase**, during which the plant cannot produce reproductive structures even if all the appropriate environmental signals are present (Lawson and Poethig 1995). The transition from the juvenile to the adult stage may require the acquisition of competence by the leaves or meristem to respond to an internal or external signal (McDaniel et al. 1992; Singer et al. 1992; Huala and Sussex 1993).

Grafting and organ culture experiments, mutant analyses, and molecular analyses give us a framework for describing the

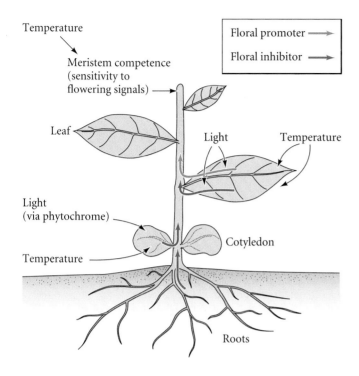

Temperature

Meristem competence
(sensitivity to
flowering signals)

Leaf

Light
(via phytochrome)

Temperature

| Floral promoter ⟶ |
| Floral inhibitor ⟶ |

Light

Temperature

Cotyledon

Roots

Figure 20.28
Regulation of the vegetative-to-reproductive transition. Internal and external factors regulate whether a meristem produces vegetative or reproductive structures. Not all of the regulatory mechanisms shown are used in all species, and some species flower independently of external environmental signals. Signals promoting or inhibiting flowering can move from the roots, cotyledons, or leaves to the shoot apex, where meristem competence determines whether or not the plant will respond to the signals. Leaves may also need to develop competence to respond to environmental signals before they can produce floral promoters.

reproductive transition in plants (Figure 20.28). Grafting experiments have identified the sources of signals that promote or inhibit flowering and have provided information on the developmental acquisition of meristem competence to respond to these signals (Lang et al. 1977; Singer et al. 1992; McDaniel et al. 1996; Reid et al. 1996). Analyses of mutants and molecular characterization of genes are yielding information on the mechanics of these signal-response mechanisms (Hempel et al. 2000; Levy and Dean 1998).

Leaves produce a graft-transmissible substance that induces flowering. In some species, this signal is produced only under specific **photoperiods** (day lengths), while other species are day-neutral and will flower under any photoperiod (Zeevaart 1984). Not all leaves may be competent to perceive or pass on photope-

Figure 20.29
The interaction of competence and signal strength. *Nicotiana tabacum* (*Nt*) is a day-neutral plant that flowers when the meristem gains competence to respond to internal signals. *N. silvestris* (*Ns*) is a long-day plant that flowers when the floral signal(s) reach a critical level. These grafting experiments illustrate that a young *Nt* shoot is less competent to respond to the *Nt* flowering signal than an older *Nt* shoot. Young *Nt* shoots respond quickly to the flowering signal from flowering *Ns* plants, but flower later when grafted to young *Ns* plants with a lower level of signal. The scion is the shoot that is grafted on to the stock; the stock is the rooted portion of the plant from which the shoot has been excised. (After Singer et al. 1992.)

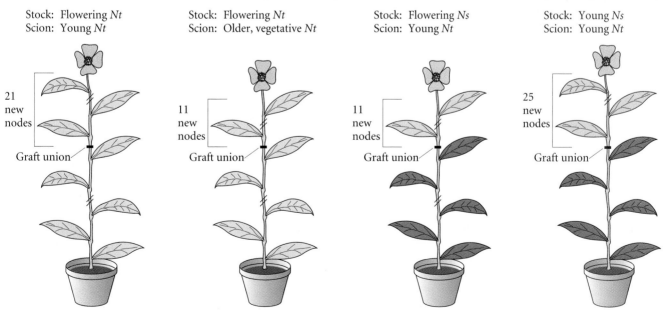

| Stock: Flowering *Nt* | Stock: Flowering *Nt* | Stock: Flowering *Ns* | Stock: Young *Ns* |
| Scion: Young *Nt* | Scion: Older, vegetative *Nt* | Scion: Young *Nt* | Scion: Young *Nt* |

21 new nodes

Graft union

11 new nodes

Graft union

11 new nodes

Graft union

25 new nodes

Graft union

riodic signals. The **phytochrome** pigments transduce these signals from the external environment. The structure of phytochrome is modified by red and far-red light, and these changes can initiate a cascade of events leading to the production of either floral promoter or floral inhibitor (Deng and Quail 1999). Leaves, cotyledons, and roots have been identified as sources of floral inhibitors in some species (McDaniel et al. 1992; Reid et al. 1996). A critical balance between inhibitor and promoter is needed for the reproductive transition.

In some species, meristems change in their competence to respond to flowering signals during development (Singer et al. 1992). **Vernalization**, a period of chilling, can enhance the competence of shoots and leaves to perceive or produce a flowering signal. The reproductive transition depends on both meristem competence and signal strength (Figure 20.29). Shoot tip culture experiments in several species (including tobacco, sunflower, and peas) have demonstrated that determination for reproductive function can occur before reproductive morphogenesis (reviewed in McDaniel et al. 1992). That is, isolated shoot tips that are determined for reproductive development but are morphologically vegetative will produce the same number of nodes before flowering in situ and in culture (Figure 20.30).

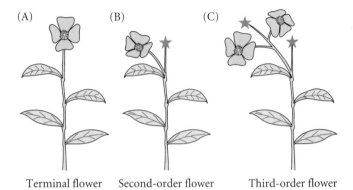

Terminal flower   Second-order flower   Third-order flower

Figure 20.31
Regulation of inflorescence branching architecture. (A) In the simplest angiosperms, a terminal flower forms directly from the terminal shoot apex. (B) *Arabidopsis* and snapdragons produce flowers on axillary branches (second-order flowers). (C) In peas, the axillary branches grow out and initiate flowers (third-order flowers), but do not directly produce flowers. Recent evidence suggests that more complex branching patterns in inflorescences are the result of suppressing the expression of flowering genes in meristems. Tthe stars indicate meristems where floral gene expression is suppressed. In *Arabidopsis* and snapdragons (A), a single gene suppresses flowering in the first-order meristem. In peas (B), two genes are necessary to suppress flowering in the first- and second-order meristems.

Figure 20.30
Determination for reproductive development. In this experiment, buds three nodes from the base of pea plants were treated when the plants had three expanded leaves. These buds were determined for reproductive development, and produced the same number of new nodes (A) when they were allowed to grow in situ by removing the terminal shoot; (B) when they were removed from the plant and cultured; and (C) when only the bud meristem was cultured. If these buds had not been committed to reproductive development, they would have developed like vegetative buds and produced many more nodes before flowering.

Seed-derived plant

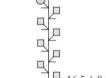

16.5 ± 0.2 nodes

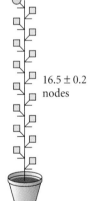

(A) Bud 3 grown in situ   (B) Bud 3 cultured   (C) Bud 3 meristem cultured

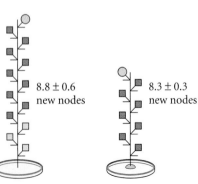

9.0 ± 0.0 new nodes   8.8 ± 0.6 new nodes   8.3 ± 0.3 new nodes

The "black box" between environmental signals and the production of a flower is vanishing rapidly, especially in the model plant *Arabidopsis*. The signaling pathways from light via different phytochromes to key flowering genes are being elucidated. Molecular explanations are revealing redundant pathways that ensure that flowering will occur. Light-dependent, gibberellin-dependent, vernalization-dependent, and autonomous pathways that regulate the floral transition have been genetically dissected.

The ancestral angiosperm is believed to have formed a terminal flower directly from the terminal shoot apex (Stebbins 1974). In modern angiosperms, a variety of flowering patterns exist in which the terminal shoot apex is indeterminate, but axillary buds produce flowers. This observation introduces an intermediate step into the reproductive process: the transition of a vegetative meristem to an **inflorescence meristem**, which initiates axillary meristems that can produce floral organs, but does not directly produce floral parts itself. The inflorescence is the reproductive backbone (stem) that displays the flowers (see Figure 20.20). The inflorescence meristem probably arises through the action of a gene that suppresses terminal flower formation. The *CENTRORADIALUS* (*CEN*) gene in snapdragons suppresses terminal flower formation (Bradley et al. 1996). It suppresses expression of *FLORICAULA* (*FLO*), which specifies floral meristem identity. Curiously, the expression of *FLO* is necessary for *CEN* to be turned on. The *Arabidopsis* homologue of *CEN* (*TERMINAL FLOWER 1* or *TFL1*) is expressed during the vegetative phase of development as well, and has the additional function of delaying the commitment to inflores-

(A)
(B)
(C)
(D)

Figure 20.32
Floral meristem identity mutants. (A) Wild-type *Arabidopsis*, (B) *leafy* mutant, (C) *apetala1*
mutant, and (D) *leafy apetala1* double mutant. (Photographs courtesy of J. Bowman.)

cence development (Bradley et al. 1997). Overexpression of *TFL1* in transgenic *Arabidopsis* extends the time before a terminal flower forms (Ratcliffe et al. 1998). *TFL1* must delay the reproductive transition. Garden peas branch one more time than snapdragons do before forming a flower. That is, the axillary meristem does not directly produce a flower, but acts as an inflorescence meristem that initiates floral meristems. Two genes, *DET* and *VEG1*, are responsible for this more complex inflorescence, and only when both are nonfunctional is a terminal flower formed (Figure 20.31; Singer et al. 1996).

The next step in the reproductive process is the specification of floral meristems—those meristems that will actually produce flowers (Weigel 1995). In *Arabidopsis*, *LEAFY* (*LFY*), *APETALA 1* (*AP1*), and *CAULIFLOWER* (*CAL*) are **floral meristem identity genes** (Figure 20.32). *LFY* is the homologue of *FLO* in snapdragons, and its upregulation during development is key to the transition to reproductive development (Blázquez et al. 1997). Expression of these genes is necessary for the transition from an inflorescence meristem to a floral meristem. Mutants (*lfy*) tend to form leafy shoots in the axils where flowers form in wild-type plants; they are unable to make the transition to floral development. If *LFY* is overexpressed, flowering occurs early. For example, when aspen was transformed with an *LFY* gene that was expressed throughout the plant, the time to flowering was dramatically shortened from years to months (Weigel and Nilsson 1995). *AP1* and *CAL*

are closely related and redundant genes. The *cal* mutant looks like the wild-type plant, but *ap1 cal* double mutants produce inflorescences that look like cauliflower heads (Figure 20.33)

Floral meristem identity genes initiate a cascade of gene expression that turns on region-specifying (**cadastral**) genes,

Figure 20.33
Arabidopsis double mutant of *ap1* and *cal*. Since *cal* alone gives a wild-type phenotype, the double mutant demonstrates the redundancy of these two genes in the flowering pathway. (Photograph courtesy of J. Bowman.)

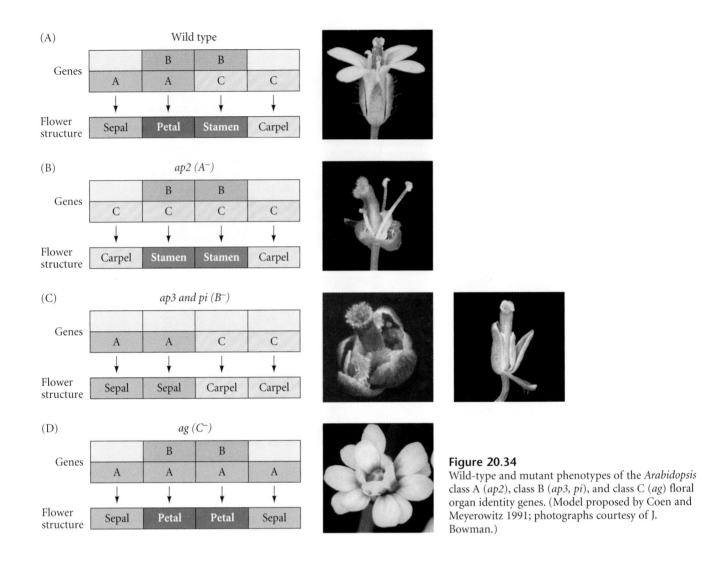

Figure 20.34
Wild-type and mutant phenotypes of the *Arabidopsis* class A (*ap2*), class B (*ap3*, *pi*), and class C (*ag*) floral organ identity genes. (Model proposed by Coen and Meyerowitz 1991; photographs courtesy of J. Bowman.)

which further specify pattern by initiating transcription of **floral organ identity genes** (Weigel 1995). *SUPERMAN* (*SUP*) is an example of a cadastral gene in *Arabidopsis* that plays a role in specifying boundaries for organ identity gene expression. Three classes (A, B, and C) of organ identity genes are necessary to specify the four whorls of floral organs (Figures 20.34 and 20.35; Coen and Meyerowitz 1991). They are homeotic genes (but not Hox genes) and include *AP2*, *AGAMOUS* (*AG*), *AP3*, and *PISTILLATA* (*PI*) in *Arabidopsis*. Class A genes (*AP2*) alone specify sepal development. Class A genes and class B genes (*AP3* and *PI*) together specify petals. Class B and class C (*AG*) genes are necessary for stamen formation; class C genes alone specify carpel formation. When all of these homeotic genes are not expressed in a developing flower, floral parts become leaflike. The ABC genes code for transcription factors that initiate a cascade of events leading to the actual production of floral parts. In addition to the ABC genes, class D genes are now being investigated that specifically regulate ovule development. The ovule evolved long before the other angiosperm floral parts, and while its development is coordinated with that of the carpel, one would expect more ancient, independent pathways to exist.

Senescence

Flowering and **senescence** (a developmental program leading to death) are closely linked in many angiosperms. Individual flower petals in some species senesce following pollination. Orchids, which stay fresh for long periods of time if they are not pollinated, are a good example. Fruit ripening (and ultimately over-ripening) is an example of organ senescence. Whole-plant senescence leads to the death of the entire sporophyte generation. **Monocarpic** plants flower once and then senesce. **Polycarpic** plants, such as the bristlecone pine, can live thousands of years (4900 years is the current record) and flower repeatedly. In polycarpic plants, death is by accident; in monocarpic plants, it appears to be genetically programmed. Flowers and fruits play a key role in the process, and their removal can sometimes delay senescence. In some legumes, senescence can be delayed by removing the developing seed—in other words, the embryo may trigger senescence in the parent plant. During flowering and fruit development, nutrients are reallocated from other parts of the plant to support the development of the next generation. The reproductive structures become a nutrient sink, and this can lead to whole-plant senescence.

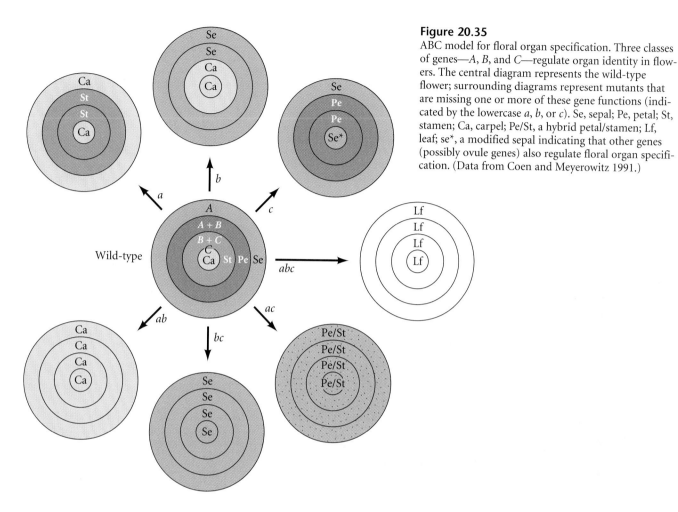

Figure 20.35
ABC model for floral organ specification. Three classes of genes—*A*, *B*, and *C*—regulate organ identity in flowers. The central diagram represents the wild-type flower; surrounding diagrams represent mutants that are missing one or more of these gene functions (indicated by the lowercase *a*, *b*, or *c*). Se, sepal; Pe, petal; St, stamen; Ca, carpel; Pe/St, a hybrid petal/stamen; Lf, leaf; se*, a modified sepal indicating that other genes (possibly ovule genes) also regulate floral organ specification. (Data from Coen and Meyerowitz 1991.)

Snapshot Summary: Plant Development

1. Plants are characterized by alternation of generations; that is, their life cycle includes both diploid and haploid multicellular generations.

2. A multicellular diploid sporophyte produces haploid spores via meiosis. These spores divide mitotically to produce a haploid gametophyte. Mitotic divisions within the gametophyte produce the gametes. The diploid sporophyte results from the fusion of two gametes.

3. The male gamete, pollen, arrives at the style of the female gametophyte and effects fertilization through the pollen tube. Two sperm cells move through the pollen tube; one joins with the ovum to form the zygote, and the other is involved in the formation of the endosperm.

4. Plant embryos develop deeply embedded in parental tissue. The parent tissue provides nutrients but only minimal patterning information.

5. Early embryogenesis is characterized by the establishment of the shoot-root axis and by radial patterning yielding three tissue systems. Pattern emerges by regulation of planes of cell division and the directions of cell expansion, since plant cells do not move during development.

6. As the embryo matures, a food reserve is established. Only the rudiments of the basic body plan are established by the time embryogenesis ceases and the seed enters dormancy.

7. Pattern is elaborated during postembryonic development, when meristems construct the reiterative structures of the plant.

8. The germ line is not reserved early in development. Coordination of signaling among leaves, roots, and shoot meristems regulates the transition to the reproductive state. Reproduction may be followed by genetically programmed senescence of the parent plant.

Literature Cited

Alessa, L. and D. L. Kropf. 1999. F-actin marks the rhizoid pole in living *Pelvetia compressa* zygotes. *Development* 126: 201–209.

Barton, M. K. and R. S. Poethig. 1993. Formation of the shoot apical meristem in *Arabidopsis thaliana*: An analysis of development in the wild type and in the *shoot meristemless* mutant. *Development* 119: 823–831.

Belanger, K. D. and R. S. Quatrano. 2000. Polarity: The role of localized secretion. *Curr. Opin. Plant Biol.* 3: 67–72.

Benfey, P. N. 1999. Is the shoot a root with a view? *Curr. Opin. Plant Biol.* 2: 39–43.

Benfey, P. N., P. J. Linstead, K. Roberts, J. W. Schiefelbein, M.-T. Hauser, and R. A. Aeschbacher. 1993. Root development in *Arabidopsis*: Four mutants with dramatically altered root morphogenesis. *Development* 119: 53–70.

Bhattacharyya, M. K., A. M. Smith, T. H. N. Ellis, C. Hedley and C. Martin. 1990. The wrinkled-seed character of pea described by Mendel is caused by a transposon-like insertion in a gene encoding starch-branching enzyme. *Cell* 60: 115–122.

Blázquez, M. A., L. N. Soowai, I. Lee and D. Weigel. 1997. *LEAFY* expression and flower initiation in *Arabidopsis*. *Development* 124: 3835–3844.

Bowman, J. 1994. *Arabidopsis: An Atlas of Morphology and Development*. Springer-Verlag, New York.

Bradley, D., R. Carpenter, L. Copsey, C. Vincent, S. Rothstein and E. Coen. 1996. Control of inflorescence architecture in *Antirrhinum*. *Nature* 379: 791–797.

Bradley, D., O. Ratcliffe, C. Vincent, R. Carpenter and E. Coen. 1997. Inflorescence commitment and architecture in *Arabidopsis*. *Science* 275: 80–83.

Brewbaker, J. L. and B. H. Kwack. 1963. The essential role of calcium ions in pollen germination and pollen tube growth. *Am. J. Bot.* 50: 859–863.

Brown, K. S. 1999. Deep Green rewrites evolutionary history of plants. *Science* 285: 990–991.

Brownlee, C. and F. Berger. 1995. Extracellular matrix and pattern in plant embryos: On the lookout for developmental information. *Trends Genet.* 11: 344–348.

Cai, G. and M. Cresti. 1999. Rethinking cytoskeleton in plant reproduction: Towards a biotechnological future? *Sex. Plant Reprod.* 12: 67–70.

Christianson, M. L. 1986. Fate map of the organizing shoot apex in *Gossypium*. *Am. J. Bot.* 73: 947–958.

Cionini, P. G., A. Bennici, A. Alpi and F. D'Amato. 1976. Suspensor, gibberellin and *in vitro* development of *Phaseolus coccineus* embryos. *Planta* 131: 115–117.

Clark, J. K. and W. F. Sheridan. 1986. Developmental profiles of the maize embryo-lethal mutants *dek22* and *dek23*. *J. Hered.* 77: 83–92.

Clark, S. E., S. E. Jacobsen, J. Z. Levin and E. M. Meyerowitz. 1996. The *CLAVATA* and *SHOOT MERISTEMLESS* loci competitively regulate meristem activity in *Arabidopsis*. *Development* 122: 1567–1575.

Coen, E. S. and E. M. Meyerowitz. 1991. The war of the whorls: Genetic interactions controlling flower development. *Nature* 353: 31–37.

Columbo, L. J. Franken, A. P. Van der Krol, P. E. Wittich, H. J. Dons and G. C. Angenent. 1997. Down regulation of ovule specific MADS box genes from petunia results in maternally controlled defects in seed development. *Plant Cell* 9: 703–715.

Cox, D. N., A. Chao, J. Baker, L. Chang, D. Qiao and H. Lin. 1998. A novel class of evolutionarily conserved genes defined by *piwi* are essential for stem cell self-renewal. *Genes Dev.* 12: 3715–3727.

Cruden, R. W. and R. M. Lloyd. 1995. Embryophytes have equivalent sexual phenotypes and breeding systems: Why not a common terminology to describe them? *Am. J. Bot.* 82: 816–825.

Deng, X. W. and P. H. Quail. 1999. Signalling in light-controlled development. *Semin. Cell Dev. Biol.* 10: 121–129.

Dodds, P. N., A. E. Clarke and E. Newbigin. 1996. A molecular perspective on pollination in flowering plants. *Cell* 85: 141–144.

Doerner, P. 1999. Shoot meristems: Intercellular signals keep the balance. *Curr. Biol.* 9: R377–R380.

Di Laurenzio, L., J. Wysocka-Diller, J. E. Malmay, L. Pysh, Y. Helariutta, G. Freshour, M. G. Hahn, K. A. Feldman, P. N. Benfey. 1996. The *SCARECROW* gene regulates an asymmetric cell division that is essential for generating the radial organization of the *Arabidopsis* root. *Cell* 86: 423–433.

Drews, G. N., D. Lee and C. A. Christensen. 1998. Genetic analysis of female gametophyte development and function. *Plant Cell* 10: 5–17.

Esau, K. 1977. *Anatomy of Seed Plants*, 2nd Ed. John Wiley and Sons, New York.

Franklin-Tong, V. E., B. K. Drobak, A. C. Allan, P. A. C. Watkins and J. Trewavas. 1996. Growth of pollen tubes in *Papaver rhoeas* is regulated by a slow-moving calcium wave propagated by inositol 1,4,5-trisphosphate. *Plant Cell* 8: 1305–1321.

Friedman, W. E. 1998. The evolution of double fertilization and endosperm: An "historical" perspective. *Sex. Plant Rep.* 11: 6–16.

Gaude, T. and S. McCormick. 1999. Signaling in pollen-pistil interactions. *Semin. Cell Dev. Biol.* 10: 139–147.

Gifford, E. M. and A. S. Foster. 1989. *Morphology and Evolution of Vascular Plants*. W. H. Freeman, New York.

Grossniklaus, U., J. Vielle-Calzada, M. A. Hoeppner and W. B. Gagliano. 1998. Maternal control of embryogenesis by *MEDEA*, a polycomb group gene in *Arabidopsis*. *Science* 280: 446–450.

Haccius, B. 1963. Restitution in acidity-damaged plant embryos: Regeneration or regulation? *Phytomorphology* 13: 107–115.

Hareven, D., T. Gutfinger, A Pornis, Y. Eshed, and E. Lifschitz. 1996. The making of a compound leaf: Genetic manipulation of leaf architecture in tomato. *Cell* 84: 735–744.

Hempel, F. D., D. R. Welch and L. J. Feldman. 2000. Floral induction and determination: Where is flowering controlled? *Trends Plant Sci.* 5: 17–21.

Hofer, J. M. I. and T. H. N. Ellis. 1998. The genetic control of patterning in pea leaves. *Trends Plant Sci.* 3: 434–444.

Huala, E. and I. M. Sussex. 1993. Determination and cell interactions in reproductive meristems. *Plant Cell* 5: 1157–1165.

Hulskamp, M., K. Schneitz and R. E. Pruitt. 1995. Genetic evidence for a long-range activity that directs pollen tube guidance in *Arabidopsis*. *Plant Cell* 7: 57–64.

Irish, V. F. and I. M. Sussex. 1992. A fate map of the *Arabidopsis* embryonic shoot apical meristem. *Development* 115: 745–754.

Jackson, D. 1996. Plant morphogenesis: Designing leaves. *Curr. Biol.* 6: 917–919.

Jaffe, L. A., M. H. Weisenseel and L. F. Jaffe. 1975. Calcium accumulation within the growing tips of pollen tubes. *J. Cell Biol.* 67: 488–492.

Janssen, B.-J., L. Lund, N. Sinha. 1998. Overexpression of a homeobox gene *LeT6* reveals indeterminate features in the tomato compound leaf. *Plant Physiol.* 117: 771–786.

Jean, R. V. and D. Barabé. 1999. *Symmetry in Plants*. World Scientific Publishing, River Edge, NJ.

Johri, B. M., K. B. Ambegaokar and P. S. Srivastava. 1992. *Comparative Embryology of Angiosperms*. Springer-Verlag, New York.

Kropf, D. L., S. R. Bisgrove and W. E. Hable. 1999. Establishing a growth axis in fucoid algae. *Trends Plant Sci.* 4: 490–494.

Lang, A., M. K. Chailakhyan and I. A. Frolova. 1977. Promotion and inhibition of flower formation in a day-neutral plant in grafts with short-day and long-day plants. *Proc. Natl. Acad. Sci. USA* 74: 2412–2416.

Laux, T. and G. Jurgens. 1994. Establishing the body plan of the *Arabidopsis* embryo. *Acta Bot. Neer.* 43: 247–260.

Lawson, E. J. R. and R. S. Poethig. 1995. Shoot development in plants: Time for a change. *Trends Genet.* 11: 263–268.

Lee, H.-S., S. Huang, and T.-H. Kao. 1994. S proteins control rejection of incompatible pollen in *Petunia inflata*. *Nature* 367: 560–563.

Levy, Y. Y. and C. Dean. 1998. The transition to flowering. *Plant Cell* 10: 1973–1989.

Liu, C.-M., Z.-H. Xu and N.-H. Chua. 1993. Auxin polar transport is essential for the establishment of bilateral symmetry during early plant embryogenesis. *Plant Cell* 5: 621–630.

Lord, E. M., L. L. Walling and G. Y. Jauh. 1996. Cell adhesion in plants and its role in pollination. *In* M. Smallwood, J. P. Knox and D. J. Bowles (eds.), *Membranes: Specialized Functions in Plants.* Bios. Sci. Pub., Oxford.

Marx, G. A. 1987. A suite of mutants that modify pattern formation in pea leaves. *Plant Mol. Biol. Rep.* 5: 311–335.

Mayer, U., R. A. Torres Ruiz, T. Berleth, S. Misera and G. Jurgens. 1991. Mutations affecting body organization in the *Arabidopsis* embryo. *Nature* 353: 402–406.

McDaniel, C. N. and R. S. Poethig. 1988. Cell-lineage patterns in the shoot apical meristem of the germinating maize embryo. *Planta* 175: 13–22.

McDaniel, C. N., S. R. Singer and S. M. E. Smith. 1992. Developmental states associated with the floral transition. *Developmental Biology* 153: 59–69.

McDaniel, C. N., L. K. Hartnett and K. A. Sangrey. 1996. Regulation of node number in day-neutral *Nicotiana tabacum*: A factor in plant size. *Plant J.* 9: 55–61.

Meyerowitz, E. M. 1997. Genetic control of cell division patterns in developing plants. *Cell* 88: 299–308.

Mogie, M. 1992. *The Evolution of Asexual Reproduction in Plants.* Chapman and Hall, New York.

Nasrallah, J. B., J. C. Stein, M. K. Kandasamy and M. E. Nasrallah. 1994. Signaling the arrest of pollen tube development in self-incompatible plants. *Science* 266: 1505–1508.

Poethig, R. S. 1987. Clonal analysis of cell lineage patterns in plant development. *Am. J. Bot.* 74: 581–594.

Preuss, D. 1999. Chromatin silencing and *Arabidopsis* development: A role for polycomb proteins. *Plant Cell* 11: 765–768.

Ratcliffe, O. J., I. Amaya, C. A. Vincent, R. Rothstein, R. Carpenter, E. S. Coen and D. J. Bradley. 1998. A common mechanism controls the life cycle and architecture of plants. *Development* 125: 1609–1615.

Ray, A. 1998. New paradigms in plant embryogenesis: Maternal control comes in different flavors. *Trends Plant Sci.* 3: 325–327.

Ray, S., G. T. Golden and A. Ray. 1996. Maternal effects of the *short integument* mutation on embryo development in *Arabidopsis*. *Dev. Biol.* 180: 365–369.

Reid, J. B., I. C. Murfet, S. R. Singer, J. L. Weller and S. A. Taylor. 1996. Physiological-genetics of flowering in *Pisum*. *Semin. Cell Dev. Biol.* 7: 455–463.

Sabatini, S., D. Beis, H. Wolkenfelt, J. Murfett, T. Guilfoyle, J. Malamy, P. Benfey, O. Leyser, N. Bechtold, P. Weisbeek and B. Scheres. 1999. An auxin-dependent distal organizer of pattern and polarity in the *Arabidopsis* root. *Cell* 99: 463–472.

SanMiguel, P., A. Tikhonov, Y. Jin, N. Motchoulskaia, D. Zakharov, A. Melake-Berhan, P. Springer, K. J. Edwards, M. Lee and Z. Avramova. 1996. Nested retrotransposons in the intergenic regions of the maize genome. *Science* 274: 765–768.

Scheres, B. and P. N. Benfey. 1999. Asymmetric cell division in plants. *Annu. Rev. Plant Physiol. Plant Mol. Biol.* 50: 505–537.

Scheres, B. and R. Heidstra. 1999. Digging out roots: Pattern formation, cell division, and morphogenesis in plants. *Curr. Top. Dev. Biol.* 45: 207–247.

Scheres, B., L. Di Laurenzio, V. Willemsen, M.-T. Hauser, K. Janmaat, P. Weisbeek and P. N. Benfey. 1995. Mutations affecting the radial organisation of the *Arabidopsis* root display specific defects throughout the embryonic axis. *Development* 121: 53–62.

Schiavone, F. M. and R. H. Racusen. 1990. Microsurgery reveals regional capabilities for pattern reestablishment in somatic carrot embryos. *Dev. Biol.* 141: 211–219.

Schwartz, B. W., E. C. Yeung and D. W. Meinke. 1994. Disruption of morphogenesis and transformation of the suspensor in abnormal suspensor mutants of *Arabidopsis*. *Development* 120: 3235–3245.

Singer, S. R., C. H. Hannon and S. C. Huber. 1992. Acquisition of competence for floral determination in shoot apices of *Nicotiana*. *Planta* 188: 546–550.

Singer, S. R., S. L. Maki, J. Sollinger, H. Mullen, J. Fick and A. McCall. 1996. Suppression of terminal flower formation in pea: Evolutionary implications. *Dev. Biol.* 175: 377.

Sinha, N. 1999. Leaf development in angiosperms. *Annu. Rev. Plant Physiol. Plant Mol. Biol.* 50: 419–446.

Smith, R. H. 1984. Developmental potential of excised primordial and expanding leaves of *Coleus blumei* benth. *Am. J. Bot.* 71: 114–1120.

Southworth, D. 1996. Gametes and fertilization in flowering plants. *Curr. Top. Dev. Biol.* 34: 259–279.

Stebbins, L. 1974. *Flowering Plants: Evolution Above the Species Level.* Belknap Press, Cambridge, MA.

Steeves, T. A. 1966. On the determination of leaf primordia in ferns. *In* E. G. Cutter (ed.), *Trends in Plant Morphogenesis.* Longman, London, pp. 200–219.

Steeves, T. A. and I. M. Sussex. 1989. *Patterns in Plant Development*, 2nd Ed. Cambridge University Press, New York.

Taylor, C. B. 1996. More arresting developments: S RNases and interspecific incompatibility. *Plant Cell* 8: 939–941.

Trewavas, A. J. and R. Malho. 1998. Ca^{2+} signalling in plant cells: The big network. *Curr. Opin. Plant Biol.* 1: 428–433.

Veit, B., S. P. Briggs, R. J. Schmidt, M. F. Yanofsky and S. Hake. 1998. Regulation of leaf initiation by the *terminal ear 1* gene of maize. *Nature* 393: 166–168.

Vielle-Calzada, J., J. Thomas, C. Spillane, A. Coluccio, M. A. Hoeppner and U. Grossniklaus. 1999. Maintenance of genomic imprinting at the *Arabidopsis* MEDEA locau requires zygotic DDM1 activity. *Genes Dev.* 13: 2971–2982.

Weigel, D. 1995. The genetics of flower development: From floral induction to ovule morphogenesis. *Annu. Rev. Genet.* 29: 19–39.

Weigel, D. and O. Nilsson. 1995. A developmental switch sufficient for flower initiation in diverse plants. *Nature* 377: 495–500.

Wilhelmi, L. K. and D. Preuss. 1996. Self-sterility in *Arabidopsis* due to defective pollen tube guidance. *Science* 274: 1535–1537.

Wilhelmi, L. K. and D. Preuss. 1999. The mating game: Pollination and fertilization in flowering plants. *Curr. Opin. Plant Biol.* 2: 18–22.

Willemsen, V., H. Wolkenfelt, G. de Vrieze, P. Weisbeek and B. Scheres. 1998. The *HOBBIT* gene is required for the formation of the root meristem in the *Arabidopsis* embryo. *Development* 125: 521–531.

Yadegari, R., G. R. de Paiva, T. Laux, A. M. Koltunow, N. Apuya, J. L. Zimmerman, R. L. Fischer, J. J. Harada and R. B. Goldberg. 1994. Cell differentiation and morphogenesis are uncoupled in *Arabidopsis raspberry* embryos. *Plant Cell* 6: 1713–1729.

Yeung, E. C. and I. M. Sussex. 1979. Embryogeny of *Phaseolus coccineus*: The suspensor and the growth of the embryo-proper *in vitro*. *Z. Pflanzenphysiol.* 91: 423–433.

Zeevaart, J. A. D. 1984. Photoperiodic induction, the floral stimulus and floral-promoting substances. *In* D. Vince-Prue, B. Thomas and K. E. Cockshull (eds.), *Light and the Flowering Process.* Academic Press, Orlando, pp. 137–142.

21

Environmental regulation
of animal development

ONE OF THE FEATURES SHARED by all the animals normally used to study developmental biology is their ability to develop regularly in the laboratory. Given adequate nutrition and temperature, these "model systems" develop independently of their environment (Bolker 1995). These animals give one the erroneous impression that "DNA provides the programme which controls the development of the embryo" (Wolpert 1991), or that everything needed to form the embryo is within the fertilized egg. Today, with new concerns about the loss of organismal diversity and about the effects of environmental pollutants, there is renewed interest in the regulation of development by the environment (see van der Weele 1999). There are numerous examples (and *Homo sapiens* provides some of the best) wherein the environment plays a critical role in determining the organism's phenotype. We discussed such environmental regulation of development in Chapters 3 and 17, where we encountered environmentally regulated sex determination and morphologies. In these and many other cases, the environment can elicit different phenotypes from the same genotype. The genetic ability to respond to such environmental factors has to be inherited, of course, but in these cases it is the environment that directs the different phenotypes from the same nuclear genotype.

In this chapter, we will discuss how organisms use environmental cues in the course of their their normal development, as well as how exogenous compounds found in the environment can divert development from its usual path and cause congenital abnormalities.

Environmental Regulation of Normal Development

Environmental Cues and Normal Development

Larval settlement

The inclusion of environmental cues into normal development occurs during the settling of marine larvae. These cues may not be constant, but they need to be part of the environment if further development is to occur (Pechenik et al. 1998). A free-swimming marine larva often needs to settle near a source of food or on a firm substrate on which it can metamorphose. Thus, if prey or substrates give off soluble molecules, these molecules can be used by the larvae as cues to settle and begin metamorphosis. Among the molluscs, there are often very specific cues for settlement

Table 21.1 Specific settlement substrates of molluscan larvae

Molluscan species	Substrate
GASTROPODA (SNAILS, NUDIBRANCHS)	
Nassarius obsoletus	Mud from adult habitat
Philippia radiata	*Porites lobata* (a cnidarian)
Adalaria proxima	*Electra pilosa* (a bryozoan)
Doridella obscura	*Electra crustulenta* (a bryozoan)
Phestilla sibogae	*Porites compressa* (a cnidarian)
Rostanga pulchra	*Ophlitaspongia pennata* (a sponge)
Trinchesia aurantia	*Tubularia indivisa* (a cnidarian)
Elysia chlorotica	Primary film of microorganisms from adult habitat
Haminoea solitaria	Primary film of microorganisms from adult habitat
Aplysia californica	*Laurencia pacifica* (a red alga)
Aplysia juliana	*Ulva spp.* (green algae)
Aplysia parvula	*Chondrococcus hornemanni* (a red alga)
Stylocheilus longicauda	*Lyngbya majuscula* (a cyanobacterium)
Onchidoris bilamellata	Living barnacles
AMPHINEURA (CHITONS)	
Tonicella lineata	*Lithophyllum sp.* and *Lithothamnion sp.* (red algae)
LAMELLIBRANCHIA (BIVALVES)	
Teredo sp.	Wood
Bankia gouldi	Wood
Mercenaria mercenaria	Clam liquor; sand
Placopecten magellanicus	Adult shell; sand; etc.
Mytilus edulis	Filamentous algae; other nonbiological silk material
Crassostrea virginica	Shell liquor; body extract; "shellfish glycogen"

(Table 21.1; Hadfield 1977). In some cases, the prey supply the cues, while in other cases the substrate gives off molecules used by the larvae to initiate settlement.*

One of the best studied cases of larval settlement is that of the red abalone, *Haliotis rufescens*. Its larvae only settle when they physically contact coralline red algae. A brief contact is all that is required for the competent larvae to stop

---

*The importance of substrates for larval settlement and metamorphosis was first demonstrated in 1880, when William Keith Brooks, an embryologist at Johns Hopkins University, was asked to help the ailing oyster industry of Chesapeake Bay. For decades, oysters had been dredged from the bay, and there had always been a new crop to take their place. But recently, each year brought fewer oysters. What was responsible for the decline? Experimenting with larval oysters, Brooks discovered that the American oyster (unlike its better-studied European cousin) needed a hard substrate on which to metamorphose. For years, oystermen had thrown the shells back into the sea, but with the advent of suburban sidewalks, the oystermen were selling the shells to the cement factories. Brooks's solution: throw the shells back into the bay. The oyster population responded, and the Baltimore wharves still sell their descendants.

swimming and begin metamorphosis. The chemical agent responsible for this change has not yet been isolated, but a receptor that recognizes an algal peptide induces metamorphosis in competent larvae. Larvae that are not competent to begin metamorphosis do not appear to have this receptor. The receptor is thought to be linked to a G protein similar to those found in vertebrates, and the activation of this G protein may be necessary for inducing larval settlement and metamorphosis (Morse et al. 1984; Baxter and Morse 1992; Degnan and Morse 1995).

Blood meals

In many mosquitoes, egg production is triggered by a blood meal. Only female mosquitoes bite, and prior to a blood meal they make no vitellogenin yolk protein. In *Aedes aegypti*, the digested products of the blood meal stimulate the brain to secrete egg development neurosecretory hormone (EDNH, also known as ovarian ecdysteroidogenic hormone, or OEH). This hormone stimulates the ovary to make ecdysteroids, which instruct the fat body cells to make vitellogenin for the oocytes (Fallon et al. 1974; Hagedorn 1983; Borovsk et al. 1990). Vitellogenin is critical for egg production. Thus, without the blood meal, there is no vitellogenin and no eggs.

In the blood-sucking bug *Rhodinus prolixus*, adult females produce a new batch of eggs each time they drink blood. This blood meal serves two functions. Blood proteins from the mammalian host supply the amino acids needed for vitellogenin synthesis, and the physical stretching of the abdomen by the blood initiates the endocrine stimuli that activates juvenile hormone secretion by the corpora allata. JH stimulates vitellogenin synthesis in the ovary and fat body (see Nijhout 1994). Moreover, a single large blood meal induces the molt. If this bug takes many small meals, it will survive, but it will not molt or grow. In these instances, mammals provide the environmental cues for part of the insect's development.

Developmental symbiosis

In some of the above examples, the development of one individual is made possible by the presence of another individual of a different species. In some organisms, this relationship has become symbiotic (Sapp 1994). In these cases, the symbionts become so tightly integrated into the host organism that the host cannot develop without them. The adult squid *Euprymna scolopes* is equipped with a light organ composed of sacs containing the luminous bacteria *Vibrio fischeri*. The juvenile squid, however, does not contain these light-emitting symbionts; nor does it have a structure to house them. Rather, the squid acquires the bacteria from the seawater pumped through its mantle cavity. The bacteria bind to a ciliated epithelium that extends into this cavity. The bacteria induce the apoptotic death of these epithelial cells, their replacement by a nonciliated epithelium, and the differentiation of the surrounding epithelial cells into storage sacs for the bacteria (Figure 21.1; McFall-Ngai and Ruby 1991; Montgomery and McFall-Ngai 1995).

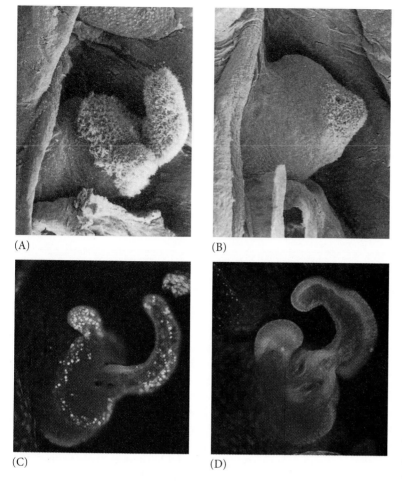

(A)

(B)

(C)

(D)

Figure 21.1
Symbiosis in the squid *Euprymna*. (A, B) Scanning electron micrographs of a light organ primordium of a 3-day-old juvenile squid *E. scolopes*. (A) Light organ in an uninfected juvenile. (B) Light organ of a juvenile infected with the symbiotic *V. fischeri* bacteria. Regression of the epithelium is obvious. (C) Bacterial-induced apoptosis shown by acridine orange staining at 12 hours after infection of the juvenile squid with the bacteria. The bright green areas indicate regions of cell death. (D) Light organ of a squid grown in the absence of *V. fischeri*. No areas of apoptosis are seen. (From Montgomery and McFall-Ngai 1995; photographs courtesy of M. McFall-Ngai.)

es of amphibian and snail eggs, photosynthesis from algal "fouling" enables net oxygen production in the light, while respiration exceeds photosynthesis in the dark (Bachmann et al. 1986; Pinder and Friet 1994; Cohen and Strathmann 1996). Thus, the algae "rescue" the eggs by their photosynthesis.

An even tighter link between morphogenesis and symbiosis is exemplified by the leafhopper *Euscelis incisus*. Here, the symbiosis occurs within the egg. Symbiotic bacteria are found within the egg cytoplasm and are transferred through the generations, just like mitochondria. These bacteria have become so specialized that they can multiply only inside the leafhopper's cytoplasm, and the host has become so dependent on the bacteria that it cannot complete embryogenesis without them. In fact, it is thought that the bacterial symbionts are essential for the formation of the embryonic gut. If the bacteria are surgically or metabolically removed from the eggs (by feeding antibiotics to larvae or adults), these symbiont-free oocytes develop into embryos that lack an abdomen (Figure 21.2; Sander 1968; Schwemmler 1974, 1989).

Symbiosis between egg masses and photosynthetic algae is critical for the development of several species. When eggs are packed together in tight masses, the supply of oxygen limits the rate of development, and the development of those embryos on the inside of the cluster is retarded compared with those near the surface (Strathmann and Strathmann 1995). While there is a steep gradient of oxygen from the outside of the cluster to deep within it, the embryos seem to get around this problem by coating themselves with a thin film of photosynthetic algae. In clutch-

WEBSITE **21.1 Developmental symbioses.** Some embryos acquire protection and nutrients by forming symbiotic associations with other organisms. The mechanisms by which these associations form are now being elucidated.

(A)

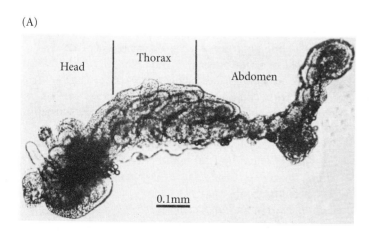

Head | Thorax | Abdomen

0.1mm

(B)

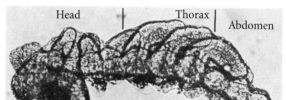

Head | Thorax | Abdomen

0.1mm

Figure 21.2
Microbial symbionts are necessary for gut formation in the leafhopper *Euscelis incisus*. (A) Control embryo, with symbionts, has normal gut formation. (B) Abnormal, gut-deficient embryo forming when antibiotics have eliminated most of the symbiotic bacteria from the egg. (From Schwemmler 1974; photographs courtesy of W. Schwemmler.)

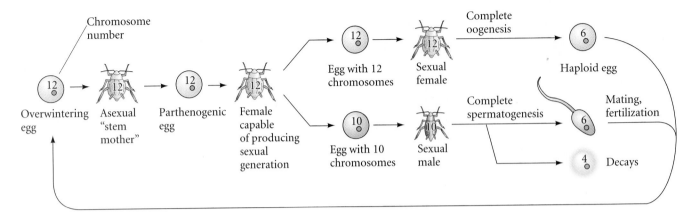

Figure 21.3
Chromosomal changes during the life cycle of the hickory aphid. Fall weather induces the production of males and females, which mate to produce the overwintering egg.

Predictable Environmental Differences as Cues for Development

If the environment contains predictable components (such as gravity) or predictable changes (such as seasons), these can become part of the development of the organism. The use of temperature and daylight length is used by numerous species to adjust their development to a changing environment. The stresses of gravitational pressure also play a role in the development of some organisms.

Seasonality and sex in aphids

Several species of aphids have a fascinating life cycle wherein an egg hatched in the spring gives rise to several generations of parthenogenetically (asexually) reproducing females. During the autumn, however, a particular type of female is produced whose eggs can give rise to both males and sexual females. These sexual forms mate, and their eggs are able to survive the winter. When the overwintering eggs hatch, each one gives rise to an asexual female.

Some of the mysteries of this type of development were solved in 1909 by Thomas Hunt Morgan (before he started working on fruit flies). Morgan analyzed the chromosomes of the hickory aphid through several generations (Figure 21.3). He found that the diploid number of the female aphids is 12. In parthenogenetically reproducing females, only one polar body is extruded from the developing ovum during oogenesis, so the diploid number of 12 is retained in the egg. This egg develops parthenogenetically, without being fertilized. In the females that give rise to eggs that become male or female, a modification of oogenesis occurs. In the female-producing eggs, 6 chromosome pairs enter the sole polar body; the diploid number of 12 is thereby retained. In male-producing eggs, however, an extra chromosome pair enters the polar body. The male diploid number is thus 10. These males and

females are sexual and produce gametes by complete meiotic divisions. The females produce oocytes with a haploid set of 6 chromosomes. The males, however, divide their 10 chromosomes to produce some sperm with a haploid number of 4 and other sperm with a haploid number of 6. The sperm with 4 chromosomes degenerate. The sperm with 6 chromosomes fertilize the eggs with their 6 chromosomes to restore the diploid chromosome number 12. These eggs overwinter, and when they hatch in the spring, females emerge.

Morgan solved one riddle. The riddle of how the autumn weather regulates whether the female reproduces sexually or parthenogenetically, however, remains unsolved. Similarly, we do not know what regulates whether the diploid oocyte gives rise to male- or female-producing eggs. Moreover, the same environmental factors are used differently by other aphid species. Figure 21.4 shows another type of life cycle found in aphids, involving an alternation of sexual and asexual generations. In *Megoura viciae*, temperature determines the sex early in development (with extreme temperatures favoring the production of females). In female development, day length and temperature determine whether the female will reproduce sexually or parthenogenetically, and a combination of temperature and population density determine whether she will be winged or wingless (Beck 1980). It appears that juvenile hormone controls the parthenogenetic/sexual switch (the addition of JH to adults producing sexual offspring causes them to have parthenogenetic offspring) and inhibits the formation of wings (Hardie 1981; Hardie and Lees 1985). But it is not known how the environmental changes become transformed into titers of JH, or how the autumn weather (or perhaps declining hours of sunlight) causes the differential movement of chromosomes into the polar body.

WEBSITE **21.2 Complex environmental effects on development**. The life cycles of certain insects are controlled by several environmental cues whose intersection provides a delicate timing mechanism.

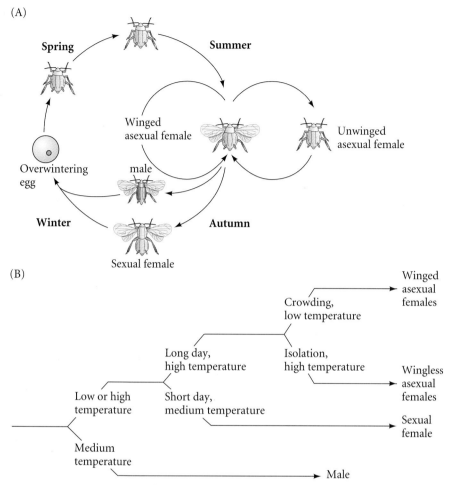

(A)

Spring

Summer

Winged
asexual female

Overwintering
egg

Unwinged
asexual female

male

Winter

Autumn

Sexual female

(B)

Crowding,
low temperature

Winged
asexual
females

Long day,
high temperature

Isolation,
high temperature

Low or high
temperature

Short day,
medium temperature

Wingless
asexual
females

Sexual
female

Medium
temperature

Male

Figure 21.4
Environmental effects on the life cycle of the aphid *Megoura viciae*. (A) Alternation of sexual and asexual generations, wherein the sexual generation is produced in the autumn. (B) Developmental alternatives provided by environmental factors in the life cycle of *Megoura*. (A after Nijhout 1994; B after Beck 1980.)

Diapause

Many species of insects have evolved a strategy called **diapause**. Diapause is a suspension of development that can occur at the embryonic, larval, pupal, or adult stage, depending on the species. The overwintering eggs of the hickory aphid provide an example of this strategy. In some species, diapause is facultative and occurs only when induced by environmental conditions; in other species, diapause has become an obligatory part of the life cycle. The latter is often seen in temperate-zone insects, in which diapause is induced by changes in the photoperiod (the relative lengths of day and night). The day length at which 50 percent of the population has entered diapause is called the critical day length, and this usually occurs quite suddenly. The critical day length is a genetically determined property (Danilevskii 1965; Tauber et al. 1986).

Diapause is not a physiological response brought about by harsh conditions. Rather, it is brought about by token stimuli that presage a change in the environment, beginning before the severe conditions actually arise. Diapause is especially important for temperate-zone insects, enabling them to survive the winter.

The silkworm moth *Bombyx mori* overwinters as an embryo, entering diapause just before segmentation. The gypsy moth *Lymantia dispar* initiates its diapause as a fully formed larva, ready to hatch as soon as diapause ends.

WEBSITE **21.3 Mechanisms of diapause**. Light and temperature are critical for the induction and maintenance of diapause. Different species use different signals for this event.

Gravity and pressure

An environmental constant that can be used in development is gravity and movement. In Chapters 10 and 11 we saw that gravity was critical for frog and chick axis formation. Moreover, there are several bones whose formation is dependent on stresses occasioned by the movement of the embryo. Such stresses have been known to be responsible for the formation of the human patella (kneecap) after birth. But more recently it has been shown that several bones in the chicken do not form if embryonic movement is suppressed in the chick egg. One of these bones is the fibular crest. This bone connects the tibia to the fibula, and it allows the force of the iliofibularis muscle to pull directly from the femur to the tibia. This direct connection is thought to be important in the evolution of birds, and the fibular crest is a universal feature of the bird hindlimb (Müller and Steicher 1989). When the bird is prevented from moving within its egg, this bone fails to develop (Figure 21.5; Wu 1996; Newman and Müller, in press). Therefore, even in the formation of important bones, the environment can play a critical role.

Phenotypic Plasticity: Polyphenism and Reaction Norms

The ability of an individual to express one phenotype under one set of circumstances and another phenotype under another set is called **phenotypic plasticity**.* There are two main

---

*The ability of environmental cues to induce phenotypic change should be considered "tertiary induction." Primary induction involves the establishment of a single field within the embryo (such that one egg gives rise to just one embryo). Secondary induction concerns those cascades of inductive events within the embryo by which the organs are formed. Tertiary induction is the induction of developmental changes by factors in the environment.

(A)

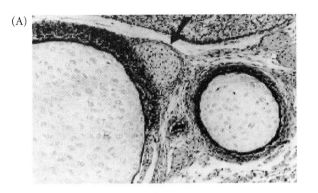

(B)   (C)

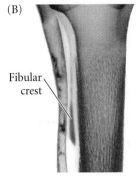

Fibular
crest

Figure 21.5
Activity-induced formation of the fibular crest. The fibular crest (syndesmosis tibiofibularis) is formed when the movement of the embryo in the egg puts stress on the tibia. (A) Traverse section through the 10-day embryonic chick limb, showing condensation (arrow) that will become the fibular crest. (B) 13-day chick embryo showing fibular crest forming between the tibia and fibular bones. (C) Absence of fibular crest in the connective tissue of a 13-day embryo whose movement was inhibited. The blue dye stains cartilage, while the red dye stains the bone elements. (Photographs courtesy of G. Müller.)

types of phenotypic plasticity: polyphenism and reaction norms. **Polyphenism** refers to discontinuous ("either/or") phenotypes elicited by the environment. Migratory locusts, for instance, exist in two mutually exclusive forms: a short-winged, uniformly colored solitary phase and a long-winged, brightly colored gregarious phase. Cues in the environment (mainly population density) determine which morphology a young locust will take (Figure 21.6; see Pener 1991). Similarly, the nymphs of planthoppers can develop in two ways, depending on their environment. High population densities and the presence of certain plant communities lead to the production of migratory insects, in which the third thoracic segment produces a large hindwing. Low population densities and other food plants lead to the development of flightless planthoppers, with the third thoracic segment developing into a haltere-like vestigial wing (Raatikainen 1967; Denno et al. 1985). The seasonal coat color changes in arctic animals are another example of polyphenism.*

In other cases, the genome encodes a range of potential phenotypes, and the environment selects the phenotype that is usually the most adaptive. For instance, constant and intense labor can make our muscles grow larger; but there is a genetically defined limit to how much hypertrophy is possible. Similarly, the microhabitat of a young salamander can cause its color to change (again, within genetically defined limits). This continuous range of phenotypes expressed by a single genotype

*Although seasonal polyphenism is usually considered adaptive, there are times when it does not increase the fitness of the organism. For instance, the photoperiod can cause a hare's color to change from brown to white, but if it doesn't snow, the hare will be conspicuous against the dark background.

(A)   Stationary   Migratory
       morph   morph

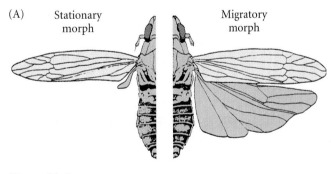

(B)

(C)

Figure 21.6
Density-induced polyphenism in planthoppers and grasshoppers. (A) Composite diagram showing the short-winged (left) and long-winged (right) forms of the planthopper *Prokelisia marginata*. The long-winged form is an excellent flier; the short-winged form is flightless. (B, C) Density-induced changes in the "plague locust" *Schistocerca gregaria*. (B) Low-density morph, showing green pigmentation and miniature wings. (C) High-density morph, showing new pigmentation and wing and leg development. (A after Denno et al. 1985; B, C from Tawfik et al. 1999, photographs courtesy of S. Tanaka).

Figure 21.7
Polyphenic variation of *Pontia* (*Pieridae*) The top row shows summer morphs: *P. protodice* female (left) and male (center); *P. occidentalis* male (right). Bottom row, spring morphs: *P. protodice* female (left) and male (center); *P. occidentalis* male (right). (Photograph courtesy of T. Valente.)

across a range of environmental conditions is called the **reaction norm** (Woltereck 1909; Schmalhausen 1949; Stearns et al. 1991). The reaction norm is thus a property of the genome and can also be selected. Different genotypes are expected to differ in the direction and amount of plasticity that they are able to express (Gotthard and Nylin 1995; Via et al. 1995).

Seasonal polyphenism in butterflies

Two dramatic examples of polyphenism were shown in Chapter 3. The two phenotypes of the butterfly *Araschnia levana* are so different that Linnaeus classified them as two different species (see Figure 3.3) and the phenotype of the moth *Nemoria arizonaria* depends on its diet (Figure 3.4). This type of polyphenism is not uncommon among insects. Throughout much of the Northern Hemisphere, one can see a polyphenism in the *Pieris* and *Colias* butterflies (the cabbage whites and sulphurs) between those that eclose (emerge from their pupal case) during the long days of summer and those that eclose at the beginning of the season, in the short, cooler days of spring. The hindwing pigments of the short-day forms are darker than those of the long-day butterflies. This has a functional advantage during the colder months of spring: The darker pigments absorb sunlight more efficiently, raising the body temperature faster than lighter pigments, so the darker short-day butterflies can use their pigments to heat themselves up between flights (Figure 21.7; Shapiro 1968, 1978; Watt 1968, 1969; Hoffmann 1973; see also Nijhout 1991).

In tropical parts of the world, there is often a hot wet season and a cool dry season. In Africa, the Malawian butterfly *Bicyclus anynana* has a polyphenism that is adaptive to seasonal changes. It occurs in two phenotypes (morphs). The dry-season morph is cryptic, resembling the dead brown leaves of its habitat. The wet-season morph is more active, and

it has ventral hindwing eyespots that deflect attacks from predatory birds and lizards (Figure 21.8). The determining factor appears to be the temperature during pupation. Low temperatures produce the dry-season morph; high temperatures produce the wet-season morph (Brakefield and Reitsma 1991). The development of butterfly eyespots begins in the late larval stages, when the transcription of the *Distal-less* gene is restricted to a small focus that will become the center of each eyespot. During the early pupal stage, *Distal-less* expression is seen in a wider area, and this expression is thought to constitute the activating signal that determines the size of the spot. Last, the cells receiving the signal determine the color they will take. The seasonal *Bicyclus* morphs appear to diverge at the later stages of signal activation and color differentiation (Figure 21.9; Brakefield et al. 1996).

WEBSITE **21.4 Polyphenisms in butterflies**. Environmental cues appear to regulate development by controlling the production of juvenile hormone or ecdysone.

Nutritional polyphenism

Not all polyphenisms are controlled by the seasons. In bees, the size of the female larva at its metamorphic molt determines whether the individual is to be a worker or a queen. A larva fed nutrient-rich "royal jelly" retains the activity of her corpora allata during her last instar stage. The juvenile hormone secreted by these organs delays pupation, allowing the resulting bee to emerge larger and (in some species) more specialized in her anatomy (Brian 1974, 1980; Plowright and Pendrel 1977). The JH level of larvae destined to become queens is 25 times that of larvae destined to become workers, and application of JH onto

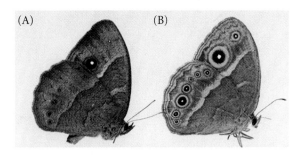

Figure 21.8
The two seasonal morphs of the *Bicyclus anynana* butterfly from Malawi. (A) The dry-season form blends into brown dead-leaf litter. (B) The wet-season form has conspicuous ventral hindwing eyespots. The wet-season form can be mimicked by raising larvae at high (23°C) temperatures, whereas larvae grown in lower temperatures (17°C, approximating the temperatures in the transition to the dry season) develop into the dry-season morph. (From Brakefield et al. 1996; photographs courtesy of S. Carroll and P. Brakefield.)

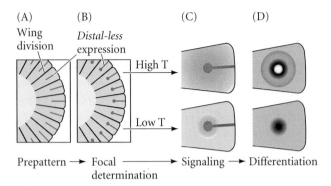

Figure 21.9
Developmental stages leading to the formation of eyespots. (A) *Distal-less* gene expression occurs in regions of the wing imaginal disc where eyespots have the potential to form. (B) Foci of *Distal-less* expression are stabilized in particular regions of the wing. (C) In the pupa, *Distal-less* foci expand. (D) The surrounding cells respond to the signal by producing particular pigments, depending on their distance from the focus and their position in the wing. In *Bicyclus*, the two morphs are indistinguishable until the signaling stage (C). (After Brakefield et al. 1996.)

worker larvae can transform them into queens as well (Wirtz 1973; Rachinsky and Hartfelder 1990).

Similarly, ant colonies are predominantly female, and the females can be extremely polymorphic (Figure 21.10A). The two major types of females are the worker and the gyne. The **gyne** is a potential queen. In more specialized species, a larger worker, the soldier, is also seen. In *Pheidole bicarinata*, these castes are determined by the levels of JH in the developing larvae. Larvae given protein-rich food have an elevated JH titer,

which causes an abrupt developmental switch that "reprograms" the size at which the larvae will begin metamorphosis. This causes a large and discontinuous size difference between the gyne, soldier, and worker castes. This reprogramming also involves changes in gene activity, since the cuticular proteins of the workers and soldiers are different (Passera 1985; Wheeler 1991).

In different ant species, caste determination can be environmental, hormonal, or a combination of both. These developmental patterns of caste determination have been analyzed by Diana Wheeler (1986 1991) and are summarized in Figures 21.10B and C. In most species, ant larvae are bipotential until near pupation. In *Myrmica rubra*, only larvae that overwinter remain bipotential. After winter, the queen stimulates workers to underfeed the last-instar larvae. This means that as long as there is a queen, no new queens can result. If the larvae are fed, they can becomes gynes. Thus, these larvae remain bipotential until late in their last instar. In other species such as *Pheidole pallidula*, the queen controls gyne formation through chemicals that act during embryogenesis, so that no new queens are formed. However, the workers remain bipotential and can become minor or major workers, depending on nutrition.

Environment-dependent sex determination

As we saw in Chapter 17, there are many species in which the environment determines whether an individual is to be male or female. The temperature-dependence of sex determination in fishes and reptiles has provided the best studied cases. Figure 17.20 displays some of the patterns of temperature-dependent sex determination in reptiles. This type of environmental sex determination has advantages and disadvantages.

One advantage is that it probably gives the species the benefits of sexual reproduction without tying the species to a 1:1 sex ratio. In crocodiles, in which temperature extremes produce females while moderate temperatures produce males, the sex ratio may be as great as 10

(A)

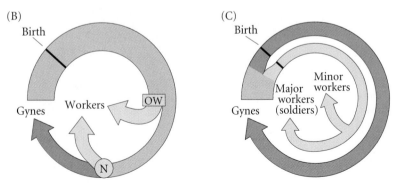

(B) Birth / Gynes / Workers / OW / N

(C) Birth / Gynes / Major workers (soldiers) / Minor workers

Figure 21.10
(A) The remarkable dimorphism of worker (left) and queen (right) ants in the species *Pheidologeton diversus*. The two are sisters, but one was fed as a larva such that she kept growing and metamorphosed into a fertile "queen." (B, C) Gyne (queen) and worker formation in ants. Lightly colored areas represent bipotentiality to become either workers or gynes. The circled N represents a nutritional switch controlled by the larva's environment. (B) *Myrmica rubra*, wherein only the larvae that overwinter (OW) remain bipotential. In the last instar, the nutritional switch determines caste. (C) *Pheidole pallidula*, wherein the queen controls gyne determination through hormones that act during embryogenesis. (A photograph © Mark W. Moffett/National Geographic Society; B, C after Wheeler 1986.)

females to each male (Woodward and Murray 1993). The major disadvantage of temperature-dependent sex determination may be its narrowing of the temperature limits within which a species can exist. This means that thermal pollution (either locally or due to global warming) could conceivably eliminate a species in a given area (Janzen and Paukstis 1991). Ferguson and Joanen (1982) speculate that dinosaurs may have had temperature-dependent sex determination and that their sudden demise may have been caused by a slight change in temperature that created conditions wherein only males or only females hatched from their eggs.

Charnov and Bull (1977) have argued that environmental sex determination would be adaptive in certain habitats characterized by patchiness—a habitat having some regions where it is more advantageous to be male and other regions where it is more advantageous to be female. Conover and Heins (1987) provide evidence for this hypothesis. In certain fishes, females benefit from being larger, since size translates into higher fecundity. If you are a female Atlantic silverside (*Menidia menidia*), it is advantageous to be born early in the breeding season, which allows you a longer feeding season and thus would allow you to grow larger. In the males, size is of no importance. Conover and Heins showed that in the southern range of *Menidia*, females are indeed born early in the breeding season. Temperature appears to play a major role in this pattern. However, in the northern reaches of its range, the species shows no environmental sex determination. Rather, a 1:1 ratio is generated at all temperatures (Figure 21.11). The researchers speculated that the more northern populations have a very short feeding season, so there is no advantage for a female to be born earlier. Thus, this species of fish has environmental sex determination in those regions where it is adaptive and genotypic sex determination in those regions where it is not. Here again, one sees that the environment can induce sexual phenotype, or sexual phenotype can be a property of the genome, as it is with most mammals.

Temperature isn't the only environmental factor that can affect sex determination in fish. The sex of the blue-headed wrasse, a Panamanian reef fish, depends on the other fish it encounters. If the wrasse larva reaches a reef where a male lives with many females, it develops into a female. When the male dies, one of the females (usually the largest) becomes a male. Within a day, its ovaries shrink and its testes grow. If the same wrasse larva had reached a reef that had no males or that had territory undefended by a male, it would have developed into a male wrasse (Warner 1984).

Figure 21.11
Relationship between temperature and sex ratio [F:(F+M)] during the period of sex determination in *Menidia menidia*. In fish collected from the northernmost portion of its range (Nova Scotia), temperature had little effect on sex determination. Among fish collected at more southerly locations (especially from Virginia through South Carolina), however, temperature had a large effect. (After Conover and Heins 1987.)

Polyphenisms for alternative conditions

Most studies of adaptations concern the roles that adult structures play in enabling the individual to survive in otherwise precarious or hostile environments. However, the developing animal, too, has to survive in its habitat, and its development must adapt to the conditions of its existence.

The spadefoot toad, *Scaphiopus couchii*, has a remarkable strategy for coping with a very harsh environment. The toads are called out from hibernation by the thunder that accompanies the first spring storm in the Sonoran desert. (Unfortunately, motorcycles produce the same sounds, causing these toads to come out from hibernation and die in the scorching Arizona sunlight.) The toads breed in temporary ponds formed by the rain, and the embryos develop quickly into larvae. After the larvae metamorphose, the young toads return to the desert, burrowing into the sand until the next year's storms bring them out.

Desert ponds are ephemeral pools that can either dry up quickly or persist, depending on the initial depth and the frequency of the rainfall. One might envision two alternative scenarios confronting a tadpole in such a pond: either (1) the pond persists until you have time to metamorphose and you live, or (2) the pond dries up before your metamorphosis is complete, and you die. These toads (and several other amphibians), however, have evolved a third alternative. The time of metamorphosis is controlled by the pond. If the pond persists at a viable level, development continues at its normal rate, and the algae-eating tadpoles eventually develop into juvenile spadefoot toads. However, if the pond is drying out and getting

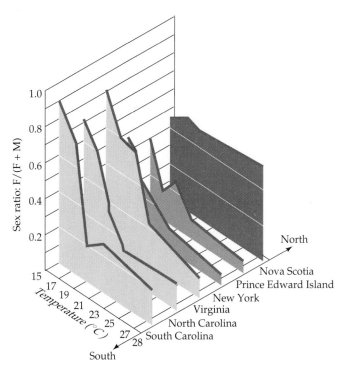

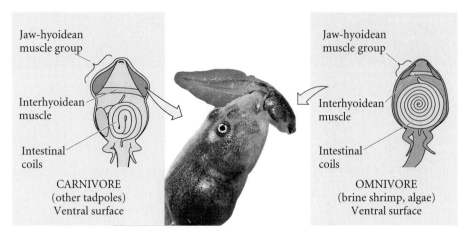

Figure 21.12
Polyphenism in the tadpoles of the spadefoot toad, *Scaphiopus couchii*. The typical morph is an omnivore, usually eating insects and algae. When ponds are drying quickly, however, the carnivorous (cannibalistic) morph forms. It develops a wider mouth, larger jaw muscles, and an intestine modified for a carnivorous diet. The center photograph shows a cannibalistic tadpole eating its smaller pondmate. (Photograph © Thomas Wiewandt; drawings courtesy of R. Ruibel.)

smaller, overcrowding occurs, and some of the tadpoles embark on an alternative developmental pathway. They develop a wider mouth and powerful jaw muscles, which enable them to eat (among other things) other *Scaphiopus* tadpoles (Figure 21.12). These carnivorous tadpoles metamorphose quickly, albeit into a smaller version of the juvenile spadefoot toad. But they survive while other *Scaphiopus* tadpoles perish, either from desiccation or ingestion by their pondmates (Newman 1989, 1992).

Such phenotypic plasticity is also seen in echinoderm larvae. When food is scarce, the ciliated arms of the pluteus larva grow longer and increase the ability of the larva to obtain food. But this is done at a cost to the adult rudiment growing within the larva, and it takes longer for those long-armed plutei (even if they can acquire food) to metamorphose (Hart and Strathmann 1994).

Phenotypic plasticity gives an individual the ability to respond to different environmental conditions. Different phenotypes are more fit in different environments. In the spadefoot toad, the faster-developing carnivorous form is more fit in quickly drying ponds, but the slower-developing tadpoles (which develop into larger, more robust toads) are more fit in wetter conditions. There is a "trade-off" in evolving this phenotypic plasticity, but it helps ensure that some animals will survive, whichever condition prevails at a given time.

Predator-Induced Defenses

One survival strategy for coping with a harsh environment is for an animal to evolve the ability to develop a new structure when confronted by a particular predator. In such cases, the development of the animal is changed by chemicals released by the predator, enabling the embryos or juveniles to better escape those same predators. This is sometimes called **predator-induced defense**, or **predator-induced polyphenism**.

To demonstrate predator-induced polyphenism, one has to show that the phenotypic change is caused by the predator (usually by soluble chemicals released by the predator) and that the phenotypic modification increases the fitness of its bearers when the predator is present (Adler and Harvell 1990; Tollrian and Harvell 1999). For instance, several rotifer species will alter their morphology when they develop in pond water in which their predators were cultured (Figure 21.13; Dodson 1989; Adler and Harvell 1990). The predatory rotifer *Asplanchna* releases into its water a soluble compound that induces the eggs of a prey rotifer species, *Keratella slacki*, to develop into individuals with slightly larger bodies, but with anterior spines 130 percent longer than they would otherwise be. These changes make them more difficult to eat. The snail *Thais lamellosa* develops a thickened shell and a "tooth" in its aperture when exposed to the effluent of the crab species that preys on it. In a mixed population, crabs will not attack the thicker snails until more than 50 percent of the normal snails are devoured (Palmer 1985). Figure 21.13 shows the typical and predator-induced morphs for several species. In each case, soluble filtrate from water surrounding the predator is able to induce these changes, and the induced morph is more successful at surviving the predator.

The predator-induced polyphenism of the parthenogenetic water flea *Daphnia* is beneficial not only to itself, but also to its offspring. When *Daphnia cucullata* encounter the predatory larvae of the fly *Chaeoborus*, their "helmets" grows to twice their normal size. This inhibits their being eaten by the fly larvae. This same helmet induction occurs if the *Daphnia* are exposed to extracts of water in which the fly larvae had been swimming (Figure 21.14). Chemicals that are released by a predator and can induce defenses in the prey are called **kairomones**. Agrawal and colleagues (1999) have shown that the offspring of such an induced *Daphnia* will be born with this same altered head morphology. It is possible that the kairomone regulates gene expression both in the adult and in the developing embryos. We still do not know how *Daphnia* evolved the ability to make receptors that bind the kairomone or to utilize the kairomone to generate an adaptive morphological change.

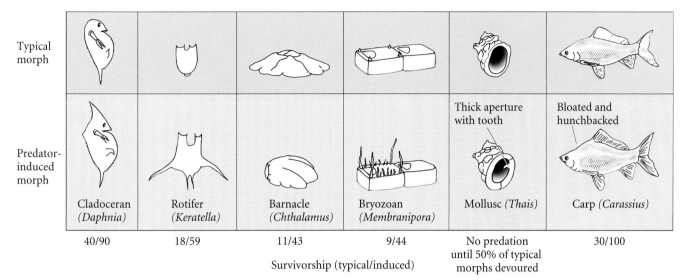

	Cladoceran *(Daphnia)*	Rotifer *(Keratella)*	Barnacle *(Chthalamus)*	Bryozoan *(Membranipora)*	Mollusc *(Thais)*	Carp *(Carassius)*

Survivorship (typical/induced)

| 40/90 | 18/59 | 11/43 | 9/44 | No predation until 50% of typical morphs devoured | 30/100 |

Figure 21.13
Predator-induced defenses. Typical (upper row) and predator-induced (lower row) morphs of various organisms are shown. The numbers beneath each column represents the percentage of organisms surviving predation when both induced and uninduced individuals were presented with predators (in various assays). (Data from Adler and Harvell 1990 and references cited therein.)

Colonial invertebrates have some spectacular developmental responses to predators. Due to their sessile growth, they cannot escape by moving away, and predators often treat them like plants, eating modules without killing off the entire colony. *Membranipora membranacea* is a widely distributed bryozoan often seen on kelp. It is grazed upon by certain nudibranch molluscs that suck in the modules at the periphery of the colony (Figure 21.15). When exposed to such predation, the modules near the nudibranch develop spines. These spines interfere with the predators establishing the suction needed to feed. Spines can also be induced within three days by treating a colony with chemical extracts of their predator (Harvell 1986; 1999.)

Predator-induced polyphenism is not limited to invertebrates. McCollum and Van Buskirk (1996) have shown that in the presence of its predators, the tail fin of the tadpole of the gray treefrog *Hyla chrysoscelis* grows larger and becomes bright red. This allows the tadpole to swim away faster and to deflect strikes toward the tail region. The carp *Carassius carassius* is able to respond to the presence of a predatory pike only if the pike had al-ready eaten other fish. The carp grows into a pot-bellied, hunchbacked morph that will not fit into the pike's jaws. However, as in most predator-induced defenses, there is a trade-off (otherwise one would expect the induced morph to become the normal phenotype). In this case, the induced morphology puts a drag on swimming efficiency, and the fatter fish cannot swim as well (Brönmark and Pettersson 1994).

WEBSITE **21.5 Inducible in ant colonies.** In some species of ants, the loss of soldier ants creates conditions that induce more workers to become soldiers.

(A) (B) (C)

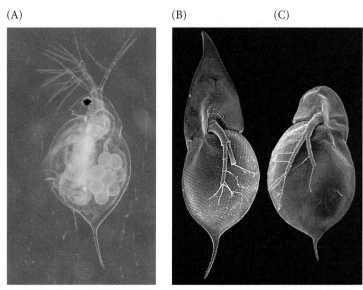

Figure 21.14
Predator-induced polyphenism in *Daphnia*. (A) *Daphnia* is an all-female species, producing eggs (seen in the picture) parthenogenetically. (B, C) Scanning electron micrographs showing predator-induced (B) and normal (C) morphs of the same clone. (A courtesy of R. Tollrian; B, C courtesy of A. A. Agrawal.)

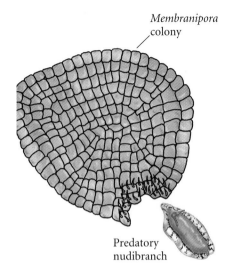

Membranipora colony

Predatory nudibranch

Figure 21.15
Predator-induced defence in the colonial bryozoan *Membranipora membranacea*. Chemicals originating from the nudibranchs grazing at the periphery of the colony cause modules close to the grazing site to develop spines. The spines prevent the nudibranchs from feeding. (After Harvell 1999.)

Mammalian Immunity as a Predator-Induced Response

If predator-induced polyphenism is an adaptive response to potential threats, the mammalian immune system is its highest achievement. The mammalian immune system is an incredibly elaborate mechanism for sensing and destroying materials that are foreign to the body. When we are exposed to a foreign molecule (called an **antigen**), we manufacture **antibodies** and secrete them into our blood serum (see Chapter 4). These antibodies combine with the antigen to inactivate or eliminate the antigen. The basis for the immune response is summarized in the five major postulates of the clonal selection hypothesis (Burnett 1959):

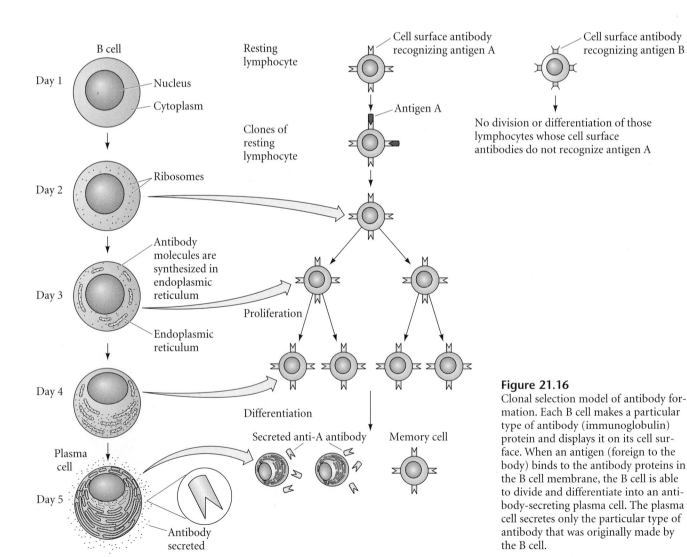

Figure 21.16
Clonal selection model of antibody formation. Each B cell makes a particular type of antibody (immunoglobulin) protein and displays it on its cell surface. When an antigen (foreign to the body) binds to the antibody proteins in the B cell membrane, the B cell is able to divide and differentiate into an antibody-secreting plasma cell. The plasma cell secretes only the particular type of antibody that was originally made by the B cell.

1. Each B lymphocyte (B cell) can make one, and only one, type of antibody. It is specific for one shape of antigen only.
2. Each B cell places the antibodies it makes into its cell membrane with the specificity-bearing side outward.
3. Antigens are presented (usually by macrophages) to the antibodies on the B cell membranes.
4. Only those B cells that bind to the antigen can complete their development into antibody-secreting **plasma cells**. These B cells divide repeatedly, produce an extensive rough endoplasmic reticulum, and synthesize enormous amounts of antibody molecules. These antibodies are secreted into the blood.
5. The specificity of the antibody made by the plasma cell is exactly the same as that which was on the cell surface of the B cells.

The type of antibody molecule on the cell surface of the B cell is determined by chance. Out of the 10 million types of antibody proteins the cell can possibly synthesize, each B cell makes only one type. That is, one B cell may be making antibodies that bind to poliovirus, while a neighboring B cell might be making antibodies to diphtheria toxin. The B cells are continually being created and destroyed. However, when an antigen binds to a set of B cells, these cells are stimulated to divide and differentiate into plasma cells (that secrete the antibody) and memory cells (that populate lymph nodes and respond rapidly when exposed to the same antigen later in life) (Figure 21.16). Thus, each person's constellation of plasma cells and memory cells differs depending on which antigens he or she has encountered. Identical twins have different populations of B cell descendants in their spleens and lymph nodes.

WEBSITE **21.6 The induction of immune cells.** The details of the mammalian immune response involve inductions between B cells, T cells, macrophages, and antigens. HIV kills the T cells responsible for allowing B cells to differentiate. The passage of antibodies through the placenta and the transfer of antibodies through milk give the newborn protection as well.

Sidelights & Speculations

Genetic Assimilation

I N DISCUSSING THE TRADE-OFFS between uninduced and induced morphs, we mentioned that if an induced morph did not have a significant trade-off, one would expect it to become the predominant form of the species. C. H. Waddington and I. I. Schmalhausen independently made this prediction to explain how some species could rapidly evolve in particular directions (see Gilbert 1994). Both scientists were impressed by the calluses of the ostrich. Most mammalian skin has the ability to form calluses on those areas that are abraded by the ground or some other surface.* The skin cells respond to friction by proliferating. While such examples of environmentally induced callus formation are widespread, the ostrich is born with calluses where it will touch the ground (Figure 21.17). Waddington and Schmalhausen hypothesized that since the skin cells are already competent to be induced by friction, they could be induced by other things as

well. As ostriches evolved, a mutation (or a particular combination of alleles) appeared that enabled the skin cells to respond to a substance within the embryo. Waddington (1942) wrote:

> *Presumably its skin, like that of other animals, would react directly to external pressure and rubbing by becoming thicker. … This capacity to react must itself be dependent upon genes. …It may then not be too difficult for a gene mutation to occur which will modify some other area in the embryo in such a way that it takes over the function of external pressure, interacting with the skin so as to "pull the trigger" and set off the development of callosities.*

By this transfer of induction from an external inducer to an internal inducer, a trait that had been induced by the environment became part of the genome of the organism and could be selected. Waddington called this phenomenon "genetic assimilation," while Schmalhausen (1949) called it "stabilizing selection." Both scientists had used orthodox embryology and orthodox genetics to explain phenomena that had been considered cases of Lamarckian "inheritance of acquired characteristics."

Figure 21.17
Ventral side of an ostrich; arrows mark the calluses. (After Waddington 1942).

*And until this century, writers were recognized by the calluses on their fingers. (Thus, from observing his fingers, Sherlock Holmes correctly surmised that the red-headed man had been hired as a scrivener.)

A shift from environmental stimulus to genetic stimulus might explain sex determination in *Menidia* and caste determination in ants. Similarly, the preexisting developmental plasticity of arm length in feeding echinoderm larvae may have bridged the transition from pluteus (feeding) larvae to larvae that lack ciliated arms. The change in the allocation of resources between larval and juvenile structures parallels that seen where the food reserves are stored in the egg. Thus, the changes already present as adaptations to external food resources may have become genetically fixed in those species whose larvae do not need to hunt their food (Strathmann et al. 1992).

If genetic assimilation is the genetic fixation of one of the phenotypes that had been adaptively expressed, then butterflies would be a good place to look for further examples. Brakefield and colleagues (1996) showed that they could genetically fix the different morphs of the adaptive polyphenism of *Bicyclus*, and Shapiro (1976) has shown that the short-day (cold-weather) adaptive phenotype of several butterflies is the same as the single genetically produced phenotype of related species or subspecies living at higher altitudes or latitudes.

Genetic assimilation may play an important role in providing a bias for evolutionary change. If an organism inherits a reaction norm, the developmental pathways leading to a particular phenotype are already in place, and all that evolution need do is supply a constant initiator of those pathways. In the next chapter, we will discuss some of the molecular evidence for genetic assimilation. ■

WEBSITE **21.7 Genetic assimilation and phenocopies.** Genetic assimilation has been documented in the laboratory, and the ability to react to environmental stimuli has been transferred to embryonic inducers.

Learning: An Environmentally Adaptive Nervous System

In Chapter 13, we saw that neuronal activity can be a critical factor in deciding which synapses are retained by the adult organism. Here we will extend that discussion to highlight those remarkable instances in which new experiences modify the original set of neuronal connections, causing the creation of new neurons, or the formation of new synapses between existing neurons. Since neurons, once formed, do not divide, the "birthday" of a neuron can be identified by treating the organism with radioactive thymidine. Normally, very little radioactive thymidine is taken up into the DNA of a neuron that has already been formed. However, if a new neuron differentiates by cell division during the treatment, it will incorporate radioactive thymidine into its DNA.

Such new neurons are seen to be generated when male songbirds first learn their songs. Juvenile zebra finches memorize a model song and then learn the pattern of muscle contractions necessary to sing a particular phrase. In this learning and repetition process, new neurons are generated in the hyperstriatum of the finch's brain. Many of these new neurons send axons to the archistriatum, which is responsible for controlling the vocal musculature (Nordeen and Nordeen 1988). These changes are not seen in males who are too old to learn the song, nor are they seen in juvenile females (who do not sing these phrases).

In white-crowned sparrows, where song is regulated by photoperiod and hormones, the exposure of adult males to long hours of light and testosterone induces over 50,000 new neurons in their vocal centers (Tramontin et al. 2000). These birds have seasonal plasticity of their brain neural circuitry. Testosterone is believed to increase the level of BDNF in the song-producing centers of the birds' brains. If female birds are given BDNF, they will also produce more neurons there (Rasika et al. 1999).

The cerebral cortices of young rats reared in stimulating environments are packed with more neurons, synapses, and dendrites than are found in rats reared in isolation (Turner and Greenough 1983). Even the adult brain continues to develop in response to new experiences. When adult canaries learn new songs, they generate new neurons whose axons project from one vocal region of the brain to another (Alvarez-Buylla et al. 1990). Studies on adult rats and mice indicate that environmental stimulation can increase the number of new neurons in the dentate gyrus (Kemperman et al. 1997a,b; Gould et al. 1999; van Praag et al. 1999). Similarly, when adult rats learn to keep their balance on dowels, their cerebellar Purkinje cell neurons develop new synapses (Black et al. 1990). Thus, the the pattern of neuronal connections is a product of inherited patterning and patterning produced by experiences. This interplay between innate and experiential development has been detailed most dramatically in studies on mammalian vision.

EXPERIENTIAL CHANGES IN INHERENT MAMMALIAN VISUAL PATHWAYS. Some of the most interesting research on mammalian neuronal patterning concerns the effects of sensory deprivation on the developing visual system in kittens and monkeys. The paths by which electrical impulses pass from the retina to the brain in mammals are shown in Figure 21.18. Axons from the retinal ganglion cells form the two optic nerves, which meet at the optic chiasm. As in *Xenopus* tadpoles, some axons go to the opposite (contralateral) side of the brain but, unlike most other vertebrates, mammalian retinal ganglion cells also send inputs into the same (ipsilateral) side of the brain (see Chapter 13). These nerves end at the two lateral geniculate nuclei. Here the input from each eye is kept separate, with the uppermost and anterior layers receiving the axons from the contralateral eye, and the middle of the layers receiving input from the ipsilateral eye. The situation becomes even more complicated as neurons from the lateral geniculate nuclei connect with the neurons of the visual cortex. Over 80 percent of the neural cells in the cortex receive input from both eyes. The result is binocular vision and depth perception.

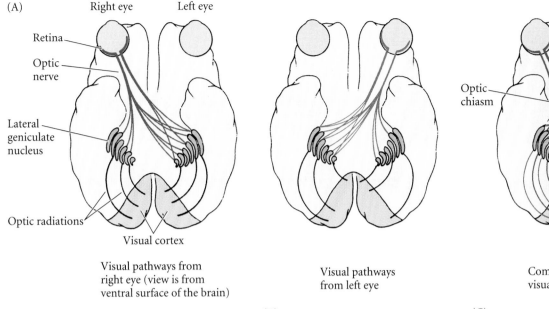

(A) Right eye   Left eye

Retina

Optic nerve

Lateral geniculate nucleus

Optic radiations

Visual cortex

Optic chiasm

Visual pathways from right eye (view is from ventral surface of the brain)

Visual pathways from left eye

Combined left and right visual pathways

Figure 21.18
Major pathways of the mammalian visual system. (A) In mammals, the optic nerve from each eye branches, sending nerve fibers to a lateral geniculate nucleus on each side of the brain. On the ipsilateral side, a particular part of the retina projects to a particular part of the lateral geniculate nucleus. On the contralateral side, the lateral geniculate nucleus receives input from all parts of the retina. Neurons from each lateral geniculate nucleus innervate the visual cortex on the same side. (B, C) Isolated (and filleted) retinas show (B) ipsilateral and (C) contralateral projections from the mouse 16-day embryonic retinal ganglion cells. The fluorescent carbocyanine dye DiI was inserted behind the optic chiasm and was allowed to enter the retinal axons. The dye diffuses along the axons, thereby labeling them as to their origin. Ipsilateral projections mostly come from a single part of the retina (in this case, the ventrotemporal region). Contralateral projections to the same site come from all over the retina. (B, C from Colello and Guillery 1990; photographs courtesy of the authors.)

Another remarkable finding is that the retinocortical projection pattern is the same for both eyes. If a cortical neuron is stimulated by light flashing across a region of the left eye 5 degrees above and 1 degree to the left of the fovea,* it will also be stimulated by a light flashing across a region of the right eye 5 degrees above and 1 degree to the left of the fovea. Moreover, the response evoked in the cortical cell when both eyes are stimulated is greater than the response when either retina is stimulated alone.

Hubel, Wiesel, and their co-workers (see Hubel 1967) demonstrated that the development of the nervous system de-

pends to some degree on the experience of the individual during a critical period of development. In other words, not all neuronal development is encoded in the genome; some is the result of learning. Experience appears to strengthen or stabilize some neuronal connections that are already present at birth and to weaken or eliminate others. These conclusions come from studies of partial sensory deprivation. Hubel and Wiesel (1962, 1963) sewed shut the right eyelids of newborn kittens and left them closed for 3 months. After this time, they unsewed the lids of the right eye. The cortical cells of such kittens could not be stimulated by shining light into the right eye. Almost all the inputs into the visual cortex came from the left eye only. The behavior of the kittens revealed the inadequacy of their right eyes: When only the left eyes of these animals were covered, the kittens became functionally blind. Because the lateral geniculate neurons appeared to be stimulated from both right and left eyes in these kittens, the physiological defect appeared to be in the connections between the lateral geniculate nuclei and the visual cortex. In rhesus monkeys, where similar phenomena are observed, the defect has been correlated with a lack of protein synthesis in the lateral geniculate neurons innervated by the covered eye (Kennedy et al. 1981).

*The fovea is a depression in the center of the retina where only cones are present and the rods and blood vessels are absent. Here it serves as a convenient landmark.

(A)

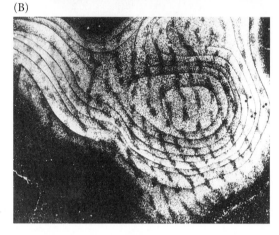

(B)

Figure 21.19
Dark-field autoradiographs of monkey striate (visual) cortex 2 weeks after one eye was injected with [³H]proline in the vitreous humor. Each retinal neuron takes up the radioactive label and transfers it to the cells with which it forms synapses. (A) Normal labeling pattern. The white stripes indicate that roughly half the columns took up the label, while the other half did not—a pattern reflecting that half the cells were innervated by the labeled eye and half were innervated by the unlabeled eye. (B) Labeling pattern when the unlabeled eye was sutured shut for 18 months. The axonal projections from the normal (labeled) eye have taken over the regions that would normally have been innervated by the sutured eye. (C, D) Drawings of axons from the lateralgeniculate nuclei of kittens in which one eye was occluded for 33 days. The terminal branching of axons receiving input from the occluded eye (C) were far less extensive than that of axons receiving input from the nonoccluded eye (D). (A and B from Wiesel 1982, courtesy of T. Wiesel; C and D after Antonini and Stryker 1993.)

(C)   (D)

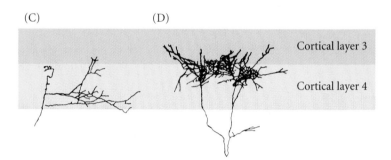

Although it would be tempting to conclude that the blindness resulting from these experiments was due to a failure to form the proper visual connections, this is not the case. Rather, when a kitten or monkey is born, axons from lateral geniculate neurons receiving input from each eye overlap extensively in the visual cortex (Hubel and Wiesel 1963; Crair et al. 1998). However, when one eye is covered early in the animal's life, its connections in the visual cortex are taken over by those of the other eye (Figure 21.19). Competition occurs, and experience plays a role in strengthening and stabilizing the connections from each lateral geniculate nucleus to the visual cortex. Thus, when both eyes of a kitten are sewn shut for 3 months, most cortical cells can still be stimulated by appropriate illumination of one eye or the other. The critical time in kitten development for this validation of neuronal connections begins between the fourth and the sixth week after birth. Monocular deprivation up to the fourth week produces little or no physiological deficit, but after 6 weeks, it produces all the characteristic neuronal changes. If a kitten has had normal visual experience for the first 3 months, any subsequent monocular deprivation (even for a year or more) has no effect. At that point, the synapses have been stabilized.

Two principles, then, can be seen in the patterning of the mammalian visual system. First, the neuronal connections involved in vision are present even before the animal sees. Second, experience plays an important role in determining whether or not certain connections remain.* Just as experience refines the original neuromuscular connections, so experience plays a role in refining and improving the visual connections. It is possible, too, that adult functions such as learning and memory arise from the establishment and/or strengthening of different synapses by experience. As Purves and Lichtman (1985) remark:

> *The interaction of individual animals and their world continues to shape the nervous system throughout life in ways that could never have been programmed. Modification of the nervous system by experience is thus the last and most subtle developmental strategy.*

WEBSITE **21.8 The phantom limb phenomenon.** Individuals who have a limb amputated sometimes feel pain in the absent appendage. This appears to be caused by a reorganization of the human cerebral cortex following the amputation.

Environmental Disruption of Normal Development

From what has been said so far in this chapter, it is clear that the instructions for development do not reside wholly in the genes or even in the zygote. The developing organism is often sensitive to cues from the environment. However, this sensitivity makes the organism vulnerable to environmental changes that can disrupt development.

If you think it is amazing that any one of us survives to be born, you are correct. It is estimated that one-half to two-thirds of all human conceptions do not develop successfully to term (Figure 21.20). Many of these embryos express their abnormality

*Recent studies (Colman et al. 1997) have shown that divergence in neurotransmitter release results in changes in synaptic adhesivity and causes the withdrawal of the axon providing the weaker stimulation. Studies in mice (Huang et al. 1999; Katz 1999) suggest that brain-derived neurotropic factor (BDNF) is crucial during the critical period.

Cortical layer 3

Cortical layer 4

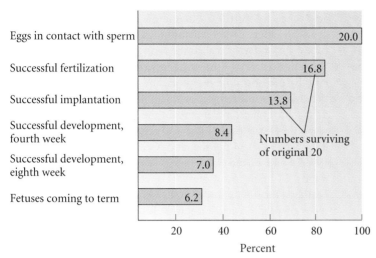

Eggs in contact with sperm — 20.0
Successful fertilization — 16.8
Successful implantation — 13.8
Successful development, fourth week — 8.4
Successful development, eighth week — 7.0
Fetuses coming to term — 6.2

Numbers surviving of original 20

Percent (20 40 60 80 100)

Figure 21.20
The fates of 20 hypothetical human eggs in the United States and western Europe. Under normal conditions, only 6.2 eggs of the original 20 would be expected to develop successfully to term. (After Volpe 1987.)

so early that they fail to implant in the uterus. Others implant but fail to establish a successful pregnancy. Thus, most abnormal embryos are spontaneously aborted, often before the woman even knows she is pregnant (Boué et al. 1985). Edmonds and co-workers (1982), using a sensitive immunological test that can detect the presence of human chorionic gonadotropin (hCG) as early as 8 or 9 days after fertilization, monitored 112 pregnancies in normal women. Of these hCG-determined pregnancies, 67 failed to be maintained.

It appears, then, that many human embryos are impaired early in development and do not survive long in utero. Defects in the lungs, limbs, face, or mouth, however, are not deleterious to the fetus (which does not depend on those organs while inside the mother), but can threaten life once the baby is born. About 5 percent of all human babies born have a recognizable malformation, some of them mild, some very severe (McKeown 1976).

Congenital ("present at birth") abnormalities and the loss of embryos and fetuses prior to birth are caused both intrinsically and extrinsically. Those abnormalities caused by genetic events (mutations, aneuploidies, translocations) are called malformations. For instance, aniridia (absence of the iris), caused by a mutation of the *PAX6* gene, is a malformation (see Chapter 4). Down's syndrome, caused by trisomy of chromosome 21, is likewise a malformation. Most early embryonic and fetal demise is probably due to chromosomal abnormalities that interfere with normal developmental processes.

Abnormalities caused by exogenous agents (certain chemicals or viruses, radiation, or hyperthermia) are called **disruptions**. The agents responsible for these disruptions are called **teratogens**.* Most teratogens produce their effects only during certain critical periods of development. The most critical time for any organ is when it is growing and forming its structures. Different organs have different critical periods, but the time from period from day 15 through day 60 of gestation is critical

for many human organs. The heart forms primarily during weeks 3 and 4, while the external genitalia are most sensitive during weeks 8 and 9. The brain and skeleton are always sensitive, from the beginning of week 3 to the end of pregnancy and beyond.

Teratogenic Agents

Different agents are teratogenic in different organisms. A partial list of agents that are teratogenic in humans is given in Table 21.2. The largest class of teratogens includes drugs and chemicals.

Some chemicals that are naturally found in the environment can cause birth defects. Even in the pristine alpine meadows of the Rocky Mountains, teratogens are found. Here grows the skunk cabbage *Veratrum californicum*, upon which sheep sometimes feed. If pregnant ewes eat this plant, their fetuses tend to develop severe neurological damage, including cyclopia, the fusion of two eyes in the center of the face (see Figure 6.25). Two products made by this plant, jervine and cyclopamine, inhibit cholesterol synthesis in the fetus and prevent Sonic hedgehog from functioning. Indeed, this resembles human genetic conditions (discussed in Chapters 6 and 12) in which the genes for either Sonic hedgehog or cholesterol synthesising enzymes are mutated (Beachy et al. 1997; Opitz 1998). The affected organism dies shortly after birth (as a result of severe brain defects, including the lack of a pituitary gland).

Quinine and alcohol, two substances derived from plants, can also cause congenital malformations. Quinine can cause deafness, and alcohol (when more than 2–3 ounces per day are imbibed by the mother) can cause physical and mental retardation in the infant. Nicotine and caffeine have not been proved to cause congenital anomalies, but women who are heavy smokers (20 cigarettes a day or more) are likely to have infants that are smaller than those born to women who do not smoke. There is controversy over whether cigarette smoking enhances the risk of facial anomalies. Smoking also significantly lowers the number and motility of sperm in the semen of males who smoke at least four cigarettes a day (Kulikauskas et al. 1985).

In addition, our industrial society produces hundreds of new artificial compounds that come into general use each year. Pesticides and organic mercury compounds have caused neurological and behavioral abnormalities in infants whose mothers have ingested them during pregnancy. Moreover,

---

*In some cases, the same condition can be caused by a disruption (from an exogenous agent) or a malformation (from the nucleus). For instance, certain axial malformations in mice can be produced either by the administration of retinoic acid or by mutations in certain Hox genes. In some instances, the mutation and the teratogen are known to affect the same enzyme. Chondroplasia punctata is a congenital defect of bone and cartilage, characterized by abnormal bone mineralization, underdevelopment of nasal cartilage, and shortened fingers. It is caused by a defective gene on the X chromosome. An identical phenotype is produced by the ingestion of the rat-killing compound, warfarin. It appears that the defective gene is normally responsible for producing an arylsulfatase protein necessary for cartilage growth. The warfarin compound inhibits this same enzyme (Franco et al. 1995).

drugs that are used to control diseases in adults may have deleterious effects on fetuses. For example, valproic acid is an anticonvusant drug used to control epilepsy. It is known to be teratogenic in humans as it can cause major and minor bone defects. Barnes and colleagues (1996) have shown that valproic acid decreases the level of *Pax1* transcription in chick somites. This causes the malformation of the somite and the corresponding malformations of vertebrae and ribs.

Retinoic acid as a teratogen

In some instances, even a compound that is involved in normal development can have deleterious effects if present in large amounts at particular times. Retinoic acid is important in forming the anterior-posterior axis of the mammalian embryo and also in forming the limbs (see Chapters 11 and 16; Morriss-Kay and Ward 1999). In these instances, retinoic acid is secreted from discrete cells and works in a small area. However, if retinoic acid is present in large amounts, cells that normally would not receive such high concentrations of this molecule will respond to it.

Inside the developing embryo, vitamin A and 13-*cis*-retinoic acid become isomerized to the developmentally active forms of retinoic acid, all-*trans*-retinoic acid and 9-*cis*-retinoic acid (Creech Kraft 1992). Retinoic acid cannot bind directly to genes. To effect gene regulation, RA must bind to a group of transcription factors called the retinoic acid receptors (RARs). These proteins have the same general structure as the steroid and thyroid hormone receptors, and they are active only when they have bound retinoic acid (Linney 1992). The RARs bind to specific enhancer elements in the DNA called retinoic acid response elements. Retinoic acid response elements contain at least two copies of the sequence GGTCA (Ruberte et al. 1990, 1991). Some of the Hox genes have retinoic acid response elements in their promoters (Yu et al. 1991; Pöpperl and Featherstone 1993; Studer et al. 1994). There are three major types of retinoic acid receptors: RAR-α, RAR-β, and RAR-γ. They each bind both forms of retinoic acid, and they each bind to the same retinoic acid response element.

Retinoic acid has been useful in treating severe cystic acne and has been available (under the name Accutane) since 1982. Because the deleterious effects of administering large amounts of vitamin A or its analogues to pregnant animals have been known since the 1950s (Cohlan 1953; Giroud and Martinet 1959; Kochhar et al. 1984), the drug carries a label warning that it should not be used by pregnant women. However, about 160,000 women of childbearing age (15 to 45 years) have taken this drug since it was introduced, and some of them have used it during pregnancy. Lammer and his co-workers (1985) studied a group of women who inadvertently exposed themselves to retinoic acid and who elected to remain pregnant. Of their 59 fetuses, 26 were born without any noticeable anomalies, 12 aborted spontaneously, and 21 were born with obvious anomalies. The malformed infants had a characteristic pattern of anomalies, including absent or defective ears, absent or small jaws, cleft palate, aortic arch abnormalities, thymic deficiencies, and abnormalities of the central nervous system.*

This pattern of multiple congenital anomalies is similar to that seen in rat and mouse embryos whose pregnant

Table 21.2 Some agents thought to cause disruptions in human fetal development[a]

DRUGS AND CHEMICALS	IONIZING RADIATION (X-RAYS)
Alcohol	
Aminoglycosides (Gentamycin)	HYPERTHERMIA
Aminopterin	INFECTIOUS MICROORGANISMS
Antithyroid agents (PTU)	Coxsackie virus
Bromine	Cytomegalovirus
Cigarette smoke	Herpes simplex
Cocaine	Parvovirus
Cortisone	Rubella (German measles)
Diethylstilbesterol (DES)	*Toxoplasma gondii* (toxoplasmosis)
Diphenylhydantoin	*Treponema pallidum* (syphilis)
Heroin	METABOLIC CONDITIONS IN THE MOTHER
Lead	Autoimmune disease (including Rh incompatibility)
Methylmercury	Diabetes
Penicillamine	Dietary deficiencies, malnutrition
Retinoic acid (Isotretinoin, Accutane)	Phenylketonuria
Streptomycin	
Tetracycline	
Thalidomide	
Trimethadione	
Valproic acid	
Warfarin	

Source: Adapted from Opitz 1991.
[a]This list includes known and possible teratogenic agents and is not exhaustive.

*This is a critical public health concern, because there is significant overlap between the population using acne medicine and the population of women of childbearing age, and because it is estimated that half of the pregnancies in America are unplanned (Nulman et al. 1997). Vitamin A is itself teratogenic when injected in megadose amounts. Rothman and colleagues (1995) found that pregnant women who took more than 10,000 international units of preformed vitamin A per day (in the form of vitamin supplements) had about a 2 percent chance of having a baby born with disruptions similar to those produced by retinoic acid.

mothers have been given these drugs. Goulding and Pratt (1986) have placed 8-day mouse embryos in a solution containing 13-*cis*-retinoic acid at a very low concentration (2×10^{-6} *M*). Even at this concentration, approximately one-third of the embryos developed a very specific pattern of anomalies, including dramatic reduction in the size of the first and second pharyngeal arches (see Figure 11.40). In normal mice, the first arch eventually forms the maxilla and mandible of the jaw and two ossicles of the middle ear, while the second arch forms the third ossicle of the middle ear as well as other facial bones.

Retinoic acid probably works by several mechanisms. One basis for the teratogenicity of retinoic acid appears to reside in the drug's ability to alter the expression of the Hox genes and thereby respecify portions of the anterior-posterior axis and inhibit neural crest cell migration from the cranial region of the neural tube (Moroni et al. 1994; Studer et al. 1994). Radioactively labeled retinoic acid binds to the cranial neural crest cells and arrests both their proliferation and their migration (Johnston et al. 1985; Goulding and Pratt 1986). The binding seems to be specific to the cranial neural crest-derived cells, and the teratogenic effect of the drug is confined to a specific developmental period (days 8–10 in mice; days 20–35 in humans). Animal models of retinoic acid teratogenesis have been extremely successful in elucidating the mechanisms of teratogenesis at the cellular level.

> WEBSITE **21.9 Mechanisms of retinoic acid teratogenesis**. Within the cell, numerous retinoid binding proteins interact to influence the ability of retinoic acid to transcribe particular genes.

> WEBSITE **21.10 Thalidomide as a teratogen**. The drug Thalidomide caused thousands of babies to be born with malformed arms and legs, and it provided the first major evidence that drugs could induce congenital anomalies. The mechanism of its action is still hotly debated.

Alcohol as a teratogen

In terms of frequency and cost to society, the most devastating teratogen is undoubtedly ethanol. In 1968, Lemoine and colleagues noticed a syndrome of birth defects in the children of alcoholic mothers. This **fetal alcohol syndrome** (**FAS**) was also noted by Jones and Smith (1973). Babies with FAS were characterized as having small head size, an indistinct philtrum (the pair of ridges that runs between the nose and mouth above the center of the upper lip), a narrow upper lip, and a low nose bridge. The brain of such a child may be dramatically smaller than normal and often shows defects in neuronal and glial migration (Figure 21.21; Clarren 1986). There is also prominent extra cell death in the frontonasal process and in cranial nerve ganglia (Sulik et al. 1988). Fetal alcohol syndrome is the third most prevalent type of mental retardation (behind fragile X syndrome and Down syndrome) and affects 1 out of every 500–750 children born in the United States (Abel and Sokol 1987).

Children with fetal alcohol syndrome are developmentally and mentally retarded, with a mean IQ of about 68 (Streissguth and LaDue 1987). FAS patients with a mean chronological age of 16.5 years were found to have the functional vocabulary of 6.5-year-olds and to have the mathematical abilities of fourth graders. Most adults and adolescents with FAS cannot handle money or their own lives, and they have difficulty learning from past experiences. Moreover, in many instances of FAS, the behavioral abnormalities exist without any gross physical changes in brain or head size (J. Opitz, personal communication). There is great variation in the ability of mothers and fetuses to metabolize ethanol, and it is thought that 30 to 40 percent of the children born to alcoholic mothers who drink during pregnancy will have FAS. It is also thought that lower amounts of ethanol ingestion by the mother can lead to fetal alcohol effect, a less severe form of FAS, but a condition that lowers the functional and intellectual abilities of the sufferer.*

A mouse model system has been used to explain the effects of alcohol on the face and nervous system. When mice are exposed to ethanol at the time of gastrulation, it induces the same range of developmental defects as in humans. As early as 12 hours after the mother ingests the alcohol, abnormalities of development are observed. The midline structures fail to form, allowing the abnormally close proximity of the medial processes of the face. Forebrain anomalies are also seen, and the more severely affected fetuses lack a forebrain entirely (Sulik et al. 1988). Studies on these mice suggest that ethanol may induce its teratogenic effects by more than one

---

*For a sensitive account of raising a child with fetal alcohol syndrome, as well as an analysis of FAS in Native American culture in the United States, read Michael Dorris's *The Broken Cord* (1989). The personal and sociological effects of FAS are well integrated with the scientific and economic data.

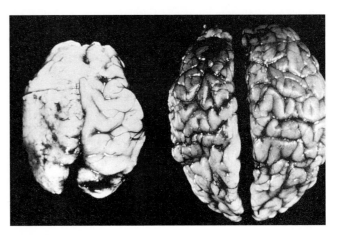

Figure 21.21
Comparison of a brain from an infant with fetal alcohol syndrome (left) with a brain from a normal infant of the same age (right). The brain from the infant with FAS is significantly smaller, and the pattern of convolutions is obscured by glial cells that have migrated over the top of the brain. (Photographs courtesy of S. Clarren.)

mechanism. First, anatomical evidence suggests that neural crest migration is severely impaired. Instead of migrating and dividing, ethanol-treated neural crest cells prematurely initiate their differentiation into facial cartilage (Hoffman and Kulyk 1999). Second, ethanol can cause the apoptosis of neurons. One way it can cause apoptosis is to generate superoxide radicals that can oxidize cell membranes and lead to cytolysis (Figure 21.22A–C; Davis et al. 1990; Kotch et al. 1995). Another pathway to apoptosis involves the activation of GABA receptors and simultaneous inhibition of glutamate receptors (Ikonomidou et al. 2000). Ethanol-induced apoptosis can delete millions of neurons from the developing forebrain, frontonasal (facial) process, and cranial nerve ganglia. Third, alcohol may directly interfere with the ability of cell adhesion molecule L1 to function in holding cells together. Ramanathan and colleagues (1996) have shown that ethanol

can block the adhesive functions of L1 proteins in vitro at levels as low as 7 mM, a concentration of ethanol produced in the blood or brain with a single drink (Figure 21.22D). Moreover, mutations in human *L1* genes cause a syndrome of mental retardation and malformations similar to that seen in severe cases of fetal alcohol syndrome.

WEBSITE **21.11 Our knowledge of alcohol's teratogenicity.** A hundred years ago, alcohol was considered dangerous to the fetus. Fifty years ago, it was considered harmless, and today it is considered very dangerous. Sociological studies look at how these assessments were made.

Other teratogenic agents

CHEMICALS. Over 50,000 artificial chemicals are currently used in our society and about 200 to 500 new materials being

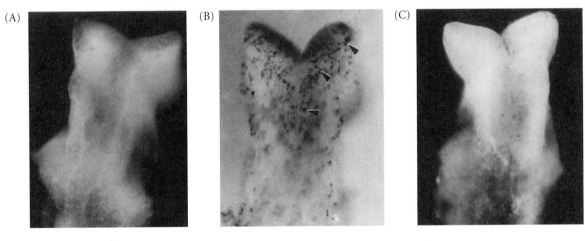

(A)   (B)   (C)

Figure 21.22
Possible mechanisms producing fetal alcohol syndrome. (A–C) Cell death caused by ethanol-induced superoxide radicals. Staining with Nile blue sulfate reveals areas of cell death. (A) Control 9-day mouse embryo head region. (B) Head region of ethanol-treated embryo, showing areas of cell death. (C) Head region of 9-day embryo treated with both ethanol and superoxide dismutase, an inhibitor of superoxide radicals. The superoxide inhibitor prevents the alcohol-induced cell death. (D) The inhibition of L1-mediated cell adhesion by ethanol. (A–C from Kotch et al. 1995; photographs courtesy of K. Sulik; D after Ramanathan et al. 1996.)

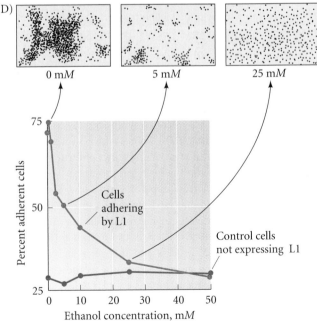

(D)

0 mM 5 mM 25 mM

Cells adhering by L1

Control cells not expressing L1

Percent adherent cells

Ethanol concentration, mM

made each year (Johnson 1980). In the former Soviet Union, the unregulated "industrial production at-all-costs" approach leaves a legacy of soaring birth defects. In some regions of Kazakhstan, teratogens such as lead, mercury, and zinc are found in high concentrations in drinking water, vegetables, and the air. In these places, nearly half the people tested have extensive chromosome breakage. In some areas, the incidence of birth defects has doubled since 1980 (Edwards 1994).

Although teratogenic compounds have always been with us, the risks increase as more and more untested compounds enter our environment each year. Most industrial chemicals have not been screened for their teratogenic effects. Standard screening protocols are expensive, long, and subject to interspecies differences in metabolism. There is still no consensus on how to test a substance's teratogenicity for human embryos.

PATHOGENS AS TERATOGENIC AGENTS. Another class of teratogens includes viruses and other pathogens. Gregg (1941) first documented the fact that women who had rubella (German measles) during the first third of their pregnancy had a 1 in 6 chance of giving birth to an infant with eye cataracts, heart malformations, or deafness. This was the first evidence that the mother could not fully protect the fetus from the outside environment. The earlier the rubella infection occurred during the pregnancy, the greater the risk that the embryo would be malformed. The first 5 weeks appear to be the most critical, because this is when the heart, eyes, and ears are being formed. The rubella epidemic of 1963–1965 probably resulted in about 20,000 fetal deaths and 30,000 infants with birth defects in the United States. Two other viruses, *Cytomegalovirus* and *Herpes simplex*, are also teratogenic. Cytomegalovirus infection of early embryos is nearly always

fatal, but infection of later embryos can lead to blindness, deafness, cerebral palsy, and mental retardation.

Bacteria and protists are rarely teratogenic, but two of them can damage human embryos. *Toxoplasma gondii*, a protozoan carried by rabbits and cats (and their feces), can cross the placenta and cause brain and eye defects in the fetus. *Treponema pallidum*, the cause of syphilis, can kill early fetuses and produce congenital deafness in older ones.

IONIZING RADIATION. Ionizing radiation can break chromosomes and alter DNA structure. For this reason, pregnant women are told to avoid unnecessary X-rays, even though there is no evidence for congenital anomalies resulting from diagnostic radiation (Holmes 1979). Heat from high fevers is also a possible teratogen.

While we know the causes of certain malformations, most congenital abnormalities are not yet able to be explained. For instance, congenital cardiac anomalies occur in about 1 in every 200 live births. Genetic causes are responsible for about 8 percent of these heart abnormalities, and about 2 percent can be explained by known teratogens. That leaves 90 percent of the them unexplained (O'Rahilly and Müller 1992). We still have a great deal of research to do.

WEBSITE **21.12 Other teratogenic agents.** Certain behavior-modifying drugs are teratogenic, while others appear not to be. This is an area of great concern, as conflicting studies debate whether caffeine, cannabis, and cocaine are teratogenic.

WEBSITE **21.13 Maternal effects on later disease.** Recent epidemiological data suggests that nutritional stress to a pregnant mother may predispose her offspring to certain diseases when they are adults.

Sidelights & Speculations

Endocrine Disruptors

ENDOCRINE DISRUPTORS are exogenous chemicals that interfere with the normal function of hormones. They can disrupt hormonal function in many ways.

1. Endocrine disruptors can mimic the effects of natural hormones by binding to their receptors. DES (diethylstilbesterol; Chapter 17), is one such example.
2. Endocrine disruptors may block the binding of a hormone to its receptor, or they can block the synthesis of the hormone. Finasteride, a chemical used to

prevent male pattern baldness and enlargement of the prostate glands, is an anti-androgen, since it blocks the synthesis of dihydrotestosterone. Women are warned not to handle this drug if they are pregnant, since it could arrest the genital development of male fetuses.
3. Endocrine disruptors can interfere with the transport of a hormone or its elimination from the body. For instance, rats exposed to **polychlorinatedbiphenyl pollutants** (PCBs; see below) have low levels of thyroid hormone. The PCBs

compete for the binding sites of the thyroid hormone transport protein. Without being bound to this protein, the thyroid hormones are excreted from the body (McKinney et al. 1985; Morse et al. 1996).

Developmental toxicology and endocrine disruption are relatively new fields of research. While traditional toxicology has pursued the environmental causes of death, cancer, and genetic damage, developmental toxicology/endocrine disruptor research has focused on the roles that environmental chemicals may have in altering development by disrupting normal endocrine function of surviving animals (Bigsby et al. 1999).

WEBSITE **21.14 Environmental endocrine disruptors**. The Wingspread Consensus Statement of 1991 began a move by scientists to influence government policy concerning potential endocrine disruptors. This site looks at that statement and at some of the policies presently being implemented.

Environmental Estrogens

There is probably no bigger controversy in the field of toxicology than whether chemical pollutants are responsible for congenital malformations in wild animals, the decline of sperm counts in men, and breast cancer in women. One of the sources of these pollutants is pesticide use. Americans use some 2 billion pounds of pesticides each year, and some pesticide residues stay in the food chain for decades. Although banned in the United States in 1972, DDT has an environmental half-life of about 100 years (Nature Genetics 1995). Recent evidence has shown that DDT (dichloro-diphenyl-trichloroethane) and its chief metabolic by-product, DDE (which lacks one of the chlorine atoms), can act as estrogenic compounds, either by mimicking estrogen or by inhibiting androgen effectiveness (Davis et al. 1993; Kelce et al. 1995). DDE is a more potent estrogen than DDT, and it is able to inhibit androgen-responsive transcription at doses comparable to those found in contaminated soil in the United States and other countries. DDT and DDE

have been linked to such environmental problems as the decrease in the alligator populations in Florida, the feminization of fish in Lake Superior, the rise in breast cancers, and the worldwide decline of human sperm counts (Carlsen et al. 1992; Keiding and Skakkebaek 1993; Stone 1994; Swan et al. 1997). Guillette and co-workers (1994; Matter et al. 1998) have linked a pollutant spill in Florida's Lake Apopka (a discharge including DDT, DDE, and numerous other polychlorinated biphenyls) to a 90% decline in the birthrate of alligators and to the reduced penis size in the young males.

Dioxin, a by-product of the chemical processes used to make pesticides and paper products, has been linked to reproductive anomalies in male rats. The male offspring of rats exposed to this planar, lipophilic molecule when pregnant have reduced sperm counts, smaller testes, and fewer male-specific sexual behaviors. Fish embryos seem particularly susceptible to dioxin and related compounds, and it has been speculated that the amount of these compounds in the Great Lakes during the 1940s was so high that none of the lake trout hatched there during that time survived (Figure 21.23; Hornung et al. 1996; Zabel and Peterson 1996; Johnson et al. 1998).

Some estrogenic compounds may be in the food we eat and in the wrapping that surrounds them, for some of the chemicals used to set plastics have been found to be estrogenic. The discovery of the estrogenic effect of plastic stabilizers was made in a frightening way. Investigators at Tufts University Medical School had been studying estrogen-responsive tumor cells. These cells require estrogen in order to proliferate. Their studies were

going well until 1987, when the experiments suddenly went awry. Then the control cells began to show the high growth rates suggesting stimulation comparable to that of the estrogen-treated cells. Thus, it as if someone had contaminated the medium by adding estrogen to it. What was the source of contamination? After spending four months testing all the components of their experimental system, the researchers discovered that the source of estrogen was the plastic tubes that held their water and serum. The company that made the tubes refused to tell the investigators about its new process for stabilizing the polystyrene plastic, so the scientists had to discover it themselves. The culprit turned out to be *p*-nonylphenol, a chemical that is also used to harden the plastic of the plumbing tubes that bring us water and to stabilize the polystyrene plastics that hold water, milk, orange juice, and other common liquid food products (Soto et al. 1991; Colburn et al. 1996). This compound is also the degradation product of detergents, household cleaners, and contraceptive creams. A related compound, 4-*tert*-pentylphenol, has a potent estrogenic effect on human cultured cells and can cause male carp (*Cyprinus carpis*) to develop oviducts, ovarian tissue, and oocytes (Gimeno et al. 1996).

Some other environmental estrogens are polychlorinated biphenyls (mentioned earlier). These PCBs can react with a number of different steroid receptors. PCBs were widely used as refrigerants before they were banned in the 1970s when they were shown to cause cancer in rats. They remain in the food chain, however (in both water and sediments), and they have been blamed for the widespread decline in the reproductive capacities of ot-

Figure 21.23 Lake trout 4 weeks after hatching. (A) Normal larva with its golden yellow yolk sac. (B) Dioxin-exposed larva exhibiting a blue yolk sac. The yolk sac has swelled with water and has numerous sites of hemorrhage. Such fish often have reduced growth, as well as heart and facial anomalies. (Photograph courtesy of R. E. Peterson.)

(A)

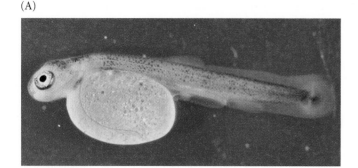

(B)

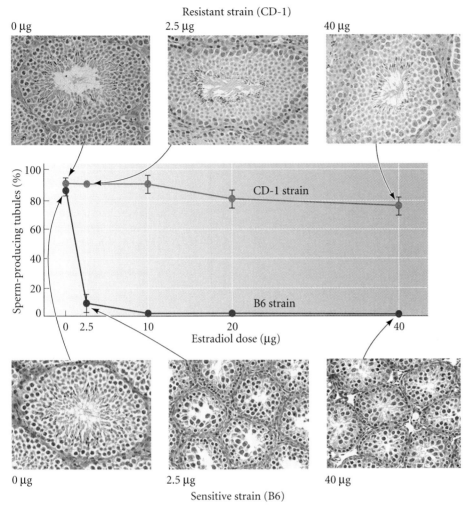

Resistant strain (CD-1)

0 µg   2.5 µg   40 µg

0 µg   2.5 µg   40 µg

Sensitive strain (B6)

Figure 21.24
Effects of estrogen implants on different strains of mice. The graph shows the percentage of seminiferous tubules containing elongated spermatozoa. (The mean standard ± error is for an average of six individuals). The photographs show cross sections of the testicles and are all at the same magnification. 40 µg of estradiol did not affect spermatogenesis in the CD-1 strain, but as little as 2.5 µg of estrogen almost completely abolished spermatogenesis in the B6 strain. (After Spearow et al. 1999; photographs courtesy of J. L. Spearow.)

found to have high affinities for the thyroid hormone serum transport protein transthyretin, and can block thyroxine from binding to this protein. This leads to the elevated excretion of the thyroid hormones. Thyroid hormones are critical for the growth of the cochlea of the inner ear, and rats whose mothers were exposed to PCBs had poorly developed cochleas and hearing defects (Goldey and Crofton in Stone 1995; Cheek et al. 1999).

Deformed Frogs: Pesticides Mimicking Retinoic Acid?

Throughout the United States and southern Canada there is a dramatic increase in the number of deformed frogs and salamanders in what seem to be pristine woodland ponds (Figure 21.26A; Oulette et al. 1997). These deformities include extra or missing limbs, missing or misplaced eyes, deformed jaws, and malformed hearts and guts. Some of these malformations (especially the limb anomalies, see Figure 3.23) may be due to trematode infestation, but other malformations do not seem to be explainable by that route. Some lakes containing high proportions of malformed frogs do not appear to be infested with trematodes, and water from these lakes is able to disrupt development in frog eggs that are placed into it. It is not known what is causing these disruptions, but there is speculation (see Hilleman 1996, Oulette et al. 1997) that pesticides (sprayed for mosquito and tick control) might be activating or interfering with the retinoic acid pathway. The spectrum of abnormalities seen in these frogs resembles those malformations caused by exposing tadpoles to retinoic acid (Crawford and Vincenti 1998; Gardiner and Hoppe 1999).

New research has focused on compounds such as methoprene, a juvenile hormone mimic that inhibits mosquito pupae from

ters, seals, mink, and fish. Some PCBs resemble diethylstilbesterol in shape, and they may affect the estrogen receptor as DES does, perhaps by binding to another site on the estrogen receptor. Another organochlorine compound (and an ingredient in many pesticides) is methoxychlor. Pickford and colleagues (1999) found that methoxychlor blocked progesterone-induced oocyte maturation in *Xenopus* at concentrations that are environmentally relevant. This would severely inhibit the fertility of the frogs, and it may be a component of the worldwide decline in amphibian populations.

Some scientists, however, say that these claims are exaggerated. Tests on mice had shown that litter size, sperm concentration, and development were not affected by environmental concentrations of environmental estrogens. However, recent work by Spearow and colleagues (1999) has shown a remarkable genetic difference in the sensitivity to estrogen among different strains of mice. The strain that had been used for testing environmental estrogens, the CD-1 strain, is at least 16 times more resistant to endocrine disruption than the most sensitive strains such as B6. When estrogen-containing pellets were implanted beneath the skin of young male CD-1 mice, very little happened. However, when the same pellets were placed beneath the skin of B6 mice, their testes shrunk, and the number of sperm seen in the seminiferous tubules dropped dramatically (Figure 21.24). This widespread range of sensitivities has important consequences for determining safety limits for humans.

Environmental Thyroid Hormone Disruptors

The structure of some PCBs resembles that of thyroid hormones (Figure 21.25), and exposure to them alters serum thyroid hormone levels in humans. Hydroxylated PCB were

Estradiol-17β
(estrogen)

Bisphenol-A

Dioxin

Thyroxine

Diethylstilbestrol

o,p′-DDT

General PCB structure

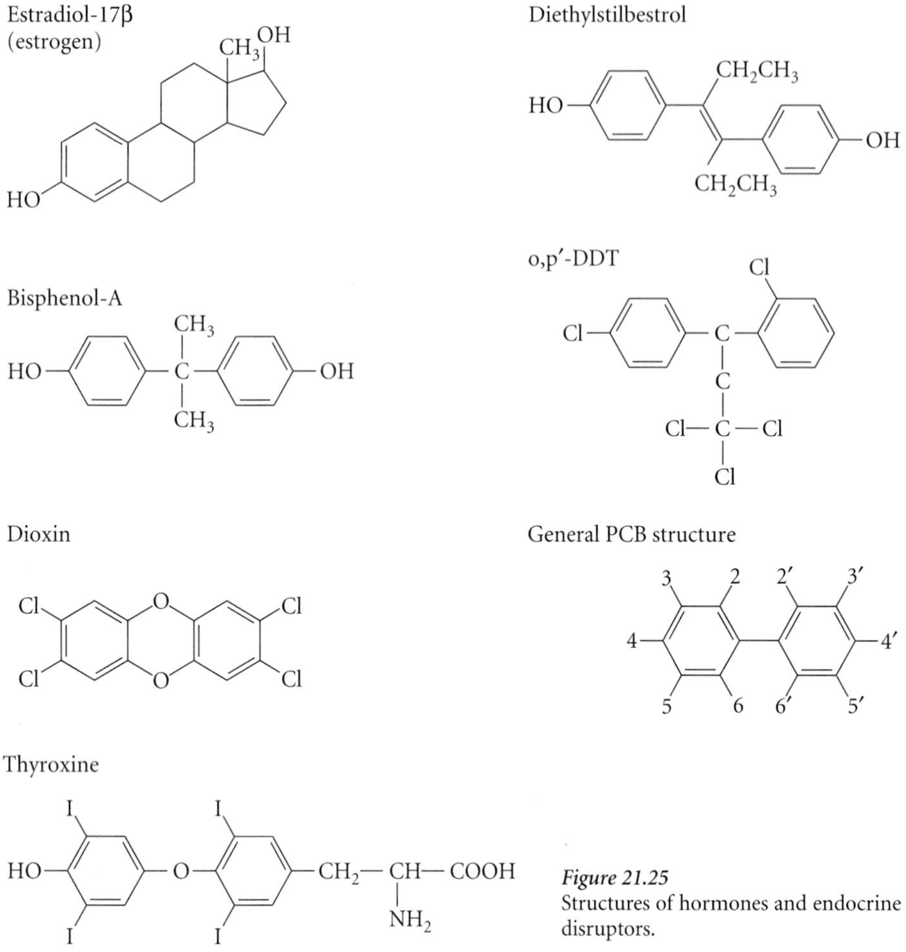

Figure 21.25
Structures of hormones and endocrine
disruptors.

endpoints must be checked, and many different levels of causation have to be established (Crain and Guillette 1998; McNabb et al. 1999). For instance, one could ask if the pollutant spill in Lake Apopka was responsible for the feminization of male alligators. To establish this, one has to ask how might the chemicals in the spill contributed to reproductive anomalies in males alligators and what would be the consequences of that happening. Table 21.3 shows the postulated chain of causation. After observing that the population level of the alligators has declined, at the organism level one discovers the unusually high levels of estrogens in the female alligators, the unusually low levels of testosterone in the males, and the decrease in the number of births among the alligators. On the tissue and organ level, the decline in birth rate can be explained by the elevated production of estrogens from the juvenile testes, the malformation of the testes and penis, and the changes in enzyme activity in the female gonads. On the cellular level, one sees ovarian abnormalities that correlate with unusually elevated estrogen levels. These cellular changes, in turn, can be explained at the molecular level by the finding that many of the components of

metamorphosing into adults. Since vertebrates do not have juvenile hormone, it was assumed that this pesticide would not harm fish, amphibians, or humans. This has been found to be the case: methoprene, itself, does not have teratogenic properties. However, upon exposure to sunlight, methoprene breaks down into products that have significant teratogenic activity in frogs (Figures 21.26B, C). These compounds have a structure similar to that of retinoic acid and will bind to the retinoid receptor (Harmon et al. 1995; La Claire et al 1998). When *Xenopus* eggs are incubated in water containing these compounds, the tadpoles are often malformed, and show a spectrum of deformities similar to those seen in the wild (La Claire et al. 1998).

Chains of Causation

Whether in law or science, establishing chains of causation is a demanding and necessary task. In developmental toxicology, numerous

Table 21.3 Chain of causation linking contaminant spill in Lake Apopka to endocrine disruption in juvenile alligators

Level	Evidence
Population	The juvenile alligator population in Lake Apopka has decreased.
Organism	Juvenile Apopka females have elevated circulating levels of estradiol-17β.
	Juvenile Apopka males have depressed circulating concentrations of testosterone.
Tissue/Organ	Juvenile Apopka females have altered gonad aromatase activity.
	Juvenile Apopka males have poorly organized seminiferous tubules.
	Juvenile Apopka males have reduced penis size.
	Testes from juvenile Apopka males have elevated estradiol (estrogen) production.
Cellular	Juvenile Apopka females have polyovular follicles that are characteristic of estrogen excess.
Molecular	Many contaminants bind the alligator estrogen receptors and progesterone receptor.
	Many of these contaminants do not bind to the alligator cytosol proteins that blockade excess hormones.

Source: After Crain and Guillette 1998.

(A)

(B)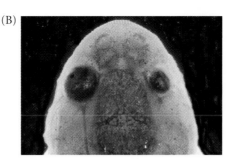

(C)

Methoprene

Sunlight

Water

Methoprenoic acid
(MA)

(D)

Retinoic acid

Figure 21.26
Teratogenesis in frogs. (A) Wild green frog (*Rana clamitans*) with an eye deformity, collected in New Hampshire in 1999 by K. Babbitt. (B) *Xenopus* tadpole with eye deformities caused by incubating newly fertilized eggs in water containing methoprenic acid, a by-product of methoprene. (C) One of several pathways by which methoprene can decay into teratogenic compounds such as methoprenic acid. (D) An isomer of retinoic acid showing the structural similarities to methoprenic acid. (A courtesy of K. Babbitt and K. Reed; B, C after La Claire et al. 1998, courtesy of J. Bantle.)

the pollutant spill bind to the alligator estrogen and progesterone receptors and that they are able to circumvent the cell's usual defenses against overproduction of steroid hormones (Crain et al. 1998).

While there is little dispute about the damage to wildlife being wrought by endocrine disrupting chemicals, it is difficult to document the effects of environmental compounds on humans. There is enormous genetic variation in the human species, and one cannot perform controlled experiments to determine the effect of any particular compound on a human population. Rather, we are exposed to "cocktails" consisting of different compounds ingested at different times. There is a great deal more research that needs to be done on the biochemistry of these compounds, their effects on development, and the epidemiology of developmental abnormalities. At the moment, evidence coming from animal studies suggests that humans and natural animal populations are at risk from these hormonal modulators, but not all the needed data are in. ■

WEBSITE **21.15 Deformed frogs and salamanders.** Considerable efforts are being made to find the causes for both the recent decline of amphibian populations and for the developmental anomalies being discovered in these animals. Parasites, fungus, ultraviolet radiation, and pesticides may all be playing a role.

Genetic-Environmental Interactions

The observation that a substance may be teratogenic in one strain of mice but not in another strongly suggests that there is a genetic component to whether a substance can produce changes in normal development. Recent evidence suggests that different alleles in the human population can influence whether a substance is benign or dangerous to the fetus. For example, among the general population, there is only a slight risk that heavy smoking by the mother will cause facial malformations in her fetus. However, if the fetus has a particular allele (A_2) of the gene for growth factor TGF-β, tobacco smoke absorbed through the placenta can raise the risk of cleft lip and palates tenfold (Shaw et al. 1996). Similarly, different alleles encoding the enzyme alcohol dehydrogenase-2 result in differing abilities to degrade ethanol. Whether heavy maternal alcohol consumption leads to fetal alcohol syndrome or fetal alcohol effect may be due to the types of alcohol dehydrogenase isozymes in the mother and fetus (McCarver-May, 1996). Thus, whether or not a compound is "teratogenic" depends on many things, including the genes of the individuals exposed to it.

Coda

Development usually occurs in a rich environmental milieu, and most animals are sensitive to environmental cues. The environment may determine sexual phenotype, may induce remarkable structural and chemical adaptations according to the season, may induce specific morphological changes that allow an individual to escape predation, and can induce caste determination in insects. The environment can also alter the structure of our neurons and the specificity of our immunocompetent cells. Unfortunately, the environment can also be the source of chemicals that disrupt normal developmental processes.

The developmental plasticity of the nervous system assures that each person is an individual. Our brain adds experience to endowment. Fears that cloning could produce "thousands of Hitlers" are unfounded. Not only have there been no genes identified for bigotry, demagoguery, or political canniness, but one would have to reconstruct Hitler's personal, social, and political milieus to even come close to replicating the dictator's personality. Wolpe (1997) has pointed out that thinking that a genetically identical clone of Hitler would become a bigoted dictator is buying into the same genetic essentialism that made Hitler so evil. Similarly, Gould (1997) points out that even Eng and Chang Bunker, the well publicized conjoined twins who most likely had both the same heredity and the same environment, became very

different people. One was cheerful and abstained from all liquor. The other was morose and alchoholic (which was a problem, since they shared the same liver). The plasticity of our nervous system enables us to be individuals and "allows us to escape the tyranny of our genes" (Childs 1999).

While development usually occurs in a complex natural environment, it can most easily be studied in the laboratory. Indeed, our "model systems" are animals that are readily domesticated and whose development is least affected by environmental factors (Bolker 1995). However, as we become aware of the complexity of development, we are realizing that development is critically keyed to the environment. It can take a community to develop an embryo. Ecologists have known about "life history strategies" of organisms for over a century. However, the proximate causes of these histories (such as how a fish becomes male in one environment and female in another) are just beginning to be understood. The exploration of environmental regulation of development is just beginning.

Snapshot Summary: The Environmental Regulation of Development

1. The environment can affect development in several ways. Development is sometimes cued to normal circumstances that the organism can expect to find in its environment. The larvae of many species will not begin metamorphosis until they find a suitable substrate. In other instances, symbiotic relationships between two or more species are necessary for the complete development of one or more of the species.

2. Developmental plasticity makes it possible for environmental circumstances to elicit different phenotypes from the same genotype. Many species have a broad reaction norm, wherein the genotype can respond in a graded way to environmental conditions.

3. Some species exhibit polyphenisms, in which distinctly different phenotypes are evoked by different environmental cues.

4. Seasonal cues such as photoperiod, temperature, or type of food can alter development in ways that make the organism more fit. Changes in temperature also are responsible for determining sex in several organisms, including many types of reptiles and insects.

5. Predator-induced polyphenisms have evolved such that the prey species can respond morphologically to the presence of a specific predator. In some instances, this induced adaptation can be transmitted to the progeny of the prey.

6. The differentiation of immunocompetent cells and the formation of synapses in the visual system are examples where experience influences the phenotype.

7. Compounds found in the environment (teratogens) can disrupt normal development. Teratogens can be naturally occurring substances or synthetic ones.

8. Alcohol and retinoic acid are two of the most intensively studied human teratogens. They may produce their teratogenic effects through more than one pathway.

9. It is possible that numerous compounds may be acting as hormone mimics or antagonists disrupt normal development by interfering with the endocrine system.

10. Genetic differences can predispose individuals to being affected by teratogens.

Literature Cited

Abel, E. L. and R. J. Sokol. 1987. Incidence of fetal alcohol syndrome and economic impact of FAS-related anomalies. *Drug Alcohol Depend.* 19: 51–70.

Adler, F. R. and C. D. Harvell. 1990. Inducible defenses, phenotypic variability, and biotic environments. *Trends Ecol. Evol.* 5: 407–410.

Agrawal A. A., C. Laforsch and R. Tollrian. 1999. Transgenerational induction of defenses in animals and plants. *Nature* 401: 60–63.

Alvarez-Buylla, A., J. R. Kirn and F. Nottebohm. 1990. Birth of projection neurons in adult avian brain may be related to perceptual or motor learning. *Science* 249: 1444–1446.

Antonini, A. and M. P. Stryker,.1993. Rapid remodeling of axonal arbors in the visual cortex. *Science* 260: 1818–1821.

Bachmann, M. D., R. G. Carlton, J. M. Burkholder and R. G. Wetzel. 1986. Symbiosis between salamander eggs and green algae: Microelectrode measurements inside eggs demonstrate effects of photosynthesis on oxygen concentrations. *Can. Zool.* 64: 1586–1588.

Barnes, G. L., B. D. Mariani and R. S. Tuan. 1996. Valproic acid-induced somite teratogenesis in the chick embryo: Relationship with *Pax-1* gene expression. *Teratology* 54: 93–102.

Baxter, G. T. and D. E. Morse. 1992. Cilia from abalone larvae contain a receptor-dependent G-protein transduction system similar to that in mammals. *Biol. Bull.* 183: 147–154.

Beachy, P. A. 1997. Multiple roles of cholesterol in hedgehog protein biogenesis and signaling. *Cold Spr. Harb. Symp. Quant. Biol.* 62: 191–204.

Beck, S. D. 1980. *Insect Photoperiodism.* 2nd Ed. Academic Press, New York.

Bigsby, R. and 8 others. 1999. Evaluating the effects of endocrine disrupters on endocrine function during development. *Environ. Health Perspect.* [Suppl.] 107: 613–618.

Black, J. E., K. R. Issacs, B. J. Anderson, A. A. Alcantara and W. T.Greenough. 1990. Learning causes synaptogenesis, whereas motor activity causes angiogenesis, in cerebellar cortex of adult rats. *Proc. Natl. Acad. Sci. USA* 87: 5568–5572.

Bolker, J. A. 1995. Model systems in developmental biology. *BioEssays* 17: 451–455.

Borovsk, D., D. A. Carlson, P. R. Griffin, J. Shabanowitz and D. F. Hunt. 1990. Mosquito oostatic factor: A novel decapeptide modulating trypsin-like enzyme biosynthesis in the midgut. *FASEB J.* 4: 3015–3020.

Boué, A., J. Boué and A. Gropp. 1985. Cytogenetics of pregnancy wastage. *Adv. Hum. Genet.* 14: 1–57.

Brakefield, P. M. and N. Reitsma. 1991. Phenotypic plasticity, seasonal climate, and the population biology of *Bicyclus* butterflies (Satyridae) in Malawi. *Ecol. Entomol.* 16: 291–303.

Brakefield, P. M. and 7 others. 1996. Development, plasticity, and evolution of butterfly eyespot patterns. *Nature* 384: 236–242.

Brian, M. V. 1974. Caste differentiation in *Myrmica rubra*: The role of hormones. *J. Insect Physiol.* 20: 1351–1365.

Brian, M. V. 1980. Social control over sex and caste in bees, wasps and ants. *Biol. Rev.* 55: 379–415.

Brönmark, C. and L. Pettersson. 1994. Chemical cues from piscivores induce a change in morphology in crucian carp. *Oikos* 70: 396–402.

Bull, J. J. 1980. Sex determination in reptiles. *Q. Rev. Biol.* 55: 3–21.

Burnett, F. M. 1959. *The Clonal Selection Theory of Immunity.* Vanderbilt University Press, Nashville.

Carlsen, E., A. Giwercman, N. Keiding and N. E. Skakkebaek. 1992. Evidence for decreasing quality of semen during past 50 years. *Brit. Med. J.* 305: 609–613.

Charnov, E. L. and J. J. Bull. 1977. When is sex environmentally determined? *Nature* 266: 828–830.

Cheek, A. O., K. Kow, J. Chen, and J. A. McLachlan. 1999. Potential mechanisms of thyroid disruption in humans: Interaction of organochlorine compounds with thyroid receptor, transthyretin, and thyroid-binding globulin. *Environ. Health Perspect.* 107: 273–278.

Childs, B. 1999. *Genetic Medicine: A Logic of Disease.* John Hopkins University Press.

Clarren, S. K. 1986. Neuropathology in the fetal alcohol syndrome. *In* J. R. West (ed.), *Alcohol and Brain Development.* Oxford University Press, New York.

Cohen, C. S. and R. R. Strathmann. 1996. Embryos at the edge of tolerance: Effects of environment and structure of egg masses on supply of oxygen to embryos. *Biol. Bull.* 190: 8–15.

Cohlan, S. Q. 1953. Excessive intake of vitamin A as a cause of congenital anomalies in the rat. *Science* 117: 535–537.

Colburn, T., D. Dumanoski and J. P. Myers. 1996. *Our Stolen Future.* Dutton, New York.

Colello, R. J. and R. W. Guillery. 1990. The early development of retinal ganglion cells with uncrossed axons in the mouse: Retinal position and axon course. *Development* 108: 515–523.

Colman, H., J. Nabekura and J. W. Lichtman. 1997. Alterations in synaptic strength preceding axon withdrawal. *Science* 275: 356–361.

Conover, D. O. and S. W. Heins. 1987. Adaptive variation in environmental and genetic sex determination in a fish. *Nature* 326: 496–498.

Crain, D. A. and L. L. Guillette Jr. 1998. Reptiles as models of contaminant–induced endocrine disruption. *Animal Reprod. Sci.* 53: 77–86.

Crair, M. C., D. C. Gillespie and M. P. Stryker. 1998. The role of visual experience in the development of columns in the cat visual cortex. *Science* 279: 566–570.

Crawford, K. and D. M. Vincenti. 1998. Retinoic acid and thyroid hormone may function through similar and competitive pathways in regenerating axolotls. *J. Exp. Zool.* 282: 724–738.

Creech Kraft, J. 1992. Pharmacokinetics, placental transfer, and teratogenicity of 13-*cis* retinoic acid, its isomer and metabolites. *In* G. M. Morriss-Kay (ed.), *Retinoids in Normal Development and Teratogenesis.* Oxford University Press, Oxford, pp. 267–280.

Danilevskii, A. S. 1965. *Photoperiodism and Seasonal Development of Insects.* Oliver and Boyd, Edinburgh.

Davis, D. L., H. L. Bradlow, M. Wolff, T. Woodruff, D. G. Hoel and H. Anton-Culver. 1993. Xenoestrogens as preventable causes of breast cancer. *Environ. Health Perspect.* 101: 372–377.

Davis, W. L., L. A. Crawford, O. J. Cooper, G. R. Farmer, D. Thomas and B. L. Freeman. 1990. Ethanol induces the generation of reactive free radicals by neural crest cells in culture. *J. Craniofac. Genet. Dev. Biol.* 10: 277–293.

Degnan, B. M. and D. E. Morse. 1995. Developmental and morphogenetic gene regulation in *Haliotis rufescens* larvae at metamorphosis. *Am. Zool.* 35: 391–398.

Denno, R. F., L. W. Douglass and D. Jacobs. 1985. Crowding and host plant nutrition: Environmental determinants of wing form in *Prokelisia marginata. Ecology* 66: 1588–1596.

Dodson, S. 1989. Predator-induced reaction norms. *BioScience* 39: 447–452.

Dorris, M. 1989. *The Broken Cord.* Harper and Row, New York.

Edmonds, D. K., K. S. Lindsay, J. F. Miller, E. Williamson and P. J. Wood. 1982. Early embryonic mortality in women. *Fertil. Steril.* 38: 447–453.

Edwards, M. 1994. Pollution in the former Soviet Union: Lethal legacy. *Natl. Geogr.* 186 (2): 70–115.

Fallon, A. M., H. H. Hagedorn, G. R. Wyatt and H. Laufer. 1974. Activation of vitellogenin synthesis in the mosquito *Aedes aegypti* by ecdysone. *J. Insect Physiol.* 26: 829–1823.

Ferguson, M. W. J. and T. Joanen. 1982. Temperature of egg incubation determines sex in *Alligator mississippiensis. Nature* 296: 850–853.

Franco, B. and 12 others. 1995. A cluster of sulfatase genes on Xp22.3: Mutations in *chondrodysplasia punctata* (*CDPX*) and implications for warfarin embryopathy. *Cell* 81: 15–21.

Gardiner, D. M. and D. M. Hoppe. 1999. Environmentally induced limb malformations in mink frogs (*Rana septentrionalis*). *J. Exp. Zool.* 284: 207–216.

Gilbert, S. F. 1994. Dobzhansky, Waddington and Schmalhausen: Embryology and the Modern Synthesis. *In* M. B. Adams (ed.), *The Evolution of Theodosius Dobzhansky: Essays on His Life and Thought in Russia and America.* Princeton University Press, Princeton, pp. 143–154.

Gimeno, S., A. Gerritsen, T. Bowmer and H. Komen. 1996. Feminization of male carp. *Nature* 384: 221–222.

Giroud, A. and M. Martinet. 1959. Teratogenese pur hypervitaminose A chez le rat, la souris, le cobaye, et le lapin. *Arch. Fr. Pediatr.* 16: 971–980.

Gotthard, K. and S. Nylin. 1995. Adaptive plasticity and plasticity as an adaptation: A selective review of plasticity in animal morphology and life history. *Oikos* 74: 3–17.

Gould, E., A. Beylin, P. Tanapat, A. Reeves and T. J. Shors. 1999. Learning enhances adult neurogenesis in the hippocampal formation. *Nature Neurosci.* 2: 260–265.

Gould, S. J. 1997. Individuality. *Sciences* (NY) 37: 14–16.

Goulding, E. H. and R. M. Pratt. 1986. Isotretinoin teratogenicity in mouse whole embryo culture. *J. Craniofac. Genet. Dev. Biol.* 6: 99–112.

Gregg, N. M. 1941. Congenital cataract following German measles in the mother. *Trans. Opthalmol. Soc. Aust.* 3: 35.

Guillette, L. J., T. S. Gross, G. R. Masson, J. M. Matter, H. F. Percival and A. R. Woodward. 1994. Developmental abnormalities of the gonad and abnormal sex hormone concentrations in juvenile alligators from contaminated and control lakes in Florida. *Environ. Health Perspect.* 102: 680–688.

Hadfield, M. G. 1977. Metamorphosis in marine molluscan larvae: An analysis of stimulus and response. *In* R.-S. Chia and M. E. Rice (eds.), *Settlement and Metamorphosis of Marine Invertebrate Larvae.* Elsevier, New York, pp. 165–175.

Hagedorn, H. H. 1983. The role of ecdysteroids in the adult insect. *In* G. Downer and H. Laufer (eds.), *Endocrinology of Insects.* Alan R. Liss, New York, pp. 241–304.

Hardie, J. 1981. Juvenile hormone and photoperiodically controlled polymorphism in *Aphis fabae*: Postnatal effects on presumptive gynoparae. *J. Insect Physiol.* 27: 347–355.

Hardie, J. and A. D. Lees. 1985. Endocrine control of polymorphism and polyphenism. *In* G. A. Kerkut and L. I. Gilbert (eds.), *Comprehensive Insect Physiology, Biochemistry, and Pharmacology*, vol. 8. Pergamon Press, Oxford, pp. 441–490.

Harmon, M. A., M. F. Boehm, R. A. Heyman and D. J. Mangelsdorf. 1995. Activation of mammalian retinoid-X receptors by the insect growth regulator methoprene. *Proc. Natl. Acad. Sci. USA* 92: 6157–6160.

Hart, M. W. and R. R. Strathmann. 1994. Functional consequences of phenotypic plasticity in echinoid larvae. *Biol. Bull.* 186: 291–299.

Harvell, C. W. 1986. The ecology and evolution of inducible defences in a marine bryozoan: Cues, costs, and consequences. *Amer. Nat.* 128: 810–823.

Harvell, C. W. 1999. Complex biotic environments: Coloniality and the hereditary variation for inducible defences. *In* R. Tollrian and C. D. Harvell (eds.), *The Ecology and Evolution of Inducible Defenses.* Princeton University Press, Princeton, NJ, pp. 231–244.

Hilleman, B. 1996. Frog deformities pose a mystery. *Chem. Engnr. News* 74: 24.

Hoffman, L. M. and W. M. Kulyk. 1999. Alcohol promotes in vitro chondrogenesis in embryonic facial mesenchyme. *Int. J. Dev. Biol.* 43: 167–174.

Hoffmann, R. J. 1973. Environmental control of seasonal variation in the butterfly *Colias eurytheme*. I. Adaptive aspects of a photoperiodic response. *Evolution* 27: 387–397.

Holmes, L. B. 1979. Radiation. *In* V. C. Vaughan, R. J. McKay and R. D. Behrman (eds.), *Nelson Textbook of Pediatrics*, 11th Ed. Saunders, Philadelphia.

Hornung, M. W., E. W. Zabel and R E. Peterson. 1996. Toxic equivalency factors of polybrominated dibenzo-*p*-dioxin, dibenzofuran, biphenyl, and polyhalogenated diphenyl ether congeners based on rainbow trout early life stage mortality. *Toxicol. Appl. Pharmacol.* 140: 227–234.

Huang, Z and 7 others. 1999. BDNF regulates the maturation of inhibition and the critical period of plasticity in mouse visual cortex. *Cell* 98: 739–755.

Hubel, D. H. 1967. Effects of distortion of sensory input on the visual system of kittens. *Physiologist* 10: 17–45.

Hubel, D. H. and T. N. Wiesel. 1962. Receptive fields, binocular interaction and functional architecture in the cat's visual cortex. *J. Physiol.* 160: 106–154.

Hubel, D. H. and T. N. Wiesel. 1963. Receptive fields of cells in striate cortex of very young, visually inexperienced kittens. *J. Neurophysiol.* 26: 944–1002.

Ikonomidou, C. and 11 others. 2000. Ethanol-induced apoptotic neurodegeneration and fetal alcohol syndrome. *Science* 287: 1056–1060.

Janzen, F. J. and G. L. Paukstis. 1991. Environmental sex determination in reptiles: Ecology, evolution, and experimental design. *Q. Rev. Biol.* 66: 149–179.

Johnson, E. M. 1980. Screening for teratogenic potential: Are we asking the proper questions? *Teratology* 21: 259.

Johnson, R. D. and 10 others. 1998. Toxicity of 2,3,7,8-tetrachlorodibenzo-*p*-dioxin to early life stage brook trout (*Salvelinus fontinalis*) following parental dietary exposure. *Environ.Toxicol. Chem.* 17: 2408–2421.

Johnston, M. C., K. K. Sulik, W. S. Webster and B. L. Jarvis. 1985. Isotretinoin embryopathy in a mouse model: Cranial neural crest involvement. *Teratology* 31: 26A.

Jones, K. L. and D. W. Smith. 1973. Recognition of the fetal alcohol syndrome. *Lancet* 2: 999–1001.

Katz, L. C. 1999. What's critical for the critical period in the visual cortex? *Cell* 99: 673–676.

Keiding, N and N. E. Skakkebaek. 1993. Are estrogens involved in falling sperm counts and disorders of the male reproductive tract? *Lancet* 341: 1392–1395.

Kelce, W. R., C. R. Stone, S. C. Laws, L. E. Gray, J. A. Kemppainen and E. M. Wilson. 1995. Persistent DDT metabolite p,p'-DDE is a potent androgen receptor antagonist. *Nature* 375: 581–585.

Kempermann, G., H. G. Kuhn and F. H. Gage. 1997a. More hippocampal neurons in adult mice living in an enriched environment. *Nature* 386: 493–495.

Kempermann, G., H. G. Kuhn and F. H. Gage.1997b. Genetic influence on neurogenesis in the dentate gyrus of adult mice. *Proc. Natl. Acad. Sci. USA* 94: 10409–10414.

Kennedy, C., S. Suda, C. B. Smith, M. Miyaoka, M. Ito and L. Sokoloff. 1981. Changes in protein synthesis underlying functional plasticity in immature monkey visual system. *Proc. Natl. Acad. Sci USA* 78: 3950–3953.

Kochhar, D. M., J. D. Penner and C. I. Tellone. 1984. Comparative teratogenic activities of two retinoids: Effects on palate and limb development. *Teratogenesis Carcinogen. Mutagen.* 4: 377–387.

Kotch, L. E., S.-Y. Chen and K. K. Sulik. 1995. Ethanol-induced teratogenesis: Free radical damage as a possible mechanism. *Teratology* 52: 128–136.

Kulikauskas, V., A. B. Blaustein and R. J. Ablin. 1985. Cigarette smoking and its possible effects on sperm. *Fertil. Steril.* 44: 526–528.

La Claire, J. J., J. A. Bantle and J. Dumont. 1998. Photoproducts and metabolites of a common insect growth regulator produce developmental deformities in *Xenopus. Environ. Sci. Tech.* 32: 1453–1461.

Lammer, E. J. and 11 others. 1985. Retinoic acid embryopathy. *N. Engl. J. Med.* 313: 837–841.

Lemoine, E. M., J. P. Harousseau, J. P. Borteyru and J. C. Menuet. 1968. Les enfants de parents alcooliques: Anomalies observées. *Oest. Med.* 21: 476–482.

Linney, E. 1992. Retinoic acid receptors: Transcription factors modulating gene expression, development, and differentiation. *Curr. Top. Dev. Biol.* 27: 309–350.

Matter, J. M. and 8 others. 1998. Effects of endocrine–disrupting contaminants in reptiles: alligators. *In* R. J. Kendall, R. L. Dickerson, J. P. Geisy and W. A. Suk (eds.), *Principles and Processes for Evaluating Endocrine Disruptors on Wildlife.* SETAC Press, Pensacola, FL, pp. 267–289.

McCarver-May, D. G. 1996. Genetic differences in alcohol dehydrogenase and fetal alcohol effects. Abstracts of the Ninth International Congress of Human Genetics. *Brazil J. Genet.* 19: 73.

McCollum, S. A. and J. Van Buskirk. 1996. Costs and benefits of a predator induced polyphenism on the gray treefrog *Hyla chrysoscelis. Evolution* 50: 583–593.

McFall-Ngai, M. J. and E. G. Ruby. 1991. Symbiont recognition and subsequent morphogenesis as early events in an animal-bacterial mutualism. *Science* 254: 1491–1494.

McKeown, T. 1976. Human malformations: An introduction. *Br. Med. Bull.* 32: 1–3.

McKinney, J. D., K. Chae, S. J. Oatley and C. C. Blake. 1985. Molecular interactions of toxic chlorinated dibenzo–*p*–dioxins and dibenzofurans with thyroxine-binding prealbumin. *J. Med Chem* 28: 375–381

McNabb, A. and 7 others. 1999. Basic physiology. *In* R. T. Di Giulio and D. E. Tillitt (eds.), *Reproductive and Developmental Effects of Contaminants in Oviparous Vertebrates.* SETAC Press, Pensacola, FL, pp. 113–223.

Montgomery, M. K. and M. J. McFall-Ngai. 1995. The inductive role of bacterial symbionts in the morphogenesis of a squid light organ. *Am. Zool.* 35: 372–380.

Morgan, T. H. 1909. Sex determination and parthenogenesis in phylloxerans and aphids. *Science* 29: 234–237.

Moroni, M. C., M. A. Vigano and F. Mavilio. 1994. Regulation of human *Hoxd-4* gene by retinoids. *Mech. Dev.* 44: 139–154.

Morriss-Kay, G. M. and S. J. Ward. 1999. Retinoids and mammalian development. *Int. Rev. Cytol.* 188: 73–131.

Morse, A. N. C., C. A. Froyd and D. E. Morse. 1984. Molecules from cyanobacteria and red algae that induce larval settlement and metamorphosis in the mollusc *Haliotis rufescens*. *Mar. Biol.* 81: 293–298.

Morse, D. C., E. K. Wehler, W. Wesseling, J. H. Koeman and A. Brouwer. 1996. Alterations in rat brain thyroid hormone status following pre- and postnatal exposure to polychlorinated biphenyls (Aroclor 1254). *Toxicol. Appl. Pharmacol.* 136: 269–279.

Müller, G. B. and J. Steicher. 1989. Ontogeny of the syndesmosis tibiofibularis and the evolution of the bird hindlimb: a caenogenetic feature triggers phenotypic novelty. *Anat. Embryol.* 179: 327–339.

Nature Genetics (editorial). 1995. Risk assessment and religion. *Nature Genet.* 11: 105–106.

Newman, R. A. 1989. Developmental plasticity of *Scaphiopus couchii* tadpoles in an unpredictable environment. *Ecology* 70: 1775–1787.

Newman, R. A. 1992. Adaptive plasticity in amphibian metamorphosis. *BioScience* 42: 671–678.

Newman, R. A. and G. B. Müller. In press. Epigenetic mechanisms of character origination. *In* G. Wagner (ed.), *The Concept Character in Evolutionary Biology*. Academic Press, San Diego.

Nijhout, H. F. 1991. *The Development and Evolution of Butterfly Wing Patterns*. Smithsonian Institution Press, Washington, D.C.

Nijhout, H. F. 1994. *Insect Hormones*. Princeton University Press, Princeton, NJ.

Nordeen, K. W. and E. J. Nordeen. 1988. Projection neurons within a vocal pathway are born during song learning in zebra finches. *Nature* 334: 149–151.

Nulman, I. and 7 others. 1997. Neurodevelopment of children exposed in utero to antidepressant drugs. *New Engl. J. Med.* 336: 258–262.

Opitz, J. M. 1998. The RSH syndrome: Paradigmatic metabolic malformation syndrome? *In* M. I. New (ed.), *Diagnosis and Treatment of the Unborn Child*. Idelson-Gnocchi, pp. 43–55.

O'Rahilly, R. and F. Müller. 1992. *Human Embryology and Teratology*. Wiley-Liss, New York.

Ouellet, M., J. Bonin, J. Rodriguez, J. L. DesGanges and S. Lair. 1997. Hindlimb deformities (ectromelia, ectrodactyly) in free–living anurans from agricultural habitats. *J. Wildlife Dis.* 33: 95–104.

Palmer, A. R. 1985. Adaptive value of shell variation in *Thais lamellosa*: Effect of thick shells on vulnerability to and preference by crabs. *Veliger* 27: 349–356.

Passera, L. 1985. Soldier determination in ants of the genus *Pheidole*. *In* J. A. L. Watson, B. M Okot-Kotber and C. Noirot (eds.) *Caste Determination in Social Insects*. Pergamon, Oxford, pp. 331–346.

Pechenik, J. A., D. E. Wendt and J. N. Jarrett. 1998. Metamorphosis is not a new beginning. *BioScience* 48: 901–910.

Pener, M. P. 1991. Locust phase polymorphism and its endocrine relations. *Adv. Insect Physiol.* 3: 1–79.

Pickford, D. B. and I. D. Morris. 1999. Effects of endocrine-disrupting contaminants on amphibian oogenesis: Methoxychlor inhibits progesterone-induced maturation of *Xenopus laevis* oocytes in vitro. *Environ. Health Perspect.* 107: 285–292.

Pinder, A. W. and S. C. Friet. 1994. Oxygen transport in egg masses of the amphibians *Rana sylvatica* and *Ambystoma maculatum*: Convection, diffusion, and oxygen production by algae. *J. Exp. Biol.* 197: 17–30.

Plowright, R. C. and B. A. Pendrel. 1977. Larval growth in bumble-bees. *Can. Entomol.* 109: 967–973.

Pöpperl, H. and M. S. Featherstone. 1993. Identification of retinoic acid response element upstream from the mouse *Hox-4.2* gene. *Mol. Cell. Biol.* 13: 257–265.

Purves, D. and J. W. Lichtman. 1985. *Principles of Neural Development*. Sinauer Associates, Sunderland, MA.

Raatikainen, M. 1967. Bionomics, enemies, and population dynamics of *Javesella pellucida* (F.) (Homoptera, Delphaidae). *Annales Agric. Fenniae* 6: 1–49.

Rachinsky, A. and K. Hartfelder. 1990. Corpora allata activity, a prime regulating element for caste-specific juvenile hormone titre in honey bee larvae (*Apis mellifera carnica*). *J. Insect Physiol.* 36: 329–349.

Ramanathan, R., M. F. Wilkemeyer, B. Mittel, G. Perides and M. E. Charness. 1996. Alcohol inhibits cell-cell adhesion mediated by human L1. *J. Cell Biol.* 133: 381–390.

Rasika, S., A. Alvarez-Buylla and F. Nottebohm. 1999. BDNF mediates the effects of testosterone on the survival of new neurons in the adult brain. *Neuron* 22: 53–62.

Reiter, H. O. and M. P. Stryker. 1988. Neural plasticity without postsynaptic action potentials: Less-active inputs become dominant when kitten visual cortical cells are pharmacologically inhibited. *Proc. Natl. Acad. Sci. USA* 85: 3623–3627.

Rothman, K. J., L. L. Moore, M. R. Singer, U.-S. D. T. Nguyen, S. Mannino and A. Milunsky. 1995. Teratogenicity of high vitamin A intake. *N. Engl. J. Med.* 333: 1369–1373.

Ruberte, E., P. Dollé, A. Krust, A. Zalent, G. Morriss-Kay and P. Chambon. 1990. Specific spatial and temporal distribution of retinoic acid γ transcripts during mouse embryogenesis. *Development* 108: 213–222.

Ruberte, E. and 7 others. 1991. Retinoic acid receptors in the embryo. *Semin. Dev. Biol.* 2: 153–159.

Sander, K. 1968. Entwicklungsphysiologische Untersuchungen am embryonalen Mycetom von *Euscelis plebejus* F. (Homoptera, Ciciadina). I. *Dev. Biol.* 17: 16–38.

Sapp, J. 1994. *Evolution by Association: A History of Symbiosis*. Oxford University Press, New York.

Schmalhausen, I. I. 1949. *Factors of Evolution: The Theory of Stabilizing Selection*. University of Chicago Press, Chicago.

Schwemmler, W. 1974. Endosymbionts: Factors of egg patterning. *J. Insect Physiol.* 20: 1467–1474.

Schwemmler, W. 1989. Insect symbiosis as a model system for egg cell differentiation. *In* W. Schwemmler and G. Gassner (eds.), *Insect Endosymbiosis*. CRC Press, Boca Raton, pp. 37–53.

Shapiro, A. M. 1968. Photoperiodic induction of vernal phenotype in *Pieris protodice* Boisduval and Le Conta (Lepidoptera: Pieridae). *Wasmann J. Biol.* 26: 137–149.

Shapiro, A. M. 1976. Seasonal polyphenism. *Evol. Biol.* 9: 259–333.

Shapiro, A. M. 1978. The evolutionary significance of redundancy and variability in phenotypic induction mechanisms of pierid butterflies (Lepidoptera). *Psyche* 85: 275–283.

Shaw, G. M., C. R. Wasserman, E. J. Lammer, C. D. O'Malley, J. C. Murray, A. M. Basart and M. M. Tolarova. 1996. Orofacial clefts, parental cigarette smoking, and transforming growth factor-alpha gene variants. *Am. J. Hum. Genet.* 58: 551–561.

Soto, A., H. Justicia, J. Wray and C. Sonnenschein. 1991. *p*-nonylphenol: An estrogenic xenobiotic released from "modified" polystyrene. *Environ. Health Perspect.* 92: 167–173.

Spearow, J. L., P. Doemeny, R. Sera, R. Leffler and M. Barkley. 1999. Genetic variation in susceptibility to endocrine disruption by estrogen in mice. *Science* 285: 1259–1261.

Stearns, S. C., G. de Jong and R. A. Newman. 1991. The effects of phenotypic plasticity on genetic correlations. *Trends Ecol. Evol.* 6: 122–126.

Stent, G. S. 1973. A physiological mechanism for Hebb's postulate of learning. *Proc. Natl. Acad. Sci. USA* 70: 997–1001.

Stone, R. 1994. Environmental estrogens stir debate. *Science* 265: 308–310.

Stone, R. 1995. Environmental toxicants under scrutiny at Baltimore meeting. *Science* 267: 1770–1771.

Strathmann, R. R. and M. F. Strathmann. 1995. Oxygen supply and limits on aggregation of embryos. *J. Mar. Biol. Assoc. UK* 75: 413–428.

Strathmann, R. R., L. Fenaux and M. F. Strathmann. 1992. Heterochronic developmental plasticity in larval sea urchins and its implication for evolution on nonfeeding larvae. *Evolution* 46: 972–986.

Streissguth, A. P. and R. A. LaDue. 1987. Fetal alcohol: Teratogenic causes of developmental disabilities. *In* S. R. Schroeder (ed.), *Toxic Substances and Mental Retardation.* American Association of Mental Deficiency, Washington, DC, pp. 1–32.

Studer, M., H. Popperl, H. Marshall, A. Kuroiwa and R. Krumlauf. 1994. Role of a conserved retinoic acid response element in rhombomere restriction of *Hoxb-1. Science* 265: 1728–1732.

Sulik, K. K., C. S. Cook and W. S. Webster. 1988. Teratogens and craniofacial malformations: Relationships to cell death. *Development* 103 [Suppl]: 213–231.

Swan, S. H., E. P. Elkin and L. Fenster. 1997. Have sperm densities declined? A reanalysis of global trend data. *Environ. Health Perspect.* 105:1228–1232.

Tawkik, A. I. and 9 others. Identification of the gregarization-associated dark–pigmentotropin in locusts through an albino mutant. *Proc. Natl. Acad. Sci USA* 96: 7083–7087.

Tollrian, R. and C. D. Harvell. 1999. *The Ecology and Evolution of Inducible Defenses.* Princeton University Press, Princeton, NJ.

Tauber, M. J., C. A. Tauber and S. Masaki. 1986. *Seasonal Adaptations of Insects.* Oxford University Press, Oxford.

Tramontin, A. D., V. N. Hartman and E. A. Brenowitz. 2000. Breeding conditions induce rapid and sequential growth in adult avian song control circuits: A model for seasonal plasticity in the brain. *J. Neurosci.* 20: 854–861.

Turner, A. M. and W. T. Greenough. 1983. Synapses per neuron and synaptic dimensions in occipital cortex of rats reared in complex, social, or isolation housing. *Acta Stereologica* 2 [Suppl. 1]: 239–244.

van der Weele, C. 1999. *Images of Development: Environmental Causes in Ontogeny.* SUNY Press, Albany.

van Praag, H., G. Kempermann and F. H. Gage. 1999. Running increases cell proliferation and neurogenesis in the adult mouse gentate gyrus. *Nature Neurosci.* 2: 266–270.

Via, S., R. Gomulkiewicz, G. De Jong, S. M. Scheiner, C. D. Schlichting and P. H. Van Tienderen. 1995. Adaptive phenotypic plasticity: Consensus and controversy. *Trends Ecol. Evol.* 10: 212–217.

Volpe, E. P. 1987. Developmental biology and human concerns. *Am. Zool.* 27: 697–714.

Waddington, C. H. 1942. Canalization of development and the inheritance of acquired characteristics. *Nature* 150: 563–565.

Warner, R. R. 1984. Mating behavior and hermaphroditism in coral reef fishes. *Amer. Sci.* 72: 382–389.

Watt, W. B. 1968. Adaptive significance of pigment polymorphism in *Colias* butterflies. I. Variation of melanin in relation to thermoregulation. *Evolution* 22: 437–458.

Watt, W. B. 1969. Adaptive significance of pigment polymorphism in *Colias* butterflies. II. Thermoregulation and periodically controlled melanin production in *Colias eurytheme. Proc. Natl. Acad. Sci. USA* 63: 767–774.

Wheeler, D. 1986. Developmental and physiological determinants of caste in social Hymenoptera: Evolutionary implications. *Am. Nat.* 128: 13–34.

Wheeler, D. 1991. The developmental basis of worker caste polymorphism in ants. *Am. Nat.* 138: 1218–1238.

Wiesel, T. N.1982. Postnatal development of the visual cortex and the influence of environment. *Nature* 299: 583–591.

Wirtz, P. 1973. Differentiation in the honeybee larva. *Meded. Landb. Hogesch. Wagningen* 73–75, 1–66.

Wolpe, P. R. 1997. If I am only my genes, what am I? *Kennedy Inst. Ethics J.* 7: 213–230.

Wolpert, L. 1991. *The Triumph of the Embryo.* Oxford University Press. Oxford.

Woltereck, R. 1909. Weitere experimentelle Untersuchungen über Artveränderung, speziell über das Wesen quantitativer Artunderscheide bei Daphniden. *Versuch. Deutsch. Zool. Ges.* 1909: 110–172.

Woodward, D. E. and J. D. Murray. 1993. On the effect of temperature-dependent sex determination on sex ratio and survivorship in crocodilians. *Proc. R. Soc. Lond.* [B] 252: 149–155.

Wu, K. C. 1996. Entwicklung, Stimulation, und Paralyse der embryonalen Motorick. *Wien Klin. Wochenschr.* 108: 303–305.

Yu, V. C. and 9 others. 1991. RXRb: A coregulator that enhances binding of retinoic acid, thyroid hormone, and vitamin D receptors to their cognate response elements. *Cell* 67: 1251–1266.

Zabel, E. W. and R. E. Peterson. 1996. TCDD-like activity of 2,3,6,7-tetrachloroxanthene in rainbow trout early life stages and in a rainbow trout gonadal cell line (RTG-2). *Environ. Toxicol. Chem.* 15: 2305–2309.

chapter 22

Developmental mechanisms of evolutionary change

How does newness come into the world? How is it born? Of what fusions, translations, conjoinings is it made? How does it survive, extreme and dangerous as it is? What compromises, what deals, what betrayals of its secret nature must it make to stave off the wrecking crew, the exterminating angel, the guillotine?
SALMAN RUSHDIE (1988)

Biology points out the individuality of every being, and at the same time reminds us of the brotherhood of all.
JEAN ROSTAND (1962)

WHEN WILHELM ROUX ANNOUNCED THE CREATION of experimental embryology in 1894, he broke many of the ties that linked embryology to evolutionary biology. According to Roux, embryology was to leave the seashore and forest and go into the laboratory. However, he promised that embryology would someday return to evolutionary biology, bringing with it new knowledge of how animals were generated and how evolutionary changes might occur. He stated that "an ontogenetic and a phylogenetic developmental mechanics are to be perfected." Roux thought that research into the developmental mechanics of individual embryos (the ontogenetic branch) would proceed faster than the phylogenetic (evolutionary) branch, but he predicted that "in consequence of the intimate causal connections between the two, many of the conclusions drawn from the investigation of individual development [would] throw light on the phylogenetic processes." A century later, we are now at the point of fulfilling Roux's prophecy and returning developmental biology to questions of evolution. This return is producing a new model of evolution that integrates both developmental genetics and population genetics.

The fundamental principle of this new evolutionary synthesis is that evolution is caused by heritable changes in the *development* of organisms. This view can be traced back to Darwin, and it is compatible with and complementary to the view of evolution based on population genetics that evolution is caused by changes in gene frequency between generations. The merging of the developmental genetic approach to evolution ("evo-devo") with the population genetic approach is creating a more complete evolutionary biology that is beginning to explain the origin of both species and higher taxa (Raff 1996; Gerhart and Kirschner 1999; Hall 1999).

"Unity of Type" and "Conditions of Existence"

Charles Darwin's synthesis

In the 1800s, debates over the origin of species pitted against each other two ways of looking at nature. One view (championed by Georges Cuvier and Charles Bell) focused on the differences among species that allowed each species to adapt to its environment. Thus, the fingers of the human hand, the flipper of the seal, and the wings of birds and bats were seen as marvelous contrivances, fashioned by the Creator, to allow these animals to adapt to their "conditions of existence." The other view, championed by Geoffroy St. Hilaire and Richard Owen, was that these adaptations were secondary, and that the "unity of type" (what Owen called "homologies") was critical. The human hand, the seal's flipper, and the wings of bats and birds are each modifications of the same basic plan (see Figure 1.13). In discovering that basic plan, one can find the form upon which the Creator designed these animals. The adaptations were secondary.

Darwin acknowledged his debt to these earlier debates when he wrote in 1859, "It is generally acknowledged that all organic beings have been formed on two great laws—Unity of Type, and Conditions of Existence." Darwin went on to explain that his theory would explain unity of type by descent from a common ancestor. Moreover, the changes creating the marvelous adaptations to the conditions of existence were explained by natural selection. Darwin called this concept "descent with modification." As mentioned in Chapter 1, Darwin found that homologies between the embryonic and larval structures of different phyla provided excellent evidence for descent with modification. He also argued that adaptations that depart from the "type" and allow an organism to survive in its particular environment develop late in the embryo. Thus, Darwin recognized two ways of looking at "descent with modification." One could emphasize the *common descent* by pointing out embryonic homologies between two or more groups of animals, or one could emphasize the *modifications* by showing how development was altered to produce structures that enabled animals to adapt to particular conditions.

WEBSITE **22.1 Haeckel's biogenetic law.** In the early 1900s, a fusion of evolution and embryology was wrongly interpreted to support a linear (as opposed to a branched) model of evolution. The interpretation of Ernst Haeckel was that every organism evolved by the terminal addition of a new stage to the end of the last "highest" organism. Thus, he saw the entire animal kingdom as representing truncated steps of human development.

WEBSITE **22.2 "Scientific creationism."** The phenomenon of creationism combines several American social traditions: fundamentalism, natural theology, and scientism. These three websites look at "scientific creationism" and provide some parables as to the nature of evolution.

E. B. Wilson and F. R. Lillie

Darwin did not attempt to construct complete phylogenies from embryological data, but his work inspired many of his contemporaries to do so. One of the first scientists to realize the evolutionary importance of von Baer's laws (see Chapter 1) was Elie Metchnikoff. Metchnikoff appreciated that evolution consists of modifying embryonic organisms, not adult ones. In 1891, he wrote:

> *Man appeared as a result of a one-sided, but not total, improvement of organism, by joining not so much adult apes, but rather their unevenly developed fetuses. From the purely natural historical point of view, it would be possible to recognize man as an ape's "monster," with an enormously developed brain, face and hands.*

But if changes in embryonic development effected evolutionary changes, how did these developmental changes take place? During the late 1800s, many investigators attempted to link development to phylogeny through the analysis of cell lineages. They meticulously observed each cell in developing embryos and compared the ways in which different organisms formed their tissues. In 1898, two eminent embryologists gave cell lineage lectures at the Marine Biological Laboratory at Woods Hole, Massachusetts, that served to emphasize the two ways in which embryology was being used to support evolutionary biology. The first lecture, presented by E. B. Wilson, was a landmark in the use of embryonic homologies to establish phylogenetic relationships. Wilson had observed the spiral cleavage patterns of flatworms, molluscs, and annelids, and he had discovered that in each case, the same organs came from the same groups of cells. For him this meant that these phyla all had a common ancestor. Indeed, modern research using DNA sequences has confirmed Wilson's conclusion and placed these three phyla together.

The other lecturer was F. R. Lillie, who had also done his research on the development of molluscan embryos and on modifications of cell lineages. He stressed the modifications, not the similarities, of cleavage. His research on *Unio*, a mussel whose cleavage pattern is altered to produce the "beartrap" larva that enables it to survive in flowing streams, was highlighted in Chapter 8. Lillie argued that "modern" evolutionary studies would do better to concentrate on changes in embryonic development that allowed for survival in particular environments than to focus on ancestral homologies that united animals into lines of descent.

In 1898, then, the two main avenues of approach to evolution and development were clearly defined: to find underlying unities that link disparate groups of animals, and to detect those differences in development that enable species to adapt to particular environments. Darwin thought these two approaches to be temporally distinguished—that is, that one would find underlying unities in the earliest stages of development, while the later stages would diverge to allow specific adaptations (see Ospovat 1981). However, Wilson and Lillie were both discussing the cleavage stage of embryogenesis. These two ways of characterizing evolution and development are still the major approaches today.

"Life's splendid drama"

Until this decade, "many invertebrate biologists saw the reconstruction of relationships among the phyla as an insoluble dilemma. … Indeed, as late as 1990, a comprehensive summary concluded that the relationships between most of the higher animal groups were entirely unresolved" (Erwin et al. 1997). However, in the 1990s, a broad consensus on the general form of a phylogenetic tree of life began to emerge among paleontologists, molecular biologists, and developmental geneticists (see Winnepenninckx et al. 1998; Davidson and

Ruvkun 1999; Erwin 1999). This consensus (one representation of which is shown in Figure 22.1A) came about from (1) improved methods of analyzing DNA, taking into account its variation within groups of animals, (2) new data on conserved regulatory gene sequences such as the Hox genes, which are usually stable within phyla but can diverge between phyla, (3) morphological evidence for the related nature of some structures that had once been thought to be distinct, and (4) computer programs that can sort out enormous amounts of data, not privileging any particular set of relationships over others. The results were surprising to many scientists.

1. The animal kingdom can be divided into Porifera (sponges), Cnidaria and Ctenophora (jellyfish and comb jellies), and the Bilateria. The Porifera lack any coherent epithelium or any symmetry. The Cnidaria and Ctenophora are diploblastic (with two epithelial layers, lacking mesoderm) and have radial symmetry.

2. The Bilateria are triploblastic (with true endoderm, mesoderm, and ectoderm), have bilateral symmetry and can be divided into two groups, the deuterostomes and the protostomes. The deuterostomes include the echinoderms and the chordates. (Echinoderm larvae are originally bilateral.) Deuterostomes form their anuses from their blastopores. They probably arose from a tornaria larva similar to that of the pluteus.

3. The protostomes can be divided into two groups, the **Ecdysozoa** (animals whose bodies are covered by an ex-oskeleton, which therefore molt as they grow) and the **Lophotrochozoa** (animals that have most or all of their soft tissues in contact with the environment and which generally use cilia in feeding or locomotion). Nematodes and flatworms (which lack true coeloms) had formerly been considered basal groups, perhaps ancestral to both the protostomes and the deuterostomes. However, the new studies have shown that the nematodes belong to the Ecdysozoa and the flatworms (as Wilson had predicted) to the Lophotrochozoa. Each of these two groups is **monophyletic**, meaning that the phyla in each of them share a common ancestor.

WEBSITE **22.3 The emergence of embryos.** How did individual cells come to sacrifice their individual potentials and generate embryos? How did gastrulation evolve? The answers may involve predation and the inability to divide and be ciliated at the same time.

WEBSITE **22.4 Why are there no new animal phyla?** It appears that the three dozen or so known phyla were all created 500 million years ago. It may be the case that no new phylum has emerged since the late Cambrian. What is the evidence for the early formation of the phyla, and are there any body plans left unused?

WEBSITE **22.5 How taxonomic groups are classified.** The advent of cladistics has put some order into the various ways of classifying animals. This does not mean, however, that there is unanimous agreement on the results.

Figure 22.1

A current phylogeny (A) and a regulatory gene whose function is conserved (B). The *Pax6* gene for eye development is an example of a gene ancestral to both protostomes and deuterostomes. The micrograph shows ommatidia emerging in the leg of a fruit fly (a protostome) in which mouse (deuterostome) *Pax6* cDNA was expressed in the leg disc. (A based on J. Garey, personal communication; B from Halder et al. 1995, photograph courtesy of W. J. Gehring and G. Halder.)

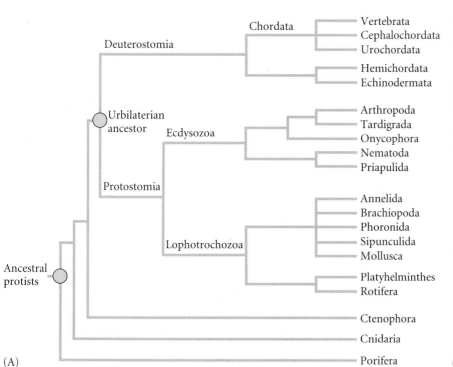

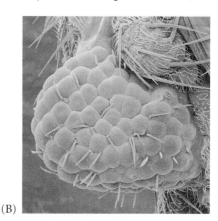

(A)

(B)

The search for the Urbilaterian ancestor

It is doubtful that we will find a fossilized representative of the ancestral phylum that gave rise to both the deuterostomes and the protostomes. Such a hypothetical animal is sometimes called the **Urbilaterian ancestor** or the **PDA** (*protostome-deuterostome ancestor*). Since such an animal probably had neither a bony endoskeleton (a deuterostome trait) nor a hard exoskeleton (characteristic of ecdysozoans), it would not fossilize well. However, we can undertake what Sean Carroll (quoted in DiSilvestro 1997) has called "paleontology without fossils." The logic of this approach is to find homologous genes that are performing the same functions in both a deuterostome (usually a chick or a mouse) and a protostome (generally an arthropod such as *Drosophila*). Many such genes have been found (Table 22.1), and their similarities of structure and function in protostomes and deuterostomes make it likely that these genes emerged in an animal that is ancestral to both groups.

Pax6, for example, plays a role in forming eyes in both vertebrates and invertebrates (see Chapters 4 and 5). Ectopic expression of *Pax6* will form extra eyes in both *Drosophila* and *Xenopus*, representatives of the protostomes and deuteros-

tomes, respectively (Chow et al. 1999; see Figure 5.14). Moreover, the ectopic expression of a deuterostome (mouse) *Pax6* gene in a fly larva will also induce ectopic fly eyes (Figure 22.1B; Halder et al. 1995). Therefore, it is a safe assumption that the same *Pax6* gene is involved in eye production in both deuterostomes and protostomes. Moreover, at least three other genes—*sine oculis*, *eyes absent*, and *dachshund*—are also used to form eyes in both *Drosophila* and vertebrates (Jean et al, 1998; Relaix and Buckingham, 1999). Since it is extremely unlikely that deuterostomes and protostomes would have evolved the *Pax6* (and other) genes independently and used it independently for the same function, it is very likely that the PDA had a *Pax6* gene and used it for generating eyes.

Another such gene shared by deuterostomes and protostomes is the homeobox-containing gene *tinman*. The Tinman protein is expressed in the *Drosophila* splanchnic mesoderm, eventually residing in the region of the cardiac mesoderm. Loss-of-function mutants of *tinman* lack a heart (hence its name, after the Wizard of Oz character) (Bodmer 1993). In mice, the homologous gene is called *Nkx2–5* and it, too, is originally expressed in the splanchnic mesoderm and then continues to be expressed in those cells that form the

Table 22.1 Developmental regulatory genes conserved between protostomes and deuterostomes

Gene	Function	Distribution
achaete-scute group	Cell fate specification	Cnidarians, *Drosophila*, vertebrates
Bcl2/Drob-1/ced9	Programmed cell death	*Drosophila*, nematodes, vertebrates
Caudal	Posterior differentiation	*Drosophila*, vertebrates
delta/Xdelta-1	Primary neurogenesis	*Drosophila*, *Xenopus*
Distal-less/DLX	Appendage formation (proximal-distal axis)	Numerous phyla of protostomes and deuterostomes
Dorsal/NFκB	Immune response	*Drosophila*, vertebrates
forkhead / Fox	Terminal differentiation	*Drosophila*, vertebrates
Fringe/radical fringe	Formation of limb margin (apical ectodermal ridge in vertebrates)	*Drosophila*, chick
Hac-1/Apaf/ced 4	Programmed cell death	*Drosophila*, nematodes, vertebrates
Hox complex	Anterior-posterior patterning	Widespread among metazoans
lin-12/Notch	Cell fate specification	*C. elegans*, *Drosophila*, vertebrates
Otx-1, Otx-2/Otd, Emx-1, Emx-2/ems	Anterior patterning, cephalization	*Drosophila*, vertebrates
Pax6/eyeless; Eyes absent/eya	Anterior CNS/eye regulation	*Drosophila*, vertebrates
Polycomb group	Controls Hox expression/ cell differentiation	*Drosophila*, vertebrates
Netrins, Split proteins, and their receptors	Axon guidance	*Drosophila*, vertebrates
RAS	Signal transduction	*Drosophila*, vertebrates
sine occulus/Six3	Anterior CNS/eye pattern formation	*Drosophila*, vertebrates
sog/chordin, dpp/BMP4	Dorsal-ventral patterning, neurogenesis	*Drosophila*, *Xenopus*
tinman/Nkx 2-5	Heart/blood vascular system	*Drosophila*, mouse
vnd, msh	Neural tube patterning	*Drosophila*, vertebrates

Source: After Erwin 1999.

heart tubes (see Chapter 15; Manak and Scott 1994). Thus, although the heart of vertebrates and the heart of insects have hardly anything in common except their ability to pump fluids, they both appear to be predicated on the expression of the same gene, *Nkx2–5/tinman*. Therefore, it is probable that the PDA had a circulatory system with a pump based on the expression of the *Nkx2–5/tinman* gene.

Another set of genes shared by the deuterostomes and protostomes are those for the transcription factors involved in head formation (Finkelstein and Boncinelli 1994; Hirth and Reichert 1999). In *Drosophila*, the brain is composed of three segments called **neuromeres**. These neuromeres are specified by three transcription factors. The genes encoding these factors are *tailless* (*tll*) and *orthodenticle* (*otd*), which are expressed predominantly in the anteriormost neuromere, and *empty spiracles* (*ems*), which is expressed in the posterior two neuromeres (Monaghan et al. 1995; Hirth et al. 1998). Loss-of-function mutations of *otd* eliminate the anteriormost neuromere of the developing *Drosophila* embryo, and loss-of-function mutations of *ems* eliminate the second and third neuromeres (Hirth et al. 1995). In frogs and mice, the homologues of these genes (*Otx-1, Otx-2, Emx-1, Emx-2*) are also expressed in the brain (Simeone et al. 1992), although the exact patterns of transcription are not identical (Figure 22.2). The *Otx-2* gene has been experimentally knocked out by gene targeting (Acampora et al. 1995; Matsuo et al. 1995; Ang et al. 1996), and the resulting mice have neural and mesodermal head deficiencies anterior to the r3 rhombomere. In humans, mutations of *EMX2* lead to a rare condition known as schizencephaly, in which there are

clefts ripping through the entire cerebral cortex (Brunelli et al. 1996). Even though the *Drosophila otd* and *ems* genes are specified by the Bicoid and Hunchback gradients and the mammalian *Otx* and *Emx* transcripts are induced by the anterior dorsal mesoderm, it appears that the same genes are used for specifying the anterior brain regions.

It appears, then, that the ancestor of all bilaterian organisms had sensory organs based on *Pax6*, a heart based on *tinman*, and a head based on *Otx*, *Ems*, and *tll*. It also had something else: an anterior-posterior polarity based on the expression of Hox genes. The analysis of Hox genes has given us critical clues as to how morphological changes could occur through alterations of development. So we return to our analysis of Hox genes.

Hox Genes: Descent with Modification

As mentioned throughout this book, the expression of Hox genes provides the basis for anterior-posterior axis specification throughout the animal kingdom. This means that the enormous variation of morphological form in the animal kingdom is underlain by a common set of instructions. Indeed,

Figure 22.2
Expression of regulatory transcription factors in *Drosophila* and in vertebrates along the anterior-posterior axis. The *Drosophila* genes *ems*, *tll*, and *otd* are expressed in the anterior regions of the brain, as are the homologous genes of vertebrates. The Hox complex genes are expressed in *Drosophila* and in vertebrates in similar patterns in the hindbrain and spinal cord. (After Hirth and Reichert 1999.)

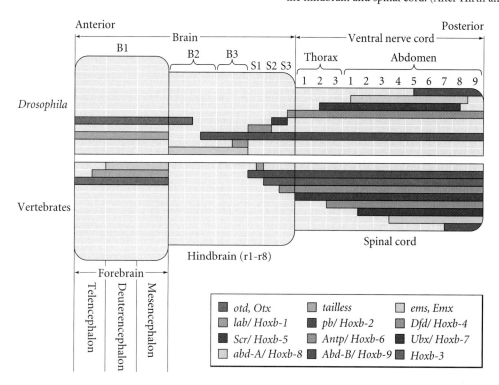

one of the most remarkable pieces of evidence for deep homologies among all the animals of the world is provided by the Hox genes. As mentioned in Chapter 11, not only are the Hox genes themselves homologous, but they are in the same order on their respective chromsomes. The expression patterns are also remarkably similar between the Hox genes of different phyla: the genes at the 3′ end are expressed anteriorly, while those at the 5′ end are expressed more posteriorly (Figure 22.2). As if this evidence of homology were not enough, Malicki and colleagues (1992) demonstrated that the human *HOXB4* gene could mimic the function of its *Drosophila* homologue, *deformed*, when introduced into *Dfd*-deficient *Drosophila* embryos. Slack and his colleagues (1993) postulated that the Hox gene expression pattern defines the development of all animals, and that the pattern of Hox gene expression is constant for all phyla.*

If the underlying Hox gene expression is uniform, how did the differences among the phyla emerge? It is thought that they arose from differences in how the Hox genes are regulated and what target genes the Hox-encoded proteins regulate. Gellon and McGinnis (1998) have catalogued four critical ways in which variation in Hox expression patterns might lead to evolutionary change (Figure 22.3):

- Changes in the Hox protein-responsive elements of downstream genes
- Changes in Hox gene transcription patterns within a portion of the body
- Changes in Hox gene transcription patterns between portions of the body
- Changes in the number of Hox genes

Changes in Hox-responsive elements of downstream genes

One of the most obvious differences between a fruit fly and a butterfly is that the fly has two halteres where the butterfly has a pair of hindwings (Figure 22.4). The expression pattern of the

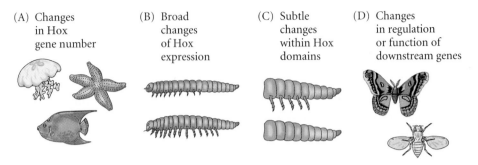

Figure 22.3

Changes in Hox genes correlate with evolutionary changes in animal morphology. (A) Changes in the number of Hox genes correlate with changes at the level of phyla, and increasing the number of Hox genes may allow increased complexity. (B) Broad changes in Hox gene expression can produce different types of structures in body segments that had been identical. (C) Hox gene expression changes within a region may be important in enabling different groups of animals to have particular differences, as in the prolegs of butterfly and moth (lepidopteran) caterpillars that are not seen in fly (dipteran) larvae. (D) Changes in the function or regulation of downstream genes may allow differences such as those distinguishing the haltere of dipterans from the hindwing of lepidopterans. (After Gellon and McGinnis 1998.)

Hox genes, however, does not differ between a butterfly larva and a fly larva. In both cases, the *Ultrabithorax* (*Ubx*) gene is expressed throughout the imaginal discs of the third thoracic segment (from which the hindwing and haltere are derived). What distinguishes a haltere from a wing is the response of the target genes. The Ubx protein downregulates several genes in the *Drosophila* imaginal discs. Many of these same genes are not regulated by Ubx in butterflies. Moreover, some other genes regulated by Ubx in *Drosophila* are regulated differently in butterflies (Carroll et al. 1995; Weatherbee et al. 1999). Thus, the wing differences between dipterans (two-winged insects such as flies) and lepidopterans (butterflies and moths) can be attributed to the different ways in which potential target genes in the imaginal discs respond to the Ubx protein.

Changes in Hox gene transcription patterns within a body portion

In Chapter 16, we saw that changes in Hox gene expression are correlated with the change in morphology from the fish fin to the tetrapod limb. Among arthropods, differences in limb morphology may also be caused by Hox gene expression differences. Insects have six legs as adults, a pair arising from each of the three thoracic segments. In *Drosophila*, the *Distal-less* (*Dll*) gene is critical for providing the proximal-distal axis of the appendages (see Figure 18.15). *Distal-less* expression occurs in the cephalic and thoracic limb-forming discs, but it is excluded in the abdomen by the abdA and Ubx proteins. Thus, the appendages grow into legs and wings in the thorax and into jaws in the head. The *Drosophila* larva never develops limbs in its abdomen.

Butterfly and moth larvae, however, are characterized by rudimentary abdominal legs called **prolegs** (see Figure 22.4).

---

*The reason for this remarkable conservation of structure in the Hox gene complex is thought to be the sharing of *cis*-regulatory regions by neighboring genes. If a Hox gene is moved to a different region within the complex, its regulation is altered. The critical regulatory regimes might be the binding sites for the Polycomb proteins. These proteins are also conserved throughout evolution, and they silence the Hox genes at specific times and places. Here, then, we see a "phyletic constraint" at the molecular level (Chiang et al. 1995; Müller et al. 1995; Kmita et al. 2000).

Figure 22.4
Differences in larval and adult morphology due to Hox gene differences. (A, B) Larva and adult of *Drosophila*, a dipteran. An arrow points to one of the halteres of the adult. The larva lacks prolegs; its anterior end is at the left. (C, D) Larva and adult of *Danaus plexippus*, the monarch butterfly, a lepidopteran. The anterior of the caterpillar is to the left, and a proleg is indicated by an arrow. The adult has hindwings rather than halteres. The regulation of Hox genes determines the presence of prolegs, and the targets of Hox genes determine whether the third thoracic segment is to generate halteres or hindwings. (A courtesy of M. Tyler; B courtesy of E. B. Lewis; C courtesy of G. Savage; D by Bill Beatty/Visuals Unlimited.)

Panganiban and her colleagues (1994) cloned the *Distal-less* homologue from the buckeye (*Precis*) butterfly and mapped its expression during development. During the early portion of *Precis* embryogenesis, *Dll* expression is the same as it is in *Drosophila*. During gastrulation, *Dll* expression is seen first in the head regions and in the thoracic regions that will give rise to the leg imaginal discs. However, as development proceeds, the *Dll* gene of *Precis* becomes expressed in the third through sixth abdominal segments (Figure 22.5). Whereas *Dll* expression is seen in both the proximal ring and the "socks" of the true thoracic legs, the expression of *Dll* in the abdomen is restricted to the proximal ring. Thus, the lepidopteran prolegs appear to be homologous to the proximal portion of the thoracic legs. The expression of *Dll* in the maxilla and labial segments in both *Drosophila* and *Precis* is interesting because it is consistent with recent paleontological evidence (Kukalova-Peck 1992) that, although these jaw structures originated from limb primordia, distal limb elements have been lost from all arthropod jaws.

The presence of larval prolegs and *Dll* expression in the *Precis* abdominal segments suggests that *Dll* is regulated dif-

ferently in dipterans and lepidopterans. Two possibilities come to the fore: first, that the *Dll* genes of *Precis* are not repressed by the abdA and Ubx homeodomain proteins, and second, that the expression of the repressing homeodomain genes is somehow abrogated in the abdominal regions of *Precis*. Warren and co-workers (1994) showed that the *Drosophila* and *Precis* embryos have the same initial pattern of *abdA* and *Ubx* gene expression. However, at about 20% of the way through *Precis* embryogenesis, the expression of these genes becomes downregulated in

Figure 22.5
Distal-less gene expression in the larva of the buckeye butterfly *Precis*. By 40% of the way through embryonic development, *Dll* expression in *Precis* has diverged significantly from that of *Drosophila* in that *Dll* expression is also seen in abdominal segments 3–6. (From Panganiban et al. 1994; courtesy of the authors.)

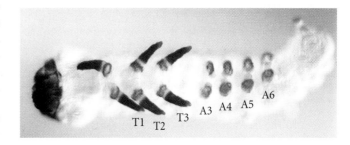

Figure 22.6

"Holes" in the expression of *abdA* and *Ubx* in the abdomen of the larval butterfly *Precis*. The abdA and Ubx proteins have been stained green. The Distal-less protein is stained red, and areas of overlap appear yellow. (A) In the early caterpillar, the thoracic limbs (in segments T1–T3) and jaws are seen to express the Distal-less protein. Some abdominal segments (A3–A6) begin to have holes (indicated by arrows) in the expression domains of abdA and Ubx. (B) In a later-stage caterpillar, these "holes" have become regions of *Distal-less* expression. (From Warren et al. 1994; photographs courtesy of B. Warren, S. Paddock, and S. Carroll.)

(A)

(B)

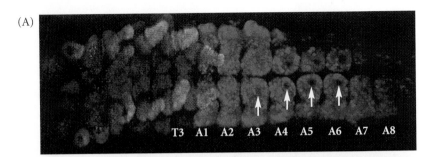

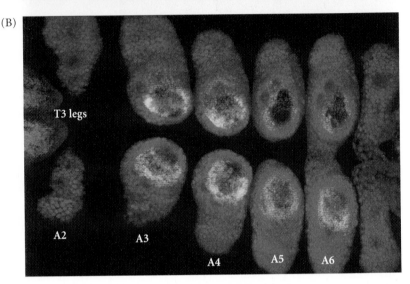

small patches of segments A3–A6, the abdominal segments that give rise to the prolegs (Figure 22.6). Shortly thereafter, the *Dll* and *Antennapedia* (*Antp*) genes are expressed in those "holes." It is not known what molecules are used to downregulate *abdA* and *Ubx* gene expression in the regions of *Dll* expression. The Polycomb group genes are the best suspects, since they can repress both genes in *Drosophila*.

Changes in Hox gene expression between body segments

THE ORIGINS OF MAXILLIPEDS IN CRUSTACEANS. There are substantial differences in Hox gene expression patterns between insects and crustaceans, and there are also significant differences in Hox gene expression patterns within the different crustacean groups. Crustaceans are characterized by a pregnathal head (similar to the insect acron), gnathal (jawed) head segments, six thoracic segments, genital segments, abdominal segments, and a telson (Figure 22.7). Each of the thoracic segments of the crustacean expresses *Antp*, *Ubx*, and

Figure 22.7

Hox gene expression and morphological change in crustaceans. At the bottom are the structures of a brine shrimp (crustacean) and a grasshopper (insect). The domains of Hox gene expression specifying the various structures are coded in color. Whereas the Hox gene expression domains segregate in the insect thorax and abdomen, they coincide in the crustacean thorax. Above them is a hypothetical model for the divergence of insect and crustacean body plans from a common ancestor. The *Antennapedia*, *Ultrabithorax*, and *abdominal A* genes are very similar and are thought to have emerged by gene duplication from a single gene in a distant ancestor of the arthropods. *Abdominal B* is expressed in the segments destined to become genitalia. Paleontological evidence suggests that the ancestral arthropod had identical thoracic segments, similar to those of extant crustaceans. In both cases, *Antennapedia*, *Ultrabithorax*, and *abdominal A* expression is seen throughout the thorax. The crustaceans later evolved tail segments. Insects, however, diversified their thoracic segments and used the Hox genes to specify the segments differently. The ancestors of the insects and crustaceans are hypothetical reconstructions based on fossils of the middle Cambrian. (After Averof and Akam 1995 and Manton 1977.)

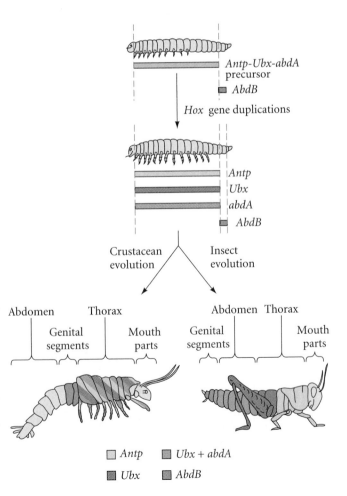

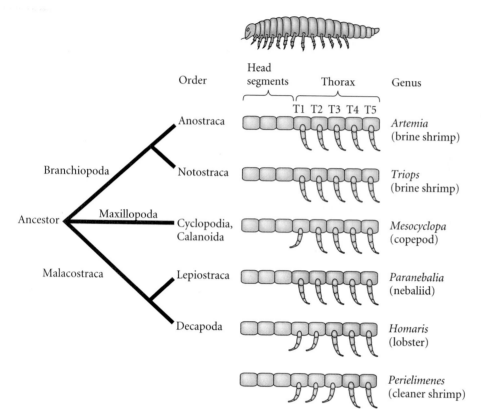

Figure 22.8
Schematic representation of the expression of *Ubx* and *abdA* (green) in the thoracic segments of different types of crustaceans. The generation of maxillipeds occurs in the thoracic segments that do not express either of these homeodomain proteins. (After Averof and Patel 1997.)

abdA, and these genes appear to be interchangeable in the crustaceans. Indeed, the thoracic segments all look alike, and there is no specialization between these segments. In the arthropod lineage that gave rise to the insects, each gene would take on different (but sometimes overlapping) functions (Averof and Akam 1995).

But within the crustacean lineages, there are interesting variations on this theme. Averof and Patel (1997) have shown that if a thoracic segment does not express *Ubx* and *abdA*, it converts its anterior locomotor limb into a feeding appendage called a **maxilliped**. Thus, brine shrimp such as *Artemia* have a uniform expression of *Ubx* and *abdA* in their thoracic segments, and they lack maxillipeds. Lobsters such as *Homaris* lack *Ubx* and *abdA* expression in their first and second thoracic segments, and these segments have paired maxillipeds (Figure 22.8). The fossil record suggests that the earliest crustaceans lacked maxillipeds and had uniform thoracic segments. This would mean that the presence of maxillipeds is a derived characteristic that evolved in several crustacean lineages.

WHY SNAKES DON'T HAVE LEGS. As shown in Chapter 11, the expression patterns of Hox genes in vertebrates determines the type of vertebral structure formed. Thoracic vertebrae, for instance, have ribs, while cervical (neck) vertebrae and lumbar vertebrae do not. The type of vertebra produced is specified by the Hox genes expressed in the somite.

One of the most radical alterations of the vertebrate body plan is seen in the snakes. Snakes evolved from lizards, and they appear to have lost their legs in a two-step process. Both paleontological and embryological evidence supports the view that snakes first lost their forelimbs and later lost their hindlimbs (Caldwell and Lee 1997; Graham and McGonnell 1999). Fossil snakes with hindlimbs, but no forelimbs, have been found. Moreover, while the most derived snakes (such as vipers) are completely limbless, more primitive snakes (such as boas and pythons) have pelvic girdles and rudimentary femurs.

The missing forelimbs can be explained by the Hox expression pattern in the anterior portion of the snake. In most vertebrates, the forelimb forms just anterior to the most anterior expression domain of *Hoxc-6* (Gaunt 1994; Burke et al. 1995). Caudal to that point, *Hoxc-6*, in combination with *Hoxc-8*, helps specify vertebrae to be thoracic. During early python development, *Hoxc-6* is not expressed in the absence of *Hoxc-8*, so the forelimbs do not form. Rather, the combination of *Hoxc-6* and *Hoxc-8* is expressed for most of the length of the organism, telling the vertebrae to form ribs throughout most of the body (Figure 22.9; Cohn and Tickle 1999).

The hindlimb buds do begin to form in pythons, but they do not make anything more than a femur. This appears to be due to the lack of *sonic hedgehog* expression by the limb bud mesenchyme. Sonic hedgehog is needed both for the polarity of the limb and for maintenance of the apical ectodermal ridge (AER). Python hindlimb buds lack the AER. Interestingly, the phenotype of the python hindlimb resembles that of mouse embryos with loss-of-function mutations of *sonic hedgehog* (Chiang et al. 1996).

(A)

(B)

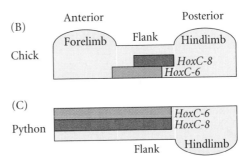

(C)

Figure 22.9
Loss of limbs in snakes. (A) Skeleton of the garter snake, *Thamnophis*, stained with alcian blue. Ribbed vertebrae are seen from the head to the tail. (B, C) Hox expression patterns in chick (B) and python (C). (Photograph courtesy of A. C. Burke; B, C after Cohn and Tickle 1999.)

HOX GENES AND ATAVISMS. As we have seen, disruptions of the Hox genes can change one type of vertebra into another type. In some instances, the mutation of a Hox gene can produce "atavistic" conditions, wherein the organism resembles an earlier evolutionary stage. Deletion or misregulation of the *Hoxa-2* genes in mice, for instance, results in a partial transformation of the second pharyngeal arch into a copy of the first pharyngeal arch. The mutant fetuses lack the stapes and styloid bones formed from the second arch, but have extra malleus, incus, tympanic, and squamosal bones. They also possess a rodlike cartilage that has no counterpart in normal mice, but looks like the pterygoquadrate cartilage thought to have been present in therapsids, the reptilian ancestors that gave rise to the mammals (Figure 22.10; Rijli et al. 1993; Lohnes et al. 1994; Mark et al. 1995). Thus, major evolutionary changes can be correlated with the alteration of Hox gene expression in different parts of the embryo.

Changes in Hox gene number

The number of Hox genes may play a role in permitting the evolution of complex structures. All invertebrates have a single Hox complex per haploid genome. In the most simple invertebrates—such as sponges—there appear to be only one or two Hox genes in this complex (Degnan et al. 1995; Schierwater and Kuhn 1998). In the more complex invertebrates, such as insects, there are numerous Hox genes in this complex. Comparing the Hox genes of chordates, arthropods, and molluscs suggests that there was a common set of seven Hox genes in the Urbilaterian ancestor of the protostomes and deuterostomes. Indeed, in invertebrate deuterostomes (echinoderms and amphioxus, an invertebrate chordate), there is only one Hox complex, which looks very much like that of the insects (Figure 22.11; Holland and Garcia-Fernández 1996).

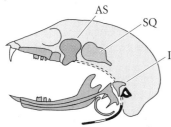

Wild-type mouse/mammal

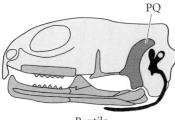

Reptile

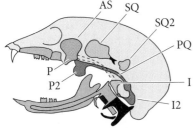

HoxA-2 null mutant

Figure 22.10
Representation of skeletal elements derived from the first pharyngeal arch (in gray) and the second pharyngeal arch (in black). (AS, alisphenoid; I, incus; I2, duplicated incus; P and P2, normal and duplicated pteroid cartilage; PQ, pterygoquadrate cartilage; SQ, squamosal; SQ2, duplicated squamosal.) (After Mark et al. 1995.)

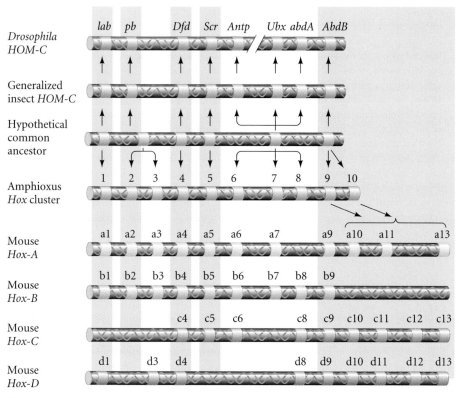

Figure 22.11
Postulated ancestry of the homeotic genes from a hypothetical ancestor of both deuterostomes and prototomes. Amphioxus has only one cluster, similar to that of insects. Vertebrates have four clusters, none of which is complete. (After Holland and Garcia-Fernández 1996.)

By the time the earliest vertebrates (agnathan fishes) evolved, there were at least four Hox complexes. The transition from amphioxus to early fish is believed to be one of the major leaps in complexity during evolution (Amores et al. 1998; Holland 1998). This transition involved the evolution of the head, the neural crest, new cell types (such as osteoblasts and odontoblasts), the brain, and the spinal cord. As we saw in Chapter 11, the regionalization of the brain and spinal cord is dependent upon Hox genes, and the regional specification of the somitic segments depends upon the paralogous members of the different Hox clusters. For instance, deletions of *Hoxa-3* (from the A cluster) affect the neural crest-derived *glands* of the neck; deletions of *Hoxd-3* (its paralogue from the D cluster) affect the somite-derived *skeleton* of the neck. This distinction may be due to the different levels of expression of these genes within the same tissues (Greer et al. 2000). Holland (1998) speculates that the generation of these new structures was allowed by the fourfold duplication of the Hox gene complex.

Assessing Homologies through Regulatory Gene Expression Patterns

What constitutes a spider's head? Once a typical pattern of expression has been determined for genes such as those of the Hox complex, one can attempt to derive the evolution of anomalous structures by looking at their patterns of gene expression. For instance, chelicerates (spiders and mites) are very derived arthropods. They do not have a distinct head, but rather a **cephalothorax**, and it is difficult to see the segmentation in their brains. While their heads and brains may look very different from those of insects, the expression pattern of Hox genes in the anterior of the spider embryo is very similar to that in the head of an insect embryo. This pattern demonstrates that spiders have the same head segments as insects and therefore supports the view that all arthropods (of which chelicerates and insects are members) have a common origin (Damen et al. 1998; Telford and Thomas 1998).

How does a new cell type form?
As mentioned earlier in the chapter, the neural crest cells were important in the origin of chordates. While we do not know how neural crest cells arose, Holland and colleagues (1996) have provided a fascinating speculation that involves the duplication and divergence of new genes. It also involves the vertebrate homologues of the *Drosophila* gene *Distal-less*. *Distal-less* is found throughout the animal kingdom, and it is expressed in those tissues that stick out from the body axis, notably limbs and antennae (Panganiban et al. 1997). But in vertebrates, *Distal-less* has acquired new functions. Amphioxus is an invertebrate chordate that has a notochord, somites, and a hollow neural tube. It lacks a brain and facial struc-

tures, and most importantly, it lacks neural crest cells. Like *Drosophila*, amphioxus has but one copy of the *Distal-less* gene per haploid genome, and as in *Drosophila*, this gene is expressed in the epidermis and central nervous system. However, whereas amphioxus has only one copy of this gene, vertebrates have five or six closely related copies of *Distal-less*, all of which probably originated from a single ancestral gene that resembles the one in amphioxus (Price 1993; Boncinelli 1994). These *Distal-less* homologues have found new functions. Some are expressed in the mesoderm, a place where *Distal-less* is not expressed in amphioxus. Other vertebrate *Distal-less* homologues are expressed in the forebrain, mimicking an expression pattern seen in the anterior of the amphioxus neural tube. These findings suggest that the vertebrate forebrain is homologous to the anterior neural tube of amphioxus. At least three of these vertebrate *Distal-less* genes function in the patterning of the neural crest cells, and deletions of these genes cause the absence or malformation of the branchial arches, face, jaws, teeth, and vestibular apparatus (Qiu et al. 1997; DePew et al. 1999). Although it remains unproved, it is possible that the new type of *Distal-less* gene could have caused the migratory ectodermal cells of amphioxus to evolve into neural crest cells.

Is the endostyle the precursor of the thyroid gland?

Similarly, there has been disagreement as to whether the endostyle of amphioxus is homologous to the thyroid gland of vertebrates. Both organs accumulate iodine, although the endocrine function of the endostyle has not been demonstrated. Recent studies (Holland and Holland 1999) have shown that in addition to its structural and functional similarities to the vertebrate thyroid, the endostyle also expresses two transcription factors that are used to specify the vertebrate thyroid. Therefore, the case for homology between these two organs is strengthened. ∎

Homologous Pathways of Development

One of the most exciting findings of the past decade has been the existence not only of homologous regulatory genes, but also of homologous signal transduction pathways (Zuckerkandl 1994; Gilbert 1996; Gilbert et al. 1996). Many of those pathways have been mentioned earlier in this book. They are composed of homologous proteins arranged in a homologous manner. In this respect, the homology is similar to that of a human forearm and a seal flipper. The parts—the proteins—are homologous, and the structures they make up–the pathways—are homologous.

Homologous pathways form the basic infrastructure of development. The targets of these pathways may differ, however, among organisms. For example, the Dorsal-Cactus pathway used in *Drosophila* for specifying dorsal-ventral polarity is also used by the mammalian immune system to activate inflammatory proteins (see Figure 9.38). This does not mean that the *Drosophila* blastoderm is homologous to the human macrophage. It merely means that there is a very ancient pathway that predates the deuterostome-protostome split, and that this pathway can be used in different systems. The pathways (one in *Drosophila*, one in humans) are homologous; the organs they form are not.

Another ancient pathway is the RTK pathway (see Figure 6.14). In *Drosophila*, the determination of photoreceptor 7 is accomplished when the Sevenless protein (on the presumptive photoreceptor 7) binds to the Bride of sevenless (Boss) protein on photoreceptor 8. This interaction activates the receptor tyrosine kinase of the Sevenless protein to phosphorylate itself. The Drk protein then binds to these newly phosphorylated tyrosines through its Src-homology-2 (SH2) region and activates the Son of sevenless (SOS) protein. This protein is a guanosine nucleotide exchanger and exchanges GDP for GTP on the Ras1 G protein. This activates the G protein, enabling it to transmit its signal to the nucleus through the MAP kinase cascade (Figure 22.12). This same system has been found to be involved in the determination of the nematode vulva, the mammalian epidermis, and the *Drosophila* terminal segments. The similarity in these systems is so striking that many of the components are interchangeable between species. The gene for human GRB2 can correct the phenotypic defects of *sem-5*-deficient nematodes, and the nematode SEM-5 protein can bind to the phosphorylated form of the human EGF receptor (Stern et al. 1993). Thus, in the ectoderm of one organism, the RTK pathway may activate the genes responsible for proliferation. But in another organism, the same pathway may activate the genes responsible for making a photoreceptor. And in a third organism, the pathway activates the genes needed to construct a vulva.

Pathways undergo descent with modification, too. This is readily seen in the Wnt pathway that we have discussed throughout the book. Figure 22.13 shows how the Wnt pathway is used in several different organisms. The pathways are homologous, but not identical. They are thought to have originated in a common ancestral pathway that predated the deuterostome-protostome split.

Earlier in this chapter, we discussed genes such as *Pax6* and *tinman* that appear to have had their developmental functions before the protostome-deuterostome split. We have also discussed homologous pathways that may or may not be used in similar structures. However, there appear to be some pathways that are used to form the same structure in all animals. When homologous pathways made of homologous parts are used for the same function in both protostomes and deuterostomes, they are said to have **deep homology** (Shubin et al. 1997).

Instructions for forming the central nervous system

One example of deep homology has already been discussed in earlier chapters. First, as seen in Chapter 10, the chordin/BMP4 pathway demonstrates that in both vertebrates and invertebrates, chordin/Short-gastrulation (Sog) inhibits the lateralizing effects of BMP4/Decapentaplegic (Dpp), thereby allowing

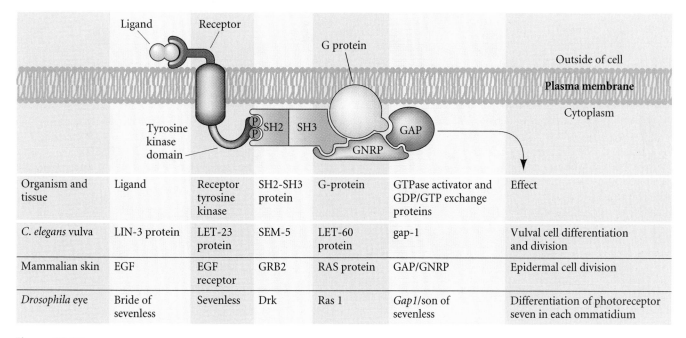

Organism and tissue	Ligand	Receptor tyrosine kinase	SH2-SH3 protein	G-protein	GTPase activator and GDP/GTP exchange proteins	Effect
C. elegans vulva	LIN-3 protein	LET-23 protein	SEM-5	LET-60 protein	gap-1	Vulval cell differentiation and division
Mammalian skin	EGF	EGF receptor	GRB2	RAS protein	GAP/GNRP	Epidermal cell division
Drosophila eye	Bride of sevenless	Sevenless	Drk	Ras 1	*Gap1*/son of sevenless	Differentiation of photoreceptor seven in each ommatidium

Figure 22.12
The widely used RTK pathway. The outline of the pathway is shown below the diagram, along with the names of its elements in different species. The ligand can be a soluble protein (as in EGF) or a membrane-bound protein on another cell (as in the Bride of sevenless protein presented to the Sevenless RTK). The cytoplasmic domains of the RTKs are autophosphorylated once they are dimerized, and this allows them to bind the adaptor protein and to stimulate the Ras G protein. The activity of the Ras G protein can be enhanced by GTPase activation or inhibited by the GAP proteins. The activated G protein initiates a cascade of phosphorylation that ends in a phosphorylated (activated) transcription factor entering into the nucleus and effecting RNA transcription.

the ectoderm protected by chordin/Sog to become the neurogenic ectoderm. These reactions are so similar that *Drosophila* Dpp protein can induce ventral fates in *Xenopus* and can substitute for the Sog protein (Holley et al. 1995).

In addition to this central inhibitory reaction of chordin/Sog inhibiting BMP4/Dpp, there are other reactions that add to the deep homology of the instructions for forming the protostome and deuterostome neural tube. For instance, the spread of Dpp in *Drosophila* is aided by Tolloid, a metalloprotease that degrades Sog. The gradient of Dpp concentration from dorsal to ventral is created by the opposing actions of

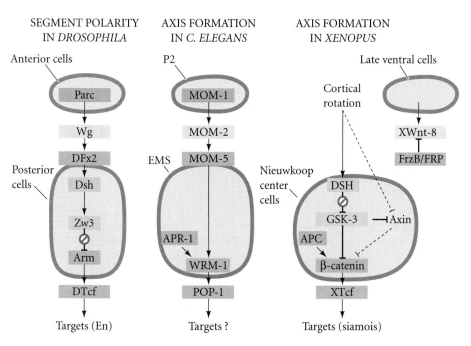

Figure 22.13
Three modifications of the Wnt pathway. Each pathway proceeds vertically. The cells secreting the Wingless protein are labeled (anterior cells in the *Drosophila* parasegment, the P2 cell in *C. elegans*). The responding cells are shown beneath them (posterior cells in the *Drosophila* parasegment, the EMS cell of *C. elegans*, and the Nieuwkoop center cells of *Xenopus*. In *Xenopus*, the Disheveled protein is thought to be brought to the Nieuwkoop center by the cortical rotation of the cytoplasm. The proteins on each level are homologous to one another. (After Cadigan and Nusse 1997.)

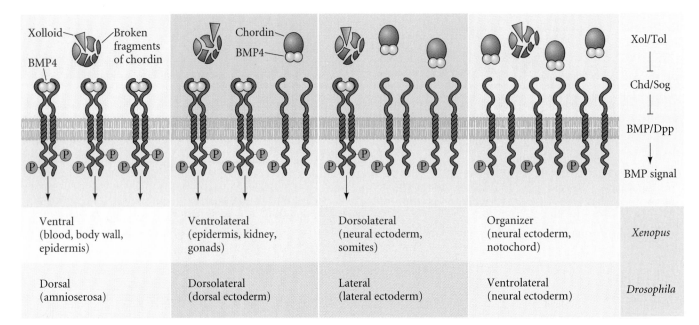

Figure 22.14
Homologous pathways specifying neural ectoderm in protostomes (*Drosophila*) and deuterostomes (*Xenopus*). Both pathways involve a source of chordin/Sog (the organizer in *Xenopus*, the presumptive neural ectoderm in *Drosophila*) and a source of BMP4/Dpp (the ventral mesoderm infrogs, the presumptive amnioserosa in flies). In both instances, the gradient is shaped by a constant supply of Tolloid/Xolloid, which degrades Sog/chordin. In both cases, the neural ectoderm forms where BMP signaling is prevented. (After Dale and Wardle 1999.)

Tolloid (increasing Dpp) and Sog (decreasing it) (Figure 22.14; Marqués et al. 1997). In *Xenopus* and zebrafish, the homologues of Tolloid—Xolloid and BMP1, respectively—have the same function. They degrade chordin. The gradient of BMP4 from ventral to dorsal is established by the antagonistic interactions of Xolloid or BMP1 (increasing BMP4) and chordin (decreasing BMP4) (Blader et al. 1997; Piccolo et al. 1997). Thus, it appears that nature may have figured out how to make a nervous system only once. The protostome and deuterostome nervous systems, despite their obvious differences, seem to be formed by the same set of instructions.

Limb formation

It is also possible that nature has only one set of instructions for forming limbs (Shubin et al. 1997). Nothing could be a better example of analogy than vertebrate and insect legs. Fly limbs and vertebrate limbs have little in common except their function. They have no structural similarities. Insect legs are made of chitin and have no inner skeleton. They are formed by the telescoping out of ectodermal imaginal discs (see Chapter 18). Vertebrate limbs, on the other hand (no pun intended), have no chitin, but possess a bony endoskeleton. These limbs are created by the interaction of ectoderm and mesoderm (see Chapter16). However, the genetic instructions to form these two distinctly different types of limbs are extremely similar.

As we have seen, Sonic hedgehog is usually expressed in the posterior part of the vertebrate limb bud. If it is expressed in the anterior part of the bud, mirror-image duplications arise (see Figure 16.19; Riddle et al. 1993). In the *Drosophila* wing or leg imaginal disc, the Hedgehog protein is expressed in the posterior portion of the disc. If it is expressed anteriorly, mirror-image duplications of the wing will form (Figure 22.15; Basler and Struhl 1993; Ingham 1994). Furthermore, certain genes regulated by the Hedgehog proteins have been conserved as well (Marigo et al. 1996). Thus, the anterior-posterior axis appears to be specified in the same way in vertebrate and in insect limbs. The dorsal-ventral axis also appears to be specified similarly. The ventral limb compartments of both insects and vertebrates appear to be specified by the expression of the *engrailed* gene (Davis et al. 1991; Loomis et al. 1996), while the dorsal compartment is defined by *apterous* (in insects) or its relative *Lmx1* (in vertebrates) (Figure 22.16).

The formation of the proximal-distal axis also proceeds similarly in vertebrates and invertebrates. In insects, the wing margin forms at the border between the dorsal cells and the ventral cells. The dorsal cells express Apterous protein, and this activates the expression of Fringe. The interaction of the Fringe protein with the ventral cells leads to the growth of the wing blade outward from the body wall. Unexpectedly, a very similar cascade induces the outgrowth of the vertebrate limb. The apical ectodermal ridge (AER) forms at the junction of the dorsal cells with the ventral cells. The dorsal cells express Radical fringe, a vertebrate homologue of the Fringe protein. This protein is critical in forming the AER. It is interesting that the ways in which these proteins become localized differ

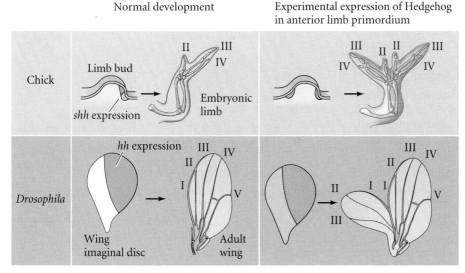

Normal development   Experimental expression of Hedgehog in anterior limb primordium

Figure 22.15
Homology of process in the formation of the anterior-posterior axes in *Drosophila* and chick appendages. A chick limb bud expresses Sonic hedgehog in its posterior region. If Sonic hedgehog is also expressed in an anterior region, the limb develops a mirror-image duplication of the anterior-posterior axis. A *Drosophila* wing disc expresses Hedgehog in its posterior compartment. If Hedgehog is expressed in the anterior compartment as well, the wing develops a mirror-image duplication of the anterior-posterior axis. (After Ingham 1994.)

Figure 22.16
Deep homology of the limbs. The same set of proteins is used to establish the polarity of limbs in both deuterostomes (chick) and protostomes (*Drosophila*). The top panels represent chick limb buds with the dorsal region on top and the apical ectodermal ridge facing the viewer. The bottom panels represent the *Drosophila* wing disc with its dorsal region upward and its anterior side to the left. (A) Proximal-distal axes are specified by the Distal-less protein in the most distal region of the limb bud or disk. This protein forms at the junction where the Fringe-containing dorsal cells meet the ventral cells. (B) Dorsal-ventral patterning is specified by the expression of a LIM protein, either Apterous (*Drosophila*) or Lmx1 (chick), in the dorsal portion of the disk or bud. A Wnt protein (Wingless in *Drosophila*, Wnt7A in the chick) induces this expression. (C) Anterior-posterior patterning is accomplished by the expression of Hedgehog in the posterior of the disk or bud. Hedgehog, in turn, activates a BMP that can relay a signal to other cells.

enormously between these phyla. Radical fringe, for instance, is induced by fibroblast growth factors, and it induces the AER to produce more fibroblast growth factors for the outgrowth of the limb. Fibroblast growth factors play no role in insect limb development. Yet, the instructions for limb outgrowth and polarity appear to be essentially the same in insects and vertebrates. Nature may have evolved the mechanism to form a limb only once, in the PDA, and both arthropods and vertebrates use modifications that process to this day.

Modularity: The Prerequisite for Evolution through Development

How can the development of an embryo change when development is so finely tuned and complex? How can such change occur without destroying the entire organism? It was once thought that the only way to promote evolution was to add a step to the end of embryonic development, but we now know that even early stages can be altered to produce evolutionary novelties. The reason why changes in development can occur is that the embryo, like the adult organism, is composed of modules (Riedl 1978; Bonner 1988).

(A) Proximal-distal patterning: Distal-less in most distal region

(B) Dorsal-ventral patterning: Lim protein in dorsal region specified by Wnt protein

(C) Anterior-posterior patterning: Hedgehog in posterior induces BMP to signal

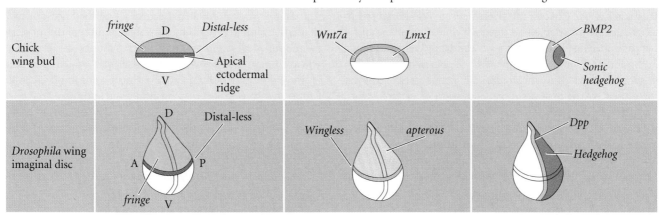

Development occurs through a series of discrete and interacting **modules** (Riedl 1978; Gilbert et al. 1996; Raff 1996; Wagner 1996). Organisms are constructed of units that are coherent within themselves and yet part of a larger unit. Thus, cells are parts of tissues, which are parts of organs, which are parts of systems, and so on. Such a hierarchically nested system has been called a level-interactive modular array (Dyke 1988). In development, such modules include morphogenetic fields (for example, those described for the limb or eye), pathways (such as those mentioned above), imaginal discs, cell lineages (such as the inner cell mass or trophoblast), insect parasegments, and vertebrate organ rudiments. Modular units allow certain parts of the body to change without interfering with the functions of other parts.

The fundamental principle of modularity allows three processes to alter development: dissociation, duplication and divergence, and co-option (Raff 1996). Since modules are found on all levels, from molecular to organismal, it is not surprising that one sees these principles operating at all levels of development.

WEBSITE **22.6 Modularity as a principle of evolution.** Complex structures are created by the assortment of pre-existing modules. It is silly to consider a protein as a collection of atoms. It is an ordered assembly of amino acids that have already formed from atoms. Modularity allows evolution to occur by forming components that can be individually modified.

Dissociation: Heterochrony and allometry

Not all parts of the embryo are connected to one another. One can dissect out the limb field of a salamander neurula, for example, and the eyes are not affected. By means of mutation or environmental perturbation, one part of the embryo can change without the other parts changing. This modularity of development can allow changes that are either spatial or temporal.

Heterochrony is a shift in the relative timing of two developmental processes from one generation to the next. In other words, one module can change its time of expression relative to the other modules of the embryo. We have come across this concept in our discussion of neoteny and progenesis in salamanders (see Chapter 18). Heterochrony can be caused in different ways. In salamander heterochronies in which the larval stage is retained, heterochrony is caused by gene mutations in the ability to induce or respond to the hormones initiating metamorphosis. Other heterochronic phenotypes, however, are caused by the heterochronic expression of certain genes. The direct development of some sea urchins involves the early activation of adult genes and the suppression of larval gene expression (Raff and Wray 1989). Thus, heterochrony can "return" an organism to a larval state, free from the specialized adaptations of the adult. Heterochrony can also give larval characteristics to an adult organism, as in the small size and webbed feet of arboreal salamanders (Figure 22.17) or the fetal growth rate of human newborn brain tissue (see Chapter 12).

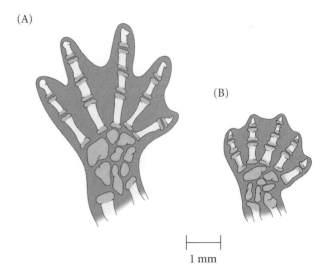

(A)

(B)

Figure 22.17
Salamander heterochrony. The progenesis of limb development in *Bolitoglossa* can create a tree-climbing salamander. (A) Foot of an adult *B. rostratus*, a terrestrial salamander. (B) Foot of an adult *B. occidentalis*, an arboreal salamander. The feet, skull, and body size of *B. occidentalis* adults resemble those of juvenile *B. rostratus*. Since the digits have not expanded past the webbing, the feet can produce suction to climb trees. (After Alberch and Alberch 1981.)

WEBSITE **22.7 Heterochrony in evolution.** Heterochrony is an important means of dissociating the development of one portion of the body from another. It has been seen to play critical roles in the evolution of direct-developing sea urchins, arboreal salamanders, and furless apes.

Another consequence of modularity is **allometry**. Allometry occurs when different parts of an organism grow at different rates (see Chapter 1). Allometry can be very important in forming variant body plans within a phylum. Such differential growth changes can involve altering a target cell's sensitivity to growth factors or altering the amounts of growth factors produced. Again, the vertebrate limb can provide a useful illustration. Local differences among chondrocytes cause the central toe of the horse to grow at a rate 1.4 times that of the lateral toes (Wolpert 1983). This means that as the horse grew larger during evolution, this regional difference caused the five-toed horse to become a one-toed horse. A particularly dramatic example of allometry in evolution comes from skull development. In the very young (4- to 5-mm) whale embryo, the nose is in the usual mammalian position. However, the enormous growth of the maxilla and premaxilla (upper jaw) pushes over the frontal bone and forces the nose to the top of the skull (Figure 22.18). This new position of the nose (blowhole) allows the whale to have a large and highly specialized jaw apparatus and to breathe while parallel to the water's surface (Slijper 1962).

Allometry can also generate evolutionary novelty by small, incremental changes that eventually cross some developmental threshold (sometimes called a bifurcation point). A change in quantity eventually becomes a change in quality

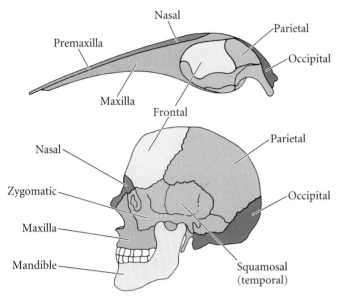

Figure 22.18
Allometric growth in the whale head. An adult human skull is shown for comparison. The whale's upper jaw (maxilla) has pushed forward, causing the nose to move to the top of the skull. (The premaxilla is present in the early human fetus, but it fuses with the maxilla by the end of the third month of gestation. The human premaxilla was discovered by Goethe, among others, in 1786.) (After Slijper 1962.)

when such a threshold is crossed. It has been postulated that this type of mechanism produced the external fur-lined "neck" pouches of pocket gophers and kangaroo rats that live in deserts. External pouches differ from internal ones in that they are fur-lined and they have no internal connection to the mouth. They allow these animals to store seeds without running the risk of desiccation. Brylski and Hall (1988) have dissected the heads of pocket gopher and kangaroo rat embryos and have looked at the way the external cheek pouch is constructed. When data from these animals were compared with data from animals that form internal cheek pouches (such as hamsters), the investigators found that the pouches are formed in very similar manners. In both cases, the pouches are formed within the embryonic cheeks by outpocketings of the cheek (buccal) epithelium into the facial mesenchyme (Figure 22.19). In animals with internal cheek pouches, these evaginations stay within the cheek. However, in animals that form external pouches, the elongation of the snout draws up the outpocketings into the region of the lip. As the lip epithelium rolls out of the oral cavity, so do the outpockets. What had been internal becomes external. The fur lining is probably derived from the external pouches' coming into contact with dermal mesenchyme, which can induce hair to form in epithelia (see Chapter 12). Such a pouch has no internal opening to the mouth. Indeed, the transition from internal to external pouch is one of threshold. The placement of the evaginations anteriorly or posteriorly determines whether the pouch is internal or not. There is no "transitional stage" hav-

ing two openings, one internal and one external.* One could envision this externalization occurring by a chance mutation or set of alleles that shifted the outpocketing to a slightly more anterior location. Such a trait would be selected for in desert environments, where dessication is a constant risk. As Van Valen reflected in 1976, evolution can be defined as "the control of development by ecology."

Duplication and divergence

Modularity also allows duplication and divergence. The duplication part of this process allows the formation of redundant structures, and the divergence part allows these structures to assume new roles. One of the copies can maintain the

---

*The lack of such transitional forms is often cited by creationists as evidence against evolution. For instance, in the transition from reptiles to mammals, three of the bones of the reptilian jaw became the incus and malleus, leaving only one bone (the dentary) in the lower jaw (see Chapter 1 and below). Gish (1973), a creationist, says that this is an impossible situation, since no fossil has been discovered showing two or three jaw bones and two or three ear ossicles. Such an animal, he claims, would have dragged its jaw on the ground. However, such a specific transitional form (and there are over a dozen documented transitional forms between reptilian and mammalian skulls) need never have existed. Hopson (1966) has shown on embryological grounds how the bones of the jaw could have divided and been used for different functions, and Romer (1970) has found reptilian fossils wherein the new jaw articulation was already functional while the older bones were becoming useless. There are several species of therapsid reptiles that had two jaw articulations, with the stapes brought into close proximity with the upper portion of the quadrate bone (which would become the incus).

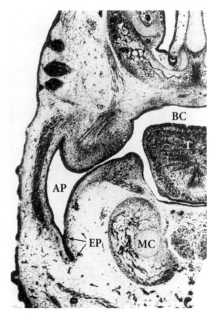

Figure 22.19
Transverse section through the anterior region of a pocket gopher (*Thomomys*) embryo, showing the anterior opening of the pouch (AP) and the continuity at this stage between the pouch and the buccal cavity (BC) across the developing lip area. (EP, epithelial cells; MC, Meckel's cartilage; T, tongue.) (From Brylski and Hall 1988; photograph courtesy of the authors.)

Figure 22.20
Recruitment of pathways in the formation of butterfly eyespots. (A) The interaction between Hedgehog (green) and Engrailed (blue) is first seen in the insect blastoderm, and is then modified during the creation of a new structure, the wing. In the wing disc, Hedgehog in the posterior compartment induces Engrailed expression in the cells immediately anterior to the anterior/posterior boundary. In the butterfly wing spot, the same pathway is utilized to create the eyespot. In the posterior of the wing, Hedgehog is locally downregulated so that re-expression can induce Engrailed. (B, C) In *Bicyclus anynana* (see Figure 21.8), the signal to produce the eyespot also activates several other genes that had been used for wing formation. The white central portion of the eyespot expresses Engrailed, Spalt, and Distal-less. The cells that will become the black region surrounding the center express Spalt and Distal-less. The peripheral ring of cells that will generate the orange pigment expresses only Engrailed. (After Keys et al. 1999 and Pennisi 2000; photograph courtesy of S. B. Carroll and S. Paddock.)

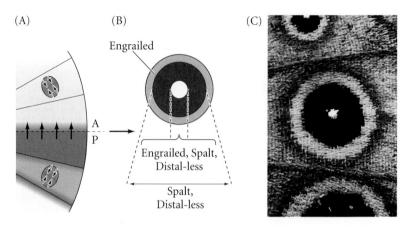

for an entirely new function. A structure originally used for walking has been recruited into a structure suitable for flying.

A famous case of co-option is the use of embryonic jaw parts in the creation of the mammalian middle ear, as explicated in Chapter 1 (see Figure 1.14; Gould 1990). First, the gill

original role while the others are free to mutate and diverge functionally. This can happen at numerous levels. The Hox genes, TGF-β family genes, MyoD family genes, and globin genes each probably started as a single gene that duplicated several times. After the duplication, mutations caused the divergences that gave the members of each family new functions. At the tissue level, one sees duplication and divergence in the somites that give rise to the cervical, thoracic, and lumbar vertebrae.

Co-option

No one structure is destined for any particular purpose. A pencil can be used for writing, but it can also be used as a toothpick, a dagger, a hole-puncher, or a drumstick. On the molecular level, the gene *engrailed* is used for segmentation in the *Drosophila* embryo, is used later to specify its neurons, and is used in the larval stages to provide an anterior-posterior axis to imaginal discs. Similarly, a protein that functions as an enolase or alcohol dehydrogenase enzyme in the liver can function as a structural crystallin protein in the lens (Piatigorsky and Wistow 1991). In other words, preexisting units can be co-opted (recruited) for new functions. Sometimes, whole pathways are co-opted from one system to another. For instance, the pathway by which Hedgehog protein induces Engrailed protein to pattern and extend the insect wing is later used within the wing blades to make the eyespots of butterflies and moths. Distal-less, another protein used to extend the wing imaginal disc, is later used to form the center of such eyespots (Figure 22.20). Co-option can also be seen on the morphological level. Wings have evolved three times during vertebrate evolution, and in each case, different forearm structures were modified

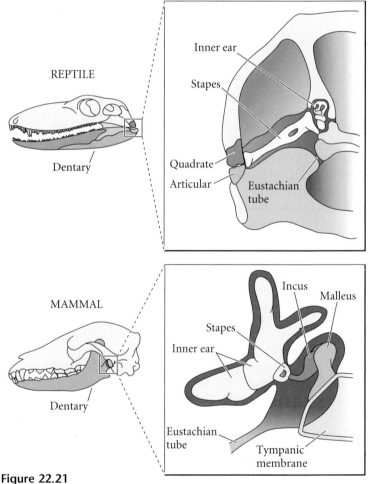

Figure 22.21
Evolution of the mammalian middle ear bones from the reptilian jaw. The quadrate and articular bones of reptiles were part of the lower jaw. Sound could be transmitted from these bones via the large stapes. When the dentary bone grew and took over the jaw functions of these two bones, the articular bone became the malleus and the quadrate bone became the incus. (After Romer 1949.)

(A)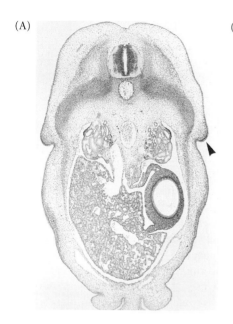

(B)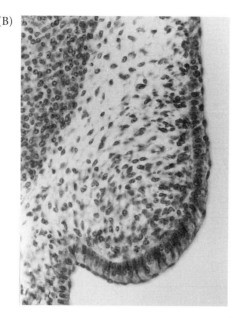

Figure 22.22
Midtrunk cross section through the embryo of the turtle *Chelydra serpentina*. (A) The carapacial ridge (arrowhead) is formed at the boundary of the somitic and lateral plate mesoderm and now represents the dorsal-ventral boundary. The thickened mesodermal bands extending from the center into the carapace area are the rib condensations. (B) Higher magnification of the carapacial ridge. (From Burke 1989a; photographs courtesy of A. C. Burke.)

arches of jawless fishes became the jaws of their descendants; then, millions of years later, the upper elements of the reptilian jawbone became the malleus and incus (hammer and stirrup) bones of the mammalian middle ear (Figure 22.21).

Scientists are looking at co-option in the formation of novel evolutionary structures. For instance, the carapace (dorsal shell) of the turtle is an evolutionary novelty that appears to form in a manner reminiscent of limbs. There is even a carapacial ridge that organizes the mesenchyme much like the apical ectodermal ridge of the limb bud (Figure 22.22; Burke 1989a). The bones themselves appear to form in the manner of skull bones. It is possible that certain developmental pathways (those used to form the limbs and those used to form the skull bones) have been recruited to form this new structure. The existence of discrete developmental modules allows the principles of dissociation, duplication and divergence, and co-option to form new types of organisms.

Developmental Correlation

Correlated progression

The modular nature of development also expects that modules will aggregate to form larger modules. One evolutionary consequence of this phenomenon is **correlated progression**, in which changes in one part of the embryo induce changes in another. Skeletal cartilage informs the placement of muscles, and muscles induce the placement of nerve axons. In such cases, if one structure changes, it will induce other structures to change with it (Thomson 1988). The dramatic changes in bone arrangement from agnathans to jawed fishes, from jawed fishes to amphibians, and from reptiles to mammals were coordinated with changes in jaw structure, jaw musculature, tooth deposition and shape, and the structure of the cranial vault and ear (Kemp 1982; Thomson 1988; Fischman 1995).

The mechanism through which the jaw apparatus has maintained its integrity from agnathans to amniotes is a remarkable example of embryonic modularity. The neural crest-derived structures of the vertebrate head include the pharyngeal arches (the precursors of the jaw, middle ear, tongue skeleton, etc.) as well as the dermal bones of the face and the facial musculature (see Chapter 13). The braincase is produced from mesodermal tissues. Köntges and Lumsden (1996) were able to map the fates of the neural crest cells associated with particular rhombomeres by replacing individual chick rhombomeres with those of quail (Figure 22.23). Antibody staining of the quail neural crest cells showed that each rhombomere gives rise to particular skeletal elements *and to the muscles attached to them*. Moreover, the muscle-and-skeleton modules from each rhombomere were found to be innervated by a particular cranial nerve. For instance, the neural crest cells from rhombomere 4 generated four skeletal tissues: the retroarticular process of the lower jaw (found in birds, but not mammals), a portion of the tongue skeleton, the stapes bone of the middle ear, and, surprisingly, the small portion of the braincase where the jaw-opening muscle attaches to the otherwise mesodermally derived skull. The muscles connecting these four skeletal elements also came from the r4 neural crest cells. These muscles are all innervated by the seventh cranial nerve. Thus, this rhombomere forms a modular unit, comprising the pharyngeal arch skeletal elements, the muscles that move them, the attachment site of the muscles to the braincase, and the nerves that innervate the muscles. Because these muscles and bones are formed from the same cells, their relationships can be maintained despite the dramatic changes in position and function that these elements might undergo over time.

One can also see correlated progression over a shorter time in domesticated animals. Humans have a great talent for selecting hereditary variants in domestic animals that involve those neural crest cells forming the frontonasal and mandibu-

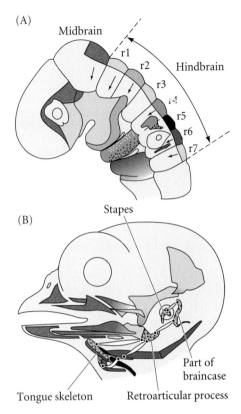

(A)

Midbrain

r1
r2   Hindbrain
r3
r4
r5
r6
r7

(B)

Stapes

Part of
braincase

Tongue skeleton   Retroarticular process

Figure 22.23
Chick embryo rhombomere neural crest cells and their musculoskeletal "packets." (A) 2-day chick embryo showing the contribution of the rhombomere neural crest cells to the pharyngeal arches.
(Most of the neural crest cells from r3 and r5 undergo apoptosis,
while the rest of these cells contribute to the larger population of r4
neural crest cells.) (B) 10-day embryo showing the bones of the
upper and lower jaws, tongue skeleton, and middle ear derived from
the rhombomeric crest cells. The muscles derived from r4 are attached to bones from the same rhombomere, and the part of the
braincase attached to the r4-derived jaw-opening muscle is also derived from r 4. (Other muscles have been omitted for clarity.) (After
Ahlberg 1997.)

lar processes. In some cases, such as that of bulldogs, the breed
is selected for a wide face with very little angle between head
and jaw. Other breeds, such as the collie, are selected for a narrow snout with a long jaw protruding away from the head. All
breeds of dogs can move their jaws, shake their heads, and bark,
despite the differences in the way their bones are shaped or positioned. Each variation is genetically determined, and it is important to note that each represents a harmonious rearrangement of the different bones with each other and with their
muscular attachments. As the skeletal elements were selected, so
were the muscles that moved them, the nerves that controlled
their movements, and the blood vessels that fed them.*

*This coordination is not quite universal, however. In dogs with greatly
shortened faces (such as bulldogs), the skin has not coordinated its development with the bones and therefore hangs in folds from the head (Stockard
1941).

Correlated progression has also been shown experimentally. Repeating the earlier experiments of Hampé (1959),
Gerd Müller (1989) inserted barriers of gold foil into the pre-
chondrogenic hindlimb buds of a 3.5-day chick embryo. This
barrier separated the regions of tibia formation and fibula formation. The results of these experiments were twofold. First,
the tibia is shortened, and the fibula bows and retains its connection to the fibulare (the distal portion of the tibia). Such relationships between the tibia and fibula are not usually seen in
birds, but they are characteristic of reptiles (Figure 22.24).
Second, the musculature of the hindlimb undergoes changes in
parallel with the bones. Three of the muscles that attach to
these bones now show characteristic reptilian patterns of insertion. It seems, therefore, that experimental manipulations
that alter the development of one part of the mesodermal
limb-forming field also alter the development of other mesodermal components. This was crucial in the evolution of the
bird hindlimb from the reptile hindlimb. As with the correlated progression seen in facial development, these changes all appear to be due to interactions within a module, in this case, the
chick hindlimb field. These changes are not global effects and
can occur independently of the other portions of the body.

WEBSITE **22.8 Correlated progression in domestic
animals.** Domestication appears to be selection for
neotenic conditions. In selecting for behavioral plasticity,
changes in skull shape and pigment patterns are also produced. This phenomenon can also be seen in current attempts to domesticate wild wolves and foxes.

Coevolution of ligand and receptor

Another example of developmental correlation involves the
ability of one tissue to interact with another. In development,
things have to fit together if the organism is to survive.
Ligands have to fit with receptors, and they have to be expressed at the right place and at the right time. Changes in the
ligand must be accommodated by complementary changes in
the receptor if the receptor is to function. If a mutation in a
gene encoding ligand (or receptor) produces too great a
change, it will not bind to its complementary receptor (or lig-
and), and development will stop. When duplications of ligand
and receptor genes occur, they can diverge and acquire new
functions. This is seen in the evolution of hormone families
and their receptors (Moyle et al. 1994).

Such separation of functions can cause reproductive isolation and the separation of species when the receptor and lig-
and are proteins on the sperm and egg. While most proteins of
closely related marine species are very similar, the proteins responsible for fertilization are often extremely different (Metz
et al. 1994). In sea urchins, the bindin of the sperm and the
complementary receptors of the egg have coevolved such that
the bindin of one species often does not recognize the bindin
receptors on the oocytes of other species. Hofmann and Glabe
(1994) have proposed a model whereby there are several distinct recognition sites on bindin and its receptor. Mutations

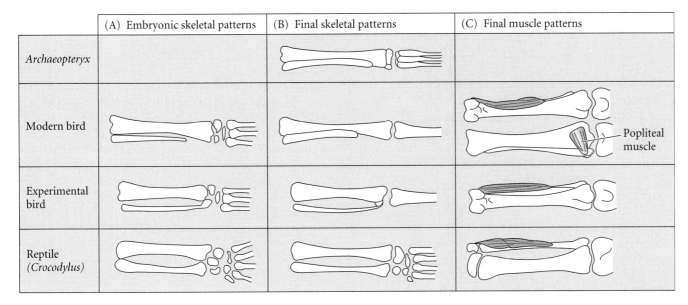

	(A) Embryonic skeletal patterns	(B) Final skeletal patterns	(C) Final muscle patterns
Archaeopteryx			
Modern bird				Popliteal muscle
Experimental bird			
Reptile (*Crocodylus*)			

Figure 22.24

Experimental "atavisms" produced by altering embryonic fields in the limb. (A–C) Results of Müller's experiments using gold foil to split the chick hindlimb field. (A, B) The embryonic and final bone patterns, indicating that the fibulare structure was retained by the experimental chick limb, as it is in extant reptiles and as it is thought to have been in *Archaeopteryx*, the earliest known bird. (C) Some of the correlated muscle patterns. The popliteal muscle is present in the normal chick limb, but is absent from reptile limbs and from the experimental limb. The fibularis brevis muscle, which normally originates from both the tibia and fibula in chicks, takes on the reptilian pattern of originating solely from the fibula in the experimental limb. (D) Fossil *Archaeopteryx* in limestone. Imprints of feathers can be clearly seen. Were it not for the feathers, this toothed organism would probably have been classified as a reptile. (A–C after Müller 1989; D, photograph courtesy of B. A. Miller/Biological Photo Service.)

(D)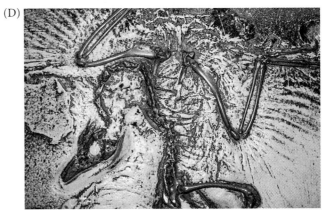

would cause some of these sites to be altered, and these alterations would select for complementary alterations on the opposite gamete. There would be a stage wherein some unaltered sperm could bind, albeit weakly, to altered eggs, but eventually, this process of alteration and accommodation would produce two reproductively isolated groups within the species (Figure 22.25). In abalones, mutations of a small region of the lysin protein and its corresponding receptor appear to be responsible for the species specificity of fertilization. Moreover, these changes in lysin and bindin proteins appear to be rapid and correlate with speciation* (Shaw et al. 1994; Metz and Palumbi 1996; Lyon and Vacquier 1999).

*Another example of a developmental mutation causing reproductive isolation involves a more mechanical function. The snail shell coiling mutations discussed in Chapter 8 are mutations that act during early development to change the position of the mesodermal organs. Mating between left-coiling and right-coiling snails is mechanically very difficult, if not impossible, in some species (Clark and Murray 1969). Because this mutation is inherited as a maternal effect gene, a group of related snails would emerge that could mate with one another but not with other members of the original population. These reproductively isolated snails could expand their range and, by the accumulation of new mutations, form a new species (Alexandrov and Sergievsky 1984).

Developmental Constraints

Another consequence of interacting modules is that these interactions limit the possible phenotypes that can be created, and they also allow change to occur in certain directions more easily than in others.[†] Collectively, these restraints on phenotype production are called **developmental constraints**.

Physical constraints

There are only about three dozen animal phyla, constituting the major body plans of the animal kingdom. One can easily imagine other types of body plans and animals that do not exist. (Science fiction writers do it all the time.) Why aren't there more major body types among the animals? To answer

[†]Leibniz, probably the philosopher who most influenced Darwin, noted that existence must be limited not only to the possible but to the *compossible*. That is, whereas numerous things can come into existence, only those that are mutually compatible will actually exist (see Lovejoy 1964). So although many developmental changes are possible, only those that can integrate into the rest of the organism (or which can cause a compensatory change in the rest of the organism) will be seen.

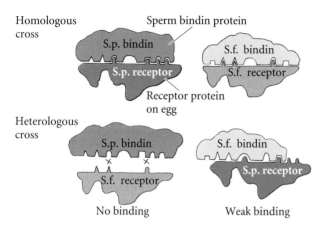

Figure 22.25
Hypothetical model for the recognition pattern of sperm and egg in two related sea urchin species. The sperm of *Strongylocentrotus purpuratus* can bind the receptor in the *S. purpuratus* egg. Similarly, the sperm of *S. franciscanus* can bind its egg receptor. *S. purpuratus* sperm will not bind to *S. franciscanus* eggs, while *S. franciscanus* sperm will bind weakly to the eggs of *S. purpuratus*. It is postulated that repeated elements in each species' bindin protein each interact with complementary sites in their species' bindin receptors. Coevolution between bindin and its receptor may have separated the two species. (After Shaw et al. 1994.)

this, we have to consider the constraints that development imposes on evolution. There are three major classes of constraints on morphogenetic evolution.

First, there are physical constraints on the construction of the organism. The laws of diffusion, hydraulics, and physical support allow only certain mechanisms of development to occur. One cannot have a vertebrate on wheeled appendages (of the sort that Dorothy saw in Oz) because blood cannot circulate to a rotating organ; this entire possibility of evolution has been closed off. Similarly, structural parameters and fluid dynamics forbid the existence of 5-foot-tall mosquitoes.

The elasticity and tensile strengths of tissues is also a physical constraint. The six cell behaviors used in morphogenisis (cell division, growth, shape change, migration, death, and matrix secretion) are each limited by physical parameters, and thereby provide limits on what structures animals can form. Interactions between different sets of tissues involves coordinating the behaviors of cell sheets, rods, and tubes in a limited number of ways (Larsen 1992).

Morphogenetic constraints

There are also constraints involving morphogenetic construction rules (Oster et al. 1988). Bateson (1894) and Alberch (1989) noted that when organisms depart from their normal development, they do so in only a limited number of ways. Some of the best examples of these types of constraints come from the analysis of limb formation in vertebrates. Holder (1983) pointed out that although there have been many modifications of the vertebrate limb over 300 million years, some modifications (such as a middle digit shorter than its surrounding digits) are not found. Moreover, analyses of natural populations suggest that there is a relatively small number of ways in which limb changes can occur (Wake and Larson 1987). If a longer limb is favorable in a given environment, the humerus may become elongated, but one never sees two smaller humeri joined together in tandem, although one could imagine the selective advantages that such an arrangement might have. This observation indicates a construction scheme that has certain rules.

The rules governing the architecture of the limb may be the rules of the reaction-diffusion model (outlined in Chapter 1; Newman and Frisch 1979). Oster and colleagues (1988) found that the reaction-diffusion model can explain the known morphologies of the limb and can explain why other morphologies are forbidden. The reaction-diffusion equations predict the observed succession of bones from stylopod (humerus/femur) to zeugopod (ulna-radius/tibia-fibula) to autopod (hand/foot). If limb morphology is indeed determined by the reaction-diffusion mechanism, then spatial features that cannot be generated by reaction-diffusion kinetics will not occur.

Evidence for this mathematical model comes from experimental manipulations, comparative anatomy and cell biology. When an axolotl limb bud is treated with the anti-mitotic drug colchicine, the dimensions of the limb are reduced. In these experimental limbs, there is not only a reduction in the number of digits, but a loss of certain digits in a certain order, as predicted by the mathematical model and from the "forbidden" morphologies. Moreover, these losses of specific digits produce limbs very similar to those of certain salamanders whose limbs develop from particularly small limb buds (Figure 22.26; Alberch and Gale 1983, 1985). The self-organization of chondrocytes into nodules can be modelled by the Turing equations, and TGF-β2 appears to have the properties of the activator molecule postulated by this hypothesis (Miura and Shiota 2000a,b). Thus, the use of reaction-diffusion mechanisms to construct limbs may constrain the possibilities that can be generated during development, because only certain types of limbs are possible under these rules.

Phyletic constraints

Phyletic constraints constitute the third set of constraints on the evolution of new types of structures (Gould and Lewontin 1979). These are historical restrictions based on the genetics of an organism's development. For instance, once a structure comes to be generated by inductive interactions, it is difficult to start over again. The notochord, for example, which is still functional in adult protochordates such as amphioxus (Berrill 1987), is considered vestigial in adult vertebrates. Yet it is transiently necessary in vertebrate embryos, where it specifies the neural tube. Similarly, Waddington (1938) noted that although the pronephric kidney of the chick embryo is considered vestigial (since it has no ability to concentrate urine), it is the source of the ureteric bud that induces the formation of a functional kidney during chick development (see Chapter 14).

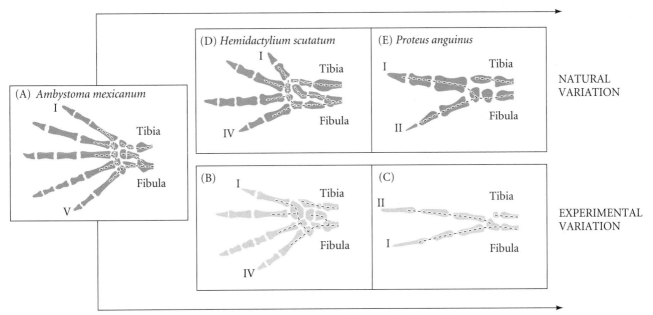

Figure 22.26
Relationship between cell number and number of digits in salamanders. (A) The hindlimb of an axolotl (*Ambystoma mexicanum*) with its five symmetrical digits. (B, C) Digits on the axolotl hindlimb after the hindlimb bud was incubated in colchicine to reduce cell number. (D, E) Hindlimbs of two wild salamanders, each having a smaller limb bud than most other salamanders: (D) *Hemidactylium scutatum* and (E) *Proteus anguinus*. The parallels between the experimental variation and the natural variation can be seen, and the common denominator is reduced cell numbers in the limb buds. (After Oster et al. 1988.)

Until recently, it was thought that the earliest stages of development would be the hardest to change, because altering them would either destroy the embryo or generate a radically new phenotype. But recent work (and the reappraisal of older work: Raff et al. 1991) has shown that alterations can be made to early cleavage without upsetting the final form. Evolutionary modifications of cytoplasmic determinants in mollusc embryos can give rise to new types of larvae that still metamorphose into molluscs, and changes in sea urchin cytoplasmic determinants can generate sea urchins that develop without larvae but still become sea urchins. In fact, while all the vertebrates arrive at a particular stage of development called the pharyngula, they do so by very different means (see Figure 1.5). Birds, reptiles, and fishes arrive there after meroblastic cleavages of different sorts; amphibians get to the pharyngula stage by way of radial holoblastic cleavage; and mammals reach the same stage after constructing a blastocyst, chorion, and amnion. The earliest stages of development, then, appear to be extremely plastic. Similarly, the later stages are very different, as the different phenotypes of mice, sunfish, snakes, and newts amply demonstrate. There is something in the middle of development, however, that appears to be invariant.

Raff (1994) argues that the formation of new body plans (*Baupläne*) is inhibited by the need for global sequences of induction during the neurula stage (Figure 22.27). Before that stage, there are few inductive events. After that stage, there are a great many inductive events, but almost all of them occur within discrete modules. During early organogenesis, however, there are several inductive events occurring simultaneously that are global in nature. At this stage, the modules overlap and interact with one another. In vertebrates, to use von Baer's example, the earliest stages of development involve specifying axes and undergoing gastrulation. Induction has not yet happened on a large scale. Moreover, as Raff and colleagues have shown (Henry et al. 1989), there is a great deal of regulative ability at these stages, so small changes in morphogen distributions or the position of cleavage planes can be accommodated. After the major

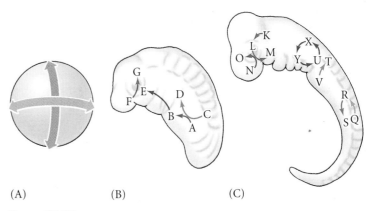

Figure 22.27
Mechanism for the bottleneck at the pharyngula stage of vertebrate development. (A) In the cleaving embryo, global interactions exist, but there are very few of them (mainly to specify the axes of the organism). (B) At the neurula to pharyngula stages, there are many global inductive interactions. (C) After the pharyngula stage, there are even more inductive interactions, but they are primarily local in effect, confined to their own modules. (After Raff 1994.)

body plan is fixed, inductions occur all over the body, but are compartmentalized into discrete organ-forming systems. The lens induces the cornea, but if it fails to do so, only the eye is affected. Similarly, there are inductions in the skin that form feathers, scales, or fur. If they do not occur, the skin or patch of skin may lack these structures, but the rest of the body is unchanged. But during early organogenesis, the interactions are more global (Slack 1983). Failure to have the heart in a certain place can affect the induction of eyes (see Figure 6.4). Failure to induce the mesoderm in a certain region leads to malformations of the kidneys, limbs, and tail. It is this stage that constrains evolution and that typifies the vertebrate phylum. Thus, once a vertebrate, it is difficult to evolve into anything else.

WEBSITE **22.9 Changing embryonic traits through natural selection.** Another factor explaining the bottleneck in developmental histories may be selection for adult traits. Just as changes in embryos can produce new phenotypes, so natural selection on adults can favor certain types of embryos that produce favorable adult phenotypes.

WEBSITE **22.10 Alternative proposals for evolutionary developmental biology.** Evolution is accomplished through heritable changes in development. In this textbook, these heritable changes are assumed to be those that alter gene expression patterns. However, other models have been proposed in which there is horizontal transmission of genetic information between phyla, or in which there is inheritance of cytoplasmic properties.

Sidelights & Speculations

Canalization and the Release of Developmental Constraints

NOT ALL MUTATIONS produce mutant phenotypes. Rather, development appears to be buffered so that slight abnormalities of genotype or slight perturbations of the environment will not lead to the formation of abnormal phenotypes (Waddington 1942). This phenomenon, called **canalization**, serves as an additional constraint on the evolution of new phenotypes.

It is difficult for a mutation to actually affect development (Nijhout and Paulsen 1997). It is the rare mutation that is 100% penetrant. Stress, however, in the form of environmental factors such as temperature, can overpower the buffering systems of development and alter the phenotype. Moreover, the altered phenotype then becomes subject to natural selection, and if selected, will eventually appear without the stress that originally induced it. Waddington called this phenomenon genetic assimilation (see Chapter 21). For instance, when Waddington subjected *Drosophila* larvae of a certain strain to high temperatures, they lost their wing crossveins. After a few generations of repeated heat shock, the crossveinless phenotype continued to be expressed in this population even without the heat shock treatment. While Waddington's results look like a case of "inheritance of acquired characteristics," there is no evidence for that view. Certainly, the crossveinless phenotype was not an adaptive response to heat. Nor did the heat shock cause the mutations. Rather, the heat shock over-

came the buffering systems, allowing preexisting mutations to result in mutant phenotypes rather than wild-type phenotypes.

In 1998, Suzanne Rutherford and Susan Lindquist showed that a major agent responsible for this buffering was the "heat shock protein" **Hsp90**. Hsp90 is a protein that binds to a set of signal transduction molecules that are inherently unstable. When it binds to them, it stabilizes their tertiary structure so that they can respond to upstream signaling molecules. Heat shock, however, causes other proteins in the cell to become unstable, and Hsp90 is diverted from its normal function (of stabilizing the signal transduction proteins) to the more general function of stabilizing any of the cell's now partially denatured peptides (Jakob et al. 1995; Nathan et al. 1997). Since Hsp90 was known to be involved with inherently unstable proteins and could be diverted by stress, it was possible that Hsp90 might be involved in buffering developmental pathways against environmental contingencies.

Evidence for the role of Hsp90 as a developmental buffer first came from mutations of *Hsp83*, the gene for Hsp90. Homozygous mutations of *Hsp83* are lethal in *Drosophila*. Heterozygous mutations increase the proportion of developmental abnormalities in the population into which they are introduced. In populations of *Drosophila* heterozygous for *Hsp83*, deformed eyes, bristle duplications, and abnormalities of legs and wings

appeared (Figure 22.28). When different mutant alleles of *Hsp83* were brought together in the same flies, both the incidence and severity of the abnormalities increased. Abnormalities were also seen when a specific inhibitor of Hsp90 (geldanamycin) was added to the food of wild-type flies, and the types of defects differed between different stocks of flies. The abnormalities observed did not show simple Mendelian inheritance, but were the outcome of the interactions of several gene products. Selective breeding of the flies with the abnormalities led over a few generations to populations in which 80–90% of the progeny had the mutant phenotype. Moreover, these mutants did not keep the *Hsp83* mutation. In other words, once the mutation in *Hsp83* had allowed the cryptic mutations to become expressed, selective matings could retain the abnormal phenotype even in the absence of abnormal Hsp90.

Thus, Hsp90 is probably a major component of the buffering system that enables the canalization of development. It provides a way to resist fluctuations due to slight mutations or slight environmental changes. Hsp90 might also be responsible for allowing mutations to accumulate but keeping them from being expressed until the environment changes. No individual mutation would change the phenotype, but mating would allow these mutations to be "collected" by members of the population. An environmental change (anything that might stress the

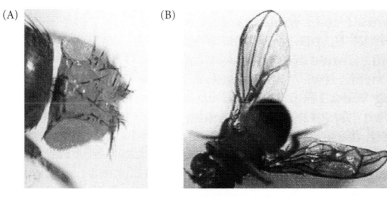

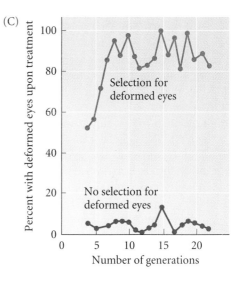

Figure 22.28
Developmental abnormalities in *Drosophila* associated with mutations in the *Hsp83* gene: (A) deformed eyes; (B) thickened wing veins. (C) The deformed eye trait (A) was selected by breeding only those individuals expressing the trait. This abnormality was not observed in the original stock, but it can be seen when individuals are mated to heterozygous *Hsp83* flies. The strong response to selection showed that even though the population was small, it contained a large amount of hidden genetic variation. (After Rutherford and Lindquist 1998; photographs courtesy of the authors.)

cells) would thereby release the hidden phenotypic possibilities of the population. In other words, transient decreases in Hsp90 (resulting from its aiding stress-damaged proteins) would uncover preexisting genetic interactions that would produce morphological variations. Most of these morphological variations would probably be deleterious, but some might be selected for in the new environment. Such release of hidden morphological variation may be responsible for the many examples of rapid speciation found in the fossil record. ■

A New Evolutionary Synthesis

In 1922, Walter Garstang declared that **ontogeny** (an individual's development) does not recapitulate **phylogeny** (evolutionary history); rather, it *creates* phylogeny. Evolution is generated by heritable changes in development. "The first bird," said Garstang, "was hatched from a reptile's egg." Thus, when we say that the contemporary one-toed horse evolved from a five-toed ancestor, we are saying that heritable changes occurred in the differentiation of the limb mesoderm into chondrocytes during embryogenesis in the horse lineage. This view of evolution as the result of hereditary changes affecting development was lost during the 1940s, when the Modern Synthesis of population genetics and evolutionary biology formed a new framework for research in evolutionary biology.

The Modern Synthesis has been one of the greatest intellectual achievements of biology. By merging the traditions of Darwin and Mendel, evolution within a species could be explained: Diversity within a population arose from the random production of mutations, and the environment acted to select the most fit phenotypes. Those animals capable of reproducing would transmit the genes that gave them their advantage. These genes included, for example, those encoding enzymes with better rates of synthesis and globins with better oxygen-carrying capacity. It was assumed that the same kinds of changes (gene or chromosomal mutations) that caused evolution within a species also caused the evolution of new species. There would need to be an accumulation of these mutations, and a mechanism of reproductive isolation to enable them to accumulate in new ways, if a new phenotype was to be produced.

Not only could the Modern Synthesis explain evolution within a species remarkably well, it also explained medically relevant questions such as why certain alleles that seem deleterious (the hemoglobin gene variant that can result in sickle cell anemia, for example) might be selected for in certain populations. The population genetic approach to evolution was summed up by one of its foremost practitioners and theorists, Theodosius Dobzhansky, when he declared, "Evolution is a change in the genetic composition of populations. The study of the mechanisms of evolution falls within the province of population genetics" (Dobzhansky (1951).

The developmental approach to evolution was excluded from the Modern Synthesis (Hamburger 1980; Gottlieb 1992; Dietrich 1995; Gilbert et al. 1996). It was thought that population genetics could explain evolution, so morphology and development were seen to play little role in modern evolutionary theory (Adams 1991). In other words, **macroevolution** (the large morphological changes seen between species, classes, and phyla) could be explained by the mechanisms of

microevolution, the "differential adaptive values of genotypes or deviations from random mating or both these factors acting together" (Torrey and Feduccia 1979).

The population genetics model contained some major assumptions that have now been called into question.

1. **Gradualism.** The supposition that all evolutionary changes occur gradually was debated by Darwin and his friends. Thomas Huxley, for instance, accepted evolution, but he felt that Darwin had burdened his theory with an unnecessary assumption of gradualism. A century later, Eldredge and Gould (1972), Stanley (1979), and others postulated **punctuated equilibrium** as an alternative to the gradualism that characterized the Modern Synthesis. According to this theory, species were characterized by their morphological stability. Evolutionary changes tended to be rapid, not gradual. At the same time, molecular studies (King and Wilson 1975) showed that 99% of the DNA of humans and chimpanzees was identical, demonstrating that a small change in DNA could cause large and important morphological changes. New findings in paleontology and molecular biology prompted scientists to consider seriously the view that mutations in regulatory genes can create large changes in morphology in a relatively short time.

2. **Extrapolation of microevolution to macroevolution.** The idea that accumulations of small mutations result in changes leading to new species has also been criticized. Richard Goldschmidt (1940) began his book *The Material Basis of Evolution* by asking the population genetic evolutionary biologists to try to explain the evolution of the following features by accumulation and selection of small mutations: hair in mammals; feathers in birds; segmentation in arthropods and vertebrates; the transformation of the gill arches into structures including aortic arches, muscles, and nerves; teeth; shells of molluscs; compound eyes; and the poison apparatus of snakes. Interestingly, both Goldschmidt and Waddington saw homeotic mutations as the kind of genetic change that could alter one structure into another and possibly create new structures or new combinations of structures. These mutations would not be in the structural genes, but in the regulatory genes. Few scientists paid attention to Goldschmidt or Waddington, however, because they were not working under the population genetics paradigm of the Modern Synthesis and because their scientific programs were suspect. (Goldschmidt did not believe in Morgan's notion of the gene as a particulate entity, and Waddington's work was misinterpreted as supporting the inheritance of acquired traits: see Gilbert 1988; 1991; Dietrich 1995.)

3. **Specificity of phenotype from genotype.** Developmental biologists have found that life is more complicated than a 1:1 relationship between genotype and phenotype. Chapter 21 documents numerous cases wherein the geno-

type can permit any of several phenotypes to form. These cases include polyphenisms induced by predators, diet, day length, or antigenic or visual experience. Moreover, development always mediates between genotype and phenotype. The same gene can produce different phenotypes depending on the other genes that are present (Wolf 1995). A mutant gene that produces limblessness in one generation can produce only a mild thumb abnormality in the next (Freire-Maia 1975). That evolution is the result of heritable changes in development (Goldschmidt 1940) is as true for whether a fly has two or three bristles on its back as for whether an appendage is to become a fin or a limb. One way of visualizing this is to use a mathematical analogy (Gilbert et al. 1996):

> Functional biology = anatomy, physiology, cell biology, gene expression
> Developmental biology = δ [functional biology]$/ \delta t$
> Evolutionary biology = $\delta /$ [developmental biology]$/ \delta t$

To go from functional biology to evolutionary biology without development is like going from displacement to acceleration without dealing with velocity.

4. **Lack of genetic similarity in disparate organisms.** We have come a long way from when Ernst Mayr (1966) could state, concerning macroevolution: "Much that has been learned about gene physiology makes it evident that the search for homologous genes is quite futile except in very close relatives." Indeed, when one considers the Hox genes, the signal transduction cascades, and the families of paracrine factors, adhesion molecules, and transcription factors, the opposite has been seen to be the case. Adult organisms may have dissimilar structures, but the genes instructing the formation of these structures are extremely similar.

The population genetics model was formulated to explain natural selection. It is based on gene differences in adults competing for reproductive advantage. The developmental genetics model is formulated to account for phylogeny—evolution above the species level. It is based on the similarities in regulatory genes that are active in embryos and larvae. We are still approaching evolution in the two ways that Darwin recognized. One can emphasize the similarities or the differences.

When the Modern Synthesis was formulated, developmental biology (and developmental genetics) were not even sciences. Embryology was left out of the Modern Synthesis, as most evolutionary biologists and geneticists felt it had nothing to contribute. However, we know now that it does. The developmental genetics approach to evolution concerns more the *arrival* of the fittest than the *survival* of the fittest.

Even critics of the Modern Synthesis (including Goldschmidt and Gould) agree that macroevolutionary change is predicated upon mutation and recombination. However, these macroevolutionary changes are in *developmental regulatory genes*, not the usual genes for enzymes and structural proteins; and these changes occur in embryos and larvae, not in adults competing for reproductive success (see Waddington 1953; Gilbert 1998).

Developmental biology brings to evolutionary biology, first, a new understanding about the relationships between genotypes and phenotypes, and second, a new understanding about the close genetic relationships between organisms as diverse as flies and frogs. In doing so, developmental biology complements the population genetics approach to evolutionary biology. It also highlights new questions. For instance, there can now be a population genetic approach to the regulatory genes (see Arthur 1997; Macdonald and Goldstein 1999; Zeng et al. 1999). One can also look at how paracrine factors, signal transduction pathways, and transcription factors have changed during the evolution of various phyla. Evolutionary developmental biology can also provide answers to classic evolutionary genetics questions such as these posed by mimicry and industrial melanism. The genes involved in these processes are being identified so the mechanisms of these phenomena can be explained (Koch et al. 1998; Brakefield 1998). To explain evolution, both the population genetics and the developmental genetics accounts are required.

Leaving developmental biology out of the population genetics model of evolution has left evolutionary biology open to attacks by creationists. According to Behe (1996), population genetics cannot explain the origin of structures such as the eye, so Darwinism is false.* How could such a complicated structure have emerged by a collection of chance mutations? If a mutation caused a change in the lens, how could it be compensated for by changes in the retina? Mutations would serve only to destroy complex organs, not create them. However, once one adds development to the evolutionary synthesis, one can see how the eye can develop through induction, and that the concepts of modularity and correlated progression can readily explain such a phenomenon (Waddington 1940; Gehring 1998). Moreover, when one sees that the formation of eyes in all known phyla is based on the same signal transduction pathway, using the *Pax6* gene, it is not difficult to see descent with modification forming the various types of eyes. This was much more difficult before the similarity of eye instructions had been discovered. Indeed, one study based in population genetics claimed that photoreceptors or eyes arose independently over forty times during the history of the animal kingdom (Salvini-Plawen and Mayr 1977).

In his review of evolution in 1953, J. B. S. Haldane expressed his thoughts about evolution with the following developmental analogy: "The current instar of the evolutionary theory may be defined by such books as those of Huxley, Simpson, Dobzhansky, Mayr, and Stebbins [the founders of the Modern Synthesis]. We are certainly not ready for a new moult, but signs of new organs are perhaps visible." This recognition of developmental ideas "points forward to a broader synthesis in the future." We have finally broken through the old pupal integument, and a new, broader, developmentally inclusive evolutionary synthesis is taking wing.

*Behe (1996) makes this point explicitly, using the example of the eye. Although he attempts to disprove the theory of evolution by using the eye as an example, he never once mentions the studies on *Pax6*. Rather, Behe mentions theories from the 1980s (based solely on population genetics) and puts them forth as contemporary science.

Snapshot Summary: Evolutionary Developmental Biology

1. Evolution is caused by the inheritance of changes in development. Modifications of embryonic or larval development can create new phenotypes that can then be selected.

2. Darwin's concept of "descent with modification" explained both homologies and adaptations. The similarities of structure were due to common ancestry (homology), while the modifications were due to natural selection (adaptation to the environmental circumstances).

3. The Urbilaterian ancestor can be extrapolated by looking at the developmental genes common to invertebrates and vertebrates and which perform similar functions. These include the Hox genes that specify body segments, the *tinman* gene that regulates heart development, the *Pax6* gene that specifies those regions able to form eyes, and the genes that instruct head and tail formation.

4. Changes in the targets of Hox genes can alter what the Hox genes specify. The Ubx protein, for instance, specifies halteres in flies and hindwings in butterflies.

5. Changes of Hox gene expression within a region can alter the structures formed by that region. For instance, changes in the expression of Ubx and abdA in insects regulate the production of prolegs in the abdominal segments of the larvae.

6. Changes in Hox gene expression between body regions can alter the structures formed by that region. In crustaceans, different Hox expression patterns enable the body to have or to lack maxillipeds on its thoracic segments.

7. Changes in Hox gene expression are correlated with the limbless phenotypes in snakes.

8. Changes in Hox gene number may allow Hox genes to take on new functions. Large changes the numbers of Hox genes correlate with major transitions in evolution.

9. Duplications of genes may also enable these genes to become expressed in new places. The formation of new cell types may result from duplicated genes whose regulation has diverged.

10. In addition to structures being homologous, developmental pathways can be homologous. Here, one has homologous proteins organized in homologous ways. These pathways can be used for different developmental phenomena in different organisms and within the same organism.

11. Deep homology results when the homologous pathway is utilized for the same function in greatly diverged organisms. The instructions for forming the central nervous system and for forming limbs are possible examples of deep homology.

12. Modularity allows for parts of the embryo to change without affecting other parts.

13. The dissociation of one module from another is shown by heterochrony (changing in the timing of the development of one region with respect to another) and by allometry (when different parts of the organism grow at different rates).

14. Allometry can create new structures (such as the pocket gopher cheek pouch) by crossing a threshold.

15. Duplication and divergence are important mechanisms of evolution. On the gene level, the Hox genes, the Distal-less genes, the MyoD genes, and many other gene families started as single genes. The diverged members can assume different functions.

16. Co-option (recruitment) of existing genes and pathways for new functions is a fundamental mechanism for creating new phenotypes. One such recruitment is the limb development pathway being used to form eyespots in butterfly wings.

17. Developmental modules can include several tissue types such that correlated progression occurs. here, a change in one portion of the module causes changes in the other portions. When skeletal bones change, the nerves and muscles serving them also change.

18. Tissue interactions have to be conserved, and if one component changes, the other must. If a ligand changes, its receptor must change. Reproductive isolation may result from changes in sperm or egg proteins.

19. Developmental constraints prevent certain phenotypes from occurring. Such restraints may be physical (no rotating limbs), morphogenetic (no middle finger smaller than its neighbors), or phyletic (no neural tube without a notochord).

20. The Hsp90 protein enables cells to accumulate genes that would otherwise give abnormal phenotypes. When the organisms are stressed during development, these phenotypes can emerge.

21. The merging of the population genetics model of evolution with the developmental genetics model of evolution is creating a new evolutionary synthesis that can account for macroevolutionary as well as microevolutionary phenomena.

Literature Cited

Acampora, D., S. Mazan, Y. Lallemand, V. Avantaggiato, M. Maury, A. Simeone and P. Brulet. 1995. Forebrain and midbrain regions are deleted in *Otx2*^{-/-} mutants due to a defective anterior neuroectoderm specification during gastrulation. *Development* 121: 3279–3290.

Adams, M. 1991. Through the looking glass: The evolution of Soviet Darwinism. *In* L. Warren and H. Koprowski (eds.), *New Perspectives in Evolution*. Liss/Wiley, New York, pp. 37–63.

Ahlberg, P. E. 1997. How to keep a head in order. *Nature* 385: 489–490.

Alberch, P. 1989. The logic of monsters: Evidence for internal constraints in development and evolution. *Geobios* (Lyon) *mémoires spécial* 12: 21–57.

Alberch, P. and J. Alberch. 1981. Heterochronic mechanisms of morphological diversification and evolutionary change in neotropical salamander *Bolitoglossa occidentalis* (Amphibia: Plethodontidae). *J. Morphol.* 167: 249–264.

Alberch, P. and E. Gale. 1983. Size dependency during the development of the amphibian foot: Colchicine induced digital loss and reduction. *J. Embryol. Exp. Morphol.* 76: 177–197.

Alberch, P. and E. Gale. 1985. A developmental analysis of an evolutionary trend: Digit reduction in amphibians. *Evolution* 39: 8–23.

Alexandrov, D. A. and S. O. Sergievsky. 1984. A variant of sympatric speciation in snails. *Malacol. Rev.* 17: 147.

Amores, A. and 12 others. 1998. Zebrafish Hox clusters and vertebrate genome evolution. *Science* 282: 1711–1714.

Ang, S. L., O. Jin, M. Rhinn, N. Daigle, L. Stevenson and J. Rossant. 1996. A targeted mouse *otx2* mutation leads to severe defects in gastrulation and formation of axial mesoderm and to deletion of rostral brain. *Development* 122: 243–252.

Arthur, W. 1997. *The Origin of Animal Body Plans: A Study in Evolutionary Developmental Biology*. Cambridge University Press, New York.

Averof, M. and M. Akam. 1995. Hox genes and the diversification of insect and crustacean body plans. *Nature* 376: 420–423.

Averof, M. and N. H. Patel. 1997. Crustacean appendage evolution associated with changes in Hox gene expression. *Nature* 388: 682–686.

Basler, K. and W. Struhl. G. 1993. Compartment boundaries and the control of *Drosophila* limb pattern by hedgehog protein. *Nature* 368: 208–214.

Bateson, W. 1894. *Materials for the Study of Variation*. Cambridge University Press, Cambridge.

Behe, M. J. 1996. *Darwin's Black Box: The Biochemical Challenge to Evolution*. Simon and Schuster, New York.

Berrill, N. J. 1987. Early chordate evolution. I. Amphioxus, the riddle of the sands. *Int. J. Invert. Reprod. Dev.* 11: 1–27.

Blader, P., S. Rastegar, N. Fischer and U. Strähle. 1997. Cleavage of the BMP4 antagonist chordin by zebrafish Tolloid. *Science* 278: 1937–1940.

Bodmer, R. 1993. The gene *tinman* is required for specification of the heart and visceral muscles in *Drosophila*. *Development* 118: 719–729.

Boncinelli, E. 1994. Early CNS development: *Distal-less* related genes and forebrain development. *Curr. Opin. Neurobiol.* 4: 29–36.

Bonner, J. T. 1988. *The Evolution of Complexity*. Princeton University Press, Princeton.

Brakefield, P. M. 1998. The evolution-development interface and advances with the eyespot patterns of *Bicyclus* butterflies. *Heredity* 80: 265–272.

Brunelli, S., A. Faiella, V. Capra, V. Nigro, A. Simeone, A. Cama and E. Boncinelli. 1996. Germline mutations in the homeobox gene *Emx2* in patients with severe schizencephaly. *Nature Genet.* 12: 94–96.

Brylski, P. and B. K. Hall. 1988. Ontogeny of a macroevolutionary phenotype: The external cheek pouches of geomyoid rodents. *Evolution* 42: 391–395.

Burke, A. C. 1989a. Development of the turtle carapace: Implications for the evolution of a novel Bauplan. *J. Morphol.* 199: 363–378.

Burke, A. C. 1989b. Epithelial-mesenchymal interactions in the development of the chelonian Bauplan. *Fortschr. Zool.* 35: 206–209.

Burke, A. C., A. C. Nelson, B. A. Morgan and C. Tabin. 1995. Hox genes and the evolution of vertebrate axial morphology. *Development* 121: 333–346.

Cadigan, K. M. and R. Nusse. 1997. Wnt signaling: A common theme in animal development. *Genes Dev.* 11: 3286–3305.

Caldwell, M. W. and M. S. Y. Lee. 1997 A snake with legs from the marine Cretaceous of the Middle East. *Nature* 386: 705–709.

Carroll, S. B., S. D. Weatherbee and J. A. Langeland. 1995. Homeotic genes and the regulation and evolution of insect wing number. *Nature* 375: 58–61.

Chiang, A., M. B. O'Connor, R. Paro, J. Simon and W. Bender. 1995. Discrete Polycomb-binding sites in each parasegment domain of the bithorax complex. *Development* 121: 1681–1689.

Chiang, C., Y. Litingtung, E. Lee, K. E. Young, J. L. Cordoen, H. Westphal and P. A. Beachy. 1996. Cyclopia and axial patterning in mice lacking *sonic hedgehog* gene function. *Nature* 383: 407–413.

Chow, R. L., C. R. Altmann, R. A. Lang and A. Hemmati-Brivanlou. 1999. Pax6 induces ectopic eyes in a vertebrate. *Development* 126: 4213–4222.

Clack, J. A. 1989. Discovery of the earliest known tetrapod stapes. *Nature* 342: 425–427.

Clark, B. and J. Murray. 1969. Ecological genetics and speciation in land snails of the genus *Partula*. *Biol. J. Linn. Soc.* 1: 31–42.

Cohn, M. J. and C. Tickle. 1999. Developmental basis of limblessness and axial patterning in snakes. *Nature* 399: 474–479.

Dale, L. and F. C. Wardle. 1999. A gradient of BMP activity specifies dorsal-ventral fates in early *Xenopus* embryos. *Semin. Cell Dev. Biol.* 10: 319–326.

Damen, W. G. M., M. Hausdorf, E.-G. Seyfarth and D. Tautz. 1998. A conserved mode of head segmentation in arthropods revealed by the expression pattern of Hox genes in a spider. *Proc. Natl. Acad. Sci. USA* 95: 10665–10670.

Darwin, C. 1859. *The Origin of Species*. John Murray, London.

Davidson, D. H. and G. Ruvkun. 1999. Themes from a NASA workshop on gene regulatory processes in development and evolution. *J. Exp. Zool.* 285: 104–115.

Davis, C. A., D. P. Holmyard, K. J. Millen and A. L. Joyner. 1991. Examining pattern formation in mouse, chicken and frog embryos with an En-specific antiserum. *Development* 111: 287–298.

Degnan, B. M., S. M. Degnan, A. Giusti and D. E. Morse. 1995. A *Hox/hom* homeobox gene in sponges. *Gene* 155: 175–177.

DePew, M. J., J. K. Liu, J. E. Long, R. Presley, J. J. Meneses, R. A. Pedersen and J. L. R. Rubenstein. 1999. *Dlx5* regulates regional development of the branchial arches and sensory capsules. *Development* 126: 3831–3846.

Dietrich, M. 1995. Richard Goldschmidt's "heresies" and the evolutionary synthesis. *J. Hist. Biol.* 28: 431–461.

DiSilvestro, R. L. 1997. Out on a limb. *BioScience* 47: 729–731.

Dobzhansky, Th. 1951. *Genetics and the Origin of Species*, 3rd Ed. Columbia University Press, New York.

Dyke, C. 1988. *The Evolutionary Dynamics of Complex Systems*. Oxford University Press, New York.

Eldredge, N. and S. J. Gould. 1972. Punctuated equilibria: An alternative to phyletic gradualism. *In* T. J. M. Schopf (ed.), *Models of Paleobiology*. Freeman, Cooper & Co., San Francisco, pp. 82–115.

Erwin, D. H. 1999. The origin of bodyplans. *Am. Zool.* 39: 617–629.

Erwin, D. H., J. Valentine and D. Jablonski. 1997. The origin of animal body plans. *Am. Sci.* 85: 126–137.

Finkelstein, R. and E. Boncinelli. 1994. From fly head to mammalian forebrain: The story of OTD and OTX. *Trends Genet.* 10: 310–315.

Fischman, J. 1995. Why mammalian ears went on the move. *Science* 270: 1436.

Freire-Maia, N. 1975. A heterozygote expression of a "recessive" gene. *Hum. Hered.* 25: 302–304.

Garstang, W. 1922. The theory of recapitulation: A critical restatement of the biogenetic law. *Zool. J. Linn. Soc.* 35: 81–101.

Gaunt, S. J. 1994, Conservation in the Hox code during morphological evolution. *Int. J. Dev. Biol.* 38: 549–552.

Gehring, W. J. 1998. *Master Control Genes in Development and Evolution: The Homeobox Story*. Yale University Press, New Haven.

Gellon, G. and W. McGinnis. 1998. Shaping animal body plans in development and evolution by modulation of Hox expression patterns. *BioEssays* 20: 116–125.

Gerhart, J. and M. Kirschner. 1997. *Cells, Embryos, and Evolution*. Blackwell Science, Oxford.

Gilbert, S. F. 1988. Cellular politics: Just, Goldschmidt, and the attempts to reconcile embryology and genetics. *In* R. Rainger, K. Benson,

and J. Maienschein, *The American Development of Biology*. University of Pennsylvania Press, Philadelphia, pp. 311–346.

Gilbert, S. F. 1991. Induction and the origins of developmental genetics. *In* S. F. Gilbert (ed.), *A Conceptual History of Modern Embryology*. Plenum Press, New York, pp. 181–206.

Gilbert, S. F. 1996. Cellular dialogues in organogenesis. *In* M. E. Martini-Neri, G. Neri and J. M. Opitz (eds.), *Gene Regulation and Fetal Development: Proceedings of the Third International Workshop on Fetal Genetic Pathology 1993*. Wiley-Liss, New York, pp. 1–12.

Gilbert, S. F. 1998. Conceptual breakthroughs in developmental biology. *J. Biosci.* 23: 169–176.

Gilbert, S. F., J. M. Opitz and R. A. Raff. 1996. Resynthesizing evolutionary and developmental biology. *Dev. Biol.* 173: 357–372.

Gish, D. T. 1973. *Evolution? The Fossils Say No!* Creation-Life Publishers, San Diego.

Goldschmidt, R. B. 1940. *The Material Basis of Evolution*. Yale University Press, New Haven.

Gottlieb, G. 1992. *Individual Development and Evolution: The Genesis of Novel Behavior*. Oxford University Press, New York.

Gould, S. J. 1990. An earful of jaw. *Nat. Hist.* 1990(3): 12–23.

Gould, S. J. and R. C. Lewontin. 1979. The spandrels of San Marcos and the Panglossian paradigm: A critique of the adaptationist programme. *Proc. R. Soc. Lond.* B 205: 581–598.

Graham, A. and I. McGonnell. 1999. Developmental evolution: This side of paradise. *Curr. Biol.* 9: R630–R632.

Greer, J. M, J. Puet, K. R. Thomas and M. Capecchi. 2000. Maintenance of functional equivalence during paralogous Hox gene evolution. *Nature* 403: 661–664.

Haldane, J. B. S. 1953. Foreword. *In* R. Brown and J. F. Danielli (eds.), *Evolution: Society of Experimental Biology Symposium 7*. Cambridge University Press, Cambridge, pp. ix–xix.

Halder, G., P. Callaerts and W. J. Gehring. 1995. Induction of ectopic eyes by targeted expression of the *eyeless* gene in *Drosophila*. *Science* 267: 1788–1792.

Hall, B. K. 1999. *Evolutionary Developmental Biology*, 2nd Ed. Kluwer, Boston.

Hamburger, V. 1980. Embryology and the Modern Synthesis in evolutionary theory. *In* E. Mayr and W. Provine (eds.), *The Evolutionary Synthesis: Perspectives on the Unification of Biology*. Cambridge University Press, New York, pp. 97–112.

Hampé, A. 1959. Contribution à l'étude du développement et la régulation des déficiences et excédents dans la patte de l'embryon de poulet. *Arch. Anat. Microsc. Morphol. Exp.* 48: 347–479.

Henry, J. J., S. Amemiya, G. A. Wray and R. A. Raff. 1989. Early inductive interactions are involved in restricting cell fates of mesomeres in sea urchin embryos. *Dev. Biol.* 136: 140–153.

Hirth, F. and H. Reichert. 1999. Conserved genetic programs in insect and mammalian brain development. *BioEssays* 21: 677–684.

Hirth, F., S. Therianos, T. Loop, W. J. Gehring, H. Reichart and K. Furukubo-Tokunaga. 1995. Developmental defects in brain segmentation caused by mutations of the homeobox genes *orthodenticle* and *empty spiracles* in *Drosophila*. *Neuron* 15: 769–778.

Hirth, F., B. Hartmann and H. Reichert. 1998. Homeotic gene action in embryonic brain development of *Drosophila*. *Development* 125: 1579–1589.

Hofmann, A. and C. Glabe. 1994. Bindin, a multifunctional sperm ligand and the evolution of new species. *Semin. Dev. Biol.* 5: 233–242.

Holder, N. 1983. Developmental constraints and the evolution of vertebrate limb patterns. *J. Theor. Biol.* 104: 451–471.

Holland, N. D. and L. Z. Holland. 1999. Amphioxus and the utility of molecular genetic data for hypothesizing body part homologies between distantly related animals. *Am. Zool.* 39: 630–640.

Holland, N. D., G. Panganiban, E. L. Henyey and L. Z. Holland. 1996. Sequence and developmental expression of *AmphiDll*, an amphioxus *Distal-less* gene transcribed in the ectoderm, epidermis, and nervous system: Insights into evolution of craniate forebrain and neural crest. *Development* 122: 2911–2920.

Holland, P. W. H. 1998. Major transitions in animal evolution: A developmental genetic perspective. *Am. Zool.* 38: 829–842.

Holland, P. W. H. and J. Garcia-Fernández. 1996. Hox genes and chordate evolution. *Dev. Biol.* 173: 382–395.

Holley, S., P. D. Jackson, Y. Sasai, B. Lu, E. M. De Robertis, F. M. Hoffmann and E. L. Ferguson. 1995. A conserved system for dorsal-ventral patterning in insects and vertebrates involving sog and chordin. *Nature* 376: 249–23.

Hopson, J. A. 1966. The origin of the mammalian middle ear. *Am. Zool.* 6: 437–450.

Ingham, P. W. 1994. Hedgehog points the way. *Curr. Biol.* 4: 345–350.

Jakob, U., H. Lilie, I. Meyer and J. Buchner. 1995. Transient interaction of Hsp90 with early unfolding intermediates of citrate synthase: Implications for heat shock in vivo. *J. Biol. Chem.* 270: 7288–7294.

Jean, D., K. Ewan and P. Gruss. 1998. Molecular regulators involved in vertebrate eye development. *Mech. Dev.* 76: 3–18.

Kemp, T. S. 1982. *Mammal-Like Reptiles and the Origin of Mammals.* Academic Press, New York.

Keys, D. N. and 8 others. 1999. Recruitment of a hedgehog regulatory circuit in butterfly eyespot evolution. *Science* 283: 532–534.

King, M. C. and A. C. Wilson. 1975. Evolution at two levels in humans, and chimpanzees. *Science* 188: 107–116.

Kmita, M. F., F. van der Hoeven, J. Zákány, R. Krumlauf and D. Duboula. 2000. Mechanisms of Hox gene colinearity: Transposition of the anterior *Hoxb-1* gene into the posterior HoxD complex. *Genes Dev.* 14: 198–211.

Koch, P. B., D. N. Keys, T. Rocheleau, K. Aronstein, M. Blackburn, S. B. Carroll and R. H. French-Constant. 1998. Regulation of dopa decarboxylase expression during color pattern formation in wild-type and melanic tiger swallowtail butterflies. *Development* 125: 2303–2313.

Köntges, G. and A. Lumsden. 1996. Rhombencephalic neural crest segmentation is preserved throughout craniofacial ontogeny. *Development* 122: 3229–3242.

Kukalova-Peck, J. 1992. The Uniramia do not exist: The ground plan of the Pterygota as revealed by Permian Diahanopterodea from Russia (Insecta: Paleodictyopteroidea). *Can. J. Zool.* 70: 236–255.

Larsen, E. W. 1992. Tissue strategies as developmental constraints: Implications for animal evolution. *Trends Ecol. Evol.* 7: 414–417.

Lillie, F. R. 1898. Adaptation in cleavage. In *Biological Lectures from the Marine Biological Laboratory, Woods Hole, Massachusetts*. Ginn, Boston, pp. 43–67.

Lohnes, D. and 7 others. 1994. Function of retinoic acid receptors in development. I. Craniofacial and skeletal abnormalities in RAR double mutants. *Development* 120: 2723–2743.

Loomis, C. A., E. Harris, J. Michaud, J. Wurst, W. Hanks and A. L. Joyner. 1996. The mouse *Engrailed-1* gene and ventral limb patterning. *Nature* 382: 360–363.

Lovejoy, A. O. 1964. *The Great Chain of Being.* Harvard University Press, Cambridge.

Lyon, J. D. and V. D. Vacquier. 1999. Interspecies chimeric sperm lysins identify regions mediating species-specific recognition of the abalone egg vitelline envelope. *Dev. Biol.* 214: 151–159.

Macdonald, S. J. and D. B. Goldstein. 1999. A quantitative genetic analysis of male sexual traits distinguishing the sibling species *Drosophila simulans* and *D. sechellia*. *Genetics* 153: 1683–1699.

Malicki, J., L. C. Cianetti, C. Peschle and W. McGinnis. 1992. Human *HOX4B* regulatory element provides head-specific expression in *Drosophila* embryos. *Nature* 358: 345–347.

Manak, J. R. and M. P. Scott. 1994. A class act: Conservation of homeodomain protein functions. *Development* [Suppl.] 61–71.

Manton, S. M. 1977. *The Arthropoda: Habits, Functional Morphology, and Evolution.* Clarendon Press, Oxford.

Marigo, V., R. L. Johnson, A. Vortkamp and C. J. Tabin. 1996. Sonic hedgehog differentially regulates expression of *GLI* and *GLI3* during development of the tetrapod limb. *Development* 180: 273–283.

Mark, M., F. M. Rijli and P. Chambon. 1995. Alteration of Hox gene expression in the branchial region of the head causes homeotic transformations, hindbrain segmentation defects and atavistic changes. *Semin. Dev. Biol.* 6: 275–284.

Marqués, G., M. Musacchio, M. J. Shimell, K. Wunnenburg-Stapleton, K. W. Y. Cho and M. B. O'Connor. 1997. The Dpp activity gradient in the early *Drosophila* embryo is established through the opposing actions of the Sog and Tld proteins. *Cell* 91: 417–425.

Matsuo, I., S. Kuratani, C. Kimura, N. Takeda and S. Aizawa. 1995. Mouse *Otx2* functions in the formation and patterning of the rostral head. *Genes Dev.* 9: 2646–2658.

Mayr, E. 1966. *Animal Species and Evolution.* Harvard University Press, Cambridge.

Metchnikoff, E. 1891. Zakon zhizni. Po- povodu nektotorykh proizvedenii gr. L. Tolstogo. *Vest. Evropy* 9: 228–260. Quoted and translated in Chernyak and Tauber, *From Metaphor to Theory: Metchnikoff and the Origin of Immunology*. Oxford University Press, New York, 1990.

Metz, E. C. and S. R. Palumbi. 1996. Positive selection and sequence rearrangements generate extensive polymorphisms in the gamete recognition protein bindin. *Mol. Biol. Evol.* 13: 397–406.

Metz, E. C., R. E. Kane, H. Yanagimachi and S. R. Palumbi. 1994. Fertilization between closely related sea urchins is blocked by incompatibilities during sperm-egg attachment and early stages of fusion. *Biol. Bull.* 187: 23–34.

Miura, T. and K. Shiota. 2000a. TGF-β2 acts as an "activator" molecule in reaction-diffusion model and is involved in cell sorting phenomenon in mouse limb micromass culture. *Devel. Dynam.* 217: 241–249.

Miura, T. and K. Shiota. 2000b. Extracellular matrix environment influence chondrogenic pattern formation in limb bud micromass culture: Experimental verification of theoretical models. *Anat. Rec.* 258: 100–107.

Monaghan, P., E. Grau, D. Bock and G. Schulz. 1995. The mouse homolog of the orphan nuclear receptor tailless is expressed in the developing brain. *Development* 121: 839–851.

Moyle, W. R., R. K. Campbell, R. V. Myers, M. P. Bernard, Y. Han and X. Wang. 1994. Co-evolution of ligand-receptor pairs. *Nature* 368: 251–255.

Müller, G. B. 1989. Ancestral patterns in bird limb development: A new look at Hampé's experiment. *J. Evol. Biol.* 1: 31–47.

Müller, J., S. Gaunt and P. Lawrence. 1995. Function of the Polycomb protein is conserved in mice and flies. *Development* 121: 2847–2852.

Nathan, D. F., M. H. Vos and S. Lindquist. 1997. In vivo functions of the *Saccharomycetes cerevisiae* Hsp90 chaperone. *Proc. Natl. Acad. Sci. USA* 94: 12949–12956.

Newman, S. A. and H. L. Frisch. 1979. Dynamics of skeletal pattern formation in the developing chick limb. *Science* 205: 662–668.

Nijhout, H. F. and S. M. Paulsen. 1997. Developmental models and polygenic characters. *Am. Nat.* 149: 394–405.

Ospovat, D. 1981. *The Development of Darwin's Theory.* Cambridge University Press, Cambridge.

Oster, G. F., N. Shubin, J. D. Murray and P. Alberch. 1988. Evolution and morphogenetic rules: The shape of the vertebrate limb in ontogeny and phylogeny. *Evolution* 42: 862–884.

Panganiban, G., L. Nagy and S. B. Carroll. 1994. The role of the *Distal-less* gene in the development and evolution of insect limbs. *Curr. Biol.* 4: 671–675.

Panganiban, G. and 13 others. 1997. The origin and evolution of animal appendages. *Proc. Natl. Acad. Sci. USA* 94: 5162–5166.

Pennisi, E. 2000. An integrative science finds a home. *Science* 287: 570–572.

Piatigorsky, J. and G. Wistow. 1991. The recruitment of crystallins: New functions precede gene duplication. *Science* 252: 1078–1079.

Piccolo, S., E. Agius, B. Lu, S. Goodman, L. Dale and E. M. De Robertis. 1997. Cleavage of chordin by the Xolloid metalloprotease suggests a role for proteolytic processing in the regulation of Spemann organizer activity. *Cell* 91: 407–416.

Price, M. 1993. Members of the *Dlx-* and *Nkx2* gene families are regionally expressed in the developing forebrain. *J. Neurobiol.* 24: 1385–1399.

Qiu, M. S. and 8 others. 1997. Role of the *Dlx* homeobox genes in proximodistal patterning of the branchial arches: Mutations of *Dlx-1*, *Dlx-2*, and *Dlx-1* and *-2* alter morphogenesis of proximal skeletal and soft tissue structures derived from the first and second arches. *Dev. Biol.* 185: 165–184.

Raff, R. A. 1994. Developmental mechanisms in the evolution of animal form: Origins and evolvability of body plans. *In* S. Bengston (ed.), *Early Life on Earth.* Columbia University Press, New York, pp. 489–500.

Raff, R. A. 1996. *The Shape of Life: Genes, Development, and the Evolution of Animal Form.* University of Chicago Press, Chicago.

Raff, R. A. and G. A. Wray. 1989. Heterochrony: Developmental mechanisms and evolutionary results. *J. Evol. Biol.* 2: 409–434.

Raff, R. A., G. A. Wray and J. J. Henry. 1991. Implications of radical evolutionary changes in early development for concepts of developmental constraint. *In* L. Warren and H. Korpowski (eds.), *New Perspectives in Evolution.* Liss/Wiley, New York, pp. 189–207.

Reichert, C. B. 1837. Entwicklungsgeschichte der Gehörknöchelchen der sogenannte Meckelsche Forsatz des Hammers. *Müller's Arch. Anat. Phys. wissensch. Med.* 177–188.

Relaix, F. and M. Buckingham. 1999. From insect eye to vertebrate muscle: Redeployment of a regulatory network. *Genes Dev.* 13: 3171–3178.

Riddle, R. D., R. L. Johnson, E. Laufer and C. Tabin. 1993. Sonic hedgehog mediates the polarizing activity of the ZPA. *Cell* 75: 1401–1416.

Riedl, R. 1978. *Order in Living Systems: A Systems Analysis of Evolution.* John Wiley and Sons, New York.

Rijli, F. M., M. Mark, S. Lakkaraju, A. Dierich, P. Dollé and P. Chambon. 1993. A homeotic transformation is generated in the rostral branchial region of the head by the disruption of *Hoxa-2*, which acts as a selector gene. *Cell* 75: 1333–1349.

Romer, A. S. 1949. *The Vertebrate Body.* Saunders, Philadelphia.

Romer, A. S. 1970. The Chanares (Argentina) Triassic reptile fauna VI. A chiniquodontid cynodont with an incipient squamosal-dentary jaw articulation. *Breviora* 344: 1–18.

Roux, W. 1894. The problems, methods and scope of developmental mechanics. In *Biological lectures of the Marine Biology Laboratory, Woods Hole.* Ginn, Boston., pp. 149–190.

Rutherford, S. L. and S. Lindquist. 1998. Hsp90 as a capacitor for morphological evolution. *Nature* 396: 336–342.

Salvini-Plawen, L. V. and E. Mayr. 1977. Evolution of photoreceptors and eyes. *Evol. Biol.* 10: 207–263.

Schierwater, B. and K. Kuhn. 1998. Homology of Hox genes and the zootype concept in early metazoan evolution. *Mol. Phylogenet. Evol.* 9: 375–381.

Shaw, A., Y.-H. Lee, C. D. Stout and V. D. Vacquier. 1994. The species-specificity and structure of abalone sperm lysin. *Semin. Dev. Biol.* 5: 209–215.

Shubin, N., C. Tabin and S. Carroll. 1997. Fossils, genes, and the evolution of animal limbs. *Nature* 388: 639–648.

Simeone, A., M. Gulisano, D. Acampora, A. Stornaiuolo, M. Rambaldi and E. Boncinelli. 1992. Two vertebrate homeobox genes related to *Drosophila empty spiracles* are expressed in the embryonic cerebral cortex. *EMBO J.* 11: 2541–2550.

Slack, J. M. W. 1983. *From Egg to Embryo: Determinative Events in Early Development.* Cambridge University Press, New York.

Slack, J. M. W., P. W. H. Holland and C. F. Graham. 1993. The zootype and the phylotypic stage. *Nature* 361: 490–492.

Slijper, E. J. 1962. *Whales.* A. J. Pomerans (trans.). Basic Books, New York.

Stanley, S. M. 1979. *Macroevolution: Pattern and Process.* W. H. Freeman, San Francisco.

Stern, M. J. and 8 others. 1993. The human *GRB2* and *Drosophila drk* genes can functionally replace the *Caenorhabditis elegans* cell signalling gene *sem-5*. *Mol. Biol. Cell.* 4: 1175–1188.

Stockard, C. R. 1941. The genetic and endocrine basis for differences in form and behaviour as elucidated by studies of contrasted pure-line dog breeds and their hybrids. *American Anatomical Memoirs No. 19.* Wistar Institute of Anatomy and Biology, Philadelphia.

Telford, M. J. and R. H. Thomas. 1998. Expression of homeobox genes shows chelicerate arthropods retain their deutocerebral segment. *Proc. Natl. Acad. Sci. USA* 95: 10671–10675.

Thomson, K. S. 1988. *Morphogenesis and Evolution.* Oxford University Press, New York.

Torrey, T. W. and A. Feduccia. 1979. *Morphogenesis of the Vertebrates.* Wiley, New York.

Van Valen, L. M. 1976. Energy and evolution. *Evol. Theor.* 1: 179–229.

Waddington, C. H. 1938. The morphogenetic function of a vestigial organ in the chick. *J. Exp. Biol.* 15: 371–376.

Waddington, C. H. 1940. *Organisers and Genes.* Cambridge University Press, Cambridge.

Waddington, C. H. 1942. Canalization of development and the inheritence of acquired characters. *Nature* 150: 563.

Waddington, C. H. 1953. Epigenetics and evolution. *In* R. Brown and J. F. Danielli (eds.), *Evolution: Society of Experimental Biology Symposium 7.* Cambridge University Press, Cambridge, pp. 186–199.

Waddington, C. H. 1956. *Principles of Embryology.* Allen and Unwin, London.

Wagner, G. P. 1996. Homologues, natural kinds, and the evolution of modularity. *Am. Zool.* 36: 36–43.

Wake, D. B. and A. Larson. 1987. A multidimensional analysis of an evolving lineage. *Science* 238: 42–48.

Warren, R. W., L. Nagy, J. Selegue, J. Gates and S. Carroll. 1994. Evolution of homeotic gene regulation and function in flies and butterflies. *Nature* 372: 458–461.

Weatherbee, S. D., H. F. Nijhout, L. W. Grunert, G. Halder, R. Galant, J. Selegue and S. Carroll. 1999. Ultrabithorax functions in butterfly wings and the evolution of insect wing patterns. *Curr. Biol.* 9: 109–115.

Wilson, E. B. 1898. Cell lineage and ancestral reminiscence. In *Biological Lectures from the Marine Biological Laboratories, Woods Hole, Massachusetts.* Ginn, Boston, pp. 21–42.

Winnepenninckx, B. M. H., Y. van de Peer and T. Backeljau. 1998. Metazoan relationships on the basis of 18S rRNA sequences: A few years later. *Am. Zool.* 38: 888–906.

Wolf, U. 1995. Identical mutations and phenotypic variation. *Hum. Genet.* 100: 305–321.

Wolpert, L. 1983. Constancy and change in the development and evolution of pattern. *In* B. C. Goodwin, N. Holder and C. C. Wylie (eds.), *Development and Evolution.* Cambridge University Press, Cambridge, pp. 47–57.

Zeng, Z. B., C. H. Kao and C. J. Basten. 1999. Estimating the genetic architecture of quantitative traits. Genet. Res. 74: 279–289.

Zuckerkandl, E. 1994. Molecular pathways to parallel evolution. I. Gene nexuses and their morphological correlates. *J. Mol. Evol.* 39: 661–678.

Appendix I — A partial list of genes active in human development

Gene	Phenotype(s) caused by mutations of the gene
I. TRANSCRIPTION FACTORS	
Androgen receptor*	Androgen insensitivity syndrome
*AZF1**	Azoospermia (male sterility due to low sperm number)
*CBFA1**	Cleidocranial dysplasia (defects in skull and shoulders due to poor osteoblast differentiation)
*EMX2**	Schizencephaly (infolding of cortical neurons due to migration defect)
Estrogen receptor*	Growth regulation problems, sterility
Forkhead-like15	Thyroid agenesis and cleft palate
*GLI3 (Ci)**	Grieg syndrome (digital malformations and skull shape)
*HOXA-13**	Hand-foot-genital syndrome
*HOXD-13**	Syndactyly (fused digits) or polysyndactyly (extra toe and webbed digits)
*LMX1B**	Nail-patella syndrome (ventralization of limb)
*MITF**	Waardenburg syndrome II (microphthalmia, deafness, pigment loss)
*MSX2**	Craniosynostoses (fusions of skull and limb bones)
PAX2	Renal-coloboma (iris) syndrome
*PAX3**	Waardenburg syndrome I (microphthalmia, deafness, pigment loss); cleiocranial dysplasia
*PAX6**	Aniridia
*PITX2**	Reiger's syndrome (dental, iris, and corneal defects)
PITX3	Congenital cataracts
POU3F4	Deafness and dystonia (mitochondrial transport protein deficiency)
*SOX9**	Campomelic dysplasia (bowed legs) and male sex reversal
*SRY**	Male sex reversal
TBX3	Schinzel syndrome (ulna, mammary, and sweat gland anomalies)
*TBX5**	Holt-Oram syndrome (hand and heart anomalies)
TCOF	Treacher-Collins syndrome (cranial facial dysplasia)
*TWIST**	Seathre-Chotzen syndrome (digit webbing, facial abnormalities)
*WT1**	Various urogenital anomalies, depending on mutation
II. PARACRINE FACTORS AND THEIR SIGNALING PATHWAYS	
*AMH**	Persistence of Müllerian duct in males
AMHR	Persistance of Müllerian duct in males due to deficiency of AMH receptor
*DHCR7**	Microcephaly or holoprosencephaly, mental retardation, hypotonia, male genital anomalies due to cholesterol deficiency

Gene	Phenotype(s) caused by mutations of the gene
EDA1*	Anhydrotic ectodermal dysplasia (poor teeth and sweat gland formation)
EDN3*	Hirschsprung's disease due to endothelin deficiency
EDNRB*	Hirschsprung's disease due to endothelin receptor deficiency
FGFR1*	Pfeiffer syndrome (digit and cranial abnormalities)
FGFR2*	Several skeletal syndromes involving digits, face, and skull, depending on type of mutation
FGFR3*	Dwarfisms: hypochondroplasia, achondroplasia, or thanatophoric dysplasia, depending on the severity of the mutation.
KIT*	Piebaldism
LIS*	Lissencephaly (mental retardation due to poor neuronal migration)
RET*	Hirschsprung's disease (pigment and gut neuron deficiency)
SHH*	Holoprosencephaly due to Sonic hedgehog deficiency

III. STRUCTURAL PROTEINS AND ENZYMES

Gene	Phenotype(s) caused by mutations of the gene
Adenosine deaminase	Combined T and B cell immunodeficiency, short stature
Arylsulfatase E	Chondrodysplasia punctata (mental retardation, deafness, cartilage nodules)
COL1A1, 2A1	Osteogenesis imperfecta (insufficient collagen) or Ehrler-Danlos syndromes, depending on severity of the mutation
COL4A5	Alport syndrome due to insufficient renal collagen
Connexin-32*	Degeneration of spinal nerve roots
Connexin-43*	Cardiac malformations, abnormal spleen and lung development
Elastin	Aortic stenosis
Fibrillin*	Marfan syndrome (tall stature, loose lenses, aortic fragility)
Glypican-3	Gigantism
Jagged*	Alagille syndrome (jaundice, abnormal vertebrae, pointed mandible)
KALIG-1*	Kallmann syndrome (anosmia and male hypogonadism)
Ketosteroid reductase A2*	Male pseudohermaphroditism
L1CAM*	(1) MASA (mental retardation, aphasia, shuffling gait, adducted thumbs) or (2) hydrocephalus, depending on mutation
Lysyl oxidase	Ehlers-Danlos syndrome VI (collagen defect)
Myosin VIIA	Usher syndrome (deafness, depigmented retina)
p57 (KIP2)	Beckwith-Wiedemann syndrome (gigantism due to failure to regulate cell cycle)
Steroid sulfatase	Ichthyosis (skin scales of cholesterol sulfate)
WRN	Premature aging—the gene probably encodes a helicase for DNA replication

*Indicates that the gene is discussed either in the text or on the website.

Sources for chapter-opening quotations

Chapter 1
Cezanne, P. Quoted in J. Gasquet, 1921. *Cezanne.* Paris, pp. 79–80.

Oppenheimer, J. M. 1955. Analysis of development: Problems, concepts, and their history. In *Analysis of Development*, B. H. Willier, P. A. Weiss and V. Hamburger (eds.). Saunders, Philadelphia, pp. 1–24.

Chapter 2
Bonner, J. T. 1965. *Size and Cycle: An Essay on the Structure of Biology.* Princeton University Press, Princeton, NJ, p. 3.

"The Circle of Life." Music by Elton John. Lyrics by Tim Rice. Copyright 1994 Wonderland Music Company, Inc. All rights reserved. Used by permission.

Chapter 3
Bard, J. 1997. Explaining development. *BioEssays* 20: 598–599.

Morgan, T. H. 1898. "Some Problems of Regeneration." Biological Lectures Delivered at the Marine Biology Laboratory, Woods Hole, 1898, p. 207.

Chapter 4
Gibson, W. 1999. Quoted in M. Peyser, "The Home of the Gay," *Newsweek* March 1, p. 50.

Ozick, C. 1989. *Metaphor and Memory.* Alfred A. Knopf, New York, p. 111.

Chapter 5
Claude, A. 1974. The coming of age of the cell. Nobel lecture, reprinted in *Science* 189: 433–435.

Waddington, C. H. 1956. *Principles of Embryology.* Macmillan, New York, p. 5.

Chapter 6
Butler, O. 1998. *Parable of the Talents.* Warner Books, New York, p. 3.

Harrison, R. G. 1933. Some difficulties of the determination problem. *Am. Nat.* 67: 306–321.

Chapter 7
Darwin, C. 1871. *The Descent of Man.* Murray, London, p. 893.

Whitman, W. 1855. "Song of Myself." In *Leaves of Grass and Selected Prose.* S. Bradley (ed.), 1949. Holt, Rinehart & Winston, New York, p. 25

Chapter 8
Hardin, G. *Exploring New Ethics for Survival: The Voyage of the Spaceship Beagle.* Viking Press, New York, p. 45.

Wolpert, L. 1986. Quoted in *From Egg to Embryo: Determinative Events in Early Development.* Cambridge University Press, Cambridge, p. 1.

Chapter 9
Kohler, R. E. 1994. *Lords of the Fly:* Drosophila *Genetics and the Experimental Life.* University of Chicago Press, Chicago, p. 33.

Schultz, J. 1935. Aspects of the relation between genes and development in *Drosophila.* Am. Nat. 69: 30–54.

Chapter 10
Rostand, J. 1960. *Carnets d'un Biolgiste.* Librairie Stock, Paris.

Spemann, H. 1943. *Forschung und Leben.* Quoted in T. J. Horder, J. A. Witkowski and C. C. Wylie, 1986. *A History of Embryology.* Cambridge University Press, Cambridge, p. 219.

Stern, C. 1936. Genetics and ontogeny. *Am. Nat.* 70: 29-35.

Chapter 11
Doyle, A. C. 1891. "A Case of Identity." In *The Adventures of Sherlock Holmes.* Reprinted in *The Complete Sherlock Holmes Treasury*, 1976. Crown, New York, p. 31.

Holub, M. 1990. "From the Intimate Life of Nude Mice." In *The Dimension of the Present Moment.* Trans. D. Habova and D. Young. Faber and Faber, London, p. 38

Chapter 12
Levi-Montalcini, R. 1988. *In Praise of Imperfection.* Basic Books, New York, p. 90.

Thomas, L. 1979. "On Embryology." In *The Medusa and the Snail,* Viking Press, New York, p. 157.

Chapter 13
Ramon y Cajal, S. 1937. *Recollections of My Life.* Trans. E. H. Craigie and J. Cano. MIT Press, Cambridge, MA, pp. 36–37.

Whitehead, A. N. 1934. *Nature and Life.* Cambridge Univesity Press, Cambridge, p. 41.

Chapter 14
Coleridge, S. T. 1885. *Miscellenea.* Bohn, London, p. 301

Whitman, W. 1867. "Inscriptions." In *Leaves of Grass and Selected Prose.* S. Bradley (ed.), 1949. Holt, Rinehart & Winston, New York, p. 1.

Chapter 15
Goethe, J. W. von. 1805. *Faust,* Part I. Trans. R. Jarrell, 1976.

Harvey, W. 1628. *Exercitio Anatomica de Motu Cordis et Sanguinis Animalibus.* Reprinted in 1928, C. C. Thomas, Baltimore, p. A2.

Chapter 16
Darwin, C. 1859. *On the Origin of Species.* New American Library, New York, p. 403.

Fuentes, C. 1989. *Christopher Unborn.* Trans. A. MacAdam. Farrar, Straus, and Giroux, New York, p. 281.

Chapter 17
Darwin, E. 1791. Quoted in M. T. Ghiselin, 1974. *The Economy of Nature and the Evolution of Sex.* University of California Press, Berkeley, p. 49.

Thomson, J. A. 1926. *Heredity.* Putnam, New York, p. 477.

Chapter 18
Schotte, O. Quoted in R. J. Goss, 1991. The natural history (and mystery) of regeneration. In *A History of Regeneration Research,* C. E. Dinsmore (ed.). Cambridge Univesity Press, Cambridge, p. 12.

Tennyson, A. 1886. *Idylls of the King,* 1958 ed. Macmillan, London, p. 292.

Williams, C. M. 1959. Hormonal regulation of insect metamorphosis. In *The Chemical Basis of Development,* W. D. McElroy and B. Glass (eds.). Johns Hopkins University Press, Baltimore, p. 794.

Chapter 19
Eliot, T. S. 1942. "Little Gidding." In *Four Quartets.* Harcourt, Brace and Company, New York, 1943, p. 39. Copyright T. S. Eliot.

Hadorn, E. 1955. *Developmental Genetics and Lethal Factors.* London 1961, p. 105.

Chapter 20
Thompson, D. W. 1942. *On Growth and Form.* Cambridge University Press, Cambridge.

Chapter 21
Haraway, D. H. 1991. "A Cyborg Manifesto." In *Simians, Cyborgs, and Women: The Reinvention of Nature.* Routledge, New York. p, 178.

Waddington, C. H. 1957. *The Strategy of the Genes.* Allen & Unwin, London, pp. 154–155.

Chapter 22
Rostand, J. 1962. *The Substance of Man.* Doubleday, Garden City, New York, p. 12.

Rushdie, S. 1989. *The Satanic Verses.* Viking, New York, p. 8.

Author index

Subject index

In-text definitions of terms are indexed in **boldfaced** type.

Italic type indicates that the information will be found in an illustration.

The designation "*n*" following a page number indicates the information is found in a footnote.

About the Book

Editor: Andrew D. Sinauer

Project Editor: Carol Wigg

Production Manager: Christopher Small

Multimedia and Ancillaries: Kathaleen Emerson, Wendy Beck, Chelsea Holabird, Gayle Sullivan

Electronic Bookbuilders: Janice Holabird, Joan Gemme

Illustration Program: J/B Woolsey Associates

Book Design: Susan Brown Schmidler

Cover Design: Jefferson Johnson

Copy Editor: Norma Roche

Subject Index: Grant Hackett

Cover Manufacture: Henry N. Sawyer Company, Inc.

Book Manufacture: Courier Companies, Inc.

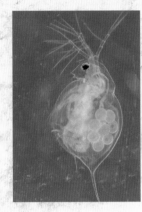
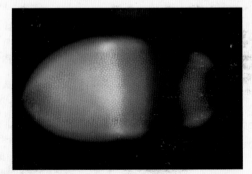